Auto Engines

Technology

Principles, Diagnosis, and Service of Engines and Related Systems

by

James E. Duffy
Automotive Writer

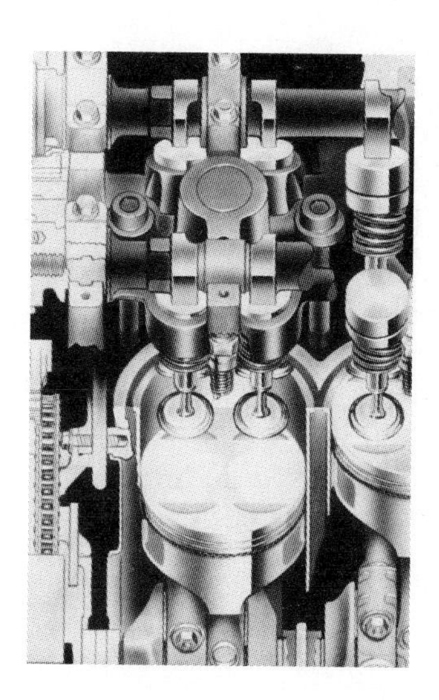

Publisher
The Goodheart-Willcox Company, Inc.
Tinley Park, Illinois

Library of Congress Catalog Card Number 96-37209
International Standard Book Number 1-56637-363-8

1 2 3 4 5 6 7 8 9 10 97 00 99 98 97 96

Library of Congress Cataloging in Publication Data

Duffy, James E.
Auto engines / by James E. Duffy.

p. cm.
Includes index.
ISBN 1-56637-363-8
1. Automobiles—Motors I. TItle.
TL210.D78 1997
629.25'04--dc21 96-37209
CIP

Introduction

Auto Engines Technology will help you learn to diagnose and repair all makes and models of gasoline and diesel engines. It shows how engines and their related systems are constructed and how they operate. The text is designed to help you use factory service manuals and specifications to complete competent engine service. It will also help you pass two of the ASE certification tests—Engine Repair and Engine Performance. The Engine Repair Test primarily includes questions on how to troubleshoot mechanical problems and how to rebuild an engine. The Engine Performance Test primarily has questions relating to performance problem diagnosis and tune-up procedures. All of these subjects receive detailed coverage. Chapter 29 summarizes the ASE voluntary testing program.

Automobile manufacturers are now using advanced engine designs to improve power, efficiency, and dependability. Today's cars are equipped with computer-controlled, high performance, 4, 6, and 8-cylinder engines. Many use four valves per cylinder, overhead camshafts, aluminum heads, aluminum blocks, turbochargers, plastic parts, and many other engineering advances. This textbook was written to help you understand and service engines with these challenging innovations.

Auto Engines Technology has 30 chapters that are grouped into six sections. Section 1 reviews fundamentals for those with no previous training in automotive technology. Section 2 details engine construction and operation. Section 3 summarizes the theory and service of engine-related systems. To be an engine technician, you must understand and be able to service starting systems, ignition systems, fuel systems, diesel injection systems, turbochargers, and more. Section 4 explains how to diagnose both mechanical and performance problems. It also summarizes engine tune-up procedures. Section 5 covers the overhauling or rebuilding of engines. Section 6 summarizes ASE certification tests and career success. Auto Engines Technology is written so you may study the chapters in the sequence presented or concentrate on certain topics. The chapters are small and cover specific engine systems or components. References help you find additional information elsewhere in the book.

Each chapter begins with learning objectives, which emphasize the important topics you will study. A summary is provided at the end of the chapter to help you review the most important concepts. The review questions at the end of each chapter include a variety of question types. ASE questions are also provided for each chapter. New words and terms are printed in *italics* and immediately defined so you can quickly learn the technical language of an engine technician.

Chapter 5 provides you with important general information about shop safety. Additional instruction regarding safety is stressed throughout the book. You can never know too much about safety.

Auto Engines Technology is a valuable guide to anyone wanting to understand or work on today's engines. It will help the student enter the trade more easily. It will help the car owner know more about the operation and servicing of engines. It will help the experienced technician become better informed about the latest engine technology.

James E. Duffy

Contents

SECTION 1—FUNDAMENTALS

SECTION 2—ENGINE CONSTRUCTION AND DESIGN

SECTION 3—ENGINE SYSTEMS

SECTION 4—ENGINE DRIVEABILITY, DIAGNOSIS, TUNE-UP

SECTION 5—ENGINE OVERHAUL

SECTION 6—ASE CERTIFICATION, CAREERS

IMPORTANT SAFETY NOTICE

Proper service and repair methods are critical to the safe, reliable operation of automobiles. The procedures described in this book are designed to help you use a manufacturer's service manual. A service manual will give the how-to details and specifications needed to do competent work.

This book contains safety precautions which must be followed. Personal injury or part damage can result when basic safety rules are not followed. Also, remember that these cautions are general and do not cover some specialized hazards. Refer to a service manual when in doubt about any service operation!

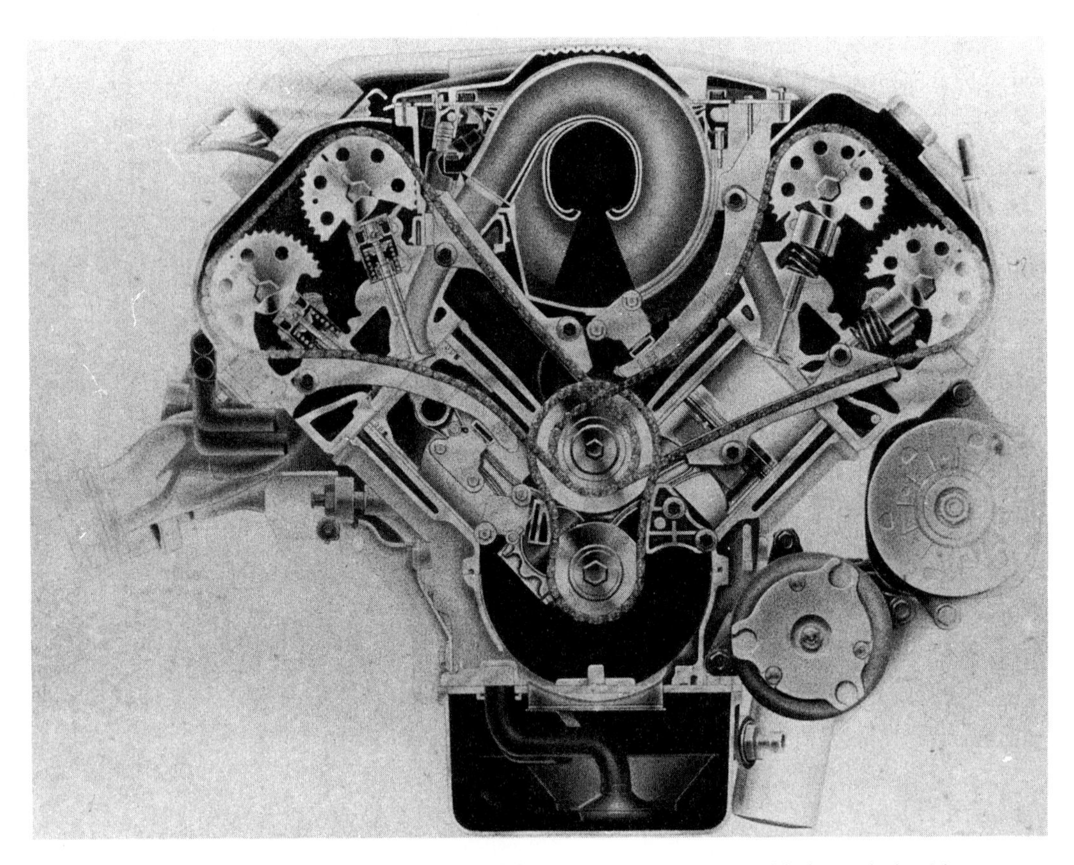

Cutaway view of a late-model V-8 engine. This engine is equipped with four chain-driven camshafts and 32 valves. (Cadillac)

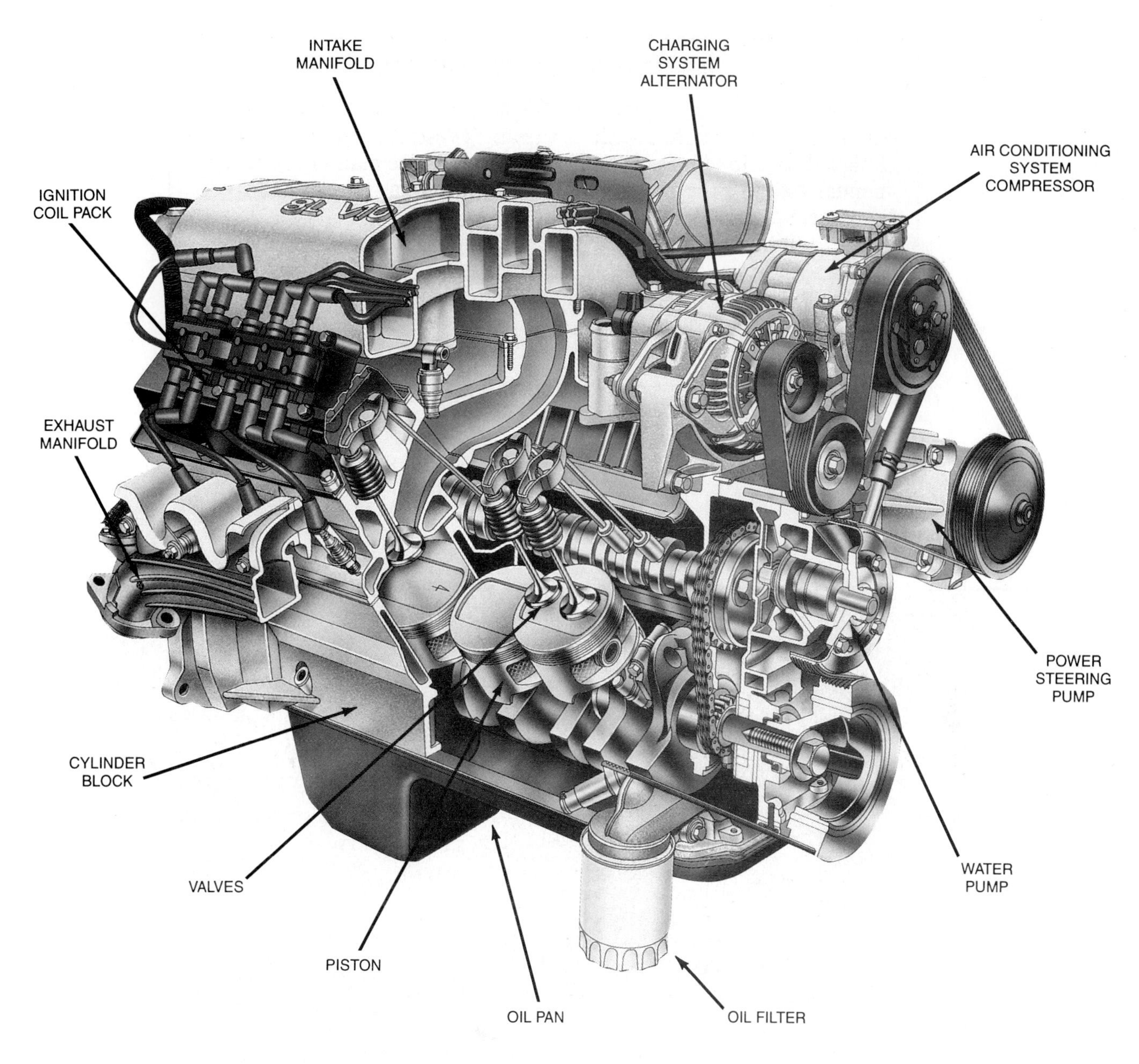

Automotive engines have changed tremendously in the past few years. This text will explain conventional engine technology as well as the latest engineering advances so that you will be prepared to service any auto engines. (Chrysler)

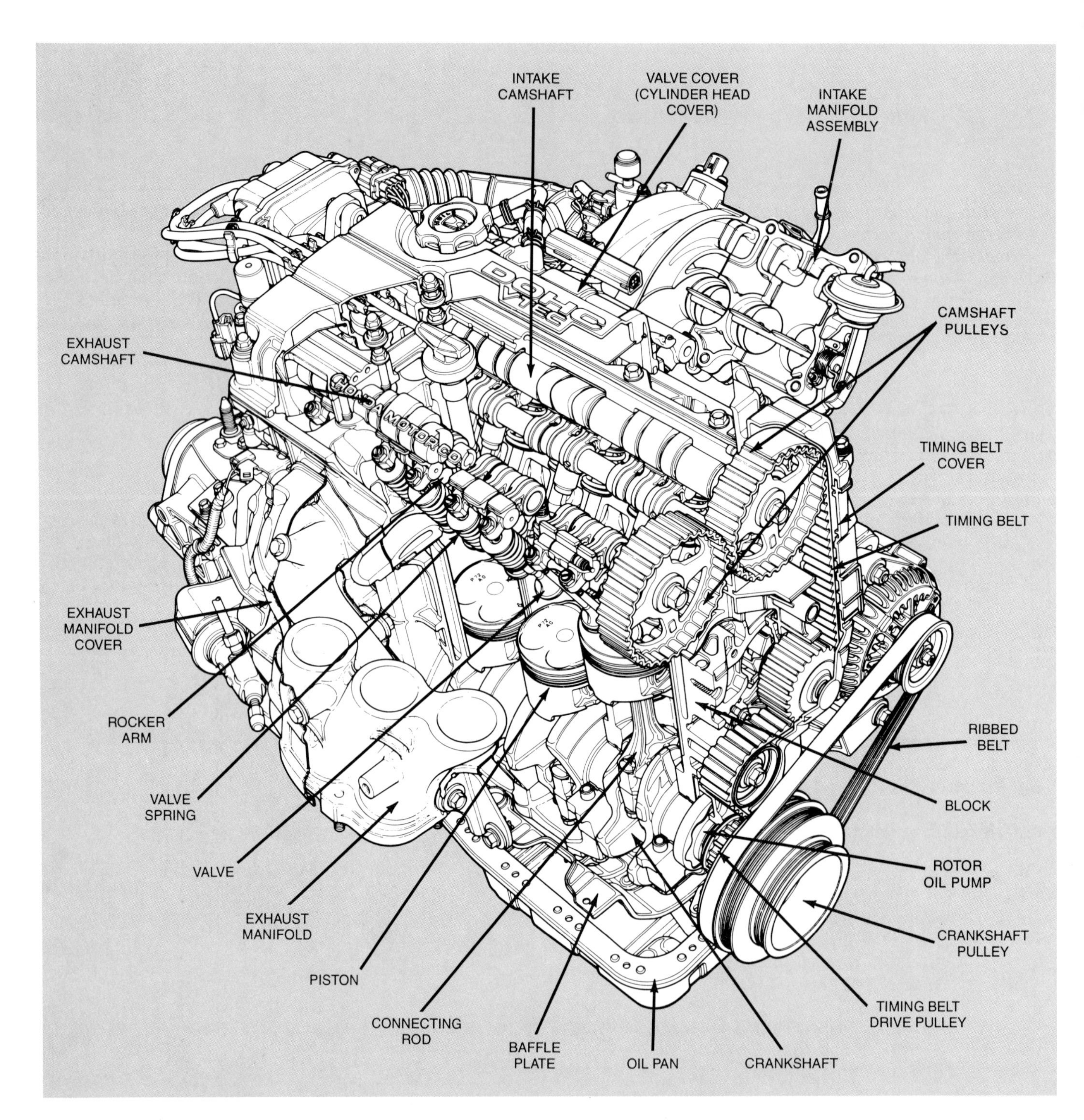

Fig. 1-1. Study general location of parts in engine. This will help you while reviewing operation of engine in chapter. (Honda)

Chapter 1

Review of Engine Operation

After studying this chapter, you will be able to:
- *List the major parts of an automotive engine.*
- *Explain the purpose of major engine parts.*
- *Describe the relationship between the components of an engine.*
- *Summarize the four-stroke cycle.*
- *List and describe the related systems of an engine.*

This chapter will quickly review the operating principles of a four-stroke cycle, piston engine. It will discuss the interaction of basic engine components and will briefly explain related systems: cooling, lubrication, fuel, computer control, and other systems. This will prepare you for later text chapters that discuss these components and systems in much more detail.

If you have already had introductory courses that covered engine operation, you should still read through this chapter to refresh your memory. If you are NOT familiar with the operation of an engine, study this chapter carefully. It will let you catch up with the students that have already had some training in engines.

This chapter will use words and illustrations to construct a basic one-cylinder engine. You will see how each part installs in a basic engine and learn how that part performs an important function. Then, near the end of the chapter, you will review the systems that supplement engine operation and protect the engine from damage.

AUTOMOTIVE ENGINE

An *engine* is the source of power for moving the car and operating the other systems. Sometimes termed *power plant,* it burns a fuel (usually gasoline, gasohol, or diesel oil) to produce heat, expansion of gases, pressure, and resulting part movement.

Since a car engine burns its fuel inside itself, it is termed an *internal combustion engine.* As you will learn, the arrangement of an engine's parts allows it to harness the energy of the burning fuel.

Fig. 1-1 illustrates the major parts of a modern, multicylinder engine. Study them as they are introduced:

1. The *block* is the supporting structure for the engine.
2. The *piston* slides up and down in the block.
3. The *piston rings* seal the space between the block and sides of the piston.
4. The *connecting rod* connects the piston to the crankshaft.
5. The *crankshaft* converts the up and down action of the piston into a usable rotary motion.
6. The *cylinder head* fits over the top of the block and holds the valves.
7. The *valves* open and close to control fuel entry and exhaust exit from the combustion chamber.
8. The *combustion chamber* is a cavity formed above the piston and below the cylinder head for containing the burning fuel.
9. The *camshaft* opens the valves at the right time.
10. The *valve springs* close the valves.
11. The *timing belt* turns the camshaft at one-half engine speed.

Engine block

The *engine block* forms the "framework" or "backbone" of an engine. Also called *cylinder block,* many of the other components of an engine fasten to the block. The block is the largest part of an engine.

Look at Fig. 1-2. It shows a cutaway view of a basic

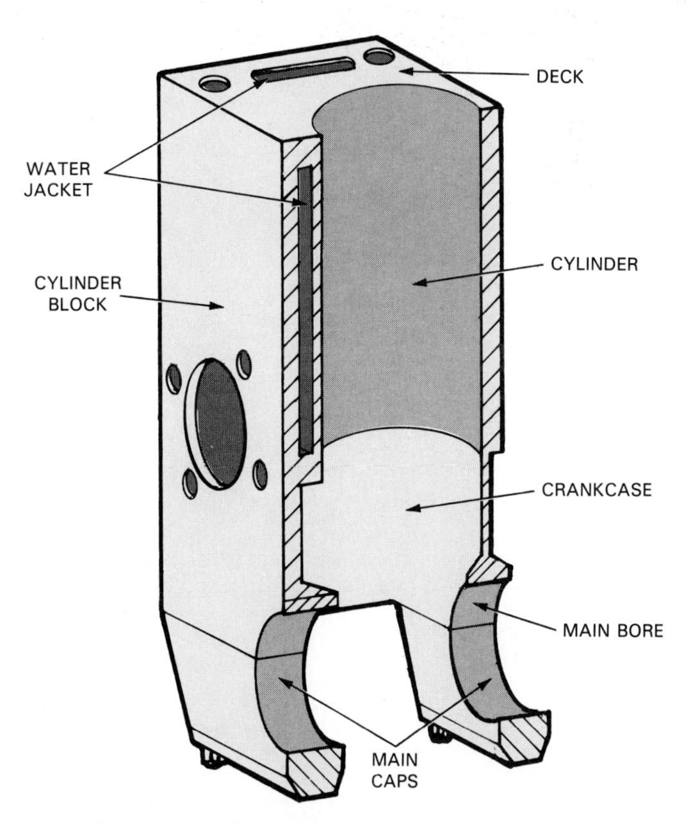

Fig. 1-2. Block is "framework" for holding other engine components. Note parts of block. This is simplified one-cylinder engine.

engine block. Note the part names.

A *cylinder* is a large hole machined through the top of the engine block. The piston fits into the cylinder. During engine operation, the cylinder, also called *cylinder bore,* guides the piston as it slides up and down in the block. The cylinder is slightly larger than the piston to produce a *clearance* (space) between the two.

Main caps bolt to the bottom of the block. They hold the crankshaft in place and form the bottom half of the main bore.

The *main bore* is a series of holes machined from the front to the rear of the block. The crankshaft fits into these holes. With the engine running, the crankshaft spins or rotates in the main bore.

The *deck* is a flat surface machined on the top of the block for the cylinder head. The head bolts to the deck. Coolant and oil passages in the deck match with openings in the cylinder head.

Water jackets surround the cylinders to carry off excess heat from the burning fuel in the cylinders. They are hollow areas inside the block and head for coolant.

The *crankcase* is the lower area of the block. The crank spins inside the crankcase.

Piston

The *piston* converts the pressure of *combustion* (burning, expanding gas) into movement. It transfers the pressure of combustion to the piston pin, connecting rod, and crankshaft. It also holds the piston rings and piston pin. See Fig. 1-3.

During engine operation, the piston slides up and down in the cylinder at tremendous speeds. At about 55 mph (88 km/h), it may go from zero to 60 miles an hour and then back to zero in one movement from top to bottom in the cylinder.

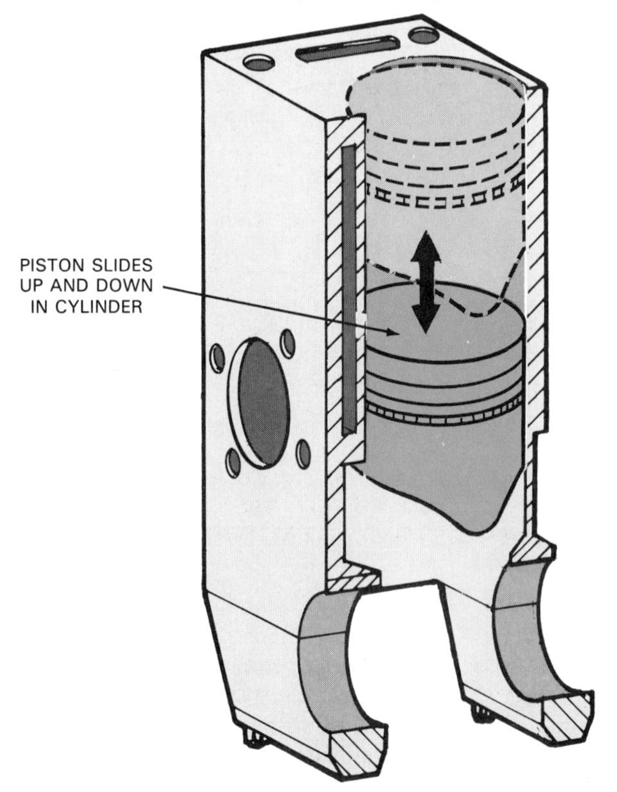

Fig. 1-3. Piston fits into cylinder bored through block. It slides up and down during engine operation. It acts as a pumping element and as a power-producing element.

Piston rings

The *piston rings* fit into grooves machined in the upper sides of the piston. They keep combustion pressure from entering the crankcase and also keep oil from entering the combustion chamber. Look at Fig. 1-4.

The *compression rings* are normally the two upper piston rings. Their job is to contain pressure formed in the combustion chamber. Without compression rings, pressure would blow past the outside diameter of the piston and into the lower area of the engine block. They seal the clearance between the block and piston.

The *oil rings* fit into the lowest groove in the piston, Fig. 1-4. They are designed to scrape excess oil from the cylinder wall and keep it from being burned in the combustion chamber. If oil were to enter the area above the piston and be burned, blue smoke would blow out the tailpipe. Figs. 1-5 and 1-6 show the action of the piston rings. Study how they work carefully.

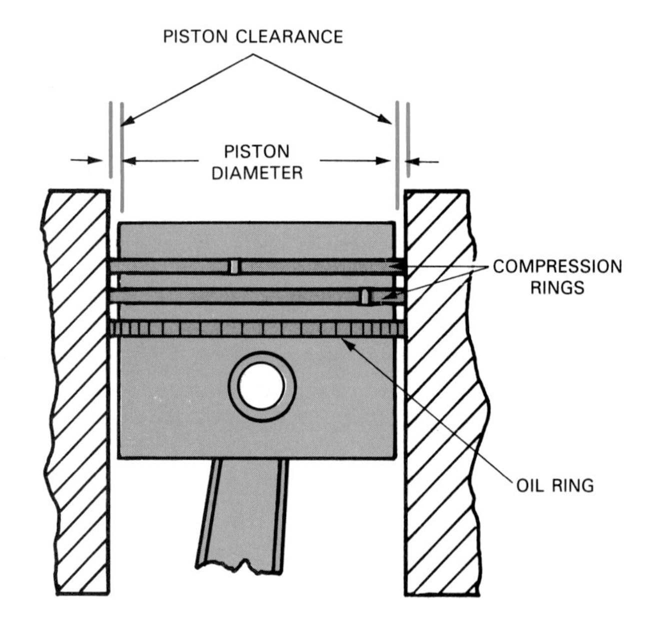

Fig. 1-4. Clearance or space between piston and cylinder allows piston to move freely in cylinder. Rings seal clearance between cylinder wall and piston.

Piston pin

A *piston pin,* also termed *wrist pin* because of its action, allows the connecting rod to swing back and forth inside the piston. The pin fits through a hole machined in the piston and through a hole in the upper end of the connecting rod. Refer to Figs. 1-7 and 1-8.

Connecting rod

The *connecting rod* links the piston to the crankshaft. It fastens to the wrist pin on top and to the crankshaft at the bottom. The connecting rod transfers the force

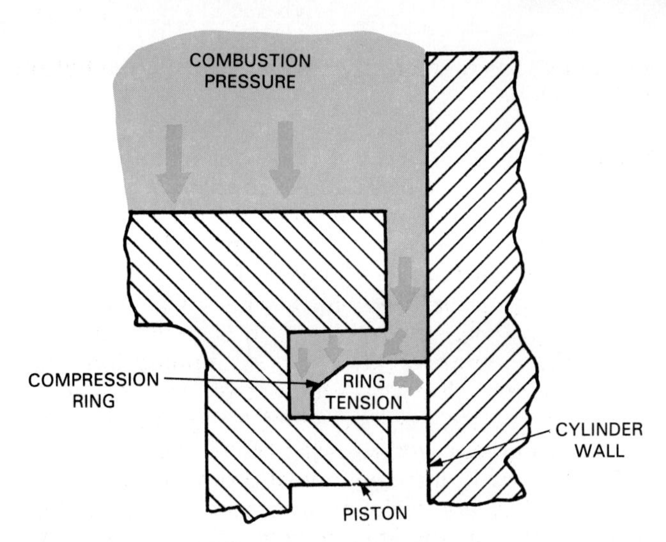

Fig. 1-5. Compression rings are two top rings. They use combustion pressure to help seal against cylinder wall. This keeps pressure up in combustion chamber and out of lower crankcase area in block.

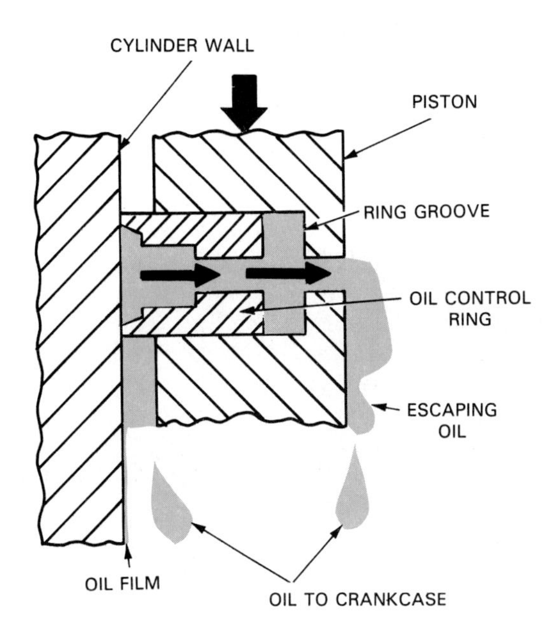

Fig. 1-6. Oil rings keep oil out of combustion chamber. They act as scraper to ''squeegee'' oil off of cylinder wall as piston slides down. (Deere & Co.)

of the piston to the crank. It also causes piston movement on nonpower producing events (up and down piston movements). See Fig. 1-9.

The *small end* of the connecting rod has a hole machined in it for the wrist pin. It extends up inside the piston.

The *big end* of the connecting rod fits around the crank. It has a removable *rod cap* that allows the installation and removal of the rod-piston assembly. *Rod bolts* hold the cap in place.

Discussed in later chapters, *bushings* normally install in the rod small end. *Rod bearings* install in the big end of the connecting rod.

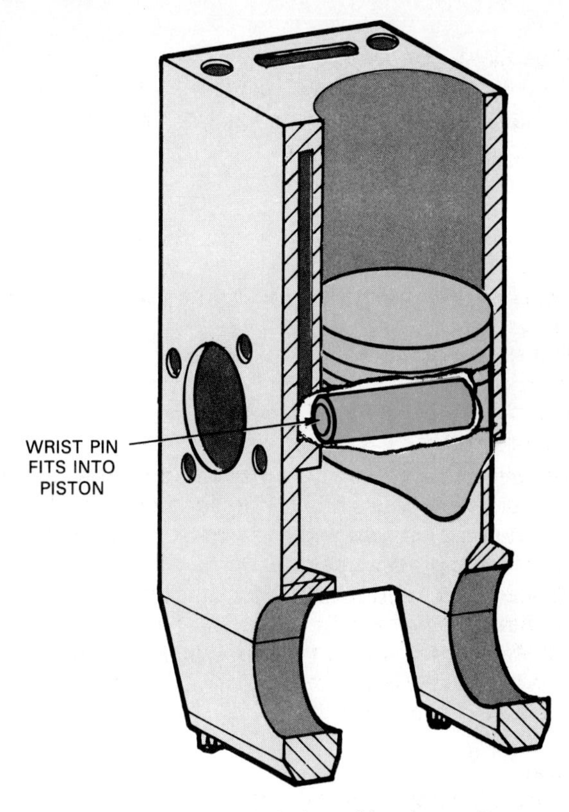

Fig. 1-7. Piston pin fits in hole bored in piston. It connects piston to connecting rod.

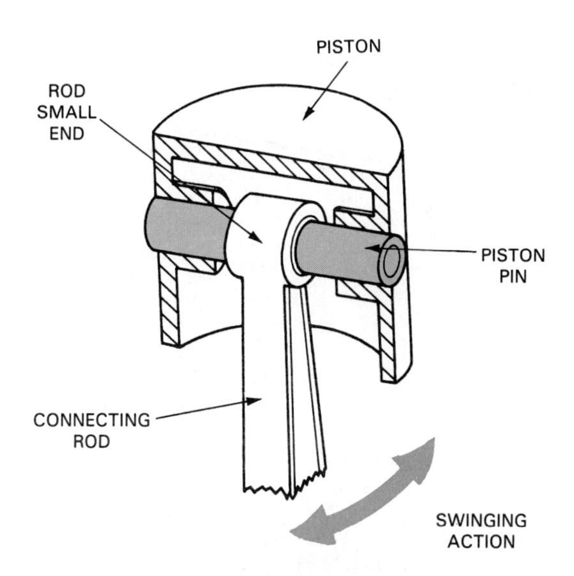

Fig. 1-8. Piston pin or wrist pin allows rod to swing inside piston. This lets bottom of rod rotate around with crankshaft as piston slides up and down in block.

Crankshaft

The *crankshaft,* as mentioned, converts the up and down movement of the connecting rod and piston into usable rotating motion. The turning motion can be used to power gears, chains, and belts for utilizing combustion energy.

The crankshaft fits into the main bore of the engine block, as shown in Fig. 1-10. It mounts on *main bear-*

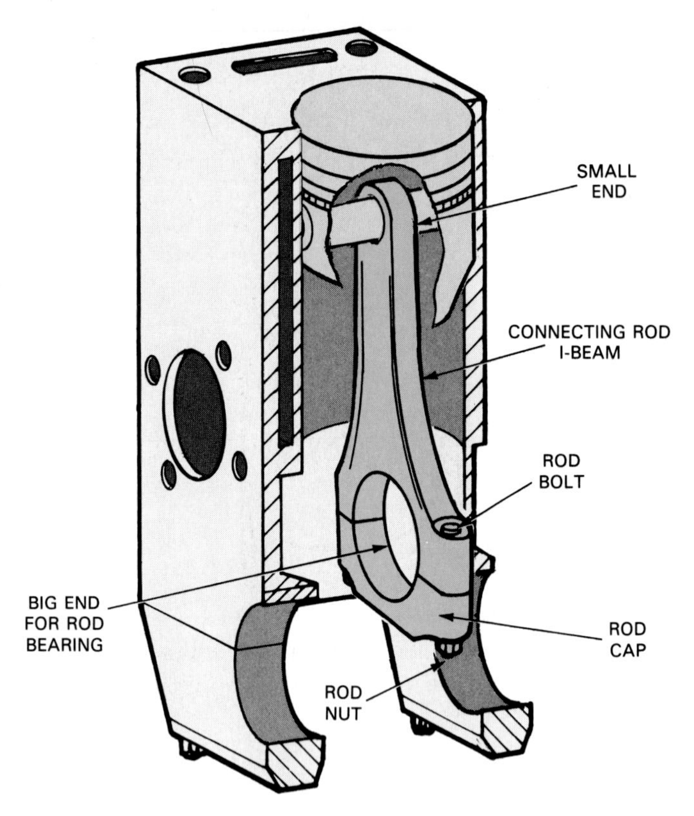

Fig. 1-9. Connecting rod links piston and crankshaft together. It has removable rod cap that allows rod to be bolted around crankshaft journal. Small end has hole for wrist pin.

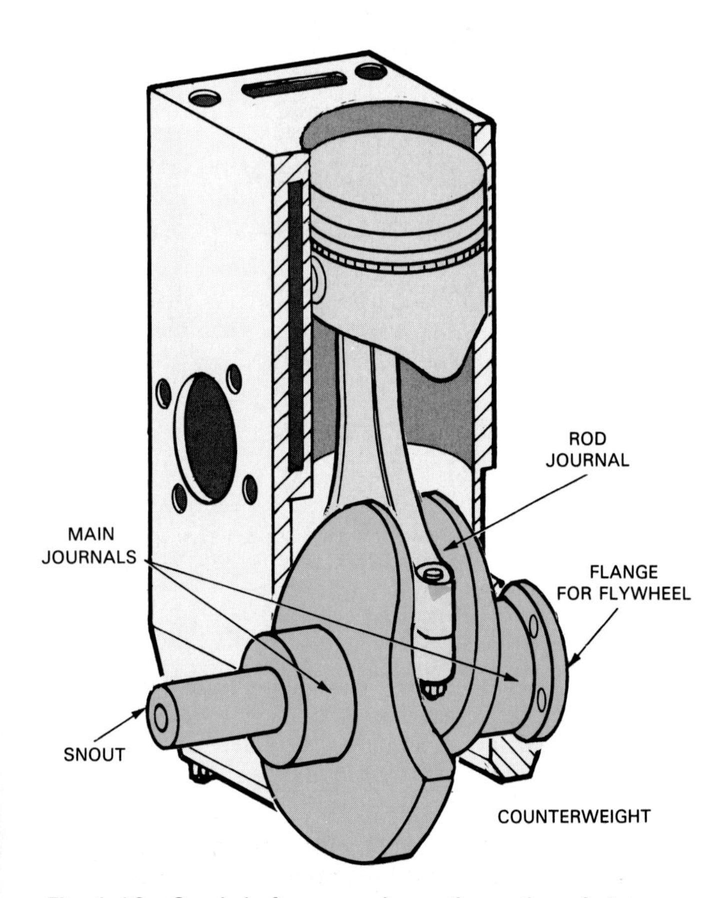

Fig. 1-10. Crankshaft uses reciprocating action of piston to produce rotating motion for car's drive line.

ings between the rod and block and is free to spin inside the block.

Fig. 1-11 shows how the crankshaft changes the *reciprocating* (up and down) action of the piston into a rotating action.

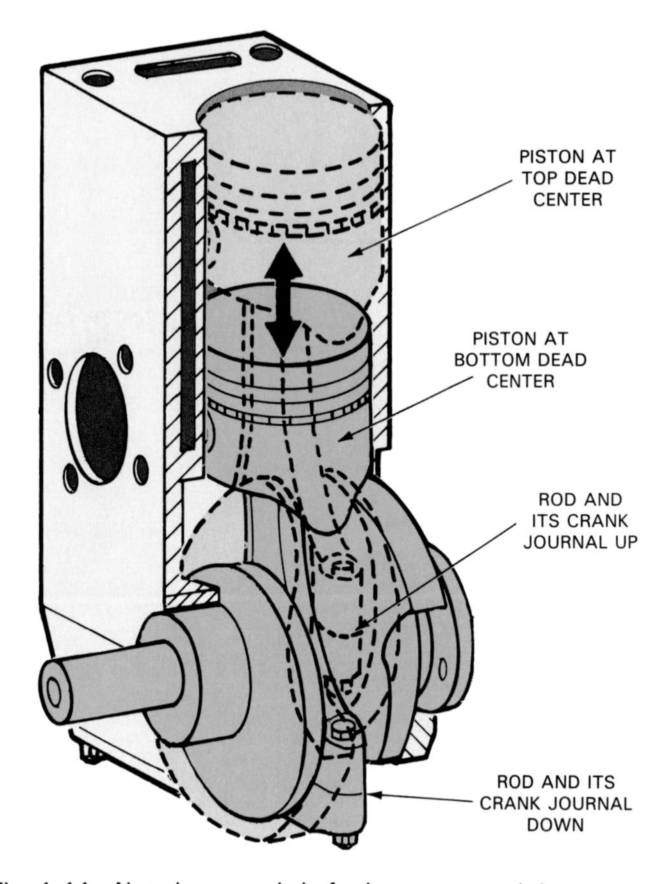

Fig. 1-11. Note how crankshaft changes up and down motion of piston into usable rotary motion.

TDC and BDC

The abbreviation TDC stands for *top dead center.* It refers to the piston being at its highest point in the cylinder. Look at Fig. 1-12.

The abbreviation BDC stands for *bottom dead center.* It means the piston is at its lowest point in the cylinder. Again, refer to Fig. 1-12.

Cylinder head

The *cylinder head* bolts to the top of the block deck to enclose the top of the cylinder, Fig. 1-13. Like the block, it contains water jackets for cooling and oil passages for lubrication of moving parts on the head.

Valve guides are machined through the top of the head for the valves. The valves slide up and down in these guides.

Cylinder head ports are passages for air and fuel to enter the combustion chamber and for burned exhaust to flow out of the engine. See Fig. 1-13.

Valve seats are machined in the opening of the *ports* (head passages), where the ports enter the combustion

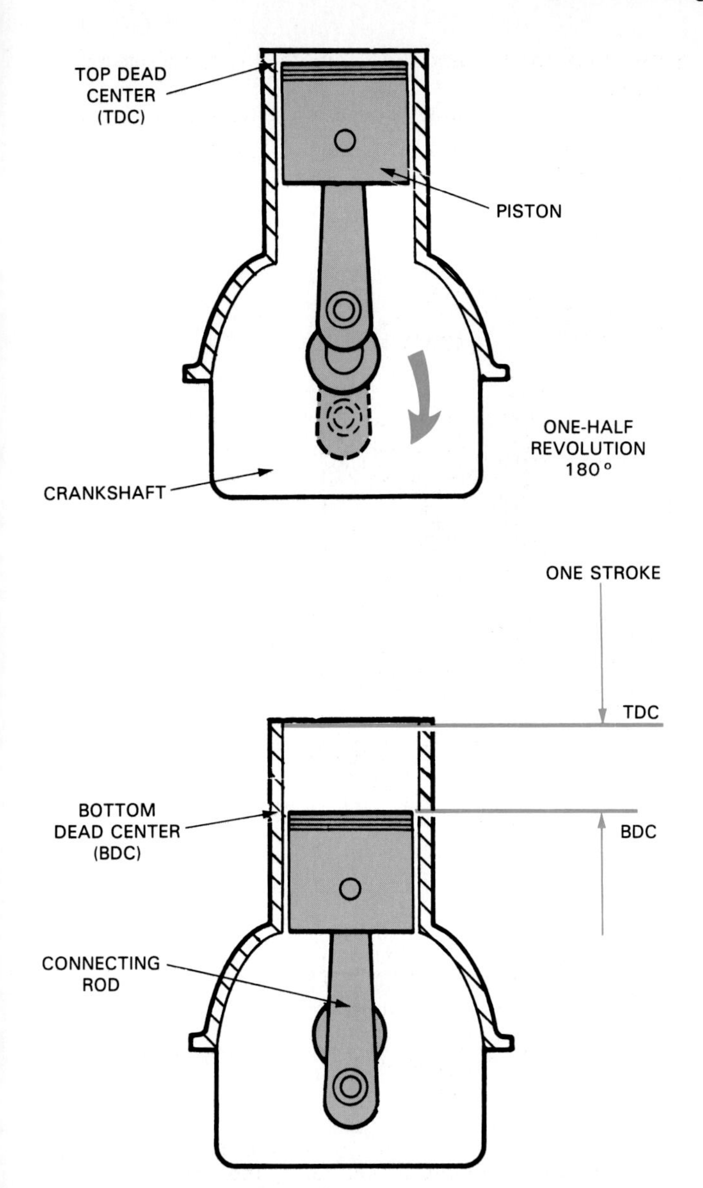

Fig. 1-12. TDC is piston at top of its stroke. BDC is piston at bottom of its stroke. One stroke is one piston movement, either up or down in cylinder. (Ford)

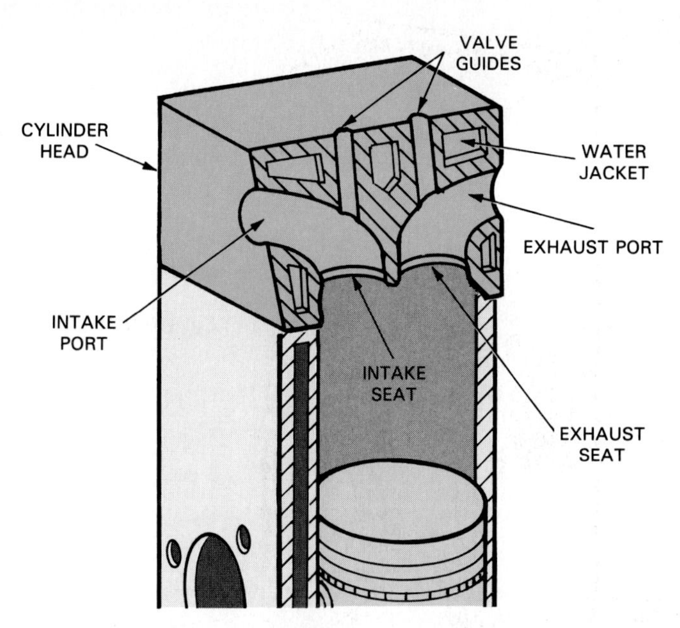

Fig. 1-13. Cylinder head bolts to top of block. It forms cover or lid over cylinder. Head also holds valves that control flow in and out of cylinder.

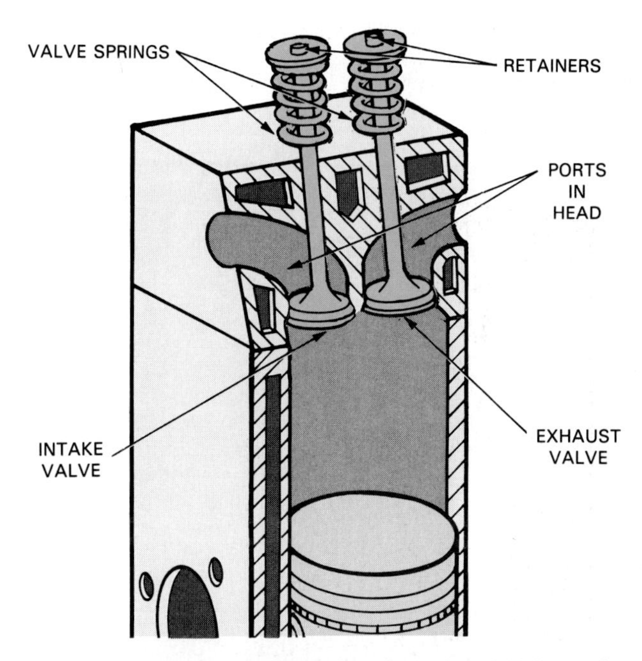

Fig. 1-14. Valves fit into guides in cylinder head. Valve springs hold valves closed. Valves seal against seats in head to close off ports and combustion chamber.

chamber. The valves close against the seats to make a leakproof seal.

Engine valves

Engine valves control the flow into and out of the engine cylinder or combustion chamber. They fit into the cylinder head and operate inside the valve guides.

Valve springs fit over the top end of the valves to keep the valves in the normally closed position, Fig. 1-14.

Fig. 1-15 shows how a valve opens and closes the port in the cylinder head. When slid down, the valve slides away from its seat and the port is opened. When slid up, the valve makes contact with its seat to seal the combustion chamber from the port.

The *intake valve* is the larger valve that allows a fuel charge to flow into the engine cylinder. The *exhaust valve* is the smaller valve that opens to let burned gases out of the engine.

Fig. 1-16 shows how the air-fuel mixture flows through the intake port, past the valve, and into the combustion chamber with the valve open.

FOUR-STROKE CYCLE

The *four-stroke cycle* needs four *strokes* (up or down piston movements) to produce a complete cycle. Every four up and down strokes of the piston results in one power-producing stroke. Two revolutions of the crank

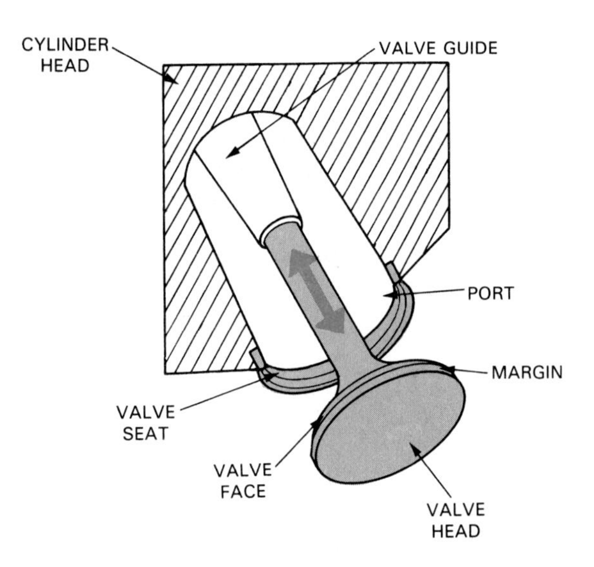

Fig. 1-15. Study valve action. When valve slides open, valve face is lifted off of valve seat. This opens port to combustion chamber. Gases are then free to enter or exit cylinder.

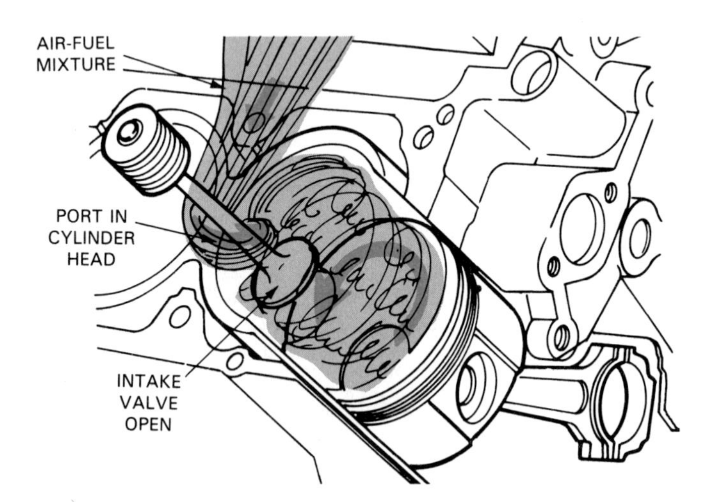

Fig. 1-16. Note action as intake valve opens. Air-fuel charge can then rush into cylinder as piston slides down. Downward movement of piston forms vacuum in cylinder. Then, atmospheric pressure or outside air pressure can push air and fuel into engine. (Ford)

are needed for one four-stroke cycle. Automotive engines, both gas and diesel, are normally four-stroke cycle engines.

The four strokes are the intake stroke, compression stroke, power stroke, and the exhaust stroke. They will be quickly reviewed. A diesel four-stroke cycle will be discussed in later textbook chapters.

Intake stroke

A gasoline engine's *intake stroke* draws air and fuel into the combustion chamber. The piston slides down to form a *vacuum* (low pressure area). The intake valve is open and the exhaust valve is closed. *Atmospheric pressure* (outside air pressure) pushes the air-fuel charge into the vacuum in the cylinder. This fills the cylinder with a burnable mixture of fuel and air.

Fig. 1-17A shows the basic action during the intake stroke. Study the position of the valves and movement of the piston.

Compression stroke

The *compression stroke* squeezes the air-fuel mixture to make the mixture more combustible. Both the intake and exhaust valves are closed. The piston slides up and compresses the mixture into the small area in the combustion chamber.

For proper *combustion* (burning), it is very important that the valves, rings, and other components do NOT allow pressure leakage out of the combustion chamber. Leakage could keep the mixture from igniting and burning on the power stroke.

Power stroke

The engine's *power stroke* ignites and burns the air-fuel mixture to produce gas expansion, pressure, and a powerful downward piston movement. This is the only stroke that does not consume (use) energy.

Both valves are still closed. The spark plug ''fires'' and the fuel mixture begins to burn. As it burns, it expands and builds pressure in the combustion chamber. Since the piston is the only part that can move, it is thrust downward with several tons of force. This downward thrust pushes on the connecting rod and crankshaft. The crankshaft is forced to turn or rotate.

Exhaust stroke

The *exhaust stroke* pushes the burned gases out of the cylinder and into the car's exhaust system. The intake valve remains closed, but the exhaust valve slides open. Since the piston is now moving up, the burned fumes are pushed out the exhaust port to ready the cylinder for another intake stroke.

With the engine operating, these strokes happen over and over very rapidly. At idle, an engine might be running at 800 *rpm* (revolutions per minute). Since it takes two complete revolutions of the crankshaft to complete a four-stroke cycle, an engine would finish 400 four-stroke cycles per minute at idle or the piston would have to slide up 800 times and down 800 times per minute. You can imagine how fast these events would be happening at highway speeds!

VALVE TRAIN

The *valve train* operates the engine valves. It times valve opening and closing to produce the four-stroke cycle. Basic valve train parts are shown in Fig. 1-18.

Camshaft

The *camshaft* opens the valves and allows the *valve springs* to close the valves at the right times. The camshaft has a series of *lobes* (egg shaped bumps) that arc on the valves (or valve train) to slide the valve down in its guide. See Fig. 1-19.

In Fig. 1-18, note how the cam lobe acts on the valve train. When the lobe moves into the lifter, it pushes up on the lifter, pushrod, and rocker arm. This opens the valve. When the lobe rotates away from the lifter, the valve spring pushes the valve and other parts into the

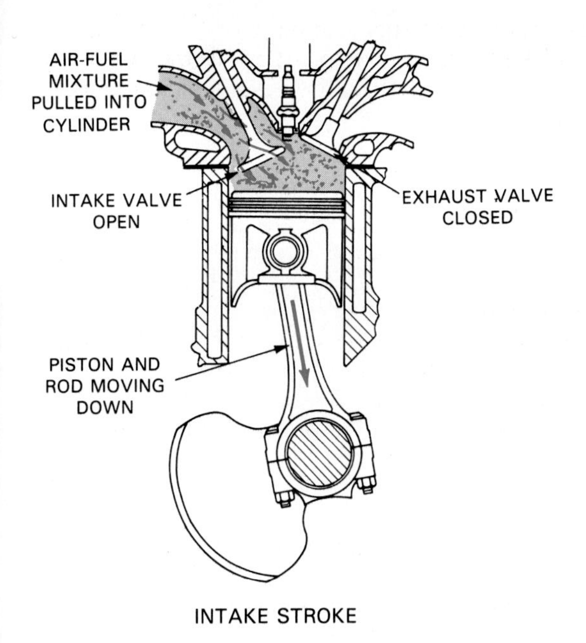

A—Intake stroke. Piston slide down when intake valve is open and exhaust valve is closed. Air-fuel charge is pulled into cylinder.

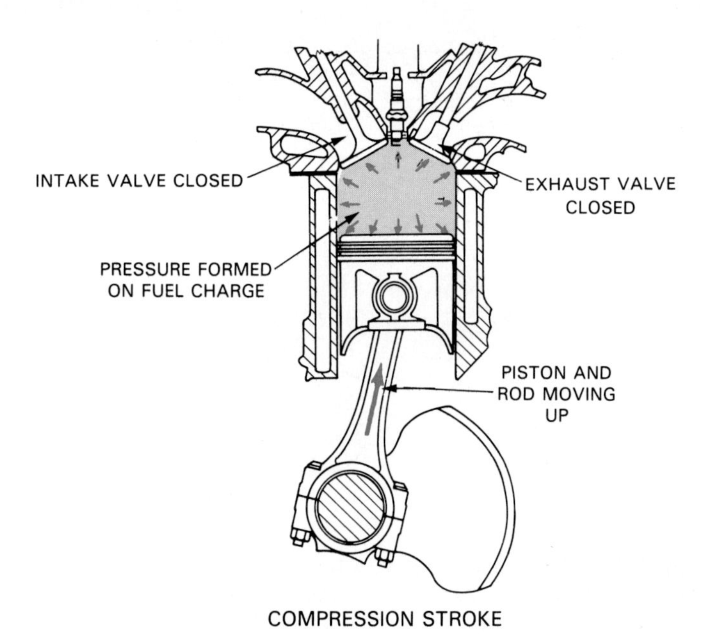

B—Compression stroke. Both valves are closed as piston slides up. This squeezes fuel charge and prepares it for combustion.

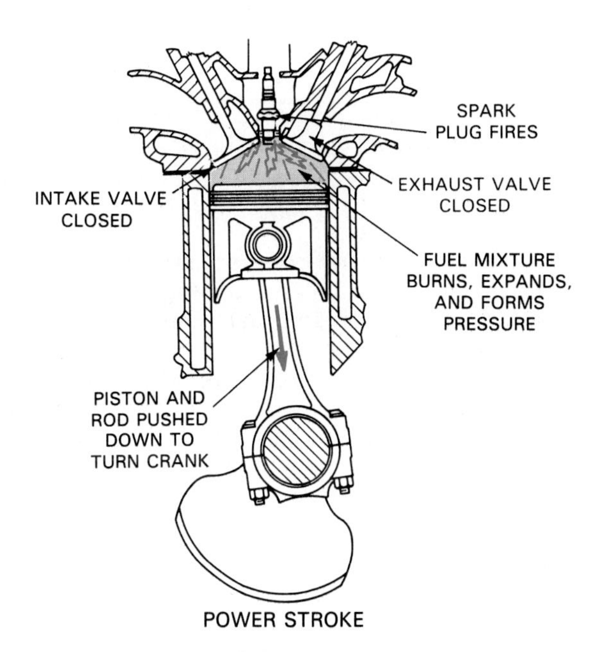

C—Power stroke. Spark plug fires and fuel begins to burn. Heat causes expansion and pressure. This pushes piston down with several tons of force to spin crankshaft.

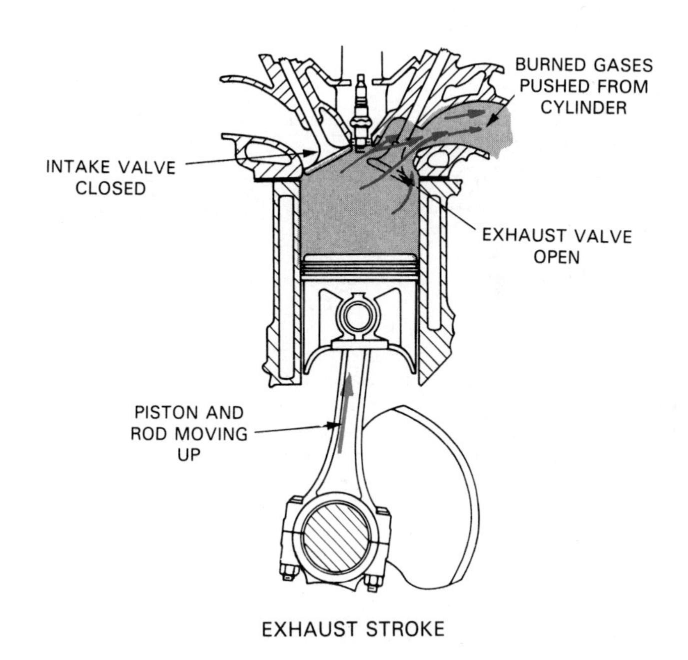

D—Exhaust stroke. Piston slides up with intake valve closed and exhaust valve open. This pushes burned fuel out of cylinder so fresh charge can enter.

Fig. 1-17. Review four-stroke cycle. You must be able to visualize these events to be a competent engine technician.

closed position.

Camshaft timing is needed to assure that the valves open and close properly in relation to the crankshaft. Either a belt, chain, or set of gears is used to turn the camshaft at one-half crankshaft speed and time the camshaft with the crankshaft. Fig. 1-20 shows how a modern timing belt is used to operate the camshaft on our basic engine. Study the part names!

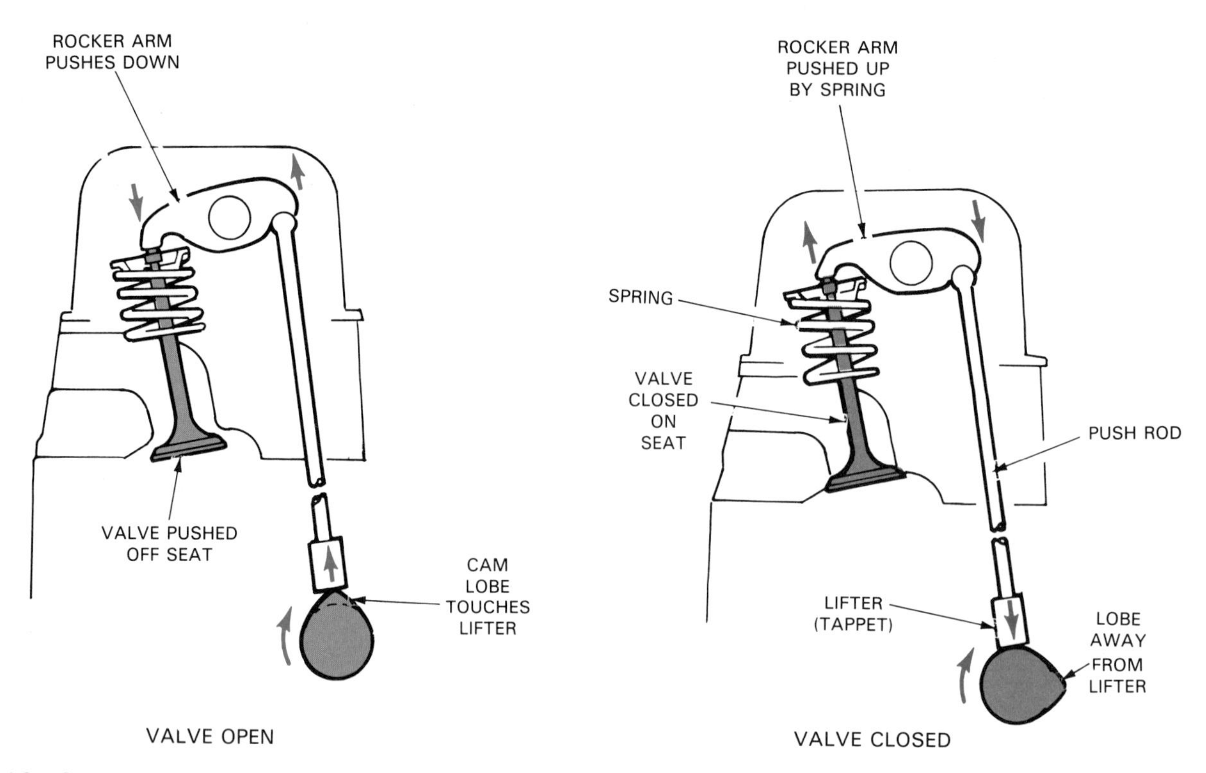

Fig. 1-18. Camshaft operates valve train. When cam lobe rotates into lifter, valve is opened. When lobe moves away from lifter, valve spring closes valve. (Ford)

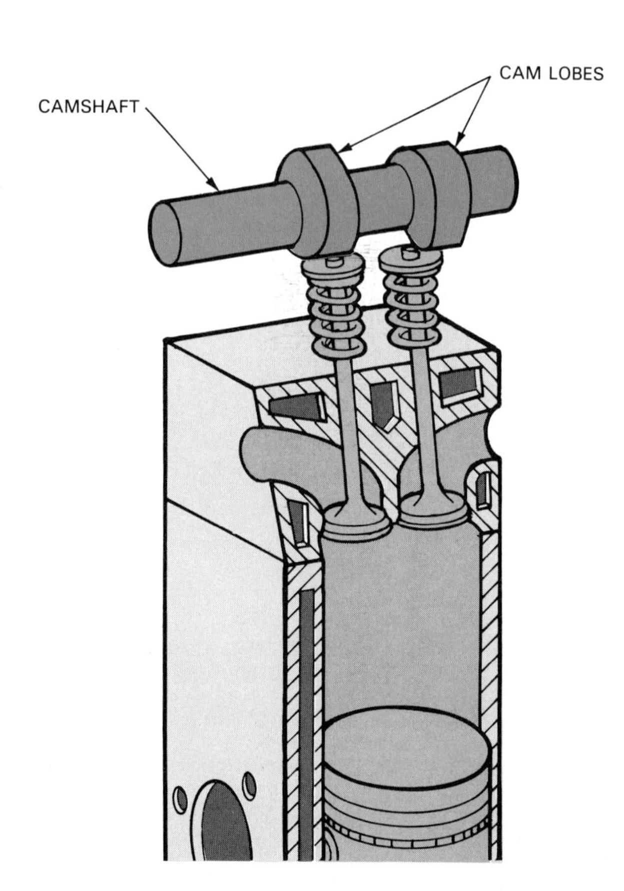

Fig. 1-19. Camshaft fits in cylinder head on many modern engines. This lets it operate directly on valves, without using pushrods.

INTAKE AND EXHAUST MANIFOLD

The *intake manifold* carries either air (diesel) or air-fuel mixture (carburetor or gasoline injection) into the cylinder head intake ports. It normally bolts to the cylinder head. Ports in the intake manifold match with the ports in the cylinder head.

The *exhaust manifold,* as its name implies, carries burned gases from the cylinder head exhaust port to the other parts of the exhaust system. Fig. 1-21 shows the basic action of intake and exhaust manifolds.

COOLING SYSTEM

The *cooling system* is needed to carry excess heat of combustion and friction away from the engine. Without a cooling system, the piston, valves, cylinder, and other parts could be ruined in a matter of minutes. The head and block could also crack from the tremendous heat.

Basically, a cooling system consists of a radiator, water pump, fan, thermostat, water jackets, and connecting hoses. This is shown in Fig. 1-22.

The *water pump* circulates an antifreeze and water solution through the water jackets, hoses, and radiator. It is driven by a *fan belt* running off of the crankshaft pulley. The *coolant* (antifreeze-water solution) picks up heat from the metal parts of the engine and carries it to the radiator.

The *radiator* transfers coolant heat into the outside air. A *fan* is used to pull air through the radiator. Large *radiator hoses* connect the radiator to the engine.

The *thermostat* is a temperature sensitive valve that controls the operating temperature of the engine. When

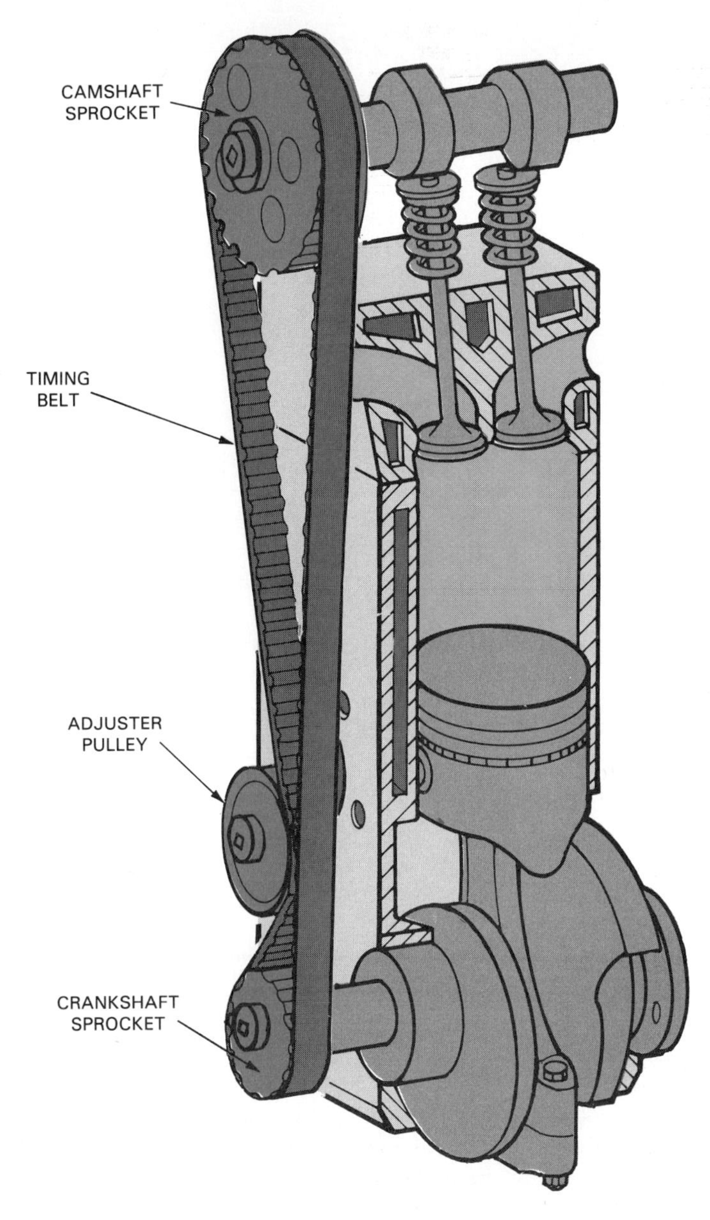

Fig. 1-20. Camshaft must be turned at one-half engine speed. Timing belt is most modern method of camshaft operation. Note part names.

the engine is cold, it blocks coolant flow through the radiator and speeds warmup. When the engine warms, it opens to allow cooling action.

LUBRICATION SYSTEM

The *lubrication system* circulates engine oil to high friction points in the engine. Without oil, friction could wear, score, and ruin parts very quickly.

The lubrication system basically consists of an oil pump, oil pickup, oil pan, and oil galleries. See Fig. 1-23.

The *oil pump* is the "heart" of the lubrication system because it pumps oil through the oil galleries. The *oil galleries* are small passages that lead to the crankshaft bearings, camshaft bearings, and valve train components. These points in the engine suffer from high friction and need oil for protection.

IGNITION SYSTEM

An *ignition system* is needed on a gasoline engine to ignite the air-fuel mixture. It must increase battery voltage enough to produce a spark in the engine combustion chamber. Refer to Fig. 1-24.

A fundamental ignition system consists of a spark plug, plug wire, ignition coil, switching device, and power source.

The *switching device* can be a set of contact points or a transistorized control unit that makes and breaks electrical current flow to the ignition coil. Its operation is timed with crankshaft rotation so that the spark occurs in the cylinder at the correct instant (end of compression stroke).

The *ignition coil* is used to step up battery voltage to over 40,000 volts. This is enough voltage to make the electricity jump the spark plug gap. The ignition coil fires every time the switching device STOPS current flow from the battery, Fig. 1-24.

The *spark plug* is the "match" that starts the air-fuel mixture burning in the combustion chamber. When the ignition coil "fires" and sends current through the spark plug wire, an electric *arc* (spark) forms at the tip of the spark plug. This makes the fuel and air start to burn, producing the power stroke.

STARTING SYSTEM

The *starting system* is needed to turn the engine crankshaft until the engine can begin running on its own power. It uses battery voltage, the ignition switch, a high current relay, and an electric motor to rotate the crankshaft, Fig. 1-25.

The *battery* stores chemical energy that can be changed into electrical energy. When the driver turns the *ignition switch* (start switch), the *solenoid* (high current relay) sends battery current to the starting motor.

The *starting motor* has a gear that engages a gear on the back of the crankshaft flywheel. The motor has enough *torque* (turning force) to spin the crankshaft through its four strokes until the engine starts and runs. Then, the driver releases the key and deactivates the starting system.

The *engine flywheel,* besides holding the large gear for the starter, smooths engine operation. It is very heavy and helps keep the crankshaft spinning between power strokes.

CHARGING SYSTEM

The *charging system* is needed to *recharge* (reenergize) the battery after starting system or other electrical system operation. The battery can become *discharged* (run down) after only a few minutes of starting motor operation.

Generally, the charging system is made up of the alternator and voltage regulator. Look at Fig. 1-26.

The *alternator* produces the electricity to recharge the battery. It is driven by a belt from the engine crankshaft pulley. The alternator sends current through the battery in reverse direction to reactivate the chemicals in the battery. This again prepares the battery for starting or other electrical loads.

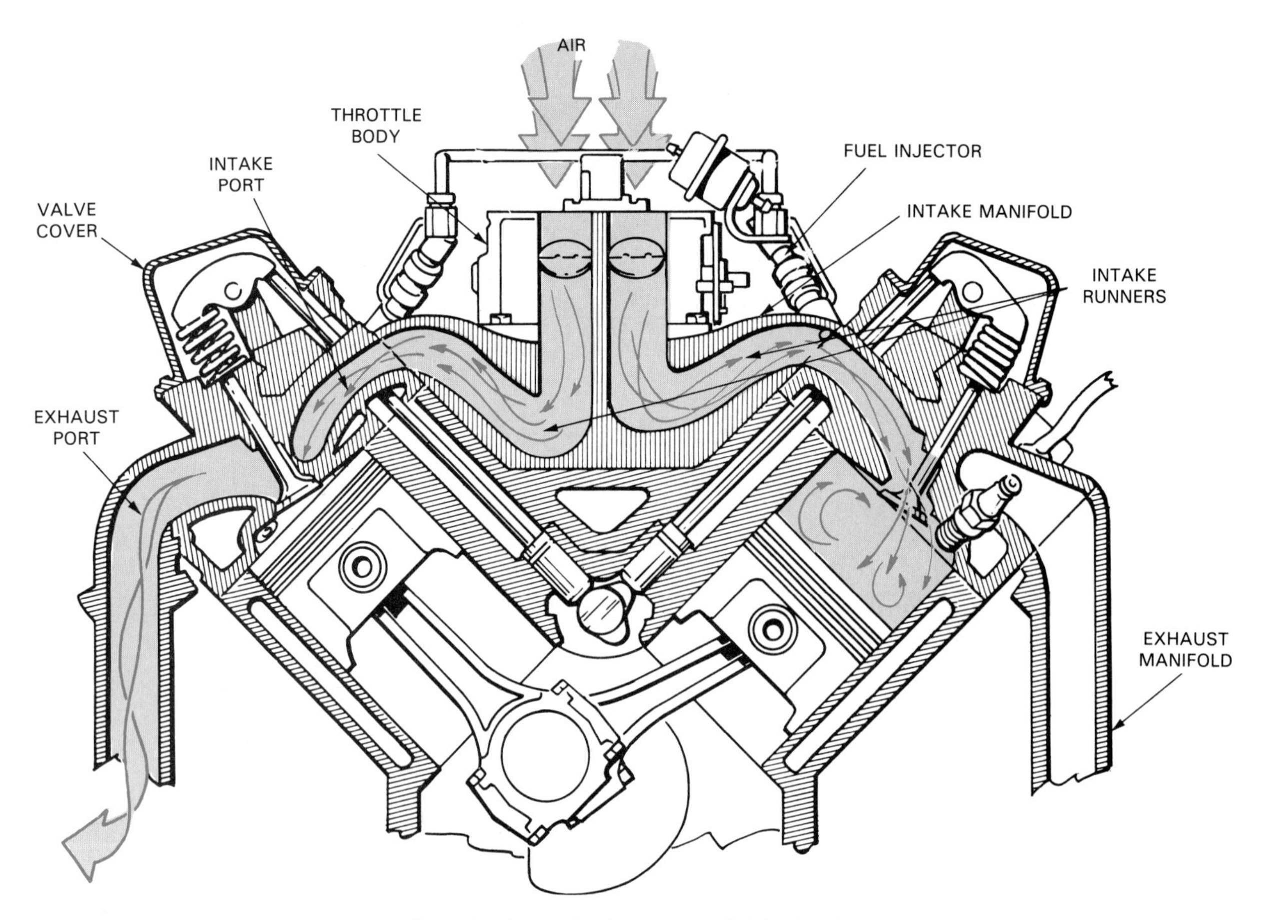

Fig. 1-21. Cutaway shows inside of engine. Note intake and exhaust manifolds. Intake manifold connects to intake ports in cylinder head. Exhaust manifolds bolt over exhaust ports in cylinder head. (General Motors)

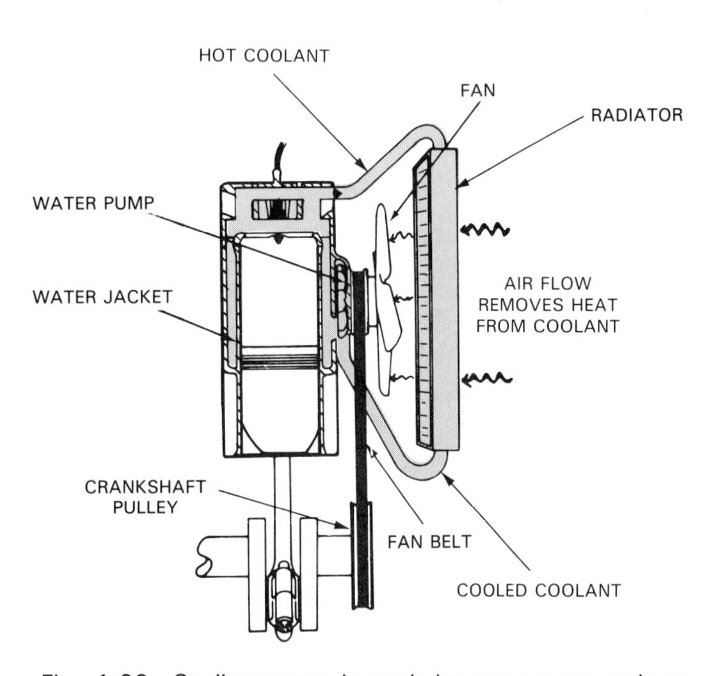

Fig. 1-22. Cooling system is needed to remove excess heat from engine and prevent severe engine damage. Water jackets allow coolant to flow around cylinders and in head. Water pump forces coolant to circulate through radiator and engine. Radiator dissipates heat into outside air. Fan pulls air through radiator. (Chrysler)

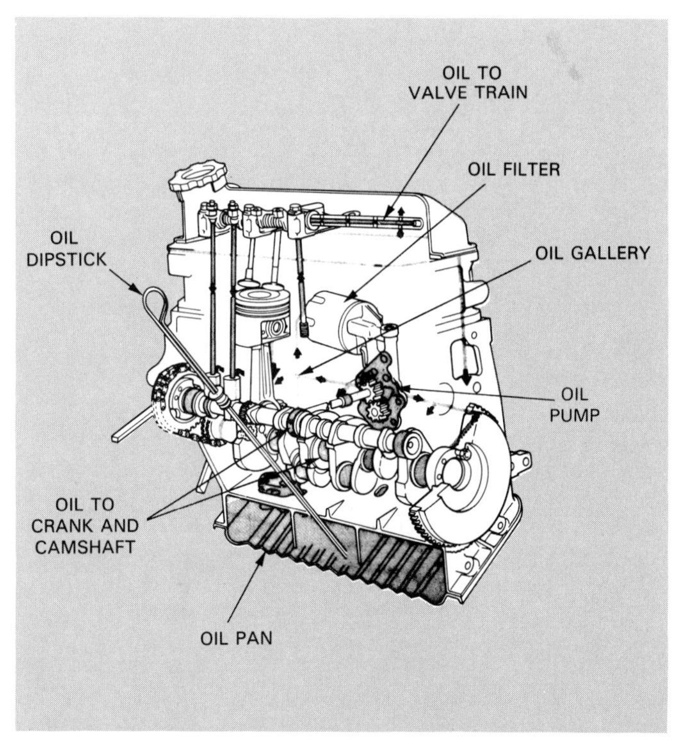

Fig. 1-23. Lubrication system prevents excess friction from damaging engine. An engine can be ruined very quickly without a lubrication system. Note part names. (Chrysler)

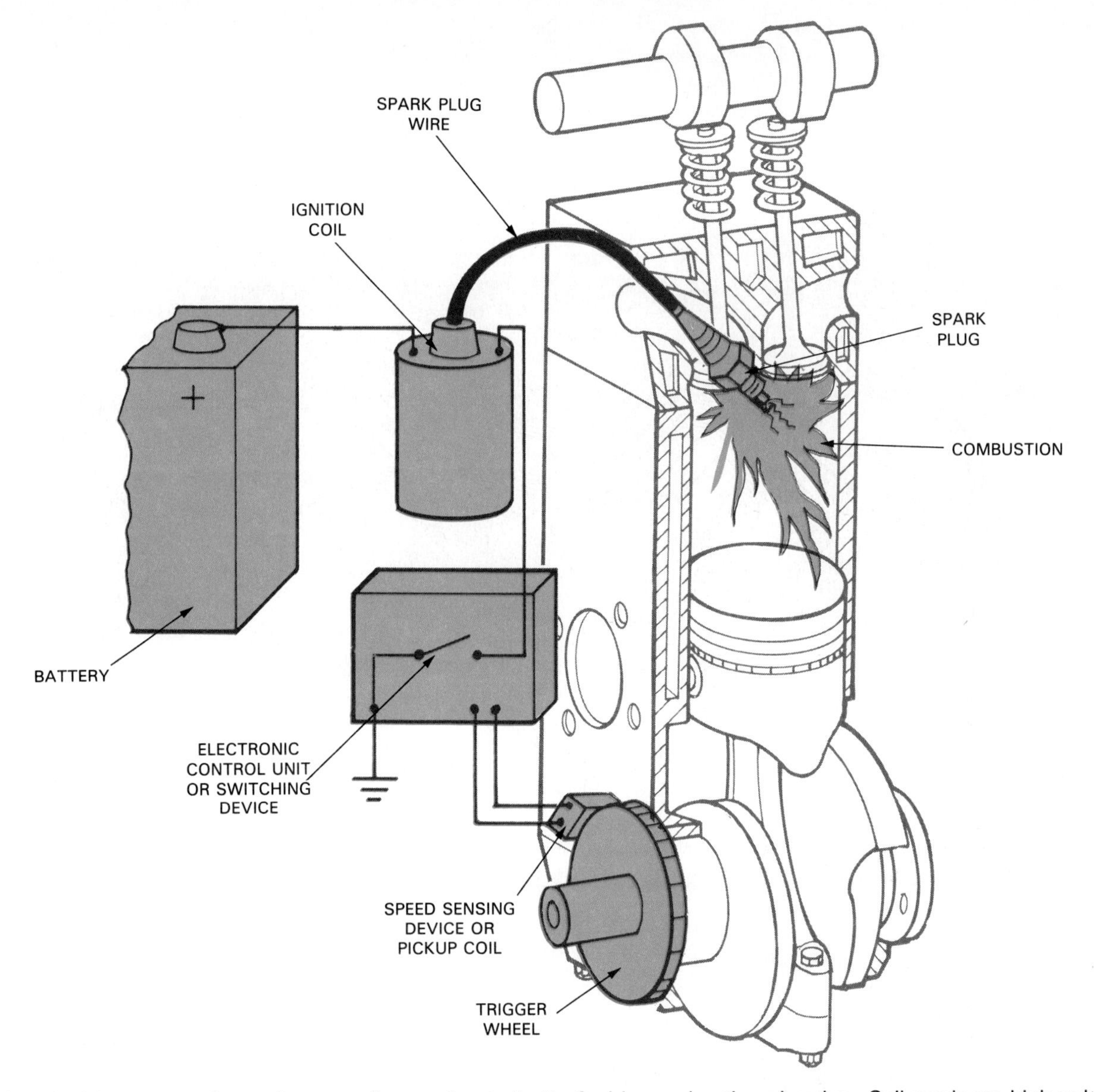

Fig. 1-24. Ignition system is used on gasoline engine to ignite fuel in combustion chamber. Coil produces high voltage for spark plug. When switching device breaks battery current to coil, coil and spark plug fire to start burning.

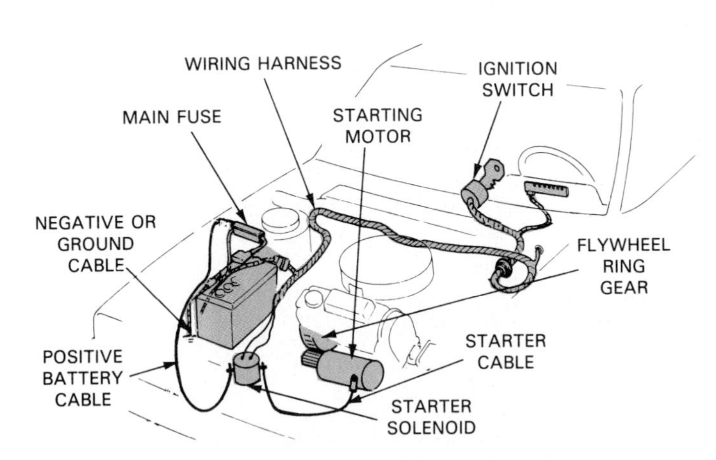

Fig. 1-25. Starting system rotates crankshaft until engine starts. Powerful electric starting motor has gear that meshes with gear on engine flywheel. Solenoid makes electrical connection between battery and starting motor when key is turned to start position.

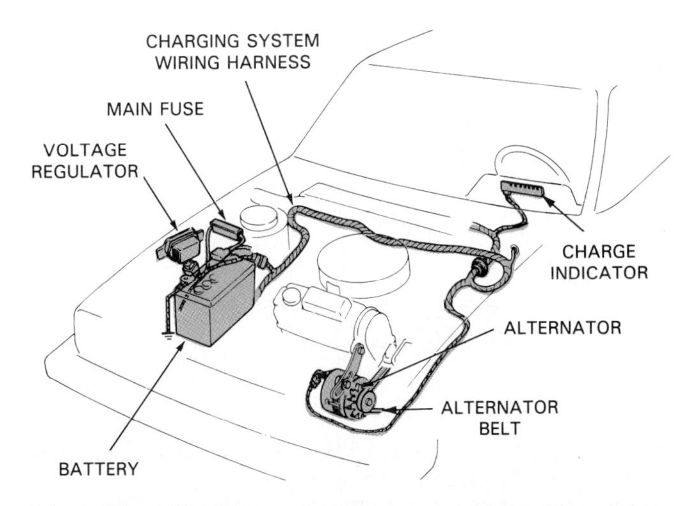

Fig. 1-26. Charging system forces current back through battery. This recharges battery and prepares it for starting motor operation and other electrical loads. (Honda)

The *voltage regulator* controls the electrical output of the alternator. It ensures that about 14.5 volts are produced by the alternator. Current then flows back into the battery, since battery voltage is only about 12.5 volts.

FUEL SYSTEM

The *fuel system* must meter the right amount of fuel (usually gasoline or diesel oil) into the engine for efficient operation under different conditions. At low speeds, it must meter a small amount of fuel and increase fuel metering as engine speed and load increases. It must also alter fuel metering with changes in engine temperature, load, and other variables.

There are three basic types of automotive fuel systems: carburetion, gasoline injection, and diesel injection.

Carburetor fuel system

A *carburetor fuel system* uses engine vacuum (suction) to pull fuel into the engine. As illustrated in Fig. 1-27, airflow through the carburetor and normal vacuum produced on the engine's intake strokes moves fuel out of the carburetor.

A *fuel pump* is used to force fuel out of the fuel tank and into the carburetor. The pump does not force fuel into the engine, however. The fuel pump is normally a mechanical pump powered by the engine, but can also be an electric pump.

The *carburetor* must control fuel and airflow into the engine. It has a *throttle valve* that is connected to the driver's "gas pedal." When the gas pedal is pressed, it opens the throttle valve. This lets more air and fuel flow into the engine for increased power. Carburetors can only be found on older vehicles.

Gasoline injection system

A *gasoline injection system* uses fuel pump pressure to spray fuel into the engine intake manifold. A basic system is pictured in Fig. 1-28.

The *gasoline injector* is simply an electrically-operated fuel valve. When energized by the computer, it opens and squirts fuel into the intake manifold. When not energized, it closes and prevents fuel entry into the engine.

An *electric fuel pump* forces fuel to the injector. A constant pressure is maintained at the injector opening.

A *computer* or *electronic control unit* is used to regulate when and how long the injector opens. It uses electrical information from various sensors to analyze the needs and operating conditions of the engine.

The *engine sensors* monitor various operating conditions: engine temperature, speed, load, etc. They send data back to the computer in the form of electrical signals.

For example, a temperature sensor may allow a large amount of electricity to flow back to the computer when the engine is cold. It may prevent electrical flow when the engine is hot. In this way, the computer can determine whether more or less fuel is needed and whether the injector should be opened for a long or short period of time.

Modern gasoline injection systems open the injector when the engine intake valve opens. Then, fuel is partially forced into the combustion chamber by pump

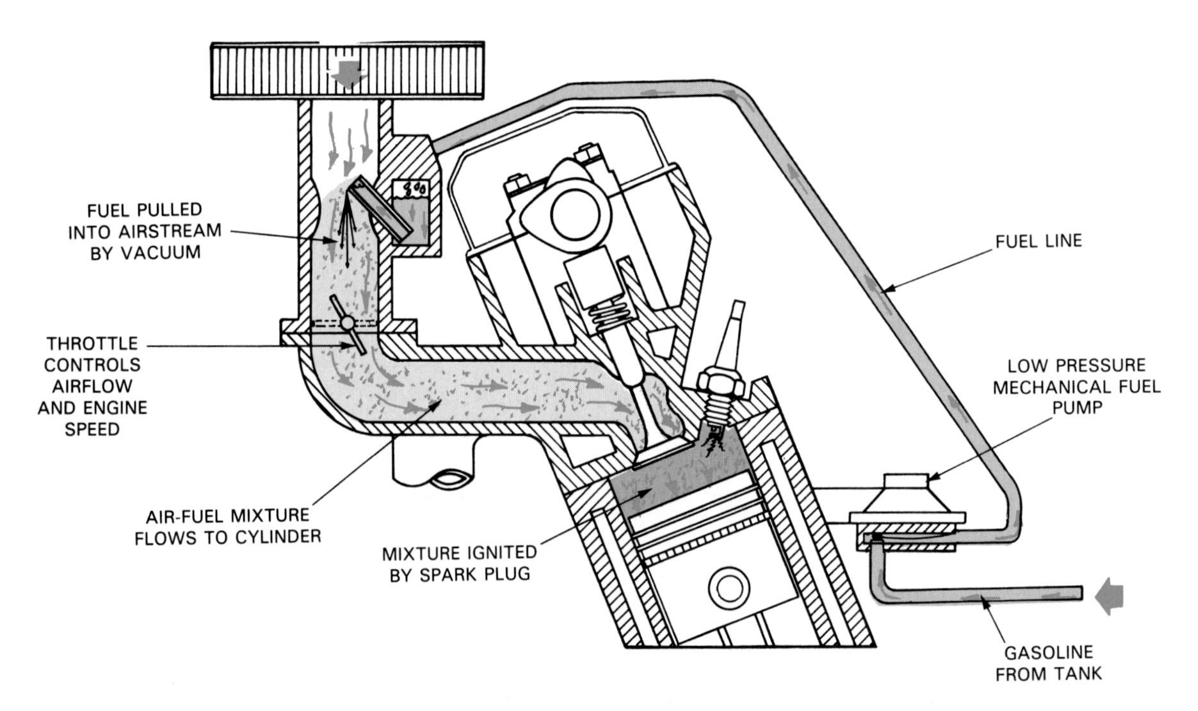

Fig. 1-27. Study basic action of carburetor type fuel system. Airflow into engine pulls fuel out of carburetor. Mechanical fuel pump fills carburetor with fuel but does not force fuel into intake manifold.

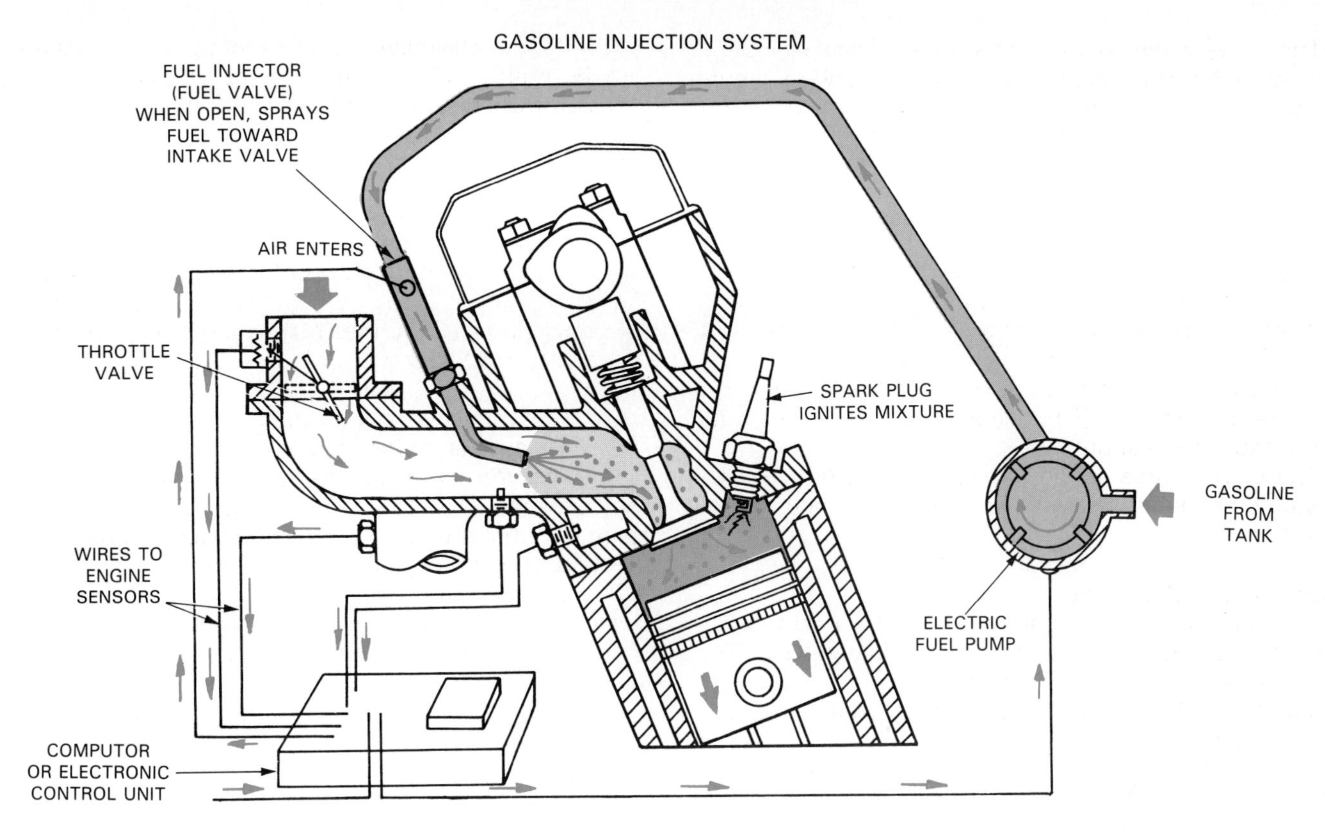

Fig. 1-28. Electronic fuel injection uses pressure from electric pump to spray fuel into engine intake manifold. Sensors monitor various conditions and send different electrical values back to computer. Computer can then open and close injector faster or slower to change amount of fuel entering engine.

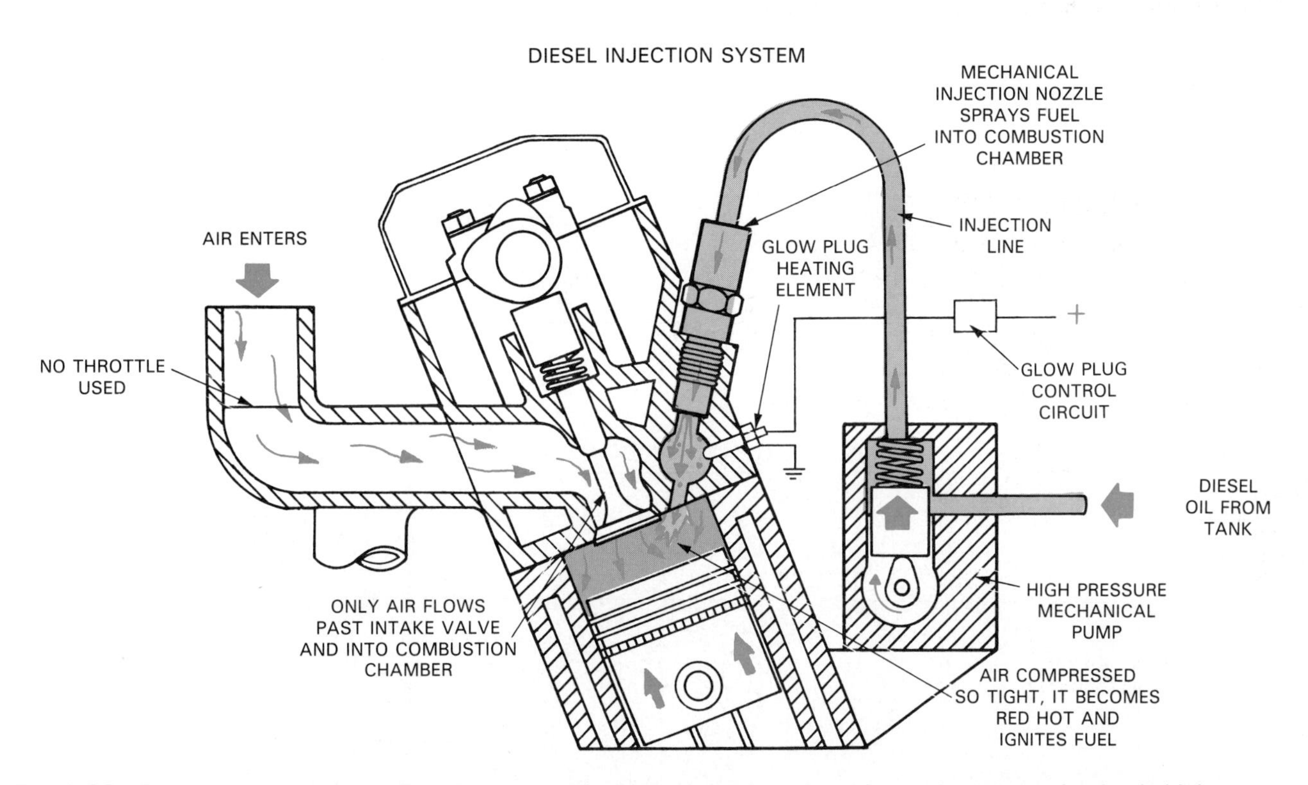

Fig. 1-29. Diesel injection is primarily mechanical. Mechanical injection pump forces fuel into spring-loaded injector nozzle. Pressure opens injector and fuel sprays directly into combustion chamber, not intake manifold as with gasoline injection. Also note that diesel does not use a throttle valve to control airflow nor a spark plug to ignite the fuel. High compression stroke pressure heats air in combustion chamber until air is hot enough to ignite fuel. Glow plug heats air for cold starting of diesel.

pressure. This helps control how much fuel enters the cylinder and also increases combustion efficiency, as you will learn in later chapters.

Diesel injection system

A *diesel injection system* forces fuel directly into the engine's combustion chamber, as shown in Fig. 1-29. The heat resulting from highly compressed air, NOT a spark plug, is then used to ignite and burn the fuel.

A diesel injection system basically consists of an injection pump, injector, and glow plug system.

The *injection pump* is primarily a high pressure, mechanical pump. It is powered by the engine and forces fuel to the diesel injector under very high pressure. A conventional fuel pump feeds fuel from the tank to the injection pump.

The *diesel injector* is simply a spring-loaded valve. It is normally closed and blocks fuel flow. However, when the injection pump forces fuel into the injector under high pressure, it opens the injector.

Note in Fig. 1-29 that a diesel does NOT use a throttle valve NOR a spark plug. When the engine valve opens, a full charge of air is allowed to flow into the cylinder. Then, on the compression stroke, the air is squeezed until it is ''red hot.'' As soon as the fuel is injected into the hot air, it begins to burn and expand.

Engine power and speed is controlled by the injection pump—more fuel increases speed and power and vice versa.

A *glow plug system* is used to aid cold starting of a diesel. The *glow plug* is an electrical heating element that warms the air in the combustion chamber. This helps the air become hot enough to start combustion.

COMPUTER SYSTEM

Modern cars use a computer system to increase vehicle efficiency. It can control the ignition system, fuel system, transmission or transaxle, emission control systems and other systems. Fig. 1-30 shows a simple diagram of a automobile's computer system.

A car's computer control system works something like the human body's central nervous system. For example, if your finger touches a hot stove, the nerves (SENSORS) in your hand send a chemical-electrical signal back to your brain. Your brain (COMPUTER) can then analyze these signals and decide that you are in pain. Your brain (computer) quickly can then activate your muscles (CONTROL DEVICES) to quickly pull your hand away from the hot stove.

A computer control system works in the same manner and for simplicity, it can be divided into three subsystems: sensing system, analyzing system, and control system.

The *sensing system,* mentioned briefly under fuel injection, checks various operating conditions using sensors. A *sensor* is a device that can change its electrical signal with a change in a condition (temperature, pressure, airflow, etc.). It can report back to the computer so that the computer can alter the operation of another component as needed.

Sensors might measure intake manifold vacuum, throttle opening, engine rpm, transmission gear position, road speed, turbocharger boost pressure, and other conditions. The sensing system sends different electrical current values back to the computer.

The *computer* must perform two functions: it must

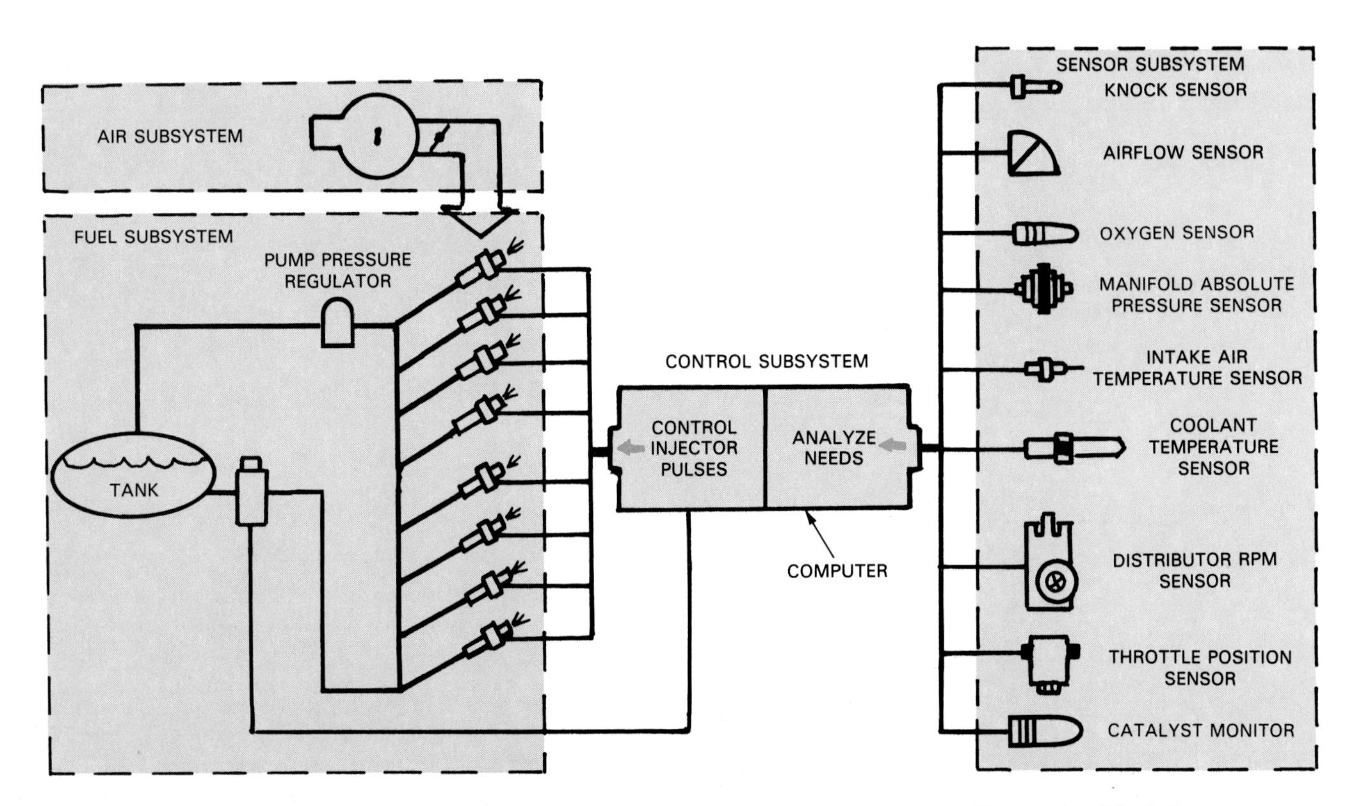

Fig. 1-30. Study basic layout and action of modern computer control system. This is simplified diagram.

analyze input from the various sensors and it must take action to control specific vehicle systems. To do this, a computer contains microscopic electrical circuits that can collect information, store information, and control output by using the collected and stored information.

The *control system* serves as the "hands" of the computer. It allows the computer to move parts, open the injectors, close the throttle, turn on the fuel pump, and do other tasks needed to increase efficiency. Electric motors, electric solenoids or relays, and switches are needed to do the work for a computer. This will be discussed in detail later in this textbook.

EMISSION CONTROL SYSTEMS

Emission control systems are designed to reduce the amounts of harmful chemicals (emissions) that enter the atmosphere from a car. There are several types of emission control systems:

1. PCV SYSTEM—The *positive crankcase ventilation system* pulls engine crankcase fumes into the intake manifold so they can be burned before entering the atmosphere.
2. HEATED AIR INLET—This is an air cleaner that uses exhaust manifold heat to warm the air entering the engine to improve combustion and reduce emissions.
3. FUEL EVAPORATION CONTROL—This system uses a charcoal-filled canister to collect and store gasoline fumes from the fuel tank and carburetor so the fumes can be burned in the engine after starting.
4. EGR SYSTEM—The *exhaust gas recirculation system* injects burned exhaust gases into the engine to lower combustion temperatures and reduce one form of pollution in the engine exhaust.
5. AIR INJECTION SYSTEM—This sytem forces air into the exhaust stream leaving the engine to help burn any fuel that enters the exhaust system.
6. CATALYTIC CONVERTER—This is an "afterburner" that burns any fuel that flows into the exhaust system.

Many of these systems work together, all reducing their share of harmful emissions. With new cars, the computer is an important part that reduces pollution. It improves the efficiency of many systems.

DRIVE LINE

The *drive line* uses power to turn the drive wheels of the car. Drive lines vary with the use of both rear- and front-wheel drive. Figs. 1-31 and 1-32 show simplified drive lines.

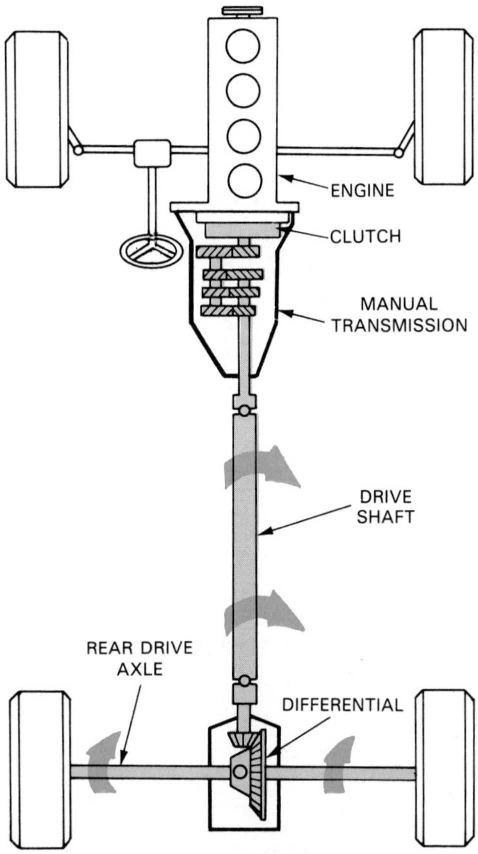

Fig. 1-31. Drive line uses engine power and crankshaft rotation to turn car's drive wheels. This is a front-engine, rear-wheel drive car. Note clutch between engine and transmission.

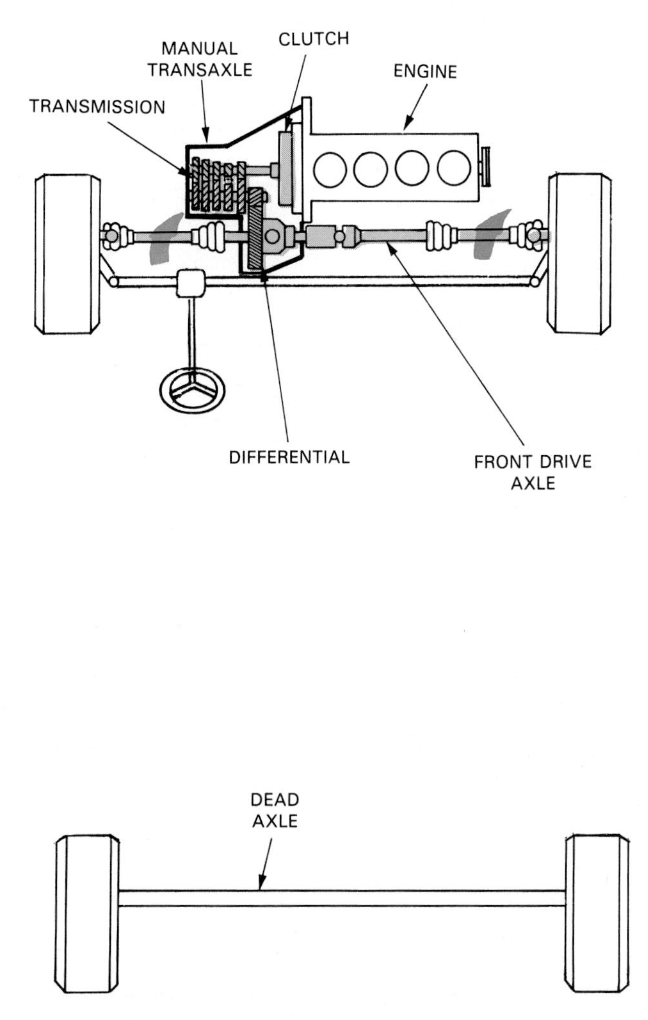

Fig. 1-32. This is a front-engine, front-wheel drive arrangement. Note differential and transmission in one housing. Swing front axles extend out from transaxle to turn drive wheels.

Clutch

The *clutch* allows the driver to engage or disengage the engine and drive line. It mounts on the engine flywheel, between the engine and transmission or transaxle. A clutch is needed when the car has a manual transmission.

Transmission

The *transmission* uses a series of gears to allow the amount of torque going to the drive wheels to vary. The driver can shift gears to change the ratio of crankshaft revolutions and drive wheel rotation. When first accelerating, more torque is needed to get the car moving. Then, at higher road speeds, less torque is needed to maintain road speed. Engine rpm also needs to be reduced at highway speeds.

A *manual transmission* is shifted by hand. Levers, cables, or rods connect the driver's shift lever to the internal parts of the transmission.

An *automatic transmission* uses a *hydraulic* (fluid pressure) system to shift gears. It uses a *torque converter* (fluid clutch) and special planetary gearsets to provide automatic operation.

A *transaxle* is a transmission and a *differential* (axle drive mechanism) combined into one housing. It is commonly used with front-wheel drive cars but may also be found on rear-engine, rear-wheel drive vehicles.

A *drive shaft* is used with a front-engine, rear-wheel drive car. It connects the transmission to the differential.

A *drive axle* connects the differential to the drive hubs or wheels. On most rear-wheel drive cars, it is a solid, steel axle. On cars with a transaxle, it is a flexible shaft extending from the transaxle to the front wheel hubs.

SUMMARY

This chapter has reviewed the basic operation of a four-stroke cycle, piston engine. A car engine is an internal combustion engine because it burns fuel inside its combustion chamber.

The block holds the other parts. The piston and rod transfer combustion pressure to the crankshaft. The crankshaft converts the up and down action of the piston into usable rotary motion.

The cylinder head bolts on top of the block. It contains the valves that control flow in and out of the cylinder. The camshaft operates the valves so that they open and close to correspond to piston action, producing the four-stroke cycle.

The intake stroke draws air-fuel mixture into the engine. The compression stroke squeezes the mixture and readies it for burning. The power stroke ignites the mixture and the expanding gases push the piston down with tremendous force. The exhaust stroke purges the burned gases to prepare the cylinder for another intake stroke.

It takes two crankshaft revolutions to complete one four-stroke cycle and one power producing stroke. The flywheel helps keep the crankshaft spinning on the non-power producing strokes.

Various systems are needed to keep the engine running. The cooling system removes excess combustion heat and prevents engine damage. The lubrication system also prevents engine damage by reducing friction between moving engine parts.

The ignition system is needed on a gasoline engine to ignite the fuel and start it burning. A diesel engine does not need a spark plug as does a gasoline engine. It uses high compression stroke pressure to heat the air in the cylinder enough to start combustion. Also unlike a gasoline engine, a diesel does not use a throttle valve to control engine speed. The amount of fuel injected into the cylinder controls power output.

The three types of fuel systems are the carburetor, gasoline injection, and diesel types. Gasoline injection has replaced the older carburetor because it can increase fuel economy by closer control of fuel use by each cylinder.

The starting system uses a powerful electric motor to turn the engine flywheel until the engine can run on its own power. The charging system pushes current back through the battery to reverse the chemical action and recharge the battery.

On-board computers are now very common. They are similar in principle to our body's central nervous system. They use sensors to monitor various conditions. Then the computer can calculate what must be done to maintain maximum efficiency. The computer can control the operation of the fuel system, ignition system, transmission, and other devices.

Emission control systems can also be computer-controlled. They reduce harmful pollutants that enter the atmosphere. Some prevent fuel vapors from evaporating into the air. Others burn any fuel that leaves the engine through the exhaust and others just increase engine efficiency to reduce air pollution.

The drive line transfers engine power to the drive wheels. Both front-wheel and rear-wheel drive types are found on today's cars. Front-wheel drive cars use a transaxle with the transmission and differential combined together.

KNOW THESE TERMS

Engine, Internal combustion, Block, Piston, Piston rings, Connecting rod, Crankshaft, Cylinder head, Valves, Camshaft, Valve springs, Timing belt, Cylinder, Main caps, Main bore, Deck, Crankcase, Combustion, Compression rings, Oil rings, Wrist pin, TDC, BDC, Valve guides, Valve ports, Valve seats, Intake valve, Exhaust valve, Four-stroke cycle, Valve train, Cooling system, Lubrication system, Ignition system, Starting system, Charging system, Carburetor, Gasoline injection, Diesel injection, Glow plug, Engine sensors, Computer, Emission control systems, Drive line, Clutch, Manual transmission, Automatic transmission, Transaxle.

REVIEW QUESTIONS—CHAPTER 1

1. The __________ is the main supporting structure of an engine.
2. These hold the crankshaft in the bottom of the engine block.
   a. Cam bearings.
   b. Main caps.
   c. Decks.

d. Rod bearings.
3. Water jackets surround the cylinders in most engines. True or False?
4. What is the basic function of a piston?
5. _________ _________ keep combustion pressure from blowing into the crankcase and they also keep _________ out of the combustion chamber.
6. Why is a piston pin or wrist pin needed?
7. The crankshaft converts the _________ and _________ motion of the piston into usable _________ motion.
8. Describe the basic parts of a cylinder head.
9. Which engine valve is larger and which is smaller?
10. In your own words, summarize the four-stroke cycle.
11. The _________ opens the valves and the _________ _________ closes the valves.
12. Why are the cooling and lubrication systems important?
13. An electric ignition system can be found on both gasoline and diesel engines. True or False?
14. Describe the three types of fuel systems.
15. In your own words, how does a computer control system operate?

ASE CERTIFICATION-TYPE QUESTIONS

1. Technician A says an engine's timing belt turns the camshaft at one-half engine speed. Technician B says an engine's timing belt turns the camshaft at one-fourth engine speed. Who is right?
   (A) A only.
   (B) B only.
   (C) Both A & B.
   (D) Neither A nor B.
2. All of the following are major parts of a modern, multicylinder automotive engine EXCEPT:
   (A) valve springs.
   (B) spool valves.
   (C) piston rings.
   (D) connecting rods.
3. An _________ is the flat surface machined on the top of the block for the cylinder head.
   (A) cam housing
   (B) engine block deck
   (C) oil gallery
   (D) None of the above.
4. Technician A says the big end of a connecting rod fits around the piston pin. Technician B says the big end of a connecting rod fits around the crankshaft. Who is right?
   (A) A only.
   (B) B only.
   (C) Both A & B.
   (D) Neither A nor B.
5. Technician A says the engine's oil rings always fit into the top groove on the pistons. Technician B says the engine's oil rings are normally located in the lowest groove on the pistons. Who is right?
   (A) A only.
   (B) B only.
   (C) Both A & B.
   (D) Neither A nor B.
6. All of the following are basic parts of an automotive engine's piston assembly EXCEPT:
   (A) connecting rod.
   (B) piston rings.
   (C) crankshaft journal.
   (D) wrist pin.
7. The intake valve is _________ than the exhaust valve.
   (A) smaller
   (B) the same size
   (C) larger
   (D) All of the above.
8. While discussing the operation of a four-stroke cycle engine, Technician A says every two up and down strokes of the engine's piston results in one power producing stroke. Technician B says every four up and down strokes of the engine's piston results in one power producing stroke. Who is right?
   (A) A only.
   (B) B only.
   (C) Both A & B.
   (D) Neither A nor B.
9. An ignition coil is used to increase battery voltage to over _________ volts.
   (A) 10,000
   (B) 15,000
   (C) 20,000
   (D) 40,000
10. All of the following are modern automotive emission control systems EXCEPT:
   (A) air injection system.
   (B) PCV system.
   (C) intake valve.
   (D) evaporative emissions control system.

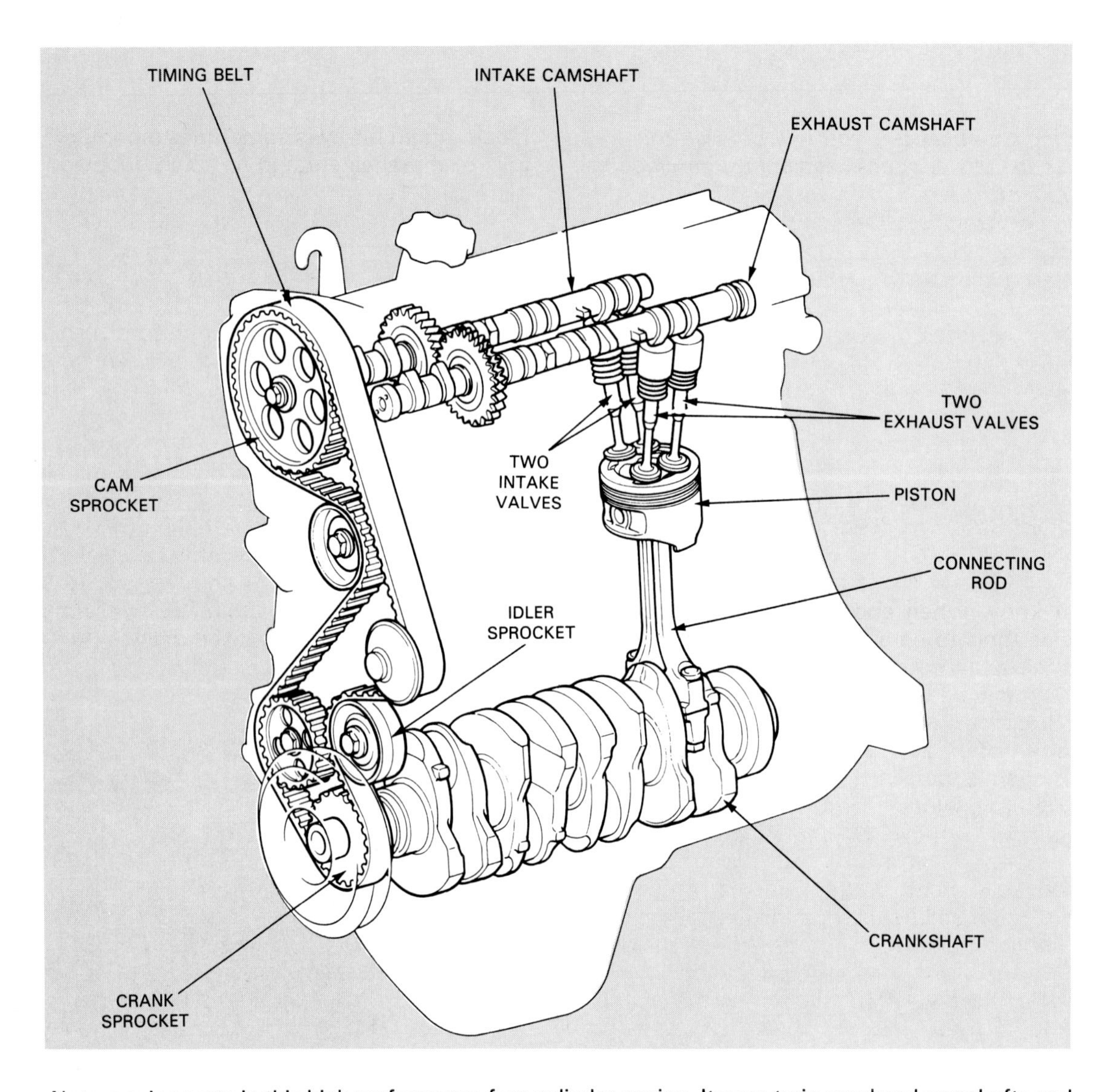

Note moving parts in this high performance four-cylinder engine. It uses twin overhead camshafts and four-valves per cylinder to increase horsepower output. Engine produces tremendous power for its small size. (Toyota)

Chapter 2

Engine Service Tools and Equipment

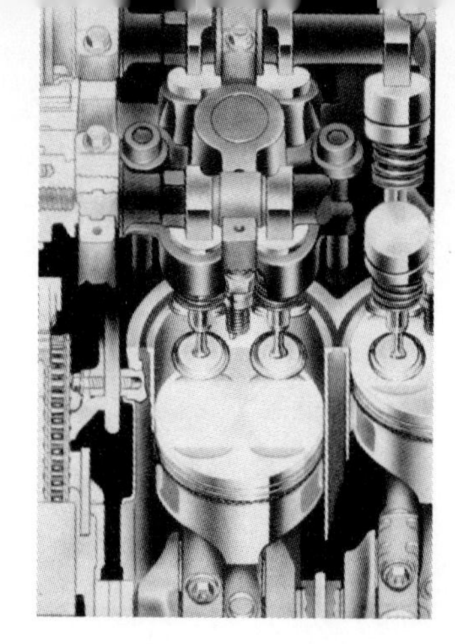

After studying this chapter, you will be able to:
- *Summarize the use of specialized engine service tools and equipment.*
- *Select the right engine service tool when doing actual engine repairs.*
- *Explain the most commonly used engine cleaning tools.*
- *Use engine measuring tools: micrometer, caliper, dial indicator, etc.*
- *Explain how to use typical engine electrical testing devices.*
- *Describe the function of commonly used engine test equipment.*
- *More fully understand later text chapters that explain the use of engine service tools and equipment in more detail.*

You must know when and how to use specialized engine service tools to be a productive technician. Today's car engines are supported by very complicated systems. To diagnose if a problem is an internal, mechanical problem or a problem with an engine system, requires the use of very specialized testing equipment. Also, with modern engine designs, specialized tools are needed to rebuild engines.

This chapter will overview the selection and utilization of engine service tools, cleaning tools, measuring tools, electrical test equipment, and engine test equipment. This will give you a solid background when learning to diagnose, service, and repair today's "high tech" engines.

Basic hand tools are NOT covered in this chapter. This text is written for students specializing in engine repair. If you need more "basics," refer to a general text, like MODERN AUTOMOTIVE TECHNOLOGY, published by Goodheart-Willcox.

ENGINE SERVICE TOOLS AND EQUIPMENT

This section of the chapter will describe the function of specialized engine service tools and equipment.

Engine stand

An *engine stand* holds an engine for easy disassembly and reassembly. It is a metal framework on small caster wheels. The cylinder block bolts to the stand.

The engine stand will hold the engine at waist level. It will also allow the engine to be rotated into different positions for access to the top or bottom of the cylinder block. A catch tray is sometimes mounted on the lower part of the engine stand to catch dripping coolant and oil, Fig. 2-1.

Engine crane

A portable *engine crane* is used to remove (raise) and install (lower) car engines, Fig. 2-2. It has a hydraulic hand jack for raising and a pressure release valve for lowering. An engine crane is also handy for lifting heavy engine parts (intake manifolds, cylinder heads), transmissions, and transaxles.

Ridge reamer

A *ridge reamer* is used to cut out the metal lip formed at the top of a worn cylinder. As shown in Fig. 2-3, it has small cutter blades that will remove a cylinder ridge. The top of the cylinder is not exposed to the rubbing, wearing action of the piston rings. It will remain unworn, while the lower section of the cylinder will wear larger

Fig. 2-1. Engine stand is a "must" when disassembling or reassembling an engine. Catch pan will keep floor clean. Engine can be rotated into different positions for easy access to crankcase, cylinders, lifter valves, cam bore, etc. (OTC)

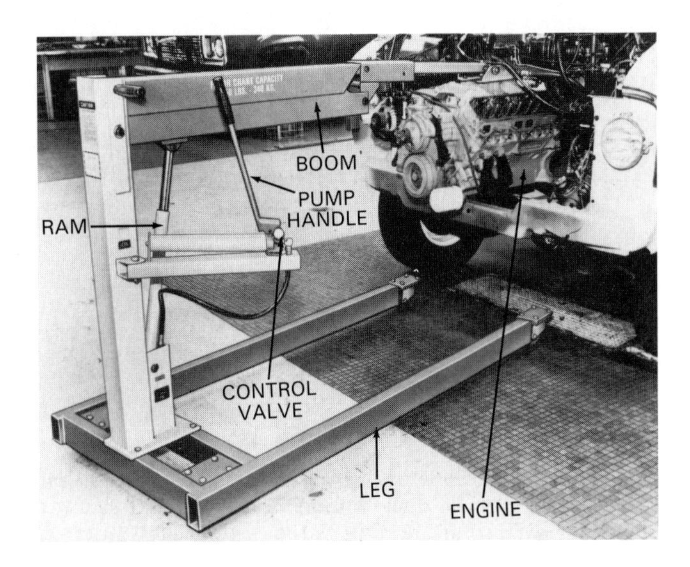

Fig. 2-2. Engine crane is for lifting engines and other heavy assemblies. Here crane is being used to replace engine in a van. Sometimes, engine can be removed through large side doors without grille and other body panel removal. (Owatonna Tools)

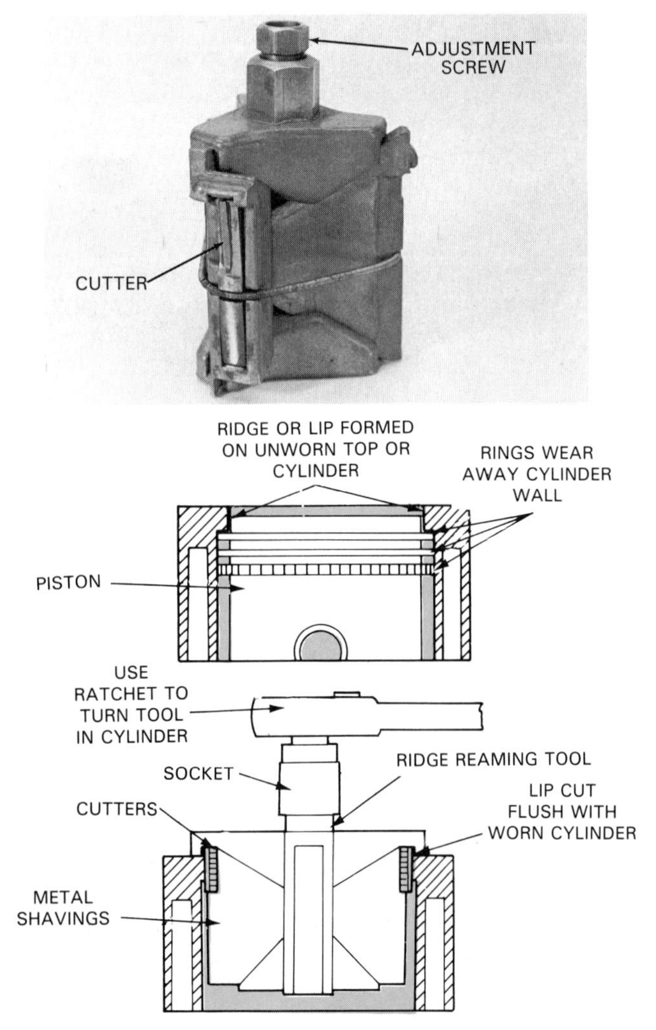

Fig. 2-3. A worn engine cylinder will have a metal lip formed at top where rings do not rub on and wear cylinder. Ridge reamer will cut this lip out and allow piston removal.

in diameter. If the ring ridge is NOT removed, the piston can be damaged when you are trying to push the piston out during engine teardown.

Cylinder hone

A *cylinder hone* is used to *deglaze* (lightly roughen or texture) a cylinder wall during an engine rebuild. It has fine grit stones that are spun inside the cylinder using an electric drill. The drill is pulled up and down to produce a crosshatched hone pattern on the cylinder wall. Then, when new piston rings are installed, the rings will wear into the cylinder wall surface and produce a good ring seal.

There are several types of cylinder hones. Each will be discussed in more detail in chapters on engine bottom end service, Fig. 2-4.

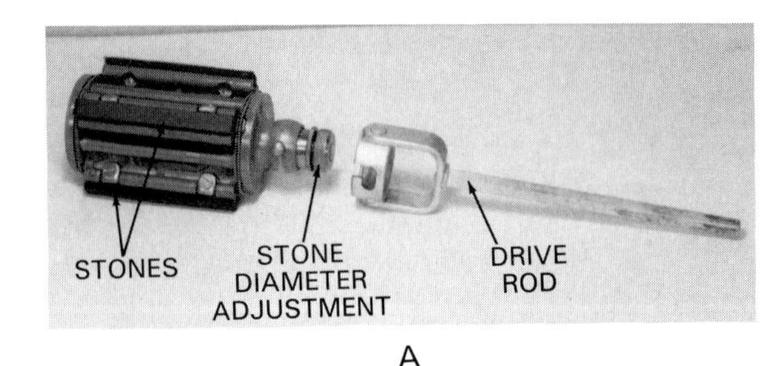

A

B

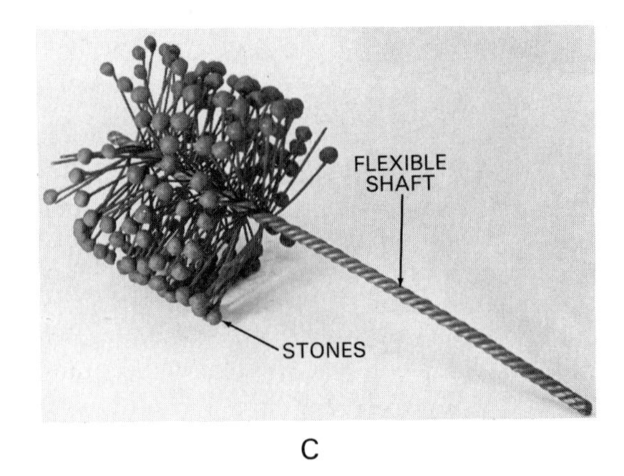

C

Fig. 2-4. Cylinder hones are needed to prepare cylinder wall surface so new rings will seal and engine will not consume oil or smoke. A—Rigid hone for heavy honing. B—Flex hone. C—Brush hone.

Ring groove cleaner

A *ring groove cleaner* will remove carbon deposits from inside piston ring grooves, Fig. 2-5. It is turned or rotated around the piston by hand. Small cutters, the same size as the ring grooves, will then scrape out carbon from the piston grooves.

If carbon is NOT removed from the piston ring grooves, the new rings could be forced out against the cylinder wall after engine operation and piston heat expansion. The rings and cylinder could be scored and ruined.

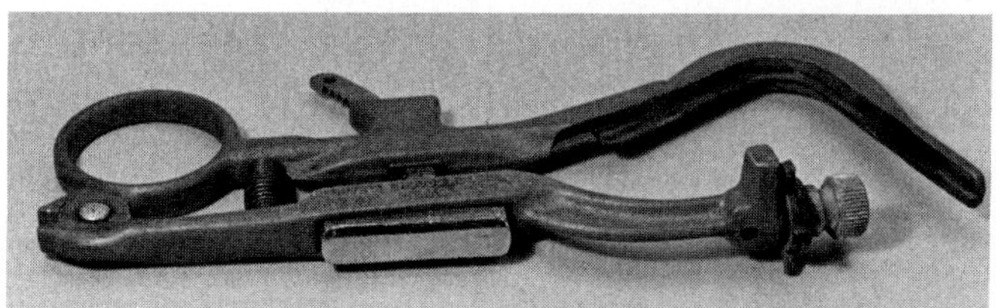

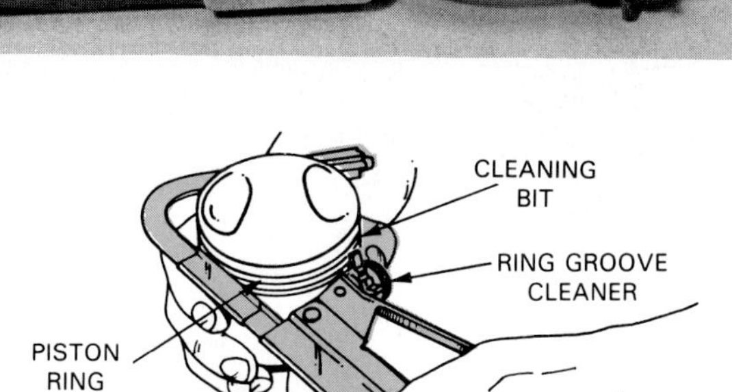

Fig. 2-5. Used pistons can have hard carbon deposits inside ring grooves. Ring groove cleaner is needed to remove these deposits from grooves so new rings fit inside grooves properly. (Lisle Tools)

Piston ring expander

A *piston ring expander* is used to remove and install piston rings. Compression rings are normally made of brittle, easy to break, cast iron. When installing new piston rings, be very careful not to overexpand and break the rings. A piston ring expander will help prevent ring breakage while speeding engine assembly, Fig. 2-6.

Piston ring compressor

A *piston ring compressor* will squeeze the piston rings into their grooves so that the piston assembly can be installed in its cylinder. It is simply a spring steel clamp that can be tightened around the piston and piston rings. A wooden hammer handle is commonly used to force the piston-rod assembly into the block while guiding the rod over the crankshaft. See Fig. 2-7.

Piston knurler

A *piston knurler* will form (not cut) grooves in the piston skirt to increase the outside diameter of the piston. Knurling is needed on slightly worn pistons to restore their size to within specifications. The knurled grooves

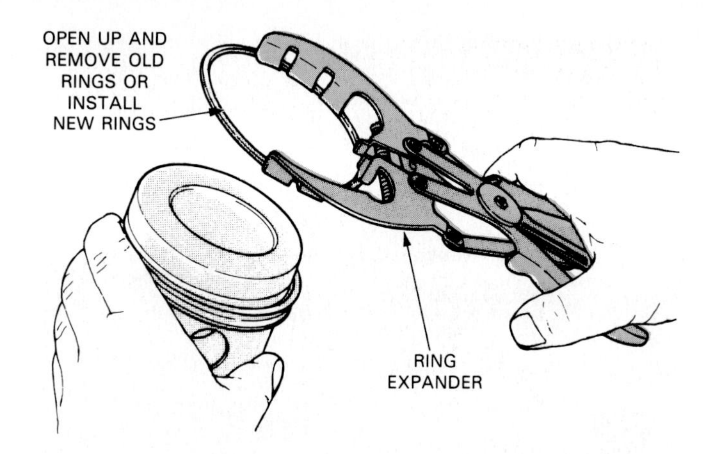

Fig. 2-6. Compression rings are very brittle and will break easily. Ring expander will make ring installation quicker and can help prevent ring breakage. (Chrysler)

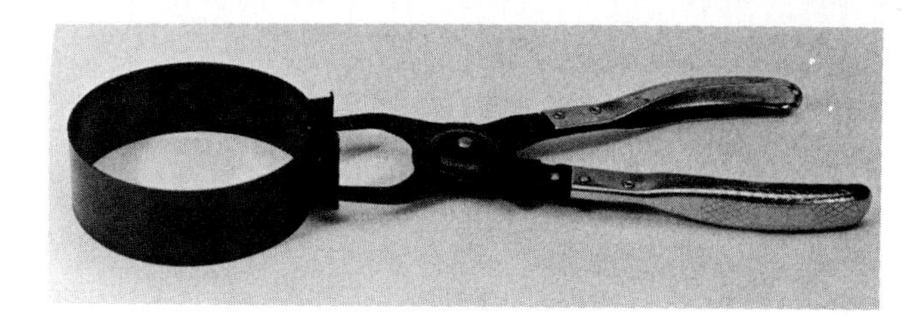

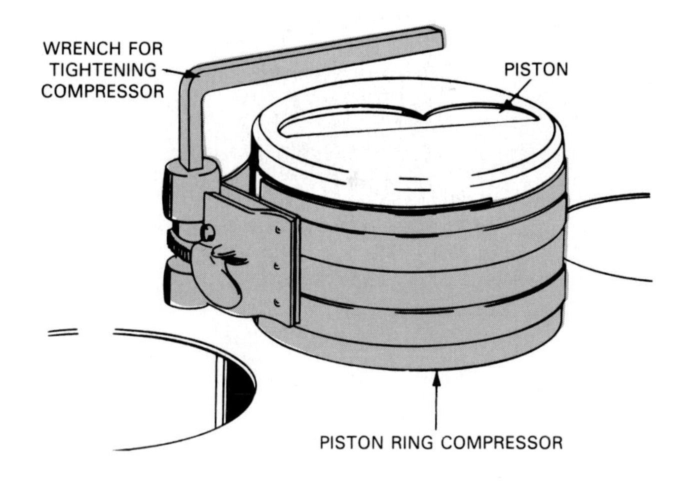

Fig. 2-7. Ring compressor is needed to hold rings in their grooves as piston assembly is forced down into cylinder block. (Cadillac)

also retain oil to prevent further piston and cylinder wear. Look at Fig. 2-8.

Flywheel lock

A *flywheel lock* will keep the engine crankshaft from turning during engine disassembly or reassembly. You might need to lock the flywheel and crankshaft when torquing flywheel bolts or the front damper bolt, for example. Refer to Fig. 2-9.

Cylinder head stand

A *cylinder head stand* will hold the head while servicing the valves and valves seats. A cylinder head can be

Fig. 2-8. Knurling tool will smash grooves in piston, making its outside diameter larger. Used pistons can then be restored to within specs. (Deere & Co.)

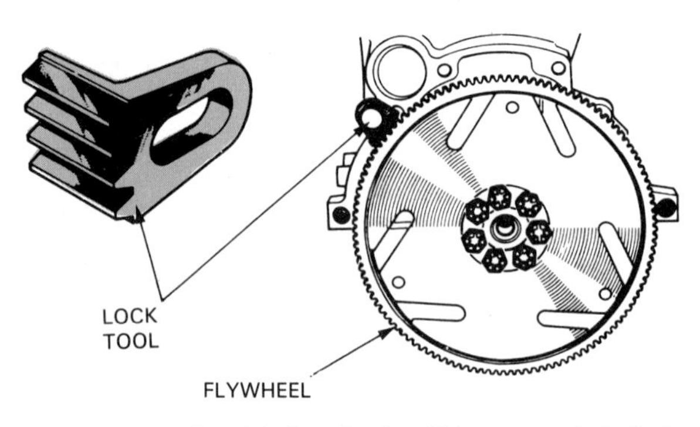

Fig. 2-9. Flywheel lock is handy. It will keep crankshaft from turning while torquing flywheel bolts or front damper. (Renault)

heavy and clumsy. The stand will make service work much easier, Fig. 2-10.

Valve spring compressor

A *valve spring compressor* allows a valve assembly to be removed from or installed in a cylinder head, Fig. 2-11. It is a clamp-like device that squeezes the valve spring together so that the valve keepers can be removed. One end of the compressor fits on the valve head. The other end fits over the spring retainer.

A valve spring compressor variation is shown in Fig. 2-12. It is for servicing modern overhead cam cylinder heads where the valve spring operates in a pocket in the cylinder head.

Valve adjusting wrench

A *valve adjusting wrench* is a special tool for turning

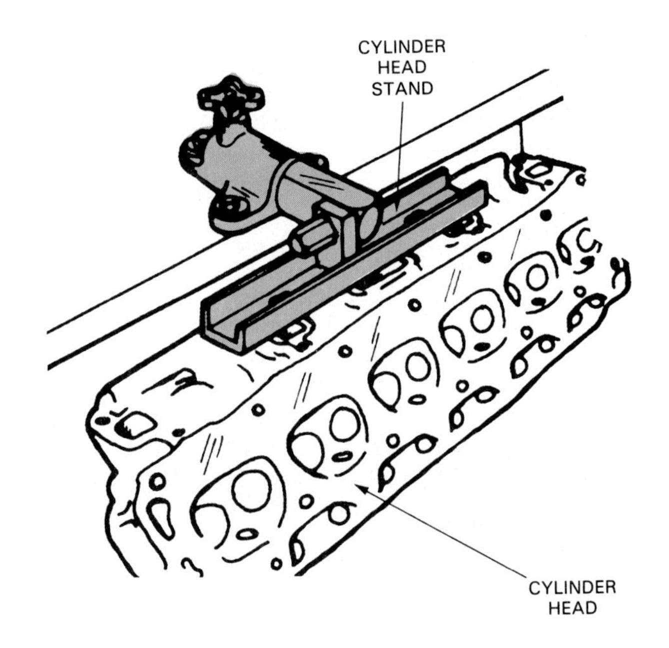

Fig. 2-10. Cylinder head stand will make valve service more convenient. Head can be swiveled into various positions. (K-Line Tools)

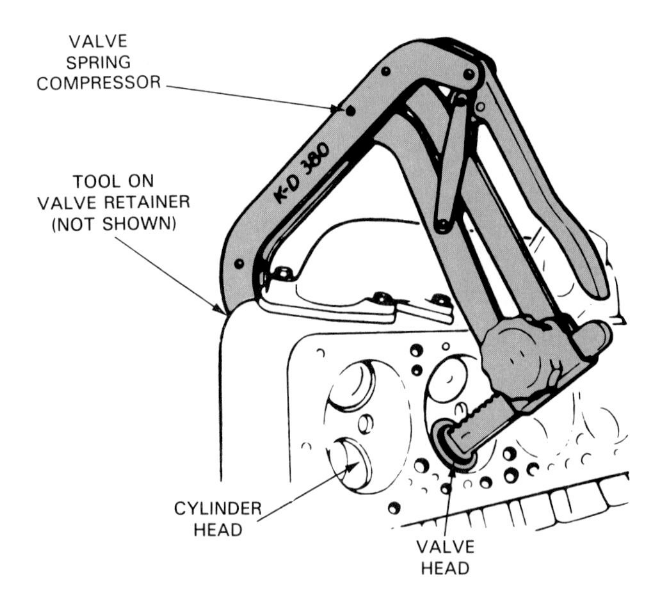

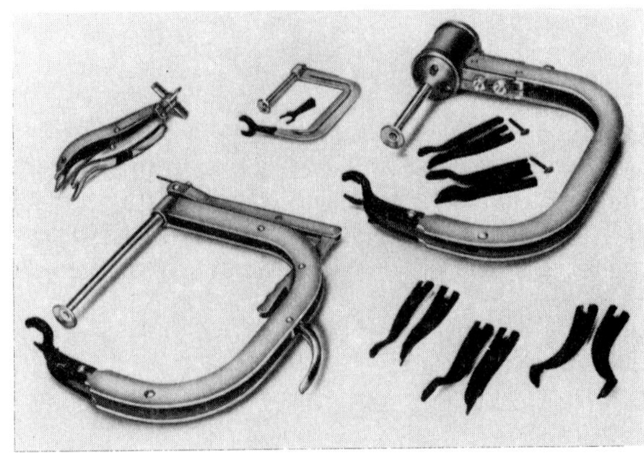

Fig. 2-11. Valve spring compressor will squeeze spring together so valve assembly can be removed from cylinder head. (K-D Tools and Snap-On)

Fig. 2-12. This valve spring compressor is for an overhead cam head with valve springs recessed into head. (K-D Tools)

the rocker arm adjusting screws. It lets you set rocker arm-to-valve stem clearance, Fig. 2-13.

Valve spring tester

A *valve spring tester* can be used to measure the tension of valve springs. As springs are used, they tend to

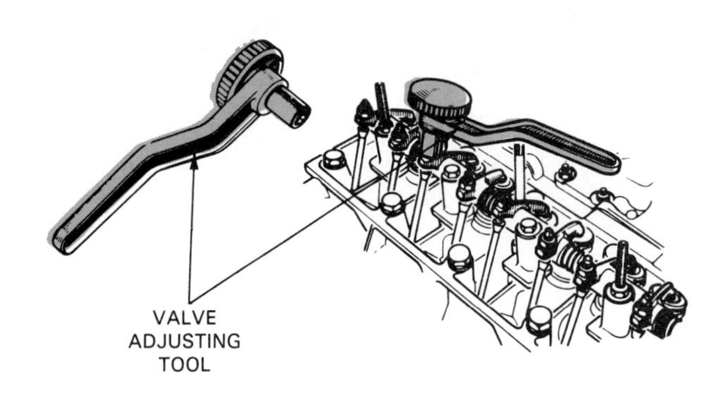

Fig. 2-13. This tool makes valve clearance adjustment easier. (Renault)

weaken and lose tension. The valve spring tester will check spring condition by compressing the spring to a specific height and showing spring pressure in pounds or kilograms. If the pressure is below specs, you would know to replace the valve springs, Fig. 2-14.

Cam bearing driver

A *cam bearing driver* will force cam bearings into or out of the cylinder block, Fig. 2-15. It is a long steel bar with a threaded end and special drivers. When screwed down, it will produce a powerful pulling action for bearing replacement.

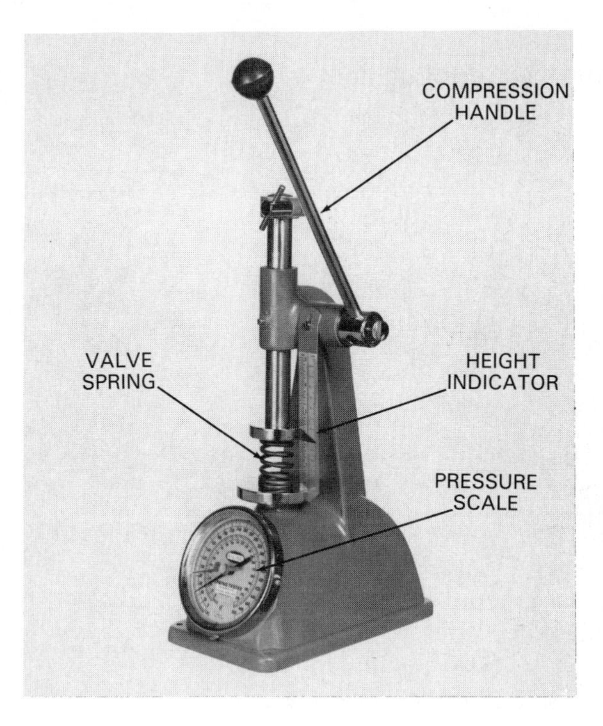

Fig. 2-14. Valve spring tester will check spring pressure to make sure springs have not lost their tension. Compress spring to specified height and check pressure. Pressure must be within specs. (K-Line Tools)

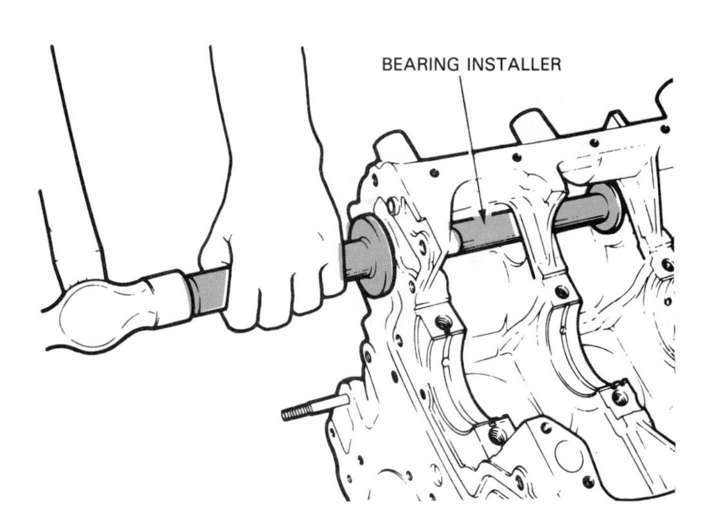

Fig. 2-15. Cam bearing driver is needed to force bearings in and out of cylinder block or head. Oil holes in bearing and part must line up. (Chrysler)

Valve guide tools

Valve guide tools are needed to restore the cylinder head valve guides when they are worn. With an integral (built-in) guide, the old guide can be machined larger and a new sleeve or insert pressed into the head. With a pressed-in guide, the old guide can be driven out and a new one driven into the head. See Fig. 2-16.

Valve grind machine

A *valve grind machine* can be used to resurface valve faces and valve stem tips. It uses grinding stones to pro-

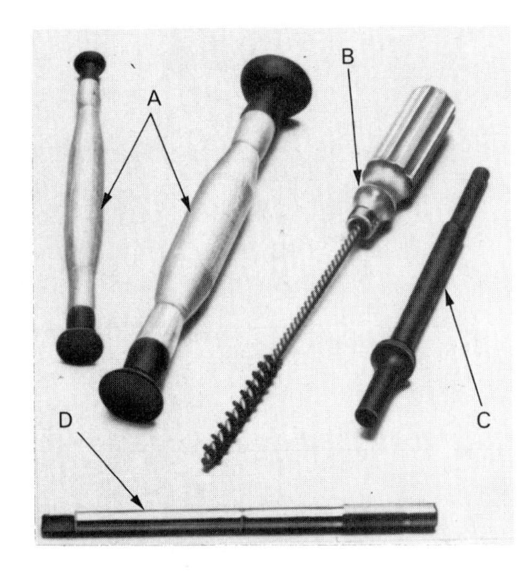

Fig. 2-16. Valve related tools. A—Lapping sticks. B, C, and D are for renewing worn valve guides in head. B—Valve guide brush. C—Guide driver. D—Guide knurling tool.

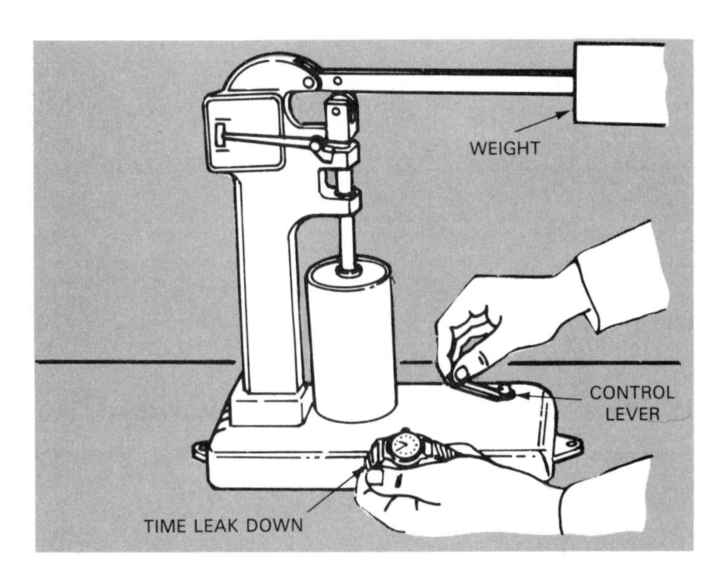

Fig. 2-18. Hydraulic lifter tester will load lifter plunger. Amount of time weight takes to push plunger down can be used to determine condition of lifter. (Buick)

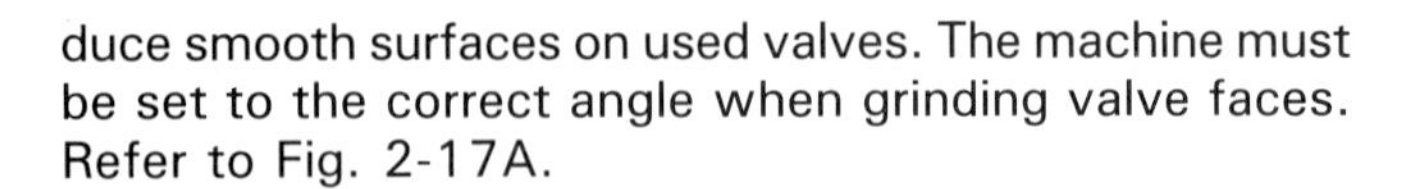
duce smooth surfaces on used valves. The machine must be set to the correct angle when grinding valve faces. Refer to Fig. 2-17A.

Valve seat grinder

A *valve seat grinder* will resurface the valve seats in a cylinder head. After prolonged use, the seats can become pitted and irregular. The seat grinder will smooth the seat and make it concentric (round) so the ground valve will form a leakproof seal when closed on the seat. One is shown in Fig. 2-17B.

Hydraulic lifter tester

A *hydraulic lifter tester* will measure the wear inside a lifter. It will preload the lifter plunger and produce a controlled collapse of the lifter. By timing how long it takes for the plunger to move to the bottom of its bore, you can find out if a lifter is good or bad. See Fig. 2-18.

Black light

A *black light* can be used to locate engine leaks. A special solution can be added to the engine oil, for example. Then, if oil is leaking, the black light will make the fluid glow and show up distinctly, Fig. 2-19.

Engine prelubricator

An *engine prelubricator* can be used to fill an engine with oil before starting. It can also be used to find worn engine bearings and other lubrication system problems during pre-teardown diagnosis, Fig. 2-20.

Boring bar

A *boring bar* is a machine shop tool for making a worn engine cylinder larger in diameter. After prolonged service, the piston rings can wear a cylinder out-of-round

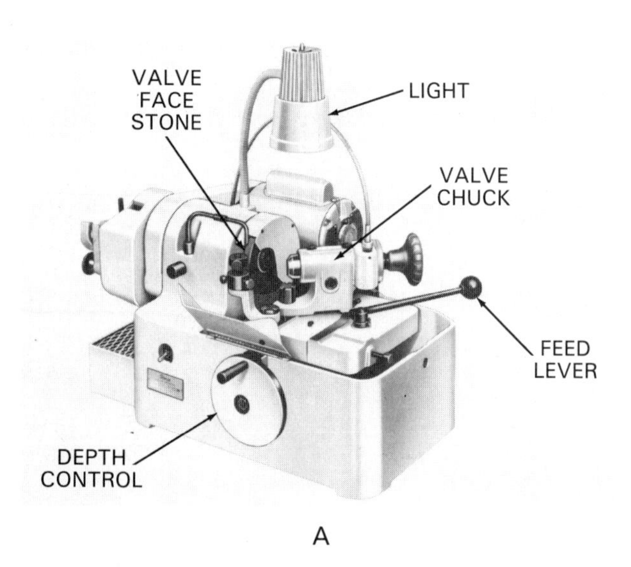

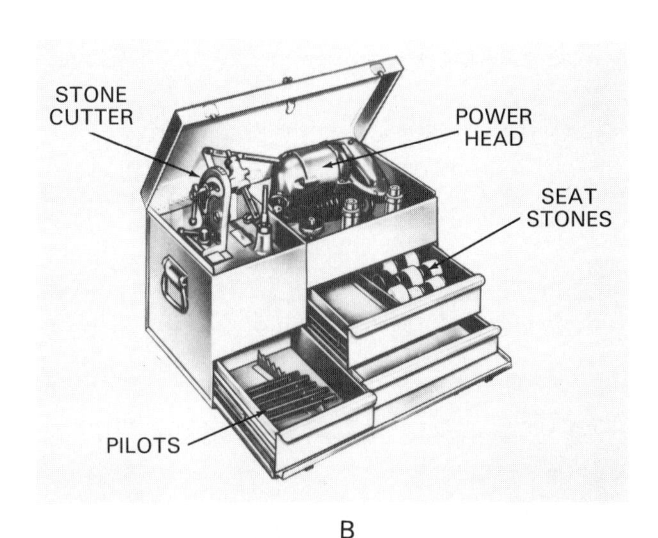

Fig. 2-17. A—Valve grind machine will resurface valve faces and valve stem tips. B—Seat grinder is for resurfacing seats in cylinder head. (Snap-on Tools)

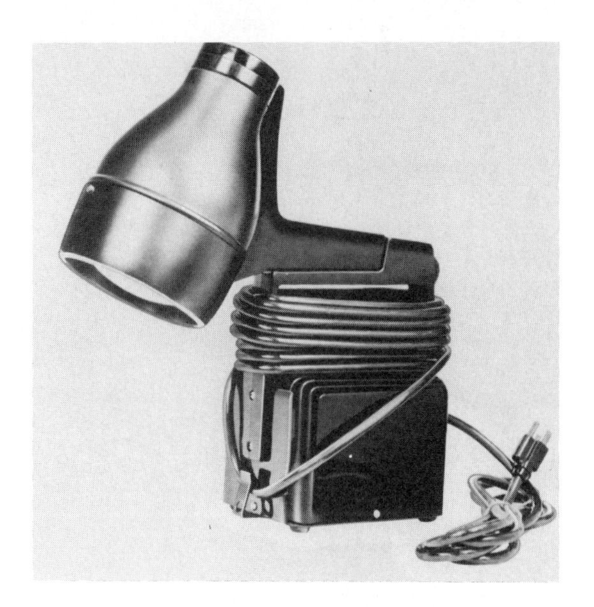

Fig. 2-19. Black light can be used to find fluid leaks and cracks in parts. Special powders or fluids are used in conjunction with black light. (Kent Moore)

Fig. 2-21. A boring bar, like this power honing machine, will remove metal from inside of cylinder bores. This will make block cylinder like new for oversize pistons. (Sunnen)

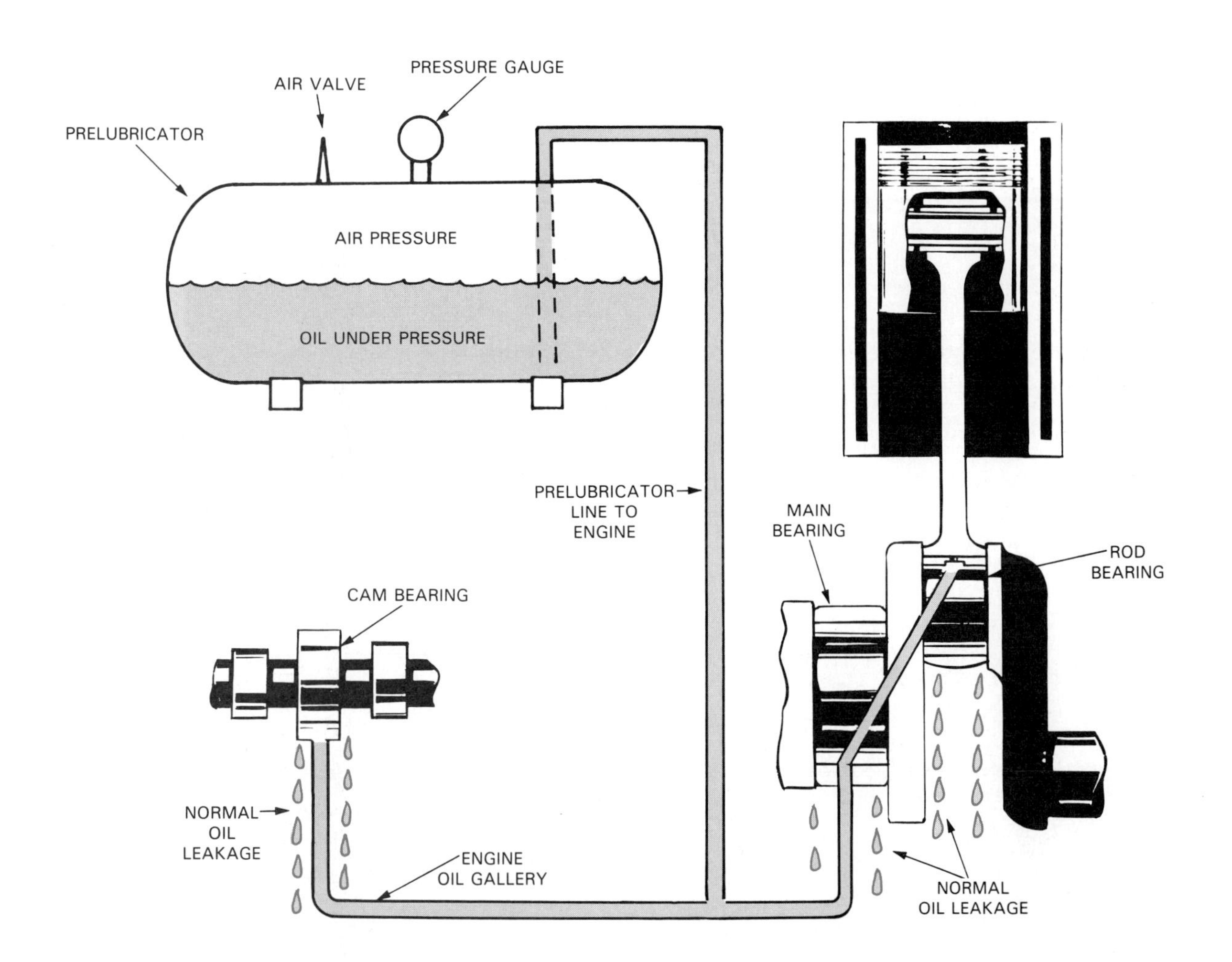

Fig. 2-20. Engine prelubricator is steel tank with air fitting and outlet hose to engine. Outlet hose is connected to engine lubrication system. Air pressure is injected into tank. Then engine oil can be forced through engine for prelubrication before starting or for finding bad bearings. Excess oil will leak out of bad bearings.

and cause taper. Part breakage can also damage a cylinder. If not too severely worn or damaged, the cylinders can be bored oversize so that new oversize pistons can be installed during the engine overhaul. A boring bar is shown in Fig. 2-21.

Crankshaft grinder

A *crankshaft grinder* will resurface worn crankshaft journals. It is a large machine tool that rotates the crankshaft against a spinning stone. This will grind the crank journals smaller in diameter. New, undersize rod and main bearings can then be used to rebuild the engine and bring bearing clearances within specs, Fig. 2-22.

Fig. 2-22. A crankshaft grinder will resurface the rod and main journals smaller. Then undersize bearings can be used to produce the proper bearing-to-crank clearance. (Storm Vulcan)

Camshaft grinder

A *camshaft grinder* will resurface the cam lobes and bearing journals on a worn camshaft. It is also a machine shop tool. The camshaft is rotated against a grinding wheel. This will restore the shape of the cam lobes. Cam lobes are prone to wear because of the high friction between the lobes and lifters or rocker arms, Fig. 2-23.

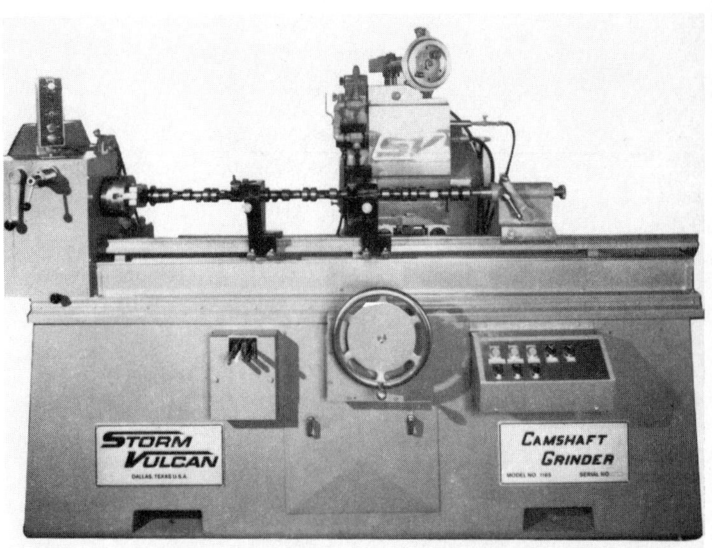

Fig. 2-23. A camshaft grinder will resurface and recontour the cam lobes. Cam lobes are one of the highest friction and wear points in an engine. (Storm Vulcan)

Fig. 2-24. Engine balancer will spin crankshaft so that weight can be added to or removed from counterweights to prevent engine vibration. It is used when weight of parts is altered.

Engine balancer

An *engine balancer* is used to spin the engine crankshaft to determine if weight must be added or removed from its counterweights, Fig. 2-24. Normally, an engine is balanced from the factory. If the same pistons, rods, and crankshaft are reused during a rebuild, balancing is NOT needed. However, if pistons, rods, or the crankshaft are changed, balancing may be required to prevent engine vibration.

Basically, the pistons, rods, rings, bearings, and piston pins are weighed on an accurate scale. Then, a formula can be used to determine how much weight must be bolted to the crankshaft journals to simulate the weight of the piston assemblies. With this weight attached, the crankshaft is spun on the balancing machine. The machine will then show where and how much weight must be changed to make the crankshaft rotate without vibrating.

Wheel puller

A *wheel puller* is for forcing a hub, damper, pulley, gear, etc. off of its shaft. The puller clamps or bolts onto the part. Then, when the puller screw is tightened, it will pull the component off its shaft, Fig. 2-25.

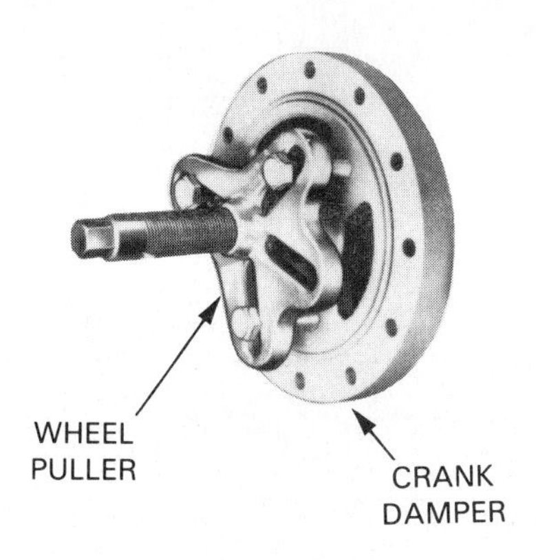

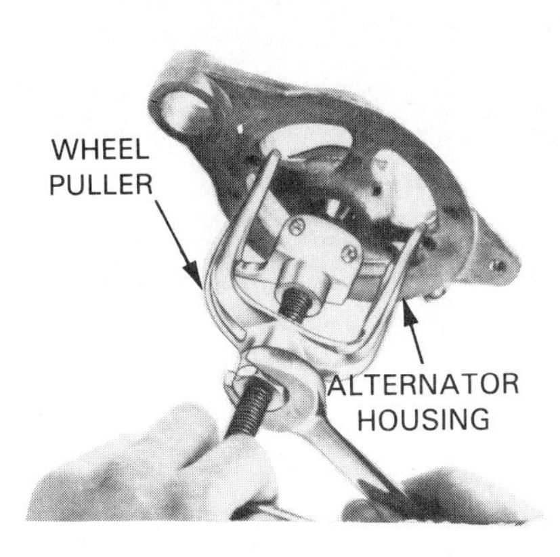

Fig. 2-25. Wheel puller will force gears, pulleys, dampers, sprockets, and other parts off of shafts. (Snap-on Tools)

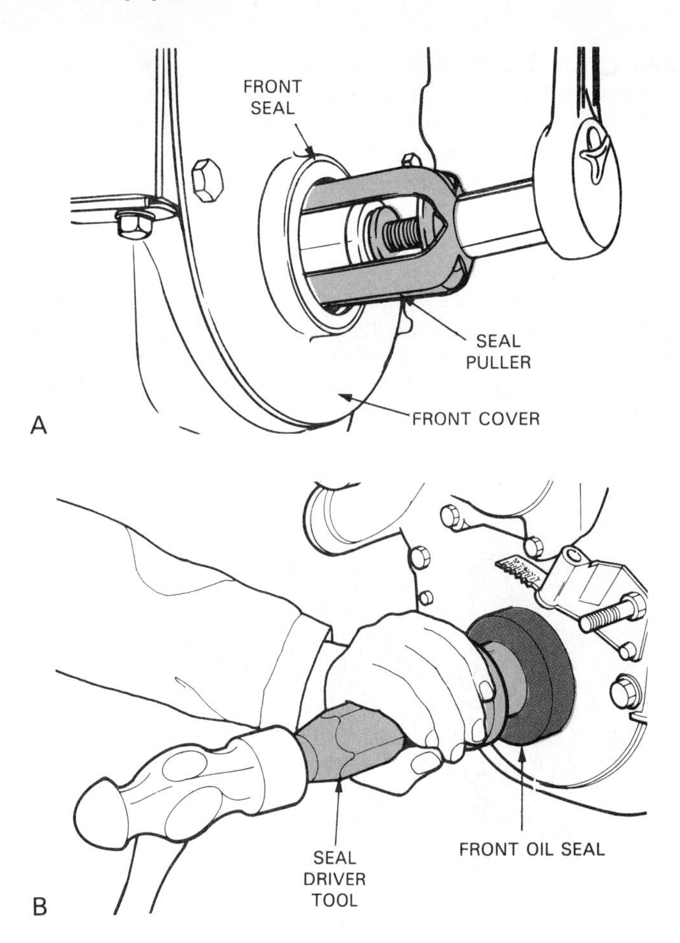

Fig. 2-26. A—Seal puller will remove old seal. B—Seal driver will produce equal force around seal body to prevent seal damage when driving in new seal. (GMC)

Danger! Wheel pullers can exert tons of force. Wear eye protection.

Seal drivers

Seal drivers are for installing and removing the front and rear crankshaft seals in an engine. As shown in Fig. 2-26, a wheel-puller like tool can be used to force out an old seal. Then, a driving tool can be used to force the new seal into place without damage to the seal body.

Stethoscope

A *stethoscope* is a listening device that will help you find abnormal noises from inside parts. The tip of the tool is touched on different engine components. When the sound is the loudest, you have found the source of the problem. Look at Fig. 2-27.

Mirror probe

A *mirror probe* will let you inspect for problems when an area is blocked from view. It is a small mirror hinged on a long handle. You can use it to find the source of hidden leaks, cracked parts, and other troubles, Fig. 2-28.

Pickup tools

A *finger pickup tool* can be used to retrieve dropped parts or tools. A special type will grasp engine lifters and pull them out of the engine. See Fig. 2-29A.

A *magnetic pickup tool* will do the same thing as a finger pickup tool. However, it will attract and hold only ferrous (metal containing iron) objects, Fig. 2-29B.

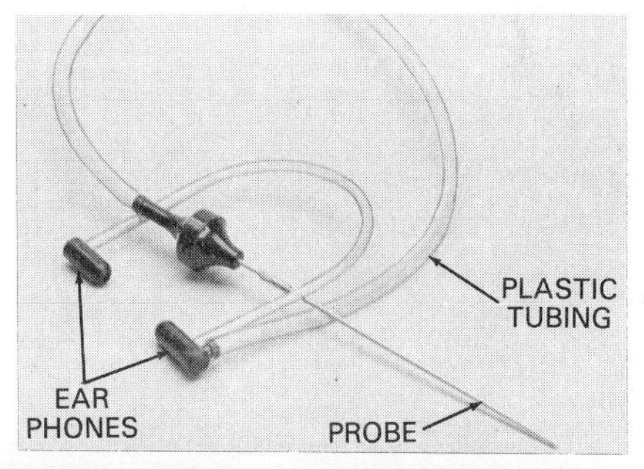

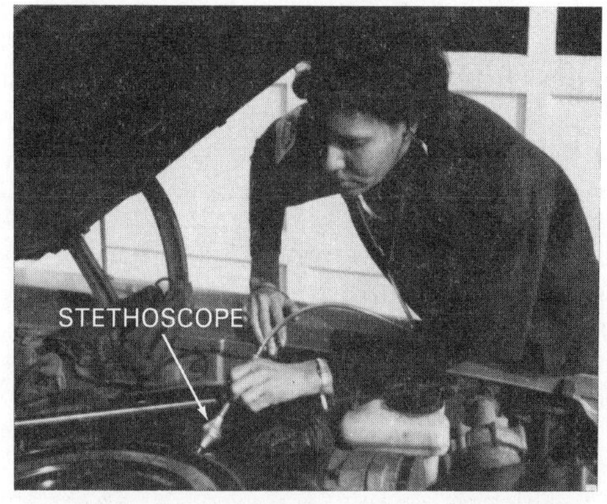

Fig. 2-27. Stethoscope can be used to find the source of abnormal engine sounds.

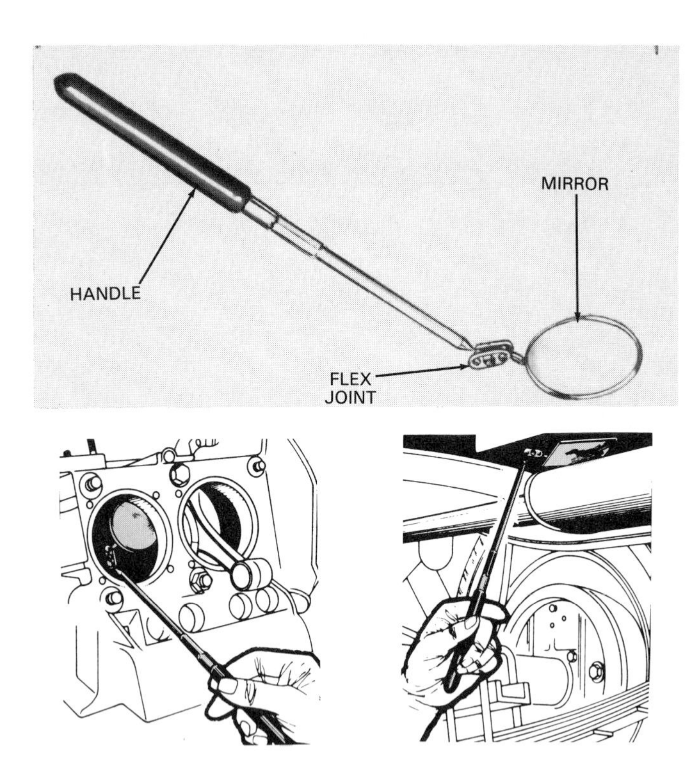

Fig. 2-28. Mirror probe will let you look in hidden areas for finding leaks, damaged parts, etc.

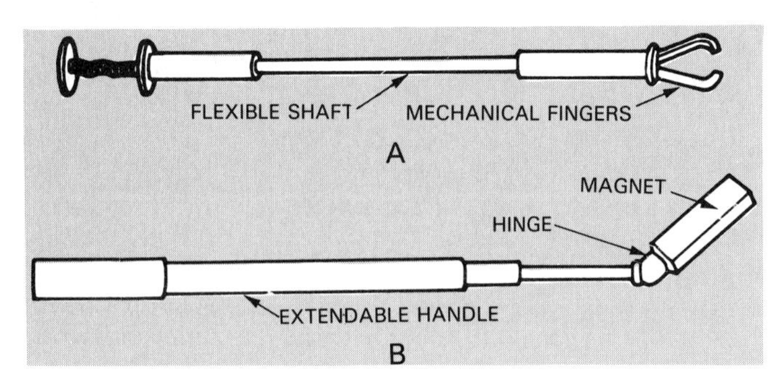

Fig. 2-29. A—Finger probe will grasp and pick up small objects when handle knob is activated. B—Magnetic pickup will retrieve dropped parts and tools. (Florida Voc. Ed.)

Bench grinder

A *bench grinder* is frequently used by the engine mechanic. The wire wheel on the grinder can be used to remove hard carbon deposits from engine valves, for example. The grinding wheel is used to resurface chisels and to do other metal removal type tasks, Fig. 2-30.

Caution! Do NOT use a wire wheel to remove carbon from soft aluminum parts—pistons, for instance. The abrasive action of the wheel can remove aluminum and ruin the part.

A few BENCH GRINDER RULES to follow are:

1. Wear eye protection and keep your hands away from the stone and brush.
2. Keep the tool rest adjusted close to the stone and brush. If NOT up close, the part can catch in the grinder.
3. Make sure the grinder shields are in place.

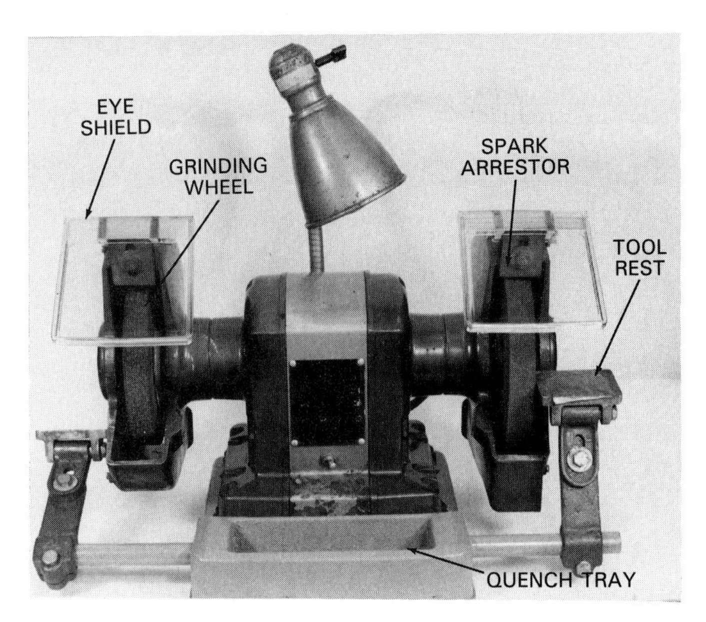

Fig. 2-30. Bench grinder is useful tool of engine mechanic. Wire brush, for example, can be used to clean carbon off of engine valves. Keep guards and shields in place!

Hydraulic press

A *hydraulic press* will produce tons of force for assembly or disassembly of pressed-together components. In engine mechanics, the hydraulic press is commonly used to force piston pins out of the pistons. A special fixture is needed to support the piston and prevent piston damage while driving, Fig. 2-31.

Danger! Respect the force generated on a hydraulic press. Always wear eye protection and stand to one side during pressing operations. If parts break, they can fly out with deadly power.

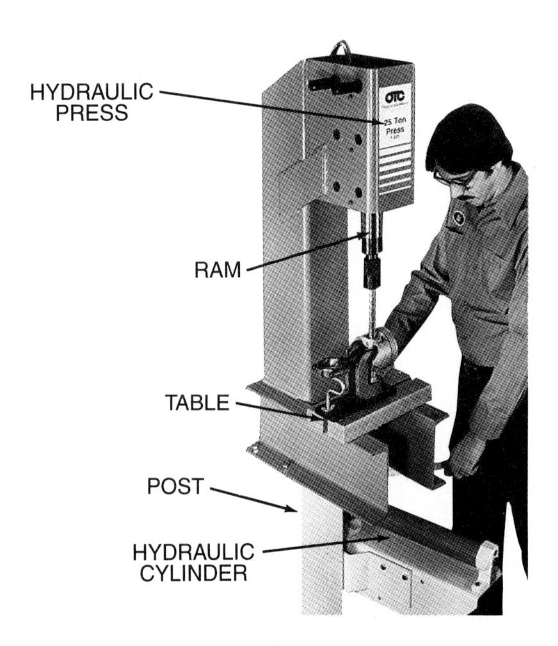

Fig. 2-31. Hydraulic press will produce powerful driving force. It is used to push piston pins, for example. Wear eye protection! (OTC Tools)

Hydraulic jacks

Hydraulic jacks are used to lift the vehicle, engine, and transmission during engine removal or installation. They can also be used to lift the engine when replacing motor mounts, exhaust manifolds, oil pans, etc. A special hydraulic transmission jack is shown in Fig. 2-32.

Never work under a car supported by a floor jack. The car must be mounted on jack stands before working. Also, keep your hands out from under an engine or transmission raised on a hydraulic jack. A hydraulic jack can fail, dropping the engine or transmission.

Fig. 2-32. Various types of hydraulic jacks are used in engine repair. This is an extension type transmission jack sometimes needed for engine removal. (OTC)

ENGINE CLEANING TOOLS AND EQUIPMENT

A great deal of time is spent cleaning parts during an engine overhaul. It is important for you to understand the most commonly used cleaning tools and equipment.

Blow guns

An air powered *blow gun* is commonly used to dry and clean engine parts after being washed in solvent. It is also used to blow dust and loose dirt off of an engine part before disassembly. See Figs. 2-33 and 2-34.

When using a blow gun, wear eye protection. Direct the blast of air away from yourself and others.

Fig. 2-33. Blow guns will remove dust, dirt, and cleaning solvent from engine parts. Do not aim nozzle towards body! (Binks)

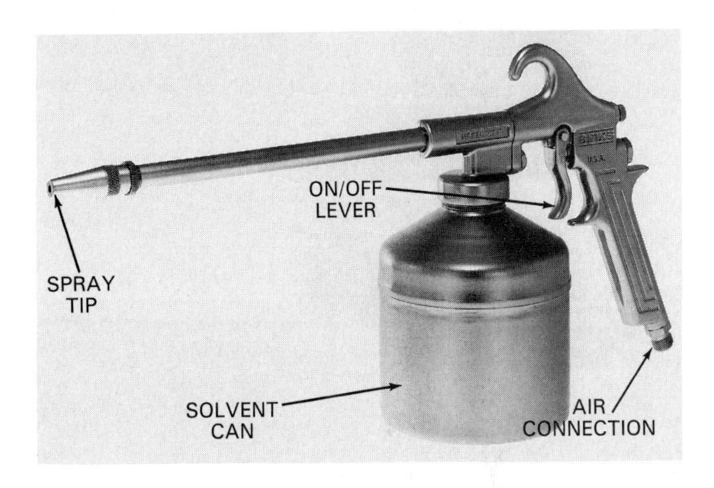

Fig. 2-34. This is a solvent cleaning gun. Solvent is pulled out of can and blown onto parts with air pressure. (Binks)

Steam cleaner and high pressure washer

A *steam cleaner* or a *high pressure washer* is used to remove heavy deposits of dirt, grease, and oil from the outside of engines. They provide an easy and rapid method of cleaning BEFORE DISASSEMBLY, Fig. 2-35.

Caution! A steam cleaner operates at relatively high pressures and temperatures. Follow all safety rules and specific operating instructions.

Cold solvent tank

A *cold solvent tank* is used to remove grease and oil from engine parts. After removing all old gaskets and scraping off excess grease, the parts can be brushed clean in the solvent, Fig. 2-36. A blow gun is normally used to remove the solvent.

Hot tank

A *hot tank* uses heat and a very powerful cleaning agent to clean engine parts. This is normally a piece of machine shop equipment.

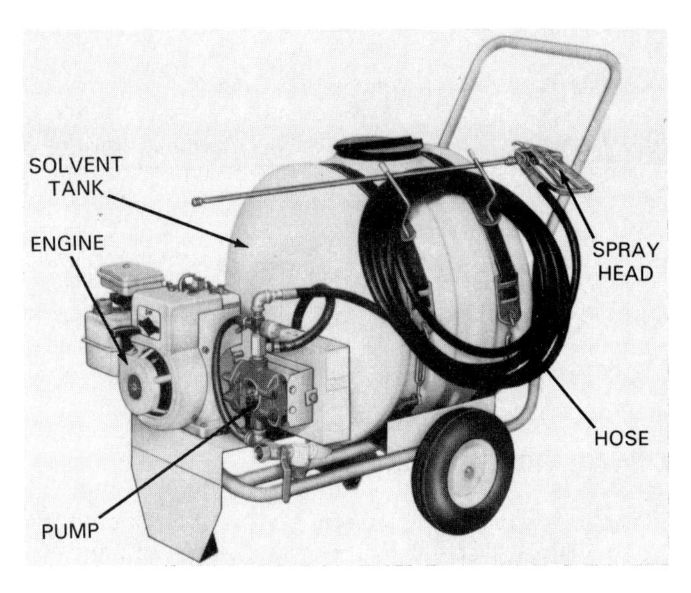

Fig. 2-35. Pressure washers, like this one, or a steam cleaner, can be used to clean engines before teardown. (Sioux Tools)

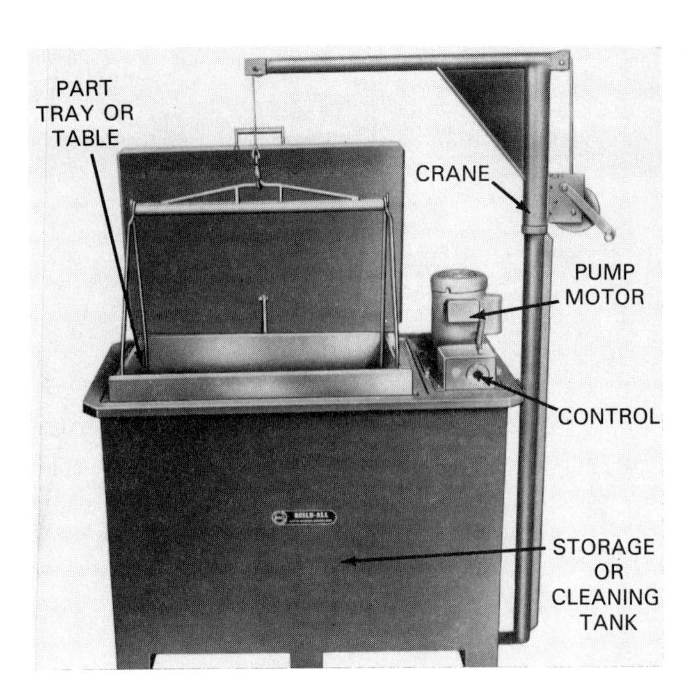

Fig. 2-37. Hot tanks use very powerful cleaning agent for removing paint, mineral deposits, etc. Wear eye protection and gloves when working around hot tank. (BAC)

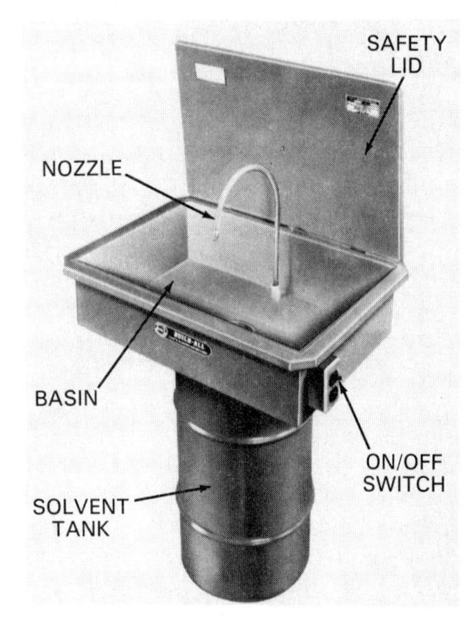

Fig. 2-36. Cold solvent tank contains mild cleaning agent for removing oil and grease from parts. (Build-All)

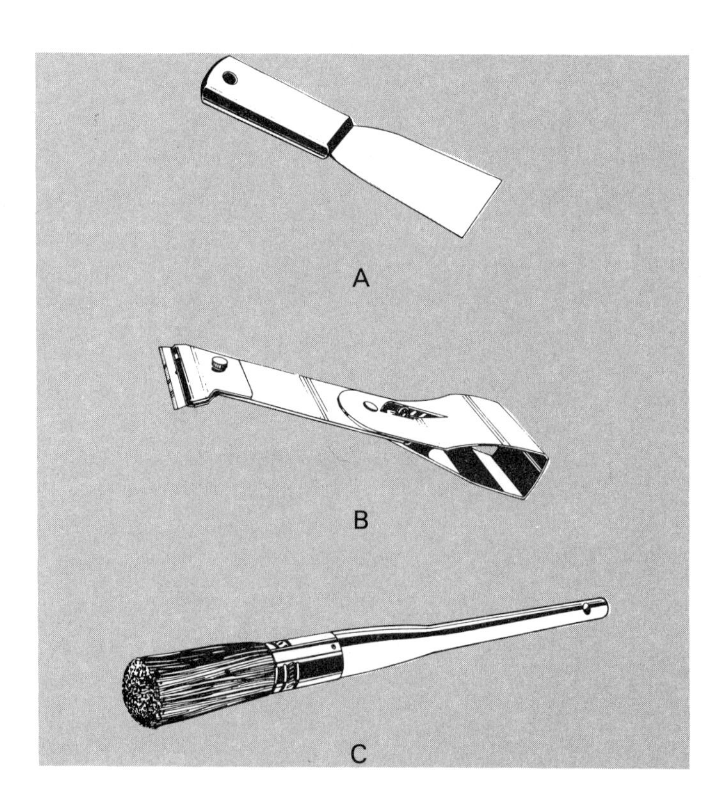

Fig. 2-38. A—Putty knife type scraper for removing old engine gaskets. B—Razor blade type scraper for removing paint and other thin coatings. C—Parts cleaning brush for use in cold solvent tank. (Plew Tools)

An engine block is frequently hot tanked during an overhaul. Hot tanking will remove all paint, grease, oil, and mineral deposits in the water jackets. Before hot tank cleaning, you should remove all core plugs. Cam bearings will be eroded and new ones must be installed after tanking. See Fig. 2-37.

Scrapers and brushes

Scrapers are commonly used by the engine technician to remove old gasket material and oil buildups. *Hand brushes* are used with solvent to remove oil and grease. A *rotary brush* is also used in a drill to remove hard carbon deposits from cylinder head combustion chambers. Refer to Figs. 2-38 and 2-39.

Tubing equipment

Tubing equipment (tube cutter, flaring tools, reamer) is frequently used to make new lines running to an

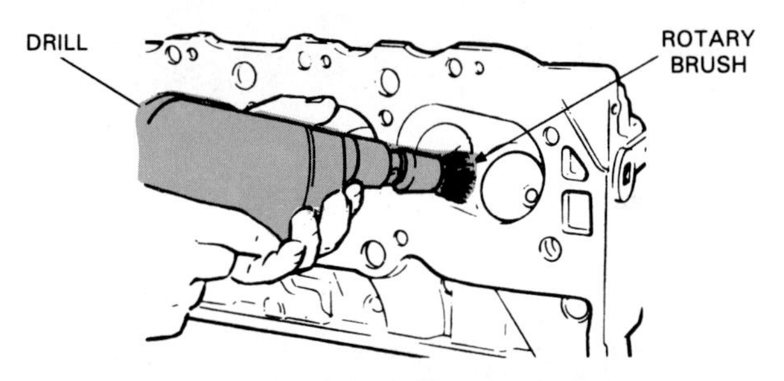

Fig. 2-39. Rotary brush is commonly used in drill to remove hard carbon deposits from inside combustion chambers in cylinder head. You must be careful not to damage valve seats.

engine. If an old fuel line, for instance, is damaged, these tools will let you produce a new line, Fig. 2-40.

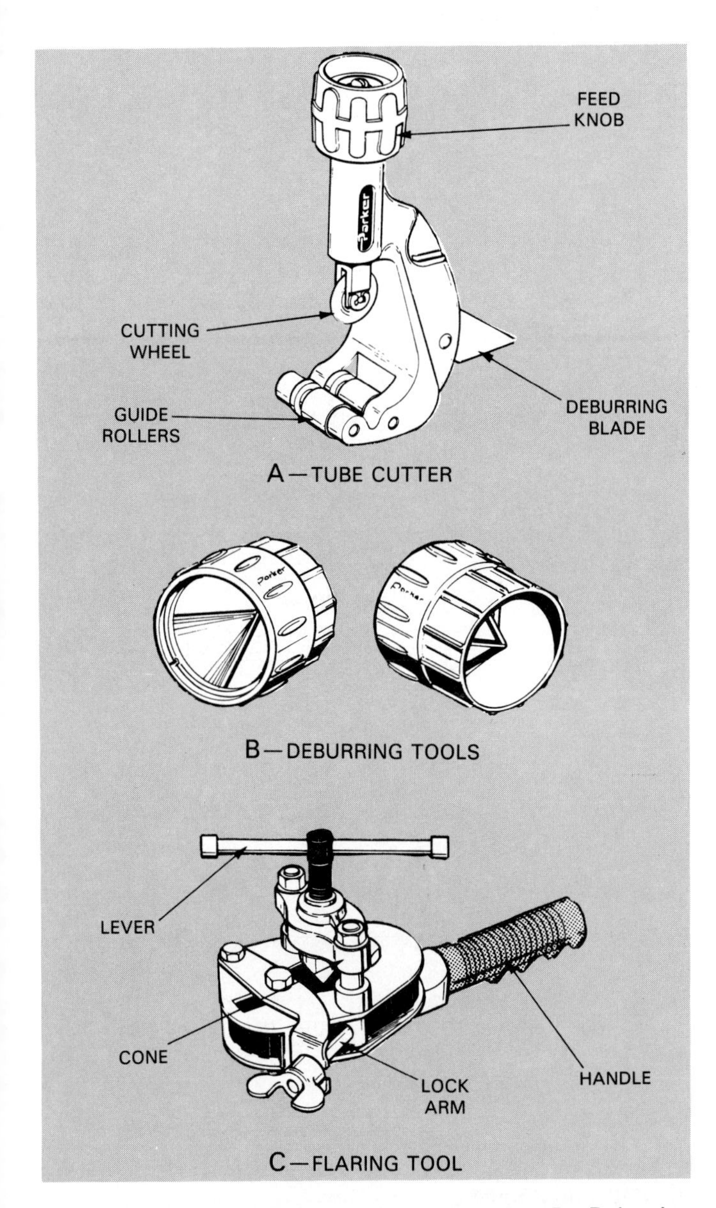

Fig. 2-40. Basic tubing tools. A—Tube cutter. B—Deburring tools. C—Flaring tool. (Parker)

MEASURING TOOLS

Numerous types of measuring tools are used by an engine technician. As an engine operates, its parts rub together and slowly wear out. Precision measurement is needed to find out if the parts are too worn to be reused in the engine.

Engine manufacturers give *specifications* (measurements or just specs) for maximum wear limits for most engine parts. If your measurements are NOT within these specs, the parts must be reconditioned or replaced.

This section of the chapter will give a brief review of engine measuring tools and equipment.

MEASURING SYSTEMS

The two *measuring systems* are the U.S. Customary Units System, also called conventional system, and the SI Metric System. Both are commonly used when working on engines.

The conventional system is mainly used in the United States. Almost all other countries use the metric system. The U.S. is slowly replacing its system with the metric system. All foreign cars and many new, American-made engines use metric bolts, nuts, and other parts. Manufacturer specifications are also given in both conventional and metric values.

Conventional measuring system

The *conventional measuring system* originated from sizes taken from parts of the human body. For example, the width of the human thumb was used to standardize the inch. The length of the human foot helped standardize the foot (12 inches). The distance between the tip of a finger and nose was used to set the standard of the yard (3 feet). Obviously, these are not very scientific standards.

Metric (SI) measuring system

The *metric* (SI) *measuring system* uses a power of 10 for all basic units. It is a simplier and more logical system than our conventional system. Computation often requires nothing more than adding zeros or moving a decimal point. For instance one meter equals 10 decimeters, 100 centimeters, or 1000 millimeters.

Conversion charts

A *measuring system conversion chart* is needed when changing from one measuring system to another: inches to centimeters, gallons to liters, liters to gallons. A conversion chart lets the technician quickly convert conventional values into equivalent metric values or metric values into equal conventional values.

A *decimal conversion chart* is commonly used by the technician to interchange and find equal values for fractions, decimals, and millimeters. Fractions are ONLY ACCURATE TO about 1/64 of an inch. For smaller measurements, either decimals or millimeters should be used. A decimal conversion chart may be needed to change a ruler measurement (fraction) into a decimal specification.

Refer to the back of this textbook for conversion charts.

Ruler (scale)

A steel *ruler,* also called *scale,* is frequently used to make low precision linear (straightline) measurements. It is accurate to about 1/64 in. (0.5 mm) in most instances, Fig. 2-41.

A conventional ruler has numbers that equal full inches. The smaller, unnumbered lines or graduations represent fractions of an inch (1/2, 1/4, 1/8, 1/16). The shortest graduation lines equal the smallest fractions of an inch.

A metric ruler normally has lines or divisions representing millimeters (mm). Each numbered line usually equals 10 mm (one centimeter).

A *pocket rule* or *pocket scale* is extremely short (typically 6 in. long). It fits in your shirt pocket and is very handy.

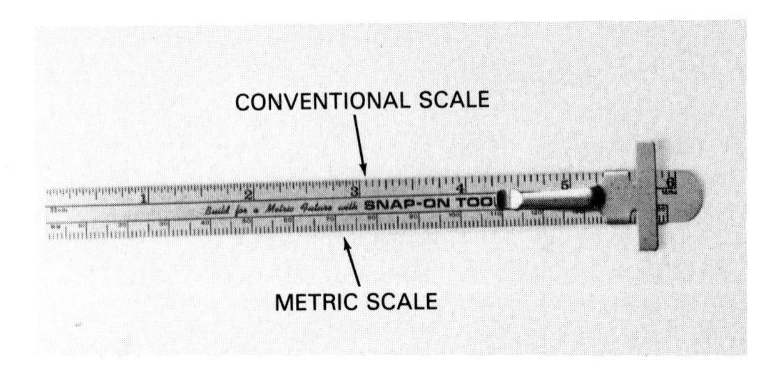

Fig. 2-41. Pocket rule is handy for making "rough measurements." This one has both conventional and metric scales.

Feeler gauges

A *feeler gauge* is used to measure small clearances or gaps between parts. The two basic types of feeler gauges are shown in Fig. 2-42.

A *flat feeler gauge* has precision ground, steel BLADES of various thicknesses. Thickness is written on each blade in thousandths of an inch (.001, .010, .017) and/or in hundredths of a millimeter (0.01, 0.06, 0.20, 0.23). A flat feeler gauge is normally used to measure small distances between PARALLEL SURFACES.

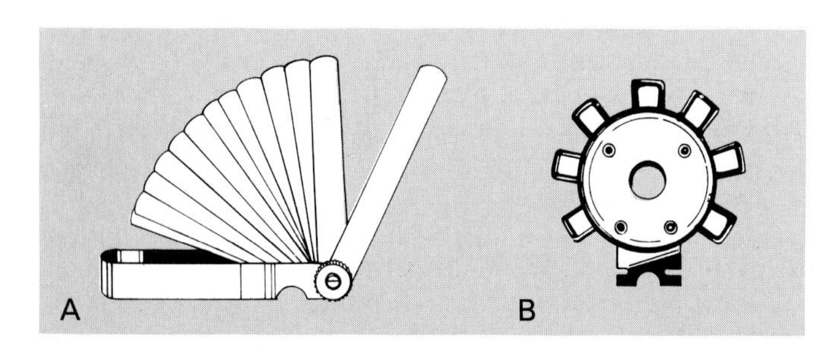

Fig. 2-42. A—Flat feeler gauge is for checking clearances between parallel surfaces. B—Wire feeler gauge is for checking clearance between unparallel surfaces. (Plew Tools-Parker)

A *wire feeler gauge* has precision sized WIRES that are also labeled by diameter or thickness. It is normally used to measure slightly larger spaces or gaps than a flat feeler gauge. The wire gauge's ground shape also makes it more accurate for measuring UNPARALLEL or CURVED SURFACES.

Using a feeler gauge

To measure with either type of feeler gauge, find the gauge thickness that just fits between the two parts being measured. The gauge should drag slightly when pulled between the two surfaces. The size given on the side of the gauge equals the clearance between the two components.

Straightedge

A *straightedge* is normally used with a flat feeler gauge to check or measure the warpage (flatness) of engine parts. The steel straightedge is placed on the part surface. Then, different size flat feeler gauge blades are slid between the straightedge and part surface. The largest blade that fits under the straightedge equals warpage.

Machining or resurfacing will usually straighten the part and allow its use on the engine. Cylinder heads, blocks, and manifolds are the most common engine parts checked with a straightedge, Fig. 2-43.

Combination square

A *combination square* has a sliding square (frame with 90 deg. angle edge) mounted on a steel ruler. It is needed when the ruler must be held perfectly square (straight) against the part being measured, Fig. 2-44.

Calipers and dividers

The *outside caliper* is sometimes helpful when making external measurements where 1/64 in. (approximately 0.40 mm) accuracy is sufficient. It can be fitted over the outside of parts and adjusted to touch the part. Then, the caliper is held against and compared to a ruler to determine part size. See Fig. 2-45A.

The *inside caliper* is designed for internal

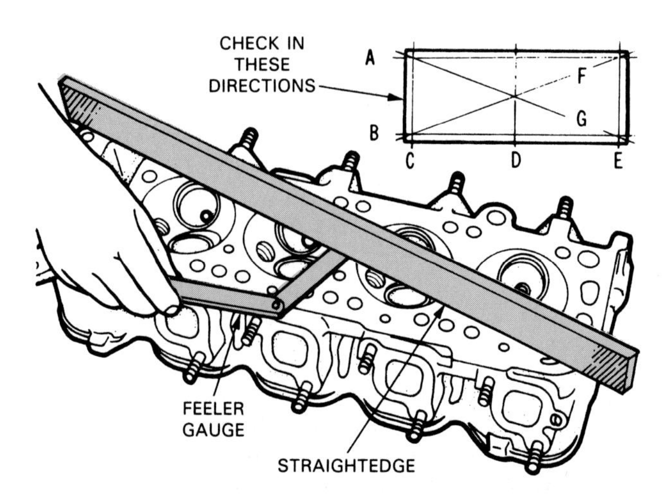

Fig. 2-43. Straightedge will show if part is warped. Feeler gauge that fits between part and straightedge equals amount of warpage. Check at various angles. (Chrysler)

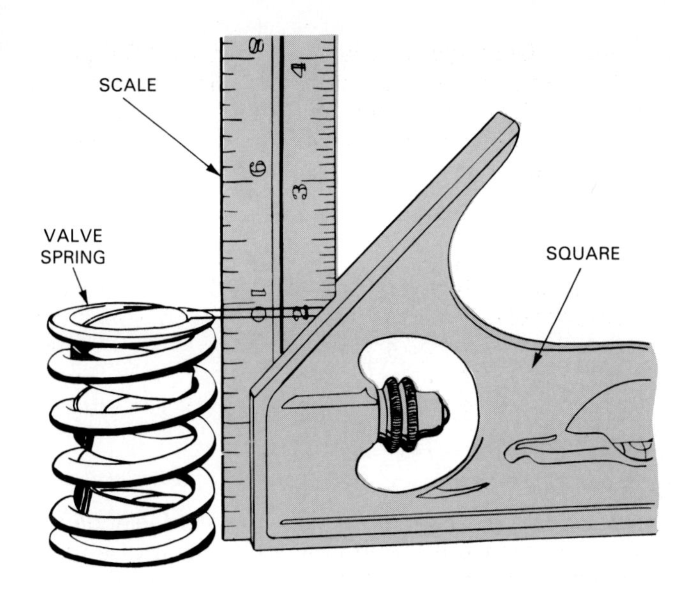

Fig. 2-44. Combination square is being used to check valve spring. If spring is not square, it should be replaced. (Cadillac)

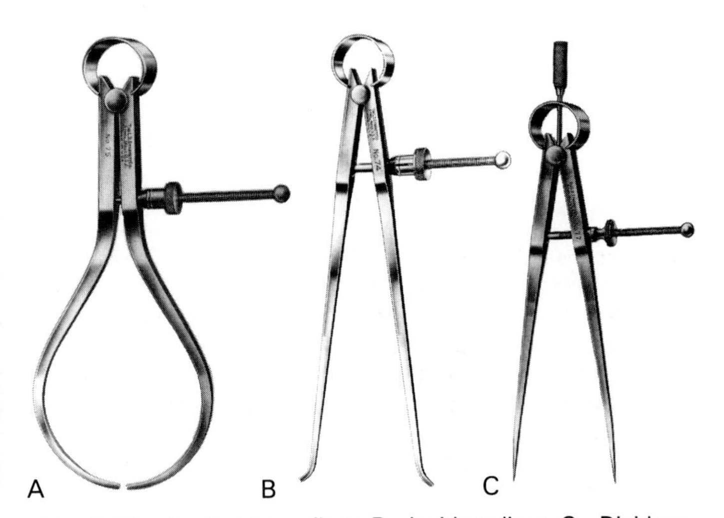

Fig. 2-45. A—Outside caliper. B—Inside caliper. C—Dividers.

measurements in holes and other openings. It must also be compared to a ruler. Look at Fig. 2-45B.

Dividers are similar to calipers but have straight, sharply pointed tips. They are commonly used for layout work on sheet metal parts. The sharp points will scribe circles and lines on sheet metal and plastic. Dividers will also transfer and make surface measurements, like calipers. Refer to Fig. 2-45C.

Dial caliper

A *dial caliper* is a very useful measuring tool with an accuracy of typically .001 in. (0.01 mm). Most types can be used to make inside, outside, and depth measurements quickly and precisely.

Fig. 2-46 shows how to use a dial caliper. Basically, read the ruler scale to get the rough measurement and then add the dial reading for thousands of an inch.

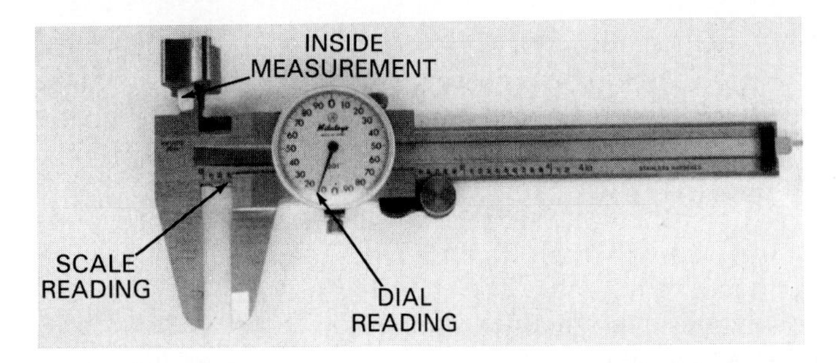

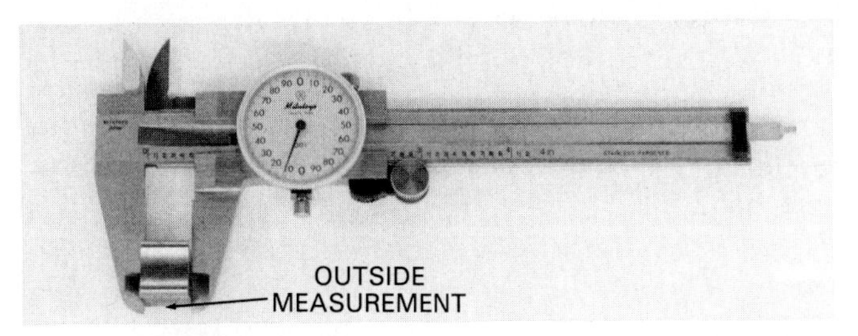

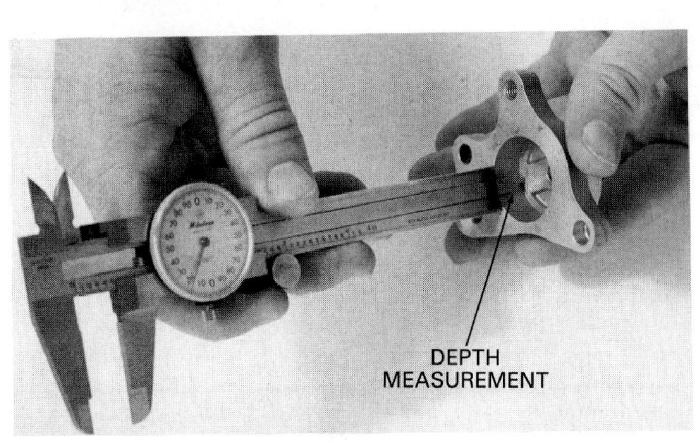

Fig. 2-46. Dial caliper is very useful because it will measure inside, outside, and depth with an accuracy of .001 in. A—Inside measurement. B—Outside measurement. C—Depth measurement. Read ruler scale and then add dial reading in thousandths of an inch.

MICROMETERS

A *micrometer,* nicknamed a "mike," is commonly used when making very accurate engine measurements. It will easily measure ONE THOUSANDTHS OF AN INCH (.001 in.) or one hundredths of a millimeter (0.01 mm). There are several types of mikes used in engine repair.

An *outside micrometer,* Fig. 2-47, is for measuring external dimensions, diameters, or thicknesses. It is fitted around the outside of the part. Then the thimble is turned until the part is lightly touching both the spindle and anvil. Measurement is obtained by reading graduations on the hub and thimble.

Reading a conventional micrometer

To read a conventional micrometer, follow the steps listed below:

1. Note the LARGEST NUMBER visible on the micrometer sleeve (barrel). Each number equals .100 in. (2 = .200, 3 = .300, 4 = .400). This is

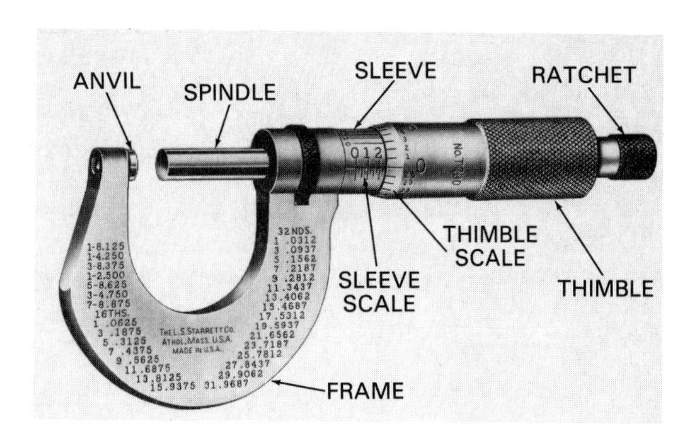

Fig. 2-47. You should know the basic parts of a micrometer. (Starret)

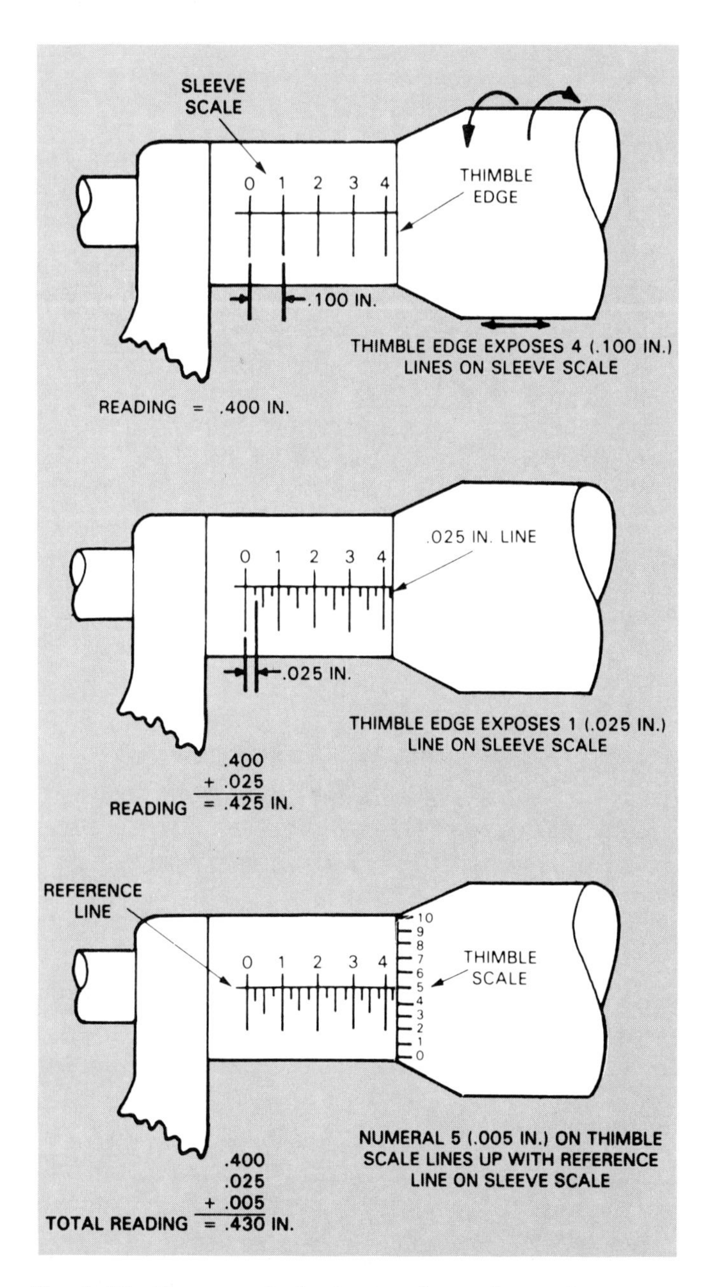

Fig. 2-48. These are the basic steps for reading a micrometer. Add to get total mike reading.

illustrated in Fig. 2-48.

2. Count the number of FULL GRADUATIONS to the right of the sleeve number. Each full sleeve graduation equals .025 in. (2 full lines = .050, 3 = .075). See Fig. 2-48.
3. Note the THIMBLE GRADUATION aligned with the horizontal sleeve line. Each thimble graduation equals .001 in. (2 thimble graduations = .002, 3 = .003). Round off when the sleeve line is not directly aligned with a thimble graduation, Fig. 2-48.
4. ADD the decimal values from steps 1, 2, and 3. Also, add any full inches. This will give you the micrometer reading, Fig. 2-48.

Reading a metric micrometer

Metric micrometers are also available. They are similar to conventional micrometers but have graduations and numbers in metric values. One revolution of the thimble equals 0.500 mm, instead of .025 in.

To read a metric micrometer, follow the four steps given below:

1. Read the SLEEVE NUMBER. Each sleeve number equals 1.00 mm (2 sleeve numbers = 2.00, 3 = 3.00). See Fig. 2-49.
2. Count and record the SLEEVE GRADUATIONS visible to the right of the sleeve number. Each line equals 0.50 mm (2 sleeve lines = 1.00, 3 = 1.50).
3. Read the THIMBLE GRADUATION lined up with the horizontal sleeve line. Each thimble graduation equals 0.01 mm (2 graduations = 0.02, 3 = 0.03).
4. ADD the values obtained in the previous 3 steps. This will give you the metric micrometer reading, as shown in Fig. 2-49.

Micrometer rules

A few micrometer rules to remember are:

1. Never drop or overtighten a micrometer. It is very delicate and its accuracy can be thrown off easily.
2. Store micrometers where they cannot be damaged by large, heavy tools. Keep them in wooden or plastic storage boxes.
3. Grasp the mike frame in your palm and turn the thimble with your thumb and finger. It should just "drag"

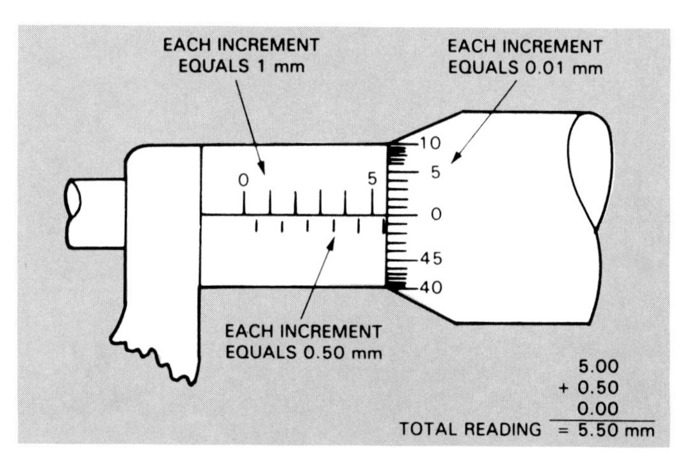

Fig. 2-49. Note scale divisions on metric micrometer. They are in millimeters. It is read like conventional "mike."

on the part being measured.

4. Hold the micrometer squarely with the work or false readings can result. Closely watch how the spindle is contacting the part.
5. Rock or swivel the mike as it is touched on round parts. This will ensure that the most accurate diameter measurement is obtained.
6. Place a thin film of oil on the micrometer during storage. This will keep the tool from rusting.
7. Always check the accuracy of a micrometer if it is dropped, struck, or after a long period of use. Tool salespeople sometimes have standardized gauges for checking micrometer accuracy.

Special micrometers and gauges

Various special micrometers and gauges are used in auto mechanics. It is important for you to learn the basic types and their uses.

A *telescoping gauge* measures internal part bores or openings. It is commonly used to measure block cylinder bores. The spring-loaded gauge is expanded to the size of the opening, Fig. 2-50A. Then, it is locked to size and measured with an outside micrometer, Fig. 2-50B.

A *hole gauge* is needed for measuring very small holes in parts. The hole gauge, like a telescoping gauge, is inserted and adjusted to fit the hole. Then, it is removed and measured with an outside micrometer.

An *inside micrometer* is a special measuring tool used for internal measurements in large holes, cylinders, or other part openings. It is read in the same manner as an outside mike, Fig. 2-51.

A *depth micrometer* is helpful when precisely measuring the depth of an opening. The base of the mike is positioned squarely on the part. Then, the thimble is turned until the spindle contacts the bottom of the opening. The depth micrometer is read in the same way as an outside micrometer. However, the hub markings are REVERSED. See Fig. 2-52.

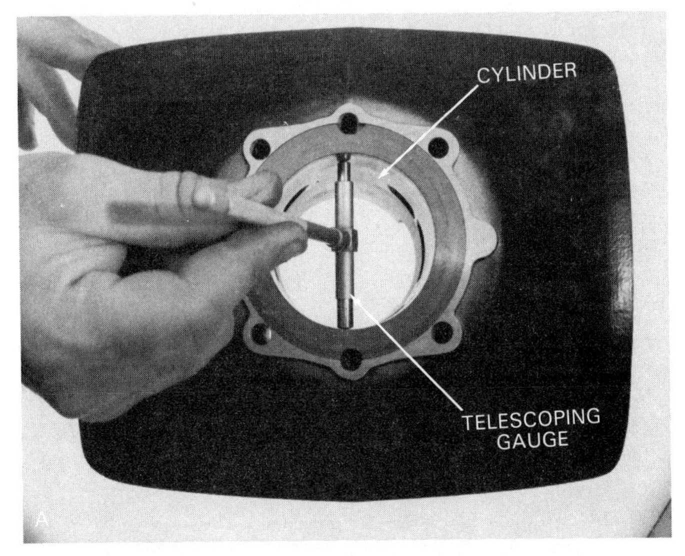

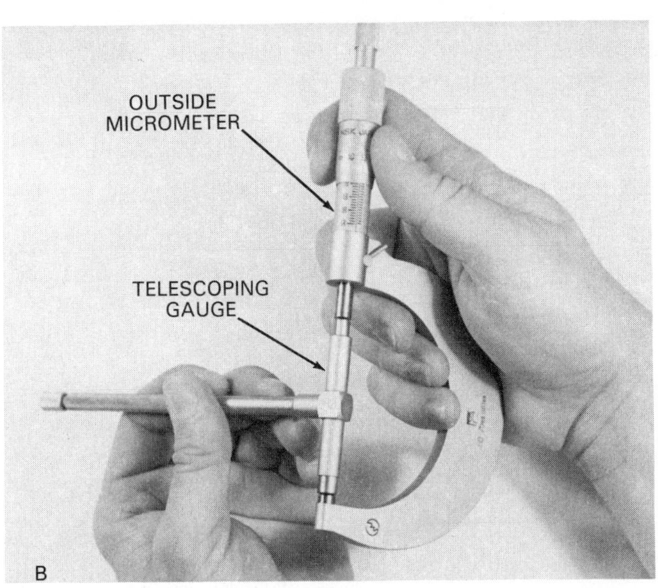

Fig. 2-50. A—Telescoping gauge is adjusted to fit inside bore. B—Then, outside mike is used to measure gauge and determine diameter of bore.

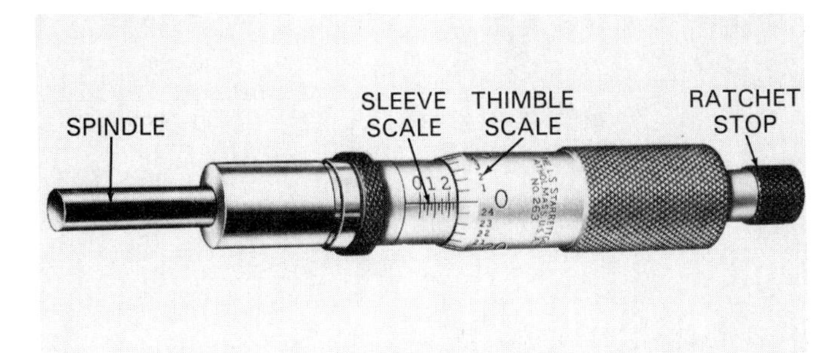

Fig. 2-51. This is an inside micrometer. It will do the same thing as a telescoping gauge but will not measure small diameters. (Starret)

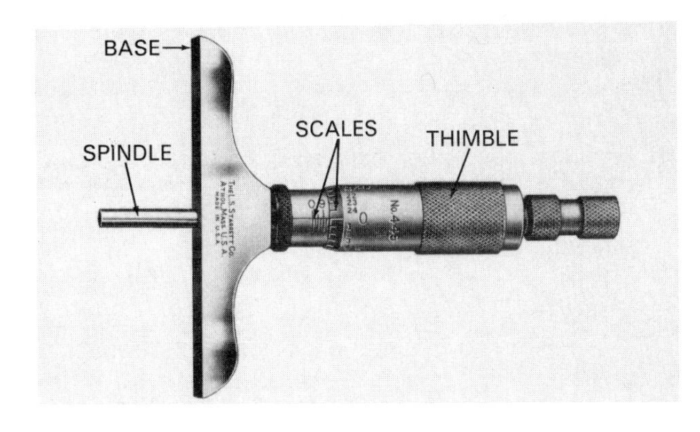

Fig. 2-52. Depth gauge is more specialized tool. (Starret)

Dial indicator

A *dial indicator* will measure part movement in thousandths of an inch (hundredths of a millimeter). The needle on the indicator face registers the amount of plunger movement, Fig. 2-53.

A dial indicator is frequently used to check gear teeth backlash (clearance), shaft end play, cam lobe lift, and other similar kinds of part movements. A magnetic

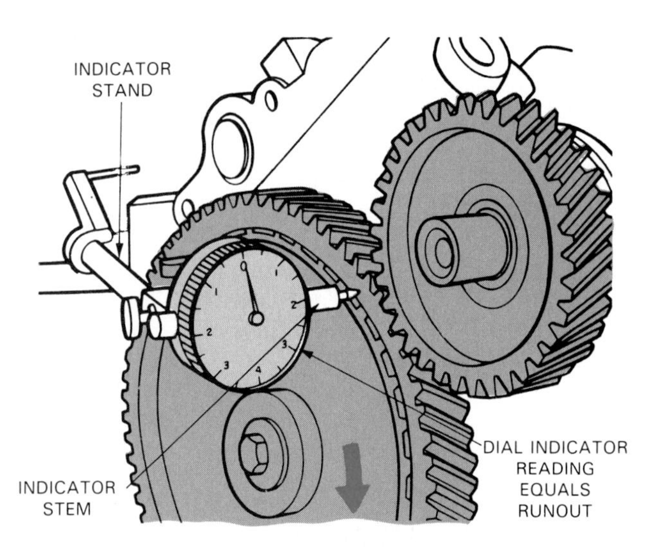

Fig. 2-53. Dial indicator will measure part movement in thousandths of an inch or hundredths of a millimeter. Here indicator is being used to measure gear runout or wobble. (Ford)

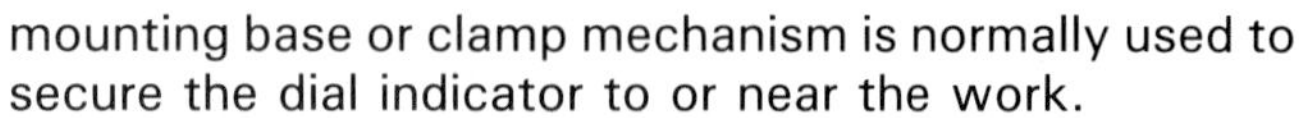
mounting base or clamp mechanism is normally used to secure the dial indicator to or near the work.

Fig. 2-54 shows a special fixture for using a dial indicator to measure valve spring installed height. Look at Fig. 2-55. It pictures a dial bore gauge for measuring cylinder wear.

Using a dial indicator

To measure with a dial indicator, follow these rules:

1. Mount the indicator securely and position the dial plunger parallel with the movement to be measured.
2. Preload or partially compress the indicator plunger before locking the indicator into place. Part movement in either direction should cause dial pointer movement.
3. Move the part back and forth or rotate the part while reading the indicator. Pointer movement equals part movement, clearance, or runout.
4. Be careful not to damage a dial indicator. It is very delicate.

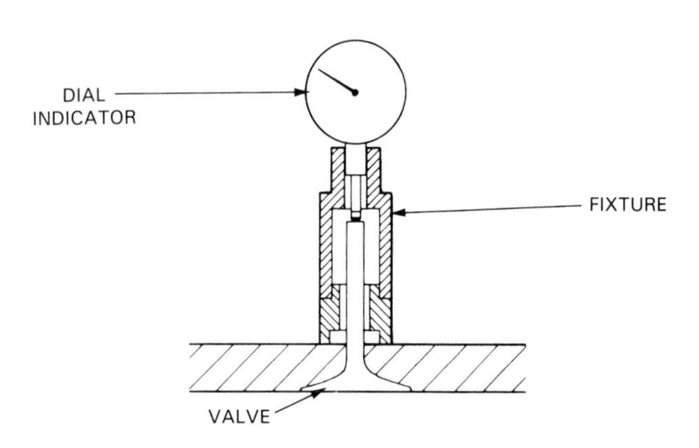

Fig. 2-54. Special fixture will let you check spring installed height with dial indicator. This is provided by one auto maker and will not work on other engines. (K-D Tools)

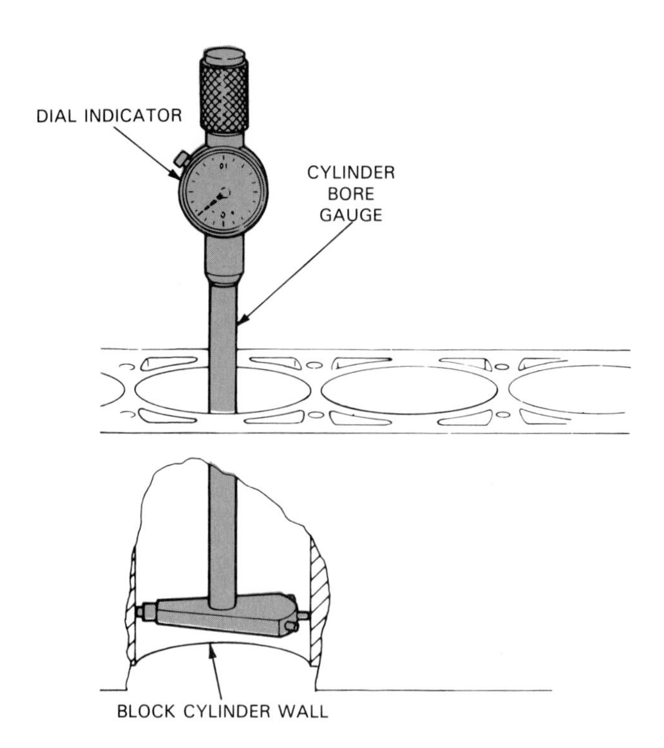

Fig. 2-55. Dial bore gauge is very fast way of checking condition of engine cylinders. It can be slid up and down in cylinder. Indicator movement shows amount of taper and out-of-roundness. (American Motors)

Plastigage

Plastigage is a special measuring device normally used to check clearances between internal surfaces. Shown in Fig. 2-56, it is commonly used to measure clearances in engine connecting rod bearings, crankshaft main bearings, and oil pumps.

To measure with Plastigage, place a small piece of the round plastic-like thread on a clean, unoiled part surface. Bolt the parts together, torque to specs and disassemble. Compare the paper covering's scale to the smashed Plastigage strip. The corresponding paper scale width indicates part clearance.

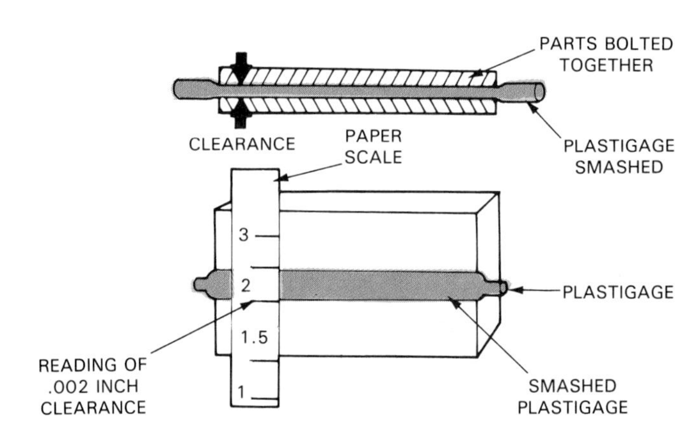

Fig. 2-56. Plastigage is placed between two parts to measure clearance. Parts will smash Plastigage wire. Then, paper scale can be compared to width of smashed Plastigage to determine clearance—the more the wire is smashed, the less the clearance.

TORQUE WRENCHES

A *torque wrench* measures the amount of turning force being applied to a fastener (bolt or nut). Conventional torque wrench scales usually read in foot-pounds (ft-lb) and metric scales in newton-meters (N•m).

The three general types of torque wrenches are the flex bar, dial indicator, and sound indicating types. These are shown in Fig. 2-57.

The flex bar torque wrench is inexpensive and adequate. The dial indicator torque wrench is very accurate, but can be hard to use in tight quarter. The sound indicating type torque wrench is very fast and easy to use. It makes a ''pop'' or ''click'' sound when a preset torque value is reached. You do not have to watch an indicating needle while torquing.

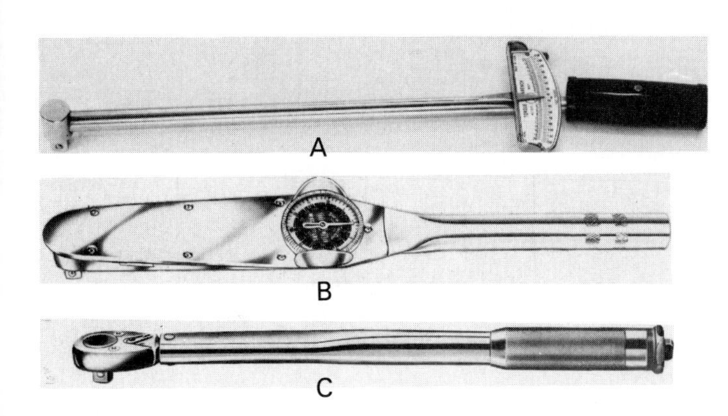

Fig. 2-57. A—Flex bar torque wrench. B—Dial indicator torque wrench. C—Click or sound indicating torque wrench. (Snap-on Tools)

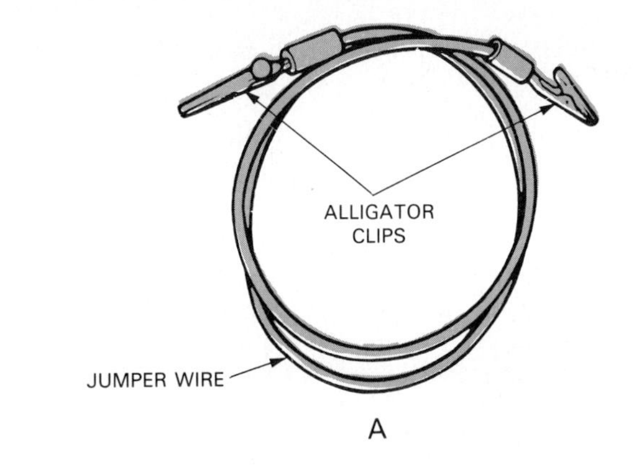

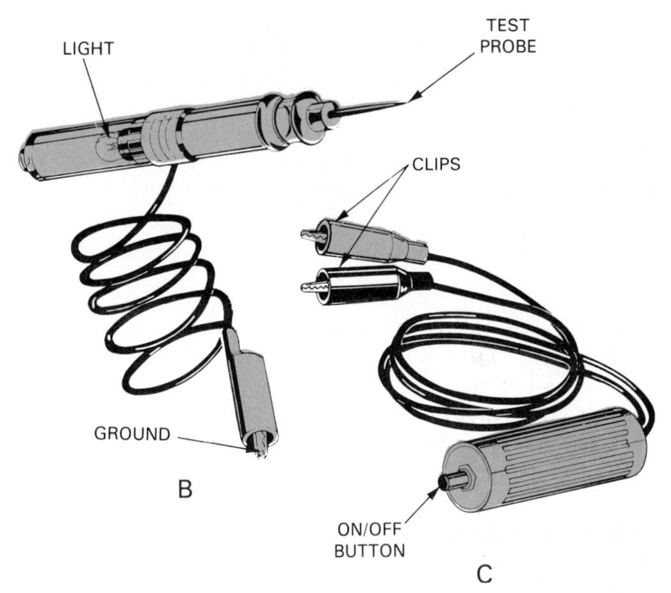

Fig. 2-58. A—Jumper wire for bypassing component or connecting voltage directly to circuit. B—Test light for making sure circuit or component is getting electricity. C—Remote starter switch is connected to starter solenoid so engine can be cranked from engine compartment or under engine. (Ford and Parker)

ELECTRICAL TEST EQUIPMENT

Today's engines are controlled by electronics. Sensors and a computer monitor various engine operating conditions and then use electrical signals to control various engine functions (ignition timing, fuel injection, idle speed, turbocharger boost, etc.)

To be a competent engine technician in a modern auto repair facility, you MUST know how to make electrical tests. This section of the chapter will review common electrical test equipment and summarize their use.

Jumper wire

A *jumper wire* is used to bypass a component in a circuit, Fig. 2-58A. For example, the jumper could be used to bypass a switch. If the circuit begins to function with the switch bypassed (jumper connected across switch terminals), then the switch is open and should be replaced. Jumper wires can also be used to connect a meter to various engine components for testing.

Test light

A *test light* is used to check for voltage in different sections of a circuit, Fig. 2-58B. One example, if you are NOT getting ''sparks'' at the spark plugs (engine not start), you could use the test light to make sure voltage is reaching the ignition coil. It will quickly check for *circuitry continuity* (completeness of circuit).

Remote starter switch

A *remote starter switch* is handy when you need to crank the engine to rotate the crankshaft, Fig. 2-58C. It is connected to the starter solenoid and to the battery. When pressed, the starter is energized to crank the engine. It is commonly used when adjusting engine valves, during engine removal for flywheel-torque converter bolt removal, and to make compression tests.

VOM (Multimeter)

A VOM (volt-ohm-milliammeter), also called multimeter, will measure voltage, current, and resistance. When you suspect a faulty electrical component on an engine, check it with your VOM. If the current, voltage, or ohms test value is NOT within specs, something is wrong with the circuit or component.

An *analog meter* uses a NEEDLE to register the electrical value, Fig. 2-59A. It is an older type design that can still be useful in engine repair. For example, the analog type is handy for finding poor electrical connections. The ohmmeter can be connected to the circuit and the suspected wires wiggled. If the needle also wiggles, you have found a loose electrical connection.

A *digital meter* uses a NUMBER DISPLAY to show the electrical values, Fig. 2-59B. It is a more modern type that is normally recommended when checking engine sensors. A digital meter normally draws less current than an analog meter. It is less likely to damage delicate electronic components.

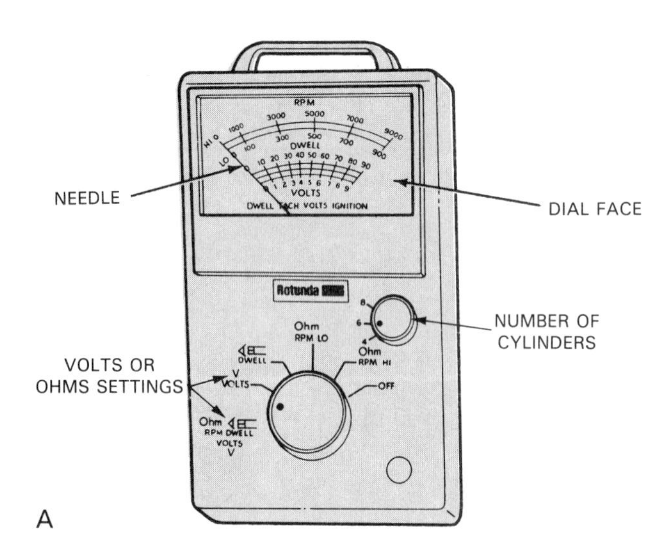

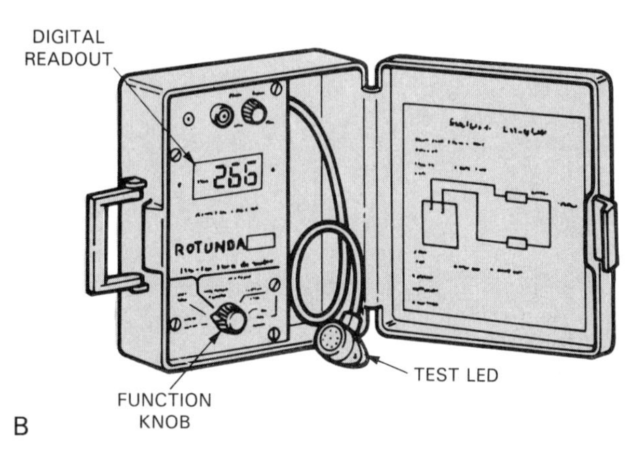

Fig. 2-59. A—Analog meter uses needle that swings across meter face. B—Digital meter uses number output. Both are needed in engine repair with modern computer controlled engines. (Ford)

Special electrical testers

Numerous types of specialized electrical testers are available. A few of these include the following:

A *charging system tester* is an electronic device for quickly checking the voltage output of the alternator. As shown in Fig. 2-60, it is connected to the electrical system. Indicator lights show whether the charging system is working normally.

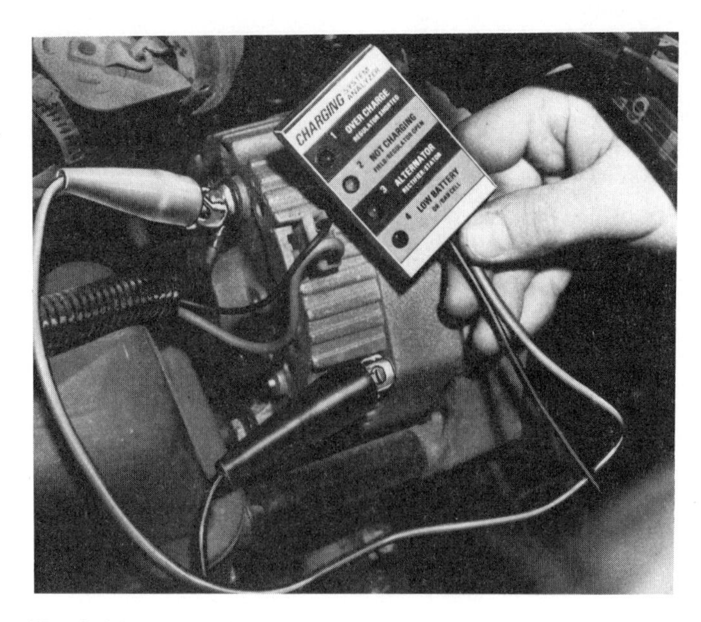

Fig. 2-60. Special charging system tester will quickly show condition of alternator and regulator with indicator lights. (Belden)

An *ignition system tester* will check the major components of an electronic ignition system. The type shown in Fig. 2-61 also uses indicator lights.

A *load tester* will check batteries, the charging system, and the starting system. It will also do other electrical tests. The load tester is commonly used because it is very fast and easy to use. One is shown in Fig. 2-62.

A *computer system tester* will analyze the condition of the on-board computer and its related components. It is normally connected to a specified test point on the

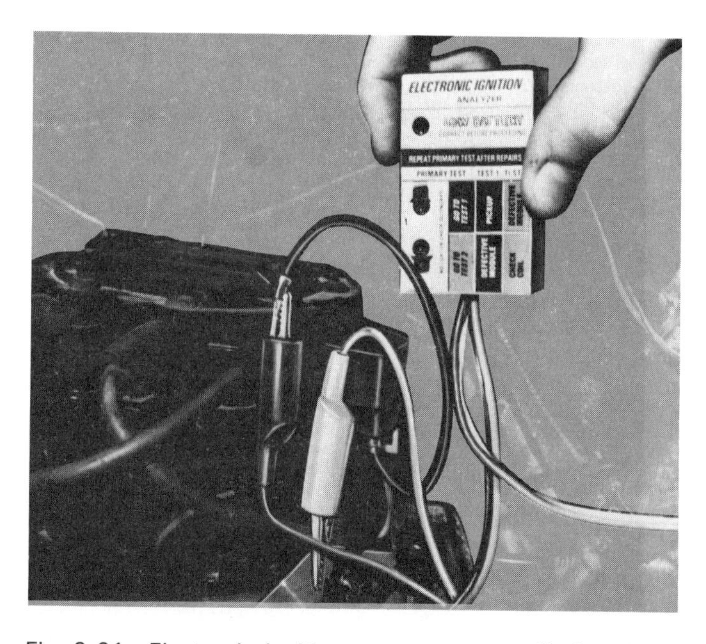

Fig. 2-61. Electronic ignition system tester will also show system condition with indicator lights. Testers vary so read operating instructions. (Belden)

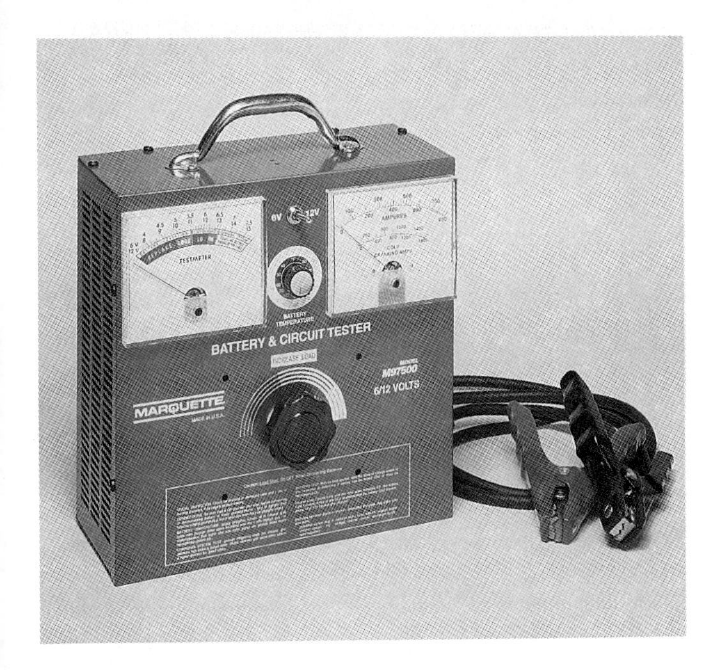

Fig. 2-62. Load tester will quickly check condition of battery, charging system, and starting system. Its use will be detailed in later chapters. (Lincoln)

system. Then, a digital code or engine trouble code (number output, flashing lights, etc.) can be used to determine the location of any faulty parts. The service manual will have a chart telling what each number code or value means. See Fig. 2-63.

Note! Many other electrical-electronic testers are used by an engine technician. These will be detailed in later textbook chapters.

Fig. 2-63. This modern tester will check computer control system. It will perform numerous tests. Variations of this tester are available so always follow operating manual. (Alltest)

Battery charger

A *battery charger* is used to recharge (energize) a discharged (deenergized) car battery. It forces current back through the battery. Normally, the red charger lead connects to the battery positive (+) terminal. The black charger lead connects the negative (−) battery terminal.

WARNING! Always connect the battery charger leads to the battery BEFORE turning the charger ON. This will prevent sparks that could ignite any battery gas. The gases around the top of a battery can EXPLODE violently.

Jumper cables

Jumper cables are used to start a car with a dead (discharged) battery. The cables can be connected between the dead battery and another car's battery. This will let you crank and start the engine.

When connecting jumper cables, connect directly to both battery terminals on the good battery. Connect the positive cable to the positive battery terminal on the discharged battery. Finally, connect the negative terminal to a good vehicle ground so as to prevent sparks from occurring around the discharged battery.

ENGINE TUNE-UP EQUIPMENT

All kinds of tune-up equipment are now needed to keep late model "high tech" engines running at maximum performance. This section of the chapter will summarize the most common types of tune-up equipment. Later chapters will explain their use in MORE DETAIL.

Pressure gauge

A *pressure gauge* measures psi (pounds per square inch) and/or kPa (kilopascals), Fig. 2-64. It is used to check various pressures such as: fuel pump pressure,

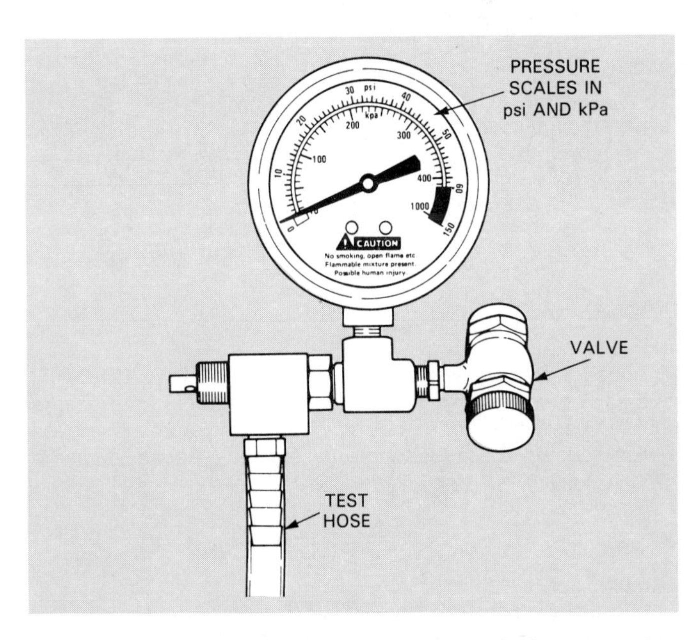

Fig. 2-64. This particular pressure gauge is for measuring pressure in fuel injection system. Engine mechanic will have several pressure gauges for checking engine oil pressure, fuel pressure, negative air pressure or vacuum air injection pump pressure, etc. (Ford)

pressure regulator pressure, engine oil pressure, or turbo boost pressure. An engine technician will usually have several pressure gauges. Each will be designed for a specific job.

Compression gauge

A *compression gauge,* Fig. 2-65, is used to measure the amount of pressure during the engine compression stroke. It provides a means of testing the MECHANICAL CONDITION of the engine. If the compression gauge readings are not within specs, something is mechanically wrong in the engine.

A compression test should be performed when symptoms (engine miss, rough idle, puffing noise in induction system or exhaust) point to major engine problems.

To do a compression test, remove all of the spark plugs (gasoline engine) or glow plugs (diesel engine). Disable the ignition or injection system. Block open the throttle valve (gasoline engine).

Screw the compression tester into a spark plug or glow plug hole. Crank the engine at least four compression strokes while reading the gauge. Measure and record the pressure for each cylinder. Repeat on other cylinders.

A normal compression reading will make the gauge increase evenly to specs. The pressure in each cylinder should NOT vary over about 10 percent.

If the pressure is low in two adjacent cylinders, the head gasket may be blown. A leak in the head gasket can allow compression pressure to blow back and forth between the two cylinders.

If compression pressure is low in only ONE cylinder, that cylinder has a mechanical problem (burned valve or piston, blown head gasket, warped or cracked head, worn piston rings, broken valve spring). There is an opening in the combustion chamber allowing compression pressure to leak out.

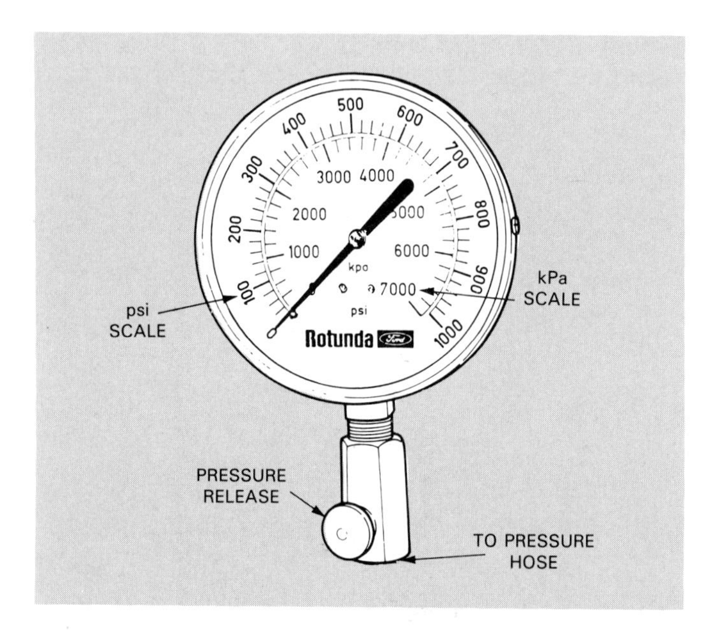

Fig. 2-65. This is a compression gauge for a diesel engine. It reads higher pressures than compression gauge for gasoline engine. Gauge is connected to combustion chamber. Engine is then cranked and compression strokes should build spec pressure for good combustion. If not, something is wrong in engine. (Ford)

If all of the pressure readings are low, you could have a "jumped" timing chain or belt. When the valve timing is off, it can reduce compression pressure in ALL of the cylinders.

Refer to the text index for more information on this subject.

Cylinder leakage tester

A *cylinder leakage tester,* Fig. 2-66, performs about the same function as a compression gauge; it measures the amount of air leakage out of the engine combustion chambers. External air pressure is forced into the cylinder with the piston at TDC on the compression stroke. Then, a pressure gauge can be used to determine the percent of air leakage out of the cylinder.

If leakage is severe enough, you will be able to hear and feel air blowing out of the engine. Air may blow out the intake manifold (bad intake valve), exhaust system (burned exhaust valve), breather (bad rings, piston, or cylinder), or into an adjacent cylinder (blown head gasket).

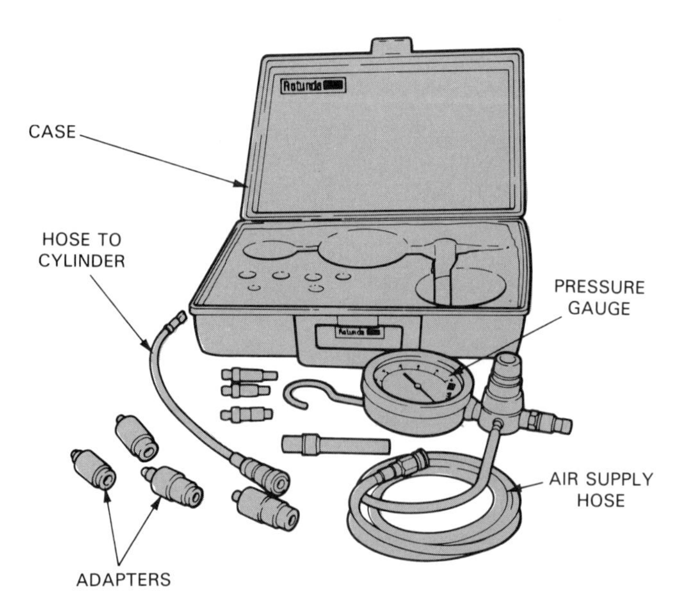

Fig. 2-66. Cylinder leakage tester will also indicate mechanical condition of engine. Air pressure is forced into cylinder. Then, pressure gauge can be used to calculate amount of leakage out of combustion chamber. (Ford Motor Co.)

Vacuum gauge

A *vacuum gauge,* Fig. 2-67, is commonly used to measure negative pressure or vacuum (suction). It is similar to a pressure gauge. However, the gauge reads in inches of mercury (in/hg) and metric kilograms per square centimeter (kg/cm^2). As one example, a vacuum gauge is used to measure the vacuum in an engine's intake manifold. If the reading is low or fluctuating, it may indicate an engine problem.

Fig. 2-68 shows a hand-vacuum pump and gauge assembly. It can be used to test many engine related

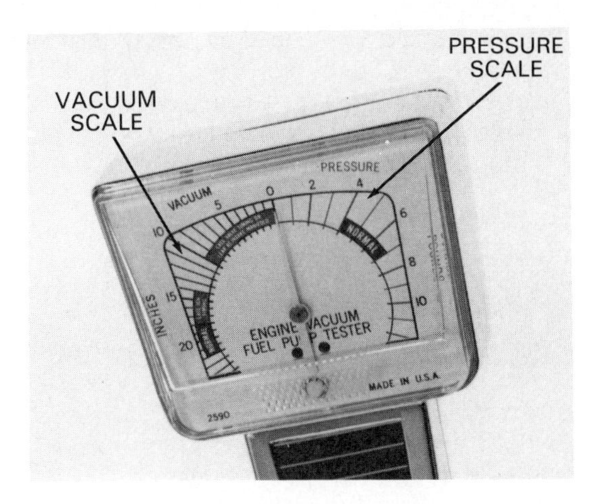

Fig. 2-67. This is a vacuum-pressure gauge. Vacuum gauge, as you will learn in later chapters, can be used to diagnose engine problems. (Sonoco)

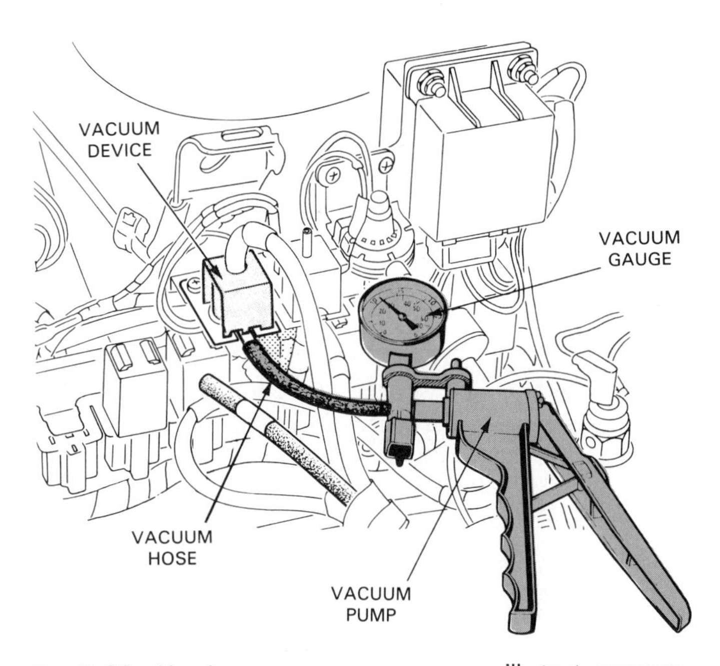

Fig. 2-68. Hand vacuum pump-gauge will test vacuum operated devices that can affect engine operation. Device should function when specific vacuum is applied. It must not leak vacuum either.

vacuum devices. If a component leaks vacuum or does not operate at its specified vacuum, it is faulty.

Spark plug cleaner

A *spark plug cleaner* will blast off deposits from spark plug tips and allow their reuse in an engine. One is shown in Fig. 2-69. It uses shop air pressure and blast compound (sand-like particles) to remove carbon deposits from a spark plug. Some manufacturers do not recommend spark plug cleaning.

Distributor tester

A *distributor tester,* also called *distributor machine,* is used to check the operation of an older ignition system

Fig. 2-69. Spark plug cleaner uses air pressure and blast compound to remove carbon from spark plug tip. This tester also produces arc at plug to test its operation. (Autolite)

distributor. The distributor is removed from the engine and mounted in the tester. After connecting the electrical and vacuum connections, the tester spins the distributor shaft to monitor distributor performance. See Fig. 2-70.

A distributor tester will check:

1. Centrifugal advance.
2. Vacuum advance.
3. Contact point or pickup coil operation.
4. Distributor shaft and cam lobe wear.
5. Condenser condition.

Adapter devices are sometimes needed to check distributors for electronic ignition systems. The adapter is an AMPLIFIER that increases the output of the pickup coil. See Chapter 14 on ignition system service.

Tach-dwell meter

A *tach-dwell meter* is a tachometer and a dwell meter combined into one instrument. The *tach* is for measuring engine rpm (speed). The *dwell meter* measures in

Fig. 2-70. Distributor machine will check operation of ignition system distributor. (Sun Electric)

degrees. For example, it is used for contact point adjustment or computer-controlled carburetor calibration. A tach-dwell is sometimes used during engine tune-ups. See Fig. 2-71.

Follow operating instructions when connecting a tach-dwell meter. Procedures vary. To measure rpm for an idle speed adjustment, you must typically connect the red lead to the negative coil terminal or tach terminal. The black tester lead connects to ground.

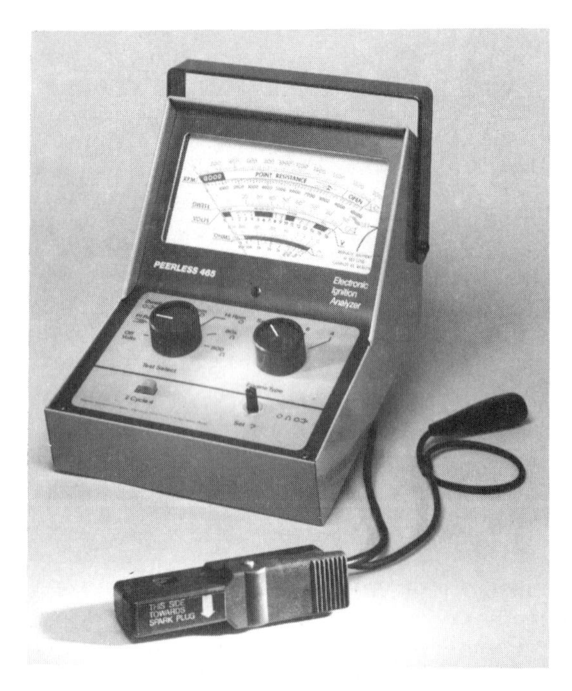

Fig. 2-71. Tach-dwell meter will measure engine speed for throttle adjustment. Dwell meter will measure in degrees for contact point and computer-controlled carburetor service. (Peerless)

Spark tester

A *spark tester* is used to check the basic operation of the engine ignition system. It is made like a spark plug but has a ground clip on it. The spark tester is connected between the end of a spark plug wire and ground. When the engine is cranked or started, a "hot" spark should jump across the tester gap. This shows ignition system operation.

Timing light

A *timing light,* Fig. 2-72, is a strobe (flashing) light used to check and adjust ignition timing. Normally, the two small timing light leads are connected to the car battery. The larger lead is connected to the NUMBER ONE spark plug wire. The timing light is then aimed on the engine timing marks. They are usually located on the front cover of the engine, crank damper, or on the flywheel.

The timing light, by flashing on and off, will make the spinning crankshaft pulley, balancer, or flywheel appear to stand still. This makes the moving timing mark visible. You can then adjust the ignition timing by turning

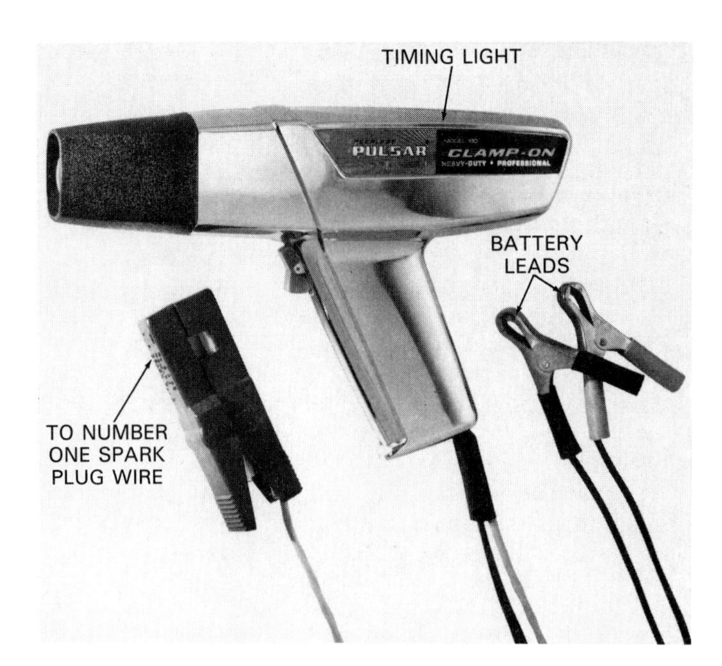

Fig. 2-72. Timing light is strobe light. It is aimed on timing marks and spinning damper or flywheel to adjust when spark plugs fire in cylinders in relation to piston position. Note lead connection points. (Peerless)

the distributor housing. When the computer controls the ignition system, there may be an adjustment on the computer or timing may NOT be adjustable.

Mag-tach, timing meter

A *mag-tach, timing meter,* Fig. 2-73, is a magnetically triggered tachometer for measuring engine rpm and for adjusting diesel injection timing. It is used on both diesel and gasoline engines. A diesel engine does NOT have an electricallly-operated ignition system to power a conventional tachometer or timing light.

The mag-tach is operated by a magnet that senses a

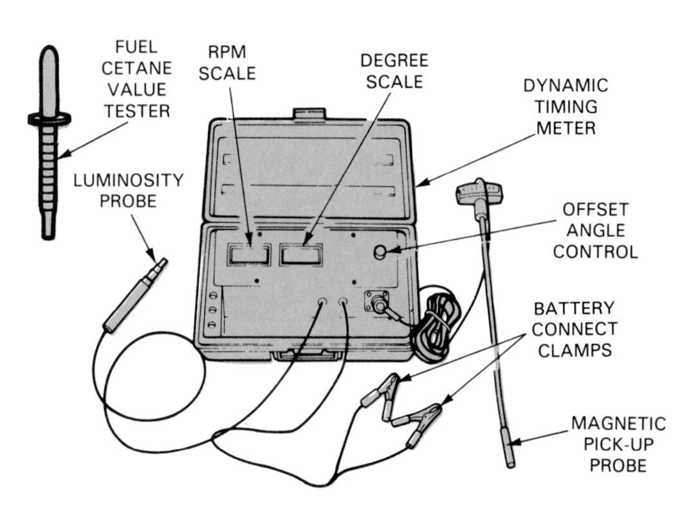

Fig. 2-73. This is a dynamic timing meter for servicing diesel engine. Magnetic probe senses crankshaft and damper position. Luminosity probe detects combustion flame in cylinder for injection timing adjustment. (Ford)

notch in the engine flywheel or damper. In this way, engine speed can be measured without a connection to the ignition system.

Some mag-tachs, also called *timing testers,* are capable of measuring ignition timing advance and injection timing advance. A luminosity probe will detect a combustion flame for setting injection timing on diesels.

Temperature probe

Special *thermometers* or *temperature gauges,* or *probes* are frequently used to measure the temperature of numerous components (radiator temperature, for example). Most temperature probes or gauges can read in either Fahrenheit (F) or Celsius (C). The temperature obtained can be compared to specs. Then, if the temperature is too low or too high, you would know that a repair or adjustment is needed. Fig. 2-74 shows one type temperature probe.

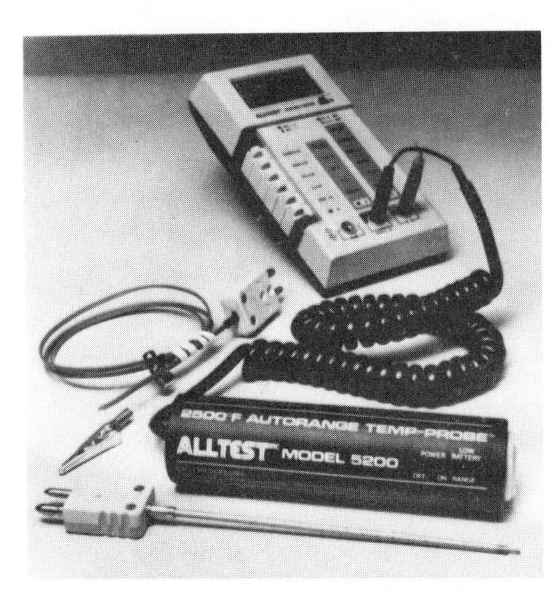

Fig. 2-74. Temperature probe will accurately measure various engine related temperatures—coolant temperature, exhaust manifold temperature, etc.

EFI tester

An *EFI tester* will check the operation of modern electronic fuel injection systems. Most EFI testers come with numerous accessory devices. The tester will usually test all of the systems sensors, circuit wiring, and computer. A pressure gauge may be provided for measuring fuel pump and pressure regulator outputs.

EFI test equipment varies. Make sure you follow the operation instructions carefully. Frequently, auto makers recommend a particular brand of tester for their vehicles. The service manual will explain the use of the exact type tester.

Exhaust gas analyzer

An *exhaust gas analyzer,* Fig. 2-75, measures the chemical content of the engine's exhaust. This allows the mechanic to check combustion efficiency and the

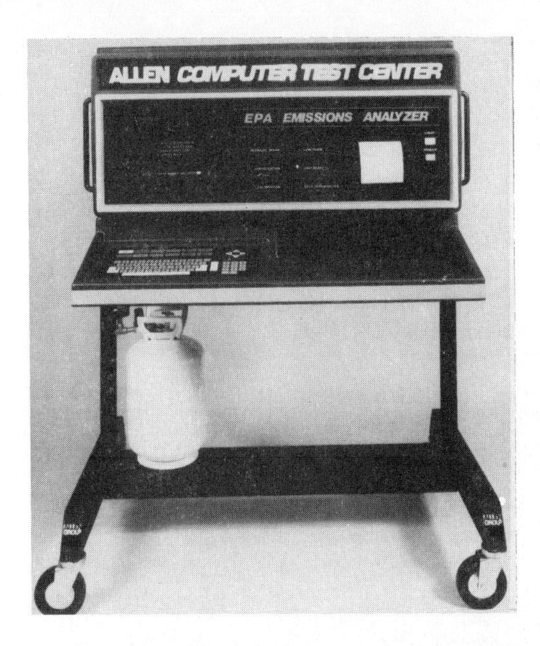

Fig. 2-75. Emissions analyzer will measure chemical content of exhaust for checking engine and combustion efficiency. It can be used to diagnose various engine problems.

amount of pollutants produced by the engine. An exhaust analyzer will check engine, fuel system, and emission control system operation.

Engine analyzer

An *engine analyzer* is a group of different testing instruments mounted in one assembly. One is pictured in Fig. 2-76. An engine analyzer may have a tachometer, dwell meter, vacuum-pressure gauge, VOM, exhaust gas analyzer, and a scope (TV screen that displays voltages) in a single roll-around cabinet.

Fig. 2-76. Engine analyzer contains various testing devices in one cabinet. (Sun Electric Corp.)

TESTING SAFETY RULES

When using test equipment, there are several rules to remember.

1. Read the operating instructions for the test equipment. Failure to follow directions could cause bodily injury and severe damage to the part or instrument.
2. If the engine is to be running during your tests, set the parking brake. Block the wheels. Place an automatic transmission in park or manual transmission in neutral. Connect an exhaust vent hose to the tailpipe if you are working in an enclosed auto shop.
3. Keep test equipment leads (wires) or hoses away from engine belts, the fan, and hot engine parts. They can be damaged easily.
4. Wear eye protection when working around an engine fan. Remove jewelry and secure long hair.
5. Never look into a carburetor or throttle body when cranking or running the engine. Do not cover the air inlet with your hand. If the engine were to backfire, it could cause serious burns.
6. Refer to the auto manufacturer's service manual for specific testing procedures.

OTHER ENGINE TOOLS AND EQUIPMENT

This chapter has only covered the most commonly used engine tools and equipment. Throughout this text, you will learn about many other more specialized tools and equipment. Refer to the text index if you need specific information on a tool not covered in this chapter.

SUMMARY

A good engine technician knows when and how to use the many tools and equipment needed to service today's engines. Selecting the right tool for the job will increase work efficiency and work quality.

Always use an engine stand when disassembling an engine. A crane is commonly used to remove and install an engine. Remove the ring ridge before piston removal. A cylinder hone is needed to deglaze cylinders to make new piston rings seal properly.

A ring expander will speed piston ring replacement and will help prevent ring breakage. A ring compressor is needed to push the rings into their grooves for piston installation into the block.

A cylinder head stand will make head work easier. A spring compressor is needed to remove the valves from the head and to install valves. A valve spring tester will make sure the springs are reusable.

A valve grind machine will resurface the faces and stem tips of valves. A valve seat grinder will resurface the cylinder head valve seats to produce a good seal.

A boring bar is a machine shop tool for making cylinders oversized to repair block wear or damage. A crankshaft grinder and camshaft grinder are also machine shop tools for restoring crankshafts and camshafts.

A cold solvent tank is for removing oil and grease deposits from parts. A hot tank is for removing more stubborn deposits: paint, grease, and mineral deposits in water jackets.

Numerous measuring tools are needed in engine repair. Some of these include: micrometers, feeler gauges, calipers, dial indicators, Plastigage, and torque wrenches. You must know how to use all of these tools properly.

Modern engines use numerous sensors to feed information back to the on-board computer. As an engine technician, you must be able to test and replace these engine sensors. Voltmeters, ohmmeters, ammeters, and special testers are normally used to diagnose electrical-electronic troubles that could upset engine operation.

Tune-up equipment is also used by the engine technician. He or she must be able to utilize pressure gauges, compression gauges, cylinder leakage testers, vacuum gauges, distributor testers, tach-dwell meters, timing lights, temperature probes, electronic fuel injection analyzers, and engine analyzers.

Always follow all safety rules and equipment operating instructions. Tools and equipment vary and so do the methods for using them!

KNOW THESE TERMS

Engine stand, Engine crane, Ridge reamer, Cylinder hone, Ring groove cleaner, Ring expander, Ring compressor, Knurling, Valve spring compressor, Valve spring tester, Cam bearing driver, Valve grind machine, Valve seat grinder, Lifter tester, Prelubricator, Boring bar, Crankshaft grinder, Camshaft grinder, Engine balancer, Wheel puller, Seal driver, Stethoscope, Hydraulic press, Blow gun, Steam cleaner, Cold solvent tank, Hot tank, Metric system, Decimal conversion chart, Feeler gauge, Straightedge, Calipers, Dividers, Torque wrench, Micrometer, Telescoping gauge, Dial indicator, Plastigage, Jumper wire, Test light, Remote starter switch, VOM, Digital meter, Analog meter, Compression gauge, Cylinder leakage tester, Vacuum gauge, Distributor tester, Tach-dwell meter, Timing light, Mag-tach, Temperature probe, EFI tester, Exhaust gas analyzer, Engine analyzer.

REVIEW QUESTIONS—CHAPTER 2

1. An __________ __________ allows for easy engine service because the cylinder block can be rotated into different positions.
2. This tool is used to cut the metal lip out of the top of a worn cylinder for piston removal.
   a. Boring bar.
   b. Ridge reamer.
   c. Hone.
   d. Ring groove cleaner
3. A cylinder hone is used to __________ a cylinder wall to let the __________ __________ seal properly.
4. What would happen if the carbon is not cleaned from inside of the piston ring grooves on a used piston?
5. Explain the difference between a piston ring expander and a piston ring compressor.
6. A valve spring compressor fits onto the cylinder head deck and onto the valve spring retainer. True or false?

7. How can you tell if valve springs should be replaced?
8. A __________ __________ is a machine shop tool for making the cylinders larger in diameter when they are worn or damaged.
9. When should you have a machine shop balance an engine?
10. You should wear eye protection when using a wheel puller because it can exert tons of force. True or false?
11. A __________ __________ is commonly used for installing and removing piston pins.
12. What is the different between a cold solvent tank and a hot tank?
13. Define the term "specs."
14. How do you use a flat feeler gauge and straightedge?
15. A dial caliper will measure to an accuracy of approximately __________ or __________.
16. Summarize the basic steps for using a conventional, outside micrometer.
17. A __________ __________ will measure part movement to within a thousandth of an inch.
18. What is a VOM?
19. Define the term "engine trouble code."
20. A car engine is missing and a puffing sound can be heard at the tailpipe. It was just tuned-up by another shop.

    Technician A says that a comparison test should be done to check the mechanical condition of the engine.

    Technician B says that the carburetor should be rebuilt because that is a common cause of a miss on this particular car.

    Who is correct?
    a. Technician A.
    b. Technician B.
    c. Both A and B.
    d. Neither A nor B.

ASE CERTIFICATION-TYPE QUESTIONS

1. A ridge reamer is used to remove the metal lip formed at the top of a worn __________.
   (A) valve guide
   (B) engine cylinder
   (C) piston
   (D) valve
2. Technician A says a cylinder hone is used to increase the diameter of an engine's cylinder wall. Technician B says a cylinder hone is used to deglaze an engine's cylinder wall. Who is right?
   (A) A only.
   (B) B only.
   (C) Both A & B.
   (D) Neither A nor B.
3. All of the following are common types of cylinder hones EXCEPT:
   (A) grinding hone.
   (B) brush hone.
   (C) flex hone.
   (D) rigid hone.
4. Technician A says piston knurling is used to restore slightly worn wrist pins to within specifications. Technician B says piston knurling is used to restore slightly worn pistons to within specifications. Who is right?
   (A) A only.
   (B) B only.
   (C) Both A & B.
   (D) Neither A nor B.
5. Technician A says a crankshaft grinder is used to resurface worn crankshaft journals. Technician B says a crankshaft grinder is used to resurface worn crankshaft counterweights. Who is right?
   (A) A only.
   (B) B only.
   (C) Both A & B.
   (D) Neither A nor B.
6. A __________ should be used to clean a cast iron engine block during an engine overhaul.
   (A) acid dip
   (B) cold solvent tank
   (C) steam cleaner
   (D) hot tank
7. All of the following cleaning equipment should be used when rebuilding an automotive engine EXCEPT:
   (A) blow gun.
   (B) rotary brush.
   (C) Plastigage.
   (D) bench grinder.
8. Technician A says an outside micrometer is for measuring external dimensions, diameters, or thicknesses. Technician B says an outside micrometer is used to measure internal depths, lengths, and diameters. Who is right?
   (A) A only.
   (B) B only.
   (C) Both A & B.
   (D) Neither A nor B.
9. Technician A says a compression gauge can be used to detect compression leakage in an engine's combustion chamber. Technician B says a cylinder leakage tester can detect compression leakage in an engine's combustion chamber. Who is right?
   (A) A only.
   (B) B only.
   (C) Both A & B.
   (D) Neither A nor B.
10. A distributor tester will check the __________.
    (A) pickup coil.
    (B) distributor shaft for wear
    (C) rotor for wear
    (D) All of the above.

Fig. 3-1. The modern engine uses a wide variety of bolts, nuts, belts, gaskets, seals, sealants, etc. (Ford)

Chapter 3

Engine Hardware

After studying this chapter, you will be able to:

- *Explain the purpose of the various fasteners used on automotive engines.*
- *Summarize the classification of threads and other bolt specifications.*
- *Properly torque fasteners to specs.*
- *Identify the various types of nuts and screw heads.*
- *Repair fastener and thread damage.*
- *Efficiently remove and install engine gaskets.*
- *Describe the use of chemical sealers.*
- *Diagnose gasket and seal failures.*
- *Replace metal lines and rubber hoses properly.*
- *Service engine belts.*

This chapter will overview the various types of hardware found on today's engines. This includes bolts, nuts, snap rings, lock pins, machine screws, studs, castle nuts, and washers. This also includes gaskets, seals, sealants, hoses, belts, and tubing. It is essential that you know how to properly handle these devices when working on an engine, Fig. 3-1.

This chapter will help you develop basic skills. It will give you important information that will prepare you for later textbook chapters on engine overhaul, cooling system service, fuel system service, etc. Study this chapter carefully and you will be taking another step toward becoming an engine technician.

FASTENERS

Fasteners include any type of holding device. Hundreds are used in the construction of an engine. The most common ones are pictured in Fig. 3-2.

Bolts are metal shafts with threads on one end and a head on the other end. When threaded into a part, without the use of a nut, they are called *cap screws.*

Nuts have internal threads and are normally hex (six-sided) in shape. Various types of nuts can be found on an engine, Fig. 3-3.

Bolt and nut dimensions

The basic dimensions of a bolt are:

1. *Bolt size* is a measurement of the outside diameter of the bolt threads, Fig. 3-4.
2. *Bolt head size* is the distance across the flats of the bolt head. It is the same as the *wrench size.*
3. *Bolt length* is measured from the bottom of the bolt head to the threaded end of the bolt.
4. *Thread pitch* is the same as thread COARSENESS.

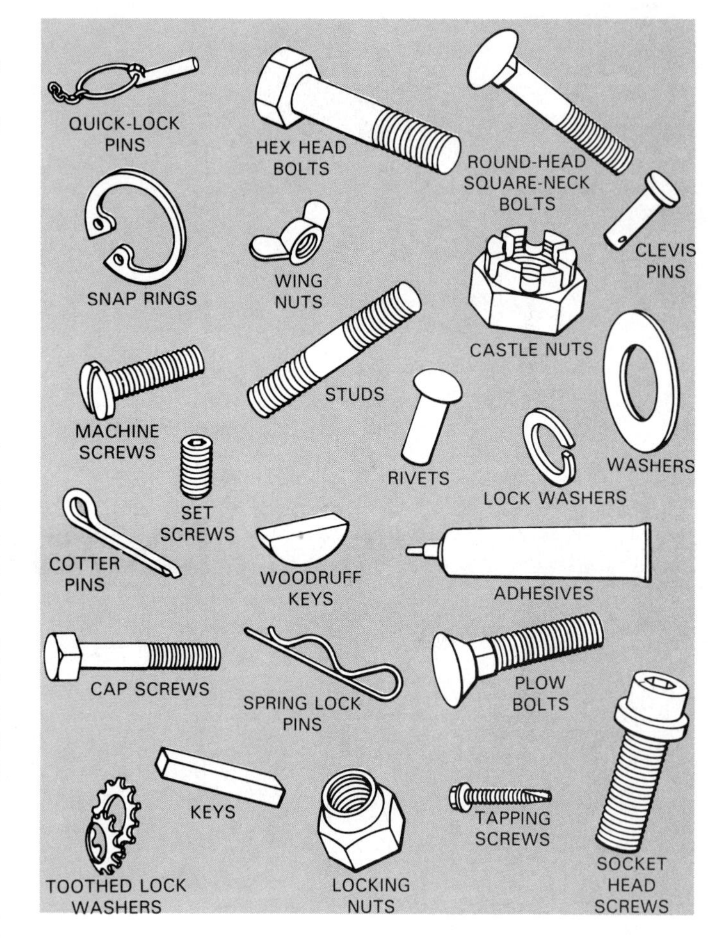

Fig. 3-2. Study names of basic types of fasteners. (Deere & Co.)

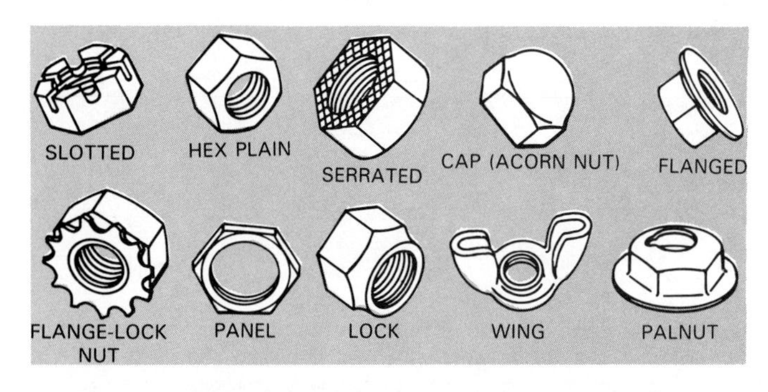

Fig. 3-3. Note various nut types found on engines. (Deere & Co.)

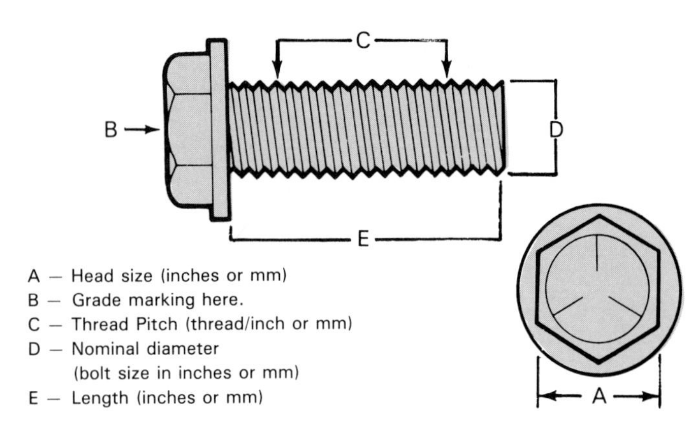

Fig. 3-4. These are basic dimensions of bolt. You need them when ordering bolts.

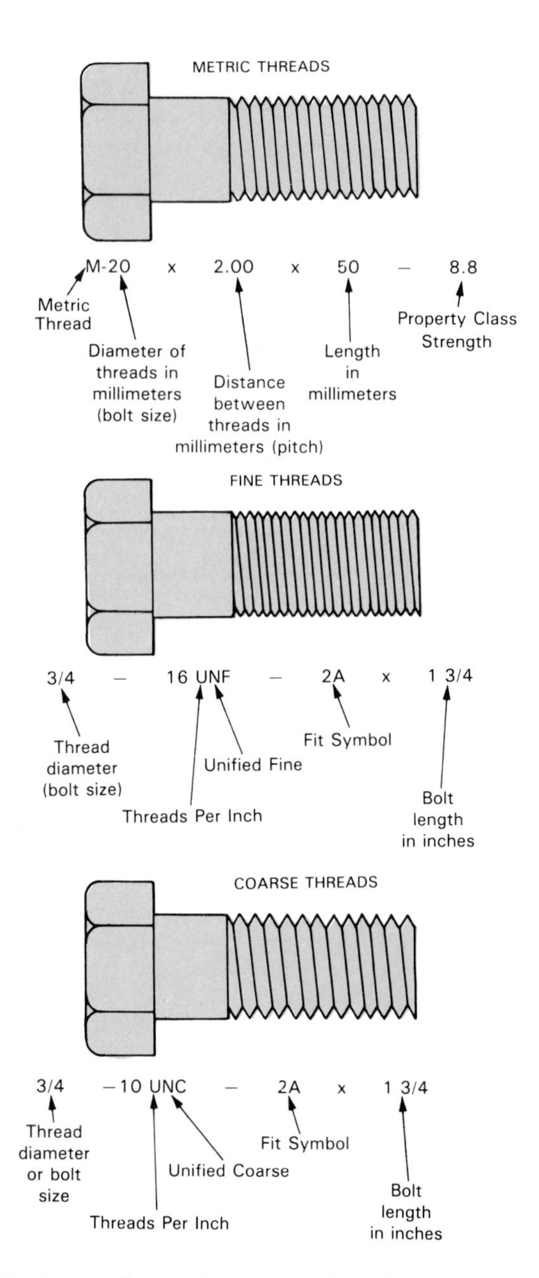

Fig. 3-5. Coarse, fine, and metric are three thread types found on engines. Never interchange them or thread damage will result. Study bolt designation numbers.

With U.S. fasteners, it is the number of threads per inch. With metric fasteners, it is the distance between each thread in millimeters.

Thread types

There are three basic types of threads used on fasteners:

1. Coarse threads (UNC or Unified National Coarse).
2. Fine threads (UNF or Unified National Fine).
3. Metric threads (SI or Scientific International).

Never interchange thread types or thread damage will result, Fig. 3-5. Metric threads could be easily confused for fine threads if not inspected carefully. If a bolt is forced into a hole with the wrong threads, either the bolt or part threads can be ruined.

A *thread pitch gauge* will check and determine thread type, Fig. 3-6. Practice, however, should make it easy for you to identify threads on sight.

Left and right-hand threads

Bolts and nuts also come in right and left-hand threads. With common *right-hand threads,* the fastener must be turned CLOCKWISE to tighten. With the less common *left-hand threads,* turn the fastener in a COUNTERCLOCKWISE direction to tighten. The letter ''L'' may be stamped on fasteners with left-hand threads.

Bolt grade (head markings)

Tensile strength or *grade* refers to the strength of a bolt. It indicates the amount of torque a fastener can withstand before breaking. Tensile strengths vary. Bolts are made of different metals; some are stronger.

Bolt head markings, also called *grade markings,* specify the tensile strength of the bolt. Conventional bolts are marked with LINES or SLASH MARKS: the more lines, the stronger the bolt. A metric bolt is marked by a NUMBERING SYSTEM: the larger the number, the stronger the bolt. Refer to Fig. 3-7.

Never replace a high grade bolt with a lower grade bolt. The weaker bolt could snap, possibly causing part failure and a dangerous condition.

Bolt description

A *bolt description* is a series of numbers and letters that give specifications for the bolt. This is also explained

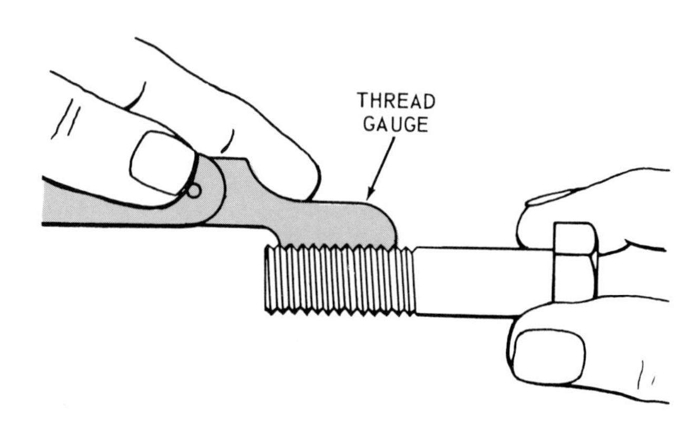

Fig. 3-6. Thread gauge will check thread type. Practice will let you tell threads by sight. (Deere & Co.)

CAUTION
The torque specifications listed below are approximate guidelines only and may vary depending on conditions when used such as amount and type of lubricant, type of plating on bolt, etc.

SAE STANDARD / FOOT POUNDS

GRADE OF BOLT	SAE 1 & 2	SAE 5	SAE 6	SAE 8		
MIN. TEN STRENGTH	64,000 P.S.I.	105,000 P.S.I.	133,000 P.S.I	150,000 P.S.I.		
MARKINGS ON HEAD					SIZE OF SOCKET OR WRENCH OPENING	
U.S. STANDARD					U.S. REGULAR	
BOLT DIA.	FOOT POUNDS				BOLT HEAD	NUT
1/4	5	7	10	10.5	3/8	7/16
5/16	9	14	19	22	1/2	9/16
3/8	15	25	34	37	9/16	5/8
7/16	24	40	55	60	5/8	3/4
1/2	37	60	85	92	3/4	13/16
9/16	53	88	120	132	7/8	7/8
5/8	74	120	167	180	15/16	1.
3/4	120	200	280	296	1-1/8	1-1/8

METRIC STANDARD

GRADE OF BOLT		5D	.8G	10K	12K	
MIN. TENSILE STRENGTH		71,160 P.S.I	113,800 P.S.I.	142,200 P.S.I.	170,679 P.S.I.	
GRADE MARKINGS ON HEAD		5D	8G	10K	12K	SIZE OF SOCKET OR WRENCH OPENING
METRIC						METRIC
BOLT DIA.	U.S. DEC EQUIV.	FOOT POUNDS				BOLT HEAD
6mm	.2362	5	6	8	10	10mm
8mm	.3150	10	16	22	27	14mm
10mm	.3937	19	31	40	49	17mm
12mm	.4720	34	54	70	86	19mm
14mm	.5512	55	89	117	137	22mm
16mm	.6299	83	132	175	208	24mm
18mm	.709	111	182	236	283	27mm
22mm	.8661	182	284	394	464	32mm

Fig. 3-7. Basic bolt torque chart can be used when engine specs are not available. Size, grade, and threads of bolt determine how tight it should be torqued.

in Fig. 3-5. When purchasing new bolts, you will need the bolt description information.

WASHERS

Washers are used under bolt heads and nuts. There are two basic types: flat washers and lock washers. Refer back to Fig. 3-2.

A *flat washer* or *plain washer increases the clamping surface under the fastener. It prevents the bolt or nut head from gouging or digging into the part.*

A lock washer prevents the bolt or nut from loosening. Under stress and vibration, a fastener could unscrew without a lock washer.

Lock tabs or *plates* perform the functions of both flat and lock washers. They increase clamping surface area

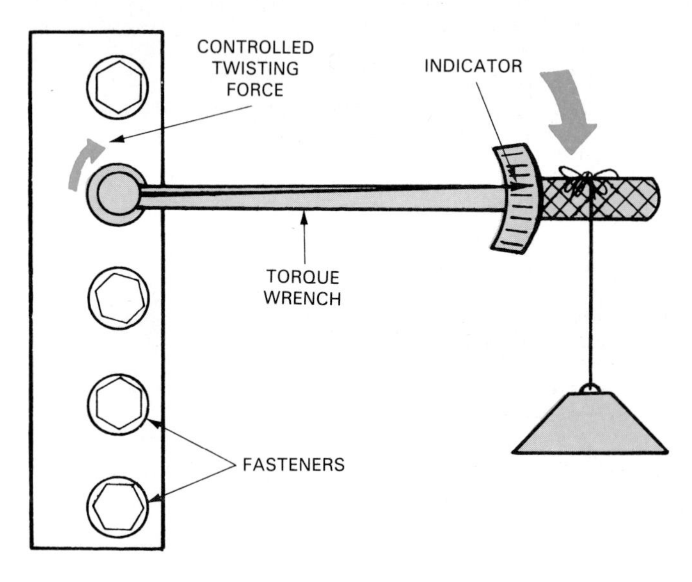

Fig. 3-8. Torque is measured in pound-feet or newton-meters. One pound-foot of torque is produced when one pound of weight is suspended from a one foot long lever arm.

and secure the fastener. They are found on engine exhaust manifolds.

TORQUING BOLTS

It is critical that engine bolts and nuts are *torqued* or tightened correctly, Fig. 3-8.

If *overtightened,* a bolt will stretch and possibly break. The threads could also fail. If *undertightened,* a bolt could work loose. Part movement could shear the fastener or break a gasket, causing leakage.

Torque specifications are tightening values given by the auto manufacturer. Torque specs are normally given for all precision assemblies (head bolts, rod bolts, oil pump bolts, etc.).

Bolt tightening sequence

A *bolt tightening sequence* or pattern makes sure the parts clamp together evenly. An incorrect or uneven pattern could break parts, cause warpage, and leakage.

A *crisscross pattern* is generally used because it goes from side to side and end to end, starting in the middle. Each bolt is tightened a little at a time, jumping back and forth from the middle outward. See Fig. 3-9.

A shop manual will show the proper sequence for critical assemblies: such as cylinder heads, intake manifolds, and exhaust manifolds. This will be detailed later in the text.

Using a torque wrench

Follow these basic rules when using a torque wrench:

1. Place a steady pull on the torque wrench. Do NOT use short, jerky motions or inaccurate readings can result.
2. Make sure that the fastener threads are clean and lightly oiled.
3. When possible, avoid using swivel joints. They can upset torque wrench accuracy.
4. When reading a torque wrench, look straight down at the scale. Viewing from an angle can give a false reading.

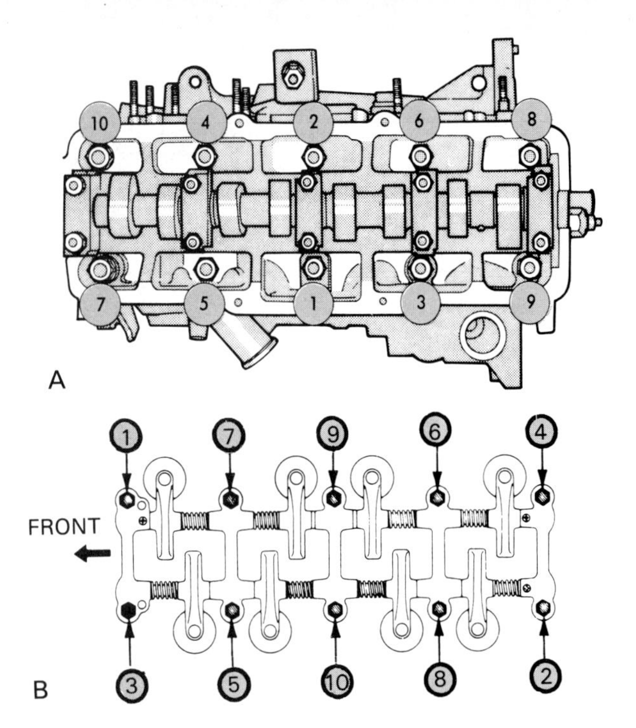

Fig. 3-9. This is a fundamental crisscross tightening sequence. Use it when several bolts hold one part. It will assure even clamping pressure between parts. Service manual will give more specific sequence for critical engine assemblies.

5. A general torque value chart should only be used when manufacturer's specs are NOT available.
6. When manufacturer torque patterns are not available, use a general crisscross pattern.
7. Only pull on the handle of the torque wrench. Do NOT allow the wrench beam to touch anything.
8. Tighten bolts and nuts in at least four steps: to one-half recommended torque, to three-fourth torque, to full torque, and to full torque a second time.
9. Retorque when required. Engine cylinder heads, intake manifold, and exhaust manifold may need to be retightened after operation and heating.
10. Replace bolts when recommended by the auto maker.

Torque-to-Yield

Service manuals sometimes recommend using new bolts because of a torque-to-yield process. *Torque-to-Yield* is a bolt tightening method that requires a specific bolt torque, followed by turning the bolt a specific number of degrees. After using a torque wrench, a *degree wheel adapter* is placed between the socket and wrench. The fastener is then turned until the degree wheel reads as specified by the manufacturer. This stretches the bolt to its correct yield point and preloads the fastener for better part clamping under varying conditions.

Torque stretch is determined by measuring the change in bolt length. For example, when building an engine, you could "mike" connecting rod bolts to measure length before and after tightening. Too much stretch would indicate bolt weakness. Not enough stretch might indicate thread problems that can affect bolt torque. If the service manual recommends new bolts because of torque-to-yield, it is necessary to discard the old bolts.

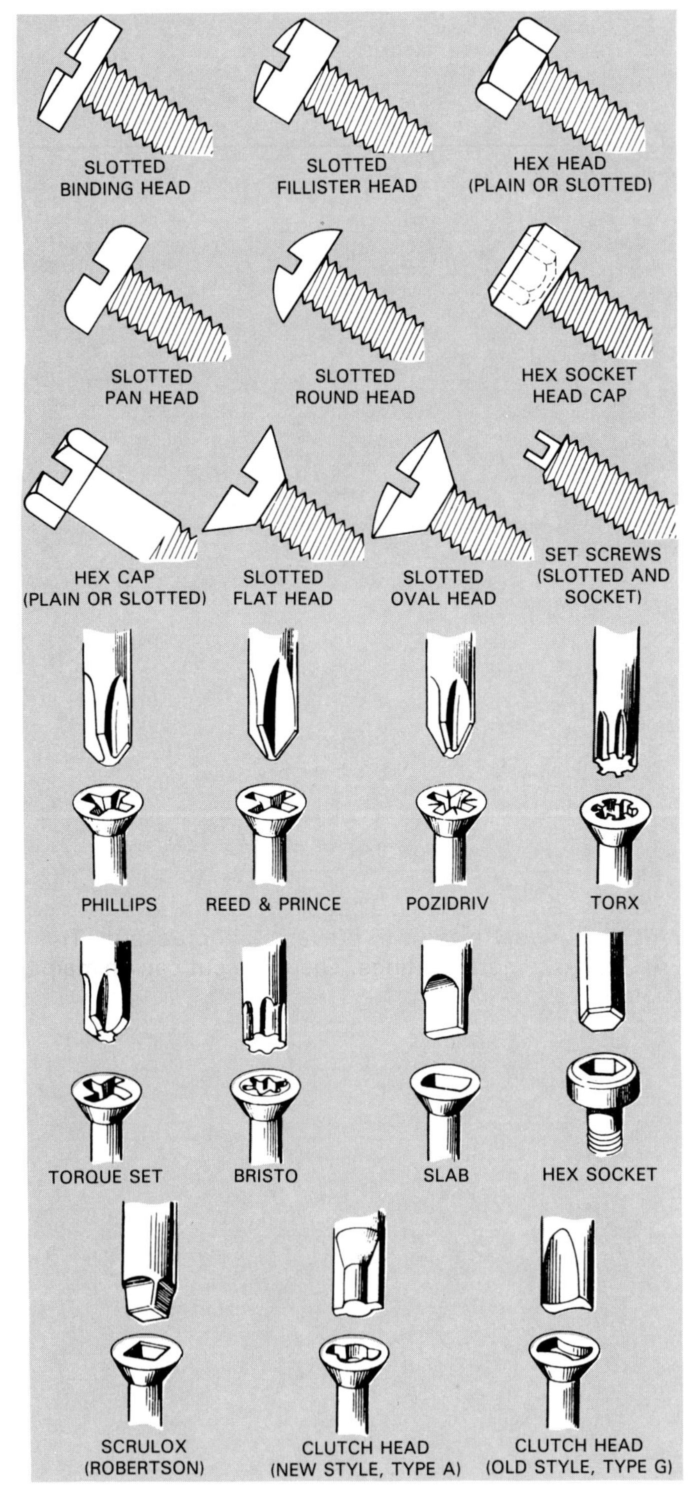

Fig. 3-10. Study machine screw or head types. (Heyco and Klein Tools)

Machine screws

Machine screws are similar to bolts, but they normally have screwdriver-type heads. They are threaded their full length and are relatively weak and small, Fig. 3-2. Machine screws are used to secure parts when clamping loads are light. They come in various head shapes, as pictured in Fig. 3-10. You might find machine screws on accessory units mounted on an engine.

THREAD REPAIRS

You must be capable of quickly and properly repairing damaged threads. Damaged threaded holes in parts may require repairs to salvage the parts. Fig. 3-11 shows several thread repair tools.

Minor thread repairs

Minor thread damage includes nicks in threads, partial flattened threads, and other less serious problems. Minor thread damage can usually be repaired with a thread chaser.

A *thread chaser* is a rethreading tool for cleaning up damaged threads. This type tool is available for fixing both internal and external threads. The chaser is run through or over the threads to restore them, Fig. 3-11.

Major thread damage generally includes badly smashed threads, stripped threads, or threads that cannot be repaired easily.

Taps and dies

A *tap* is a tool for cutting internal threads in holes, Fig. 3-12. Various tap shapes are provided. Some are for starting the threads. Others are for cutting the threads all the way to the bottom of a hole.

A *die* is for cutting external threads. It is used to cut threads on rods, bolts, shafts, and pins, Fig. 3-12.

Taps and dies are mounted in special handles called *tap handles* and *die handles.* Hold the handle squarely while rotating it into the work.

As soon as the tap or die begins to bind, back the tool off about a quarter turn to prevent tool breakage. This will clean out metal cuttings. Then, the cut can be made another half turn deeper. Keep rotating a half turn in and a quarter turn out until completing the cut.

A few rules for using taps and dies are:

1. Never force a tap handle or the tool may break. Back off the handle as described to clean out metal shavings.
2. Keep the tap and die well oiled to ease cutting.
3. Always use the right size tap in a correctly sized and drilled hole.

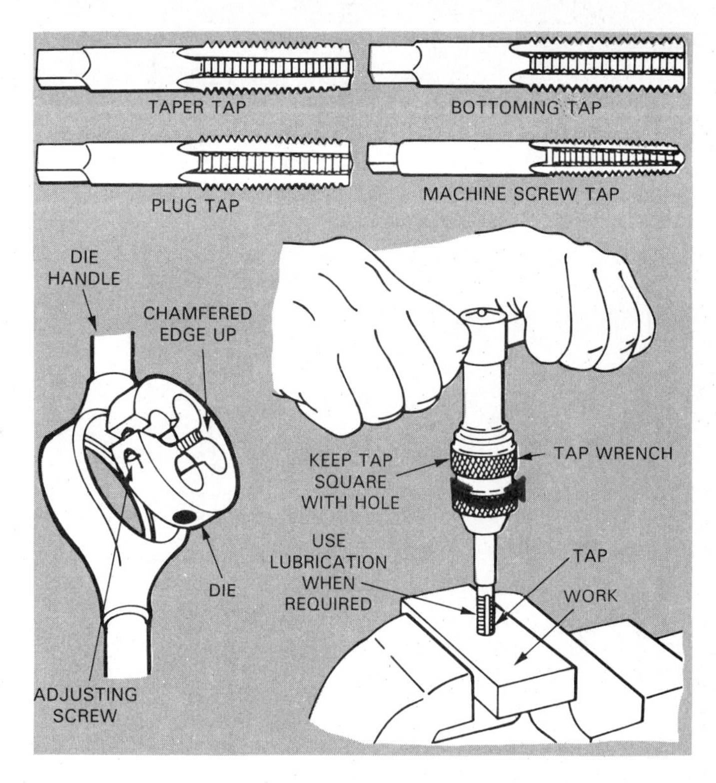

Fig. 3-12. Tap is for cutting threads in hole. Die is for cutting threads on shaft or bolt. When using tap or die, rotate clockwise until you feel a slight resistance to rotation. Then back off handle to clean threads before turning in a little deeper.

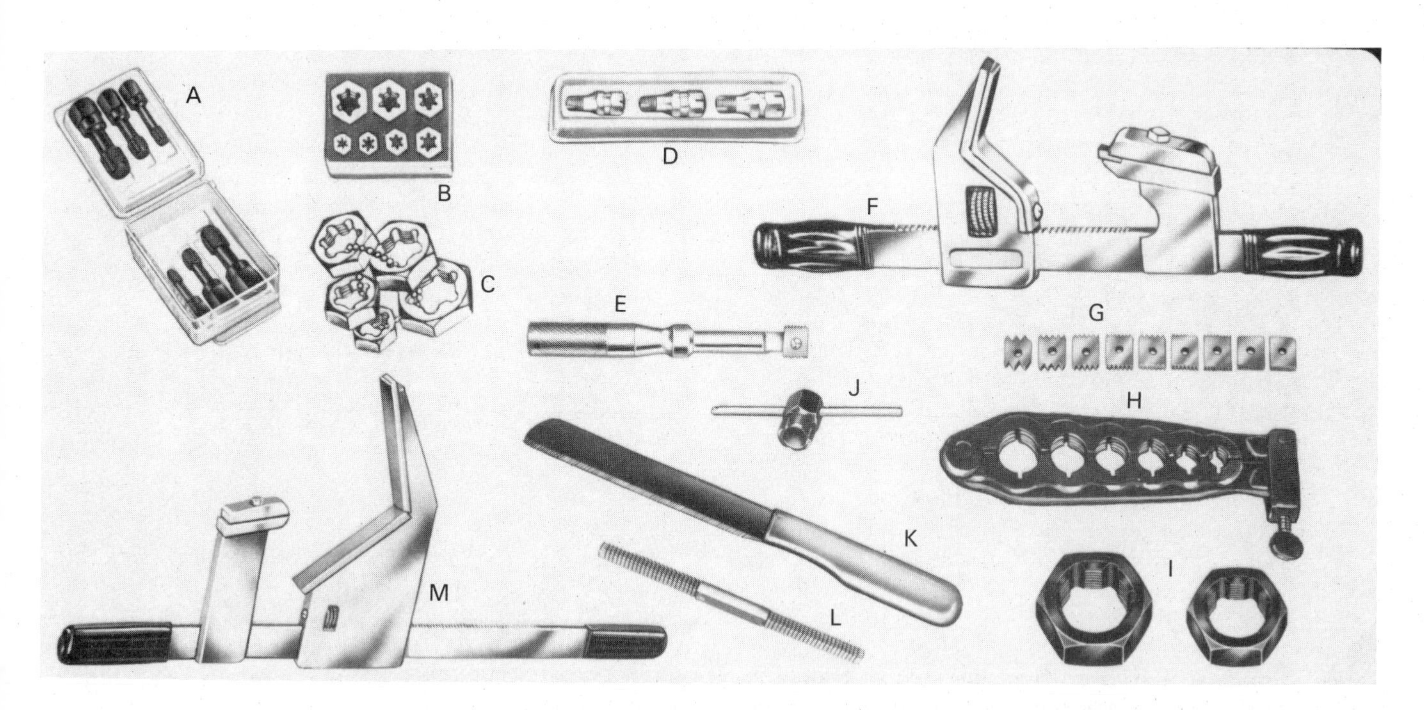

Fig. 3-11. A—Spark plug hole thread chasers. B & C—Nut type chasers. D—Metric chasers. E through J—External chasers. K & L—Thread files. M—External thread chaser. (Snap-On).

Tapping oversize

When a thread chaser cannot be used to clean up damaged threads, the hole can be drilled and tapped oversize. First, drill out the hole one diameter or size larger. Then, cut new threads in the drilled hole with the correct size tap. A larger bolt can then be installed.

A *drill and tap size chart, given in Fig. 3-13, will help you select the right size drill bit and tap. One example: the chart shows that when a 1/2 in. hole is drilled, a 27/64 in. coarse tap is required.*

Removing damaged fasteners

An engine technician must be able to remove broken bolts and bolts with rusted or rounded-off heads. This is an important skill that is not covered in service manuals.

Certain tools and methods are needed for removing problem fasteners:

1. *Vise grip pliers* can sometimes be used to remove broken bolts or bolts with rounded heads. Lock vise grip pliers tightly on bolt for removal, Fig. 3-14A.
2. A *stud puller* or stud wrench will remove studs and bolts broken off above the surface of the part. It will also install studs. Position the stud puller so that it will not clamp on and damage the threads.
3. In some cases, broken fasteners are too short to grasp with any tool. Either cut a *screwdriver slot* in the bolt with a hacksaw or weld on another BOLT HEAD. Then a screwdriver or wrench can be used to unscrew the broken bolt.
4. When the fastener is broken FLUSH with the part surface, a *hammer* and *punch* can be used for removal. Angle the punch so that blows from the hammer can drive out and unscrew the broken bolt.
5. A *screw extractor* or "easy-out" can also be used to remove bolts broken flush or below the part surface. To use a screw extractor, drill a hole in the exact center of the broken fastener. Then, lightly drive the extractor into the hole using a hammer. Unscrew the broken bolt using a wrench, Fig. 3-14B.

CAUTION! Be extremely careful not to break a tap or a screw extractor. They are case hardened and cannot be easily drilled out of a hole. You will compound your problems if you overtwist and break one of these tools.

6. On some broken bolts, you may have to drill a hole almost as large as the inside diameter of the threads. Then, use a tap or punch to remove the *thread shell* (thin layer of threads remaining in hole).

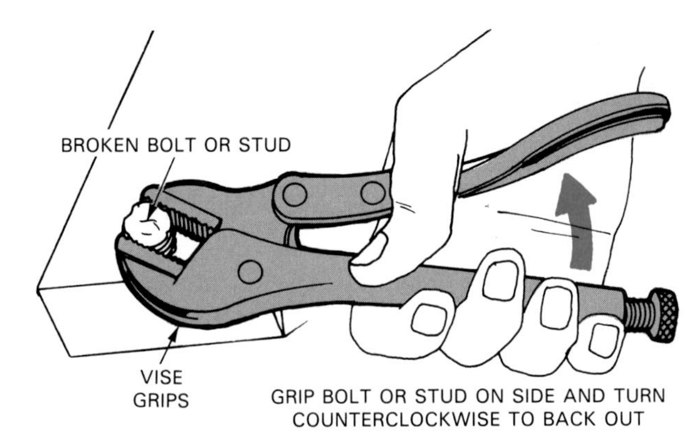

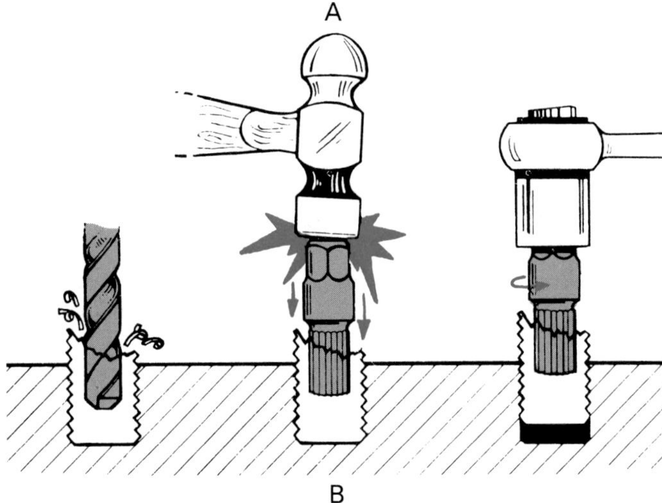

Fig. 3-14. A—Vise grip pliers will sometimes remove broken bolt or stud when part of fastener is sticking up above part. Clamp vise grips tightly on first try. (Florida Voc. Ed.) B—Screw extractor is method of removing broken bolt when below surface of part. Drill hole in center of broken bolt. Tap extractor into hole. Use wrench to unscrew and remove broken bolt. (Florida Voc. Ed. and Lisle Tools)

AMERICAN NATIONAL SCREW THREAD PITCHES

COARSE STANDARD THREAD (N. C.)
Formerly U. S. Standard Thread

Sizes	Threads Per Inch	Outside Diameter at Screw	Tap Drill Sizes	Decimal Equivalent of Drill
1	64	.073	53	0.0595
2	56	.086	50	0.0700
3	48	.099	47	0.0785
4	40	.112	43	0.0890
5	40	.125	38	0.1015
6	32	.138	36	0.1065
8	32	.164	29	0.1360
10	24	.190	25	0.1495
12	24	.216	16	0.1770
1/4	20	.250	7	0.2010
5/16	18	.3125	F	0.2570
3/8	16	.375	5/16	0.3125
7/16	14	.4375	U	0.3680
1/2	13	.500	27/64	0.4219
9/16	12	.5625	31/64	0.4843
5/8	11	.625	17/32	0.5312
3/4	10	.750	21/32	0.6562
7/8	9	.875	49/64	0.7656
1	8	1.000	7/8	0.875
1 1/8	7	1.125	63/64	0.9843
1 1/4	7	1.250	1 7/64	1.1093

FINE STANDARD THREAD (N. F.)
Formerly S.A.E. Thread

Sizes	Threads Per Inch	Outside Diameter at Screw	Tap Drill Sizes	Decimal Equivalent of Drill
0	80	.060	3/64	0.0469
1	72	.073	53	0.0595
2	64	.086	50	0.0700
3	56	.099	45	0.0820
4	48	.112	42	0.0935
5	44	.125	37	0.1040
6	40	.138	33	0.1130
8	36	.164	29	0.1360
10	32	.190	21	0.1590
12	28	.216	14	0.1820
1/4	28	.250	3	0.2130
5/16	24	.3125	I	0.2720
3/8	24	.375	Q	0.3320
7/16	20	.4375	25/64	0.3906
1/2	20	.500	29/64	0.4531
9/16	18	.5625	0.5062	0.5062
5/8	18	.625	0.5687	0.5687
3/4	16	.750	11/16	0.6875
7/8	14	.875	0.8020	0.8020
1	14	1.000	0.9274	0.9274
1 1/8	12	1.125	1 3/64	1.0468
1 1/4	12	1.250	1 11/64	1.1718

Fig. 3-13. Tap drill chart will give size of hole to drill for each tap.

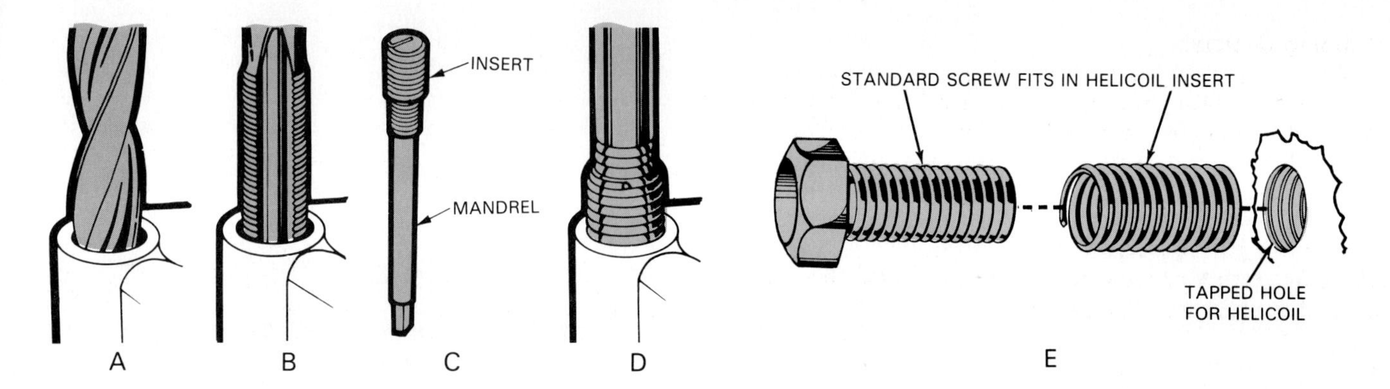

Fig. 3-15. Helicoil will let you repair stripped threads in part and use same size bolt. A—Drill hole oversize. B—Tap out hole. C and D—Use mandrel to screw helicoil into hole. D—Original size bolt will then thread into helicoil. (Buick and Chrysler)

Helicoil (thread repair insert)

A *helicoil* or *thread repair insert* should be used when an oversize hole and fastener is NOT acceptable. A helicoil will repair damaged internal threads and allow the use of the ORIGINAL SIZE HOLE.

To use a helicoil, drill the hole oversize as described in the helicoil instructions. Tap the hole. Then, screw the helicoil into the threaded hole. The internal threads of the helicoil will act as threads the same size as the damaged hole, Fig. 3-15.

NONTHREADED FASTENERS

Numerous types of *nonthreaded fasteners* are utilized in the assembly of an automobile. It is essential to learn the most common types.

Snap rings

A *snap ring* fits into a groove in a part and commonly holds shafts, bearings, gears, pins, and other similar components in place.

Several types of snap rings are given in Fig. 3-16.

Snap ring pliers are needed to remove and install snap rings. They have special jaws that grasp the snap ring.

WARNING! Wear eye protection when working with snap rings. When flexed, the ring can shoot into your face with considerable force.

Keys and keyways

A metal *key* fits into a *keyway* (groove) cut into a shaft and part (gear, pulley, collar). The key prevents the part from turning on its shaft. Refer to Fig. 3-17.

Setscrews are normally used to lock a part onto a shaft. They may be used with or without a key and keyway. A setscrew is a headless fastener normally designed to accept an *allen wrench* (hex wrench) or screwdriver.

Pins

A *cotter pin* is a safety device normally used in a slotted nut, Fig. 3-18. It fits through a hole in the bolt or part. This keeps the slotted nut from turning, and

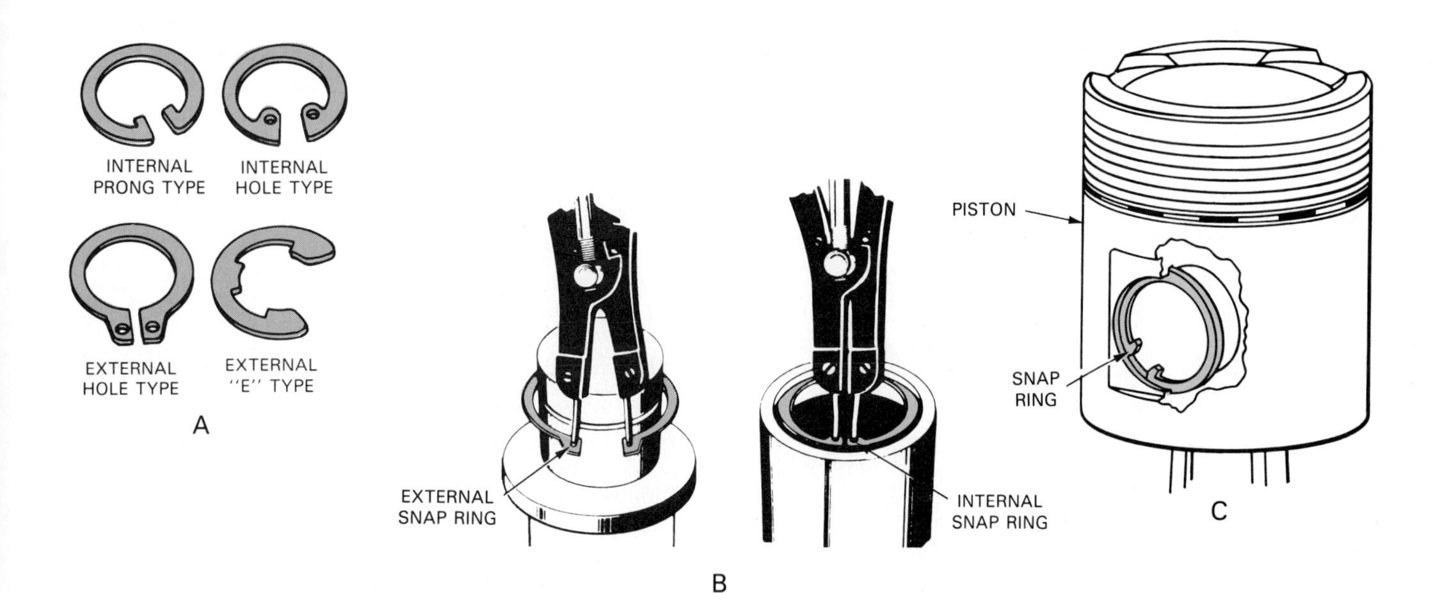

Fig. 3-16. Snap rings. A—Snap rings types. B—Using snap ring pliers. C—Snap rings are sometimes used to hold piston pin inside piston. They fit into small groove machined in pin bore. (Deere & Co.)

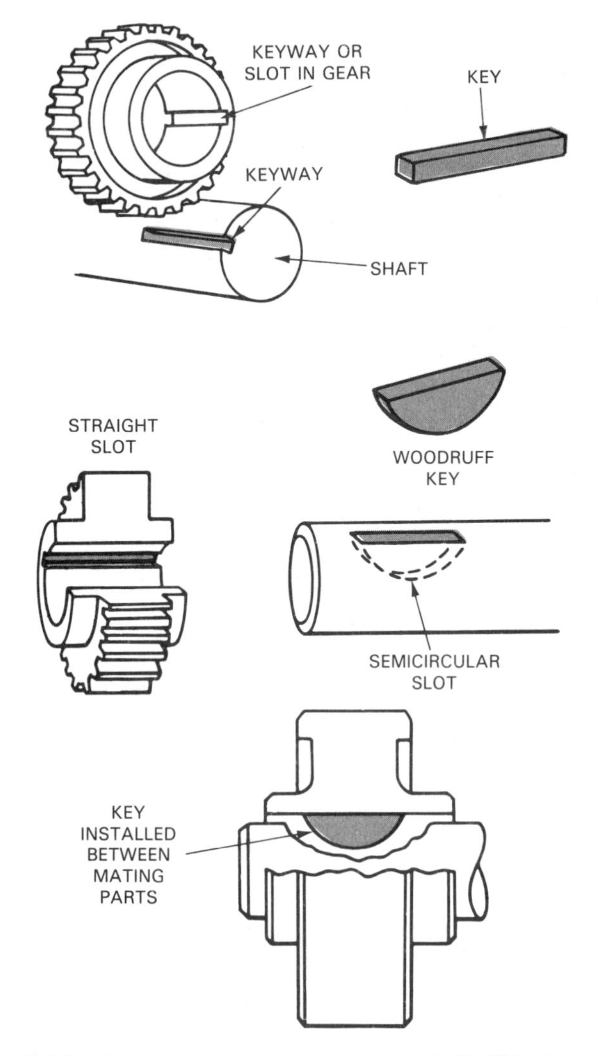

Fig. 3-17. Key and keyway lock part to shaft. This is commonly done on front of crankshaft to hold damper and pulley. (Deere & Co.)

possibly coming off. Cotter pins are also used with pins, shafts, and linkages.

Pins (straight, tapered, and split) are frequently used to secure a gear or pulley to its shaft. Several are shown in Fig. 3-18.

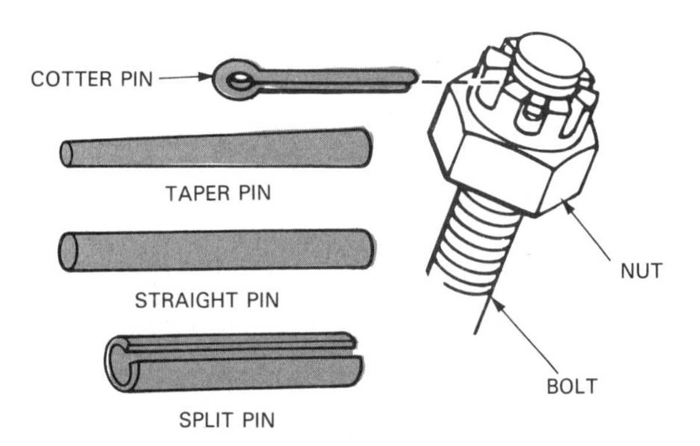

Fig. 3-18. Note various pin types. (Florida Voc. Ed. and Deere & Co.)

One example, a split pin, is commonly used to hold the drive gear on the distributor shaft. Since the oil pump may be driven by the distributor shaft, the pin serves as a safety device. If the oil pump locks up, the pin will shear off and stop the engine from running (no distributor rotation). This prevents severe engine damage from lack of lubrication.

GASKETS AND SEALS

Gaskets and seals are used between engine parts to prevent fluid leakage. It is important to understand a few principles about gaskets and seals. If they are serviced improperly, serious customer complaints and mechanical failures can result.

Gaskets

A *gasket* is a soft, flexible, material placed between parts, Fig. 3-19. It can be made of fiber materials, rubber, neoprene (synthetic rubber), cork, treated paper, or thin steel.

When the parts are fastened tightly together, the gasket is compressed and deformed. This forces the gasket material to fill small gaps, scratches, or other imperfections in the part surfaces. A leakproof seal is produced between the parts.

Gasket rules

When working with gaskets, remember the following:

1. Inspect for leaks before engine disassembly. If the two parts are leaking, the part surface should be inspected closely for problems.
2. Avoid part damage during disassembly. Be careful not to nick, gouge, or dent mating surfaces while removing parts. The slightest unevenness could cause leakage.
3. Clean off old engine gasket carefully. All of the old gasket material must be scraped or wire brushed from the parts. Use care especially on aluminum

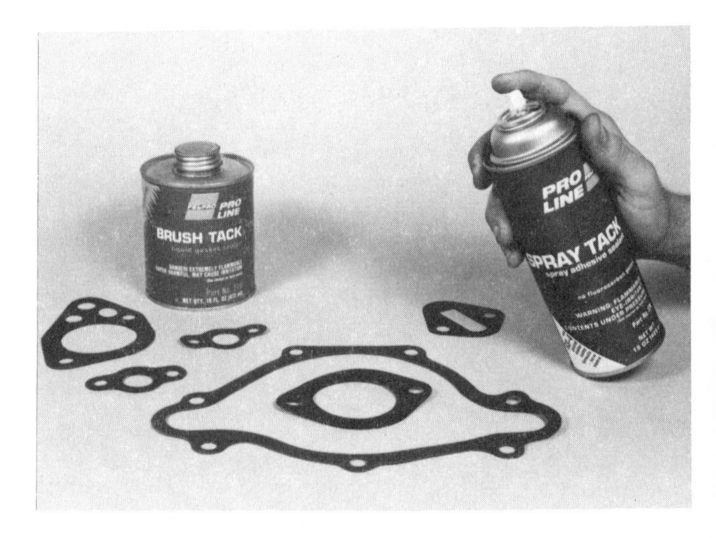

Fig. 3-19. Gaskets prevent leakage between parts. Various sealers and adhesives can be used with gaskets, primarily to hold the gaskets in place during part assembly. (Fel-Pro Gaskets)

and other soft metals. These soft metals are easily damaged. Use a dull scraper and wire brush lightly.

Note! Some engine parts (valve covers for example) are plastic. They can be damaged easily during removal and cleaning.

4. Wash and dry parts thoroughly! After gasket removal, wash parts in solvent. Blow dry with compressed air. Then wipe mating surfaces with a clean shop rag.
5. Check new gasket shape! Compare the new gasket to the shape of the mating surface. Lay the new gasket into place. All holes and sealing surfaces must match perfectly.
6. Use sealer if needed! Some gaskets require sealer, Fig. 3-20. Sealer is normally used where two different gaskets come together. It will prevent leakage where gaskets overlap. Check a service manual for details. Use sparingly. Too much sealer could clog internal passages in the assembly. See Figs. 3-21 and 3-22.
7. Hand start all fasteners before tightening! After fitting the gasket and parts in place, screw all bolts in by hand. This will assure proper part alignment and threading of fasteners. It also lets you check bolt lengths.
8. Tighten in steps! When more than one bolt is used to hold a part, tighten each bolt a little at a time. Tighten one to about half of its torque spec, then the others. Tighten to three-fourth torque and to full torque. Then, RETORQUE each fastener again.
9. Use a crisscross tightening pattern! Either a basic crisscross or factory recommended torque pattern should be used when tightening parts having multiple fasteners. This will assure even gasket compression and sealing.
10. Do not overtighten fasteners! It is very easy to tighten the bolts enough to dent sheet metal parts and smash or break gaskets. Apply only the specified torque.

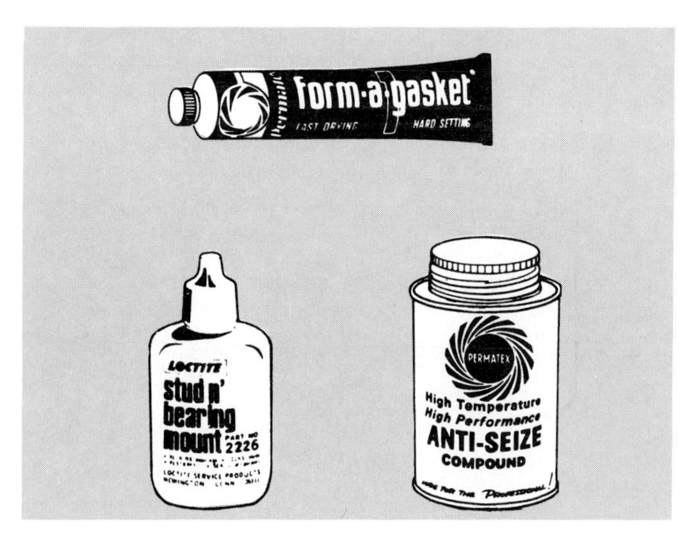

Fig. 3-20. Hard setting sealer may be used on threaded fittings. Anti-seize compound is used when bolt rusting and corrosion could prevent removal. Locking compound can be used when fastener must not unscrew or loosen in service. (Permatex)

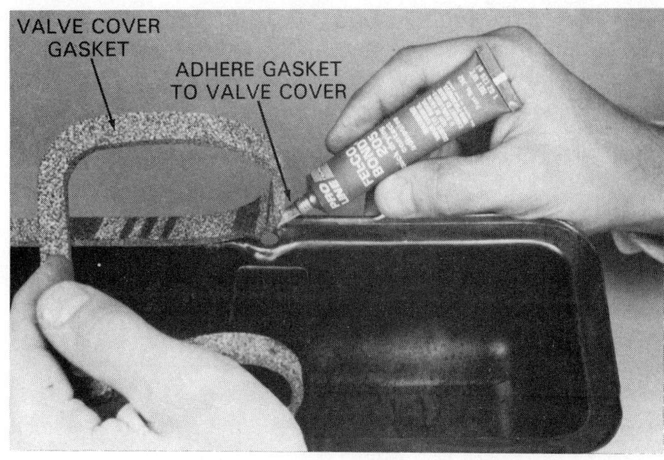

Fig. 3-21. Technician is using adhesive type sealer to hold valve cover gasket in place. This will keep gasket from shifting while fitting cover on engine. (Fel-Pro Gaskets)

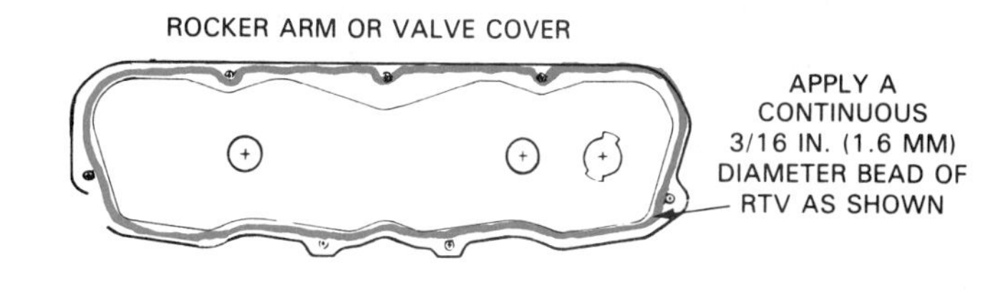

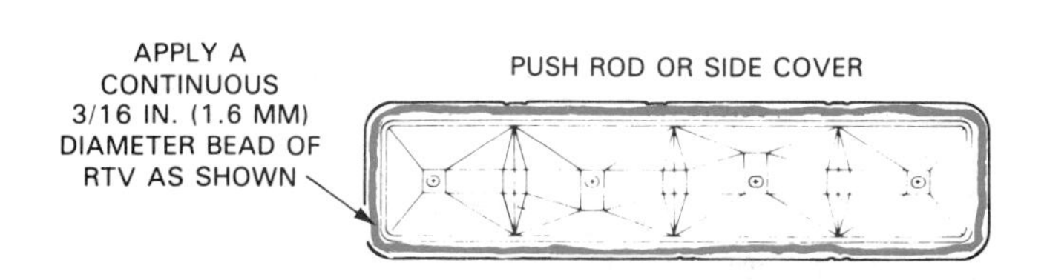

Fig. 3-22. RTV or silicone sealer can be used instead of a cork or rubber gasket. It will seal properly if parts are clean before assembly. Note size of bead run around covers. (Chevrolet)

SEALERS

A *sealer* is commonly coated on a gasket to help prevent leakage, and to hold the gasket during assembly. There are numerous kinds of sealers. They have different properties and are designed for different uses. Always read the manufacturer's label and a service manual before selecting a sealer, Fig. 3-20.

Hardening sealers are used on permanent assemblies such as fittings and threads, and for filling uneven surfaces. They are usually resistant to most chemicals and heat.

Nonhardening sealers are for semipermanent assemblies: cover plates, flanges, threads, hose connections, and other applications. They are also resistant to most chemicals and moderate heat, Fig. 3-21.

Shellac is a nonhardening sealer. It is a gummy, sticky substance that remains pliable. It is frequently used on fiber gaskets as a sealer and to hold the gasket in place during assembly.

Form-in-place gaskets

Form-in-place gasket refers to a special sealer that is used INSTEAD of a conventional fiber or rubber gasket. Two common types of form-in-place gaskets are: RTV (room temperature vulcanizing) sealer, also called silicone sealer, and anaerobic sealer. Note: Use only oxygen sensor safe sealers.

RTV or silicone sealer

RTV or *silicone sealer* cures (dries) from moisture in the air. It is used to form a gasket on thin, flexible flanges.

RTV sealer normally comes in a tube. Depending upon the brand, it can have a shelf life from one year to two years. Always inspect the package for the expiration date before use. If too old, RTV sealer will NOT CURE (harden) and seal properly.

Using RTV sealer

RTV or silicone sealer should be applied in a continuous bead approximately 1/8 in. (3 mm) to 3/16 in. (1.6 mm) in diameter, Fig. 3-22. All mounting holes must be circles. Uncured RTV may be removed with a rag.

Components should be torqued in place while the RTV is still wet to the touch (within about 10 minutes). The use of locating dowels is often recommended to prevent the sealing bead from being smeared. If the continuous bead of silicone is broken, a leak may result.

Anaerobic sealer

Anaerobic sealer cures in the absence of air and is designed for tightly fitting, thick parts. It is used between two smooth, true surfaces, NOT on thin, flexible flanges.

Anaerobic sealer should be applied sparingly. Use 1/16 to 3/32 in. (1.5 to 2 mm) diameter bead on one gasket surface. Be certain that the sealer surrounds each mounting hole. Typically, bolts torque within 15 minutes.

Sealing engine parts

When selecting a form-in-place gasket, refer to a manufacturer service manual. Scrape or wire brush gasket surfaces to remove all loose material. Check that all gasket rails are flat. Using a shop rag and solvent, wipe off oil and grease. The sealing surfaces must be CLEAN and DRY before using a form-in-place gasket.

A few gasket manufacturers sell pre-cut gaskets designed to replace form-in-place gaskets. When working on an engine installed in a car, it can be difficult to properly clean the sealing surfaces. Or, it may be almost impossible to fit a part on the engine without hitting and breaking the bead of sealant. When this is the case, a pre-cut gasket might work better than sealer alone.

Note! Sealer is sometimes recommended on bolts that extend into water jackets. Fig. 3-23 shows sealer on a front cover bolt.

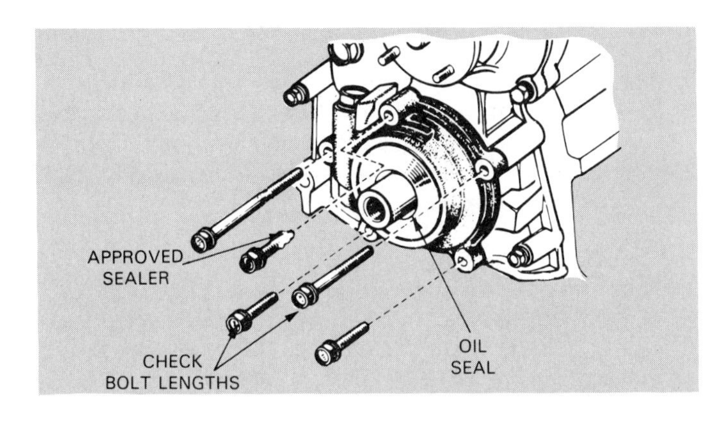

Fig. 3-23. Some bolts extend into water and oil passages in engine. Approved sealer should be applied to threads on these bolts to prevent leakage. Check service manual for details when in doubt. (Toyota)

SEALS

Seals prevent leakage between a stationary part (engine front cover for example) and a moving part (crankshaft for example). A seal allows the shaft to spin or slide inside the nonmoving part without fluid leakage, Fig. 3-24. Seals are normally made of synthetic rubber molded onto a metal body.

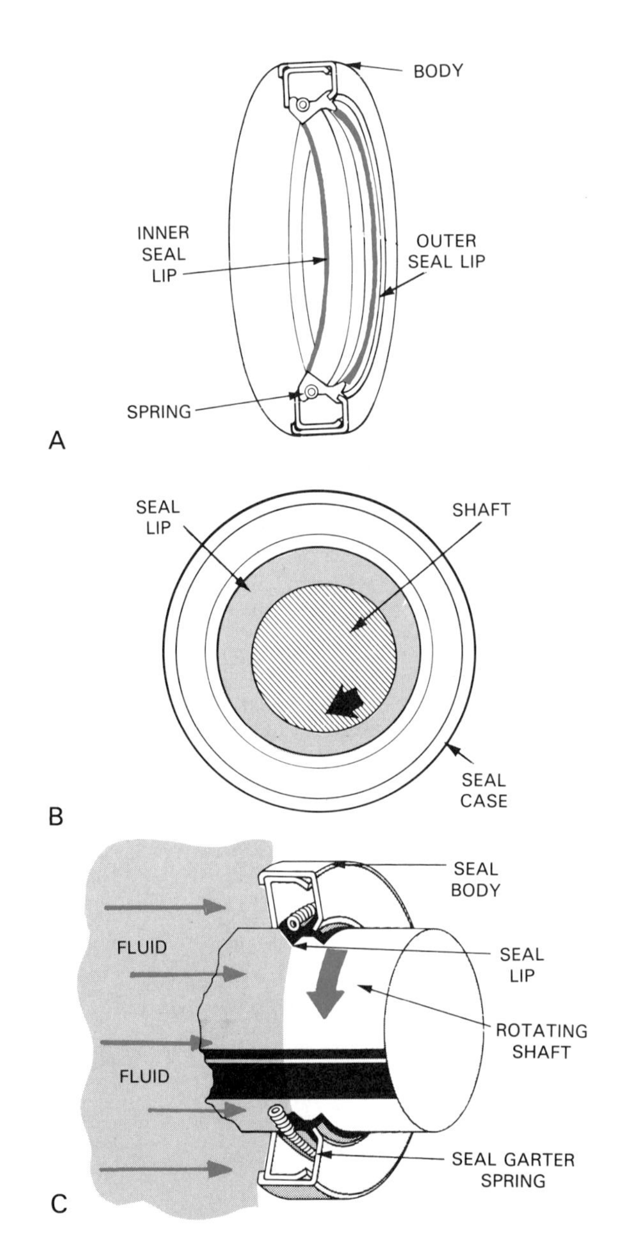

Fig. 3-24. Seal prevents leakage between rotating shaft and stationary housing or cover. A—Seal parts. B—Seal on shaft. C—Seal action. (Chrysler and Caterpillar)

Servicing seals

There are several procedures to remember when working with seals.

1. Inspect seal for leakage before disassembly! If a seal is leaking, there may be other problems beside a defective seal. Look for a scored shaft, misaligned seal housing, or damaged parts. Leakage requires close inspection after disassembly.
2. Remove old seal carefully! Pry out the old seal without scratching the seal housing. Sometimes, a special puller is required for seal removal. This will be discussed in other chapters.
3. Inspect shaft for wear and burrs! Look at the shaft closely where it contacts the seal. It should be smooth and flat. File off any burrs that could cut the new seal. A badly worn shaft will require polishing, a shaft sleeve repair kit, or replacement, Fig. 3-25.
4. Check new seal size! Compare the old seal to the new seal. Hold them next to each other. Both the inside diameter (ID) and outside diameter (OD) must be the same. To double-check the inside diameter, slip the seal over the shaft. It should fit snugly.
5. Install new seal correctly! Coat the outside of the seal housing with approved sealer. Lubricate the inner lip of the seal with system fluid.

   Install the seal with the sealing LIP FACING THE INSIDE OF THE PART. If installed backwards, a large oil leak will result. Also, check that the seal is squarely and fully seated in its bore.

O-ring seals

An *O-ring seal* is a stationary type seal that fits into a groove between two parts, Fig. 3-26. When the parts are assembled, the synthetic rubber seal is partially compressed and forms a leakproof joint.

Normally, O-ring seals should be coated with system fluid (engine oil, diesel fuel, or type fluid used in component). This will help the parts slide together without scuffing or cutting the seal.

Usually, sealants are NOT used on O-ring type seals. When in doubt about any seal installation, check in a shop manual.

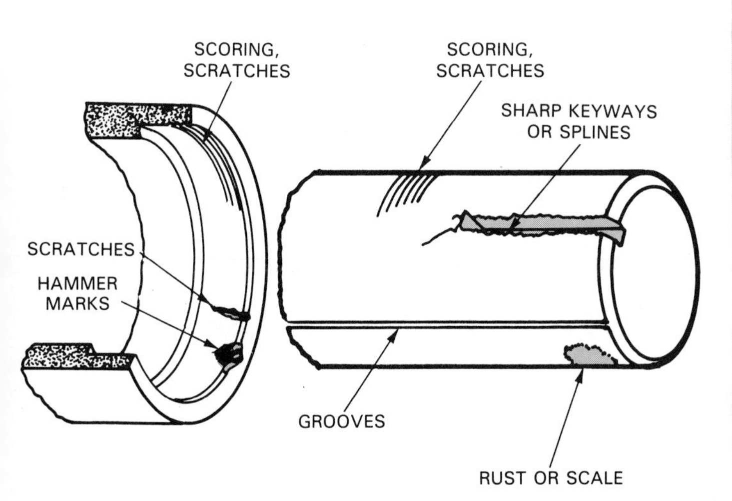

Fig. 3-25. When servicing a seal, check for these kinds of problems. (Federal Mogul)

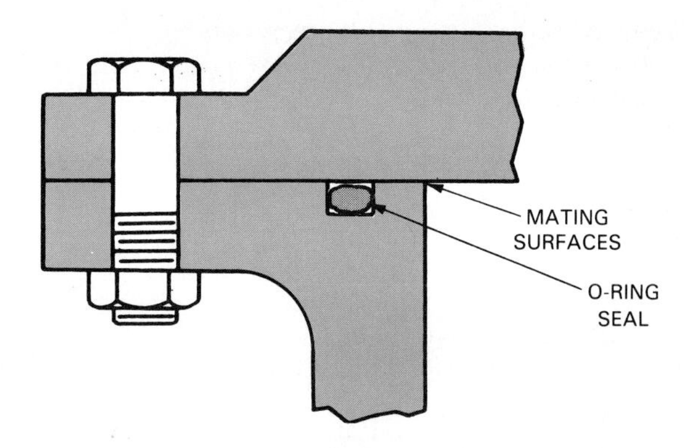

Fig. 3-26. O-ring seal is usually stationary type seal between parts.

OTHER INFORMATION

Special gaskets and seals sometimes require other installation techniques. These special situations will be discussed in later chapters where they apply.

HOSES AND LINES

When working on engines, you will have to service numerous types of hoses and lines. It is very important that you handle them properly. If not, dangerous or damaging leaks can result. Fig. 3-27 shows some of the types of lines and hoses found in an engine compartment.

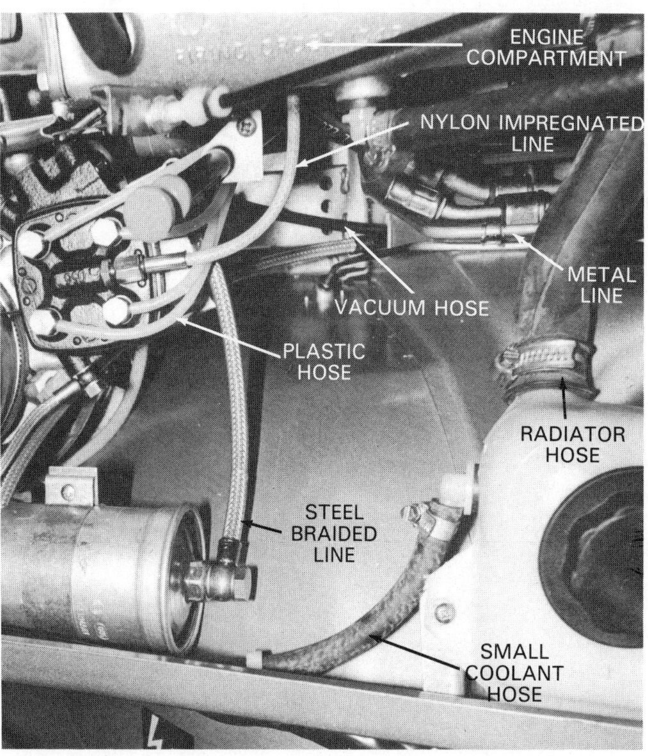

Fig. 3-27. Note a few types of hoses found in engine compartment. You must service all of these properly when working on an engine. (Saab)

DANGER! Diesel injection lines can contain over 8000 psi (56 000 kPa) pressure. Never attempt to remove a diesel injection component with the engine running. Fuel could squirt out, blinding you, or puncturing your skin and causing blood poisoning or death! Wear eye protection when working around running diesel engines.

Metal lines

Metal lines or tubing are used as fuel lines, especially from the fuel tank to the engine. They are usually double-wall steel tubing. Fig. 3-28 shows some of the tools needed to replace a section of metal line.

To service a fuel line, for example, follow the steps illustrated in Fig. 3-29A through G. Cut off the old line and a section of new line with a tubing cutter. Rotate the cutter around the line while progressively tightening the thumbscrew.

Use a reamer to remove metal burrs from inside the line. Bend the line to the shape of the old line using a bending spring or other type of bending tool.

To form a flare on the line, use a flaring tool, as shown in Fig. 3-29E and F. Form the tubing inward with the flaring tool adapter. Then force the cone into the end of the line to produce a double-lap. A fitting nut can then be used on the line. Blow line clean before use.

Hoses

Hoses are used on an engine to make a flexible connection between two fittings. They carry fuel, coolant, oil, air, and other fluids to the engine. Like metal lines, they must be serviced properly to prevent engine failure. Fig. 3-30 shows the most common ones.

Hose clamps secure hoses tightly to their fittings. The three basic types are illustrated in Fig. 3-31.

Hose inspection

To check a hose, visually inspect and squeeze it. If you see cracking of the outer rubber layer or if the hose has

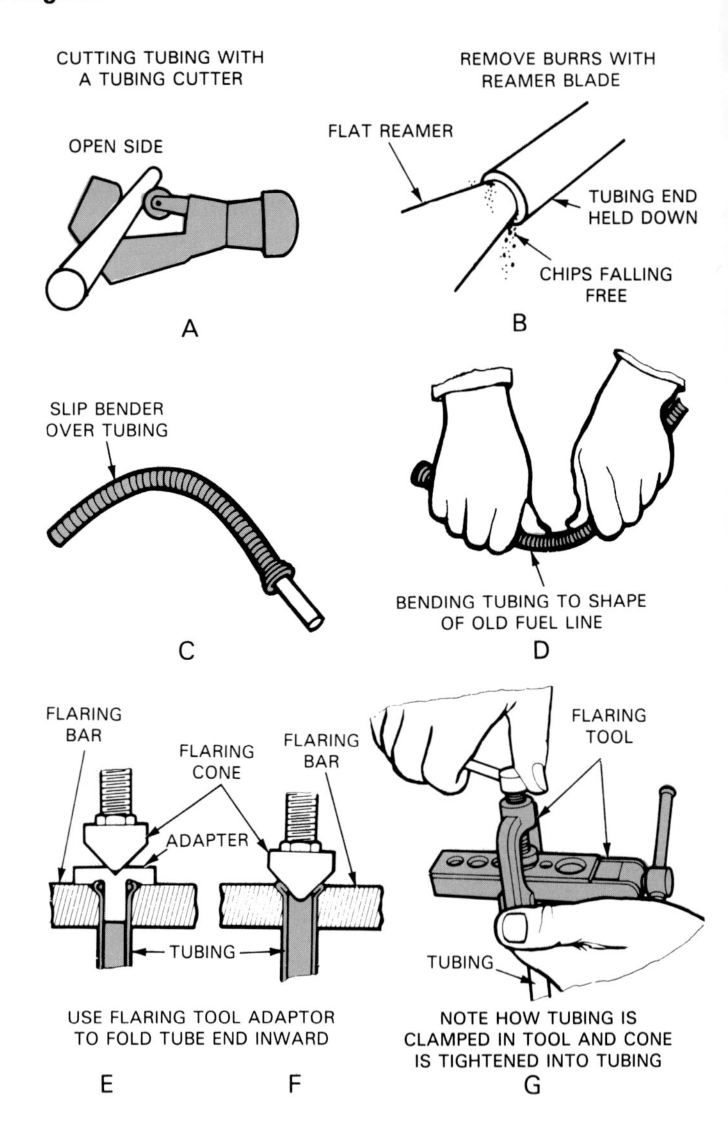

Fig. 3-29. Study basic steps for forming flare on end of line. This will allow use of tube fitting. (Chrysler)

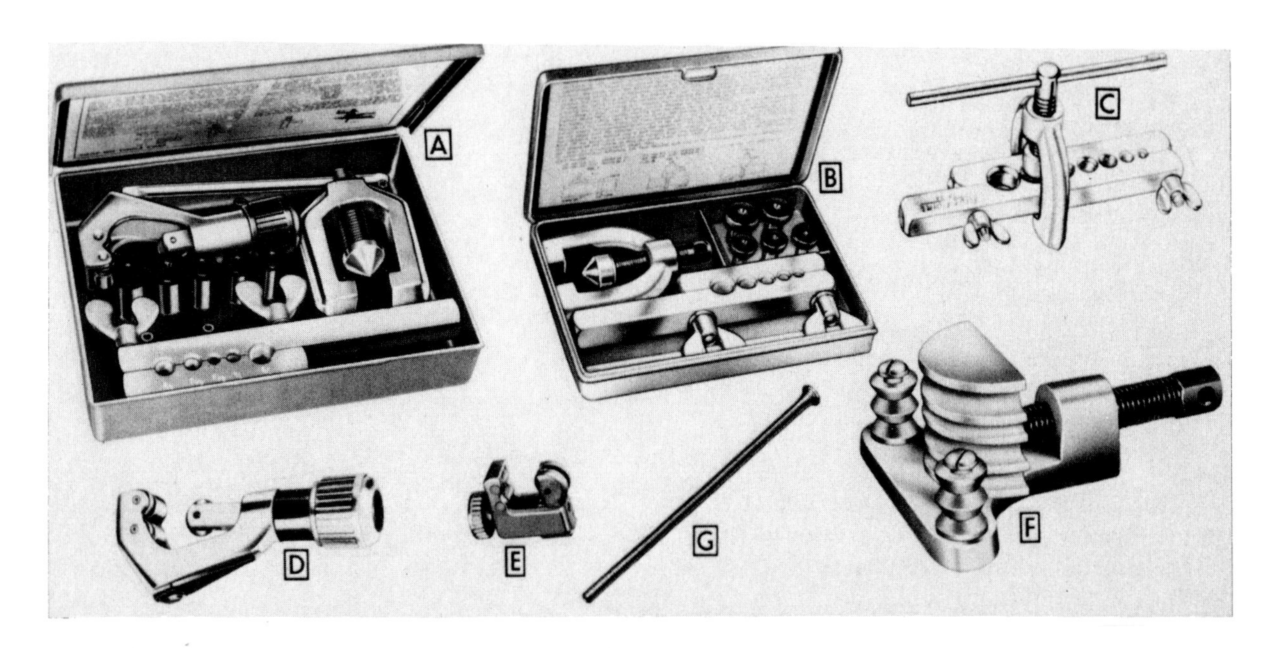

Fig. 3-28. A & B—Tube flaring sets. C—Flaring tool. D & E—Tube cutters. F & G—Tube benders.

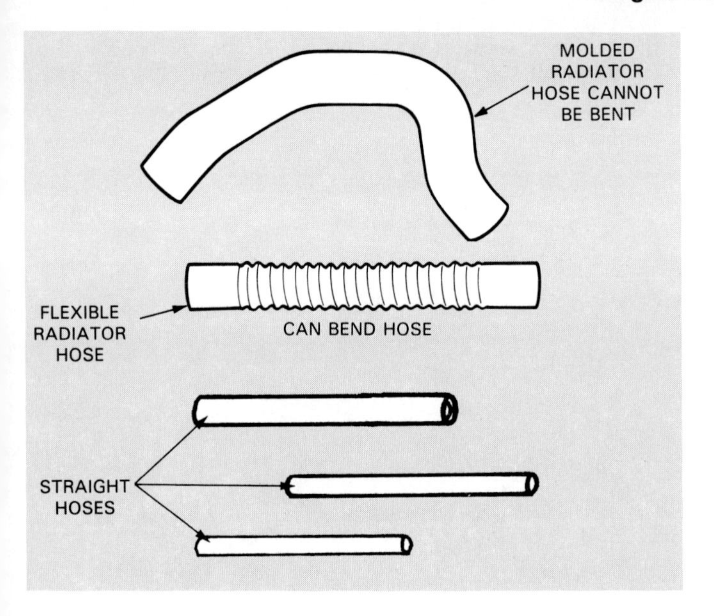

Fig. 3-30. Note hose types found in engine compartment.

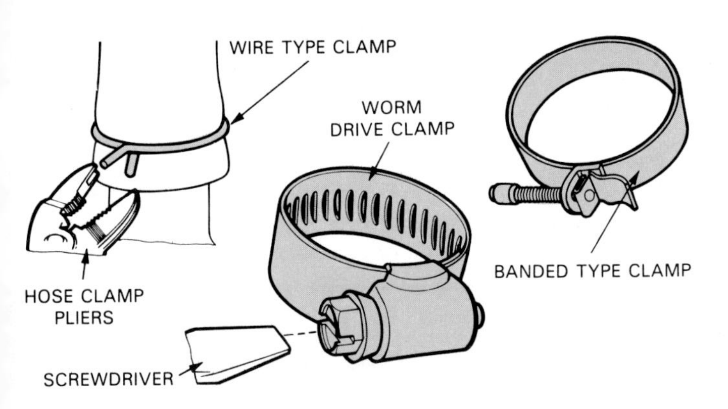

Fig. 3-31. Study clamp types for hoses. (Mopar)

hardened (aged) or softened (fluid contamination), install a new hose. Do this for the radiator hoses, heater hoses, fuel system hoses, vacuum hoses, etc.

Hose replacement

To remove a hose, loosen the clamps. If to be replaced, you can cut a slit in the hose with a razor blade knife to ease removal. Twist and pull the hose off. Obtain a new hose and compare it to the old one.

Clean the hose fittings. Install the new hose, following recommended procedures for the type hose being serviced. Then, check the hose for leaks.

Line and hose service rules

Remember these rules when working with lines and hoses:

1. Place a shop rag around the fuel line fitting during removal. This will keep fuel from spraying on you or on the hot engine. Use a flare nut or tubing wrench. Two wrenches may be needed where two fittings join, Fig. 3-32. Do not bend or kink plastic fuel lines.
2. Only use approved double-wall steel tubing for fuel lines. Never use copper tubing.
3. Make smooth bends when forming a new line. Use a bending spring or bending tool.
4. Form double-lap flares on the ends of fuel lines. A single-lap flare is NOT approved for fuel lines.
5. Reinstall all line hold-down clamps and brackets. If not properly supported, the line can vibrate and fail.
6. Route all lines and hoses away from hot or moving engine parts. Double-check clearance after installation.
7. Torque fittings when recommended in the service manual, Fig. 3-33.
8. Only use hoses approved for the application. If the wrong type is accidentally used, the fluid can chemically attack and rapidly ruin the hose. A dangerous or damaging leak could result.
9. Check the condition of all hoses closely before reinstallation, Fig. 3-34.
10. Cut hoses off squarely, Fig. 3-35.

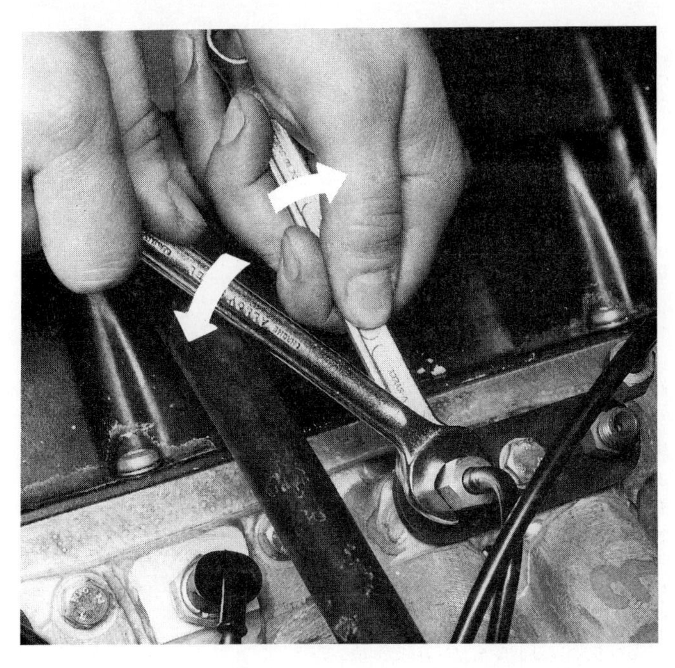

Fig. 3-32. Sometimes, two wrenches are needed to remove fitting when two fittings are joined. This keeps both fittings from turning and twisting off line. (Saab)

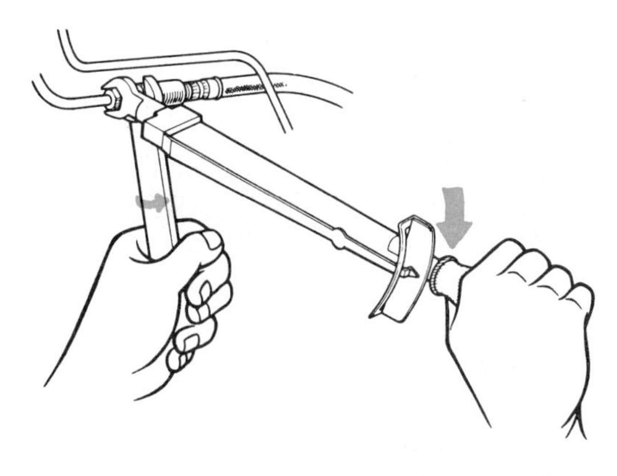

Fig. 3-33. When told to do so in service manual, torque line fittings. This is especially important on high pressure injection lines.

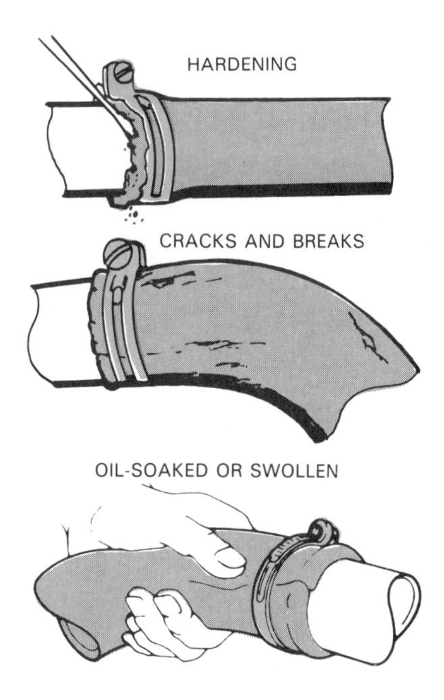

Fig. 3-34. Inspect hoses for these problems when working on engine. (Gates Rubber Co. and Ford)

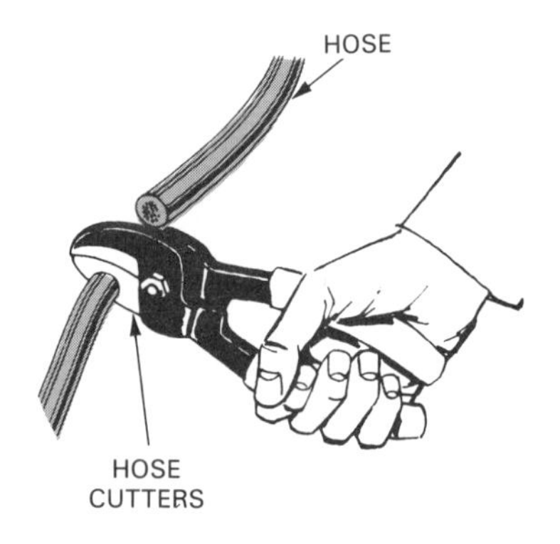

Fig. 3-35. Cut hose off squarely. These special cutters will help. (Belden, NAPA)

11. Make sure a hose fully covers its fitting or line before installing the clamps. Pressure in the system could force the hose off if not installed properly.
12. Be careful not to break plastic fittings when removing vacuum hoses. See Fig. 3-36A.
13. Label hoses if their routing is confusing, Fig. 3-36B.
14. Double-check all fittings for leaks. Start the engine and inspect the connections closely.

DANGER! Many gasoline injection systems retain pressure after the engine has been shut off. Relieve system pressure at the pressure test fitting before disconnecting fuel lines.

BELTS

Engines use several types of belts to operate the water pump, alternator, power steering pump, air conditioning compressor, and sometimes the air injection pump, and auxiliary shaft. Either one or several belts can be used to operate these components. See Figs. 3-37 and 3-38.

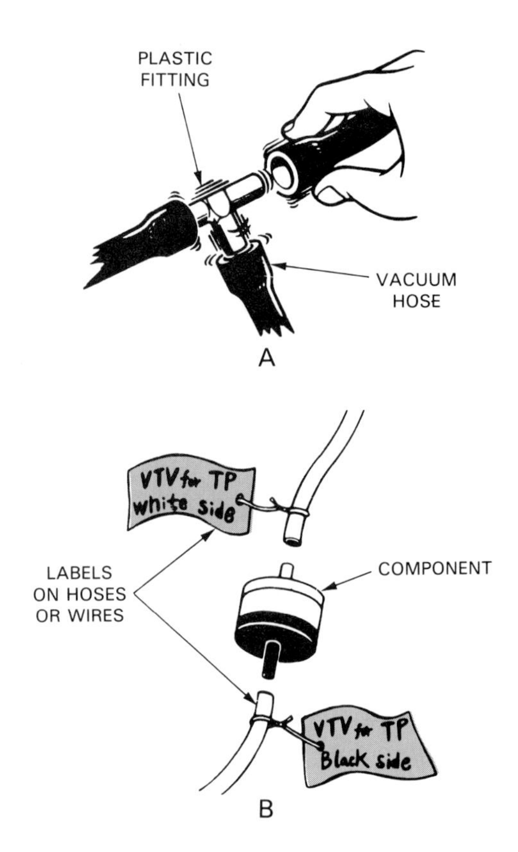

Fig. 3-36. A—Be careful not to break plastic fittings when disconnecting vacuum hoses from engine. Grasp fitting and twist hose while pulling. B—Tag or label hoses and wires with tape on engine to avoid confusion upon reassembly. (Toyota)

Fig. 3-37. Note arrangement of drive belts on front of this engine. (Ford)

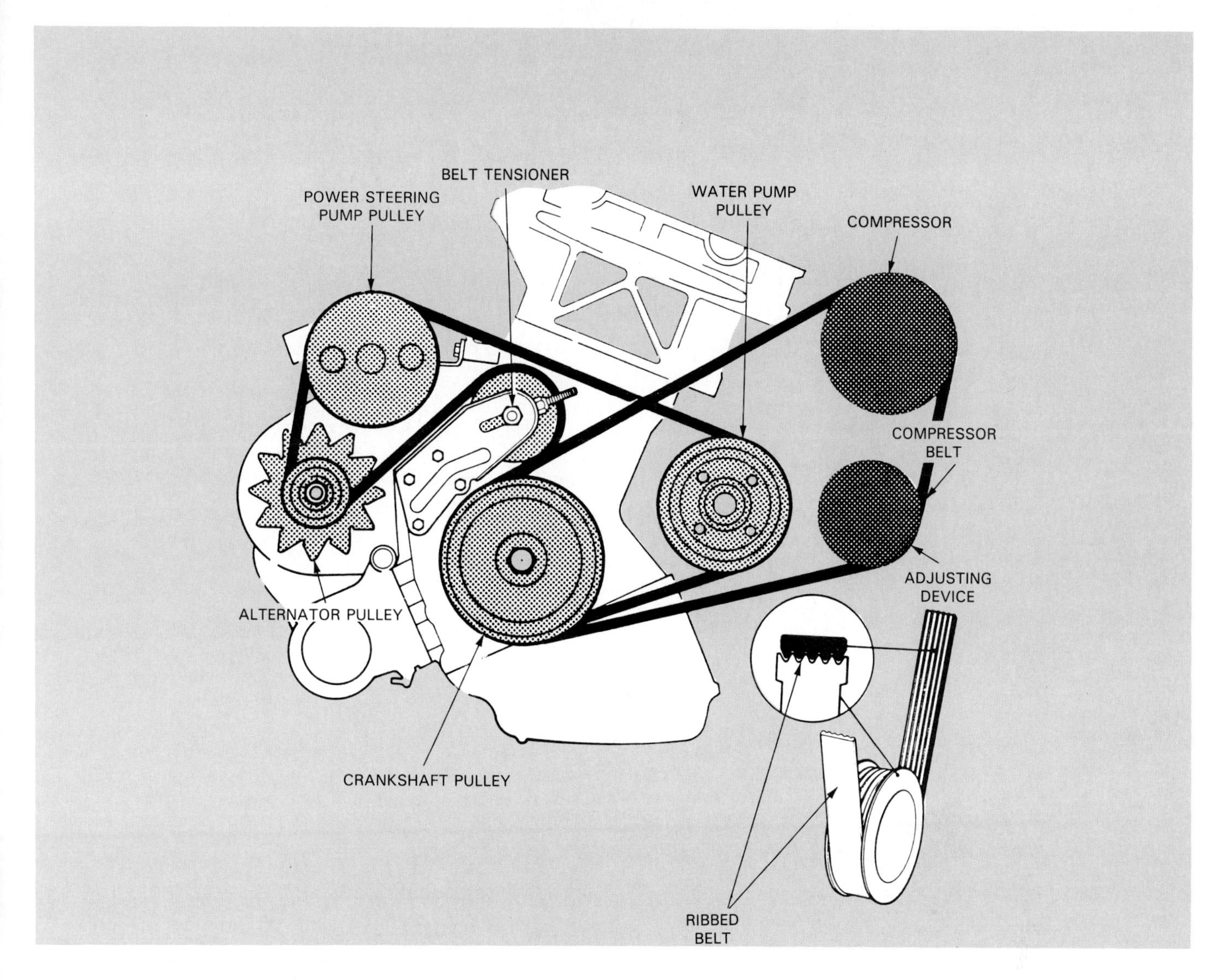

Fig. 3-38. Ribbed belt is newer design that will handle more load. Note how only two belts are needed. (Saab)

A *V-belt,* when cross-sectioned, looks like the letter ''V.'' It is an older, conventional design. Normally, two, three, or four V-belts are used on an engine.

A *ribbed belt* is more flat in shape and has small ridges that are formed on the inside diameter of the belt. This is shown in Fig. 3-38. This is a more recent design. A ribbed belt can handle more torque than a comparable V-belt. For this reason, only one or two ribbed belts are needed to operate all engine accessory units.

A COGGED BELT has teeth or ribs that operate in matching ribs in the belt sprockets. This type is commonly used to drive the engine camshaft and auxiliary shaft. Refer to Fig. 3-39.

Belt service

Belt service typically includes periodic inspection and replacement. To inspect a belt, look for cracking and fraying. Twist the belt over and check its underside. Look for the troubles shown in Figs. 3-39 and 3-40.

If any belt problems are found, install a new belt. Follow service manual procedures and tighten the belt to specifications.

Generally, the alternator belt should be tightened just enough to prevent slippage. Overtightening could cause premature failure of the alternator bearings. The other units (power steering pump, air conditioning compressor, auxiliary drive) operate in oil and can tolerate more belt tightening. Look at Fig. 3-41.

BEARINGS

There are basically four types of bearings that can be used in or on an engine: friction bearing, roller bearing, ball bearing, and needle bearings. Needle and friction bearings are commonly used in an engine. Roller and ball bearings are used elsewhere in the car. These are shown in Fig. 3-42.

A *friction bearing* is a plain bearing that has two smooth, sliding surfaces. This type is commonly used as engine main, connecting rod, and camshaft bearings.

Antifriction bearings use balls, rollers, or needle bearings between the two moving surfaces. A rolling, instead of sliding, action reduces friction. This type bearing (needle type) can be used on roller lifters and sometimes

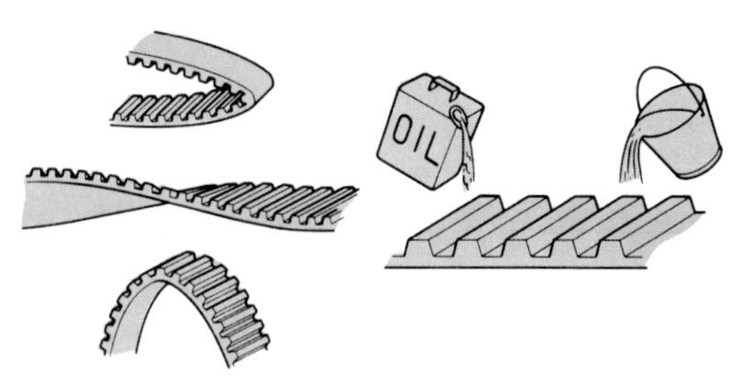

A — Do not bend, twist, or turn belt excessively. Oil and water will deteriorate belt. Fix all engine leaks.

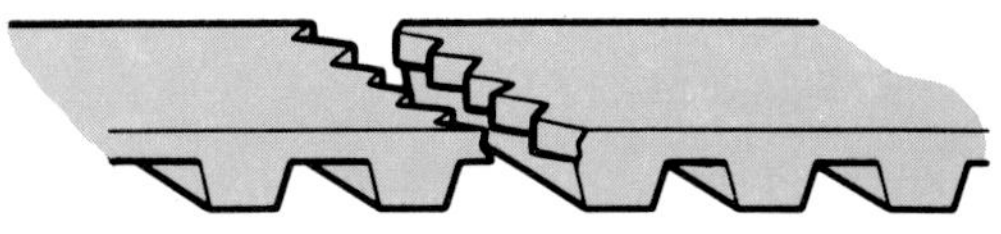

B — Belt breakage may be caused by sprocket or tensioner problem. Check these parts before installing new belt.

C — If timing belt teeth are missing, check for locked component. Oil pump, injection pump, etc., could be frozen.

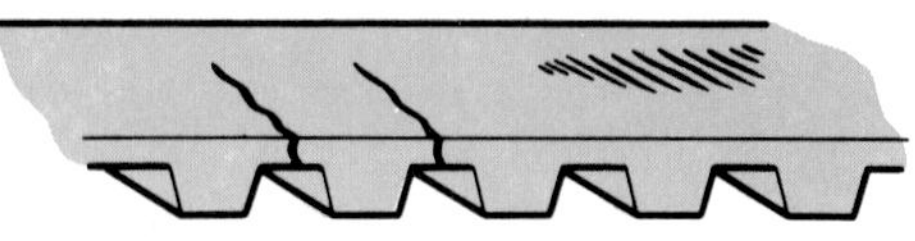

D — If there is wear or cracks on smooth side of belt, check idler or tensioner pulley.

E — With damage or wear on one side of belt, check belt guide and pulley alignment.

F — Wear on timing belt teeth may be caused by timing sprocket problem. Inspect sprockets carefully. Oil contamination will also cause this trouble.

Fig. 3-39. Check cogged belts for these problems. (Toyota)

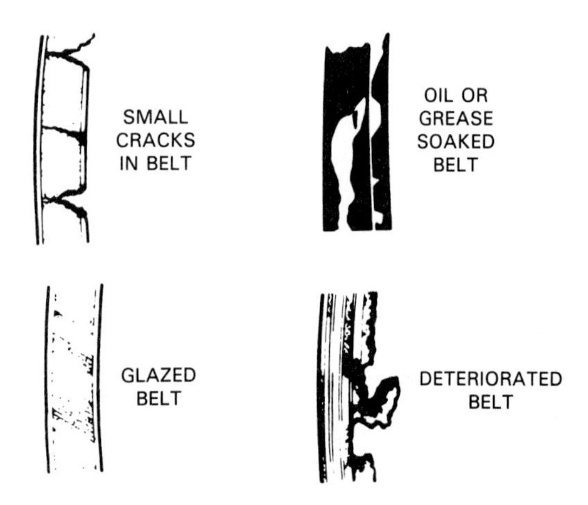

Fig. 3-40. Inspect V-belts closely for trouble. (Sun Electric)

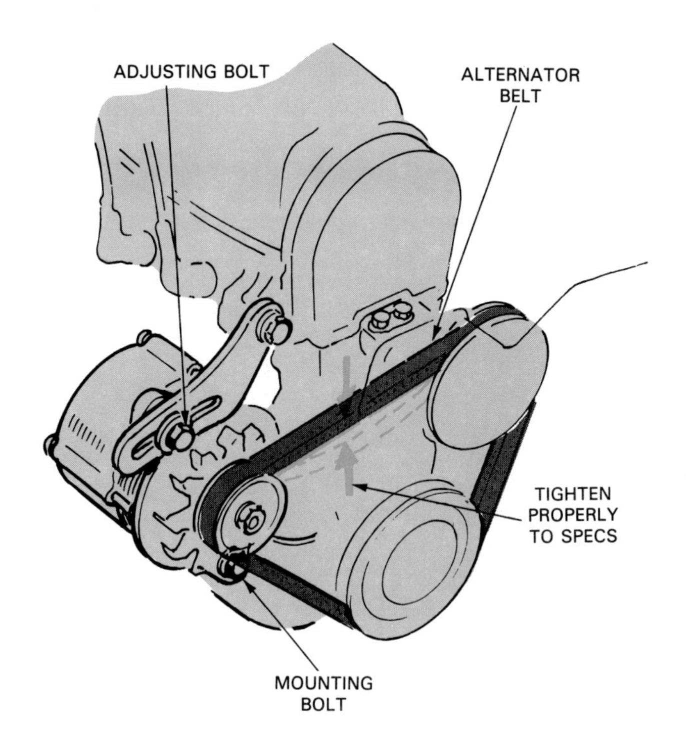

Fig. 3-41. Tighten belts to specs. Alternator belt should not be as tight as other belts because bearings could fail from overtightening. (Honda)

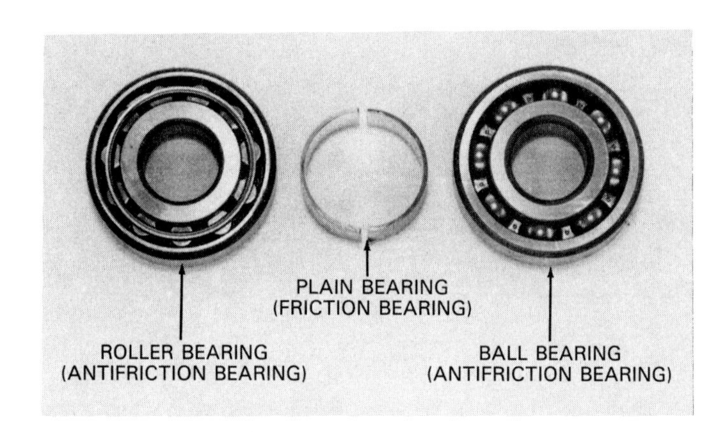

Fig. 3-42. Study basic types of bearings. Friction bearings are most common ones found in engine. A few parts, however, use needle bearings, as you will learn in later textbook chapters. Needle bearing is roller bearing with tiny rollers.

the rocker arms. They are also found in engine accessory units (alternator, pumps, etc.).

SUMMARY

To be a competent engine mechanic or technician, you must know how to work with bolts, nuts, snap rings, belts, hoses, lines, seals, sealants, and other similar hardware.

Fasteners include any type of holding device, from a cap screw to glue. Bolts are the most common fasteners used on engines. Bolt size is measured across the threads. Bolt length is from the bottom of the head to the tip of the threads. The three thread types are coarse, fine, and metric. Thread types must never be mixed or damage will result.

Flat washers increase bolt or nut clamping surface and keep the head from digging into the part. Lock washers keep the bolt or nut from loosening in service.

Bolts and nuts should be torqued to specs in a crisscross pattern when required. Many engine components must be torqued during assembly.

Many nonthreaded fasteners are found in engines. Snap rings can be used to secure piston pins, for example. Lock pins can hold the distributor drive gear. A key and keyway are normally used on the front of the engine crankshaft.

Gaskets, seals, and sealants also play an important role in engine operation—their failure could cause major engine damage. Proper service methods are essential.

Metal lines and hoses carry fuel, coolant, and other fluids. They must be well maintained to assure engine dependability. Remember the danger of fuel spray from both gasoline and diesel injection systems. Replace any hose that shows signs of deterioration. Also keep the engine belts in good condition.

KNOW THESE TERMS

Fastener, Bolt, Nut, Cap screw, Hex, Bolt size, Head size, Bolt length, Pitch, Thread types, Left and right-hand threads, Tensil strength, Grade markings, Bolt description, Flat washer, Lock washer, Bolt torque, Tightening sequence, Torque-to-Yield, Degree wheel adapter, Machine screw, Minor and major thread damage, Tap, Die, Helicoil, Screw extractor, Nonthreaded fastener, Key and keyway, Gasket, Seal, RTV, Anaerobic sealer, Line, Hose, Hose clamp, V-belt, Ribbed belt, Cogged belt, Antifriction bearing, Friction bearing.

REVIEW QUESTIONS—CHAPTER 3

1. Define the term "fastener."
2. When a bolt threads into a part, it is called a ______ ______.
3. Bolt size is measured across the inside diameter of the threads. True or False?
4. List the three types of bolt threads.
5. Turn ______ ______ clockwise to tighten.
6. This refers to the strength of a bolt.
   a. Tensile strength or grade
   b. Tensile strength and static strength.
   c. Static strength or grade.
   d. None of the above are correct.
   e. All of the above are correct.
7. Explain the purpose of a flat washer and a lock washer.
8. Why is a crisscross pattern normally recommended on parts with multiple fasteners?
9. What is a helicoil?
10. Slip joint pliers are needed to remove and install snap rings. True or False?
11. A ______ fits in a slot and prevents a part from rotating on its shaft.
12. How is RTV sealer used?
13. An ______ ______ is a stationary type seal that usually fits into a groove in a component.
14. Diesel injection lines can contain over ______ psi (______ kPa) pressure.
15. This is the most common type of bearing found in an automotive engine.
   a. Antifriction bearing.
   b. Friction bearing.
   c. Needle bearing.
   d. Ball bearing.

ASE CERTIFICATION-TYPE QUESTIONS

1. All of the following are basic dimensions of a bolt EXCEPT:
   (A) bolt length.
   (B) thread pitch.
   (C) thread shape.
   (D) bolt head size.
2. Technician A says that a Woodruff key is a common fastener used on an automotive engine. Technician B says that a snap ring is a common fastener used on an automotive engine. Who is right?
   (A) A only.
   (B) B only.
   (C) Both A & B.
   (D) Neither A nor B.
3. Technician A says that metric threads can be easily confused for fine threads if not inspected carefully. Technician B says that metric threads can be easily confused for coarse threads if not inspected carefully. Who is right?
   (A) A only.
   (B) B only.
   (C) Both A & B.
   (D) Neither A nor B.
4. Technician A says that you should make sure that a fastener's threads are clean and lightly oiled before torquing. Technician B says that you should view the torque wrench scale from an angle when torquing a fastener. Who is right?
   (A) A only.
   (B) B only.
   (C) Both A & B.
   (D) Neither A nor B.
5. A die can be used to cut threads on all of the following, EXCEPT:
   (A) rods.
   (B) holes.
   (C) bolts.
   (D) shafts.
6. Technician A says that a helicoil will repair damaged external threads. Technician B says that a helicoil will repair damaged internal threads. Who is right?
   (A) A only.
   (B) B only.
   (C) Both A & B.
   (D) Neither A nor B.

7. When installing a new gasket, you should ________.
   (A) tighten fasteners in a crisscross pattern
   (B) never overtighten the engine component fasteners
   (C) coat the gasket with grease
   (D) Both A & B.
8. All of the following are basic rules to follow when working with automotive seals EXCEPT:
   (A) check the size of the new seal.
   (B) use a hammer and screwdriver to remove the old seal.
   (C) check the shaft or seal housing for damage.
   (D) coat the outside of the housing with sealer before installing a new seal.
9. Metal fuel lines are made of ________ tubing.
   (A) single-wall steel
   (B) double-wall copper
   (C) double-wall steel
   (D) single-wall copper
10. Technician A says that antifriction bearings are commonly used as engine main bearings. Technician B says that friction bearings are commonly used as engine main bearings. Who is right?
   (A) A only.
   (B) B only.
   (C) Both A & B.
   (D) Neither A nor B.

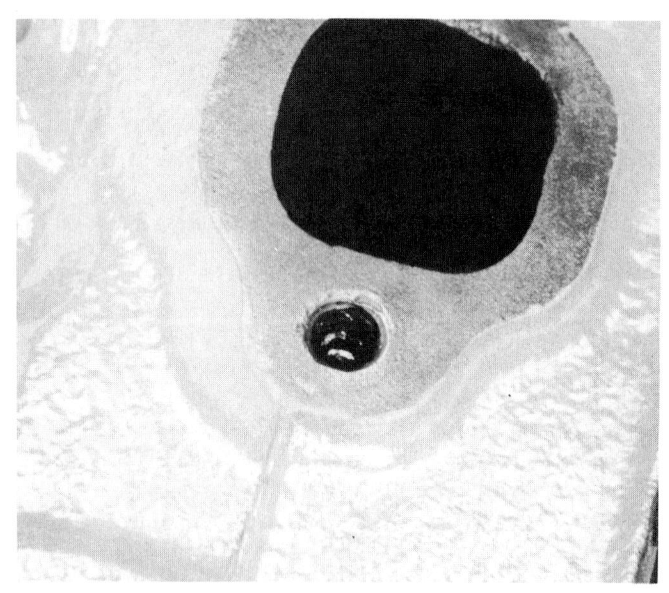

A—Threads in part have been pulled out by excess bolt tightening.

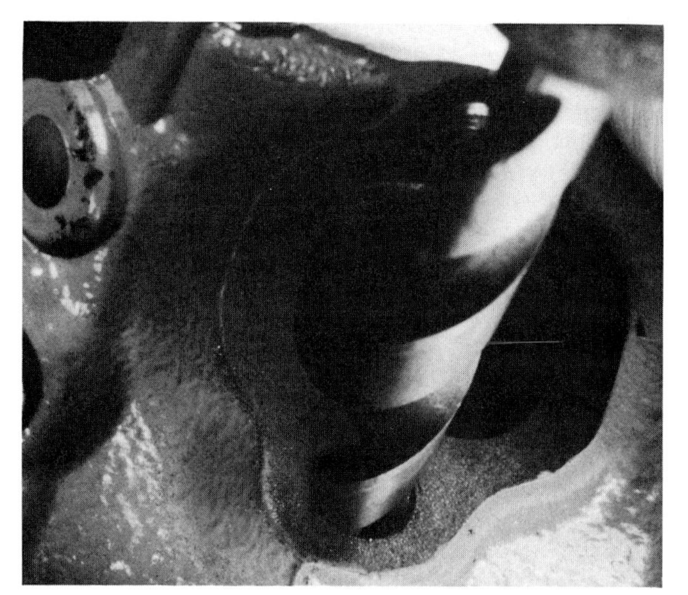

B—Drill and tap hole oversize using the recommended bit and tap size.

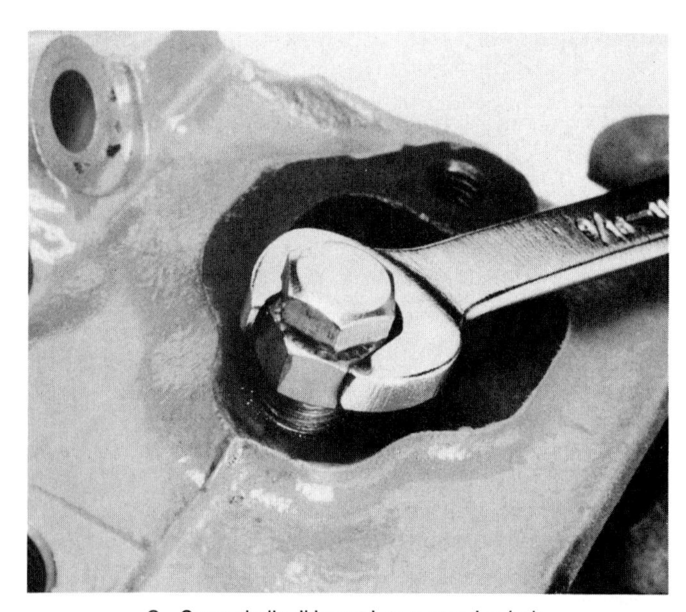

C—Screw helicoil insert into part using bolt and nut or special tool.

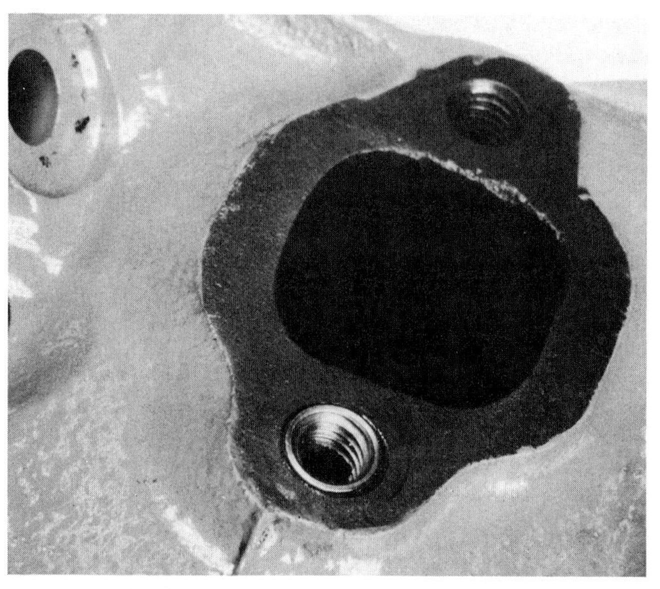

D—Repair will allow use of original bolt diameter because helicoil has internal threads.

These are the basic steps for repairing a stripped hole in a part with a helicoil insert. (The Eastwood Company)

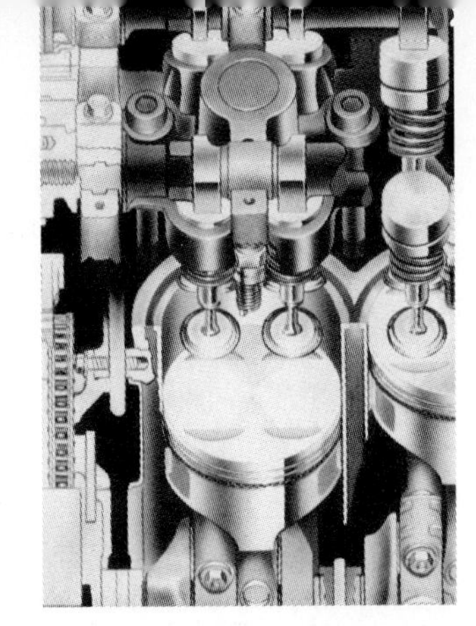

Chapter 4

Electricity-Electronics for Engines

After studying this chapter, you will be able to:
- *Describe a simple electrical circuit.*
- *Explain the types of automotive circuits.*
- *Describe the function of basic electrical components.*
- *Explain the operation of electronic components.*
- *Use Ohm's Law.*
- *Properly handle and repair automotive wiring.*
- *Identify electrical symbols and use wiring diagrams.*
- *Troubleshoot and repair fundamental electrical problems.*
- *Use electrical-electronic test instruments.*

Today's engines use a wide variety of electrical-electronic devices. Some monitor engine conditions. Others control engine related systems. In general, electrical-electronic devices help make an engine more efficient because they are much more precise and faster acting than mechanical devices. For this reason, it is impossible to service almost any part of an engine without working with some type of wiring or electrical device.

This chapter will cover the most general and basic information related to engine electrical and electronic components. It is an important chapter that will help you understand test equipment, ignition systems, charging and starting systems, computer control systems, and other subjects covered in this book. Study this chapter carefully!

WHAT IS ELECTRICITY?

Atoms are the "building blocks" of everything around us. You are made of atoms, this book, air, the sun—everything! An atom is made of tiny particles called protons, neutrons, and electrons.

The *electrons* are negatively charged particles that circle around the center or core of the atom. The protons and neutrons are in the center core of the atom. An atom works something like the planets circling the sun in our solar system.

Different substances have different atomic structures (arrangement of particles in atom). Some atoms have extra, *free electrons* that are NOT bound tightly to the atom. They can move from atom to atom. This movement of free electrons is what we call "electricity."

Insulators and conductors

An *insulator* is a substance without free electrons; it blocks electrical flow. Plastic, rubber, ceramic and air are examples of insulators. Wires are covered with insulation to prevent unwanted electrical flow.

A *conductor* is a substance having free electrons and allows the flow of electricity. Metal is the most common conductor and it is used in wiring of electrical devices. Electrons can easily move through a conductor.

Current, voltage, and resistance

The three basic elements of electricity are: CURRENT (amps), VOLTAGE (volts), and RESISTANCE (ohms). It is important to understand each value.

Current (abbreviated I or A) is the FLOW of electrons through a conductor. Just as water molecules flow through a garden hose, electrons flow through a wire in a circuit.

As an example, when current flows through a light bulb, the electrons rub against the atoms in the bulb *filament* (resistance wire inside bulb). This produces an "electrical friction" that heats the filament, making it glow "red hot."

Voltage (abbreviated V or E) is the force or ELECTRICAL PRESSURE that causes current flow. Similarly, water pressure causes water to squirt out the end of a garden hose.

An increase in voltage (pressure) causes an increase in current. A decrease in voltage causes a decrease in current. Automobiles normally use a 12 V (actually 12.5 V) electrical system.

Resistance (abbreviated R or Ω) is the OPPOSITION to current flow. Resistance is needed to control the flow of current in a circuit.

Just as the on/off valve on a garden hose can be opened or closed to control water flow, circuit resistance can be increased or decreased to control the flow of electricity. High resistance reduces current. Low resistance increases current.

CIRCUIT TYPES

There are various types of fundamental circuits. It is important for you to understand the configuration of each when working with engine-related electrical devices.

Simple circuit

A *simple circuit* is made up of a power source, conductors, and a load. These are the minimum requirements for a circuit, Fig. 4-1.

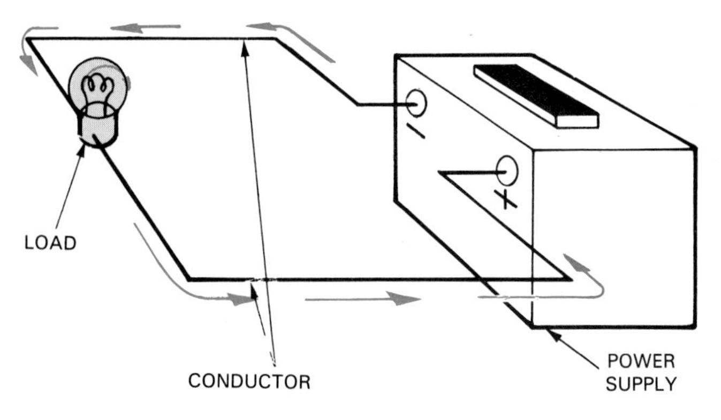

Fig. 4-1. To make basic electric circuit, you need a power source, conductors or wiring, and a load. Current can then flow through circuit and back to power source.

A circuit *power source* could be the battery or the alternator (charging system). It provides the force that moves electrons through the circuit. See Fig. 4-2.

The *load* could be any resistance element in the circuit: light bulb, electric motor, etc. The load performs the function or purpose of the circuit. For example, it may be an electric cooling fan that draws air through the radiator and protects the engine from overheating.

The *conductors* in the simple circuit are metal wires or the metal parts of the car. They carry current from the power source to the load.

Frame ground circuit

A *frame ground circuit,* also called *one wire circuit,* uses the car's metal body or frame as a conductor to complete an electrical circuit. This is commonly used in an automobile because it saves wiring. Wires do not have to return to the battery. A frame ground circuit is shown in Fig. 4-3.

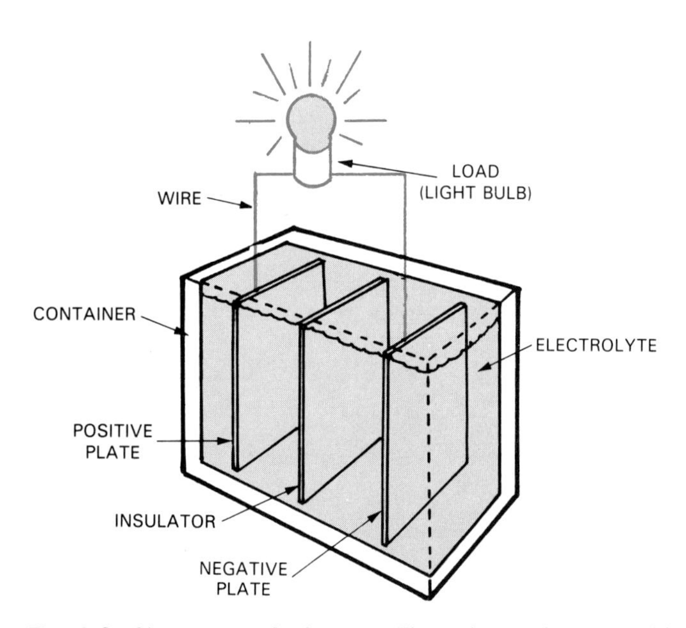

Fig. 4-2. Note parts of a battery. Electrolyte or battery acid produces chemical reaction that make electrons flow between plates.

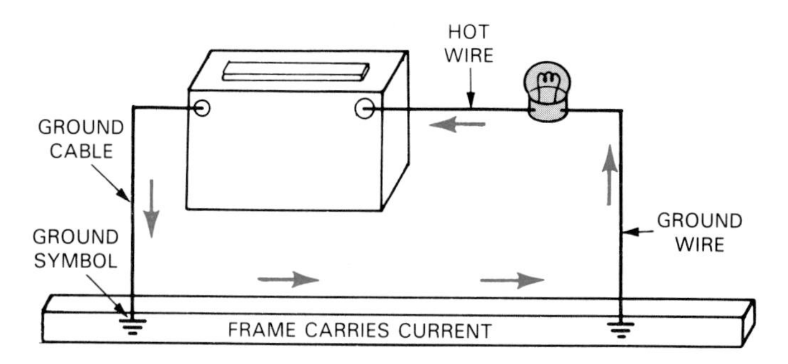

Fig. 4-3. Frame ground or one-wire circuit is commonly used in automobiles. Metal structure of car is used to complete the circuit.

Series and parallel circuits

A *series circuit* only has one path for current flow. Two or more loads are wired into this single path, Fig. 4-4A. When two loads are wired in series, they are dependent upon each other. If two light bulbs are in series, one burning out will break the circuit and both bulbs will stop glowing.

A *parallel circuit* has more than one path for current flow, Fig. 4-4B. The loads are wired into separate *legs* (paths) and can work independently of each other. If one bulb burns out, the other will still glow.

A *series-parallel circuit* is a combination series and parallel circuit. Part of the load is in series and part of the load is wired in parallel. Refer to Fig. 4-4C.

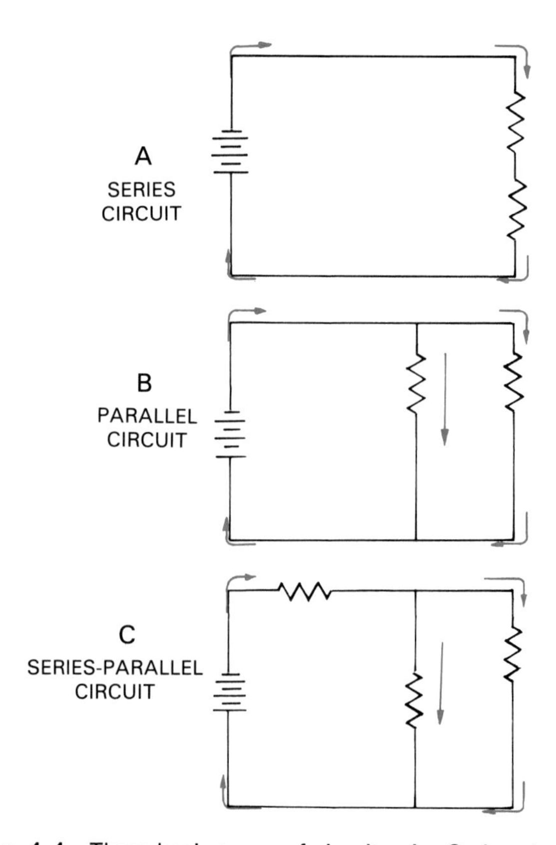

Fig. 4-4. Three basic types of circuits: A—Series circuit has only one electrical path. B—Parallel circuit has more than one path providing voltage to each load. C—Series-parallel combines these two circuit types.

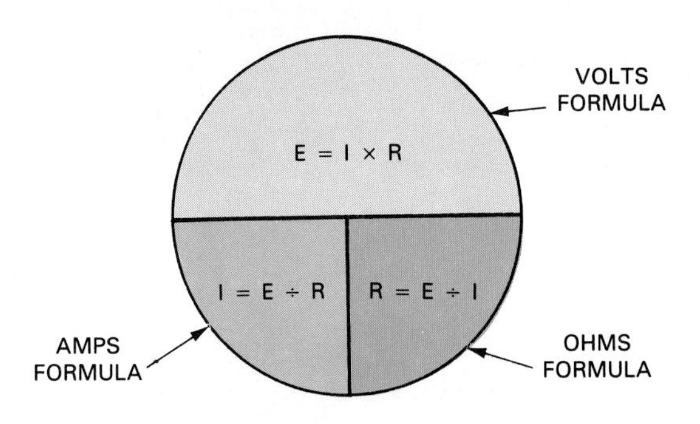

Fig. 4-5. Ohm's Law Pie Chart represents the formulas for calculating either voltage, current, or ohms in a circuit.

All of these circuit types can be found in engine related wiring. You must remember how series circuit loads affect each other and how a parallel circuit has independent legs that have little affect on each other.

OHM'S LAW

Ohm's Law is a basic formula for calculating unknown values in electrical circuits. If you know two values, you can use Ohm's Law and basic math to find the other.

Fig. 4-5 shows an *OHM'S LAW PIE CHART* that represents the use of the formula. Note the location of E(volts), I(amps), and R(ohms) in the pie chart. Volts is above amps and ohms. Amps and ohms are next to each other. This should help you remember—Volts equals amps times ohms; amps equals volts divided by ohms; and ohms equals volts divided by amps.

One example, if you know that a circuit has 12 volts applied and its load has 10 ohms, simply divide 12 by 10 to get 1.2 amps. See Fig. 4-6.

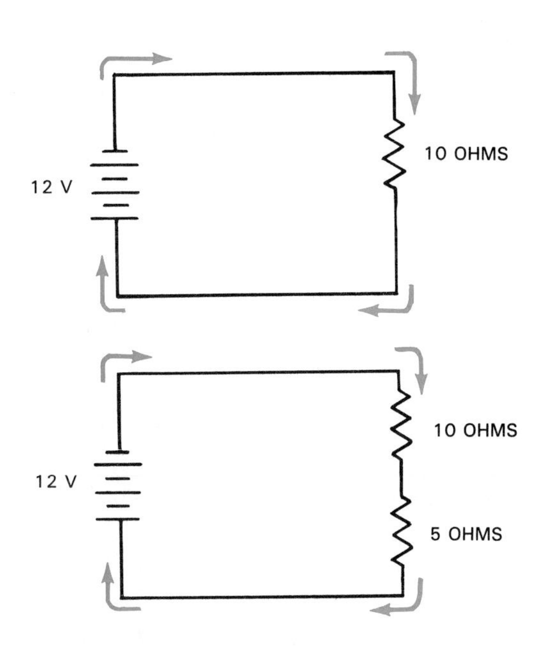

Fig. 4-6. Simple circuits show relationship of resistance, current, and voltage. As resistance increases, current drops. A lowered resistance would increase current. Increased voltage would also increase current and vice versa. How much current would flow in each of these circuits?

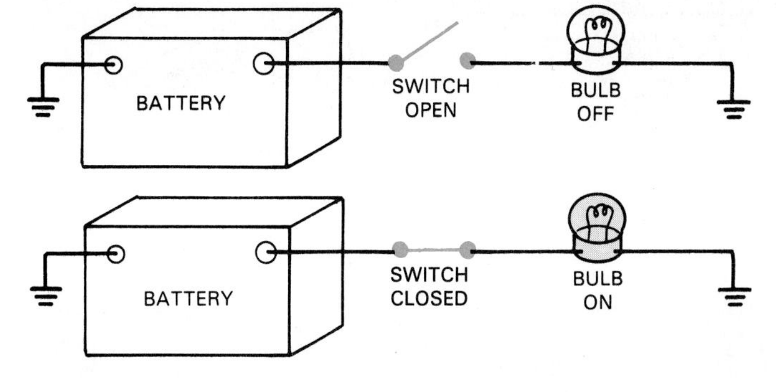

Fig. 4-7. Switch provides means of breaking path for current in circuit.

ELECTRICAL COMPONENTS

Electrical components include electrical devices that were used before the advent of electronics. Mechanical switches, fuses, and circuit breakers are examples of electrical components. Since these devices are still common, you must understand them fully.

Switch

A *switch* allows an electric circuit to be turned on or off manually (by hand), Fig. 4-7. When the switch is CLOSED (on), the circuit is *complete* (fully connected) and will operate. When the switch is OPEN (off), the circuit is *broken* (disconnected) and does NOT function.

Fuse

A *fuse* protects a circuit against damage caused by a "short circuit." As shown in Fig. 4-8, the fuse link will

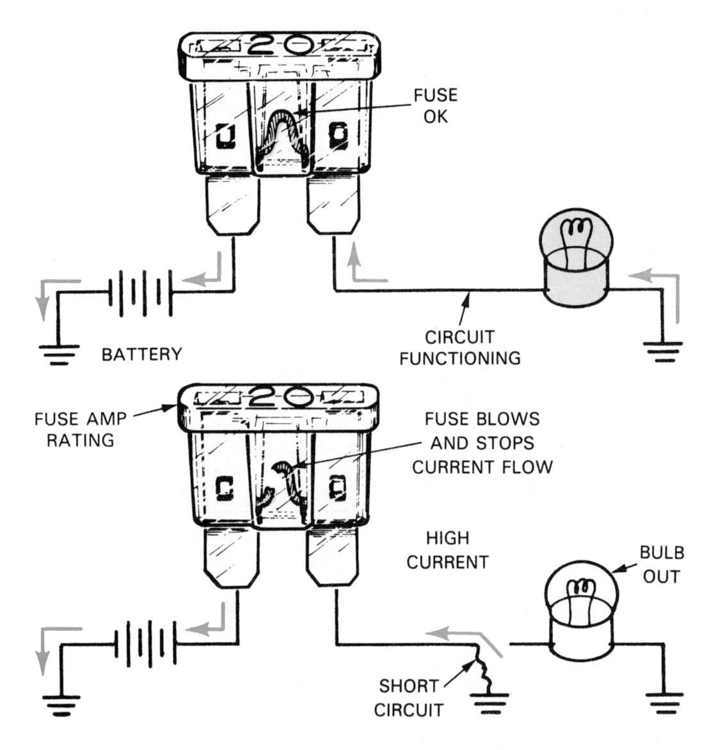

Fig. 4-8. Note how fuse will burn and break circuit if a short causes too much current in circuit. (General Motors)

melt and burn in half to stop current and any further circuit damage.

A *fuse box* is often located under the car's dashboard. It contains fuses for the various circuits. One is pictured in Fig. 4-9.

A *short circuit* or "short" is caused when a defective wire or component touches ground and causes excess current flow. See Fig. 4-9. If a short to ground exists between the battery and load, unlimited current flow can cause an *"electrical fire"* (melting and burning of wire insulation).

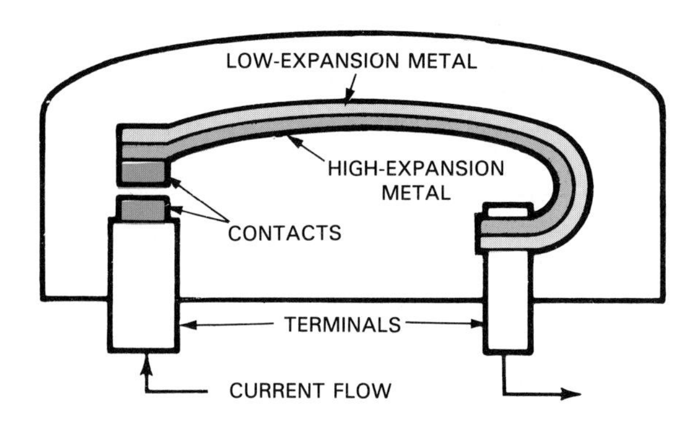

Fig. 4-10. Circuit breaker protects circuit from short as does a fuse. However, it will reset itself and allow circuit to function again if short is fixed. Excess current heats dissimilar metals and makes arm bend to disconnect contacts. (Ford)

Circuit breaker

A *circuit breaker* performs the same function as a fuse. It disconnects the power source from the circuit when current becomes too high. Excess current and heat bends a bimetal strip to open a set of breakers (contact points). Normally, a circuit breaker will automatically reset itself when current returns to normal levels and the bimetal strip cools. Circuit breaker action is shown in Fig. 4-10.

Magnetic field

You are probably familiar with magnetism from using a simple permanet magnet. It produces an invisible *magnetic field* (lines of force) that will attract ferrous (metal containing iron) objects.

A magnetic field can also be created using electricity. A long piece of wire can be wound into a *coil* (spiral of wire). The ends of the wire can be connected to a battery or other power source. Then, when current passes through the wire, a magnetic field is produced.

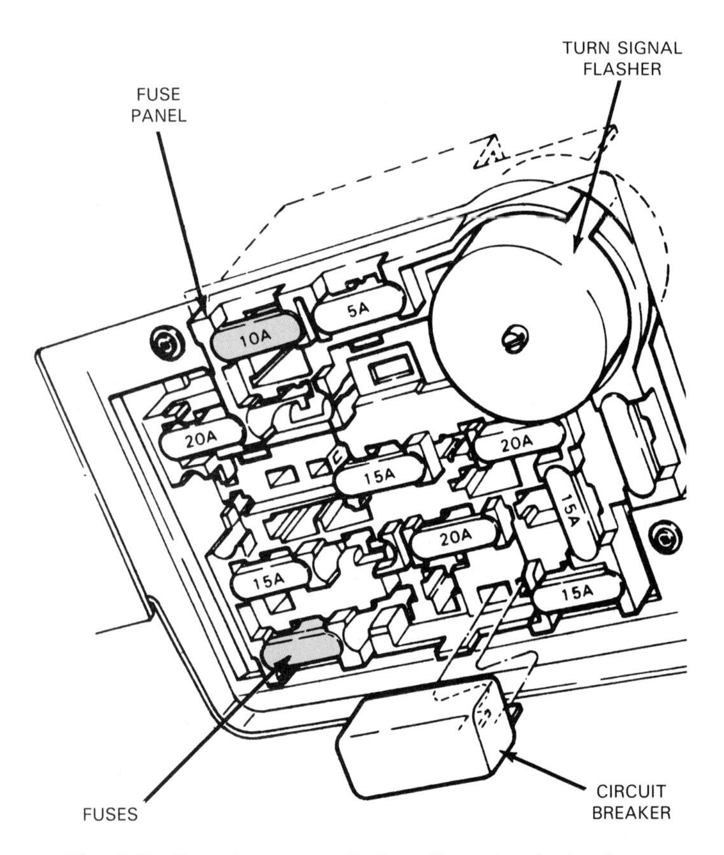

Fig. 4-9. Fuse box normally installs under dash of car. (Ford)

To make the field or lines of force STRONGER, a soft iron bar or *core* can be inserted into the center of the coil. The iron core will become magnetized, making an *electromagnet (electric magnet).*

If a magnetic field is passed over a wire, an electric current is INDUCED or generated in the wire. The wire cutting the lines of force causes a tiny amount of electricity to flow through the wire. This action is called *induction* (current generated in wire by a magnetic field).

Many engine related components use the characteristics of a magnetic field. Electronic fuel injection, electric motors, relays, ignition systems, and on-board computers are just a few examples.

Electric motor

To build a basic electric motor, you would start by bending a piece of wire into a loop. When current is passed through the wire loop or *winding* a magnetic field forms around the wire.

A magnet or *pole piece* would be needed to make the loop of wire move. A magnetic field is set up between the pole pieces, also called *pole shoes.* This is illustrated in Fig. 4-11.

Relay

A *relay* is an electrically operated switch, Fig. 4-12. It allows a small dash switch to control another circuit by remote control. It also allows very small wires to be used behind the dash, while larger wires are needed in the relay-operated circuit. See Fig. 4-13.

Solenoid

A *solenoid* is similar to a relay but has much larger coil and can handle much higher current values. One is shown in Fig. 4-14. Note how a small input current controls a much higher current.

A solenoid is used to complete a high current circuit and can also be used to move components. The plunger in the solenoid moves a large metal disc against two terminals to complete a circuit. The plunger can be used to move a gear for example. A solenoid, as you will learn in later chapters, is commonly used on an engine starting motor.

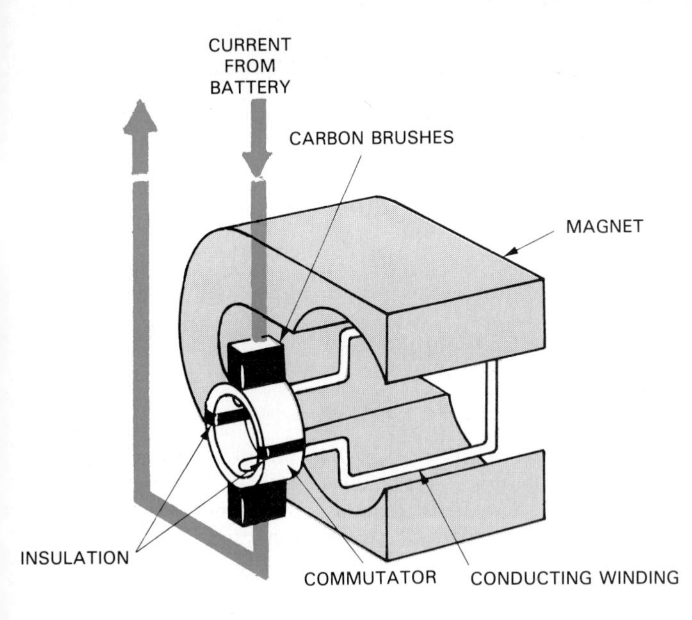

Fig. 4-11. Study principle of an electric motor. Magnet forms field that acts on loop of wire. Current flow through loop causes loop to spin as commutator reverses polarity of field around loop. Loop rapidly spins in magnet because of attraction and repulsion of magnetic fields. (Robert Bosch)

Fig. 4-13. Relays can be located under dash or in engine compartment. Service manual will tell you the function of each relay. (Saab)

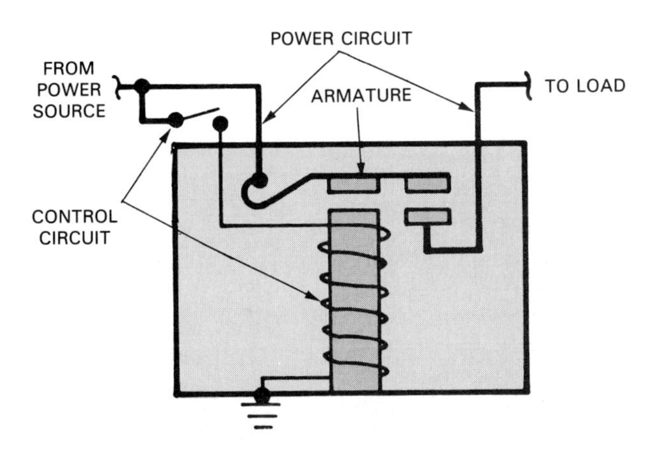

Fig. 4-12. Relay allows small control current to connect circuit with slightly higher current. Current through coil produces magnetic field that pulls armature down. This completes load circuit. (Ford)

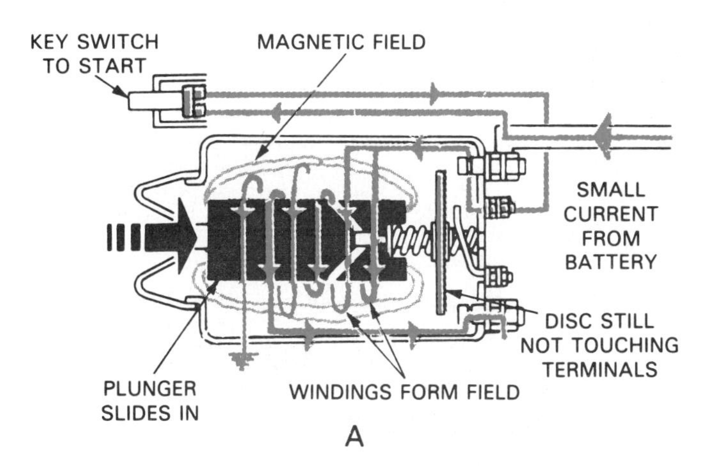

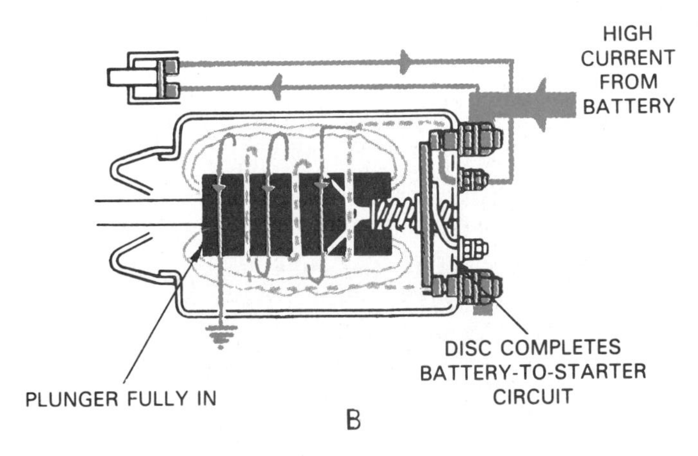

Fig. 4-14. Solenoid is powerful relay and can also move other components. A—Small control current activates coil. B—Magnetic field around coil pulls plunger to make disc touch two terminals and complete high load circuit.

AC AND DC CURRENT

AC or *alternating current* rapidly changes from positive to negative and current flows back and forth through the circuit.

DC or *direct current* only flows in one direction and is commonly used throughout the automobile.

A comparison of AC and DC current is given in Fig. 4-15. This illustration represents what you would see on a *scope* (television tube type electronic testing device for measuring voltage). When voltage is zero, the line would stay on the base or zero level. When voltage goes positive, the line would move up on the screen. The line or trace would move down for negative.

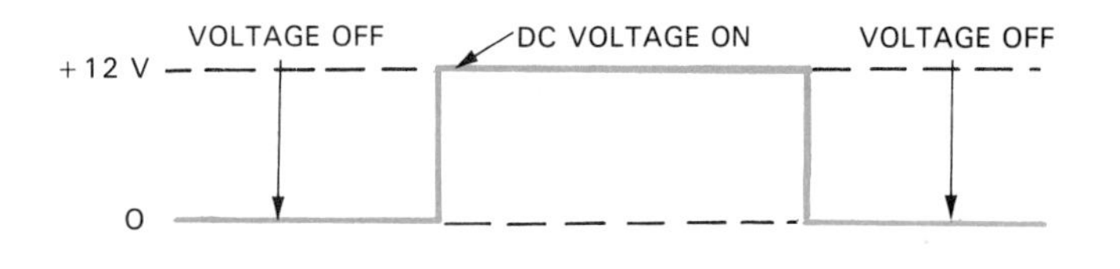

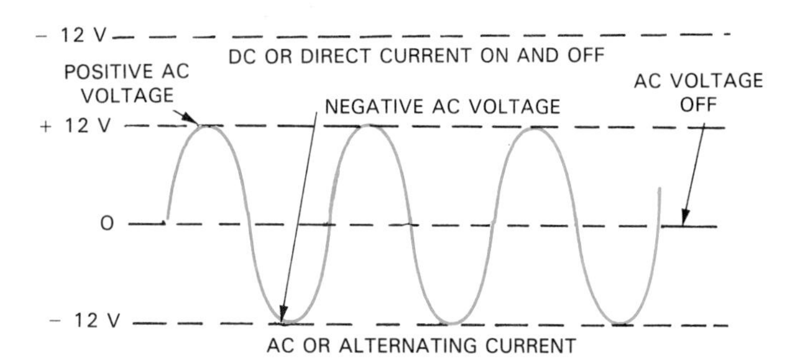

Fig. 4-15. Diagram shows difference between DC and AC. A—Note how direct current stays at relatively constant voltage when on. Line moves up from zero line to 12 V line. B—Alternating current voltage rapidly changes from positve to negative. AC is produced by speed sensing types of engine sensors.

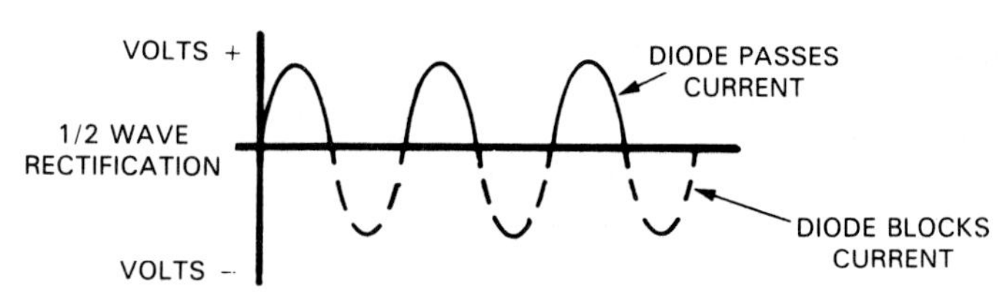

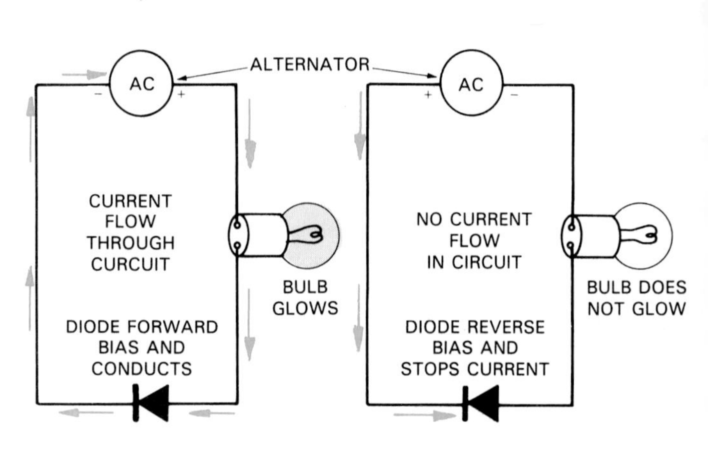

Fig. 4-16. Note how diode will only pass current in one direction. Diode can be used to change AC into DC.

Note! This subject will be covered fully in later chapters on charging systems and testing for engine performance problems.

ELECTRONIC DEVICES

Electronic devices do NOT use moving parts as do many electrical components (relays, solenoids, etc.). Termed *solid state,* they use special semiconductor substances that can change from a resistor to a conductor with electrical stimulation. To be a competent engine technician, you must be familiar with the operation of common electronic devices.

Diode

A *diode* is an ''electronic check valve'' that will only allow current to flow in one direction. It is commonly used to change alternating current into direct current. See Fig. 4-16.

When *forward bias* (current entering in right direction), a diode acts as a CONDUCTOR and allows flow.

When *reverse bias* (current tries to enter wrong way), the diode changes into an INSULATOR. It stops current from passing through the circuit.

Transistor

A *transistor* performs the same basic function as a relay; it acts as a remote control switch. It is much more efficient than a relay, however. A transistor can sometimes turn on and off faster than 2000 times a second. It does this without using moving parts which can wear and deteriorate.

Look at Fig. 4-17. A transistor *amplifies* (increases) a small control or base current. The small base current energizes the semiconductor material, changing it from an insulator to a conductor. This allows the much larger circuit current to pass through the transistor.

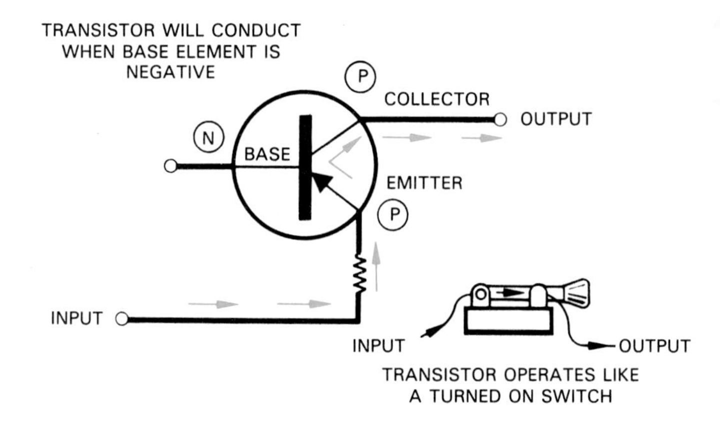

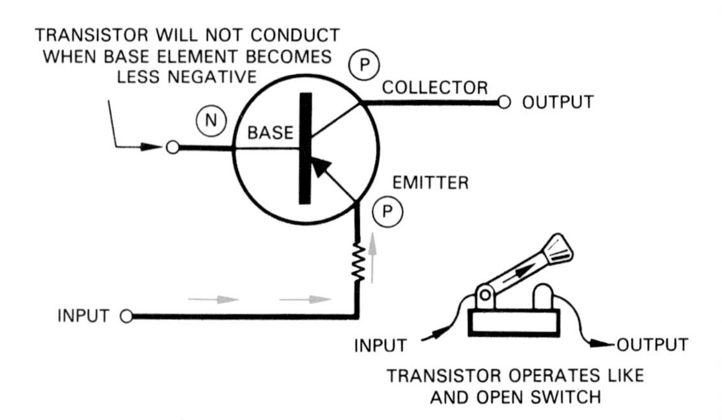

Fig. 4-17. Transistor acts as a relay. Small base current can be used to turn transistor on to control much higher current to load. (Motorola)

Condenser or capacitor

A *condenser* or *capacitor* is a device used to absorb unwanted electrical pulses (AC voltage fluctuations) in a circuit. They are used in various types of electrical and electronic circuits.

Integrated circuit

An *integrated circuit,* abbreviated IC, contains almost microscopic diodes, transistors, resistors, and capacitors in a wafer-like *chip* (small plastic housing with metal terminals). Integrated circuits are used in very complex electronic circuits; computers for example.

Printed circuit

Printed circuits do NOT use conventional, round wires; they use flat conductor strips mounted on an insulating board. Printed circuits are normally used instead of wires on the back of the instrument panel. This eliminates the need for a bundle of wires going to the engine indicators, gauges, and instrument bulbs.

Amplifier

An *amplifier* is an electronic circuit designed to use a very small current to control a very large current. Its function is much the same as a transistor. However, higher output currents and smaller control currents are possible.

A good example of an amplifier is an ignition system control unit (amplifier), introduced in Chapter one. It uses small electrical pulses from the distributor to produce strong on/off cycles to operate the ignition coil.

Computer

A *computer* or *electronic control unit* (ECU) uses very complex electronic circuits to perform various functions, Fig. 4-18. It uses input from engine sensors to gather information about engine efficiency. It is programmed to then send out electrical signals to relays, solenoids, and other components to affect engine operation.

An integrated computer system uses feedback data from several smaller computer systems (engine, brakes, suspension, traction control for example) to better control each individual system. All systems are controlled from a central processor instead of individual ones. This allows the computer to monitor more functions to better decide how to control the engine, drivetrain, braking, traction control, and ride stiffness for optimum efficiency.

Sensors

Sensors are electrical devices that change their resistance or voltage output with a change in a condition, Fig. 4-18. They are mounted in various locations on the engine and other systems. A few of the most common engine sensors are:

1. *Oxygen sensor* or *exhaust gas sensor* that measures the amount of oxygen in the engine exhaust. It changes voltage output with a change in oxygen. The amount of oxygen in the exhaust indicates combustion efficiency. This data is used by the computer to determine if the mixture is too rich or too lean.

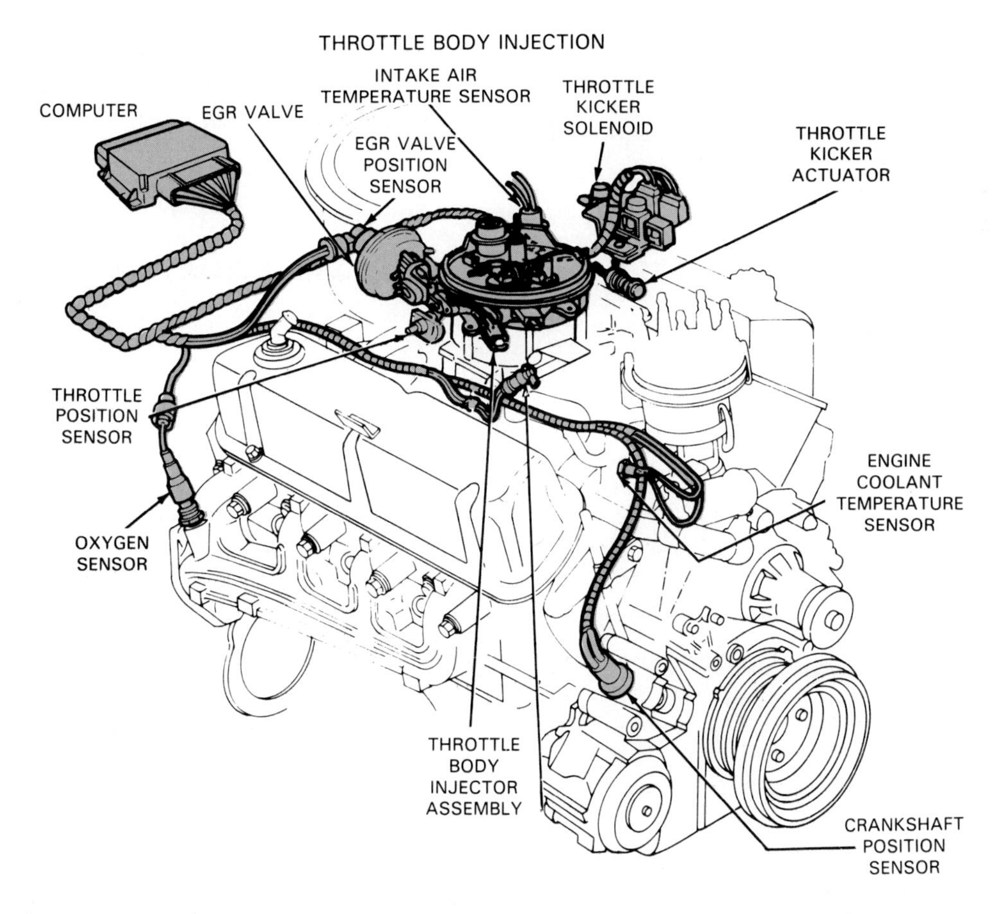

Fig. 4-18. Computer and engine sensors are now used to increase engine efficiency. To service an engine, you will be working with engine sensors. (Ford)

2. *Manifold pressure sensor* that monitors the vacuum in the engine intake manifold. Engine manifold vacuum changes with engine load and this information is also used by the computer to increase efficiency.
3. *Throttle position sensor* that changes resistance as the engine throttle valve is opened and closed. This reports to the computer about engine speed, acceleration, or deceleration. This lets the computer alter ignition timing, fuel mixture, and other variables as needed.
4. *Engine temperature sensor* that checks coolant temperature and changes its resistance as engine temperature changes. This allows the computer to richen the mixture when the engine is cold.
5. *Airflow sensor* measures the amount of air flowing into the engine. This lets the computer process data on engine speed, air density, and air temperature.
6. *Crankshaft position sensor* produces an electrical signal corresponding to engine rpm (revolutions per minute). This tells the computer how fast the engine is running for altering various systems.

Other engine sensors are also used: knock sensors, oil pressure sensors, intake air temperature sensors, etc. They all use similar principles and will be explained in more detail later in this book.

AUTOMOTIVE WIRING

An automobile uses various types of wiring in its many electrical systems. To be an engine technician, it is important to learn the different types, how they are used, and how to service them.

Wire types

Primary wire is small and carries battery or alternator voltage, Fig. 4-19. Primary wire normally has plastic insulation to prevent shorting. The insulation is usually *color coded* (different wires are marked with different colors). This lets you trace (follow) wires that are partially hidden.

Primary wires are often enclosed in a wiring harness. A *wiring harness* is a plastic or tape covering that helps protect and organize the wires. See Fig. 4-20.

Wire size is determined by the diameter of the wire's metal conductor. The diameter is stated in *gauge size* which is a number system. The larger the gauge number is, the smaller the diameter of the wire conductor.

When replacing a section of wire, always use wire of equal size. If a smaller wire is used, the circuit could malfunction due to high resistance. Undersize wire could heat up and melt its protective insulation. An electrical fire could result.

Note! When disconnecting wiring harness connectors, make sure you use approved methods. Most modern electrical connectors lock together with plastic or metal clips. You must release the clips to disconnect the wiring. Fig. 4-21 shows examples.

Secondary wire, also called high tension cable, spark plug wire, or coil wire, is only used in a car's ignition system. It has extra thick insulation for carrying high voltage from the ignition coil to the spark plugs. The conductor, however, is for very small currents, Fig. 4-22.

Battery cable is extremely large gauge wire capable of carrying high currents from the battery to the starting motor. Usually, a starting motor draws more current than all of the other electrical components combined (normally over 100 amps).

Ground wires or *ground straps* connect electrical components to the chassis or ground of the car. Since they connect circuits or parts to ground, insulation is NOT needed.

CODE	
B	BLACK
Br	BROWN
G	GREEN
Gy	GRAY
L	BLUE
Lb	LIGHT BLUE
Lg	LIGHT GREEN
O	ORANGE
R	RED
W	WHITE
Y	YELLOW

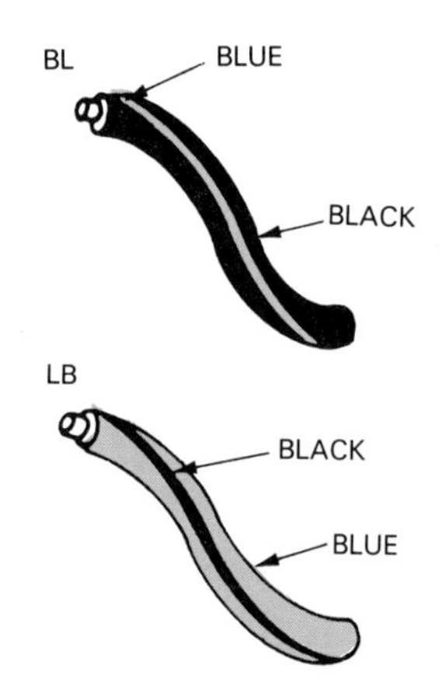

Fig. 4-19. Primary wires are normally color coded. This will let you trace or follow a wire through car. Wiring diagrams will give abbreviations for color coding of wires.

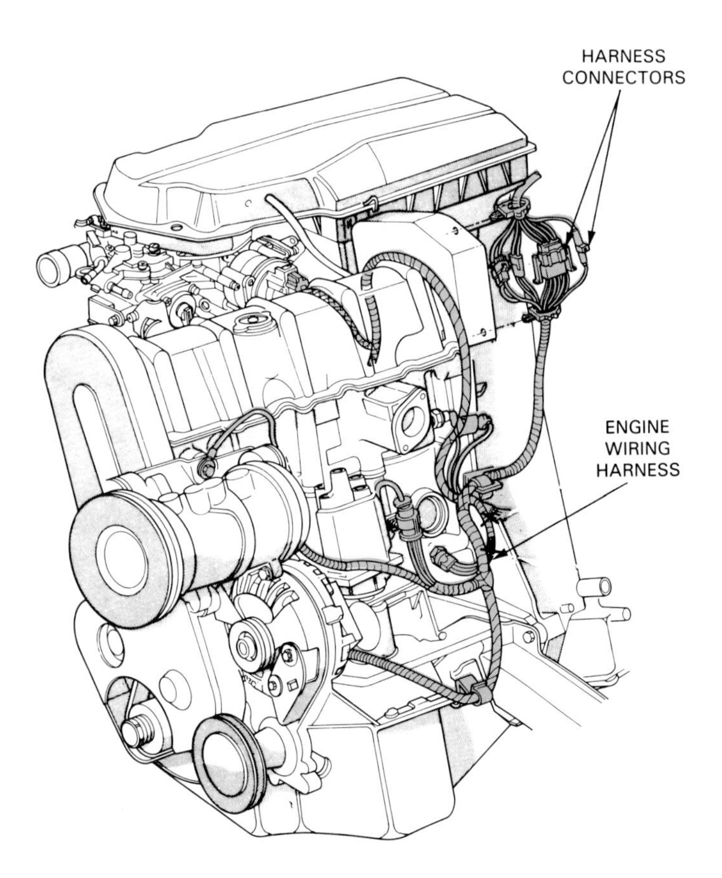

Fig. 4-20. Engine wiring harness surrounds and protects wires for engine sensors, ignition system, and other components. Plastic connectors are used to join sections of wires. (Chrysler)

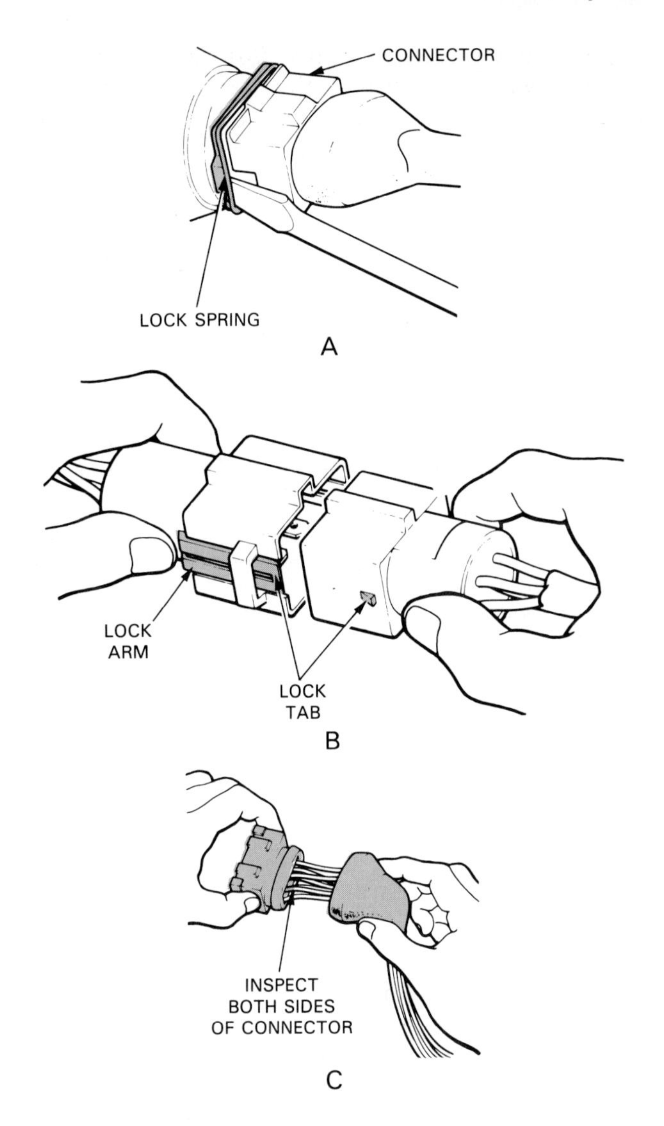

Fig. 4-21. Always handle wiring harness connectors properly. A—This connector has a metal lock spring that keeps connector from pulling apart. It must be released before disconnecting connector. B—This connector has plastic arms that lock around tabs. You must squeeze arms to free connector. C—When inspecting connector, check the corrosion of terminals and for moisture in either side of connector. (Honda)

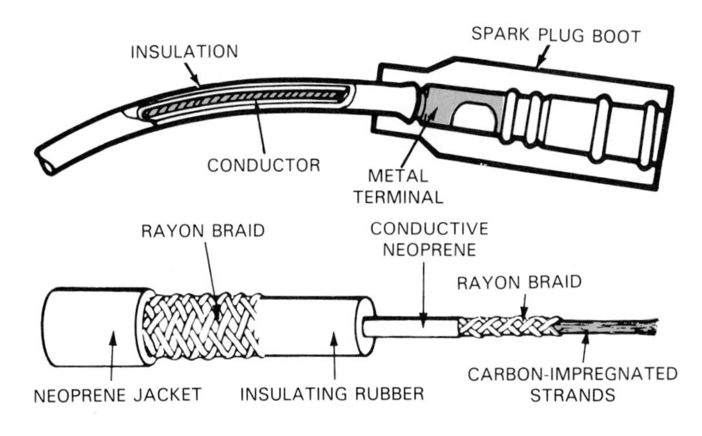

Fig. 4-22. Secondary wire has thick insulation to carry high voltage. Most secondary wires have carbon-impregnated strands that provide some internal resistance to prevent radio noise. (Champion Spark Plugs)

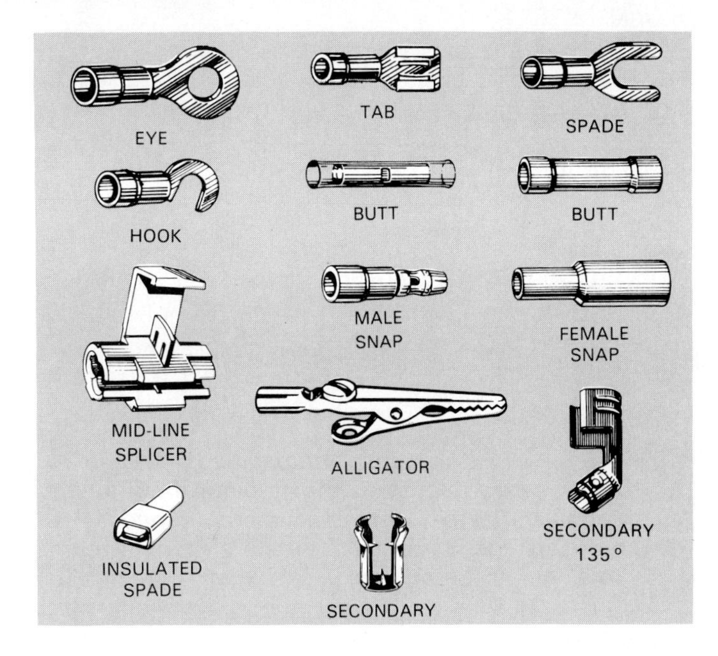

Fig. 4-23. Study types of wire terminals and connectors. (Belden)

Wiring repairs

Numerous methods can be used to fix automotive wiring. The most important general methods will be introduced.

Crimp connectors and *terminals* can be used to quickly repair automotive wiring. Several are given in Fig. 4-23. *Terminals* allow a wire to be connected to an electrical component. *Connectors* or *splicers* allow a wire to be connected to another wire.

Crimping pliers are used to deform the connector or terminal around the wire. Fig. 4-24A shows a technician installing a crimp terminal.

A *soldering gun* or iron can also be used to permanently fasten wires to terminals or to other wires, Fig. 4-25B. The soldering gun produces enough heat to melt *solder* (lead and tin alloy). The soldering gun is touched to the wire and other component to preheat them. Then, the solder is touched to the heated joint and it melts. When cooled, the solder makes a solid connection between the electrical components.

Note! *Rosin core solder* should be used on all electrical repairs. *Acid core solder* can cause corrosion of electrical components. It is recommended for nonelectrical repairs (radiator repairs, for example).

Electrical component location chart

Today's cars have dozens of electrical-electronic components in the engine compartment. As mentioned, many can upset engine operation if faulty.

A *component location diagram* illustrates where each type electrical-electronic part is positioned in the engine compartment. One is shown in Fig. 4-25. This type chart will let you quickly find a component that is suspected of causing trouble.

Wiring diagram

A road map shows how various cities are connected by roads and highways. Similarily, a *wiring diagram*

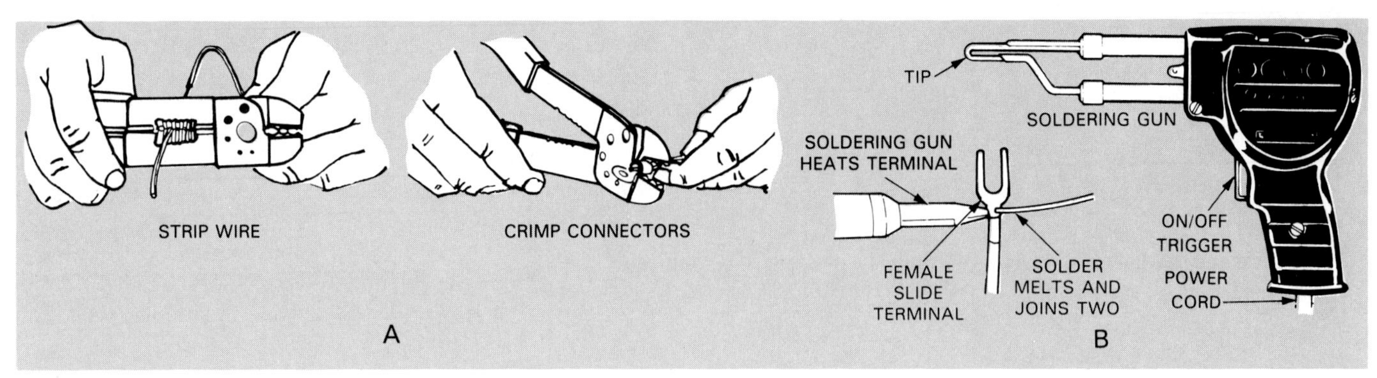

Fig. 4-24. Wire repairs can be done with crimp connectors or soldering gun. A—Strip wire so that connector can be slid over metal end of wire. Use crimping pliers to form connector around wire. Pull on wire lightly to make sure connector is tight. B—Soldering gun can be used to heat wire and terminal. Then, solder is melted onto preheated connection. (Klein Tools and Florida Dept. of Voc. Ed.)

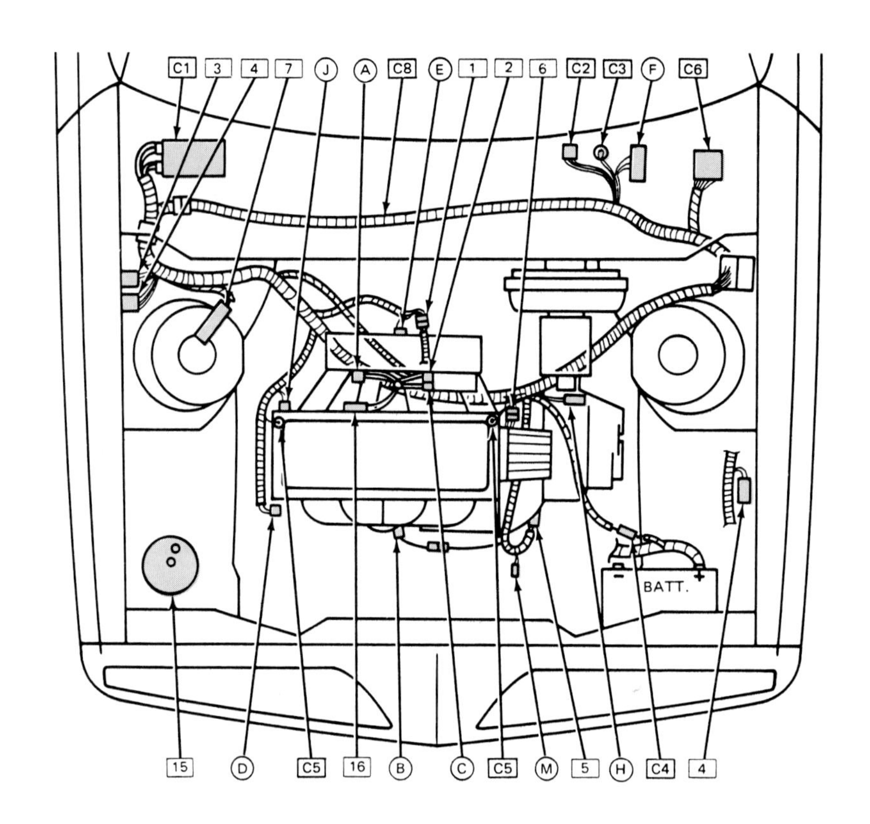

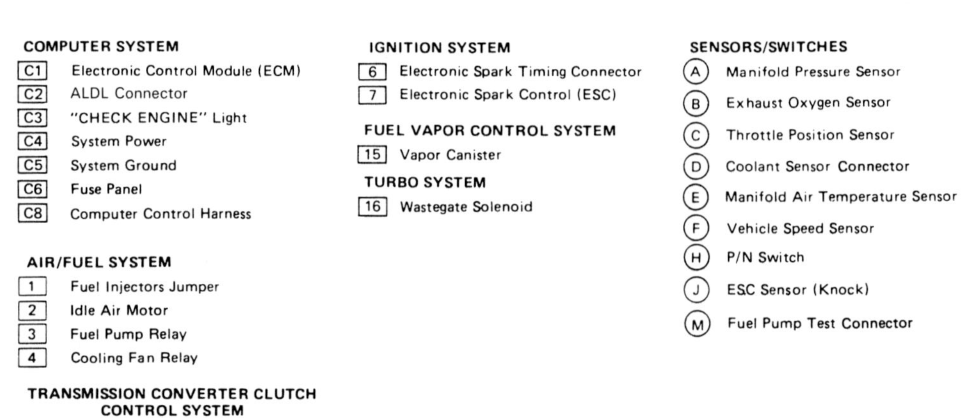

Fig. 4-25. This is an example of a service manual electrical component location chart for one car's engine compartment. Can you find the fuse panel, oxygen sensor, and fuel pump test connector? (Buick)

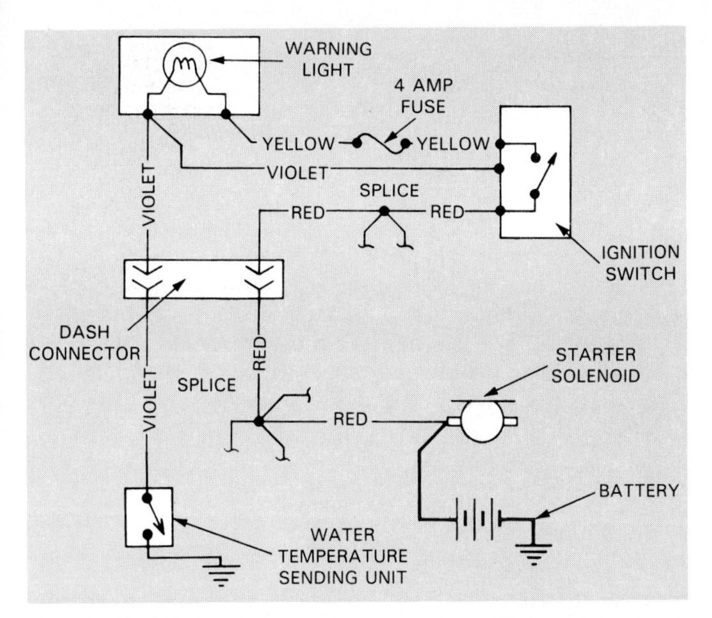

Fig. 4-26. Wiring diagram uses symbols and lines to represent electrical components and wires. It lets you quickly see how wires connect to components, where connections in the wiring harness are located, etc.

shows how electrical components are connected by wires, Fig. 4-26. A wiring diagram serves as an "electrical map" that helps the technician with difficult repairs.

Electrical symbols

Wiring diagrams use *electrical symbols* to represent the electrical components in a circuit. The lines on the diagram represent the wires and connect to the symbols. In this way, you can trace each wire and see how it connects to each component. Fig. 4-27 gives several electrical symbols.

BASIC ELECTRICAL TESTS

Various electrical tests and testing devices are used by an engine technician. To be prepared for many later chapters, you should have a general understanding of these tools and how to use them.

Jumper wire

A *jumper wire* is handy for testing switches, relays, solenoids, wires, and other nonresistive components (switch, connector, wire). The "jumper" can be

Symbol	Meaning
+	POSITIVE
−	NEGATIVE
	GROUND
	FUSE
	CIRCUIT BREAKER
	CAPACITOR
Ω	OHMS
	RESISTOR
	VARIABLE RESISTOR
	SERIES RESISTOR
	COIL
	STEP UP COIL
	OPEN CONTACT
	CLOSED CONTACT
	CLOSED SWITCH
	OPEN SWITCH
	CLOSED GANGED SWITCH
	OPEN GANGED SWITCH
	TWO POLE SINGLE THROW SWITCH
	PRESSURE SWITCH
	SOLENOID SWITCH
	MERCURY SWITCH
	DIODE OR RECTIFIER
	BY-DIRECTIONAL ZENER DIODE

Symbol	Meaning
	CONNECTOR
	MALE CONNECTOR
	FEMALE CONNECTOR
	MULTIPLE CONNECTOR
	DENOTES WIRE CONTINUES ELSEWHERE
	SPLICE
J2 2	SPLICE IDENTIFICATION
	OPTIONAL WIRING WITH / WIRING WITHOUT
	THERMAL ELEMENT (BI-METAL STRIP)
	"Y" WINDINGS
88:88	DIGITAL READOUT
	SINGLE FILAMENT LAMP
	DUAL FILAMENT LAMP
	L.E.D.-LIGHT EMITTING DIODE
	THERMISTOR
	GAUGE
TIMER	TIMER
	MOTOR
	ARMATURE AND BRUSHES
	DENOTES WIRE GOES THROUGH GROMMET
#36	DENOTES WIRE GOES THROUGH 40 WAY DISCONNECT
#19 STRG COLUMN	DENOTES WIRE GOES THROUGH 25 WAY STEERING COLUMN CONNECTOR
INST PANEL #14	DENOTES WIRE GOES THROUGH 25 WAY INSTRUMENT PANEL CONNECTOR

Fig. 4-27. These are the symbols most commonly found on automotive wiring diagrams. Study them carefully! (Chrysler Corp.)

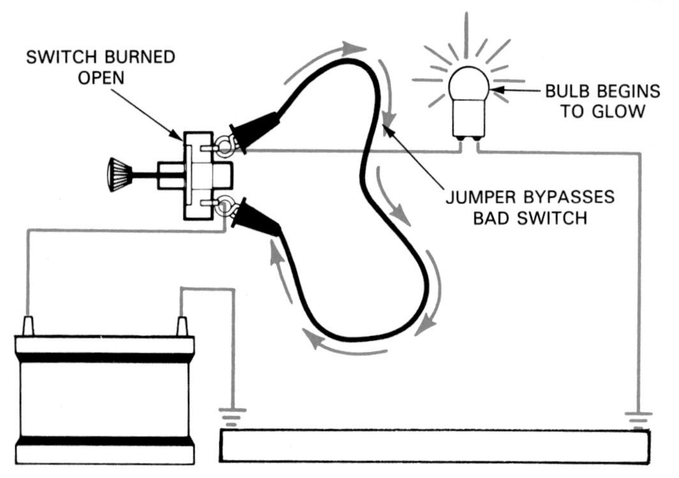

Fig. 4-28. Jumper wire will let you bypass voltage around or to a component. For example, if you think a switch is bad, jumper around the switch. If the circuit then begins to function, the switch is bad. (Ford)

substituted for the component, as shown in Fig. 4-28. If the circuit begins to function with the jumper in place, then the component being bypassed is defective.

Test light

A *test light* is a fast method of checking a circuit for power or voltage, Fig. 4-29. It has an alligator clip that connects to ground. Then, the pointed tip can be touched to the circuit to check for power. If there is voltage, the light will glow. If it does NOT glow, there is an open or break between the power source and the test point.

A *self-powered test light* is similar to a flashlight with a lead attached. It contains batteries and is used to check for *circuit continuity* (whether circuit is complete). To use this type test light, the normal source of power (car battery or feed wire) must be disconnected. If the light glows, the circuit or part has continuity (low ohms). If it does NOT glow, there is an open or break (high ohms) between the two test points.

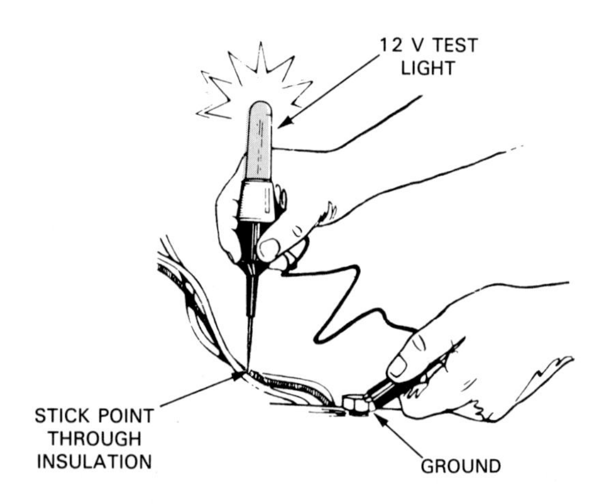

Fig. 4-29. Test light is fast way of making sure power is being fed to section of circuit. Sharp probe can be touched on metal connections or pierced through insulation to check for voltage. Do not pierce insulation on sensor wires carrying low voltage output. (Lisle Tools)

Using a voltmeter

A *voltmeter* is used to measure the amount of voltage (volts) in a circuit. It is normally connected across or in PARALLEL to the circuit. The voltmeter reading can be compared to specifications to determine whether an electrical problem exists. See Figs. 4-30A and 4-31.

An *analog meter* uses a needle that swings across the face of the meter, Fig. 4-31. It is an older type but is still useful for checking small fluctuations in voltage.

A *digital meter* uses a number readout instead of a needle, Fig. 4-32. It is a modern type frequently recommended for testing computer system components. It will normally draw less current and not damage delicate electronic parts.

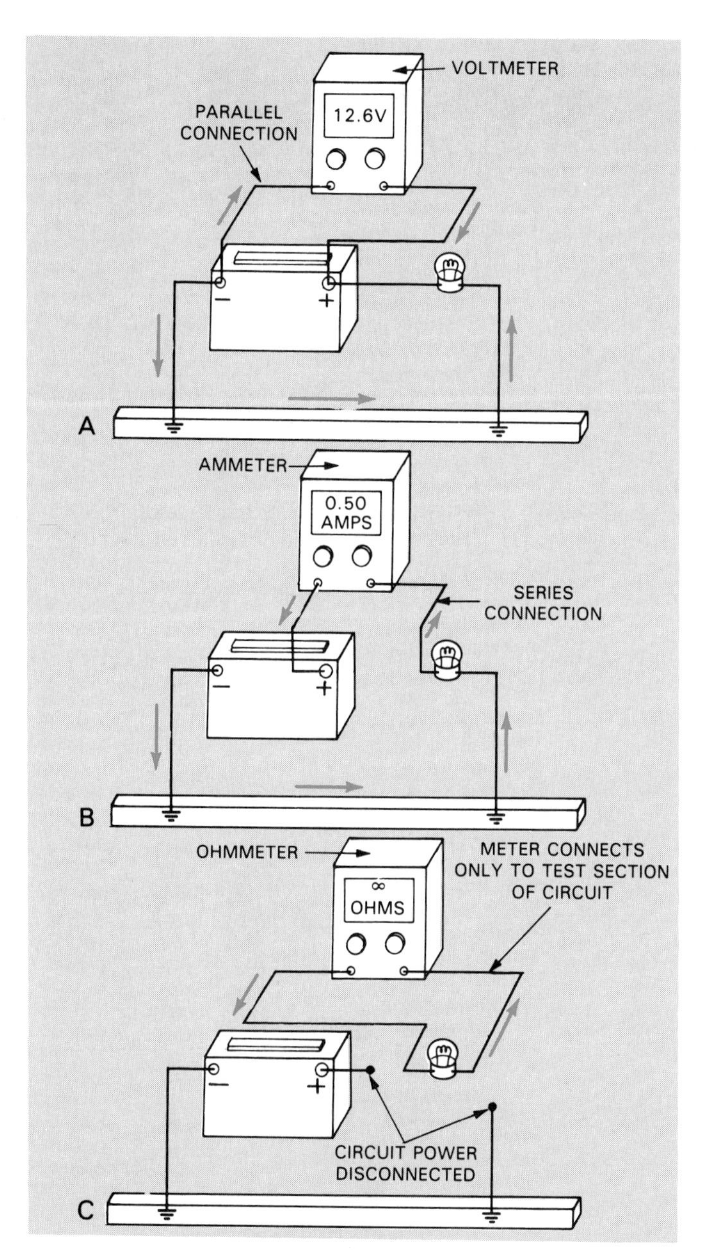

Fig. 4-30. Study basic meter connections. A—Voltmeter is connected across or in parallel to measure "electrical pressure." B—Ammeter connects in series. You must break circuit so that current flows through meter. Inductive ammeters, however, have a clip that fits over or around insulation to measure current. C—Ohmmeter must not be connected to power source or meter damage may result.

Fig. 4-31. Analog meter uses a needle to indicate reading. It is handy when you have a bad connection. Wiggle wires while watching needle. If needle fluctuates, there may be a poor electrical connection. This type meter should not be used to check some engine sensors. It can draw too much current and burn the sensors. (Saab)

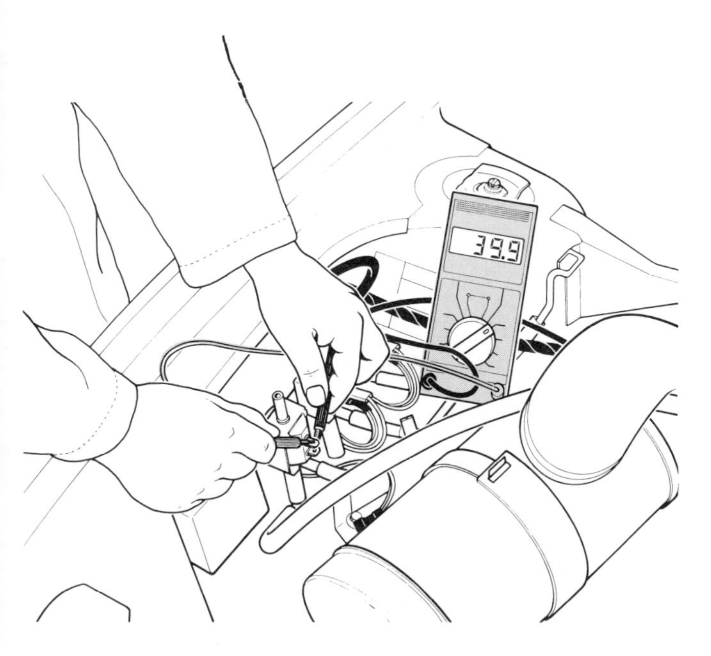

Fig. 4-32. Digital meter has number display instead of needle. This is a more modern type that is usually recommended for testing sensors. Mechanic is measuring resistance of vacuum valve windings to check for open or short. Readings must be within specs. (Peugeot)

Using an ammeter

An *ammeter* measures the amount of current (amps) in a circuit, Fig. 4-30B. Conventional types must be connected in SERIES with the circuit. All of the current in the circuit must pass through the ammeter.

A modern *inductive* or *clip-on ammeter* is simply slipped over the outside of the wire insulation. It uses the magnetic field around the outside of the wire to determine the amount of current in the wire. An inductive ammeter is very fast and easy to use.

Using an ohmmeter

An *ohmmeter* will measure the amount of resistance (ohms) in a circuit or component.

As in Fig. 4-30C, the ohmmeter is connected across the wire or component being tested with the power source disconnected. Then, the ohmmeter reading can be compared to specifications. If too high or low, the part is defective.

Multimeter or VOM

A *multimeter,* also called a *VOM,* is an ohmmeter, ammeter, and voltmeter combined into one case. As pictured in Fig. 4-31, a *function knob* (control knob) can be turned to select the type measurement to be made (volts, amps, or ohms). A VOM must be connected to the circuit as described for each individual meter.

A few rules to remember when using a multimeter are:

1. Do NOT connect the meter, when set on ohms, to a source of voltage. This could damage the meter or blow its fuse. Disconnect the battery from the circuit before measuring resistance.
2. Use a high impedance (high resistance) meter when checking electronic components, especially in an engine computer control system. Some meters with low impedance will draw too much current through the electronic device, ruining it.
3. Disconnect electrical terminals carefully or you could damage the terminal and cause high circuit resistance and an engine malfunction.
4. When using a pointed test probe, be careful not to stick the sharp probe into your hand. It is very easy for the tip to slip off of a wire and stab through the skin. Do not stick holes in wires going to some engine sensors (oxygen sensor for example). They produce such a small voltage that moisture entering through the hole in the insulation could affect sensor signals to the computer.
5. Preset the meter to a high range before measuring an unknown electrical quantity. For example, when measuring current, set the meter to a high range. This will prevent excess needle deflection or current flow through the meter that could damage the meter.
6. Follow service manual procedures when making electrical tests. The wrong meter connection could cause serious damage to today's computer systems.
7. Position the meter so that it cannot be damaged from a fall or hot or spinning engine parts. Keep the leads away from the hot exhaust manifolds and spinning engine fan.
8. Always read the meter's operating instructions before use. Meter ranges, specifications, and capabilities will vary.

SUMMARY

An engine technician, working in a modern garage, must have a sound understanding of electrical-electronic components and how they are serviced. The inside of the engine compartment on a late model car is full of engine sensors, solenoids, relays, and other devices. As an engine technician, you must be trained in diagnosing and replacing these parts.

Electricity is the movement of free electrons through a conductor or wire. Insulation is used to prevent electron flow.

A simple circuit consists of a power source, conductors, and a load. Cars commonly use a frame ground circuit to save on wiring.

Voltage is the electrical pressure that causes current flow. Resistance is needed to limit and control current. Ohm's Law can be used to find an unknown electrical value when two values are given.

A fuse or circuit breaker is used to protect a circuit from a short. Either will break the circuit and protect the circuit from excess current and heat.

A relay is an electrically operated switch. A solenoid is similar to a relay but it handles much higher loads.

Alternating current changes direction while direct current only flows in one direction. A car primarily uses DC.

A diode is an electrical check valve that only allows current flow in one direction. A transistor is a solid state switch. A small current can stimulate the semiconductor material in the transistor, changing it from an insulator to a conductor. An integrated circuit contains microscopic components in a small chip. IC's are commonly used in computers.

A computer and sensors are now used to monitor and control engine operation. The sensors convert a condition (temperature, movement, etc.) into an electrical signal. The computer uses these electrical changes to determine the needs of the engine. It can then use relays, solenoids, and small DC motors to affect engine operation and improve efficiency.

Primary wire is the small wire carrying battery voltage to electrical components. Secondary wire has thicker insulation for carrying high voltage in the ignition system.

Wiring can be repaired with crimp connectors or soldering. Wiring diagrams are used to make complex electrical repairs. They are "road maps" that show how all of the components in a circuit connect to each other.

Test lights, jumper wires, and a multimeter are commonly used to make electrical tests. A digital, high impedance meter is normally recommended for tests on computer control systems.

KNOW THESE TERMS

Free electrons, Insulator, Conductor, Current, Voltage, Resistance, Ohms, One-wire circuit, Ohm's Law, Switch, Fuse, Circuit breaker, Short circuit, Magnetic field, Electromagnet, Induction, Relay, Solenoid, AC, DC, Diode, Transistor, Capacitor, Printed circuit, Integrated circuit, Amplifier, Computer, Integrated computer systems, Oxygen sensor, Pressure sensor, Throttle position sensor, Engine temperature sensor, Airflow sensor, Crank position sensor, Primary wire, Secondary wire, Wiring harness, Crimp connectors, Rosin core solder, Component location diagram, Wiring diagram, Electrical symbol, Jumper wire, Test light, Analog meter, Digital meter, VOM.

REVIEW QUESTIONS—CHAPTER 4

1. Explain the difference between an insulator and a conductor.
2. __________, abbreviated ____ or ____, is the flow of electrons through a circuit.
3. Why is resistance needed in a circuit?
4. A frame ground circuit uses the car body or metal structure as a return wire to the battery negative. True or False?
5. If a circuit has 12.5 volts applied and 25 ohms of resistance, how much current will flow in the circuit?
6. What is the purpose of a fuse or circuit breaker?
7. Magnetism is used in the following components:
   a. Relay.
   b. Solenoid.
   c. Motor.
   d. All of the above.
   e. None of the above.
8. A __________ has no moving parts and performs the same function as a relay by acting as a remote control switch.
9. In your own words, how does a computer control system increase engine efficiency?
10. __________ __________ is smaller wire that carries battery voltage and __________ __________ has much thicker insulation for carrying high voltage to the spark plugs.
11. To disconect modern wiring harness connectors, simply pull on the wires next to the connector. True or False?
12. __________ __________ solder should be used during electrical repairs.
13. When would a wiring diagram be useful?
14. An engine does not crank when the key is turned to start. The battery is new and has been tested by the customer and another garage.
    Technician A says that a test light should be used to determine if voltage is being applied to the starter solenoid from the ignition switch.
    Technician B says that a jumper wire can be used to connect battery voltage directly to the solenoid to check its operation.
    Who is correct?
    a. Technician A.
    b. Technician B.
    c. Both A and B.
    d. Neither A nor B.
15. Two technicians disagree on how to test several sensors in a computer control system.
    Technician A says a test light is the fastest way of checking for voltage to and from various sensors.
    Technician B says a digital, high impedance meter should be used to check the sensors.
    Which technician is correct?
    a. Technician A.
    b. Technician B.
    c. Both A and B.
    d. Neither A nor B.

ASE CERTIFICATION-TYPE QUESTIONS

1. Technician A says voltage controls the flow of current in a circuit. Technician B says that voltage is the electrical pressure that causes current flow. Who is right?
   (A) A only.
   (B) B only.
   (C) Both A & B.
   (D) Neither A nor B.
2. All of the following are basic types of electrical circuits EXCEPT:
   (A) frame ground circuit.
   (B) series circuit.
   (C) series-ground circuit.
   (D) parallel circuit.
3. Technician A says that if a circuit has 12 volts with a 2 ohm load, the circuit has 6 amps flowing through it. Technician B says that if a circuit has 12 volts with a 2 ohm load, the circuit has 3 amps flowing through it. Who is right?
   (A) A only.
   (B) B only.
   (C) Both A & B.
   (D) Neither A nor B.
4. A circuit breaker performs the same function as a __________.
   (A) relay
   (B) solenoid
   (C) fuse
   (D) None of the above.
5. Technician A says that alternating current flows in one direction and is commonly used throughout an automobile. Technician B says that direct current flows back and forth through a circuit and is commonly used throughout an automobile. Who is right?
   (A) A only.
   (B) B only.
   (C) Both A & B.
   (D) Neither A nor B.
6. Technician A says that an electrical relay can handle much higher current values than a solenoid. Technician B says that a solenoid can handle much higher current values than a relay. Who is right?
   (A) A only.
   (B) B only.
   (C) Both A & B.
   (D) Neither A nor B.
7. Technician A says that electronic devices use moving parts to conduct electricity through a circuit. Technician B says electronic devices do not use moving parts to perform their function in a circuit. Who is right?
   (A) A only.
   (B) B only.
   (C) Both A & B.
   (D) Neither A nor B.
8. All of the following are solid state electronic devices EXCEPT:
   (A) condenser.
   (B) relay.
   (C) diode.
   (D) transistor.
9. Technician A says that primary wire is small and carries battery or alternator voltage. Technician B says that primary wire is used to carry high voltage from the ignition coil to the spark plugs. Who is right?
   (A) A only.
   (B) B only.
   (C) Both A & B.
   (D) Neither A nor B.
10. An ammeter measures the amount of __________ in a circuit.
    (A) current
    (B) resistance
    (C) voltage
    (D) wattage

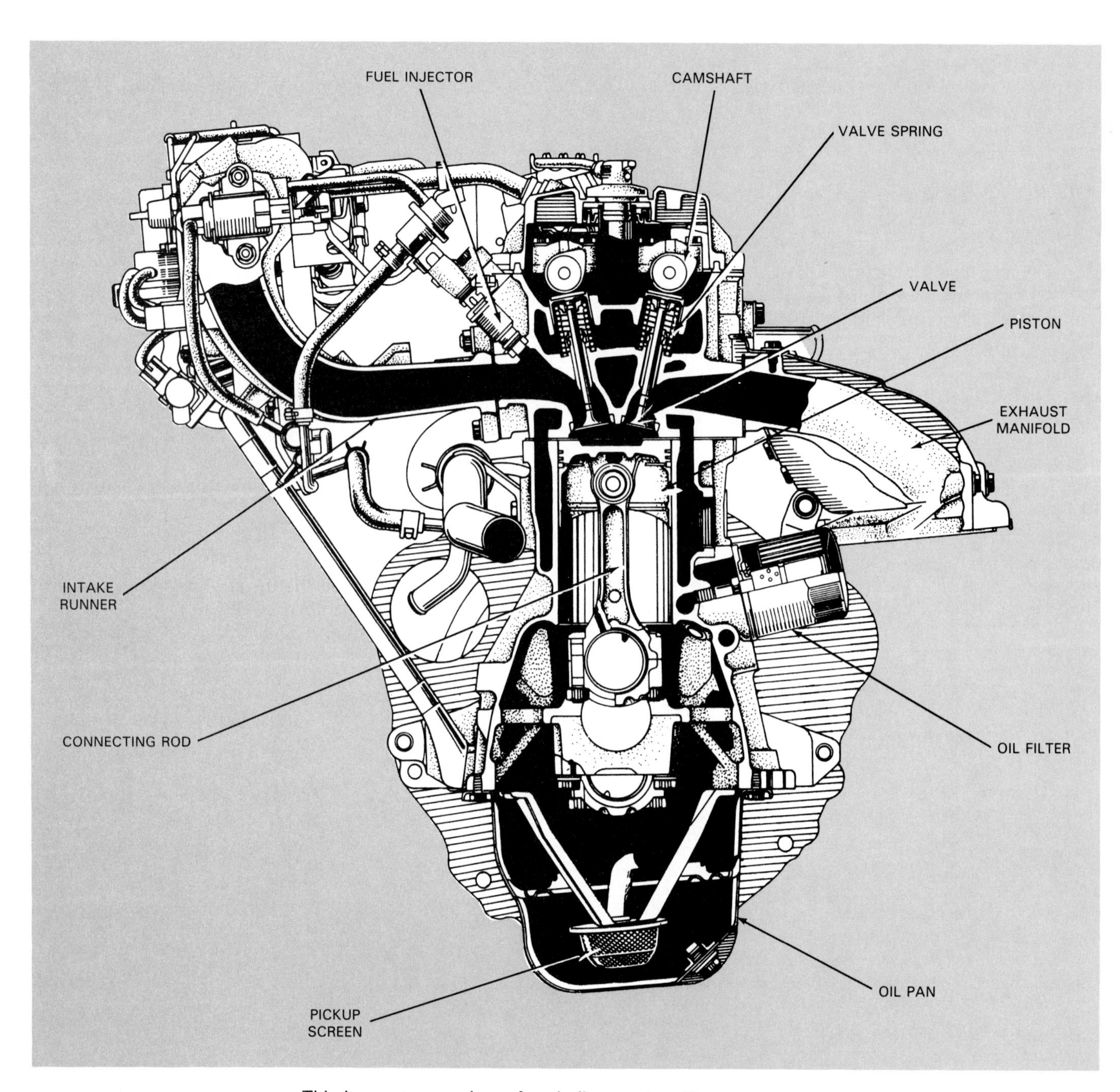

This is a cutaway view of an in-line engine. Note part names.

Chapter 5

Shop Safety

After studying this chapter, you will be able to:

- *Prevent shop accidents.*
- *Explain the causes of typical shop accidents.*
- *List the most important safety rules.*
- *Work around auto engines with a minimum of danger.*
- *Describe the proper procedures for disposing of hazardous wastes.*

This chapter will review shop safety. Hundreds of engine technicians are injured or killed while on the job each year. A majority of these injuries or deaths resulted from broken safety rules.

As an engine technician, you wil be exposed to many potentially dangerous situations. You will be working around running engines, engines suspended in the air on a hoist, around gasoline, and numerous other possible hazards. You must learn to control these dangerous situations and keep your job safe and enjoyable. This chapter will help you gain basic ''safety skills.''

TYPES OF ACCIDENTS

There are six kinds of accidents you must prevent: fires, explosions, asphyxiation, chemical burns, electric shock, and physical injuries. Each could cripple or kill you or someone else!

Fires

Fires can cause horrible destruction and death. When working on engines, gasoline, diesel fuel, cleaning solvents, and oily rags are just a few of the many possible causes of a shop fire.

Gasoline is the most dangerous flammable in the shop. Just a cup of gasoline can engulf a whole engine compartment in flames. The flames can then consume the rest of the car and maybe the complete shop.

A few rules for handling gasoline include:

1. NEVER use gasoline as a cleaning agent. Cleaning solvents are not as flammable as gasoline.
2. Keep sources of heat (welding and cutting equipment for example) away from the engine's fuel system.
3. Wipe up gasoline spills right away. Do not spread ''quick dry'' (oil absorbent) on gasoline because the absorbent will become even more flammable.
4. Disconnect the car battery when disconnecting a fuel line on an engine.
5. Wrap a rag around any fitting when disconnecting a fuel line. This will keep fuel from leaking or spraying out and possibly starting a fire when

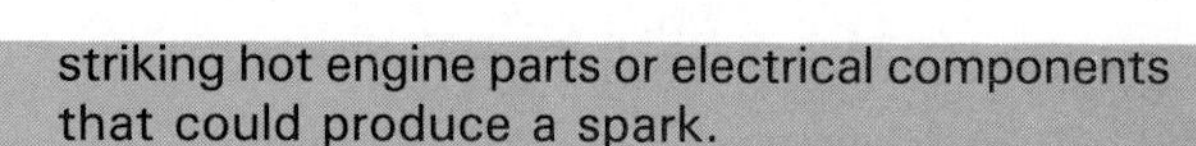

striking hot engine parts or electrical components that could produce a spark.

6. Store gasoline and other flammables in approved containers with a lid.

Electrical fires occur when a current carrying wire ''shorts'' to ground. The wire can then carry too much current. The excess current will make the wire glow ''red hot,'' melting its insulation. The insulation can then catch on fire and other wires can also short. Wiring through much of the car can quickly begin to burn. See Fig. 5-1.

Disconnect the car battery to prevent electrical fires. Then, if you accidentally short a hot wire to ground, nothing will happen.

Explosions

Several types of explosions are possible when servicing an engine or its systems. You should be aware of these sources of sudden death and injury.

Car BATTERIES can EXPLODE! As shown in Fig. 5-2, hydrogen gas can surround the top of car batteries being charged or discharged. This gas is highly explosive. The slightest spark or flame can ignite and cause the battery to explode. Chunks of battery case and acid can blow into your eyes and face. Blindness, facial cuts, acid burns, and scars can result.

Various other sources can cause shop explosions. For example, special sodium-filled engine valves, welding tanks, propane-filled bottles, and fuel tanks can all explode if mishandled. These hazards, if engine related, will be discussed in later chapters.

Asphyxiation

Asphyxiation is caused by breathing toxic or poisonous substances in the air. Mild cases of asphyxiation will cause dizziness, headaches, and vomiting. Severe asphyxiation can cause death.

The most dangerous source of asphyxiation in an auto shop is an automobile engine. An engine's EXHAUST GASES are DEADLY POISON. Connect an exhaust hose to the tailpipe of any car being operated in the shop, Fig. 5-3. Also, make sure the exhaust system is turned ON.

Discussed in related chapters, other shop substances are harmful if inhaled. One of these harmful substances is *asbestos* clutch disc dust for instance.

Respirators (filter masks) should be worn when working around any kind of airborne impurities.

Keep in mind that a filter mask or respirator will NOT block some poisonous fumes.

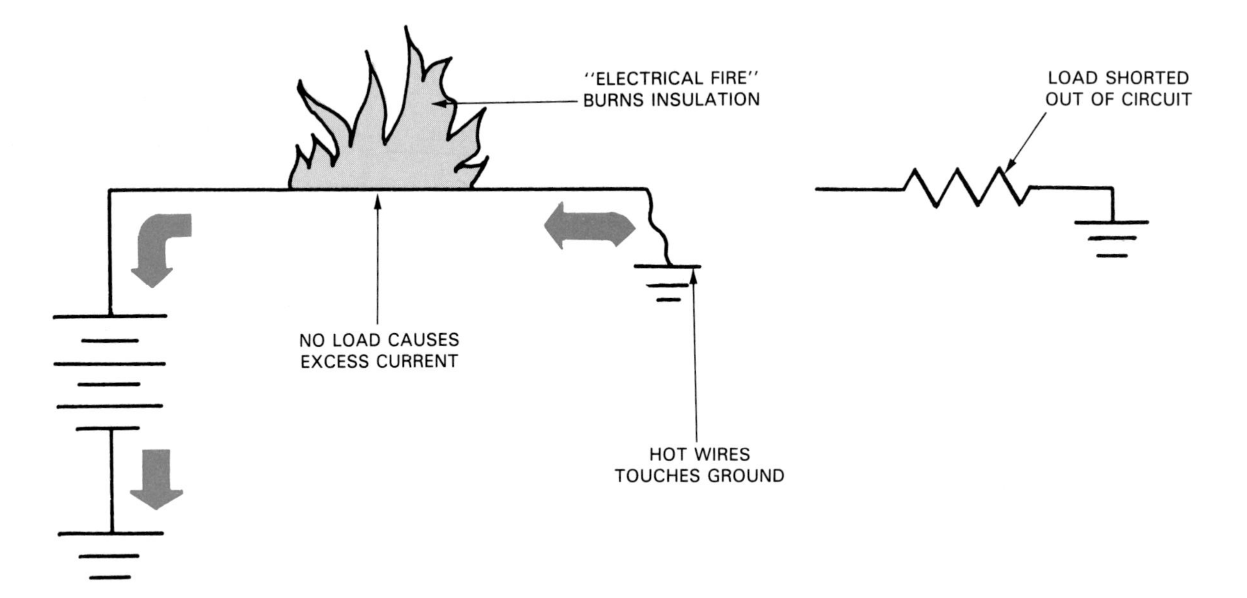

Fig. 5-1. Electrical fire can cause tremendous damage to car in a matter of seconds. Hot wire touches ground and high current causes enough heat to make wire insulation burn. Always disconnect battery when wiring is disconnected to service engine.

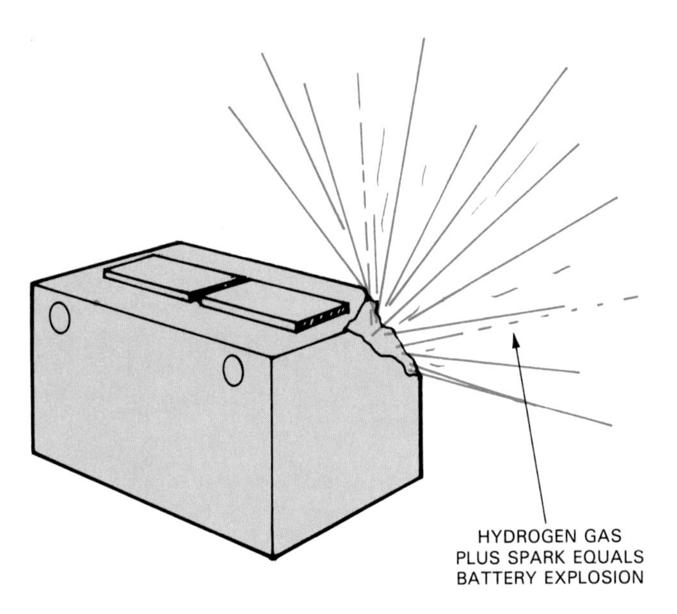

Fig. 5-2. Car batteries can explode with blinding or lethal force. Hydrogen gas can collect around top of battery. Spark or flame can ignite gases. Then, gases inside battery can ignite, blowing chunks of battery case and battery acid into shop.

Fig. 5-3. Use exhaust hose when engine is running in enclosed shop. Poisonous gases will then be drawn out of shop. (Chrysler)

Chemical burns

Various solvents (parts cleaners), battery acid, and a few other shop substances can cause *chemical burns* to the skin. Always read the directions and warnings on chemicals.

Carburetor cleaner (decarbonizing type cleaner), for example, is super powerful and can severely burn your hands in a matter of seconds, Fig. 5-4. Wear rubber gloves and eye protection when using carburetor cleaner. If a skin burn occurs, follow label directions.

Fig. 5-4. Carburetor cleaner is very powerful and can cause severe chemical burns. Wear rubber or plastic gloves and eye protection when working with carburetor cleaner. (Penray)

Electric shock

Electric shock can occur when using improperly grounded electric power tools. Never use an electric tool unless it has a functional GROUND PRONG (third, round prong on plug socket). This prevents current from accidentally passing through your body. Also, never use an electric tool on a wet shop floor.

Physical injury

Physical injuries (cuts, broken bones, strained backs) can result from hundreds of different accidents. As an engine mechanic, you must constantly think and evaluate every repair technique. Decide whether a particular operation is safe or dangerous and take action as required.

One example: why move an engine block by hand when a crane is handy? You and a friend may be strong enough to lift an engine block but why risk injuring your or someone else's back? Once your back is injured, it will NEVER be the same!

GENERAL SAFETY RULES

Listed are several general safety rules that should be remembered and followed at all times.

1. Wear eye protection during any operation that could endanger your eyes! This would include operating power tools, working around a spinning engine fan, and carrying batteries.
2. Avoid anyone who does not take shop work seriously.
3. Keep your shop organized! Return all tools and equipment to their proper storage areas. Never lay tools, creepers, or parts on the floor, Fig. 5-5.
4. Dress like a technician! Remove rings, bracelets, necklaces, watches, and other jewelry. They can get caught in engine fans, belts, etc.—tearing off flesh, fingers, and ears. Also, roll up long sleeves and secure long hair, they too can get caught in spinning parts.
5. Never carry sharp tools or parts in your pockets. They can easily puncture your skin.
6. Wear eye protection when grinding, welding, and during other operations where eye hazards are present, Fig. 5-6.
7. Work like a professional! When learning to be an engine technician, it is easy to get excited about your work. However, avoid working too fast. You could overlook a repair procedure or safety rule and cause an accident.
8. Use the right tool for the job! There is usually a ''best tool'' for each repair task. Always ask yourself this question—is there another tool that will work better?
9. Keep equipment guards or shields in place! If a power tool has a safety guard, use it!
10. Lift heavy engine parts with your legs, NOT with your back. There are many assemblies that are very heavy. When lifting, bend at your knees while keeping your back as straight as possible. On extremely heavy assemblies (transmissions, engine blocks, transaxles) use a portable crane.
11. Use adequate lighting! A portable shop light

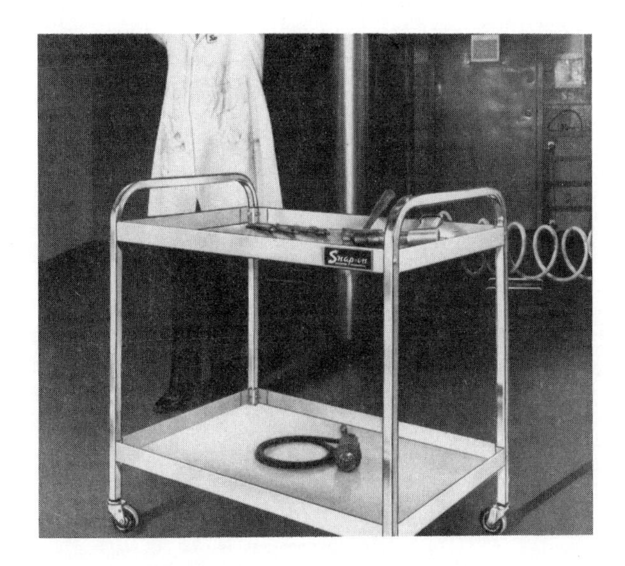

Fig. 5-5. Always keep tools organized while working. Do not lay tools on shop floor. This roll around cart will let you bring all needed tools to job. (Snap-On Tools)

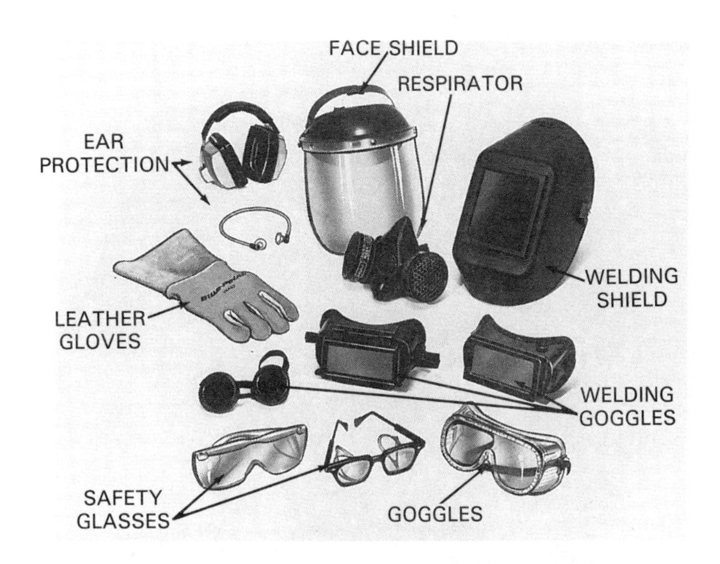

Fig. 5-6. Always use a protection device when needed. Remember to think of potential dangers and the action that must be taken to prevent injury.

increases working safety and it also increases work speed and precision.

12. Ventilate your work area when needed! Turn on the shop ventilation fan or open the shop doors anytime fumes are present in the shop.
13. Never stir up asbestos dust! Asbestos dust (particles found in engine clutch assemblies) are powerful CANCER-CAUSING AGENTS. Do NOT use compressed air to blow the dust off these parts.
14. Jack up or raise a car slowly and safely! A car can weigh between one and two tons. Never work under a car unless it is supported by jack stands, Fig. 5-7. It is NOT safe to work under a car held by a floor jack. Also, chock wheels when the car is on jack stands.

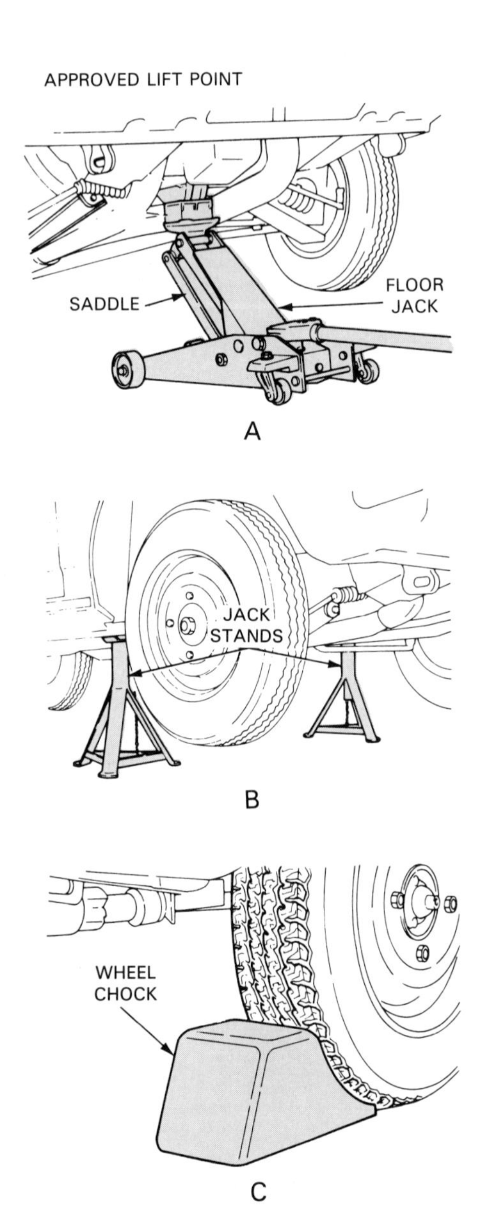

Fig. 5-7. A—Use floor jack to raise car slowly. Place jack under an approved lift point on car. Place transmission in neutral while raising. B—Secure car on jack stands before working under car. C—Chock wheels to make sure car cannot roll off of jack stands. (Subaru)

15. Drive slowly when in the shop area! With other students and cars in the shop, it is very easy to have an accident.
16. Report unsafe conditions to your instructor! If you notice any type of hazard, inform your instructor right away.
17. Keep away from engine fans! The fan on a car engine is like a SPINNING KNIFE. It can inflict serious injuries. Also, if a part or tool is dropped into the fan, it can fly out and hit someone or damage the radiator.
18. Respect running engines! When a car engine is running, make sure that the transmission or transaxle is in park. Check that the emergency brake is set, and that the wheels are blocked. If the transmission were knocked into gear, it could run over you or a friend.
19. No smoking! No one should smoke in an auto shop. Smoking is a serious fire hazard considering fuel lines, cleaning solvents, and other flammables may be exposed.
20. Lead absorbed through the skin and lungs can cause anemia, nerve damage, and brain disorders. Lead can be found in exhaust gases and combustion chamber deposits of older cars. Avoid contact with lead!
21. Chemical pneumonia can result from inhaling oil mist. Dermatitis can result from skin contact with oil.
22. Obtain instructor permission before using any new or unfamiliar power tool, lift, or other shop equipment! Your instructor will need to give a demonstration.
23. ATTENTION! If an accident or injury ever occurs in the shop, notify your instructor immediately. Use common sense on deciding to get a fire extinguisher or to take other action.

ENGINE SERVICE SAFETY

Specialized safety rules that apply to specific service or repair tasks on an engine will be covered in later chapters. Refer to the index for more safety-related information if needed.

DISPOSING OF AUTO SHOP WASTES

Automotive service and maintenance facilities frequently generate hazardous wastes, which are regulated by the *Resource Conservation and Recovery Act.* This federal act covers businesses that generate, transport, or manage hazardous wastes. Any business that maintains or repairs vehicles, heavy equipment, or farm equipment should comply with the regulations of the act.

Draining automotive fluids and replacing non-repairable components are the most common automotive repair activities that produce hazardous wastes. Automotive fluid and solid wastes include:

- Used motor oil (combustible and may contain toxic chemicals).
- Other discarded lubricants, such as transmission and differential fluids (may contain toxic chemicals).

- Used parts cleaners and degreasers (combustible and may contain toxic chemicals).
- Carburetor cleaners (contain flammable or combustible liquids).
- Worn out batteries (contain lead and toxic chemicals).
- Fan belts, mufflers, and catalytic converters.

One of the best ways to deal with hazardous wastes is to minimize the quantity of wastes produced. This can be accomplished by practicing good housekeeping, improving inventory control, and following proper spill containment techniques.

When hazardous wastes are produced, they must be disposed of properly. Regulations require that these wastes be collected by a registered hazardous waste hauler. Several major companies offer pick up and recycling services. Repair or maintenance facilities (such as service stations, dealerships, or independent repair garages) that generate 220 lb. (100 kg) of hazardous waste monthly must fill out a *Uniform Hazardous Waste Manifest* before shipping the wastes to a disposal or recycling site. The manifest is simply a tracking document that must accompany hazardous wastes when they are shipped from the work facility. It contains detailed information about the origin, character, and destination of the wastes. When shipping certain wastes, the proper Department of Transportation shipping descriptions must be listed on the manifest. Tables listing these descriptions are available from each state's hazardous waste management agency or from a regional EPA office.

EPA regulations state that no manifest is needed for used oil or lead-acid batteries if they sent off for recycling. In such cases, the material is not regarded as hazardous. Your state might have its own requirements. Check with your state hazardous waste management agency.

Unless recycled for scrap metal, used oil filters are considered hazardous waste. If not recycled, they must be listed on the manifest as hazardous. Before disposal, oil filters should be gravity drained so that they do not contain free-flowing oil. Store them in a closed, labeled container for pickup by a recycler.

Antifreeze

In the past, used antifreeze was not considered hazardous and, therefore, was not regulated by the Resource Conservation and Recovery Act. However, antifreeze has been reclassified as a hazardous waste due to the heavy metal and chlorinated solvents that it picks up when circulating through an engine's cooling system. Used antifreeze should never be mixed with used oil. If this occurs, th entire mixture must be classified as a hazardous waste, even though the oil may not be hazardous. (Some states classify used oil as a hazardous waste.)

Refrigerants

Refrigerants, such as R-12 and R-134a, removed from automotive air conditioning systems during service must not be vented to the atmosphere. State regulations require that they be recovered and recycled. Recovery systems are available for a cost of about $3500. Technicians servicing automotive air conditioning systems must be certified in refrigerant recycling. Certification is offered through several EPA-approved testing organizations.

Material Safety Data Sheets

For each chemical that they produce, chemical manufacturers are required to provide a *Material Safety Data Sheet (MSDS).* These sheets list all the known dangers and treatment procedures for a specific chemical. There should be an MSDS for each chemical used in the auto shop. Be sure to read the MSDS for any chemical that you are not familar with.

Hazardous Waste

Hazardous waste is a solid, liquid, or gas that can harm people and the environment. There are several criteria for determining if a substance is hazardous:

- A material is considered an *ignitable hazard* if it will ignite and burn easily. Gasoline, diesel oil, solvents, and other chemicals in an auto shop are considered an ignitable hazard.
- A material or waste is a *corrosive hazard* if it dissolves metals and other materials or burns human skin. Battery acid and many parts cleaners are considered a corrosive hazard.
- Anything that reacts violently with other materials or releases poisonous gases is considered *reactivity hazard.* Materials that generate toxic mists, fumes, vapors, and flammable gases are also a reactive hazard.
- Materials like lead, cadmium, chromium, arsenic, and other heavy metals that can pollute and make water and soil harmful are considered a *toxicity hazard.* Used motor oil, solvents, and other chemicals in the auto shop are a toxic hazard that must be disposed of properly.

Always dispose of hazardous wastes properly. Used motor oil, cleaning solvent, and other materials should be sent to a established waste disposal facility. They can treat and recycle this material to help prevent environmental pollution. Never pour used motor oil or other wastes down a drain because this will contaminate water, soil, and air.

SUMMARY

An auto shop can be a safe and enjoyable place to work if safety rules are followed. If safety regulations are NOT followed, it can be a very dangerous place to work. Always make sure you are using approved practices when working on an engine.

You must prevent fires, explosions, chemical burns, electrocution, and other physical injuries. Constantly think about what you are doing and take corrective action when needed!

Hazardous wastes are often generated in the auto shop. These wastes are regulated by the Resource Conservation and Recovery Act and must be disposed of properly.

KNOW THESE TERMS

Gasoline spill, Electrical fire, Battery explosion, Asphyxiation, Asbestos, Respirator, Carburetor cleaner, Electric shock, Ground prong, Physical injury, Equipment guard, Resource Conservation and Recovery Act,

Uniform Hazardous Waste Manifest, Material Safety Data Sheet (MSDS), Hazardous waste, Ignitable hazard, Corrosive hazard, Reactivity hazard, Toxicity hazard.

REVIEW QUESTIONS—CHAPTER 5

1. List six kinds of accidents.
2. __________ is the most common source of shop fires.
3. Apply "quick dry" or absorbent to gasoline spills. True or False?
4. What should you do when disconnecting an engine fuel line?
5. How does an electrical fire occur?
6. Car batteries can explode! True or False?
7. Why are engine exhaust gases so dangerous?
8. This can cause severe chemical burns.
   a. Gasoline.
   b. Diesel fuel.
   c. Carburetor cleaner.
   d. All of the above.
   e. None of the above.
9. Why is a ground prong provided on electrical equipment?
10. Why should you NOT wear jewelry when on the job?

ASE CERTIFICATION-TYPE QUESTIONS

1. Technician A says that you should disconnect the battery before removing a fuel line from an engine. Technician B says that you should wrap a shop rag around the fitting before disconnecting a fuel line. Who is right?
   (A) A only.
   (B) B only.
   (C) Both A & B.
   (D) Neither A nor B.
2. The most dangerous source of asphyxiation in an auto shop is a(n) __________.
   (A) gasoline can
   (B) oily shop rag
   (C) engine
   (D) None of the above.
3. Technician A says that a battery can produce gases that can cause an explosion. Technician B says that special sodium-filled engine valves, if mishandled, can cause an explosion. Who is right?
   (A) A only.
   (B) B only.
   (C) Both A & B.
   (D) Neither A nor B.
4. Technician A says that a respirator will block all poisonous fumes that can be produced in an auto shop. Technician B says that a respirator will not block all poisonous fumes that can be produced in an auto shop. Who is right?
   (A) A only.
   (B) B only.
   (C) Both A & B.
   (D) Neither A nor B.
5. Technician A says that battery acid can cause chemical burns to the skin. Technician B says that some carburetor cleaners can cause chemical burns to the skin. Who is right?
   (A) A only.
   (B) B only.
   (C) Both A & B.
   (D) Neither A nor B.
6. Technician A says that an electric power tool that has a faulty ground prong can cause electrocution. Technician B says that if a power tool's faulty ground prong is removed, the power tool is safe to use. Who is right?
   (A) A only.
   (B) B only.
   (C) Both A & B.
   (D) Neither A nor B.
7. Technician A says that eye protection should be worn when carrying an automotive battery. Technician B says that eye protection should be worn when working around an engine fan. Who is right?
   (A) A only.
   (B) B only.
   (C) Both A & B.
   (D) Neither A nor B.
8. Technician A says that used batteries are considered a hazardous waste. Technician B says that used carburetor cleaner is considered a hazardous waste. Who is right?
   (A) A only.
   (B) B only.
   (C) Both A & B.
   (D) Neither A nor B.
9. All of the following are safety rules to follow when working in an auto shop EXCEPT:
   (A) never carry sharp parts or tools in your pocket.
   (B) always lift heavy engine parts with your back, not with your legs.
   (C) keep the auto shop organized.
   (D) always keep equipment guards or shields in place.
10. Technician A says that refrigerants can be vented into the atmosphere as long as proper filtering systems are utilized. Technician B says that refrigerants must never be vented into the atmosphere during air conditioning service. Who is right?
   (A) A only.
   (B) B only.
   (C) Both A & B.
   (D) Neither A nor B.

Chapter 6

Engine Types, Classifications

After studying this chapter, you will be able to:
- *Explain the different engine cylinder arrangements.*
- *Compare two and four-stroke cycle engines.*
- *Describe combustion chamber designs.*
- *Classify an engine by its valve location.*
- *Compare overhead valve and overhead cam engine designs.*
- *Describe alternate types of engines.*

As an engine technician, you must be able to differentiate between engine types. Understanding how an engine is designed and constructed will help you when troubleshooting problems. An experienced technician can usually glance into an engine compartment and instantly describe numerous facts about how the engine is constructed and operates.

One example, you might hear a technician say—"This is a dual overhead cam, four-cylinder with four valves per chamber, a distributorless ignition, and an intercooled turbocharger." This information would help the technician if he or she had to work on the engine.

This chapter will introduce the many classifications and designs of modern engines. This information will prepare you for many later chapters that discuss engine construction and service in more detail. In a sense, the chapter will help you develop the "language" of an engine technician; so study carefully!

CLASSIFYING ENGINES

There are many ways to classify an engine, even though the fundamental parts (block, pistons, crankshaft, camshaft) are basically the same. These design differences, however, can greatly affect how the engine performs and how it is serviced.

Modern automotive engines are normally classified by:

1. Arrangement of cylinders.
2. Number of cylinders.
3. Crankshaft design.
4. Firing order.
5. Cooling system type.
6. Method of fuel entry into engine.
7. Type of fuel burned (gas or diesel for example).
8. Combustion chamber shape.
9. Cylinder head port design.
10. Number of valves per cylinder.
11. Method of driving camshaft.
12. Valve location.
13. Camshaft location.
14. Method of balancing engine.
15. Location of combustion.
16. Two or four-stroke cycle.
17. Reciprocating or rotary design.
18. Method of using combustion (alternate engines).

CYLINDER ARRANGEMENT

Cylinder arrangement refers to the position of the cylinders in the engine block in relation to the crankshaft. There are four basic cylinder arrangements found in automobiles: in-line, V-type, slant, and opposed.

In-line engine

An *in-line engine* has its cylinders positioned one after the other in a straight line. The cylinders are located vertically in a line parallel with the crankshaft centerline. This is shown in Fig. 6-1A.

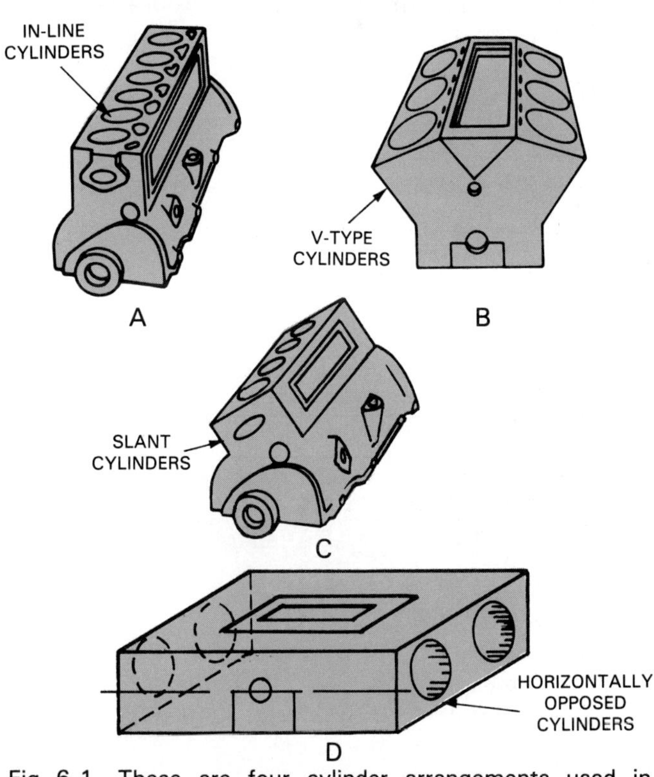

Fig. 6-1. These are four cylinder arrangements used in automobiles. In-line and V-type are more common than slant and opposed.

In-line engines are very common today. They are well suited to small *displacement* (size) engines. Three, four, five, and six cylinder engines are frequently an in-line design. Because of their small size and good fuel economy, in-line four-cylinder engines are one of the most common types found in today's cars. It is used by both American and foreign auto makers.

V-type engine

A *V-type engine* looks like the letter "V" when viewed from the front or rear, Fig. 6-1B. The two *banks* (sets) of cylinders lay at an angle from vertical on each side of the crankshaft. A V-type design reduces the length and the height of the engine for fitting in a small engine compartment.

Four, six, and eight cylinder engines can be a V-type. An even number of cylinders is needed on each bank.

V-eight engines were very common in the past. They were compact, very powerful, and idled very smoothly because of all of the power strokes per crank revolution. However, the V-6 engine is now becoming more popular because of its increased efficiency. It will get better gas mileage than a larger, V-8 engine. A V-type engine is primarily an American design.

Slant engine

A *slant engine* only has one set of in-line cylinders but they are angled to one side. The block's crankcase is vertical while the cylinders lean over or slant, Fig. 6-1C.

Like a V-type, a slant engine enables the car hood to be sloped or lowered for increased aerodynamics (reduced drag as the car cuts through the air).

A slant engine is less common than a straight in-line or a V-type engine. It was more common a few years ago when slant-six cylinder engines were produced in quantity by Chrysler.

Opposed engine

An *opposed engine,* also termed "pancake engine" has two banks of cylinders that lay flat or horizontal on each side of the crankshaft. It is usually placed in the rear of the car. This arrangement is in Fig. 6-1D.

Opposed four and twelve cylinder engines are fairly rare. However, an opposed or flat-twelve is now being used by Ferrari in their top of the line performance car. Porsche also uses a flat-six in many of their performance cars. This design produces a very low center of gravity (weight near ground) that helps cornering. In addition, with it located in the rear, the weight of the engine is over the drive wheels for increased traction upon acceleration. Several years ago, Volkswagen used an air-cooled, rear-mounted, opposed-four engine in their famous "Beetle" or "Bug" economy car.

NUMBER OF CYLINDERS

Automotive engines normally have either 4, 6, or 8 cylinders. A few car engines, however, have 3, 5, 12, or 16 cylinders.

A greater number of cylinders generally increases engine smoothness and power. For instance, an 8-cylinder engine would produce twice as many power strokes per crank revolution as a 4-cylinder engine. This would reduce power stroke pulsations and roughness, especially at idle.

Four-cylinder engines are usually in-line, slant, and sometimes opposed. Six-cylinder engines can be in-line, slant, or a V-type. Five-cylinder engines are normally in-line. Eight-cylinder engines are commonly a V-type.

Smaller four-cylinder and six-cylinder engines are the trend of the future. With high-tech designs and turbocharging, smaller displacement engines are producing as much power as many larger V-8 engines. Yet they can produce much higher fuel economy and less *emissions* (air pollution).

Cylinder numbers

Cylinder numbers identify the cylinders, pistons, and connecting rods of the engine. Cylinder numbers can be cast into the intake manifold; corresponding numbers are normally stamped into the sides of the connecting rods. Cylinder numbers vary with the engine design and cylinder arrangement.

Engine manufacturers use cylinder numbers so the engine technician can make repairs and do tune-up operations. When rebuilding an engine, you will normally have to return each piston and rod to the same cylinder. When doing a tune-up, for example, you will need to now which is the number one cylinder for connecting a timing light.

Look at Fig. 6-2. It shows typical cylinder numbers for V-6 and in-line 4-cylinder engines.

On V-type engines, you can normally tell the NUMBER ONE CYLINDER because it is located slightly ahead or in front of the cylinder on the other side of the block, Fig. 6-3. With a V-type engine, two connecting rods bolt to the same crankshaft rod journal, so one must be located in front of the other, Fig. 6-4.

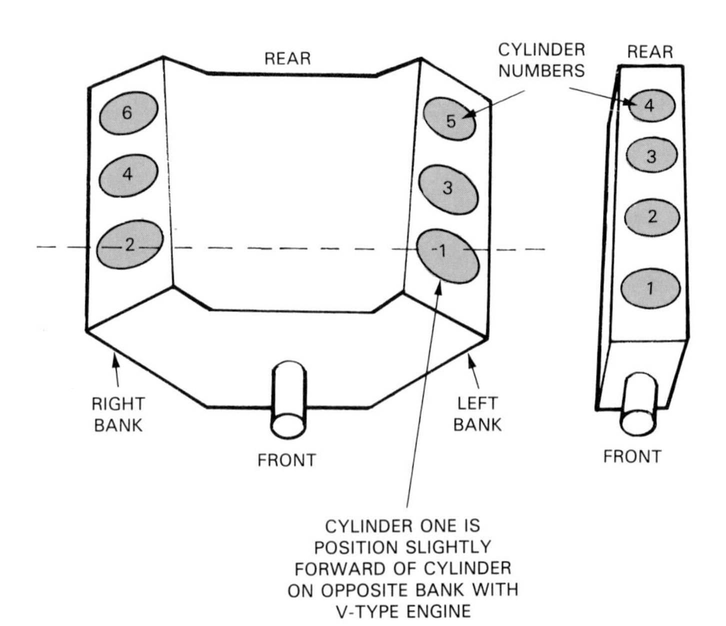

Fig. 6-2. Cylinder numbers vary with V-type engine. Numbers usually run in sequence, from front to rear, on in-line engines. Note how you can tell number one cylinder on V-type engines. Number one cylinder is slightly ahead of cylinder on opposite bank.

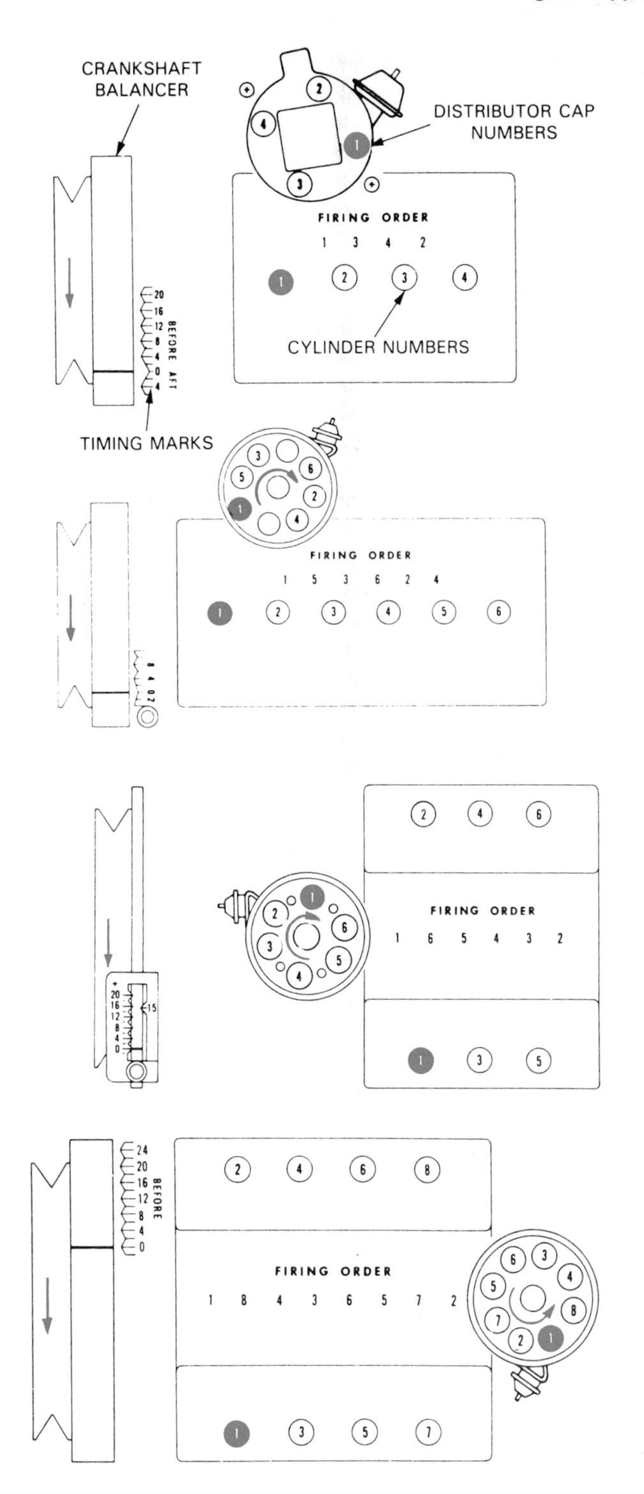

Fig. 6-3. These are service manual illustrations showing cylinder numbers and firing orders. Cylinder numbers are needed during engine repairs. Both are needed when doing tune-up type tasks. (Mitchell Manuals)

Four-cylinder engines are usually numbered 1-2-3-4 from the front to the rear.

You should keep these rules in mind, but there are exceptions to any rule. Always refer to a service manual to make sure of cylinder numbers. Sometimes with a V-engine, one bank will have odd numbers (1, 3, 5) and the other even (2, 4, 6). The cylinders can also be numbered in sequence (1, 2, 3) on one bank and in sequence on the other (4, 5, 6). The service manual will usually provide a simple drawing of the engine showing cylinder numbers and firing order, Fig. 6-3.

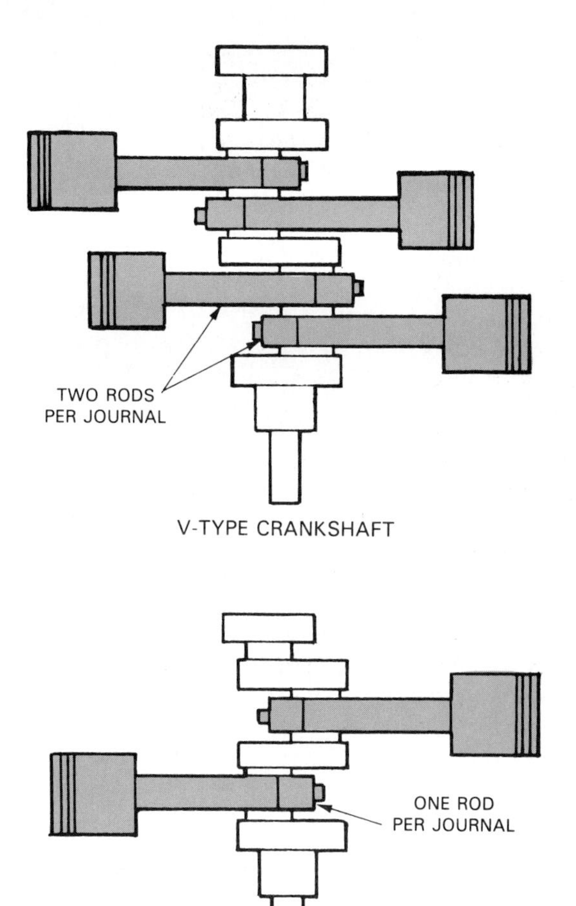

Fig. 6-4. A V-type crankshaft has two connecting rods per rod journal. This is why number one cylinder on V-type engine is ahead of one on opposite side. Note how in-line engine only has one rod per journal.

CRANKSHAFT CLASSIFICATION

There are several different ways to classify an engine crankshaft. The most common are:

1. An *in-line crankshaft* only has one connecting rod fastened to each rod journal, Fig. 6-4.
2. A *V-type crankshaft* has two connecting rods bolted to the same journal, Fig. 6-5.
3. A *cast iron crankshaft* is made by pouring molten iron into a mold. It is a common type for passenger cars.
4. A *forged steel crankshaft* is made by hammering semi-molten steel into a mold. It is much stronger and more rigid than a cast iron crank and is used in high performance applications.
5. A *splayed crankshaft* has the individual rod journals on the same crank journal machined off-center. This design is commonly found on V-6 engines to help smooth engine operation.
6. An *internally balanced crankshaft* uses the

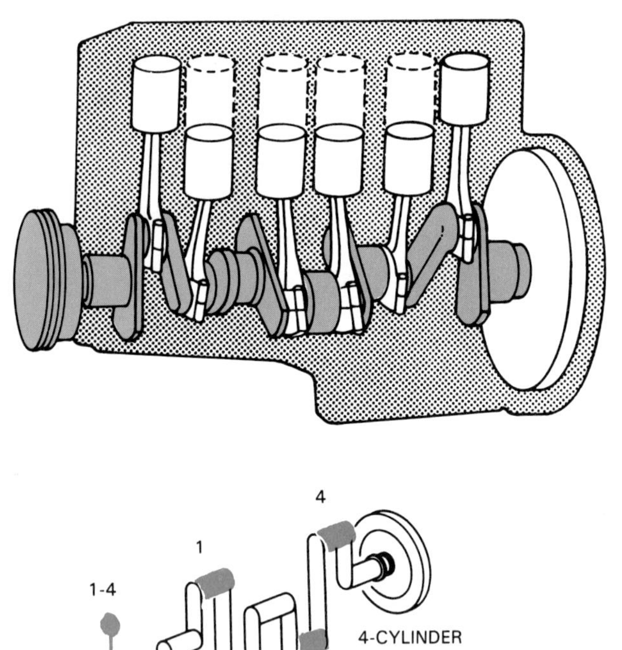

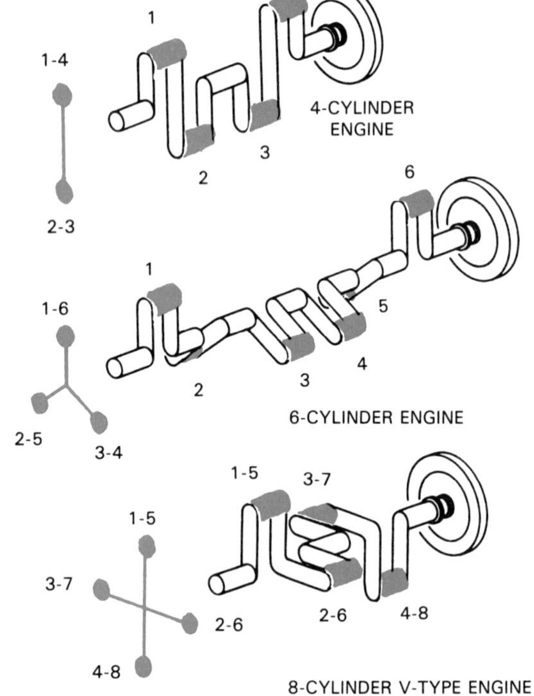

Fig. 6-5. Crankshaft determines when each piston will reach TDC. (Caterpillar Inc.)

counterweights to offset the weight of the rod and piston assemblies. Metal is added or removed from the counterweights to balance the engine and prevent vibration.

7. An *externally balanced crankshaft* uses both the counterweights and weights on the flywheel and harmonic balancer to prevent vibration. A small metal pad is added to the flywheel or balancer to counteract a longer stroke machined on the rod journals.

FIRING ORDER CLASSIFICATION

Firing order refers to the sequence that combustion occurs in each engine cylinder. The position of the crankshaft rod journals in relation to each other determines engine firing order.

The service manual will have a drawing showing the firing order for the engine. This information may also be given on the engine intake manifold. Refer to Fig. 6-3.

Two similar engines can have completely different firing orders. For example, a 4-cylinder, in-line engine may fire 1-3-4-2 or 1-2-4-3. Firing orders for 6 and 8-cylinder engines also vary.

A technician needs to know firing order when working on the engine's ignition system. It can be used when installing spark plug wires, a distributor, or when doing other tune-up related operations.

COOLING SYSTEM CLASSIFICATION

The two types of cooling systems are the liquid cooling system and the air cooling system. Almost all automobiles now use a liquid type.

Liquid cooling system

The *liquid cooling system* surrounds the cylinders with *coolant* (water and antifreeze solution). The coolant carries combustion heat out of the cylinder head and engine block to prevent engine damage. See Fig. 6-6.

The liquid cooling system is very efficient because it will let the engine warm up quickly and can closely control engine operating temperature. This increases engine performance and reduces exhaust emissions.

Air cooling system

An *air cooling system* circulates air over cooling fins on the cylinders and cylinder heads. This removes heat from the cylinders and heads to prevent overheating damage. Look at Fig. 6-7.

Air-cooled engines are not commonly used in modern passenger cars. They can be found on motorcycles and lawnmowers. With strict exhaust emission regulations, auto makers have phased out almost all air-cooled engines. They cannot maintain as constant a temperature as a liquid type cooling system. This reduces engine efficiency and increases exhaust pollution.

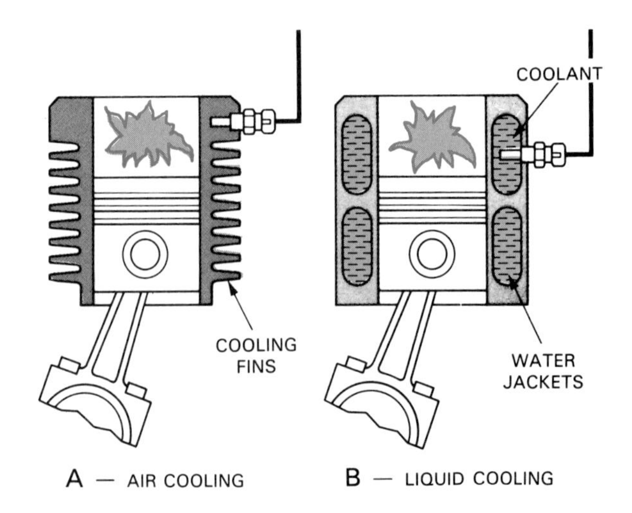

Fig. 6-6. A—Air cooling system has fins that transfer engine heat into surrounding air. B—Liquid cooling system has pockets around cylinder to hold coolant and collect heat. Liquid cooling is more common.

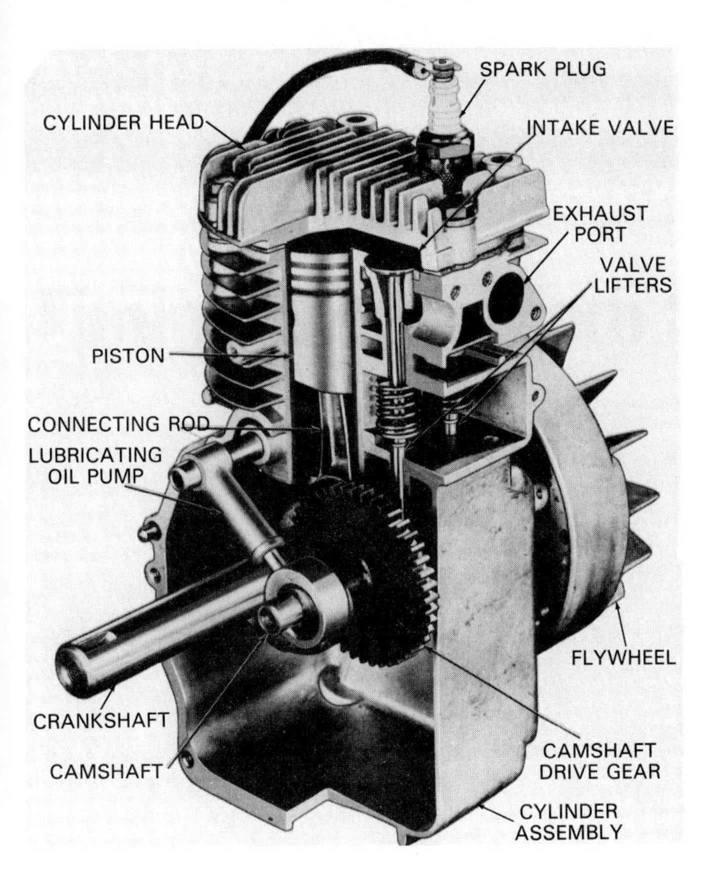

Fig. 6-7. Cutaway of this small, one-cylinder engine shows cooling fins. Air cooling system is seldom used on cars any more but can be found on lawnmowers and motorcycles.

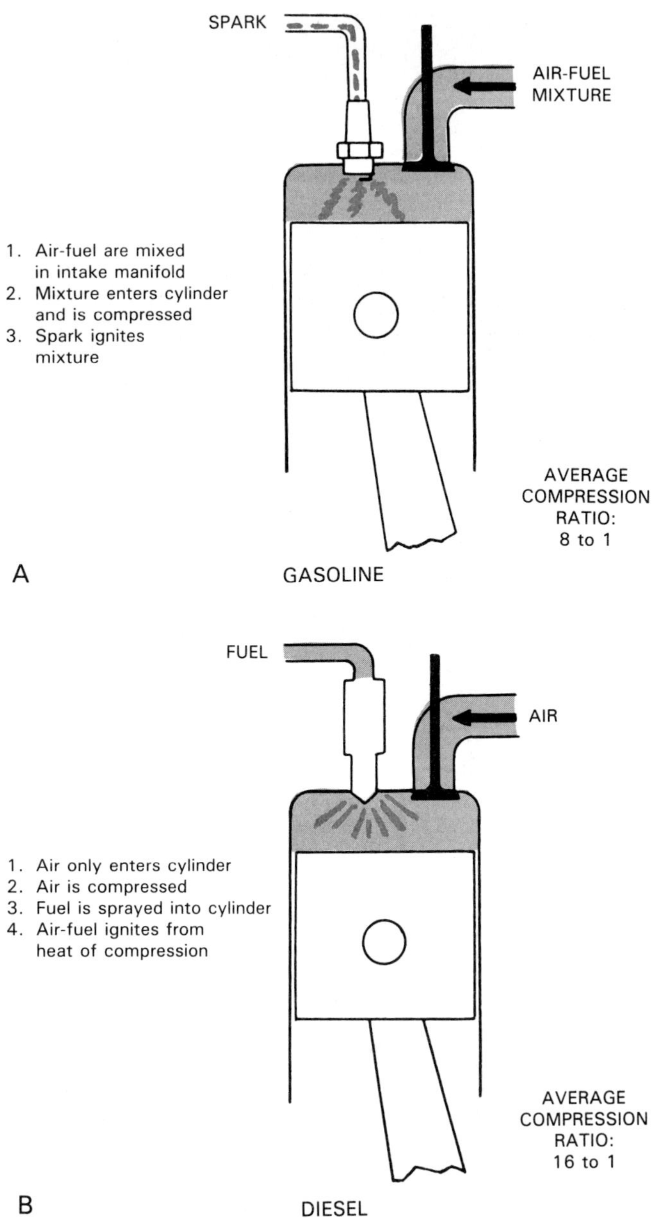

Fig. 6-8. A—Gasoline engine meters fuel into intake manifold. Spark plug ignites fuel. B—Diesel engine is direct injection engine because fuel is sprayed right into combustion chamber. Hot air ignites fuel. (Deere & Co.)

FUEL SYSTEM CLASSIFICATIONS

An automotive engine can also be classified by the type of fuel it burns, how it ignites the fuel, and how fuel is injected into the engine.

Gas and diesel engines

A *gasoline engine* burns gasoline and the fuel is metered into the intake manifold. A throttle valve is used to meter air and fuel into the engine to control engine speed and power. Spark plugs ignite the fuel, Fig. 6-8A.

A *diesel engine* burns diesel oil (thicker fraction of crude oil) and fuel is injected right into the engine combustion chambers. Only air flows through the intake manifold and a throttle valve is NOT used to control airflow. Super-high compression stroke pressure heats the air in the cylinder. When fuel is injected into the cylinder, it ignites and burns. No spark plugs are needed, Fig. 6-8B.

Fig. 6-9 illustrates the operating differences of a gasoline engine and a diesel engine. Review their operation carefully.

Automotive diesels commonly used a *precombustion chamber* to house the tip of the injector and the tip of the glow plug (heating element for cold starting). This is shown in Fig. 6-10. Fuel is injected into the prechamber where it begins to burn and expand into the main chamber. The precombustion chamber smooths engine operation and helps the engine start when cold.

Carburetor vs gasoline injection

A gasoline engine can also be classified depending upon whether it has a carburetor or fuel injection.

A *carbureted engine* uses engine vacuum to draw fuel out of the carburetor and into the engine intake manifold. Carbureted engines are being phased out for more efficient electronic gasoline injection.

A *fuel injected gasoline engine,* more properly termed *gasoline injected engine* because a diesel is also an injected engine, sprays fuel into the intake manifold. Computer controlled gasoline injection is the most common type found on the modern car because it can closely match the fuel fed into the engine with engine needs.

There are two major classifications of gasoline injection: throttle body (single-point) injection and multiport

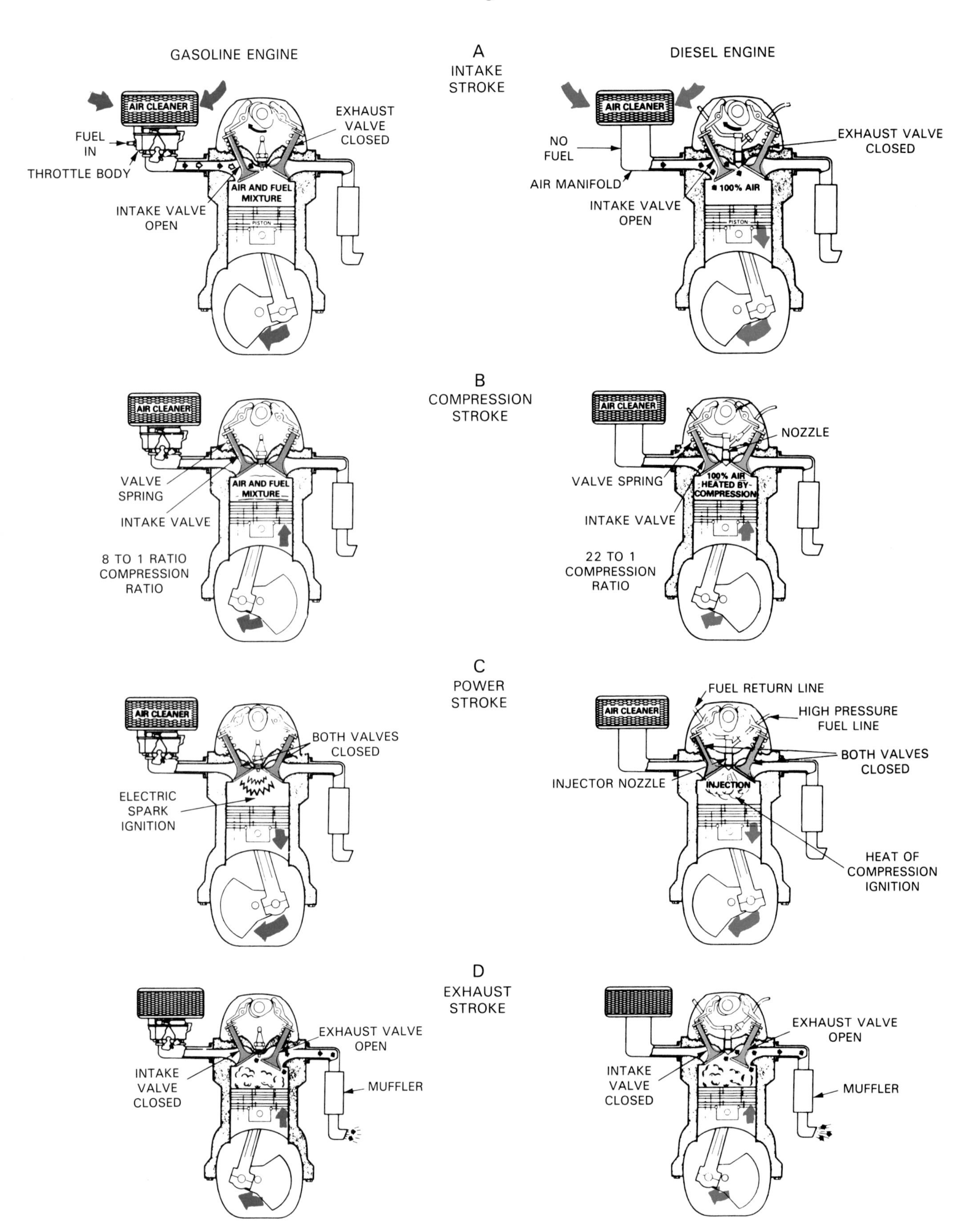

Fig. 6-9. Compare operation of gasoline and diesel engines. A—Air-fuel mixture is drawn into gasoline engine. Diesel only draws air into cylinder. B—Gasoline engine compresses air-fuel mixture. Diesel compresses only air but it squeezes air so much air becomes red hot. C—On power stroke, gasoline engine has arc at spark plug to start fuel burning and expanding. Diesel injects fuel into combustion chamber and fuel instantly begins to burn. D—Exhaust strokes in both engines are similar. (General Motors)

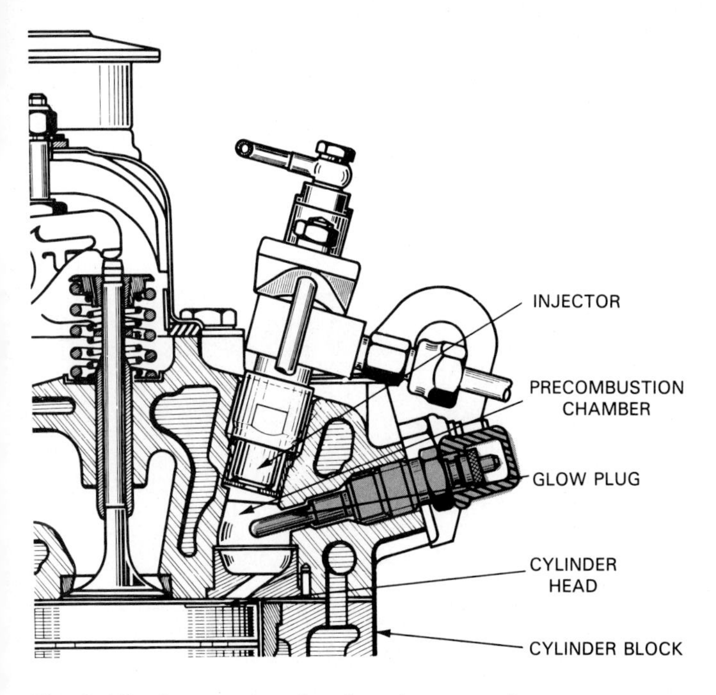

Fig. 6-10. Automotive diesel engine normally uses precombustion chamber. It is small pocket for tip of injector and tip of glow plug.

injection. Both are illustrated in Fig. 6-11. This subject will be covered in detail in later textbook chapters.

Other fuel system classifications

There are many other fuel system classifications besides carburetion, gasoline injection, and diesel injection. Each of these types can be sub-divided depending upon its particular design. This will be explained later.

An LPG or *liquefied petroleum gas* fuel system burns a very light fraction of crude oil. It stores the gas in a high pressure, thick metal tank. The high pressure keeps the gas vapor in a liquid form. Then, a special converter and carburetor are used to change the liquid back into a gas and meter the gas into the engine intake manifold.

An *alcohol fuel system* is similar to a carburetor fuel system but twice as much fuel must be metered into the engine. Also, for maximum efficiency, the compression stroke pressure or compression ratio of the engine must be increased to efficiently burn and use alcohol.

IGNITION CLASSIFICATION

Two methods are commonly used to ignite (light) the fuel in the engine combustion chamber: an ELECTRIC ARC (spark plug) and HOT AIR (compressed air). Both were mentioned earlier under gasoline and diesel engines, Figs. 6-8A and 6-8B.

A *spark ignition engine* uses an electric ignition and a spark plug to start the combustion of the fuel. Gasoline, LPG, and alcohol-fueled engines use this method of ignition.

A *compression ignition engine* uses super-high compression stroke pressure to heat the air and ignite the fuel. Diesel engines are compression ignition engines but not gasoline engines.

COMBUSTION CHAMBER CLASSIFICATIONS

An engine combustion chamber can also be used to classify an engine. Its shape, number of valves per cylinder, port configuration, etc., can all be used to label how an engine is constructed. You should understand these differences if you are going to be a ''top notch'' engine technician.

The three basic combustion chamber shapes for gasoline engines are: pancake, wedge, and hemispherical. These are pictured in Fig. 6-12.

Pancake combustion chamber

The *pancake combustion chamber,* also called ''bath tub'' chamber, has valve heads almost parallel with the top of the piston. The chamber forms a flat pocket over the piston head, Fig. 6-12A.

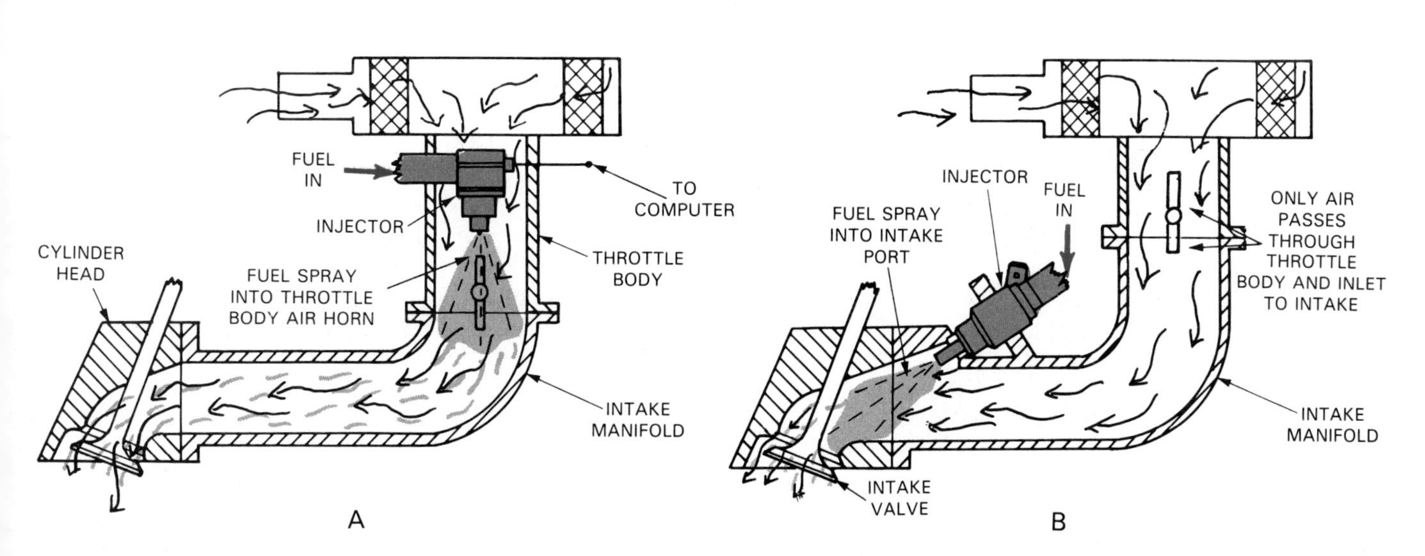

Fig. 6-11. Two types of gasoline injection can also be used to classify an engine. A—Single-point or throttle body injection. B—Multiport or port injection.

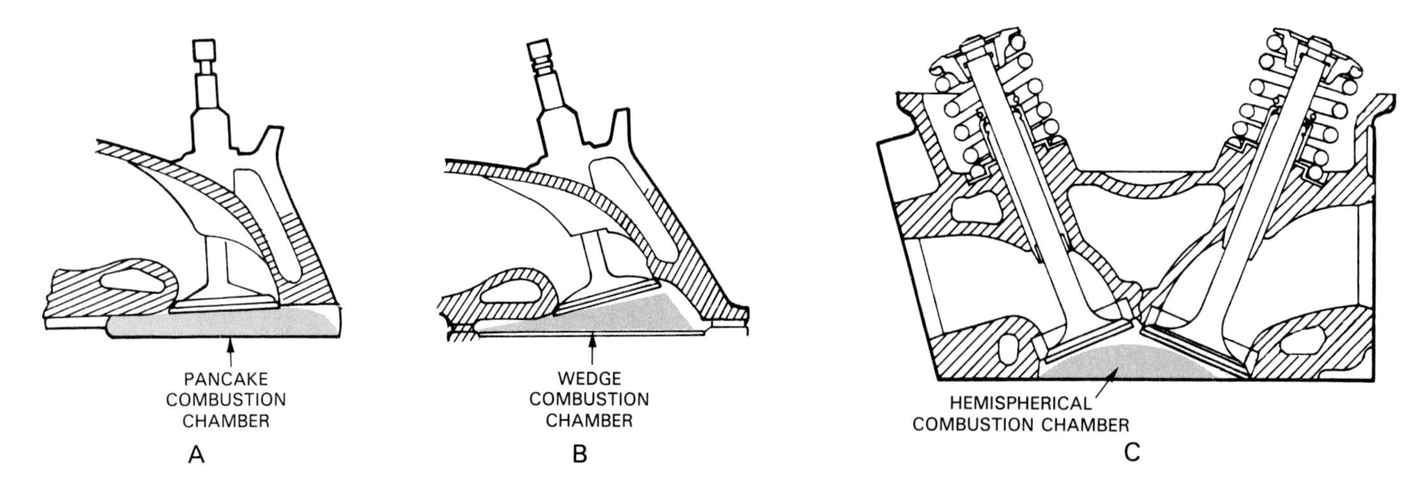

Fig. 6-12. There are three basic combustion chamber shapes. A—Pancake combustion chamber. B—Wedge combustion chamber. C—Hemispherical combustion chamber. (Chrysler)

Wedge combustion chamber

A *wedge combustion chamber,* called a *wedge head,* is shaped like a triangle or a wedge when viewed as in Fig.. 6-12B. Valves are placed side-by-side with the spark plug next to the valves.

A *squish area* is commonly formed inside a wedge type cylinder head. When the piston reaches TDC, the piston comes very close to the bottom of the cylinder head. This squeezes the air-fuel mixture in that area and causes it to squirt or squish out into the main part of the chamber. Squish can be used to improve air-fuel mixing and burning at low engine speeds.

Hemispherical combustion chamber

A *hemispherical combustion chamber,* nicknamed *hemi head,* is shaped like a dome. The valves are *canted* (tilted) on each side of the chamber. The spark plug is located near the center. A hemi is shown in Fig. 6-12C. Compare it to the others.

A hemi combustion chamber is extremely efficient. There are no hidden pockets for incomplete combustion. The surface area is very small, reducing heat loss from the chamber. The centrally located spark plug produces a very short flame path for fast and complete combustion. The canted valves help increase breathing ability.

The hemi head was first used in high horsepower, racing engines. It is now used in many OHC passenger car engines. It allows the engine to operate at high rpms and makes it very fuel efficient. It also produces complete burning of the fuel to reduce emissions.

Semi-hemi combustion chamber

A *semi-hemi* is a combustion chamber that is not quite spherical in shape. Two sides are usually flat and not curved into a dome. This chamber design is a cross beween a wedge and a full hemi. The flat areas are used to help low speed turbulence by producing a squish area. The hemi area is desirable because of its high speed combustion efficiency.

Swirl combustion chamber

A *swirl combustion chamber* uses the shape of the intake and exhaust ports and the shape of the combustion chamber roof to mix the air-fuel mixture. As shown in Fig. 6-13, a curve is provided in the intake port right before the intake valve and seat. Sometimes, a mask area is used to also control the movement of the mixture through the port and into the combustion chamber.

Just as stirring up a pile of smoking leaves in your lawn will make the leaves burn faster, swirling the air-fuel charge will improve combustion efficiency. This is the principle of a swirl combustion chamber.

Crossflow combustion chamber

A *crossflow combustion chamber* has the intake ports on one side of the head and the exhaust ports on the other side. This is pictured in Fig. 6-14.

During engine operation, the fuel charge enters the combustion chamber on the intake stroke from one side of the head. Then, on the exhaust stroke, the burned gases leave on the opposite side. This tends to make the exiting exhaust gases help pull more fuel charge into the combustion chamber to increase power. See Fig. 6-15.

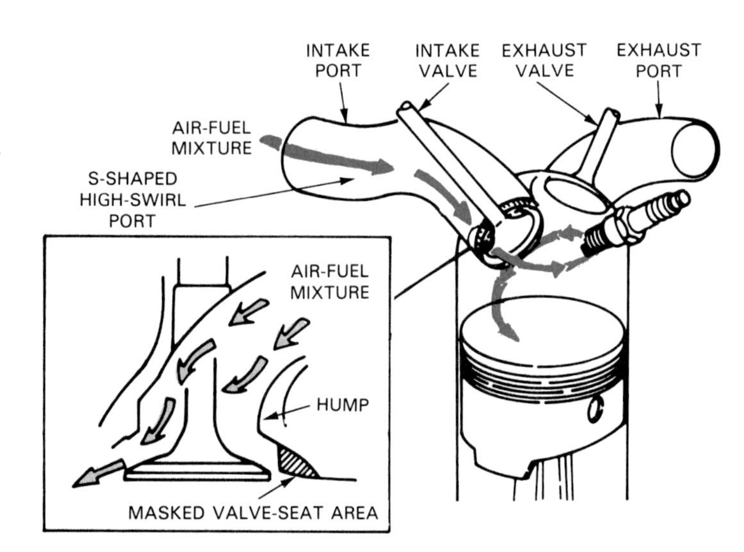

Fig. 6-13. Swirl combustion chamber uses shape of intake port and its entry into combustion chamber to mix air-fuel mixture for more efficient combustion.

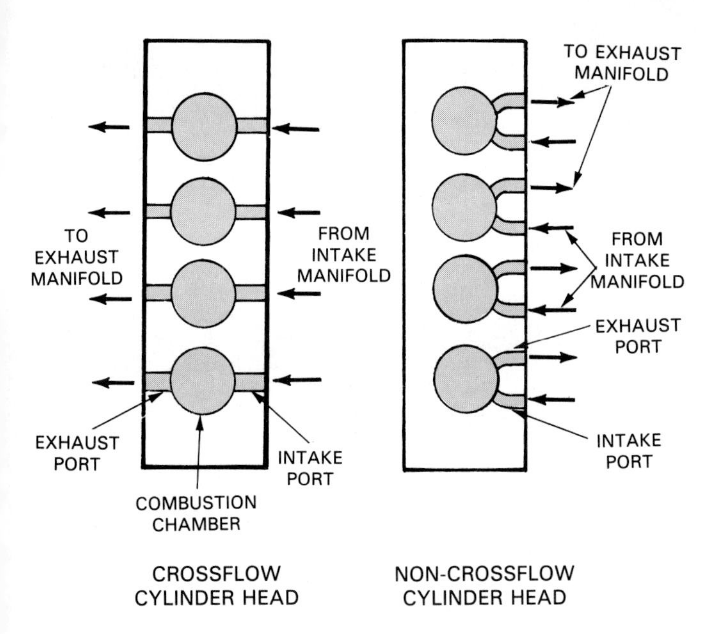

Fig. 6-14. Crossflow cylinder head has intake ports on one side and exhaust ports on other. This is a better design for engine performance.

A *noncrossflow* or *backflow combustion chamber* is also illustrated in Fig. 6-14. It is an older design that has been phased out for the more efficient crossflow combustion chamber.

Two-valve combustion chamber

A *two-valve combustion chamber* has one intake valve and one exhaust valve per cylinder. This is a conventional design that has been used for years. The intake valve is slightly larger than the exhaust valve. One is pictured in Fig. 6-15.

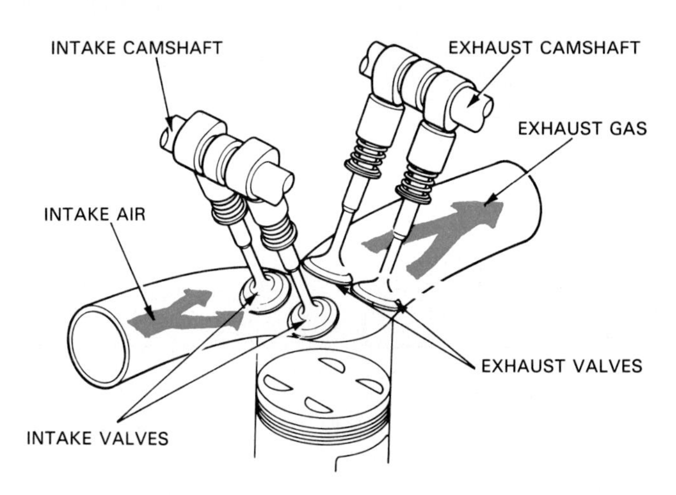

Fig. 6-16. Many modern high performance engines use four-valves per cylinder to increase engine power. Extra valves allow engine to breathe. Less valve lift can be used to obtain same flow and this can be used to increase engine compression ratio because piston reliefs can be reduced. (Toyota)

Four-valve combustion chamber

A *four-valve combustion chamber* has two intake valves and two exhaust valves for each engine cylinder. This is a relatively new design used on several high performance cars. A four-valve chamber is in Fig. 6-16.

Siamese ports are normally used with a four-valve combustion chamber. One large port divides into two smaller ports right before the engine valves.

The valves in a four-valve chamber are generally smaller than the valves in a two-valve chamber. However, they have a larger cross section when open and will increase flow in and out of the chamber. This increases engine rpm capability and the resulting power.

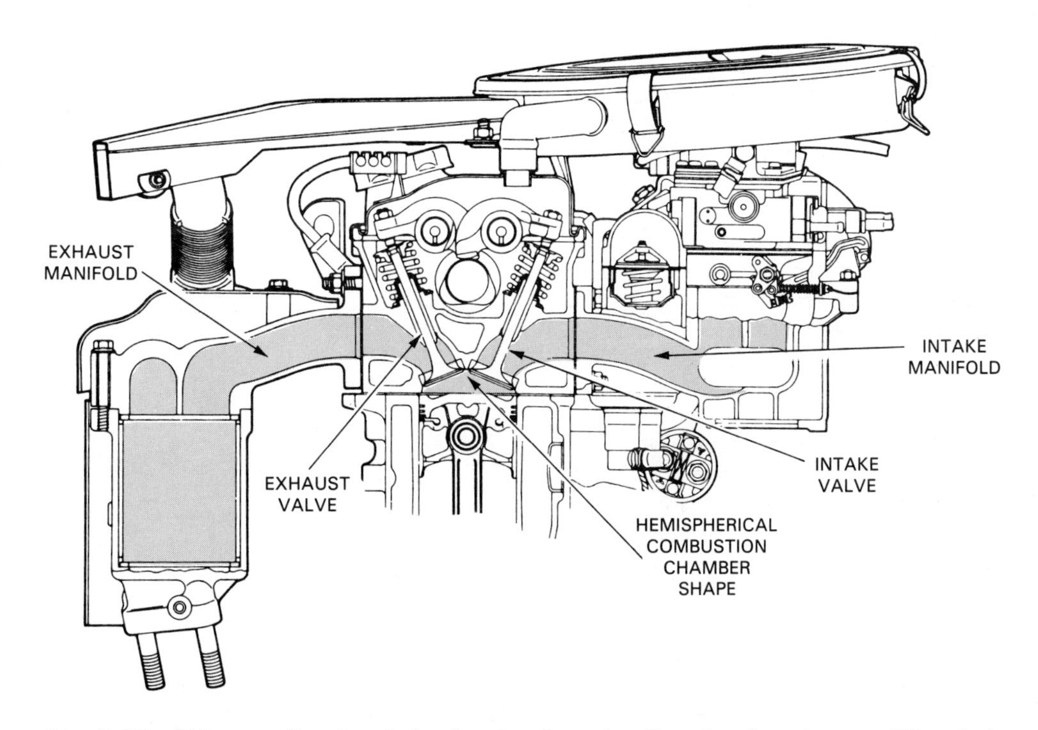

Fig. 6-15. This crossflow head also has hemi combustion chamber shape. (Chrysler)

Twin camshafts are normally used with a four-valve chamber. Too many rocker arms and pushrods would be needed to operate so many valves. A view of a four-valve, twin-cam cylinder head is given in Fig. 6-17.

A higher compression ratio can also be obtained with a four-valve combustion chamber. The valves do not have to be opened as wide to obtain the same flow; therefore, the piston can be designed with less valve relief to compress the air-fuel charger tighter. This also increases combustion efficiency.

Mixture jet combustion chamber

A *mixture jet combustion chamber* uses an extra passage running to the intake valve port. The mixture jet is used to aid swirl in the port and combustion chamber to help burning. The mixture jet mainly helps engine efficiency at low speeds. Refer to Fig. 6-18.

Air jet combustion chamber

An *air jet combustion chamber* has a small, extra valve that allows a stream of air to enter the combustion chamber to aid swirl and combustion efficiency. Shown in Fig. 6-19, two conventional valves are provided. A third, smaller valve is also used. It opens to admit a gush of air that mixes the air-fuel charge to speed burning.

A passage runs from the carburetor to the combustion chamber and jet valve. During the intake stroke, the

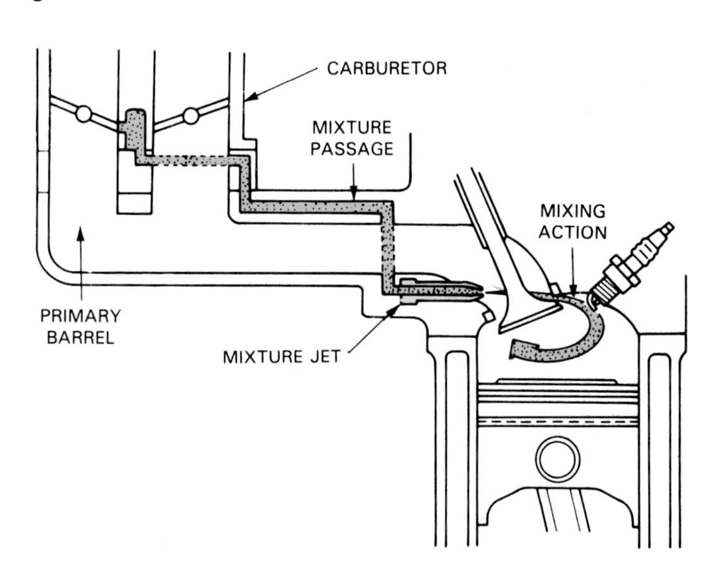

Fig. 6-18. Mixture jet injects charge into intake port right before intake valve. This increases mixing action and helps burning. (General Motors)

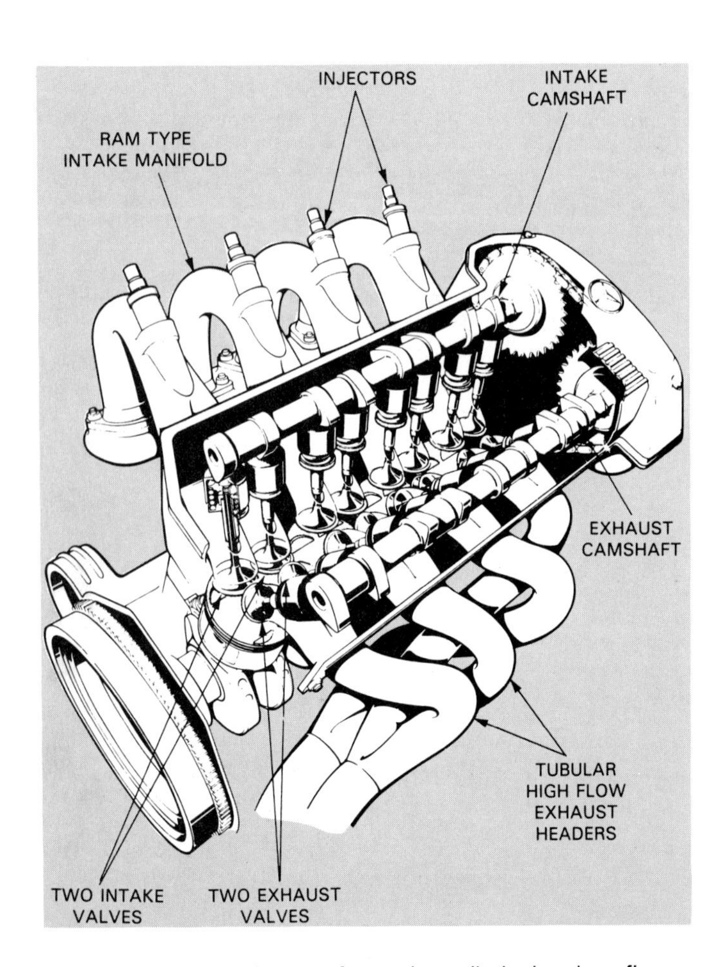

Fig. 6-17. Note twin-cam, four-valve cylinder head configuration. Intake runners and exhaust are tubular in shape to increase flow through cylinder head at high speeds. (Mercedes-Benz)

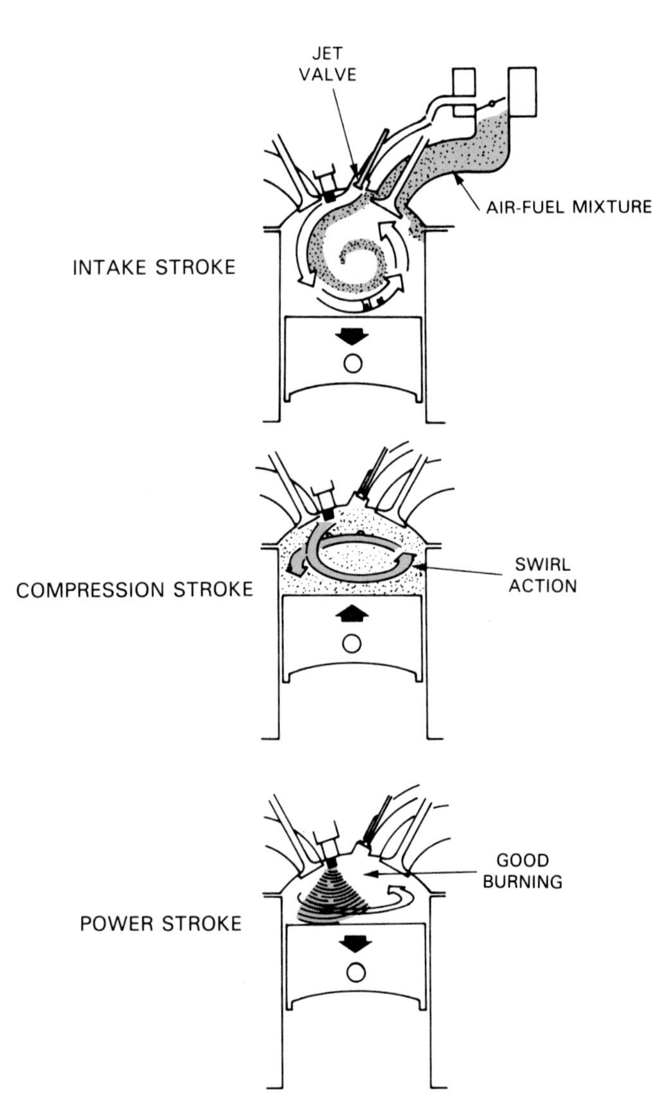

Fig. 6-19. Air jet chamber uses small, third valve to inject air stream directly into combustion chamber. This helps mixing and burning at low speeds. (Chrysler)

engine camshaft opens both the conventional intake valve and the air jet valve. This allows fuel mixture to flow into the cylinder past the conventional intake valve. At the same time, a stream of air flows into the cylinder through the jet valve.

The jet valve only works at idle and low engine speeds. At higher rpm, normal air-fuel mixing is adequate for efficient combustion.

Stratified charge combustion chamber

A *stratified charge combustion chamber* uses a small combustion chamber flame to ignite and burn the fuel in the main, large combustion chamber, Fig. 6-20.

A very *lean* mixture (high ratio of air to fuel) is admitted into the main combustion chamber. The mixture is so lean that it will not ignite and burn easily.

A *richer mixture* (higher ratio of fuel to air) is admitted into the small chamber by an extra valve. When the fuel mixture in the small chamber is ignited, flames blow into and ignite the hard to burn lean fuel mixture in the main combustion chamber.

A stratified charge chamber allows the engine to operate on a lean, high efficiency air-fuel ratio. Fuel economy is increased and exhaust emission output is reduced.

Precombustion chamber

A *precombustion chamber,* mentioned earlier, is commonly used in an automotive diesel engine. It houses the tip of the injector and tip of the glow plug. This small chamber helps the glow plug warm the air so that the fuel will ignite when injected into a cold engine. Refer back to Fig. 6-10.

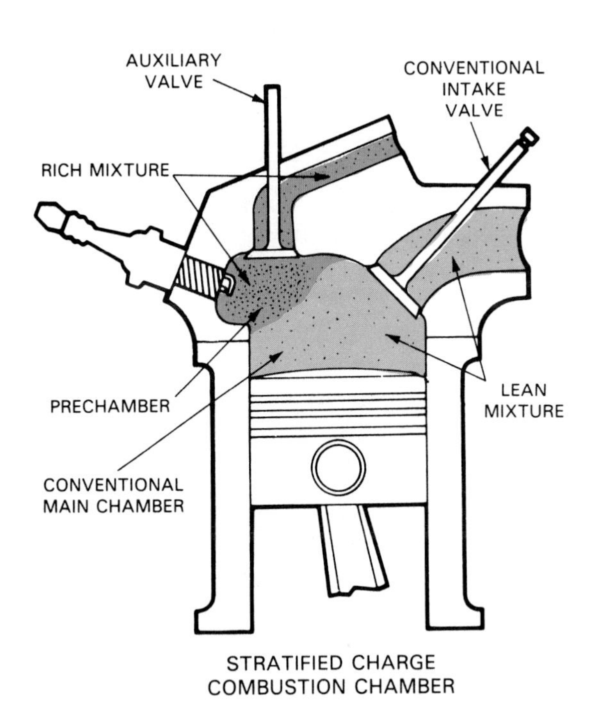

Fig. 6-20. Stratified charge combustion chamber uses small chamber to hold rich fuel mixture. Lean fuel mixture enters main chamber. Rich mixture is ignited by spark plug and blows flames into lean mixture. This ignites hard to burn lean mixture. (Ford Motor Co.)

CAMSHAFT DRIVE CLASSIFICATIONS

An engine can be classified by the method used to operate the camshaft. Three techniques are common: belt drive, gear drive, and chain drive, Fig. 6-21.

A *camshaft belt drive* uses a cogged rubber belt to turn the camshaft at one-half crankshaft speed. This type drive is normally used when the camshaft is located in the cylinder head. It can span the long distance from the crank to the cam. A belt drive is a relatively new design that is very smooth, quiet, and trouble-free.

A *camshaft gear drive* uses two or more gears to turn the cam. This type drive is a heavy duty arrangement found on a few diesel engines and severe service gasoline engines. The gears are extremely dependable but are heavy and noisy. They are only used with push rod type engines.

A *camshaft chain drive* is a common method of driving the camshaft when the cam is located in the engine block. However, a chain can also be found on a few overhead cam engines.

VALVE LOCATION CLASSIFICATIONS

The location of the engine valves is still another way to describe the type of automotive engine.

L-head engine

An *L-head engine* has both the intake and exhaust valves in the block, Fig. 6-22. Also called a *flat head* engine, its cylinder head simply forms a cover over the cylinders and valves. The camshaft is in the block and pushes upward to open the valves. Most four-stroke cycle, lawnmower engines are L-head types. Car engines are no longer L-head types. See Fig. 6-23.

I-head engine

In an *I-head engine,* both valves are in the cylinder head. Another name for this design is *overhead valve engine* (OHV), Fig. 6-22.

The OHV engine has replaced the flat head engine in automobiles. Numerous variations of the overhead valve engine are now in use, Fig. 6-24 and 6-25.

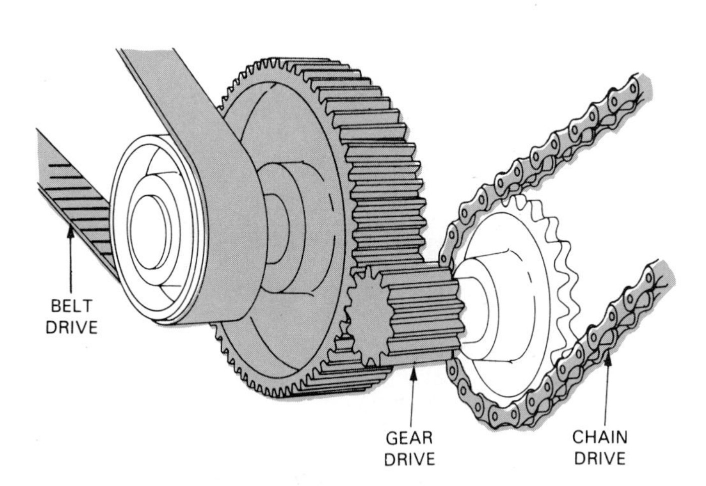

Fig. 6-21. Three methods of driving an engine camshaft: belt, gears, or a chain. (Deere & Co.)

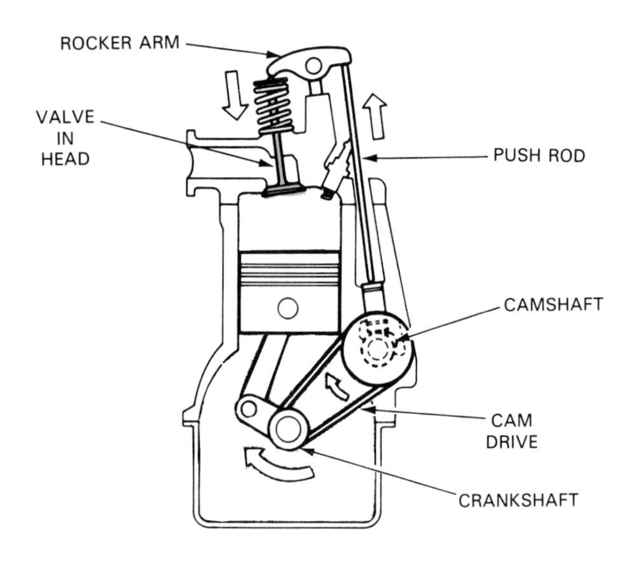

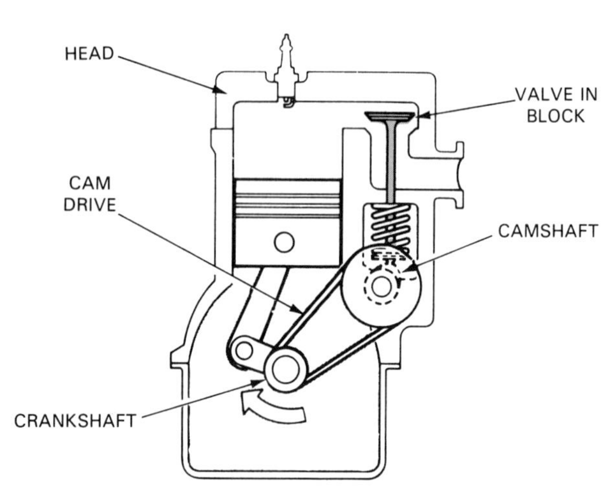

Fig. 6-22. Older engines had valve in cylinder block. This can still be found in one-cylinder lawnmower engines. Today's car engines have valves in cylinder head. (Ford)

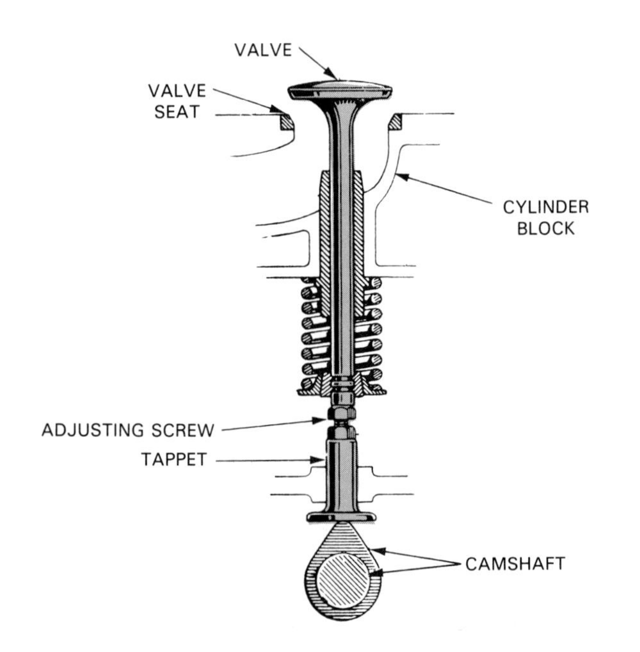

Fig. 6-23. L-head has valve in block. Cam lobe pushes valve up to open port. This design is no longer used in automobiles.

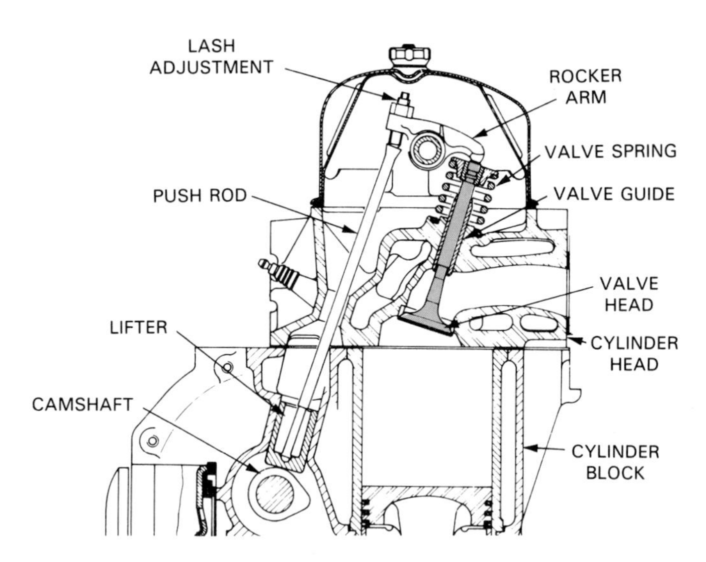

Fig. 6-24. I-head has valve in cylinder head. Note how camshaft is in block and push rod transfers motion to rocker arm and valve. (Renault)

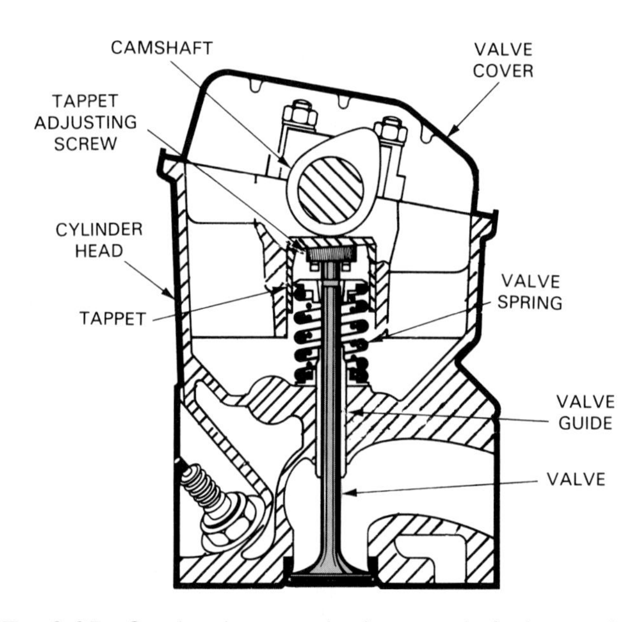

Fig. 6-25. Overhead cam engine has camshaft above valves in cylinder head. (American Motors)

Other valve configurations have been used in the past. However, they are so rare that their mention is not important.

CAMSHAFT LOCATION CLASSIFICATIONS

There are two basic locations for the engine camshaft; in the block and in the cylinder head. Both locations are common. Look at Figs. 6-24 and 6-25.

Cam-in-block engine

A *cam-in-block engine* uses push rods to transfer motion to the rocker arms and valves, Fig. 6-26A. The term overhead valve (OHV) is sometimes used instead of cam-in-block.

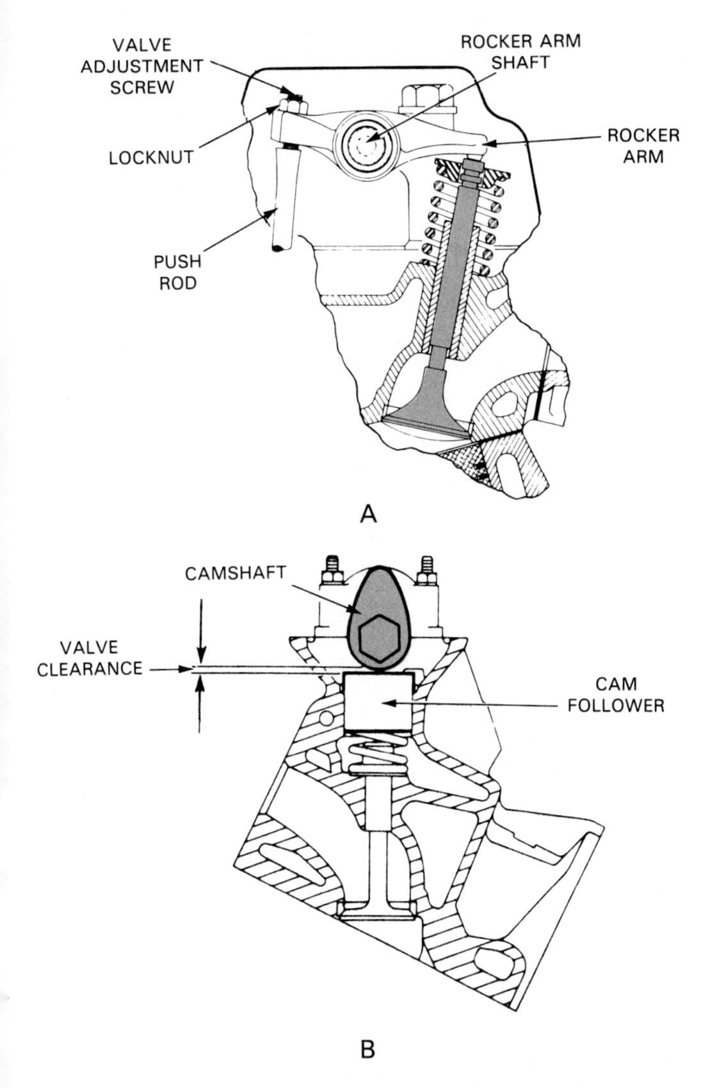

Fig. 6-26. Compare overhead valve and overhead cam configurations. (Federal Mogul)

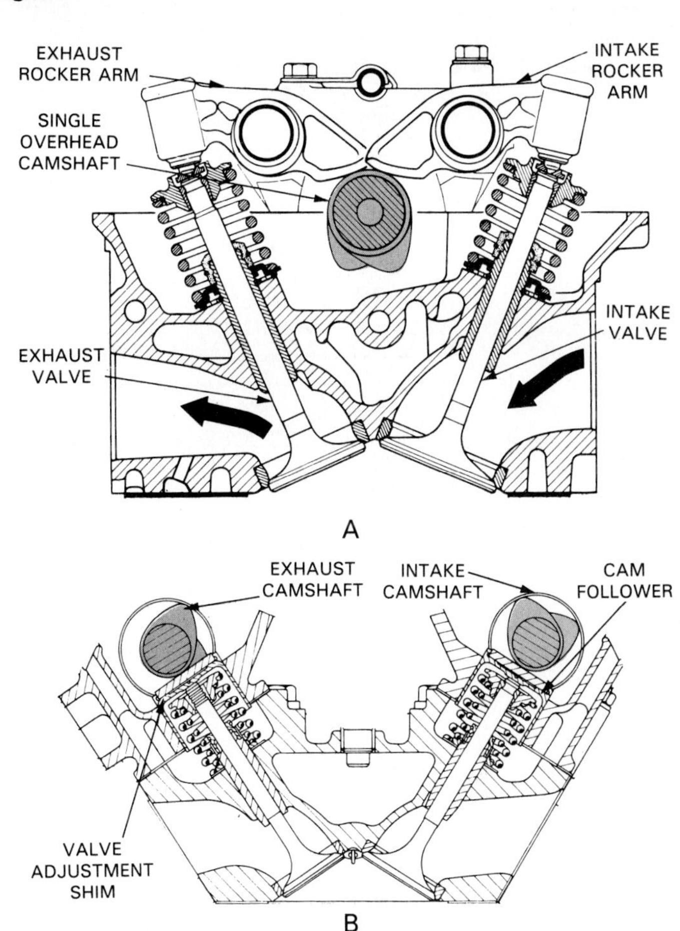

Fig. 6-27. A—SOHC or single overhead cam engine has one camshaft in cylinder head. Cam lobes can act directly on valves or rocker arms can be used. B—DOHC or dual overhead cam engine has two camshafts in cylinder head. Cams act directly on valves without rocker arms. (Fiat and Mercedes-Benz)

Overhead cam engine

In an *overhead cam engine* (OHC), the camshaft is located in the top of the cylinder head. Push rods are NOT needed to operate the rockers and valves. This type engine is a refinement of the overhead valve engine. Refer to Fig. 6-26B.

With the cam in the head, the number of valve train parts is reduced. This cuts the weight of the valve train. Also, the valves can be placed at an angle to improve *breathing* (airflow through cylinder head ports).

OHC engines were first used in racing cars because of their high rpm (revolutions per minute) efficiency. Now they are commonly used in small, high rpm, economy car engines. Without push rods to flex, less valve train weight, and improved valve positioning, the OHC increases high speed efficiency and power output.

Single overhead cam engine

A *single overhead cam engine,* abbreviated SOHC, has only one camshaft per cylinder head. The cam may act directly on the valves, or rocker arms may transfer motion to the valves. This is a very common design found on today's cars. See Figs. 6-27A and 6-28.

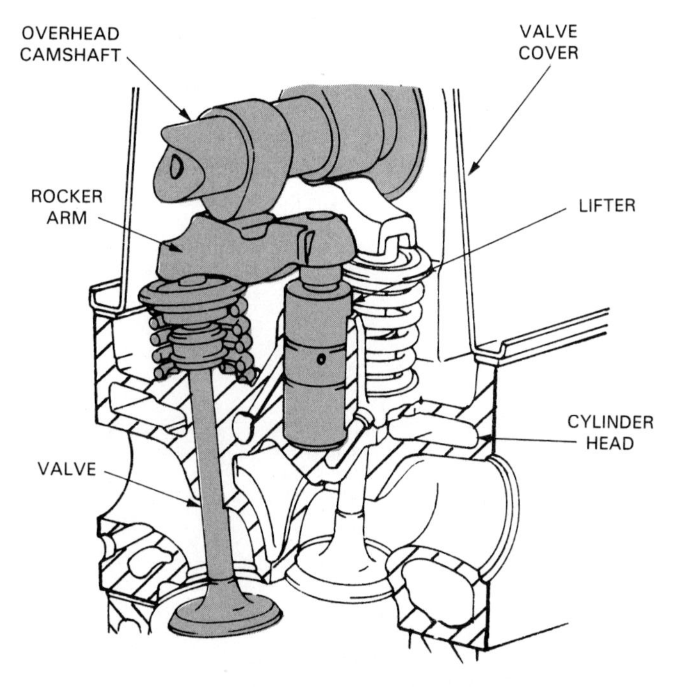

Fig. 6-28. This SOHC engine uses hydraulic lifter to quiet engine operation. (Ford)

Dual overhead cam engine

A *dual overhead cam* engine, abbreviated DOHC, has two cams per cylinder head. One cam operates the intake valves and the other cam operates the exhaust valves. This design is currently being used in several late-model engines. It can be used with a two-valve, three-valve, or four-valve combustion chamber. Look at Fig. 6-27B.

NORMAL ASPIRATION VS TURBOCHARGING

An engine can be classified as being normally aspirated or turbocharged (supercharged), Fig. 6-29.

A *normally aspirated engine* uses atmospheric pressure to force air or air-fuel mixture into the combustion chambers. This is a limiting factor of engine power because atmospheric pressure is only 14.7 psi (101 kPa) at sea level. Only so much air or fuel charge can enter the combustion chamber on each intake stroke. Usually, the chamber is not completely filled.

A *turbocharged* (exhaust driven blower) or *supercharged* (mechanically driven blower) engine forces air or air-fuel mixture into the cylinders under pressure. Discussed fully in later chapters, this more completely fills the chambers and theoretically raises the compression ratio of the engine for more power output. A turbocharger or supercharger is capable of increasing engine power up to 50%. Several engine modifications are needed, however.

VARIABLE DISPLACEMENT ENGINE

A *variable displacement engine* uses solenoid-operated rocker arms to alter the number of engine cylinders that function during engine operation. In a way, by disabling rocker arms, valves, and cylinders, the number of cylinders and resulting displacement of the engine is reduced.

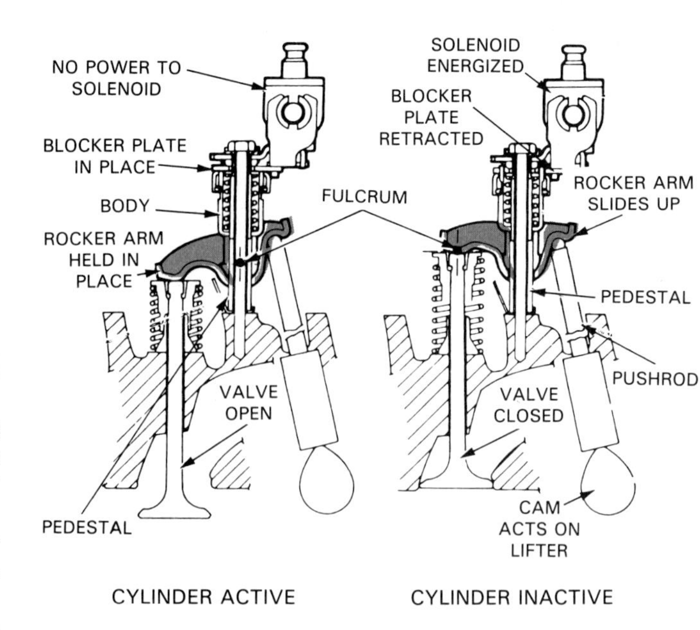

Fig. 6-30. Variable displacement engine simply disables rocker arms to keep certain cylinders from working and consuming fuel. These engines are rare. (Cadillac)

Fig. 6-30 shows the basic parts of one type variable displacement engine. This is not a very common design. Even with the valves deactivated, the rings, bearings, and other moving parts in the "dead" cylinders still contribute to frictional losses and reduced efficiency.

ENGINE BALANCING CLASSIFICATION

Mentioned briefly under crankshafts, an engine can be classified by its method of balancing. Just as you balance a wheel and tire to keep it from vibrating, you must also balance an engine crankshaft and its related parts to make the engine run smoothly.

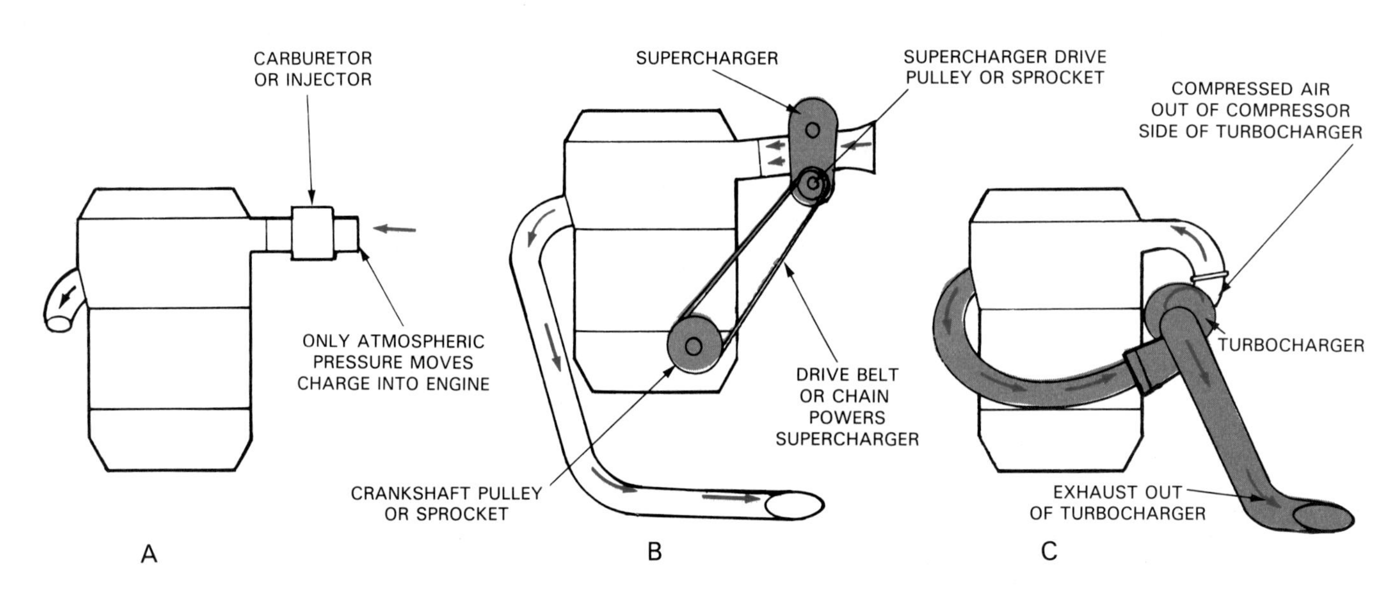

Fig. 6-29. A—Normally aspirated engine only uses atmospheric pressure to force charge into cylinders. B—Supercharged engine uses belt or chain driven blower to force charge into engine under pressure. C—Turbocharged engine uses exhaust driven blower to pressurize intake manifold and increase power output.

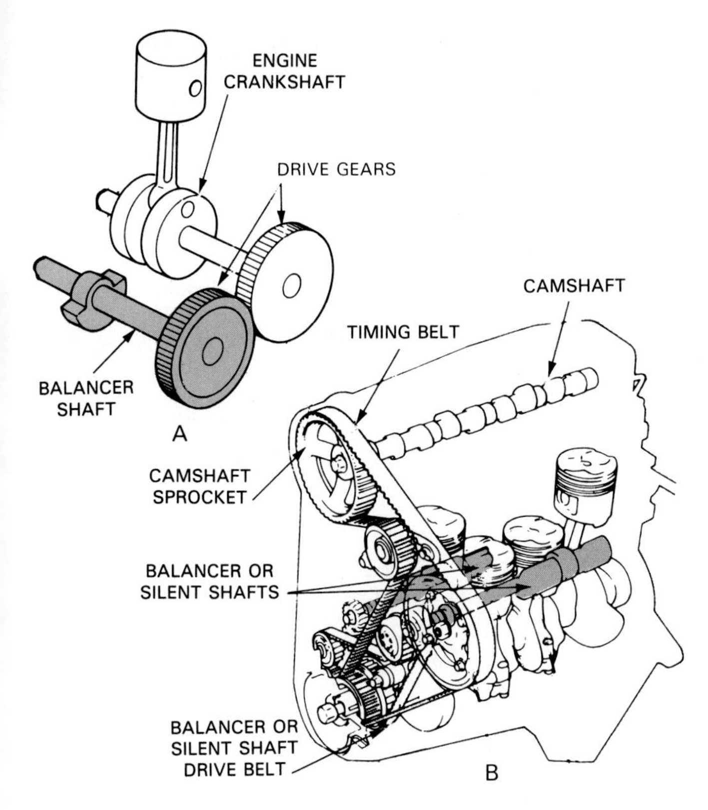

Fig. 6-31. Balancer shafts can be found on some modern engines. Counterweights on the shafts help keep crank spinning on non-power producing strokes. This makes engine run smoother at idle. A—Simple balancer shaft. B—Turbo diesel balancer shafts. (Ford)

Most engines are internally balanced. The flywheel and front damper are *neutral balanced* which means that they are equal in weight around their circumference. A few engines, however, are *externally balanced* and have extra weight added to certain points on the flywheel and balancer.

A *balancer shaft* is sometimes used to help smooth the operation of modern automobile engines. It is an extra shaft(s) with counter weights. Gears, a chain or a belt from the crankshaft is used to spin the balancer shaft(s) usually at two times engine speed. Two balancing shafts are frequently used in one engine, Fig. 6-31.

The weights on the balancer shaft are positioned to counteract the non-power producing strokes. A four-cylinder engine normally runs very rough at idle and vibration can be felt at idle in the passenger compartment. The balancer shafts help keep the crankshaft spinning between power strokes. With the balancing shafts, low speed smoothness is increased tremendously.

Only a small amount of energy is needed to spin the shafts so there is little reduction of fuel economy.

INTERNAL AND EXTERNAL COMBUSTION CLASSIFICATIONS

Engines can also be classified as internal and external combustion engines.

An *internal combustion engine* burns its fuel on the inside. An automotive engine is an internal combustion engine because its fuel is ignited and burns in the cylinder. Most other engines (rotary, turbine, etc.) are also internal combustion engines.

An *external combustion engine* burns its fuel on the outside. As shown in Fig. 6-32, a *steam engine* is the most typical example of an external combustion engine. The fuel is burned in an external chamber. The resulting heat is used to produce steam. The steam is then piped into the engine cylinder to produce pressure to operate the engine piston.

ALTERNATE ENGINES

As you have learned, automobiles generally use internal combustion, 4-stroke cycle, piston engines.

Alternate engines include all other engine types that may be used to power a vehicle. Various engine types have been developed but few have been placed into production in automobiles.

Two-stroke cycle engine

A *two-stroke cycle engine* is similar to an automotive four-stroke engine, but it only requires one revolution of the crankshaft for a complete power-producing cycle. Two piston strokes (one upward and one downward) complete the intake, compression, power, and exhaust events. Fig. 6-33 illustrates the basic operation of a two-stroke cycle engine.

As the the piston moves up, it compresses the air-fuel

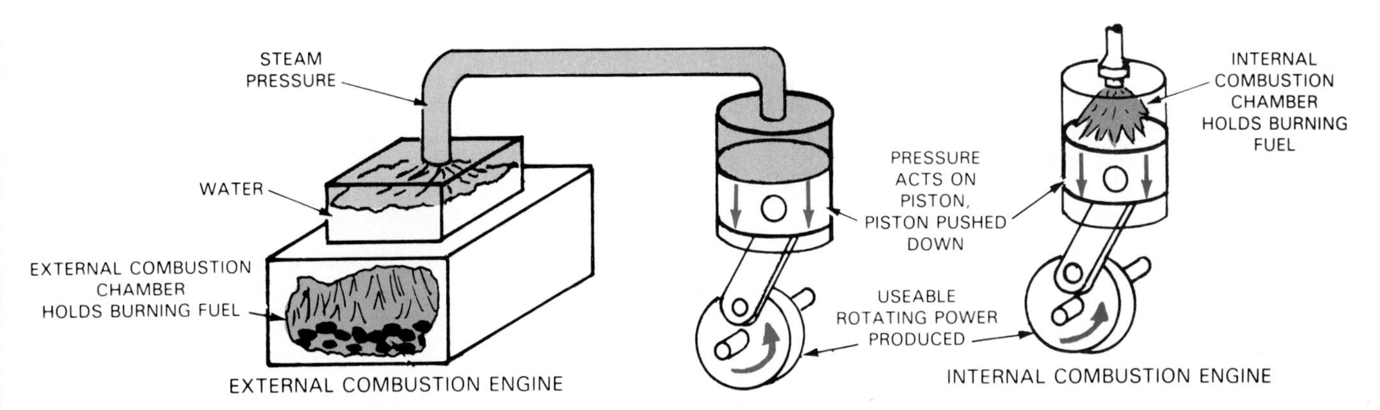

Fig. 6-32. External combustion engine burns fuel outside cylinder. A steam engine is an external combustion engine. Internal combustion engine, like car engine, burns fuel inside cylinder.

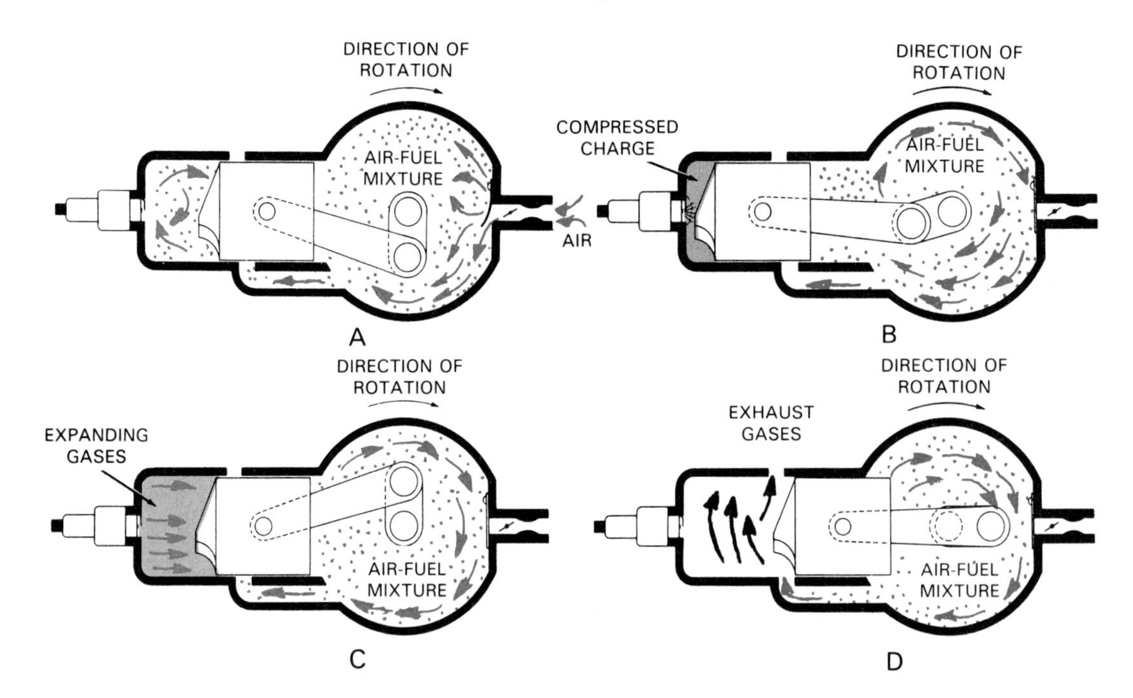

Fig. 6-33. Study operation of two-stroke cycle engine. A—Piston slides up and charge is pulled into crankcase. B—As piston moves up further, charge in cylinder is compressed and charge in crankcase lubricates moving parts. C—Mixture is ignited, forcing piston down. This compresses and pressurizes mixture in crankcase. Burned gases can leave cylinder when piston uncovers exhaust port. Pressurized crankcase can refill cylinder when piston uncovers transfer port. Either reed valve or rotary valve controls flow into crankcase.

mixture in the combustion chamber. At the same time, the vacuum created in the crankcase by the piston movement draws fuel and oil into the crankcase. Either a *reed valve* (flexible metal, flap valve) or a *rotary valve* (spinning, disc-shaped valve) can be used to control flow into the crankcase.

When the piston reaches the top of the cylinder, ignition occurs and the burning gases force the piston down. The reed valve or rotary valve closes. This compresses and pressurizes the fuel mixture in the crankcase.

As the piston moves far enough down in the cylinder, it uncovers an exhaust port in the cylinder wall. Burned gases leave the engine through the exhaust port.

As the piston continues downward, it uncovers the transfer port. Pressure in the crankcase causes a fresh fuel charge to flow into the cylinder. Upward movement of the piston again covers the transfer and exhaust ports, compression begins and the cycle is repeated.

Since the crankcase is used as a storage chamber for each successive fuel charge, the fuel and lubricating oil are premixed and introduced into the engine crankcase. Inside the crankcase, some of the oil separates from the gasoline. The oil mist lubricates and protects the moving parts inside the engine, Fig. 6-33.

Generally, two-stroke cycle engines are NOT used in automobiles because they:

1. Produce too much exhaust pollution.
2. Have poor power outout at low speeds.
3. Require more service than a four-stroke.
4. Must have motor oil mixed into the fuel.
5. Are not as fuel efficient as a four-stroke engine.

Wankel (rotary) engine

A *Wankel engine,* also termed a *rotary engine,* uses a triangular rotor instead of conventional pistons. The rotor turns or spins inside a specially shaped housing, as shown in Fig. 6-34.

While spinning on its own axis, the rotor orbits around a mainshaft. This eliminates the normal reciprocating (up and down) motion found in piston engines.

One complete cycle (all four strokes) takes place every time the rotor turns once. Three rotor faces produce three power strokes per revolution. Fig. 6-35 illustrates the basic operation of a rotary engine.

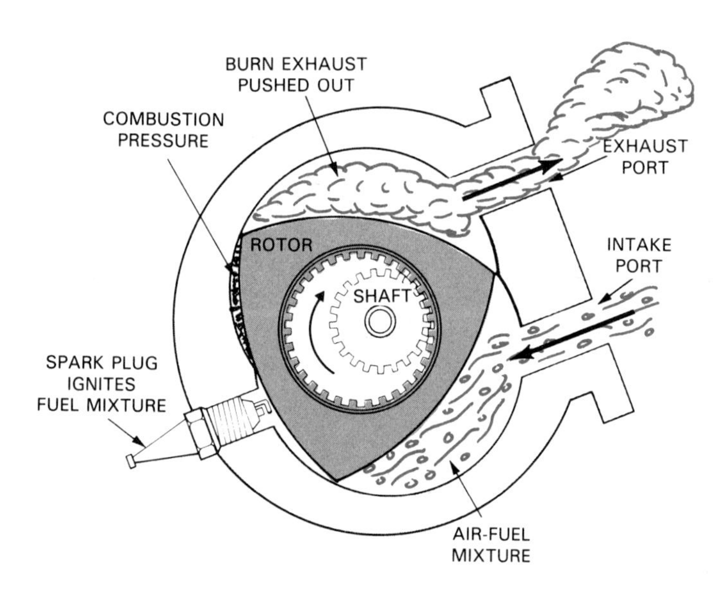

Fig. 6-34. Note chambers in rotary engine. Rotor spins inside housing for smooth operation. Pistons are not sliding up and down and changing direction.

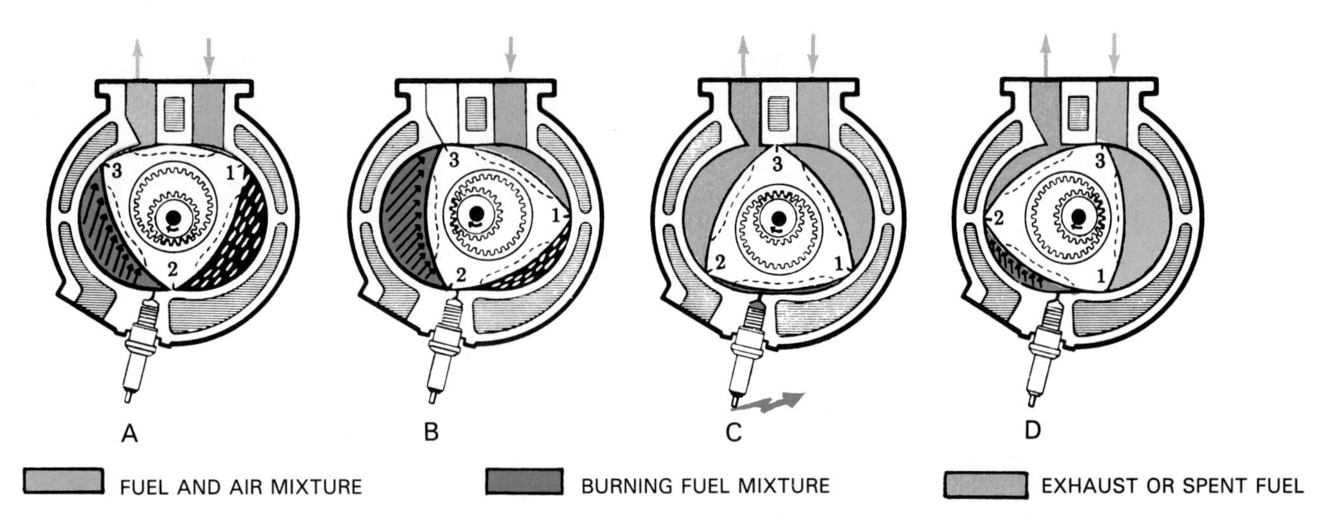

Fig. 6-35. Wankel rotary engine sequence of events: A—Intake is starting between 1 and 3; compression is occurring between 1 and 2; power is being produced between 2 and 3; exhaust is finishing between 3 and 1. B—Intake continues between 1 and 3; compression continues between 1 and 2; power is finishing between 2 and 3. D—Intake is finished between 1 and 3; power is being produced between 1 and 2; exhaust is continuing between 2 and 3.

A rotary engine is very powerful for its size. Also, because it spins—rather than moves up and down—engine operation is very smooth and vibration free.

A complicated emission control system is needed to make the rotary engine pass emission standards. This has limited its use. A Wankel engine is one of the few alternate engines to be mass produced and installed in production vehicles.

Fig. 6-36 shows a cutaway view of an actual Wankel rotary with the transmission attached. Study the parts.

Gas turbine

The *gas turbine* uses burning and expanding fuel vapor to spin fan type blades. The ''fan blades'' are connected to a shaft that can be used for power output. Fig. 6-37 illustrates a basic gas turbine.

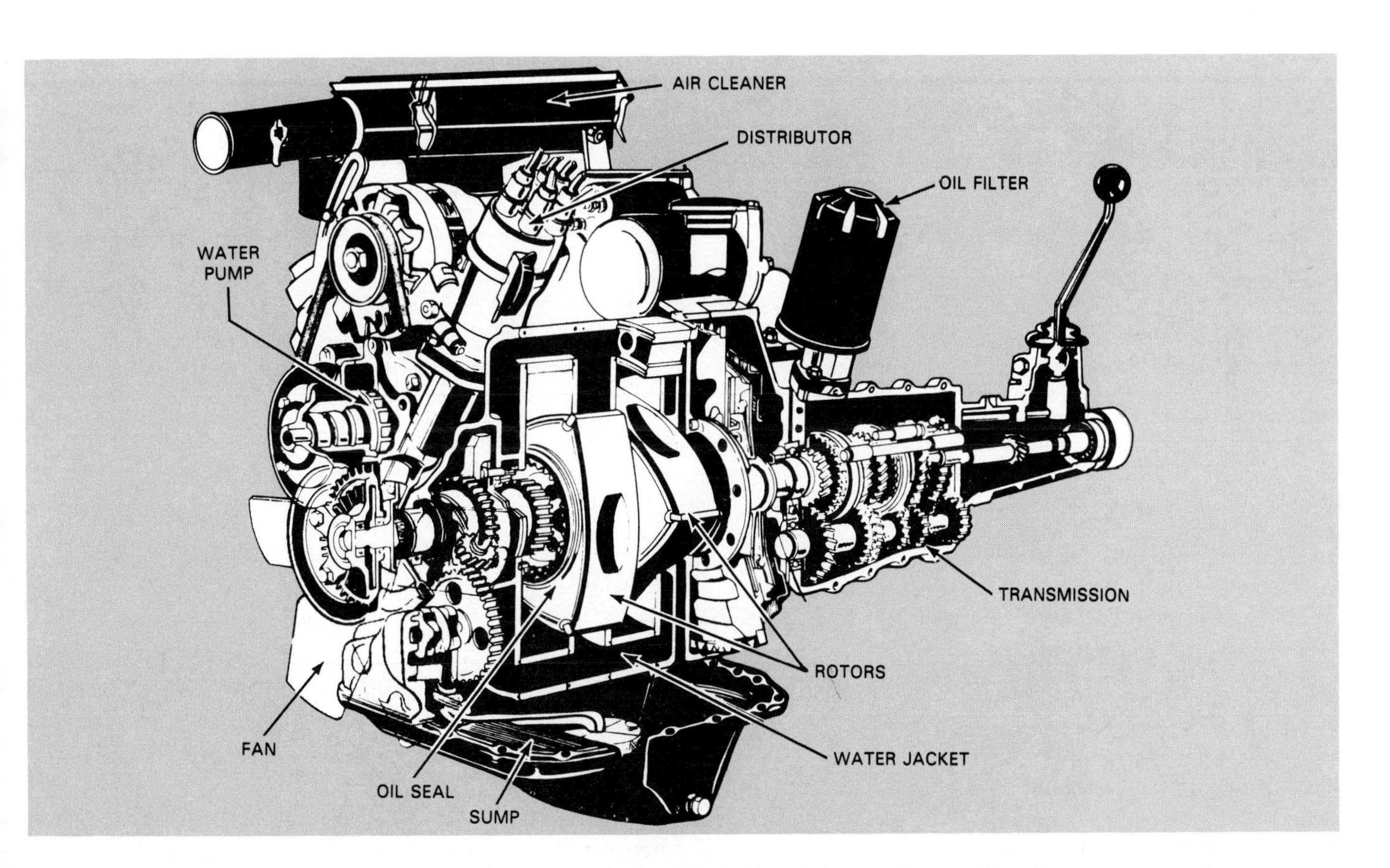

Fig. 6-36. Study parts of complete, automotive rotary engine. (Sun/Mazda)

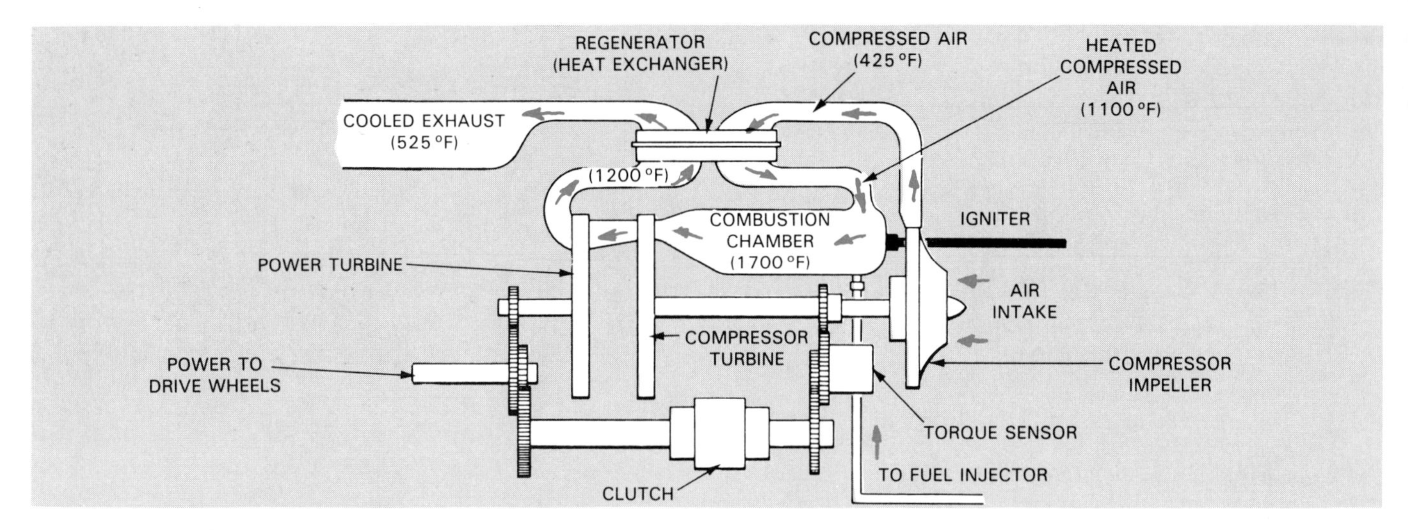

Fig. 6-37. This is a simple schematic of a gas turbine engine. Burning, expanding gas spins turbine wheels which are like large fan blades. Spinning turbines can then be used to produce usable power.

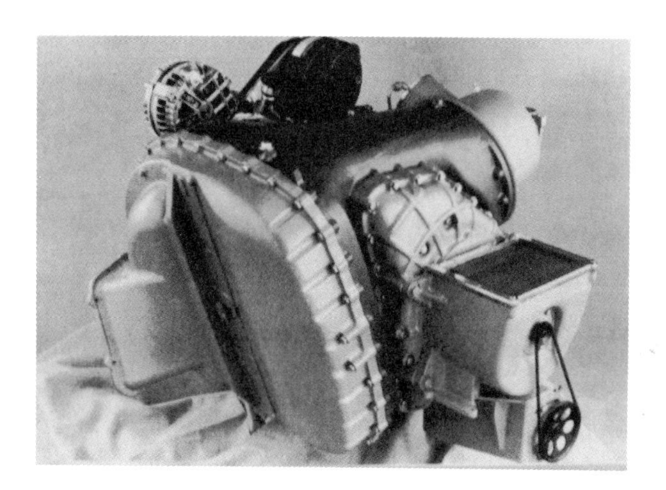

Fig. 6-38. This is an experimental gas turbine engine for automotive application. (Chrysler)

A gas turbine is a very promising alternate type of engine. It is capable of extremely high efficiency—much higher than a conventional piston engine. It can burn many types of fuel: gasoline, kerosene, or oil. A gas turbine can also produce tremendous power for its size. Because of the spinning action, its power output is very smooth, Fig. 6-38.

The gas turbine is not presently in use in automobiles because of its high manufacturing costs. It requires many special metals, ceramic parts, and precision machining and balancing. Some day the gas turbine may be a very common automotive engine.

Stirling engine

The *Stirling engine* is an experimental piston engine that uses the heating (expansion) and cooling (contraction) of a confined fluid (gas) to produce pressure and piston movement. Fig. 6-39 shows a Stirling engine.

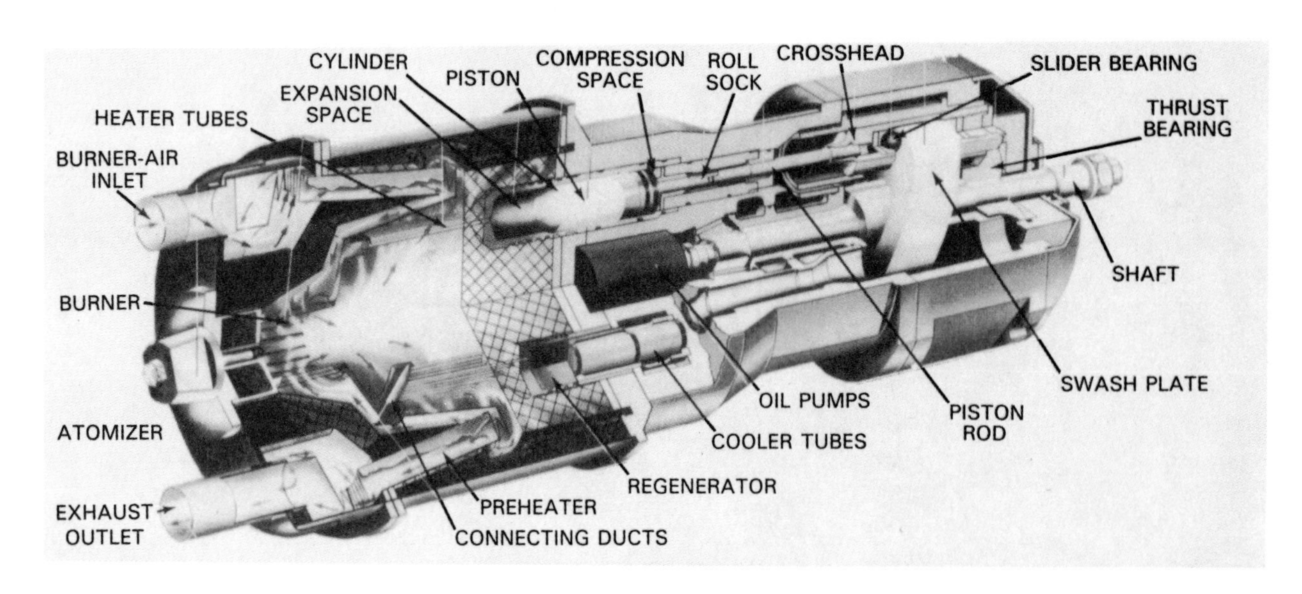

Fig. 6-39. Experimental Stirling engine uses change in temperature, and resulting expansion or contraction of confined fluid to operate pistons. (Ford)

A Stirling engine uses the principle—as a gas is heated, it will expand and as a gas is cooled, it will contract and take up less space.

When the gas in a Stirling engine is heated, it expands and pushes the piston down. On the other piston, the gas is cooled, contracts, and pulls the piston down because of vacuum (suction).

A *regenerator* or heat exchanger is used between the hot and cold sides of the piston. It is similar in construction to a car radiator. It returns heat to the gas before the gas returns to the hot side of the piston.

Free piston engine

The *free piston engine* combines the principles of diesel combustion and the power output of a turbine. It has very few moving parts, no connecting rods, crankshaft, nor ignition system. However, it suffers from excessive noise, hard starting, and poor speed control.

Look at Fig. 6-40. It illustrates the operation of a free piston engine.

Electric motor (electric car)

An *electric motor* and large storage batteries can be used to power an automobile. Electric cars have been produced in limited numbers by some companies. They have seen some success as a means of transportation for short trips. Speed and driving distance is limited with today's technology. Fig. 6-41 shows a modern electric car produced by one company.

Fig. 6-41. Electric cars are being developed by manufacturers to comply with future mandates for zero emissions vehicles. (General Motors)

Regenerative braking can be used on an electric car to help recharge the storage batteries. When the brakes

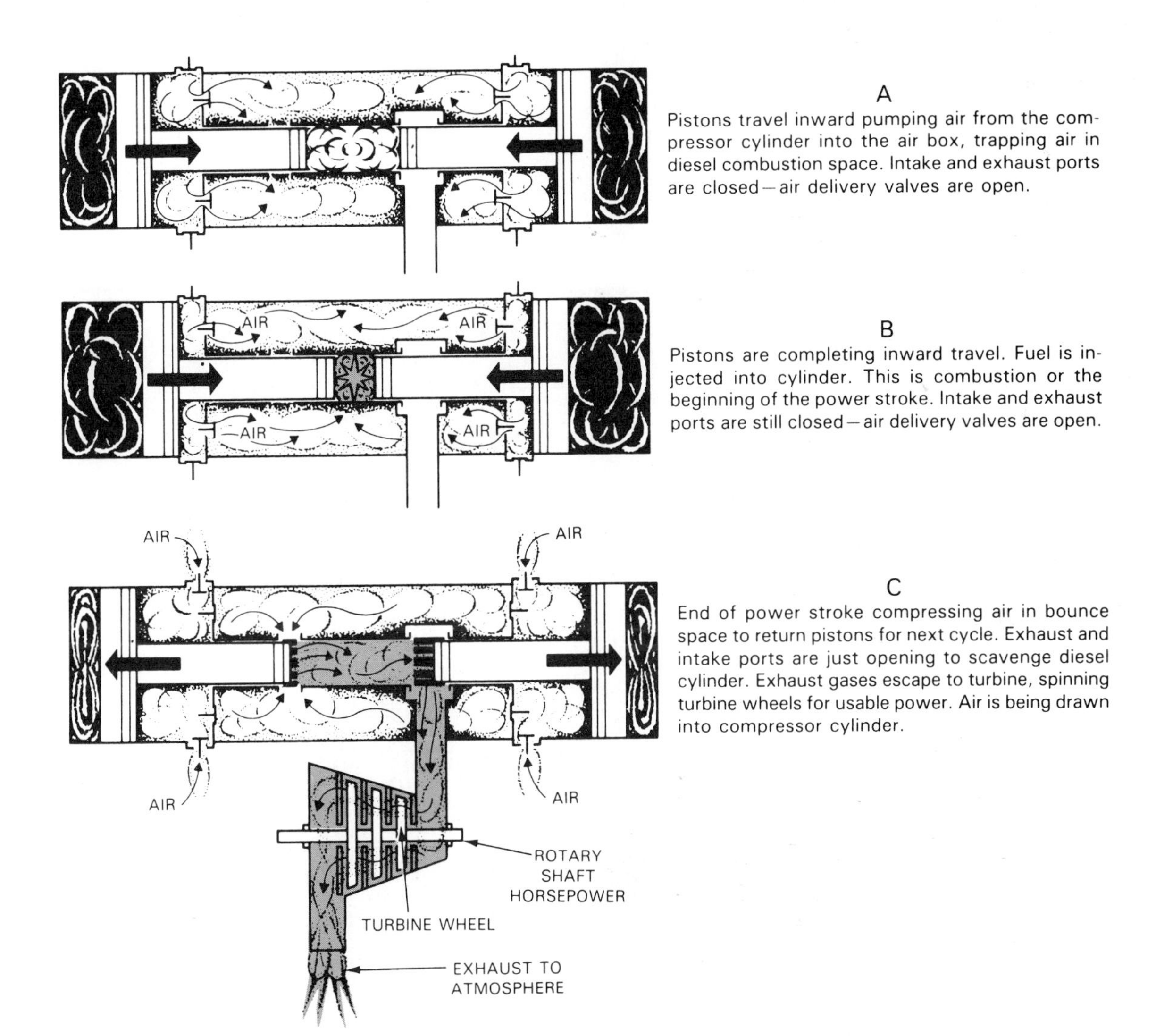

Fig. 6-40. Study operation of experimental, free-piston engine.

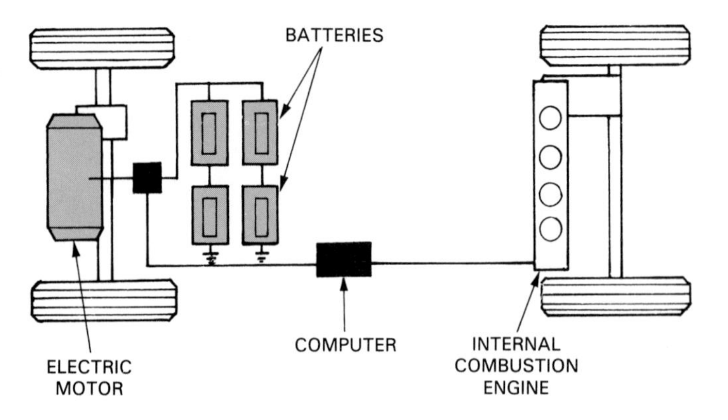

Fig. 6-42. Hybrid car uses two methods of developing power. Here batteries and electric motor help car accelerate from standstill. Then, very small, fuel efficient, gasoline engine can be used to keep car traveling at highway speeds. Engine can also be used to recharge batteries.

are applied, the electric motors are used as powerful generators to produce electricity. Since it takes considerable energy to turn the motors as generators, the car is slowed, as if conventional brakes were applied.

This principle would be beneficial if the car is used in stop and go traffic. Regenerative braking would keep the batteries charged for longer periods of time.

Hybrid power source

A *hybrid power source* uses two different methods to power the car. For example, a small gasoline engine and an electric motor and storage batteries may both be used to propel the car. This is illustrated in Fig. 6-42.

The batteries and electric motor supply power when the car first accelerates. This provides enough energy to accelerate the car quickly. Once cruising speeds are reached, the gasoline engine takes over. It is a very small engine that can supply adequate power to keep the car moving. The gasoline engine also provides enough energy to recharge the batteries.

TYPICAL AUTOMOTIVE ENGINES

Figs. 6-43 through 6-59 give typical automotive engines. Study each carefully. Note the design variations between the different types. Also, study the names of all of the parts. This will help you in later chapters as you continue to learn more about engines.

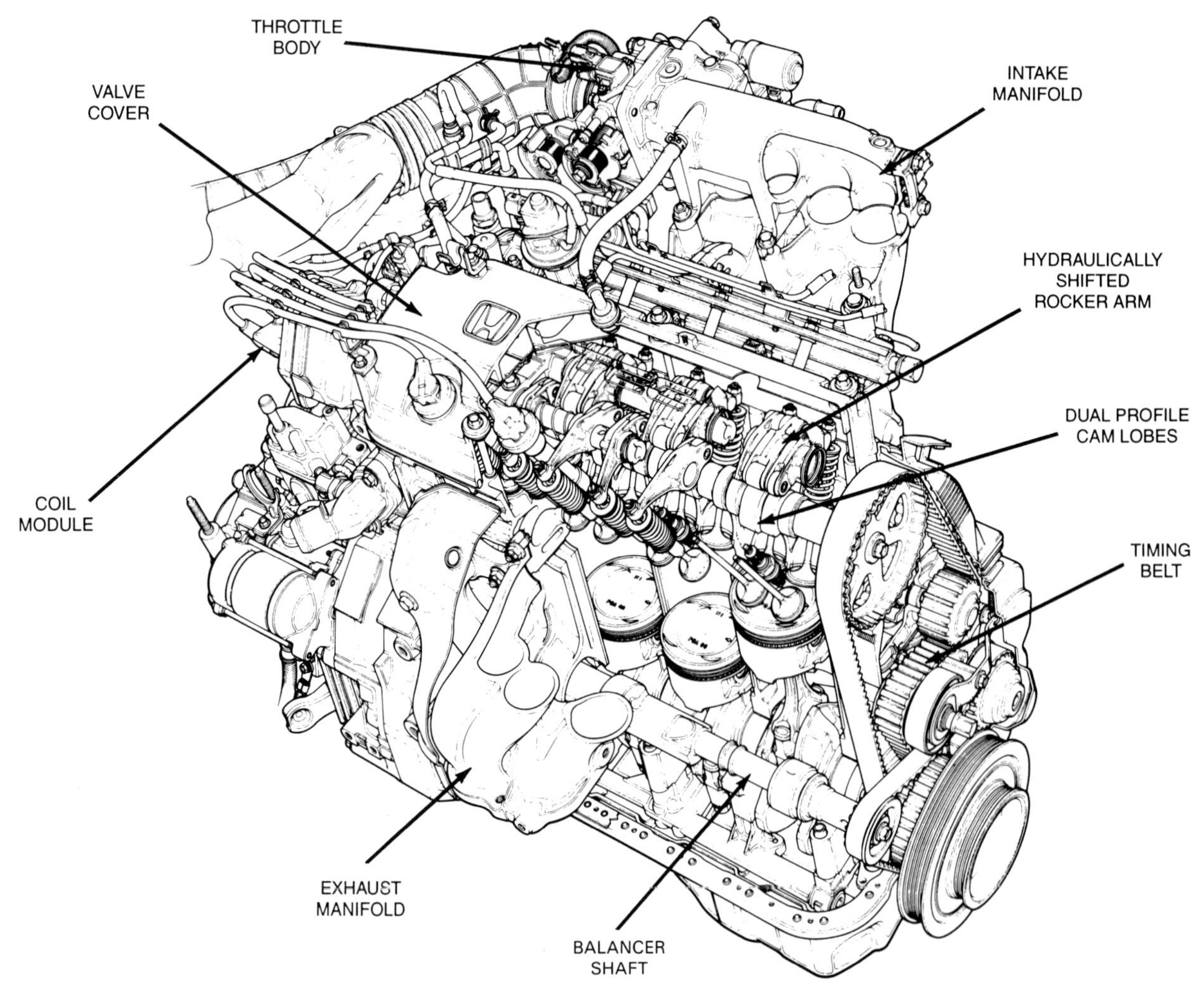

Fig. 6-43. This is a modern inline, four-cylinder 16-valve engine. It has hydraulically controlled rocker arms that can shift to use different cam lobes on the intake camshaft. One set of cam lobes has profiles for low speed efficiency, the other set has profiles for high speed power and fuel economy. (Honda Motor Co.)

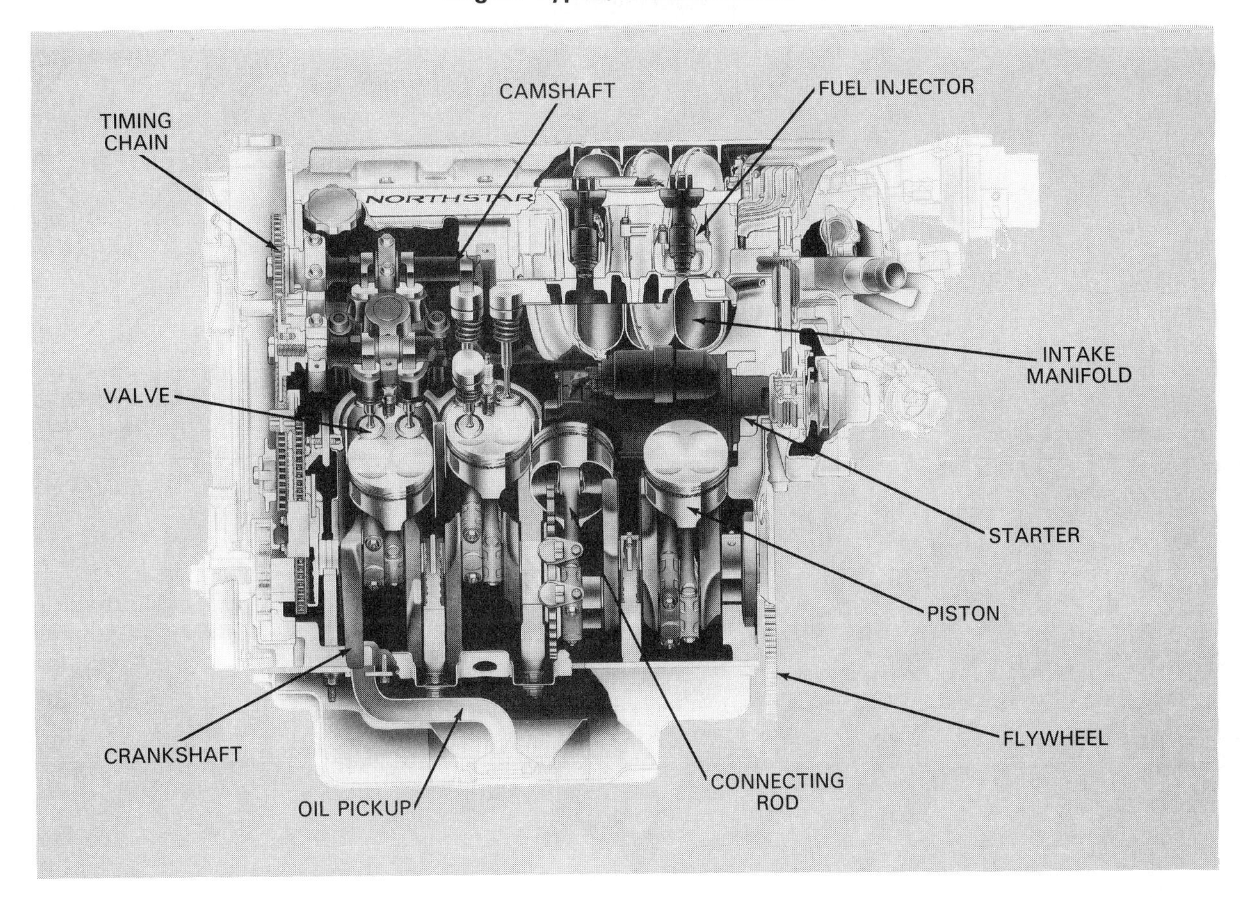

Fig. 6-44. Cutaway of a late-model V-8 engine. This engine is equipped with four camshafts and 32 valves. Note position of starter. (Cadillac)

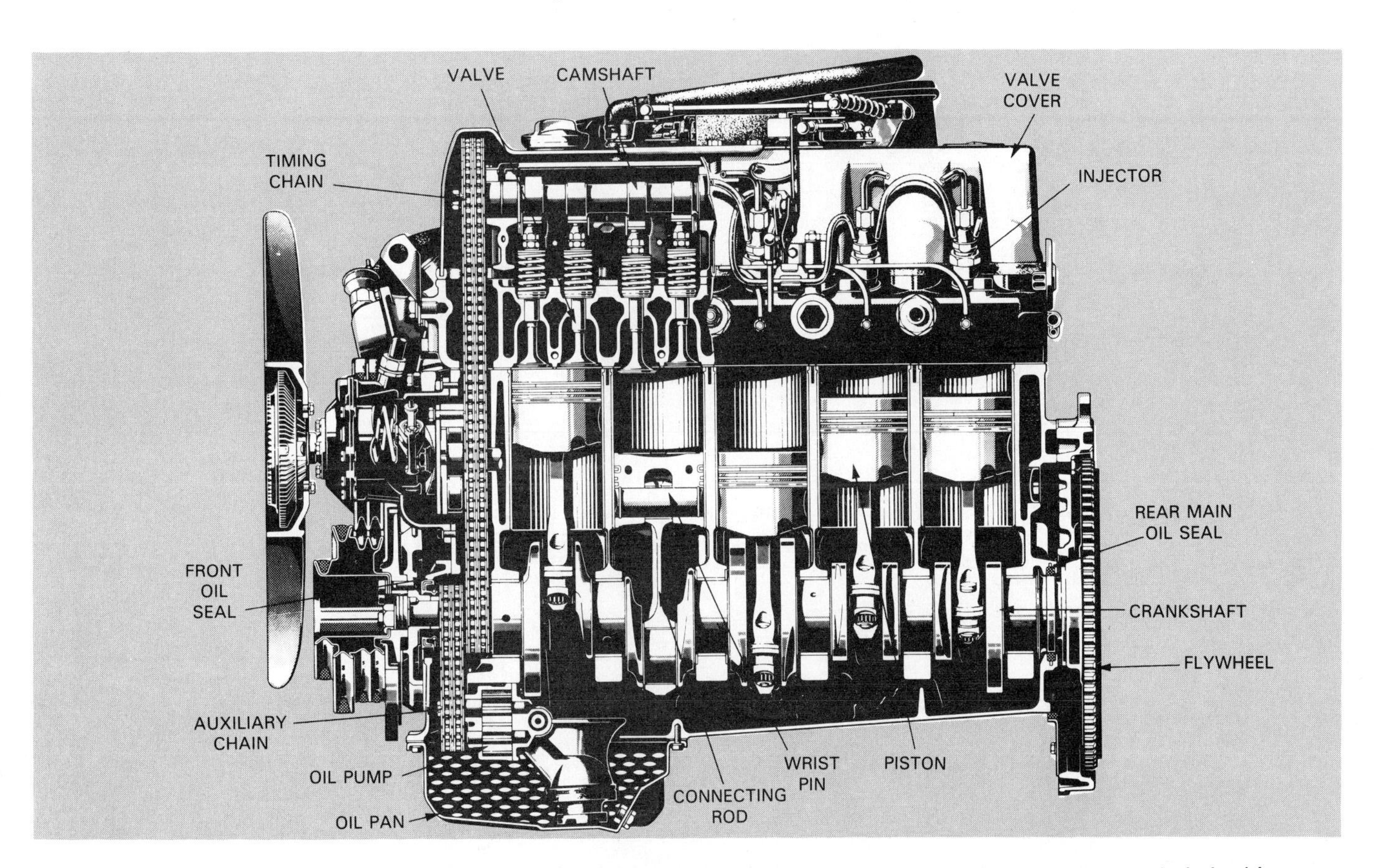

Fig. 6-45. This is a five-cylinder diesel engine with single overhead camshaft. Note gear oil pump and chain drive. (Mercedes-Benz)

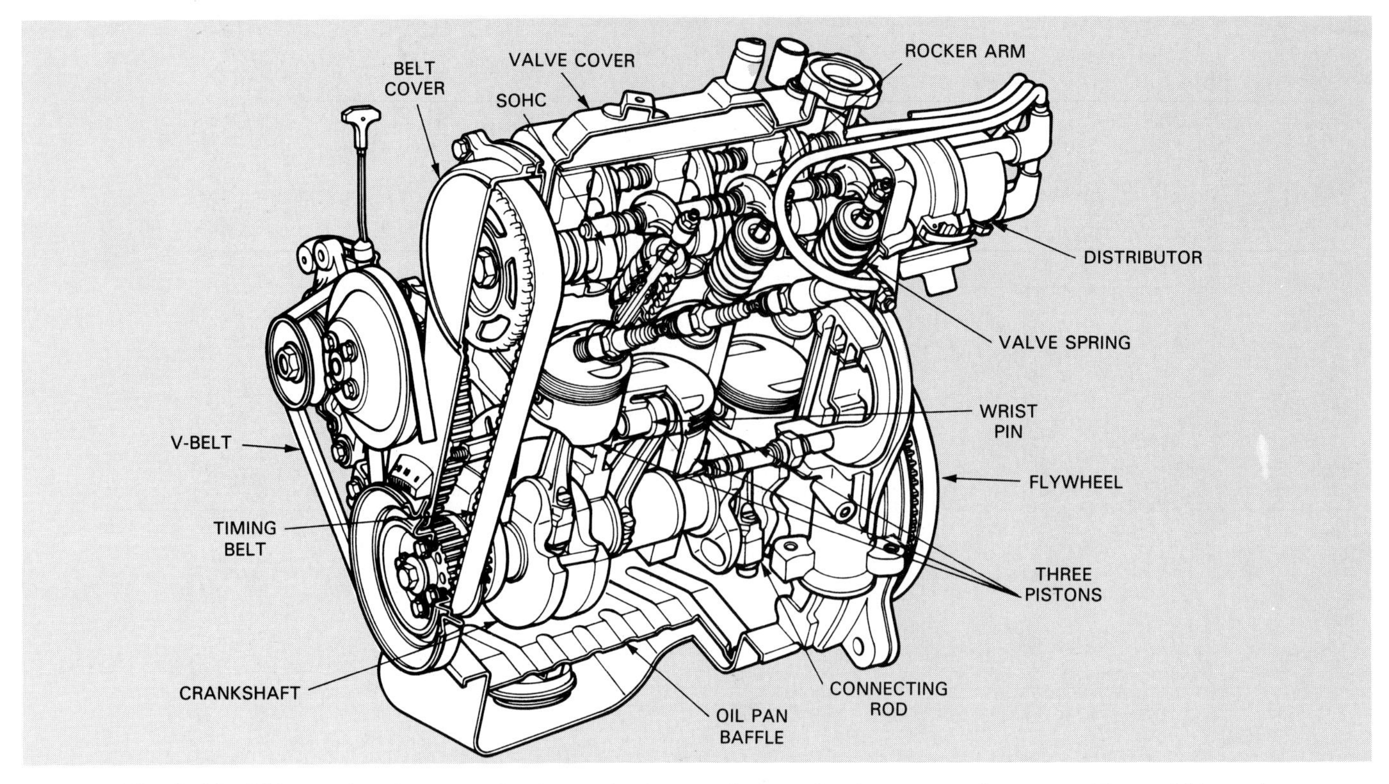

Fig. 6-46. This overhead cam engine has only three cylinders. Study names of parts. (General Motors)

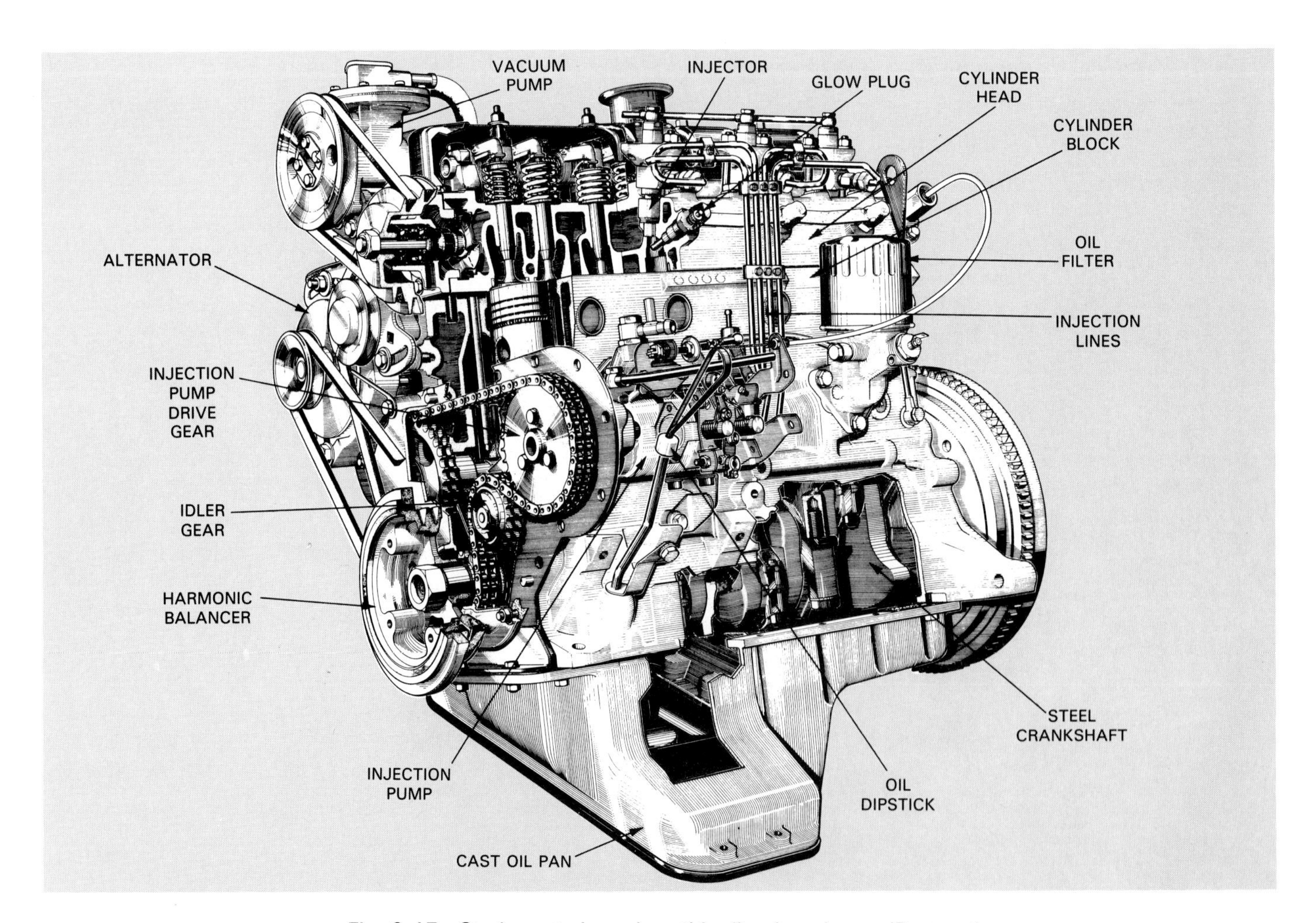

Fig. 6-47. Study parts in and on this diesel engine. (Peugeot)

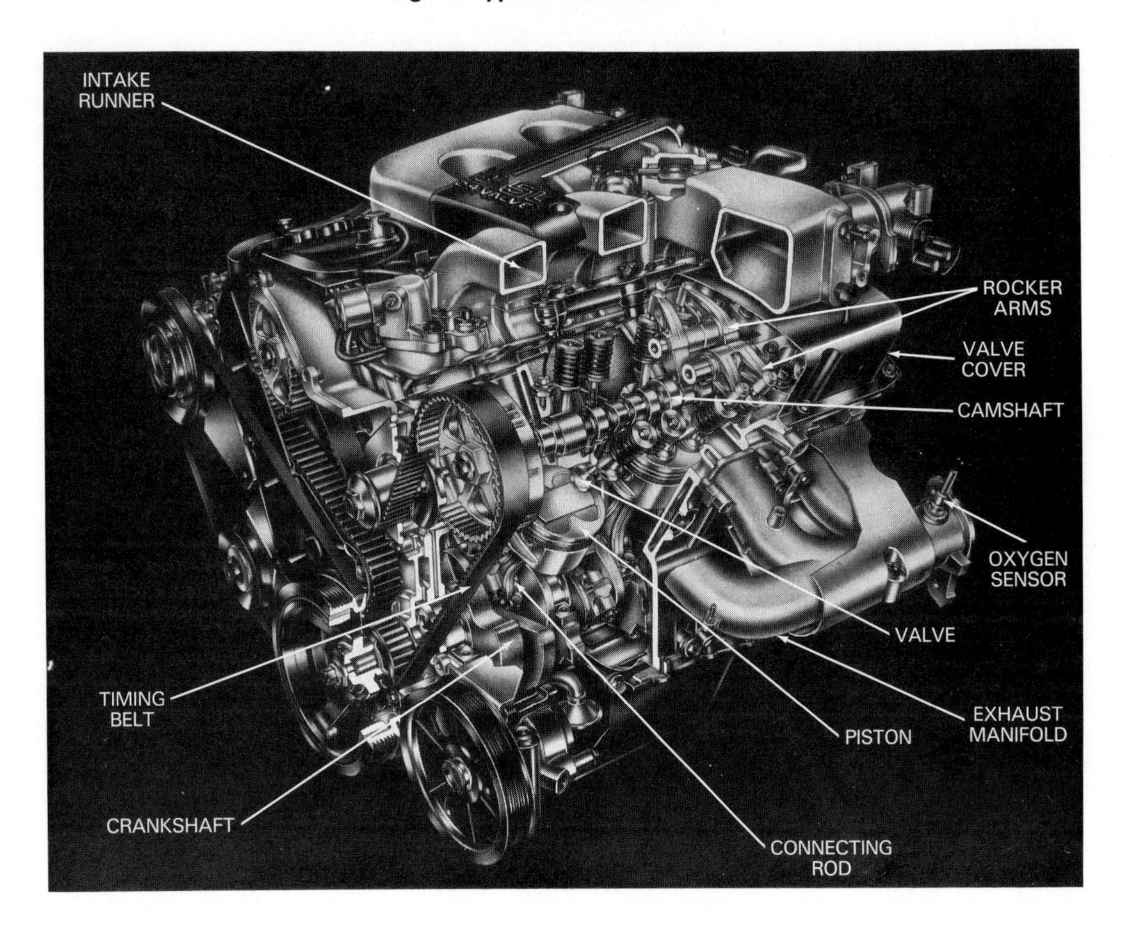

Fig. 6-48. Cutaway view of a 24-valve, V-6 engine. Study arrangement of major components. (Chrysler)

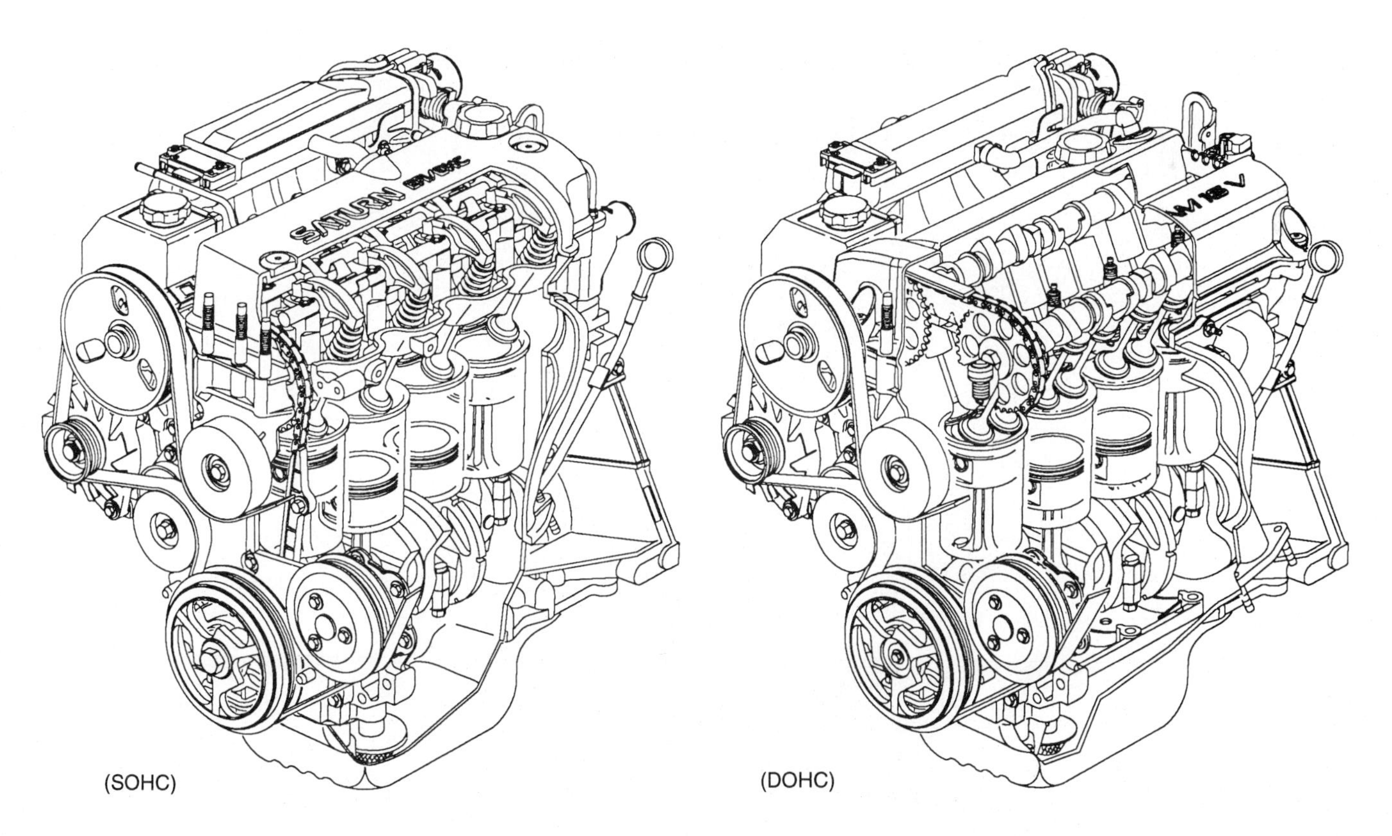

Fig. 6-49. Compare single overhead camshaft engine with dual overhead camshaft engine. How many differences can you find? (Saturn)

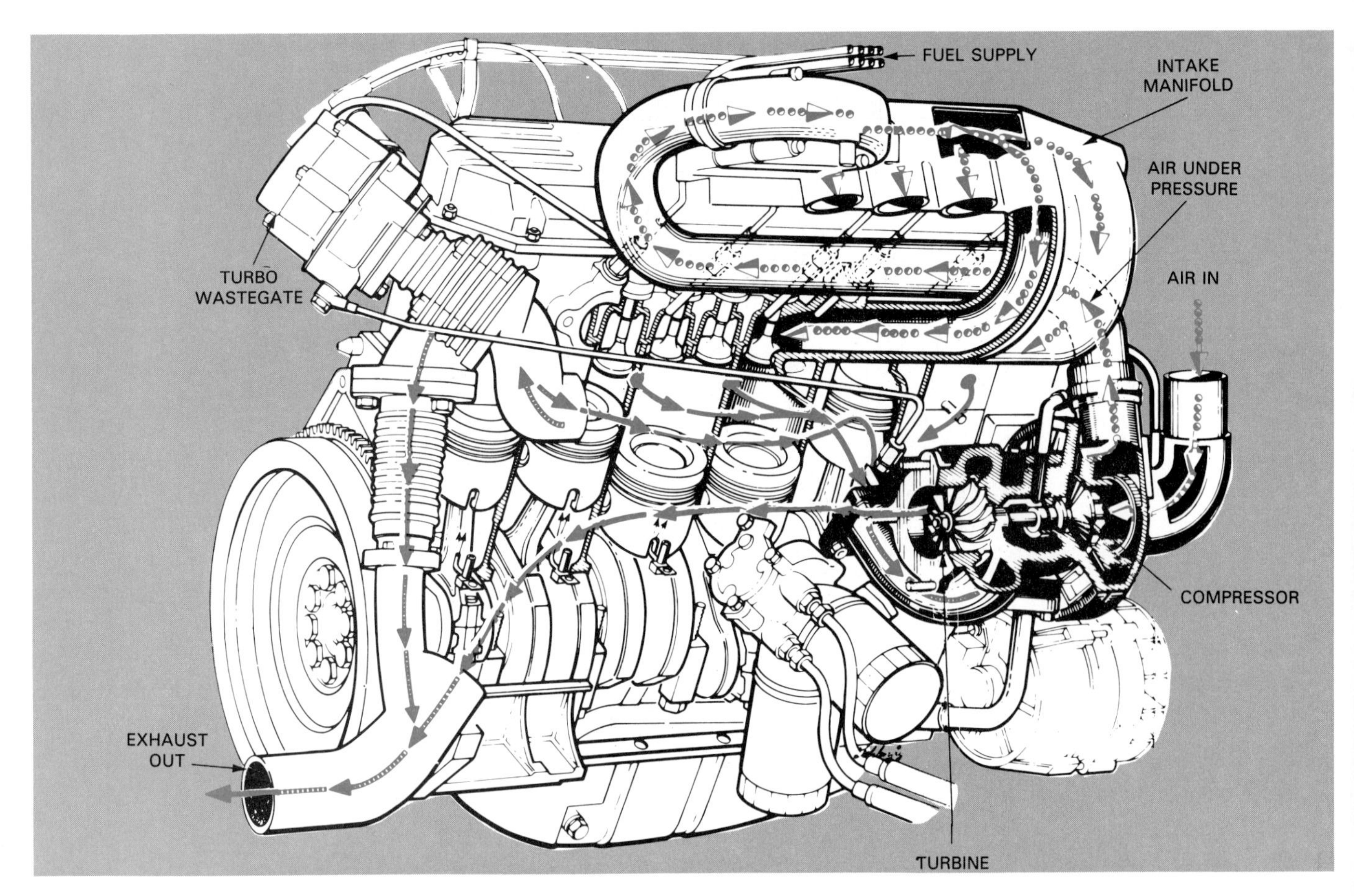

Fig. 6-50. Turbocharged, in-line engine uses wastegate to limit boost pressure. Note flow through engine and system. (Audi)

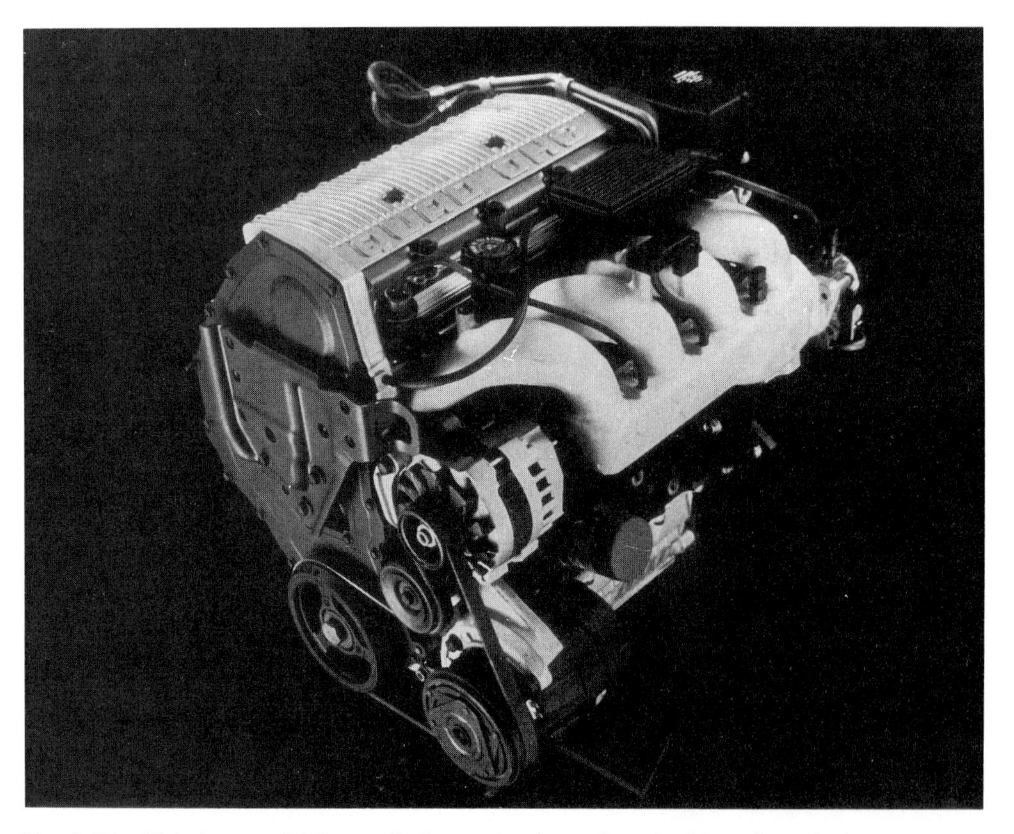

Fig. 6-51. This late-model four-cylinder engine is equipped with a direct ignition system. Spark plug wires are not used on this engine. (Buick)

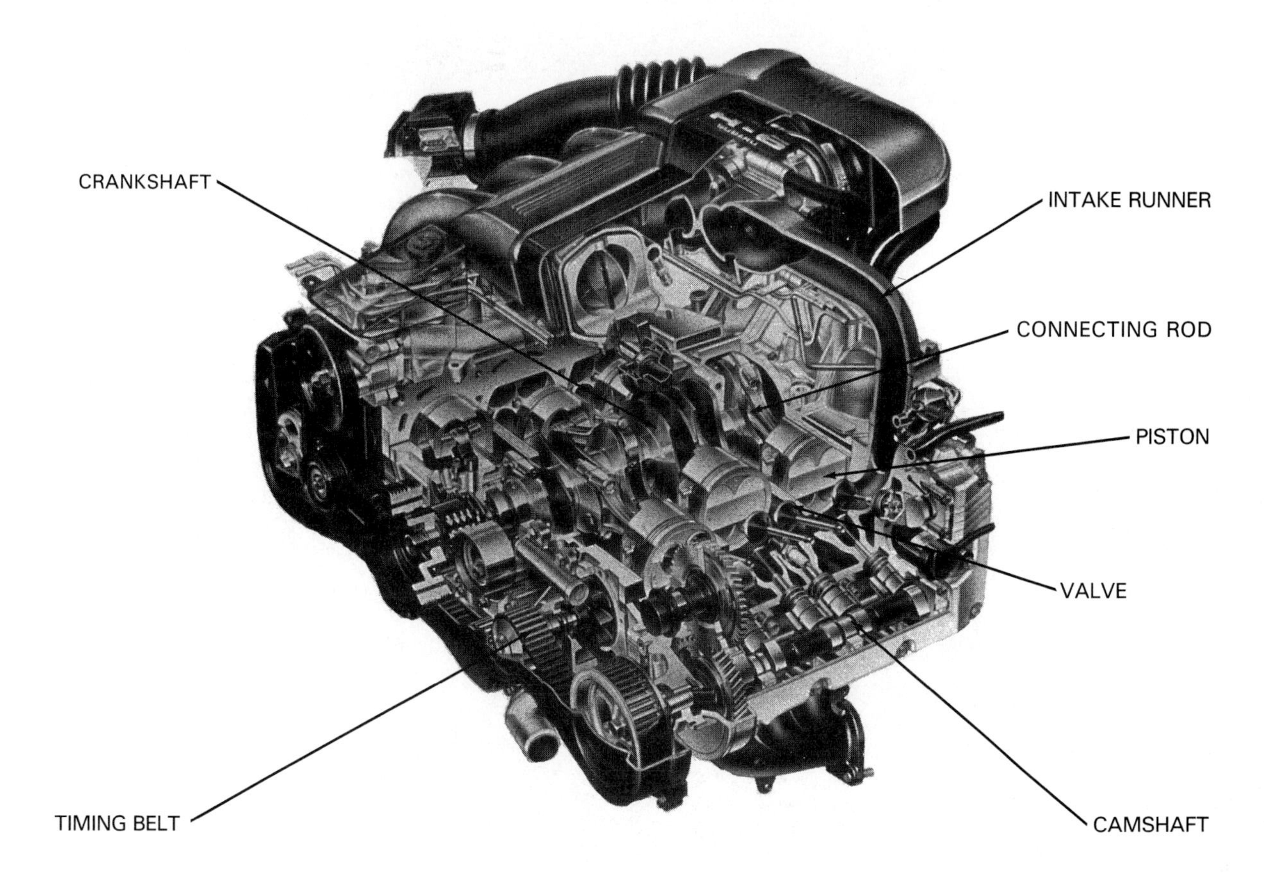

Fig. 6-52. Cutaway of a horizontally opposed, 24-valve, six-cylinder engine. This engine is equipped with dual overhead cams and fuel injection. (Subaru)

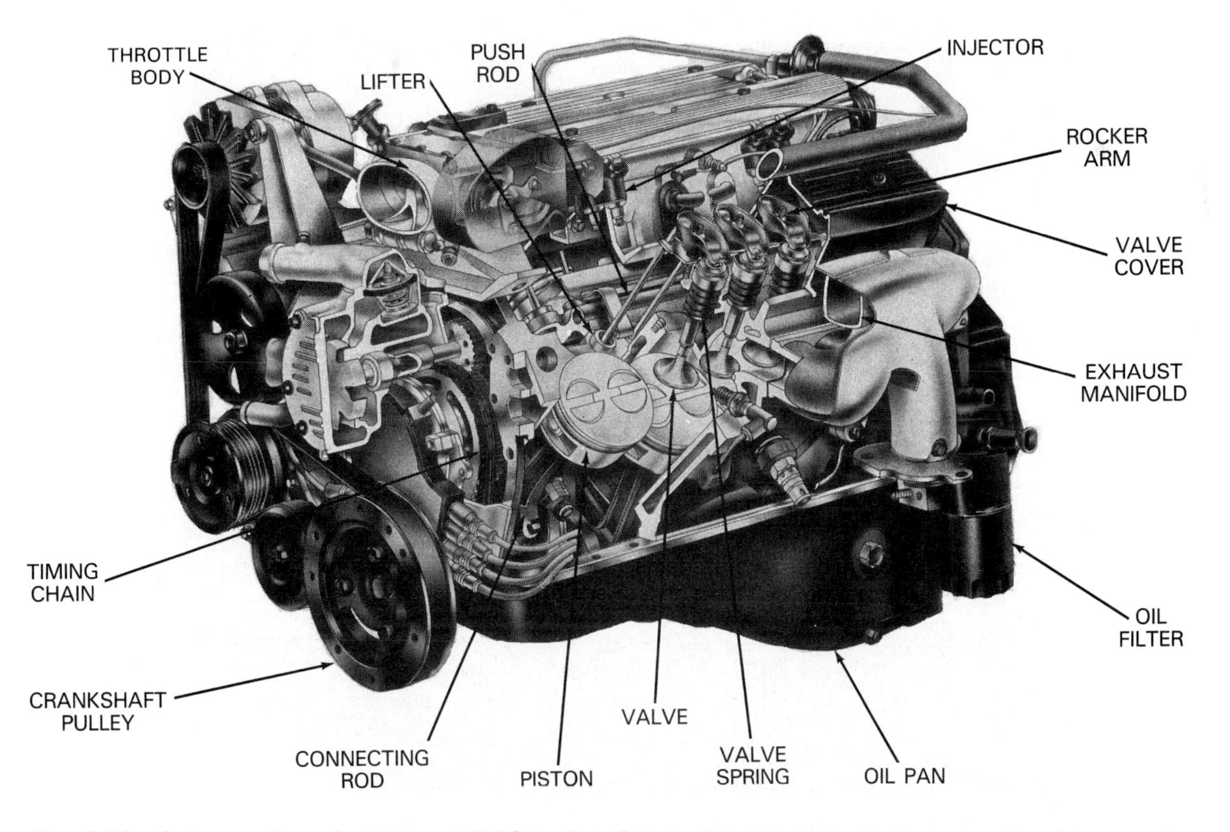

Fig. 6-53. Cutaway view of a late-model V-8 engine. Camshaft is located in block, and push rods are used to transfer motion to the rocker arms. (Pontiac)

Fig. 6-54. Cutaway view of an overhead cam, V-8 engine. This engine is equipped with a distributorless ignition system and chain-driven camshafts. (Ford)

Fig. 6-55. Term 16-valve engine refers to total number of valves in engine. This engine has four cylinders with four valves per cylinder for a total of 16. It also has twin, overhead camshafts and produces tremendous power for its size. (Toyota)

Fig. 6-56. Another 16-valve, DOHC engine, but this one also has turbocharger and intercooler. (Saab)

Fig. 6-57. How many classifications can you make about this engine—number of cylinders, cylinder arrangement, fuel system type, cam location, etc.? (Porsche)

Fig. 6-58. How many parts can you identify on this late-model engine? (BMW)

Fig. 6-59. This is a racing engine. Note tubular exhaust headers, twin cams, aluminum head, and external oil pump. (Saab)

KNOW THESE TERMS

In-line, V-type, Slant, Opposed, Cylinder numbers, Splayed crankshaft, Internally balanced, Externally balanced, Firing order, Liquid cooling, Air cooling, Gasoline engine, Diesel engine, Precombustion chamber, Carbureted engine, Injected engine, LPG, Spark ignition, Compression ignition, Pancake combustion chamber, Wedge combustion chamber, Hemi combustion chamber, Squish area, Semi-hemi, Swirl combustion chamber, Crossflow chamber, Two-valve chamber, Four-valve chamber, Air jet chamber, Stratified charge, L-head, I-head, Overhead valve, Overhead cam, SOHC, DOHC, Normally aspirated, Turbocharged, Variable displacement, Neutral balanced, Balancer shaft, Internal combustion, External combustion, Alternate engines, Two-stroke cycle engine, Wankel engine, Gas turbine, Stirling engine, Free piston engine, Regenerative braking, Hybrid power.

REVIEW QUESTIONS—CHAPTER 6

1. Explain the four common types of automotive engine cylinder arrangements.
2. Auto engines normally have ____, ____, or ____ cylinders.
3. ____________ ___________ identify the cylinders, pistons, and connecting rods of an engine.
4. How can you normally tell which cylinder is number one in a V-type engine.
5. Define the term "firing order."
6. An air cooling system is just as common as a liquid cooling system on automobiles. True or false?
7. Describe the major differences between a gasoline engine and a diesel engine.
8. A diesel engine is a compression ignition engine. True or false?
9. Which of the following is NOT a common combustion chamber shape?
   a. Wedge.
   b. Symmetrical.
   c. Hemi.
   d. Pancake.
10. How does swirl help combustion efficiency?
11. Explain the advantages of a four-valve combustion chamber.
12. What is a stratified charge combustion chamber?
13. In an ____________ or ____________ valve engine, both valves are located in the cylinder head.
14. A __________ engine has one camshaft in the cylinder head and a __________ engine has two camshafts per cylinder head.
15. Describe the operation of a rotary engine.

ASE CERTIFICATION-TYPE QUESTIONS

1. The term cylinder arrangement refers to the position of the cylinders in the engine block in relation to the __________.
   (A) camshaft
   (B) crankshaft
   (C) pistons
   (D) valves
2. All of the following are basic cylinder arrangements found in modern automobiles EXCEPT:
   (A) horizontally opposed.
   (B) V-type.
   (C) vertically opposed.
   (D) inline.
3. Technician A says that modern automotive engines can be classified by firing order. Technician B says that modern automotive engines can be classified by intake valve size. Who is right?
   (A) A only.
   (B) B only.
   (C) Both A & B.
   (D) Neither A nor B.
4. Technician A says that an air cooling system can maintain a constant engine temperature better than a liquid cooling system. Technician B says that a liquid cooling system can maintain a constant engine temperature better than an air cooling system. Who is right?
   (A) A only.
   (B) B only.
   (C) Both A & B.
   (D) Neither A nor B.
5. All of the following are examples of modern fuel systems EXCEPT:
   (A) multiport gasoline injection
   (B) throttle body gasoline injection
   (C) diesel multiport injection
   (D) throttle body diesel injection.
6. Technician A says that some gasoline engines use compression ignition to ignite the fuel in their combustion chambers. Technician B says that there are no gasoline engines that use compression ignition to ignite the fuel in their combustion chambers. Who is right?
   (A) A only.
   (B) B only.
   (C) Both A & B.
   (D) Neither A nor B.
7. All of the following are basic gasoline engine combustion chamber shapes EXCEPT:
   (A) pancake.
   (B) oval.
   (C) wedge.
   (D) hemispherical.
8. Technician A says that some automotive engine camshafts are belt-driven. Technician B says that chain and gear driven camshafts are the only types used on automotive engines. Who is right?
   (A) A only.
   (B) B only.
   (C) Both A & B.
   (D) Neither A nor B.
9. Technician A says that push rods are not needed to operate the rocker arms and valves in an overhead cam engine. Technician B says that push rods are used to operate the rocker arms and valves in an overhead cam engine. Who is right?
   (A) A only.
   (B) B only.
   (C) Both A & B.
   (D) Neither A nor B.
10. Technician A says that two-stroke cycle engines are not used in automobiles because they have poor power output at low speeds. Technician B says that two-stroke cycle engines are not used in automobiles because they are not as fuel efficient as a four-stroke cycle engine. Who is right?
   (A) A only.
   (B) B only.
   (C) Both A & B.
   (D) Neither A nor B.

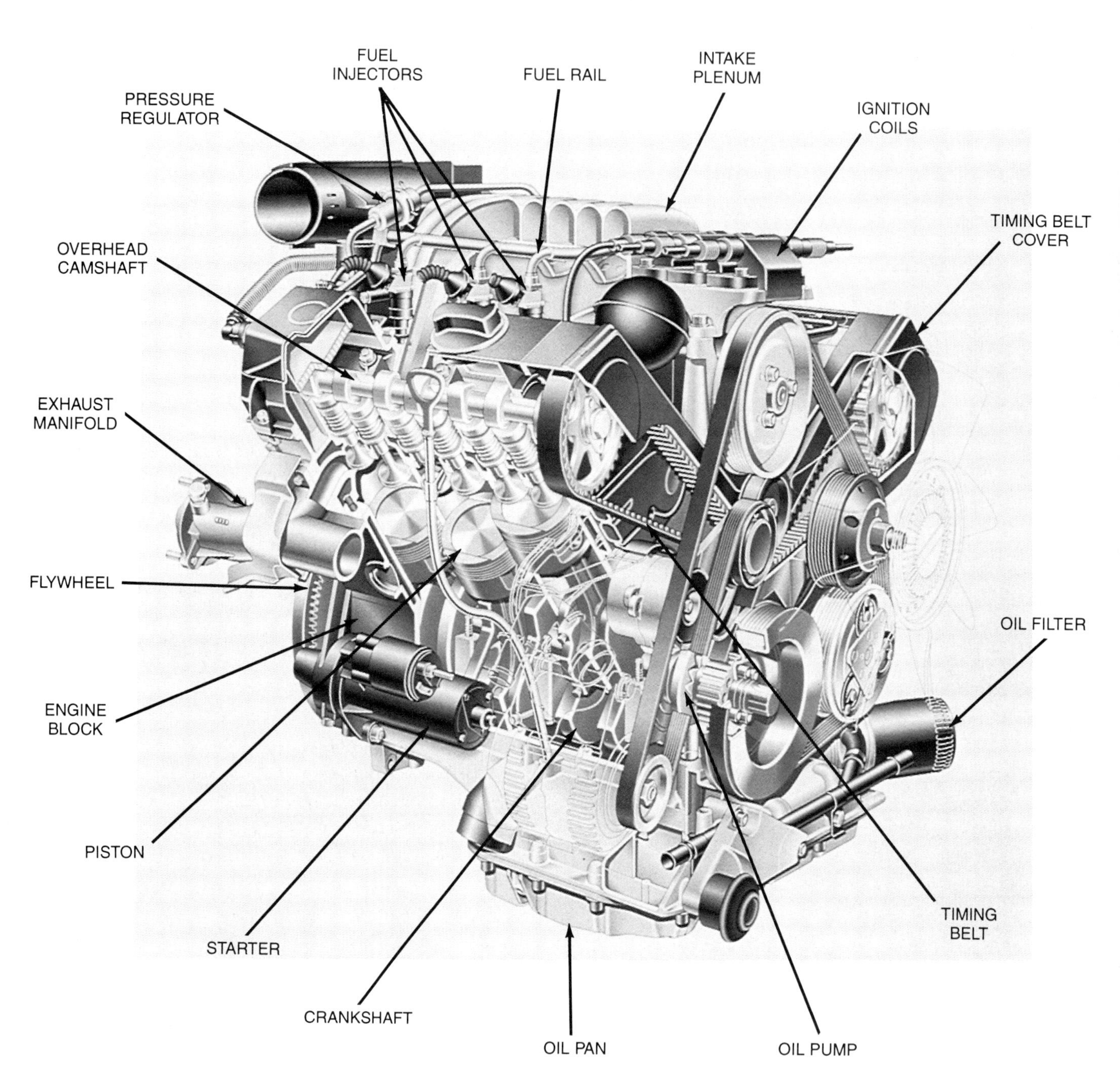

Study this cutaway view of this late model overhead camshaft engine. This engine uses belts to drive the camshafts. (Audi)

Chapter 7

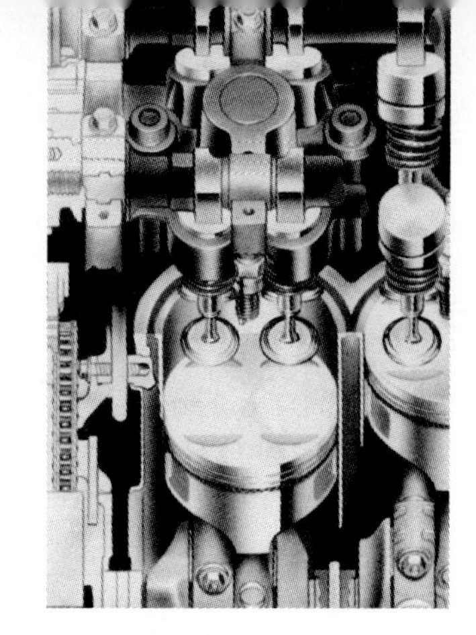

Engine Measurements

After studying this chapter, you will be able to:

- *Compare engine bore, stroke, and displacement measurements.*
- *Calculate CID.*
- *Explain bore/stroke ratio.*
- *Define the terms oversquare and undersquare engine.*
- *Describe why piston protrusion measurement is critical, especially with a diesel engine.*
- *Explain compression ratio measurements.*
- *Summarize rotational and angle measurement in degrees.*
- *Explain how atmospheric pressure affects engine performance.*
- *Define the term "volumetric efficiency."*
- *Describe horsepower and torque measurements.*
- *Explain the operation and use of engine and chassis dynamometers.*

This chapter will briefly discuss the many measurements or ratings given by engine manufacturers. Some give the physical size of the engine and others rate the power output of an engine.

To be a specialized engine technician, you should understand engine measurements. You might use this information in an auto shop when explaining something to a customer or when trying to analyze a problem. You could also continue your training into automotive engineering and find this subject matter even more essential.

ENGINE SIZE MEASUREMENTS

Engine size is determined by cylinder diameter, amount of piston travel per stroke, and number of cylinders. Any of these three variables can be changed to alter engine size. Engine size information is commonly used when ordering parts or when measuring wear during major engine repairs.

Bore and stroke

Cylinder bore is the diameter of the engine cylinder wall. It is measured across the cylinder, parallel with the top of the block. Cylinder bores in automotive engines vary in size from 3 to 4 in. (76 to 102 mm), Fig. 7-1.

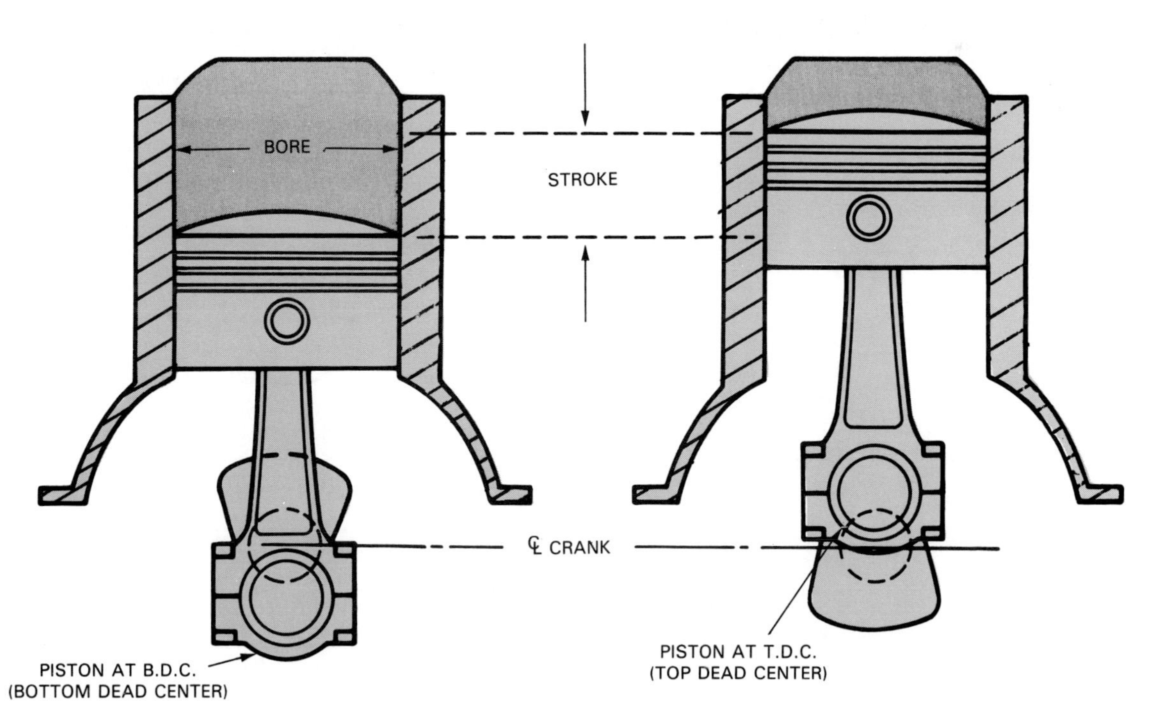

Fig. 7-1. Bore is diameter of cylinder. Stroke is distance piston travels up and down in cylinder. TDC stands for top dead center or upper point of piston travel. BDC stands for bottom dead center or piston at lowest point in stroke. (Ford)

Piston stroke is the distance the piston moves from TDC (top dead center) to BDC (bottom dead center). The amount of offset built into the crankshaft connecting rod journals or throws control piston stroke. Piston stroke also varies from about 3 to 4 in. (76 to 102 mm). Look again at Fig. 7-1.

A shop manual normally gives bore and stroke specs together. For instance, suppose a spec for bore and stroke is given as 4.00 in. × 3.00 in. This means that the engine cylinder is 4 inches in diameter and the piston stroke is 3 inches. Bore is the first value given and stroke the second.

Generally, a larger bore and stroke makes the engine more powerful. It can pull more fuel and air in on each intake stroke. Then, more pressure is exerted on the head of the piston during the power stroke.

Piston displacement

Piston displacement is the volume the piston displaces (moves) from BDC to TDC. It is determined by comparing cylinder diameter and piston stroke. A large cylinder diameter and large piston stroke would produce a larger piston displacement and vice versa.

The formula for finding piston displacement is:

$$\text{Piston Displacement} = \frac{\text{Bore squared} \times 3.14 \times \text{Stroke}}{4}$$

If an engine has a bore of 3 in. and a stroke of 3 in., what is its piston displacement?

$$\text{Piston Displacement} = \frac{3 \times 3 \times 3.14 \times 3}{4}$$

$$\text{Piston Displacement} = \frac{84.78}{4} = 21.195 \text{ cu./in.}$$

Engine displacement

Engine displacement or *engine size* equals piston displacement times the number of engine pistons or cylinders, Fig. 7-2. For example, if one piston in an engine displaces 22 cubic inches and the engine has six cylinders, what would the engine displacement be?

Engine displacement = piston displacement × no. of cylinders

Engine displacement = 22 × 6 = 132 CID

CID (cubic inch displacement), CC (cubic centimeters), and L (liters) are used to state engine displacement. For example, a V-8 engine might have a 350 CID. A V-6 could be a 3.3 L engine. A four-cylinder engine might have a displacement of 2 300 cc. Since one liter equals 1 000 cc, a 2 liter engine would have 2 000 ccs.

Engine displacement is usually matched to the weight of the car. A heavier car, truck, or van needs a larger engine that produces more power. A light, economy car only needs a small, low power engine for adequate acceleration.

As an engine technician, you might need to find engine displacement after boring an engine during a rebuild. Then you can tell the customer the new engine size or displacement. Use the formula for finding PISTON DISPLACEMENT and then the formula for ENGINE DISPLACEMENT.

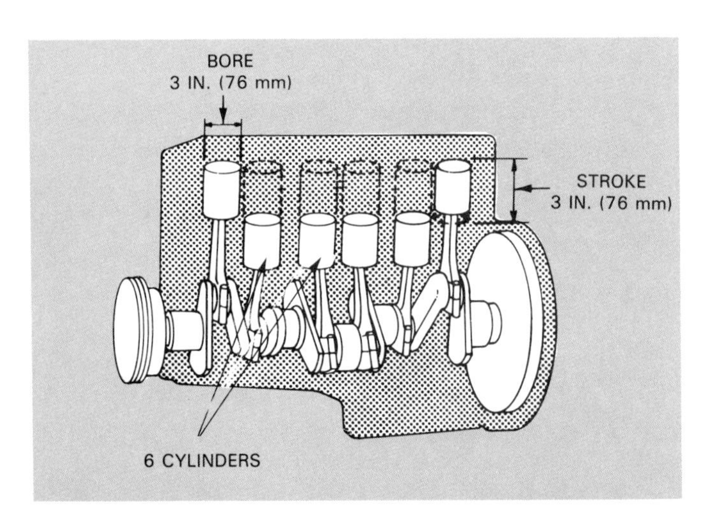

Fig. 7-2. To find engine displacement, find displacement for one cylinder and then multiply it by number of cylinders. (Deere & Co.)

Bore/stroke ratio

An engine's *bore/stroke ratio* is the relationship between the bore and the stroke. It is calculated by simply dividing the bore by the stroke, Fig. 7-3.

CID	LITERS	BORE & STROKE	B/S RATIO
140	2.3	3.78 x 3.13	1.20:1
153	2.5	3.68 x 3.30	1.11:1
177	2.9	3.66 x 2.83	1.29:1
183	3.0	3.50 x 3.14	1.11:1
232	3.8	3.81 x 3.39	1.12:1
245	4.0	3.94 x 3.31	1.19:1
281	4.6	3.55 x 3.54	1.01:1
302	5.0	4.00 x 3.00	1.33:1
351	5.8	4.00 x 3.50	1.14:1
460	7.5	4.36 x 3.85	1.13:1

Fig. 7-3. Chart shows cubic inch displacement, bore, stroke, and bore/stroke ratio. (Ford)

An *oversquare engine* has a bore that is larger than the stroke. The bore/stroke ratio would be larger than one. This is the most common engine design. A larger bore and shorter stroke allows for higher engine speeds, as is needed in an automotive engine.

An *undersquare engine* has a stroke that is larger than the bore. The bore/stroke ratio would be smaller than one. This is less common in cars. Large industrial and tractor engines are sometimes undersquare because they operate at low rpm. See Fig. 7-4.

A *square engine* has the same bore and stroke dimensions.

Piston protrusion

Piston protrusion is a measurement of how far the piston moves up in the block when at TDC. Frequently, the piston will actually protrude or stick out the top of

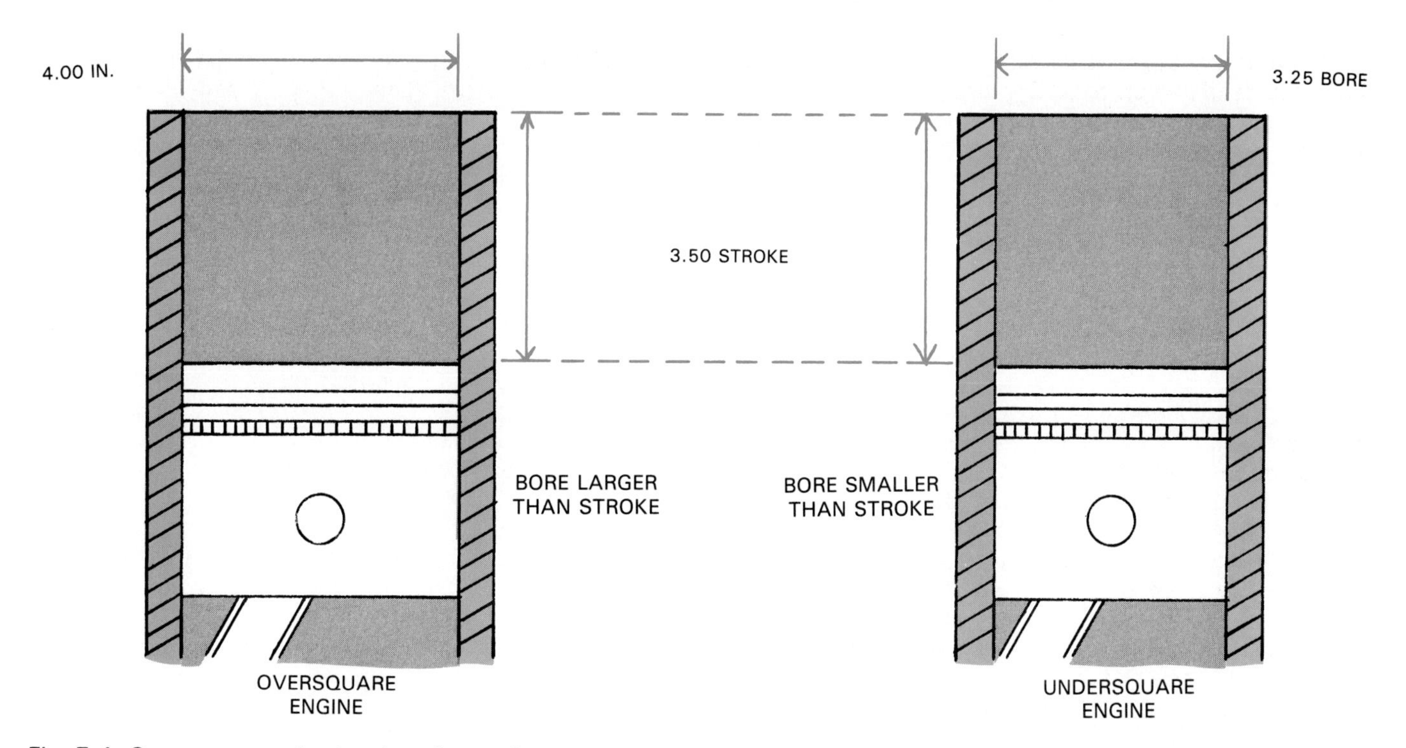

Fig. 7-4. Oversquare engine has bore larger than stroke. Undersquare engine has stroke larger than bore. Oversquare is more common.

the block, Fig. 7-5.

Piston protrusion is especially critical with diesel engines and their high compression ratio. As you will learn in later chapters, you must commonly measure piston protrusion to select the correct diesel head gasket thickness.

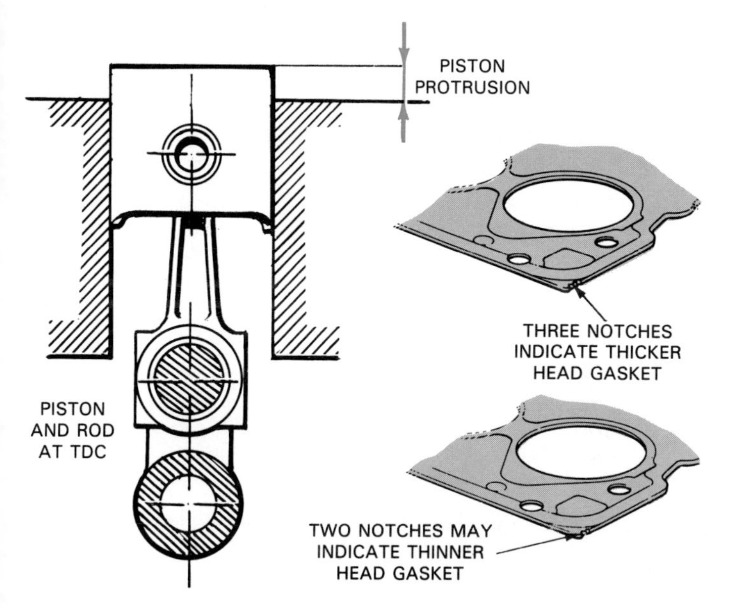

Fig. 7-5. Piston protrusion is comparison of piston head location at TDC and location of block deck. This is critical with a diesel because correct head gasket thickness must be used because of small clearance between cylinder head and piston head. (Peugeot)

FORCE, WORK, AND POWER

Force is a pushing or pulling action. When a spring is compressed, an outward movement or force is produced. Force is measured in pounds or newtons.

Work is done when force causes movement. If the compressed spring moves another engine part, work has been done. If the spring does NOT cause movement, no work has been done. Work is measured in foot-pounds or Newton meters (N•m).

The formula for work is:

$$\text{Work} = \text{distance moved} \times \text{force applied}$$

If you use a hoist to lift a 400 lb. engine 3 ft. in the air, how much work has been done?

$$\text{Work} = 3 \text{ ft.} \times 400 \text{ lb.}$$

$$= 1200 \text{ foot-pounds (ft.-lb.) of work completed.}$$

Power is the rate or speed at which work is done. It is measured in foot pounds per second or horsepower. The metric unit for power is the watt or kilowatt.

High power output can do a large amount of work. Low power can only do a small amount of work. The formula for power is:

$$\text{Power} = \frac{(\text{Distance} \times \text{Force})}{\text{Time in minutes}}$$

If an engine moves a 3000 lb. car 1000 ft. in one minute, how much power is needed?

$$\text{Power} = \frac{1000 \times 3000}{1}$$

$$= 3{,}000{,}000 \text{ foot-pounds per minute}$$

COMPRESSION RATIO

Engine *compression ratio* compares cylinder volumes with the piston at the extreme top and bottom of its travel. An engine's compression ratio controls how tight the air-fuel mixture (gas engine) or just air (diesel engine) is squeezed on the compression stroke, Fig. 7-6.

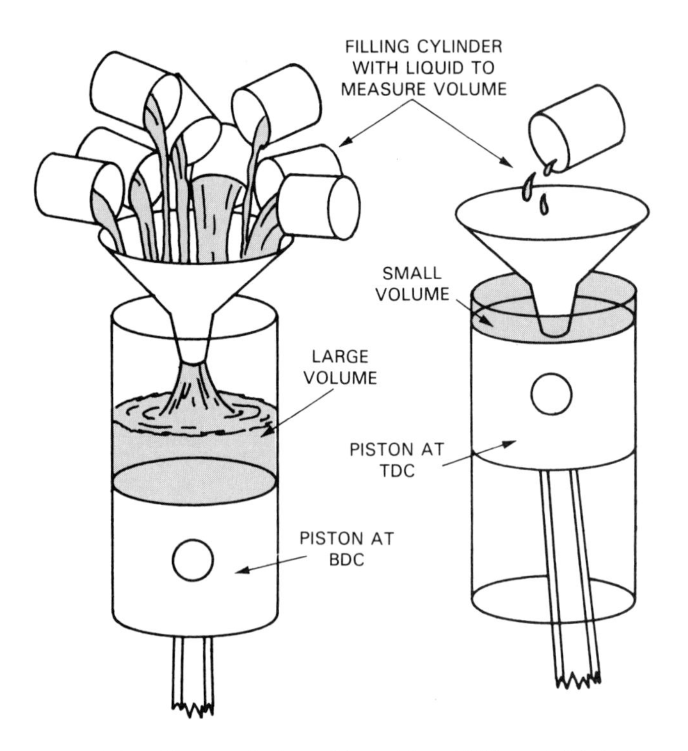

Fig. 7-6. Filling cylinder with fluid at TDC and BDC shows compression ratio of engine.

A compression ratio is given as two numbers. For example, an engine may have a compression ratio of 9:1 (9 to 1). This means the maximum cylinder volume is nine times as large as the minimum cylinder volume. At BDC, a cylinder has maximum volume. At TDC, a cylinder has minimum volume.

When a gasoline engine piston is at BDC, the cylinder volume might be 40 cubic inches (0.65 L). When the piston slides to TDC, the volume may reduce to five cubic inches (0.08 L). Dividing 40 by 5 (0.65 by 0.08), the compression ratio for this engine would be 8:1.

With the diesel (compression ignition engine), BDC cylinder volume is 17 to 20 times as large as TDC cylinder volume. The compression ratio would then be 17:1 to 20:1.

Fig. 7-7 compares gasoline and diesel engine compression ratios.

Older gasoline engines designed for leaded gasoline had higher compression ratios. Up to a point, a high compression ratio increases engine fuel efficiency and power. However, it also causes higher exhaust emission levels.

Today's gasoline engines use a lower compression ratio (about 8 or 9 to 1). This allows the use of cleaner burning, unleaded fuel. There is a slight reduction in engine power and efficiency, however.

Diesel engines have a very high compression ratio. The diesel compresses the air in the cylinder tight enough for it to heat up and ignite the fuel. Automotive fuels and combustion will be explained in detail in the next chapter.

Compression pressure

Compression pressure is the amount of pressure produced in the engine cylinder on the compression stroke. Compression pressure is normally measured in pounds per square inch (psi) or Kilopascals (kPa).

A gasoline engine may have compression pressure

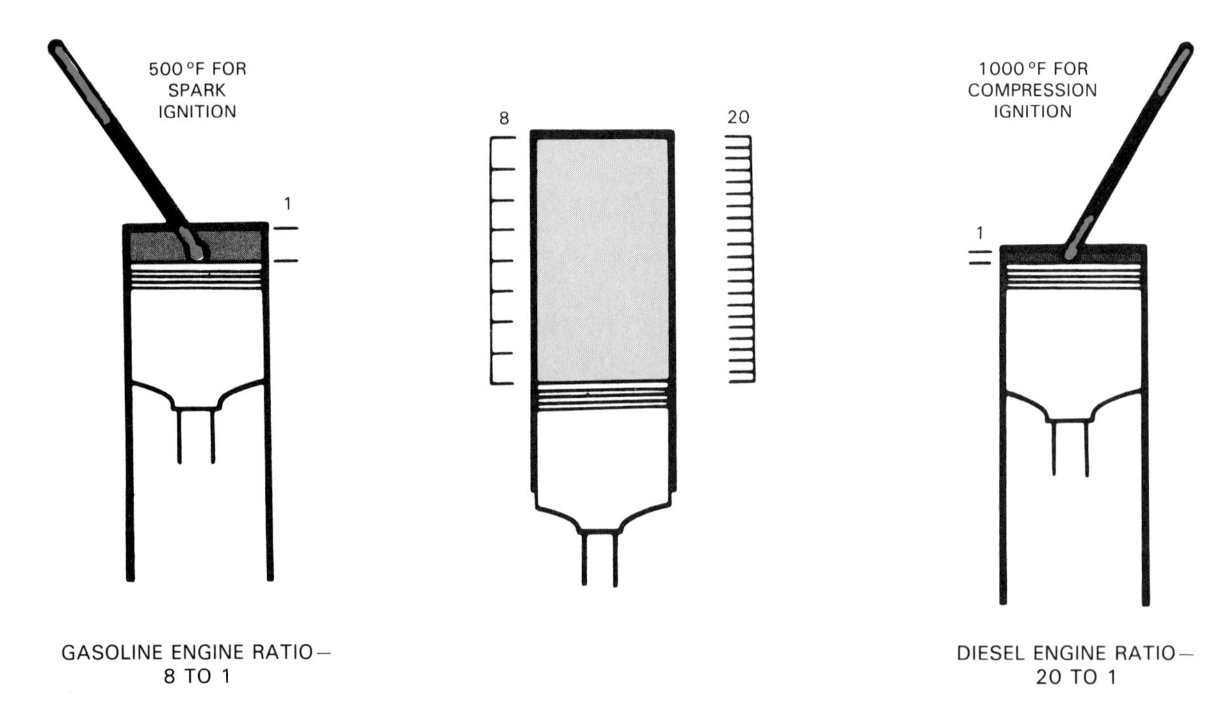

Fig. 7-7. Diesel has much higher compression ratio than gas engine. It must squeeze air until hot enough to ignite fuel. (Chrysler)

from 130 to 180 psi (896 to 1 240 kPa). A diesel engine has a much HIGHER compression pressure of about 250 to 400 psi (1 723 to 2 756 kPa).

A *compression gauge* is used to measure compression stroke pressure. It is screwed into the spark plug or glow plug hole. The ignition or injection system is disabled. Then, the engine is cranked over with the starting motor. The gauge will read compression stroke pressure.

Discussed in later chapters, compression stroke pressure is an indicator of engine condition. If low, there is a problem allowing air to leak out of the cylinder. The engine might have bad rings, burned valves, or a blown head gasket.

ROTATION OR ANGLE MEASUREMENT

Part *rotation* or *angle measurements* are made in degrees. A circle is divided into 360 equal parts or degrees. This is shown in Fig. 7-8. One-half of a complete rotation would equal 180 deg. One-fourth of a revolution would equal 90 deg. and one-eighth 45 deg.

This measuring technique is commonly used in engine repair. For example, crankshaft rotation for ignition or injection timing adjustment is given in degrees. Throttle plate angle adjustment can also be given in degrees.

ATMOSPHERIC PRESSURE

Atmospheric pressure is air pressure produced by the weight of the air above the surface of the earth. It is measured in pounds per square inch (psi) or kilopascals (kPa).

At sea level, atmospheric pressure is 14.7 psi (101 kPa). As altitude above sea level increases, atmospheric pressure decreases because there is less air above, pushing down, Fig. 7-9.

Atmospheric pressure pushes equally in all directions, Fig. 7-10. As a result, an engine uses atmospheric pressure to force the air-fuel mixture or air into the cylinders.

VACUUM

A *vacuum* is an area of negative pressure or pressure below that of normal atmospheric pressure. For example, when the engine piston slides down, it forms a vacuum. Then, outside air pressure can rush in to fill the cylinder on the intake stroke.

Vacuum is measured in inches of mercury (in./Hg.) or kilograms per square centimeter (kg/cm^2). A vacuum gauge, discussed in Chapter 2, is commonly used to measure intake manifold vacuum. Manifold vacuum can be used to help diagnose several engine problems, as will be explained in later chapters.

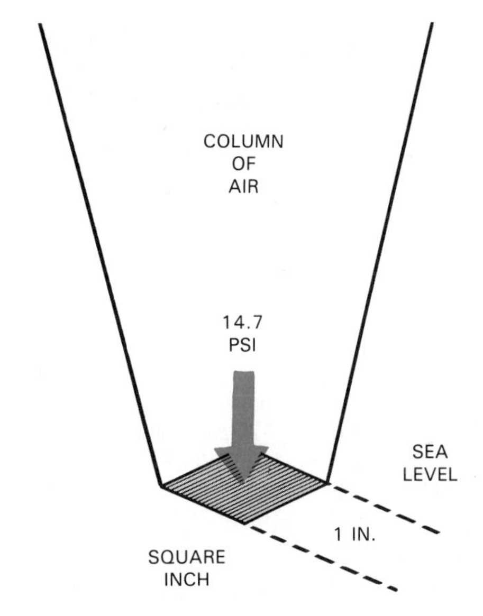

Fig. 7-9. Imaginary column of air over earth produces atmospheric pressure of 14.7 psi at sea level. (Ford)

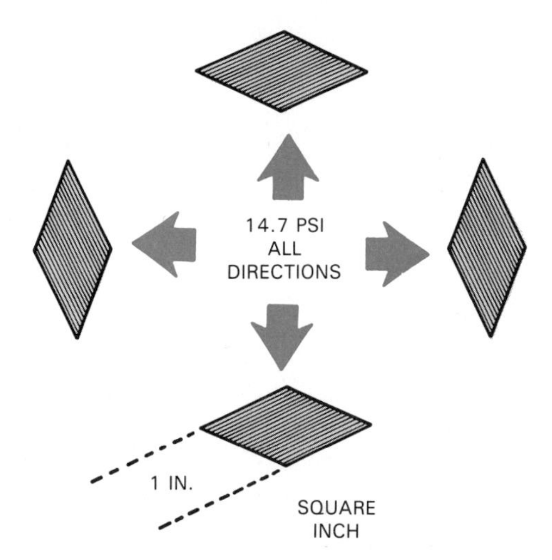

Fig. 7-10. Atmospheric pressure pushes equally in all directions and causes flow into engine cylinders on intake strokes. (Ford)

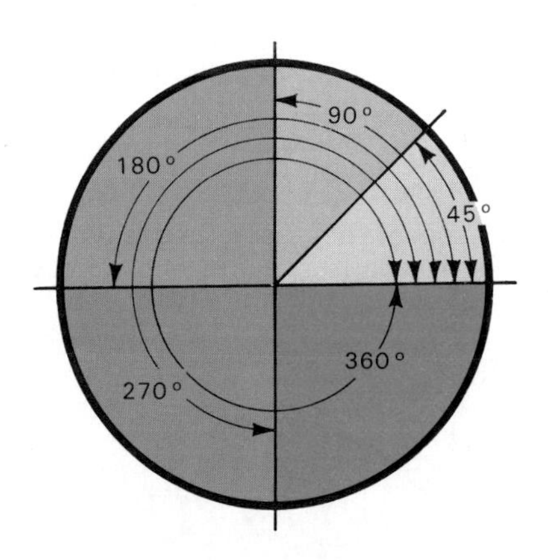

Fig. 7-8. Note how angles or rotation are measured in degrees. Crank rotation is one example of degree measurement.

ENGINE TORQUE

Engine torque is a rating of the turning force at the engine crankshaft. When combustion pressure pushes the piston down, a strong rotating force is applied to the crank. This turning force is sent to the transmission or transaxle, drive line or drive axles, and drive wheels, propelling the car.

Engine torque specs are given in a shop manual. One example, 78 lb.-ft. @3000 (at 3000) rpm is given for one particular engine. This engine would be capable of producing a maximum of 78 pound-feet of torque when operating at 3000 revolutions per minute. In metrics, engine torque is often stated in newton-meters (N•m).

HORSEPOWER

Horsepower, abbreviated hp, is a measure of an engine's ability to perform work. At one time, one horsepower was the average strength of a horse, Fig. 7-11. A 300 hp engine could, theoretically, do the work of 300 horses.

In metrics, horsepower is measured in kilowatts. Kilowatts is abbreviated kW.

One horsepower equals 33,000 ft.-lb. of work per minute. To find engine hp, use the following formula:

$$hp = \frac{\text{Distance (ft.)} \times \text{Weight (lb.)}}{33{,}000 \times \text{Time in minutes}}$$

or

$$\frac{\text{Work (ft.-lb.)}}{33{,}000 \times \text{Minutes}}$$

For a small engine to lift 500 pounds a distance of 700 feet in one minute, about how much horsepower would be needed? To calculate, apply these values to the formula:

$$hp = \frac{\text{500 Pounds} \times \text{700 Feet}}{33{,}000 \times \text{1 Minute}}$$

$$= \text{10.6 Horsepower}$$

ENGINE HORSEPOWER RATINGS

Automobile makers rate engine hp output at a specific engine rpm. For instance, a high performance, turbocharged engine might be rated at 300 hp at 5000 rpm. This engine power rating is normally stated in a service manual. There are several different methods of calculating engine horsepower.

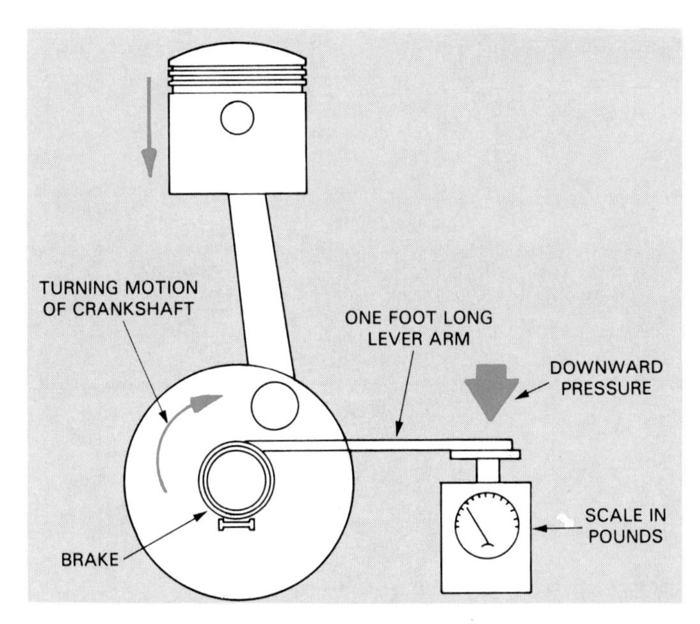

Fig. 7-12. Prony brake shows how engine horsepower is theoretically measured.

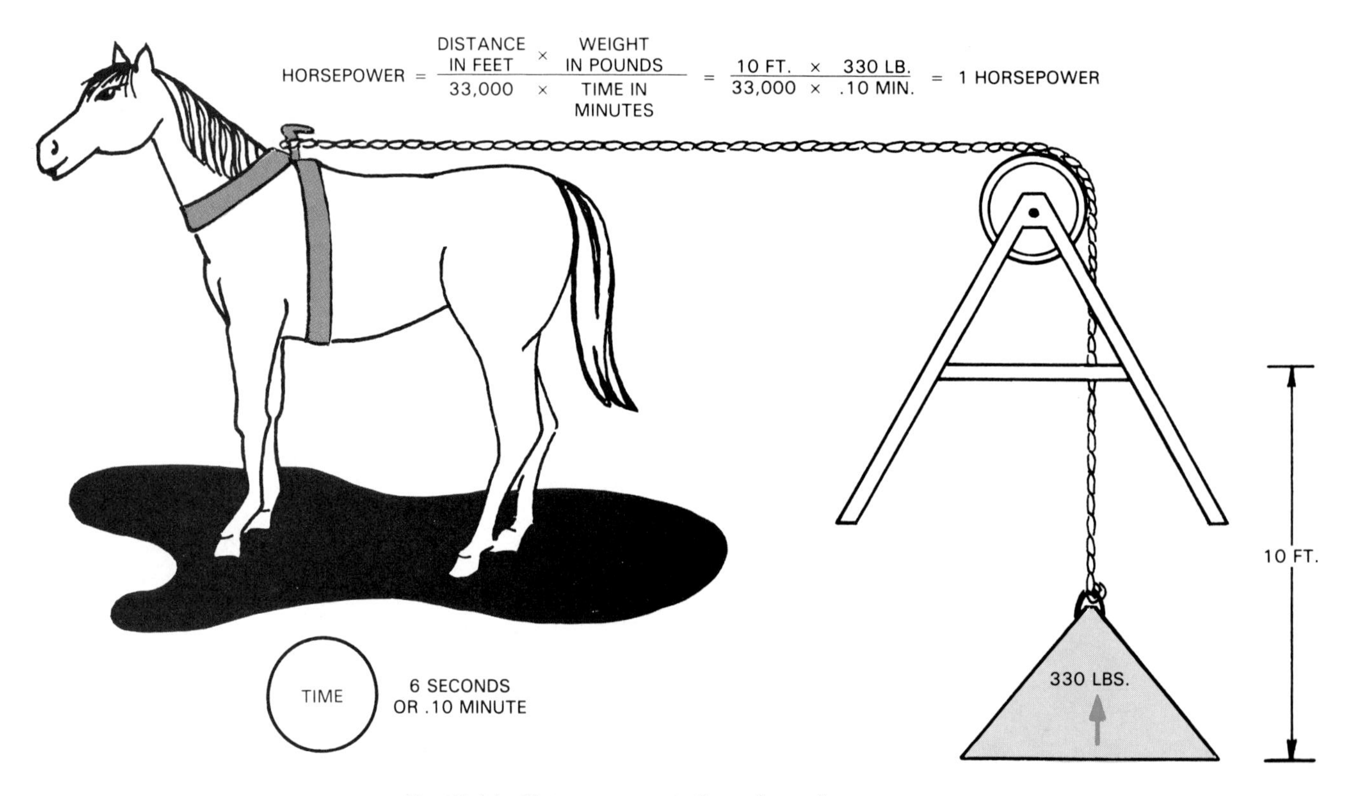

Fig. 7-11. Note representation of one horsepower.

Brake horsepower

Brake horsepower (bhp) measures the usable hp at the engine crankshaft. Shown in Fig. 7-12, a prony brake was first used to measure brake hp. The engine crank spun the prony brake while the braking mechanism was applied. The resulting amount of pointer deflection could then be used to find brake hp.

Dynamometers

An *engine dynamometer* (dyno) is now used to measure the brake hp of modern car engines, Fig. 7-13. It functions in much the same way as a prony brake. Either an electric motor or fluid coupling is used to place a drag on the engine crankshaft. Then, power output can be determined.

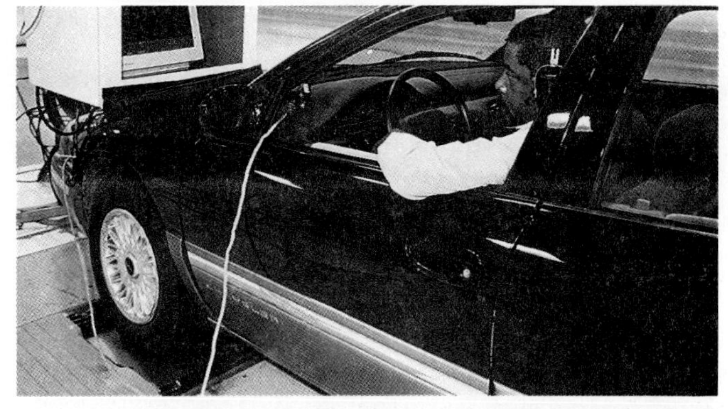

Fig. 7-14. A chassis dynamometer is used to determine the horsepower available to power the vehicle. This rating is sometimes referred to as net horsepower. (Chrysler)

Fig. 7-13. Engine dynamometer or dyno is used to measure actual engine brake horsepower. This engine is being tested before installation into a car. (Oldsmobile)

A *chassis dynamometer* measures the horsepower delivered to the car's drive wheels. See Fig. 7-14. It indicates the amount of horsepower available to propel the car. Fig. 7-15 illustrates chassis dyno operation.

Indicated horsepower

Indicated horsepower (ihp) refers to the amount of power or pressure formed in the engine combustion chambers. A special pressure sensing device is placed in the cylinder. The pressure readings are used to find indicated hp.

Frictional horsepower

Frictional horsepower (fhp) is the power needed to overcome engine friction. It is a measure of the resistance to movement between engine parts.

Frictional hp is POWER LOST to friction. It reduces the amount of power left to propel the car.

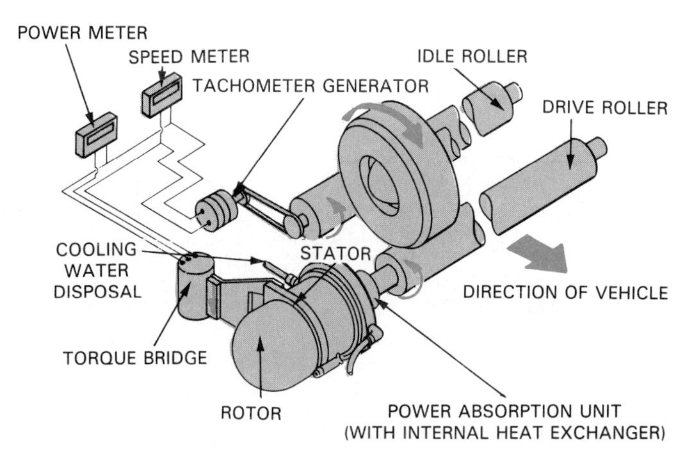

Fig. 7-15. Study operating principles of a chassis dynamometer. (Clayton)

Net horsepower

Net horsepower is the maximum power developed when an engine is loaded down with all accessories (alternator, water pump, fuel pump, air injection pump, air conditioning, and power steering pump). It is the amount of useful power with the engine installed in the vehicle. Net hp indicates the amount of power available to move the car. See Figs. 7-16 and 7-17.

Gross horsepower

Gross horsepower (ghp) is similar to Net hp, but it is the engine power available with only basic accessories installed (alternator, water pump).

Gross hp does NOT include the power lost to the power steering pump, air injection pump, air conditioning compressor, or other extra units. For the same engine, Gross hp is higher than Net hp.

Fig. 7-16. This prototype DOHC V-6 engine features variable valve timing and lift electronic control system. (Acura)

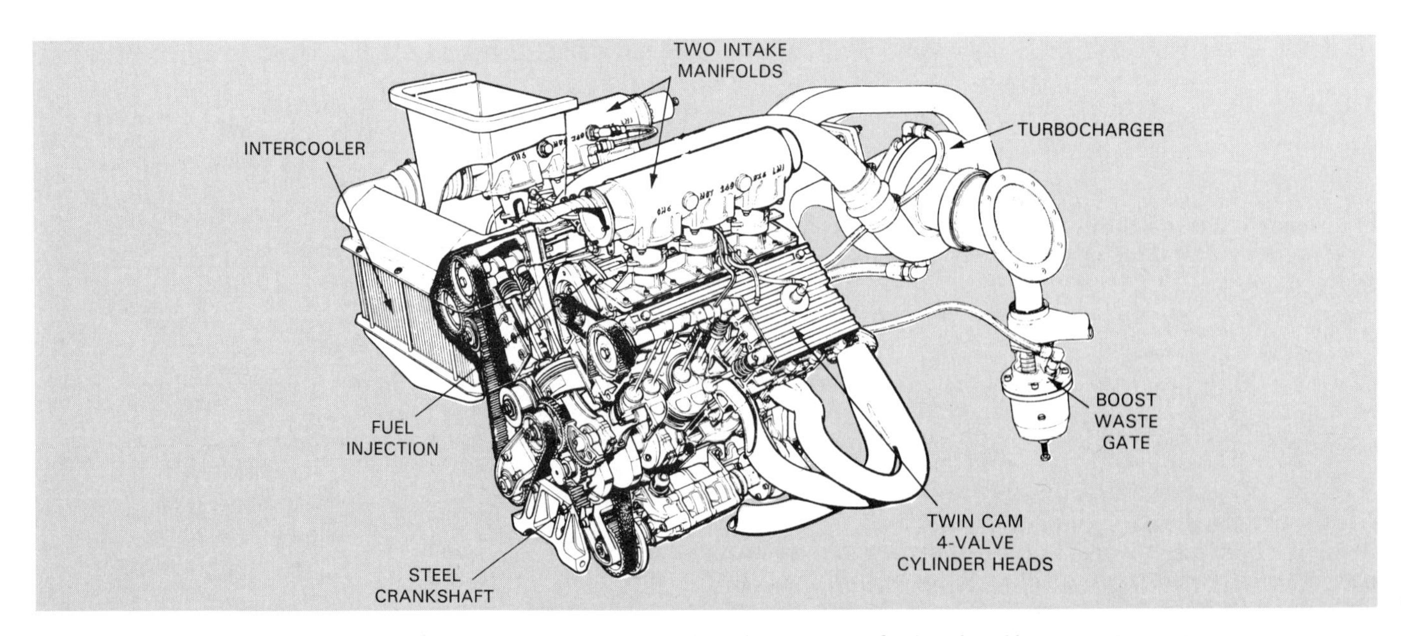

Fig. 7-17. This high performance engine will produce tremendous horsepower for its size. However, it would not be practical in a passenger car because its low speed efficiency is poor.

Taxable horsepower

Taxable horsepower (thp) is simply a general rating of engine size. In many states, it is used to find the tax placed on a car. The formula for taxable horsepower is:

thp = Cylinder Bore × Cylinder Bore × Number of cylinders × .4

Horsepower and torque curves

Horsepower and *torque curves* are used to show how engine horsepower and torque change with engine rpm. Refer to Fig. 7-18.

In a passenger car, it is important that an engine makes adequate torque and horsepower over a wide range of engine speeds. Then, the car will accelerate properly in all transmission gears.

A racing engine or performance engine is usually designed to make maximum power at high rpm. It does NOT operate well at lower rpm. Low speed power is not important to this type engine.

Horsepower and torque curves will show the operating characteristics of an engine. A passenger car engine would have more of a level, gradual curve. A performance engine would generally have a steeper curve that peaked at a higher rpm.

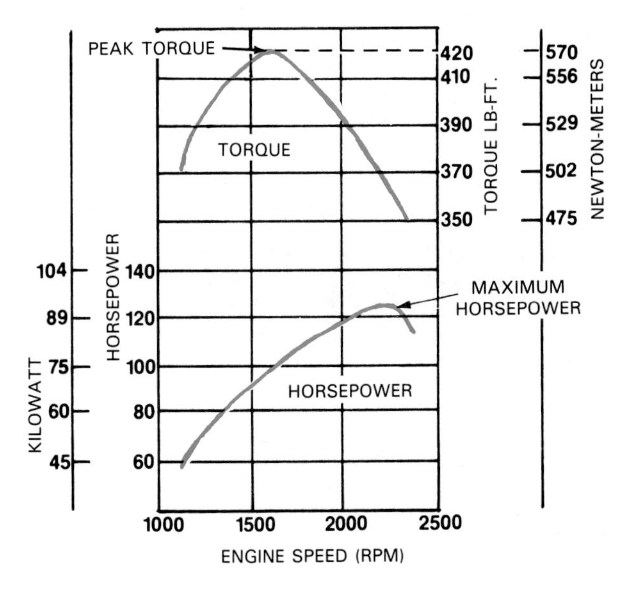

Fig. 7-18. Torque and horsepower curves show operating characteristics of an engine. What is this engine's torque and horsepower at 2000 rpm?

ENGINE EFFICIENCY

Engine efficiency is the ratio of power produced by the engine and the power supplied to the engine (heat content of fuel). By comparing fuel consumption to engine power output, you can find engine efficiency.

If all of the heat energy in the fuel were converted into useful work, an engine would be 100 percent efficient. This much efficiency is not possible with a piston engine, however. Modern piston engines are only about 25 percent efficient.

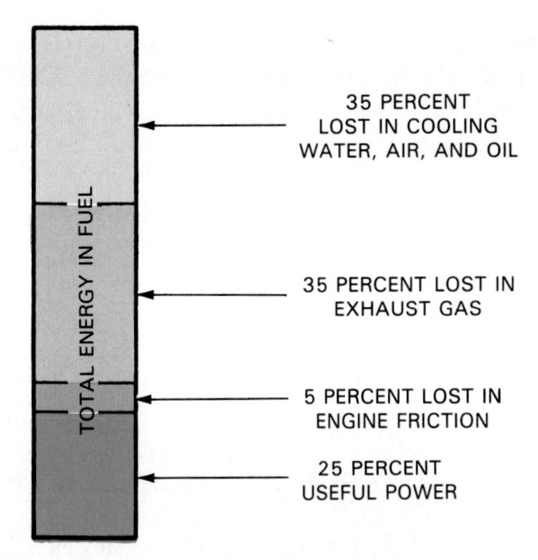

Fig. 7-19. A piston engine is not very efficient. Note that only about 25 percent of the fuel energy is used. (Ford)

Fig. 7-19 illustrates how the heat energy of the fuel is used by a piston engine. About 70 percent of the fuel's heat energy enters the cooling and exhaust systems. About 5 percent is lost to friction. This leaves little heat energy to produce heat expansion and pressure on the engine pistons.

Volumetric efficiency

Volumetric efficiency is the ratio of actual air drawn into the cylinder and the maximum possible amount of air that could enter the cylinder. It refers to how well an engine can "breathe" on its intake stroke. Volumetric efficiency is illustrated in Fig. 7-20.

If volumetric efficiency were 100 percent, the cylinder would completely fill with air on the intake stroke.

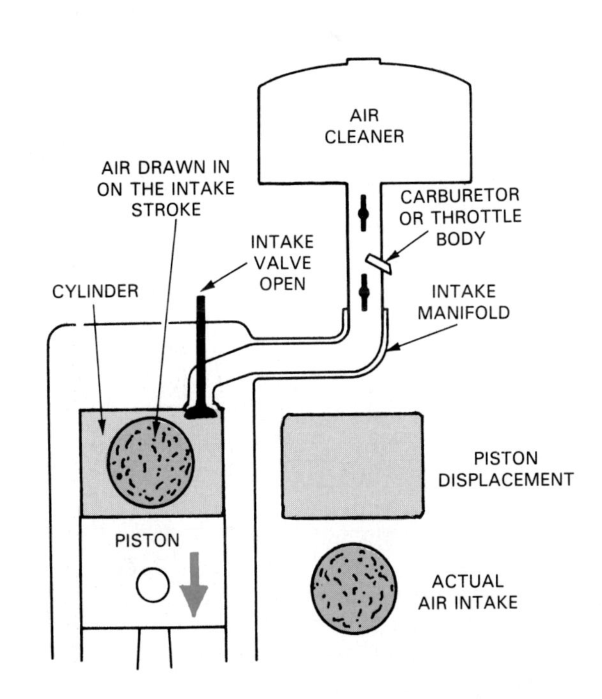

Fig. 7-20. Volumetric efficiency is comparison of actual air intake and theoretical potential for air intake. (Deere & Co.)

Engines are only capable of about 80 to 90 percent volumetric efficiency. Airflow restriction in the ports and around the valves limit airflow.

High volumetric efficiency INCREASES ENGINE POWER because more fuel and air can be burned in the combustion chambers. Turbocharging increases volumetric efficiency because air is forced into the engine under pressure.

Volumetric Efficiency =

$$\frac{\text{Volume of air taken into cylinder}}{\text{Maximum possible volume in cylinder}}$$

Mechanical efficiency

Mechanical efficiency compares brake hp and indicated hp. It is a measurement of mechanical friction. Remember that ihp equals theoretical power produced by combustion. Brake hp is the actual power at the engine crankshaft. The difference between the two is due to frictional losses.

Mechanical efficiency of around 75 to 95 percent is normal. This means that about 5 to 25 percent of the engine's power is lost to friction (frictional hp loss). The friction between the piston rings and cylinder walls accounts for most of this loss.

Thermal efficiency

Thermal efficiency is heat efficiency found by comparing fuel burned and horsepower output. It indicates how well an engine can use the fuel's heat energy. Thermal efficiency measures the amount of heat energy converted into crankshaft rotation.

A gallon of gasoline, for example, has about 19,000 Btu (British thermal units) of heat energy. One horsepower equals about 42.4 Btu of heat energy per minute. With this information, you can find engine thermal efficiency.

Thermal efficiency =

$$\frac{\text{Brake hp} \times 42.2 \text{ Btu minute}}{19{,}000 \text{ Btu/gallon} \times \text{gallons per minute}}$$

Generally, engine thermal efficiency is around 20 to 30 percent. The rest of the heat energy is absorbed into the metal parts of the engine, blown out the exhaust, or lost to friction.

SUMMARY

Engine size is determined by cylinder bore, piston stroke, and number of cylinders. Cylinder bore is the diameter of the cylinder. Stroke is piston movement from TDC to BDC and is controlled by crank rod journal offset.

Piston displacement is the volume removed as the piston moves up in the cylinder. Engine displacement adds all of the cylinders. CID stands for cubic inch displacement. CID is usually matched to the weight of the car.

An oversquare engine has a bore larger than its stroke. An undersquare engine is just the opposite.

Piston protrusion is a measurement of how far the piston extends up in the block at TDC. It is a critical measurement on diesel engines with their high compression ratio.

Compression ratio is a comparison of cylinder volumes with the piston at TDC and BDC. A gasoline engine typically has a compression ratio of about 8 or 9:1. A diesel has a compression ratio of about 17:1 because it is a compression ignition engine.

Compression pressure is a result of the upward piston movement on the compression stroke—the higher the compression ratio, the higher the compression pressure.

Atmospheric pressure is the air pressure all around us. It is used to fill the engine cylinders on the intake stroke. Vacuum is the opposite; it is pressure below atmospheric.

Engine torque is a rating of the turning force at the engine crankshaft. An engine with high torque will maintain rpm more when a load is placed on the engine. With a low torque spec, engine rpm would drop with a load or drag.

Engine horsepower is a measurement of the engine's ability to perform work. There are several horsepower ratings. Brake hp measures the usable power at the engine crankshaft and it is measured on a prony brake or engine dynamometer. A chassis dyno measures brake hp at the car's drive wheels.

Indicated hp refers to the amount of power or pressure formed in the combustion chambers. Frictional hp is the power needed to overcome engine friction. Net hp is the maximum power left when an engine is loaded with all of its accessory units. Taxable hp is a rating based on engine size. Curves can be used to show how hp and torque change with engine rpm.

Engine efficiency is the ratio of power produced by the engine and the power supplied to the engine in the form of heat energy in the fuel. Volumetric efficiency is the ratio of actual air intake and potential air intake. It rates the engine's "breathing ability." Mechanical efficiency compares brake hp and indicated hp or it rates power loss to friction. Thermal efficiency is heat efficiency found by comparing fuel used with horsepower output.

KNOW THESE TERMS

Cylinder bore, Piston stroke, Piston displacement, Engine displacement, CID, cc, L, Bore/stroke ratio, Oversquare, Undersquare, Square, Piston protrusion, Force, Work, Power, Compression ratio, Compression pressure, Atmospheric pressure, Vacuum, Torque, Engine torque, Horsepower, Brake hp, Dyno, Chassis dyno, ihp, fhp, Net hp, Gross hp, Taxable hp, Horsepower and torque curves, Engine efficiency, Volumetric efficiency, Mechanical efficiency, Thermal efficiency.

REVIEW QUESTIONS—CHAPTER 7

1. Engine size is determined by which of the following?
   a. Main journal offset.
   b. Rod journal offset.
   c. Cylinder bore.
   d. Piston stroke.
2. How do you measure cylinder bore?
3. Which of the following is NOT a possible bore and stroke for an automobile engine?
   a. 4.00 × 3.00
   b. 4.23 × 3.25
   c. 6.50 × 5.50
   d. 3.50 × 3.00
4. If an engine has a bore of 4 inches and a stroke of

3 inches, what is its piston displacement?
5. If the engine in the previous question has 6 cylinders, what is its displacement?
6. If you were to bore the cylinders .030 in. oversize in the engine in questions 5 and 6, how much would you increase its displacement or size?
7. A 3.5 × 3.0 bore/stroke ratio would be an example of _________ _________.
8. Why is diesel piston protrusion critical?
9. When an engine is rated with a compression ratio of 8.5:1, what does that mean?
10. Compression pressure for a gasoline engine might be ______ to ______ psi (______ to ______ kPa). A diesel might have ______ to ______ psi (______ to ______ kPa).
11. Atmospheric pressure 2000 feet above sea level would be 14.9 psi. True or false?
12. Vacuum is measured in ______ ______ ______ or ______ ______ ______ ______.
13. What is the difference between an engine dyno and a chassis dyno?
14. Explain four engine horsepower ratings.
15. _________ efficiency is a measurement of frictional losses in an engine.

ASE CERTIFICATION-TYPE QUESTIONS

1. Technician A says that one of the factors that determines an automotive engine's size is the amount of piston travel per stroke. Technician B says that one of the factors that determines an automotive engine's size is cylinder diameter. Who is right?
   (A) A only.
   (B) B only.
   (C) Both A & B.
   (D) Neither A nor B.
2. If one piston in a V-8 engine displaces 30 cubic inches, the engine displacement would be _________.
   (A) 180 CID
   (B) 240 CID
   (C) 248 CID
   (D) 3.75 CID
3. Technician A says that the term piston protrusion is a measurement of how far the piston travels downward in the cylinder. Technician B says that piston protrusion refers to the distance the piston moves up in the block when at TDC. Who is right?
   (A) A only.
   (B) B only.
   (C) Both A & B.
   (D) Neither A nor B.
4. If an engine moves a 2000 lb. car 500 ft. in one minute, how much power is needed?
   (A) 1,000,000 foot-pounds per minute.
   (B) 2,000,000 foot-pounds per minute.
   (C) 2500 foot-pounds per minute.
   (D) None of the above.
5. A cylinder's compression pressure is normally measured in _________.
   (A) foot-pounds
   (B) Newton-meters
   (C) pounds per square inch
   (D) None of the above.
6. When a particular engine piston is at BDC, the cylinder volume is 30 cubic inches. When this piston reaches TDC, the cylinder volume is 5 cubic inches. What is the compression ratio of this engine?
   (A) 6:1.
   (B) 5:1.
   (C) 150:1.
   (D) None of the above.
7. Technician A says that as altitude above sea level increases, atmospheric pressure increases. Technician B says that as altitude above sea level increases, atmospheric pressure decreases. Who is right?
   (A) A only.
   (B) B only.
   (C) Both A & B.
   (D) Neither A nor B.
8. Engine torque is a rating of the turning force at the engine's _________.
   (A) camshaft
   (B) connecting rods
   (C) crankshaft
   (D) flywheel
9. If a small engine lifts 600 lb. a distance of 1200 ft. in one minute, approximately how much horsepower is needed?
   (A) 6.5 horsepower.
   (B) 21.8 horsepower.
   (C) 218 horsepower.
   (D) 65 horsepower.
10. Technician A says that turbocharging increases an engine's volumetric efficiency. Technician B says that turbocharging decreases an engine's volumetric efficiency. Who is right?
    (A) A only.
    (B) B only.
    (C) Both A & B.
    (D) Neither A nor B.

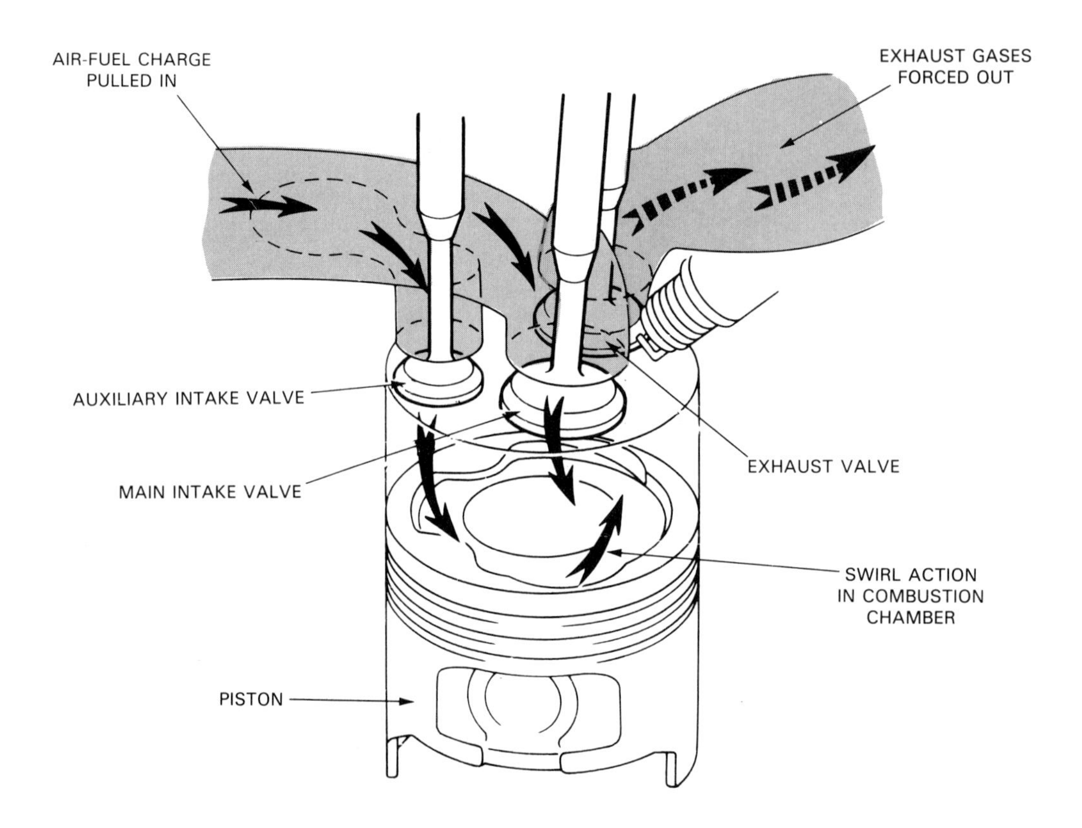

This engine design uses three valves per cylinder—two intake valves and one exhaust valve. Twin intake ports and intake valves reduce resistance to flow into combustion chamber to increase volumetric efficiency. Flow from both intake valves also increases swirl and mixing for better combustion. One exhaust valve can handle flow out of cylinder since piston pushes on and pressurizes gases. (Toyota)

Chapter 8

Engine Combustion, Fuels

After studying this chapter, you will be able to:
- *Summarize how crude oil is converted into gasoline, diesel fuel, and other fuels.*
- *Explain gasoline octane ratings.*
- *Describe what happens during normal and abnormal combustion.*
- *Explain the difference between preignition and detonation.*
- *Summarize the damage that can be caused by detonation.*
- *List the factors contributing to preignition and detonation.*
- *Describe the operation of a knock sensing system.*
- *Explain dieseling and spark knock.*
- *Summarize the factors that affect engine combustion.*
- *Describe the rating and operating characteristics of diesel fuel.*
- *Explain normal and abnormal diesel combustion.*
- *Summarize the use of alternate fuels for automotive engines.*

This chapter will help you grasp what is happening inside an automotive engine when its fuel burns and produces pressure on the pistons. It is a theoretical chapter that can help you when trying to diagnose dozens of engine performance problems. If you can understand what is occurring inside an engine, you will be better prepared to troubleshoot internal engine problems.

As you will learn, improper combustion or a fuel-related problem can reduce engine efficiency or even cause severe engine damage. As an engine technician, you must be able to analyze all types of troubles. Study this chapter carefully!

PETROLEUM

Crude oil or *petroleum* is oil in its natural state. It is a mixture of semisolids, liquids, and gases. Crude oil must be heated and broken down into lighter liquids and gases such as: gasoline, diesel fuel, and LP gas.

Crude oil is largely made up of flammable hydrocarbons. *Hydrocarbons* are chemical mixtures of hydrogen (light gas vapor) and carbon (heavy black solid).

The search for oil deposits is known as *exploration.* The oil may lie over ten miles below the earth's surface. Various tests are used to locate it.

MAKING ENGINE FUELS

Refining breaks crude oil down into different substances. Refining begins with *distillation.* This process uses heat, pressure, and a huge device called a ''fractionating tower.'' This enables the crude oil to be separated into natural gas, gasoline, motor oil, asphalt, and other products.

The heated petroleum enters the base of the fractionating tower where they separate by weight or thickness, Fig. 8-1. The temperature of the *fractionating tower* or *pipe still* is hottest near the bottom and becomes progressively cooler at higher levels. When the hot oil enters the tower, it instantly ''flashes'' or vaporizes and begins to rise in the tower. As the oil vapors rise, they condense (return to a liquid) and settle out on the trays in the tower by weight. The lighter fractions are able to rise to the higher trays in the tower.

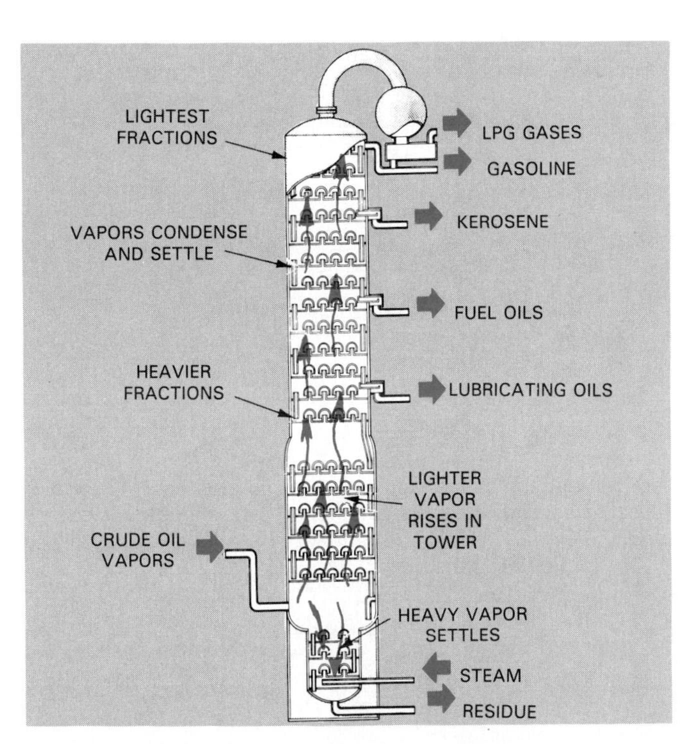

Fig. 8-1. Fractionating tower or pipe still is needed to break crude oil into its parts. Heavier fractions, like motor oil, settle on lower trays. Lighter fractions, like gasoline, settle on upper trays and flow out of tower. (Ford)

OCTANE RATINGS

The *octane rating* of a gasoline indicates its ability to resist knocking or pinging. For example, if the fuel mixture ignites too soon (excessive spark advance), the burning and rapidly expanding gases may actually try to force the piston backwards (against normal direction of piston movement). This violent action of the piston slamming against the expanding gases can cause a loud knocking noise.

In general, the higher the fuel's octane-rating or number, the greater the amount of heat and compression the gasoline can withstand before abnormal combustion or knocking occurs.

Gasoline grades

Gasoline is sold in two general grades—PREMIUM and UNLEADED. Naturally, premium fuel, often called high-test, super-test, or ethyl, has the highest octane rating. It would have the higher antiknock number and should be used in high compression engines. For example, if an engine knocks or pings on one fuel, a fuel with a higher octane rating may be needed.

It is very important to use fuel with the manufacturer's specified octane rating. There are many engine design variables (compression, combustion chamber shape, ignition timing, operating temperature, etc.) that affect engine knock and octane ratings. For this reason, every automobile engine has been factory tested in order to determine its minimum and most efficient octane requirement.

A lower fuel octane number will hurt overall engine performance. It can reduce power, lower fuel economy, cause run-on or dieseling (engine continues to run with the ignition key off), and spark knock. Under extreme conditions, a low-octane fuel can actually cause physical damage to the engine. Detonation, the most severe and harmful type of engine knock, can burn or knock holes in the top of pistons, bend connecting rods, burn valves. and finally destroy the engine.

UNLEADED GASOLINE

Unleaded gasoline does NOT contain lead antiknock additives. Also, lead-free gasoline must meet Environmental Protection Agency (EPA) standards of minimum octane.

GASOLINE COMBUSTION

Combustion is the burning action above the engine piston that produces enough heat to cause expansion of the gases, Fig. 8-2. Proper combustion of any automotive fuel requires the fuel to be mixed with air and compressed in the cylinder before ignition.

The fuel must be broken into tiny droplets and thoroughly *mixed* with air (oxygen). Liquid fuel will burn, but too slowly for adequate engine combustion, Fig. 8-3.

Atomizing fuel

Atomization refers to how the fuel injector or carburetor breaks up liquid gasoline into tiny droplets. For instance, an insect sprayer atomizes liquid insecticide. When the handle is pumped, Fig. 8-3, the sprayer sends out a fine mist of tiny drops.

Combustion temperature

At the start of the compression stroke, the combustion chamber temperature may only be around 575 °F (302 °C). After combustion, the temperature may rise to as high as 4000 ° or 5000 °F (2 204 ° or 2 760 °C). This temperature increse will expand the gases in the cylinder to form pressure.

For instance, the cylinder pressure during the compression stroke may be around 150 psi (1 033 kPa); after combustion, the cylinder pressure can increase to almost

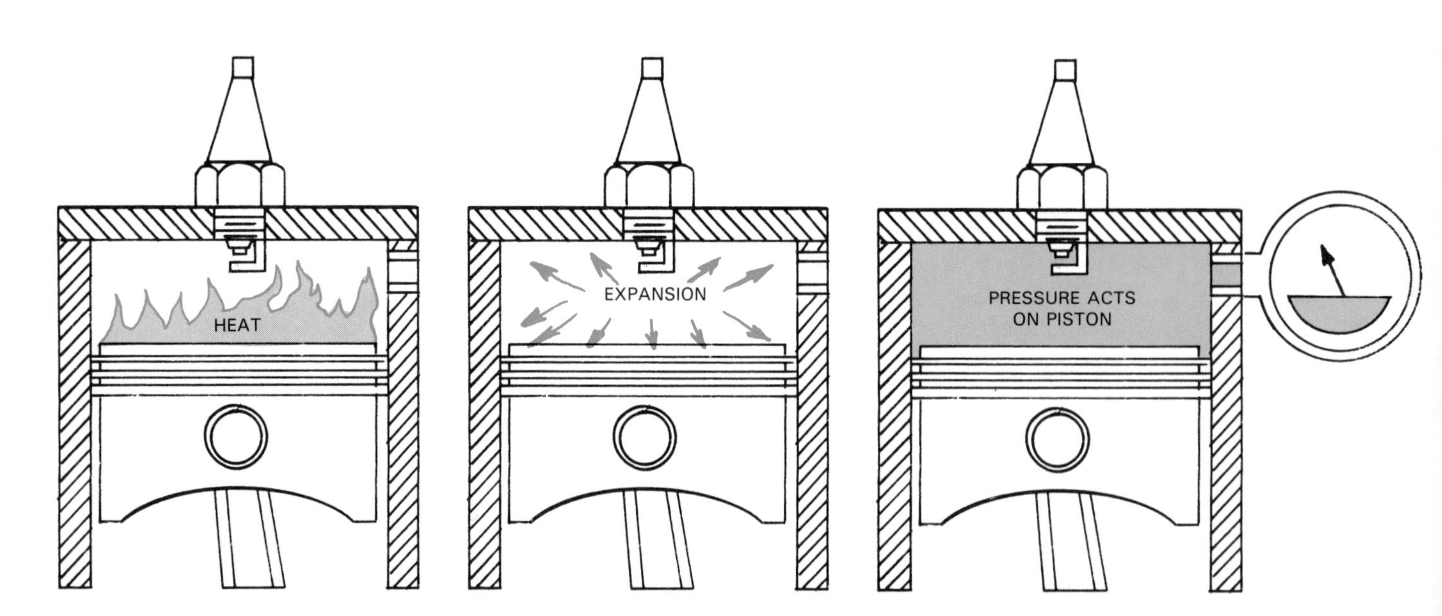

Fig. 8-2. During engine operation, burning fuel first produces heat. Heat then causes expansion of gases. Since gases are confined in chamber, they produce pressure on head of piston.

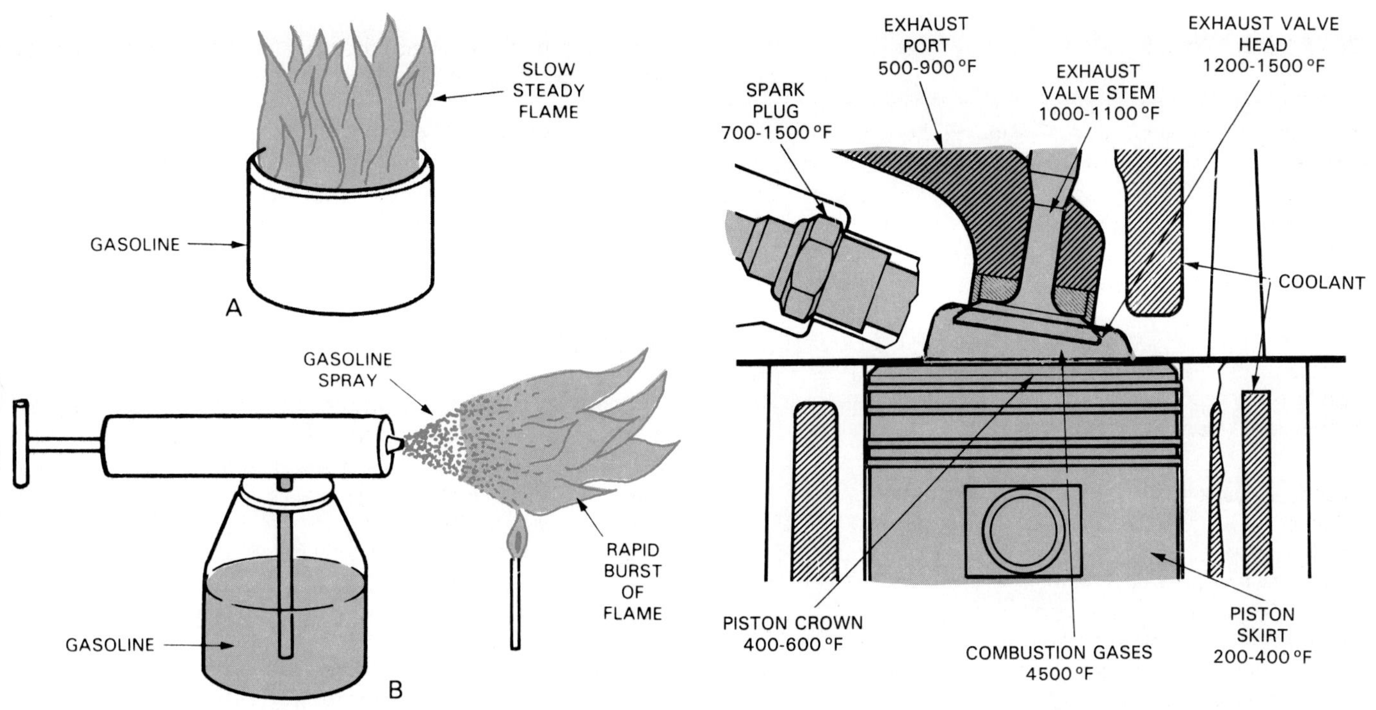

Fig. 8-3. A—Gasoline burns too slowly when in liquid form. B—If it is atomized into a mist, it will burn better.

Fig. 8-4. Combustion produces tremendous heat in engine. Combustion flame can reach 4500 °F. Note other temperatures of engine parts.

800 psi (5 512 kPa). See Fig. 8-4.

This is a tremendous amount of force pushing on the top of the piston. Every square inch has 800 pounds of pressure pushing on it. If a piston has an area of 12.6 sq. in. (around a 4 in. cylinder bore), the combustion pressure would be like having a 5 ton (4 525 kg) weight resting on the piston. This tremendous force rams the piston down the cylinder and spins the crankshaft. This pressure, as you will learn in later chapters, can go much higher in supercharged or turbocharged engines.

COMBUSTION CHEMISTRY

Combustion is a chemical reaction (union of fuel and oxygen) that produces useful heat. It is a rapid oxidation (burning) of carbon and hydrogen. Rapid oxidation or combustion of gasoline, in an ideal situation, occurs when the hydrogen in gasoline combines with oxygen to form steam, and the carbon in gasoline combines with oxygen to form harmless carbon dioxide gas.

Under ideal situations, combustion forms harmless water (H_2O) and carbon dioxide (CO_2). Carbon dioxide can be seen as bubbles in soda water.

Actual combustion in an engine is NOT this perfect nor complete. Air is NOT pure oxygen, Fig. 8-5. As a consequence, real combustion produces other undesirable chemicals or pollutants. As you can see in Fig. 8-5, several harmful gases are created because of the other substances in air and because some of the gasoline does NOT burn.

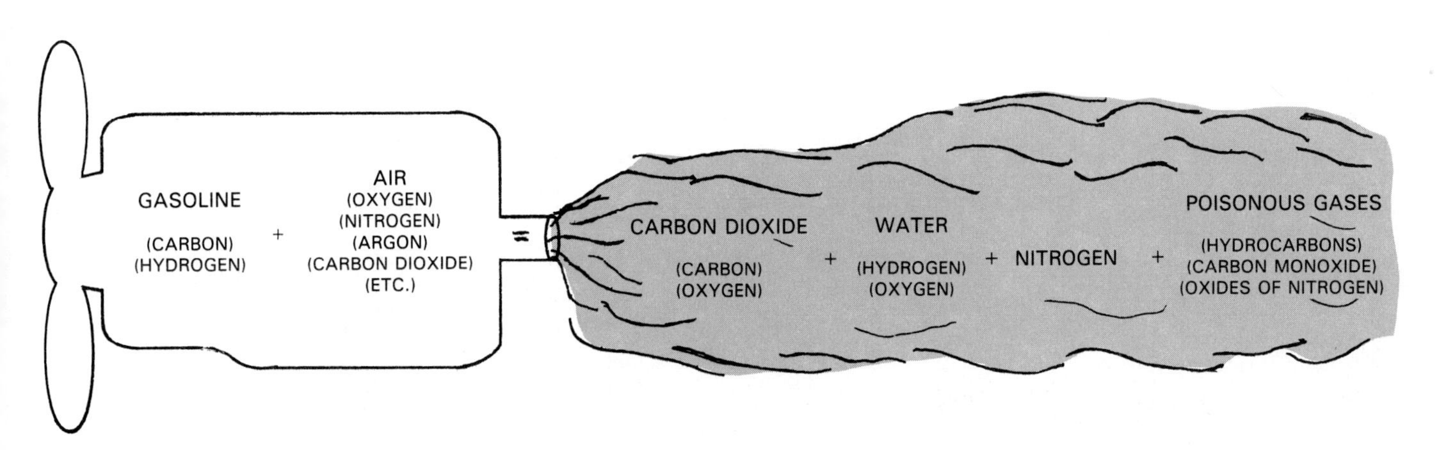

Fig. 8-5. Combustion in an engine produces several poisonous substances. Controlling combustion can help reduce these emissions.

NORMAL COMBUSTION

If all of the factors affecting combustion are favorable (air/fuel ratio, compression, spark timing, fuel characteristics, etc.), the fuel burning process will be normal. *Normal combustion* occurs in a gasoline engine when a single flame spreads evenly and uniformly away from the spark plug electrodes. Fig. 8-6 illustrates normal combustion.

Combustion speed and power

Normal combustion takes arround 3/1000 of a second, which is comparatively slow. Normal combustion is NOT an explosion. An explosion, like that of dynamite,

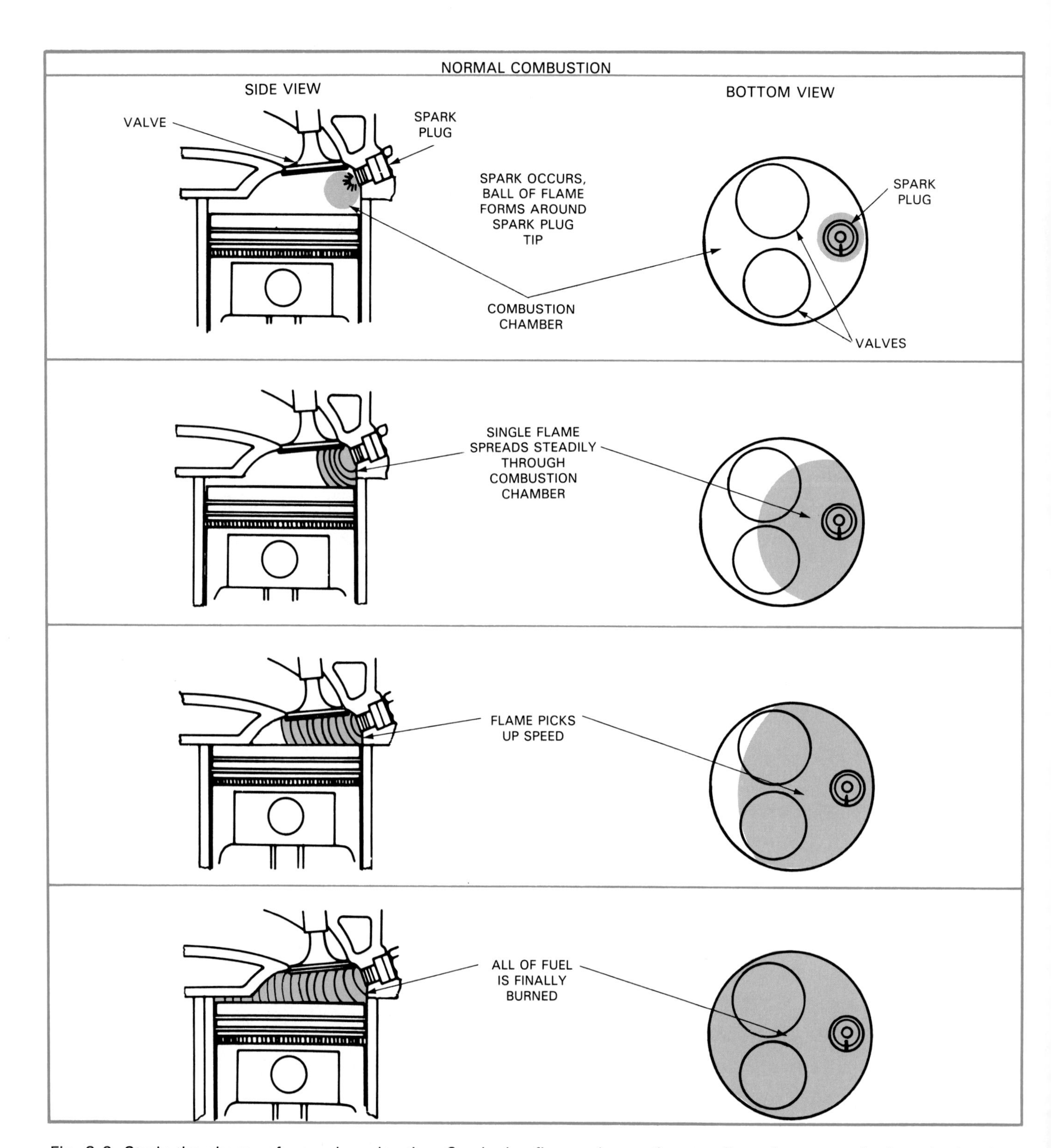

Fig. 8-6. Study the phases of normal combustion. Spark plug fires and smooth, even flame front spreads through chamber. Peak cylinder pressure is developed a few degrees past TDC on power stroke.

would take only around 1/50,000 of a second: a fraction of the time it takes for the combustion of gasoline.

The combustion of gasoline, pound for pound, produces more total energy than dynamite, however. Just a thimble full of gasoline has enough energy to produce tons of force. Under some conditions, gasoline can burn too fast, making combustion like an explosion.

PREIGNITION

Preignition, also known as *"ping,"* occurs when a hot spot in the combustion chamber ignites the fuel mixture before the spark plug "fires."

Fig. 8-7 shows what happens during preignition. Study each phase carefully.

Note difference from detonation in Fig. 8-8.

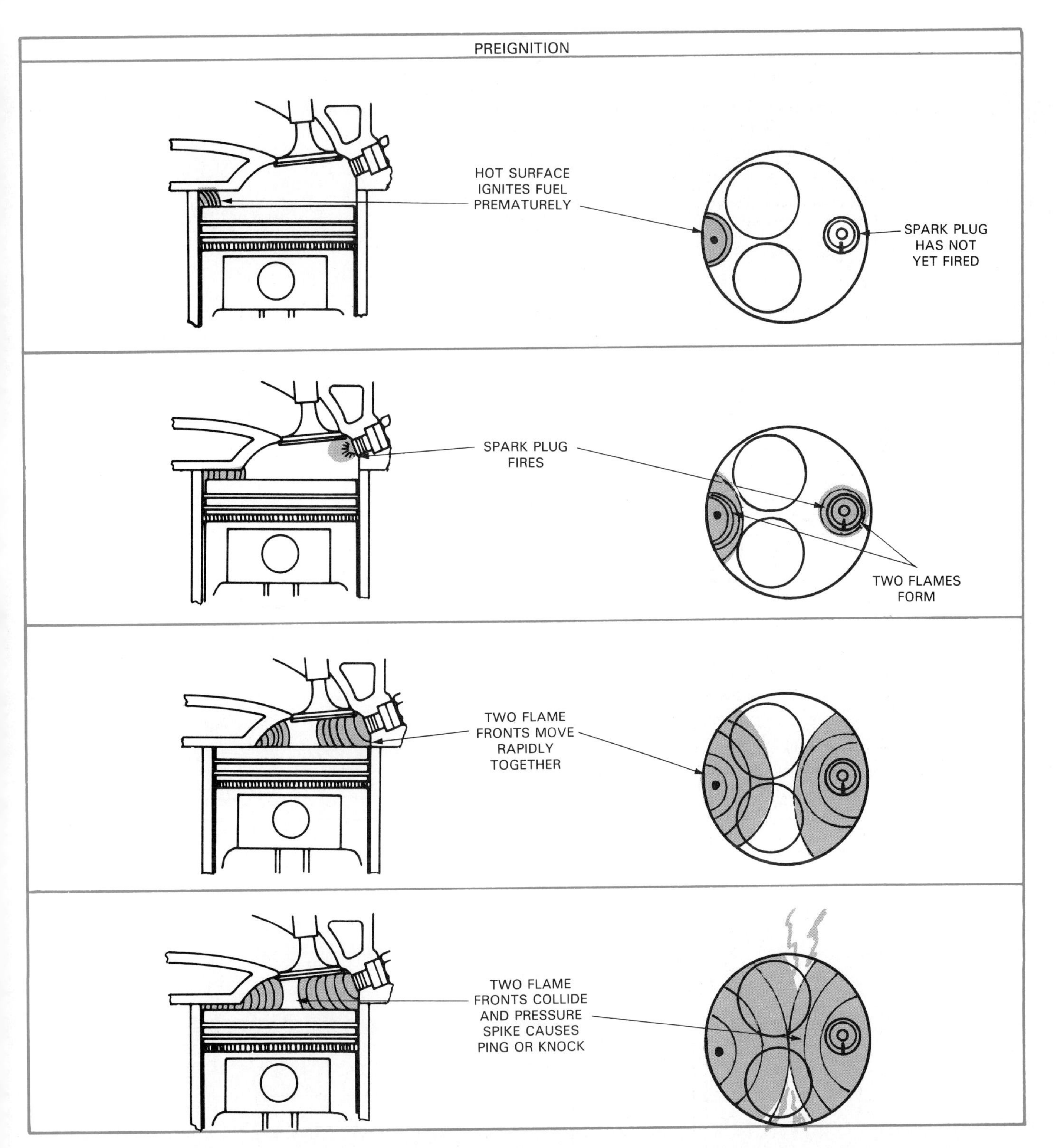

Fig. 8-7. Preignition is result of hot spot, glowing bit of carbon for instance, igniting mixture before spark plug fires. When plug fires, two flame fronts spread rapidly through chamber. When they collide, pressure spike rattles parts and produces pinging noise.

The effects of preignition are usually a slight loss of power and fuel economy. You can hear preignition as a mild pinging, knocking, or rattling.

Severe or prolonged cases of preignition can cause more serious detonation and part damage. Remember their differences, however.

Surface ignition

To fully understand preignition, you must first understand its cause—surface ignition. *Surface ignition* is caused by:

1. Incandescent (glowing hot) piece of carbon in the combustion chamber.

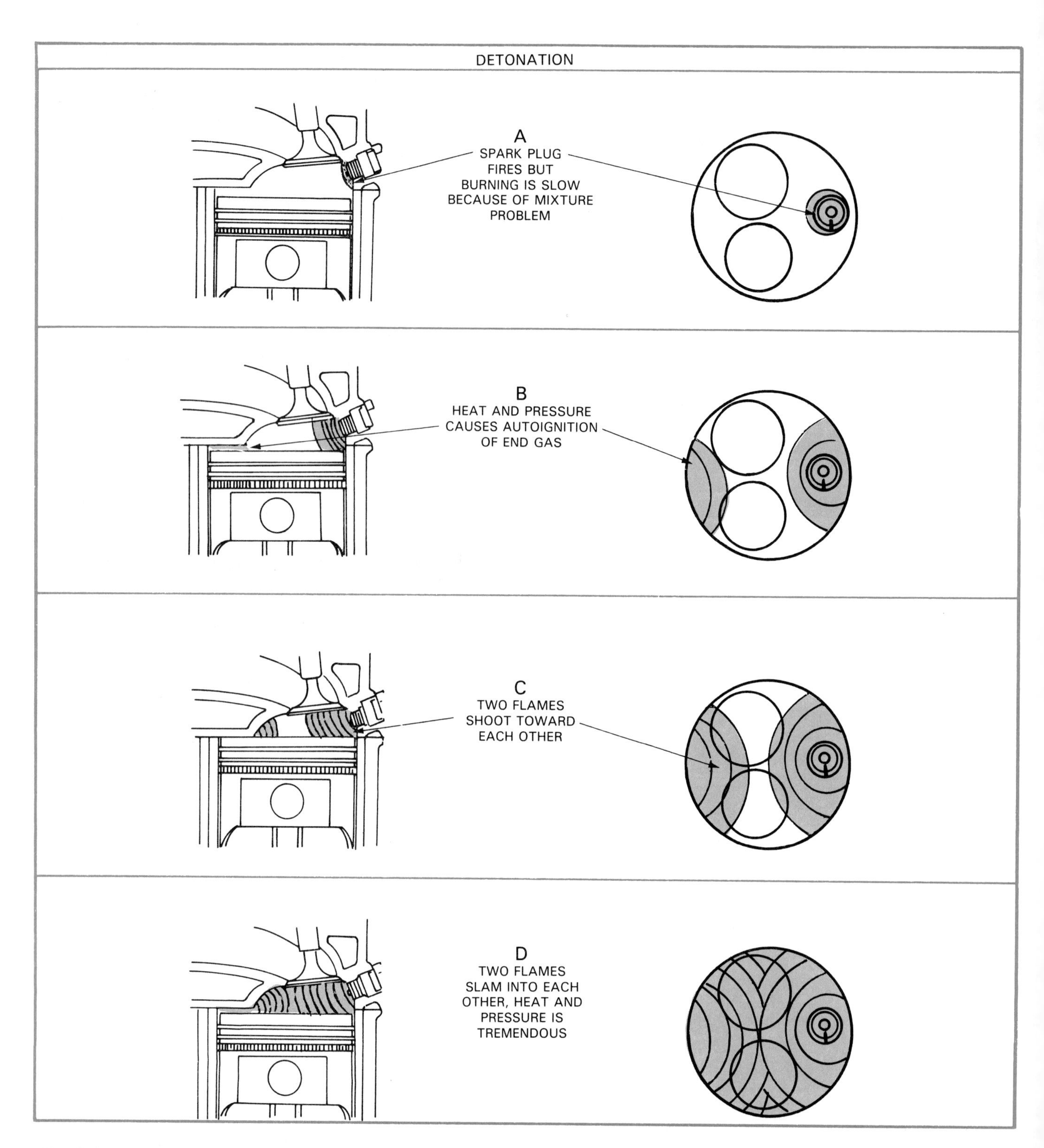

Fig. 8-8. Detonation is more damaging to engine than preignition. Spark plug fires but normal combustion flame is too slow because of mixture problem. Mixture on other side of chamber is heated and pressurized. If this end gas ignites, a violent explosion occurs and can damage piston, head, and other components.

2. Overheated engine from improper operation of cooling system.
3. Exhaust valve overheated by lean fuel mixture (lean carburetor setting, clogged injector strainer, vacuum leak, stuck EGR valve, computer system malfunction, etc.).
4. Overheated spark plug (heat range too high).
5. Exhaust valve overheated by leakage (insufficient tappet clearance, weak valve spring, sticking valve, etc.).
6. Sharp edges in the combustion chamber (overheated threads on spark plug, edge of head gasket, sharp, machined surface, etc.).
7. Excessively dry and hot atmospheric conditions or an air filter door stuck shut.

Postignition

Surface ignition can also happen after the spark plug fires. Since it occurs after normal spark ignition, it is termed *postignition.* Its affects are minimal.

Wild knock

When carbon breaks loose from the combustion chamber wall, it can cause *wild knock.* When this happens, the carbon particle bounces around in the cylinder. While suspended in the burning fuel, it will heat up and ignite the fuel mixture each time fuel enters that cylinder. This can cause a pinging sound that can last for a few moments or come and go for no apparent reason.

Detonation

Detonation is abnormal combustion where the last unburned portion of the fuel mixture almost explodes in the combustion chamber. A part of detonation is, in fact, almost as fast as an explosion of dynamite. The total process of engine detonation is about six times faster than normal combustion. See Fig. 8-8.

Detonation can cause a very rapid pressure rise in the engine cylinder. It can be so fast and extreme that parts of the engine (cylinders, cylinder heads, etc.) actually flex and vibrate. This vibration is extremely high pitched and can be heard as a loud knock or ping. A detonation knock will sound something like a ball peen hammer rapidly striking the pistons or engine block. It is much louder than preignition ping.

Detonation is a very serious engine combustion problem. Under severe detonation, connecting rods may bend, cylinder heads crack, pistons melt, spark plugs shatter, and bearings can be ruined. The extreme heat can partially melt and soften the top of a piston. Then the pressure can force a hole in the softened aluminum, Fig. 8-9.

Detonation-caused knocking will usually be heard when you push the gas pedal all the way to the floorboard while the car is only traveling at moderate speeds (20 to 40 mph). Normally, at high engine speeds there will not be enough time for the end gas to heat up, ignite without a spark, and detonate.

Detonation factors

You can think of detonation as a "race"—a race between the normal flame front and the autoignition (burning without spark) of the end gas. If normal combustion is faster than the heating and squeezing on the unburned fuel, normal combustion will win and detonation will be prevented.

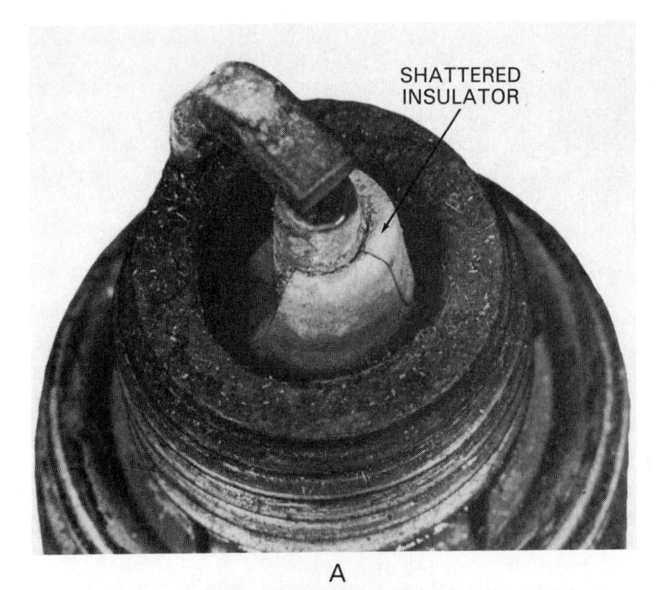

A

B

Fig. 8-9. A—Detonation has shattered this spark plug insulator. B—Prolonged detonation has overheated head of piston and blown hole through it. (Champion Spark Plugs)

Conditions causing detonation are:

1. Slow-burning lean fuel mixture (faulty carburetor or injector, fuel pump, blocked fuel filter or line, vacuum leak at higher engine speeds caused by bad PCV valve or EGR valve, computer troubles).
2. Gasoline with low octane or antiknock rating.

This is more common with unleaded gasoline.
3. Carbon deposits increasing compression ratio. This is the result of poor detergent action of gasoline or oil entering cylinders.
4. Engine operating at above-normal temperature due to low coolant level, water jacket blockage, or other trouble.
5. Ignition timing too advanced (improper setting of initial ignition timing, inaccurate distributor advance curve, computer malfunction, etc.).
6. Bad rings and/or valve seals allowing oil to be burned in cylinders.
7. Air cleaner door stuck allowing too much hot exhaust manifold air to enter engine.
8. Excessive turbocharger boost pressure from a bad pressure-limiting valve or faulty knock sensor.

Preignition versus detonation

Preignition is similar to, yet quite different from, detonation; the processes of each are reversed. Detonation occurs after the start of normal combustion and preignition begins before normal combustion. However, one can cause the other and both are capable of causing engine damage.

KNOCK SENSING SYSTEM

Many late model turbocharged engines are equipped with a knock sensing system. The *knock sensing system* uses a sensor and the computer to retard ignition timing or limit boost pressure to prevent damaging detonation and annoying preignition. Fig. 8-10 shows a simple diagram of a knock sensing system.

The *knock sensor* works like a "microphone" to listen for the sound of engine knock. When turbocharger boost is high, heat and pressure in the combustion chambers are also very high. This can cause ping or knock.

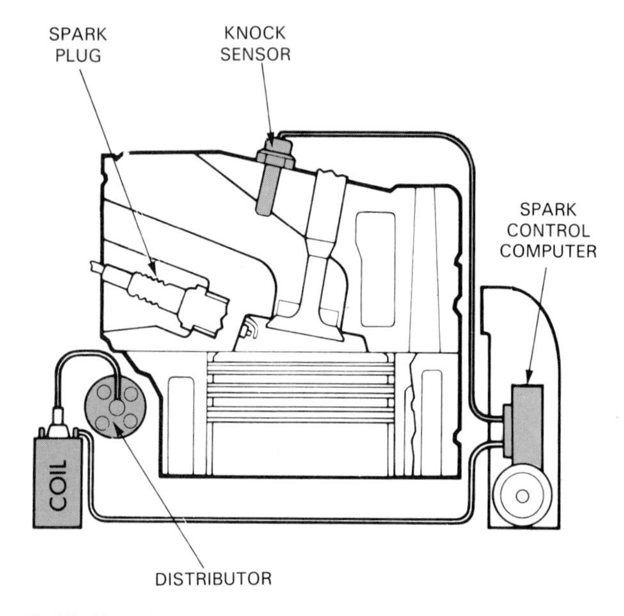

Fig. 8-10. Knock sensor system. Sensor sends electrical pulse to computer if it "hears" engine knock. Then, computer can retard ignition timing or open turbo wastegate and stop knocking. (Chrysler Corporation)

When the sensor "hears" engine knock, an electrical pulse is sent to the on-board computer. The computer is programmed to retard the timing or open the wastegate (pressure limiting valve) until the knock sound is eliminated.

To check the action of a knock sensor, tap on it with a wrench. This should make the computer retard the ignition timing or activate the wastegate and engine speed should drop slightly. Follow service manual instructions for testing details.

DIESELING (AFTER RUNNING, RUN-ON)

Surface ignition (preignition or postignition) can also cause an engine to run after the ignition key switch is turned off. A knocking, coughing, or fluttering noise can be heard. In some cases, an unpleasant "rotten egg" odor may also be evident.

The term *dieseling* is used to identify this condition because, like a diesel engine, the gasoline engine runs without sparks. The ignition switch is off and no voltage is present in the ignition system or at the plugs. The terms "run-on" and "after running" are also used to describe dieseling.

Usually, the engine continues to run because "hot spots" in the combustion chambers are igniting the fuel. The fluttering or coughing sound usually occurs right before the engine stops. This coughing sound can be a result of the engine actually "kicking" backwards. Air can blow back through the intake ports.

After running places undue strain on pistons, cylinders, valves, throttle body or carburetor gaskets, and the air cleaner.

Dieseling can be stopped by leaving the automatic transmission in drive, or by dragging a manual clutch to lug the engine. However, for a permanent fix, the source of the problem must be found and corrected.

The reasons an engine may diesel include:

1. Engine idle speed set too high.
2. Carbon deposits increasing engine compression ratio.
3. Inadequate or low gasoline octane rating.
4. Engine overheating (low coolant level, bad thermostat, etc.).
5. Spark plug heat range too high.
6. Idle fuel mixture adjustment incorrect (usually too lean), which results in improper idle speed screw setting.
7. Sticking throttle or gas pedal linkage.
8. Engine needs a tune-up (plugs, timing, carburetor adjustments, etc.).
9. Driver not allowing engine enough time to return to curb idle speed before turning off the ignition switch.
10. Oil entry into cylinders from engine mechanical problem.

Spark knock

Spark knock is an engine ping caused by the electric arc happening too early at the spark plugs. It is an ignition timing problem. The timing is advanced too far and is causing combustion pressures to slam into the upward moving piston (maximum cylinder pressure before TDC).

Sometimes, as mentioned earlier, early spark timing can cause detonation. It is hard to distinguish the various types of engine knocks by sound alone. If retarding the ignition timing cures or corrects a pinging problem in an engine, your problem was primarily spark knock.

FACTORS AFFECTING COMBUSTION

Numerous factors control combustion and determine whether burning will take place normally or abnormally. In general, the objective of good combustion is to utilize as much of the fuel's heat energy as possible, to burn the fuel as fast as possible (short of detonation), and to produce as little exhaust emissions as possible. As you will learn, there are many interacting factors affecting combustion.

Spark timing

Spark or ignition timing is an important condition controlling combustion efficiency. Generally, when ignition timing is advanced without harmful knock, engine power will be at MAXIMUM.

At the other extreme, too little spark advance, called *spark retard,* will waste energy by letting combustion heat leave the engine unused, as is shown in Fig. 8-11.

With late spark timing, much of the heat will be passed into the engine block, cylinder head, and exhaust system instead of being used to power the engine. In fact, late ignition timing can cause an engine to overheat so much that the exhaust manifolds crack. Spark timing should be set to factory specifications.

Spark intensity and duration

The size and duration of the spark at the spark plug electrodes is very important to combustion, especially with today's lean fuel mixtures. Such mixtures are hard to ignite and keep burning. This is the main reason for wider spark plug gaps and high ignition voltages. A wider plug gap requires more secondary ignition voltage (40,000 V plus) to sustain an arc across the spark plug electrodes. The resulting spark has more energy or intensity to start and maintain the combustion of a lean fuel mixture.

Spark plugs gapped too narrow could cause the engine to have a *"lean misfire"* (commonly a miss at moderate to high engine speeds). The smaller plug gap could not produce a "hot" enough spark to ignite the lean mixture.

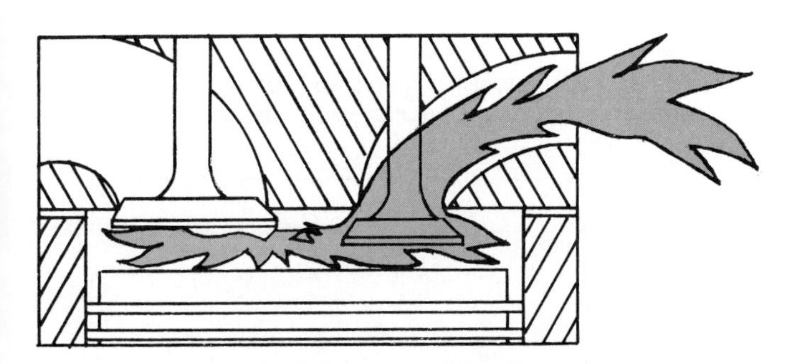

Fig. 8-11. Retarded ignition timing lowers engine efficiency because less pressure is developed and more combustion heat can blow out open exhaust valve. Too much retard can even overheat and crack engine exhaust manifold.

For complete fuel burning, it is important to "hold" or sustain a spark for a definite period of time (about 1/3 of the time it takes for combustion). Normally, a spark should occur at a spark plug for at least one millisecond (1/1000 of a second).

Note! Some engine analyzers are capable of measuring the duration of the sparks at the spark plugs.

Air-fuel ratio

The amount of fuel compared to air entering an engine is called the *air-fuel ratio,* Fig. 8-12. The proportions can be adjusted to meet the engine's changing needs.

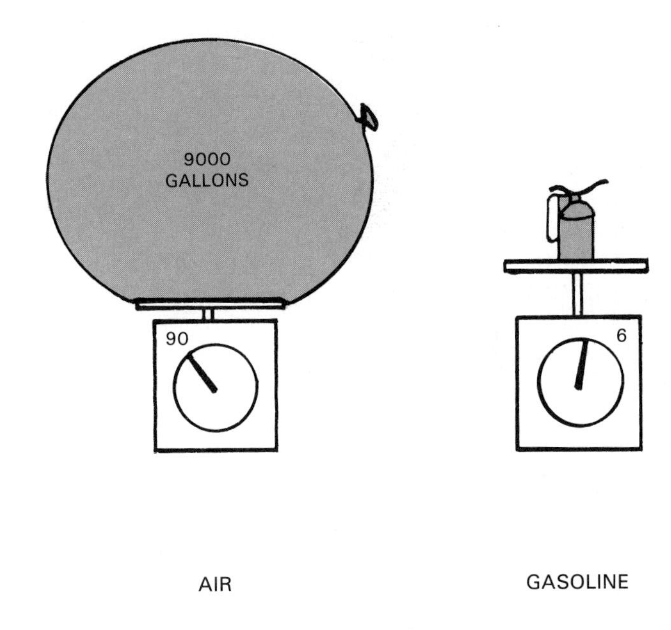

Fig. 8-12. Combustion requires a tremendous amount of air. (General Motors)

A *lean fuel mixture,* containing a lot of air compared to fuel, will give better fuel economy and fewer exhaust emissions, Fig. 8-13.

A *rich fuel mixture,* with a larger percentage of fuel, improves engine power and cold engine starting. However, it will increase emissions and fuel consumption, Fig. 8-13.

Fig. 8-14 shows a graph comparing engine power, fuel consumption, and air/fuel ratio. Note how, for high fuel economy, the mixture must be leaner.

Air-fuel ratios affect the speed of flame travel during combustion. Flame travel is quite slow whenever the mixture is extremely lean or rich.

A chemically correct fuel ratio is called a *stoichiometric mixture.* It is a theorectically ideal ratio of around 14:7:1 (14.7 parts air to 1 part fuel). Under steady-state engine conditions (unchanged engine load, temperature, fuel distribution, etc.), this ratio of fuel to air would assure that all of the fuel will blend with all of the air and be burned.

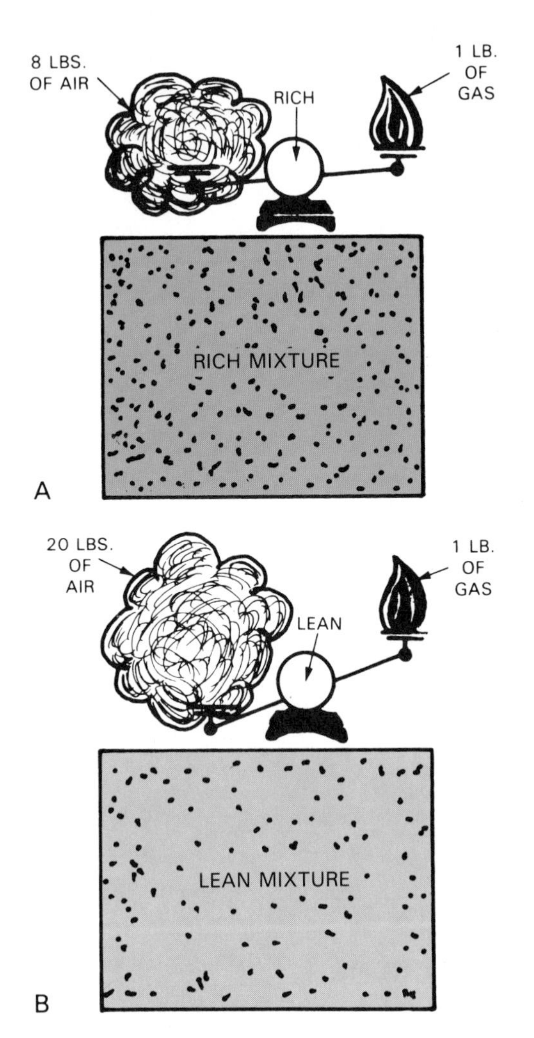

Fig. 8-13. A—Rich fuel mixture has more fuel mixed with air. B—Lean fuel mixture has less fuel in air.

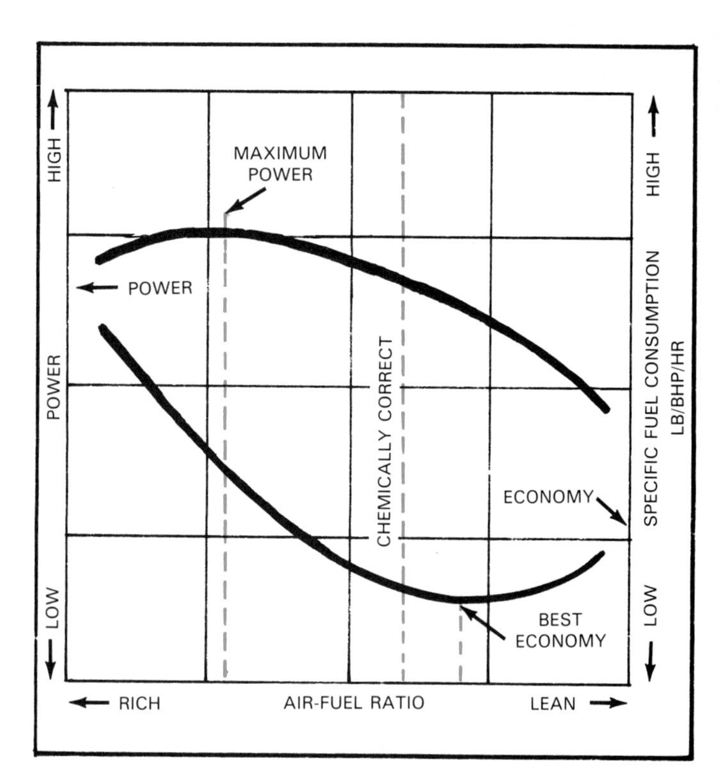

Fig. 8-14. Note graph showing relationship of engine power, fuel consumption, and air-fuel ratio. Note how stoichiometric mixture or ideal mixture is between maximum power and best economy. (Ethyl Corporation)

Multicylinder automobile engines, however, do not operate under ideal conditions. They suffer from variations in fuel delivery from cylinder to cylinder, as well as differences in compression, temperature, and timing. For this reason, a stoichiometric mixture is not always desirable, as will be explained later.

Compression ratio

The compression pressure or compression ratio of an engine is very important to combustion. Generally, a high compression ratio will increase engine power and fuel economy, but will increase harmful exhaust emission (NOx) levels. (See the text index for more information on compression ratios.)

Combustion chamber turbulence (swirl)

Air and fuel movement or *turbulence* in the cylinder will normally speed up and improve combustion efficiency. Turbulence will mix or stir the fuel mixture and expose more of the unvaporized (unbroken) fuel droplets to the combustion flame. With more fuel area touching the hot flame, combustion will be faster and cleaner (less exhaust pollution). Refer to Fig. 8-15.

For instance, if you stir up a pile of smoking leaves (slow combustion), a flame can reappear and speed up the burning process. You have exposed more of the unburned leaves to the hot materials and oxygen.

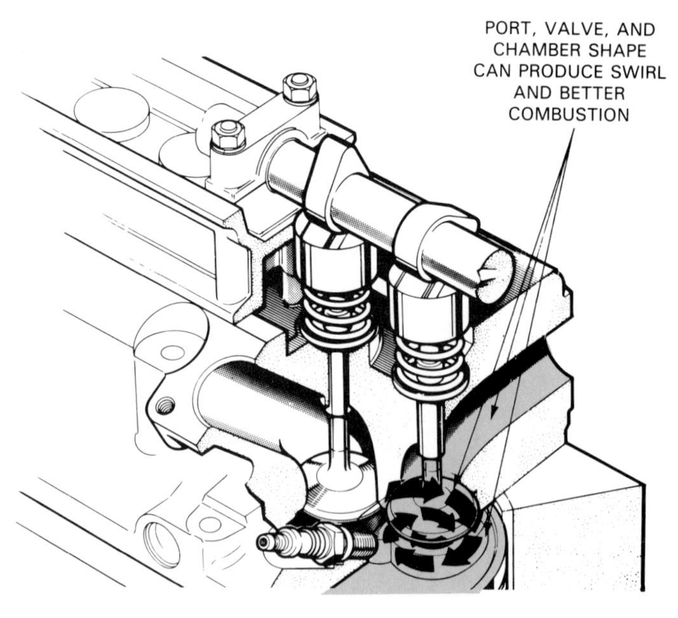

Fig. 8-15. Engine design is very critical to combustion efficiency. Ports, valves, camshaft, combustion chamber shape, and operating temperature all interact and affect how well fuel burns. (Jaguar)

Port, valve, piston, and combustion chamber shape all

affect combustion. They affect flow and turbulence of the incoming fuel charge.

Piston head shape

Since the piston head serves as one surface of the combustion chamber, it is very important that it be shaped to work with the cylinder head. A critical period of combustion occurs while the piston is at TDC. If the piston head gets too close to the cylinder head, it can restrict combustion by either blocking off areas or by quenching (cooling) a portion of the fuel mixture.

Any area not easily reached by the flame can lower combustion efficiency. Also, these areas can get an uneven mixing of fuel. Thus, they may ignite and burn poorly, Fig. 8-16.

Intake manifold pressure

The amount of *negative pressure* (vacuum) inside the intake manifold of an engine also affects combustion. It regulates how much fuel enters the engine. A high intake manifold pressure forces large amounts of fuel into the combustion chamber. This increases power output. On the other hand, low intake manifold pressure reduces the power developed by decreasing the quantity of fuel and air entering the cylinders.

A high intake manifold or inlet pressure occurs when the gas pedal of a car is pushed to the floor. This opens the throttle or air valves completely permitting full atmospheric pressure (14.7 psi at sea level) to push the fuel into the engine cylinders.

As will be discussed later, supercharging (turbocharging) raises the maximum intake manifold pressure higher than atmospheric pressure. This increases the fuel charge entering the engine and raises the compression pressure before combustion. As a result, supercharging can increase engine power by as much as 50 percent.

Cylinder bore and stroke

To a certain degree, a smaller cylinder bore and short stroke will improve combustion efficiency. With a smaller bore, the distance the combustion flame must travel is reduced. As a result, combustion speed increases, heat loss is cut, and efficiency is improved.

With a short stroke, less energy is lost to friction (rings rubbing against cylinder walls). More power is available to spin the crankshaft.

Fuel delivery system

The type of fuel delivery system will have a slight effect on combustion. Fuel injection improves combustion for two basic reasons. Injection systems allow more air to enter the cylinders, and have more control over the amount of fuel entering each cylinder. Gasoline injection is covered in later chapters.

Valve timing

Valve timing (when valves open in relation to piston position) is another important factor controlling combustion. For instance, the intake valve is sometimes held open a little past BDC.

Called *valve overlap,* both the intake and exhaust valves are open at the same time. The inertia of the fuel mixture in the intake manifold port will help to force more fuel and air into the cylinder. This increases combustion power, Fig. 8-17.

This principle is also used by the exhaust valve to draw out a little more of the burned gases. This is called *scavenging.* It allows more room for the fresh fuel mixture. Also, valve timing can be used to "clean up" combustion by retaining some of the exhaust gases in the engine cylinders where they can limit peak combustion temperature and limit Nox pollution.

Valve lift and duration

Valve lift (how far engine valves open), and *valve duration* (how long valves stay open) can have a pronounced effect on combustion, engine power, fuel economy, and drivability. Generally, lift and duration determine (at what engine speed) best performance will be developed. Fig. 8-18 shows some basic cam lobe *profiles* (shapes).

Valve lift is determined by the height of the camshaft lobe (centerline of cam to lobe tip). A tall lobe will have

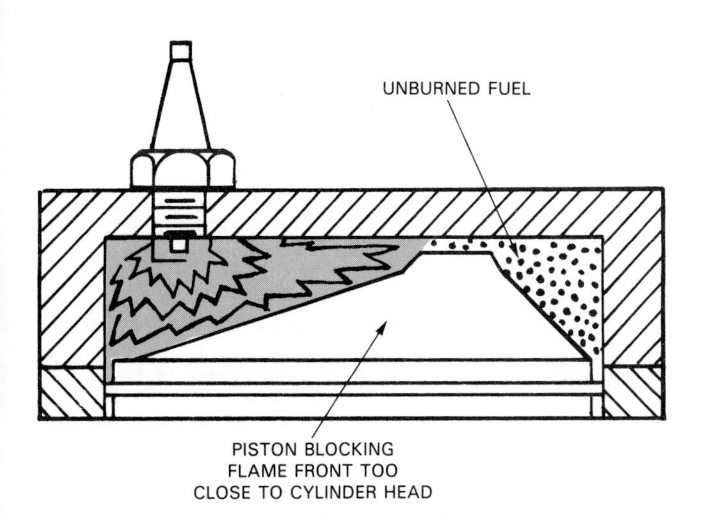

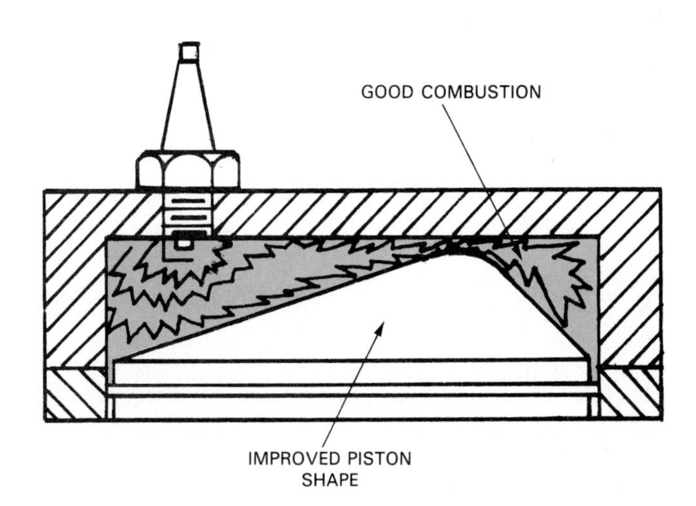

Fig. 8-16. Piston must be designed to work with shape of cylinder head. Piston and head surfaces both form combustion chamber walls. Note how piston on left is blocking flame from reaching some of the mixture. This would reduce power and increase emissions.

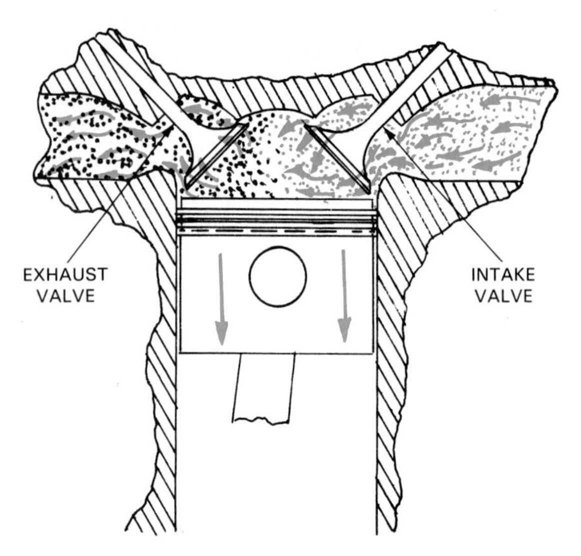

Fig. 8-17. Valve overlap has both intake and exhaust valves open at same time to increase power. Exhaust gas flow is helping to scavenge a little extra fuel mixture into combustion chamber.

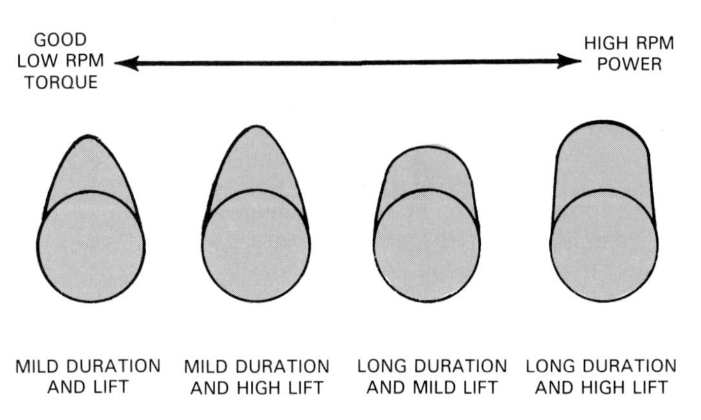

Fig. 8-18. Study basic camshaft profiles and how they affect engine operation.

a high lift. Valve duration is controlled by the width of the cam lobe; the wider the lobe, the longer it will keep the valve open. This will be covered in Chapter 10.

In most stock, unmodified engines, lift and duration specs are designed to give maximum combustion efficiency (combustion power, fuel economy, and cleanliness) between idle speed (around 800 rpm) and highway speed (around 55 mph or 3000 rpm).

In a race engine designed for maximum power and maximum engine speeds, valve lift and duration will be increased. This will open the valves wider and longer to help the engine "breathe" (take in more fuel mixture) at high speeds (6000 to 10,000 rpm). It will increase high speed horsepower, but at the cost of fuel economy and emissions.

Note that a race cam will usually DECREASE engine power, fuel economy, and exhaust cleanliness at normal engine speeds. The engine will operate inefficiently from idle up to normal highway speed. In fact, if a race cam is installed without other engine modifications (exhaust headers, increased compression ratio, etc.), usable engine power will normally be reduced at normal operating rpm.

Intake air temperature

The temperature of the air entering the air cleaner has some effect upon combustion. Generaliy, cool air increases engine performance. Cool air is more dense or compact than hot air. Therefore, it carries more power-giving oxygen into the cylinders.

Some automobiles use a thermostatically controlled air cleaner which maintains a constant inlet air temperature of between 80 °F and 110 °F (27 ° and 43 °C). This intake air temperature will provide a happy medium between engine power and low exhaust emissions.

Spark plug location

Ideally, a spark plug should be centrally located in the combustion chamber. With the plug in the center of the chamber, the distance from the plug electrodes (start of combustion) to the farthest point in the chamber will be at a minimum. This speeds up combustion because the flame will have the shortest possible distance to travel.

Fig. 8-19 shows that in an engine with a 4 in. cylinder

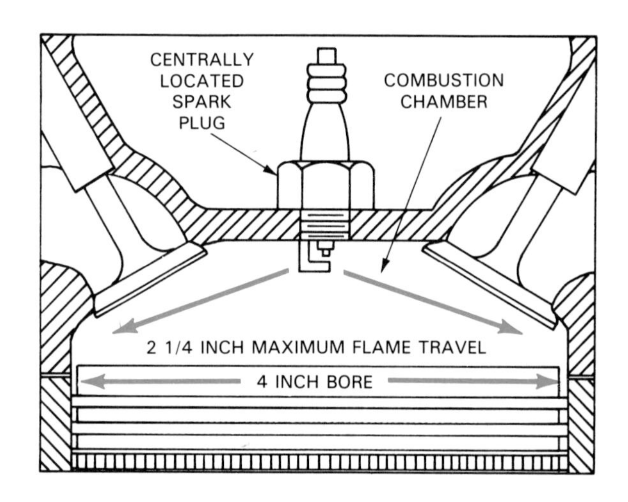

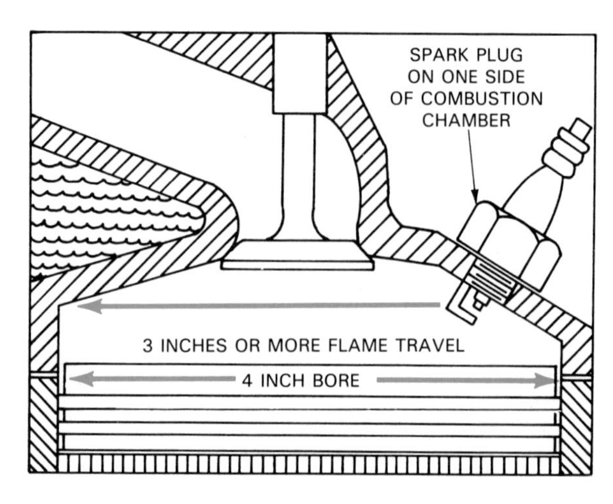

Fig. 8-19. Upper chamber has shorter path for flame travel and would be more efficient than lower chamber.

bore, a centrally located spark plug would have a maximum flame travel distance of around 2 1/4 in. If the plug were located on one side of the combustion chamber, the distance would be increased up to 3 in. As a consequence, combustion would take longer and more heat energy would be absorbed into the metal walls of the cylinder, combustion chamber, and piston head.

Combustion chamber surface area

A *small surface area* in a combustion chamber, even with a constant compression ratio, will improve the efficiency of combustion. A sphere or dome would be an excellent combustion chamber shape because it would have a small surface area. By having less area to absorb heat, there will be more heat remaining in the burning fuel to cause expansion, pressure, and piston movement. Notice in Fig. 8-20 that the irregular chamber has more surface area than the smoother, domed chamber.

Combustion chamber temperature

Combustion is improved, in general, as the temperature of the combustion chamber is increased. This is true so long as the temperature is not high enough to cause surface ignition. If the chamber wall is cool, fuel near the cold metal surface will not vaporize and burn properly. This unburned fuel can leave the engine as wasted energy and pollution. Also, a cool chamber wall will absorb too much combustion heat and produce less cylinder pressure.

Combustion chamber temperatures are affected by compression ratio, cooling system design, air-fuel ratios, and ignition timing. Normally, a combustion chamber will average a little under 1000 °F (538 °C) during moderate engine operation.

DIESEL FUEL

Next to gasoline, *diesel fuel* is the most popular automotive fuel. Once used mainly in trucks and trains, diesel engines are growing in popularity in automobiles because of their high fuel economy.

Diesel fuel, gallon for gallon, delivers more heat energy than gasoline. It is a thicker fraction of crude oil. Diesel fuel, therefore, requires a different engine design and fuel delivery system. It has different burning characteristics than gasoline.

Diesel fuel will not vaporize as readily as gasoline. If it were to enter the intake manifold of an engine, it would collect as a liquid on the manifold walls. Therefore, diesel engines have an injection system that puts the diesel fuel directly into the combustion chambers, Fig. 8-21.

Diesel fuel grade

Diesel fuel is classified into *grades.* These grades assure that the diesel fuel sold all over the country has uniform standards of service, including weight, cleanliness, ignition qualities, etc.

There are three diesel fuel grades: Number 1-D (D stands for diesel), Number 2-D, and Number 4-D. Number 3-D has been discontinued.

Usually, No. 2-D fuel is recommended for automotive diesel engines. It is the only diesel fuel available in some areas.

CAUTION! Do not try to use home heating oil or gasoline in a diesel engine. Major engine or injection system damage can occur.

Gasoline should never be used in a diesel engine. It would require a complete drain and flushing of the fuel system. Always use the grade of diesel fuel oil recommended by the engine manufacturer!

Heat value

The *heat value* of a fuel indicates the amount of heat and power developed during combustion. It is an important property that indicates the heat content in a fuel.

Diesel fuel heat content is approximately 12 percent higher than gasoline. In general, a heavy diesel fuel will have more heat energy per gallon. Commonly, heat value is measured in BTUs (British Thermal Units).

Diesel fuel viscosity

The *viscosity* of a diesel fuel refers to the fuel's ability to flow under varying conditions. A high viscosity fuel oil is thick and resists flow. A low viscosity fuel is thin and runny, like water.

Viscosity has an important effect on how diesel fuel

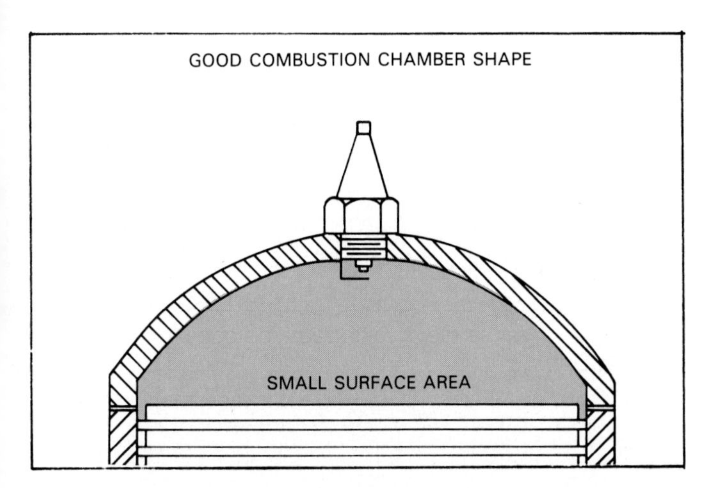

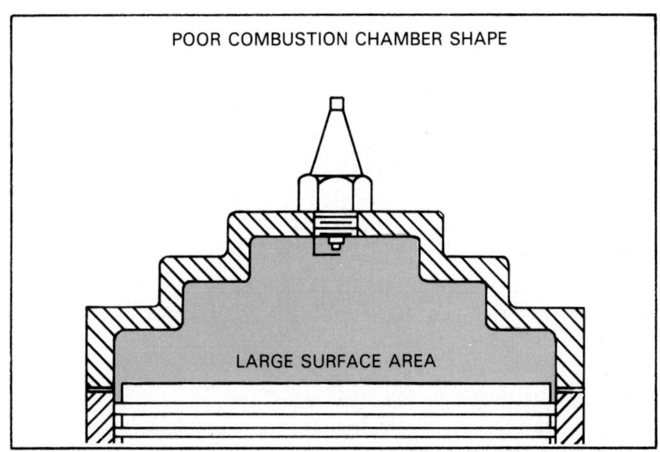

Fig. 8-20. Smooth, small combustion chamber is more efficient. It will absorb less heat and leave more heat to produce expansion and pressure on piston. Irregular shape can also block flame from burning fuel in pockets.

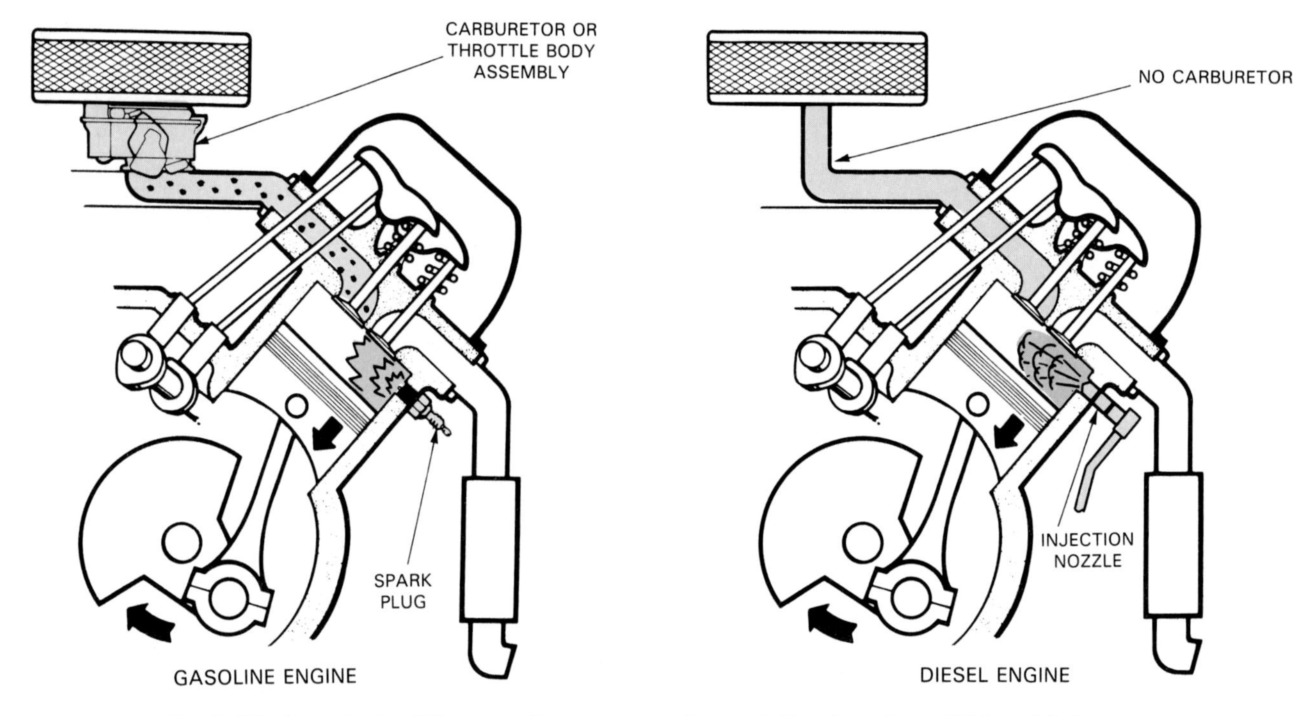

Fig. 8-21. Note basic differences between gasoline and diesel engines. (Oldsmobile)

burns and how well it sprays into the combustion chamber. It also affects the fuel's ability to lubricate the injection system and engine components. It must be thick enough to protect parts from friction.

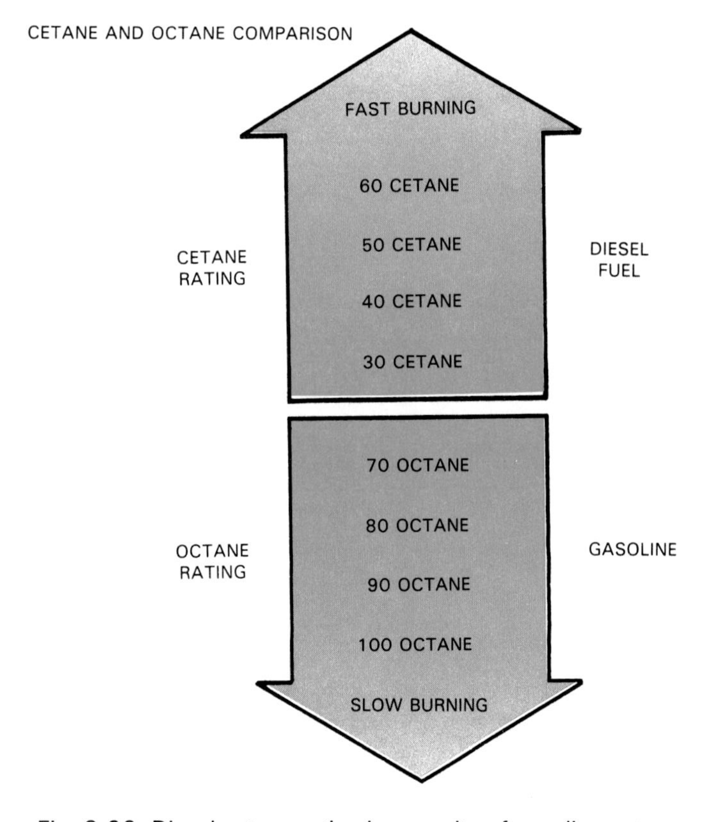

Fig. 8-22. Diesel cetane rating is opposite of gasoline octane rating.

Cetane rating

All diesel fuels have a *cetane number* or *rating* that indicates the starting quality of the diesel fuel. The HIGHER the cetane number, the better the COLD STARTING characteristics and the quieter the engine will operate when cold.

Most auto makers recommend a cetane rating of about 45 for diesel engines. This is the average cetane value of No. 2-D fuel. The cetane rating of a diesel fuel is similar, in some ways, to gasoline octane ratings. Both relate to the KNOCK tendencies of a fuel, Fig. 8-22.

In general, high cetane fuels permit a diesel engine to start properly at low temperatures and provide faster warm-up without misfiring. They limit or reduce the emission of white smoke from the exhaust during warm-up. They also reduce diesel knock and roughness.

Cloud point

Diesel fuel contains a wax called paraffin. At very cold temperatures, this wax separates from the other components in the fuel. As the wax separates, the fuel turns cloudy or milky. The temperature at which this occurs is the *cloud point* of the fuel.

When a fuel clouds, the wax can clog fuel filters. The wax buildup can restrict or stop the flow of fuel. To avoid fuel flow or clouding problems, diesel fuel with a low temperature cloud point is sold during winter in some areas.

Sulfur content

Diesel fuel *sulfur* contamination can cause a wide range of problems including engine corrosion and air pollution. Combined with water (in fuel, or formed by combustion), it will produce sulfuric acid. This acid can corrode parts of the engine, especially those made of brass, bronze, or copper. Mixed with engine oil, it will also speed the formation of engine *sludge*—thick emulsion of oil and contaminants. Sludge looks somewhat like "chocolate pudding." Sulfuric acid also adds to air pollution as part of the engine exhaust.

Water and sediment contamination

Contamination of diesel fuel by water and sediment (rust, dirt, and other particles) is a very common problem. Either can clog fuel filters and increase engine wear. In cold weather, water can freeze (jell) and block fuel flow into the engine. Water mixed with diesel fuel is also very corrosive.

CAUTION! Always make sure that the fuel cap, fuel cans, and fuel nozzles on gas station pumps are clean when fueling a diesel engine. A gasoline engine can tolerate and pass a certain amount of water but a diesel engine may not.

Diesel *fuel analysis kits* are available for checking the quality of the fuel in the car's tank. Basically, a sample of the fuel is mixed with a test chemical. Then, you can tell whether impurities (water for example) are mixed into the diesel fuel.

DIESEL COMBUSTION

In many respects, a four-stroke cycle automotive diesel engine is very similar to a four-stroke cycle gasoline engine. Since diesel engines burn a diesel oil, their fuel must be ignited by the heat of extreme air compression. A spark plug CANNOT adequately ignite diesel fuel.

Air alone is compressed in the cylinder until it is hotter than the ignition point of the fuel. Then, the fuel is sprayed into the hot air. Termed *compression ignition,* this assures thorough combustion of the diesel fuel. Gasoline cannot be ignited by compression ignition because it would ignite, detonate, and knock. Fig. 8-23 illustrates normal combustion in a diesel engine.

Diesel knock

The combustion process in a diesel must be carefully controlled. If all of the fuel were ignited in the cylinder at one time, a powerful explosion and knock would result.

Ignition lag is the time it takes the fuel to heat and vaporize before combustion. It is one factor controlling knocking and detonation in a diesel engine.

If the ignition lag is TOO LONG and the fuel vaporizes too slowly, a large amount of fuel can be present in the cylinder at one time. As a consequence, when the fuel does ignite, detonation and a violent explosion can occur. A high cetane fuel with a short lag period helps prevent diesel knock. Fig. 8-24 shows what happens during diesel knock. Study the phases.

Diesel knock is not as damaging as in a gasoline engine because of the heavy design of a diesel engine. However, detonation will lower engine power, fuel economy, and shorten engine life.

Diesel combustion factors

Many factors affect combustion in a diesel engine. Some are very similar to gasoline combustion while others are not.

Diesel engine power

The power and speed of a diesel engine is controlled by the amount of fuel injected into the cylinders—more fuel produces more power and speed.

The amount of air entering the engine stays about the same at all engine speeds. This allows an excess amount of air to enter the cylinders. Since oxygen is one of the limiting factors of power and combustion, the extra supply of air is one of the factors increasing diesel engine efficiency over a gasoline engine.

A diesel engine, however, does not produce as much horsepower as a gasoline engine of equal size. Operating speeds are much lower and parts must be made heavier and stronger to withstand compression ignition.

Diesel efficiency

Other factors that increase the fuel economy of a diesel engine are its "super lean" air-fuel ratio, extremely high compression ratio, and the high heat value of the fuel.

Air-fuel ratios in a diesel engine vary from around 100:1 (100 parts air to 1 part fuel) at idle speed to 20:1 at full load. By comparison, a gasoline engine has an air-fuel ratio of between 18:1 and 12:1 which is much richer and less fuel efficient.

This large quantity of air in a diesel assures that all of the fuel is consumed. It also aids combustion by absorbing some of the heat of the burning fuel mixture. The extra air, when heated, expands to help form more cylinder pressure, engine power, and fuel economy.

The exhaust emissions of a diesel engine are also quite different from a gasoline engine. As will be discussed under emissions, a diesel engine produces few hydrocarbon and carbon monoxide emissions but a large quantity of smoke and oxides of nitrogen. Many of the emission control systems used on gasoline engines are NOT needed on a diesel.

LIQUIFIED PETROLEUM GAS (LPG)

Natural gas is one of the lightest fractions (parts) of crude oil. Chemically, however, it is very similar to gasoline. Natural gas contains several impurities. For it to be used as an automotive fuel, it must be refined into *liquified petroleum gas* (LPG), a more pure fuel.

LPG is mainly propane and butane, along with small amounts of other gases. It has combustion qualities equal to or better than high-test gasoline.

LPG is a very good fuel. It produces good power, economy, and LOW exhaust pollution levels.

LPG is a vapor or gas at normal room temperature and pressure. This poses some design problems. In fact, since LPG is not a liquid, the entire fuel system must be redesigned to handle it.

LPG has operating characteristics almost identical to those of gasoline. However, because it is already a gas, the problem of breaking up liquid fuel is eliminated. Since the fuel enters the intake manifold and combustion

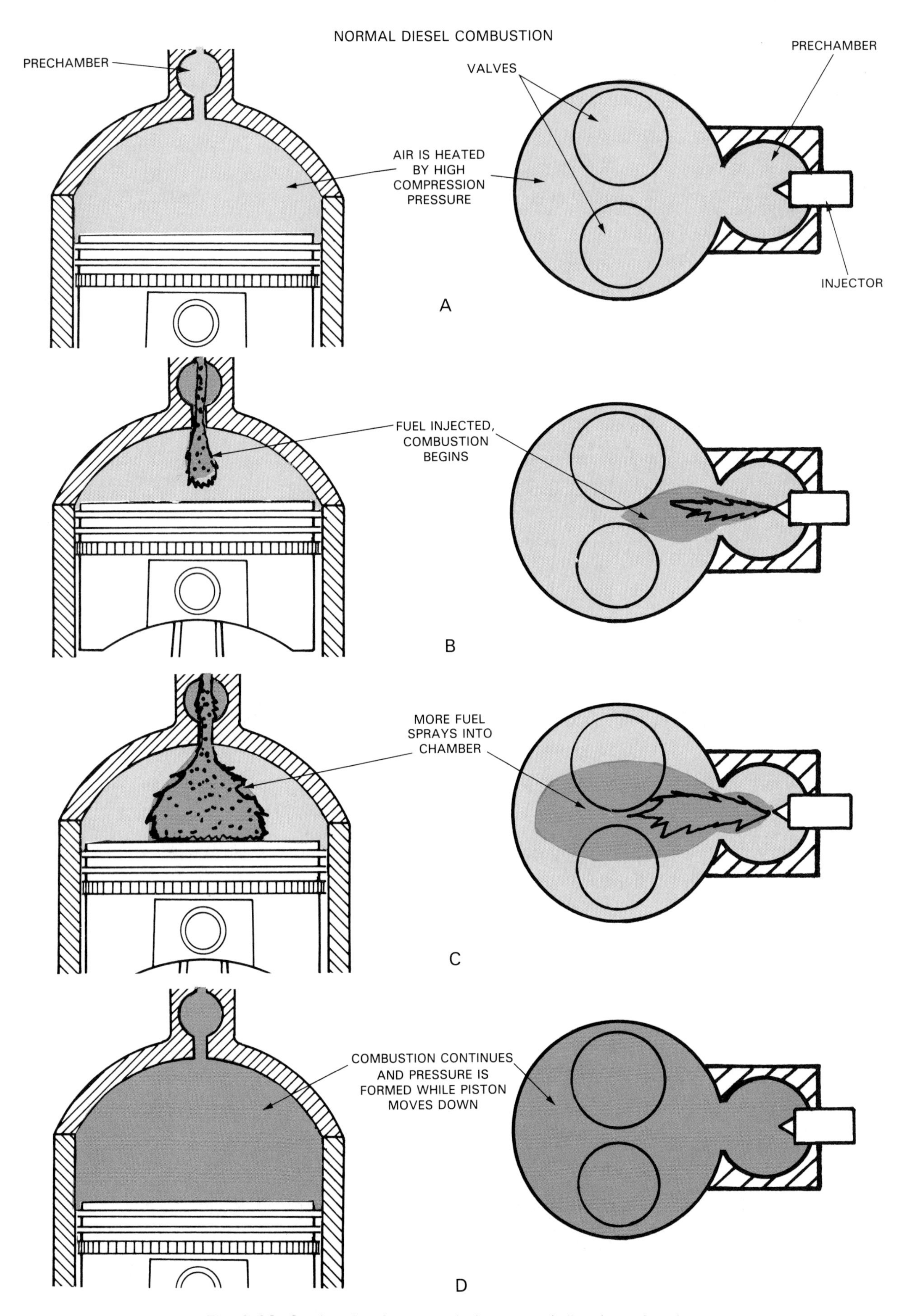

Fig. 8-23. Study what happens during normal diesel combustion.

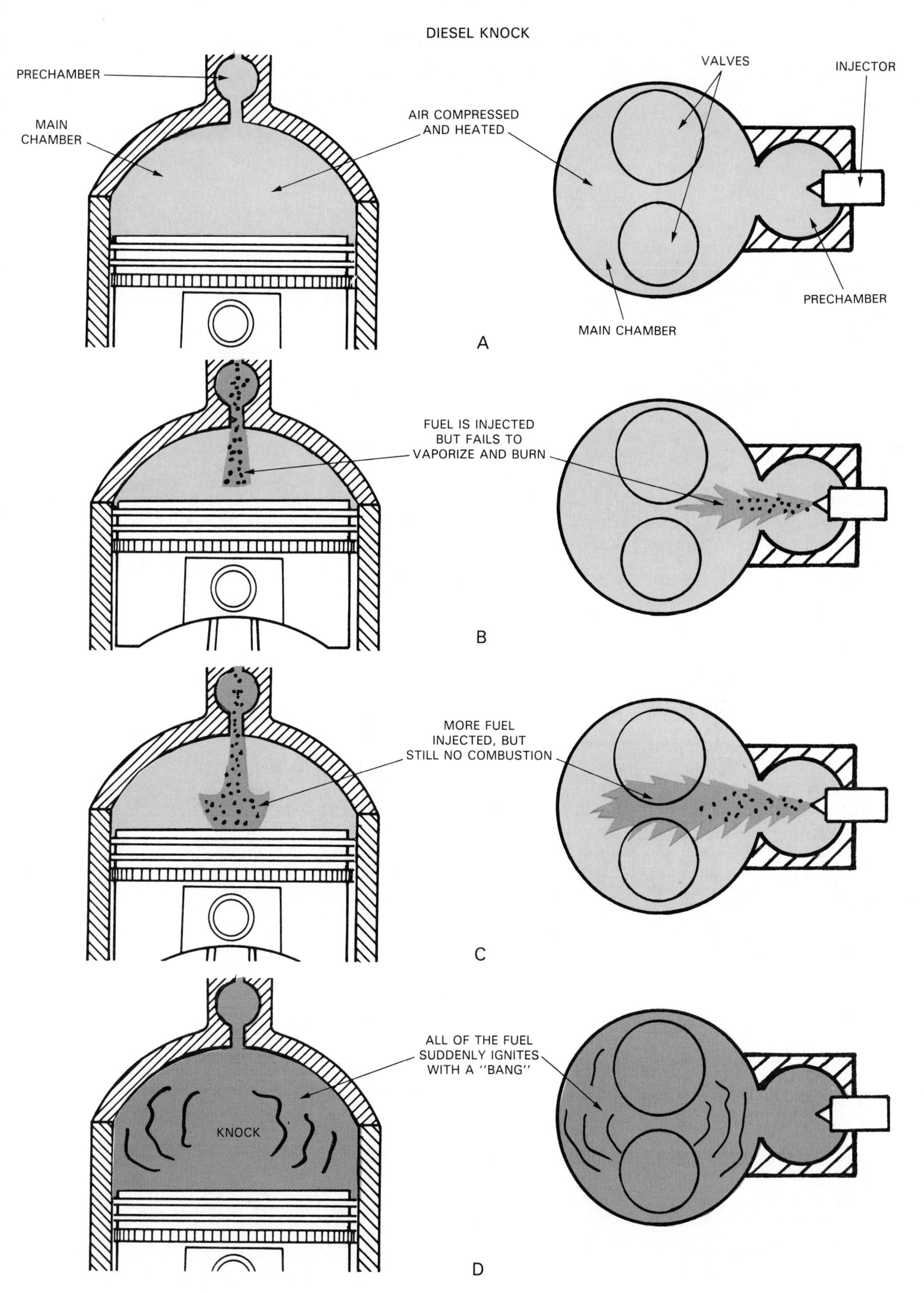

Fig. 8-24. Note how ignition lag or time when fuel does not start burning causes diesel knock.

chambers as a vapor, combustion is much more efficient. For this reason, LPG is an excellent automotive fuel.

ALCOHOL

Alcohol is an excellent alternate fuel for automobiles. The two types used to power internal combustion engines are *ethyl alcohol* (ethanol) and *methyl alcohol* (methanol). Alcohols are especially desirable as an automotive fuel because they can be manufactured from sources other than crude oil. Alcohol, intended as an automotive fuel, must be almost pure. Quite often, several refining steps are needed to approach this purity.

Grain alcohol

Ethanol or "grain alcohol" can be made from numerous farm crops, such as: wheat, corn, sugar cane, potatoes, fruit, oats, soy beans, or any crop rich in carbohydrates. Crop wastes are also a source. Ethanol is colorless, harsh tasting, and highly flammable.

Methanol

Methanol or "wood alcohol," can be made from wood chips, coal, oil shale, tar sands, corn stalks, garbage, or even manure. Like ethanol, methyl alcohol is a colorless, smelly, flammable liquid.

CAUTION! Methanol is highly poisonous and can be fatal if taken internally. Do not confuse it with ethanol!

Generally, both alcohols are similar in performance and operating characteristics. They have a very high octane or antiknock rating. In fact, they perform best in a high compression engine. Alcohol must be tightly compressed to produce maximum power.

Alcohol design considerations

With an alcohol fuel, engine compression ratios of between 11:1 and 13:1 are not unusual. Remember, today's engines use a compression ratio of around 7:1 or 9:1, much too low for pure alcohol. Power and economy would suffer.

On the other hand, in a properly designed engine and fuel system, alcohol produces fewer harmful exhaust emissions. Alcohol contains about half the heat energy of gasoline per gallon. The alcohol-air mixture must be much richer (about two times richer) than for gasoline. To provide a proper fuel-air mix, a carburetor's or injector's fuel passages should be DOUBLED in diameter to allow extra fuel flow.

Alcohol does not vaporize as easily as gasoline. This affects cold weather starting. The alcohol liquifies in the engine and will not burn properly. Thus, the engine may be difficult or even impossible to start in extremely cold weather. One solution is to introduce gasoline into the engine until the engine starts and warms up. Once the engine warms, the alcohol will vaporize and burn normally.

Alcohol burns at about HALF THE SPEED of gasoline. For this reason, timing or the advance curve must be changed so that more spark advance is provided. This will give the slow-burning alcohol more time to develop pressure and power in the cylinder.

Gasohol

Gasohol, as the name indicates, is usually a mixture of 87 octane unleaded gasoline and grain alcohol. While the mixture can range from 2 to 20 percent alcohol, most blends are 10 percent alcohol. This mixture requires little or no engine modification or adjustment.

Gasohol has a lower heat content than gasoline but usually has a higher octane rating. The alcohol acts like an antiknock additive. It can make low octane fuel perform like high octane premium. For example, 10 percent ethyl alcohol added to 87 octane unleaded gasoline can increase the fuel's octane value to more than 90 octane. This is equal to the octane value of premium gasoline.

SYNTHETIC FUELS

Automotive fuels made from coal, shale oil, and tar sands, are known as *synthetic fuels* because they are *synthesized* (changed) from a solid hydrocarbon state to a liquid or gaseous state. Synthetic fuels are presently being developed to help replace petroleum as a major source of motor fuel.

Coal synthesized fuel

Through various processes, coal can be converted into either a hydrocarbon gas or a liquid, suitable for use in internal combustion engines. Changing coal into a gas, is called *gasification.* Changing coal into a liquid is known as *liquification.* A ton of coal normally yields around 2 1/2 barrels of synthetic oil suitable for refinement into gasoline.

Oil shale

Oil shale is another natural resource that can be altered and made into oil and automotive fuels. Oil shale is a sedimentary type rock that contains a tar-like substance called KEROGEN. The shale looks very much like chocolate marble ice cream, the "chocolate" being the thick kerogen hydrocarbon and the "vanilla" being the sedimentary rock. As does coal, oil shale contains a relatively large amount of sulfur and other impurities that must be removed through refinement.

Tar sand

Tar sands are heavy hydrocarbons mixed with sand and dirt. There are several methods of removing these tars from the sand. One is to submerge the tar sand in hot water and steam. This forms a hot slurry that melts and liquifies the tar. Since oil is lighter than water, the petroleum based oils float to the top of the slurry where it can be easily removed.

Usually, tar sands contain about 12 1/2 percent petroleum compounds by weight. Around 4 to 5 tons of tar sand will produce a barrel of oil.

HYDROGEN

Many fuel and energy experts believe that, in the long run, *hydrogen* is the most promising fuel. Most important, it is the most common element in the universe. When burned, it produces little or no pollutants. In this respect, it is superior to petroleum-based fuels.

One of the best sources of hydrogen is water. Each

molecule of water is made up of two atoms of hydrogen and one of oxygen. One simple method used to separate the hydrogen is by sending electric current through water. The process is called *electrolysis.*

When burned with pure oxygen, hydrogen produces nothing but pure water. When burned in an engine using air instead of oxygen, its combustion releases water and a very small amount of pollution (NOx), but this is only a fraction of the pollutants formed by the combustion of most other fuels.

When technology overcomes the high cost of production, hydrogen may very well become an excellent automotive fuel. Solar energy could some day make hydrogen production feasible.

SUMMARY

Automotive fuels and combustion play an important role in determining the performance and service life of an engine. Combustion must be a controlled burning in the cylinders so that maximum pressure is developed on the head of the piston a few degrees past TDC.

Crude oil is converted into gasoline, diesel fuel, and other products through a refinement process. Gasoline is graded by an octane rating system that indicates how well the fuel resists knock or ping. A higher octane number means the fuel burns more slowly and steadily under heat and pressure. It would be ideal for a higher compression or turbocharged engine.

Combustion produces poisonous by-products. For this reason, combustion must be highly controlled in the engine combustion chambers. Spark timing, fuel mixing, fuel distribution to all cylinders, valve action, etc. must be perfect for combustion to be efficient.

Normal combustion occurs in a gasoline engine when the spark plug fires and a single flame burns through the mixture.

Preignition results when a "hot spot" in the combustion chamber ignites the fuel prematurely. Then, the spark plug fires and the two flames slam into each other, producing a pressure spike and a ping noise.

Detonation is more damaging and it occurs when normal combustion is too slow. The end gas or unburned fuel autoignites and almost explodes in the combustion chamber. Pistons, connecting rods, cylinder heads, etc. can be ruined by prolonged detonation.

There are many factors that contribute to preignition and detonation. These factors must be considered when trying to fix an engine suffering from abnormal combustion.

Knock sensing systems are used to detect and correct ping or knock. Used on gasoline engines, especially turbocharged engines, the system uses a knock sensor. The knock sensor is similar to a microphone installed in the engine. Any knocking or pinging sound will make the sensor produce an electrical signal for the computer. The computer can then retard the ignition timing or open the turbo wastegate to prevent knock.

Dieseling is a condition where the engine keeps running when the driver turns the key and engine off. Surface ignition is igniting fuel in the cylinders and forcing the pistons down, as in a diesel engine.

Diesel fuel is a heavier fraction of crude oil. It requires compression ignition for proper burning. Diesel fuel contains more heat energy per gallon than gasoline. This partially explains the better fuel economy of a diesel.

A cetane rating system is used with diesel fuel—the higher the cetane number, the better the cold starting ability. The cloud point of diesel fuel is also important. It indicates at what temperature wax separates from the fuel and blocks fuel filters. This is critical in cold climates.

Diesel knock usually occurs when the engine is cold. Fuel is injected into the combustion chambers but does not ignite quickly. An excess amount of fuel enters the chamber and when combustion does occur, it is too powerful and produces a louder than normal knock.

LPG is a lighter fraction of crude oil and can be used to power car engines. A special fuel system is needed since LPG is a vapor at room temperature and pressure. It burns very cleanly, producing little emissions.

Alcohol is another potential fuel for automobiles. Twice as much alcohol must be metered into the engine as gasoline, so fuel consumption about doubles. Gasohol is commonly used and it is a mixture of usually 10 percent grain alcohol and 90 percent unleaded gasoline. The alcohol serves as an octane booster to prevent ping or knock.

Synthetic fuels can be made from coal, oil shale, and tar sands. Hydrogen may be the fuel of the future. It can be made through electrolysis or sending electric current through water.

KNOW THESE TERMS

Crude oil, Hydrocarbons, Distillation, Fractionating tower, Octane rating, Unleaded, Atomization, Combustion, Preignition, Surface ignition, Postignition, Wild knock, Detonation, Knock sensor, Dieseling, Spark knock, Lean mixture, Rich mixture, Stoichiometric, Valve timing, Valve overlap, Scavenging, Valve lift, Valve duration, Diesel fuel grade, Cetane rating, Cloud point, Water contamination, Diesel knock, LPG, Ethanol, Methanol, Gasohol, Synthetic fuels.

REVIEW QUESTIONS—CHAPTER 8

1. Crude oil is a mixture of:
   a. Semisolids.
   b. Liquids.
   c. Gases.
   d. All of the above.
   e. None of the above.
2. How is crude oil changed into gasoline, diesel fuel, and other products?
3. High octane gasoline would ignite and burn faster than a low octane fuel. True or false?
4. ________ refers to how the carburetor or injector breaks fuel into tiny droplets.
5. Normal gasoline combustion occurs when a ________ ________ spreads evenly away from the ________ ________ ________.
6. Normal combustion takes about ________ of a second.
7. Define the term "preignition."
8. Define the term "detonation."
9. List eight factors contributing to detonation.

10. How does a knock sensing system work?
11. An older car engine diesels when the ignition key is shut off.
Technician A says that the problem could be a high idle speed setting.
Technician B says that low-octane fuel could be the trouble.
Who is correct?
    a. Technician A.
    b. Technician B.
    c. Both A and B.
    d. Neither A nor B.
12. Explain five factors that affect engine combustion.
13. How does air-fuel ratio affect the operation of a gasoline engine?
14. What causes excessive knock in a diesel engine?
15. Gasohol is usually a mix of _____ percent grain alcohol and _____ percent unleaded gasoline.

ASE CERTIFICATION-TYPE QUESTIONS

1. Technician A says that natural gas is a by-product of crude oil. Technician B says that asphalt is a by-product of crude oil. Who is right?
(A) A only.
(B) B only.
(C) Both A & B.
(D) Neither A nor B.
2. Technician A says that gasoline with a low octane rating can reduce an automotive engine's power efficiency. Technician B says that gasoline with a low octane rating can lower an automotive engine's fuel economy. Who is right?
(A) A only.
(B) B only.
(C) Both A & B.
(D) Neither A nor B.
3. All of the following are general grades of gasoline sold in the U.S. EXCEPT:
(A) unleaded.
(B) leaded.
(C) premium.
(D) regular.
4. Technician A says that normal combustion occurs in a gasoline engine when a single flame spreads evenly and uniformly away from the spark plug electrodes. Technician B says that normal combustion occurs in a gasoline engine when dual flames spread evenly and uniformly away from the spark plug electrodes. Who is right?
(A) A only.
(B) B only.
(C) Both A & B.
(D) Neither A nor B.
5. Technician A says that preignition occurs when the last unburned portion of the fuel mixture almost explodes in the combustion chamber. Technician B says that preignition occurs when a hot spot ignites the fuel mixture in the combustion chamber before the spark plug fires. Who is right?
(A) A only.
(B) B only.
(C) Both A & B.
(D) Neither A nor B.
6. Surface ignition can be caused by _________.
(A) a defected exhaust valve
(B) an overheated spark plug
(C) a glowing piece of carbon in the combustion chamber
(D) All of the above.
7. Technician A says that under certain circumstances, engine detonation can crack an engine's cylinder head. Technician B says that under certain circumstances engine detonation can actually melt an engine's piston. Who is right?
(A) A only.
(B) B only.
(C) Both A & B.
(D) Neither A nor B.
8. Technician A says that spark timing can affect the quality of engine combustion. Technician B says that intake manifold pressure can affect the quality of engine combustion. Who is right?
(A) A only.
(B) B only.
(C) Both A & B.
(D) Neither A nor B.
9. All of the following are diesel fuel grades sold in the U.S. EXCEPT:
(A) Number 2-D.
(B) Number 1-D.
(C) Number 3-D.
(D) Number 4-D.
10. Technician A says that the term ignition lag refers to the time it takes air to vaporize in a diesel engine's combustion chamber. Technician B says that the term ignition lag refers to the time it takes diesel fuel to heat and vaporize before combustion. Who is right?
(A) A only.
(B) B only.
(C) Both A & B.
(D) Neither A nor B.

Chapter 9

Short Block Construction

After studying this chapter, you will be able to:

- *Explain the construction of a cylinder block.*
- *Compare engine block designs.*
- *Summarize the use of cylinder sleeves.*
- *Explain engine crankshaft designs.*
- *Describe crankshaft related components: rear oil seal, pilot bearings, etc.*
- *Describe the construction of connecting rods.*
- *Summarize the operation of piston pins, bushings, pistons, and piston rings.*
- *Explain the characteristics of engine bearings.*

Late model engines are made using "high tech" materials and machining techniques. To work on today's engines, you must have a full understanding of how engine components are constructed and designed. This knowledge will help you solve problems and make important decisions concerning how to service engines. If you do NOT understand engine construction fully, it would be very easy to overlook a repair task and make it very difficult to diagnose many engine troubles.

Study the information in this chapter carefully! It will provide a base for more fully comprehending many later chapters on engine troubleshooting and repair.

SHORT BLOCK (BOTTOM END)

A *short block* includes the cylinder block with all of its internal parts installed. The pistons, rods, crankshaft, and bearings would be in the block. The heads, oil pan, and front-end parts would be removed. The term *bottom end* is often used to mean the same thing as short block.

Fig. 9-1 shows the parts of a typical short block. Study which parts are NOT considered in the short block.

Frequently, a technician will replace the short block with a factory new or rebuilt short block. The reconditioned old heads, the front cover, water pump, motor mounts, flywheel, etc., would be cleaned and used over. They would be installed on the new short block.

The engine technician may also rebuild or overhaul the short block by replacing or reconditioning worn parts. This will be explained in detail in later chapters.

LONG BLOCK

Long block is a technician's term that refers to the short block with just the heads installed. Parts like the valve covers, front cover, flywheel, mounts, etc., are NOT included in the long block. Long blocks (short block and heads) can be purchased from major auto manufacturers and specialized engine rebuilders to replace a high mileage, badly worn engine.

Can you visualize what parts are considered part of a long block or a short block in Fig. 9-2?

BARE BLOCK

A *bare block* is a cylinder block with ALL parts removed. There would not be pistons, rods, a crankshaft, nor other parts in the block.

CYLINDER BLOCK CONSTRUCTION

A *cylinder block* is the main foundation for a car engine. It supports or holds all of the other engine components. Various surfaces and holes are machined in the block for attaching these other parts.

Fig. 9-3 shows a typical cylinder block for a four-cylinder, in-line, OHC (overhead cam) engine.

Cylinder block materials

Cylinder blocks are normally made of cast iron or an aluminum alloy. Both materials are used in engines for modern automobiles.

A *cast iron block* is heavy and strong. Older engines normally used cast iron because of its dependability. Nickel, chromium, and other metals can be added to form an iron alloy that is more wear resistant and stronger than plain gray iron.

An *aluminum block* is much lighter and almost as strong as a cast iron. The aluminum must be made much thicker than cast iron, however. Aluminum blocks are becoming more popular on late model cars because the reduced weight increases fuel economy. Aluminum also dissipates heat much faster than cast iron.

Casting a cylinder block

Cylinder block casting is done by pouring hot, molten metal (iron or aluminum) into a mold. The outside of the mold forms the external shape of the block. The inside of the mold is made of sand and forms the water jackets of the block. After the metal solidifies, the mold is removed and the sand is shaken out of the block to make the block hollow.

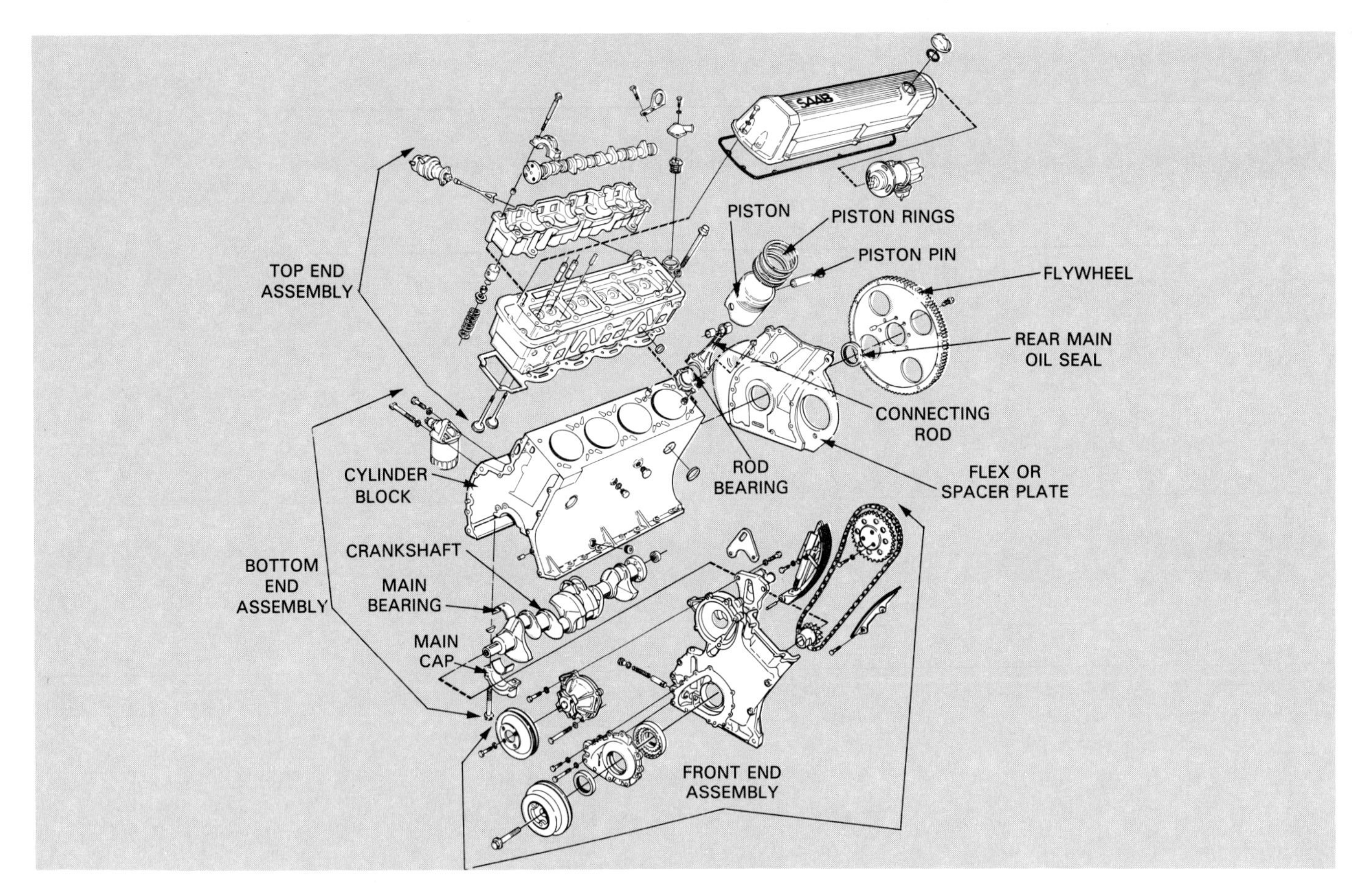

Fig. 9-1. Engine bottom end consists of major parts that install in cylinder block. Top end is primarily cylinder head and valve train. Front end is camshaft and its drive mechanism. (Saab)

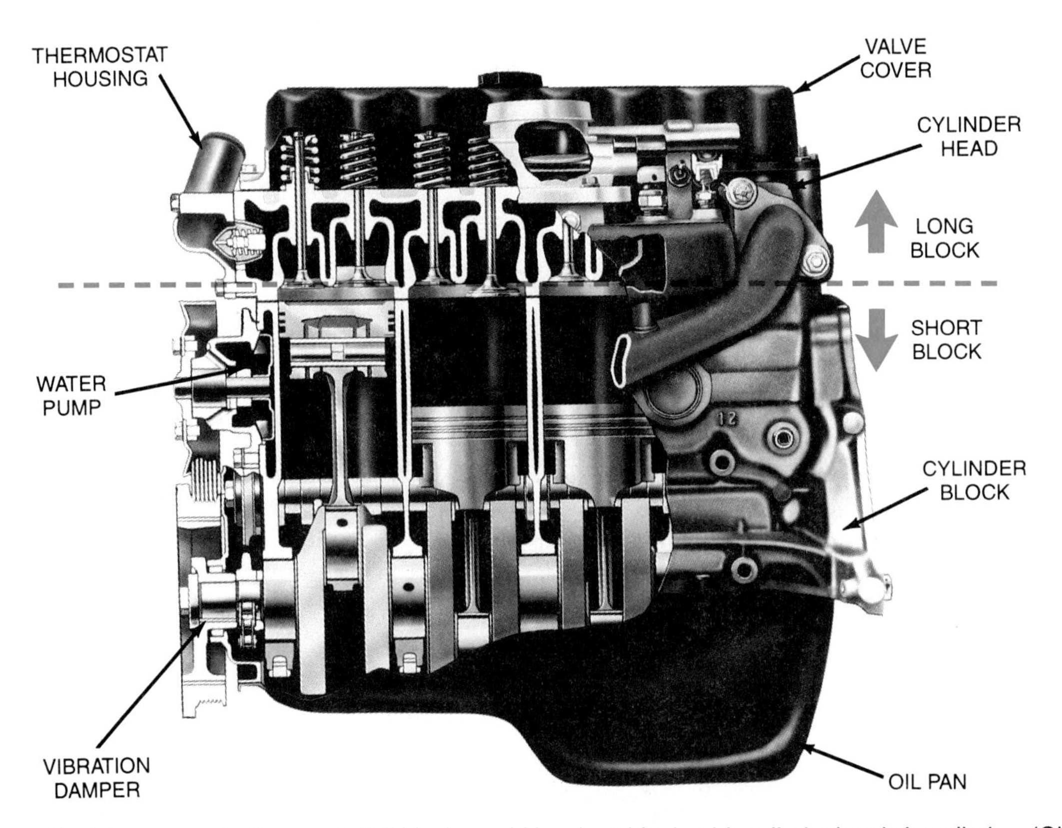

Fig. 9-2. Short block is same as bottom end. Tall block would be short block with cylinder heads installed. (Oldsmobile)

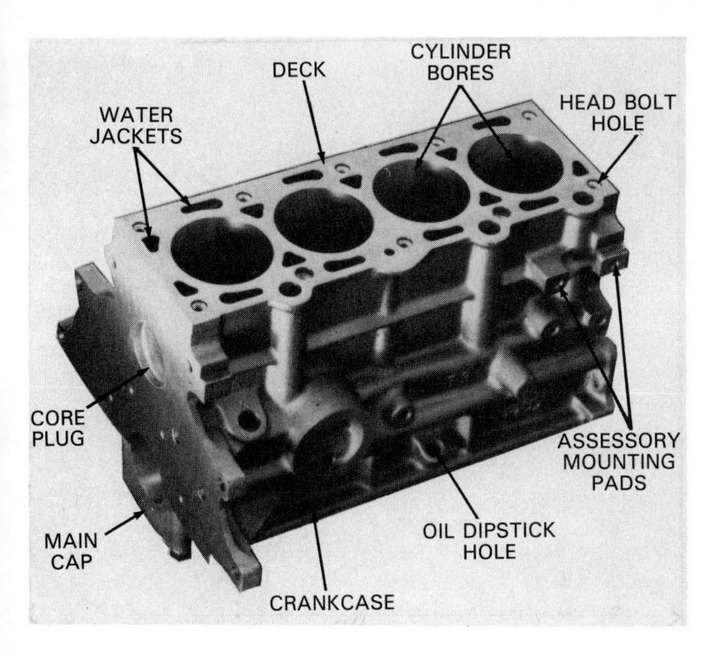

Fig. 9-3. Cylinder block is made by pouring molten metal into a mold. Internal sand mold hollows out major areas for cylinders, water jackets, etc. (Ford)

Core plugs are small metal plugs pressed or screwed into the holes in the sides of the block. These holes allow the internal sand mold to be removed from inside the block.

The term *freeze plug* is commonly used to refer to the core plugs because they pop out if the water inside the block freezes in cold weather. The ice expands and forces the plugs out. Note however, core or freeze plugs do NOT protect the block from damage when a weak coolant solution freezes. If the plugs pop out, the cylinder block will frequently be cracked by the tremendous pressure of ice expansion inside the block.

Fig. 9-4 shows a cutaway cylinder block showing the internal water jackets. Also note the core or freeze plugs on the side of the block.

Note! Various *oil passages* are also machined in the block so motor oil can reach high friction areas. This is discussed in the chapter on lubrication systems.

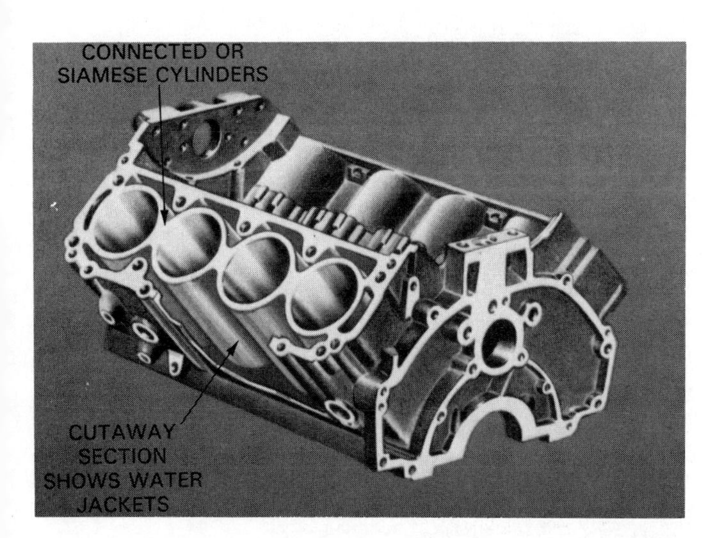

Fig. 9-4. Water jackets allow coolant to remove excess heat from engine. Note how cutaway exposes internal jackets and connected, siamese cylinders in this block. (Detroit Diesel)

Cylinders

Cylinders, also called *cylinder walls,* are large holes machined in the cylinder block for the pistons. An *integral cylinder* is part of the block, Fig. 9-5. A *sleeved cylinder* is a separate part pressed into the block, Fig. 9-6.

A cast iron block normally has integral, machined cylinders. Cast iron has excellent wear characteristics and can withstand the rubbing action of the piston rings very well.

An aluminum block commonly uses cast iron sleeves. Aluminum does NOT have good wear resistance so iron liners are commonly fitted into the block.

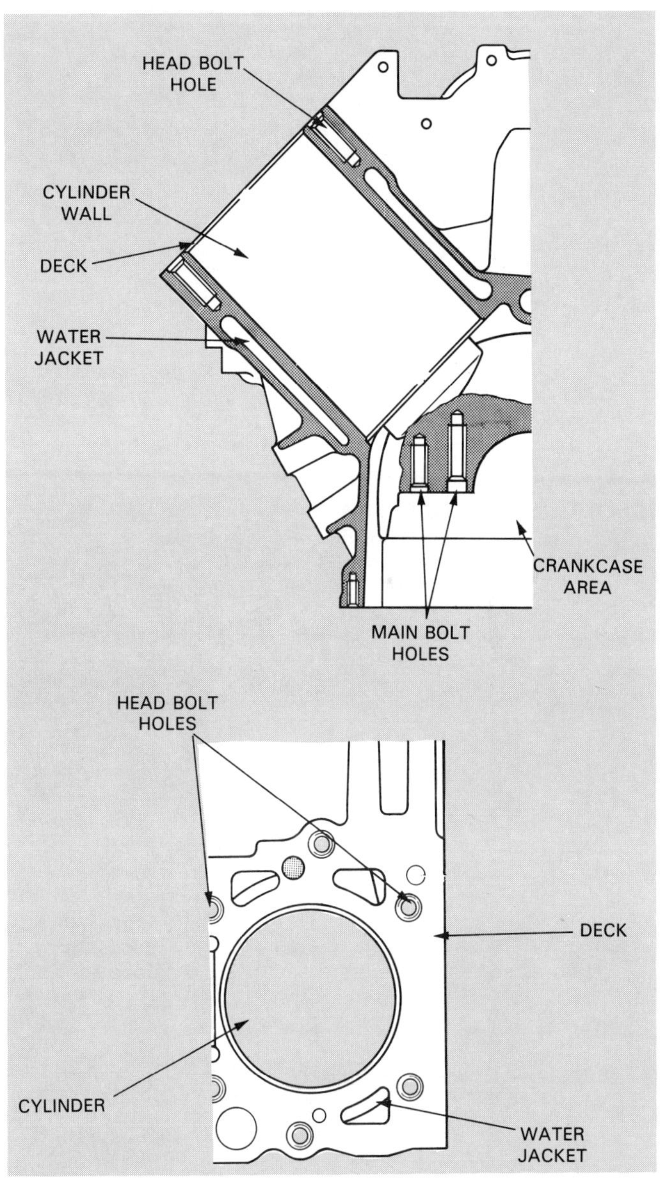

Fig. 9-5. Side and top views of block show bolt holes, openings for coolant, crankcase area, and other parts of block. (Mercedes Benz)

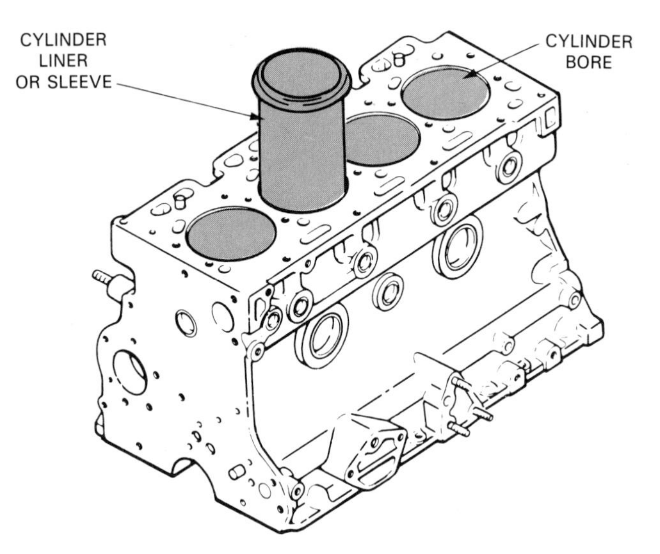

Fig. 9-6. Cylinder sleeves or liners are pressed into cylinder block. (Ford)

A few aluminum blocks, however, do NOT use cast iron sleeves. Termed a *sleeveless aluminum block,* a special process saturates the aluminum with silicone, a much harder metal than aluminum. An acid-like etching solution is coated on the cylinders to ''eat'' away a thin layer of aluminum, leaving the silicone exposed. Then, the rings rub against the harder silicone and not the soft aluminum.

Cylinder block deck

The *cylinder block deck* is a flat, machined surface for the cylinder head. The head gasket and head fit onto the deck surface. See Fig. 9-5.

Bolt holes are drilled and tapped in the deck for *head bolts. Coolant* and *oil passages* allow fluids to circulate through the block, head gasket, and cylinder heads.

Cylinder block sleeves

Cylinder block sleeves or *liners,* just mentioned, are metal pipe-shaped inserts that fit into the block. They act as removable cylinder walls. The pistons slide up and down in them. Sleeves can be installed to repair badly worn, cracked, or scored cylinder walls. Look at Fig. 9-6.

There are two basic types of cylinder sleeves: dry sleeves and wet sleeves, as illustrated in Fig. 9-7.

Dry sleeves

A *dry sleeve* presses into a cylinder that has been bored (machined) oversize. A dry sleeve is relatively thin and is NOT exposed to engine coolant. The outside of a dry sleeve touches the existing walls of the block. This provides support for the thin dry sleeve. See Fig. 9-8.

When a cylinder becomes badly worn or damaged, a dry sleeve can be installed to repair the cylinder block. The original cylinder must be bored almost as large as the outside diameter of the sleeve. Then, the sleeve is pressed into the oversized hole. Next, the inside of the sleeve is machined to the original bore diameter. This allows the use of the *standard* (original) *piston size* (diameter).

Wet sleeves

A *wet sleeve* is exposed to the engine coolant. Therefore, it must be made thicker than a dry sleeve. It must withstand combustion pressure and heat without the added support of the cylinder block, Fig. 9-8.

A wet sleeve will normally have a flange at the top. When the head is installed, the clamping action pushes down on and holds the sleeve into position. The cylinder head gasket keeps the top of the sleeve from leaking.

A rubber or copper O-ring is used at the bottom, and sometimes the top, of a wet sleeve to prevent coolant leakage into the crankcase. The O-ring seal is pinched between the block and liner to form a leakproof joint.

Today's trend is towards light aluminum cylinder blocks with cast iron, wet sleeves. The thick, wet

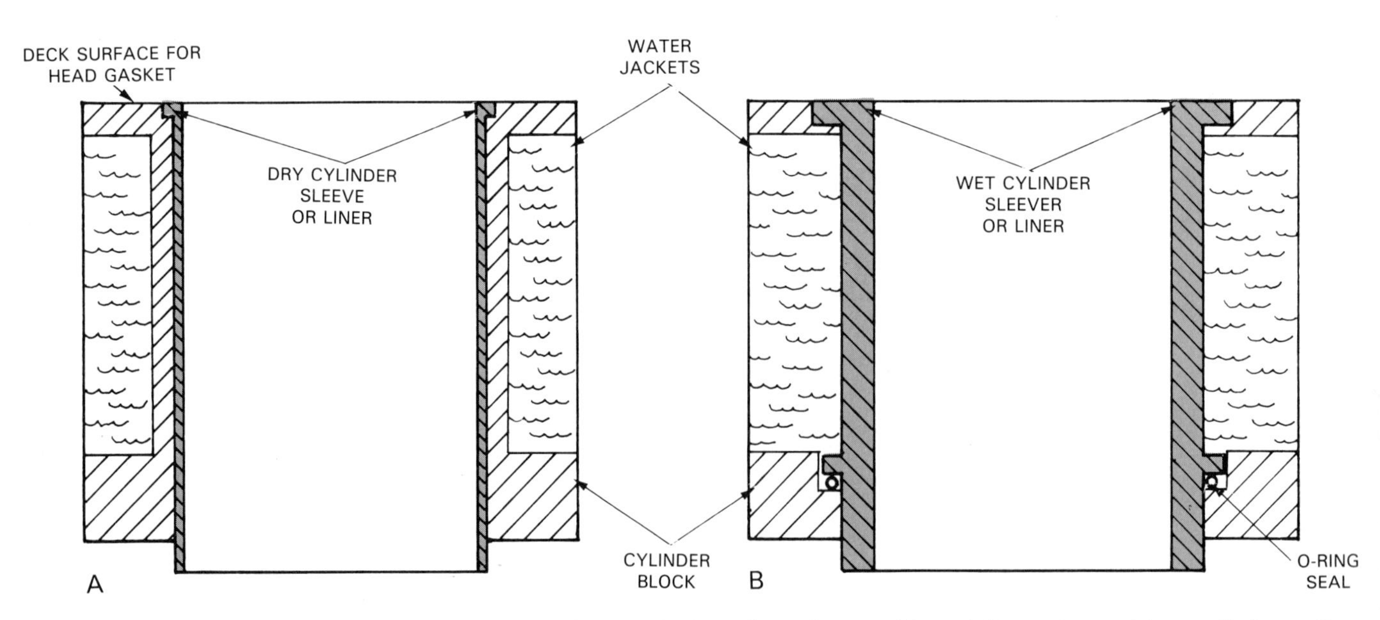

Fig. 9-7. Study basic illustration of wet and dry sleeves. A—Dry sleeve is very thin and fits against exisitng cylinder wall. B—Wet sleeve is thicker and is not supported by wall of block.

Fig. 9-8. Wet sleeves are commonly installed by factory in aluminum or diesel engine. Dry sleeves are used by machine shops to repair damaged, integral cylinder in block so block can be salvaged. (Dana)

sleeves keep a uniform temperature during engine operation. The cast iron sleeves also wear very well. Both charcteristics help increase engine service life.

Fig. 9-9 shows a cutaway view of a modern engine block. Note the use of the aluminum block, cast iron liner, and O-ring. The *head gasket* seals the connection between the cylinder block, cylinder head, and sleeve.

Siamese cylinders

A block with *siamese cylinders* has the cylinders connected with cast metal ribs. Conventional blocks have water jackets between the cylinders. Siamese cylinders do NOT because the extra metal is needed to strengthen the block and hold the cylinders and deck surface in place. Refer to Fig. 9-10.

Siamese cylinders are used when the cylinders are extremely close together or when the block is aluminum and needs reinforcement. Sometimes, grooves are machined in the top of the siamese section to allow limited coolant circulation.

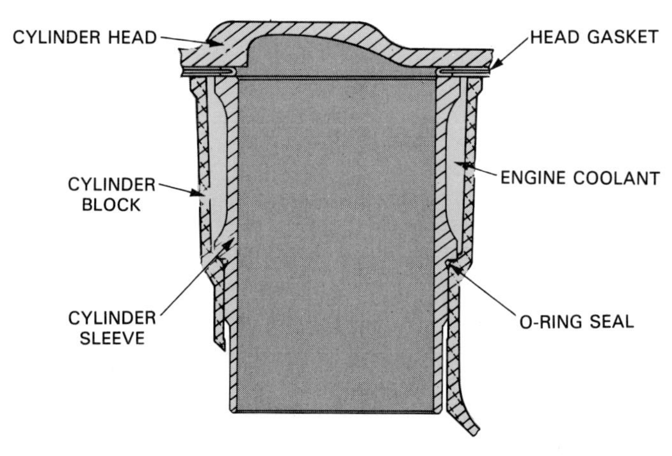

Fig. 9-9. Wet sleeve has O-ring seal at bottom to prevent coolant leakage into crankcase. Head gaskets seals top of cylinder but sometimes O-ring seal is also used at top of sleeve. (Cadillac)

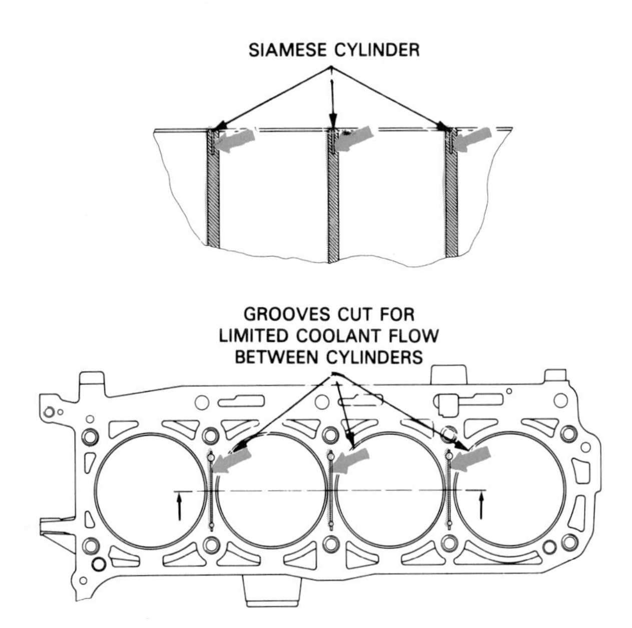

Fig. 9-10. Siamese cylinders have metal rib between each cylinder for added strength. Siamese cylinders were also shown in Fig. 9-4. (Mercedes Benz)

Lifter bores

Lifter bores are precision machined and honed holes in cylinder block on OHV (overhead valve) engines. They are located in the lifter valley.

The *lifter valley* is the area in the block for the lifters and push rods. Lifter bores and the lifter valley area are shown in Figs. 9-11 and 9-12. Some engine designs place the starter motor in the lifter valley.

Cam bore

The *cam bore* is a long hole machined through the cylinder block for the camshaft bearings. One-piece, bushing-type cam bearings are press-fit into the cam bore holes and are locked in place. Oil holes in the block line up with holes in the cam bearings for cam journal lubrication. See Fig. 9-11 again.

Main bore

The *main bore* of a cylinder block is a series of holes for the crankshaft main bearings. They are machined through the lower area of the block, below the cam bore. Split shell main bearings fit into these holds, Fig. 9-13.

Line boring is a term that refers to the machining of the main bore and the cam bore.

A long *boring bar* is rotated through the bore to cut a precisely straight and equal diameter hole for each bearing insert. In an OHC engine, the cam bore would be line bored in the cylinder head, NOT the block.

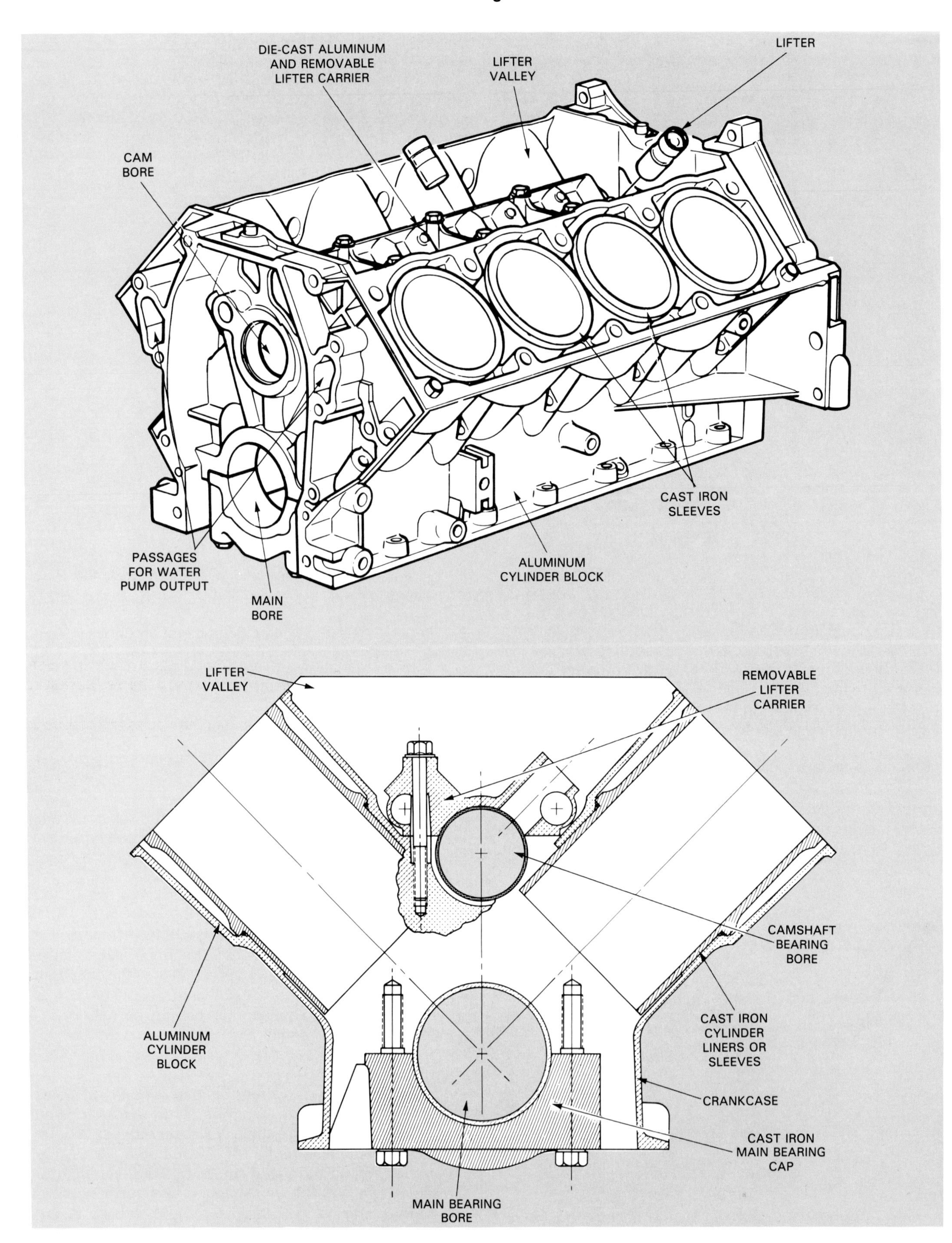

Fig. 9-11. This is a modern aluminum cylinder block with cast iron wet sleeves. It is very light and the iron sleeves wear very well. Also note removable lifter carrier and iron main caps. (Cadillac)

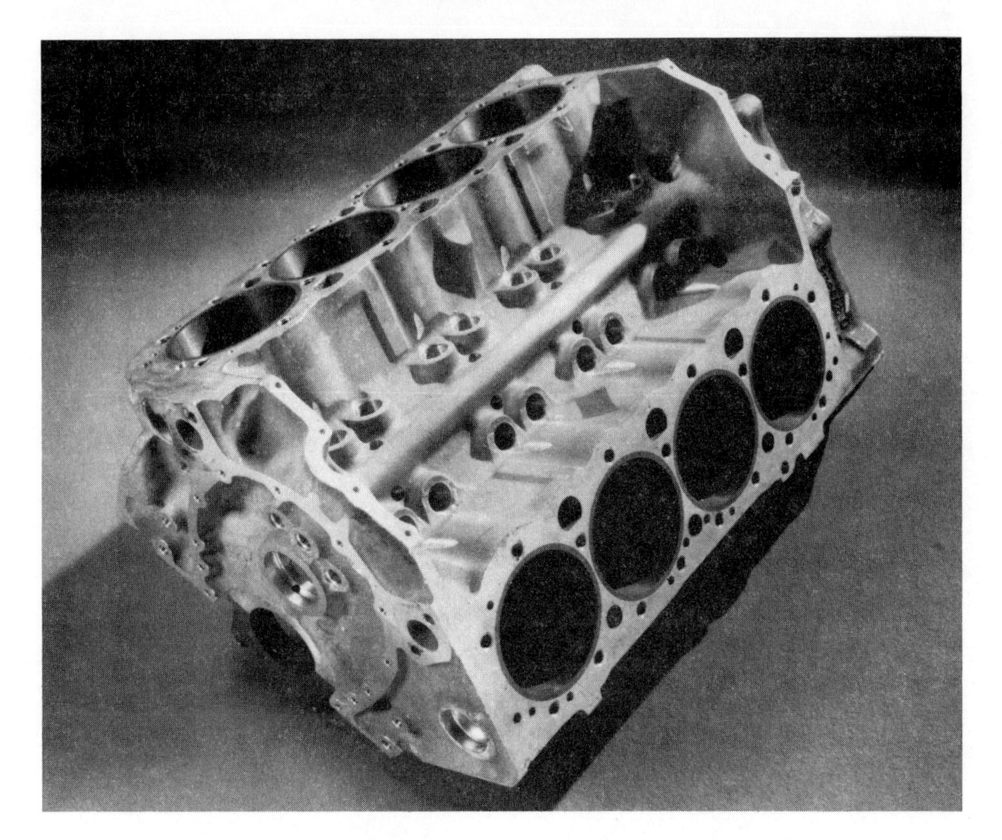

Fig. 9-12. Top view of another aluminum cylinder block with iron sleeves. Aluminum walls in block must be made thicker than iron for strength. Aluminum still reduces weight of engine considerably.

Main caps

Main caps bolt to the bottom of the cylinder block and form one-half of the main bore. Large *main cap bolts* screw into holes in the block to secure the caps to the block. The crankshaft and main bearings are held in place by the main caps. Therefore, they must be very strong, Fig. 9-13.

When being line bored, the main cap bolts are torqued.

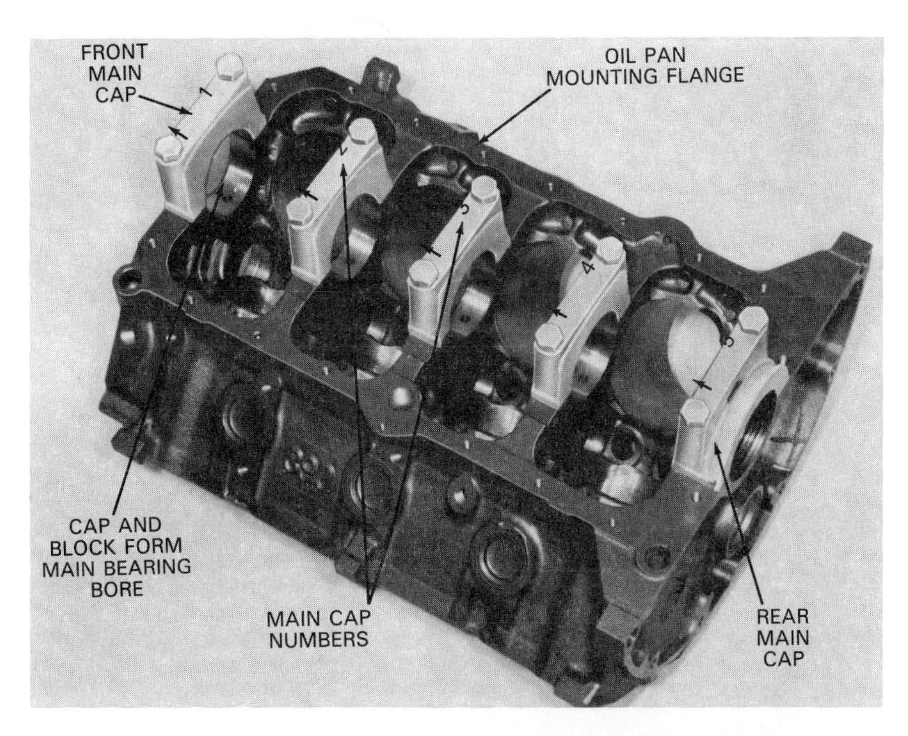

Fig. 9-13. Bottom view of cast iron block shows main caps. Caps are numbered and may have arrows so that they can be reinstalled correctly. Punch mark caps if needed before disassembling block.

to specs. Then the boring bar is run through the bore.

For this reason, you must never mix up the main caps. They CANNOT be interchanged or the bore diameters will become out-of-round. This could lock the main bearings against the crank and damage the bearings and crankshaft. Main caps are usually numbered 1-2-3-4, etc. from front to rear of the block so they will not be mixed up. Arrows are sometimes on the caps to show the front.

There are several types of main cap designs. The most common ones include:

1. TWO-BOLT MAINS (Only two large cap screws hold main caps on cylinder block). See Fig. 9-14A.
2. FOUR-BOLT MAINS (Two large and two small bolts per main cap are used. The bolts screw straight into the block from the bottom), Fig. 9-14B.
3. CROSS-BOLT MAINS (Two conventional bolts screw in through the bottom of the caps. Two more bolts fit through the sides of the block horizontally and screw into the sides of the main caps), Fig. 9-14C.

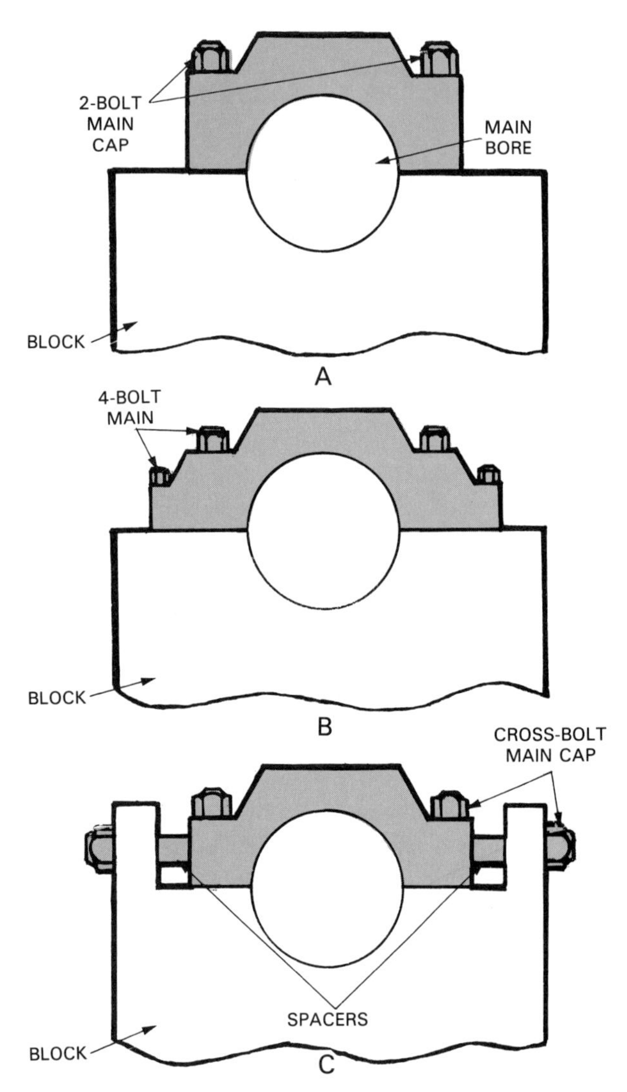

Fig. 9-14. Study basic main cap designs. Two-bolt design is for normal duty. Four-bolt and cross-bolt designs increase holding power of caps for heavy duty or high performance applications.

4. GIRDLE CAP, BEDPLATE, or UNIT CAP (One large main cap is used to hold all the lower main bearings. This is a very strong design because more surface area of the block is used to hold the crankshaft, Fig. 9-15.

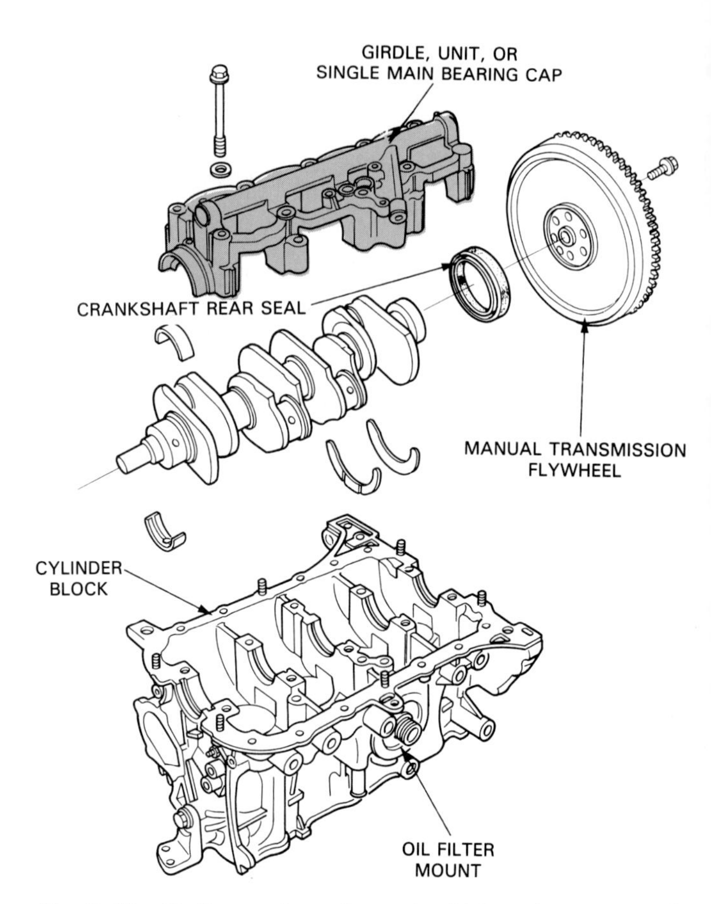

Fig. 9-15. Girdle or unit cap is another high performance main cap design. It provides tremendous strength to hold crank and main bearings in place under high rpm, high power conditions. (Honda)

Main bearings

Main bearings snap-fit into the cylinder block and main caps to provide an operating surface for the crankshaft main journals. The crank main journals slide on the lubricated main bearings as the crank spins.

A *plain main bearing* is a split-type and is used on all but one of the main bore holes. They simply limit up and down movement of the crankshaft in the block.

One *thrust main bearing* is also needed to limit *crankshaft endplay* or crankshaft front-to-rear movement in the block. The main bearing has thrust surfaces on its sides that have a small clearance between the cheeks or thrust surfaces machined on the crankshaft. In this way, clutch action, for example, will only make the crank move a few thousandths of an inch forward in the block.

Thrust washers serve the same purpose as a conventional main bearing. However, they are NOT made as part of the main bearing, Fig. 9-16. The washers fit into the block and main cap, next to the plain main bearing.

Note! Bearing construction is discussed in more detail at the end of the chapter.

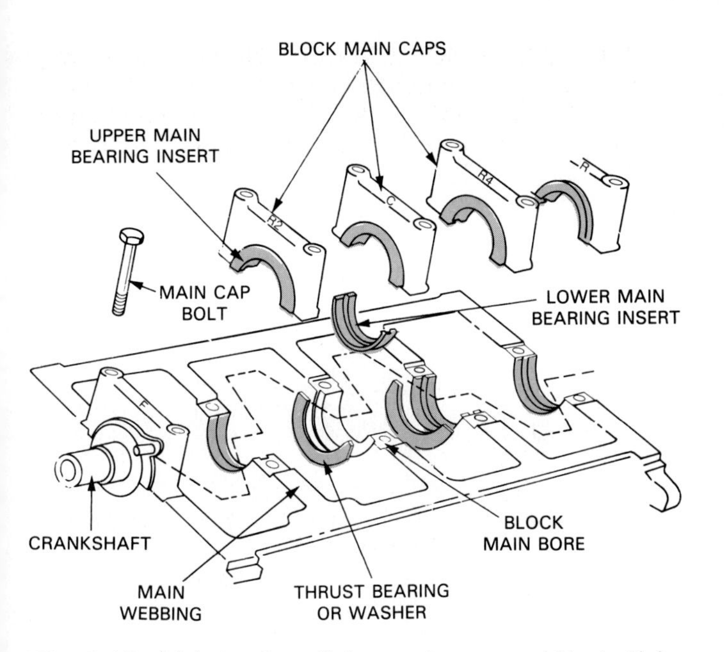

Fig. 9-16. Main bearings fit into main caps and block. Holes machined through caps and block form main bore for these bearings. Main bearings are also called main bearing inserts. (Ford)

Engine crankshafts are usually made of cast iron or forged steel. Forged steel crankshafts are needed for heavy duty applications, such as turbocharged or diesel engines. A steel crankshaft is stiffer and stronger than a cast iron crankshaft. It will withstand greater forces without flexing, twisting, or breaking. Look at Fig. 9-17.

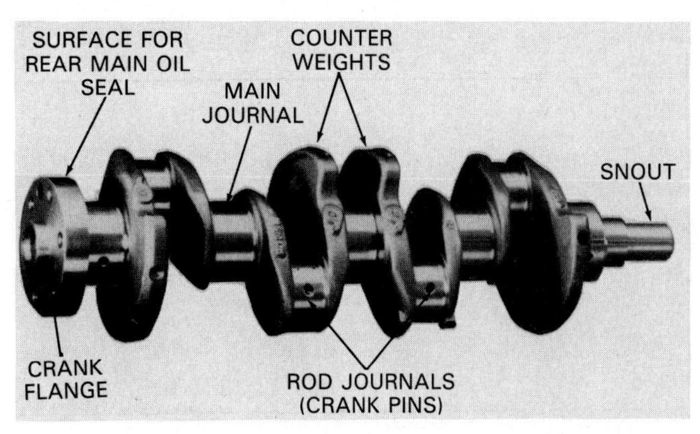

Fig. 9-17. Crankshaft can be made from cast iron or forged steel. Forged crank is for high performance gasoline engines or diesel engines because of its increased rigidity and strength. (Ford)

CRANKSHAFT CONSTRUCTION

The *crankshaft* harnesses the "tons" of force produced by the downward thrust of the pistons on their power strokes. It converts the *reciprocating* (up and down motion) of the pistons into a *rotating* (spinning) motion. The crankshaft fits into the main bore of the block, Fig. 9-16.

Crank main journals

Crankshaft main journals are precision machined and polished surfaces that ride on the main bearings. They are machined along the centerline of the crankshaft. They spin but remain stationary, in relation to the rods and pistons, when the engine is running, Fig. 9-18.

Grooved main journals have a small notch cut around

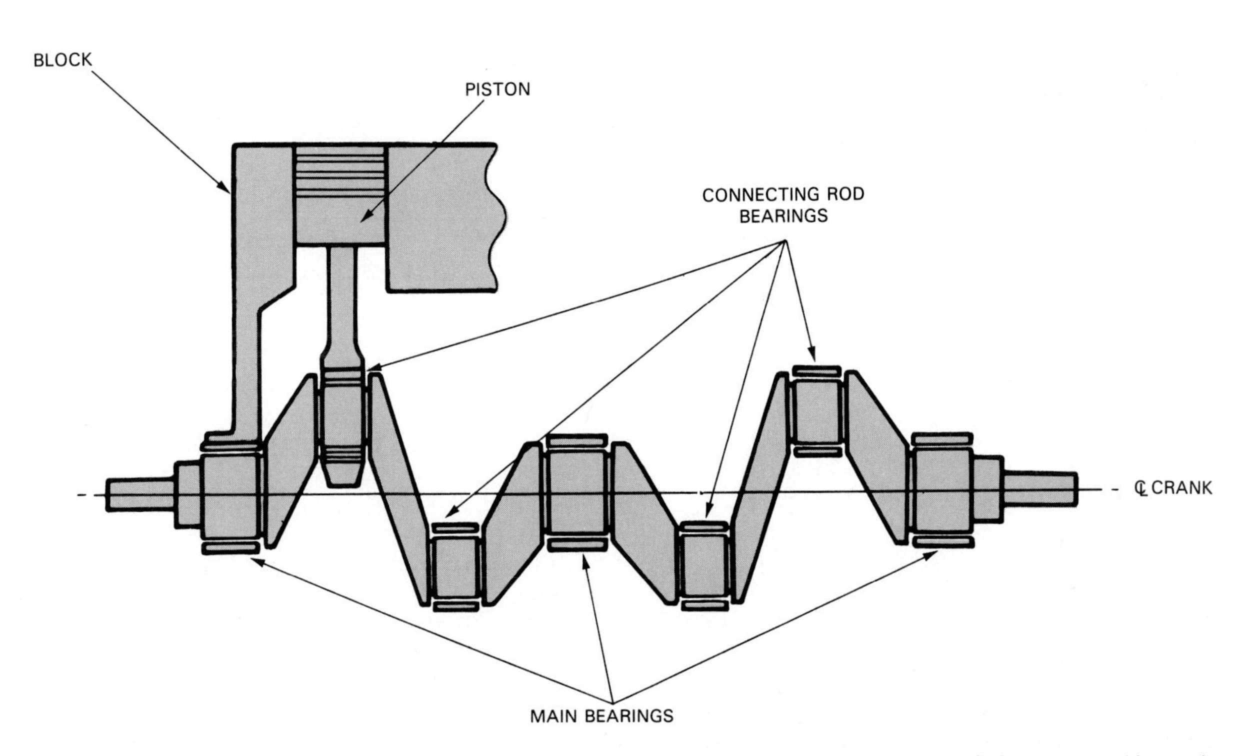

Fig. 9-18. Main or crankshaft bearings fit between main journals and block. Rod bearings fit between rod journals on crank and block.

their surface. The main groove is used to help distribute oil evenly around the journal and main bearing.

Main bearing clearance

Main bearing clearance is the space between the crankshaft main journal and the main bearing insert. The clearance allows lubricating oil to enter and separate the journal and bearing. This allows the two to rotate without rubbing on each other and wearing. A typical bearing clearance is .001 in. (0.25 mm).

Crank rod journals

Crankshaft rod journals, also termed *crankpins,* are also machined and polished surfaces, but they are for the connecting rod bearings. The connecting rods bolt around these journals. The rod journals are offset from the main journals so that the rods and pistons will slide up and down in the block when the crankshaft rotates. Normally, they are NOT grooved like main journals. Look at Figs. 9-19 and 9-20.

Since automobile engines have multiple cylinders, the crankshaft rod journals are arranged so that there is always at least one cylinder on a power stroke. Then, force is always being transmitted to the crankshaft to smooth engine operation. The trend in engine design is to reduce crankpin diameter, which reduces bearing surface velocity and friction in the bearings.

Splayed rod journal

A splayed rod journal has two separate rod journals machined offset from each other on the same crankpin. This design is used on some V-8 and V-10 engines to help smooth engine operation. It staggers the combustion strokes away from each other so the cylinders fire more evenly. See Fig. 9-21.

Roll hardened journal

A *roll hardened journal* has been specially treated during manufacturing. Basically, a large roller is driven around the journal to compress the freshly forged metal and make it harder and more wear resistant.

Explained in later chapters, a machine shop must not

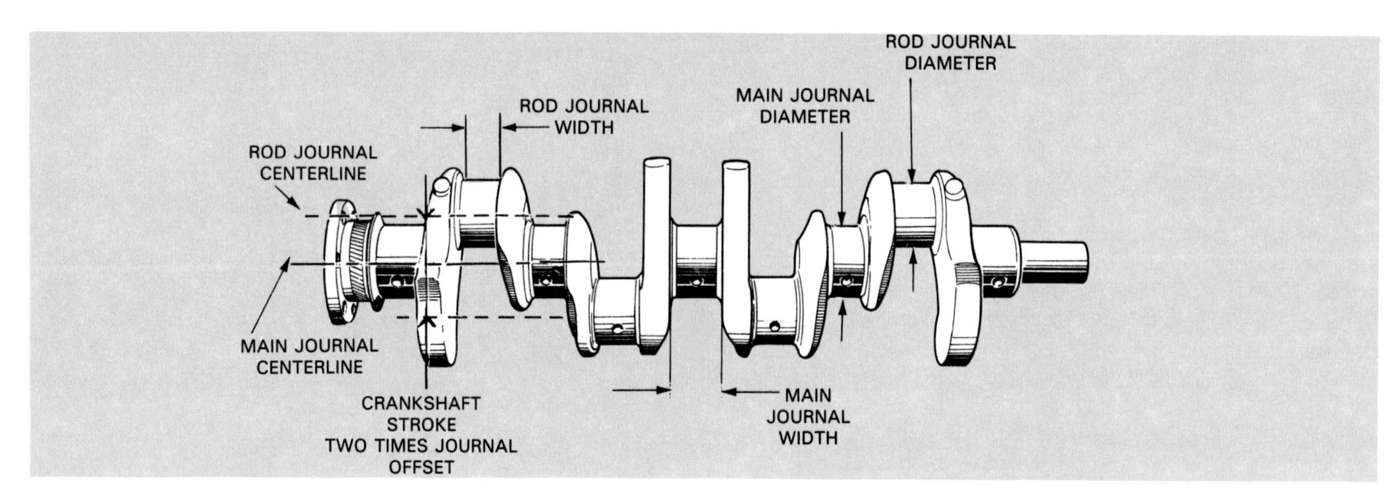

Fig. 9-19. Study basic dimensions of crankshaft. Rod journal offset times two equals stroke of crank.

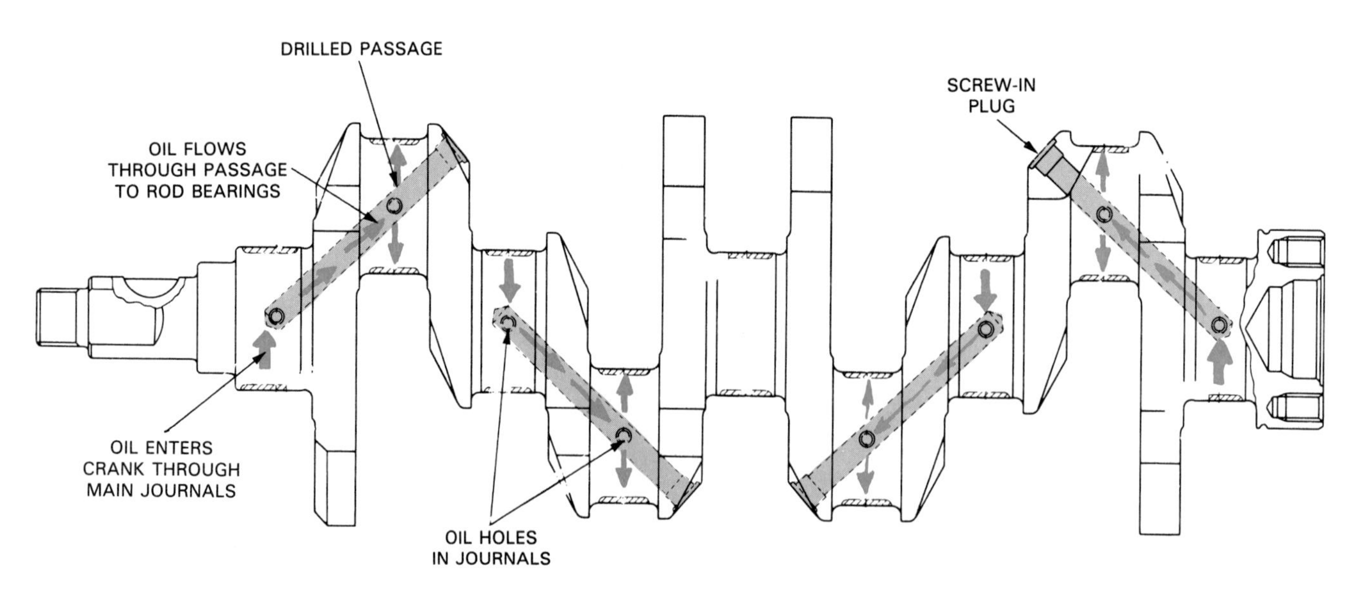

Fig. 9-20. Oil passages through crank allow oil to pass through block, through main bearings, through crank, and into rod bearing clearance.

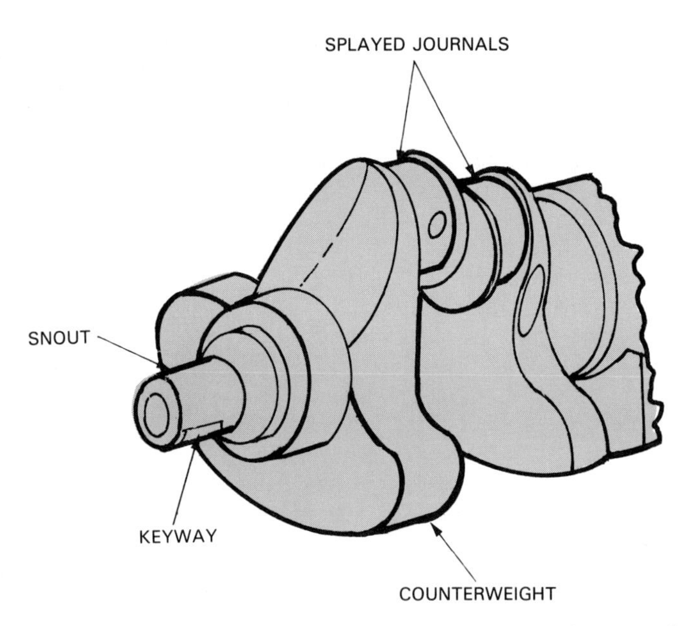

Fig. 9-21. This is a splayed rod journal. It is sometimes used in V-6 engines to help smooth operation.

grind through this hardened layer of metal when reconditioning a crankshaft or the softer, more wear prone metal will be exposed. Crankshaft service life can then be reduced.

Crankshaft cheeks

The *crankshaft cheeks* or thrust surfaces are located on each side of the main and rod journals. This is the area where the sides of the grinding stone rub on the crankshaft during its initial machining. The distance between cheeks determines *journal width,* Fig. 9-19.

Crankshaft oil passages

Crankshaft oil passages are cast or machined holes through the crank for lubricating oil. The oil pump forces oil through the block, main bearings, and into the crank. Then the oil can flow through the crank oil passages and to the engine bearings.

Fig. 9-20 shows the oil passages in one crankshaft. Note how the holes have been drilled. Screw-in or press-in plugs are installed on the ends of the drilled oil passages.

Crankshaft counterweights

Counterweights are heavy lobes on the crankshaft to help prevent engine vibration. They are used to counteract the weight of the pistons, piston rings, connecting rods, and rod bearings. They are normally formed as part of the crankshaft. However, on a few older engines, they bolt to the crankshaft. Refer to Figs. 9-19 through 9-21. A *fully counterweighted crankshaft* has weights formed opposite every crankpin. A *partially counterweighted crankshaft* only has weights formed on the center areas of the crankshaft. A fully counterweighted crankshaft will operate with less vibration.

Crankshaft snout

The *crankshaft snout* is a shaft machined on the front of the crankshaft to hold the damper, camshaft drive sprocket or gear, and crank pulleys. It sticks through the front of the block and front cover. A keyway is cut into the snout to keep the sprocket, gear, damper, or pulley from spinning or slipping, Fig. 9-21.

Crankshaft flange

The *crankshaft flange* is machined on the rear of the crank to hold the engine flywheel. It is a round disc with threaded holes for the flywheel bolts. Look at Fig. 9-22.

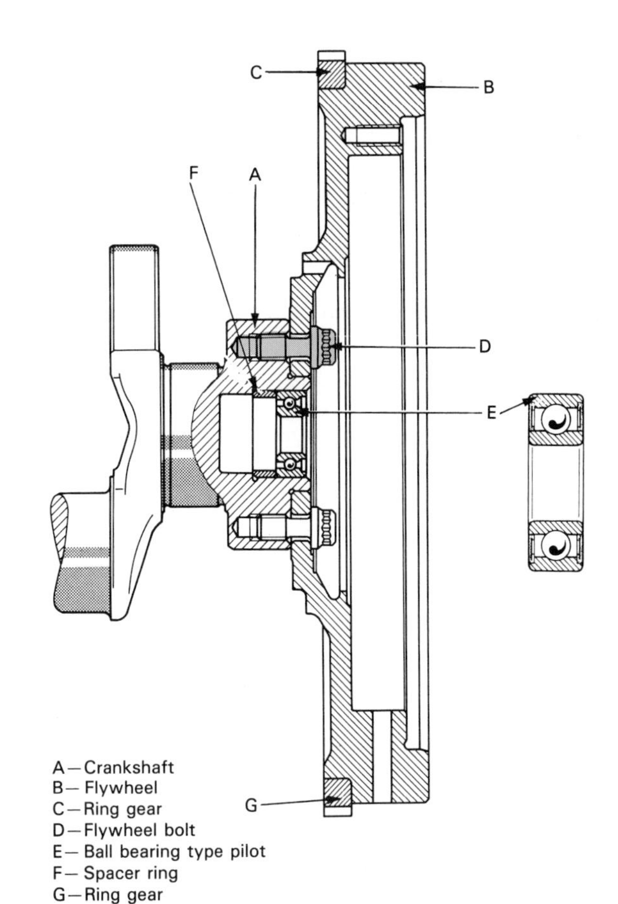

Fig. 9-22. Flywheel bolts to crank flange. It helps smooth engine operation by helping to keep crank spinning on non-power producing strokes. It may also hold ring gear for starting motor engagement. (Mercedes Benz)

Flywheel

A *flywheel* is a large steel disc mounted on the rear flange of the crankshaft. It can have several functions:

1. A flywheel for a car with a manual transmission is heavy and helps smooth engine operation, Fig. 9-22.
2. The flywheel connects the engine crankshaft to the transmission or transaxle. Either the manual clutch or the automatic transmission torque converter bolts to the flywheel.
3. The flywheel commonly has a large *ring gear* that allows engine starting. A small gear on the starting motor engages the flywheel ring gear and turns it.

Hardened *flywheel bolts* secure the flywheel to the crankshaft. The holes in the crank and flywheel are usually staggered unevenly so that the flywheel will only install in ONE POSITION. This assures that the engine stays in balance when the flywheel is removed and then reinstalled. The flywheel is centered on the crank by a hub or lip machined on the crank flange.

Pilot bearing

A *pilot bearing* is mounted in the rear of the crankshaft to support the manual transmission input shaft, Fig. 9-22. It can be a bushing type bearing or a roller type bearing. It is press-fit into a pocket machined in the crank.

An engine coupled to an automatic transmission does NOT need a pilot bearing. The large hub on the torque converter normally fits directly into the crankshaft.

Crankshaft oil seals

Crankshaft oil seals keep oil from leaking out the front and rear of the engine. The oil pump forces oil into the main and rod bearings. This causes oil to spray out of the bearings. Seals are placed around the front and rear of the crank to contain this oil, Fig. 9-23.

The *rear main oil seal* fits around the rear of the crankshaft to prevent oil leakage. It can be a one-piece or a two-piece seal. The seal lip rides on a smooth, machined and polished surface on the crank. The front seal is covered later, Fig. 9-24.

A *two-piece neoprene rear oil seal* usually installs in a groove cut into the block and rear main cap. The seal

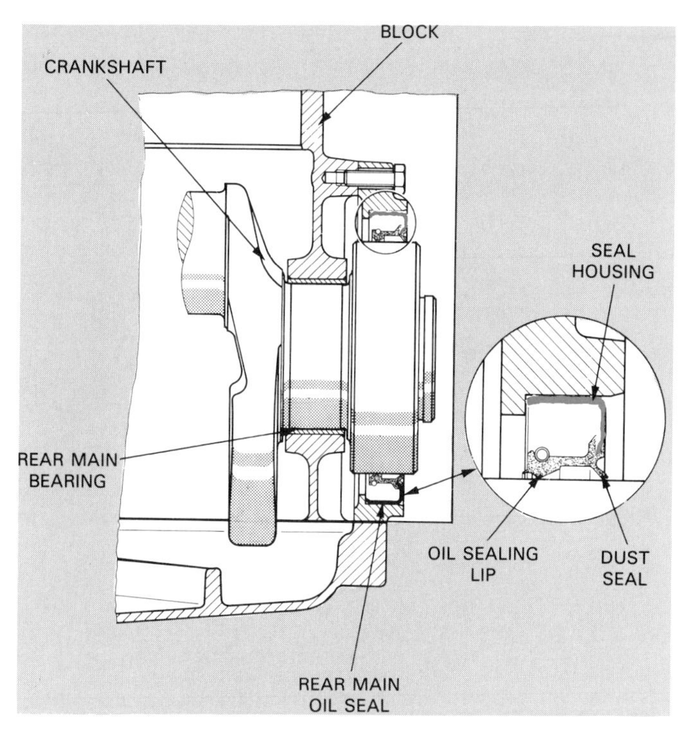

Fig. 9-23. Rear main oil seal keeps oil from spraying out rear main bearing. Oil pressure tends to push sealing lip tight against crank. (Mercedes Benz)

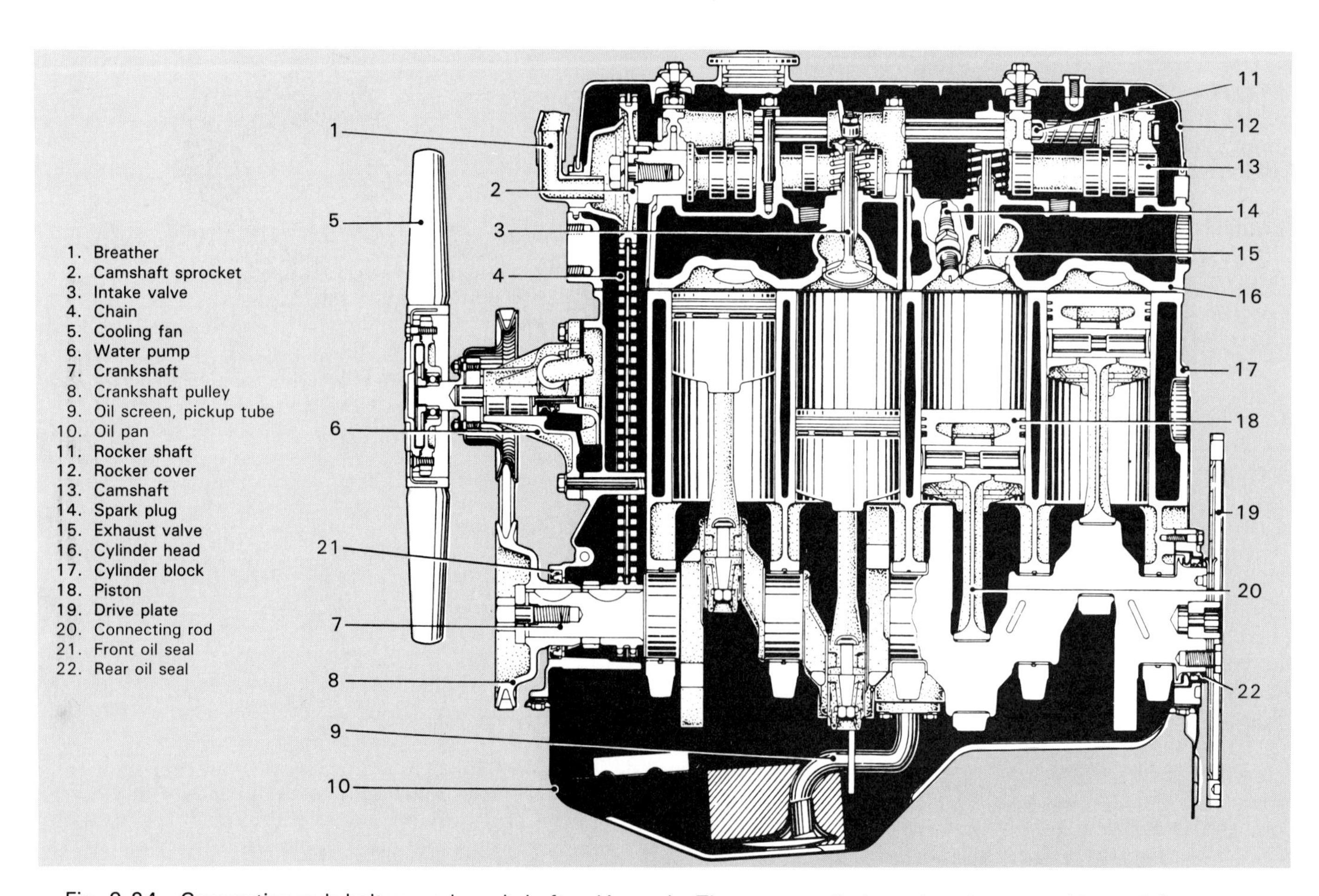

Fig. 9-24. Connecting rods bolt around crankshaft rod journals. They are usually forged so they can withstand the tremendous loads as pistons change direction in cylinder. (Toyota)

has a lip that traps oil and another lip that keeps dust and dirt out of the engine. The sealing lips ride on a machined surface of the crankshaft. *Spiral grooves* may be used on this crank surface to help throw oil inward to prevent leakage.

A *rope* or *wick rear oil seal* is simply a woven rope impregnated (filled) with graphite. One piece of the rope seal fits into a groove in the block. Another piece fits a groove in the main cap. This type seal is being replaced by one and two-piece neoprene seals.

A *one-piece neoprene rear oil seal* fits around the rear flange on the crankshaft. It has sealing lips similar to a two-piece neoprene seal.

CONNECTING ROD CONSTRUCTION

The *connecting rod* fastens the piston to the crankshaft. It transfers piston movement and combustion pressure to the crankshaft rod journals. The connecting rod also causes piston movement during the non-power producing strokes (intake, compression, and exhaust). See Fig. 9-25.

Most connecting rods are made of steel. The rod must withstand tremendous force as the piston moves up and down in the cylinder. Connecting rods normally have an *I-beam shape* because of this shape's high strength-to-weight ratio. Refer to Fig. 9-26.

At 55 mph, a rod and piston will be stationary at TDC. They will then accelerate to about 60 mph halfway down in the cylinder and decelerate to zero at BDC. This metal shearing action happens in about 3 in. of travel and is repeated millions of times during the life of an automotive engine.

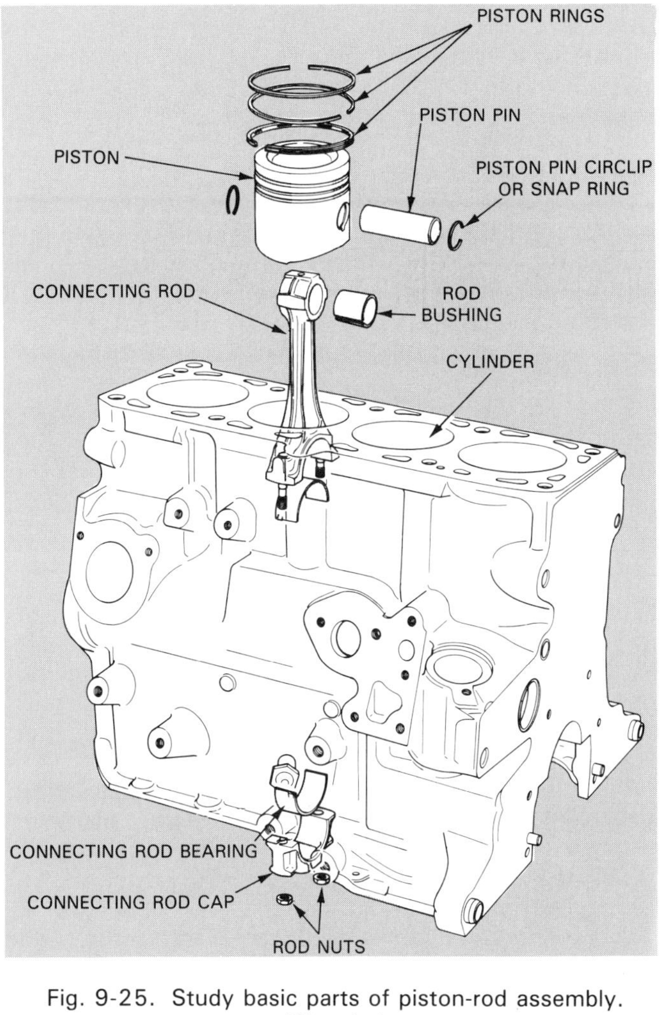

Fig. 9-25. Study basic parts of piston-rod assembly. (Chrysler)

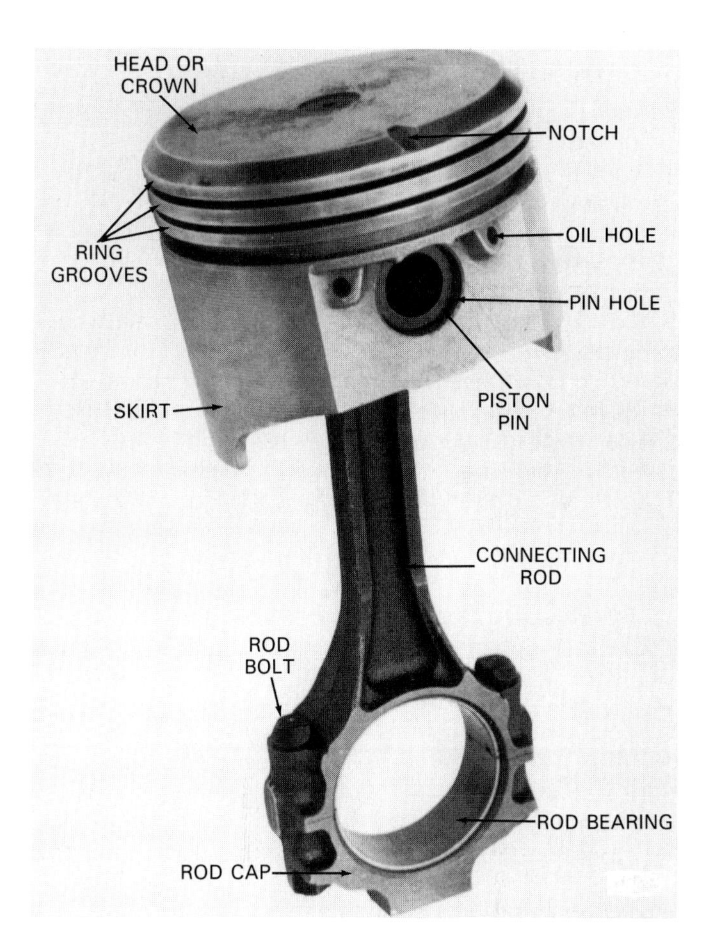

Fig. 9-26. Rod bearing fits snugly in rod big end. Small end of rod holds piston pin. (Chevrolet)

Rod oil holes

Some connecting rods have an *oil spurt hole* that provides added lubrication for the cylinder walls, piston pin, and other surrounding parts. Oil from inside the crankshaft will spray out when holes align in the crank journal, bearing, and spurt hole, Fig. 9-27A.

A *drilled connecting rod* has a machined hole its entire length. Its purpose is to supply oil to the piston pin. Refer to Fig. 9-27B.

Connecting rod cap

The *connecting rod cap* bolts to the bottom of the connecting rod body. It can be removed for disassembly of the engine. Refer again to Fig. 9-26.

The connecting *rod big end* or *lower end* is the hole machined in the rod body and cap. The connecting rod bearing fits into the big end.

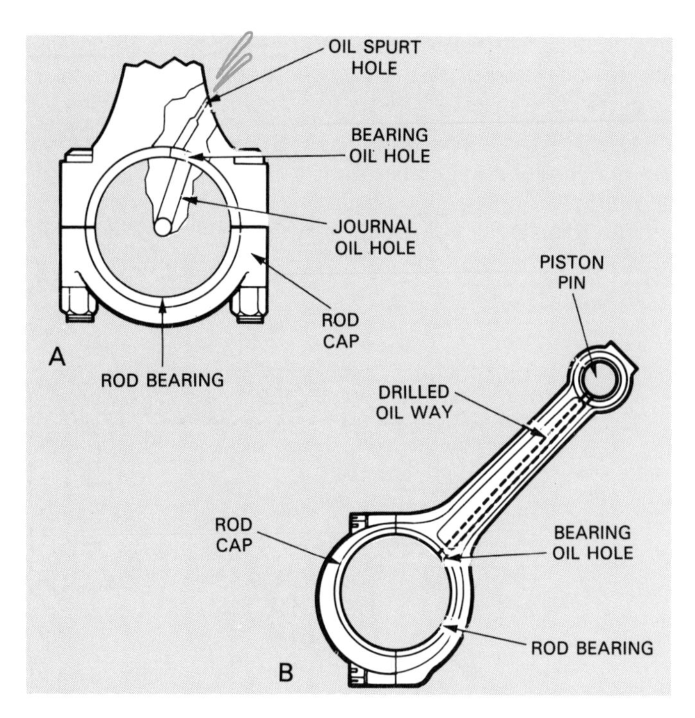

Fig. 9-27. A—When holes line up, oil can squirt out hole to provide extra lubrication for cylinder walls. B—Drilled connecting rod provides positive lubrication for piston pin. (Federal-Mogul)

Connecting *rod bolts* and *nuts* clamp the rod cap and rod together. They are special high tensile strength fasteners. Some rods use cap screws without a nut. The cap screw threads into the rod itself.

The connecting rod and the rod cap are generally produced as a unit. In most cases, the cap is cut (machined) from the rod during manufacture, creating a smooth mating surface. However, the cap on a *broken-surface connecting rod* is scribed and cracked off to produce a rough, irregular mating surface, Fig. 9-28. This surface helps lock the rod and cap in perfect alignment. It also prevents the components from shifting during engine operation. Although a broken-surface connecting rod cannot be rebuilt, oversize rod bearings can be installed during an engine overhaul.

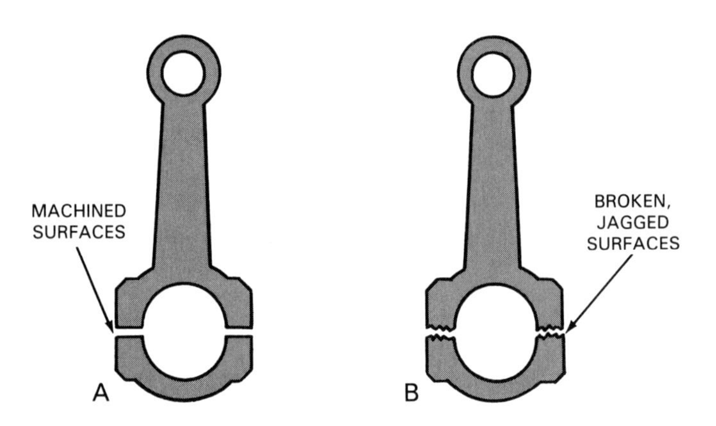

Fig. 9-28. Connecting rod assemblies. A—Conventional connecting rod. B—Broken-surface connecting rod.

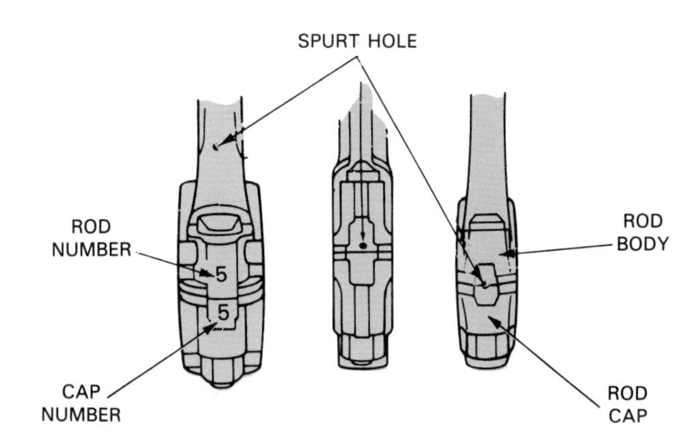

Fig. 9-29. Rod cap is bolted to rod body. Then big end bore is machined. Rod numbers must match to assure same cap is on rod. Also note oil spurt holes. (Chrysler)

Connecting rod numbers

Connecting rod numbers are used to assure proper location of each connecting rod in the engine. They also assure that the rod cap is installed on the rod body correctly. Look at Fig. 9-29.

When being manufactured, connecting rod caps are bolted to the connecting rods. Then, the big end holes are machined in the rods. Since the holes may NOT be perfectly centered, rod caps must NOT be mixed up or turned around.

If the cap is installed without the rod numbers in alignment, the bore will NOT be perfectly round. Severe rod, crankshaft, and bearing damage will result.

Connecting rod bearings

The *connecting rod bearings* ride on the crankshaft rod journals. They fit between the connecting rods and the crankshaft, as was shown in Figs. 9-25 and 9-26. The rod bearings are also removable inserts, as are main bearings.

Rod bearing clearance is the small space between the rod bearing and the crankshaft journals. As with the main bearings, it allows oil to enter the bearing. The oil prevents metal-to-metal contact that would wear out the crank and bearings.

A *rod bushing* is normally used in the small end bore. It is a one-piece type pressed into the rod.

Connecting rod dimensions

Connecting rod length is measured from the center of the pin hole to the center of the big end bore. Special machine shop fixtures are available for checking rod length. Rod length is important when doing engine service because two connecting rods may look the same but may have slightly different lengths.

Small end diameter is the distance across the bushing in the top of the rod. *Big end diameter* is the distance across the big end bore with the bearing removed. As you will learn, both of these are checked during an engine overhaul to make sure the rod is in good condition.

Connecting rod width is the distance across the sides of the big end. Rod width partially determines *rod side play* (distance rod can move sideways when installed on crankshaft). Look at Fig. 9-30.

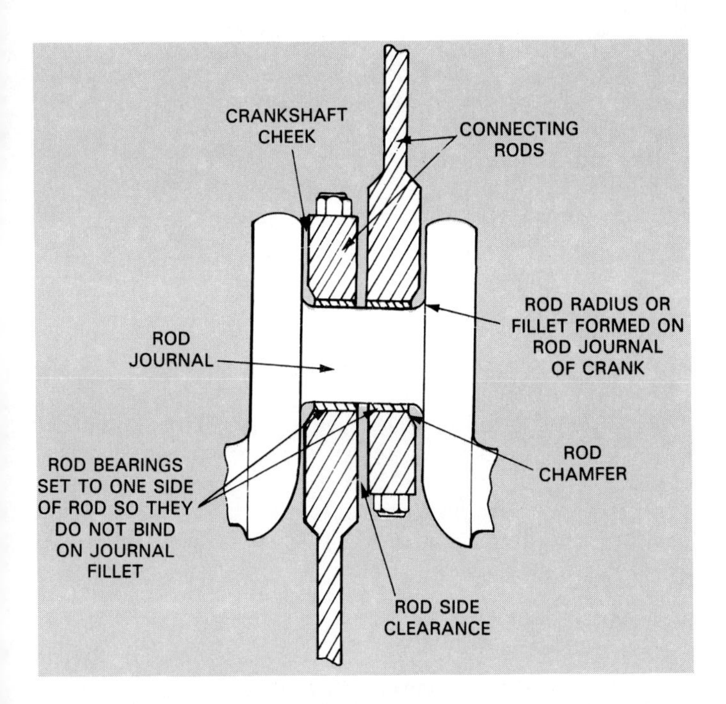

Fig. 9-30. On V-type engines, two connecting rods bolt to same journal. Rod bearings are offset to one side of rods so that they will not bind up on round fillet formed on outer edges of journal. Rod side clearance is determined by difference between thickness of two rods and width of journal.

PISTON PIN CONSTRUCTION

The *piston pin,* also called *wrist pin,* allows the piston to swing on the connecting rod. The pin fits through the hole in the piston and the connecting rod small end. This is pictured in Figs. 9-25 and 9-26.

Piston pins are normally made of casehardened steel. The hollow piston pin is also machined and polished to a very precise finish.

Casehardening is a heating and cooling process that increases the wear resistance of the piston pin. It hardens the outer layer of metal on the pin. The inner metal remains unhardened so the pin is not too brittle.

Piston pins in modern engines are normally held in the piston by one of two means: snap rings or a press-fit.

Floating piston pin

A *full floating piston pin* is secured by snap rings and is free to rotate in both the rod and piston. The pin is free to "float" in both the piston pin bore and the connecting rod small end. See Fig. 9-31.

A bronze bushing is usually used in the connecting rod. The piston pin hole serves as the other bearing surface for the pin. The *snap rings* fit into grooves machined inside the piston pin hole. Full floating piston pins are better than press fit pins since they reduce friction and wear.

Press-fit piston pin

A *press-fit piston pin* is forced tightly into the connecting rod's small end. It can rotate freely in the piston pin hole. The pin is NOT free, however, to move in the connecting rod. This holds the pin inside the piston and prevents it from sliding out and rubbing on the cylinder.

The press-fit piston pin is a very dependable design. It is also inexpensive to manufacture. The trend is towards the press-fit piston pin in today's engines.

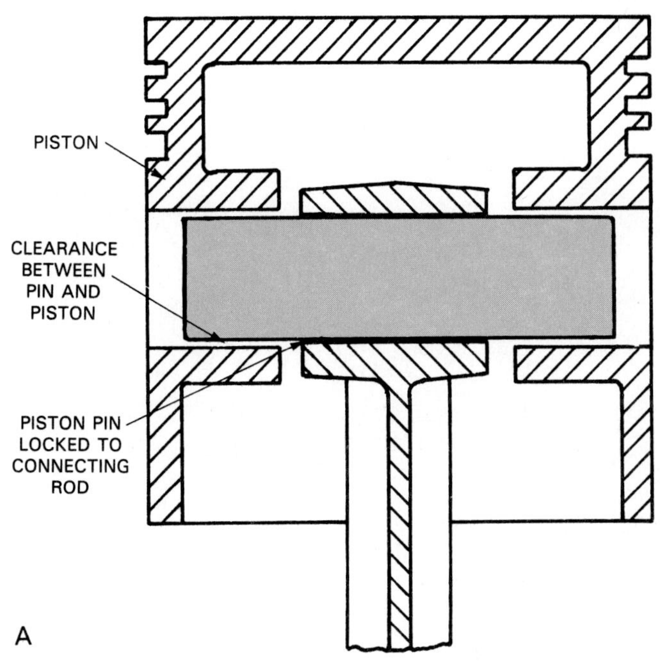

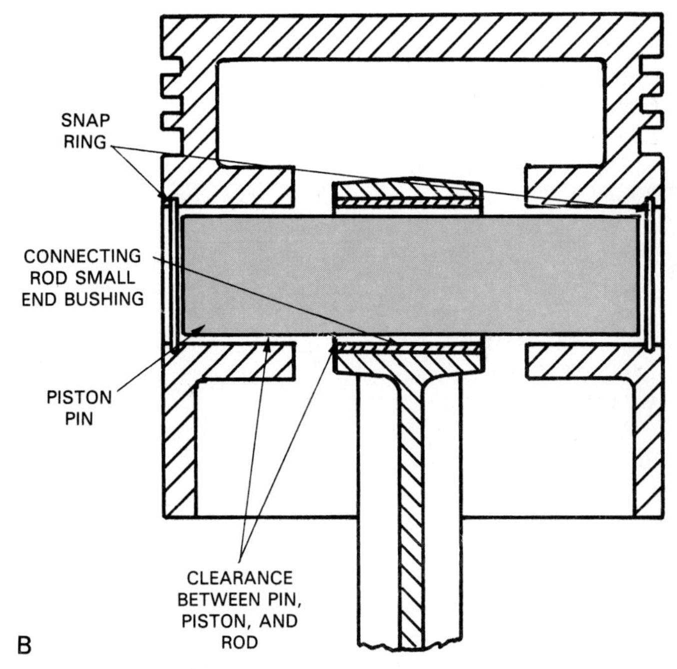

Fig. 9-31. Two modern methods of holding piston pin in piston. A—Press-fit pin is locked in connecting rod small end. This keeps pin from sliding out and ruining cylinder wall. B—Floating piston pin is free to rotate in both piston and rod. Snap rings keep pin inside piston.

Piston guided connecting rod

A *piston guided connecting rod* has thrust surfaces formed next to the piston pin to limit axial movement of the rod small end. Most engines use only the rod big end and crank to limit rod side-to-side movement. As shown in Fig. 9-32, a few high performance engines have two thrust surfaces at the inner faces of the wrist pin bore. This keeps the rod from cocking sideways and sliding to one side of the pin under a heavy load.

Piston pin offset

Piston pin offset locates the piston pin hole slightly to one side of the piston centerline to quiet piston operation. The pin hole is moved toward the piston's *major thrust surface* (piston surface pushed tightly against cylinder during power stroke).

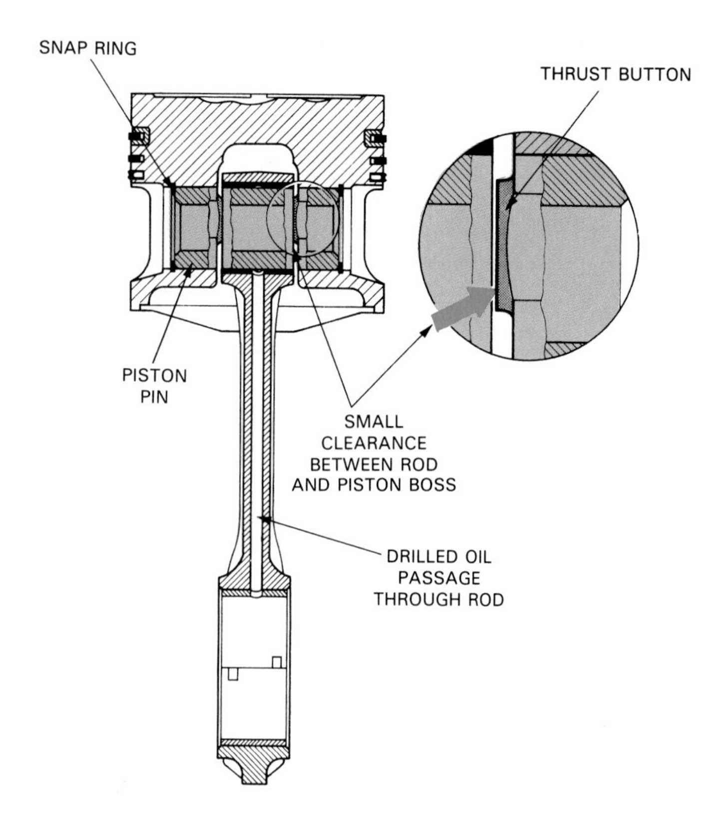

Fig. 9-32. Piston-guided connecting rod is excellent design. Boss thrust surface will only allow rod small end to move slightly from side-to-side under heavy load. Most engines use only rod-to-crank side clearance to try to control side movement at top of rod. (Mercedes Benz)

If the pin hole is centered in the piston, the piston could *slap* (knock) in the cylinder. As the piston moved up in the cylinder, it could be positioned opposite the major thrust surface. Then, during combustion, the piston could be rapidly pushed to the opposite side of the cylinder, producing a KNOCK SOUND.

With pin offset, the piston tends to be pushed against its major thrust surface. This reduces its tendency to slap sideways in the cylinder.

A *piston notch,* arrow, etc., on the head of the piston is used to indicate piston pin offset and the front of the piston. The piston may also have the word ''front'' stamped on it. This information lets you know how to position the piston in the block for correct location of piston pin offset.

PISTON CONSTRUCTION

The engine *piston* transfers the pressure of combustion (expanding gas) to the connecting rod and crankshaft. It must also hold the piston rings and piston pin while operating in the cylinder.

Fig. 9-33 shows a cutaway view of a piston. Study this illustration as the piston is described.

Forged and cast pistons

An engine piston is normally cast from aluminum. However, forged aluminum pistons may be found in turbocharged, fuel injected, diesel, or other engines that expose the pistons to severe stress.

Since aluminum is very light and relatively strong, it is an excellent material for engine pistons. When an engine is running, the piston must withstand tremendous heat and pressure.

Parts of a piston

The *piston head* or *crown* is the top of the piston. It is exposed to the heat and pressure of combustion. This area must be thick enough to withstand these forces. It must also be shaped to match and work with the shape

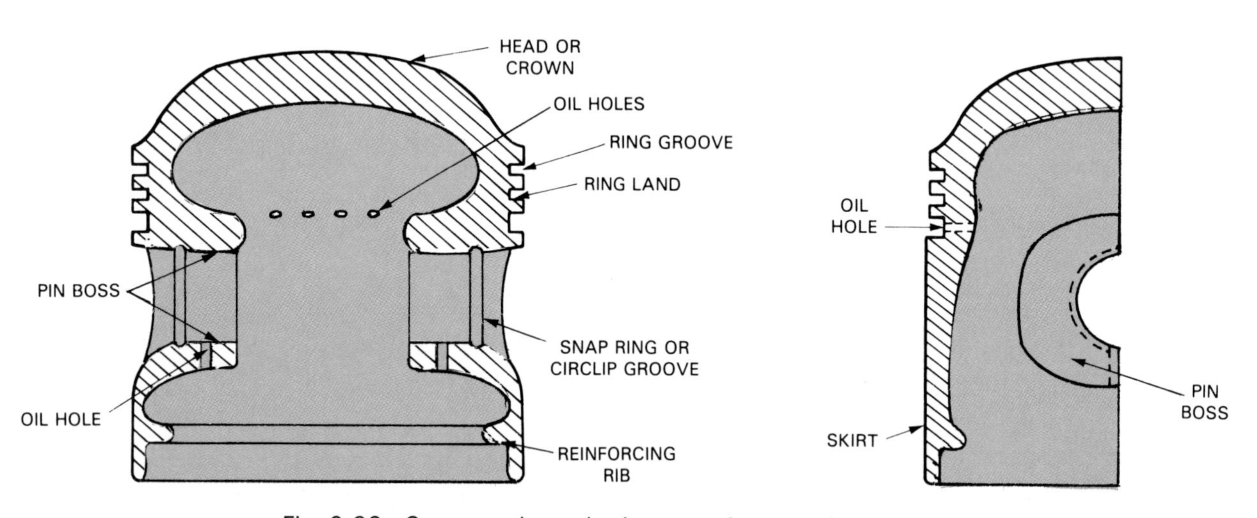

Fig. 9-33. Cutaway shows basic parts of piston. Study them.

of the combustion chamber for complete combustion.

Piston ring grooves are slots machined in the piston for the piston rings, Fig. 9-33. The upper two grooves hold the compression rings. The lower piston groove holds the oil ring.

Oil holes in the bottom groove allow the oil to pass through the piston. The oil then drains back into the crankcase.

The *ring lands* are the areas between and above the ring grooves. They separate and support the piston rings as they slide on the cylinder.

A *piston skirt* is the side of the piston below the last ring. It keeps the piston from tipping in its cylinder. Without a skirt, the piston could cock and jam in the cylinder, Fig. 9-33.

The *piston boss* is a reinforced area around the piston pin hole. It must be strong enough to support the piston pin under severe loads.

A *pin hole* is machined through the pin boss for the piston pin. It is slightly larger than the pin, Fig. 9-33.

Piston shapes

Piston shape generally refers to the shape of the piston head or crown. Usually, a piston head is shaped to match and work with the shape of the cylinder head combustion chamber.

A *flat top piston* has a crown that is almost flat and parallel with the block's deck surface, Fig. 9-34A. A flat top piston is commonly used with a wedge or pancake type cylinder head.

A *domed* or *pop-up piston* has a head that is curved upward, Fig. 9-34B. This type is normally used with a hemi type cylinder head. The piston crown must be enlarged to fill the domed combustion chamber and produce enough compression pressure.

A *valve relief piston* has small indentations either cast or machined in the piston crown, Fig. 9-34C. They provide ample piston-to-valve clearance. Without valve reliefs, the valve heads could strike the pistons.

A *dished piston* has most of the piston head recessed to lower compression. The crown is concave, as shown in Fig. 9-34D. It might be found in a turbocharged engine.

Other special piston shapes are also available. For example, Fig. 9-35 shows a piston that is partially flat top, partially dished, and has a small valve relief. It is designed for a diesel engine. Also note the steel insert on both sides of the top compression ring.

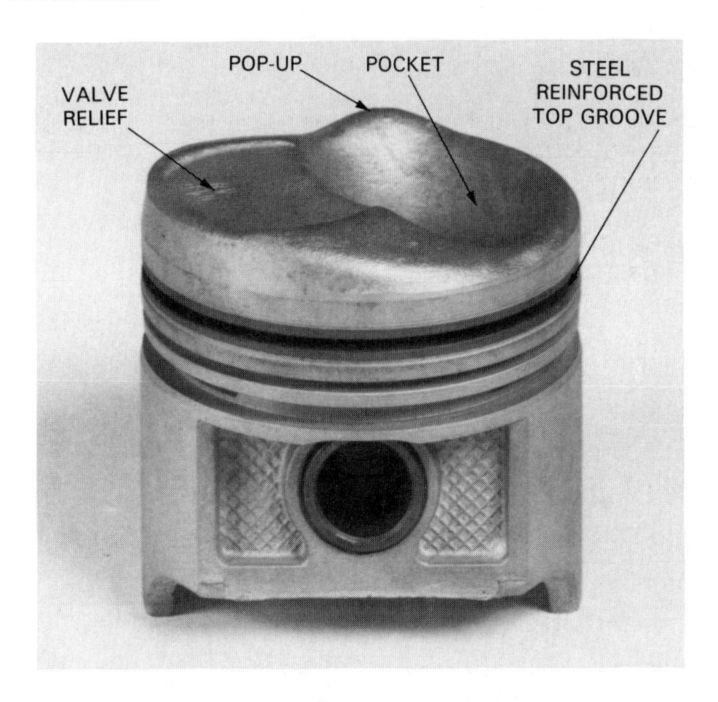

Fig. 9-35. This diesel piston has steel reinforced top ring groove, valve relief, and pocket for fuel under precombustion chamber and injector. (Dana)

Skirt design

A *slipper skirt* is produced when the portion of the piston skirt below the piston pin ends are removed. A slipper skirt provides clearance between the piston and crankshaft counterweights. The piston can slide farther down in the cylinder without hitting the crankshaft.

A *straight skirt* is flat across the bottom, as in Fig. 9-33. This style is no longer common in auto engines that are oversquare.

Piston dimensions

Fig. 9-36 illustrates several piston dimensions. These include:

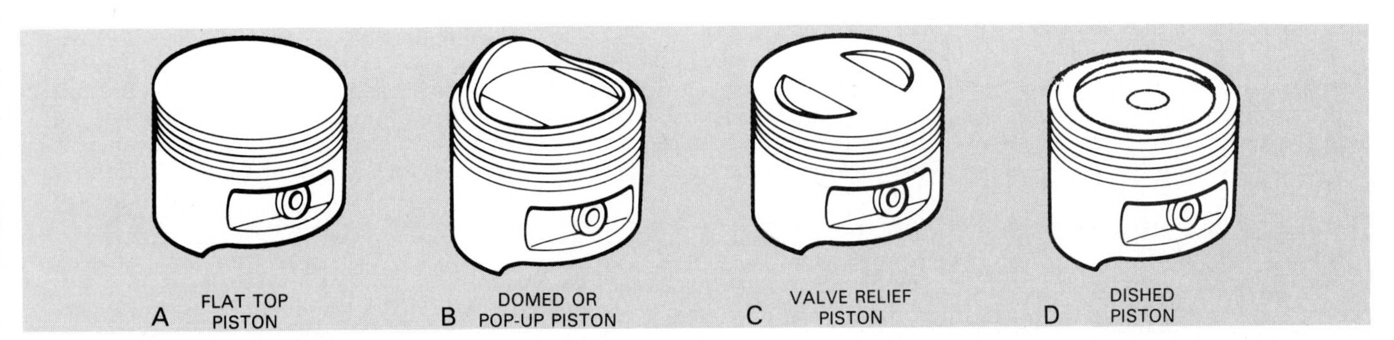

Fig. 9-34. Shape of piston head is common way of classifying piston. A—Flat top piston head is almost smooth and flat. B—Domed piston has bump on its head to raise compression ratio and power output. It is also used with hemi combustion chamber to fill chamber and maintain compression. C—Valve reliefs are used to prevent valve heads from striking piston heads. D—Dished piston caves in and lowers compression ratio. Dish also provides more valve clearance. (Ford)

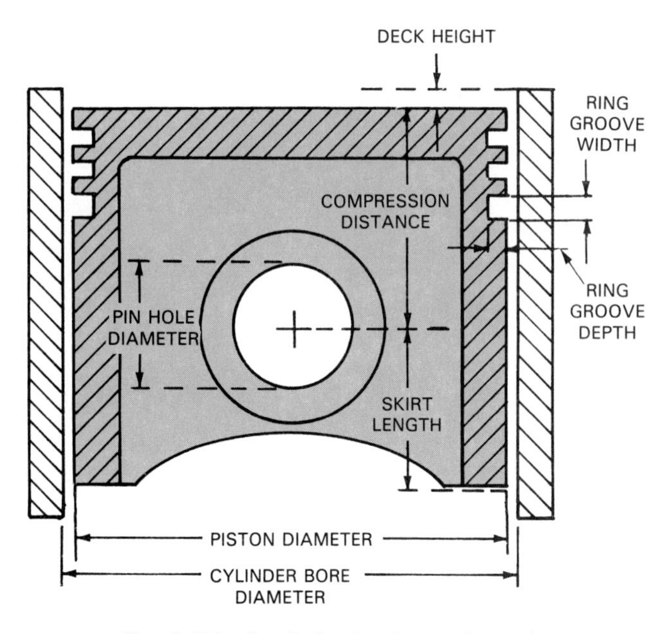

Fig. 9-36. Study basic piston dimensions.

1. PISTON DIAMETER (distance measured across sides of piston).
2. PIN HOLE DIAMETER (distance measured across inside of hole for piston pin).
3. RING GROOVE WIDTH (distance measured from top to bottom of ring groove).
4. RING GROOVE DEPTH (distance measured from ring land to back of ring groove).
5. SKIRT LENGTH (distance from bottom of skirt to centerline of pin hole).
6. COMPRESSION DISTANCE (distance from centerline of pin hole to top of piston).

As you will learn, these dimensions affect how the piston functions in the cylinder. Many are also important when working on an engine.

Cam ground piston

A *cam ground piston* is machined slightly out-of-round when viewed from the top. The piston is a few thousandths of an inch (hundredths of a millimeter) larger in diameter perpendicular (opposite) to the piston pin centerline. See Fig. 9-37.

Cam grinding is done to compensate for different rates of piston expansion (enlargement) due to differences in thickness. As the piston is heated by combustion, the thickness around the pin boss causes the piston to expand more in a line parallel with the piston pin. The cam ground or oval shaped piston then becomes ROUND WHEN HOT.

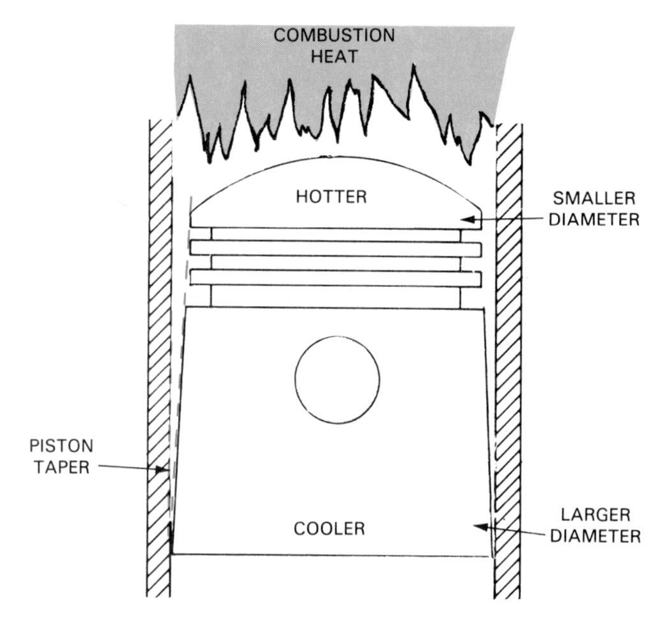

Fig. 9-38. Piston taper is commonly used because head of piston gets hotter and expands more than skirt. Combustion flame heats head and makes piston almost equal in size when hot.

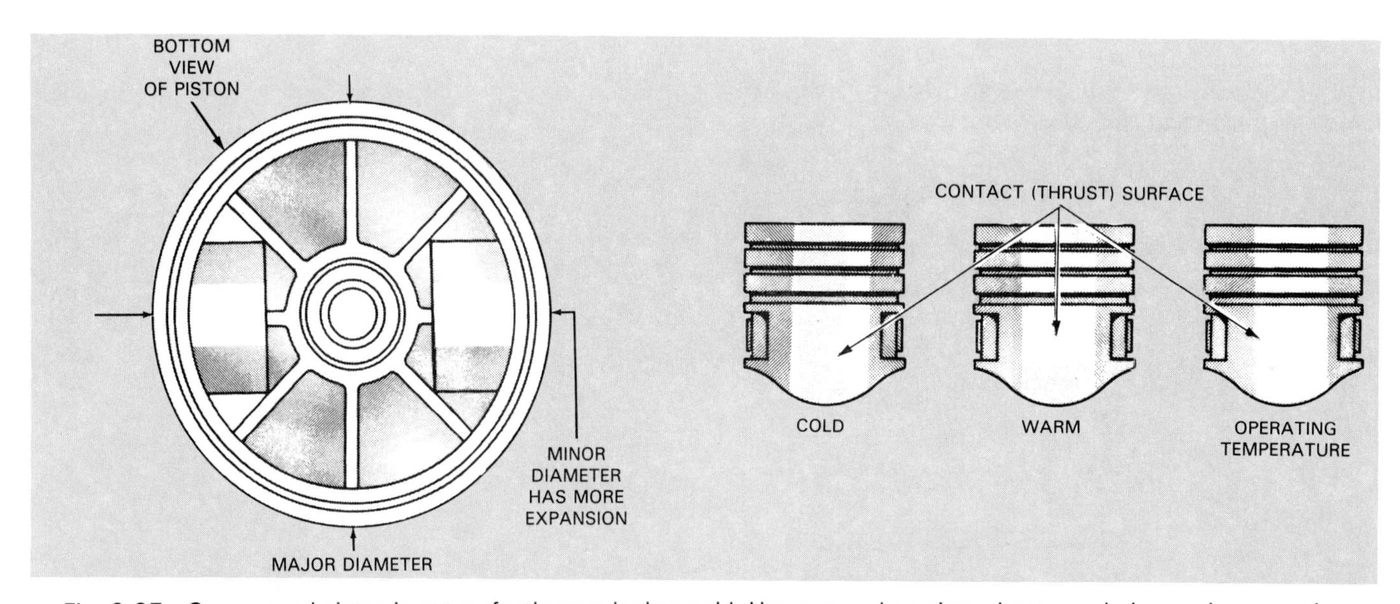

Fig. 9-37. Cam ground piston is not perfectly round when cold. However, when piston heats up during engine operation, boss area expands more and piston becomes round. Note how more of skirt area touches cylinder wall as piston warms. (Ford)

With cam grinding, a cold piston will have the correct piston-to-cylinder clearance. The unexpanded piston will NOT slap, flop sideways, and knock in the cylinder because of too much clearance. Yet, the cam ground piston will not become too tight in the cylinder when heated to full operating temperature.

Piston taper

Like cam grinding, *piston taper* is normally used to maintain the correct piston-to-cylinder clearance, Fig. 9-38. The top of the piston is machined slightly smaller than the bottom.

Since the piston head gets hotter than the skirt and expands more, piston taper makes the piston almost equal in size at the top and bottom when in operation.

Fiber reinforced piston

A *fiber reinforced piston* uses a special alloy with metal fibers to increase piston dependability under severe operating conditions. As shown in Fig. 9-39, the area above the top compression ring may be fiber reinforced to increase wear resistance, heat resistance, and cooling ability. This area of the piston is exposed to extreme loads and is prone to failure.

Piston head coatings

A *piston head coating* refers to the aluminum piston head being covered with another substance (ceramic, stainless steel, etc.) to improve durability, Fig. 9-40.

For instance, one piston manufacturer bonds a stainless steel cap to the aluminum piston. A steel mesh is used between the aluminum and stainless steel to form a good bond. The thin coating or cap on the piston head can resist heat, carbon buildup, and pressure better than plain aluminum.

Teflon-coated pistons

Teflon-coated pistons reduce friction, especially during hot restart conditions, by reducing drag. When an engine is shut down, heat can soak into aluminum pistons. This can cause slow cranking and hard starting problems.

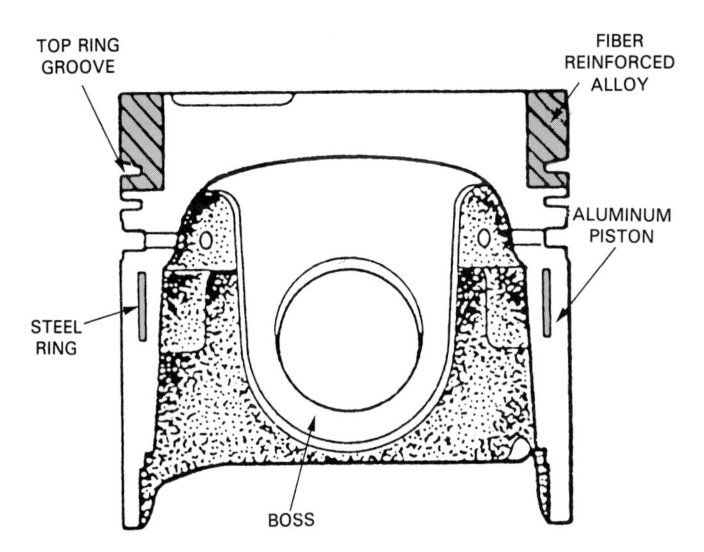

Fig. 9-39. Fiber reinforced piston is new design that uses special alloy to increase strength and heat resistance to critical top land of piston. (Toyota)

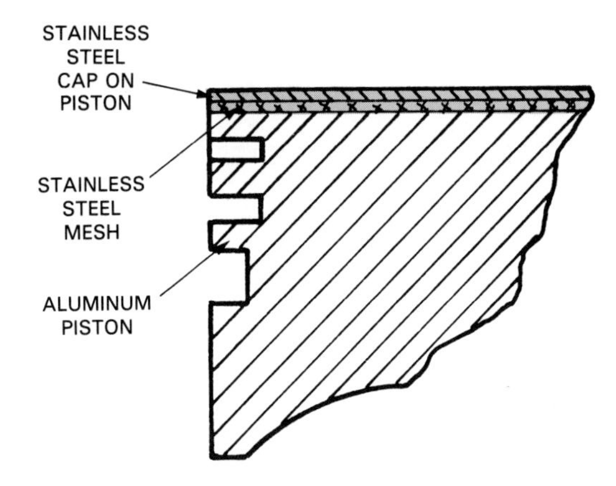

Fig. 9-40. A few high performance pistons are designed with steel cap bonded to the aluminum piston. Steel can withstand combustion heat better than aluminum, increasing efficiency.

Two-piece piston

A *two-piece piston* is constructed in two parts; the crown and the skirt. As pictured in Fig. 9-41, the piston pin holds the two together. This is an experimental design that has proven very successful in racing engines. It also allows the use of different substances in the crown and skirt. The crown could be made of fiber reinforced aluminum and stainless steel. The skirt could be made of lightweight plastic.

Oil cooled piston

An *oil cooled piston* uses pump oil pressure to direct a stream of oil through and on the piston to help cool the piston head. A simple drawing of this type piston is shown in Fig. 9-42.

Variable compression piston

A *variable compression piston* is a two-piece design controlled by engine oil pressure. The piston head fits and slides over the piston main body. Engine oil pressure fed between the halves forms a hydraulic cushion. With normal driving, the piston top extends out for maximum compression ratio and power. When engine speed increases, combustion pressure pushes the piston head down to lower compression and prevent engine knocking.

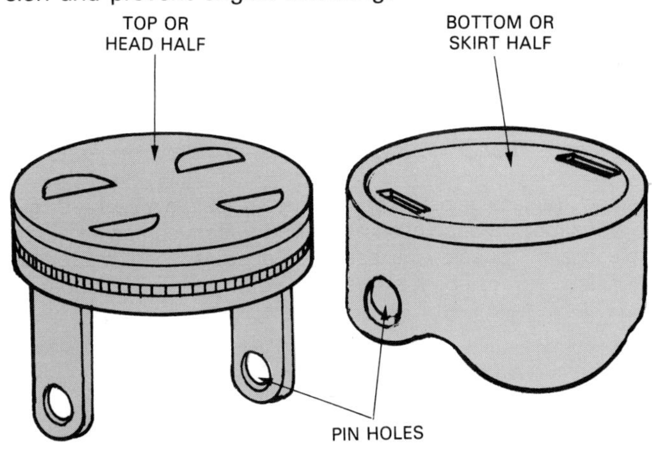

Fig. 9-41. Two-piece piston is another racing design. Top half of piston can be made of one metal alloy and bottom another or bottom can be made of space age plastic. This design reduces weight and piston can be made to run cooler.

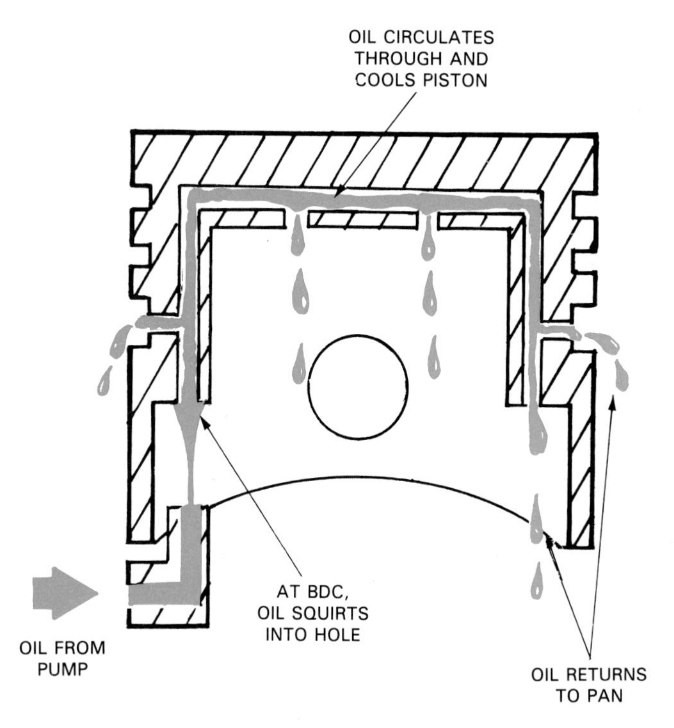

Fig. 9-42. Another advanced piston design allows oil pump to circulate oil through passages in piston. This cools piston head to increase reliability at high rpm.

Piston clearance

Piston clearance is the amount of space between the sides of the piston and the cylinder wall. Clearance is needed for a lubricating film of oil and to allow for expansion when the piston heats up. The piston must always be free to slide up and down in the cylinder block.

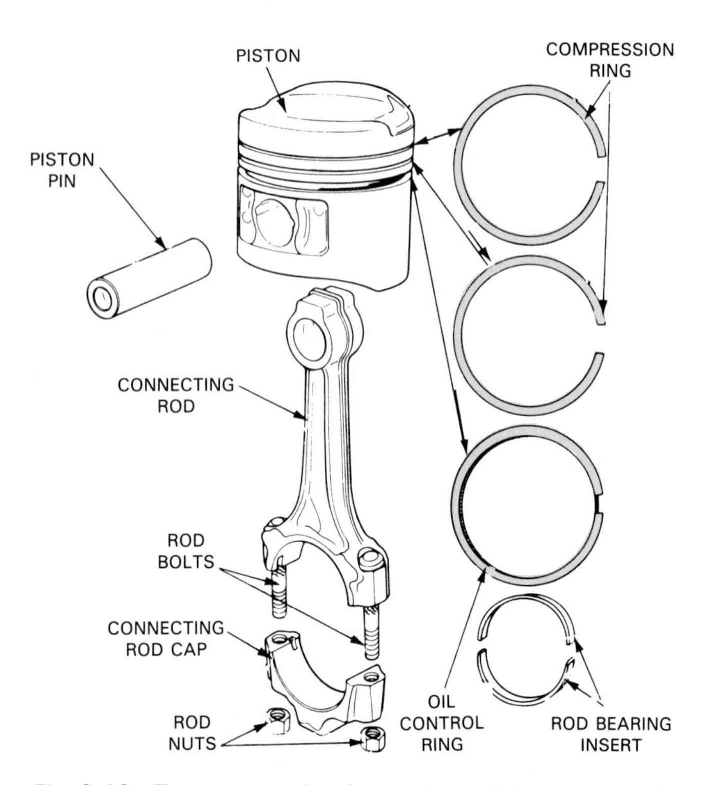

Fig. 9-43. Two compression rings and one oil ring are normally used on passenger car pistons. (Ford)

PISTON RING CONSTRUCTION

As you learned earlier, automotive pistons normally use three rings—two compression rings and one oil ring. It is important for you to understand how variations in ring construction provide different operating characteristics. See Fig. 9-43.

The *piston rings* seal the clearance between the outside of the piston and the cylinder wall. They must keep combustion pressure from entering the crankcase. They must also keep oil from entering the combustion chambers. They must do this when very hot and when traveling a high speed, Fig. 9-44.

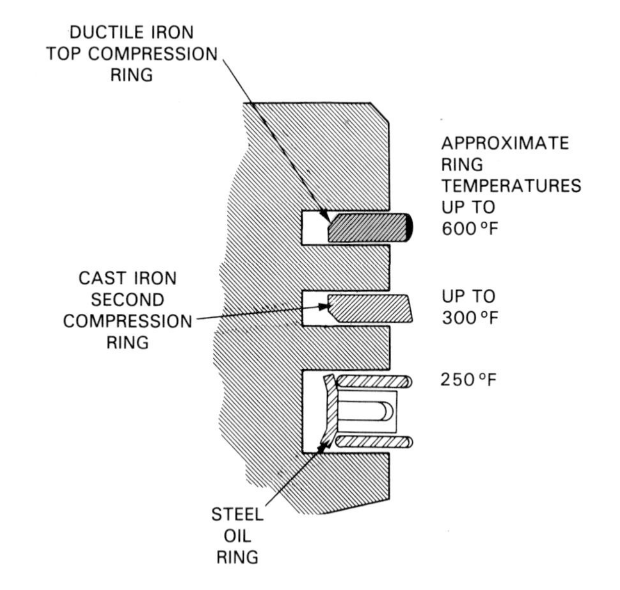

Fig. 9-44. Top compression ring operates much hotter than other rings. It is frequently made of ductile iron to withstand this heat. Second compression ring can be made of normal cast iron. Oil ring is commonly steel to maintain good contact and resist wear on cylinder wall. (Sealed Power)

Compression rings

The *compression rings* prevent *blow-by* (combustion pressure leaking into engine crankcase). Fig. 9-45 shows ring action.

On the compression stroke, pressure is trapped between the cylinder and piston grooves by the compression rings. Combustion pressure pushes the compression rings down in their grooves and out against the cylinder wall. This produces an almost leakproof seal, Fig. 9-46.

Compression rings are usually made of cast iron. An outer layer of chrome, molybdenum, or other metal may be coated on the face of the ring to increase wear resistance and oil absorption. See Fig. 9-47.

The face of the compression rings may also be GROOVED to speed *ring seating* (initial ring wear that makes ring match surface of cylinder perfectly).

Piston ring dimensions

Basic piston ring dimensions include ring width, ring wall thickness, and ring gap. These dimensions affect

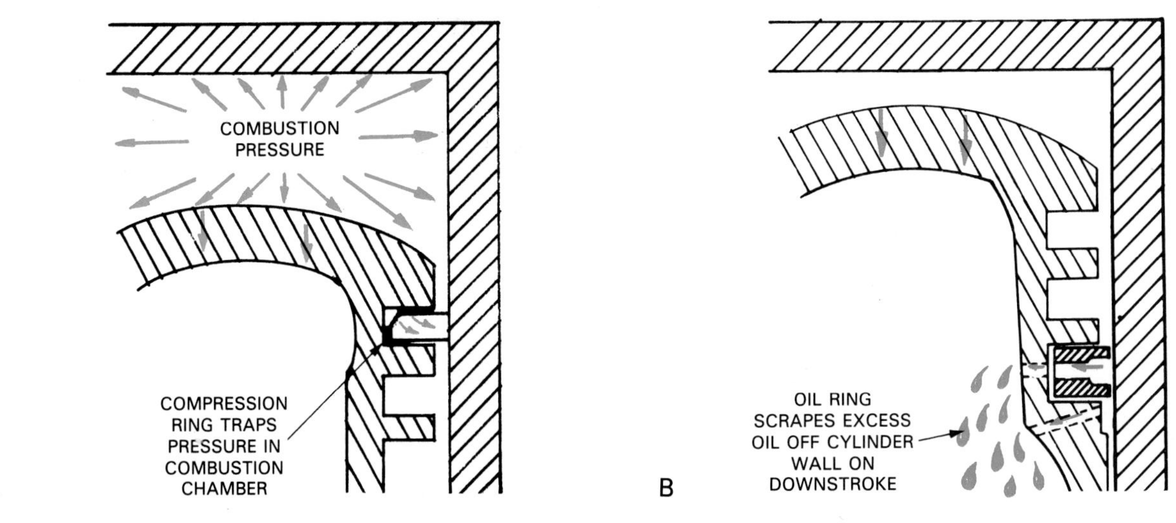

Fig. 9-45. Note basic action of compression and oil rings. A—Combustion pressure flows down between piston and cylinder. It actually helps push ring out against cylinder. B—Oil ring scrapes excess oil from cylinder. Oil flows through ring and piston and back into crankcase.

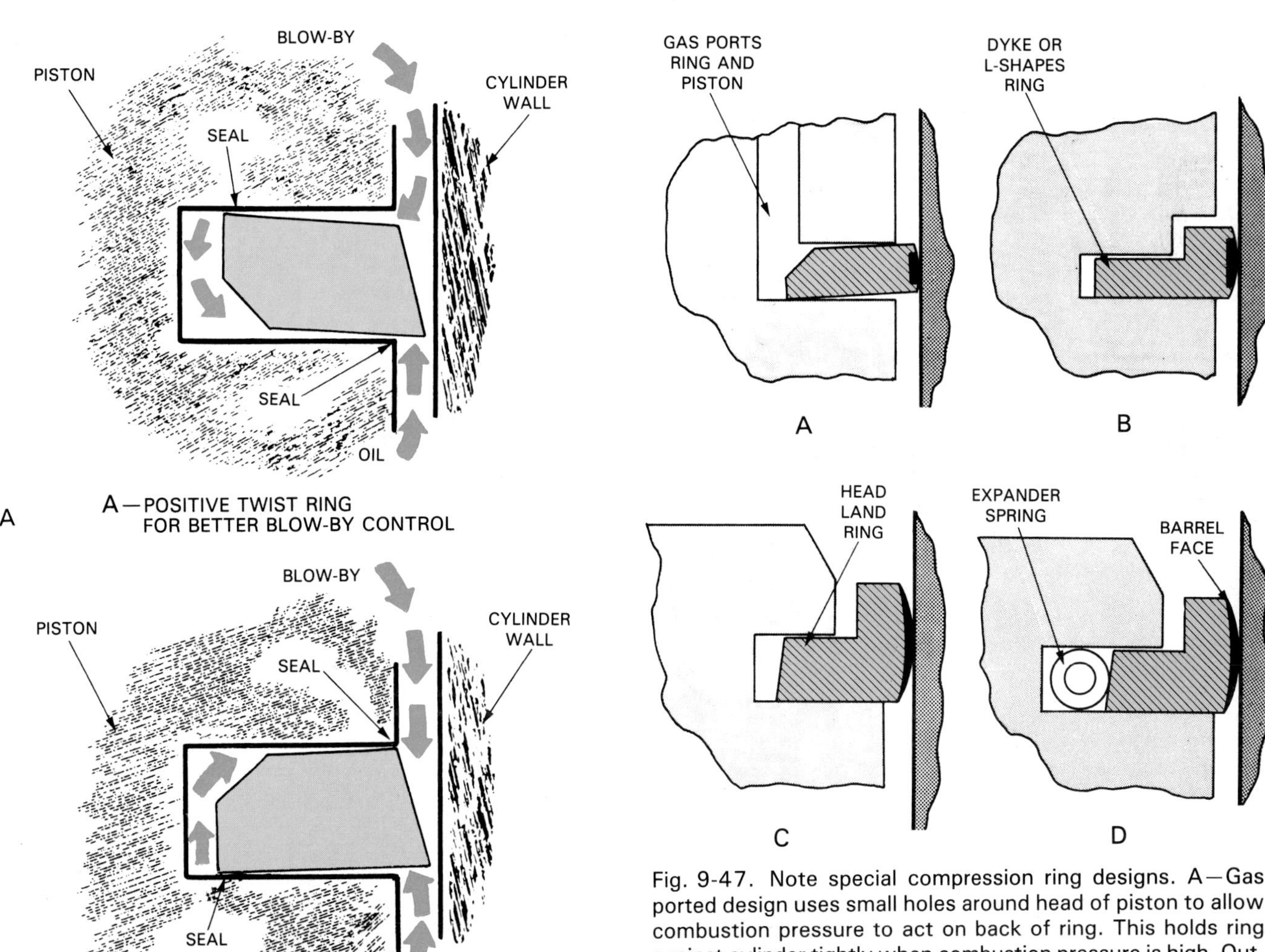

Fig. 9-46. A—Positive twist ring is better for controlling blow-by. B—Reverse twist ring is for improved oil control and it is more common. (Sealed Power)

Fig. 9-47. Note special compression ring designs. A—Gas ported design uses small holes around head of piston to allow combustion pressure to act on back of ring. This holds ring against cylinder tightly when combustion pressure is high. Outward force is reduced when pressure in combustion chamber is low to reduce ring drag or friction. B—Dyke or L-shaped ring also uses combustion pressure to force ring out against cylinder wall. C—Head land ring has L-shape but is partially outside groove. D—Spring expander can be used to increase tension forcing compression ring out against cylinder. It increases friction, however. (Sealed Power)

the operation of the engine.

Ring width is the distance from the top to the bottom of the ring. The difference between the ring width and the width of the piston ring groove determines *ring side clearance.*

Ring radial wall thickness is the distance from the face of the ring to its inner wall. The difference between the ring wall thickness and ring groove depth determines *ring back clearance.*

Ring gap is the split or space between the ends of a piston ring when installed in the cylinder. The ring gap allows the ring to be spread open and installed on the piston. It also allows the ring to be made slightly larger in diameter than the cylinder. When squeezed together and installed, the ring spreads outward and presses on the cylinder wall to aid ring sealing. Refer to Fig. 9-48.

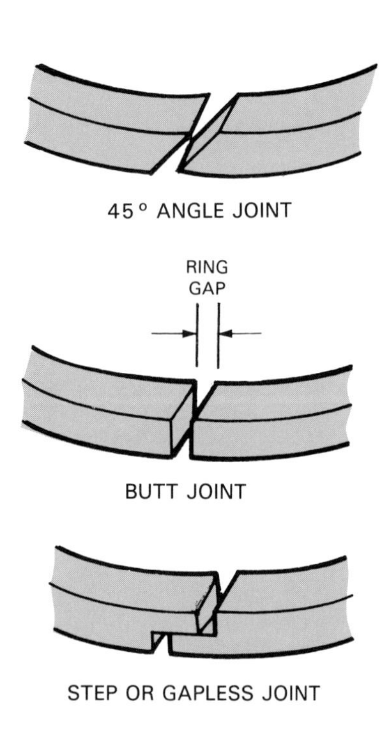

Fig. 9-48. Note basic ring gap designs. Butt joint is most common. Step or gapless ring provides improved seal.

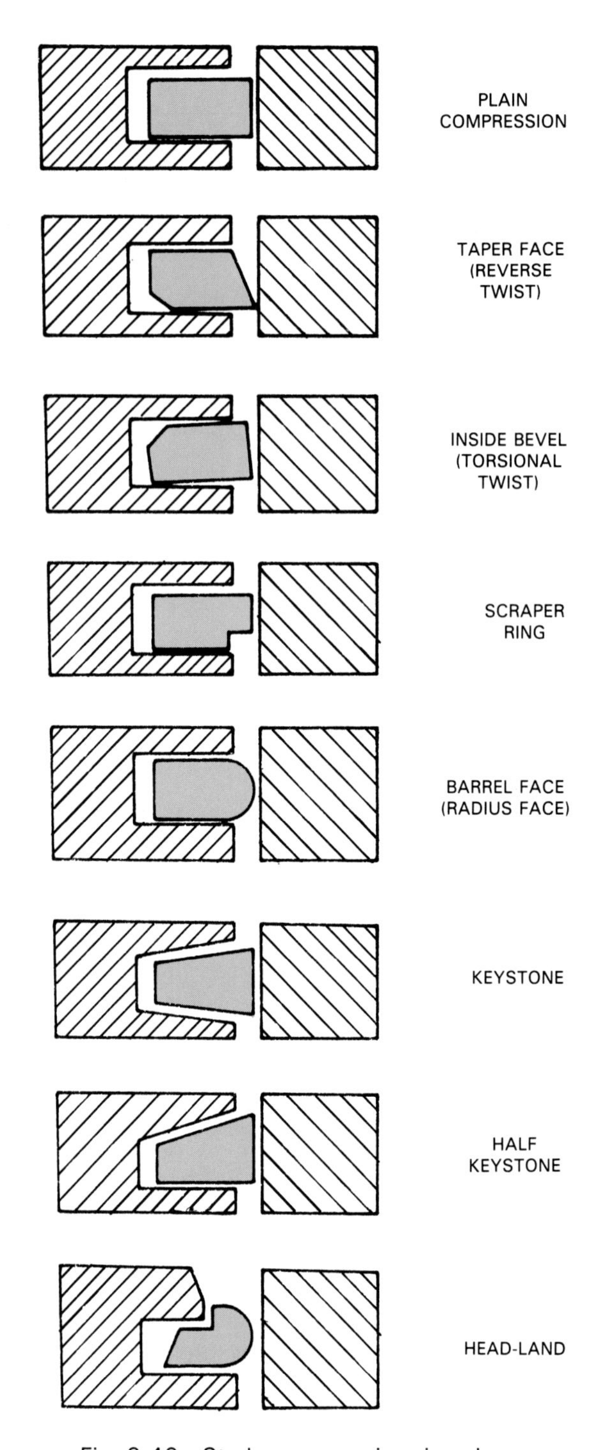

Fig. 9-49. Study compression ring shapes.

Compression ring shapes

Piston ring shape refers to the cross-section of a piston ring. The shape is also used to control how the piston ring operates in its groove and on the cylinder wall. Fig. 9-49 shows several piston ring shapes. The shape of a ring also determines the top or how the ring installs in the piston.

Oil rings

Primarily, an *oil ring* must prevent engine oil from entering the combustion chamber. It scrapes excess oil off the cylinder wall. If too much oil got into and was burned in the combustion chamber, blue smoke would come out of the car's exhaust pipe. Refer again to Fig. 9-45B.

Oil rings are available in two basic designs: rail-spacer type and one-piece type. An oil ring consisting of two rails and a spacer is the most common, Fig. 9-50. Fig. 9-51 shows some one-piece oil ring designs.

Ring expanders

A *ring expander* can be placed behind a one-piece oil ring to increase ring tension. It can also be used behind the second compression ring. The expander helps push the ring out against the cylinder wall, increasing the ring's sealing action, Fig. 9-47.

A *ring expander-spacer* is part of a three-piece oil ring. It holds the two steel oil ring scrapers apart and helps push them outward, Fig. 9-50.

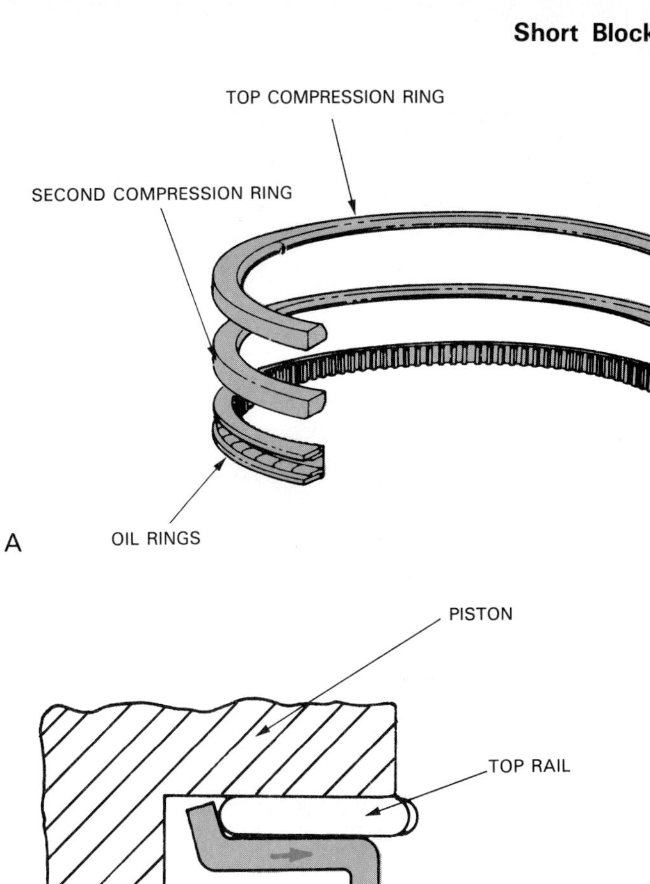

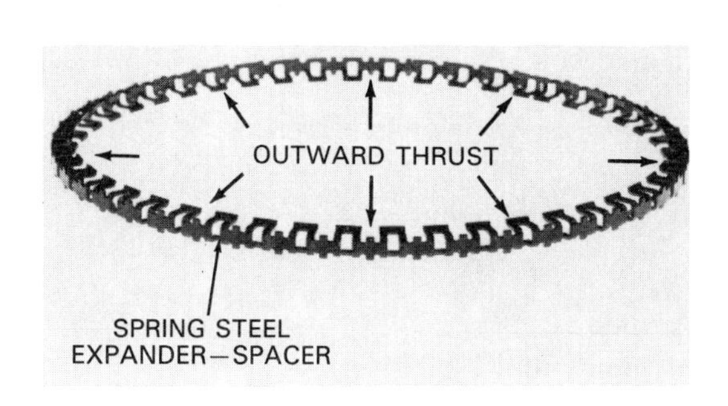

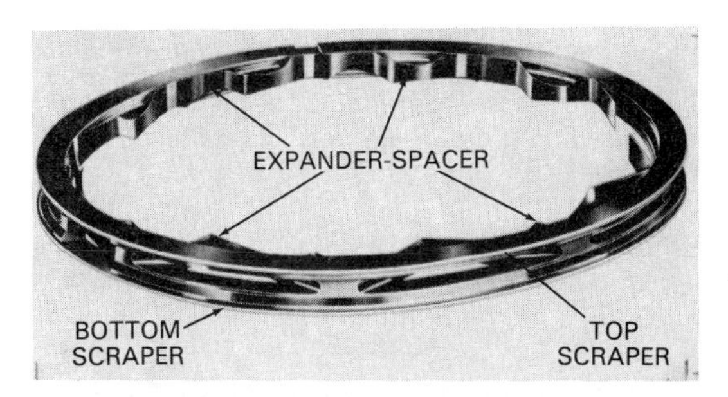

Fig. 9-50. A—Oil ring is bottom ring on piston. B—Expander-spacer holds and pushes scraper rails outward. C—Expander-spacer is made of spring steel. D—Complete 3-piece oil ring. (Dana, Perfect Circle)

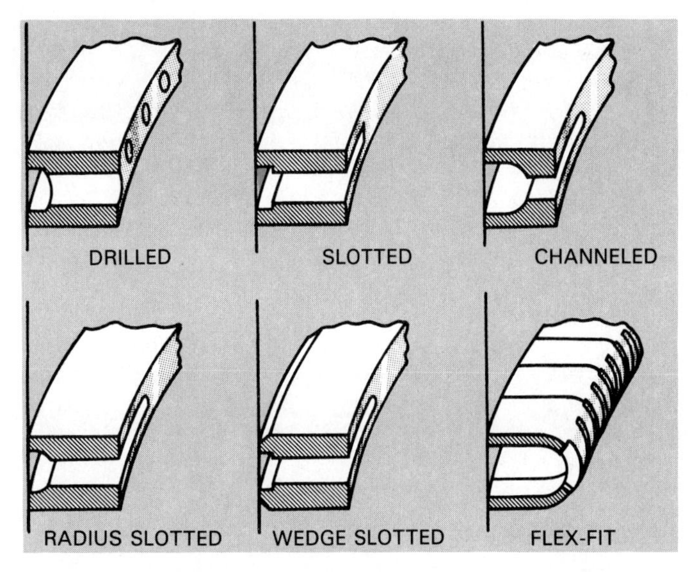

Fig. 9-51. These are variations of one-piece oil ring. They are not as common as three-piece oil ring. (Ford)

Piston ring coatings

Mentioned briefly, the face of piston rings can be coated with chrome or other metals.

Soft ring coatings (usually iron or molybdenum) help the ring wear-in quickly, forming a good seal. The soft, grooved outer surface will wear away rapidly so the ring conforms to the shape of the cylinder. Also called *quick seal rings,* they are commonly recommended for USED CYLINDERS that are slightly worn.

Hard ring coatings, such as chrome, are used to increase ring life and reduce friction. They are used in new or BORED CYLINDERS that are perfectly round and NOT worn. To aid break-in, chrome plated rings usually have ribbed faces. The ribs hold oil and wear quickly to produce a good seal.

Porous ring coatings help break-in and service life because they hold oil and reduce friction. Molybdenum is a porous metal sometimes used on compression rings.

ENGINE BEARING CONSTRUCTION

As discussed earlier, there are three basic types of engine bearings: connecting rod bearings, crankshaft main bearings, and camshaft bearings. This is illustrated in Fig. 9-52.

Steel is normally the main *backing material* (body of bearing that contacts stationary part) for engine bearings. Softer alloys are bonded over the backing to form the bearing surface.

Any one of three basic types of metal alloys can be plated over the top of the steel backing: babbitt (lead-tin alloy), copper, or aluminum. These three metals may be used in different combinations to design bearings for either light, medium, heavy duty, or extra-heavy duty applications. See Fig. 9-53.

Bearing characteristics

Engine bearings must operate under tremendous loads, severe temperature variations, abrasive action, and corrosive surroundings. Essential bearing characteristics

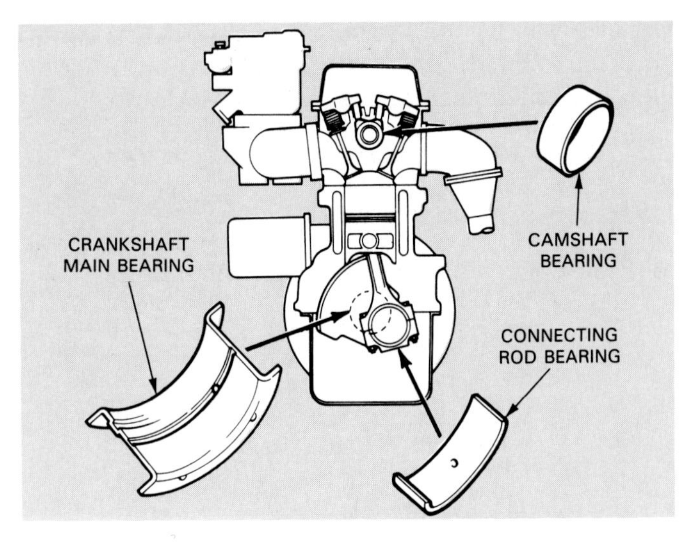

Fig. 9-52. Engine uses three basic types of bearings. (Ford)

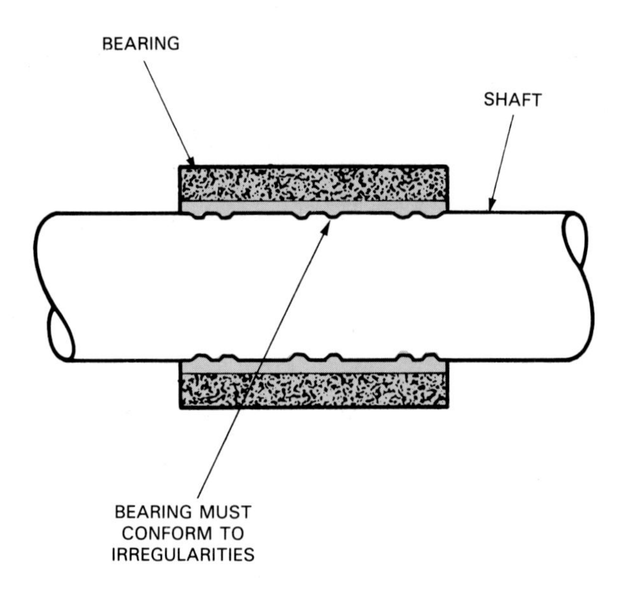

Fig. 9-54. Bearing must conform to irregularities in journal.

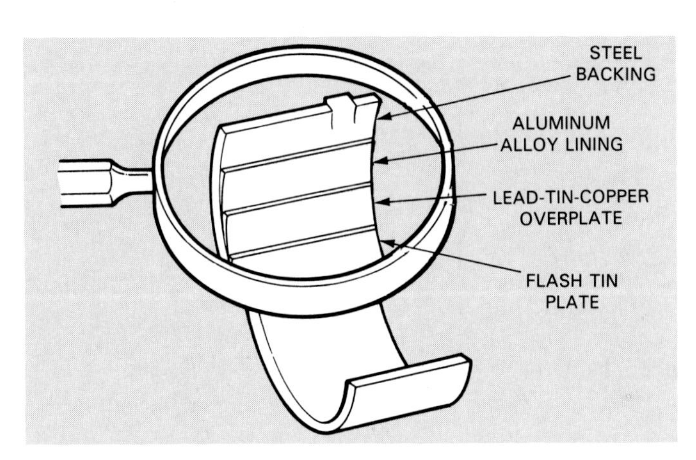

Fig. 9-53. Engine bearings are made with several different layers of metal. Steel backing is hard. Top layer is very thin and soft. (Federal Mogul)

include:

1. *Bearing load strength* is the bearing's ability to withstand pounding and crushing during engine operation. The piston and rod produce several TONS of downward force. The bearing must not fatigue, flatten, or split under these loads. If bearing load resistance is too low, the bearing can smash, fail, and spin in its bore. This can ruin the bore or journal.
2. *Bearing conformablity* is the bearing's ability to move, shift, or adjust to imperfections in the journal surface. Usually, a soft metal will be plated over a hard steel. This lets the bearing conform to any defects in the journal. See Fig. 9-54.
3. *Bearing embedability* refers to the bearing's ability to absorb dirt, metal, or other hard particles. Dirt or metal are sometimes carried into the bearings. The bearing should allow the particles to sink beneath the surface and into the bearing material. This will prevent the particles from scratching, wearing, and damaging the surface of the crankshaft or camshaft journal.
4. *Bearing corrosion resistance* is the bearing's ability to resist corrosion from acids, water, and other impurities in the engine oil. Combustion blow-by gases cause oil contamination that could corrode the engine bearings. Aluminum-lead and other alloys are now commonly used because of their excellent corrosion resistance.
5. *Bearing crush* is used to help prevent the bearing from spinning inside its bore during engine operation, Fig. 9-55. The bearing is made slightly larger in diameter than the bearing bore. The end of the bearing is slightly above the bore.

   When the rod or main cap is tightened, the bearing ends hit each other. This jams the backsides of the bearing inserts tightly against the bore, locking them in place.

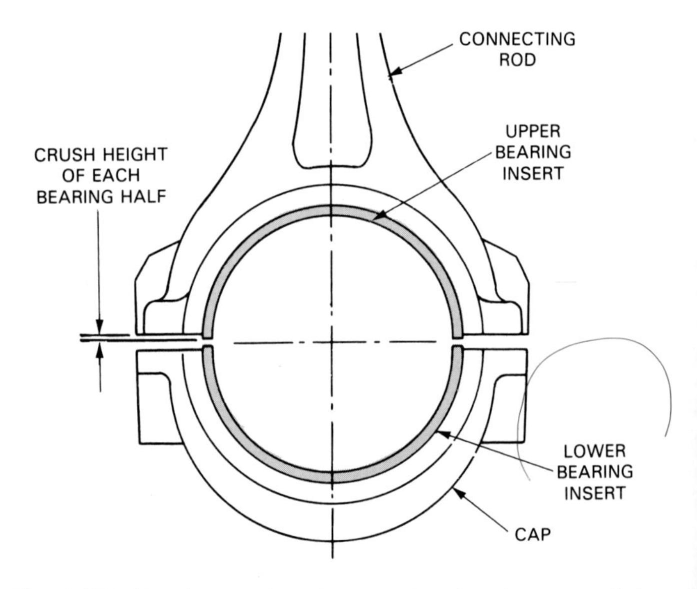

Fig. 9-55. Bearing crush makes sure bearing does not spin inside its bore. (Deere & Co)

6. *Bearing spread* is used on split type engine bearings to hold the bearing in place during assembly. The distance across the parting line of the bearing is slightly wider than the bearing bore, Fig. 9-56. This causes the bearing insert to stick in its bore when pushed into place with your fingers. Tension from bearing spread keeps the bearing from falling out of its bore as you assemble the engine.

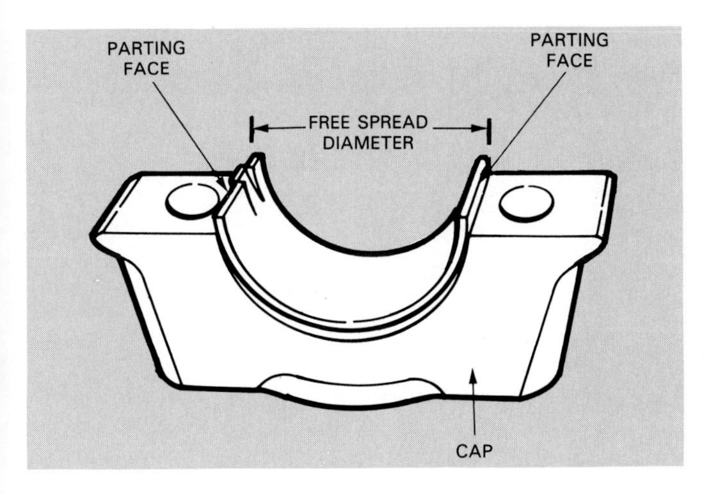

Fig. 9-56. Bearing spread holds bearing in place during engine assembly. Bearing diameter is slightly larger than bore. (Federal Mogul)

Standard and undersize bearings

A *standard bearing* has the original dimensions specified by the engine manufacturer for a new, unworn, or unmachined crankshaft. A standard bearing may have the abbreviation "STD" stamped on the back.

An *undersize bearing* is designed to be used on a crankshaft journal that has been machined to a smaller diameter. If the crank has been worn or damaged, it can be ground undersize by a machine shop. Then, undersize bearings would be needed, Fig. 9-57.

Connecting rod and main bearings are available in undersizes of .010, .020, .030, and sometimes .001 and .040 in. The undersize will normally be stamped on the back of the bearing. The crankshaft may also have an undersize number stamped on it by the machine shop.

Bearing locating lugs and dowels

Locating lugs or *dowels* position the split bearings in their bores, Fig. 9-58. The bearing usually has a lug that fits into a recess machined in the bearing bore or cap.

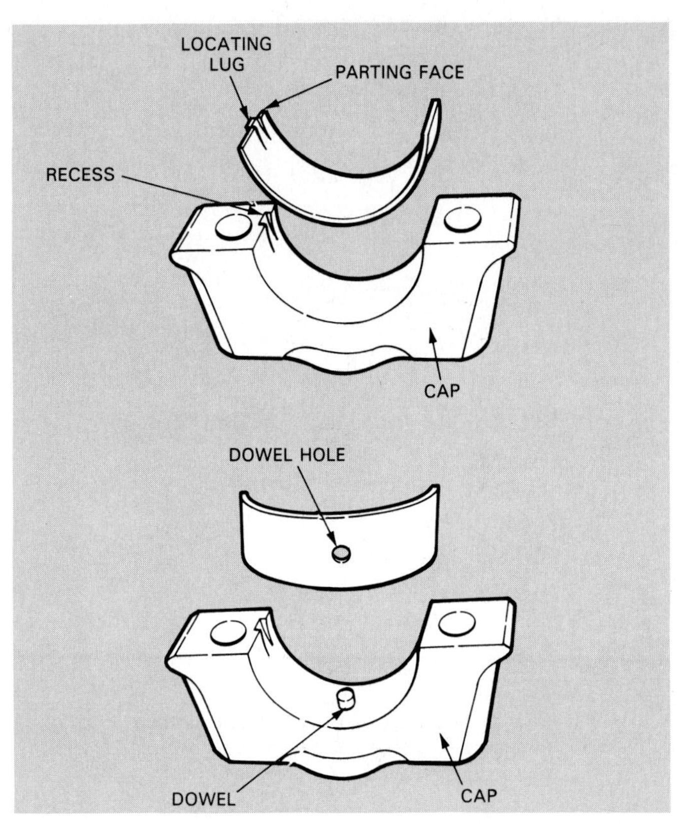

Fig. 9-58. Lug and dowel are two methods of securing bearing in bore. Lug is more common. (Federal Mogul)

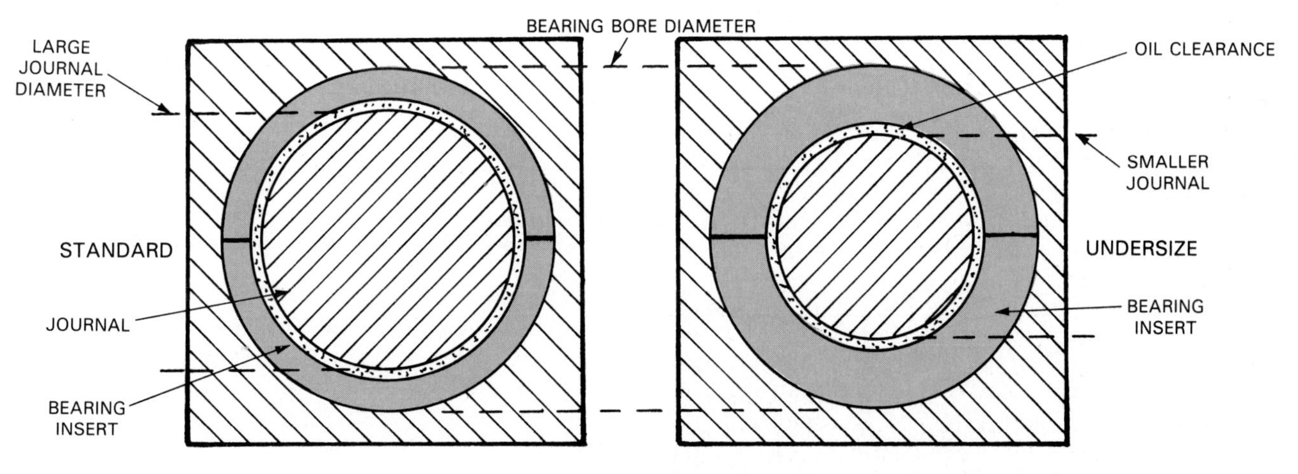

Fig. 9-57. Undersize bearing is thicker than standard bearing. It is used when crank journals have been machined smaller in diameter by machine shop.

Sometimes however, a dowel in the cap or bore fits in a hole in the bearing insert. Either method helps keep the insert from shifting or turning during crankshaft rotation.

Bearing oil holes and grooves

Oil holes and *grooves* in the engine bearings permit bearing lubrication. The holes allow oil to flow through the block and into the clearance between the bearing and crankshaft journal. The grooves provide a channel so oil can completely circle the bearing before flowing over and out of the bearing. See Fig. 9-59.

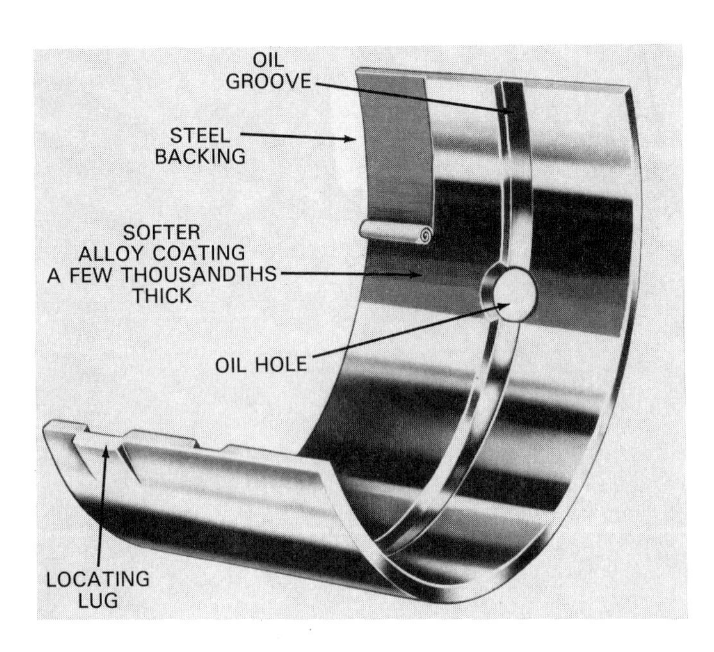

Fig. 9-59. Main bearings are frequently grooved so oil can circle bearing and distribute evenly. Grooves also let constant supply of oil enter crankshaft for rod bearings. Note thin layer of soft alloy over steel backing. (Clevite)

SELECT FIT PARTS

Select fit means that some engine parts are selected and installed in a certain position to improve the fit or clearance between parts. For example, pistons are commonly select fit into their cylinders, Fig. 9-60. The engine manufacturer will measure the diameter of the cylinders. If one cylinder is machined slightly larger than another, a slightly larger piston will be installed (select fit) in that cylinder. Because of select fit parts, it is important that you reinstall parts in their original locations when possible.

BALANCER SHAFTS

Balancer shafts are used in some engines to reduce vibration. These counterweighted shafts are generally mounted in the left and right sides of the engine block and are driven by a belt or chain, Fig. 9-61. Most balancer shafts are designed to rotate at twice crankshaft speed. The rotating shafts counteract the vibrations produced by crankshaft, piston, and connecting rod movement.

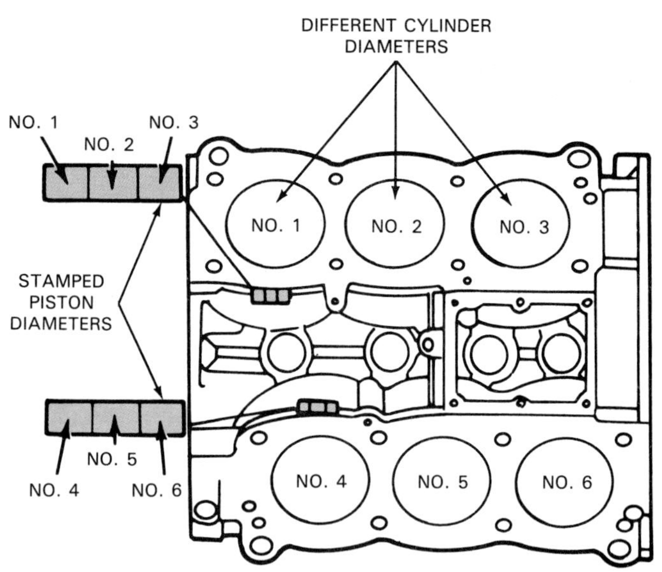

Fig. 9-60. Pistons are often select fit into specific cylinders. Note that required piston diameter for each cylinder is stamped on engine block. Piston size will vary with cylinder diameter. (Ford)

Fig. 9-61. Balancer shafts are used in this engine to minimize vibration. (Chrysler)

SUMMARY

It is very important for an engine mechanic to understand how today's "high tech" engines are designed and constructed. This information will be useful when troubleshooting problems or when doing engine repairs. Modern engines use complex materials and assembly methods.

A short block includes everything installed in the cylinder block. A tall block also includes the cylinder heads.

Blocks can be made of cast iron or aluminum. Aluminum blocks normally have cast iron sleeves. Core plugs are in the sides of the block from manufacturing. A wet sleeve is thick and serves as the cylinder wall. A dry sleeve normally installs in a bored cylinder.

Siamese cylinders are found in aluminum blocks to strengthen the area between the cylinders. The crankshaft installs in the main bore on main bearing inserts. With an OHV engine, the cam bore is also in the block. Main caps secure the crank and main bearings in the block. There are several main cap designs: two-bolt mains, four-bolt mains, cross-bolt mains, and girdle cap. A thrust bearing limits crankshaft end play.

Journals are machined and polished surfaces on the crankshaft for the main and rod bearings. A steel crankshaft is stronger than a cast iron crank and can be found in high performance or diesel engines. A splayed rod journal has the journals offset from each other to smooth V-6 engine operation. A rear main oil seal keeps engine oil from leaking out the rear of the engine from the rear main bearing.

Connecting rods are normally made of steel. Bolts or cap screws hold the cap on the rod body. Rod numbers are used so the caps can be reinstalled correctly on the same rods. Connecting rod length is measured from the center of the small end to the center of the big end.

Piston pins are casehardened. A full floating pin is free to rotate in both the rod and piston. It is held in place with snap rings. A press-fit pin is locked in the connecting rod. A piston-guided connecting rod fits closely in the piston boss and the piston limits side movement of the rod small end. Piston pin offset is used to prevent piston slap. Several piston head shapes are used in today's engines.

A fiber reinforced piston has a special alloy with steel or carbon fibers to increase durability. Piston head coatings of stainless steel are also used to increase durability.

Automotive pistons normally have three rings: two compression rings and a single, lower oil ring. Piston ring shape can be used to help increase sealing efficiency. A ring expander is used to force the ring out against the cylinder wall with more pressure. Soft ring coatings are recommended for worn cylinders. Harder coatings are for new or freshly bored cylinders that are perfectly round and not tapered.

An undersize bearing is needed for a turned crankshaft journal. A standard bearing fits on an unmachined crank.

KNOW THESE TERMS

Short block, Bottom end, Long block, Cylinder block, Core or freeze plugs, Integral cylinder, Cylinder sleeve, Deck, Dry sleeve, Wet sleeve, Siamese cylinders, Cam bore, Main bore, Main cap, Main bearing, Thrust bearing, Main bearing clearance, Rod journal, Splayed journal, Roll hardened, Counterweights, Fully counterweighted crankshaft, Partially counterweighted crankshaft, Snout, Flywheel, Pilot bearing, Rear main oil seal, Rod oil spurt hole, Rod cap, Broken-surface connecting rod, Rod numbers, Full floating piston pin, Piston-guided connecting rod, Piston pin offset, Piston head, Piston skirt, Piston boss, Valve reliefs, Cam ground piston, Piston taper, Fiber reinforced piston, Piston head coating, Teflon coated piston, Variable compression piston, Two-piece piston, Oil cooled piston, Piston clearance, Blow-by, Ring gap, Ring expander, Soft ring coating, Hard ring coating, Standard bearing, Undersize bearing, Select fit parts, Balancer shaft.

REVIEW QUESTIONS—CHAPTER 9

1. An aluminum cylinder block normally has:
   a. Aluminum cylinder walls.
   b. Cast iron sleeves.
   c. Silicone impregnated cylinders.
   d. Chrome cylinder walls.
2. Freeze plugs are used because they prevent engine block damage if a weak coolant solution freezes inside the block. True or false?
3. Explain the difference between a wet sleeve and a dry sleeve.
4. What are siamese cylinders?
5. Why must you NEVER mix up main caps?
6. Describe four main cap designs.
7. A _______ _______ _______ is used to limit crankshaft end play.
8. Main bearing clearance is the space between the crankshaft main _______ and the main _______ _______.
9. Why is a splayed journal sometimes used in V-6 engines?
10. This part is usually installed in the rear of the crankshaft when the car has a manual transmission.
    a. Snout.
    b. Flange.
    c. Pilot bearing.
    d. Throw-out bearing.
11. Describe three tyypes of rear main seals.
12. Passenger car engine connecting rods are frequently made of aluminum. True of false?
13. Connecting rod length is measured from the center of the _______ _______ to the center of the _______ _______ _______.
14. A full floating piston pin uses snap rings to keep the pin inside the piston. True or false?
15. Why is piston pin offset used?
16. Why are valve reliefs formed in some piston heads.
17. Why are pistons usually cam ground?
18. _______ _______ is used because the head or top of the piston runs hotter and expands more than the skirt.
19. When would you use a soft ring coating?
20. This makes it important for many engine parts to be reinstalled in the same location during repairs.
    a. Oversize parts.
    b. Undersize parts.
    c. Worn parts.
    d. Select fit parts.

ASE CERTIFICATION-TYPE QUESTIONS

1. Technician A says that the piston assemblies are included in an engine's short block. Technician B says that the valve train components are also included in an automotive engine's short block. Who is right?
   (A) A only.
   (B) B only.
   (C) Both A & B.
   (D) Neither A nor B.

2. Another term used to refer to an engine's short block is ________.
   (A) bare block
   (B) cylinder block
   (C) bottom end
   (D) None of the above.
3. Technician A says that an aluminum cylinder block dissipates heat much faster than a cast iron cylinder block. Technician B says that a cast iron cylinder block dissipates heat much faster than an aluminum cylinder block. Who is right?
   (A) A only.
   (B) B only.
   (C) Both A & B.
   (D) Neither A nor B.
4. Technician A says that an engine's lifter bores are located in the cylinder block deck. Technician B says that an engine's lifter bores are located in the lifter valley. Who is right?
   (A) A only.
   (B) B only.
   (C) Both A & B.
   (D) Neither A nor B.
5. All of the following are main cap designs EXCEPT:
   (A) cross-bolt mains.
   (B) girdle cap.
   (C) four-bolt mains.
   (D) three-bolt mains.
6. Technician A says that forged steel crankshafts are normally used in turbocharged automotive gasoline engines. Technician B says that forged steel crankshafts are normally used in diesel engines. Who is right?
   (A) A only.
   (B) B only.
   (C) Both A & B.
   (D) Neither A nor B.
7. The flywheel is used to ________.
   (A) smooth engine operation.
   (B) connect the crankshaft to the transmission or transaxle
   (C) hold the starter motor
   (D) Both A & B.
8. Technician A says that a floating piston pin is press fit into the connecting rod's small end. Technician B says that a floating piston pin is held in place by snap rings. Who is right?
   (A) A only.
   (B) B only.
   (C) Both A & B.
   (D) Neither A nor B.
9. Technician A says that the term piston clearance refers to the amount of space between the piston head and cylinder head. Technician B says that the term piston clearance refers to the amount of space between the sides of the piston and the engine's cylinder wall. Who is right?
   (A) A only.
   (B) B only.
   (C) Both A & B.
   (D) Neither A nor B.
10. Technician A says that bearing crush is used to help the bearing from spinning inside its bore during engine operation. Technician B says that bearing crush is used to hold the bearing in place during engine assembly. Who is right?
   (A) A only.
   (B) B only.
   (C) Both A & B.
   (D) Neither A nor B.

Chapter 10

Top End Construction

After studying this chapter, you will be able to:
- *Explain the parts of a top end assembly.*
- *Compare OHV and OHC valve trains.*
- *Describe the construction of a cylinder head.*
- *Explain how the various components install in a cylinder head.*
- *Compare cylinder head design variations.*
- *Summarize the construction of engine valves.*
- *Explain the construction of valve seats and valve guides.*
- *Describe the purpose of valve stem seals.*
- *Explain the construction of valve keepers, retainers, and valve springs.*
- *Describe why valve spring shims, valve rotators, and stem caps are sometimes needed.*
- *Summarize the parts of a camshaft.*
- *Describe the construction of camshaft bearings.*
- *Explain lifter or tappet design variations.*
- *Compare rocker arm design variations.*

Engine top end refers to the parts that fasten on top of the short block (cylinder heads, valves, camshaft, and other related components). These parts work together to control the flow of air and fuel into the engine cylinders. These components also control the flow of exhaust out of the engine. Refer to Fig. 10-1.

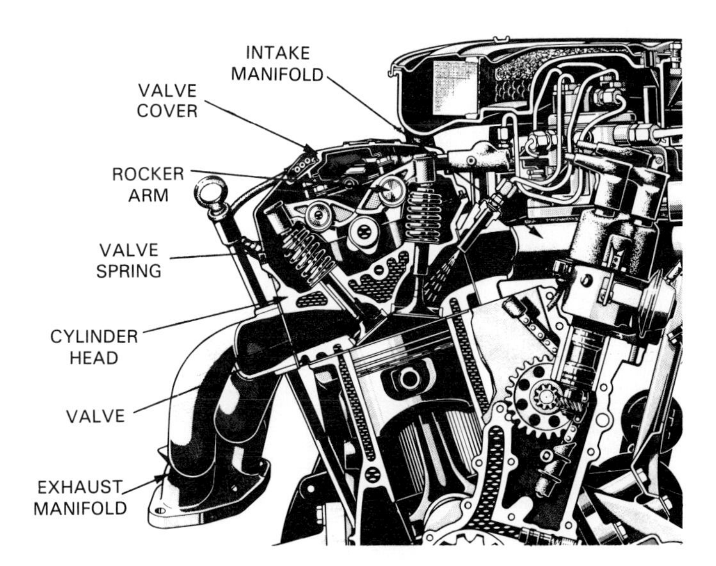

Fig. 10-1. Cutaway view of engine shows major parts of engine top end. (Mercedes Benz)

This chapter will help you learn more about how the parts of an engine top end are constructed and designed. As a result, you will be better prepared to understand following textbook chapters and to diagnose, test, and repair engine problems. See Fig. 10-2.

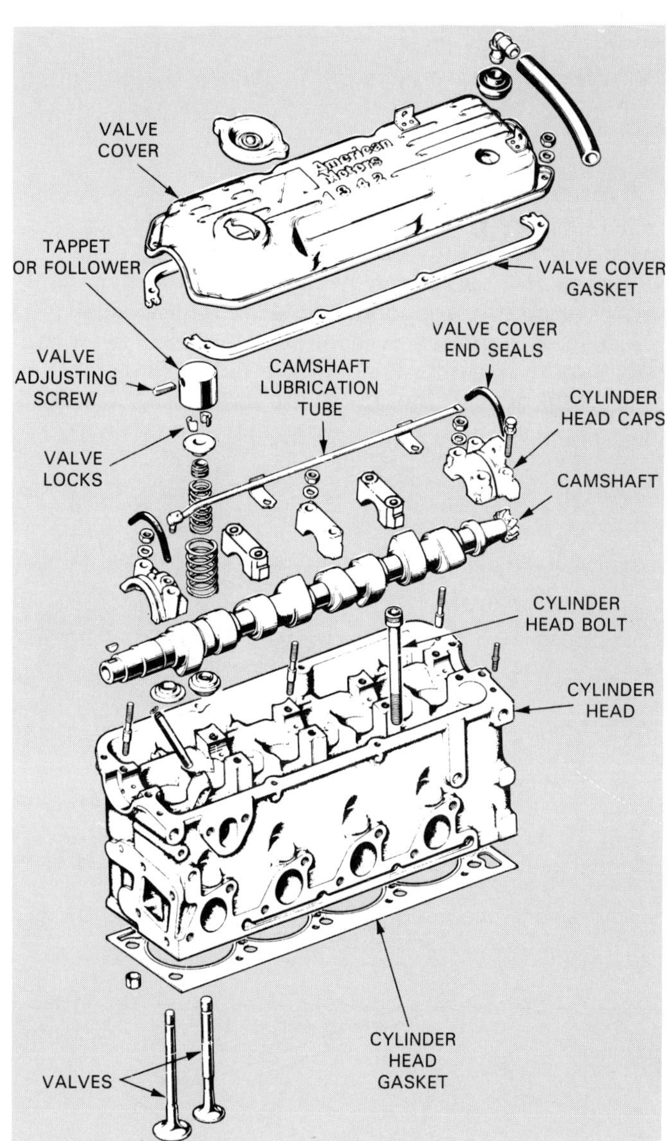

Fig. 10-2. Exploded view of engine top end shows typical components for OHC engine. (AMC)

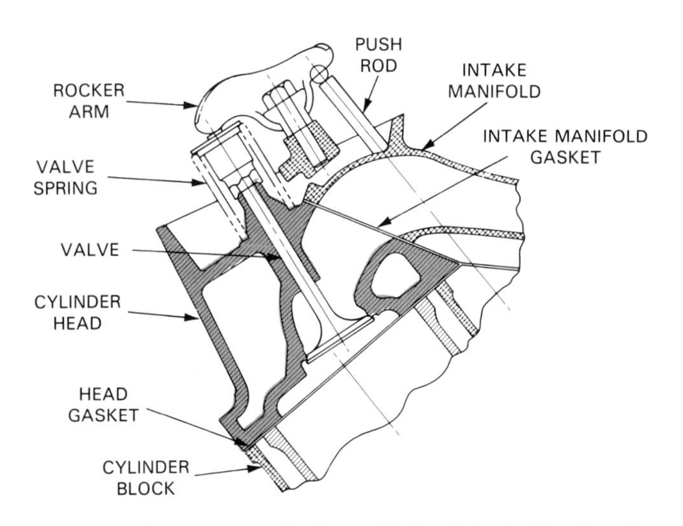

Fig. 10-3. Cylinder head bolts to deck of block. Head gasket seals surfaces between head and block. (Cadillac)

CYLINDER HEAD CONSTRUCTION

The *cylinder head* bolts to the deck of the block and covers the top of the cylinders. A *head gasket* seals the block and head surfaces to prevent oil, coolant, and pressure leakage as shown in Fig. 10-3.

Cylinder head construction can vary considerably, depending upon whether the engine is an OHV or OHC configuration. With an OHV engine, the head holds the valves and rocker arms. However, with an OHC engine, it must also hold the camshaft, Fig. 10-4.

Fig. 10-5 pictures a cylinder head for a dual overhead cam engine. Note how *cam housings,* for holding the camshafts, bolt to the top of the heads.

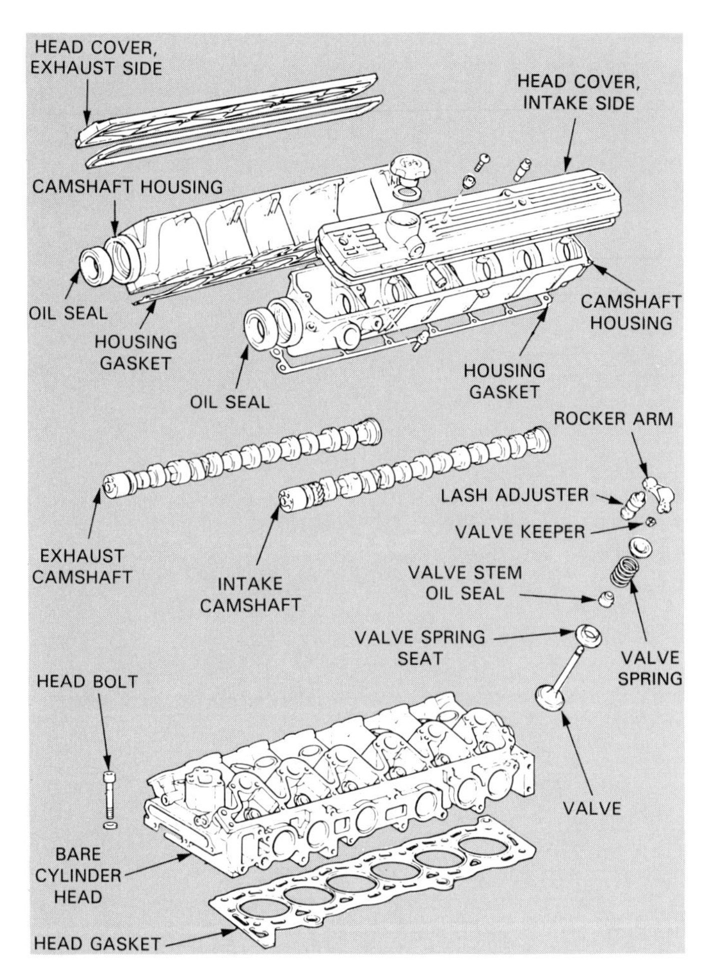

Fig. 10-5. This cylinder head is for a dual overhead cam engine. Note how cam housings bolt to top of head. This head uses four valves per cylinder. (Toyota)

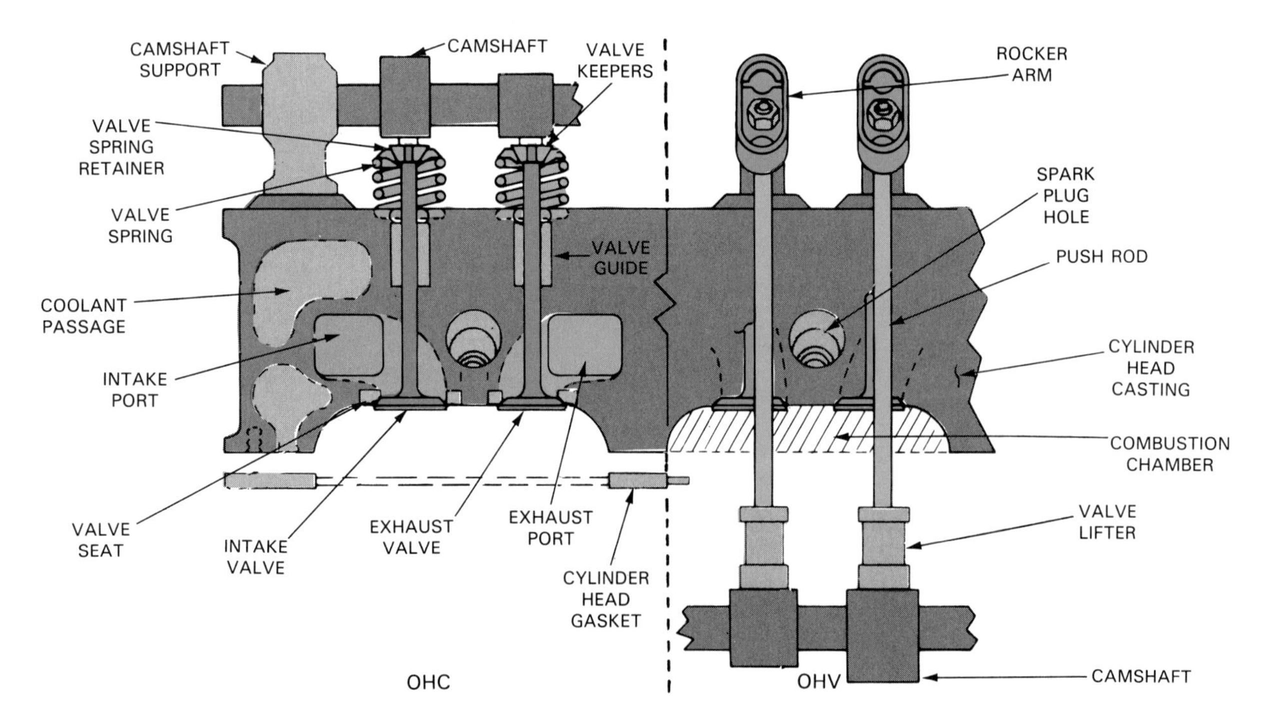

Fig. 10-4. These are the two major variations of engine top ends—overhead valve and overhead cam configurations. (Ford Motor Co.)

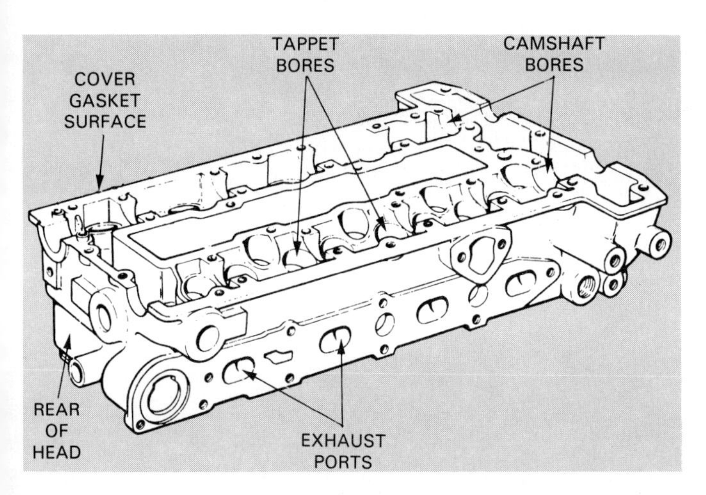

Fig. 10-6. Bare cylinder head has all of its parts removed. It can be cast from iron or aluminum. (Saab)

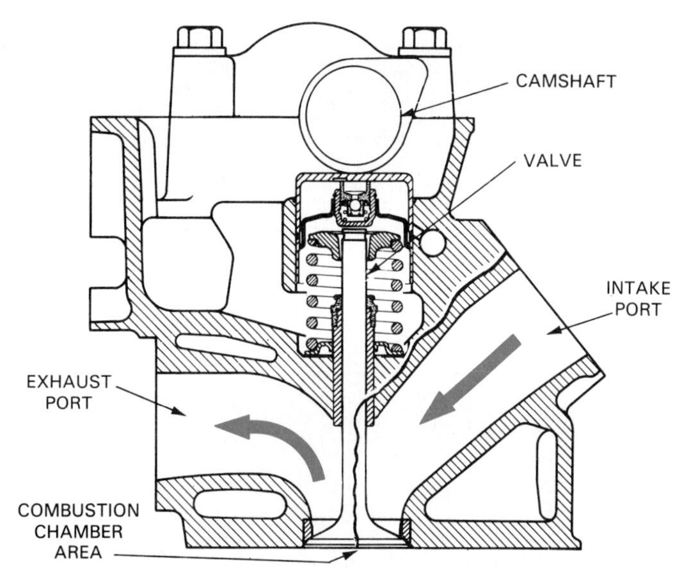

Fig. 10-8. Cutaway of side of head shows intake and exhaust ports. (Mercedes Benz)

Bare cylinder head

A *bare cylinder head* is a head casting, with all of its parts (valves, keepers, retainers, springs, seals, and rocker arms) removed. It is commonly made of cast iron or a lighter alloy of aluminum, Fig. 10-6.

If a cylinder head becomes badly damaged (cracked, warped) in service, the technician may need to install another bare head. All of the old, reusable parts would be removed and installed in the new head.

Parts of cylinder head

The major parts of a bare cylinder head include:

1. COMBUSTION CHAMBERS—Cavities or pockets formed in head deck surface above cylinders. They are located directly over the pistons. The burning of the fuel mixture occurs in these chambers. Spark plugs (gasoline engine) or glow plugs and injectors (diesel engine) protrude into these chambers. Refer to Fig. 10-7.
2. INTAKE PORTS—Passages through cylinder head from intake manifold runners to combustion chambers. These ports carry air-fuel mixture (gasoline engine) or just air (diesel engine) to the cylinders, Fig. 10-8.
3. EXHAUST PORTS—Other passages, usually on opposite side of head, leading from combustion chambers, to exhaust manifold. These ports carry burned gases out of cylinders, Fig. 10-8.
4. OIL PASSAGES—Holes through head so oil can enter from block deck surface. Matching holes align in head gasket. Oil is needed in head to lubricate moving parts of valve train. See Fig. 10-9.
5. WATER JACKETS—As in the cylinder block, coolant passages are formed in the cylinder head. They allow coolant to circulate around the backside of the combustion chambers and carry off excess heat from the burning fuel.
6. INTAKE DECK—This is a machined surface on the head for bolting on the intake manifold. The intake ports and intake bolt holes are in the intake deck.

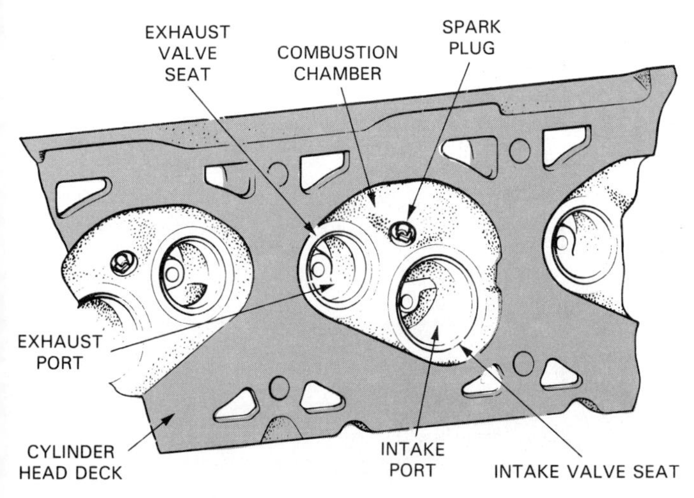

Fig. 10-7. Bottom view of cylinder head shows combustion-chamber for gasoline engine. Note part names. (Cadillac)

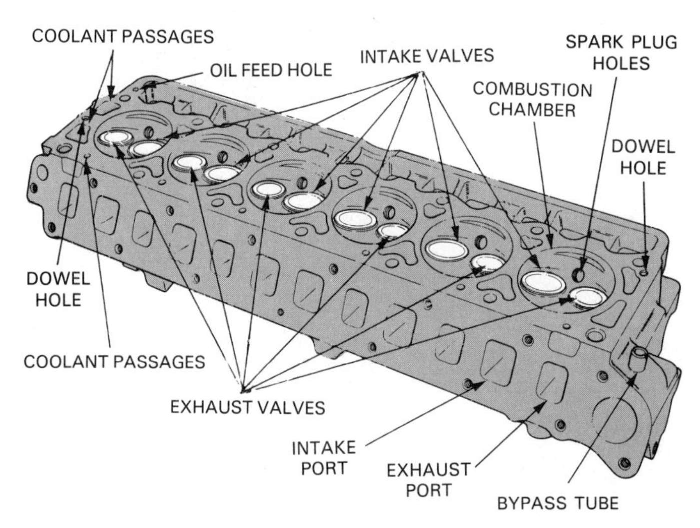

Fig. 10-9. This cylinder head still has the valves installed. Note passages in deck for coolant and oil. (Chrysler)

7. EXHAUST DECK—This is another machined surface on the head for the exhaust manifold. The exhaust ports and exhaust manifold bolt holes are on this surface. With some older designs, both the intake and exhaust ports can be on the same surface, Fig. 10-9.
8. DOWEL HOLES—These are sometimes provided so that the cylinder head can be accurately aligned when installing the head on the block. Dowels in the block fit into the dowel holes in the head. This also holds the head gasket in alignment.

Other machined surfaces and holes are provided on the cylinder head for mounting accessory units (alternator, air conditioning compressor, power steering pump, etc.).

Fig. 10-10 shows a cutaway view of a typical OHC cylinder head for a gasoline engine. Note how the spark plug installs in the head. Note how the *cylinder head caps* bolt to the top of the head. They hold the cam bearings and camshaft in place.

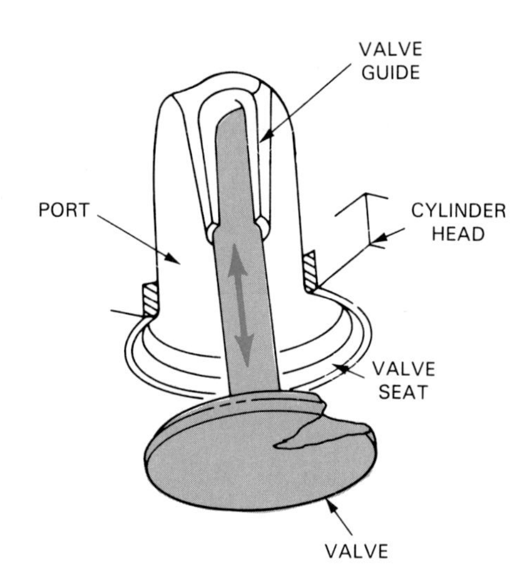

Fig. 10-11. Valve slides open and closed to block and unblock cylinder head port. Valve touches and seals on seat. Valve guide supports valve stem.

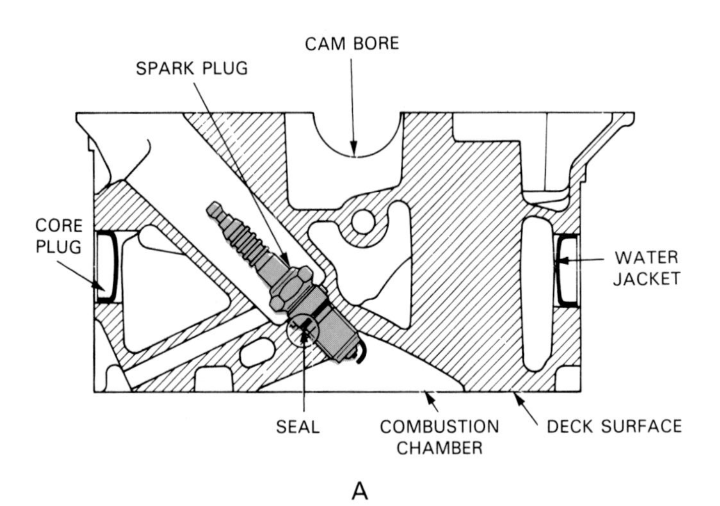

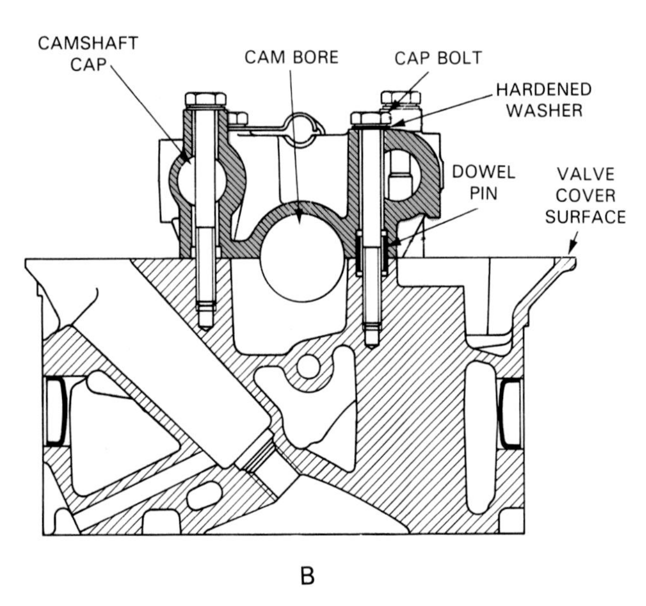

Fig. 10-10. A—Cutaway of head shows how spark plug screws into head and extends into combustion chamber. Also note coolant passages or water jackets. B—This view of head for OHC engine shows cylinder head caps that hold camshaft and cam bearings in place. (Mercedes Benz)

VALVE GUIDES

Valve guides are small holes machined through the top of the head, down into the intake and exhaust ports. The engine valves slide up and down in these guides, Fig. 10-11. The two basic types of valve guides are the integral and the pressed-in. Both are commonly used in modern passenger car engines.

Integral valve guides

Integral valve guides are made as part of the cylinder head casting. The guide is simply drilled and reamed in the head itself. This is shown in Fig. 10-12.

Integral valve guides are inexpensive to produce and are very common in cast iron cylinder heads. They are NOT used in aluminum heads because the soft aluminum would wear too quickly.

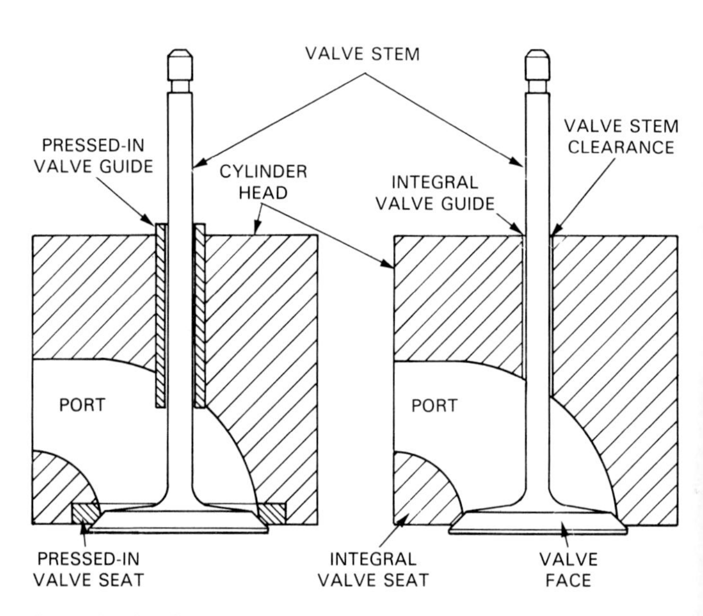

Fig. 10-12. Compare integral valve guides and valve seats and pressed-in or insert type guides and seats.

Pressed-in valve guides

Pressed-in valve guides or *valve guide inserts* are separate sleeves made of iron, steel, or bronze. They are force-fitted into the cylinder head. A hole, slightly smaller than the guide outside diameter, is precisely machined in the head. Then, a press is used to drive the guide into the smaller hole, locking the guide in place, Fig. 10-12.

Pressed-in guides can be found in both cast iron and aluminum cylinder heads. Some cast iron heads have these type guides while ALL aluminum heads use them.

When in cast iron heads, the pressed-in guide may have been installed by a machine shop to recondition worn guides. This will be discussed in later chapters. Refer to Fig. 10-13 and find the valve guides.

Knurled valve guides

Knurled valve guides have spiral grooves pressed into the inside diameter of the guide. *Knurling* is a machine shop process that decreases the inside diameter of the guide. It is often used to restore slightly worn guides to specifications. This process is also discussed in a later chapter.

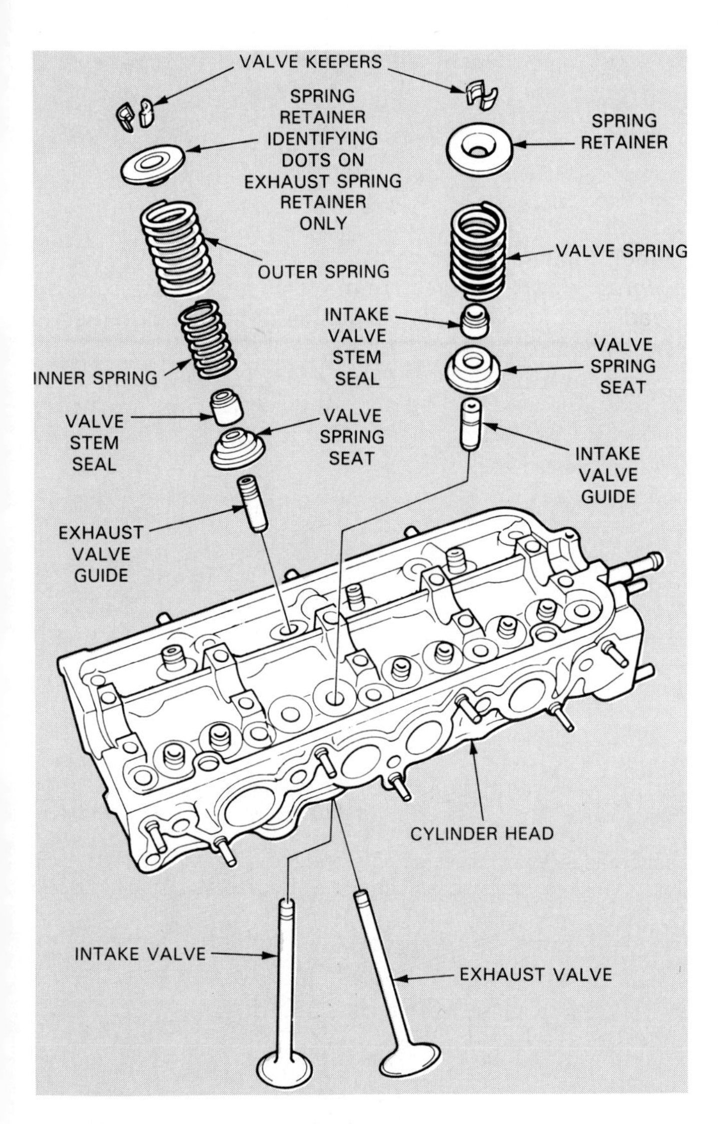

Fig. 10-13. These are the major parts of valve assemblies. Study part names and relationships carefully. (Honda)

VALVE SEATS

Valve seats are round, machined surfaces in the port openings to the combustion chambers. When the engine valve closes, the valve touches the seat to close off the port. Refer back to Fig. 10-11.

As with valve guides, valve seats can be part of the head or a separate pressed-in component. Both can be found in today's engines. See Fig. 10-12.

Integral valve seat

An *integral valve seat* is made by using cutters to machine a precise face on the port opening into the combustion chamber. The seat is aligned with and centered around the valve guide so the valve centers on the seat. Look at Fig. 10-14.

Pressed-in valve seat

A *pressed-in valve seat* or a *seat insert* is a separate machined part force-fitted into the cylinder head. A recess is cut into the combustion chamber slightly smaller than the OD of insert, Fig. 10-15. Then a press is used to drive the insert into the head. Friction keeps the seat from falling out.

Steel valve seat inserts are used in aluminum cylinder heads. The steel is needed to withstand the high operating temperatures produced by combustion.

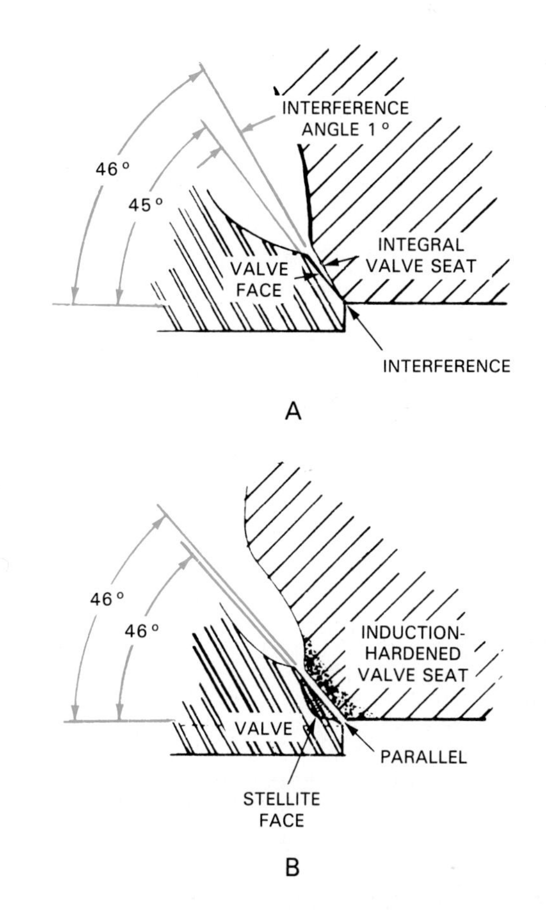

Fig. 10-14. A—Interference angle is one degree angle difference between valve face and valve seat. B—Induction hardened seat and stellite valve face coating are used to increase service life. They can withstand unleaded fuel better.

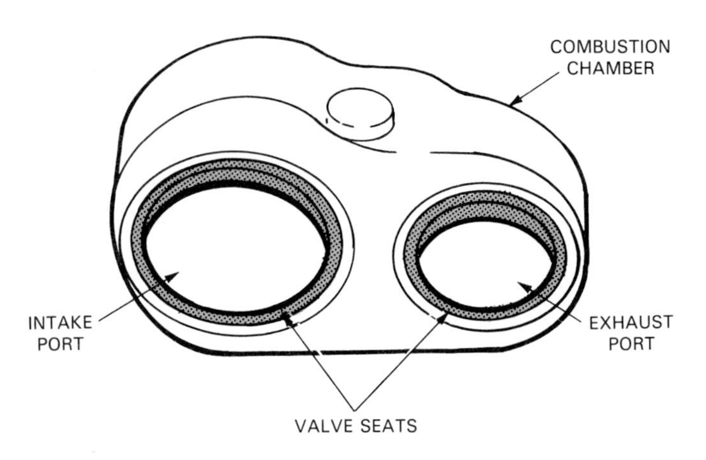

Fig. 10-15. Valve seats are formed at openings of ports into combustion chamber. They can be integral or pressed-in. (Ford)

When inserts are found in cast iron heads, they have normally been installed to repair worn or damaged seats. A seat insert is not commonly used in cast iron heads because they will not dissipate heat as quickly as integral seats. This will make the valves run hotter.

Induction hardened seats

Induction hardened valve seats are commonly used on late model engines to increase service life. *Induction hardening* is a electric-heating operation that makes the surface of the metal much harder and more resistant to wear. This is shown in Fig. 10-14.

In the past, lead additives in fuel helped to lubricate the contact point between the valves and the valve seats. The lead acted as a high temperature lubricant between the two surfaces. However, unleaded fuel of today does NOT contain lead type lubricants. Also, engine operating temperatures are higher. Valve and valve seat wear is more pronounced. Hardened valve faces and valve seats, especially on the exhaust seats, are needed to withstand these severe conditions.

Valve seat angles

The *valve seat angle* is the angle formed by the face or operating surface of the seat. A 45 ° seat angle is commonly used on passenger car engines on both the intake and exhaust seats. However, some intake seats are 30 °. The 30 ° angle intake seats are found on a few high performance engines. Look at Fig. 10-14.

Interference angle

An *interference angle* is a one-half to one degree angle difference between the valve seat face and the face of the valve. This is illustrated in Fig. 10-14.

The interference angle reduces the amount of contact area between the seat and valve. This increases the pressure between the two and speeds valve *seating* (sealing) during engine break-in.

VALVE CONSTRUCTION

As you learned earlier, *engine valves* open and close to control flow in and out of the combustion chamber. They are usually made of steel. See Fig. 10-16.

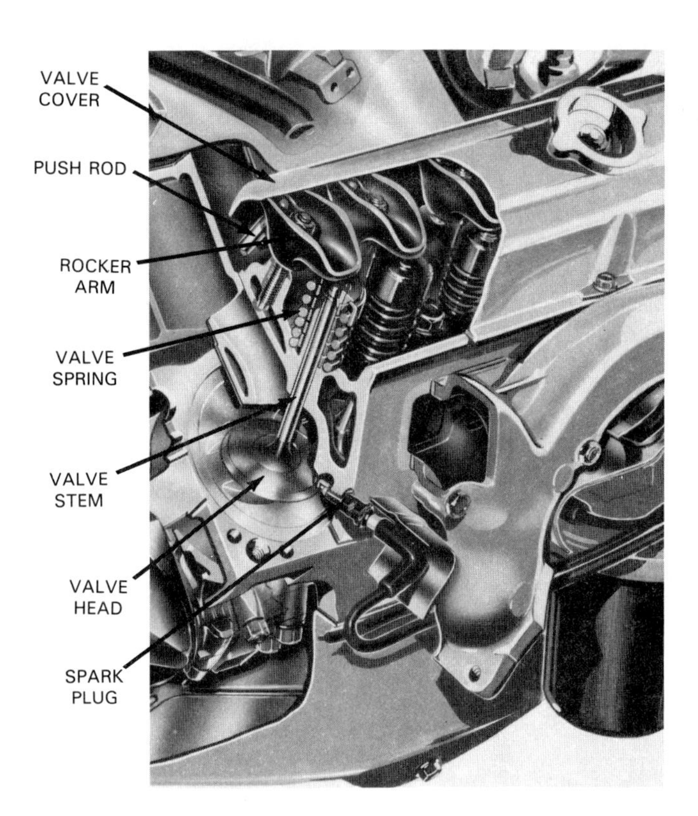

Fig. 10-16. Another cutaway of engine shows rocker arms and other valve train parts. (Chevrolet)

Most engines have one intake and one exhaust valve per cylinder. However, a few high performance engines have four valves, or even six valves, per cylinder, Fig. 10-17. This was discussed in detail in Chapter 6.

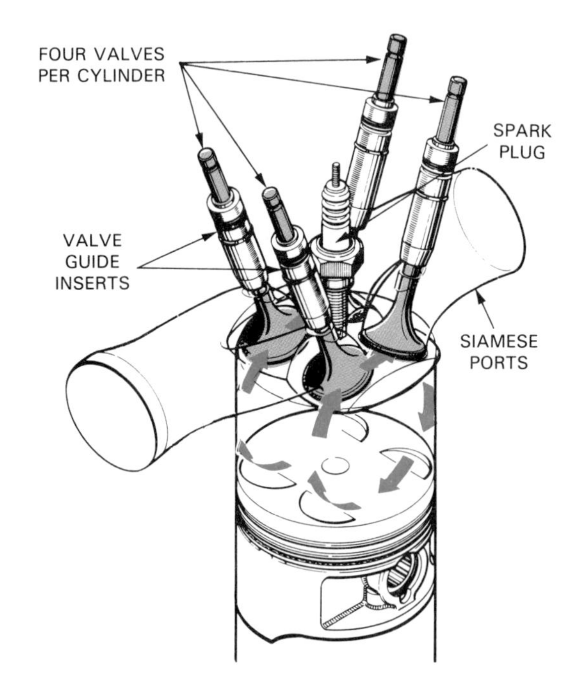

Fig. 10-17. This engine uses four valves per cylinder. Note valve guide inserts around valve stems.

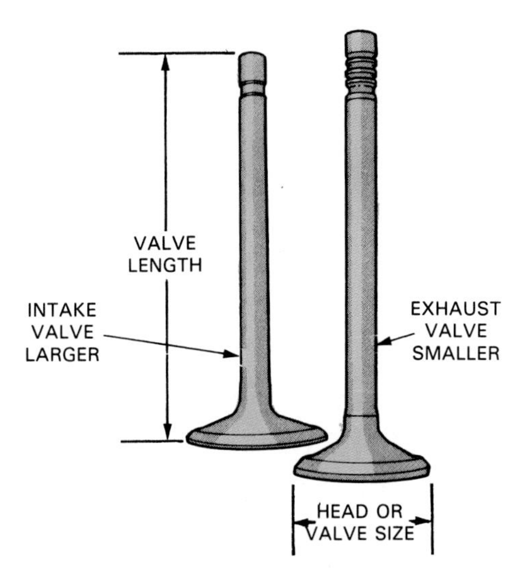

Fig. 10-18. Intake valve is normally larger than exhaust valve. Valve size is determined by diameter of valve head. (Chrysler)

An *intake valve* controls the flow of air-fuel mixture or just air through the cylinder head intake port. It is the LARGER VALVE, Fig. 10-18.

An *exhaust valve* controls the flow of burned gases out of the combustion chamber. It is located over the cylinder head exhaust port. The exhaust valve is SMALLER, Fig. 10-18.

Automotive engines commonly are *poppet* or *mushroom valves.* These terms were derived from the valve's shape (looks like a mushroom) and action (valve ''pops'' open).

Parts of a valve

The basic parts of a valve are shown in Fig. 10-19 and they include:

1. VALVE HEAD—It is the large, disc shaped surface exposed to the combustion chamber. Its outside diameter determines the *valve size.*
2. VALVE FACE—It is a machined surface on the back of the valve head. It touches and seals against the seat in the cylinder head, Fig. 10-23.
3. VALVE MARGIN—This is the flat surface on the outer edge of the valve head. It is located between the valve head and face. The margin is needed to allow the valve to withstand the high temperatures of combustion. Without a margin, the valve head would melt and burn.
4. VALVE STEM—It is a long shaft extending out of the valve head. The stem is machined, polished, and sometimes chrome plated. It fits into the guide machined through the cylinder head, Fig. 10-23.
5. KEEPER GROOVES or LOCK GROOVES—They are machined into the top of the valve stem. They accept small keepers or locks that hold the spring retainer on the valve.
6. STEM TIP—This is the upper end of the valve stem. It is ground perpendicular to the stem. A small chamfer is formed around the edge of the stem tip. The tip can be hard-alloy coated to resist rocker arm-caused wear.

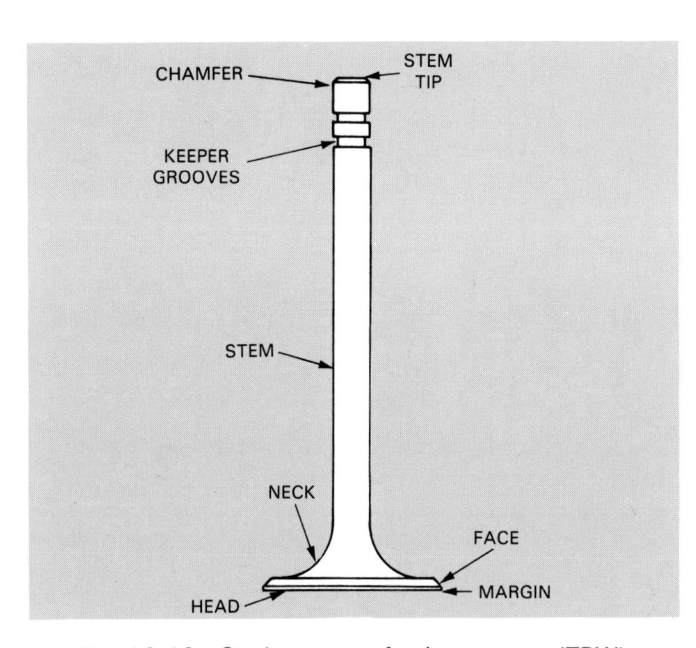

Fig. 10-19. Study names of valve parts. (TRW)

Almost every surface of a valve is machined. The stem must accurately fit the guide. The face must contact the seat perfectly. The margin must be thick enough to prevent valve burning. Grooves are accurately cut into the valve stem for the keepers.

Valve face angle

The valve face angle is the angle formed between the valve face and valve head. Normal valve face angles are 45° and 30°, 45° being more common. This was illustrated in Fig. 10-14.

Valve stem-to-guide clearance

Valve stem-to-guide clearance is the small space between the valve stem and the valve guide. It lets the valve slide freely up and down in the guide. It also allows a small amount of engine oil to lubricate the valve stem.

Valve shapes

Valve shape refers to the configuration of the valve head, Fig. 10-20. The head of the valve can be recessed,

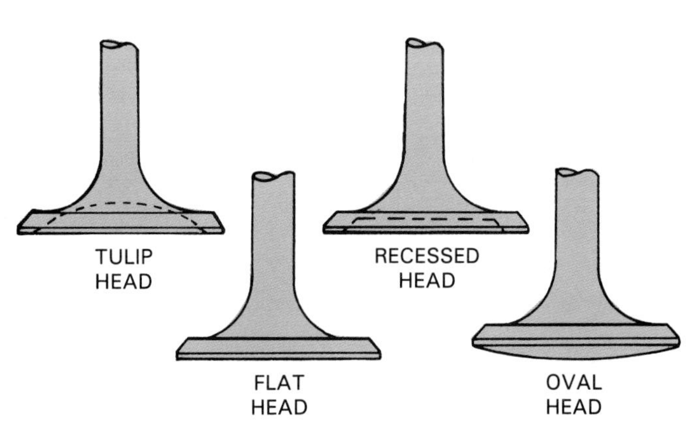

Fig. 10-20. Various valve shapes can be used. Tulip or recessed head is generally for high performance. Flat or oval head would be found in low rpm engines. (TRW)

flat, or oval in shape. The shape of the valve head partially controls the flexibility of the valve. A high rpm engine needs a flexible valve so that the valve will not bounce off its seat when closing and so that the valve can conform to any irregularities in the seat.

A recessed valve, sometimes termed *tulip valve,* is very flexible. It is also lighter than an oval or flat headed valve because metal has been removed from the head. It is ideal for high performance engines.

Sodium filled valves

A *sodium filled valve* is used when extra cooling action and lightness are needed.

During engine operation, the sodium inside the hollow valve melts. When the valve opens, the sodium splashes down into the valve head and collects heat. When the valve closes, the sodium splashes up into the valve stem. Heat transfers out of the sodium, into the stem, valve guide, and engine coolant. In this way, the valve is cooled, Fig. 10-21.

Sodium filled valves are used in a few high performance engines. They are very light and allow high engine rpm for prolonged periods without valve overheating.

Valve operating temperatures

Fig. 10-22 illustrates the typical operating temperatures of an exhaust valve. The exhaust valve is exposed to higher temperatures than the intake valve. The intake valve has cooler outside air flowing over it. Hot combustion gases blow over the exhaust valve. For the exhaust valve not to burn, it must transfer heat into the cylinder head.

DANGER! Never break open a sodium filled valve. The sodium is very reactive and can cause serious burns. If sodium is dropped into water, a reaction will occur and the sodium will burst into flames violently. When removing a damaged sodium filled valve from an engine, use extreme caution. Wear gloves and eye protection!

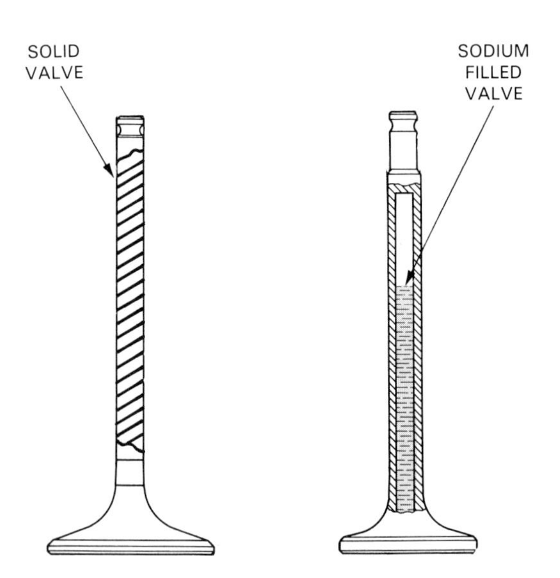

Fig. 10-21. Sodium filled valve is hollow. It will be lighter and run cooler than solid stem valve. Do not break open valve stem. Sodium could cause severe burns. (Mercedes Benz)

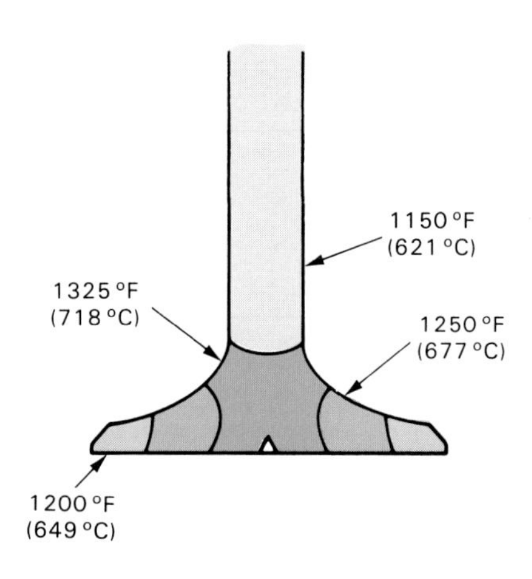

Fig. 10-22. Note operating temperatures for various surfaces of exhaust valve.

Stellite valve

A *stellite valve* has a special hard metal coating on its face. A stellite coating is often used in engines designed to burn unleaded fuel. Refer back to Fig. 10-14.

VALVE SEAL CONSTRUCTION

Valve seals prevent oil from entering the cylinder head ports through the valve guides. See Fig. 10-23.

The valve seals fit over the valve stems and keep oil from entering through the clearance between the stems and guides.

Without valve seals, oil could be drawn into the engine cylinders and burned during combustion. Oil consumption and engine smoking could result.

Valve seals come in two basic types: umbrella and O-ring types. Both are commonly used on modern engines.

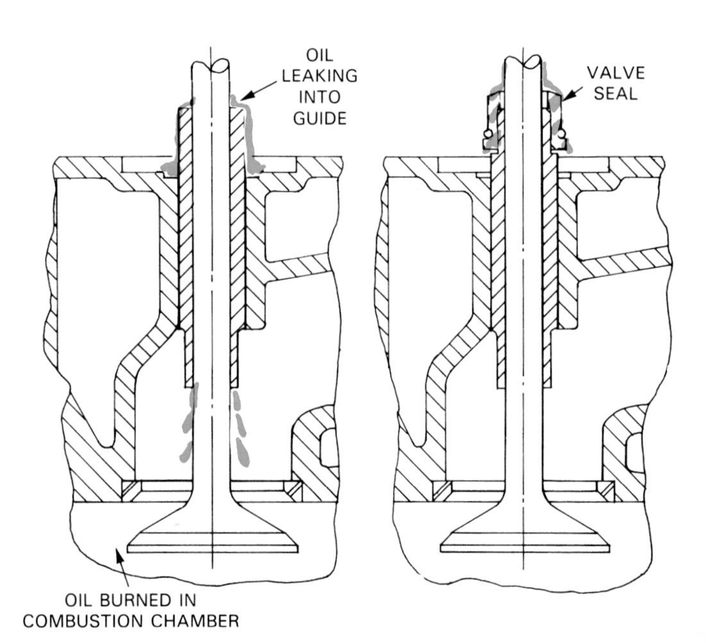

Fig. 10-23. Oil seal keeps too much oil from entering valve guide. (American Hammered Piston Rings)

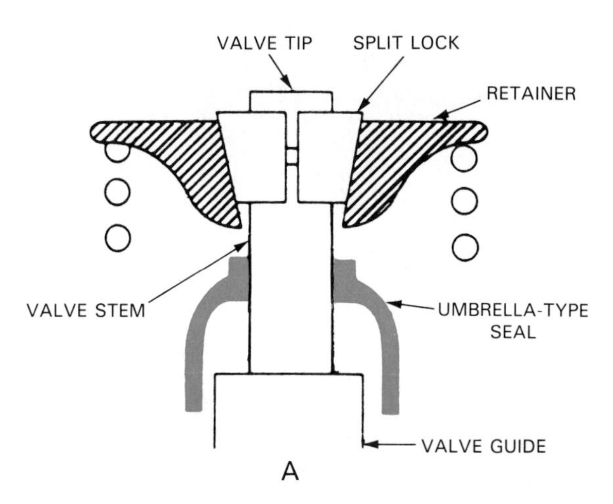

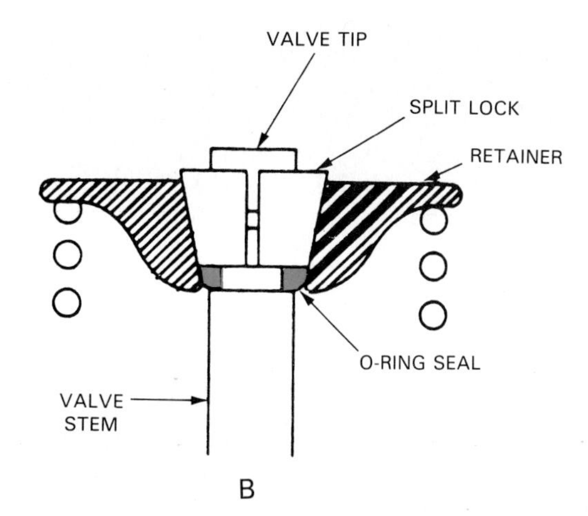

Fig. 10-24. A—Umbrella type valve seal surrounds opening into top of valve guide. B—O-ring type valve seal keeps oil from flowing down through retainer and into guide. Note part names. (Chrysler Corporation)

Umbrella valve seals

An *umbrella valve seal* is shaped like a cup and can be made of neoprene rubber or plastic. This type is shown in Fig. 10-24A.

An umbrella valve seal slides down over the valve stem before the spring and retainer. It covers the small clearance between the stem and guide. This keeps too much oil from splashing into the guide.

Three umbrella seals are pictured in Fig. 10-25.

O-ring valve seals

An *O-ring valve seal* is a small round seal that fits into an extra groove cut in the valve stem, Fig. 10-24B.

Unlike the umbrella type, it seals the gap between the retainer and valve stem, not the guide and stem. It stops oil from flowing through the retainer, down the stem, and into the guide.

An O-ring valve seal fits onto the valve stem AFTER the spring and retainer. It is made of soft synthetic rubber which allows it to be stretched over the valve stem and into its groove.

Valve oil shedder

A *valve oil shedder* is a variation of an umbrella type oil seal. It is made of hard plastic or nylon, Fig. 10-26. The shedder simply keeps excess oil from splashing on the valve stem and flowing down through the guide.

Fig. 10-25. These are three variations of umbrella type valve seals. They lock down onto valve guide. A—Seal with one lock ring and nylon insert. B—All synthetic rubber seal with two lock rings. C—Hard plastic seal with lock strap. (Dana)

Valve stem caps

Valve stem caps can be used on the tips of the valve stems. They are normally used to reduce rocker arm and stem tip wear. They sometimes can be used to provide a means of adjusting valve-to-rocker arm clearance. The caps can come in different thicknesses or be ground to provide for adjustment.

Fig. 10-27 shows a valve stem cap. It is simply a hardened steel cup that installs over the valve stem. It is free to rotate on the tip to reduce friction and wear.

VALVE SPRING ASSEMBLY CONSTRUCTION

The *valve spring assembly* is used to close the valve. It basically consists of a valve spring, retainer, and two keepers. Look at Fig. 10-26.

Valve spring construction

Valve spring construction is basically the same for all engines. Spring steel is wound into a coil. However, the number and types of coils can vary.

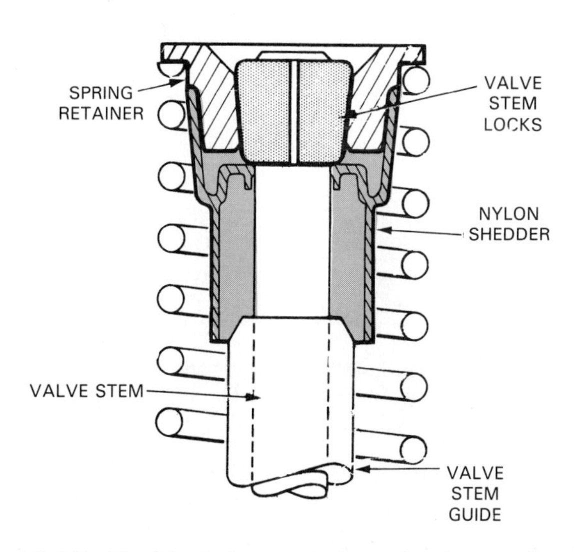

Fig. 10-26. Shedder is large nylon cup that keeps oil out of guide. (Buick)

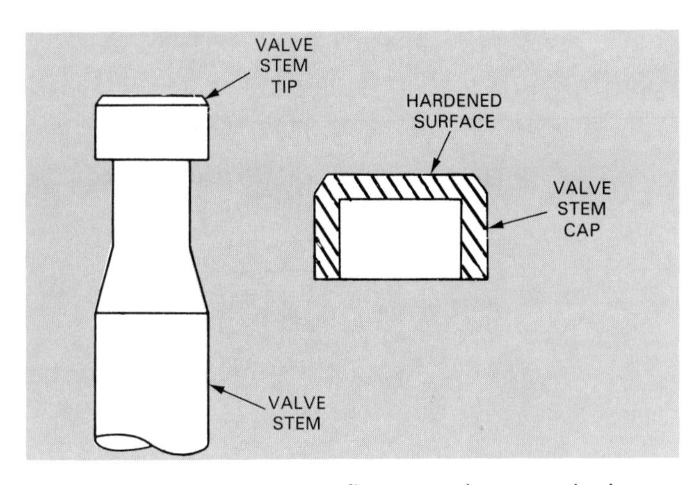

Fig. 10-27. Valve stem cap fits over valve stem tip. It cuts friction and can provide for valve clearance adjustment. (Deere & Co.)

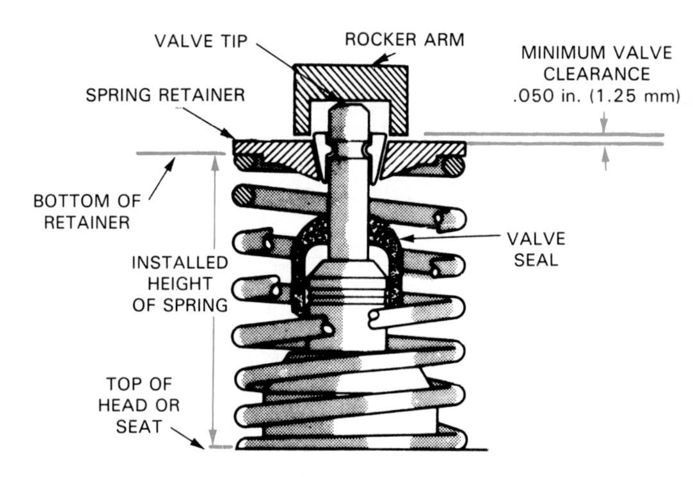

Fig. 10-29. Installed height of valve spring determines spring tension. When valves are ground, it will increase this dimension and reduce spring pressure. (Chrysler)

Fig. 10-28 shows single and multiple coil valve springs. The extra coils increase the amount of pressure holding the valve closed.

Multiple coils are needed for high rpm engines. *Single coils* are adequate for low rpm engines.

Valve spring specifications

Various specifications are given for valve springs. These include:

1. *Valve spring tension* refers to the stiffness of a valve spring. Spring tension is usually stated for both opened and closed valve positions. The service manual will give the tension in pounds or kilograms for specific compressed lengths.
2. *Valve spring free length* is the length of the valve spring when removed from the engine.
3. *Valve spring open length* is the length of the valve spring when installed on the engine with that valve fully open. It is measured from the bottom of the spring to the bottom of the spring retainer.
4. *Valve spring closed length* or *installed height* is the length of the valve spring when installed on the engine with the valve closed. Measurement is the same as for open length. See Fig. 10-29.

As you will learn in later chapters, these spring specifications are important. They affect valve action. Low spring tension can cause *valve float* (spring too weak to close valve at high rpm) which reduces engine performance.

Valve spring shims

Valve spring shims can be used to increase or restore valve spring tension. They are precise thickness flat washers that fit under the valve springs to reduce installed height. When a shim is placed under a valve spring, it reduces the opened and closed length of the spring. This compresses the spring more to increase its closing pressure. See Fig. 10-30.

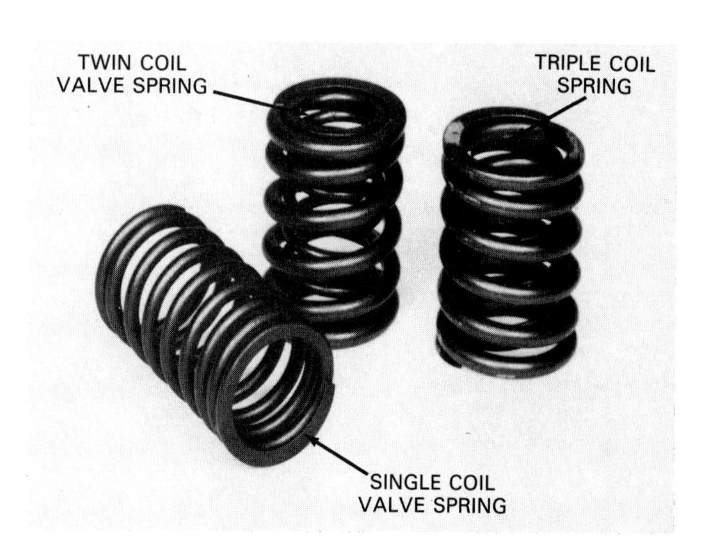

Fig. 10-28. Engine valve springs can have different number of coils. More coils increase engine rpm capabilities. (Dana)

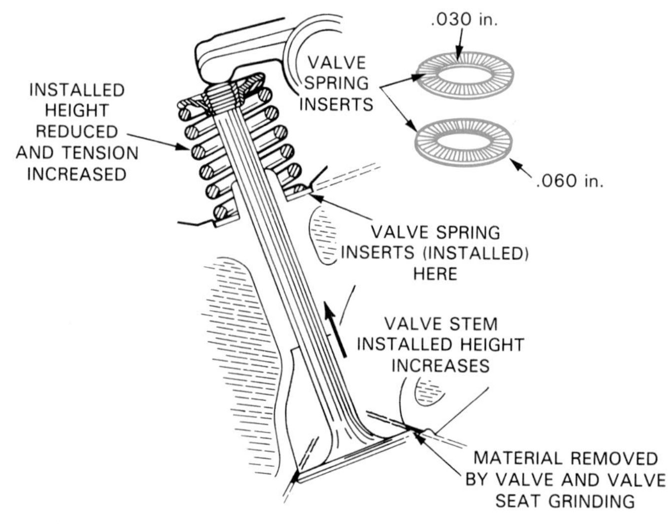

Fig. 10-30. Valve spring shims provide means of restoring correct spring tension. If installed height is increased from valve grinding or head reconditioning, you can install shims to bring spring pressure to within specs. (Ford)

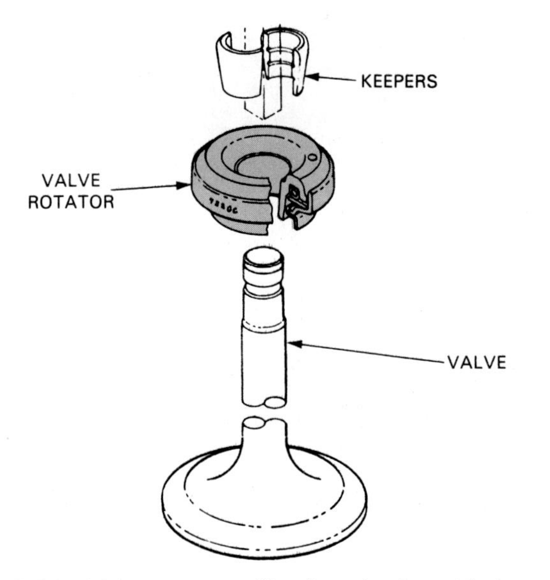

Fig. 10-31. Valve rotator will spin valve in guide to reduce localized hot spots and carbon buildup. Either could reduce valve service life. Rotators can fit under or over valve spring. (Sealed Power Corporation)

A used valve spring may weaken and lose some of its tension. Also, valve and seat grinding (head reconditioning) increases spring installed height. Valve spring shims provide a means of restoring full spring tension and pressure without spring replacement.

Selection and installation of valve spring shims are covered under cylinder head service.

Valve rotators

Valve rotators are used on some engines to spin or turn the valves in their guides. The turning action helps prevent a carbon buildup on the valve. It also prevents a localized hot spot or wear pattern on the valve or seat. Look at Fig. 10-31.

A valve rotator may be located under the valve spring (seat type rotator) or it may be on top of the valve spring (retainer type rotator). Rotators are commonly used on engine exhaust valves. Exhaust valves are exposed to more heat than intake valves.

Valve retainer construction

Valve retainers fit over the top of the valve springs, as shown in Fig. 10-29. They have a machined lip so that they center on the spring. Valve retainers are usually made of steel but can sometimes be made of lighter alloys.

Valve keeper construction

Valve keepers fit between the valve retainers and valve stems to hold the springs on the valves. See Figs. 10-24 and 10-31. Two usually install on each valve. The keepers are hardened steel to resist the pounding action of valve motion.

Valve spring seat

A *valve spring seat* is a cup-shaped washer installed between the cylinder head and bottom of the valve spring, Fig. 10-32. It provides a pocket to hold the bottom of the valve spring squarely on the head.

Valve spring shield

A *valve spring shield* is normally used in conjunction with an O-ring type oil seal. The thin metal shield surrounds the top and upper sides of the spring and helps keep excess oil off of the valve stem.

VALVE TRAIN CONSTRUCTION

As you learned earlier, the *valve train* controls the opening and closing of the valves and cylinder head ports. Although the basic function of these parts is the same, their construction can vary.

Fig. 10-32 shows the valve train parts for an overhead

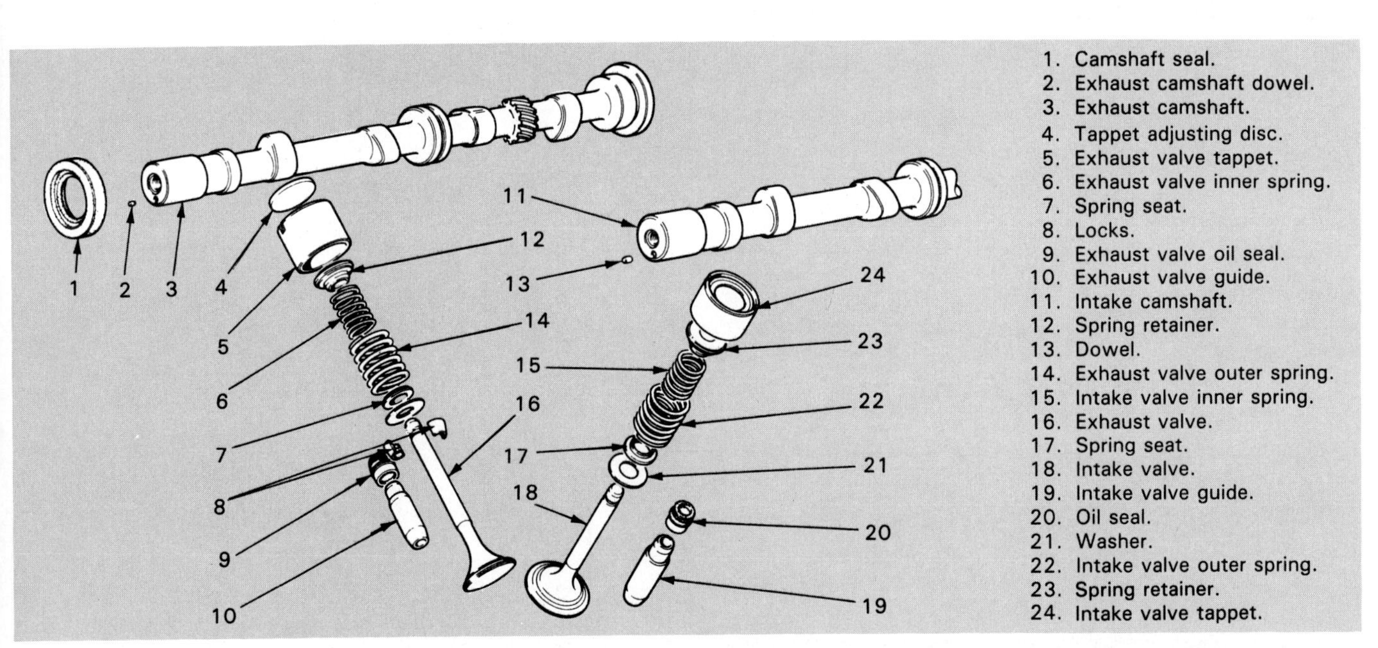

Fig. 10-32. These are basic parts of valve train for overhead camshaft engine. Study them. (Fiat)

cam engine. Fig. 10-33 illustrates the components of a valve train for an overhead valve or push rod engine. Compare their similarities and differences.

These are the two major valve train variations. You will find both when working in today's automotive repair facility. You must understand each type!

Fig. 10-34 shows the top view of a V-type engine with the intake manifold and valve covers removed.

CAMSHAFT CONSTRUCTION

A *camshaft,* mentioned in earlier chapters, opens the engine valves at the right time during each stroke. It can be powered by gears, a chain, or a belt running off of the crankshaft. The camshaft turns at one-half crankshaft speed. See Fig. 10-35.

The camshaft, sometimes called just "cam," is usually made of cast iron. The cam can be located in the block or in the cylinder head.

Although most engines have only one camshaft, some have two or four camshafts. This was explained in the chapter on engine classifications.

Cam lobe construction

The *cam lobes* are egg-shaped protrusions on the camshaft that change rotary motion into an up and down motion. They are machined, polished, and hardened to prevent friction and wear from the lifters, Fig. 10-35.

Although the cam lobes are an integral part of most camshafts, a *built-up camshaft* consists of individual lobes that are mounted on a separate hollow shaft. A built-up camshaft is light and extremely strong.

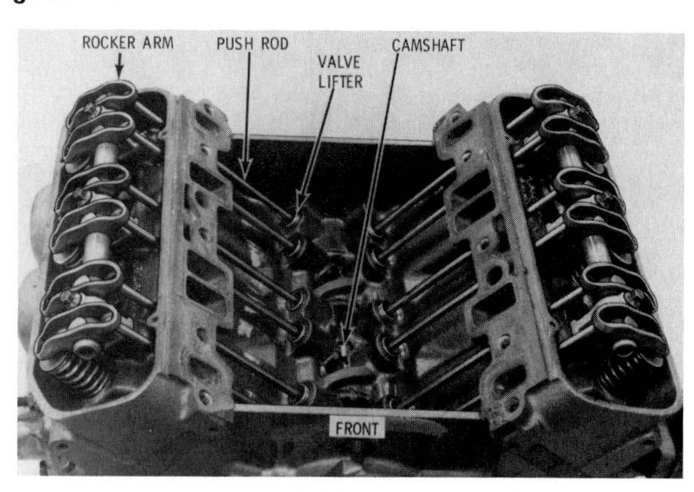

Fig. 10-34. Top view of engine with intake manifold and valve covers removed shows valve train. (Oldsmobile)

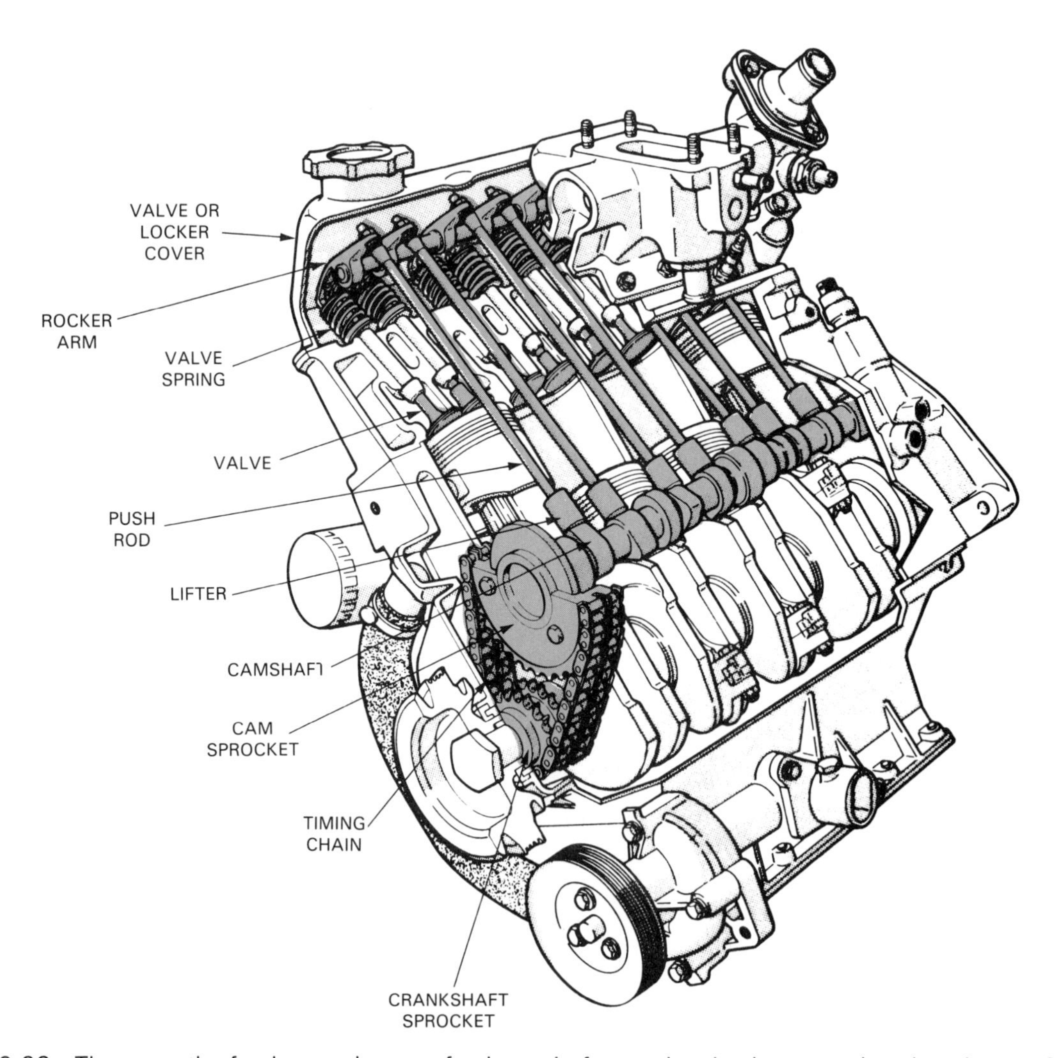

Fig. 10-33. These are the fundamental parts of valve train for overhead valve or push rod engine. (Chrysler)

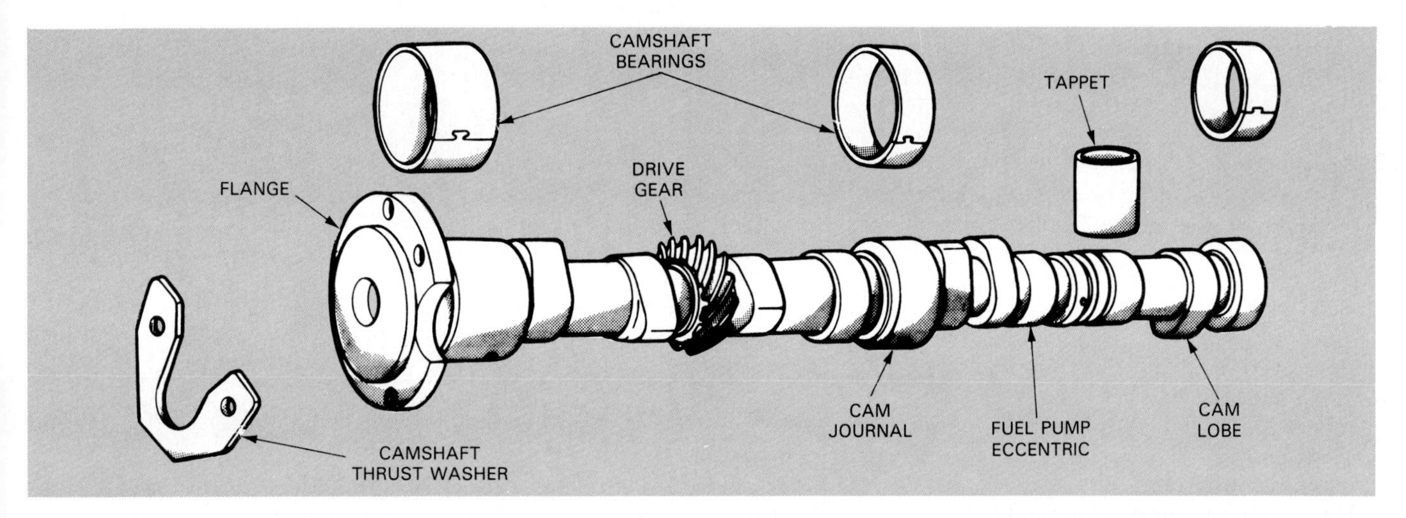

Fig. 10-35. Study parts of camshaft and its related components. Note that oil pump drive shaft is driven by cam drive gear. (Chrysler)

Usually, one cam lobe is provided for each engine valve. If the cylinder has two valves, there would be an intake lobe and exhaust lobe for that cylinder. A four-valve cylinder would usually require two more lobes and a second camshaft.

The shape of the cam lobes, Fig. 10-36, affect:

1. When each valve opens in relation to piston position.
2. How long each valve stays open.
3. How far each valve opens.

Some camshafts are machined with dual cam lobes with different profiles (shapes). One cam lobe is designed for good low speed efficiency. The other lobe profile provides power at high engine speeds. The ECU operates a solenoid valve that controls oil flow to shift the rocker arms from one lobe profile to the other.

Camshaft lift

Camshaft lift is the amount of valve train movement produced by the cam lobe. It partially determines how far the valve opens. Shown in Fig. 10-36, camshaft lift is found by subtracting the cam base circle diameter from the height of the cam lobe. Cam lift is given in inches or millimeters. A typical camshaft lift would be .450 in. (11.4 mm).

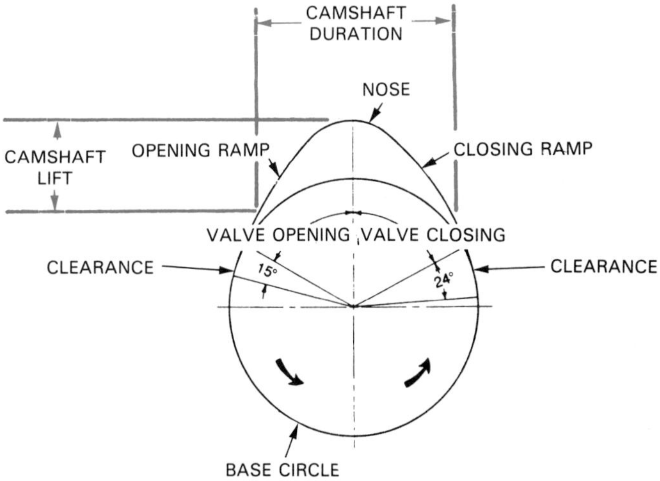

Fig. 10-36. Note parts of camshaft lobe. (Chrysler)

Camshaft duration

Camshaft duration determines how long the valves stay open, Fig. 10-36. The shape of the cam lobe nose and flank regulates camshaft duration. For instance, a pointed cam lobe would have a shorter duration than a more rounded lobe, Fig. 10-37.

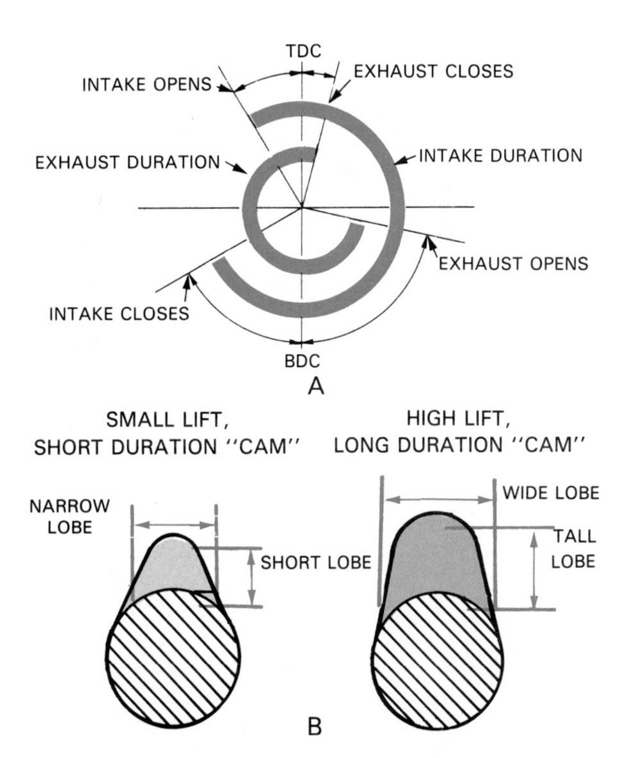

Fig. 10-37. Shape of cam lobe determines when and how long valves will open. A—Diagram shows cam action for one grind. B—Pointed lobe would generally be short duration cam. Rounder lobe would usually provide more duration. Height of lobe determines lift.

Cam duration is given in degrees of rotation—the larger the number, the longer the duration. An example, a camshaft might have 250° duration.

Valve timing

Valve timing refers to when the valves open and close in relation to the position of the pistons in the cylinders. Valve timing is designed into the camshaft and drive sprockets or gears by the manufacturer.

Detailed later, timing marks are placed on the timing gears or sprockets to assure correct valve timing. Aligning the marks assures that valve timing is within specifications.

Valve overlap

Valve overlap is the time when both of the valves in a cylinder are opened at the same time, Fig. 10-38.

Valve overlap is used to help *scavenge* (draw or suck) burned gases out of the cylinder. It also helps pull more fresh fuel charge into the cylinder.

With both the intake and exhaust valve open at the same time, the inertia (movement) of the gases through one cylinder head port and cylinder act on the gases in the other port. This results in slightly more flow into and out of the cylinder. Valve overlap helps engine breathing, especially at higher engine speeds.

Stock and high performance camshafts

The term *"stock camshaft"* refers to the original camshaft design installed by the engine manufacturer. Usually, it has mild lift and duration to provide both good performance and low exhaust emissions. See Fig. 10-37.

A high performance or "race cam" is a camshaft with increased lift and longer duration. Generally, higher lift helps the engine breath at all engine speeds. Long duration and overlap generally are for increasing maximum engine rpm capabilities.

A race cam will usually decrease engine power, fuel economy, and exhaust cleanliness at normal engine speeds. If duration is extreme, the engine will idle very roughly and not have power brakes (vacuum assist) at idle. The engine will be very inefficient from an idle up to normal highway speeds. In fact, a race cam installed without other engine modifications (exhaust headers, increased compression ratio, etc.) can reduce usable engine power. A race cam only increases power at higher engine speeds.

Warning! It is against the law to change camshaft specifications. The EPA will not allow engine modifications that increase exhaust pollution. Avoid modifying engines to be operated on city streets!

Hydraulic and mechanical camshafts

A *hydraulic camshaft* is designed to be used with hydraulic lifters. Its cam lobes are shaped to initially open the valve more quickly, called an *accelerated cam lobe.* This makes the hydraulic lifters operate properly.

A *mechanical camshaft* is made to work with solid lifters. The lobe is shaped to produce more constant opening of the valve, termed *constant-velocity lobe.*

Note! Do not use a hydraulic cam with solid lifters. Also, do not use a mechanical cam with hydraulic lifters. Either mistake could increase valve train wear and noise.

Camshaft gear

A camshaft sometimes has a *drive gear* for operating the distributor and oil pump. A gear on the ignition system distributor may mesh with this gear.

Camshaft eccentric

An *eccentric* (oval) may be machined on the camshaft for a mechanical (engine driven) fuel pump. It is similar to a cam lobe but is more round. As the cam turns, the eccentric moves the fuel pump arm up and down.

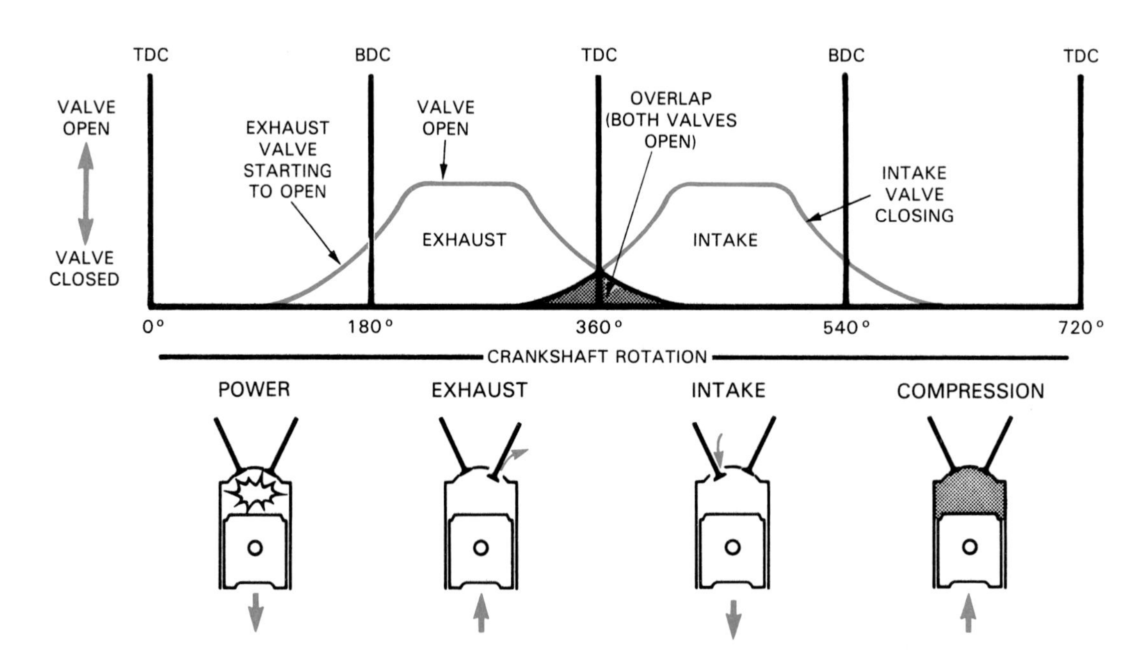

Fig. 10-38. Overlap is time when both valves in same cylinder are open. Graph shows valve action through complete cycle. Study events. (VW)

Camshaft thrust plate

A *camshaft thrust plate* is used to limit *camshaft end play* (front-to-rear movement). Refer back to Fig. 10-35.

The thrust plate bolts to the front of the block or cylinder head. When the drive gear or sprocket is bolted in place, the thrust plate sets up a predetermined clearance.

Camshaft bearings

Camshaft bearings are usually one-piece inserts pressed into the block or cylinder head, Fig. 10-39. They can also be two-piece inserts when mounted in the head. The camshaft journals ride in the cam bearings.

Cam bearings are usually constructed like engine main and connecting rod bearings. For more information on bearing construction, refer to Chapter 9.

VALVE LIFTER CONSTRUCTION

Valve lifters, also called *tappets,* ride on the cam lobes and transfer motion to the rest of the valve train, Fig. 10-40. The lifters can be located in the block or cylinder head. They fit into machined holes, termed *lifter bores.*

When the cam lob turns into the lifter, the lifter is pushed up in its bore. This pushes on the push rod and rocker arm to open the valve. Then, when the lobe rotates away from the lifter, the lifter is pushed down in its bore by the action of the valve spring. This keeps the lifter in constant contact with the camshaft.

The bottom of a lifter is *crowned* (slightly curved). Also, the lifter bore is offset from the cam lobe. As a result, one side of the lifter touches the cam lobe. This rotates the lifter in its bore to reduce wear, Fig. 10-41.

There are four basic types of lifters: hydraulic lifter, mechanical lifter, roller lifter, and OHC follower.

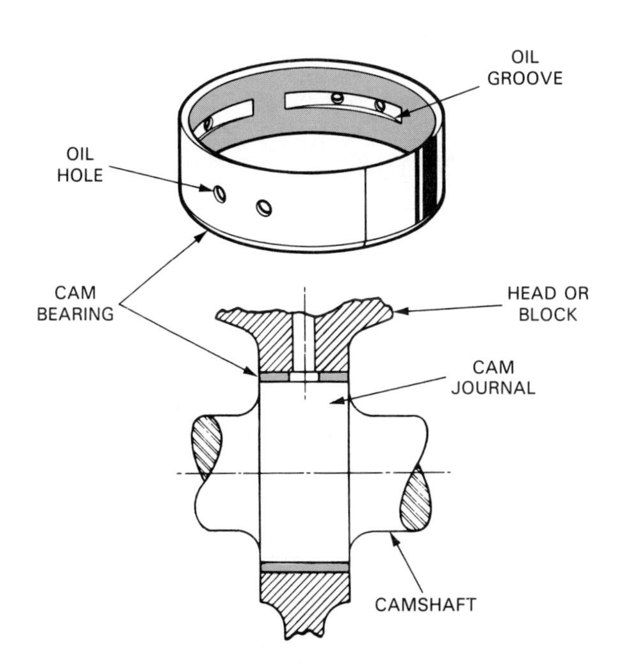

Fig. 10-39. Cam bearing is made like engine rod or main bearings. Cam bearings can be installed in block or head. Oil pump forces oil through bearing to prevent friction between cam journal and bearing surface. (Federal Mogul)

Hydraulic lifter construction

Hydraulic lifters operate quietly by maintaining zero valve clearance. They are filled with engine oil and can take up all of the clearance in the valve train when the engine is running. The oil-filled lifter adjusts itself automatically with any change in engine temperature and part wear. Hydraulic lifters are the most common type found on push rod engines, Fig. 10-42.

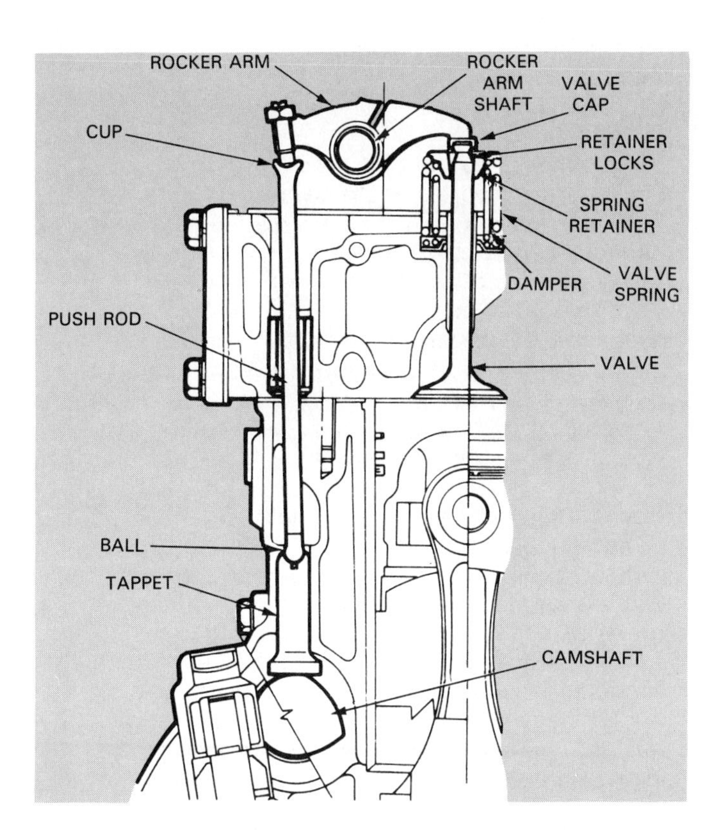

Fig. 10-40. Lifter rides on camshaft. It transfers motion to push rod and rocker arm. This is a solid lifter. Note valve lash or clearance adjustment on rocker arm. (Ford)

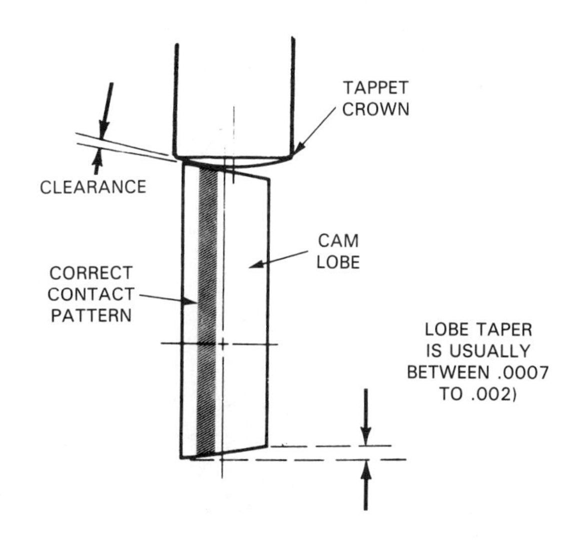

Fig. 10-41. Bottom of lifters are crowned. Cam is offset to one side of lifter. This produces contact point on one side of lifter. As a result, lifter rotates in its bore to limit wear. (Sealed Power)

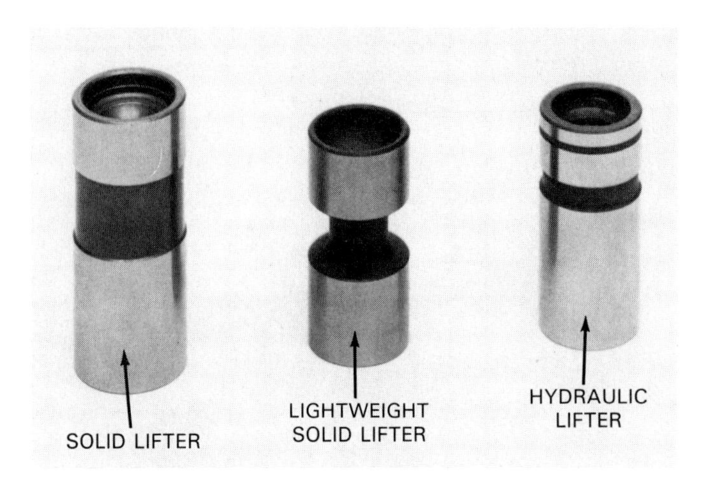

Fig. 10-42. Photo shows two solid and one hydraulic lifter. (Dana)

Fig. 10-43 shows the basic parts of a hydraulic lifter. A plunger fits into the lifter body. A check valve is provided to control oil flow into the lifter from the engine oil pump. A plunger spring keeps a constant upward pressure on the plunger.

During engine operation, oil pressure fills the inside of the hydraulic lifter with motor oil. The pressure pushes the lifter plunger up in its bore until all of the play is out of the valve train. See Fig. 10-44.

As the camshaft pushes on the lifter, the lifter check valve closes to seal oil inside the lifter. Since oil is NOT compressible, the lifter acts as a solid unit to open the valve. Look at Fig. 10-44.

If any oil leaks past the plunger or if parts wear, the oil pressure and spring push the plunger up. This compensates for any change in part clearance and maintains zero lash.

Mechanical lifter construction

Mechanical lifters, also called *solid lifters,* do NOT contain oil and are noisy during engine operation. They simply transfer camshaft lobe action to the push rod. Mechanical lifters are NOT self-adjusting and require periodic setting, Fig. 10-42.

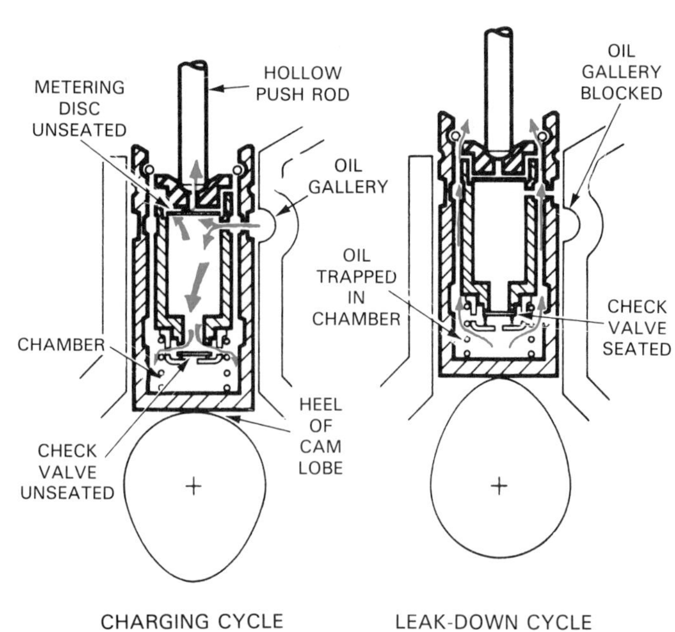

Fig. 10-44. Operation of hydraulic lifter. During charging cycle, cam lobe is away from lifter. Oil flows into lifter and gently moves plunger up to take up clearance. Then, during leak-down cycle, lobe pushes lifter up. Oil inside lifter is trapped and acts as solid unit. There is not enough time for lifter to leak down enough to affect valve opening. (American Motors)

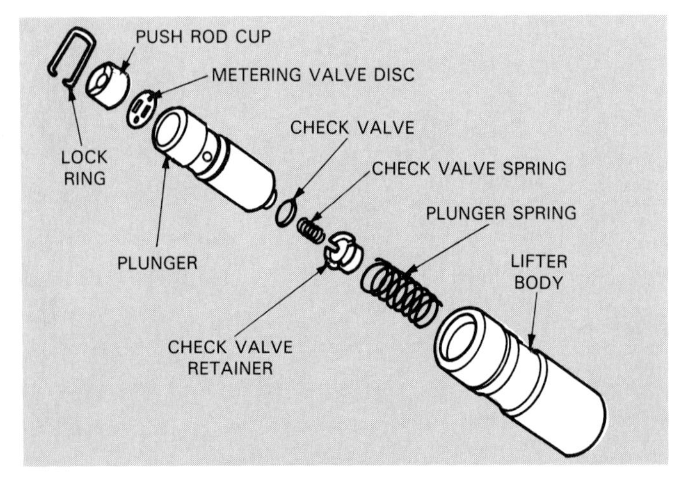

Fig. 10-43. Study internal parts of hydraulic lifter. Oil pressure acts on plunger to maintain zero clearance in valve train. (Ford Motor Co.)

A screw adjustment is normally provided at the rocker arm when solid lifters are used. Turning the adjustment screw down will reduce valve clearance (play in valve train).

The small clearance, needed with solid lifters, causes valve train noise. A clattering or clicking noise is produced as the valves open and close. This is the main reason hydraulic lifters are more common.

Roller lifter construction

A *roller lifter,* either mechanical or hydraulic, has a small roller that operates on the camshaft lobe, Fig. 10-45. The roller reduces friction and wear between the lifter and camshaft, one of the highest friction points in the engine.

A roller lifter is sometimes used in diesel engines and high performance gasoline engines to reduce frictional power loss. See Fig. 10-46.

Cam followers

Cam follower is a term frequently used to denote a valve tappet mounted in the cylinder head. The camshaft lobe is located directly above the follower and pushes down to open the valve. The tappet usually fits into a pocket machined in the cylinder head. The pocket serves as a bore to guide the tappet as it slides up and down in the head. Cam followers can be hydraulic or mechanical.

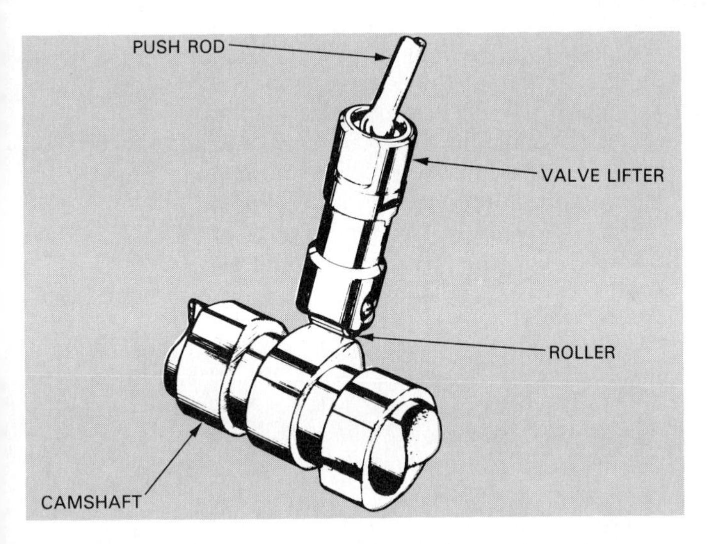

Fig. 10-45. Roller lifter is used in some high performance engines and many diesels. It cuts friction between cam lobe and bottom of lifter. This increases cam service life, fuel economy, and engine power. (Oldsmobile)

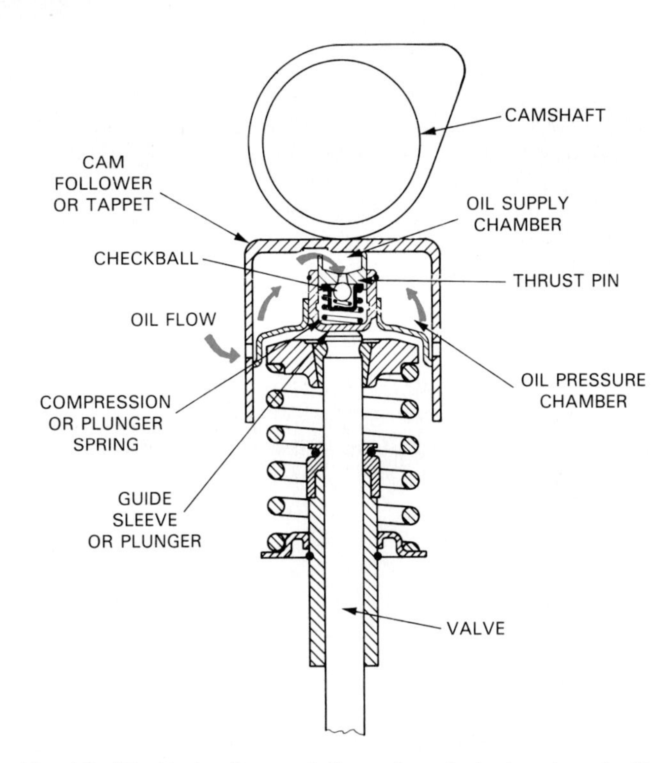

Fig. 10-47. Hydraulic cam follower in cylinder head works like conventional hydraulic lifter. Oil fills chambers in follower to maintain zero clearance. When lobe acts on follower, oil is trapped and transfers motion to valve. (Mercedes Benz)

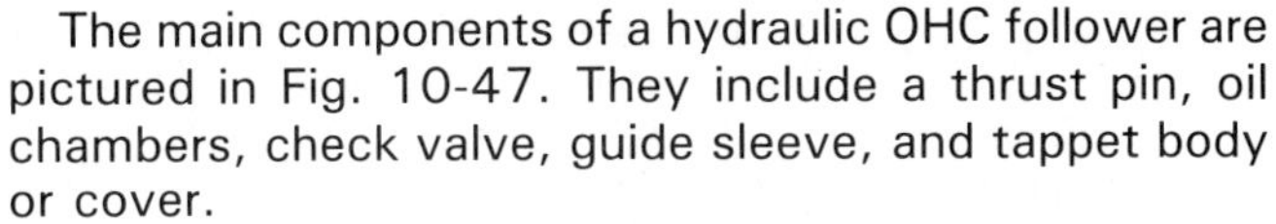

The main components of a hydraulic OHC follower are pictured in Fig. 10-47. They include a thrust pin, oil chambers, check valve, guide sleeve, and tappet body or cover.

A hydraulic OHC follower works like a hydraulic lifter for a push rod engine. The engine oil pump forces oil into the pressure chamber, Fig. 10-47. Then, when the cam lobe pushes on the tappet, the oil acts as a solid element to transfer motion to the valve. However, check valve action allows the tappet to take up any clearance between the parts to provide silent operation.

A mechanical cam follower will normally have a screw or disc (shim) to provide for valve adjustment. You can turn the screw or install a disc of a different thickness to maintain the correct valve clearance.

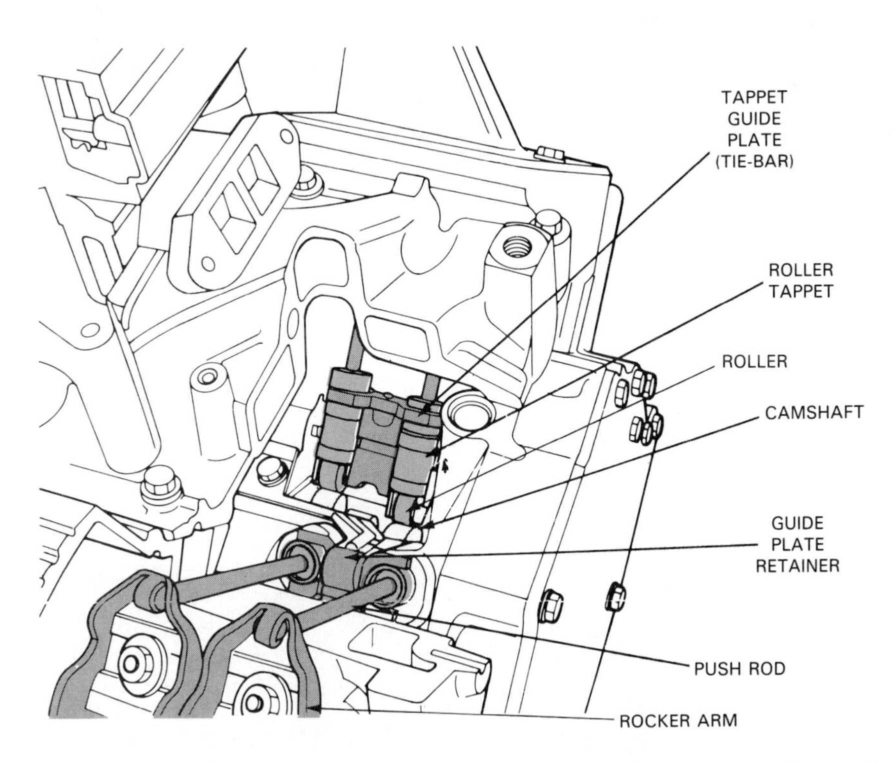

Fig. 10-46. Cutaway shows how roller lifters install in this high performance gasoline engine. (Ford)

Rocker-mounted tappets

A *rocker-mounted tappet* has the hydraulic lifter installed inside the end of the rocker arm. One is shown in Fig. 10-48. This is not a very common design but can be found on some cars. Its function and operation is similar to a hydraulic lifter or cam follower.

PUSH ROD CONSTRUCTION

Push rods transfer motion between the lifters and the rocker arms. They are needed when the camshaft is located in the cylinder block. See Fig. 11-46.

Push rods are hollow metal tubes with specially formed ends. Some push rods have a ball on both ends. Others have a ball on one end and a female socket on the other end.

Hollow push rods with holes in the ends can be used to feed oil from the lifters to the rocker arms. This prevents wear on the tip of the push rod and on the rocker arm.

Push rod guide plates

In some engines, *push rod guide plates* are used to limit side movement on the push rods. The guides hold the push rods in alignment with the rocker arms. When the push rods pass through holes in the cylinder head or intake manifold, guide plates are NOT needed.

Rocker arm construction

Rocker arms can be used to transfer motion from the push rods to the valves. They can be used in both OHC and OVH engines. In any engine, the rockers mount on top of the cylinder head, Fig. 10-48.

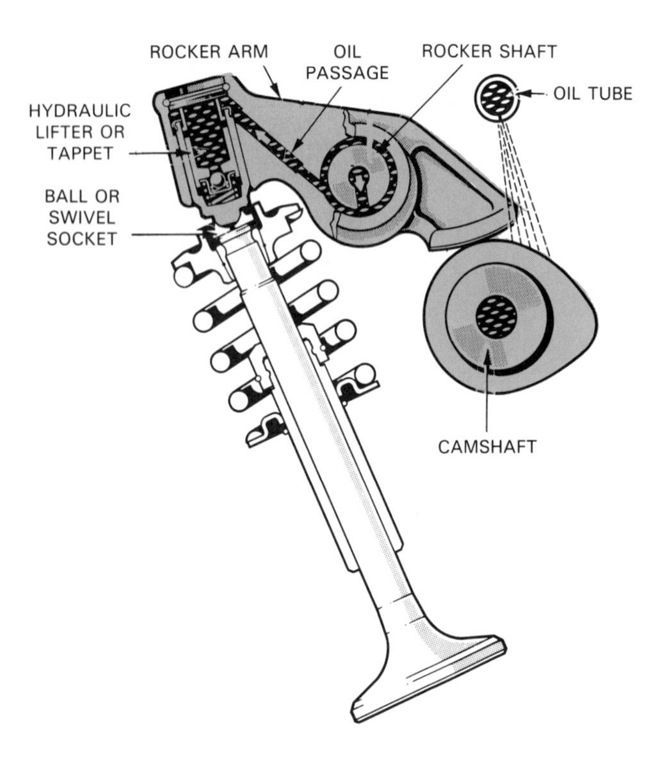

Fig. 10-48. Rocker-mounted lifter is simply a rocker arm with a hydraulic tappet mounted in its end. Operation is the same as a hydraulic lifter or follower. (Mercedes Benz)

Rocker arms are usually made of cast iron or stamped steel. They are normally drilled to allow lubrication of the valve stem tip and to allow oil entry from the push rod or rocker arm shaft. Some rocker arms are forked so that they can actuate two valves at once.

Various methods are used to support the rocker arm on the cylinder head: rocker shaft, rocker stud, or rocker pedestal.

Rocker arm shaft

A *rocker arm shaft* is a long steel tube that allows the rocker arms to swivel up and down. The shaft is held in place with metal stands. It is hollow and fills with oil during engine operation. Holes in the bottom of the shaft allow oil to lubricate the rocker arm bushings, Fig. 10-49.

Rocker arm stud

A *rocker arm stud* is simply a threaded stud in the head for supporting the rocker arm. A *half-ball* or *fulcrum* fits between the rocker and stud to let the rocker swivel. Refer to Fig. 10-50.

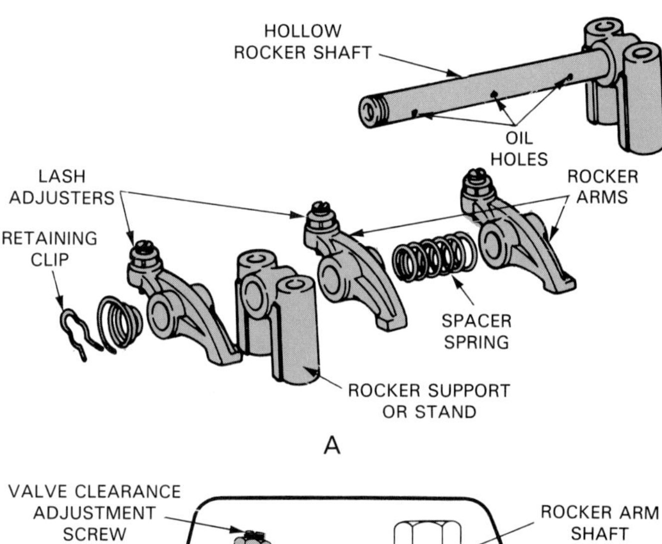

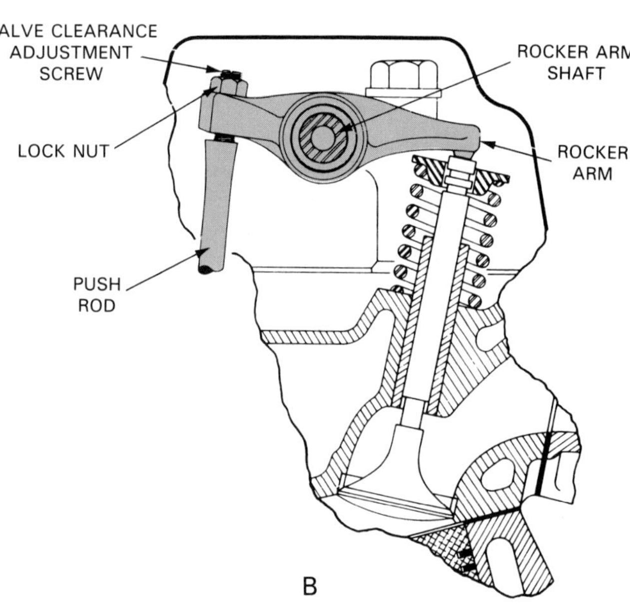

Fig. 10-49. Shaft type rocker arm assembly. A—Exploded view shows basic parts of assembly. B—Side view shows how shaft supports rocker as it swivels to open valve. (Chrysler and Federal Mogul)

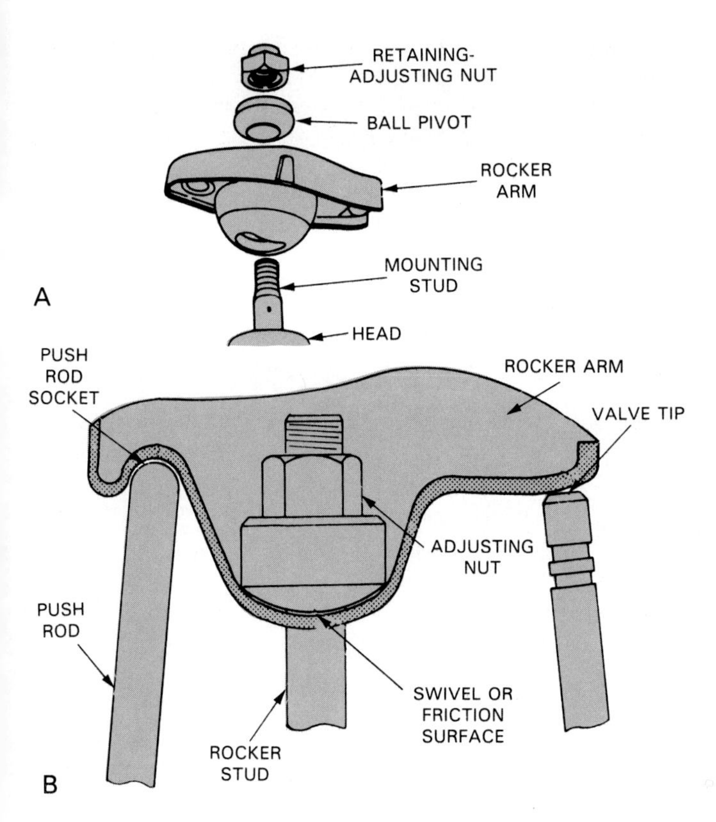

Fig. 10-50. Stud type rocker arm assembly. A—Exploded view. B—Side view. (Chrysler and TRW)

The half-ball pivot fits inside a friction surface in the rocker arm. Oil drips down into this area to provide lubrication.

Rocker arm pedestal

A *rocker arm pedestal* is a variation of a rocker arm stud type design. Raised pads are formed in the cylinder head. An aluminum pivot and bridge bolt to pedestal to support the rocker arm, as shown in Fig. 10-51. Capscrews secure the assembly to the head.

Nonadjustable rocker arms

Nonadjustable rocker arms provide no means of changing valve clearance. They are only used with some hydraulic lifters. The rocker arm assembly is tightened to a specific torque. This presets the lifter plunger halfway in its travel. Then, during engine operation, the hydraulic lifter automatically maintains zero clearance.

Push rod lengths can be changed to adjust nonadjustable rocker arms. Special push rod lengths can be purchased for unusual situations (heads milled bringing rockers closer to lifters for example).

Adjustable rocker arms

Adjustable rocker arms provide a means of changing valve train clearance. Either a screw is provided on the rocker arm or the rocker arm pivot point (stud nut for instance) can be screwed up or down.

Adjustable rocker arms MUST be used with mechanical lifters. Adjustable rockers are sometimes used with hydraulic lifters for a more accurate initial setting and for adjustment with severe part wear.

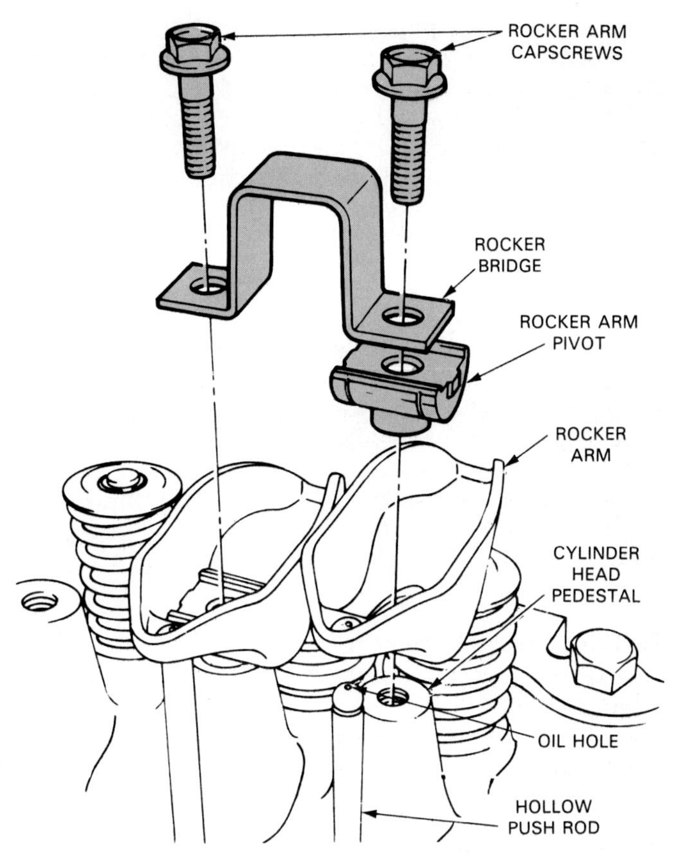

Fig. 10-51. Pedestal type rocker assemblies have pads formed on cylinder head. Bridge extends over two pivots. Capscrews pass through bridge pivots, rockers, and into pedestals. (Buick)

Rocker arm ratio

Rocker arm ratio refers to the ratio between the length of each end of the rocker arm. It compares the length from the pivot point of the rocker on the valve and push rod sides. Look at Fig. 10-52.

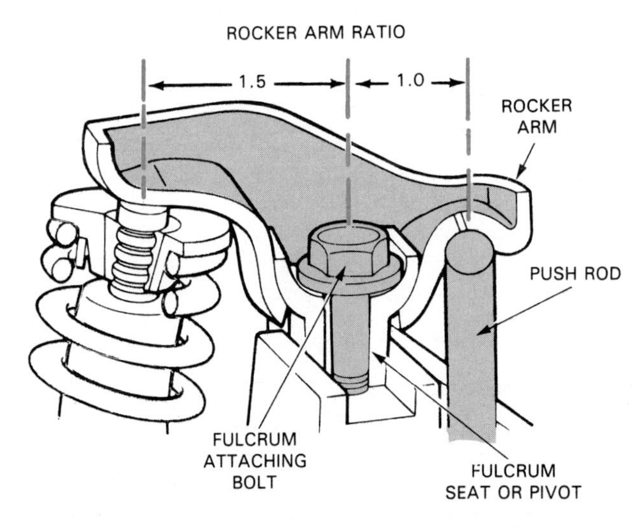

Fig. 10-52. Rocker arm ratio is determined by distance from center of rocker pivot to contact points of push rod and valve stem tip. (Ford Motor Co.)

A typical rocker ratio might be 1.5:1. This means that cam lobe lift would be multiplied by a factor of 1.5 by the action of the rocker arm. The valve will open 1 and one-half more than the lift of the camshaft.

Rocker arm ratios vary. A larger ratio would open the valves wider for the same camshaft lift. Normally, ratios cannot be changed without upsetting rocker geometry. With four-valve cylinder heads, some engine rocker arms open one of the intake valves more than the other. This is done to produce a swirling action on the air-fuel mixture entering the combustion chamber. The mixing action helps reduce low speed emissions.

Solenoid-operated rockers

Solenoid-operated rocker arms are used on variable displacement (size) engines. The solenoids can be turned on or off to deactivate or activate some of the engine valves. In this way, a cylinder can be deactivated (shut down) to reduce the effective number of cylinders.

A V-8 engine, for example, can be convereted to a 6 or 4-cylinder. This tends to increase fuel economy. However, this design has been phased out because the engine still suffers from frictional losses in the dead cylinders.

Variable Valve Timing

Variable valve timing systems alter valve timing as engine speed changes. This is done to optimize engine power and efficiency at all operating speeds. Generally, the valves in these systems function with a normal duration at low speeds. As speed increases, however, the variable timing mechanism engages to hold the valves open longer, increasing volumetric efficiency. See Fig. 10-53.

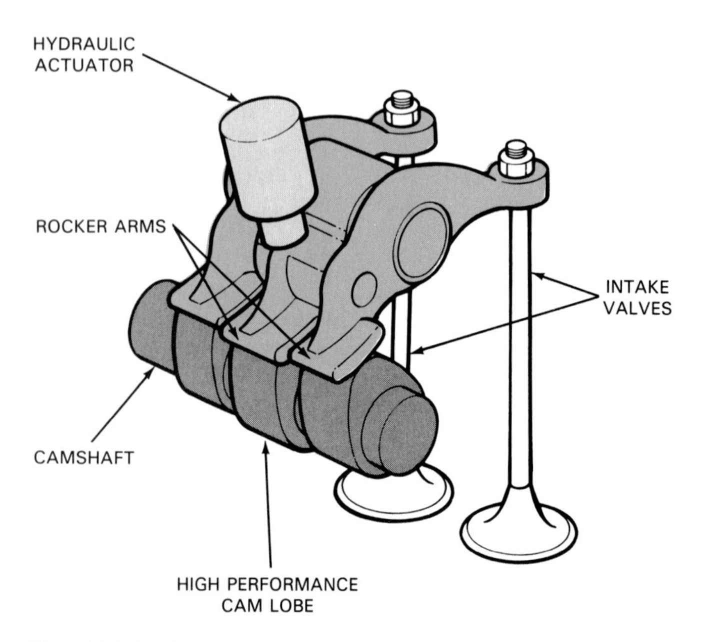

Fig. 10-53. One type of variable valve timing system. At low engine speeds, valves are controlled by two outer cam lobes. As engine speed increases, actuator causes center rocker arm to lock with outer rocker arms. Consequently, valves are controlled by center cam lobe. (Honda)

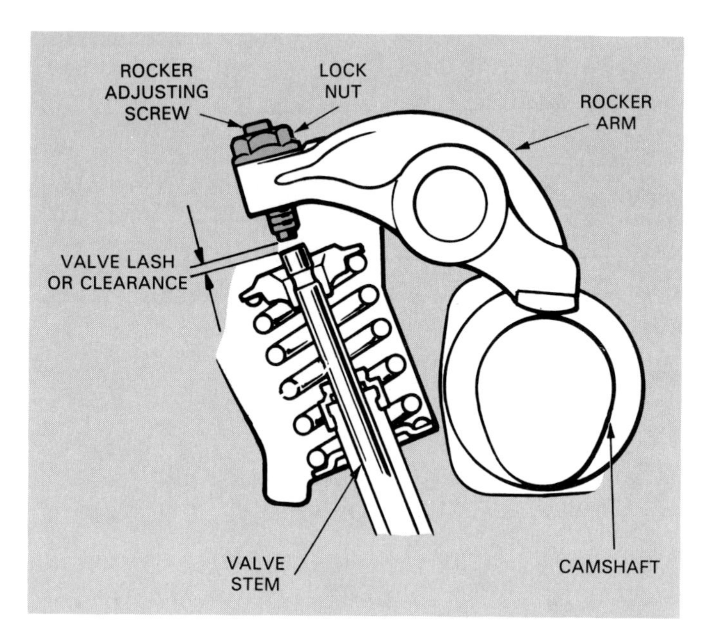

Fig. 10-54. Mechanical valve train requires method of adjusting clearance. Normally, screw and lock nut are provided on rocker arm. Turning screw down reduces clearance. (General Motors)

Valve lash

Valve lash or *valve clearance* is the space between the rocker arm tip and the valve stem tip with the VALVE CLOSED. Valve lash is needed with a mechanical valve train to assure complete valve closing, Fig. 10-54.

As mentioned, an adjustment screw is located on the rocker arm. Turning the screw down reduces valve clearance. Turning it up increases lash.

Engine manufacturers give valve lash specifications. Usually, the exhaust valve lash is slightly more than the intake valve lash. This is because the exhaust valve runs hotter and will expand more.

A typical valve lash spec might be .010 in. (0.25 mm) for the intake and .012 in. (0.30 mm) for the exhaust. Lash specs may be given HOT (engine must be warm during adjustment) or COLD (allow engine to cool during adjustment).

SUMMARY

The engine top end controls the flow in and out of the engine cylinders. Today's engine uses very complex materials and construction techniques. To service these engines properly, you must understand engine design differences.

A bare cylinder head has all of its parts removed. Heads can be made of cast iron or aluminum. Aluminum is becoming more popular because of its reduced weight.

Combustion chambers provide a place for the burning of the fuel mixture above the piston. The ports are passages in and out of the head for intake and exhaust. Valves open and close these ports.

Valve guides can be integral or pressed into the head, so can valve seats. Induction hardened seats are now used to resist the action of unleaded fuels. Most valve seats have a 45° angle on their face. An interference angle of one degree is used between the valve and seat for better initial sealing.

The valve head is the bottom of the valve. The face is the part that touches the seat. The margin is the lip between the two. A retainer and two keepers normally lock the valve spring on the valve. Valve shape is determined by the shape of the valve head.

A sodium filled valve is used in some high performance engines. It is lighter and will transfer heat into the head quicker than a solid valve. A stellite valve has a hard coating on its face. This helps it resist heat and wear from the valve seat.

Valve seals come in umbrella and O-ring designs. The umbrella seal shrouds the opening to the valve guide. The O-ring type seals the gap between the retainer and the valve stem. Both keep excess oil from flowing through the stem-to-guide clearance and into the head port.

Valve spring shims can be used to increase spring tension. This can help prevent valve float where the spring is too weak to close the valve at high rpm. Valve rotators can be used to turn the valve in its guide. This prevents localized hot spots and wear patterns that could reduce valve service life.

The camshaft is usually made of cast iron. It has lobes that operate the valve train. Camshaft lift determines how wide the valves open. Camshaft duration controls how long the valves stay open. A high performance cam would have more lift and duration.

Valve overlap is the time when both the intake and exhaust valves are open together. It is used to help scavenge burned gases out of the cylinder and pull more air-fuel mixture into the cylinder.

Valve lifters or tappets transfer cam lobe action to the rest of the valve train. A hydraulic lifter is for silent operation. It maintains zero valve train clearance. A mechanical cam is for severe or high performance and it makes a clicking sound during operation. A cam follower is used in some overhead cam engines. Its function is the same as a conventional lifter.

Push rods are hollow tubes that connect the lifters to the rocker arms. Sometimes they carry oil to the rockers.

Rocker arms swivel to transfer push rod motion to the valves. They can be made of steel or cast iron. The rockers can be mounted on a steel shaft, studs, or bridge. Rocker arm ratio refers to the ratio between the length of each end of the rocker.

Valve lash or clearance is the space between the rocker arm tip and valve stem with the valve closed. It is needed with mechanical lifters to assure complete closing of the valve.

KNOW THESE TERMS

Top end, Bare cylinder head, Intake port, Exhaust port, Intake deck, Exhaust deck, Camshaft caps, Integral valve guide, Valve guide insert, Knurled guides, Integral seat, Seat insert, Induction hardened, Valve seat angle, Interference angle, Valve face, Valve head, Valve margin, Valve stem, Keepers, Stem tip, Face angle, Guide clearance, Sodium filled valve, Stellite valve, Umbrella valve seal, O-ring valve seal, Valve oil shedder, Valve stem cap, Valve spring tension, Valve spring free length, Valve spring installed height, Valve float, Valve spring shims, Valve rotators, Valve retainers, Valve keepers, Valve spring seat, Valve spring shield, Cam lobe, Cam lift, Cam duration, Valve timing, Valve overlap, Scavenge, Hydraulic cam, Mechanical cam, Cam drive gear, Cam eccentric, Cam thrust plate, Cam bearing, Valve tappet, Hydraulic lifter, Mechanical lifter, Roller lifter, Cam followers, Rocker-mounted tappet, Push rod, Rocker arm, Rocker arm shaft, Rocker arm stud, Rocker arm bridge, Nonadjustable rocker, Adjustable rocker, Rocker arm ratio, Solenoid-operated rocker, Variable valve timing, Valve lash.

REVIEW QUESTIONS—CHAPTER 10

1. If a cylinder head becomes badly damaged and cannot be repaired, the technician would usually install a new or reconditioned ________ cylinder head and all of the reusable parts would be installed on this head.
2. ________ ________ are usually provided in the cylinder head and block for aligning the head and ________ ________ on the block.
3. On an OHC engine, what is the purpose of cylinder head caps?
4. Integral valve guides are very common in aluminum heads. True or false?
5. Why could valve guide inserts be found in used, cast iron heads?
6. What are knurled valve guides?
7. Steel valve seat ________ are used in aluminum heads to withstand the high ________ ________.
8. Induction hardened valve seats are used to help withstand the heat and friction aggravated by unleaded fuel. True or false?
9. Valve seat angles can be either ________ or ________, but the ________ angle is more common.
10. What is a valve interference angle?
11. The intake valve is larger than the exhaust valve. True or false?
12. Why is a valve margin important?
13. Some valve stem tips are ________ coated to resist wear.
14. How does a sodium filled valve operate?
15. A ________ ________ has a special hard coating on its face to resist heat and friction.
16. How does an O-ring valve seal keep too much oil from entering the valve guide?
17. What is the main or normal function of valve stem caps?
18. Describe four valve spring specifications.
19. Valve spring ________ can be used to restore spring tension in used valve springs.
20. What is the difference between a valve retainer and a valve keeper?
21. The shape of a camshaft lobe affects which of the following?
    a. When valve opens.
    b. How long valve opens.
    c. How far valve opens.
    d. All of the above are correct.

22. Explain valve overlap.
23. A young customer complains of a rough idle, no power brakes, decreased engine power at low speeds, and increased fuel consumption. A new high performance cam has just been installed in the engine.
    Technician A says that the cam probably has a long duration, which can cause all of these problems.
    Technician B says that there must be a vacuum leak and a carburetor problem. It will take a few minutes labor to check out the conditions.
    Who is correct?
    a. Technician A.
    b. Technician B.
    c. Both A and B.
    d. Neither A nor B.
24. The main purpose of hydraulic lifters over mechanical lifters is quiet operation. True or false?
25. A customer complains of clicking noises from inside the valve cover of an OHC engine. The engine has mechanical cam followers.
    Technician A says that the camshaft lobes must be worn and major repairs may be needed.
    Technician B says that valve cover removal is needed to check whether the followers are out of adjustment.
    Who is correct?
    a. Technician A.
    b. Technician B.
    c. Both A and B.
    d. Neither A nor B.

ASE CERTIFICATION-TYPE QUESTIONS

1. Technician A says that a cylinder head for an OHV engine holds the valves and rocker arms. Technician B says that a cylinder head for an OHV engine holds the rocker arms, valves, and camshaft. Who is right?
   (A) A only.
   (B) B only.
   (C) Both A & B.
   (D) Neither A nor B.
2. Technician A says that integral valve guides are very common in cast iron cylinder heads. Technician B says that integral valve guides are very common in both cast iron and aluminum cylinder heads. Who is right?
   (A) A only.
   (B) B only.
   (C) Both A & B.
   (D) Neither A nor B.
3. All of the following are parts of a bare cylinder head EXCEPT:
   (A) intake deck.
   (B) valve keepers.
   (C) water jackets.
   (D) oil passages.
4. Technician A says that a 30° seat angle is commonly used on both the intake and exhaust seats for most vehicle engines. Technician B says that a 45° seat angle is used on both the intake and exhaust seats for most vehicle engines. Who is right?
   (A) A only.
   (B) B only.
   (C) Both A & B.
   (D) Neither A nor B.
5. Technician A says that umbrella type valve seals are used on modern automotive engines. Technician B says that O-ring type valve seals are used on modern automotive engines. Who is right?
   (A) A only.
   (B) B only.
   (C) Both A & B.
   (D) Neither A nor B.
6. Technician A says that valve spring shims can be used to increase or restore valve spring free length. Technician B says that valve spring shims can be used to increase valve spring tension. Who is right?
   (A) A only.
   (B) B only.
   (C) Both A & B.
   (D) Neither A nor B.
7. All of the following are basic components of a overhead camshaft engine valve train EXCEPT:
   (A) camshaft seal.
   (B) intake valve tappet.
   (C) exhaust valve push rod.
   (D) intake valve inner spring.
8. Technician A says that camshaft lift determines how long the valves stay open. Technician B says that camshaft lift is the amount of valve train movement produced by the cam lobe. Who is right?
   (A) A only.
   (B) B only.
   (C) Both A & B.
   (D) Neither A nor B.
9. Technician A says that the term cam follower refers to a valve tappet mounted in the cylinder head. Technician B says that the term cam follower refers to a valve tappet mounted in the engine block. Who is right?
   (A) A only.
   (B) B only.
   (C) Both A & B.
   (D) Neither A nor B.
10. Technician A says that valve lash is adjusted when the engine valve is open. Technician B says that valve lash is adjusted when the engine valve is in the closed position. Who is right?
    (A) A only.
    (B) B only.
    (C) Both A & B.
    (D) Neither A nor B.

Chapter 11

Front End, Manifold, Gasket Construction

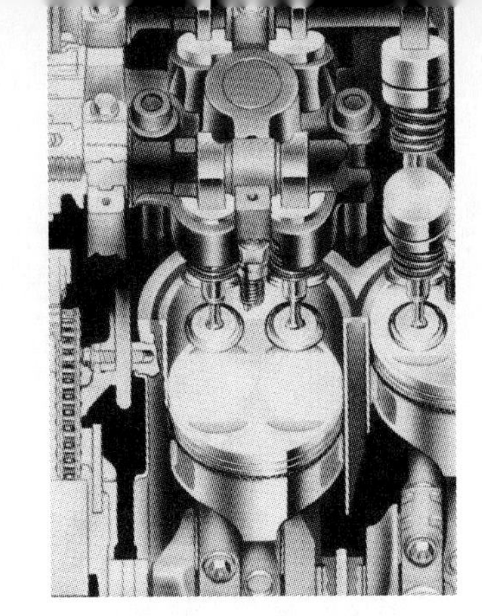

After studying this chapter, you will be able to:

- *Describe the three types of camshaft drives.*
- *Explain why the camshaft turns at one-half crankshaft speed.*
- *Summarize the construction of timing gears.*
- *Explain how timing chains and their related parts are constructed.*
- *Describe the use of chain tensioners and chain guides.*
- *Explain how crank sprockets and cam sprockets are locked in place with keys or dowels.*
- *Summarize the construction of auxiliary drive sprockets.*
- *Describe the difference between an engine front cover and a timing belt cover.*
- *Explain the construction of a timing belt mechanism.*
- *Describe the construction of crankshaft dampers.*
- *Summarize the construction of intake and exhaust manifolds.*
- *Explain the construction of engine gaskets.*
- *Identify major external parts fastened to an engine.*

Engine front end construction has changed considerably in the past few years. Today's engines frequently have overhead cams with a chain or belt drive. Auxiliary drive sprockets, chain tensioners, belt tensioners, chain guides, and balancer shaft drives can all be found on modern passenger car engines. Only a few years ago OHV engines were predomenant and many of these mechanisms were limited to exotic engines.

This chapter will help you grasp the many design variations found on present-day car engines. It will help you to understand how these parts operate and are constructed. As a result, you will be better prepared to diagnose, test, adjust, and repair these components.

The chapter will also cover intake manifold design, crankshaft balancer construction, and engine gasket construction. This too will help you in later chapters on service and repair. Study carefully!

CAMSHAFT DRIVE CONSTRUCTION

The *camshaft drive* mechanism, also called just *timing mechanism,* must turn the engine camshaft and keep it in time with the engine crankshaft and pistons. Sometimes, it must also power other units (balancer shaft, oil pump, distributor, etc.). It must drive the camshaft and other units smoothly and dependably.

There are three basic types of camshaft drives: gear drive, chain drive, and belt drive, Fig. 11-1.

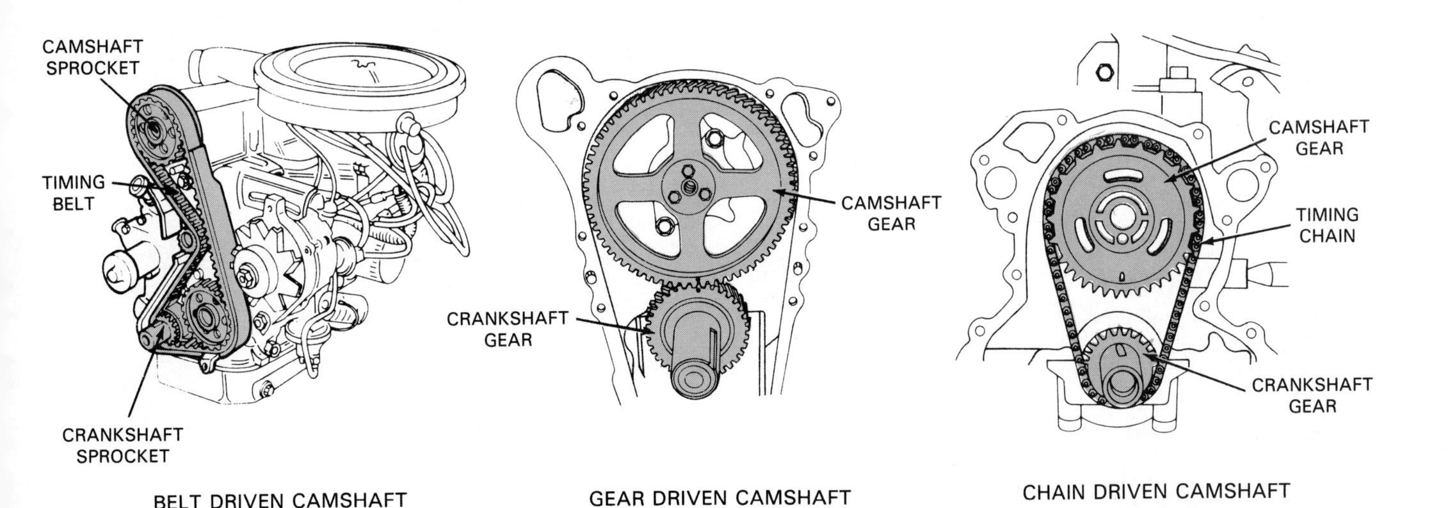

Fig. 11-1. These are the three types of camshaft drive mechanisms found on today's engines. (Ford Motor Co.)

Cam drive ratio

The *cam drive ratio* is two to one (2:1) which means the crankshaft must turn twice to turn the camshaft once. As shown in Fig. 11-2, the cam gear or sprocket turns at one-half engine rpm (revolutions per minute). The cam gear has twice as many teeth as the crank gear and is twice the diameter.

Timing gear construction

Timing gears are two helical gears on the front of the engine that operate the engine camshaft. A *crank gear* is keyed to the crankshaft snout. It turns a *camshaft gear* on the end of the camshaft, Fig. 11-3.

Timing marks on the two gears show the mechanic how to install the gears properly. The marks may be circles, indentations, or lines. The marks must line up for the cam to be in time with the crank, Fig. 11-4.

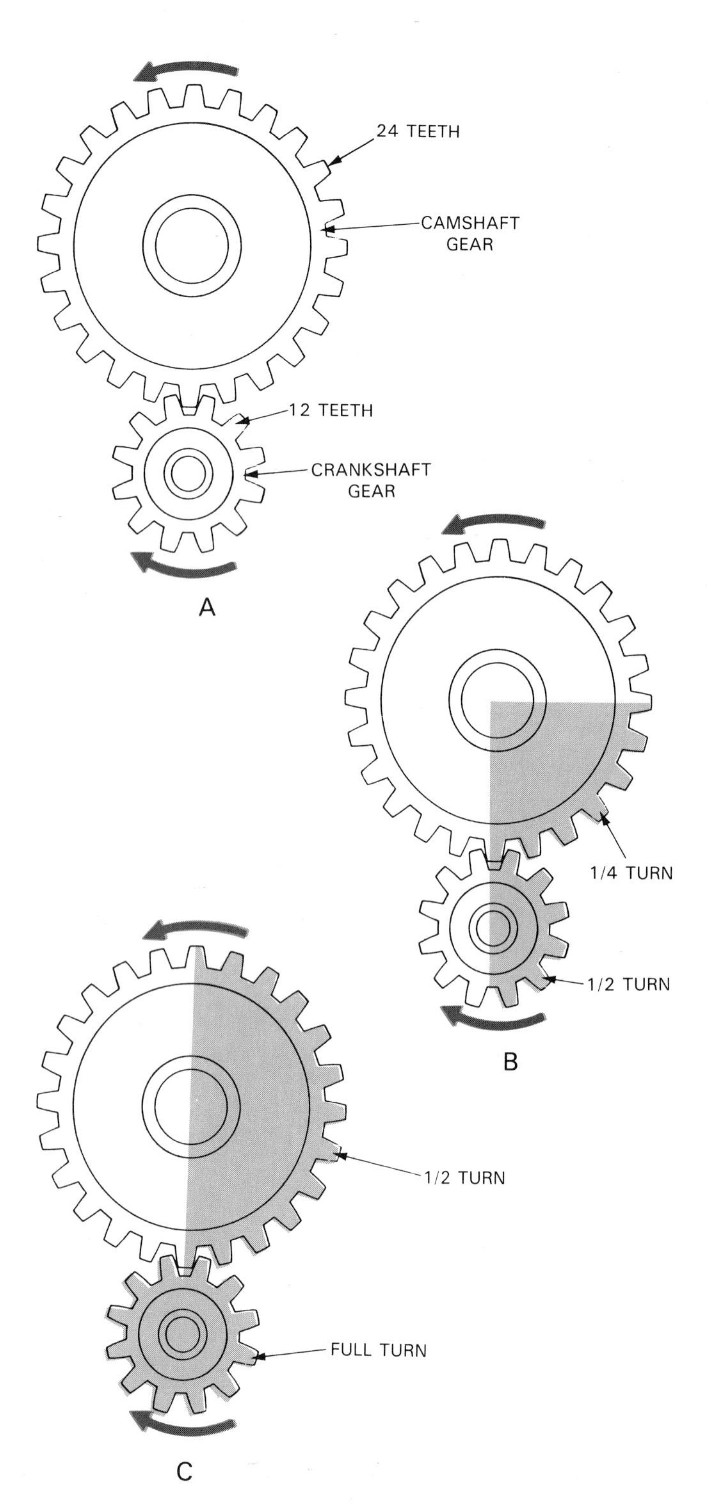

Fig. 11-2. Cam drive must turn camshaft at one-half engine speed. Crankshaft gear or sprocket must revolve twice to turn camshaft gear or sprocket once.

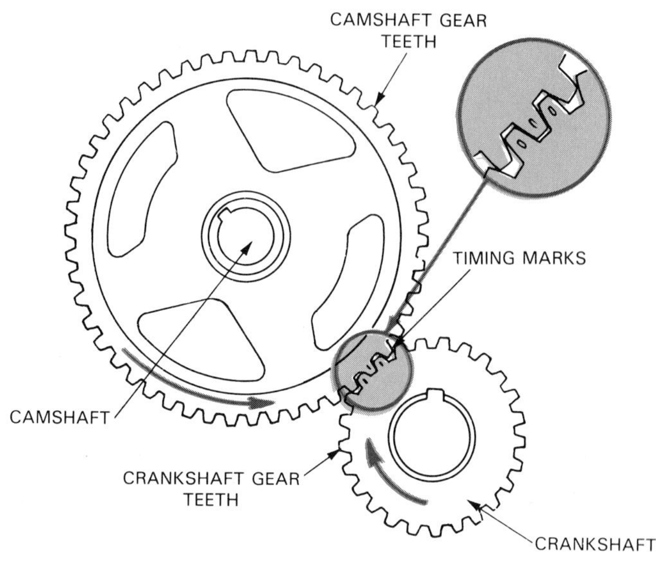

Fig. 11-3. Timing gears have timing marks so camshaft and crankshaft can be positioned correctly during engine assembly. (Deere & Co.)

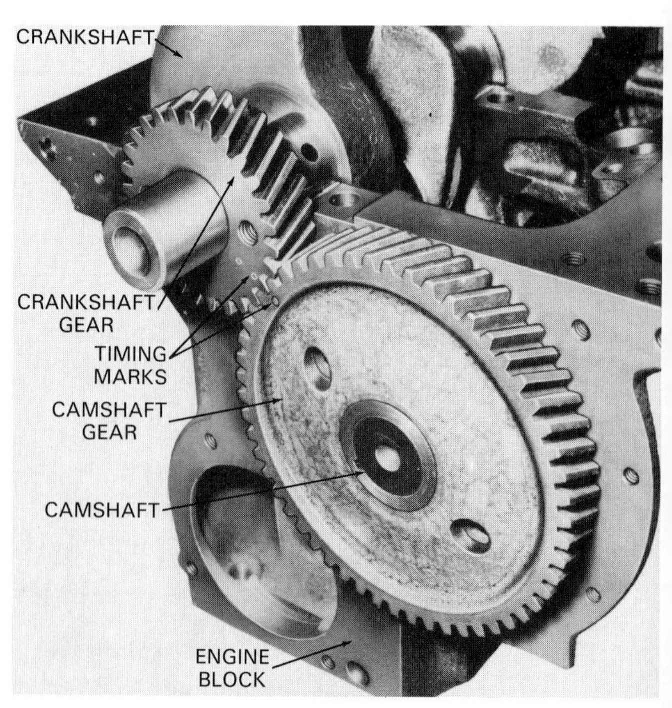

Fig. 11-4. Photo of front of this heavy-duty gasoline engine shows timing gears. Note how gears are helical cut for quieter, more dependable service. (Pontiac Motor Div.)

Timing gears are commonly used on heavy-duty applications (taxicabs or trucks). They are very dependable and long lasting. However, they are noisier than a chain or belt drive. Gears are primarily used in cam-in-block engines where the crank is close to the cam, Fig. 11-5.

Timing gear backlash is the clearance between the teeth of the timing gears. Backlash is needed so that the gears can expand slightly when hot and still maintain clearance for oil. Oil is needed between the gear teeth to prevent friction and wear.

Explained in later chapters, a dial indicator is commonly used to mesure timing gear backlash.

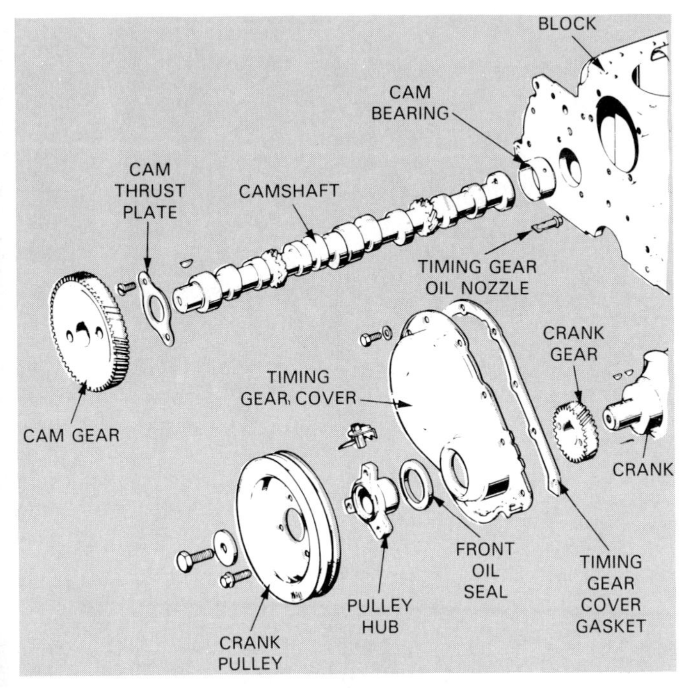

Fig. 11-5. Timing gears fit inside timing cover, also cailed engine front cover, timing case cover, etc. (American Motors)

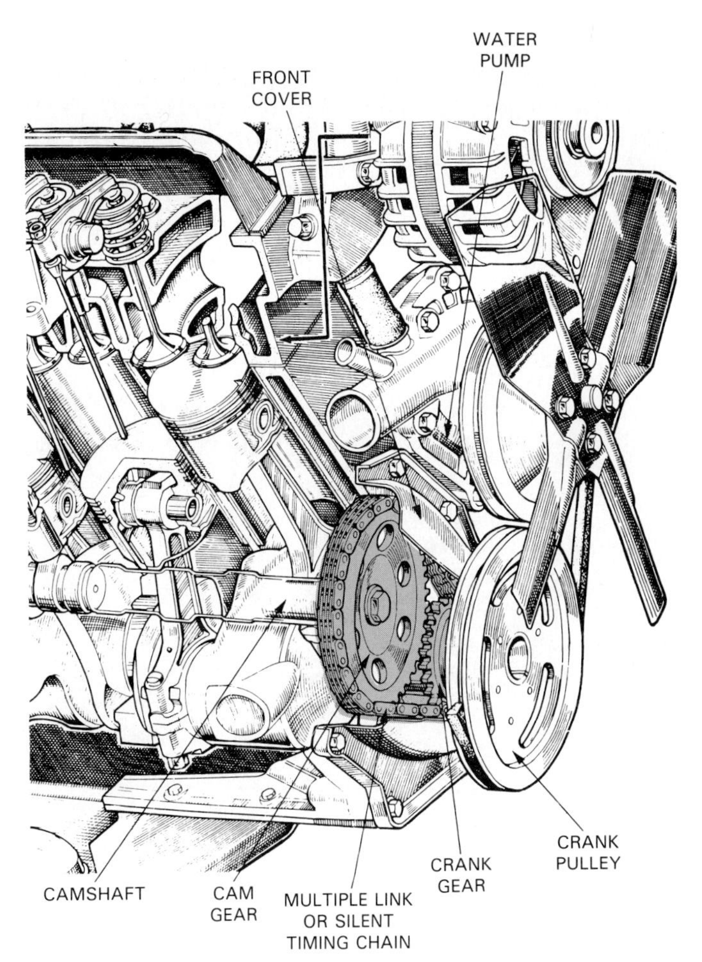

Fig. 11-6. Cutaway of this engine shows silent or multiple-link chain. (Chrysler)

Timing chain and sprocket construction

A *timing chain* and two *sprockets* can also be used to turn the camshaft. This is the most common type of drive on cam-in-block engines. It may also be used on OHC engines, however. Look at Fig. 11-6.

A *crank sprocket* is mounted on the crankshaft snout. A larger *cam sprocket,* with either metal or plastic teeth, bolts to the camshaft.

The *timing chain* transfers power from the crank sprocket to the cam sprocket. It can be a multiple link (silent) type, Fig. 11-6, or a double-roller type, Fig. 11-7.

Like timing gears, the chain sprockets have timing marks. The marks must line up to place the camshaft in time with the pistons and crankshaft, Fig. 11-7.

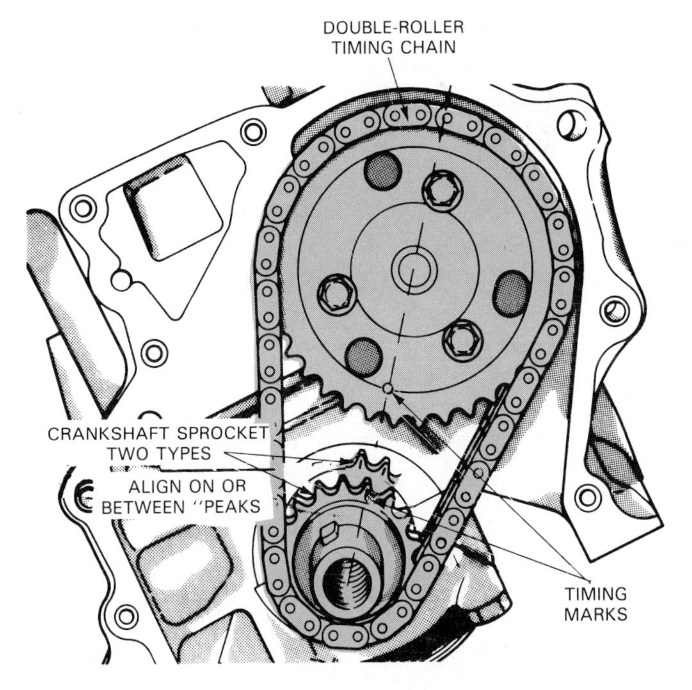

Fig. 11-7. Timing sprockets also have timing marks. They must be aligned so pistons do not hit valves. This is a double-roller type chain. (Chrysler)

Crankshaft key

A *crankshaft key* is used to lock the crankshaft sprocket (or gear) to the snout of the crank. This is shown in Figs. 11-8 and 11-9A. The key allows the

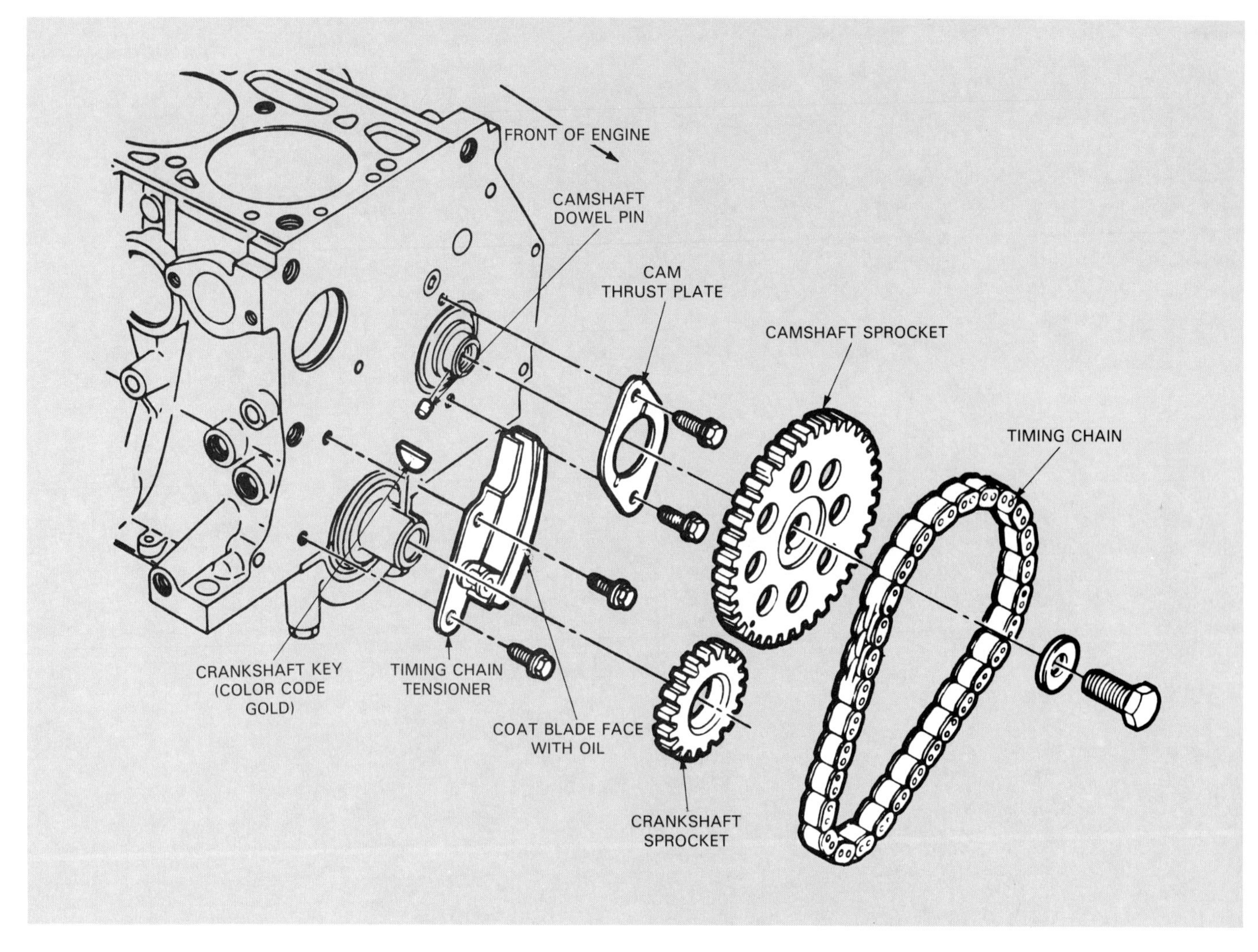

Fig. 11-8. Thrust plate is commonly used to hold cam in block. This OHV engine uses a chain tensioner. (Ford)

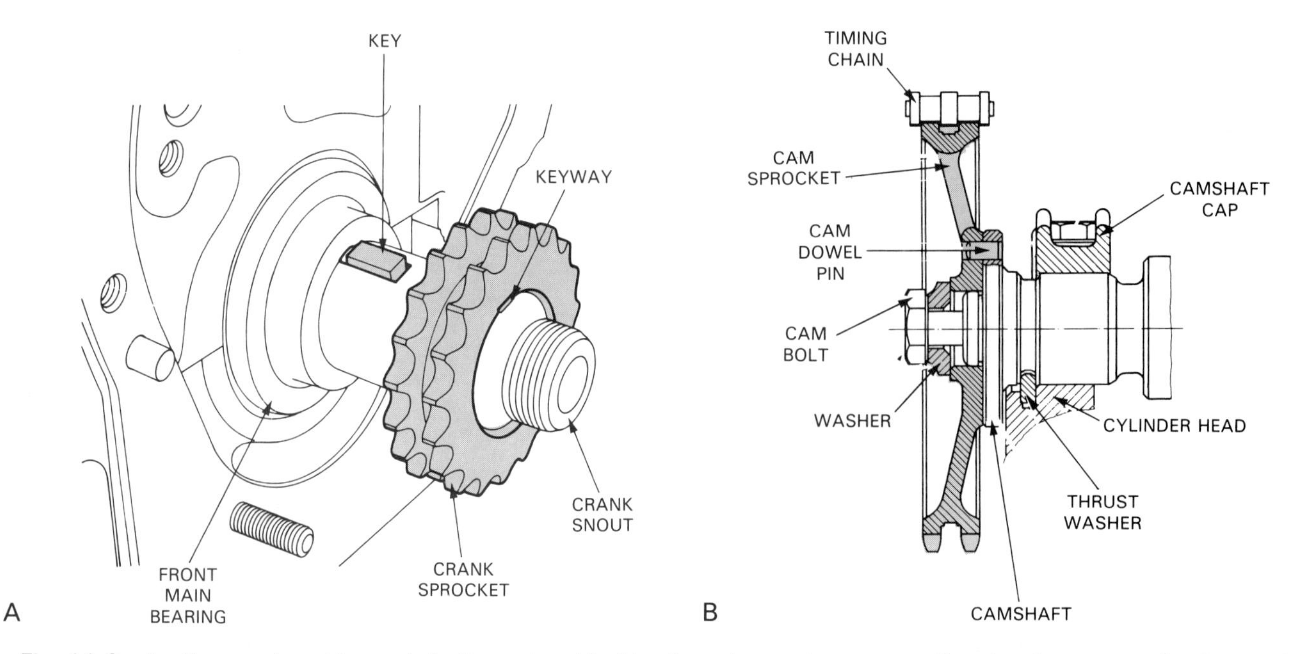

Fig. 11-9. A—Keyway is cut in crankshaft snout and inside of crank sprocket or gear. Key then keeps sprocket from spinning on crankshaft. B—Cam sprocket is frequently locked to camshaft with steel dowel pin. Bolt holds sprocket on cam. Note thrust washer that limits cam endplay. (Peugeot and Mercedes Benz)

sprocket to slide over the crank but keeps it from turning on the snout. The key also positions the timing mark properly.

Camshaft key or dowel

A *camshaft key* or *dowel* is used to secure the camshaft sprocket (or gear) on the cam. It assures that the sprocket does not spin on the camshaft and go out of time. Figs. 11-8 and 11-9B show common methods of locking the cam sprocket to the cam.

Camshaft eccentric

A *camshaft eccentric* may bolt on the front of the cam sprocket to power a mechanical fuel pump. The eccentric is offset (not centered) on the sprocket. This lets it push the fuel pump arm up and down as it rotates.

Chain tensioner

A *chain tensioner* can be used to take up excess slack as the chain and sprockets wear. It is a spring-loaded device that pushes on the unloaded side of the chain to keep the chain tight. A chain tensioner is commonly used on OHC engines but can sometimes be found on OHV engines. See Fig. 11-10.

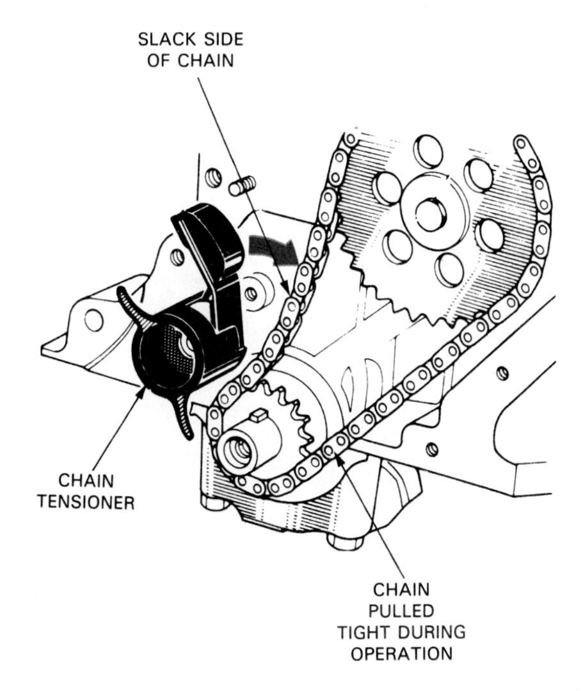

Fig. 11-10. Chain tensioner is used to keep chain tight on sprockets. It presses on slack side of chain to counteract chain and sprocket wear. (Renault)

There are several designs of chain tensioners. These are shown in Fig. 11-11. The most common classifications are the lever type tensioner and the plunger type tensioner. Both types perform the same function and are used on the engine in Fig. 11-12.

A plastic, teflon, or fiber foot normally contacts the chain. It can rub on the chain with a minimum amount of wear.

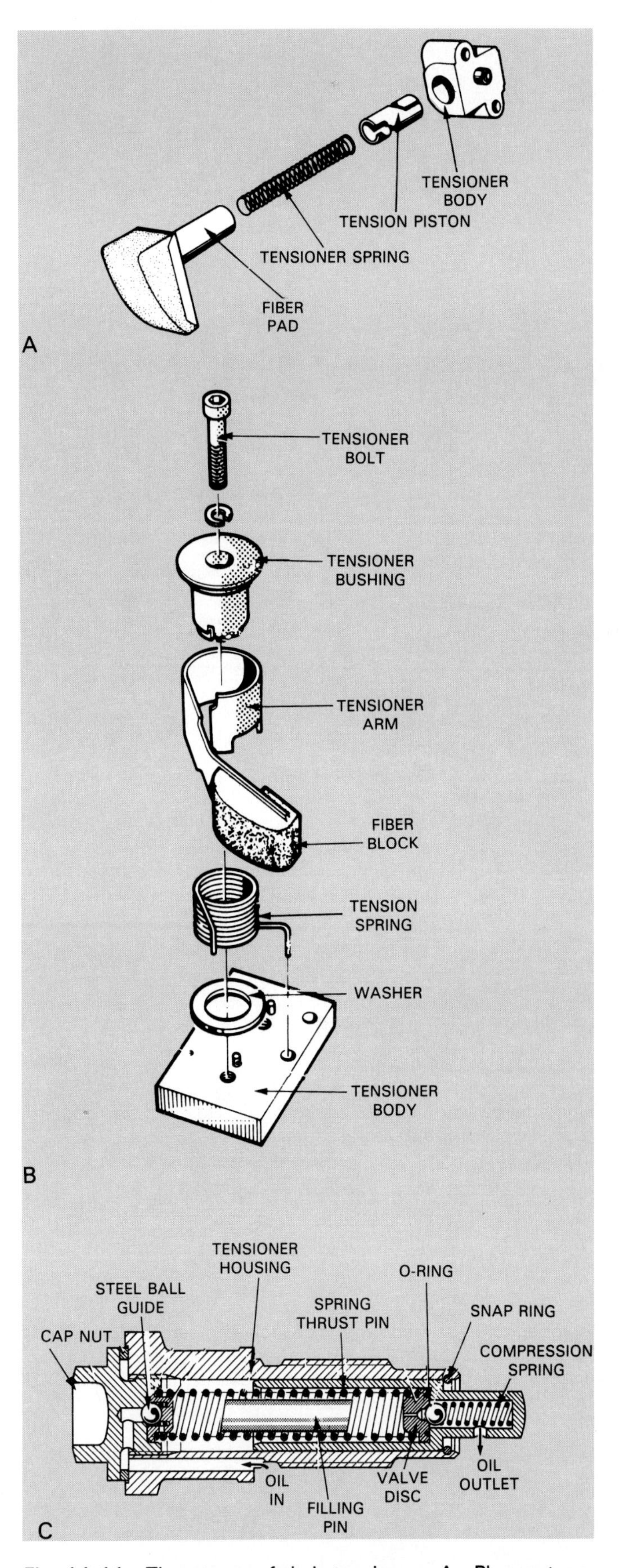

Fig. 11-11. Three types of chain tensioners. A—Plunger type pushes straight out against chain. B—Arm type swings out. C—Oil-filled plunger type increases tension with engine rpm because oil pressure increases with engine speed.

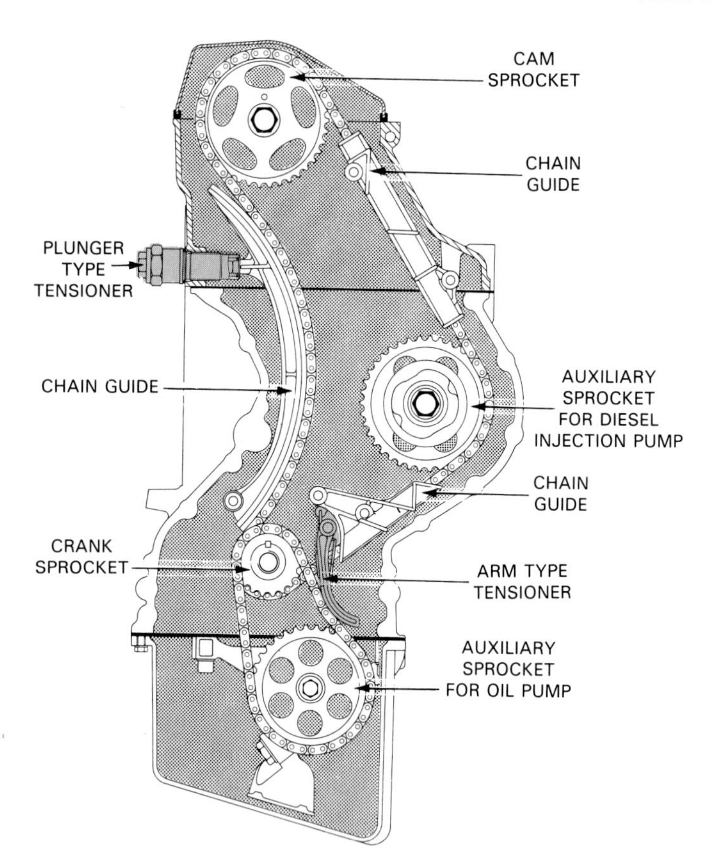

Fig. 11-12. This engine has both arm and oil-filled plunger types of chain tensioners. Plunger type is on cam drive chain. Arm type is on auxiliary chain for engine oil pump. (Mercedes Benz)

Chain guide

A *chain guide* may be needed to prevent *chain slap* (slack lets chain flap back and forth) on long lengths of chain. The guides have a metal body with either a plastic, nylon, or teflon face. This allows the chain to slide with minimum wear, Figs. 11-12 and 11-13.

A chain guide is especially needed when there is a great distance between the chain sprockets. The guide pushes in slightly on the chain to keep it running smoothly. Slots in the bolt holes allow the guides to be correctly adjusted, Fig. 11-14.

Auxiliary chain and sprockets

An *auxiliary chain* may be used to drive the engine oil pump, balancer shafts, and other units on the engine. Look at Fig. 11-15.

An auxiliary chain is driven by an extra sprocket, usually placed in front of the crankshaft timing chain. Do NOT confuse an auxiliary chain with a timing chain.

An *auxiliary drive sprocket* powers the accessory shaft. It is an extra sprocket mounted on the crankshaft. The *auxiliary driven sprocket* mounts on the auxiliary shaft, Fig. 11-16.

Fig. 11-17 shows the view of an OHC engine with the front cover removed. Study the names and relationships of the parts.

Oil slinger

An *oil slinger* helps spray oil on the timing chain to prevent wear. It also helps prevent oil from leaking out the front seal. An oil slinger is a washer-shaped part that fits in front of the crankshaft sprocket. See Fig. 11-18.

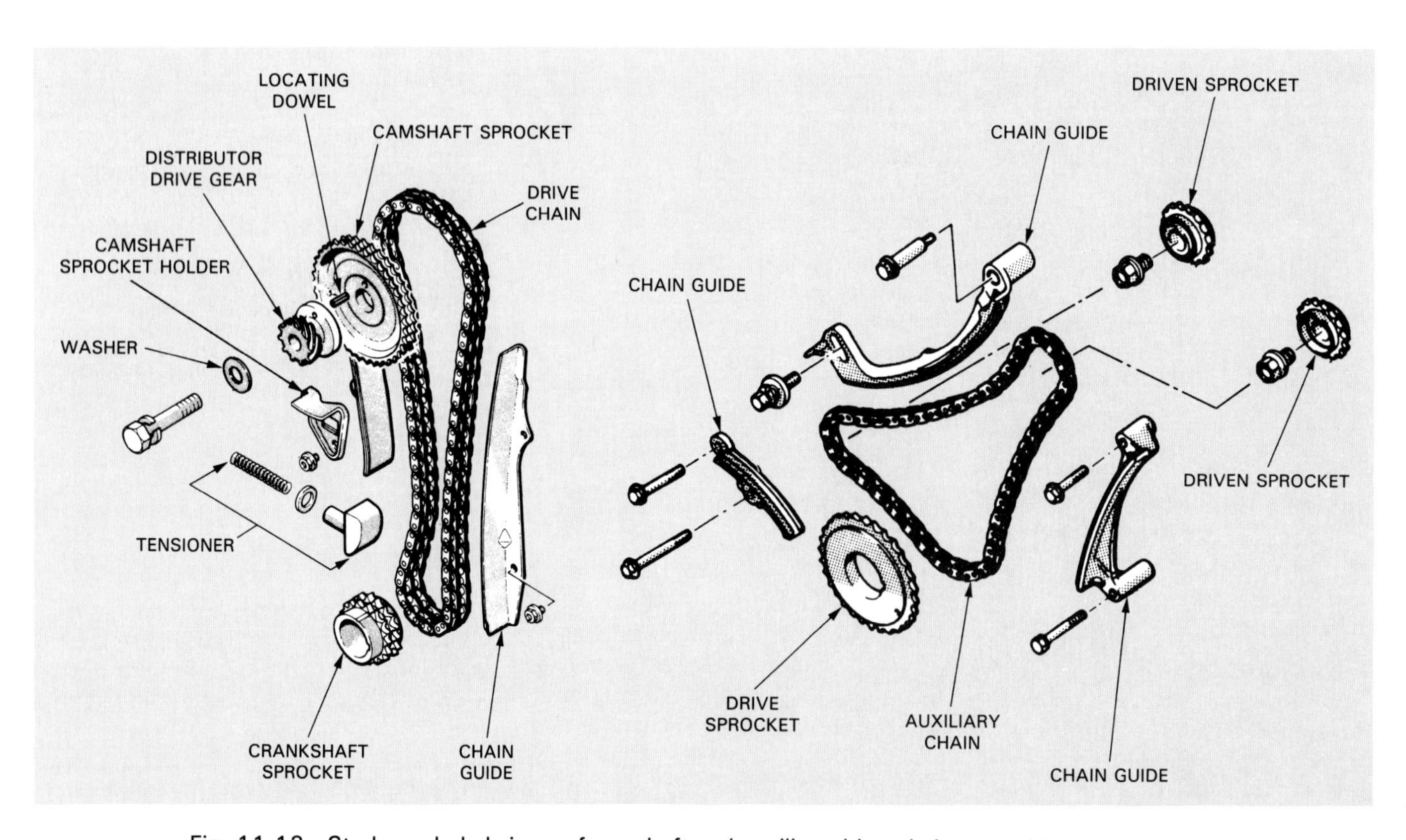

Fig. 11-13. Study exploded views of camshaft and auxiliary drive chain assemblies. (Chrysler)

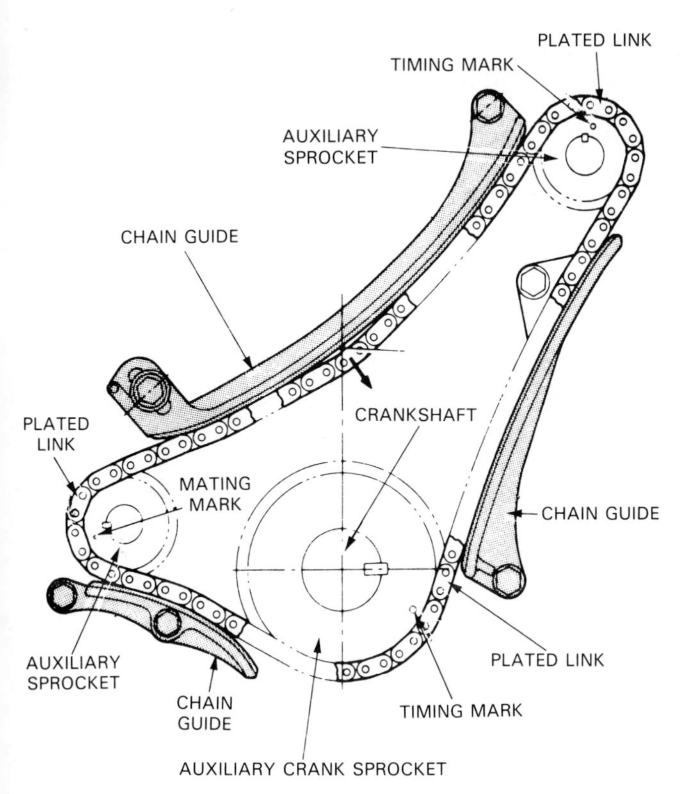

Fig. 11-14. Chain guides lightly touch chain to keep chain from slapping or flapping back and forth. This auxiliary chain for balancer shafts uses three guides.

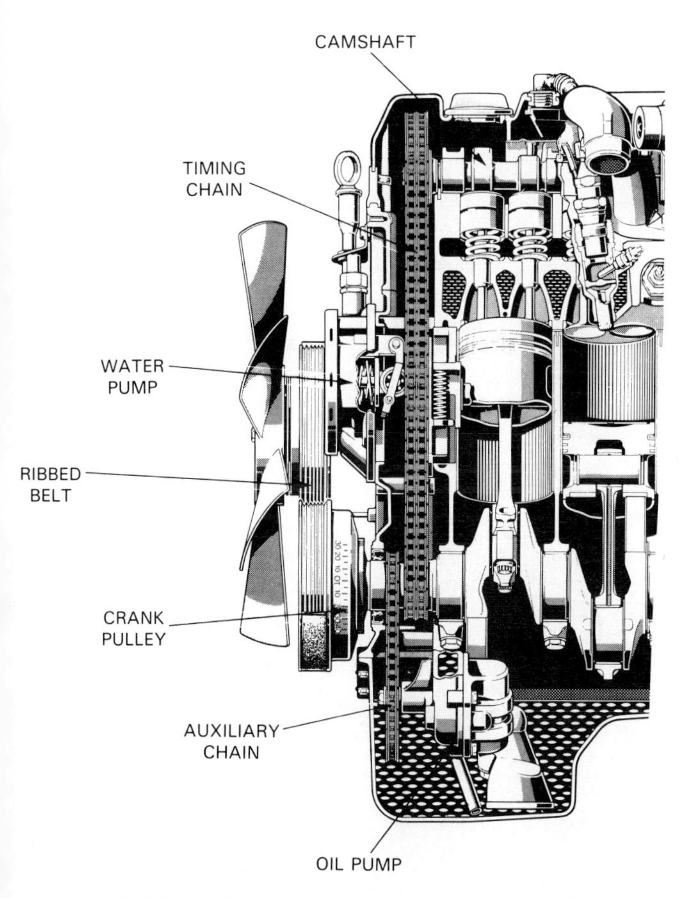

Fig. 11-15. Side view of engine shows chain drives. (Mercedes Benz)

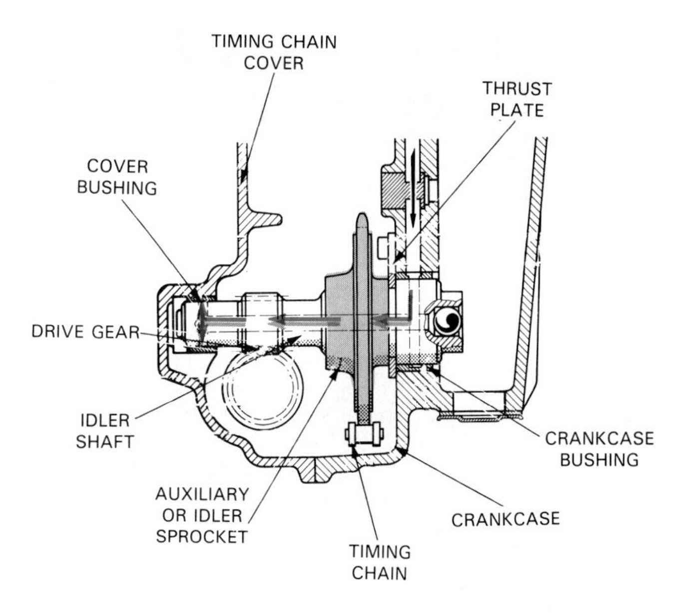

Fig. 11-16. Study construction of this idler or auxiliary sprocket. It mounts inside front cover to drive ignition system distributor. (Mercedes Benz)

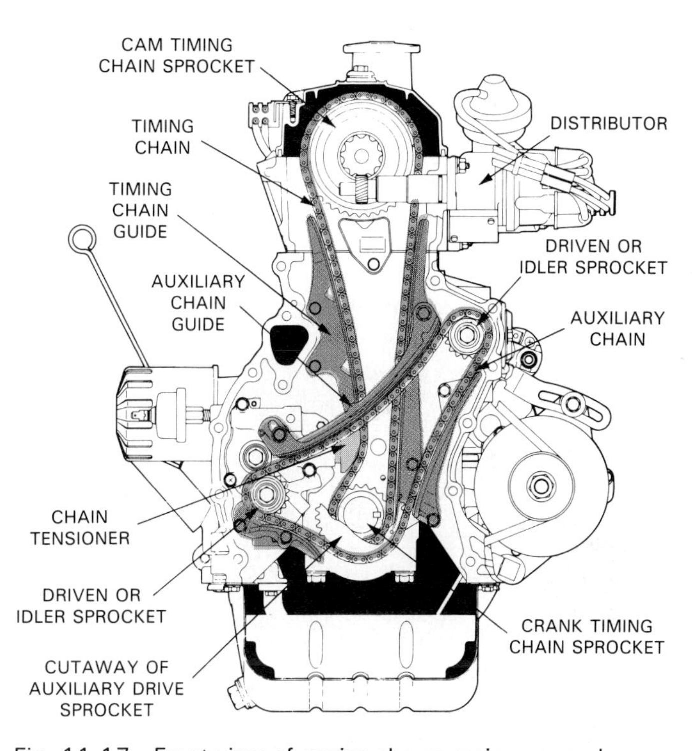

Fig. 11-17. Front view of engine shows major arm and auxiliary drive components. Study them. (Plymouth)

When the engine is running, oil squirts out of the front main bearing. Centrifugal force on the spinning slinger throws oil outward, lubricating the timing chain or gears. Since it is in front of the oil seal, only oil vapors reach the seal.

Engine front cover construction

An *engine front cover,* also called *timing chain* or *timing gear cover,* is a metal housing that bolts on the front of the engine. It encloses the timing chain or gears to keep oil from spraying out. It can be made of thin, stamped steel or cast aluminum. A gasket seals the joint between the cover and engine block.

As shown in Fig. 11-18, the front cover holds the crankshaft oil seal. Other engine parts can sometimes fasten to the front cover. It can hold a mechanical fuel pump, water pump, oil pump, distributor, and other units. When it supports other devices, the cover will usually be made of cast aluminum.

Timing belt mechanism construction

A *timing belt mechanism* is commonly found on late model overhead cam engines. This system consists of a cogged belt, belt sprockets, belt tensioner, and a belt cover. These parts are pictured in Fig. 11-19.

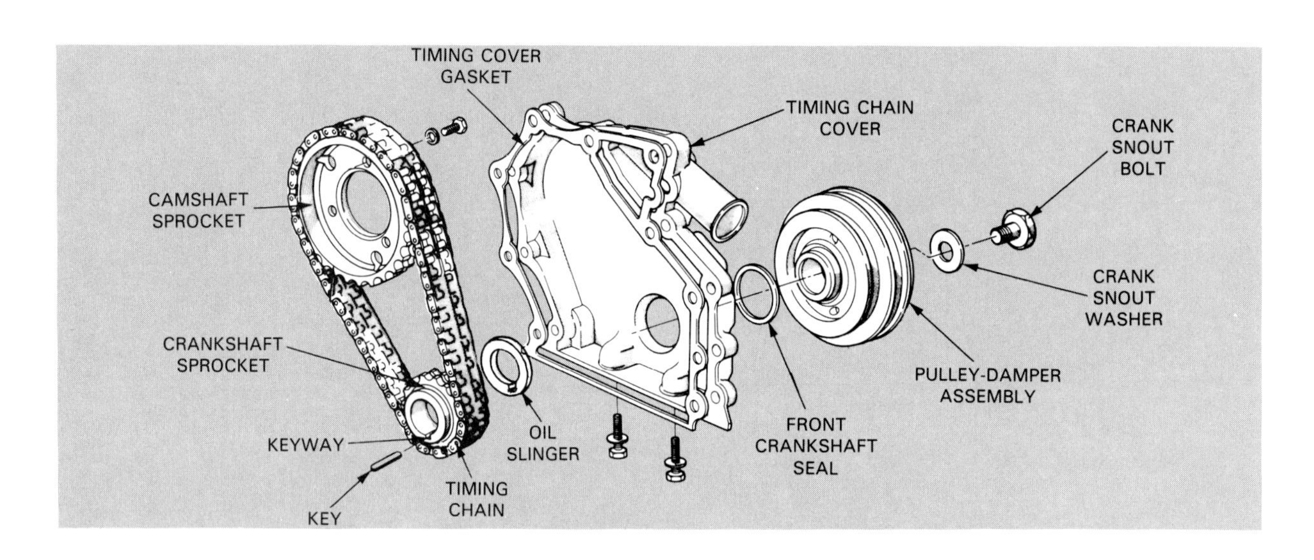

Fig. 11-18. Oil slinger fits in front of crankshaft sprocket. Its spinning action throws oil out onto chain and keeps too much oil from contacting and leaking out of front cover seal. (Chrysler)

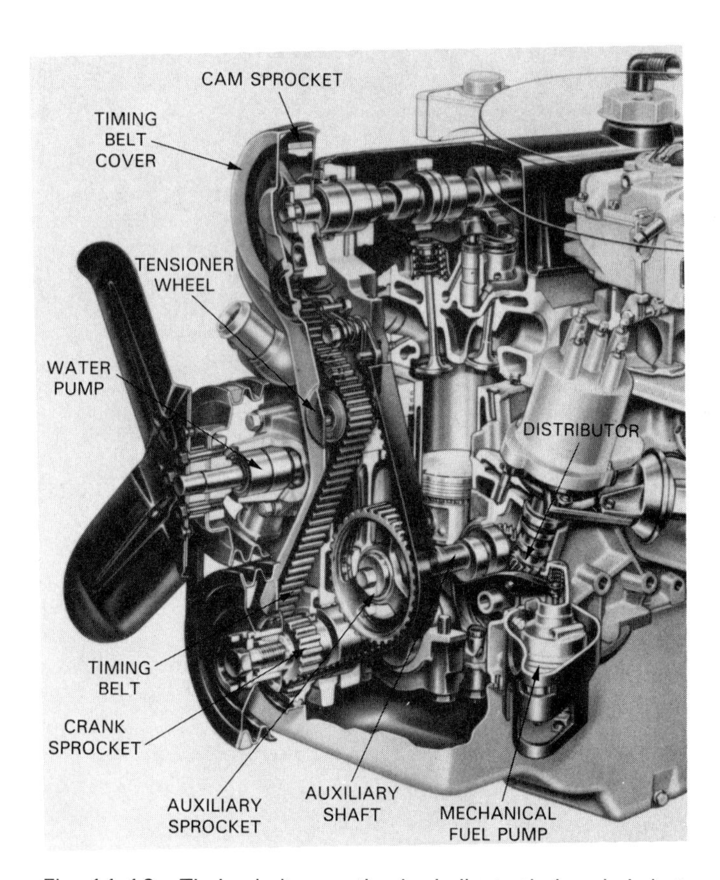

Fig. 11-19. Timing belt operation is similar to timing chain but belt runs dry, not in oil. Note part names. (Ford)

A timing belt provides an excellent way of driving the camshaft. The belt runs very smoothly and without gear backlash nor chain slack. It is very quiet and dependable.

With an OHC engine, the crankshaft and camshaft are located far away from each other. It is impractical to use gears to transfer power to the cam. Although a chain is frequently used, it requires friction-producing chain guides and a tensioner to prevent chain slack and slap. Look at Fig. 11-20.

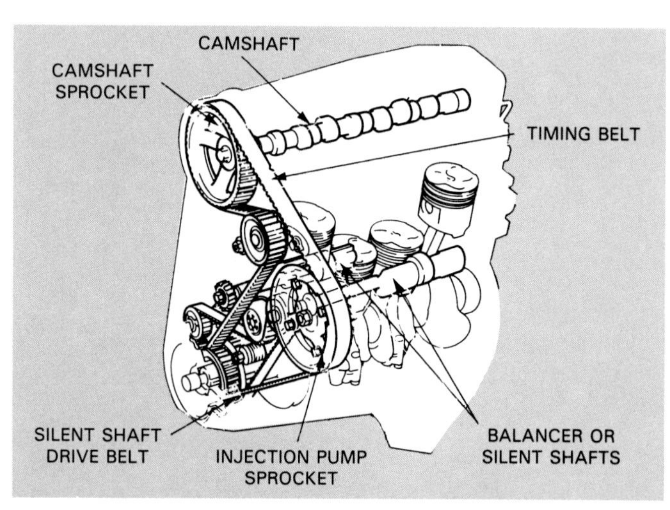

Fig. 11-20. Two cogged belts are used on the front of this engine. One drives camshaft; other drives balancer shaft. (Ford)

Timing belt construction

A *timing belt* is a cogged (toothed) belt with internal reinforcing strands to limit stretch. Teeth are formed in the inside of the belt. They mesh with teeth on the outside of the crank and cam sprockets. This provides a positive, slip-free drive mechanism, Fig. 11-21. Some late model timing belts are made of fiberglass-reinforced nitril rubber for increased strength and durability. Some timing belts are designed to last the life of the engine.

Cam belt sprocket

The *camshaft belt sprocket* performs the same function as a cam sprocket for a timing chain. However, it normally has square teeth, as shown in Fig. 11-22A.

The belt sprocket can be locked to the camshaft with a key or dowel pin. A bolt normally secures the sprocket to the front of the camshaft. An oil seal, usually in the cylinder head, keeps oil off the sprocket and belt. Oil will ruin a timing belt.

Crank belt sprocket

A *crankshaft belt sprocket* fits over the snout of the crankshaft. It is normally keyed to the crank. The crank sprocket has teeth shaped like the cam sprocket and belt. However, the crank sprocket is half the size of the cam sprocket so the camshaft turns at one-half engine speed. See Fig. 11-22B.

The crank sprocket for a timing belt fits in front of the crankshaft oil seal. A large bolt screws into the crankshaft snout to hold the sprocket in place. Belt sprockets can be made of cast iron, steel, or aluminum.

Timing belt tensioner

A *belt tensioner* is a wheel that keeps the timing belt tight on its sprockets, Fig. 11-23. The tensioner pushes inward on the back side of the belt. This prevents the belt teeth from moving away from the sprockets and slip-

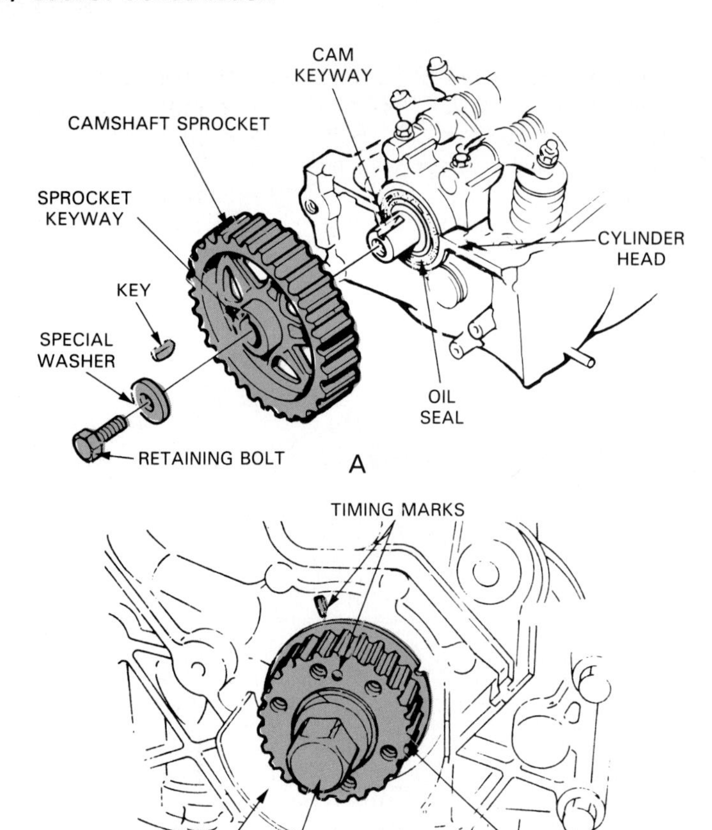

Fig. 11-22. A—Camshaft sprocket fits over small snout on this camshaft. Key locks sprocket to cam. Oil seal in head keeps oil from spraying onto belt. B—Crank sprocket is also keyed to crank. Large bolt threads intosnout to hold sprocket. Also note timing marks on front cover and sprocket. (Honda and Ford)

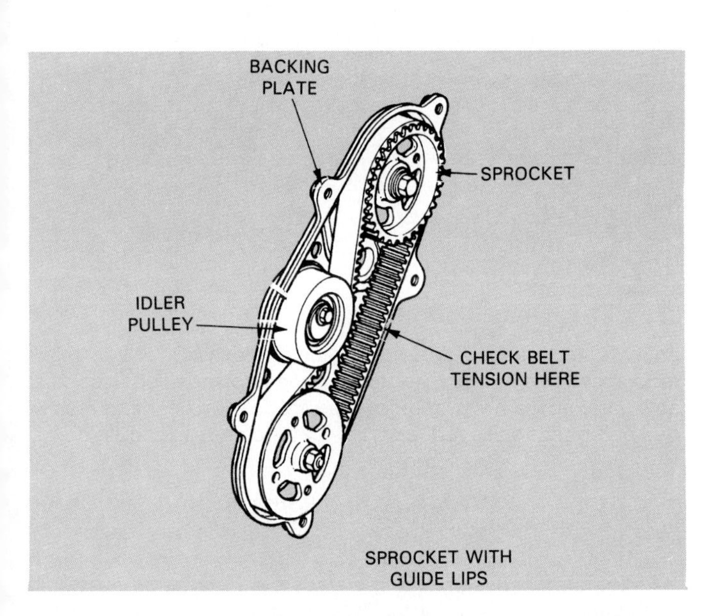

Fig. 11-21. Crankshaft sprocket turns cogged belt. Timing belt can then turn camshaft or other units. Idler or tensioner wheel pushes on back of belt to keep it on sprockets. One sprocket on the tensioner wheel must have lips that guide belt and center it on sprockets.

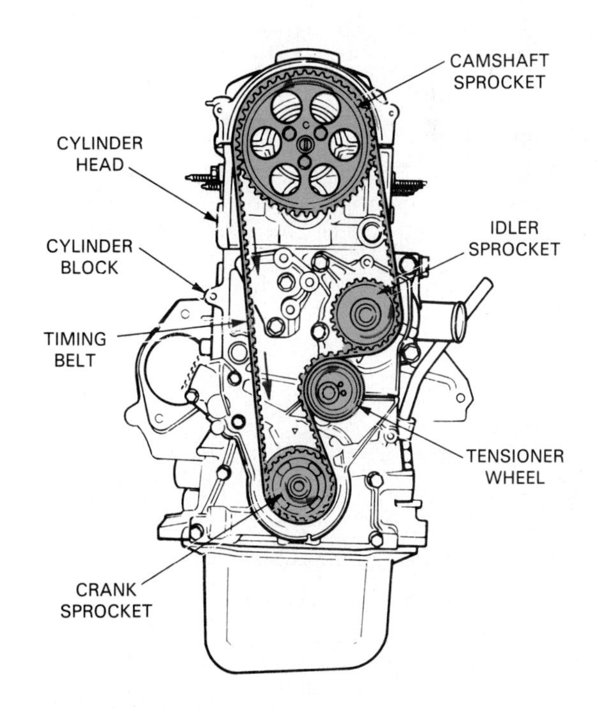

Fig. 11-23. Note rotation of timing belt, sprockets, and tensioner wheel. Idler or auxiliary sprocket can be used to power water pump, distributor, oil pump, etc.

ping on the sprocket teeth. It also keeps the belt from possibly flying off the sprockets.

The belt tensioner wheel is usually mounted on an *adjustment bracket* that allows it to be moved into or away from the belt. The wheel rides on an antifriction bearing that is *permanently sealed* (filled with grease and sealed at factory). Fig. 11-24 shows the tensioner bracket. Some belt tensioners use both spring and hydraulic pressure to maintain belt tightness. The spring tensioner keeps the belt tight when the engine is shut off. The hydraulic tensioner adjusts belt tension with engine speed. At higher rpm's, belt tension is increased to keep the belt from slipping or flying off.

Timing Belt Sensor

A *timing belt sensor* detects excessive tensioner extension and timing belt wear and stretch. When the sensor detects belt stretch, an indicator of possible belt failure, it signals the ECU. The ECU can then illuminate a dash light to warn the driver of the problem. This will allow belt replacement before failure. On many engines, if the timing belt breaks or slips off its sprockets, the pistons will hit and bend the valves.

Some computer control systems can detect probable belt failure by using data from the crankshaft and camshaft position sensors. If the timing belt jumps a tooth or two, the ECU can warn the driver of the timing belt problem by illuminating the Check Engine or MIL light before major engine damage occurs.

Auxiliary belt sprocket

An *auxiliary belt sprocket,* also termed *intermediate sprocket,* can be used to operate the oil pump, diesel injection pump, water pump, distributor, balancer shaft, or other units requiring crankshaft power. It is a third sprocket mounted to one side of the engine, Fig. 11-24.

The timing belt simply extends around this extra sprocket. The diameter of the sprocket varies, depending upon what device it is used to rotate. Its construction is similar to the cam and crank sprockets, Fig. 11-25.

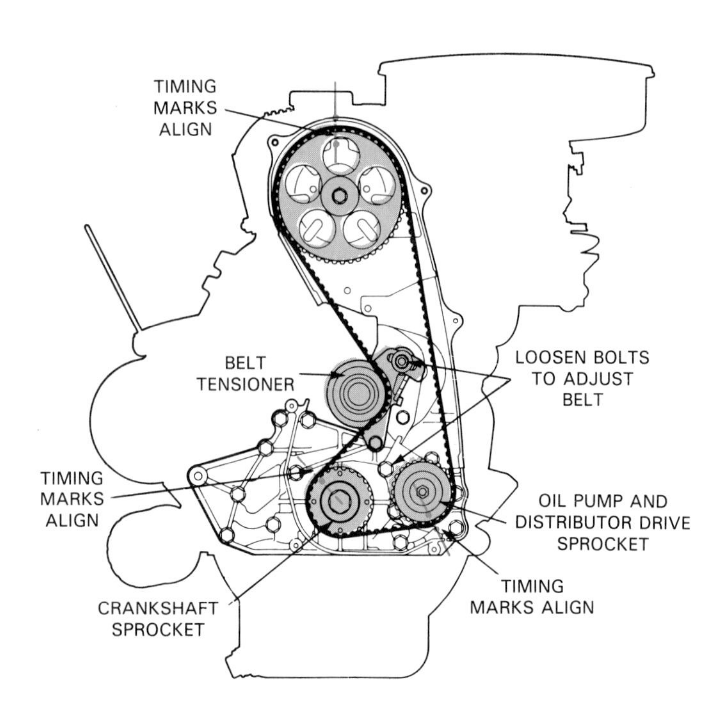

Fig. 11-24. Setting cam timing can be a little more "tricky" with OHC engine and timing belt. There are timing marks on three sprockets. This will be covered in detail in a later chapter. (Chrysler)

Belt timing marks

Belt timing marks are placed on the sprockets so the cam, crank, and intermediate sprockets can be synchronized. The intermediate sprocket must be kept in time when it turns the distributor or diesel injection pump.

The timing marks, as with chain sprockets, can be small dots, indentations, circles, triangles, holes, or other markings on the belt sprockets. The timing marks must align with marks on the engine when the number one piston is at TDC on the compression stroke. This is shown in Fig. 11-24.

Timing belt cover

A *timing belt cover* is simply a sheet metal or plastic shroud around the cam drive belt. Its function is to protect the belt from debris (rocks, tree branches, snow, etc.) that could force the belt off its sprockets. It also protects people from possible injury. See Fig. 11-25.

A timing belt cover should NOT be confused with an engine front cover. An engine front cover contains oil and houses the crankshaft oil seal. A timing belt cover fits over the belt and is in front of the engine cover. Look at Fig. 11-26.

Fig. 11-27 shows the difference between an engine front cover and a belt cover. Note that an OHC engine using a belt still uses a front cover for the oil seal.

Fig. 11-28 shows a cutaway view of a modern engine with a camshaft drive belt. Study the parts.

VIBRATION DAMPER CONSTRUCTION

A *harmonic balancer,* also called a *vibration damper* or *crankshaft damper,* is a heavy wheel mounted on rubber to help control engine vibration. The balancer is keyed to the crankshaft snout. This makes the damper spin with the crankshaft, Fig. 11-29.

Fig. 11-30 illustrates the basic construction of a typical harmonic balancer. Note how a rubber ring separates the outer inertia ring and the inner sleeve. The inertia and rubber rings set up a damping action on the crankshaft as it tries to twist and untwist. This DEADENS vibrating action. A *dual-mass harmonic balancer* has one weight mounted on the outside of the pulley and another on the inside. The extra rubber mounted weight helps reduce vibration at high engine speeds.

Harmonic vibration

Harmonic vibration is a high frequency vibration resulting from twisting and untwisting of the crankshaft. Each piston and rod assembly can exert over a ton of downward force on its journal. This can actually flex (bend) the crank throws in relation to each other.

If harmonic vibration is not controlled, the crankshaft could vibrate like a musician's tuning fork or string type musical instrument. Serious engine damage (usually crankshaft breakage) could result.

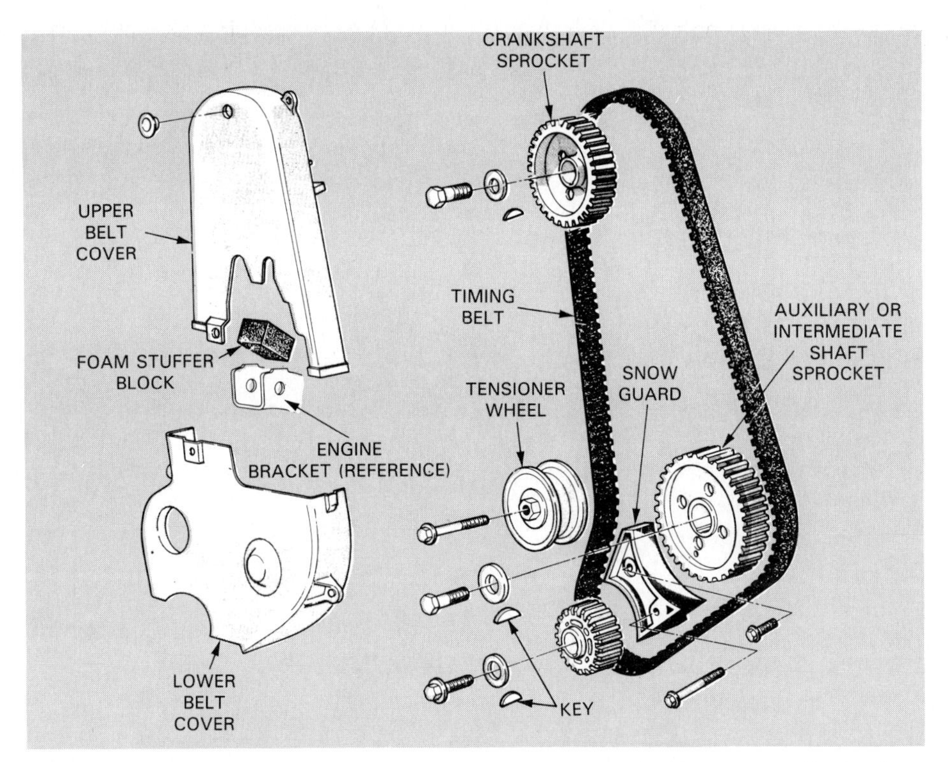

Fig. 11-25. Timing belt cover should not be confused with engine front cover. Engine front cover encloses oil and has crankshaft front seal. Belt cover simply surrounds belt and sprockets. Note snow guard that prevents buildup of snow from forcing belt off of sprockets. (Chrysler)

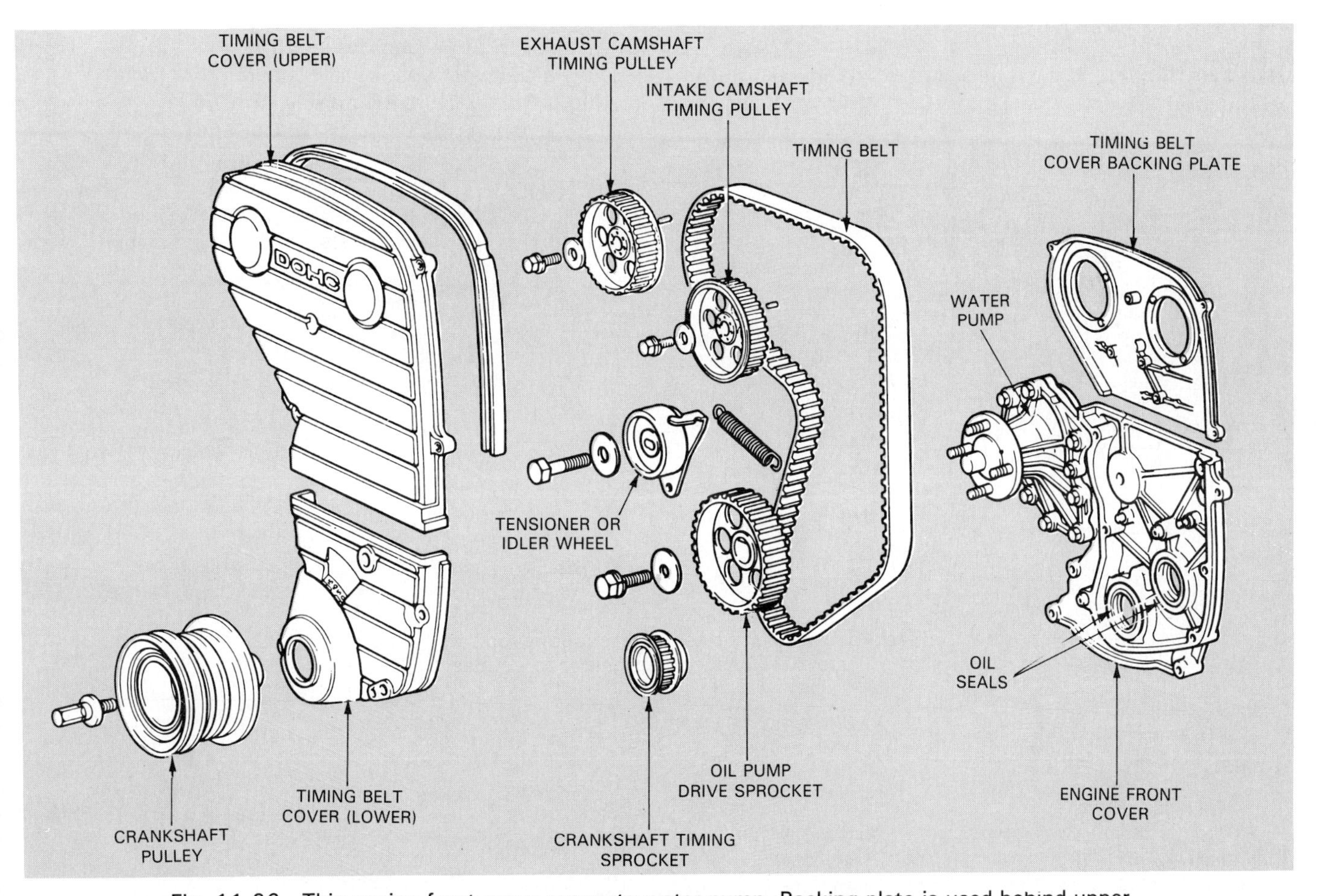

Fig. 11-26. This engine front cover supports water pump. Backing plate is used behind upper section of timing belt cover. (Toyota)

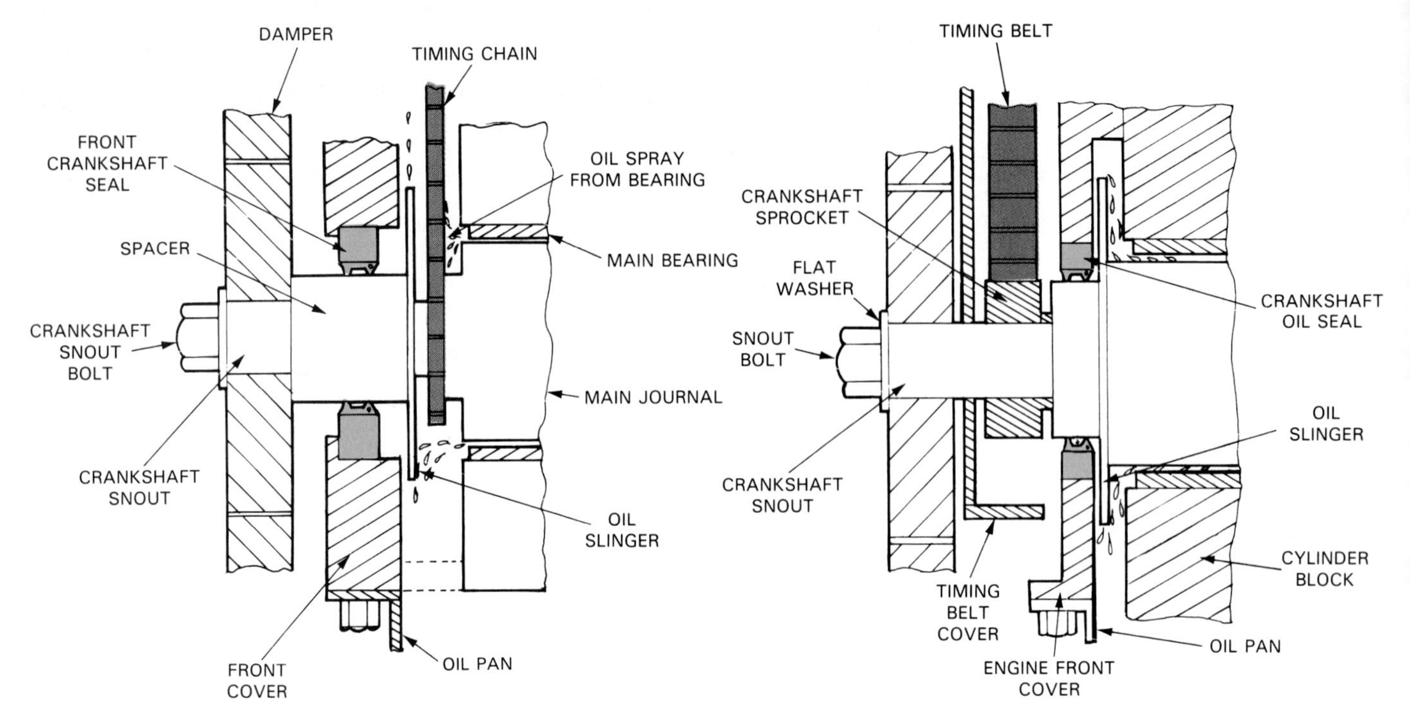

Fig. 11-27. Note differences between location of parts with timing chain and timing belt setups. Oil seal is in front of timing chain but behind timing belt.

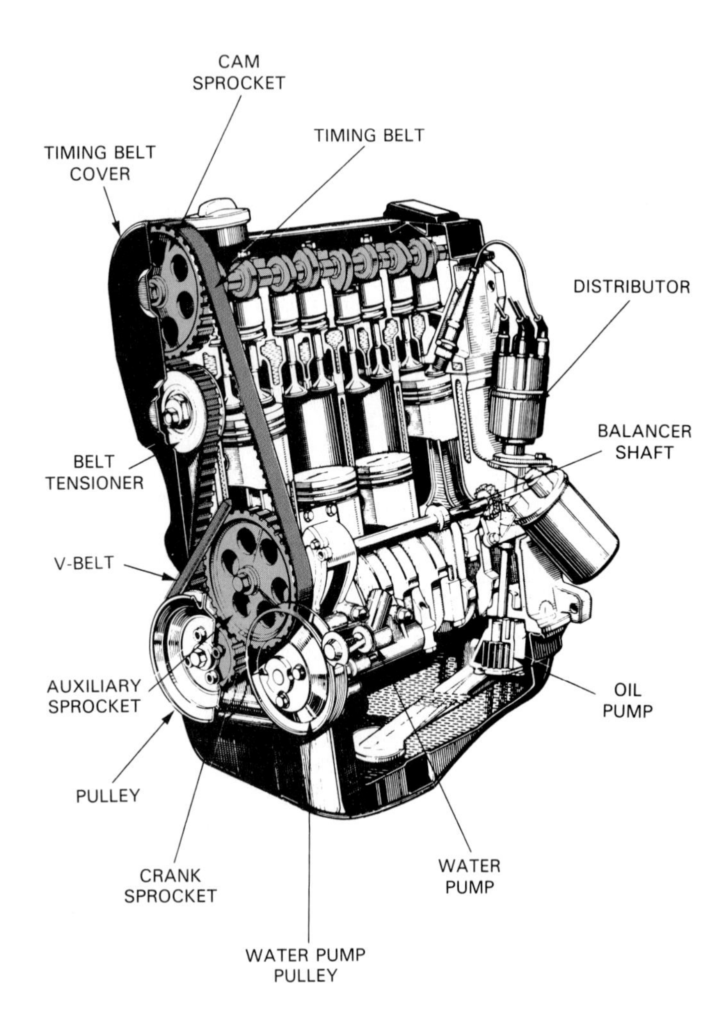

Fig. 11-28. Can you explain the operation of these components? (Chrysler)

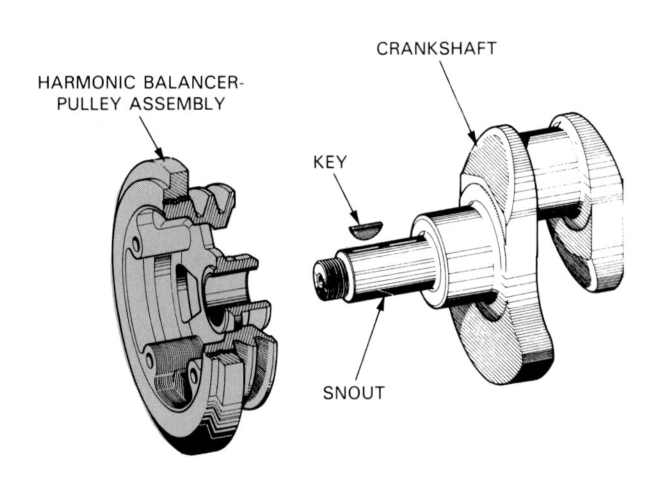

Fig. 11-29. Harmonic balancer, also termed crank damper or vibration damper, mounts on tip of crankshaft snout.

Fig. 11-31 shows a cutaway of a vibration damper assembly. Note how the assembly fastens to the crankshaft and how pulleys bolt to the damper.

Crankshaft pulleys

Crankshaft pulleys are needed to operate the alternator, power steering pump, air conditioning compressor, air injection pump, and other devices. The pulleys can be formed as part of the crankshaft damper or they can bolt to the damper, Fig. 11-32.

A crank pulley can be a V-type for use with V-belts or a more modern ribbed design for a ribbed belt. Belt pulleys are usually made of stamped steel. A hub on the center of the damper centers the pulley on the crankshaft.

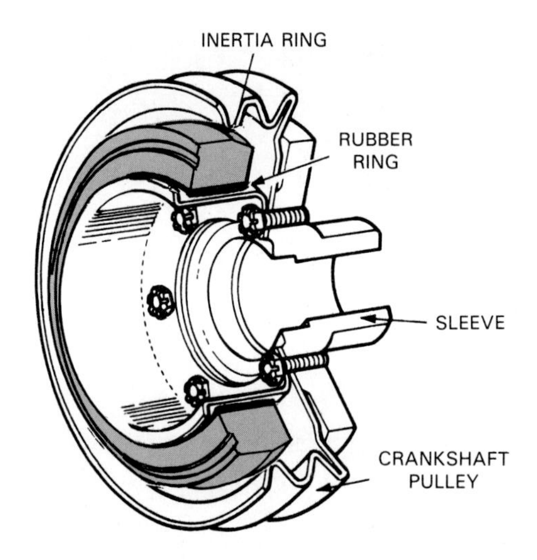

Fig. 11-30. Vibration damper is heavy wheel mounted on rubber. Flexing action of rubber ring and inertia of heavy wheel help counteract twisting and untwisting action of crankshaft at higher speeds. This prevents harmonic vibration that could break crankshaft. (Chrysler)

Engine balance shafts

Some modern engines use an internal balance shaft that is geared to the crankshaft or camshaft. The balance shaft has bob weights that spin in the opposite direction of crankshaft rotation. This cancels out torsional vibrations created by the crankshaft, providing a smoother engine idle.

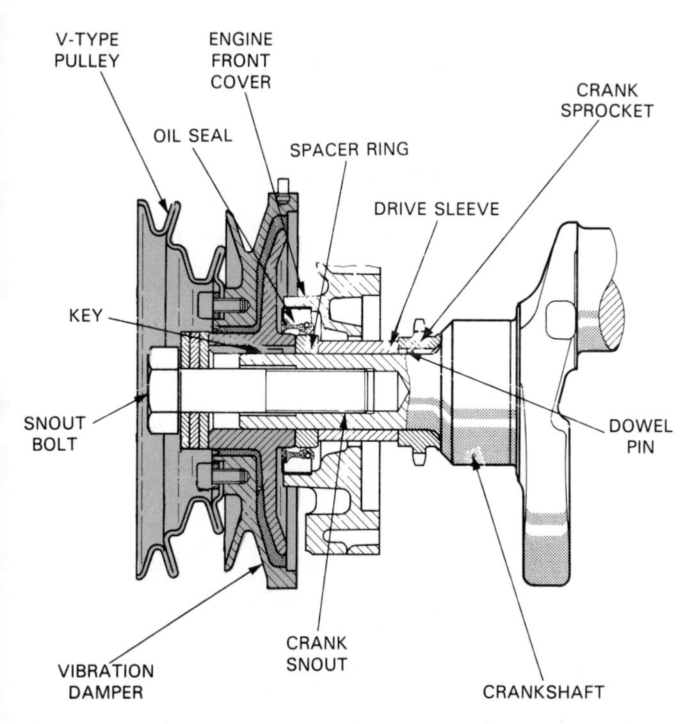

Fig. 11-31. Cutaway view shows how crankshaft pulleys bolt to harmonic balancer or vibration damper. Also note location and construction of other parts. (Mercedes Benz)

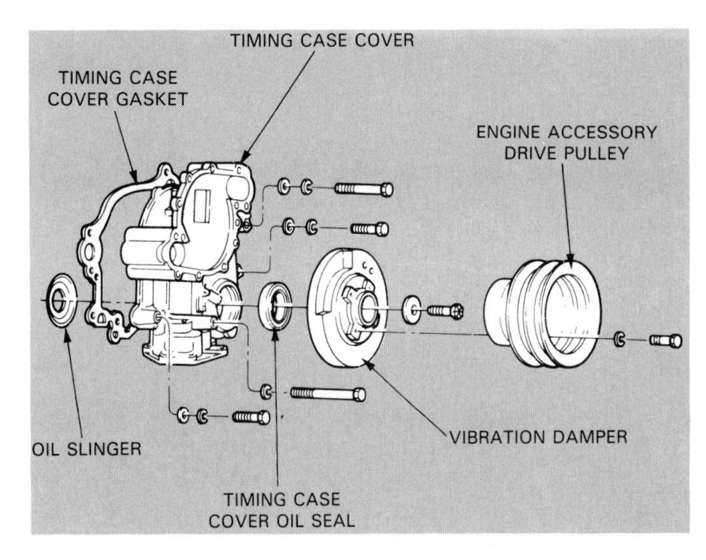

Fig. 11-32. Three or four bolts secure V or ribbed pulley to damper. Pulleys can also be formed as part of damper.

INTAKE MANIFOLDS

An engine *intake manifold* is a metal casting or plastic part casting that bolts over and covers the intake ports on the cylinder head(s). See Fig. 11-33.

An intake manifold has several functions:

1. Carry air-fuel mixture (gasoline engine) or air (diesel engine) into the cylinder head ports.
2. Evenly distribute air or air-fuel mixture to each engine cylinder.
3. Prevent the fuel from separating from the air.
4. Assist fuel vaporization and atomization.
5. Help warm the fuel charge when the engine is cold.
6. Assist in filling the engine combustion chambers with a fuel charge.

These functions vary with engine and intake manifold design. To troubleshoot and repair today's engines, you should have a sound understanding of intake manifold design and construction. Look at Fig. 11-34.

Intake manifold construction

Engine *intake manifolds* are usually made of lightweight aluminum, plastic, or a composite material. Older intake manifolds were made from heavy cast iron. Lighter materials are now used to reduce engine weight to help improve vehicle fuel economy and acceleration. The inside of the plastic runners are also smoother than metal to improve airflow into the engine. Surfaces are machined where the manifold attaches to the engine, throttle body or carburetor, and other components.

Fig. 11-35 shows cutaway views of two engine intake manifolds. Study their construction!

Fig. 11-36 pictures intake manifolds for in-line and V-type engines. Note the differences. The inline manifold positions the inlet to one side and passages run out to meet the ports in the cylinder head. The V-type has the inlet in the middle with passages feeding out to each side for both cylinder heads.

Parts of an intake manifold

The exact parts of an intake manifold vary with design.

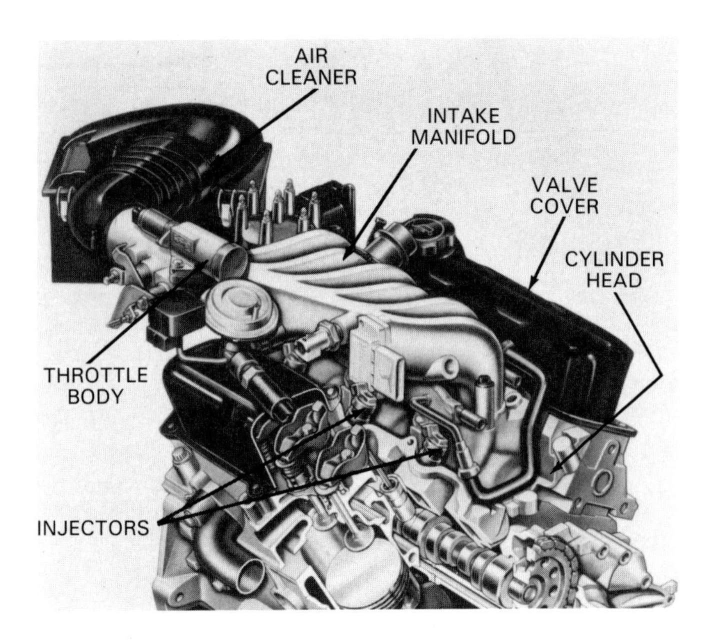

Fig. 11-33. Intake manifold routes air or air-fuel mixture to cylinder head ports. This is a modern design with long runners for increased performance. (Ford Motor Co.)

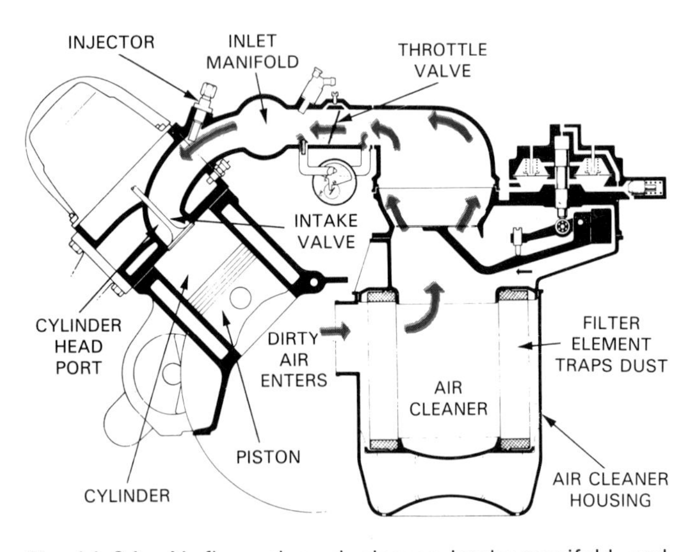

Fig. 11-34. Air flows through cleaner, intake manifold, and then into cylinder head. Today's engines frequently have injectors right before intake valves. Only air is routed through intake manifold.

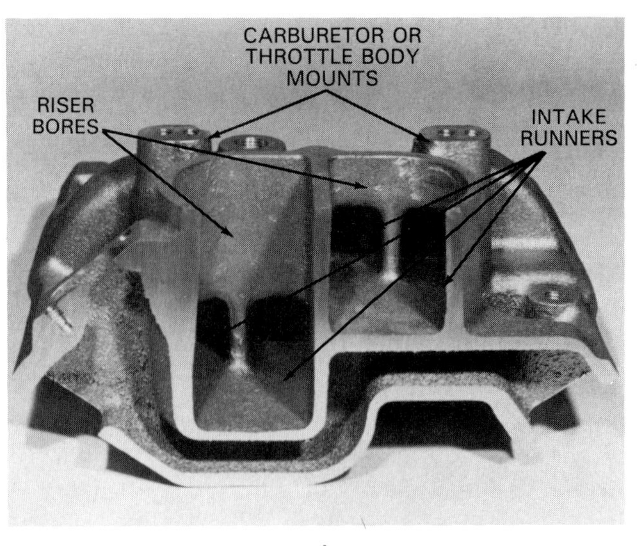

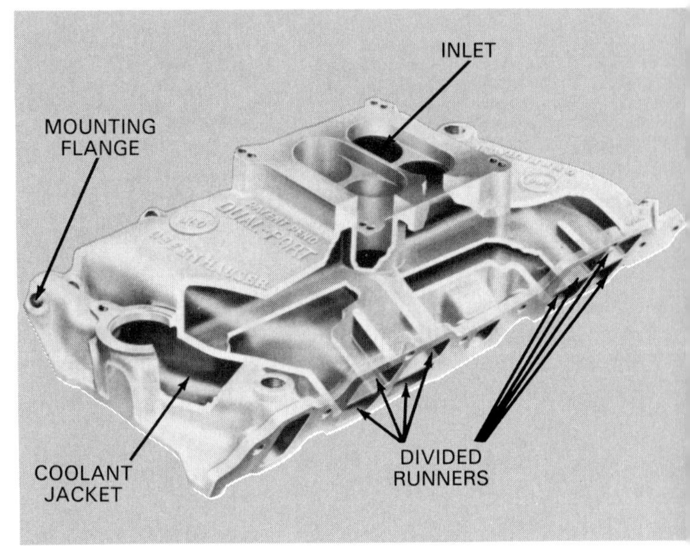

Fig. 11-35. Cutaway and phantom views of intake manifolds show construction. Study part names. (Edelbrock and Offenhauser)

However, most intake manifolds have:

1. RUNNERS (passages for air-fuel mixture or air) as shown in Fig. 11-37.
2. INLET or RISER BORE (opening for incoming air), Fig. 11-37.
3. PLENUM (chamber below inlet or bore for equalizing pressure at openings to runners), Fig. 11-37.
4. EXHAUST PASSAGE (separate passage for allowing hot engine exhaust gases to warm base of plenum), Fig. 11-39.
5. EGR PASSAGE (opening that allows small amount of exhaust gases to enter plenum to reduce NO_X emissions), Fig. 11-39.
6. WATER JACKETS (water passages for removing excess heat from intake manifold), Fig. 11-35A.
7. HEAD MATING SURFACE (machined flange that seals against cylinder head). See Fig. 11-36 again.
8. OTHER PROVISIONS (openings for vacuum fittings, EGR valve, fuel injectors, etc.).

Fuel distribution

Fuel distribution refers to how well an intake manifold routes an equal amount of air and fuel to each combustion chamber. Equal fuel distribution is needed for maximum engine performance. Poor fuel distribution occurs when cylinders receive an UNEQUAL amount fuel or air. This can upset engine efficiency.

Shown in Fig. 11-38, multiport fuel injection eliminates the problem of poor fuel distribution. Fuel is injected on the head of each intake valve. Only air flows through the intake manifold runners. This provides for equal fuel distribution and peak efficiency.

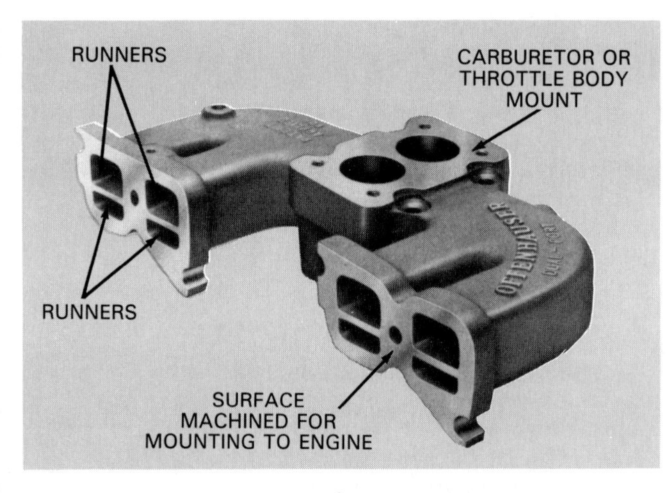

A

B

Fig. 11-36. Two basic intake manifold types: A—In-line type. B—V-type. Both are made of lightweight aluminum. (Offenhauser and Edelbrock)

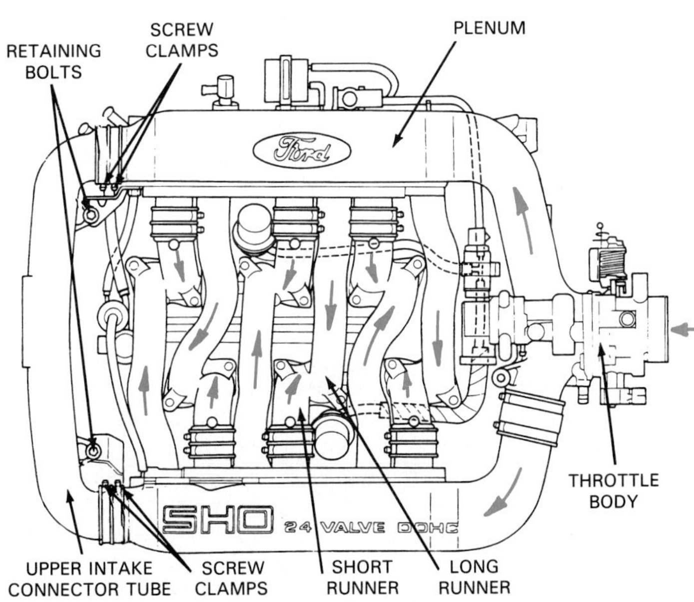

Fig. 11-37. Top view of intake shows how runners feed out from inlet and plenum. Blue arrows indicate airflow. (Ford)

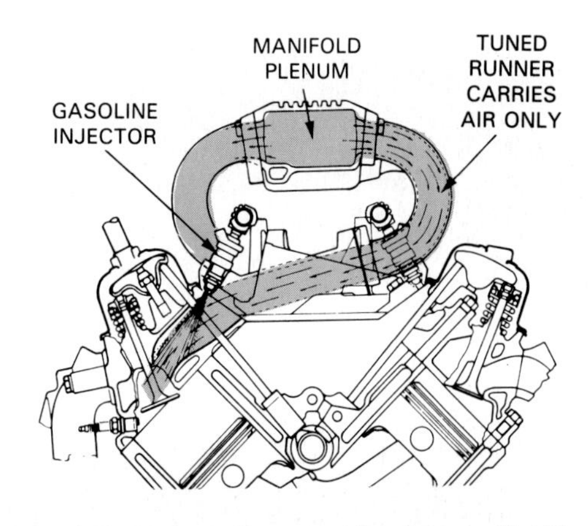

Fig. 11-38. This high performance intake manifold has large central plenum and long, equal length runners. Injectors are mounted to spray fuel directly in engine intake valves. Equal length runners helps balance pressure pulses in intake manifold as valves open and close. This results in better engine "breathing." (Chevrolet)

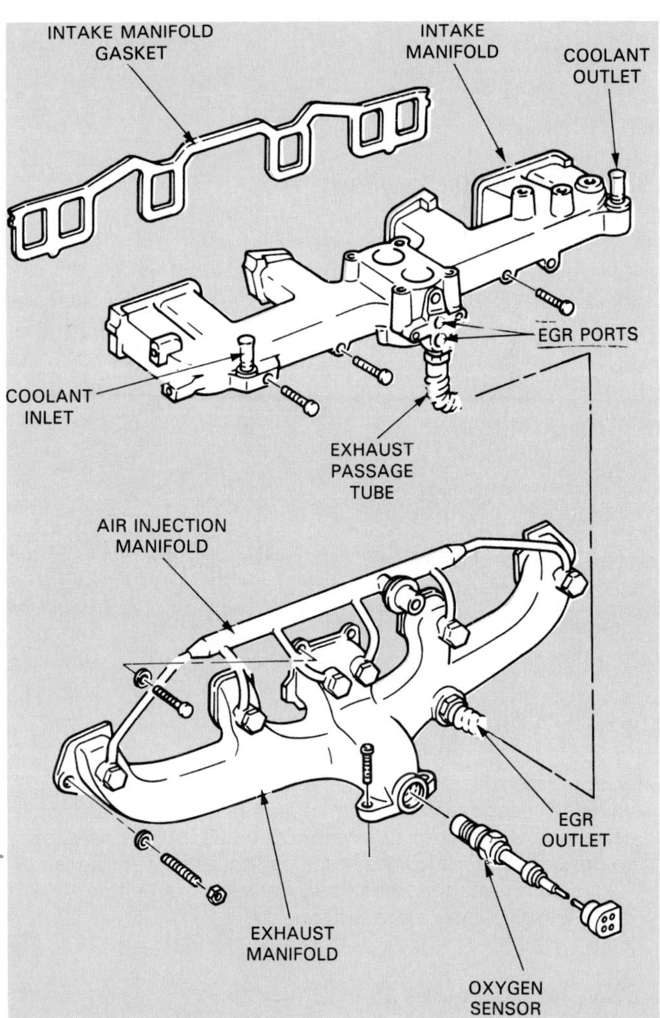

Fig. 11-39. This in-line intake manifold has metal tube running to it from exhaust manifold. Hot exhaust gases help warm intake and are used by EGR valve to reduce NO_X emissions. Note oxygen sensor that screws into exhaust manifold. It "sniffs" exhaust to check combustion efficiency. (American Motors)

Intake manifold types

Intake manifold types differ in plenum design, runner length, port shape, runner configuration, and method of construction. There are other considerations, but these are the most important.

Dual plane intake manifold

A *dual plane* intake manifold has a two-compartment plenum. This type is common on V-type passenger car engines with carburetors. It provides good fuel vaporization and distribution at low engine speeds. This makes it very efficient for everyday driving. Refer back again to Fig. 11-36B.

Single plane intake manifold

In a *single plane* intake manifold, one common plenum feeds all runners. This type is used primarily on high performance engines. It is not as efficient as the dual plane type at lower engine rpm. A single plane, however, is capable of allowing greater flow at higher engine speeds, increasing engine horsepower capabilities.

Port configuration

An intake manifold can also be classified by *port shape.* The three common port shapes are round (oval) port, D-port, and square (rectangular) port. These are pictured in Fig. 11-40.

Split runner intake manifold

A *split-runner intake manifold* has two separate runners leading to each intake valve. Usually, the ECU operates a butterfly valve over the shorter of the two runners. At low engine speeds, the long runner forces air into the cylinders. When a specific engine speed is reached, the computer opens the butterfly valve over the short intake runner. This increases air flow into the engine for added power and torque.

Tuned runner intake manifold

A *tuned runner* or *ram intake manifold* has very long runners of equal length, Fig. 11-41. It is used on high performance engines.

The long runners use the inertia of the moving air to ram more charge into the engine combustion chambers. Even at low rpm, a large mass of air flows through the runners. This air wants to keep flowing even when the cylinders are almost full. As a result, the inertia forces a little more air and fuel into the engine to increase horsepower.

Fig. 11-42 shows another tuned runner or ram type intake manifold. Note the location of the fuel injectors.

Sectional intake manifold

A *sectional intake manifold* is constructed in two or more pieces. The pieces bolt together to form the intake manifold assembly. A sectional manifold allows for a more exotic shape for improved efficiency.

A sectional intake manifold is shown in Fig. 11-43. The upper manifold or plenum bolts to the lower manifold or runners. Injectors mount in the lower manifold. A throttle body for controlling airflow bolts to the upper manifold. Gaskets seal the connections.

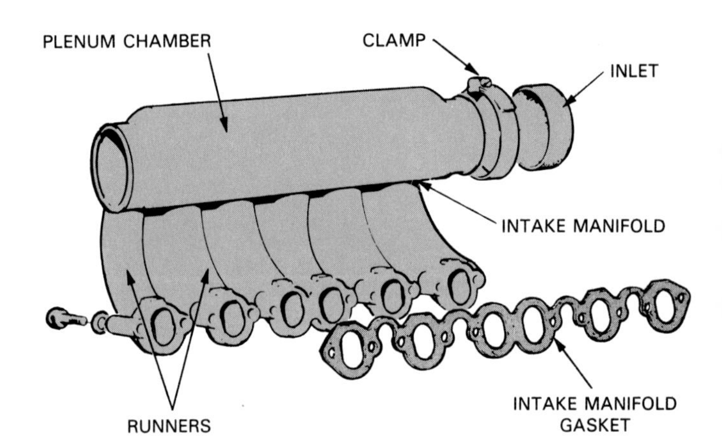

Fig. 11-41. Intake manifold gasket seals joint between cylinder head and manifold.

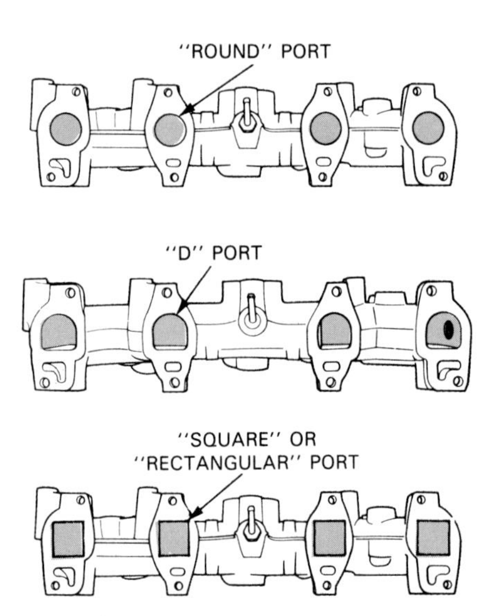

Fig. 11-40. Intake manifolds commonly have either round, D-shaped, or square ports.

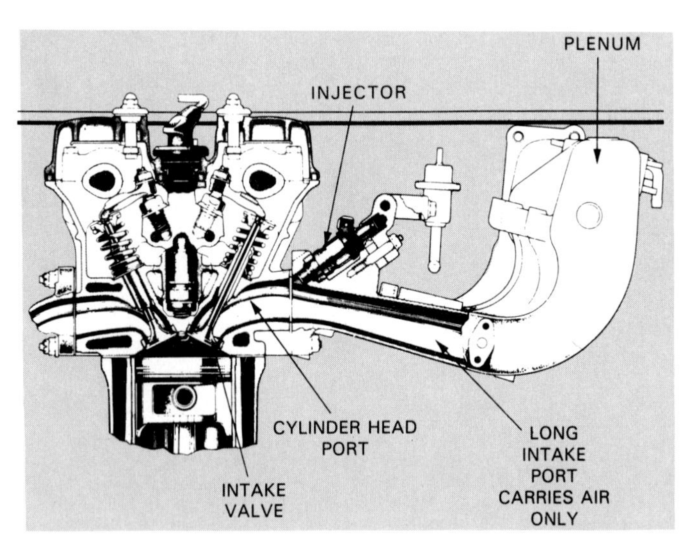

Fig. 11-42. This high performance intake manifold has extremely long runners. Called a ram manifold, it helps force extra air and fuel into cylinders by using inertia of moving air in runners. (Honda)

Variable induction sysems

Some engines are equipped with a *variable induction system,* which is designed to increase engine efficiency. Variable induction systems generally consist of two sets of intake runners that are controlled by butterfly valves.

At low engine speeds, the valves in the secondary runners are closed. Consequently, only the primary runners provide airflow to the engine. This arrangement increases low-speed power and fuel economy by improving fuel atomization and mixing. At higher speeds, the vehicle's computer or control system opens the valves in the secondary runners so that all runners feed air into the combustion chambers. This improves volumetric efficiency and engine power. Because variable induction systems are relatively complex, they are primarily used on high-performance engines.

Intake manifold reed valves

Intake manifold reed valves can also be used in the intake runners of four-stroke cycle engines. They can be used to help prevent fuel flow back into the intake manifold during valve overlap. They snap shut and only allow flow into the engine combustion chambers.

EXHAUST MANIFOLD CONSTRUCTION

The *exhaust manifold* also bolts to the cylinder head, but over the exhaust ports, Fig. 11-44. During the exhaust strokes, hot gases blow into this manifold before entering the rest of the exhaust system. An exhaust manifold is normally made of cast iron or lightweight steel tubing.

Note! Exhaust manifolds are also discussed in Chapter 19, Turbocharging Systems.

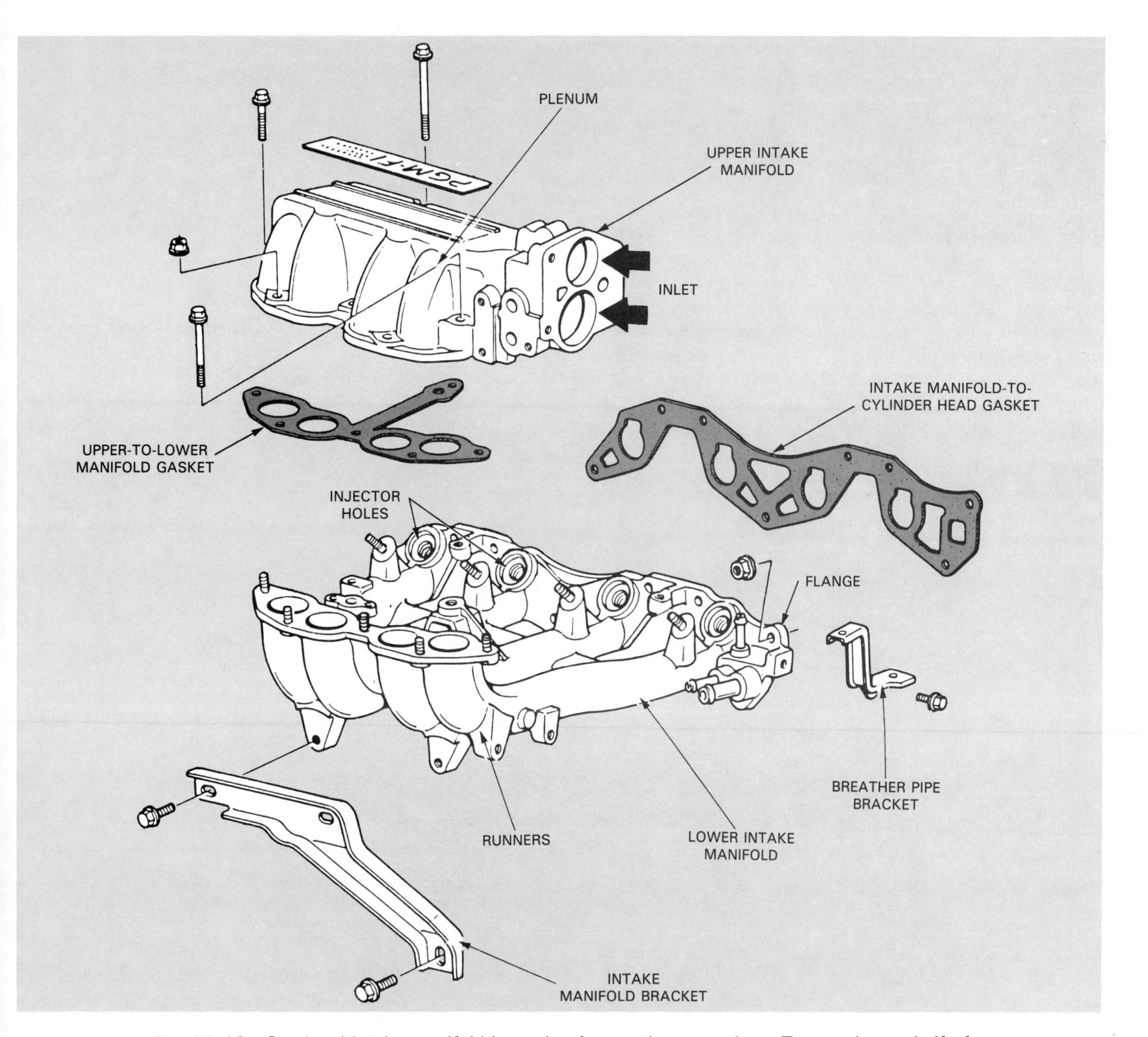

Fig. 11-43. Sectional intake manifold is made of more than one piece. Top or plenum half of intake bolts to bottom or runner half. This allows for more exotic shape. (Honda)

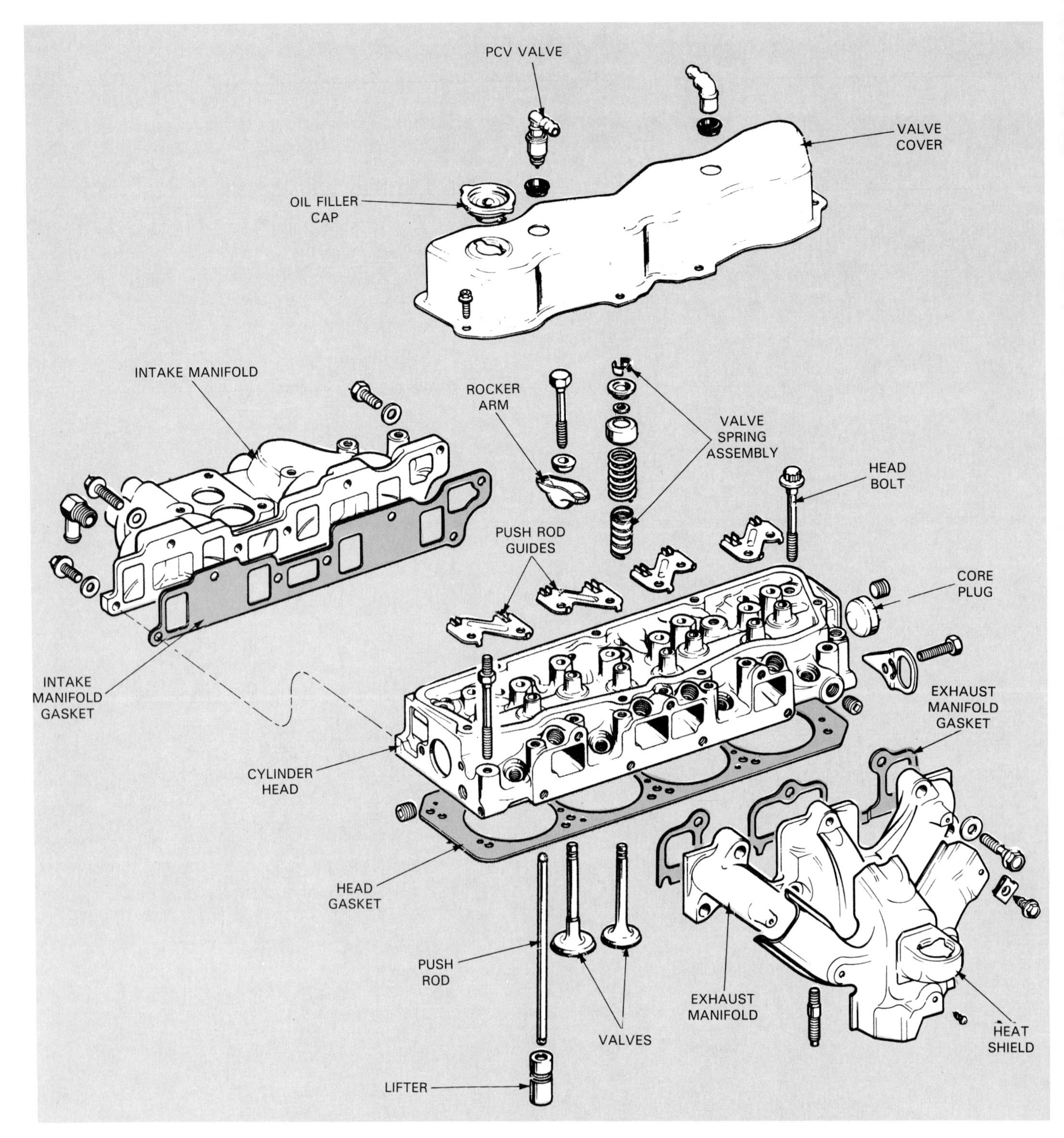

Fig. 11-44. Exhaust manifold bolts to cylinder head so that it can accept hot gases and route them into exhaust system header pipe. Note other part names. (AMC)

VALVE COVER CONSTRUCTION

The *valve cover,* also called *rocker cover* or *cam cover* on OHC engines, is a thin housing over the top of the cylinder head. It simply keeps valve train oil spray from leaking out of the engine. The valve cover can be made of thin, stamped steel, cast aluminum, or plastic. The cover is sealed by a gasket or sealant.

Fig. 11-44 shows a valve cover.

Remember! If a gasket leaks, it can cause serious engine damage and unsatisfied customers.

Some valve covers and other parts are designed to reduce the amount of engine noise or acoustic dampened. Acoustic dampened construction means the part is designed to help reduce the transmission of sound. For example, some head or valve covers are acoustically dampened by making them out of two sheets of steel formed around a center layer of mastic material.

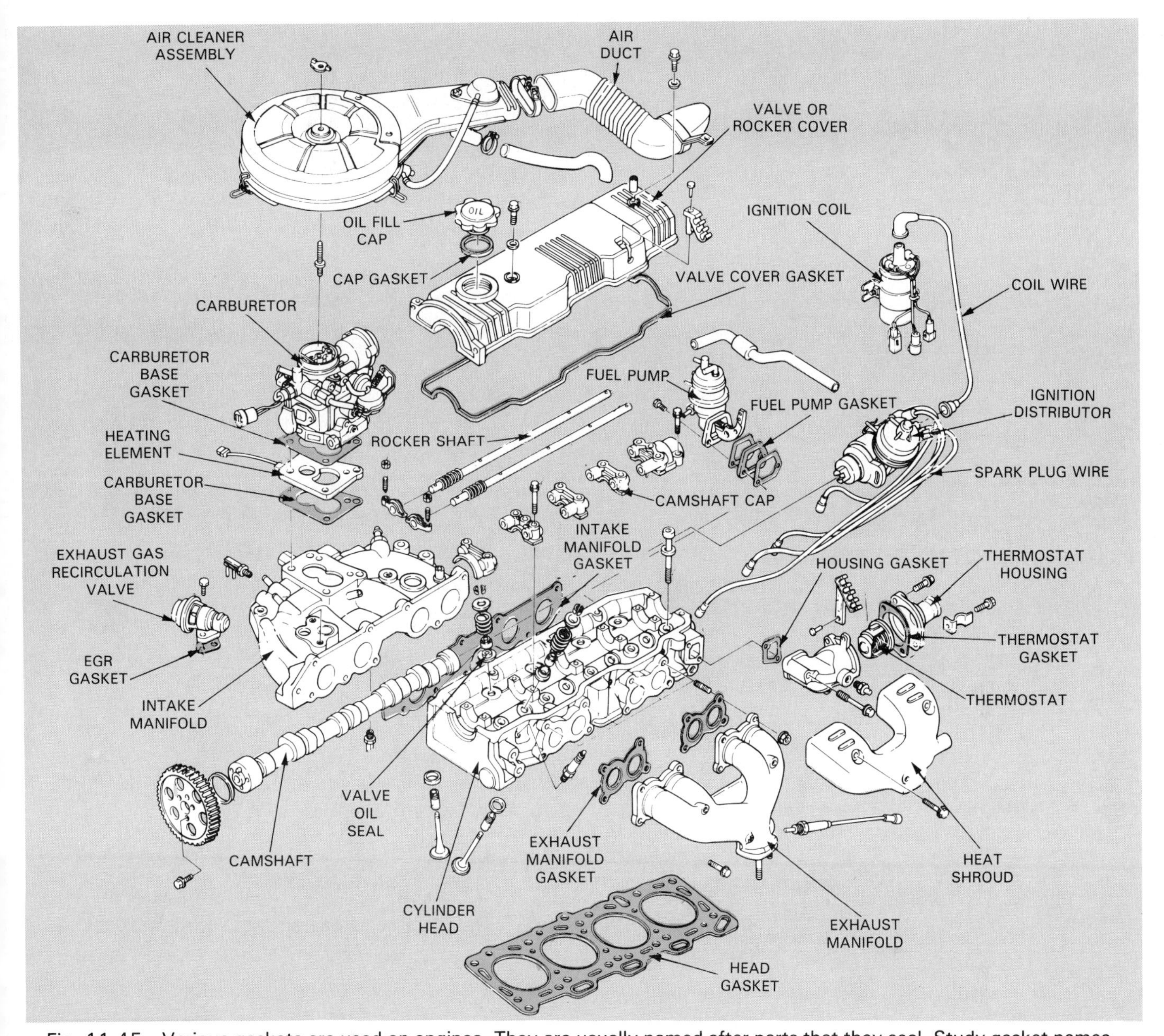

Fig. 11-45. Various gaskets are used on engines. They are usually named after parts that they seal. Study gasket names.

ENGINE GASKET CONSTRUCTION

Engine gaskets prevent pressure, oil, coolant, and air leakage between engine components. They are very important to the performance and dependability of an engine. Fig. 11-45 shows major engine gaskets.

Cylinder head gasket construction

A *cylinder head gasket* must seal the pressure of combustion and the flow of oil and coolant between the head and block. It must withstand tremendous pressure, chemical action, expansion and contraction, and severe change in temperature. This is illustrated in Fig. 11-46.

Modern cylinder head gaskets are made of several materials. A metal O-ring is normally formed around the cylinder opening. It is needed to withstand combustion heat and pressure. The body of the gasket is a composite material that can withstand the compression and relaxation.

When the engine warms, its parts expand and smash the head gasket tighter. Then, when the engine cools, the parts contract and less clamping pressure is produced on the head gasket. The head gasket must be resilient so that it can seal through thousands of these heating-cooling cycles.

Many modern head gaskets are coated with non-stick teflon. This is especially important when the cylinder head is aluminum and the block is cast iron. Aluminum expands more than iron and can produce a scrubbing action. The teflon will let the surface of the head move without gasket failure.

Also, many late model head gaskets have a bead of sealer formed around water and oil passages. the rubber-like bead helps assure that the gasket does not leak. This is illustrated in Fig. 11-47. Some late model cylinder head gaskets have *compression limiters* or metal rings that surround bolt holes to help control gasket crush for better sealing.

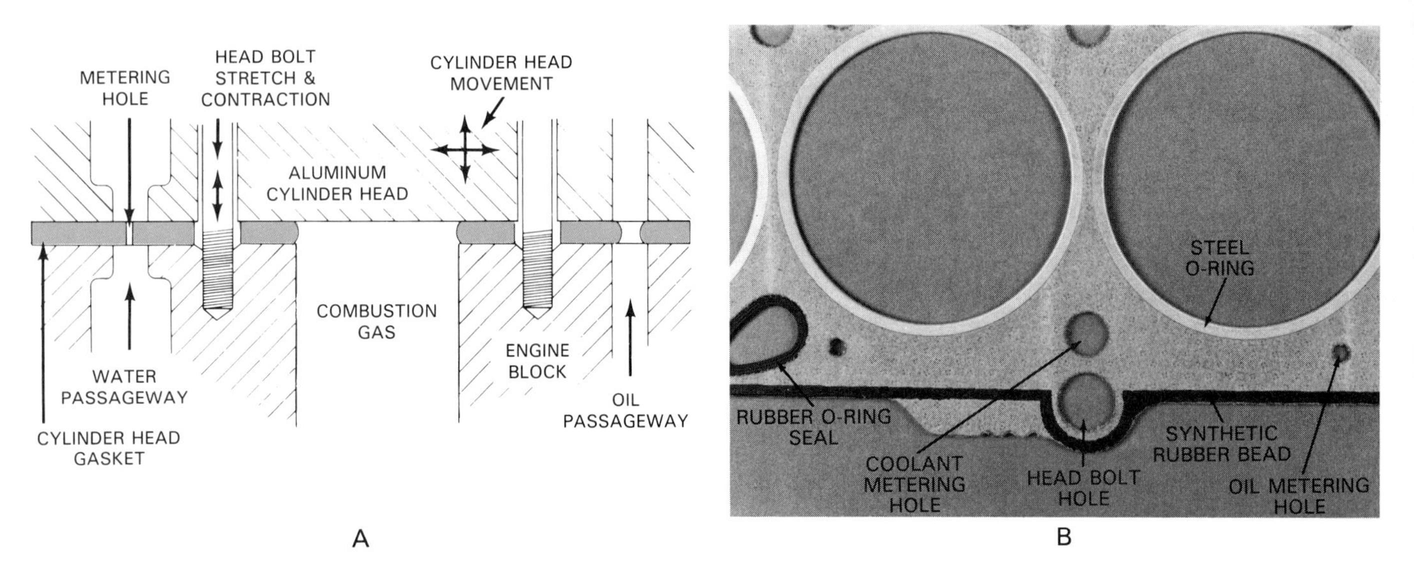

Fig. 11-46. Cylinder head gasket must operate under severe conditions. Combustion heat, pressure, oil, coolant, and head movement all try to break seal. Modern head gaskets use advanced construction methods to prevent leakage. (Fel-Pro)

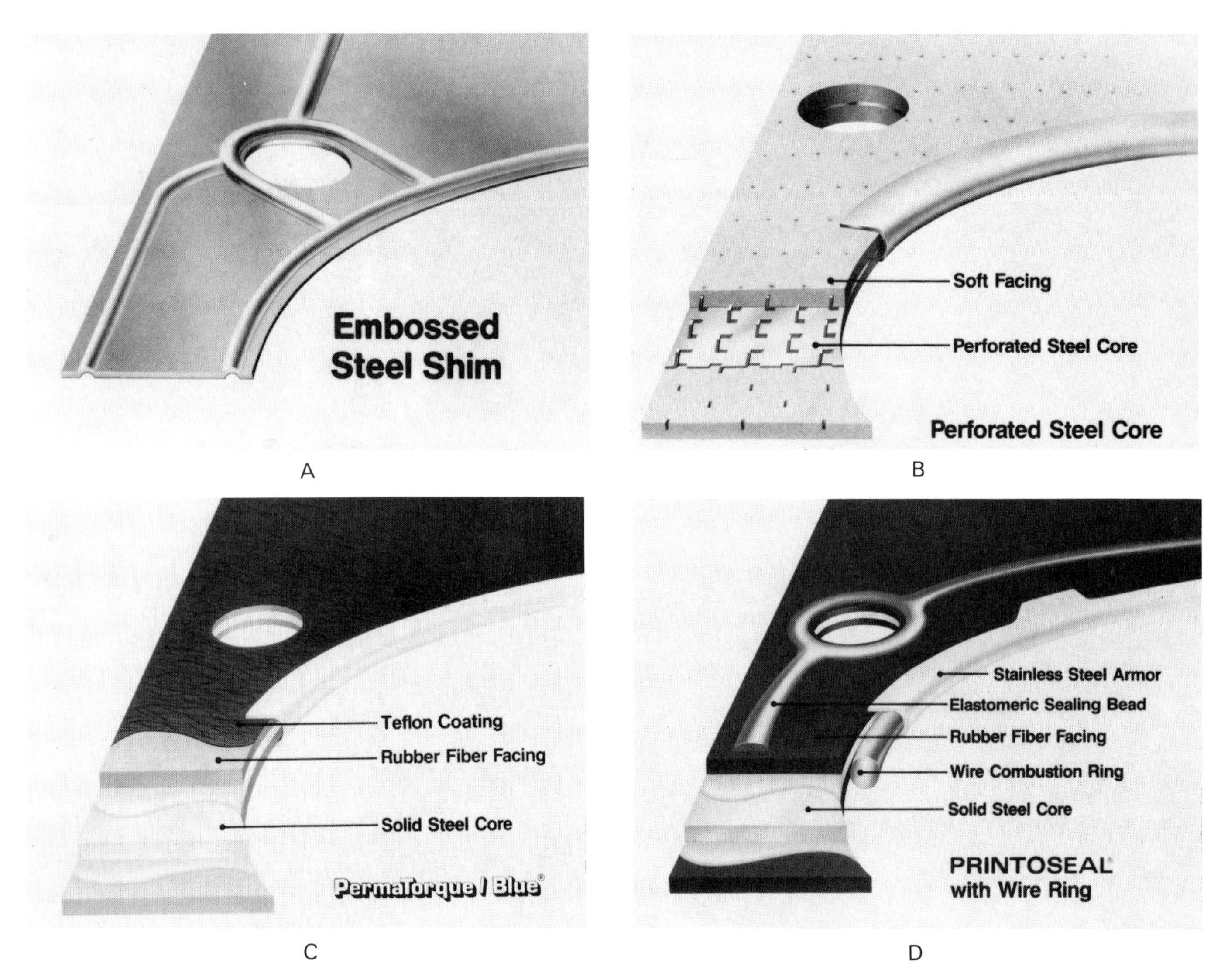

Fig. 11-47. Note various types of head gasket construction. A—Steel shim head gasket. B—Perforated steel core. C—Permanent torque head gasket. D—Permanent torque with steel ring and extra sealing bead. (Fel-Pro)

High quality head gaskets are *permanent torque* and do NOT have to be retorqued after a period of engine operation. They are resilient enough to maintain the initial torque. Many older steel shim gaskets or low quality gaskets should be torqued a second time after engine operation.

Cover and housing gasket construction

Cover and *housing gaskets* include the valve cover gasket, oil pan gasket, front cover gasket, thermostat housing gaskets, intake and exhaust manifold gaskets, etc. These gaskets are normally made of synthetic rubber, cork, asbestos, treated paper, or they may be a chemical gasket (RTV sealer). Their construction depends on their operating conditions.

When the engine cover or housing is made of thick metal, the gasket is made thinner or an anaerobic sealer may be used. However, with a thin, stamped cover, the gasket must be thicker or RTV (silicone sealer) is needed to conform to the irregular shape of the cover.

In Fig. 11-45, note how the names of the gaskets relate to the parts that they seal. This makes it easy to remember gasket names.

For more information on gaskets, seals, and sealers, refer to the text index. It will give pages for more coverage of this subject.

EXTERNAL ENGINE COMPONENTS

Figs. 11-48 through 11-50 show photos of actual engines. Study the names of the parts that fasten to the outside of the engines.

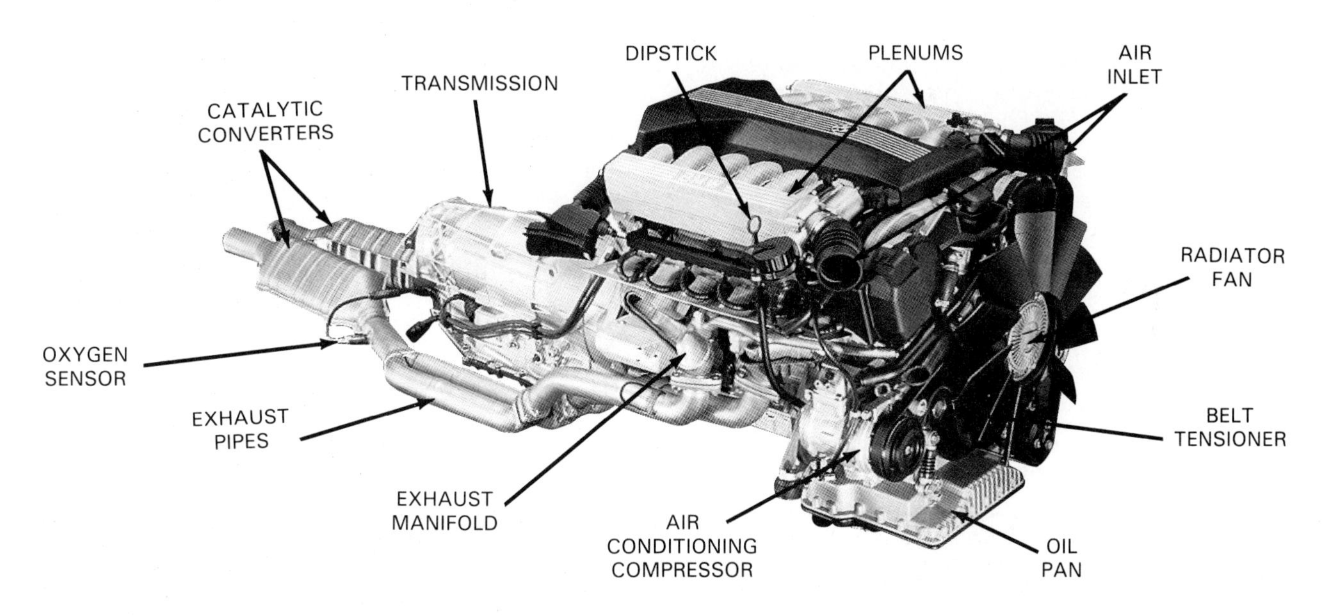

Fig. 11-48. Note parts on this engine. (BMW)

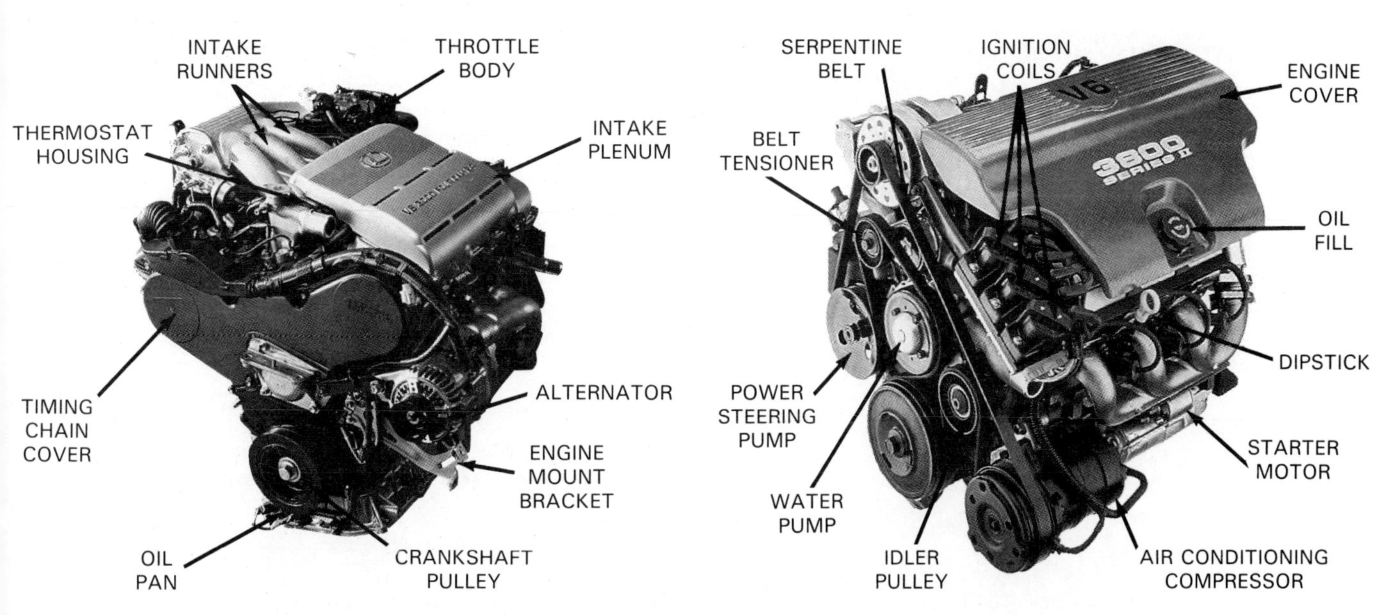

Fig. 11-49. Study parts on outside of this engine. (Lexus)

Fig. 11-50. Can you explain these components?

SUMMARY

Various designs are now used to drive the engine camshaft and other accessory units. It is important for you to understand these differences before attempting to service today's engine front ends.

The camshaft drive mechanism must rotate the camshaft at one-half crankshaft speed. A cam drive ratio of 2:1 is needed. Either gears, a chain, or a cogged belt can operate the camshaft.

Timing gears are for heavy-duty applications. They are very dependable but are heavy and noisy. Timing marks on the gears allow you to synchronize the crank and cam. Timing gear backlash is the small clearance between the gear teeth for expansion and lubricating oil.

A timing chain and sprockets are found on both OHV and OHC engines. Either a double-roller or a silent type chain can be used. The crank sprocket is one-half as large as the cam sprocket.

A crankshaft key locks the crank sprocket or gear to the crankshaft. A key or dowel pin secures the cam gear or sprocket. An eccentric can be mounted on the cam sprocket to operate a mechanical fuel pump.

A chain tensioner is a spring-loaded device that presses on the chain. It takes up excess slack as the chain and sprockets wear out. A plastic, teflon, or fiber block contacts the chain to reduce friction.

A chain guide may be needed on long lengths of chain to prevent chain slap. Chain slap is a back and forth flapping motion resulting from too much unsupported length of chain. The body of the guide is metal and the face is plastic, nylon, or teflon.

An auxiliary chain may be used to drive the engine oil pump, balancer shafts, or other units. Extra auxiliary sprockets are also needed.

The engine front cover surrounds the timing gears or timing chain and sprockets. It contains a crankshaft oil seal that keeps oil from leaking out around the crank snout. The cover can be made of stamped steel or cast aluminum.

A timing belt mechanism is normally used on OHC engines because of the great distance between the crank and cam. However, timing chains are also used on OHC engines.

A timing belt is very quiet and smooth. Teeth of the cogged belt mesh with teeth on the belt sprockets. A belt tensioner keeps the belt tight on its sprockets. A belt cover surrounds the belt and sprockets to keep debris away from the rubber belt.

A harmonic balancer or vibration damper is a heavy wheel mounted on rubber. It fits over the crankshaft snout. The damper is used to reduce harmonic vibration produced by the twisting-untwisting action of the crankshaft. This protects the crankshaft from potential breakage.

Either V-type or a ribbed pulley is used on the crankshaft to operate the alternator, power steering pump, air conditioning compressor, etc. The pulley either bolts to the damper or can be made as part of the damper.

Modern engines use exotic intake manifold designs to increase engine performance and efficiency. The plenum and runners are shaped to route the same amount of air and fuel to each cylinder. With multiport fuel injected engines, the intake only carries air. The fuel is injected into the runners right before the engine intake valves.

A tuned runner or ram manifold has long, equal length runners. They use the inertia of the airflow to force extra air and fuel into the cylinders as the valves open at higher engine speeds. This increases engine power because of a supercharging type effect.

A sectional intake manifold is made in two or more pieces. It allows a more complicated shape. An upper manifold bolts to a lower manifold to form the complete intake manifold assembly.

Gaskets are used to prevent pressure oil, coolant, and air leakage. The head gasket must contain the tremendous heat and pressure of combustion. It must also seal coolant and oil flowing between the head and block. Other gaskets are named after the parts that they seal.

KNOW THESE TERMS

Camshaft drive ratio, Timing gears, Timing marks, Timing gear backlash, Timing chain, Crankshaft key, Camshaft dowel, Camshaft eccentric, Chain tensioner, Chain guide, Auxiliary chain, Oil slinger, Engine front cover, Timing belt, Crank sprocket, Cam sprocket, Belt tensioner, Permanently sealed bearing, Timing belt sensor, Timing belt cover, Harmonic balancer, Harmonic vibration, Dual-mass harmonic balancer, Crankshaft pulleys, Intake manifold, Runners, Rise bore, Plenum, Fuel distribution, Split-runner intake manifold, Dual plane intake manifold, Tuned runner intake manifold, Sectional intake manifold, Variable induction system, Exhaust manifold, Valve cover, Cylinder head gasket.

REVIEW QUESTIONS—CHAPTER 11

1. Engine front end construction is about the same now as it was 50 years ago. True or False?
2. What are the three methods used to drive the engine camshaft?
3. The camshaft drive ratio is:
   a. 3:1.
   b. 1:1.
   c. 2:1.
   d. 4:1.
4. ________ ________ are usually used on OHV, heavy-duty application because of their dependability.
5. Define the term "timing gear backlash."
6. A timing chain can be used in cam-in-block engines and overhead cam engines. True or false?
7. The two types of timing chains are the multiple-link or ________ and the ________ ________ type.
8. What is the purpose of a crankshaft key?
9. Why is a chain tensioner sometimes used?
10. Why is a chain guide sometimes used?
11. An auxiliary chain may be used to drive:
    a. Balancer shafts.
    b. Camshaft.
    c. Alternator.
    d. None of the above.
12. How does an oil slinger work?

13. Which of these parts is NEVER fastened to the engine front cover?
    a. Mechanical fuel pump.
    b. Oil pump.
    c. Water pump.
    d. Starting motor.
14. A timing belt runs very ________ and without gear ________ nor chain ________.
15. Why are timing marks needed on timing gears, chain sprockets, or belt sprockets?
16. What is the difference between an engine front cover and a timing belt cover?
17. Crankshaft harmonic vibration is a result of the ________ and ________ of the crankshaft due to the tons of force produced by the ________ and ________ assemblies.
18. List some of the units that the crankshaft pulleys operate.
19. List five basic parts of an engine intake manifold.
20. Most modern, high quality head gaskets have to be retorqued after a short period of engine operation. True or false?

ASE CERTIFICATION-TYPE QUESTIONS

1. An engine's timing mechanism is sometimes used to power the ________.
   (A) ignition system distributor
   (B) engine's oil pump
   (C) engine balance shaft(s)
   (D) All of the above.
2. Technician A says that timing gears are normally used in OHC engines. Technician B says that timing gears are primarily used in cam-in-block automotive engines. Who is right?
   (A) A only.
   (B) B only.
   (C) Both A & B.
   (D) Neither A nor B.
3. Technician A says that the timing sprockets can be adjusted to compensate for timing chain wear. Technician B says that a chain tensioner is sometimes provided on some engines to compensate for timing chain wear. Who is right?
   (A) A only.
   (B) B only.
   (C) Both A & B.
   (D) Neither A nor B.
4. Technician A says that a timing belt cover houses the crankshaft oil seal. Technician B says that a timing belt cover contains oil and protects the timing belt mechanism. Who is right?
   (A) A only.
   (B) B only.
   (C) Both A & B.
   (D) Neither A nor B.
5. Which of the following is another name for the harmonic balancer?
   (A) Intermediate pulley.
   (B) Vibration damper.
   (C) Crankshaft damper.
   (D) Both B & C.
6. Technician A says that an engine's intake manifold helps warm the fuel charge when the engine is cold. Technician B says that an engine's intake manifold carries the air-fuel mixture or air to the engine's cylinder block. Who is right?
   (A) A only.
   (B) B only.
   (C) Both A & B.
   (D) Neither A nor B.
7. Technician A says that cast iron intake manifolds are commonly used on modern engines because they increase the fuel efficiency. Technician B says that aluminum intake manifolds are normally used on modern automobile engines because they help lower fuel consumption. Who is right?
   (A) A only.
   (B) B only.
   (C) Both A & B.
   (D) Neither A nor B.
8. Technician A says that an inline intake manifold is commonly used on V-type passenger car engines. Technician B says that a ram intake manifold is normally used on V-type passenger car engines. Who is right?
   (A) A only.
   (B) B only.
   (C) Both A & B.
   (D) Neither A nor B.
9. All of the following are basic parts of an engine's intake manifold EXCEPT:
   (A) water jackets.
   (B) riser bore.
   (C) cylinder mating surface.
   (D) runners.
10. Technician A says that several modern engine head gaskets are aluminum coated to improve engine warm-up. Technician B says that many modern head gaskets are Teflon® coated to help prevent head gasket failure. Who is right?
    (A) A only.
    (B) B only.
    (C) Both A & B.
    (D) Neither A nor B.

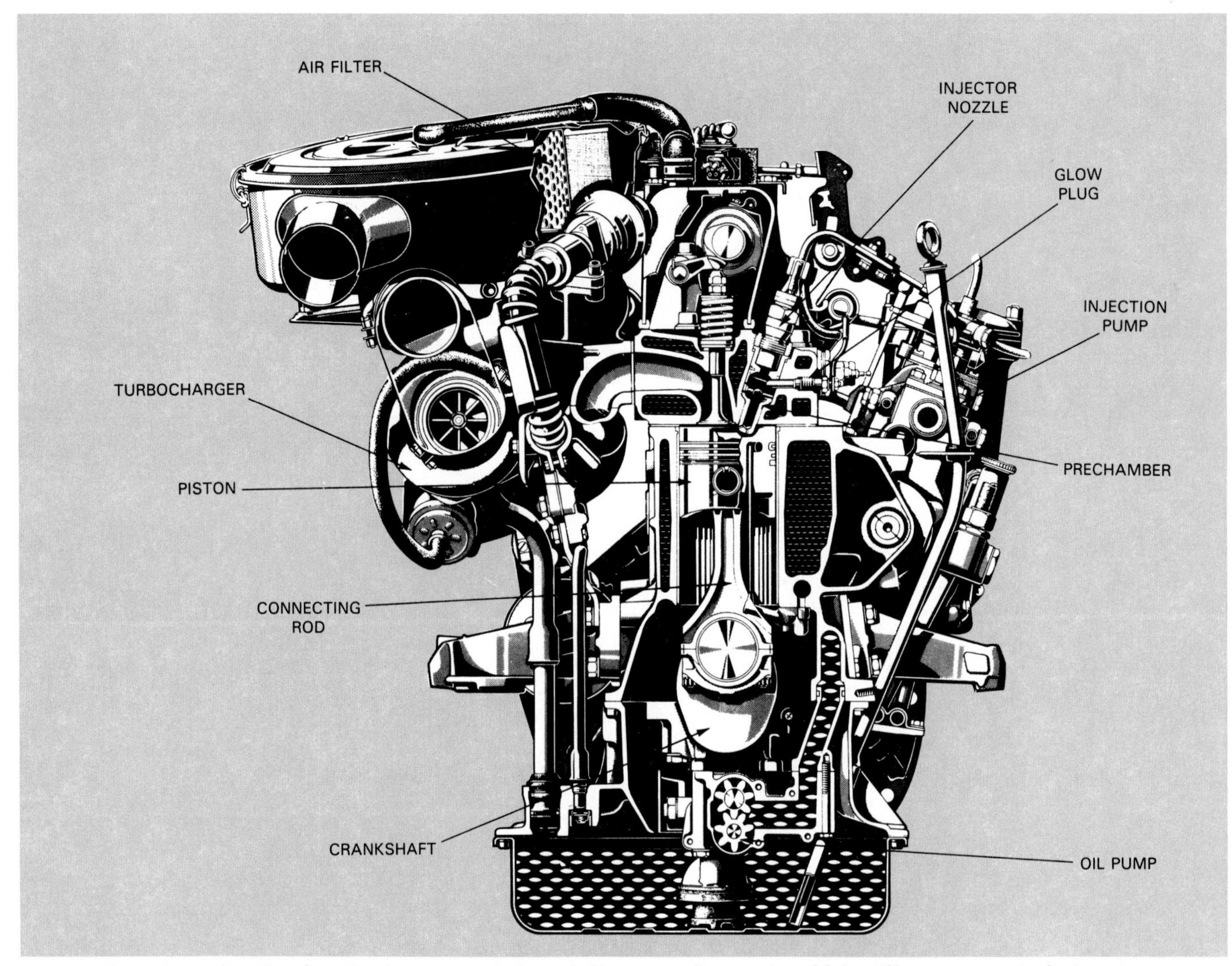

An automotive diesel engine is constructed much like a gasoline engine. Main differences are in fuel system. (Mercedes Benz)

Chapter 12

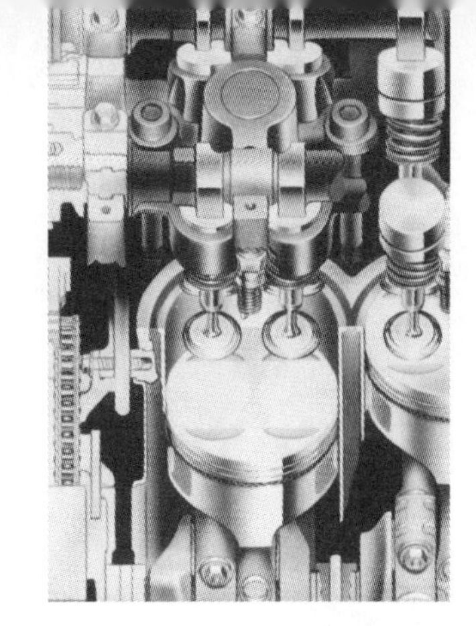

Diesel Engine Construction

After studying this chapter, you will be able to:
- *Compare the operation and construction of gasoline and diesel engines.*
- *Explain the need for the higher compression ratio of a diesel.*
- *Describe why a diesel does not need a throttle valve.*
- *Compare gasoline and diesel engine cylinder heads.*
- *Summarize why various diesel engine components must be made stronger than gasoline engine components.*
- *Explain the engine provisions needed for the injection pump.*
- *Be better prepared to understand later text chapters on diesel injection and engine service.*

This chapter will discuss the construction of automotive diesel engines. It will do this by comparing gasoline and diesel engines. By understanding design and construction differences, you will be better prepared to service a diesel engine and comprehend later textbook chapters.

Note! This chapter concentrates on engine construction, NOT fuel system components. Refer to Chapter 18 for more information on diesel injection.

DIESEL ENGINE OPERATION

Covered briefly earlier, automotive diesel engines are four-stroke cycle engines. Just like gasoline engines, they use four movements of the piston to complete one cycle or series of events. Two-cycle diesel engines are limited to larger truck and industrial applications.

Fig. 12-1 shows a basic four-stroke cycle for a simplified diesel engine.

No throttle valve

A diesel engine does NOT use a throttle valve to control airflow into the cylinders. A full charge of air is allowed to enter the intake manifold at all engine speeds.

The amount of fuel injected into the cylinders controls engine power and speed. The driver's gas pedal is linked to the injection pump. The injection pump, by controlling how much fuel enters the cylinder, lets the driver control diesel engine operation.

As you learned, a gasoline engine needs a throttle valve. The throttle valve(s) can be in the carburetor, or the throttle body of a fuel injected engine. When the throttle is closed, air and fuel flow is limited to reduce rpm, Fig. 12-2.

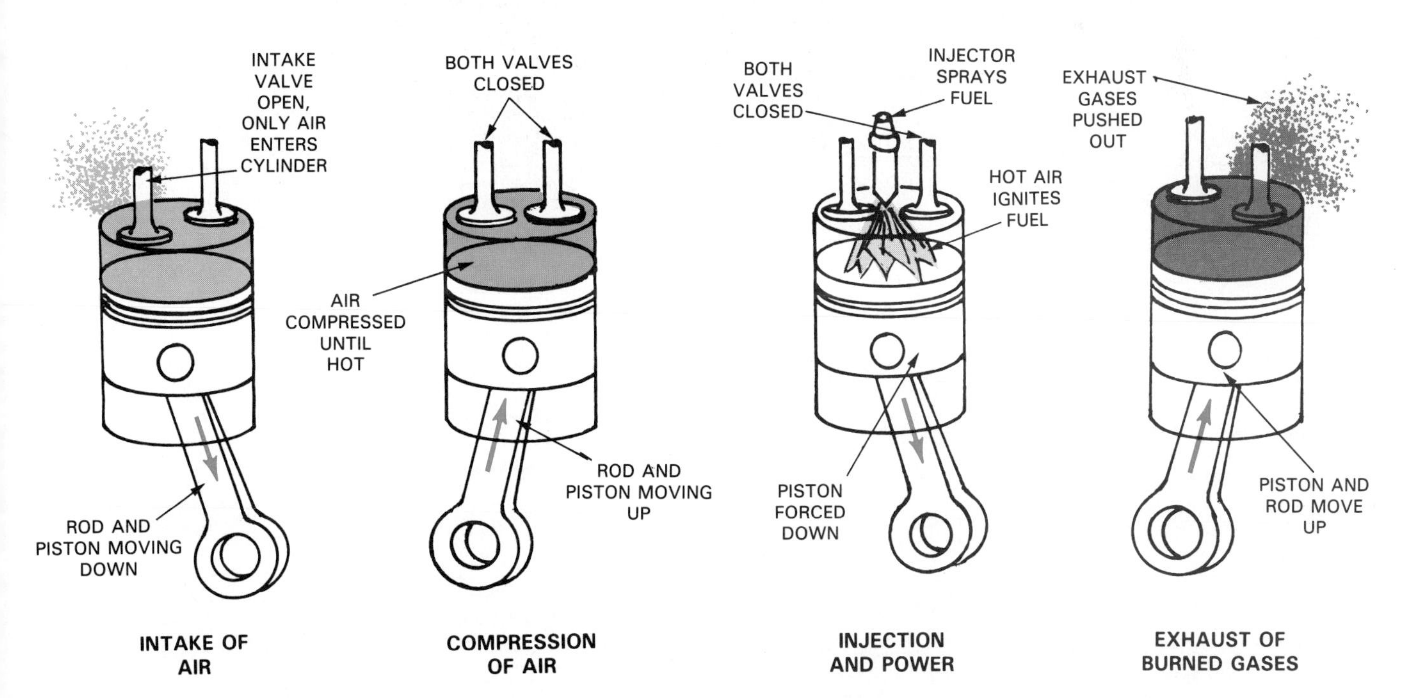

Fig. 12-1. Review series of events during diesel engine operation. They are almost the same as a gasoline engine but only air is pulled into the cylinder and fuel is injected directly into the cylinder.

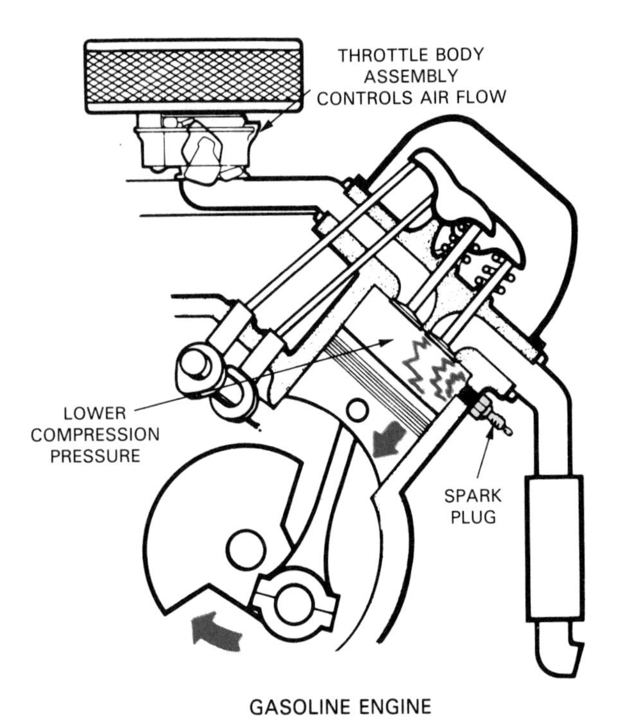

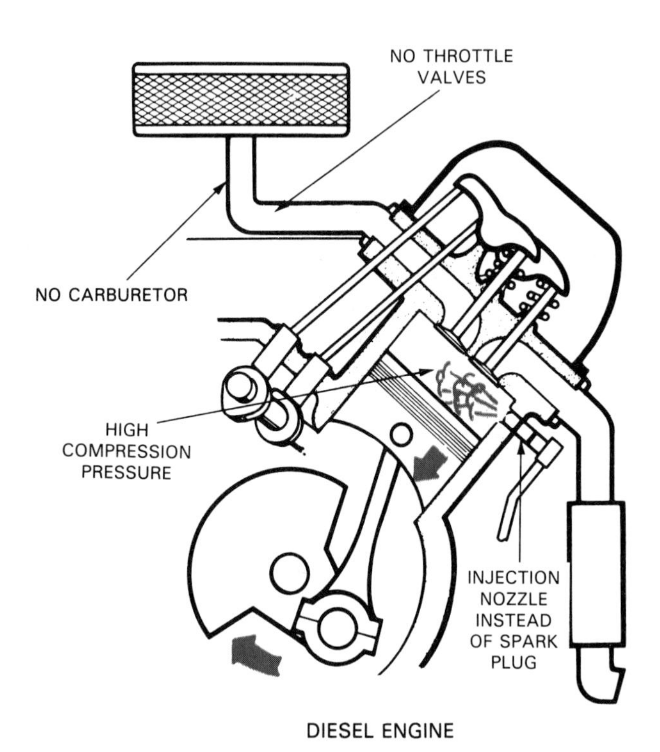

Fig. 12-2. Gasoline engine on top uses spark plug to ignite air-fuel mixture. Throttle valves in carburetor or throttle body control airflow and engine power output. Diesel engine, bottom, only draws air into cylinder on intake stroke. Air is compressed until it is hot enough to ignite fuel. No throttle valve is needed since amount of fuel injected into cylinder controls diesel engine power. (General Motors)

No intake manifold vacuum

A diesel engine does NOT produce intake manifold vacuum as does a gasoline engine. This is because it does not have a throttle valve to limit airflow into the engine. The large opening at the inlet of the intake manifold allows free flow of outside air into the engine. If you were to connect a vacuum gauge to the intake, it would read almost zero.

For this reason, a vacuum pump is commonly needed on diesel engines. The vacuum pump produces the suction needed to operate vacuum devices (power brake booster, vacuum switches, etc.).

No spark plugs

A diesel engine also does NOT use spark plugs as does a gasoline engine. An electrical-electronic ignition system is NOT needed because compression stroke heat ignites the fuel, Fig. 12-2.

Fig. 12-3 compares the operation of gasoline and diesel engines. Note the similarities and differences between the two.

A diesel is considered a *compression ignition* engine because the hot air in the cylinder makes the fuel burn. A gasoline engine is a *spark ignition* engine because it uses an electric arc to ignite its fuel.

Direct injection

A diesel is also a *direct injection* engine because it sprays fuel right into the combustion chambers, Fig. 12-4. With a gasoline engine, fuel is injected into the intake manifold (indirect gasoline injection) or pulled into the intake manifold (carburetion) by engine vacuum.

Compresses only air

A diesel only compresses air on its compression stroke. This is unlike a gasoline engine that compresses both air and fuel before the power stroke.

Higher compression ratio

A diesel engine uses a much higher compression ratio that a gasoline engine. A diesel might have a compression ratio of 17:1 to 23:1. A gasoline engine has a lower compression ratio of about 8 or 9:1. See Fig. 12-5.

Leaner air-fuel mixture

A diesel engine uses a much leaner air-fuel mixture than a gasoline engine. A diesel might have an air-fuel ratio of about 20:1 to 100:1. A gasoline engine typically has an air-fuel ratio of 13 to 17:1. This is illustrated in Fig. 12-6.

A diesel's leaner fuel mixture is partially why it has higher fuel economy than a gasoline engine. On the other hand, a gasoline engine will normally produce more power than an equal size diesel.

DIESEL ENGINE CONSTRUCTION

Diesel engines are constructed much like gasoline engines. However, they are made much heavier to withstand the higher operating pressures produced by compression ignition.

The compression stroke pressure of a gasoline engine is only about 125 to 175 psi (860 to 1200 kPa). A diesel can have as much as 275 to 400 psi (1895 to 2755 kPa) on its compression stroke.

Since the fuel *autoignites* (burns from heat and pressure) in a diesel, pressures are also much higher on

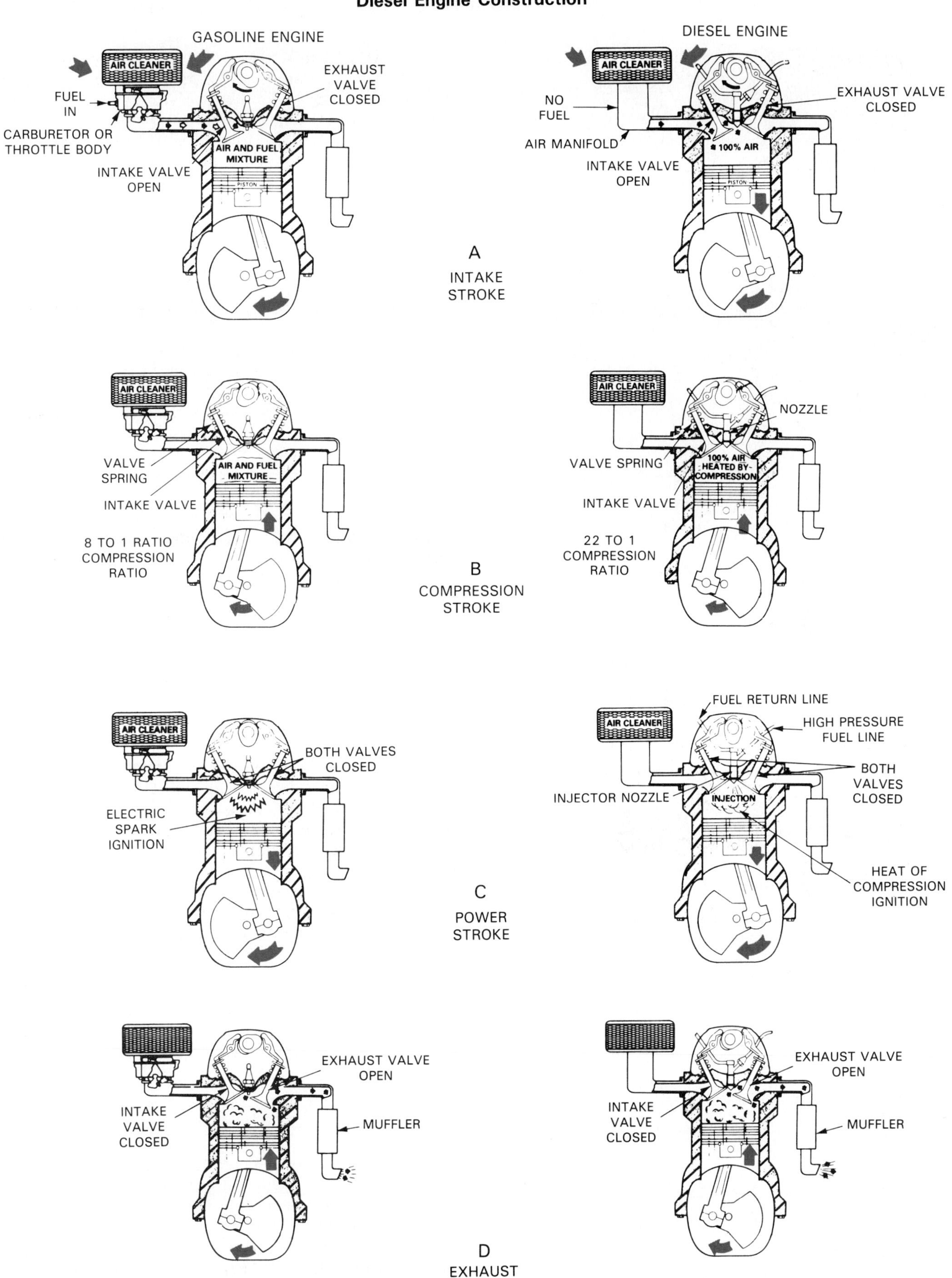

Fig. 12-3. Contrast and compare diesel engine and gasoline engine operation through four-stroke cycle. A—Intake stroke. B—Compression stroke. C—Power stroke. D—Exhaust stroke. (GM)

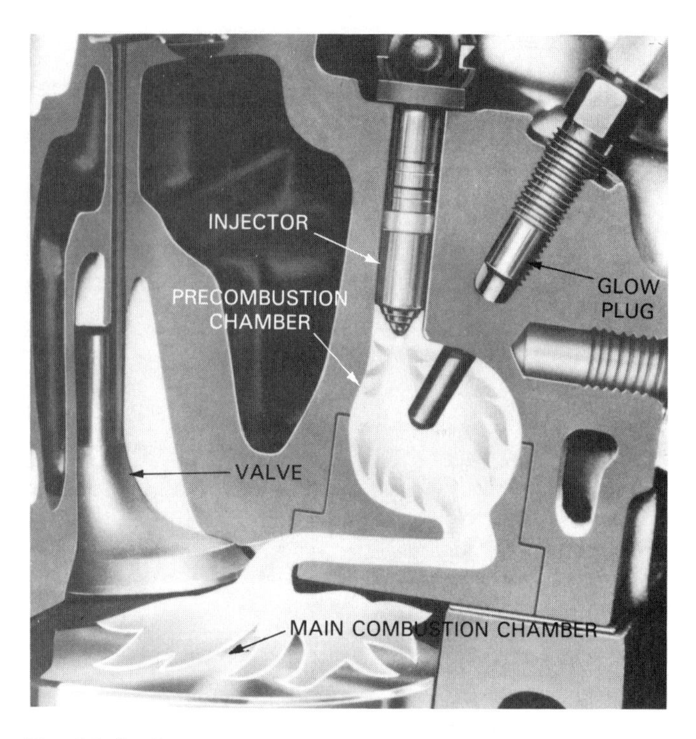

Fig. 12-4. Cutaway of diesel shows how injector sprays fuel right into combustion chamber. Hot air in chamber makes fuel ignite and burn. Glow plug simply heats air when engine is cold to aid starting. (Oldsmobile)

GASOLINE ENGINE

MINIMUM CYLINDER VOLUME

MAXIMUM CYLINDER VOLUME

8.5 TO 1

DIESEL ENGINE

SMALLER MINIMUM CYLINDER VOLUME

MAXIMUM CYLINDER VOLUME

17 TO 1

Fig. 12-5. Higher compression ratio of diesel is how fuel ignites. By squeezing air in cylinder into a tiny space, air heats up beyond the ignition point of diesel oil. Gasoline engine has much lower compression ratio so spark plug initiates combustion at right time. (Ford)

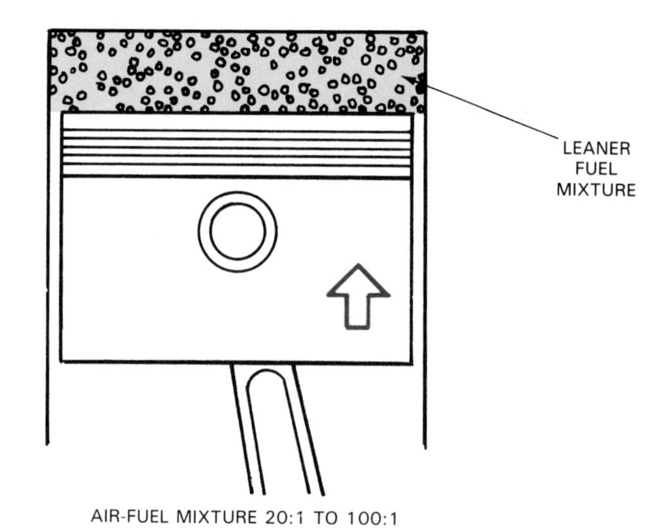

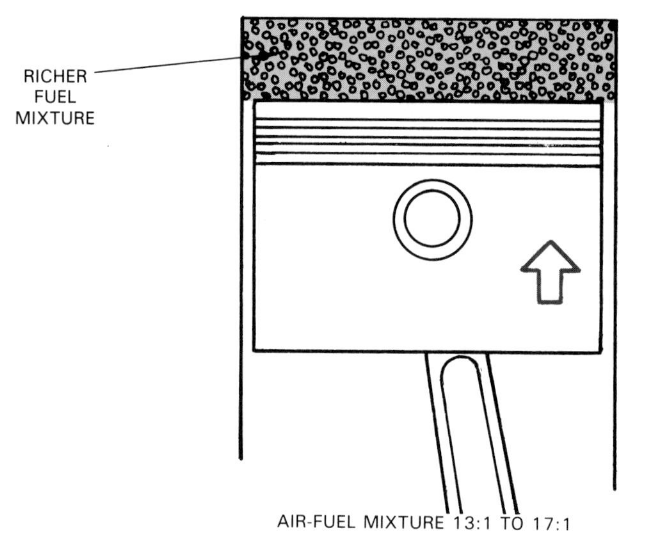

Fig. 12-6. Diesel also uses much leaner air-fuel mixture than a gasoline engine. This contributes to better fuel economy.

the diesel power stroke. This is illustrated in Fig. 12-7.

This section of the chapter will show how diesel engine components are designed to withstand these higher operating pressures. Most diesels are designed to give an extremely long service life before needing major repairs—well over 100,000 miles (160 000 km).

Diesel pistons

Diesel pistons, especially the piston heads, are thicker and heavier than pistons for gasoline engines. The top of the piston must withstand the heat and pressure as the diesel oil rapidly burns.

Fig. 12-8 compares typical pistons for gasoline and diesel engines. Note how the diesel piston has a pocket. This pocket would be located directly below the precombustion chamber outlet. It allows the combustion flame to spread into the main chamber even with the piston at TDC.

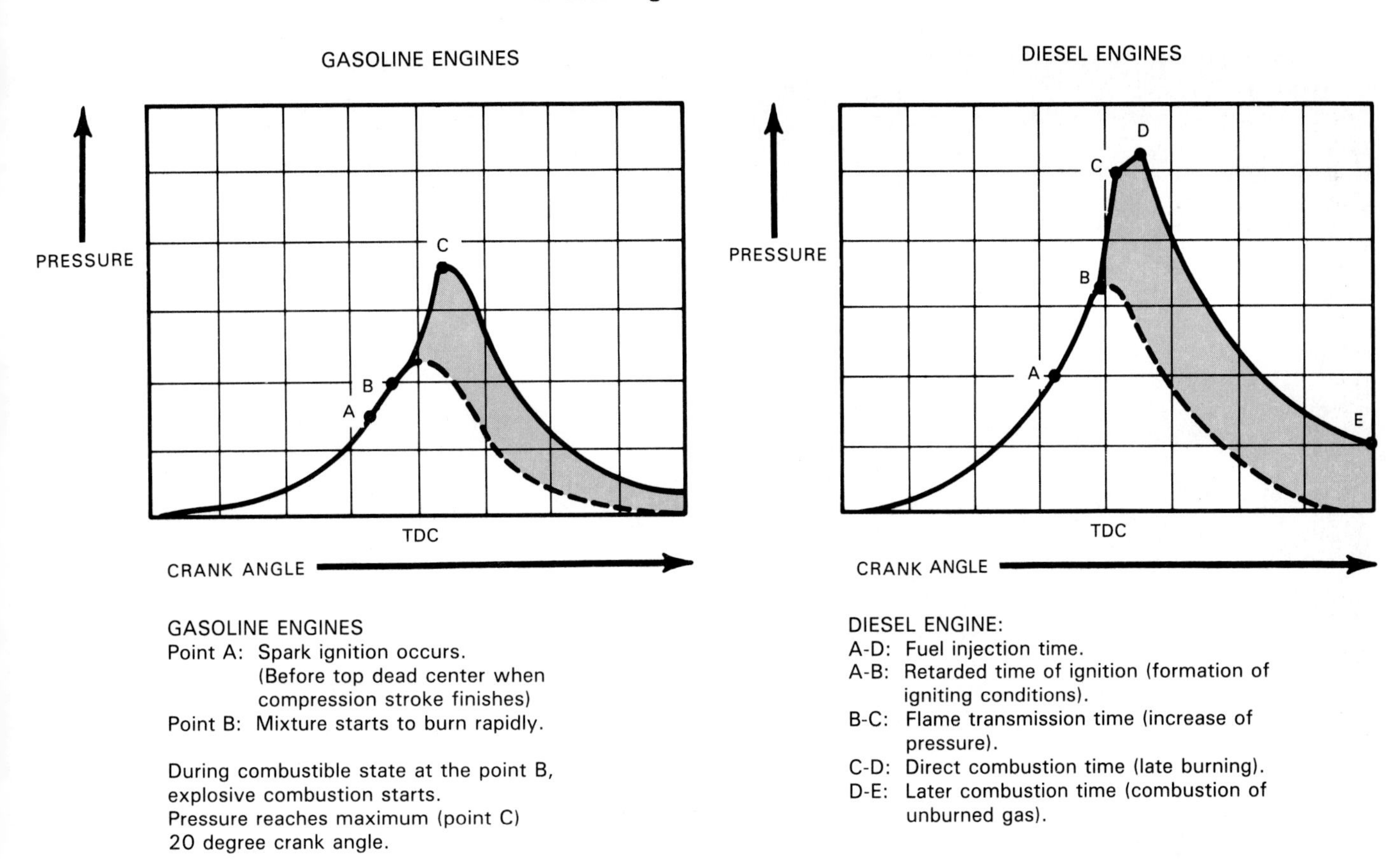

Fig. 12-7. Graphs show how cylinder pressures differe in gasoline and diesel engines. High pressures in diesel require engine construction to be much stronger. (Ford)

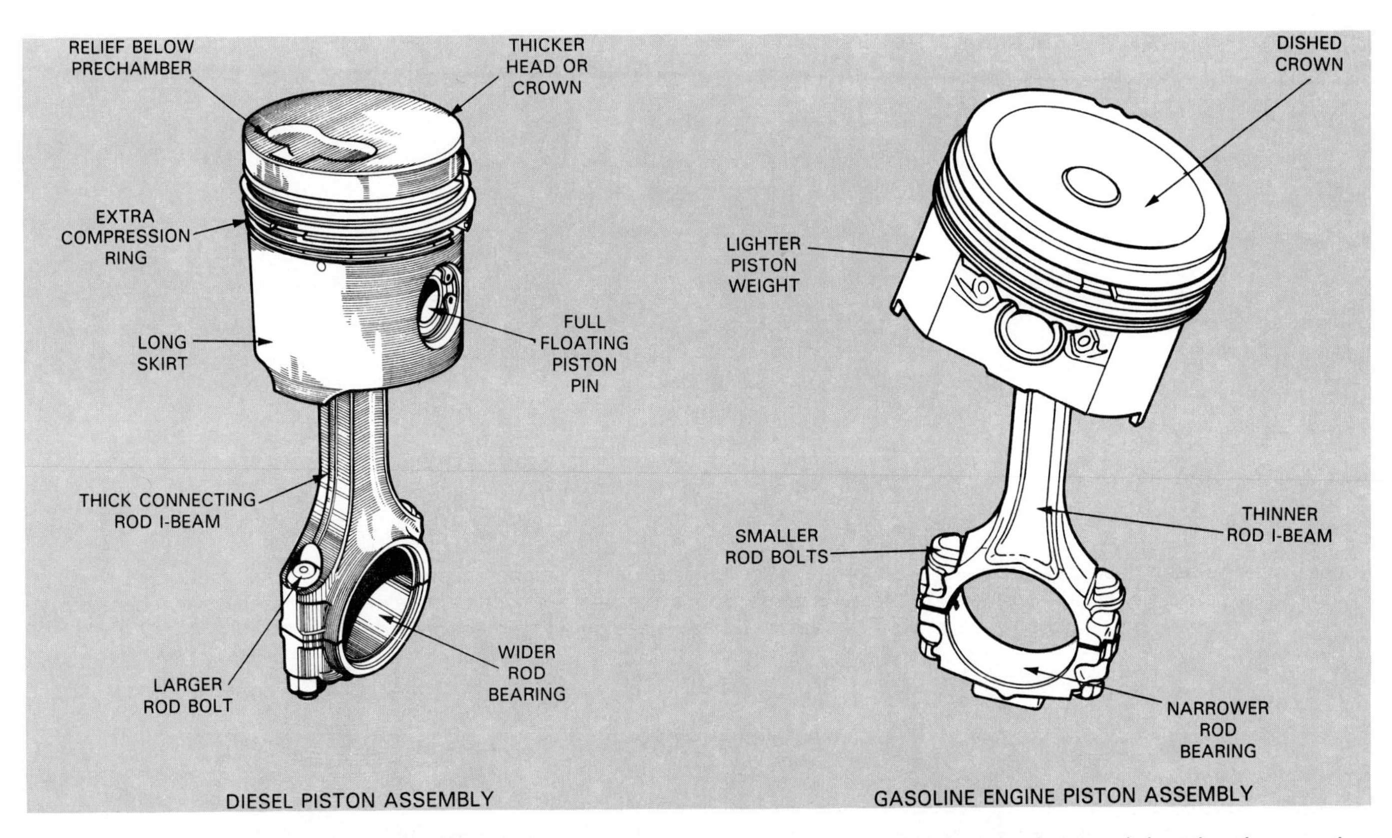

Fig. 12-8. Compare diesel and gasoline engine piston assemblies. Note how diesel engine has much heavier piston and connecting rod. Diesels operate at much lower rpm so extra reciprocating weight is acceptable. Diesel piston is also designed to produce high compression stroke pressure. (Mercedes Benz)

The piston for the gasoline engine is more dished in this example, Fig. 12-8. This would result in less compression stroke pressure so that the gasoline does not ping or detonate.

Diesel connecting rods

Diesel connecting rods are also thicker and heavier than gasoline engine connecting rods. A diesel engine operates at much lower speeds. A governor is used to prevent excessive engine rpm and resulting damage.

For this reason, the connecting can be made heavier to withstand higher loads. A lighter connecting rod is needed in a gasoline engine so that the engine can run at higher rpm. A heavy connecting rod, in a higher rpm engine, could pound out and ruin the rod bearings. Since a diesel runs at low rpm, the large reciprocating mass of the piston and rod does not damage the rod bearings. Refer to Fig. 12-8.

Diesel piston protrusion

Diesel piston protrusion occurs at TDC when the head of the piston actually moves slightly out the top of the block. This can be used in a diesel to produce the needed compression stroke pressure. Look at Fig. 12-9.

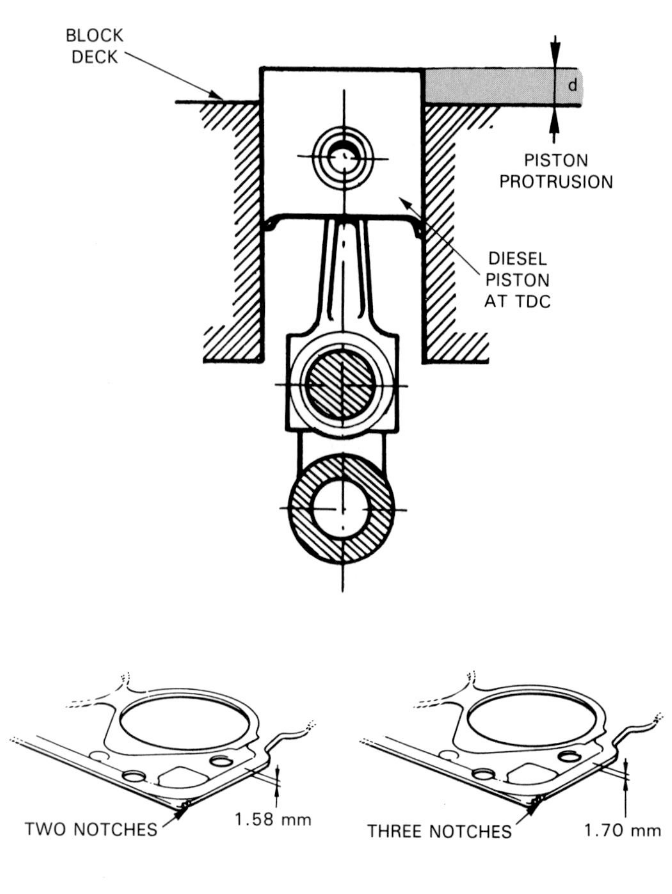

Fig. 12-9. Piston protrusion is common with automotive diesel engines. Piston head actually sticks out of block deck at TDC. For this reason, head gasket thickness is very critical. Different thicknesses of head gaskets are provided. Notches are frequently used on head gasket to denote thickness. (Peugeot)

Diesel engine head gaskets

Diesel engine pistons come very close to the cylinder head at TDC. This compresses the air enough for combustion of the diesel oil.

Diesel engine head gaskets frequently come in different thicknesses. Notches or other markings on the gaskets can be used to denote head gasket thickness. A thinner head gasket would raise compression pressure and reduce piston-to-head clearance. A thicker head gasket would reduce compression pressure and increase piston-to-head clearance, Fig. 12-9.

Different head gasket thicknesses allow the mechanic to adjust the clearance between the top of the piston and the cylinder head. By measuring piston protrusion, the mechanic can make sure the piston does not come too close or is not too far away from the cylinder head.

Diesel head gasket selection is covered in detail in later textbook chapters. Refer to the index for more information.

Diesel air inlet

As mentioned, an unrestricted supply of air is allowed to enter a diesel at all speeds. As shown in Fig. 12-10, air flows through the air cleaner, sometimes an air crossover, and then through the intake manifold. Neither a throttle body nor a carburetor are mounted on the top of the intake manifold.

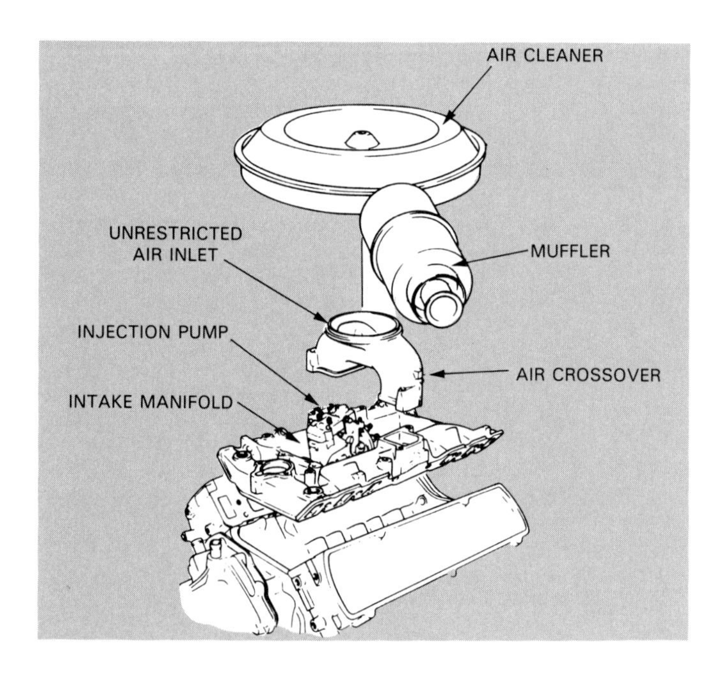

Fig. 12-10. A V-6 or V-8 diesel uses intake system similar to this one. Muffler on air cleaner is to reduce noise as large amounts of air flow into engine. Air crossover allows unrestricted flow of outside air to enter intake manifold at all speeds. (Oldsmobile)

Diesel cylinder heads

Diesel cylinder heads vary considerably from those used on gasoline engines. A spark plug screws into a threaded hole in a gasoline engine cylinder head. With

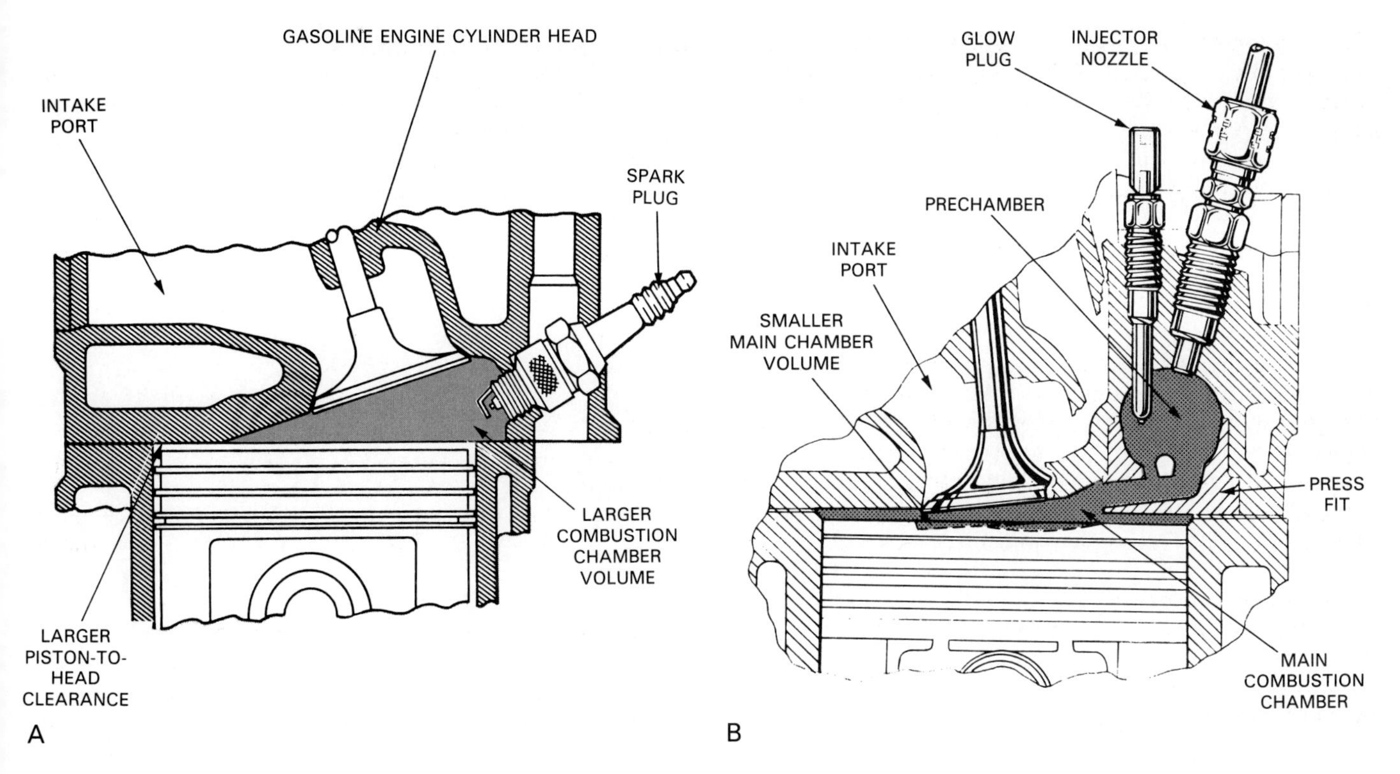

Fig. 12-11 Compare cylinder heads for gasoline and diesel engines: A—Gasoline engine uses spark plug and has fairly large combustion chamber to hold air-fuel mixture that enters from intake port. B—Diesel has much smaller combustion chamber area. Prechamber fits into head and encloses tips of injector nozzle and glow plug. (Ford and Oldsmobile)

a diesel, glow plugs, injectors, and precombustion chambers are installed in the cylinder head.

Fig. 12-11 shows cutaway views of gasoline and diesel engine cylinder heads. Compare them.

Note that in a diesel, the glow plugs and injectors usually screw into threaded holes in the head. Both the tips of the glow plugs and injectors protrude into the precombustion chamber.

Precombustion chambers, also called prechambers or just cylinder head cups, press-fit into the deck surface of the head, Fig. 12-12. Precision holes are machined in the head. The cups are slightly larger than these holes. Therefore, they lock in place when pressed into the cylinder head, Fig. 12-13.

Fig. 12-14 shows the top view of a diesel cylinder head. Note the threaded holes for the injectors and glow plugs.

Fig. 12-15 shows a cutaway view of a diesel engine. Locate the glow plug and injector.

Diesel engine cylinder block

A diesel engine cylinder block is constructed like a gasoline engine cylinder block. However, it is made thicker in high stress areas. Frequently, four-bolt mains will be used to secure the crank in the block, Fig. 12-16.

The sound of a running diesel engine gives an indication of why the block must be made stronger. The clattering sound of a diesel is caused by the fuel igniting or

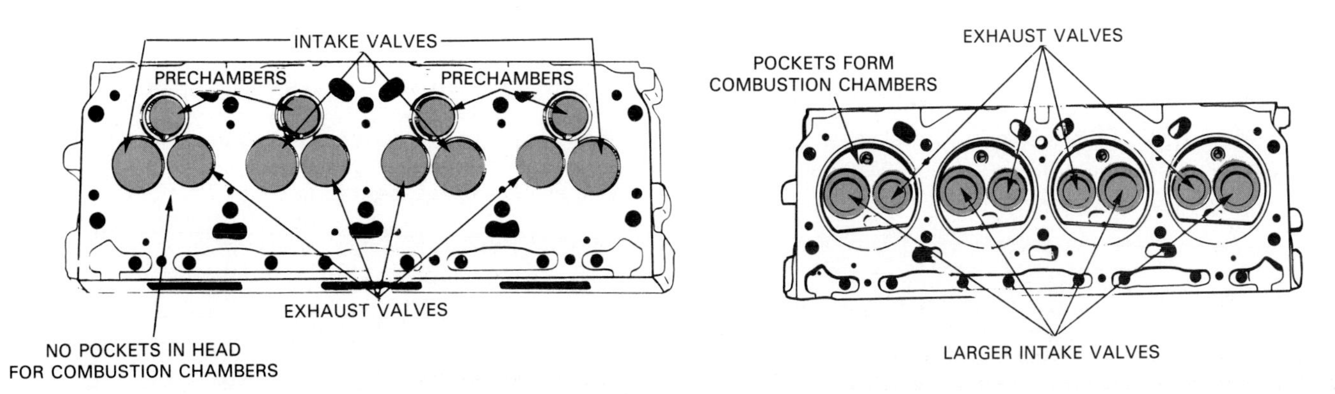

Fig. 12-12. Compare deck sides of gasoline and diesel cylinder heads. Note on diesel cylinder head how prechambers install in head. Intake and exhaust valves are also closer to the same size than in gasoline engine. Bottom of diesel head is also flat with absence of pockets forming combustion chambers. (Oldsmobile)

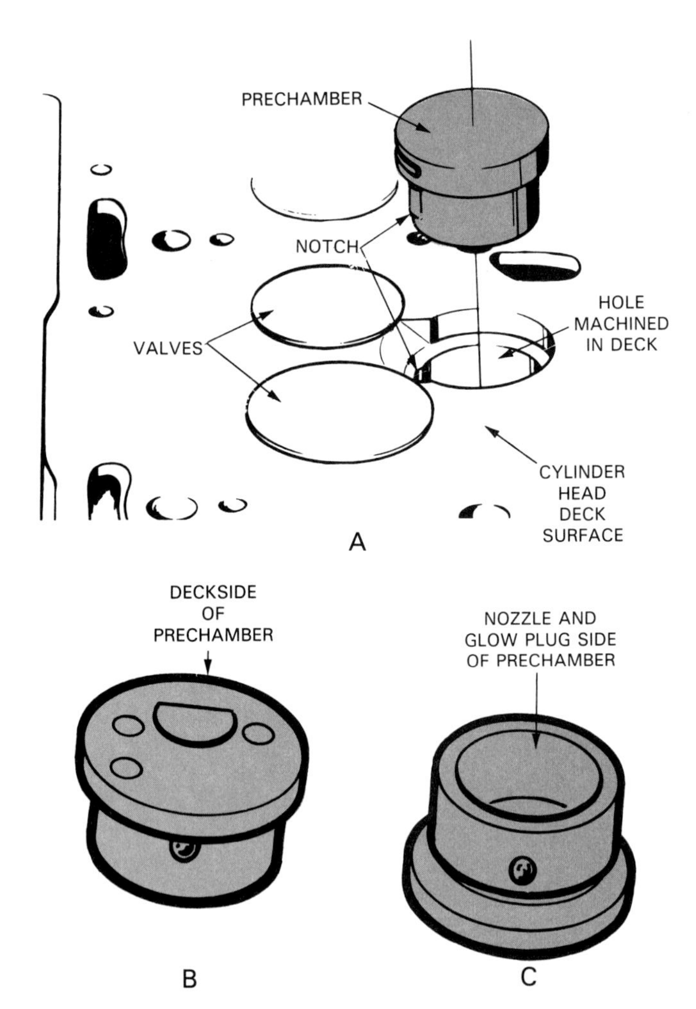

Fig. 12-13. A—Precombustion chamber is press-fit in deck of diesel cylinder head. Notch assures correct alignment. B—Deck side of prechamber has small opening so burning fuel can blow out into main combustion chamber. C—Other side has larger opening for tips of injector and glow plug. (General Motors and Peugeot)

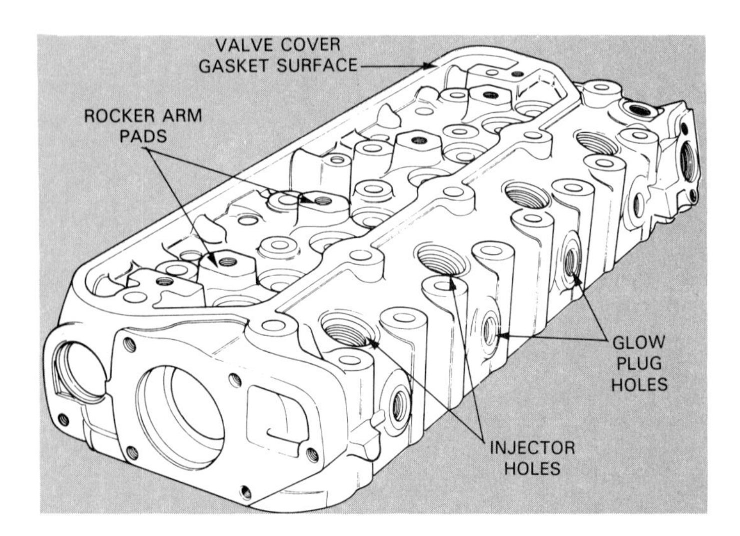

Fig. 12-14. Top view of diesel cylinder head shows threaded holes for injectors and glow plugs. It will have thicker construction to prevent combustion leakage and damage from extreme pressures in cylinders. (Peugeot)

Fig. 12-15. Phantom view of diesel engine shows major components. (Ford)

"almost exploding" in the cylinders. Much higher cylinder pressures result. The cylinder walls must be thicker to withstand the added pressure.

Diesel roller lifters

Many automotive diesel engines use roller lifters to cut friction between the lifter and camshaft lobes. They are almost identical to roller lifters used in some high performance gasoline engines. The engine in Fig. 12-16 has roller lifters.

Diesel engine valves

Diesel engine valves are also similar to gasoline engine valves. However, the intake and exhaust valves are closer to the same size. The intake valve in a gasoline engine must be much larger than the exhaust valve because both fuel and air flow through the intake port and because the engine operates at higher rpm. Refer back to Fig. 12-12.

DIESEL LUBRICATION SYSTEM

A diesel lubrication system may use piston cooling jets. These are small tubes that direct a spray of engine oil onto the bottom of the pistons. This helps cool the pistons and increases engine dependability.

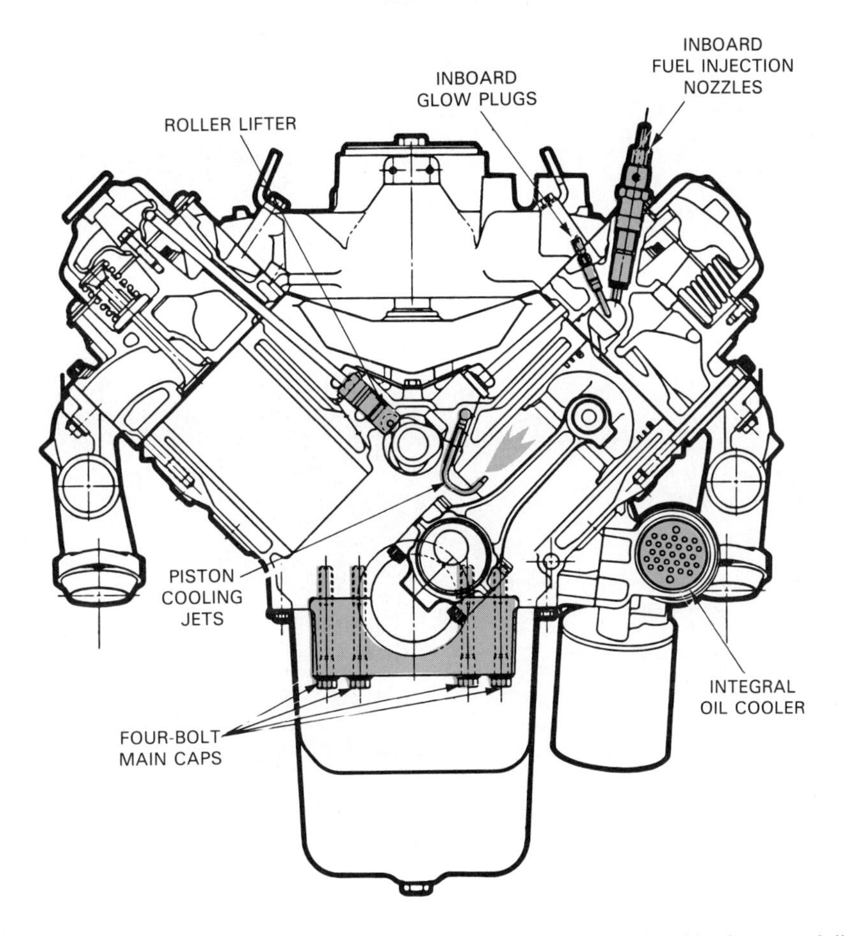

Fig. 12-16. Diesel cylinder block is also made much heavier than gasoline engine block, especially the cylinder walls. Note oil jet for cooling pistons, four-bolt mains, roller lifters, and oil cooler. (Ford Motor Company)

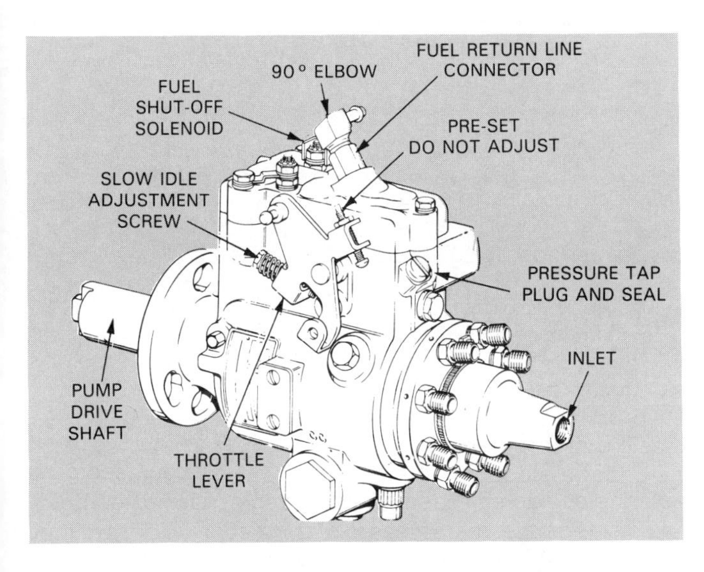

Fig. 12-17. Injection pump is primary control of diesel engine, as will be detailed in a later chapter. (Oldsmobile)

An oil cooler is also found on many automotive diesel engines, Fig. 12-16. It is smaller to oil coolers found on high performance gasoline engines. The oil cooler helps maintain oil viscosity (thickness) to protect engine components from friction and damage.

DIESEL STARTING SYSTEM

A diesel engine starting system must be more powerful than a starting system for a gasoline engine. The higher compression ratio makes it much more difficult for the starting motor to spin the crankshaft. Also, the diesel relies on high cranking speeds for engine starting. The crankshaft must be turned at a minimum rpm to produce enough compression stroke heat to start combustion.

A diesel starting system normally uses a high torque starting motor. Either a high power battery or twin batteries are needed to feed enough current to the diesel starting motor.

OTHER DIESEL SYSTEMS

Other systems (charging, cooling, etc.) are about the same as those found on gasoline engines. They will be explained in later chapters.

DIESEL INJECTION PUMP

A diesel injection pump is normally mounted on and driven by the engine, Fig. 12-17. Provisions are made on the engine for the injection pump. The injection pump can be driven by gears, a chain, or by a cogged belt.

Fig. 12-18 shows how an injection pump drive gear

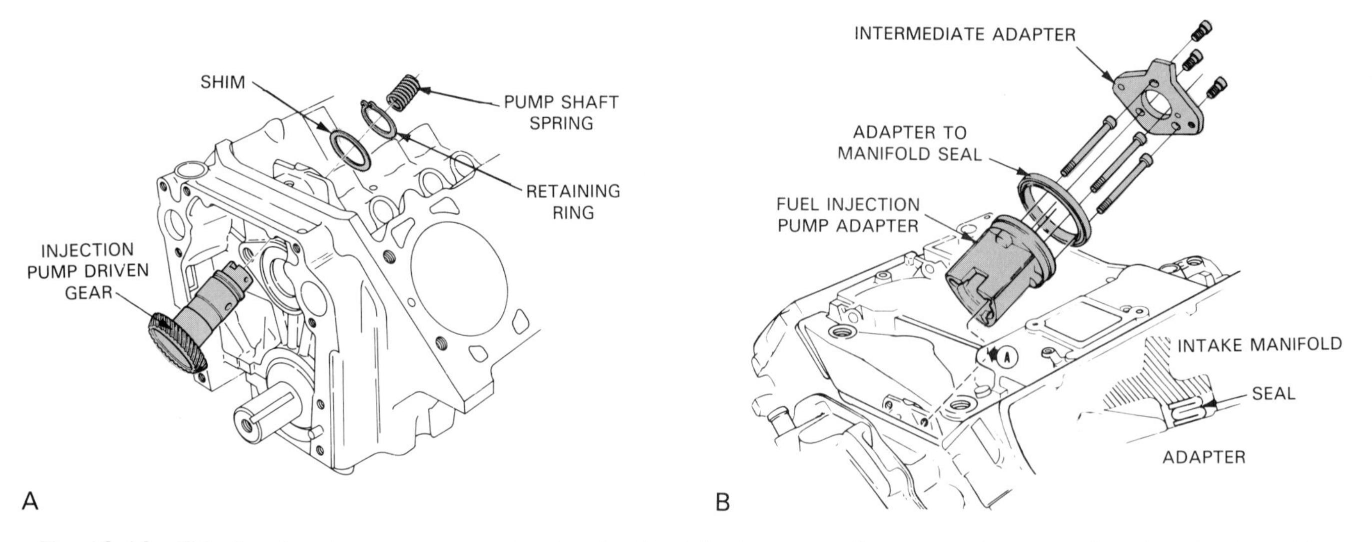

Fig. 12-18. This diesel engine uses gear mechanism to drive injection pump. Pump mounts on top of engine. A—Gear drive mechanism. B—Adapter for injection pump. (Oldsmobile)

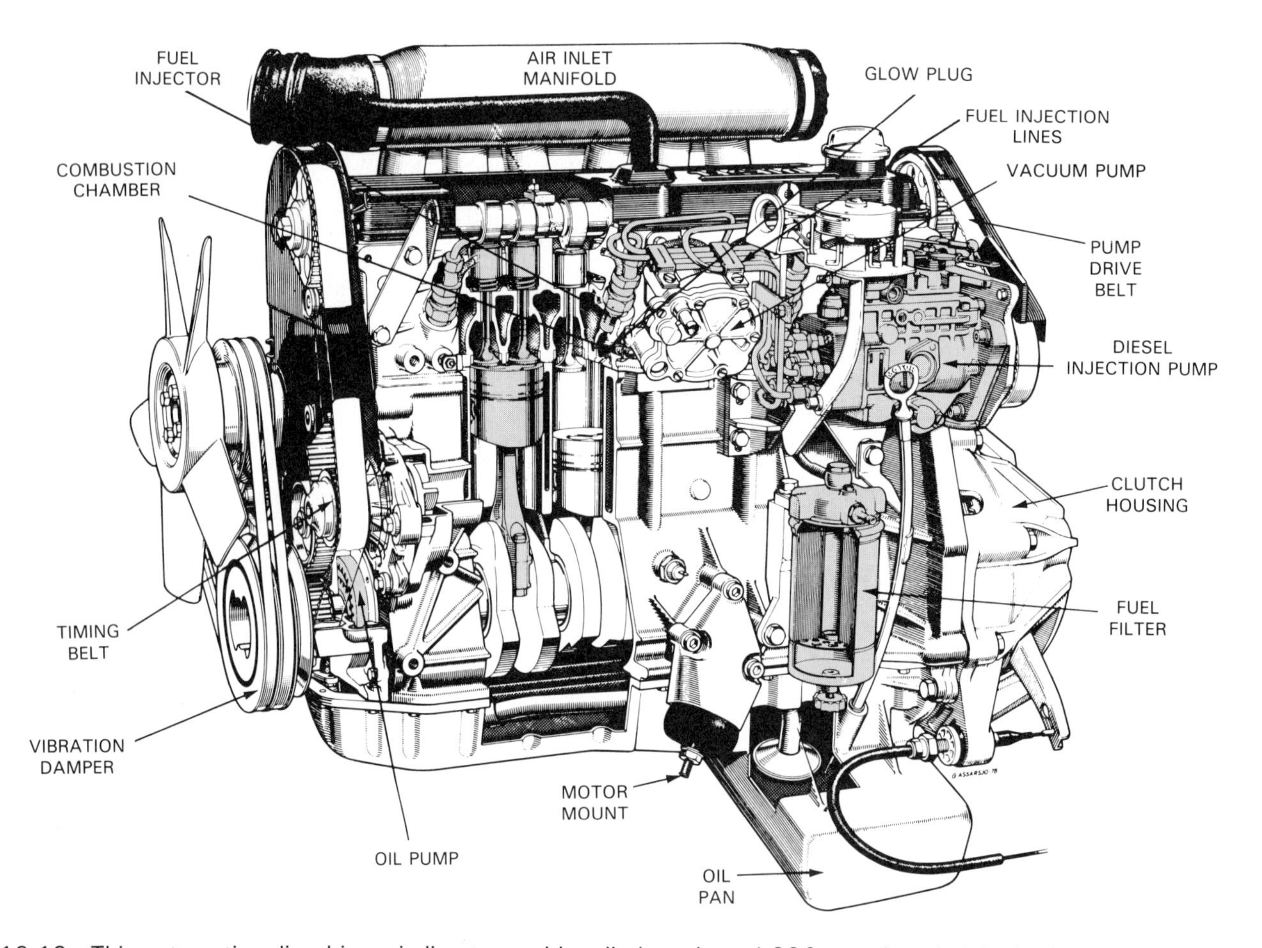

Fig. 12-19. This automotive diesel is an in-line type with cylinders slanted 20° to reduce height for hood clearance. It uses cast iron cylinder block and aluminum cylinder head. (Volvo)

is installed on the front of one type diesel engine. A shaft extends up through the block. The injection pump, mounted on top of the engine, is turned by this gear and shaft. Adapters are frequently used for pump mounting.

DIESEL ENGINE DESIGN VARIATIONS

Numerous design variations are found on today's automotive diesel engines. Figs. 12-19 and 12-20 show

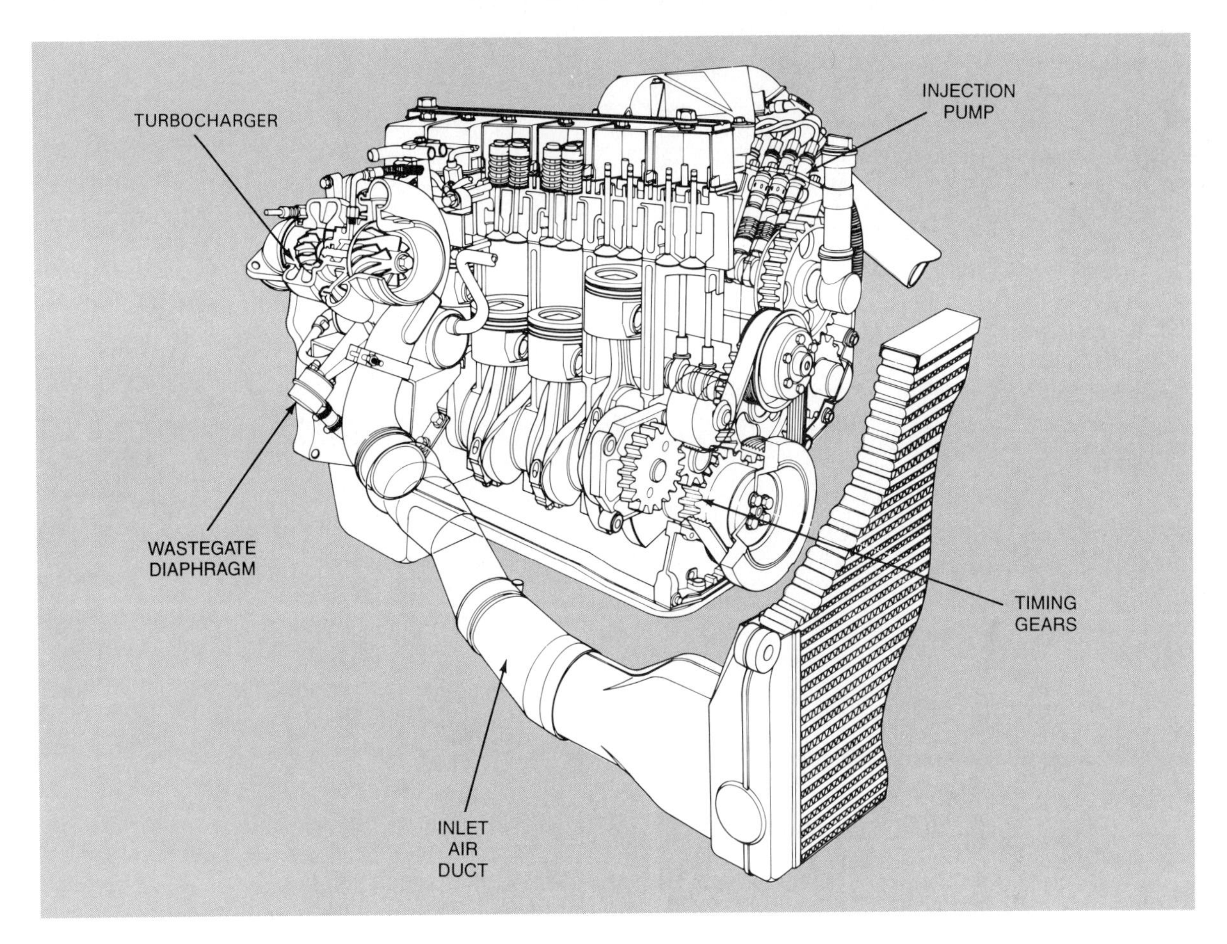

Fig. 12-20. Many automotive diesel engines now use turbocharging to increase power output. This is a 5.9L, inline, six-cylinder diesel for a pickup truck. (Dodge)

different diesel engines. Study the part names and note their construction.

Later textbook chapters will explain diesel engine construction and operation in more detail. As you will learn, if you can work on a gasoline engine, it does not take that much more knowledge to service diesel engines. Most of the components are the same and require similar repair or rebuild techniques.

SUMMARY

Auto diesel engines are four-stroke cycle engines. Two-stroke cycle diesels are only used in larger industrial and truck engine applications.

A diesel does not use a throttle valve nor spark plugs. The injection pump controls engine speed and power. High compression stroke pressure and the resulting heat starts combustion in a diesel.

Fuel is injected directly into the engine cylinders. The hot air in the cylinders makes the diesel oil begin to burn and expand. A diesel has a compression ratio of about 17:1 to 23:1. It also uses a very lean air-fuel mixture.

Generally, many of the parts of a diesel engine are the same as those in a gasoline engine. However, they are constructed for heavy duty operation. The pressures in the diesel cylinders are much higher. The block, pistons, connecting rods, crankshaft, etc. must be able to withstand the increased loads.

Diesel pistons are designed to help produce a high compression ratio. The heads of the pistons are almost flat and they can protrude out of the top of the block. Piston heads must be very thick to withstand combustion heat and pressure.

Diesel connecting rods are thicker and heavier than those in a gasoline engine. A diesel operates at much lower speeds and a heavier rod can be used without adverse effects.

Diesel engine cylinder heads are quite different from gasoline engine cylinder heads. The head of a diesel must hold the prechambers, glow plugs, and fuel injectors. It too must be thick to withstand combustion pressure.

Diesel head gaskets come in various thicknesses. This lets the technician make sure the correct compression ratio rresults after engine reassembly.

Provisions must be provided on the engine for mounting and driving the diesel injection pump. Either gears, a chain, or cogged belt can be used to operate the injection pump.

KNOW THESE TERMS

Compression ignition, Spark ignition, Direct injection, Diesel compression ratio, Diesel air-fuel mixture, Compression stroke pressure, Piston protrusion, Head gasket thickness markings, Precombustion chamber.

REVIEW QUESTIONS—CHAPTER 12

1. Two-cycle diesel engines are NOT commonly used in automotive applications. True or false?
2. Which of the following are NOT used on automotive diesel engines?
   a. Spark plugs.
   b. Throttle valves.
   c. Ignition system.
   d. Glow plugs.
3. How is diesel engine speed and power controlled?
4. Why is diesel engine intake manifold vacuum almost zero?
5. A diesel engine is considered a:
   a. Spark ignition engine.
   b. Arc ignition engine.
   c. Compression ignition engine.
   d. Hydraulic ignition engine.
6. The compression ratio for a diesel is about ________ while a gasoline engine only has a compression ratio of about ________.
7. A diesel uses a richer air-fuel mixture than a gasoline engine. True or false?
8. Why is diesel engine construction much heavier than a gasoline engine?
9. Define the term "piston protrusion."
10. Why do diesel engine head gaskets come in different thicknesses?
11. Summarize the differences between the cylinder heads for diesel and gasoline engines.
12. List three methods of driving the diesel injection pump.

ASE CERTIFICATION-TYPE QUESTIONS

1. Technician A says that a full charge of air enters the engine's intake manifold while the engine is operating at high speeds. Technician B says that a full charge of air enters the diesel's intake manifold at all engine speeds. Who is right?
   (A) A only.
   (B) B only.
   (C) Both A & B.
   (D) Neither A nor B.
2. Technician A says that a gasoline engine uses a leaner air-fuel mixture than a diesel engine. Technician B says that a diesel engine uses a much leaner air-fuel mixture than a gasoline engine. Who is right?
   (A) A only.
   (B) B only.
   (C) Both A & B.
   (D) Neither A nor B.
3. Technician A says that most diesel engines are designed to give a service life of 50,000 miles, (80 000 km) before major repairs are needed. Technician B says that most diesel engines are designed to give a service of 100,000 miles (160 000 km) or more before major repairs are required. Who is right?
   (A) A only.
   (B) B only.
   (C) Both A & B.
   (D) Neither A nor B.
4. Technician A says that a thick head gasket reduces compression pressure and decreases piston-to-head clearance. Technician B says that a thick head gasket raises compression pressure and increases piston-to-head clearance. Who is right?
   (A) A only.
   (B) B only.
   (C) Both A & B.
   (D) Neither A nor B.
5. A diesel engine produces ________ intake manifold vacuum than a gasoline engine.
   (A) higher
   (B) lower
   (C) does not produce vacuum
   (D) the same
6. Technician A says that diesel engine piston heads are thinner than gasoline engine piston heads. Technician B says that a diesel engine piston normally has one more compression ring than a gasoline engine piston. Who is right?
   (A) A only.
   (B) B only.
   (C) Both A & B.
   (D) Neither A nor B.
7. Technician A says that two-bolt main caps are always used to secure the diesel crankshaft in the block. Technician B says that four-bolt main caps are sometimes used to secure the diesel crankshaft in the block. Who is right?
   (A) A only.
   (B) B only.
   (C) Both A & B.
   (D) Neither A nor B.
8. All of the following are basic components of a V-8 diesel intake system EXCEPT:
   (A) air crossover.
   (B) muffler.
   (C) throttle valve.
   (D) air cleaner.
9. Technician A says that a diesel engine's exhaust valve is much larger than the intake valve. Technician B says that a diesel engine's intake and exhaust valves are normally about the same size. Who is right?
   (A) A only.
   (B) B only.
   (C) Both A & B.
   (D) Neither A nor B.
10. Technician A says that a diesel engine injection pump is sometimes driven by a cogged belt. Technician B says that a diesel engine injection pump is sometimes chain driven. Who is right?
   (A) A only.
   (B) B only.
   (C) Both A & B.
   (D) Neither A nor B.

Chapter 13
Charging and Starting Systems—Operation and Service

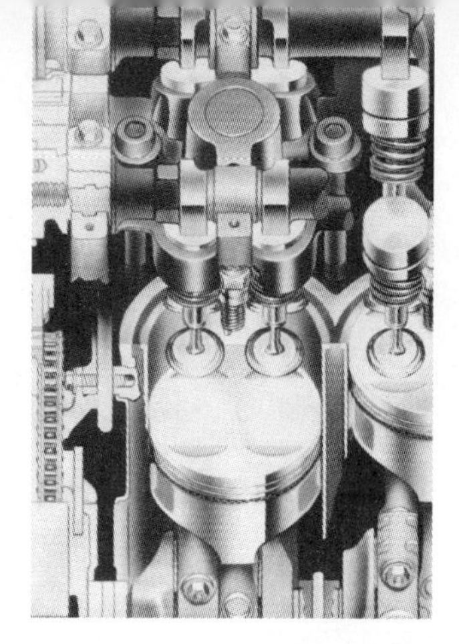

After studying this chapter, you will be able to:
- *Summarize the interactions of the battery, charging system, and starting system.*
- *Explain battery construction.*
- *Summarize the operation of a starting system.*
- *Describe starting motor and solenoid construction.*
- *Explain how a charging system works.*
- *Summarize the construction and operation of an alternator.*
- *Diagnose battery, starting system, and charging system problems.*
- *Perform common battery tests, charging system tests, and starting system tests.*
- *Jump start a car.*
- *Inspect for battery, starting system, and charging system troubles.*
- *Use a load tester.*
- *Replace a battery, starting motor, and alternator.*

As an engine technician, you must be able to do fundamental service operations on batteries, starting motors, alternators, and other engine related components. Today's engine technician must do more than just rebuild and repair the mechanical components inside an engine. He or she is frequently required to service the many systems and components that supplement engine operation.

Without a properly operating battery, alternator, and starting motor, an engine is useless. It will not crank and start! Also, during engine removal, the technician must remove the alternator and starter and properly install them when the engine is refitted in the vehicle.

The first half of this chapter covers the theory and construction of charging and starting systems. The last half of the chapter summarizes the service procedures most commonly done by an engine technician.

Note! Following text chapters will explain the operation and repair of other vital engine systems—cooling system, lubrication system, fuel system, computer systems, ignition system, etc. As a result, this section of the book will help prepare you to become a COMPLETE ENGINE TECHNICIAN.

ENGINE STARTING

Engine starting is dependent upon two main factors—the battery and starting system (starting motor, solenoid, ignition switch, and wiring). However, the charging system must assure proper battery charge and it also affects whether the engine will crank and start when the ignition key is turned.

As shown in Fig. 13-1, the battery feeds a huge amount of current to the starting motor during engine cranking. This drains power out of the battery at a tremendous rate. It takes considerable torque for the starting motor to rotate the engine crankshaft.

The alternator must feed current back into the battery after the engine starts and runs on its own power. It must ready the battery for another engine start-up.

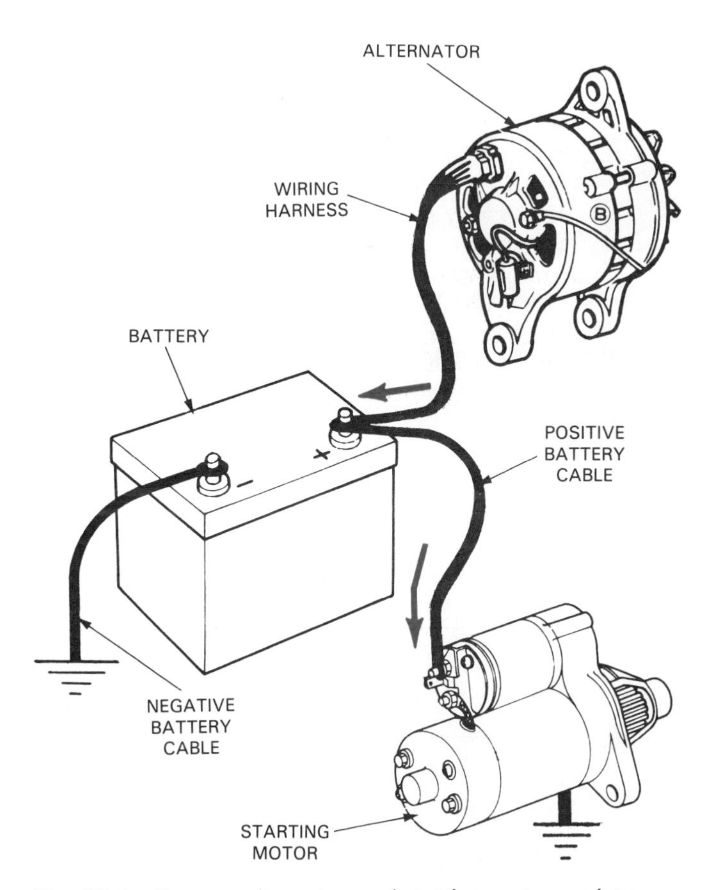

Fig. 13-1. Battery, alternator, and starting motor are interacting. One can affect performance of another component or system.

As you can see, these three main components or systems are interacting. The starting system relies on the battery for power. The battery relies on the charging system for energization.

When the driver turns the ignition switch, battery current flows to the starting motor. A gear on the starter then engages the flywheel ring gear. This turns the flywheel and the engine goes through its four-stroke cycle until combustion power takes over. Then the engine runs on its own and the alternator recharges the battery.

BATTERY

A *car battery* must store energy and produce electricity. It is an electro-chemical device that uses chemical action to make electrical energy.

A battery produces DC (direct current) that only flows in one direction. Note that this is unlike the current in a house. A home wall outlet has AC (alternator current) that flows one way and then the other at 60 Hertz (60 cycle per second).

When *discharging* (current flowing out of battery), the battery changes chemical energy into electrical energy. In this way, it releases stored energy.

During *charging* (current flowing into battery from charging system), electrical energy is converted into chemical energy. The battery can then store energy until needed.

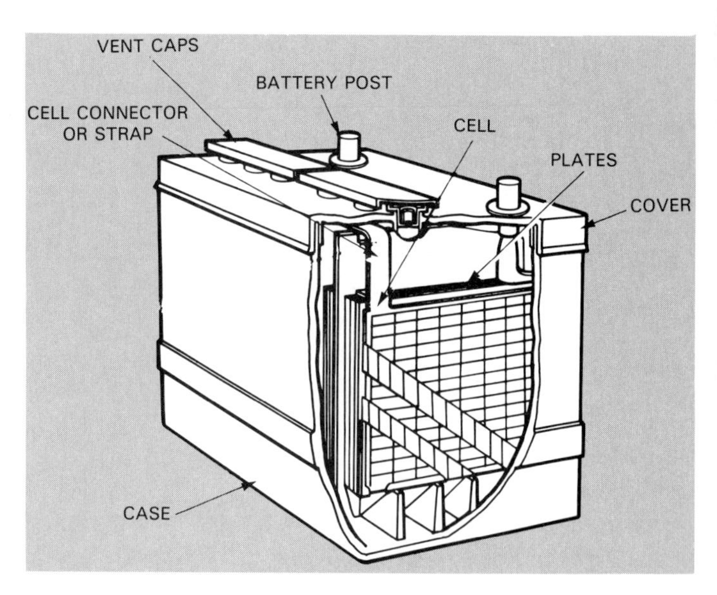

Fig. 13-2. Cutaway shows inside of battery. Note part names. (Chrysler)

BATTERY CONSTRUCTION

A car battery must be made to withstand the severe conditions found in an engine compartment. It must operate under severe vibration, severe cold, severe heat, around corrosive chemicals, during high current discharge, and prolonged periods of storage.

Parts of a battery

The major parts of an automotive battery are:

1. CELL (compartment in battery that contains dissimilar plates, plate separators, and battery acid), Fig. 13-2.
2. BATTERY CASE (plastic housing that forms cell compartments and holds other parts).
3. BATTERY COVER (plastic lid bonded to battery case).
4. BATTERY CAPS (removable plastic covers over cells).
5. BATTERY TERMINALS (metal connectors, either posts or side terminals, for feeding current from plates to battery cables).
6. ELECTROLYTE (battery acid that reacts with lead plates to produce electron flow).
7. BATTERY STRAPS (lead connections between plates in each battery cell).
8. BATTERY PLATES (porous sheets of lead, several plates are positive and several are negative because of types of metal in the lead), Fig. 13-2.

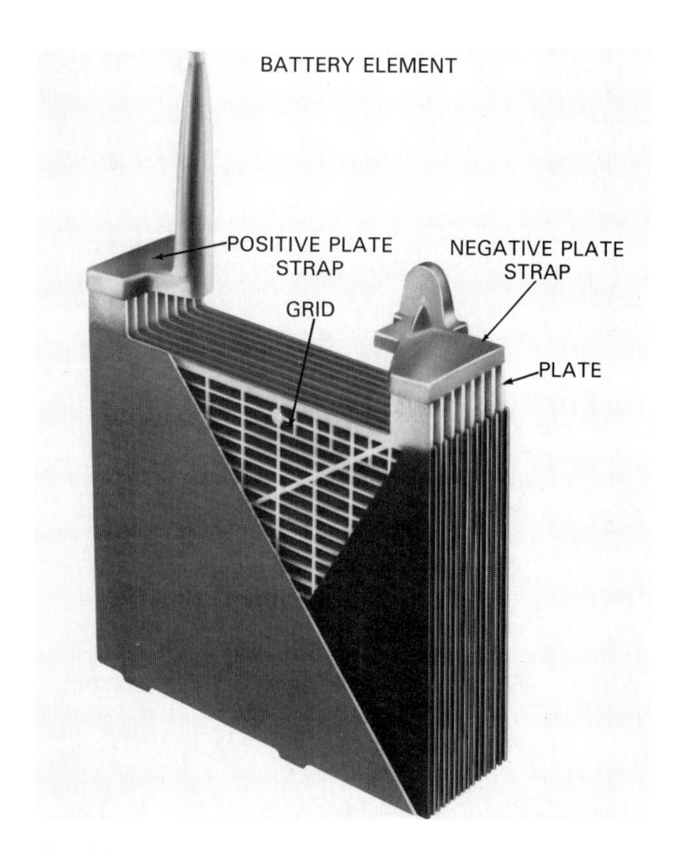

Fig. 13-3. Element is made of dissimilar plates, plate separators or insulators, and straps. When submersed in electrolyte, it is called a battery cell and produces 2.1 volts. (Gould)

Battery elements

A *battery element* consists of the positive plates, negative plates, straps, and separators. Each element fits into a separate compartment in the battery case. When acid surrounds the element, it produces a *battery cell.*

Fig. 13-3 shows an actual battery element. Study the names of the components.

Fig. 13-4 is a cutaway showing how the electrolyte covers the battery plates.

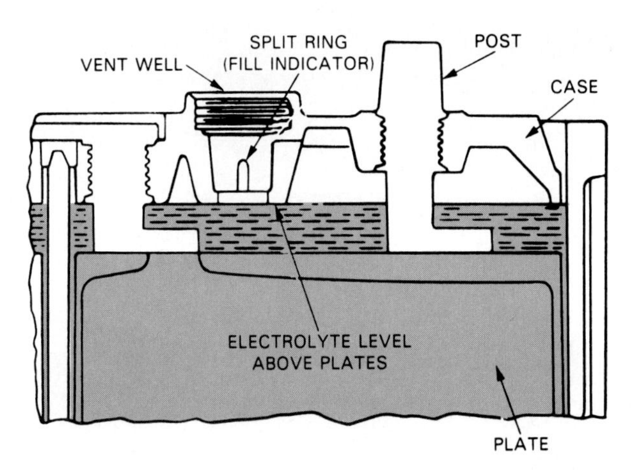

Fig. 13-4. Electrolye or battery acid covers plates. Chemical reaction between plates and acid results in electron flow or electricity. (GMC)

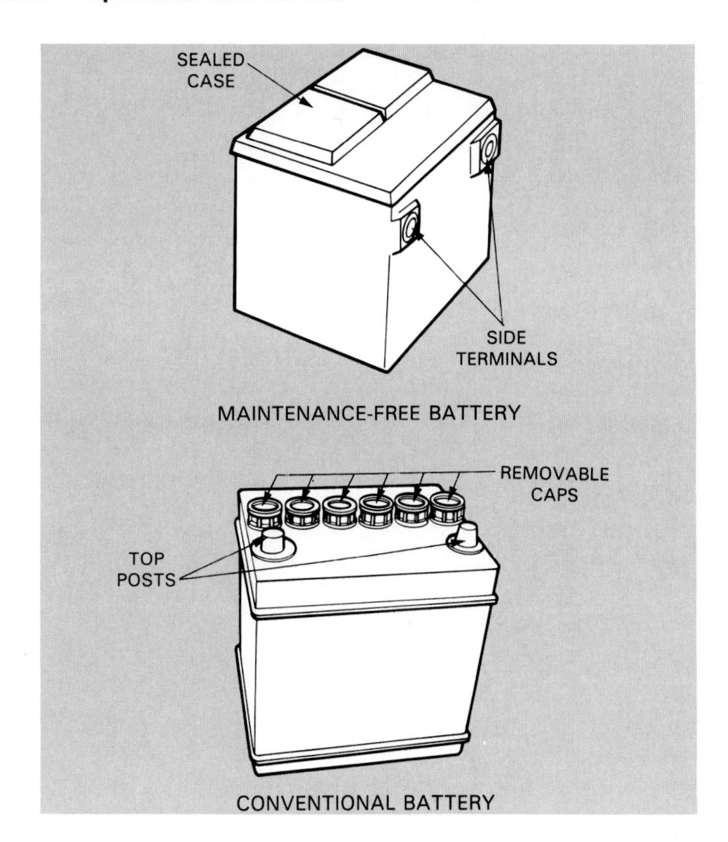

Fig. 13-5. Maintenance-free battery does not have removable caps for adding water to electrolyte. Maintenance-free batteries can have either side terminals or top posts, however. (Chrysler Corporation)

Battery voltage

Battery open circuit (no load) *cell voltage* is 2.1 volts, often rounded off to 2.0 volts. Since the cells in a car battery are connected in series, battery voltage depends upon the number of cells.

A 12-*volt battery* has 6 cells that produce an actual open circuit voltage of 12.6 volts. Modern autos use a 12 V (12.6 V actual) battery and 12 V electrical system.

Some cars with diesel engines use TWO 12 V batteries connected in parallel. When two batteries are in parallel, their output voltage stays the same but current output increases. Dual batteries may be needed to crank and start a compression ignition, diesel engine.

Battery posts and side terminals

Battery posts are round metal terminals sticking out of the top of the battery cover. They serve as male connections for female battery cable ends, Fig. 13-5.

The *positive post* will be larger than the negative post. It may be marked with red paint and a positive (+) symbol. The *negative post* is smaller and may be black or green in color. It normally has a negative (−) symbol on or near it.

Side terminals are electrical connections on the side of the battery, Fig. 13-5. They have female threads that accept a special bolt on the battery cable end. Side terminal polarity is identified by positive and negative symbols on the case. The side terminal design is slowly replacing the older post configuration.

Maintenance-free battery

A *maintenance-free battery* is easily identified because it does NOT use removable filler caps, Fig. 13-5. The top of the battery cells are covered with a large, snap-in cover. Since calcium is used to make the battery plates, water does not have to be added to the electrolyte periodically as with older battery designs.

Battery charge indicator

A *charge indicator,* also called an *eye* or *test indicator,* shows the general charge on the battery. The charge indicator changes color with changes in battery charge. For example, the indicator may be green with the battery fully charged. It may turn black when discharged, clear when low on acid, or yellow when the battery needs replacement. See Fig. 13-6.

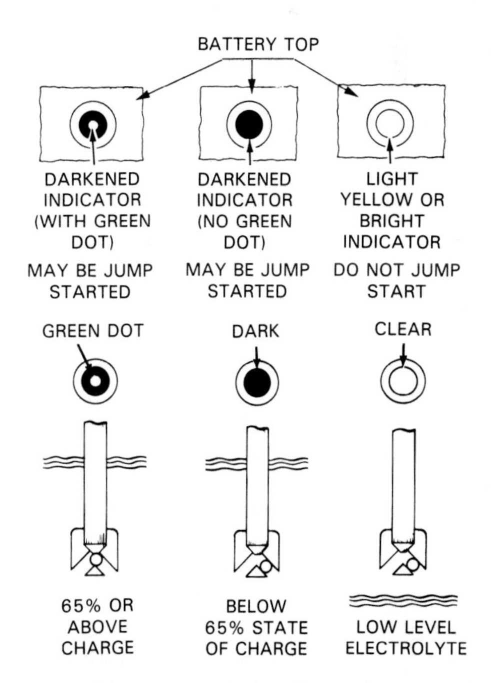

Fig. 13-6. Charge indicator eye on top of battery will show general condition of battery. (GMC)

Battery temperature and efficiency

As battery temperature drops, becomes colder, battery power is reduced. At low temperatures, the chemical action inside the battery is slowed. It will not produce as much current as when warm. This affects a battery's ability to start an engine in extremely cold weather. Look at Fig. 13-7.

Battery cables

Battery cables are large wires that connect the battery terminals to the electrical system of the car. The *positive cable* is normally red and fastens to the starter solenoid. The *negative battery cable* is usually black and connects to ground on the engine block.

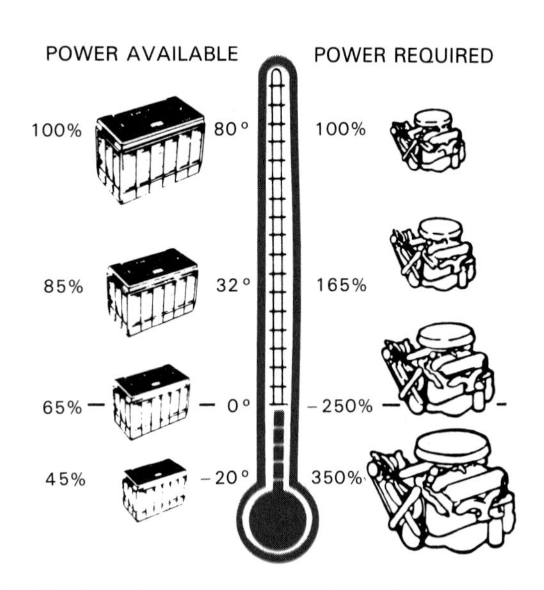

Fig. 13-7. As battery temperature drops, battery performance or current output also drops. This is one reason engine is difficult to crank in cold weather. Also, thickened engine oil increases friction making starting motor work harder. (GMC)

STARTING SYSTEM PRINCIPLES

The car's *starting system* uses battery power and an electric motor to turn the engine crankshaft for engine starting. See Fig. 13-8.

The major parts of a starting system include:

1. BATTERY (source of energy for starting system).
2. IGNITION SWITCH (allows driver to control starting system operation).
3. SOLENOID (high current relay [switch] for connecting battery to starting motor).
4. STARTING MOTOR (high torque electric motor for turning gear on engine flywheel).

Starting system action

When you turn the ignition key to start, current flows through the solenoid coil, as shown in Fig. 13-8. This closes the solenoid contacts, connecting battery current to the starting motor. The motor turns the flywheel ring gear until the engine starts and runs on its own power.

When the engine starts, you release the ignition key. This breaks the current flow to the solenoid and starter. The starter stops turning and the starter gear moves away from the flywheel gear.

STARTING MOTOR CONSTRUCTION

The *starting motor* converts electrical energy into mechanical or rotating energy. It is similar to other electric motors but is more powerful.

Starting motor parts

The construction of starting motors is similar but there are slight design variations. If you understand the construction of a basic motor, however, you can relate it to any starting motor type.

Illustrated in Fig. 13-9, the basic parts of a starting motor include:

1. POLE PIECES (large blocks of iron mounted inside

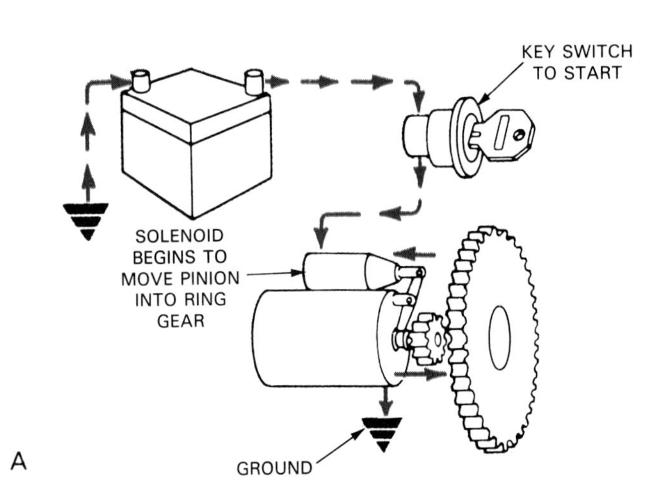

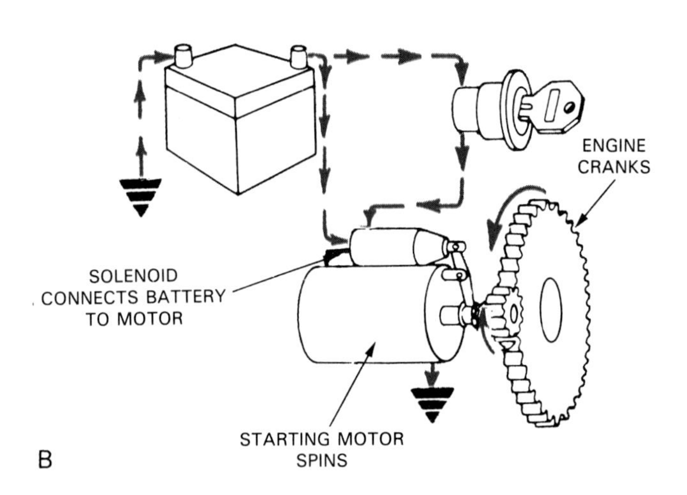

Fig. 13-8. A—When ignition key is turned to start, small current activates starter solenoid. Contacts in solenoid close and conduct larger current. B—Solenoid connects battery to starter and slides pinion gear into flywheel gear to crank engine. (Deere & Co.)

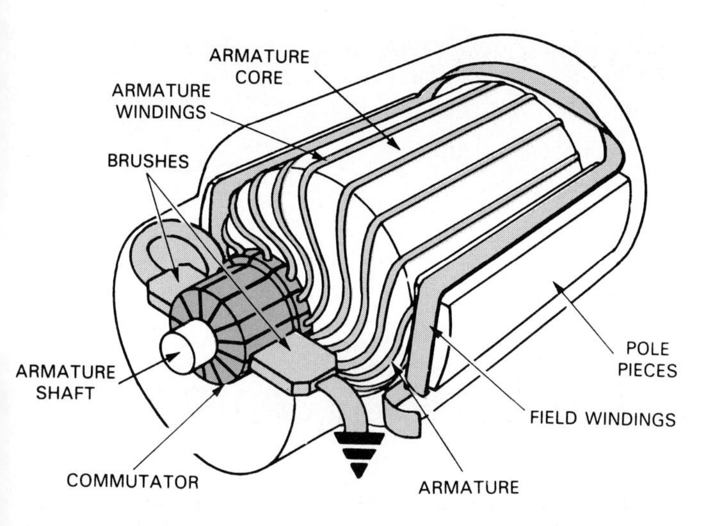

Fig. 13-9. Study basic construction of a starting motor. Pole pieces and field windings set up strong, stationary magnetic field. Current through brushes, commutator, and armature windings sets up another strong field around armature. Two fields then attract and repel each other to spin armature.

windings and next to outside of armature).

2. FIELD WINDINGS (coils of wire that develop strong, stationary magnetic field for rotating armature).
3. ARMATURE (set of coils or windings mounted on a shaft assembly, it is turned by interaction of magnetic fields).
4. COMMUTATOR (rotating connection for feeding current into armature windings).
5. BRUSHES (sliding contacts that ride on commutator and connect voltage to each segment [part] of commutator).

When voltage is fed to the starting motor, current flows through the brushes and into the armature windings. This sets up a strong magnetic field around one set of windings. As a result, the magnetic field around the field windings acts on the magnetic field around the armature. The attraction and repulsion (like between different poles of magnets) then rotates the armature.

As the armature turns, a new set of windings are energized because the brushes touch another set of segments on the commutator. This makes the armature rotate a little more. The process is repeated on each segment so that the armature keeps spinning smoothly.

The major housings of a starting motor are:

1. COMMUTATOR END FRAME (end housing for brushes, brush springs, and shaft bushing).
2. FIELD FRAME (center housing that holds field coils and shoes).
3. DRIVE END FRAME (end housing around pinion gear; has bushing for armature shaft).

Starter pinion gear

A *starter pinion gear* is a small gear on the armature shaft that engages a large gear on the engine flywheel, Fig. 13-10. It moves into and meshes with the flywheel ring gear anytime the starter is energized.

Most starter pinion gears are made as part of a pinion drive mechanism. The pinion drive unit slides over the end of the armature shaft.

The starter *overrunning clutch* locks the pinion gear in one direction and releases it in the other. This allows the pinion gear to turn the flywheel ring gear for start-

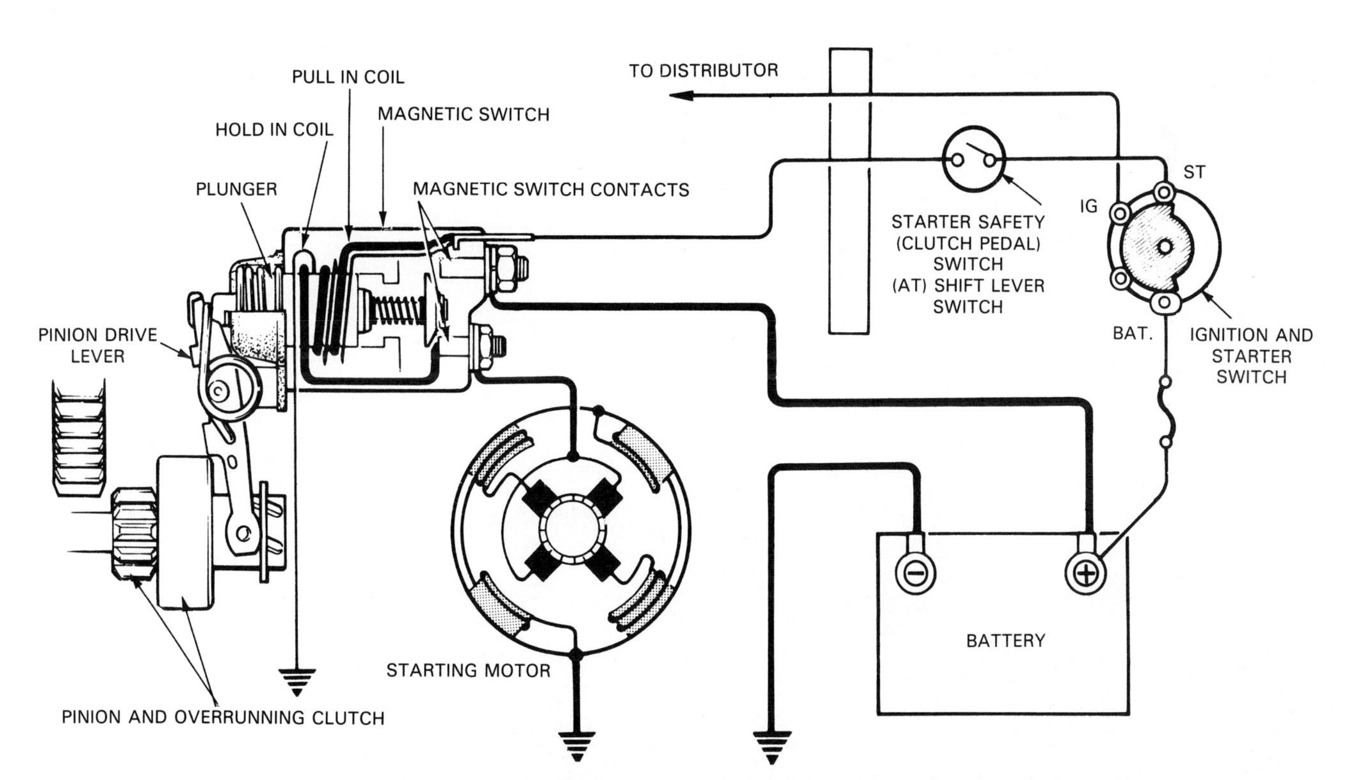

Fig. 13-10. Schematic shows how solenoid completes starting motor circuit and also engages pinion gear. Study connections. (Chevrolet)

ing. It also lets the pinion gear freewheel when the engine begins to run.

Starting Motor Locations

Starting motors are bolted to the engine. They can be located in one of two positions on or in the engine. With many designs, the starting motor bolts to the lower, rear, side of the engine block. With other designs, the starting motor is under the intake manifold.

STARTER SOLENOID

The *starter solenoid* is a high current relay. As shown in Fig. 13-10, it makes an electrical connection between the battery and starting motor. The starter solenoid is an electro-magnetic switch (switch using electricity and magnetism). It is similar to other relays but is capable of handling much higher current levels. A cutaway view of a starter solenoid is given in Fig. 13-11.

A starter solenoid, depending upon starter design, may have three functions:

1. Complete battery-to-start circuit.
2. Push starter pinion gear into mesh with flywheel.
3. Bypass resistance wire in ignition circuit.

The starter solenoid may be located away from or on the starting motor. When mounted on a body panel (away from starter), the solenoid simply makes and breaks electrical connections. When mounted on the starter, it also slides the pinion gear into mesh.

Fig. 13-12 shows an exploded view of a starting motor.

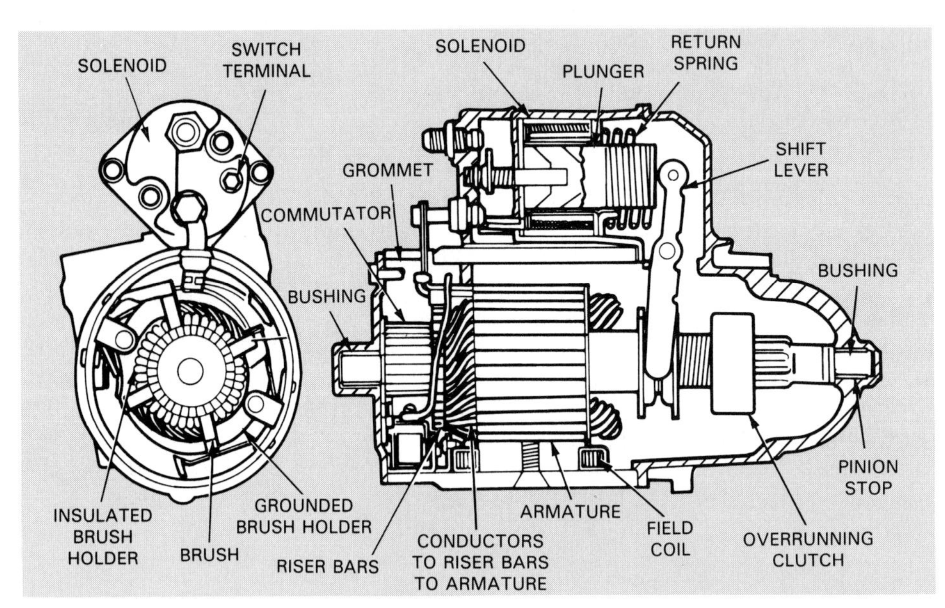

Fig. 13-11. Cutaway shows major components of starting motor. (Chevrolet)

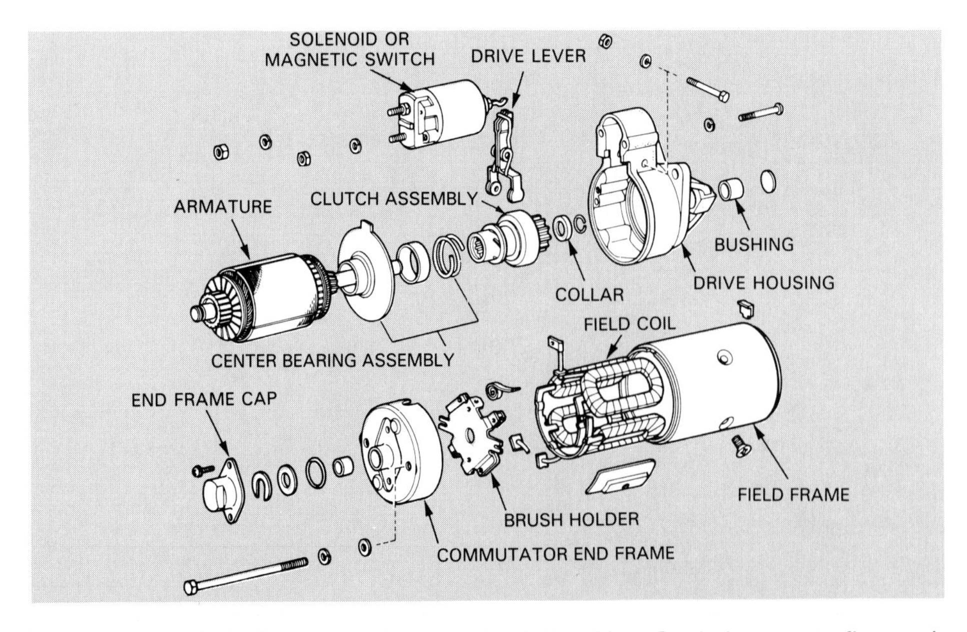

Fig. 13-12. Exploded view of starting motor shows part relationships. Study how parts fit together. (Toyota)

NEUTRAL SAFETY SWITCH

A *neutral safety switch* prevents the engine from cranking unless the shift selector is in neutral or park. It stops the starting system from working when the transmission is in gear.

Cars with automatic transmissions commonly have a neutral safety switch. The switch may be mounted on the shift lever mechanism or on the transmission. One is illustrated in Fig. 13-10.

STARTER RELAY

A *starter relay* uses a small current from the ignition switch to control a slightly larger current to the starter solenoid. This further reduces the load on the ignition key switch. Refer to Fig. 13-10.

Look at Fig. 13-13. It shows a wiring diagram for a typical starting circuit.

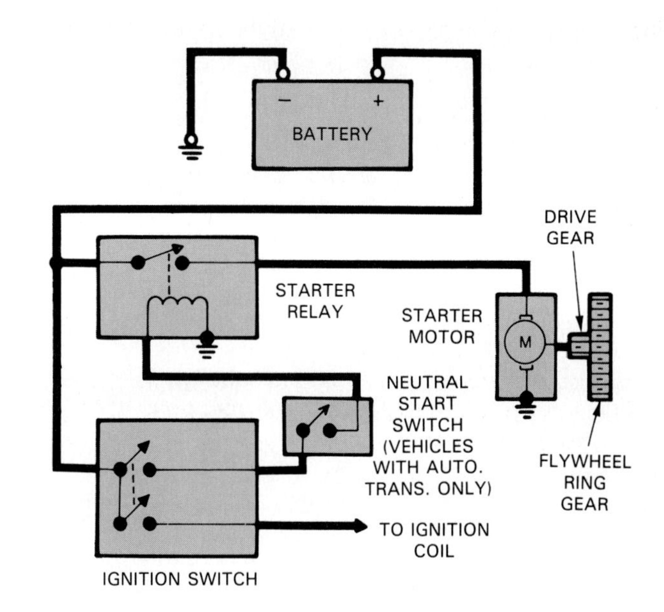

Fig. 13-13. Simplified diagram shows how relay and neutral safety switch wire into starting circuit. (Ford Motor Co.)

CHARGING SYSTEM PRINCIPLES

This section of the chapter will summarize the construction and operation of a charging system.

Charging system functions

The *charging system* has several basic functions:

1. Recharges battery after engine cranking or after use of electrical accessories with engine shut off.
2. Supplies all of the car's electricity when engine is running.
3. Provides voltage output slightly higher than battery voltage.
4. Change current output to meet electrical loads.

Charging system parts

The basic parts of a charging system are shown in Fig. 13-14 and they include:

1. ALTERNATOR (generator that uses mechanical [engine] power to produce electricity).
2. VOLTAGE REGULATOR (electrical device for controlling output voltage and current of alternator).
3. ALTERNATOR BELT (connects engine crankshaft pulley to alternator pulley for driving alternator).

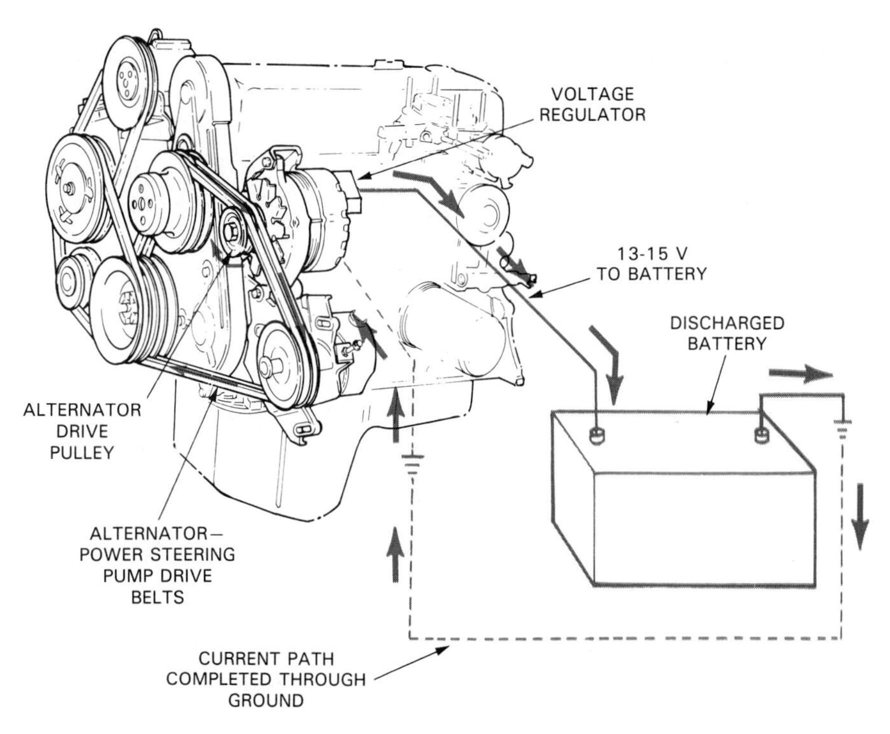

Fig. 13-14. Charging system uses engine power to spin belt and alternator pulley. Alternator converts this motion into electricity to recharge battery. Voltage regulator controls alternator output. (Ford)

4. BATTERY (provides current to initially energize or excite alternator and also helps stabilize alternator output).

Charging system operation

During engine cranking, the battery supplies all electricity. Then, when the engine starts running, the charging system takes over to provide current to the car's electrical system.

The engine crankshaft pulley and alternator belt spins the alternator pulley. This powers the alternator and causes it to produce electricity.

The voltage regulator keeps alternator output at a preset *charging voltage* (approximately 13 to 15 volts). Since this is HIGHER than battery voltage (12.6 volts), current flows back into the battery and recharges it. Current also flows to the ignition system, electronic fuel injection system, on-board computer, radio, or any other device using electricity.

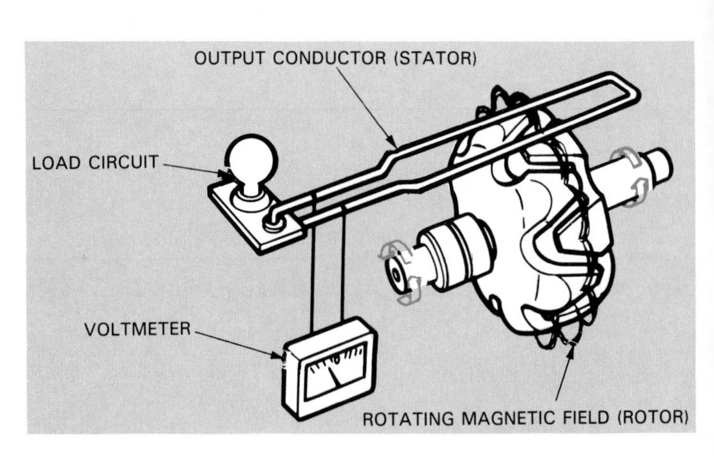

Fig. 13-16. Rotor is spinning magnetic field. Its magnetic field cuts across stator or output conductor. Current is induced into stator windings and fed to car's electrical system. (Chrysler)

ALTERNATOR

Pictured in Fig. 13-15, the main components of a typical alternator are:

1. ROTOR ASSEMBLY (field windings, claw poles, rotor shaft, and slip rings).
2. STATOR ASSEMBLY (three stator windings or coils, stator core, and output wires).
3. BRUSH ASSEMBLY (brush housing, brushes, brush springs, and brush wires).
4. RECTIFIER ASSEMBLY (diodes, heat sink or diode plate, and electrical terminals).
5. HOUSING (drive end frame, slip ring and frame, end frame bolts).
6. FAN AND PULLEY ASSEMBLY (fan, spacer, pulley, lock washer, and pulley nut).
7. INTEGRAL REGULATOR (electronic voltage regulator mounted in or on rear of modern alternators).
8. ALTERNATOR BEARINGS (anti-friction bearing on each end of rotor shaft).

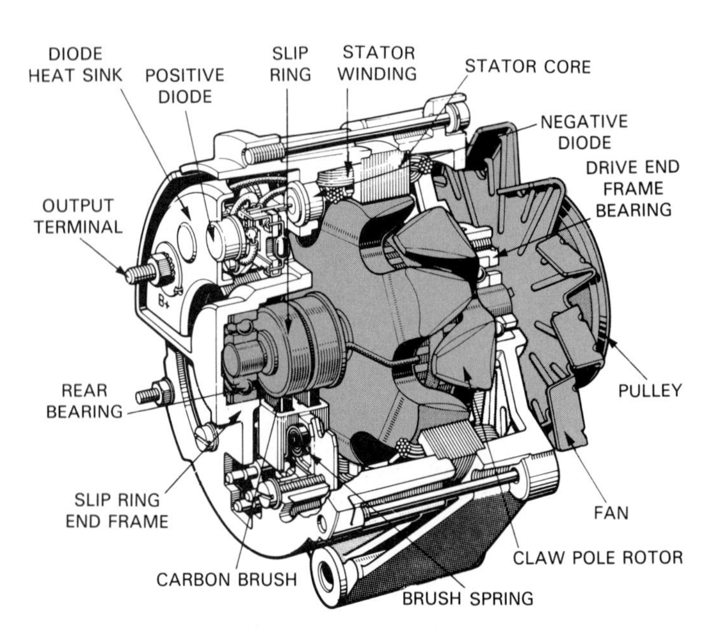

Fig. 13-15. Study internal parts of this alternator. (Bosch)

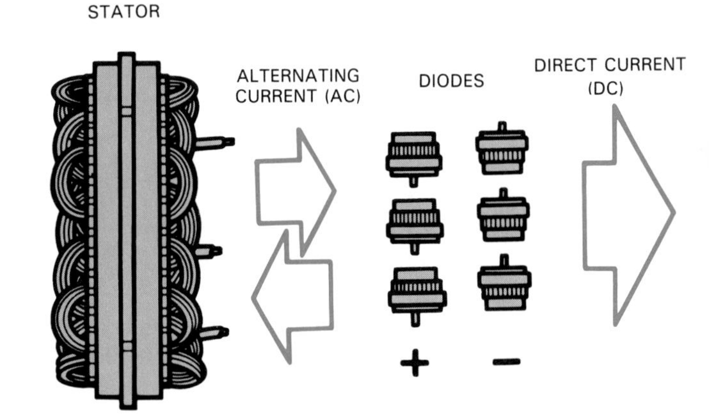

Fig. 13-17. Stator output is AC. Diodes are needed to convert alternating current into direct current for use by vehicle electrical system. Six rectifying diodes are commonly used in alternator. (Mopar)

Alternator operation

Basically, an alternator produces electricity by spinning a magnetic field (rotor) over a stationary set of windings (stator). As pictured in Fig. 13-16, a magnetic field is produced around the rotor. As the rotor is turned by engine power, the conductor or stator has current induced into it, Fig. 13-17.

Fig. 13-18 shows the rear of a typical alternator. Study the electrical connections.

Alternator pulley and belt

An *alternator pulley* is secured to the front of the rotor shaft by a large nut. It provides a means of spinning the rotor. See Fig. 13-19.

An *alternator belt,* running off the engine crankshaft pulley, turns the alternator pulley and rotor. One of four types of belts may be used: V-belt, cogged V-belt, ribbed belt, and serpentine belt.

VOLTAGE REGULATORS

A *voltage regulator* controls alternator output by changing the amount of current flowing through the rotor

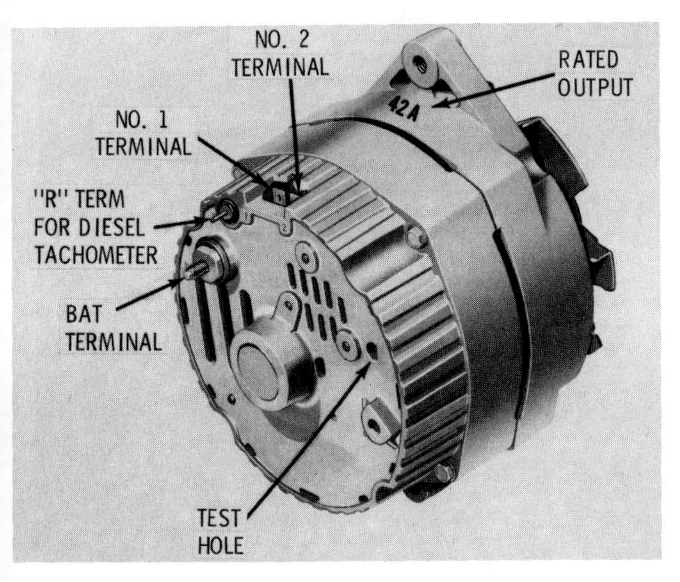

Fig. 13-18. Rear view of modern alternator shows electrical connections. (Pontiac)

windings. Any change in rotor current changes the field strength acting on the stator windings. In this way, the regulator can maintain a preset charging voltage.

There are three basic types of voltage regulators:

1. Electronic voltage regulator mounted inside or on back of alternator, Fig. 13-19.
2. Electronic regulator mounted away from alternator in engine compartment.
3. Contact point regulator mounted away from alternator in engine compartment. This type is only found on older cars.

Computerized regulation

In some cases, a vehicle's computer is used to regulate the alternator's output. This arrangement eliminates the need for a conventional regulator.

Basically, the computer controls the duty cycle of the alternator's field windings. The *duty cycle* is the percentage of time that current is fed to the windings. When the electrical load is heavy, the duty cycle is long. This causes the alternator to produce a high output. When the electrical load is light, the duty cycle is short and the alternator's output is low. A duty cycle of about 50% is normal.

CHARGING AND STARTING SYSTEM SERVICE

Now that you understand the operation and construction of batteries, alternators, starting motors, and related components, you are ready to review basic testing, service, and repair of these parts.

Always keep in mind that the battery, charging system, and starting system are interacting. A malfunction in one component can adversely affect the operation of another component.

Fig. 13-20 shows a diagram for a modern electrical diagnostic connector circuit. The *diagnostic connector* can be used to measure voltages in various parts of the circuit without disconnecting wires. Also, special testers can be plugged into the diagnostic connector to analyze troubles. Study the connections in this circuit.

BATTERY PROBLEMS

A *"dead battery"* (discharged battery) is a very common problem. The engine will usually fail to crank and start. Even though the lights and horn may work, there

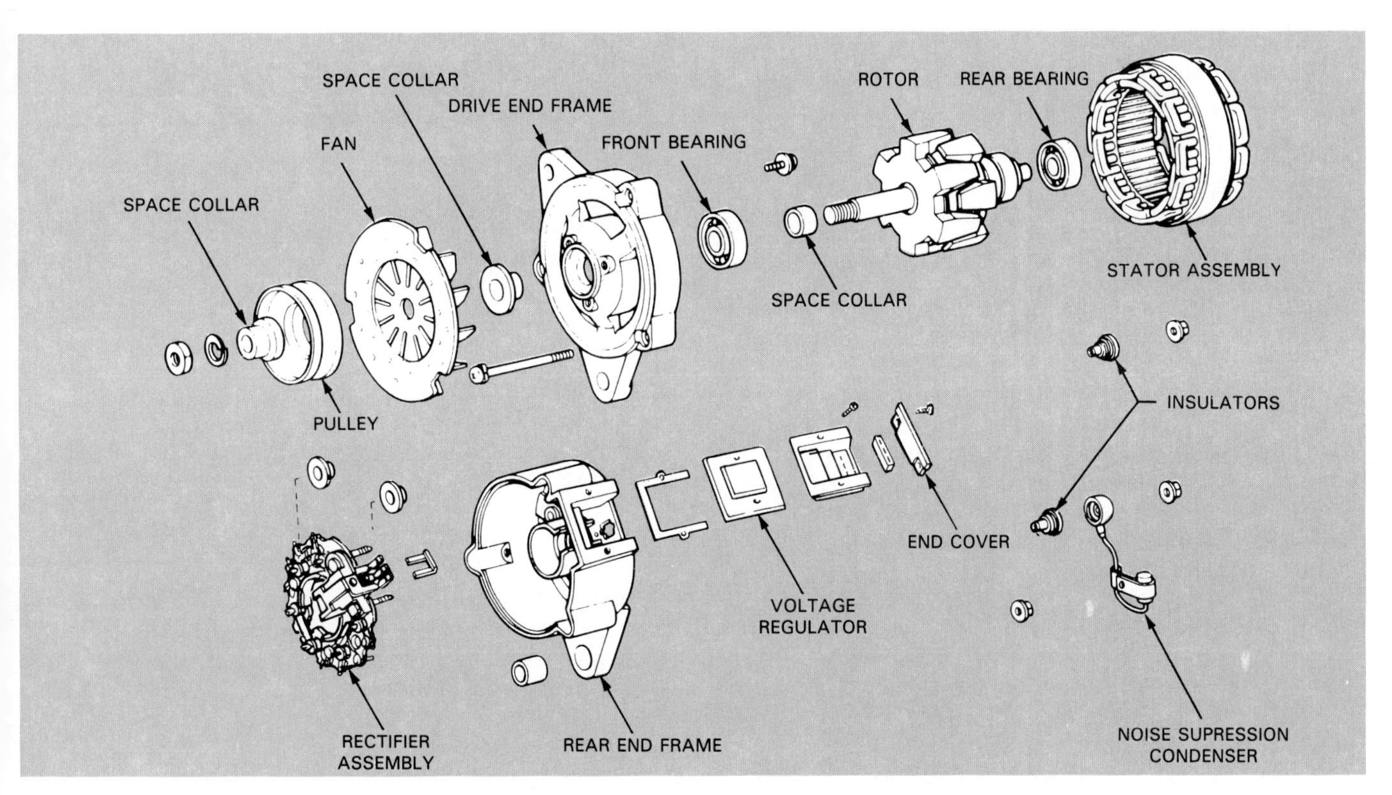

Fig. 13-19. Study exploded view of typical alternator with built-in voltage regulator. (Toyota)

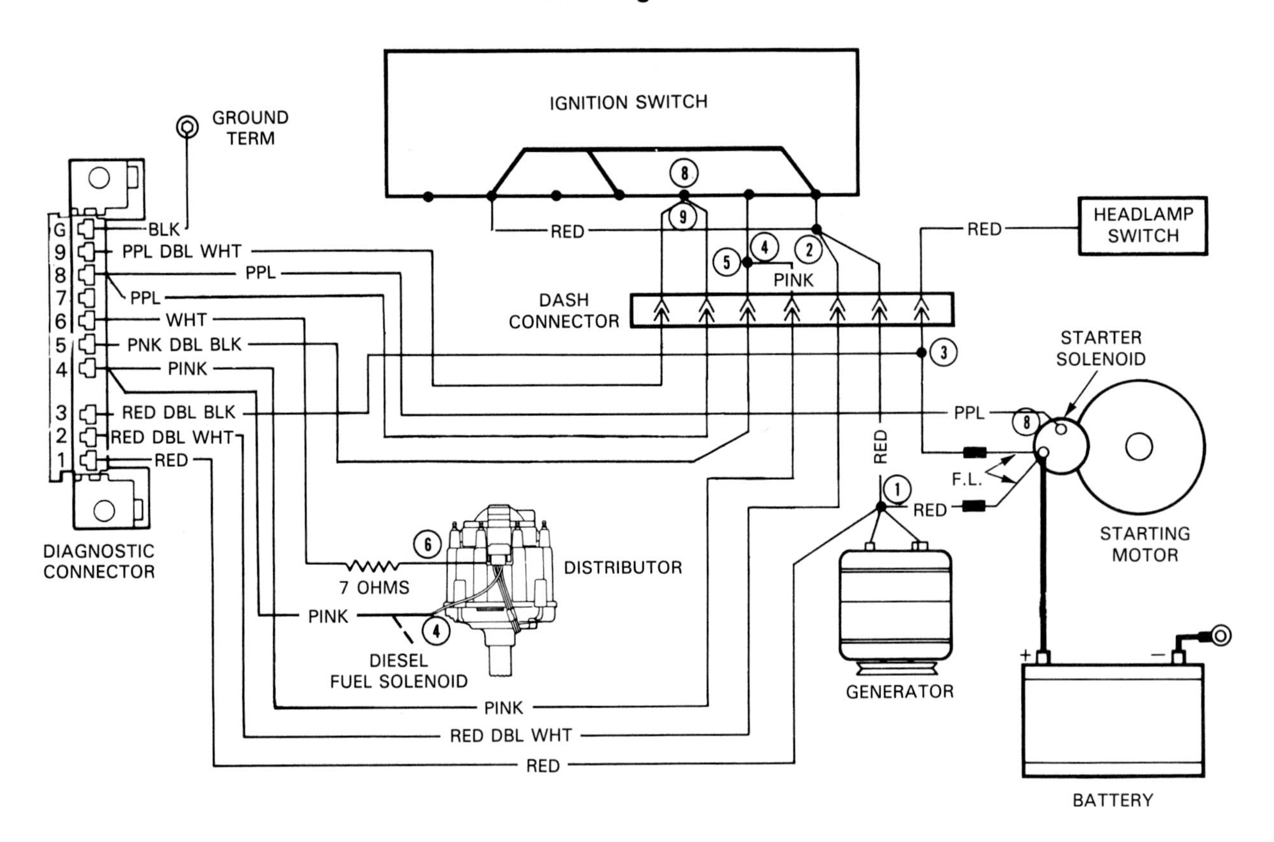

Fig. 13-20. This wiring diagram shows how special test connector is wired. Special tester or voltmeter can be used to quickly check circuit voltage at connector, without disconnecting wires. (Buick)

is not enough "juice" (current) in the battery to operate the starting motor.

A *discharged battery* could be caused by:

1. Defective battery (sulfated plates).
2. Charging system problem, loose alternator belt for example.
3. Starting system problem.
4. Poor cable connections.
5. Engine performance problem requiring excessive cranking time.
6. Electrical problem drawing current out of battery with ignition key OFF.

Battery maintenance

If a battery is not maintained properly, its service life will be reduced. Battery maintenance should be done periodically—during tune-ups, grease jobs, or anytime symptoms indicate battery problems.

Battery maintenance typically includes:

1. Checking electrolyte level or indicator eye.
2. Cleaning battery terminal connections.
3. Cleaning battery top.
4. Checking battery hold-down and tray.
5. Inspecting for physical damage to case and terminals.

Inspecting battery condition

Inspect the battery anytime the hood is opened, Fig. 13-21. Look for a buildup on the battery top, case damage, loose or corroded connections, or any trouble that could upset battery operation.

DANGER! Wear eye protection when working around batteries. Batteries contain acid that could cause blindness. Even the film buildup on a battery can contain acid.

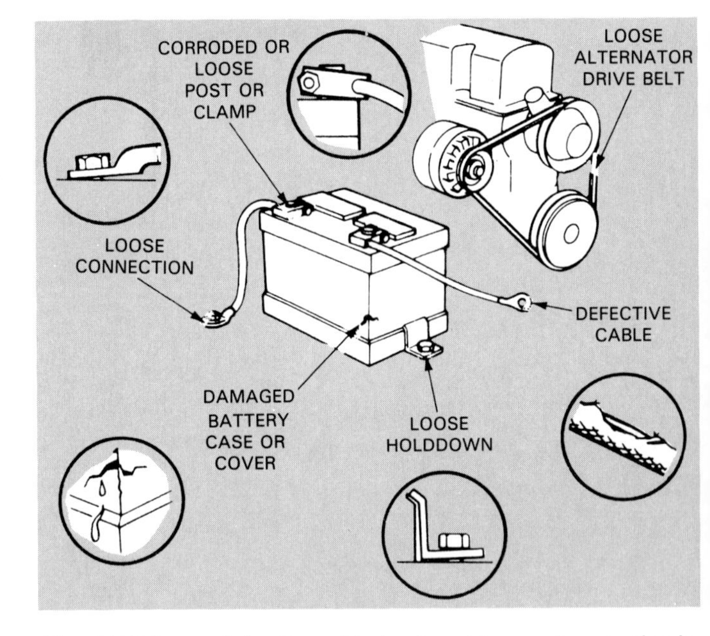

Fig. 13-21. Look for these kinds of problems when engine is difficult to start or will not start. (GM)

Battery terminal test

A *battery terminal test* quickly checks for a poor electrical connection between the battery cables and terminals. A voltmeter is used to measure voltage drop across the cables and terminals when cranking the engine, Fig. 13-22.

If the voltmeter shows OVER about .5 volt, there is a high resistance at the cable connection. This would tell you to clean the battery connections.

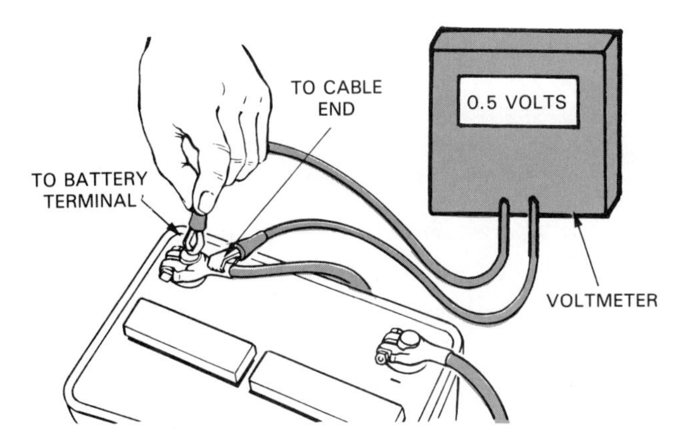

Fig. 13-22. Voltmeter can be used to quickly check for poor battery cable connections. Measure voltage drop across cable and terminal while cranking engine. Voltage drop should be under .5 volt. (NAPA)

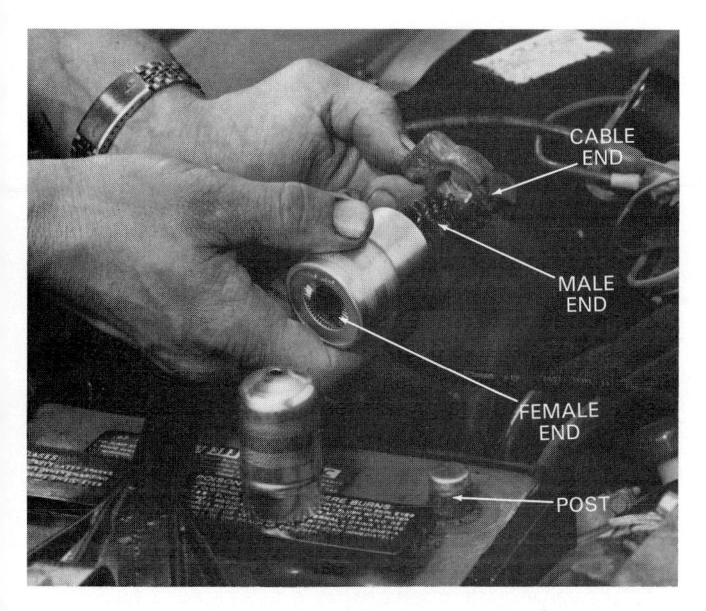

Fig. 13-23. Technician is using special post type cleaner to remove deposits from surface of metal on post and cable end. (Peerless)

Cleaning battery terminals

To clean post type terminals, use a cleaning tool like the one in Fig. 13-23. Use the female end to clean the post. Use the male end on the terminal. Twist the tool to remove the oxidized outer surface on the connections.

To clean side terminals, use a small wire brush. Polish both the cable end and the mating surface on the battery terminal.

Do NOT use a knife or scraper to clean battery terminals. This removes too much metal and can ruin the connection.

When reinstalling the cables, coat the terminals with petroleum jelly or white grease. This will keep acid fumes off of the connections and keep them from corroding again. Tighten the fasteners just enough to secure the connection.

Jump starting

In emergency situtations, it will be necessary to *jump start* the car by connecting another battery to the discharged battery. The two batteries are conected POSITIVE to POSITIVE and NEGATIVE to NEGATIVE.

Connect the red jumper cable to the positive terminal on both batteries. Then, connect the black jumper cable to any ground on both vehicles as shown in Fig. 13-24.

WARNING! Do not short the jumper cables together or connect them backwards. This could cause serious damage to the charging or computer systems.

BATTERY CHARGERS

When tests show that the battery is discharged, a battery charger may be used to re-energize the battery. The *battery charger* will force current back into the battery to restore the charge on the battery plates.

DANGER! Before connecting a battery charger to a battery, make sure the charger is turned OFF. Also, check that the work area is well ventilated. If a spark ignites any battery gas, the battery could EXPLODE. Wear eye protection!

To use a battery charger, connect the RED charger lead to the positive terminal. Connect the BLACK charger lead to the negative terminal of the battery. With side terminal batteries, use adapters.

Make sure you do NOT reverse the charger connections or the charging system could be damaged. Set the charger controls and turn on the power.

BATTERY LOAD TEST

A *battery load test, also termed a battery capacity test,* is one of the BEST methods of checking battery

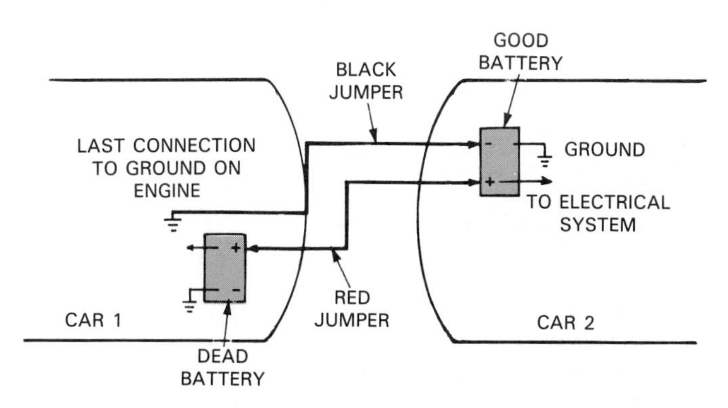

Fig. 13-24. Connect jumper cables as shown. The last connection, black ground lead, should be made away from battery on any good ground—a metal engine part for example.

condition. It tests the battery under full current load. The battery load test actually measures the current output and performance of the battery. It is one of the most common and informative battery tests used in modern automotive garages. Refer to Fig. 13-25.

Connecting load tester

Connect the load tester to the battery terminals. If the tester is an *inductive type* (clip-on ammeter lead senses field around outside of cables), use the connections shown in Fig. 13-26. If the tester is NOT inductive, connect the ammeter in series.

Control settings and exact procedures vary. Follow the directions provided with the testing equipment.

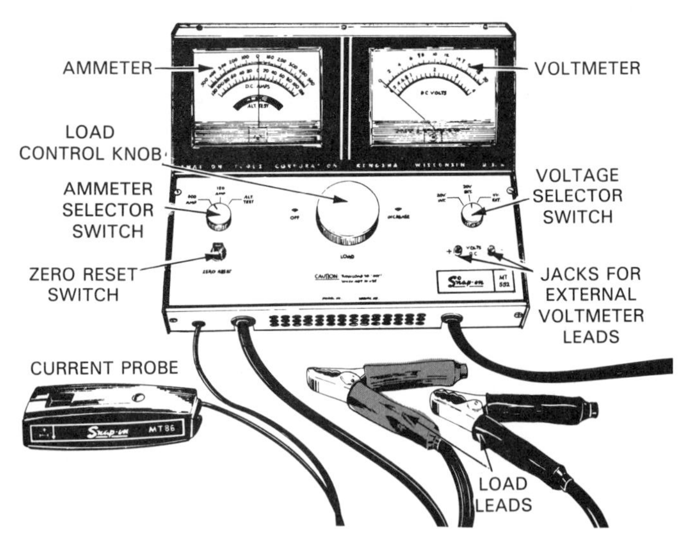

Fig. 13-25. Note controls on this modern load tester. (Snap-On Tools)

Determine battery load

Before loading a battery, calculate how much current draw should be applied to the battery.

If the *amp-hour rating* is given, load the battery to THREE TIMES its amp-hour rating. For example, if the battery is rated at 60 amp-hours, test the battery at 180 amps (60 × 3 = 180). Fig. 13-27 gives a chart showing load test amps.

A load conversion chart will normally be provided with the load testing equipment. Refer to this material when in doubt.

BATTERY RATINGS			LOAD TEST AMPS
Cold Cranking Current	**Amp-Hour (Approx.)**	**Watts**	
200	35-40	1800	100 amps
250	41-48	2100	125 amps
300	49-62	2500	150 amps
350	63-70	2900	175 amps
400	71-76	3250	200 amps
450	77-86	3600	225 amps
500	87-92	3900	250 amps
550	93-110	4200	275 amps

Fig. 13-27. This chart shows how different ratings affect how much current you should draw during battery load test.

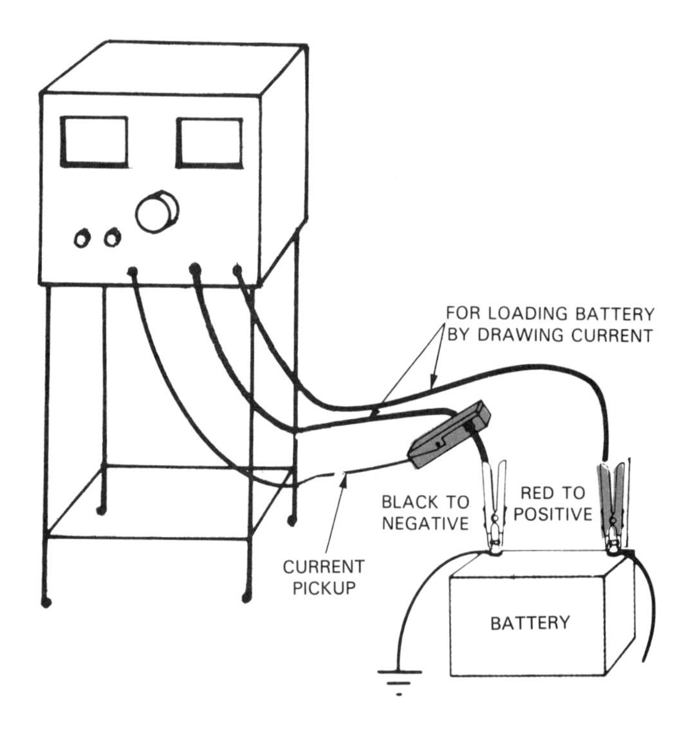

Fig. 13-26. To load test battery, connect tester as shown or use operating instructions. Basically, use tester to load battery and draw specified current depending upon battery rating. Battery voltage must stay above 9.5 volts for battery condition to be acceptable.

Loading the battery

Double-check that the tester is connected properly. Then, turn the load control knob until the ammeter reads the correct load for your battery.

Hold the load for 15 seconds. Then, read the voltmeter while the load is applied. Turn the load control completely OFF so the battery will not be discharged.

Load test results

If the voltmeter reads 9.5 volts or MORE at room temperature, the battery is good. This voltage is based on a battery temperature above 70 °F (21 °C). A cold battery may show a lower voltage. You will need a temperature compensation chart like the one in Fig. 13-28. It allows for any reduced battery performance caused by a low temperature.

If the voltmeter reads below 9.5 volts at room temperature, battery performance is POOR. This would show that the battery is not producing enough current to properly run the starting motor.

STARTING SYSTEM SERVICE

A starting system is easy to work on compared to the car's other electrical systems. It only has about five major components that cause problems.

APPROXIMATE ELECTROLYTE TEMPERATURE	MINIMUM ACCEPTABLE VOLTAGE UNDER LOAD FOR GOOD BATTERY
60 °F (16 °C)	9.5
50 °F (10 °C)	9.4
40 °F (4 °C)	9.3
30 °F (−1 °C)	9.1
20 °F (−7 °C)	8.9
10 °F (−12 °C)	8.7
0 °F (−18 °C)	8.5

Fig. 13-28. Temperature should also be considered when testing battery condition. Note how temperature affects voltage reading during load test.

Fig. 13-29 shows the most common starting system troubles. If any of these parts have high resistance, low resistance, damage, or wear, the engine may not crank normally.

Common starting system problems

A *no-crank problem* results when the engine crankshaft does NOT rotate properly with the ignition key at start. The most common causes are a dead battery, poor electrical connection, or faulty system component.

A *slow cranking problem* occurs when the engine crankshaft rotates at lower than normal speed. It is usually caused by the same kind of problems producing a no-crank problem.

A *buzzing* or *clicking sound* from the solenoid, without cranking, is commonly due to a DISCHARGED BATTERY or poor battery cable connections. Low current flow is causing the solenoid plunger to rapidly kick in and out, making a clattering sound.

A *single click sound,* without cranking, may point to a bad starting motor, burned solenoid contacts, dead battery, or engine mechanical problems. The click is probably the solenoid closing or the pinion gear contacting the flywheel gear.

A *humming sound,* after momentary engine cranking, may be due to a bad starter overrunning clutch or pinion gear unit. Pinion gear wear can make the gear disengage from the flywheel gear too soon. This can let the motor armature spin rapidly, making a humming sound.

A *metallic grinding noise* may be caused by broken flywheel teeth or pinion gear teeth wear. The grinding may be the gears clashing over each other.

Normal *cranking, without starting,* is usually NOT caused by the starting system. There may be trouble in the fuel or ignition systems. With a diesel engine, check engine cranking speed. If cranking rpm is low, the diesel may not start.

Fig. 13-29. These are the types of problems that you can find in starting system. (Echlin)

Check the battery first

A *dead* or *discharged battery* is one of the most common reasons the starting system fails to crank the engine properly. A starting motor draws several times the amount of current (over 200 amps) as any other electrical component.

If needed, load test the battery as described earlier. Make sure the battery is good and fully charged. A starting motor will NOT function without a fully charged and well connected battery.

STARTER CURRENT DRAW TEST

A *starter current draw test* measures the number of amps used by the starting system. It will quickly tell you about the condition of the starting motor and other system parts. If current draw is higher or lower than specs, there is a problem.

To do a starter current draw test, connect meters or a load tester to measure battery voltage and current flow out of the battery. See Fig. 13-30.

To keep the engine from starting during the test, disconnect the coil supply wire or ground the coil wire. With a diesel engine, disable the injection system. You may have to unhook the fuel shut-off solenoid. Check in a shop manual for details.

WARNING! Do NOT crank the engine for more than 15-30 seconds or starter damage may result. If cranked too long, the starter could overheat. Allow the

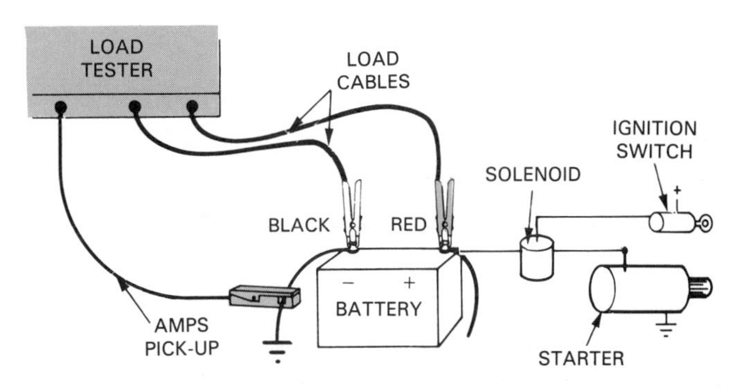

Fig. 13-30. Load tester will also measure current draw of starting motor. High or low current draw when cranking will indicate possible troubles.

starter to cool a few minutes if more cranking time is needed.

Crank the engine and note the current readings. If not within specs, something is wrong in the starting system or engine. Further tests would be needed. Fig. 13-31 gives the average current draw values for various engine sizes.

ENGINE DISPLACEMENT	12-VOLT SYSTEM MAX. CURRENT
Most 4-6 Cylinders	125-175 Amps Max.
Under 300 C.I.D.	150-200 Amps Max.
300 C.I.D. or Over	175-250 Amps Max.

CRANKING CIRCUIT TROUBLESHOOTING CHART

Cranking Voltage	Cranking Amps	Possible Problem
Voltage Within Specs	Current Within Specs	System OK
Voltage OK	Current Low Engine Cranks Slowly	Starter Circuit Connections Faulty
Voltage Low	Current Low Engine Cranks Slowly	Battery Low
Voltage Low	Current High	Starter Motor Faulty

Fig. 13-31. Chart shows how test voltages and current values may indicate potential problems in starting system.

STARTER SERVICE

A *faulty starting motor* can cause a wide range of symptoms: slow cranking, no cranking, overheating of starter cables, and abnormal noises while cranking. If the battery, cables, solenoid, and other starting system parts are good but the engine does not crank properly, the starter may be bad.

A current draw test and other tests would suggest when the starter motor should be removed for further inspection, testing, or replacement.

Starting motor removal

Before deciding to remove the starting motor, inspect it closely for problems. Check that the starter-to-engine bolts are tight. Loose starter bolts can upset motor operation by causing a poor ground or incorrect pinion gear meshing. Make sure all wires on the motor and solenoid are tight.

To remove the starting motor, first DISCONNECT THE BATTERY. Then, unbolt the battery cable and solenoid wires (starter-mounted solenoid type) and any braces on the motor. Unscrew the bolts while holding up on the motor. Refer to Fig. 13-32.

CAUTION! Be careful not to drop the starting motor when removing it. The starter is heavy enough to cause injury if it falls.

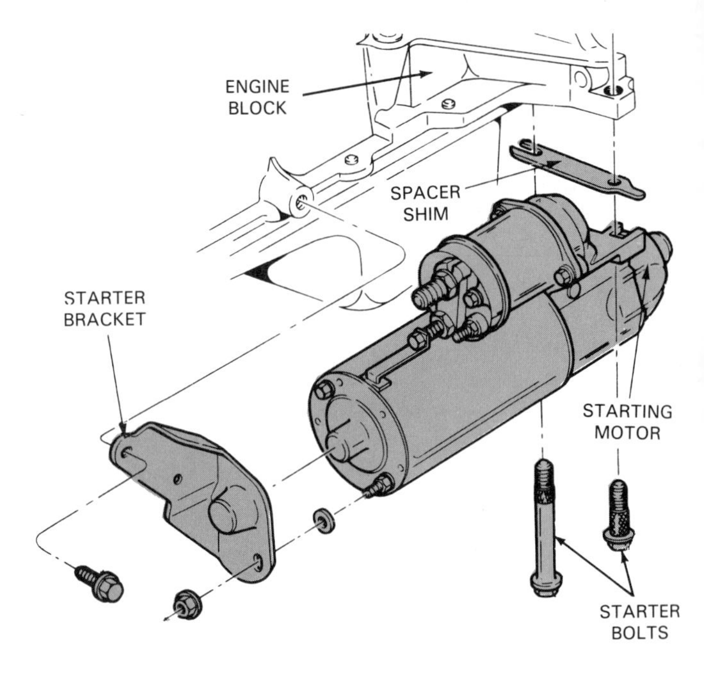

Fig. 13-32. During starting motor installation, make sure you connect wires properly, replace any shims, and secure all brackets. When removing motor, hold onto it tightly because motor is heavy for its small size. (General Motors)

Starter shims may be used to adjust the space between the pinion gear and flywheel ring gear, Fig. 13-32. Shims position the starter farther away from or closer to the ring gear.

During starter removal, always check for shims. They must be put in the same place during reassembly. If not replaced, the pinion and flywheel gears may not mesh. A grinding noise will result.

When the starting motor is located inside the block valley, you must remove the intake manifold to access the motor. First, disconnect the battery and remove wires, fuel lines, and cables between the manifold and chassis. Then, remove the intake manifold bolts, brackets, and other parts inhibiting manifold removal. Lift off the manifold. This will let you access the bolts that secure the motor to the cylinder block.

Overrunning clutch (pinion gear) service

Normally, the overrunning clutch or pinion gear assembly is replaced anytime the starting motor is disassembled. The pinion gear is subjected to extreme wear and tear when engaged and disengaged from the engine flywheel. It is usually wise to replace the pinion gear during starter service, Fig. 13-33.

Installing starting motor

Install the starter in the reverse order of removal. Make sure that any spacer shims are replaced between the motor and the engine block.

If the starter has a solenoid on it, connect the wires on the solenoid before bolting the starter to the engine. Tighten the starter bolts to the recommended torque. Replace any bracket or shields and reconnect the battery. Crank the engine several times to check starting motor operation.

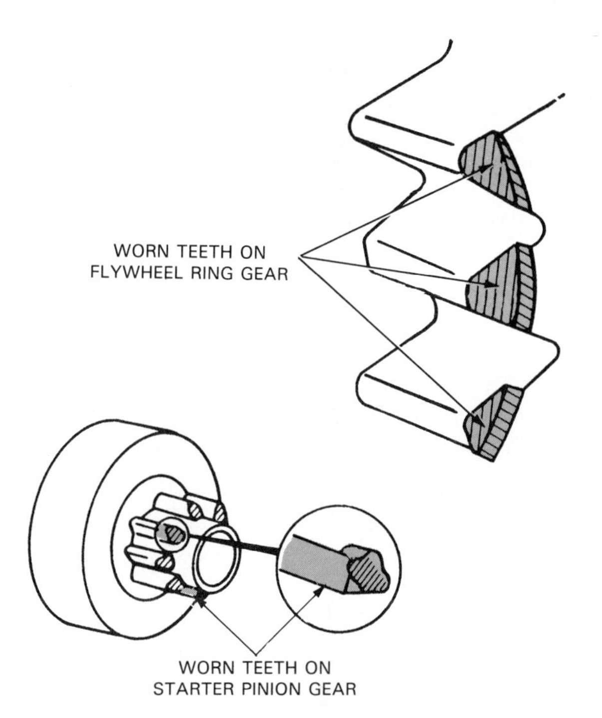

Fig. 13-33. Common problem is worn teeth on starter pinion gear and sometimes flywheel ring gear. Worn teeth can make starter disengage prematurely. (Ford)

CHARGING SYSTEM SERVICE

Even though a charging system only has two major parts (alternator and regulator), be careful when troubleshooting. Sometimes, another fault, (bad starting motor, battery, or wiring) will appear to be problems in the charging system.

There are four common types of trouble symptoms caused by the charging system:

1. Dead battery (slow or no cranking).
2. Overcharged battery (water must be added to battery frequently).
3. Abnormal noises (grinding, squealing, buzzing).
4. Indicator shows problem (light glows all the time or incorrect indicator reading).

Visual inspection

Open the hood and visually inspect the parts of the charging system. Check for obvious troubles like the ones given in Fig. 13-34.

Alternator belt

Check for *alternator belt problems.* Make sure the belt is adjusted properly. A loose belt may squeal or flap up and down. Also inspect the condition of the alternator belt. Check for cracks, glazing (hard, shiny surface), grease or oil contamination, and deterioration.

If needed, adjust alternator belt tension, Fig. 13-35. Loosen the alternator mounting and adjusting bolts. Then pry on a strong surface of the end frame. Pull hard enough to produce proper tension. Tighten the adjusting bolt while holding the pry bar. Tighten the other mounting bolts. Recheck tension.

WARNING! Tighten an alternator belt only enough to prevent belt slippage or belt flap. Overtightening is a COMMON MISTAKE that quickly ruins alternator bearings because of excess load.

CHARGING SYSTEM TESTS

Charging system tests should be done when symptoms point to low alternator voltage and current. These tests will quickly determine the operating condition of the charging system.

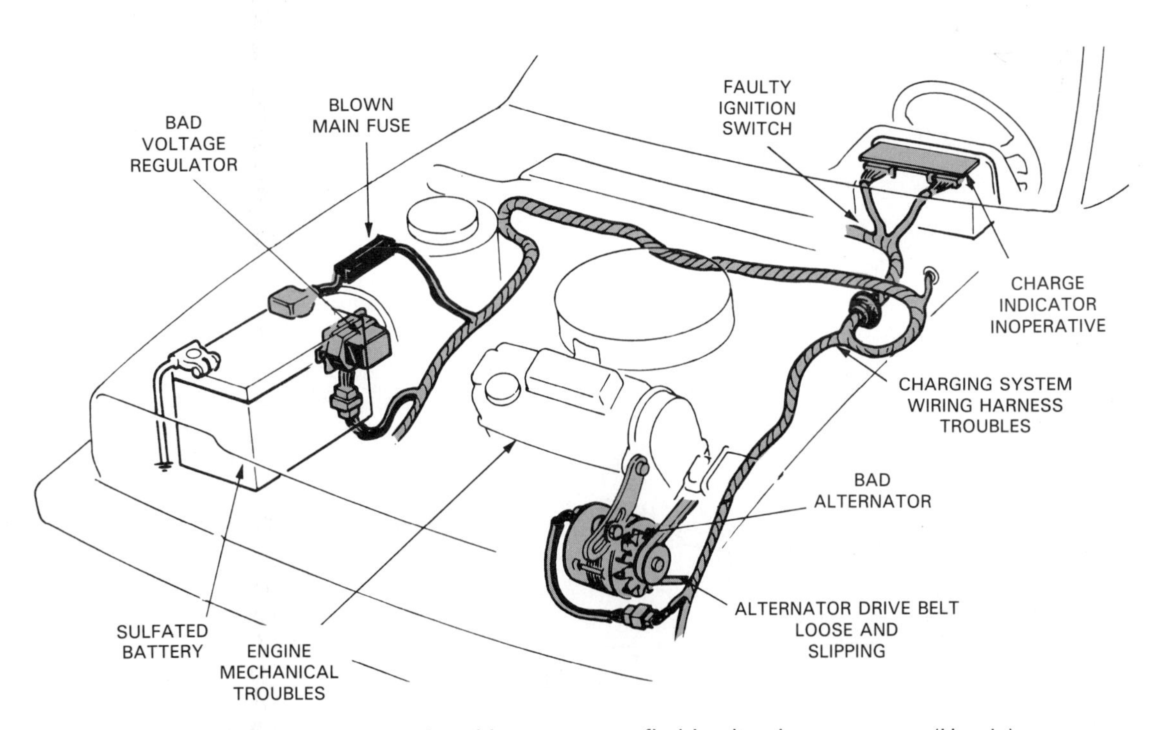

Fig. 13-34. Study types of problems you can find in charging system. (Honda)

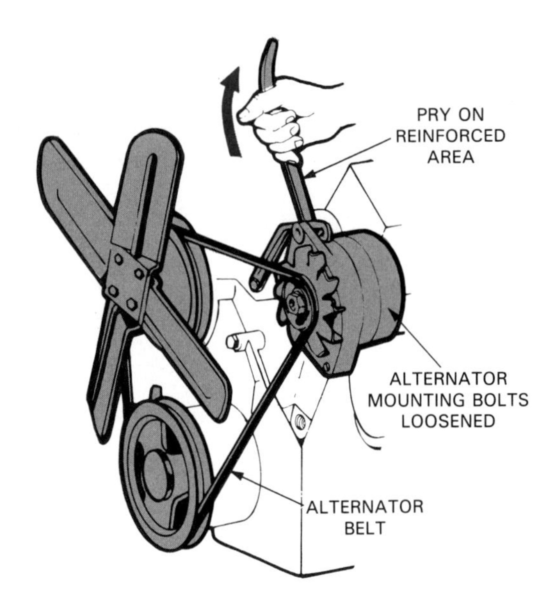

Fig. 13-35. Adjust alternator belt by prying on reinforced area of housing. Do not tighten belt too much or alternator bearings can fail. (Florida Dept. of Voc. Ed.)

There are four common tests made on a charging system:

1. CHARGING SYSTEM OUTPUT TEST (measurement of current and voltage output of charging system under a load).
2. REGULATOR VOLTAGE TEST (measurement of charging system voltage under low output, low load conditions).
3. REGULATOR BYPASS TEST (connect full battery voltage to alternator field, leaving regulator out of circuit).
4. CIRCUIT RESISTANCE TESTS (measure resistance in insulated and ground circuits of system).

Charging system output test

A *charging system output test* makes sure the alternator can produce enough current to operate the electrical devices and still have enough current to recharge the battery. This test can be done with a load tester or an accurate voltmeter.

Basically, to use a LOAD TESTER, connect the tester to the battery as described in the operating manual. Start the engine and load the system to the current specs of the alternator. The system should produce spec current and maintain spec voltage. If not, something is wrong in the system.

To use a VOLTMETER to check the charging system, connect the voltmeter to the battery cables. Then use three basic steps:

1. Measure battery voltage with the engine off. Write this value down. See Fig. 13-36A. Battery voltage should be around 12.3 to 12.5 V.
2. Start and fast idle the engine while all electrical devices are off. Again, read the voltmeter and record the voltage value. Voltage should increase above battery voltage by 2 or 3 volts. This checks regulator no-load voltage, Fig. 13-36B.
3. Now, turn on all electrical accessories (headlights, wipers, radio, air conditioning, etc.) with the engine running. Read the voltmeter, Fig. 13-36C.

Voltage will drop, but it must NOT drop below battery voltage (reading in step one or above 12.5 volts).

If voltage stays above battery voltage, there is enough charging output to run the accessories AND recharge the battery. If the voltage across the battery is equal to or below battery voltage, the charging system is NOT working properly. Other tests are needed to pinpoint the trouble.

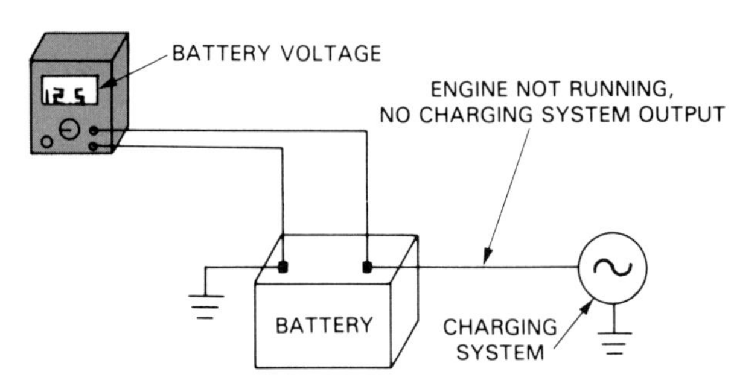

A—BATTERY VOLTAGE IS MEASURED TO CHECK BATTERY CONDITION AND ALLOW FOR METER INACCURACY.

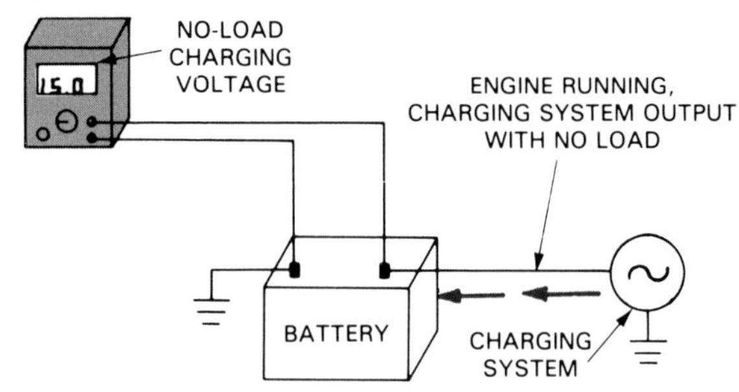

B—NO-LOAD TEST CHECKS ACTION OF VOLTAGE REGULATOR AND ALTERNATOR. VOLTAGE SHOULD INCREASE OVER BATTERY VOLTAGE BUT NOT BEYOND SPECS.

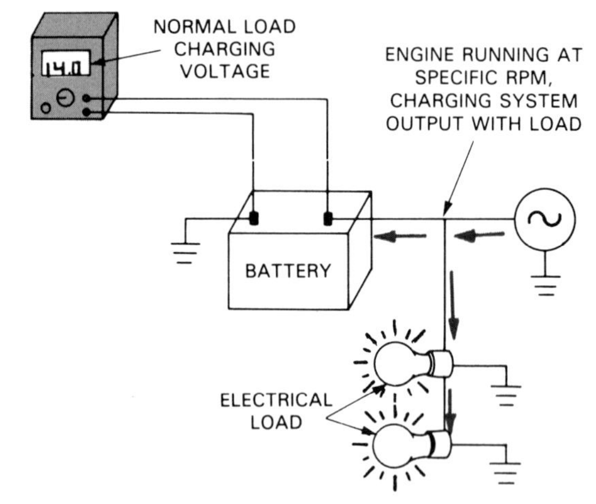

C—TURN ON HEADLIGHT HIGH BEAM AND TAILLIGHTS. THE VOLTAGE WILL DROP BUT SHOULD STAY ABOVE BATTERY VOLTAGE. IF VOLTAGE IS HIGHER THAN STEP A, CHARGING SYSTEM SHOULD BE RECHARGING BATTERY.

Fig. 13-36. This is simple method of using a voltmeter to check charging system output. A—Measure battery voltage. B—Measure charging voltage with a load. C—Measure charging voltage with accessory load. No-load voltage should be around 13 to 15 volts. Load voltage must stay above battery voltage to assure recharging.

Regulator by-pass test

A *regulator by-pass test* should be done when charging system output is below specs. It will help find out whether the alternator OR the voltage regulator is at fault.

To bypass the regulator, you must do one of the following:

1. Insert screwdriver into small hole on back of alternator.
2. Use jumper wire to connect battery voltage to alternator field connection on back of alternator.
3. Use jumper wire to connect voltage regulator harness terminals so direct battery voltage is applied to alternator field.

When the regulator is by-passed, the alternator should produce maximum current and voltage. If the voltage and current is NOT within specs, the ALTERNATOR IS BAD. If output is within specs, then the REGULATOR may be faulty or you could have WIRING TROUBLES.

Other tests, like circuit resistance tests, would be needed to find the exact trouble. Refer to a service manual for details.

Alternator service

A *bad alternator* will show up during your tests as a low voltage and current output problem. Even when the regulator is bypassed and full voltage is applied to the alternator field, charging voltage and current will NOT be up to specs.

Before unbolting the alternator, first DISCONNECT THE BATTERY to prevent damage to parts if wires are shorted. Shown in Fig. 13-37, most alternators are held to the front of the engine with two bolts. Loosen the bolts and remove the belt. Then remove the alternator.

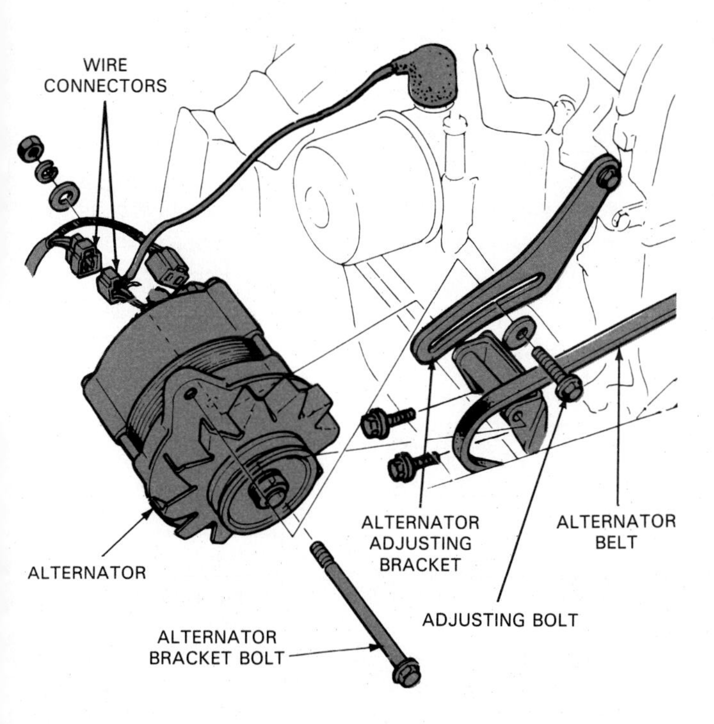

Fig. 13-37. Two or three bolts usually hold alternator on engine. Disconnect battery to avoid shorts that could damage electrical system when replacing alternator. (Honda)

When removing the wires from the back of the alternator, note their location and whether special insulating washers are used. If you make a mistake reattaching the wires to the alternator, system damage can occur. After reinstallation, adjust belt tension properly.

SUMMARY

A battery stores the energy needed for engine cranking. It is made to withstand severe vibration, temperature change, and other abuses.

Cars normally use 12 volt batteries that actually have an open circuit voltage of 12.6 volts. A maintenance-free battery has a sealed top and usually a charge indicator eye.

The starting system uses battery voltage to rotate the engine crankshaft. Current flows through the ignition switch, starter solenoid, and motor windings. Magnetic fields act upon each other to turn the starter armature.

The charging system must recharge the battery after engine cranking and after the use of electrical accessories. It uses an alternator, driven by the engine, to force current back into the battery. The voltage regulator maintains a charge voltage of about 13 to 15 volts, which is higher than the 12.6 battery voltage.

When troubleshooting and repairing a charging system or a starting system, remember that these systerms interact.

A car battery should be properly maintained to assure dependable engine starting. Inspect the battery for problems whenever you are working under the hood.

Jump-start a car by connecting the jumper cables positive to positive and negative to ground. The negative cable should be connected to any ground (negative) away from the batteries.

Make sure a battery charger is shut off before connecting the cables to the battery.

A load test is the best way to check battery condition. After connecting the load tester, turn the load control knob until the ammeter reads the output specs for the specific battery. Hold the load for 15 seconds while reading the voltmeter. The voltmeter should stay above 9.5 volts at room temperature if the battery is good. A lower voltage reading might indicate a bad battery.

A starter current draw test will check the general condition of the starting motor.

Do not tighten an alternator belt as tight as other engine belts. Too much belt tension could overheat the alternator bearings and cause their failure.

KNOW THESE TERMS

Discharging, Charging, Cell, Plate, Terminal, Electrolyte, Element, Battery voltage, Post, Side terminal, Maintenance-free battery, Charge indicator, Battery cable, Ignition switch, Solenoid, Starting motor, Pole pieces, Field windings, Armature, Commutator, Brushes, Pinion gear, Ring gear, Neutral safety switch, Relay, Alternator, Voltage regulator, Rotor, Stator, Rectifier, Diodes, Diagnostic connector, Dead battery, Battery terminal test, Jump starting, Battery charger, Hydrogen gas, Load tester, Battery load test, Starter current draw test, Starter shims, Charging system output test, Regulator by-pass test.

REVIEW QUESTIONS—CHAPTER 13

1. When discharging, a battery changes __________ energy into __________ energy.
2. Battery cell voltage is __________ volts and a 12 volt battery actually has __________ volts.
3. The positive battery post is normally smaller and is painted red. True or false?
4. How does temperature affect battery performance?
5. What is a starter solenoid?
6. Explain starting motor operation.
7. A ________ ________ ________ prevents engine cranking unless the transmission shift selector is in park or neutral.
8. Define the term "charging voltage."
9. Describe the operation of an alternator.
10. To quickly check for poor battery cable connections, you can measure the voltage drop across the cable-to-terminal connections of the battery. When cranking the engine, voltage drop should be:
    a. Above .5 volt.
    b. Below .5 volt.
    c. Around .9 volt.
    d. Below .3 volt.
11. What is the correct way of connecting jumper cables from one car to the next?
12. How do you load test a battery?
13. A customer complains of a rapid clicking or buzzing sound and no engine cranking when the key is turned to start.
    Technician A says that the starting motor must be bad. High current draw is overloading the solenoid.
    Technician B says that a battery load test should be performed because a weak battery or poor cable connections could let the solenoid kick in and out.
    Which technician is correct?
    a. Technician A.
    b. Technician B.
    c. Both A and B.
    d. Neither A nor B.
14. A humming sound, after momentary cranking, is commonly caused by:
    a. Faulty solenoid.
    b. Bad armature.
    c. Worn pinion gear unit.
    d. Loose flywheel bolts.
15. How can you use a voltmeter to check charging system operation?

ASE CERTIFICATION-TYPE QUESTIONS

1. When an automotive battery discharges during engine starting, the battery changes chemical energy into __________.
   (A) heat energy
   (B) electrical energy
   (C) mechanical energy
   (D) All of the above.
2. All of the following are basic parts of an automotive battery EXCEPT:
   (A) aluminum plates.
   (B) battery straps.
   (C) electrolyte.
   (D) battery case.
3. Technician A says that the starter pole pieces are located next to the commutator. Technician B says that the starter pole pieces are mounted inside the field windings. Who is right?
   (A) A only.
   (B) B only.
   (C) Both A & B.
   (D) Neither A nor B.
4. The starter solenoid is used on some automobiles to __________.
   (A) push the starter pinion gear into mesh with the flywheel gear
   (B) bypass the resistance wire in the ignition circuit
   (C) energize the alternator stator
   (D) Both A & B.
5. Technician A says that one of the functions of an automotive charging system is to supply all the car's electricity when the engine is running. Technician B says that one of the functions of an automotive charging system is to provide voltage output slightly higher than battery voltage. Who is right?
   (A) A only.
   (B) B only.
   (C) Both A & B.
   (D) Neither A nor B.
6. A charging system voltage regulator normally keeps alternator output at approximately __________.
   (A) 16-18 volts
   (B) 13-15 volts
   (C) 11-13 volts
   (D) 13-14 amps
7. All of the following are basic components of a typical alternator EXCEPT:
   (A) rotor assembly.
   (B) armature shaft.
   (C) rectifier assembly.
   (D) brush assembly.
8. Technician A says that when performing a battery load test on a battery rated at 80 amp-hours, the battery should be tested at 240 amps. Technician B says that when performing a battery load test on a battery rated at 80 amp-hours, the battery should be tested at 160 amps. Who is right?
   (A) A only.
   (B) B only.
   (C) Both A & B.
   (D) Neither A nor B.
9. When performing a starter current draw test, you should crank the engine for no more than __________ seconds while reading the test results.
   (A) 45-60
   (B) 15-30
   (C) 30-45
   (D) 60-90
10. Technician A says that a regulator bypass test can detect an alternator problem. Technician B says that a regulator bypass test can detect a faulty voltage regulator. Who is right?
    (A) A only.
    (B) B only.
    (C) Both A & B.
    (D) Neither A nor B.

Chapter 14

Ignition Systems—Operation and Service

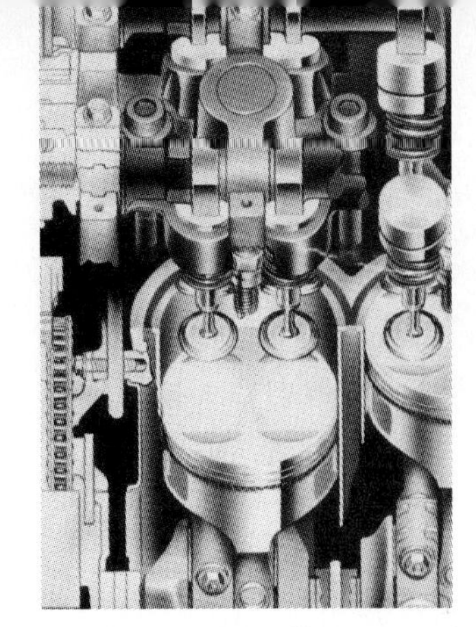

After studying this chapter, you will be able to:
- *Explain how an ignition system operates.*
- *Compare ignition system design variations.*
- *Sketch the primary and secondary circuits of an ignition system.*
- *Troubleshoot basic ignition system problems.*
- *Perform basic ignition system tests.*
- *Gap and replace spark plugs.*
- *Adjust ignition timing.*
- *Adjust contact points and pickup coil gaps.*
- *Replace major components of an ignition system.*

A gasoline engine's ignition system is very important to engine power and efficiency. The ignition system must make the fuel start burning in each cylinder at exactly the right time. Combustion must begin a few degrees before TDC on the compression stroke. This assures that combustion pressure acts on the piston properly.

This chapter will briefly summarize the most essential information needed to become a competent engine technician. First, it will review the theory and construction of ignition system components. Then, the chapter will summarize the most common tests, adjustments, and repairs. As a result, you will be better prepared to correct engine and ignition system problems when on the job.

IGNITION SYSTEM FUNDAMENTALS

An *ignition system* changes battery voltage into extremely high voltage for operating the spark plugs. It must also send these high voltage surges to the spark plugs at exactly the right time.

Today's engines can have one of two basic types of ignition systems:

1. DISTRIBUTOR IGNITION SYSTEM (mechanical-electrical device for operating system and feeding voltage to spark plug wires), Fig. 14-1A.
2. DISTRIBUTORLESS IGNITION (computer and sensors operate multiple ignition coils), Fig. 14-1B.

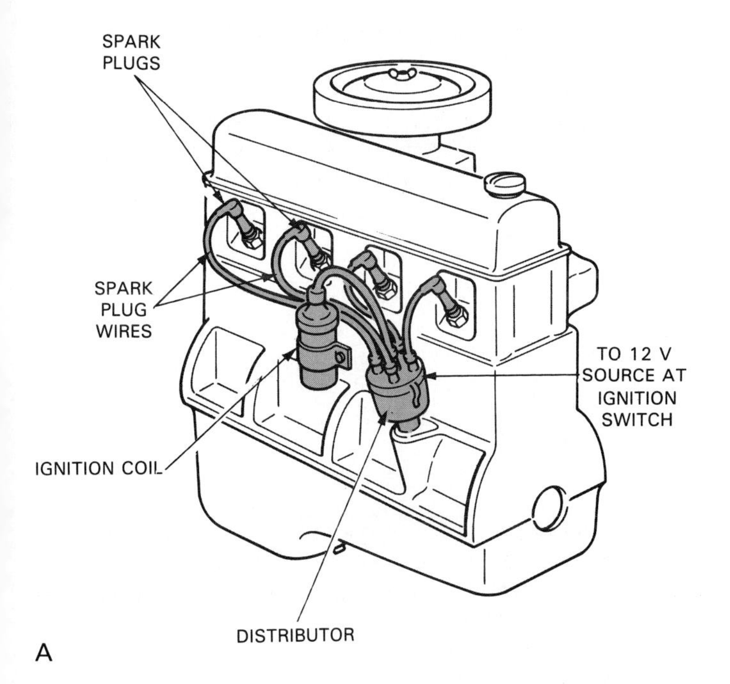

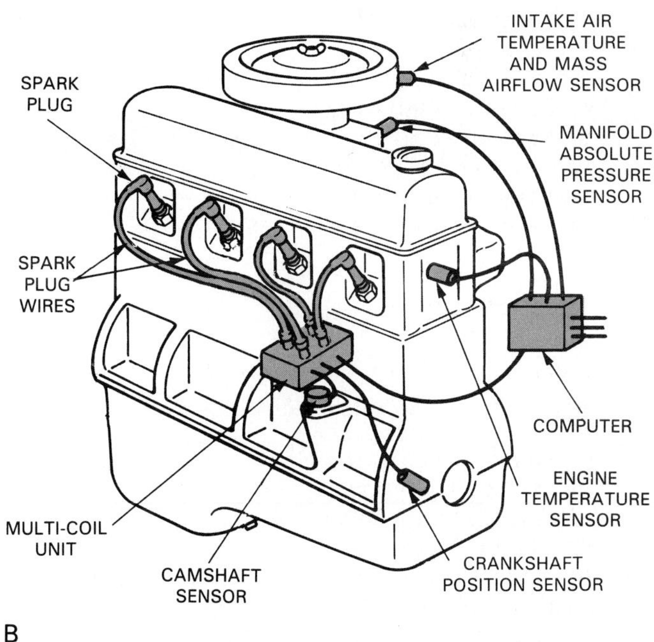

Fig. 14-1. Today's ignition systems can be distributor or computer controlled. A—Older distributor ignition is still found on millions of engines. B—Distributorless ignition uses sensors, electronic control unit, and multiple coil unit to fire plugs. (Ford)

Functions of ignition system

With either type, an ignition has several critical functions to perform. These include:

1. Provide a method of turning a spark ignition or gasoline engine ON and OFF.
2. Be capable of operating on various supply voltages (battery or alternator voltage).
3. Produce a high voltage arc at the spark plug electrodes.
4. Distribute high voltage pulses in the correct sequence.
5. Time the spark so that it occurs as the piston nears TDC on the compression stroke.
6. Vary spark timing with engine speed, load, and other conditions.

Primary and secondary circuits

The two main sections of an ignition system are the primary and secondary circuits.

The *primary circuit* includes all of the components and wires operating on low voltage (battery or alternator voltage). This is shown in Fig. 14-2A.

The *secondary circuit* is the high voltage (30,000 volt) section, Fig. 14-2B. It consists of the wires and parts

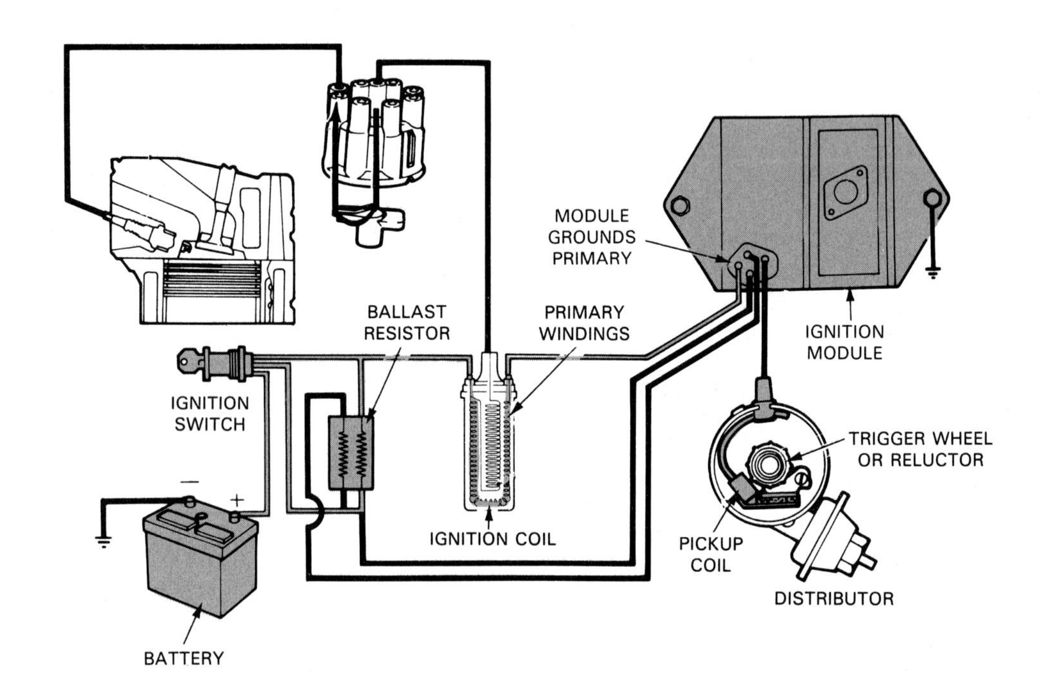

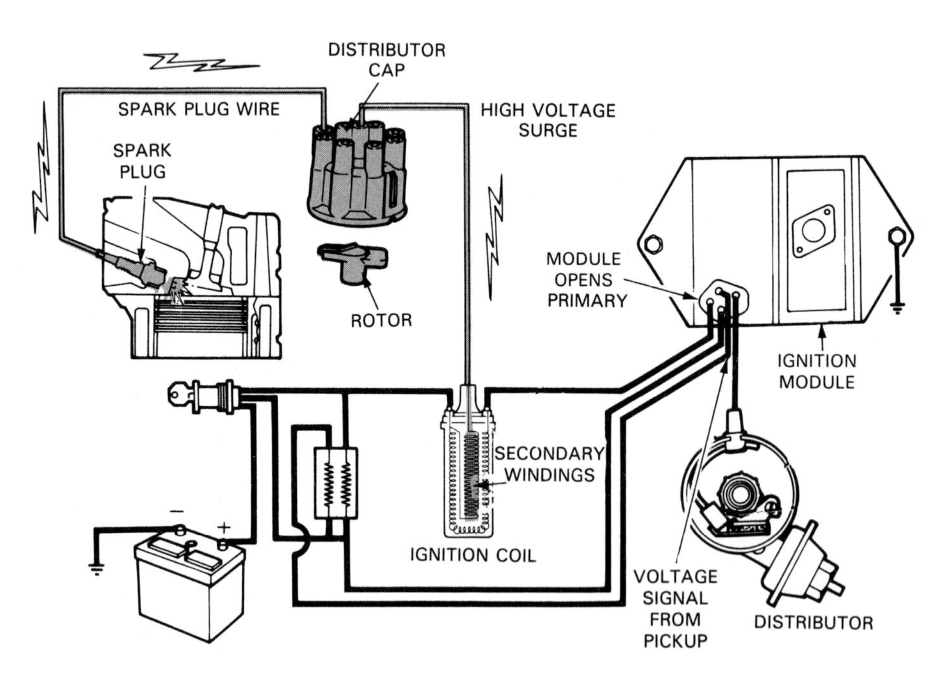

Fig. 14-2. A—Primary section of ignition system is low voltage circuit. B—Secondary of ignition is high voltage circuit. (Chrysler)

between the ignition coil output and the spark plug ground.

The primary circuit of the ignition system uses conventional wire, similar to the wire used in the other electrical systems of the car. The secondary wiring (spark plug wire and coil wire), however, must have much THICKER INSULATION to prevent *leakage* (arcing) of the high voltage.

Parts of ignition system

The basic parts of a distributor type ignition system are illustrated in Fig. 14-2 and they include:

1. BATTERY (source of electrical energy for system when engine is not running).
2. IGNITION SWITCH (switch for connecting and disconnecting voltage to circuit so driver can turn gasoline engine on and off).
3. BALLAST RESISTOR or RESISTANCE WIRE (reduces amount of battery voltage in circuit to about 9.5 to 10 volts).
4. IGNITION COIL (step-up transformer for changing battery voltage into high voltage for spark plugs).
5. SWITCHING MECHANISM (either mechanical contact points or electronic circuit for interrupting current flow through ignition coil so that coil fires).
6. SPARK PLUGS (provide air gaps in combustion chambers so that electric arcs can start combustion).
7. WIRING (primary and secondary wire for connecting components of ignition circuit).

Ignition system operation

With the ignition switch ON, current flows to the parts of the ignition system. When the switching device is closed (conducting current), current flows through and energizes the ignition coil.

When the piston is nearing TDC on the compression stroke, the switching device opens. This causes high voltage to shoot out of the ignition coil and to the spark plug.

The electric arc at the plug ignites the fuel mixture. The mixture begins to burn, forming pressure in the cylinder for the engine's power stroke.

When the ignition key is turned to OFF, the battery-to-coil circuit is broken. Without current to the ignition coil, sparks are NOT produced at the spark plugs and the engine stops running.

IGNITION COIL

An *ignition coil* produces the high voltage (30,000 volts or more) needed to make current jump the gap at the spark plugs. It is a pulse type transformer capable of producing a short burst of high voltage.

The ignition coil consists of two sets of *windings* (insulated wire wrapped in circular pattern). The coil has two primary terminals (low voltage connections), an *iron core* (long piece of iron inside windings), and a high voltage terminal (output or coil wire connection). This is shown in Fig. 14-3.

The coil's *primary windings* are several hundred turns of heavy wire. Low voltage flows through these windings to produce a magnetic field.

The *secondary windings* in the ignition coil are several

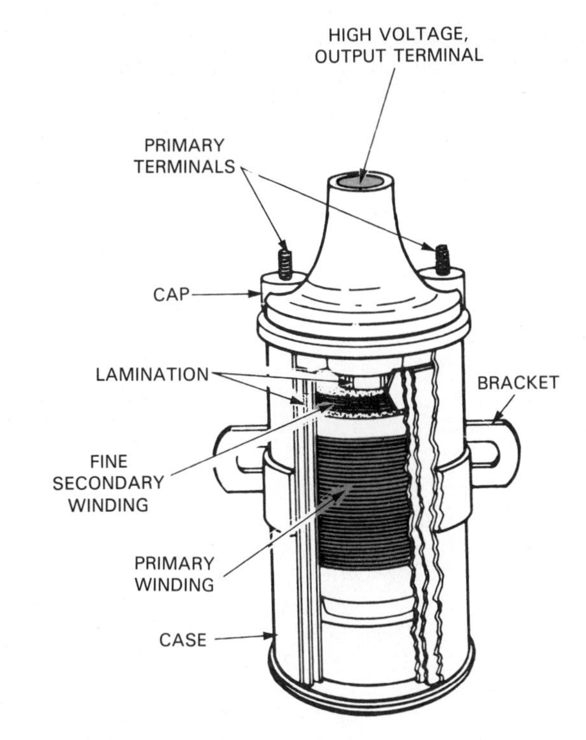

Fig. 14-3. Ignition coil produces high voltage for spark plugs. Primary windings surround secondary windings. Primary terminals carry current to primary windings. Secondary windings connect to output terminal.

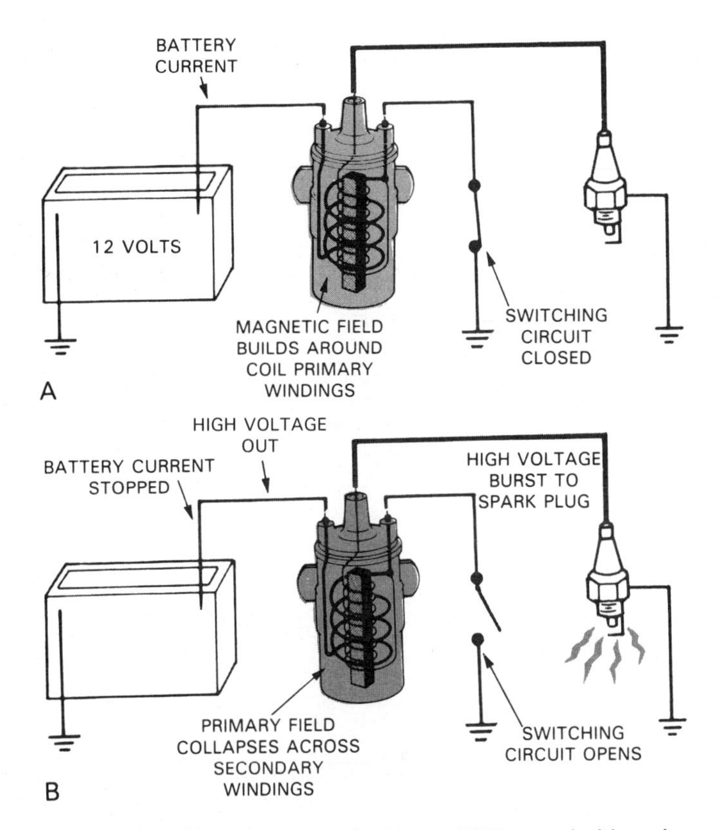

Fig. 14-4. Basic coil operation. When ECU or switching circuit is closed, current flows through primary of coil. Magnetic field builds around windings. When ECU or switching circuit opens, primary field collapses across secondary. High voltage is then induced into secondary windings and shoots out top of coil. (Saab)

thousand turns of fine wire. Both windings are wrapped around an iron core.

When the switching device opens, current flow is stopped in the primary windings, Fig. 14-4. This makes the strong magnetic field collapse across the secondary coil windings. The magnetic field movement induces voltage into the secondary windings. Since there are more turns or windings in the secondary, much high voltage is produced.

SPARK PLUGS

The *spark plugs* use ignition coil high voltage to ignite the fuel mixture, Fig. 14-5. Somewhere between 4000 and 10,000 volts are needed to make current jump the gap at the plug electrodes. This is much lower than the coil's output potential.

The parts of a spark plug are:

1. CENTER ELECTRODE (metal rod in center of plug that conducts current to air gap), Fig. 14-5.
2. SHELL (metal housing with hex for tightening plug in cylinder head).
3. INSULATOR (ceramic enclosure around center electrode to prevent electrical arcing to cylinder head). See Fig. 14-5.
4. TERMINAL BOLT (upper half of center electrode that also provides electrical connection to plug wire).
5. SIDE ELECTRODE (metal prong on tip of plug that produces air gap with center electrode), Fig. 14-5.

Spark plug gap

Spark plug gap is the distance between the center and side electrodes, Fig. 14-5. Normal gap specs range from .030 to .060 inch (0.76 mm to 1.52 mm).

Smaller spark plug gaps are used on older engines equipped with contact point ignition systems. Larger spark plug gaps are now used with modern electronic ignition systems.

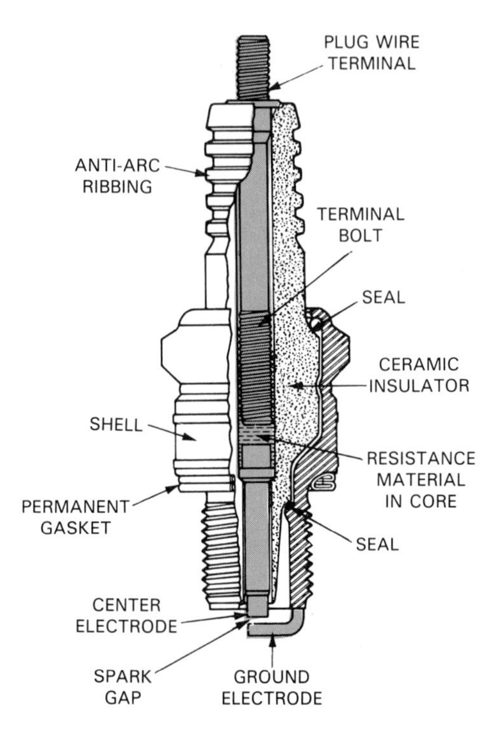

Fig. 14-5. Cutaway shows basic parts of spark plug. (VW)

Spark plug heat range

Spark plug heat range is a rating of the operating temperature of the plug tip. Plug heat range is basically determined by the length and diameter of the ceramic insulator tip and the ability of the plug to transfer heat into the cooling system.

A *hot spark plug* has a long insulator tip and will tend to burn off deposits and self-clean itself. See Fig. 14-6.

A *cold spark plug* has a shorter insulator tip; its tip operates at a cooler temperature. A cold plug is used in engines operated at high speeds. The cooler tip will help prevent tip overheating and preignition, Fig. 14-6.

Auto manufacturers normally recommend a specific spark plug heat range for their engines. The heat range will normally be coded and given as a number on the plug insulator.

Generally, the larger the number on the plug, the hotter the spark plug tip will operate. For instance, a 52 plug would be hotter than a 42 or 32.

The only time you should deviate from plug heat range specs is when abnormal engine or driving conditions are encountered. For example, a hotter plug may be installed in an old, worn out, oil-burning engine. The hotter plug will help burn off oil deposits and prevent oil fouling of the plug.

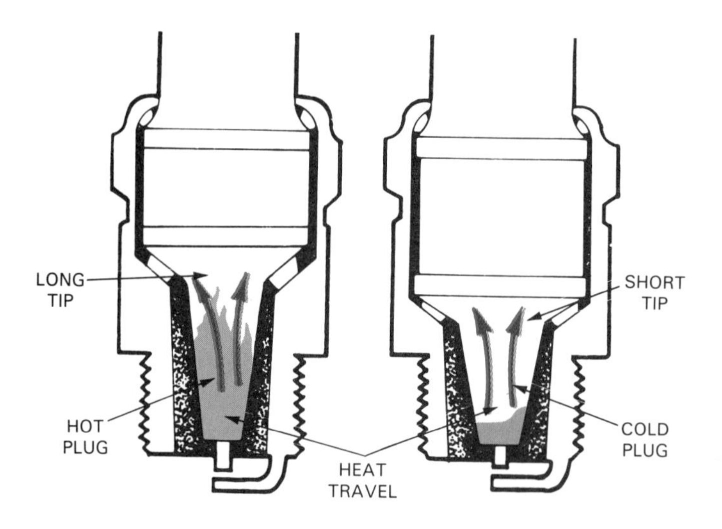

Fig. 14-6. Hot plug would have longer tip. More heat can build in tip during engine operation to clean off deposits. Cold plug's shorter tip can dissipate heat into head easier for high speed operation. (Champion)

Spark plug reach

Spark plug reach is the distance between the end of the plug threads and the seat or sealing surface on the plug. Plug reach determines how far the plug extends through the cylinder head, Fig. 14-7.

If spark plug reach is too long, the plug electrode may be struck by the piston at TDC. If reach is too short, the

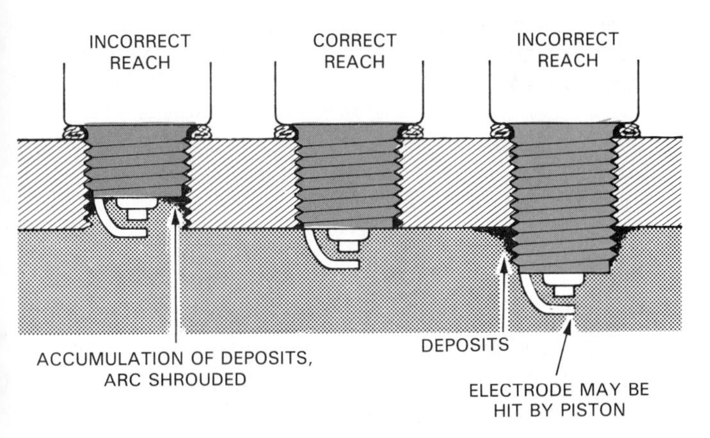

Fig. 14-7. Reach is determined by length of plug threads. Note problems with wrong reach. (WGK)

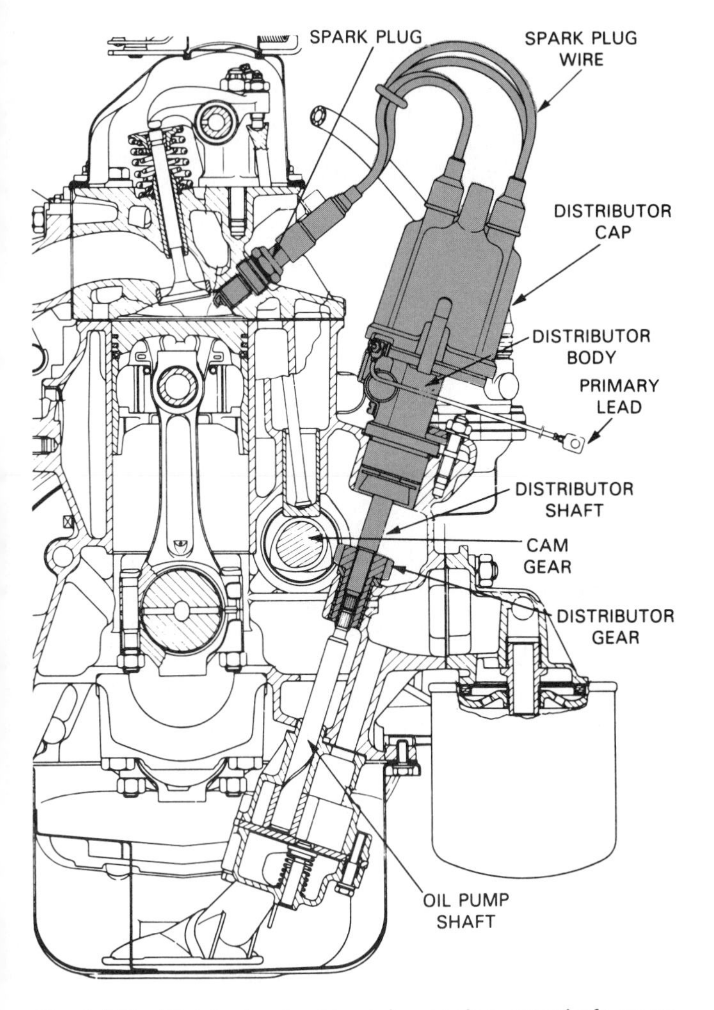

Fig. 14-8. Distributor is driven by engine camshaft or accessory shaft. It turns at one-half engine speed. (Fiat)

plug electrodes may not extend far enough into the chamber and combustion efficiency may be reduced.

Spark plug resistance

A *resistor spark plug* has built-in internal resistance to help prevent radio noise or static. The resistance is about 10,000 ohms. Almost all new cars come equipped with resistor spark plugs.

A *non-resistor spark plug* does not have internal resistance. The center electrode is a solid piece of metal from one end to the other. This type is seldom used on passenger cars. Off-road and racing engines frequently use non-resistor spark plugs.

IGNITION DISTRIBUTORS

An *ignition distributor* can have several functions:

1. Actuate ON/OFF cycles of current flow through ignition coil primary windings.
2. Distribute coil's high voltage pulses to each spark plug wire.
3. Cause spark to occur at each plug earlier in compression stroke as engine speed increases and vice versa.
4. Change spark timing with changes in engine load. As more load is placed on engine, spark timing must occur later in the compression stroke to prevent spark knock (abnormal combustion).
5. Sometimes, bottom of the distributor shaft powers the engine oil pump.
6. Some distributors house the ignition coil and electronic switching circuit internally.

There are two basic types of distributors. Modern types have a pickup coil for operating the electronic switching circuit. Older types have mechanical breakers or contact points that operate the ignition coil directly.

Fig. 14-8 shows how the camshaft can be used to drive the distributor. Small primary wires connect to the electronic control unit (electronic switching circuit) or to the ignition coil when equipped with breaker points. Spark plug wires extend out from the distributor cap to the plugs.

Contact point distributor

Before going on to study today's electronic ignition systems, you should have a basic understanding of older contact point systems. The operation of each is similar in many ways.

The *distributor cam* is the lobed part on the distributor shaft that opens the contact points, Fig. 14-9. The cam turns with the shaft at one-half engine speed. One lobe is normally provided for each spark plug.

The *contact points,* also called *breaker points,* is a spring-loaded electrical switch in the distributor. Small screws hold the contact points on the distributor advance plate. A *rubbing block,* of fiber material, rides on the distributor cam. Wires from the condenser and ignition coil primary connect to the points, Fig. 14-9.

The *condenser* or *capacitor* prevents the contact points from arcing and burning, Fig. 14-9. It also provides a storage place for electricity as the points open. This electricity is fed back into the primary when the points reclose.

Contact point operation

With the engine running, the distributor shaft and distributor cam rotate. This causes the cam to open and close the points.

Since the points are wired to the primary windings of the ignition coil, the points make and break the ignition

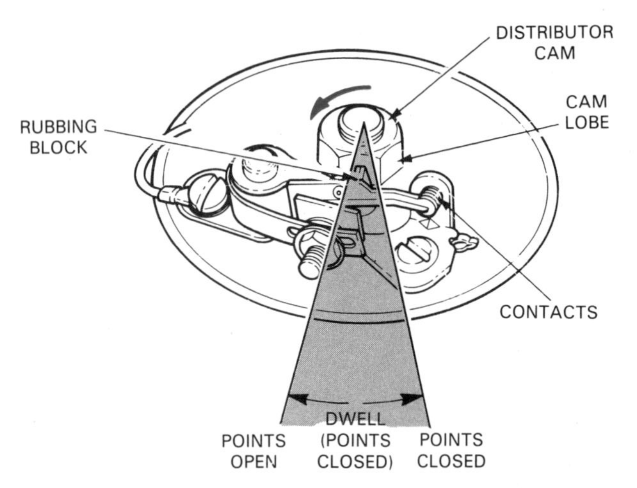

Fig. 14-9. Dwell or cam angle is time when points are closed. Distributor cam has small lobes that push points open. Lobes act on fiber rubbing block. (Chrysler)

coil primary circuit. When the points are closed, a magnetic field builds in the coil. When the points open, the field collapses and voltage is sent to one of the spark plugs.

Dwell (cam angle)

Dwell or *cam angle* is the amount of time, given in degrees of distributor rotation, that the points remain closed between each opening. A typical dwell for a V-8 engine is 30°. A dwell period is needed to assure that the coil has enough time to build up a strong magnetic field, Fig. 14-9.

Contact point gap

Contact point gap is the distance between the opened points. As you will learn later, it is adjusted during a tune-up and it affects dwell.

ELECTRONIC IGNITION SYSTEM

An *electronic ignition system,* also called a *solid state* or *transistor ignition system,* uses an electronic control circuit and a distributor pickup coil to operate the ignition coil.

An electronic ignition is more dependable than a contact point type. There are no mechanical breakers to wear or burn. This helps avoid trouble with ignition timing and dwell.

An electronic ignition is also capable of producing much higher secondary voltages. This is an advantage because wider spark plug gaps and higher voltages are needed to ignite today's lean air-fuel mixtures. Lean mixtures are now used for reduced exhaust emissions and fuel consumption.

Trigger wheel

The *trigger wheel,* also called *reluctor* or *pole piece,* is fastened to the upper end of the distributor shaft. With new systems, the trigger wheel can be mounted on the crankshaft and/or camshaft. The trigger wheel induces a signal into the sensor or pickup so the computer can determine engine speed and crankshaft/camshaft positions. Sometimes one or more trigger wheel teeth represent each cylinder, Fig. 14-10.

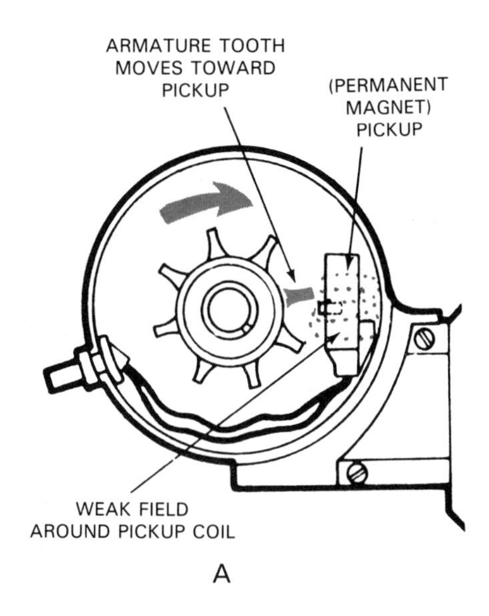

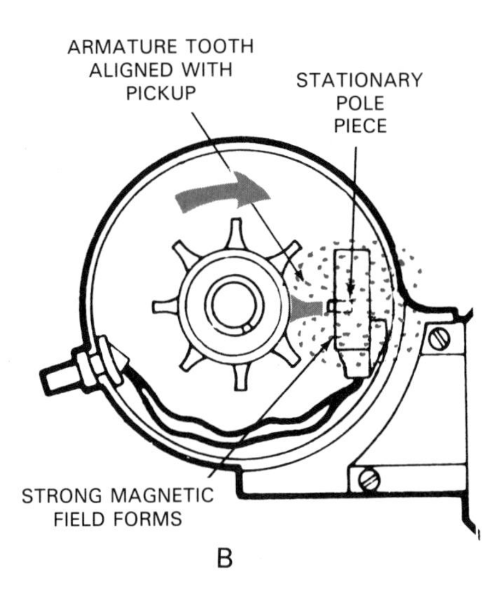

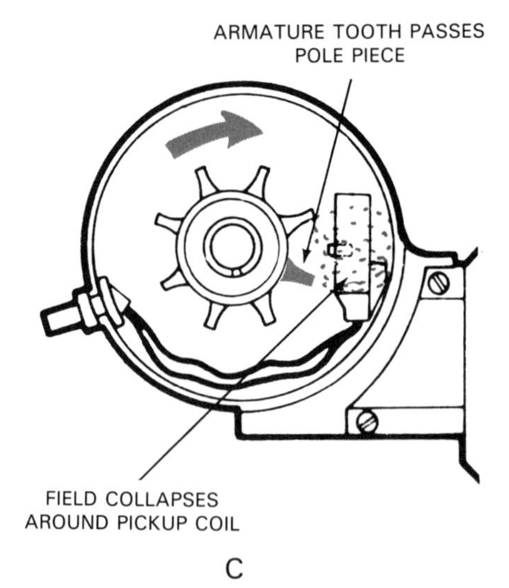

Fig. 14-10. Electronic distributor operation. A—Trigger wheel tooth moves toward pickup coil. Field around coil is still weak. B—Trigger wheel tooth aligns with pickup coil and field is strengthened. C—Tooth passes pickup and field again weakens. Change of field strength around pickup produces signal for electronic switching circuit. (Florida Dept. of Voc. Ed.)

Pickup coil

The *pickup coil,* also called *sensor assembly* or *sensor coil,* produces tiny voltage pulses for the ignition system's electronic control unit. The sensor assembly is a small set of windings forming a coil, Fig. 14-10.

As a trigger wheel tooth passes the pickup coil, it strengthens the magnetic field around the coil. This causes a change in the current flow through the coil. As a result, an electrical pulse (voltage or current change) is sent to the electronic control unit as the trigger wheel teeth pass the pickup, Fig. 14-10.

Ignition system ECU

The ignition system *electronic control unit, amplifier* or *control module,* is an "electronic switch" that turns the ignition coil primary current on and off. It looks like a small box with wires coming out. The ECU does the same thing as contact points, Fig. 14-11.

An ignition module is a network of transistors, resistors, capacitors, and other electronic components. The circuit is sealed in a plastic or metal housing.

The module can be located:

1. On the side of the distributor.
2. Inside the distributor.
3. In the engine compartment.
4. Under the car dash.

Electronic ignition operation

With the engine running, the trigger wheel spins with engine operation, Fig. 14-11. As the teeth pass the pickup, a change in the magnetic field causes a change in output voltage or current. This results in engine rpm electrical signals entering the module.

The module increases these tiny pulses into ON/OFF current cycles for the ignition coil. When the module is ON, current flows through the primary windings of the

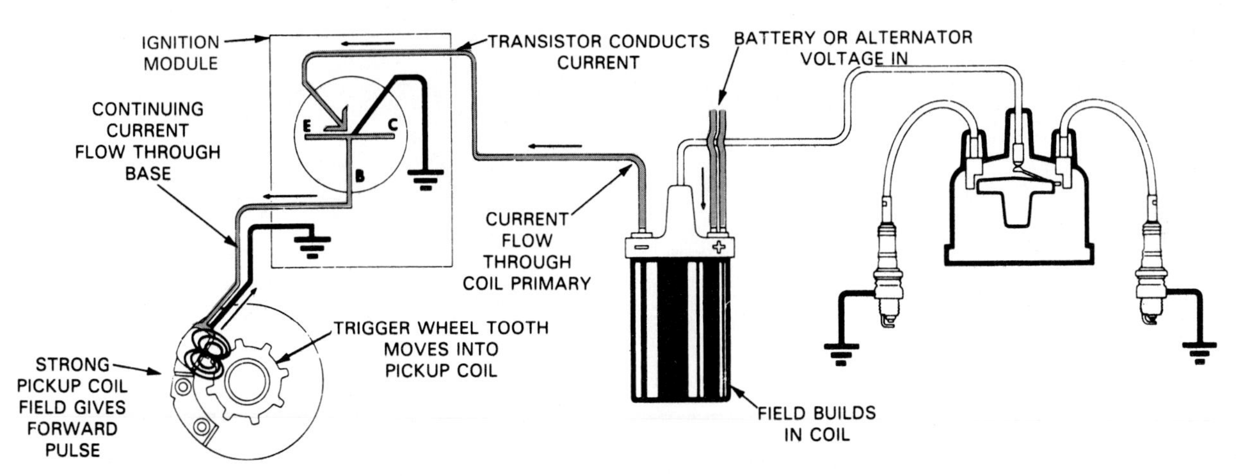

A — As trigger wheel tooth aligns with pickup coil, current flow through base of transistor turns transistor on. Current flows through ignition coil primary and through emitter-collector of transistor. Strong field builds in ignition coil primary windings.

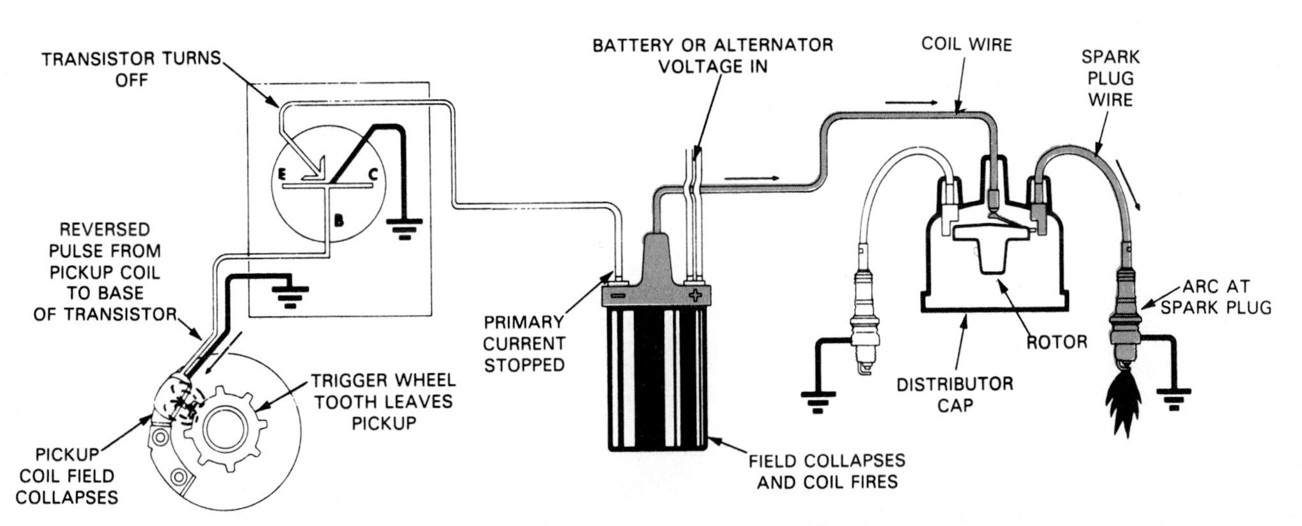

B — Just as trigger wheel passes pickup coil, current pulse flows out of pickup and to base of transistor. Electrical pulse is opposite the polarity of emitter-base voltage. This turns transistor off. Without current flow through emitter and collector, field collapses in ignition coil and 30,000 volts are induced into coil secondary windings. Spark plug fires.

Fig. 14-11. Electronic ignition system operation. A—Trigger wheel moves past pickup coil and produces signal. Transistor conducts current through ignition coil. B—Tooth moves away from pickup coil and transistor shuts off. This fires ignition coil and spark plug. (Echlin)

ignition coil, developing a magnetic field. Then, when the trigger wheel and pickup turn the module OFF, the ignition coil field collapses and fires a spark plug.

COIL AND SPARK PLUG WIRES

The *coil wire* carries high voltage from the high voltage terminal of the ignition coil to the center terminal of the distributor cap, Fig. 14-11. It is constructed like a very short spark plug wire.

With a unitized distributor or distributorless ignition, a coil wire is NOT needed.

Spark plug wires carry coil voltage from the distributor cap side terminals to each spark plug, Fig. 14-11. In more modern computer-coil (distributorless) ignitions, the spark plug wires carry coil voltage directly to the plugs.

Spark plug wire boots protect the metal connectors from corrosion, oil, and moisture. Boots fit over both ends of the secondary wires.

Solid spark plug wires are used on racing engines and very old automobiles. The wire conductor is simply a stranded metal wire.

Resistance spark plug wires are now used because they contain internal resistance that prevents radio noise. They use carbon-impregnated strands of rayon braid. Also called *radio suppression wires,* they have about 10,000 ohms per foot. This avoids high voltage induced popping or cracking in the radio speakers.

DISTRIBUTOR CAP AND ROTOR

The *distributor cap* is an insulating, plastic part that fits over the top of the distributor housing. Its center terminal transfers voltage from the coil wire to the rotor. The distributor cap also has outer or side terminals that send electric arcs to the spark plug wires. Metal terminals are molded into the plastic cap. Sometimes the cap also holds the ignition coil, Fig. 14-12.

The *distributor rotor* transfers voltage from the distributor cap center terminal to the distributor cap outer terminals.

The rotor is mounted on top of the distributor shaft. It is a spinning electrical switch that feeds voltage to each spark plug wire, Fig. 14-12.

Voltage is high enough that it can jump the air space between the rotor and cap. About 3000 volts is used as the spark jumps this rotor-to-cap gap. See Fig. 14-13.

IGNITION TIMING

Ignition timing, also called *spark timing,* refers to when the spark plugs fire in relation to the position of the engine pistons. Ignition timing must vary with engine speed and load.

Timing advance

Timing advance occurs when the spark plugs fire sooner on the engine's compression strokes. The timing is set several degrees before TDC. More advance is needed at higher engine speeds to give combustion enough time to develop pressure on the power stroke. See Fig. 14-14.

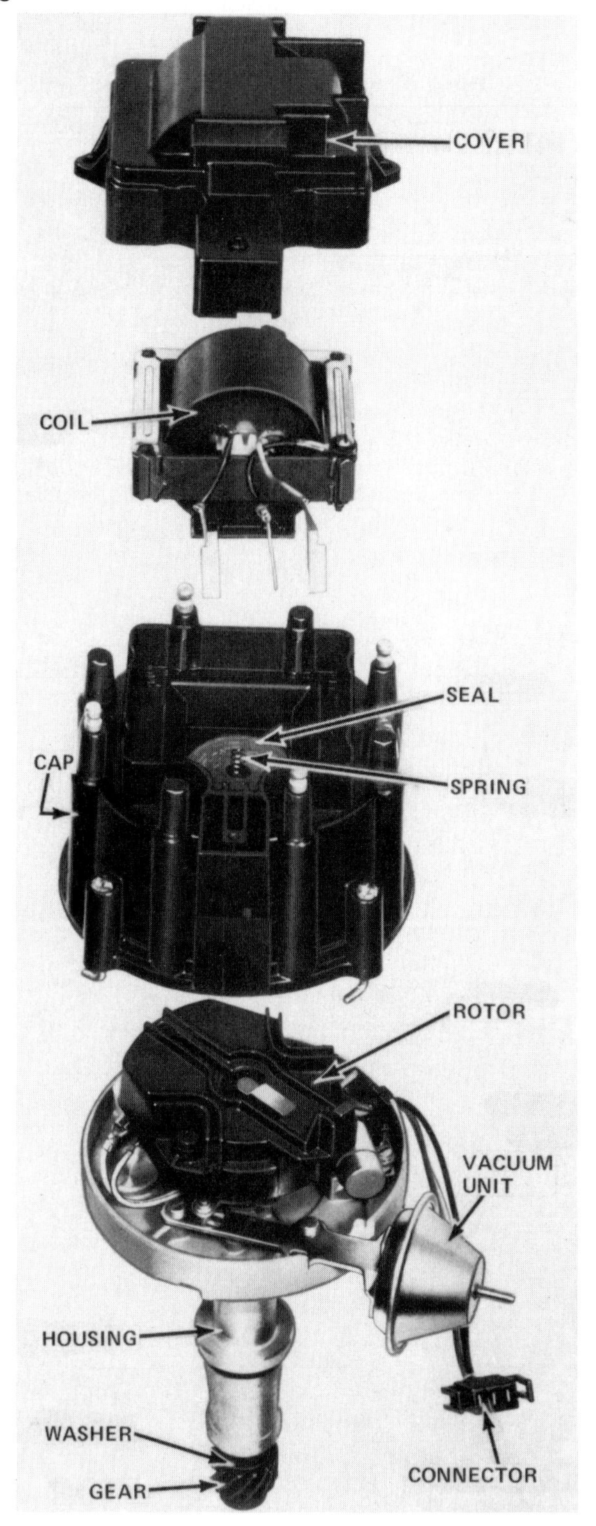

Fig. 14-12. This is a unitized distributor. Ignition coil-pickup coil, and module are all inside distributor. (Chevrolet)

Timing retard

Timing retard occurs when the spark plugs fire later on the compression strokes. It is the opposite of timing advance. Spark retard is needed at lower engine speeds and under high load conditions. Timing retard prevents the fuel from burning too much on the compression stroke, causing spark knock or ping (abnormal combustion).

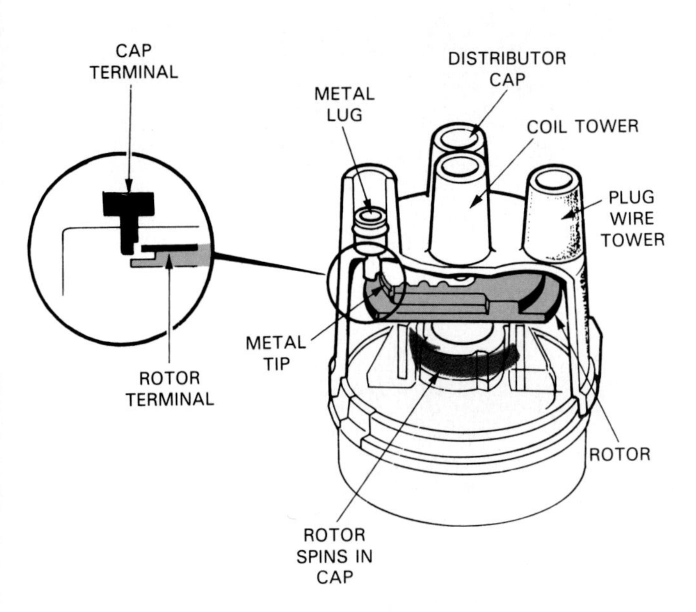

Fig. 14-13. Rotor spins inside distributor cap. It sends coil voltage from center cap terminal to each outer, spark plug wire terminal. Electric arcs jump gap between rotor tip and other cap terminals. (Honda)

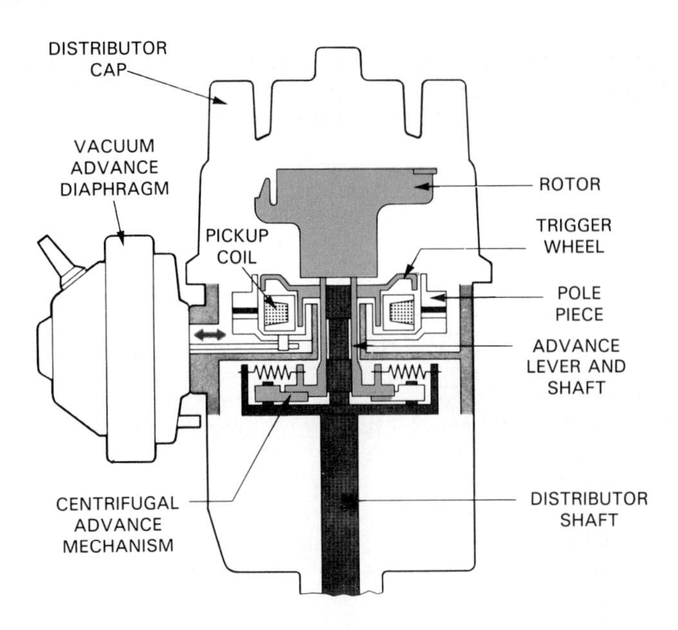

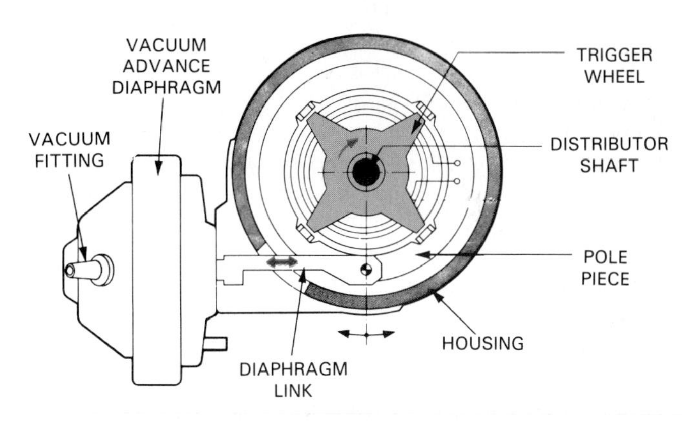

Fig. 14-15. Note parts of distributor. Centrifugal advance uses spring-loaded weights to rotate trigger wheel against shaft rotation as speed increases. Vacuum advance also turns plate and pickup coil against shaft rotation as needed. (Bosch)

Controlling ignition timing

There are three basic methods used to control spark timing:

1. DISTRIBUTOR CENTRIFUGAL ADVANCE (controlled by engine speed), Fig. 14-15.
2. DISTRIBUTOR VACUUM ADVANCE (controlled by engine intake manifold vacuum and load), Fig. 14-15.
3. ELECTRONIC (COMPUTER) ADVANCE (controlled by computer and various engine sensors: engine rpm, temperature, intake manifold vacuum, throttle position, etc.)

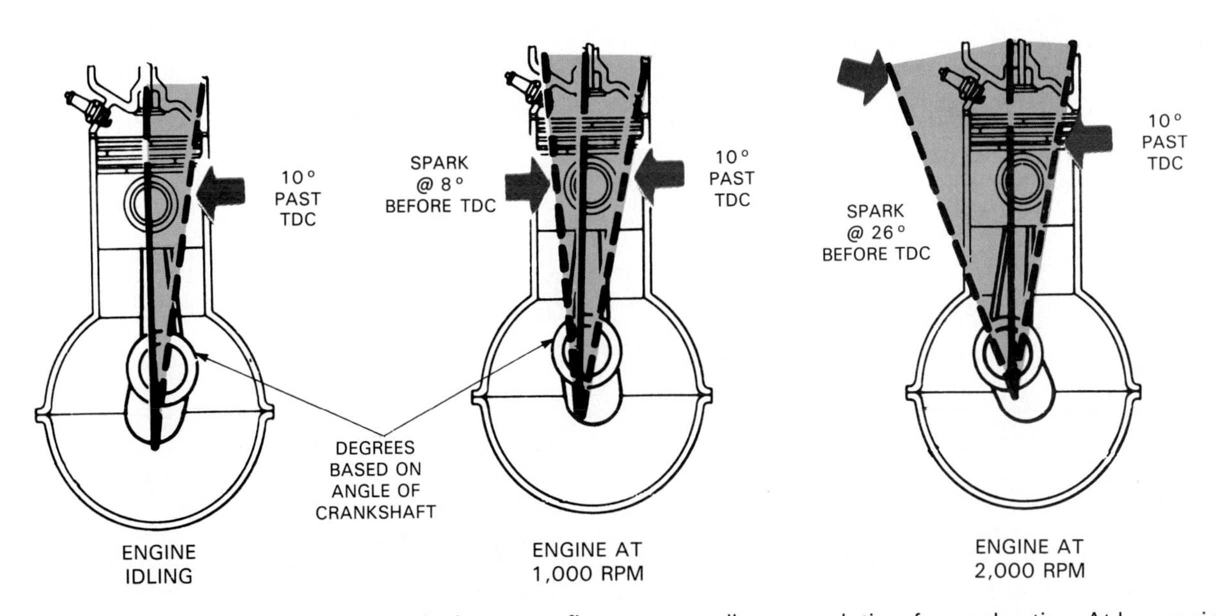

Fig. 14-14. As engine speed increases, spark plugs must fire sooner to allow enough time for combustion. At low engine speeds, only a slight timing advance is needed. Centrifugal advance changes timing with engine speed. (GM Trucks)

Distributor centrifugal advance

The *distributor centrifugal advance* makes the ignition coil fire sooner as engine speed increases. It uses spring-loaded weights, centrifugal force, and lever action to rotate the distributor cam, or trigger wheel on the distributor shaft. By rotating the cam or trigger wheel against distributor shaft rotation, spark timing is advanced, Fig. 14-15.

Distributor vacuum advance

The *distributor vacuum advance* provides additional spark advance when engine load is low at part throttle. It is a method of matching ignition timing with load.

The vacuum advance mechanism increases FUEL ECONOMY because is helps maintain ideal spark advance at all times. Look at Fig. 14-16.

Fig. 14-17 shows the top view of a distributor. Note the vacuum advance diaphragm, pickup coil, and electronic module.

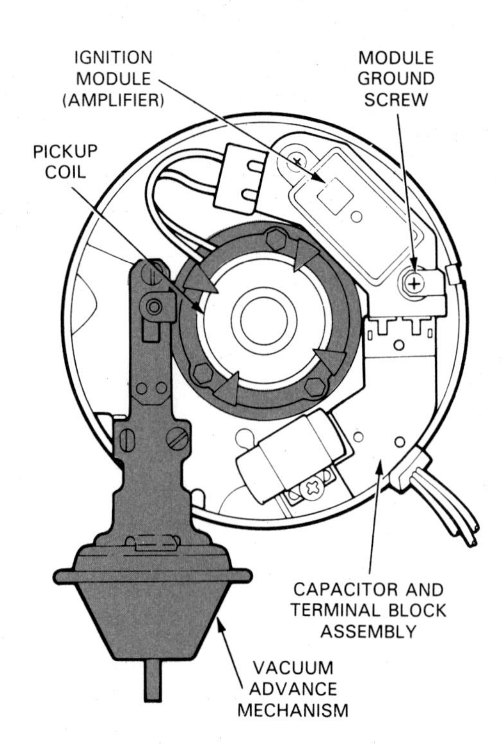

Fig. 14-17. Cap and rotor are removed from this electronic distributor. (Chrysler)

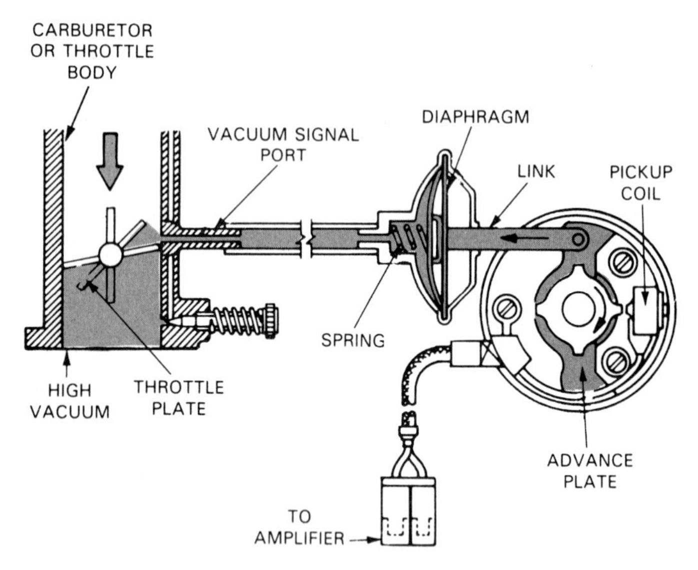

Fig. 14-16. Basically, vacuum hose is attached between carburetor or throttle body fitting and diaphragm. When throttle opens, it provides vacuum signal to diaphragm. Vacuum pulls diaphragm to advance timing under cruising, low load conditions. (Fiat)

Electronic spark advance

Electronic spark advance uses engine sensors and a computer to control ignition timing. A distributor may be used but it does NOT contain centrifugal or vacuum advance mechanisms, Fig. 14-18.

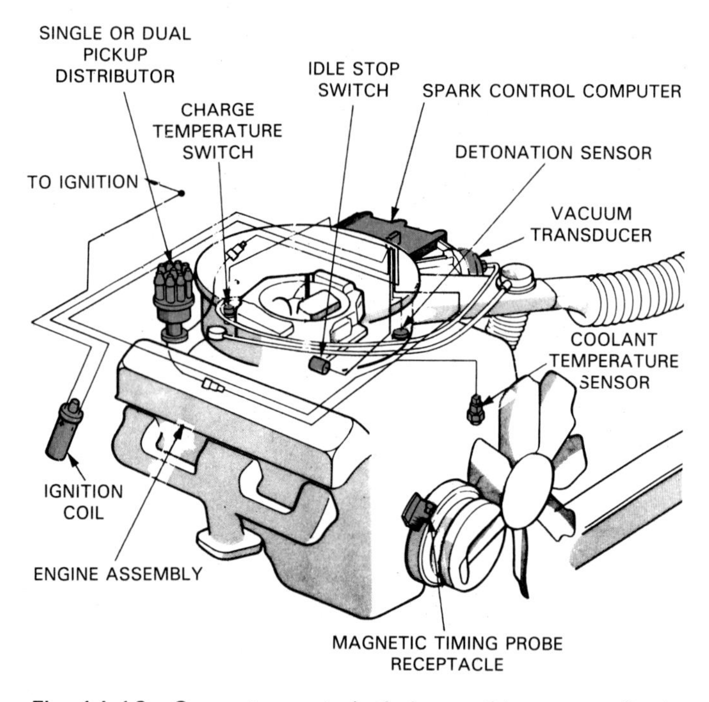

Fig. 14-18. Computer controls timing on this system. Study sensor names and locations. (Chrysler)

Engine sensors

The *engine sensors* check various engine operating conditions and send electrical data to the computer. The computer can then change ignition timing for maximum engine efficiency.

Ignition system engine sensors typically include:

1. ENGINE SPEED SENSOR (reports engine rpm to computer).
2. CRANKSHAFT POSITION SENSOR (reports piston position).
3. MAP SENSOR (measures engine vacuum, an indicator of load).
4. INTAKE AIR TEMPERATURE SENSOR (check temperature of air entering engine).
5. ENGINE COOLANT TEMPERATURE SENSOR (measures operating temperature of engine).
6. THROTTLE POSITION SWITCH (notes position of throttle valves).

7. KNOCK SENSOR (allows computer to retard timing when engine pings or knocks), Fig. 14-19.

An ionization knock sensing system is used on some engines to detect abnormal combustion and knocking. In this type of system, the engine's computer triggers the ignition coil to send a low-voltage discharge across the spark plug gap immediately following combustion. The quality of combustion affects the resistance across the gap by varying the degree of ionization (the process by which atoms lose or gain electrons). Therefore, the quality of combustion influences the strength of the discharge across the gap. The computer uses feedback from the discharge to determine if the turbo boost or spark advance should be lowered to reduce knocking. A conventional detonation sensor is not needed in an ionization knock sensing system.

Computer

The *spark control computer* receives input signals (different current or voltage levels) from the sensors. It is *programmed* (preset) to adjust ignition timing to meet different engine conditions. The computer may be mounted on the air cleaner, fender panel, under the car dash, or under a seat, Figs. 14-18 and 14-19.

Electronic spark advance operation

Imagine a car traveling down the highway at 55 mph (88 km/h). The speed sensor would detect moderate engine rpm. The throttle position sensor would detect part throttle. The intake air and coolant temperature sensors would report normal operating temperatures. The manifold pressure sensor would send high vacuum signals to the computer.

The computer could then calculate that the engine would need maximum spark advance. The timing would occur several degrees before TDC on the compression stroke. This would ensure that the engine attained high fuel economy on the highway.

If the driver began to pass a car, engine intake manifold vacuum would drop to a very low level. The vacuum sensor signal would be fed to the computer. The throttle position sensor would detect WOT (wide open throttle). Other sensor outputs would stay about the same. The computer could then retard ignition timing slightly to prevent spark knock or ping.

Since computer systems vary, refer to a service manual for more information. The manual will detail its operation.

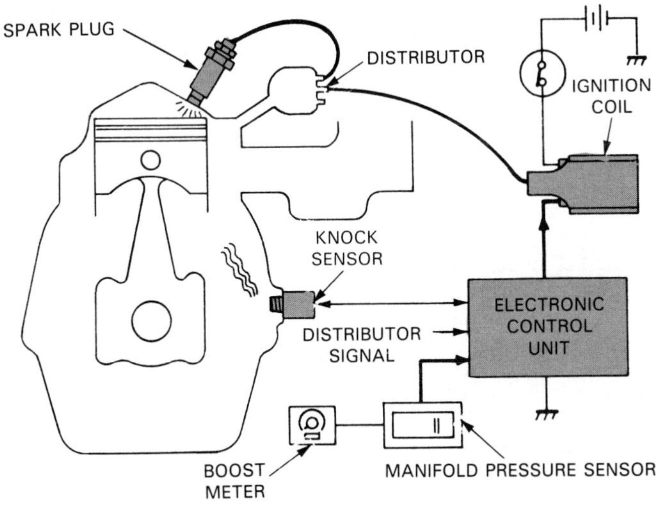

SPARK PLUG
DISTRIBUTOR
IGNITION COIL
KNOCK SENSOR
ELECTRONIC CONTROL UNIT
DISTRIBUTOR SIGNAL
BOOST METER
MANIFOLD PRESSURE SENSOR

Fig. 14-19. Detonation sensor detects ping or spark knock and sends signal to electronic control unit. Control unit can then retard timing enough to stop pinging. (Chrysler)

CRANKSHAFT TRIGGERED IGNITION

A *crankshaft triggered ignition* system places the pickup coil and trigger wheel (pulse ring) unit on the front or rear of the engine. These parts are NOT located inside the distributor.

A *pulse ring* is usually mounted on the crakshaft damper to provide engine speed information to the pickup unit. It performs the same function as the trigger wheel in a distributor for an electronic ignition. The teeth on the pulse ring correspond to the number of engine cylinders. Most crankshaft position sensors are missing one tooth. The missing tooth represents top dead center or a reference for measuring crankshaft position.

The *crankshaft position sensor* is mounted next to the crank pulse ring and sends electrical pulses to the system computer. It does the same thing as a distributor pickup coil. Other sensors are commonly used to also feed data to the computer.

The distributor for a crankshaft triggered ignition is simply used to transfer high voltage to each spark plug wire.

A crank triggered ignition can maintain more precise ignition timing than a system with a distributor-mounted pickup coil. There is no backlash or play in the distributor drive gear, timing chain, or gears to upset ignition timing. Crank and piston position is "read" right off the crankshaft.

DISTRIBUTORLESS IGNITION

A *computer-coil ignition,* also called a *distributorless* (no distributor) *ignition,* uses multiple ignition coils, a coil control unit, engine sensors, and a computer to operate the spark plugs. A distributor is NOT needed, Fig. 14-20.

Fig. 14-20. Note coil modules on this engine. No distributor is needed. (Buick)

Distributorless ignition coil and module

A *distributorless ignition coil* consists of multiple ignition coils. These coils can be part of one coil pack or individual coils that can be replaced. A four-cylinder engine uses two coils, a six-cylinder three coils, and a V-8, four coils. The coils are wired so that they fire TWO SPARK PLUGS at once. One cylinder is on the power stroke and the other is on the exhaust stroke. This is called the *waste spark* and has no effect on engine operation.

A *distributorless ignition module* operates in the same manner as a distributor's ignition module. However, it is more complex, since it must analyze data from input sensors and the ECU and provide outputs to the ignition coils and information back to the ECU, Fig. 14-21.

Sensors

A *cam sensor* is commonly installed in place of the ignition distributor. It sends pulses to the coil module giving data on camshaft and valve position, Fig. 14-21.

The *crank sensor,* as discussed, feeds pulses to the module which show engine speed and piston position.

A knock sensor may be used to allow the system to retard timing if the engine begins to ping or knock.

Distributorless ignition operation

The on-board computer monitors engine operating conditions and controls ignition timing. Some sensor data is also fed to the ignition module, Fig. 14-21.

When the computer and sensors send correct electrical pulses to the module, the module fires one of the ignition coils.

Since each coil secondary output is wired to two spark plugs, both spark plugs fire. One produces the power stroke. The other spark plug arc does nothing because that cylinder is on the exhaust stroke. Burned gases are simply being pushed out of the cylinder.

When the next pulse ring tooth aligns with the crank sensor, the next ignition coil fires. Another two spark plugs arc for one more power stroke. This process is repeated over and over as the engine runs.

IGNITION SYSTEM SERVICE

This section of the chapter will explain the most common ignition system tests, adjustments, and repairs done by an engine technician. For example, after a major ENGINE OVERHAUL, the technician must install spark plugs, plug wires, the distributor, ignition coil—which are all part of the ignition system. Study carefully!

TROUBLESHOOTING THE IGNITION SYSTEM

The ignition system and several other systems (fuel, emission, electrical systems) all work together. A problem is one system may affect, or appear to affect, the operation of another system. As a consequence, problem diagnosis can be challenging.

One example, a shorted gasoline injector can cause an engine miss or rough idle. An oil fouled spark plug will also cause an engine miss. The symptoms for each will be almost identical. You must use proper testing methods to find the faulty component.

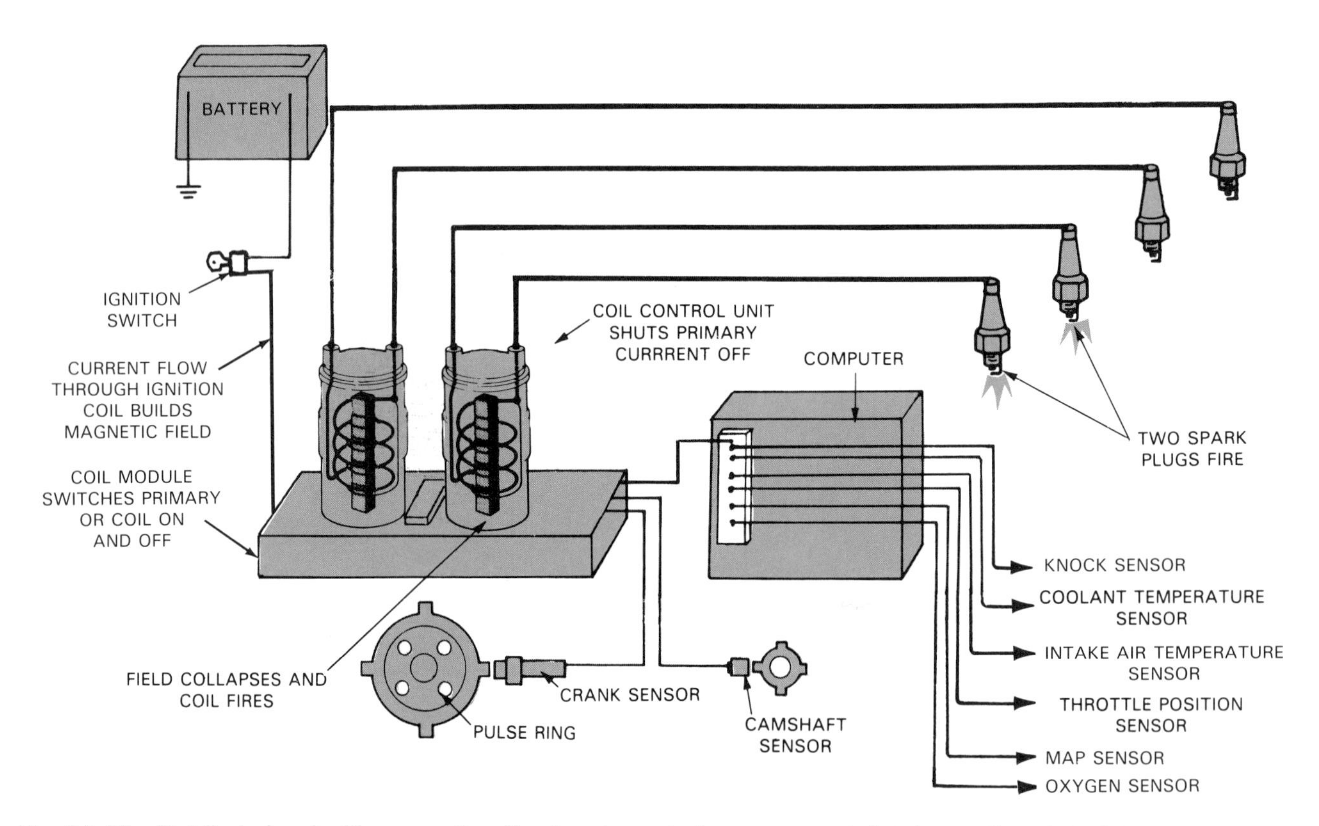

Fig. 14-21. Distributorless ignition operation. Crank and camshaft sensors send signals to coil module. Other sensors report to computer. Computer then signals coil module when to fire two plugs. This system eliminates wear problems found with distributors.

Preliminary checks

Inspect the ignition with and without the engine running. Look for obvious problems: loose primary connections, spark plug wire pulled off, deteriorated wire insulation, cracked distributor cap, or other troubles. Look over other engine systems. Try to find anything that could upset engine operation. Fig. 14-22 shows some typical problems.

Spark test

A *spark test,* also called a *spark intensity test,* measures the brightness and length of the electric arc (spark) produced by the ignition system. It is a quick way of checking the condition of the ignition system.

The spark test is commonly used when an engine cranks, but will NOT start. The test will help show whether the trouble is in the fuel system (no fuel problem) or in the ignition (no spark problem). It may also be used to check the spark plug wires, distributor cap, and other secondary components.

A *spark tester* is a device with a very large air gap for checking ignition output voltage. It is like a spark plug with a wide gap and a ground wire.

Remove one of the wires from a spark plug. Insert the spark tester into the wire. Ground it on the engine. Crank or start the engine. Observe the spark at the tester air gap, Fig. 14-23.

WARNING! Only run the engine for a short time with a spark plug wire off. Unburned fuel from the dead cylinder could foul and ruin the catalytic converter.

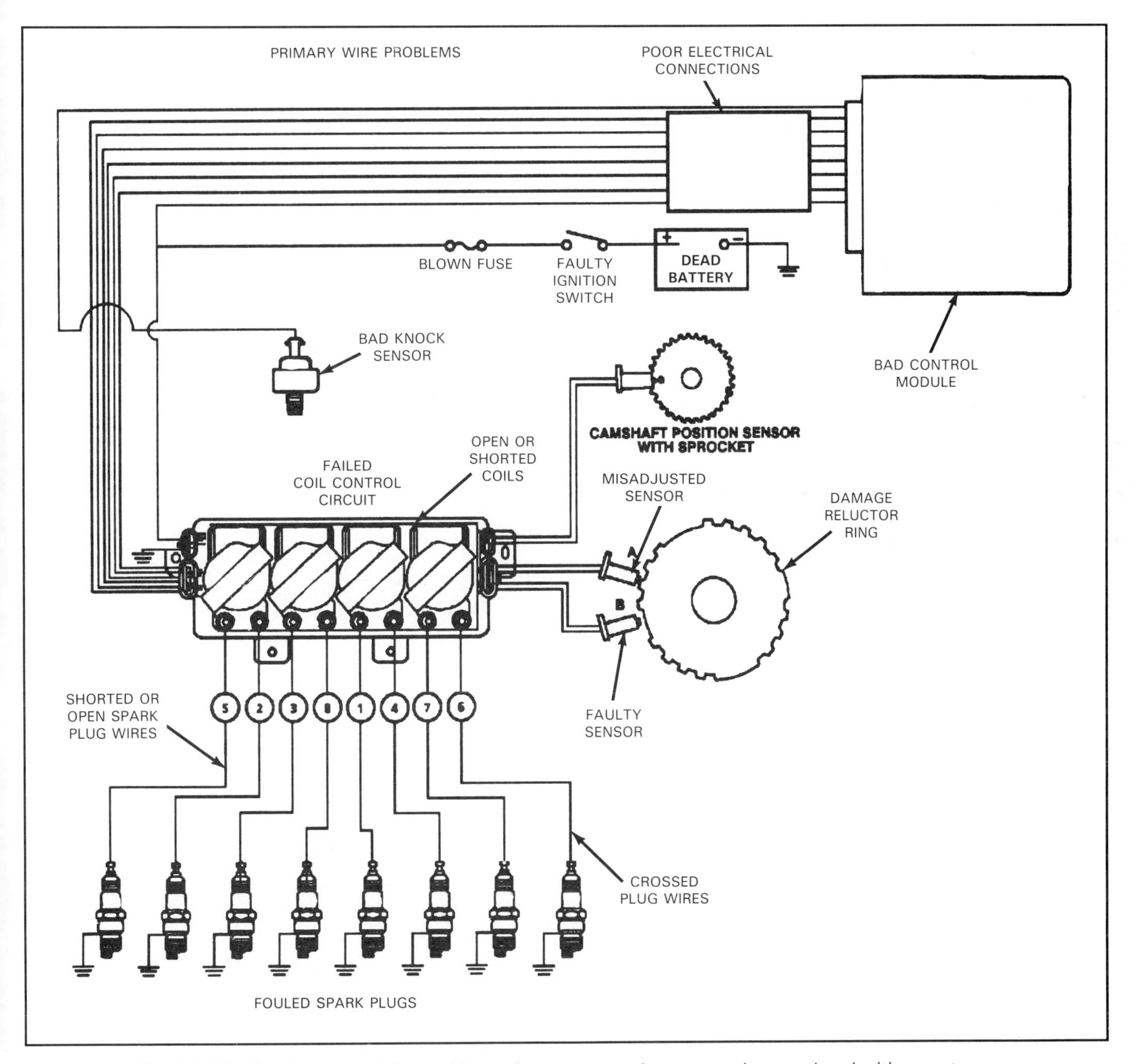

Fig. 14-22. Study some of the problems that can sometimes occur in a modern ignition system.

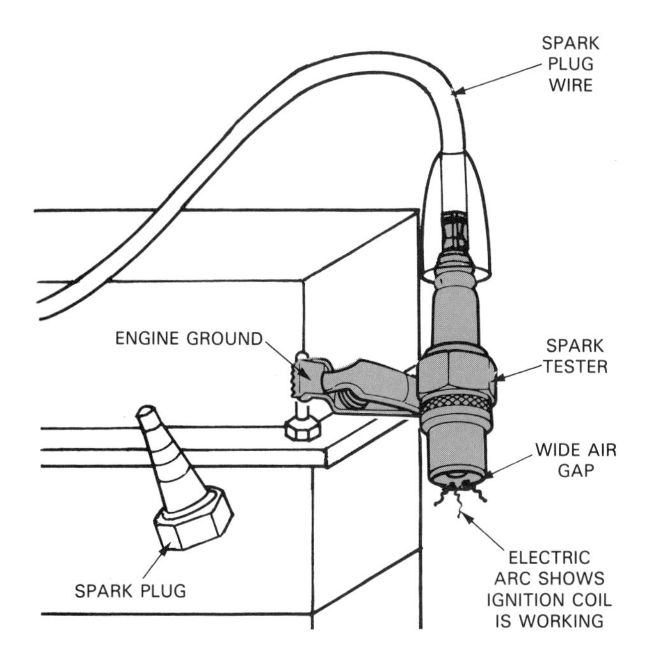

Fig. 14-23. Spark test should be done when engine cranks, but does not start and run. It will make sure ignition system is producing hot spark.

A *strong spark* shows that ignition voltage is good. The engine ''no-start'' problem might be due to fouled spark plugs, fuel system problem, or engine trouble. A strong spark indicates that the ignition coil, pickup coil, electronic control unit, and other parts are working.

A *weak spark* or *no spark* shows that something is wrong in the ignition system.

If spark is weak at all spark plug wires, the problem is common to all of the cylinders (bad ignition coil, rotor, coil wire). Other tests (covered shortly) would be needed to pinpoint the trouble.

Dead cylinder

A *dead cylinder* is one NOT burning fuel on the power stroke. It could be due to ignition system troubles or problems in the engine, fuel system, or another system. A very rough idle and a puffing noise in the engine exhaust may indicate a dead cylinder.

To check for a dead cylinder, pull off one spark plug wire at a time. Hold it next to ground. Engine rpm and idle smoothness should decrease with the wire removed.

If idle smoothness and rpm stay the same with the plug wire off, that cylinder is DEAD. It is not producing power. You need to check for spark at the wire, spark plug condition, and possible low cylinder compression. Special test instruments may be needed to diagnose ignition system problems. These tools are discussed in Chapter 21.

SPARK PLUG SERVICE

Bad spark plugs can cause a wide range of problems: misfiring, lack of power, poor fuel economy, and hard starting. After prolonged use, the spark plug tip can become coated with ash, oil, and other substances. Also, the plug electrodes can burn and widen the gap. This can make it more difficult for the ignition system to produce an arc between the electrodes.

Platinum spark plugs

Platinum spark plugs use precious metal on their electrodes to make them more resistant to electric arc wear or corrosion. This is especially important with today's distributorless ignition systems. Platinum plugs are now commonly used to increase spark plug service life up to 100,000 miles (160 000 km) before replacement is needed.

Spark plug removal

To remove the spark plugs, first check that the spark plug wires are numbered or located correctly in their clips. Grasping the spark plug wire boot, pull the wire off. Twist the boot back and forth, Fig. 14-24A.

WARNING! Never remove a spark plug wire by pulling on the wire. If you pull on the wire, you can break the strand type conductor.

Blow debris away from the spark plug holes with compressed air. Then, using a spark plug socket, extension, swivel, and ratchet, as needed, unscrew each plug, Fig. 14-24B. As you remove each spark plug, lay it in order on the fender cover or workbench. Do not mix up the plugs. After all of the plugs are out, inspect them to diagnose the condition of the engine. To prevent thread damage, it is best to remove spark plugs from aluminum heads with the engine cold.

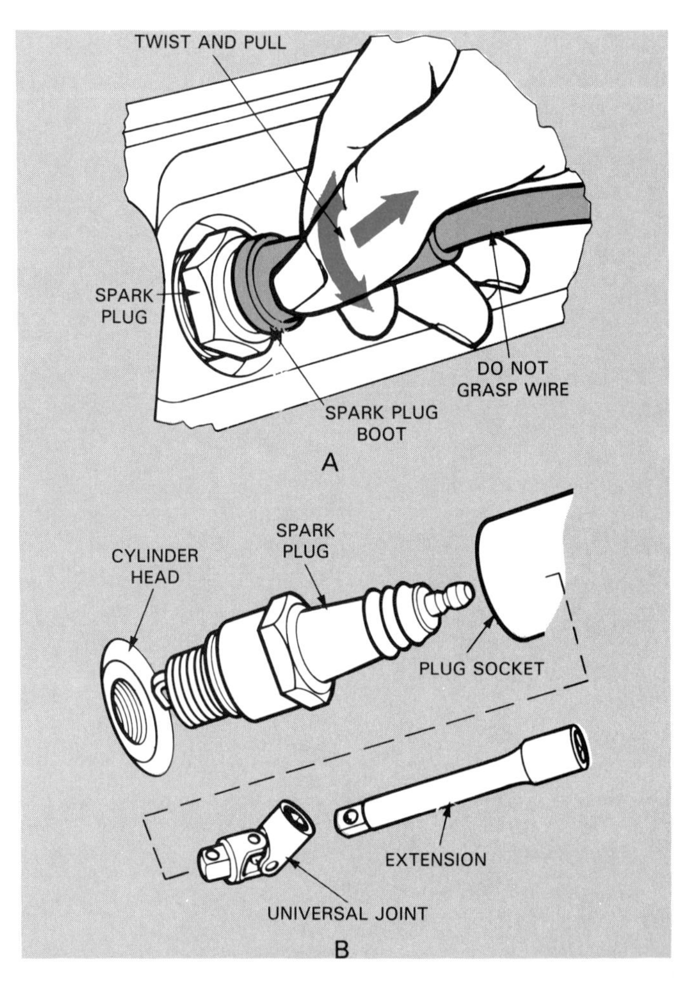

Fig. 14-24. A—Remove spark plug wire by pulling and twisting on boot, not wire itself. B—Spark plug socket and ratchet are used to unscrew spark plugs. (Chrysler)

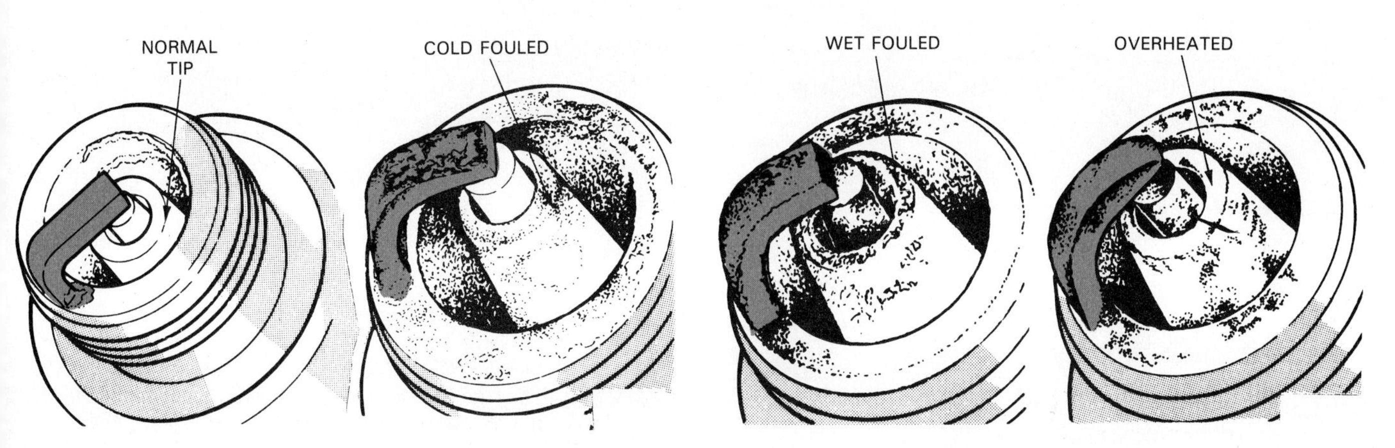

Fig. 14-25. When removing plugs, read them to check ignition, fuel system, and engine condition. (Chrysler)

Reading spark plugs

To *read spark plugs,* closely inspect and analyze the condition of each spark plug tip and insulator. This will give you information on the condition of the engine, fuel system, and ignition system.

For example, a properly burning plug should have a *brown* to *greyish-tan* color. A *black* or *wet plug* indicates that the plug is NOT firing or that there is an engine problem (worn piston rings, leaking valve stem seals, low engine compression, or rich fuel mixture) in that cylinder.

Study the spark plugs in Fig. 14-25. Learn to read the condition of used spark plugs. They can provide valuable information when troubleshooting problems.

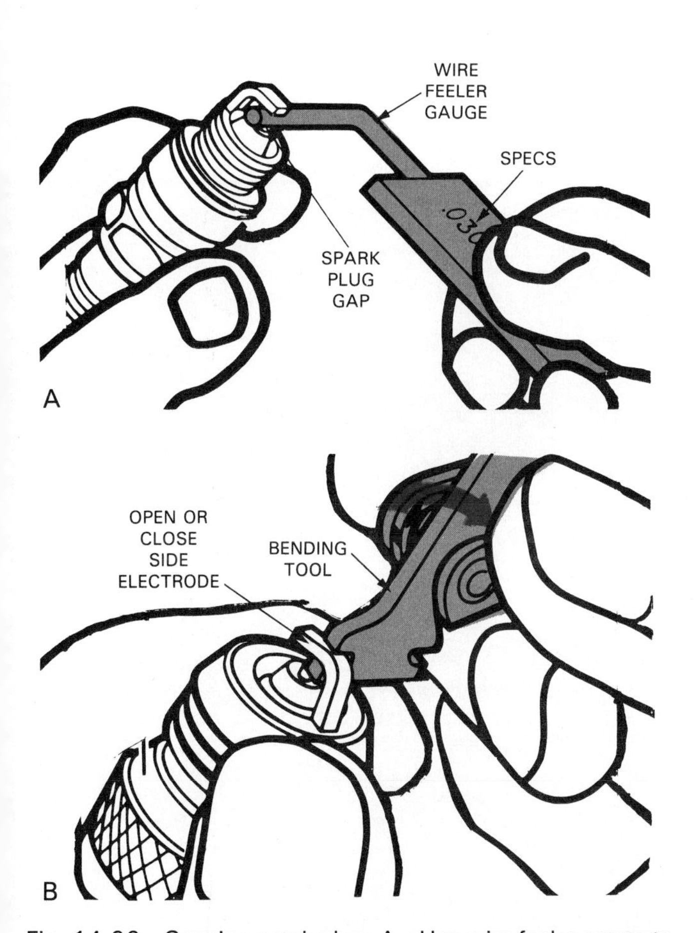

Fig. 14-26. Gapping spark plug. A—Use wire feeler gauge to measure space between electrodes. It should just drag when pulled through gap. B—Use tool to bend side electrode opened or closed as needed. (Sun Electric Corp.)

Gapping spark plugs

Obtain the correct replacement plug recommended by the manufacturer. Then, set *spark plug gap* by spacing the side electrode the correct distance from the center electrode. If the new spark plugs have been dropped or mishandled, the gap may NOT be within specs.

A *wire feeler gauge* should be used to measure spark plug gap, Fig. 14-26. Slide the feeler gauge between electrodes. If needed, bend the side electrode until the feeler gauge fits properly. The gauge should drag lightly as it is pulled in and out of the gap.

Spark plug gaps vary from approximately .030 inch (0.76 mm) on contact point ignitions to over .060 inch (1.52 mm) on electronic systems. Use manual specs for spark plug gaps.

Installing spark plugs

Use your fingers, a spark plug socket, or a short piece of vacuum hose to START the plugs in their holes. Do NOT use the ratchet because the plug and cylinder head threads could be crossthreaded and damaged.

With the spark plugs threaded into the head a few turns by hand, spin them in the rest of the way with your ratchet.

Tighten the spark plugs to specs. Some auto makers give a spark plug torque. Others recommend bottoming the plugs on the seat and then turning an additional one-quarter to one-half turn. Refer to a service manual for exact procedures.

SECONDARY WIRE SERVICE

A *faulty spark plug wire* can have a burned or broken conductor or it could have deteriorated insulation. Most spark plug wires have a resistance conductor that is easily separated. If the conductor is broken, voltage and current cannot reach the spark plug.

If the spark plug insulation is faulty, sparks may leak through the insulation to ground or to another wire, instead of reaching the spark plug.

Secondary wire resistance test

A *secondary wire resistance test* will check the condition of a spark plug or coil wire conductor. To do this test, connect an ohmmeter across each end of the spark plug or coil wire, as in Fig. 14-27. The meter will read internal resistance in ohms. Compare your reading to specs.

Typically, spark plug wire resistance should NOT be over approxiamtely 5000 ohms per inch or 100 K (100,000) ohms total. Since specs vary, always check in a service manual for an exact value. A bad spark plug wire will often have almost infinite (maximum) resistance.

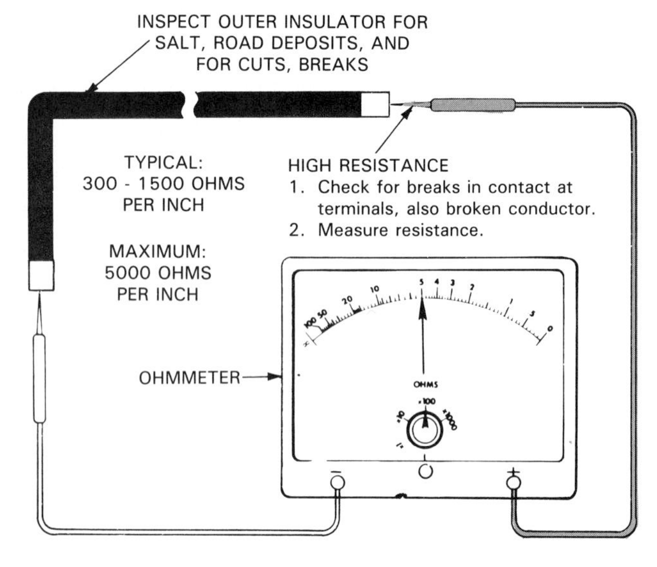

Fig. 14-27. Ohmmeter can be used to check for internal conductor breakage in coil and spark plug wires. If resistance is too high, replace wire. (Ford)

Secondary wire insulation test

A *secondary wire insulation test* checks for sparks arcing through the insulation to ground. An ohmmeter test will NOT detect bad insulation.

To check spark plug and coil wire insulation, place fender covers over the sides of the car hood to block out light. Start the engine. Inspect each wire for leakage.

A special test light or grounded screwdriver can be moved next to the wire insulation. If sparks jump through the insulation and onto the tool, the wire is bad.

Spark plug wire leakage is a condition in which electric arcs pass through the wire insulation. This problem requires wire replacement.

Replacing spark plug wires

Installing new spark plug wires is simple if one wire at a time is replaced. Compare old and new wire lengths. Replace each wire with one of equal length. Make sure the wire is fully attached on the plug and in the distributor or coil module.

Spark plug wire replacement is more complicated if all of the wires are removed at once. Then, you must use the engine firing order and cylinder numbers to route each wire correctly.

Fig. 14-28 gives *firing order illustrations* similar to the types found in a service manual. Each can be used to trace the wires from a distributor cap tower to the correct spark plug or cylinder.

Engine firing order.

Engine firing order refers to the sequence in which the spark plugs fire to cause combustion in each cylinder. A four-cylinder engine may have one of two firing orders: 1-3-4-2 or 1-2-4-3. The cylinders are numbered 1-2-3-4 starting at the front of the engine. In this way, you can tell which cylinders will fire in sequence. Firing orders and cylinder numbers for V-6 and V-8 engines vary.

The engine firing order is sometimes cast into the top of the intake manifold. When not on the manifold, the firing order can be found in a service manual, Fig. 14-28. Engine firing order is commonly used when installing spark plug wires and when doing other tune-up tasks.

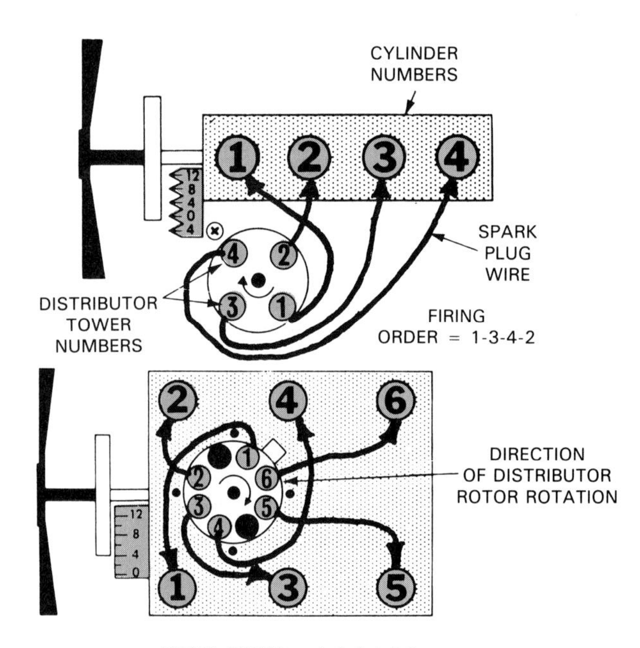

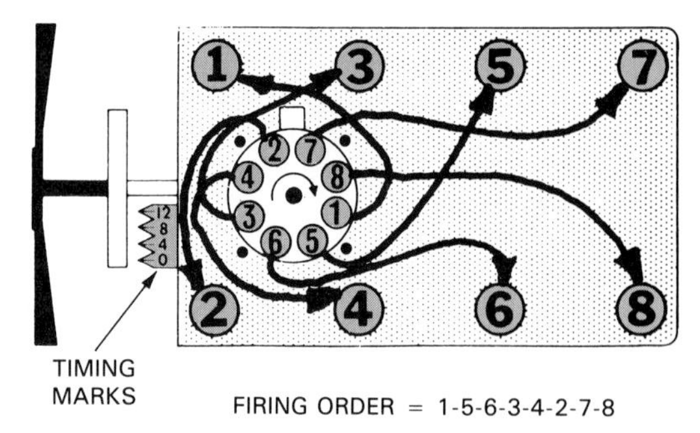

Fig. 14-28. These are service manual illustrations showing firing order, distributor rotor rotation, and spark plug wire routing. They would be needed during several engine tune-up type tasks.

DISTRIBUTOR SERVICE

A distributor is critical to the operation of an ignition system. It senses engine speed, alters ignition timing, and distributes high voltage to each spark plug wire. If any part of the distributor is faulty, engine performance suffers.

Distributor cap and rotor service

A *bad distributor cap* or *rotor* can cause engine missing, BACKFIRING, (popping noise in induction system), and other engine performance problems. In some cases, the engine may not start.

A common trouble arises when a *carbon trace* (small line of carbon-like substance that conducts electricity) forms on the distributor cap or rotor. The carbon trace will short coil voltage to ground or to a wrong terminal in the cap. A carbon trace can cause the spark plugs to fire poorly, out of sequence, or not at all.

When problems point to possible distributor cap or rotor troubles, remove and inspect them. Using a drop light, check the inside of the cap for cracks and carbon traces. A carbon trace is black, making it difficult to see on a black colored distributor cap.

If a crack or carbon trace is found, replace the cap or rotor. Also check the rotor tip for excessive burning, damage, or looseness. Make sure the rotor fits snugly on the distributor shaft.

Distributor cap and rotor replacement

Distributor caps may be secured by either screws or spring type metal clips. Normally, turn the screws counterclockwise for removal. With clips, pry on the top of the spring clips, being careful not to crack the cap. The clip should pop free. Wiggle and pull upward to remove the cap from the distributor body.

Rotors may be held by screws or they may be force-fitted around the distributor shaft. Pulling by hand will usually free a press-fit rotor. However, if stuck, carefully pry under the rotor.

With some ignition systems, the ignition coil is housed inside the distributor cap. In this case, the coil must be taken out of the old cap and installed in the new one.

WARNING! If a distributor cap is NOT installed correctly, rotor and distributor damage can result. The whirling rotor can strike the sides of the cap.

Pickup coil service

Most electronic ignition distributors use a pickup coil to sense trigger wheel rotation and engine speed. If the distributor fails to produce these tiny signals, the complete ignition system and engine can stop functioning.

A *bad pickup coil* can produce a wide range of engine problems: stalling, missing, no-start troubles, and loss of power at specific speeds. If the tiny windings in the pickup coil break, they can cause intermittent problems that only occur under certain conditions. Also, because of vacuum advance movement, the thin wire leads going to be pickup coil can break internally.

With the ignition key on, try moving a screwdriver or electric soldering gun over the pickup coil. This should make the ignition coil fire. See Fig. 14-29.

A *pickup coil ohmmeter test* compares actual pickup

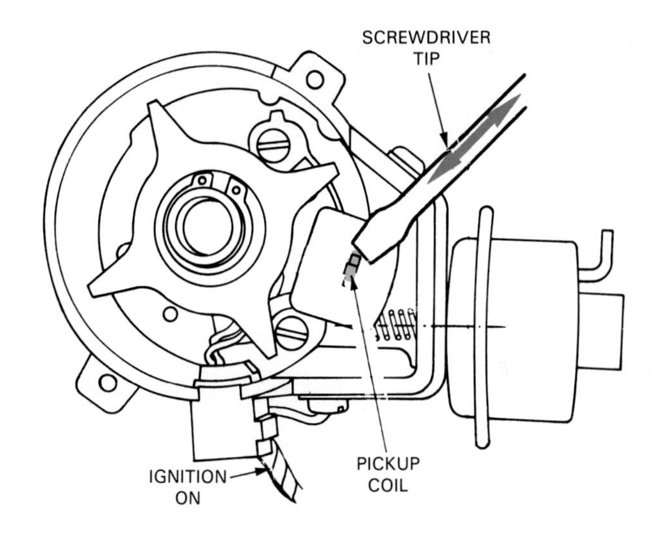

Fig. 14-29. With coil wire removed and held next to ground and with ignition on, pass screwdriver tip over pickup coil. This should make ignition coil fire and spark jump out of coil wire. Electric soldering gun will also fire ignition when held near pickup. This will show you that the module, ignition coil, etc., are working.

coil resistance with specs. If resistance is too high or low, the unit is bad.

Connect an ohmmeter across the pickup coil output leads, Fig. 14-30. Observe the meter reading. WIGGLE THE WIRES to the pickup coil while watching the meter. This will help locate a break in the leads to the pickup.

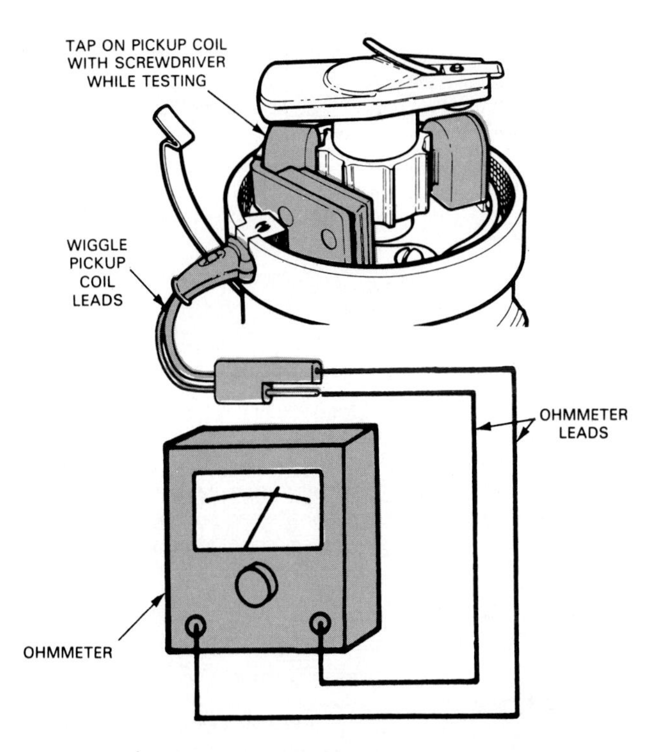

Fig. 14-30. Most common way of checking pickup coil is ohmmeter. Pickup resistance must be within specs. Wiggle wires to pickup because they can be broken internally. (Chrysler)

Also, lightly tap on the coil with the handle of a screwdriver. This could uncover any break in the coil windings.

Pickup coil resistance will usually vary between 250 and 1500 ohms. Check a service manual for specs. If the meter reading changes when the wire leads are moved or when the coil is tapped, replace the pickup coil.

A low-reading AC *voltmeter* can also be used to test a pickup coil. With the engine cranking, a small ac voltage of about 3 to 8 volts should be produced by the pickup coil. Refer to specs for exact voltage values and procedures.

A distributor pickup coil can usually be replaced by removing the distributor cap, rotor, and the screws on the advance plate. Sometimes the pickup coil is mounted around the distributor shaft. Since procedures vary, find detailed directions in a service manual.

The *pickup coil air gap* is the space between the pickup coil and a trigger wheel tooth. With some designs, it must be set after installing the pickup coil, Fig. 14-31.

To obtain an accurate reading, slide a NONMAGNETIC FEELER GAUGE (plastic or brass gauge) between the pickup coil and trigger wheel. One of the trigger wheel teeth must point at the pickup coil. Move the pickup coil in or out until the correct size gauge fits in the gap. Tighten the pickup screws and double-check the air gap setting.

Contact point service

Bad contact points (points having burned, pitted, misaligned contacts, or worn rubbing block) cause a wide range of engine performance problems: high speed missing, no-start problem, and many other ignition related troubles.

A *faulty condenser* could leak (allow some DC current to flow to ground), be shorted (direct electrical connection to ground), or be opened (broken lead wire to condenser foils). If leaking or open, a condenser will cause POINT ARCING and BURNING. If the condenser is shorted, primary current will flow to ground and the engine will NOT start.

The distributor points and condenser are held in place by small screws. To prevent dropping the screws, use a magnetic or clip-type screwdriver that firmly holds the screws. Use a small wrench to disconnect the primary wires from the points. Lift the points and condenser out of the distributor.

Wipe the distributor cam and breaker plate clean. If recommended, apply a small amount of lubrication to points on the distributor (wick in center of shaft, cam wick, or oil hole in side of distributor housing).

To prolong service life, place a small amount of grease on the side of the breaker rubbing block. This will reduce friction between the distributor cam and fiber block.

Fit the points and condenser into the distributor. Then, install their screws. If the distributor has a window (square metal plate) in the distributor cap, tighten the point hold-down screws. If the distributor does NOT have a window, only partially tighten the screws so that the points can be adjusted.

Adjusting distributor points

Distributor points can be adjusted using either a *feeler gauge* (metal blade ground to precision thickness) or a *dwell meter* (meter that electrically measures point setting in degrees of distributor rotation).

To use a feeler gauge to gap (set) distributor points, crank the engine until the points are FULLY OPEN. The point rubbing block should be on top of a distributor cam lobe. This is illustrated in Fig. 14-32.

Distributor point gap is the recommended distance between the contacts in the fully open position. Look up this spec in the service manual. It may also be given on the emission sticker in the engine compartment.

Typical point gap settings average around .015 in. (0.38 mm) for eight-cylinder engines to .025 in. (0.53 mm) for six and four-cylinder engines.

With the distributor points open, slide the specified thickness feeler gauge between the points. Adjust the points so that they just touch the blade. If needed,

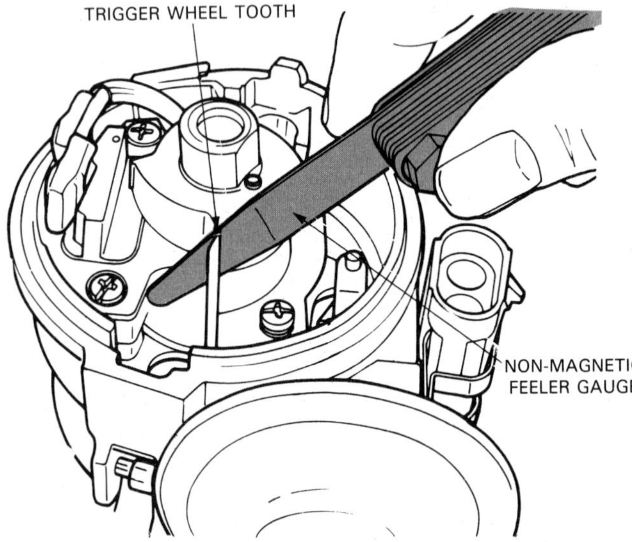

TRIGGER WHEEL TOOTH

NON-MAGNETIC FEELER GAUGE

Fig. 14-31. Some pickup coil gaps must be adjusted. Use a non-magnetic feeler gauge for accurate reading. (Pontiac)

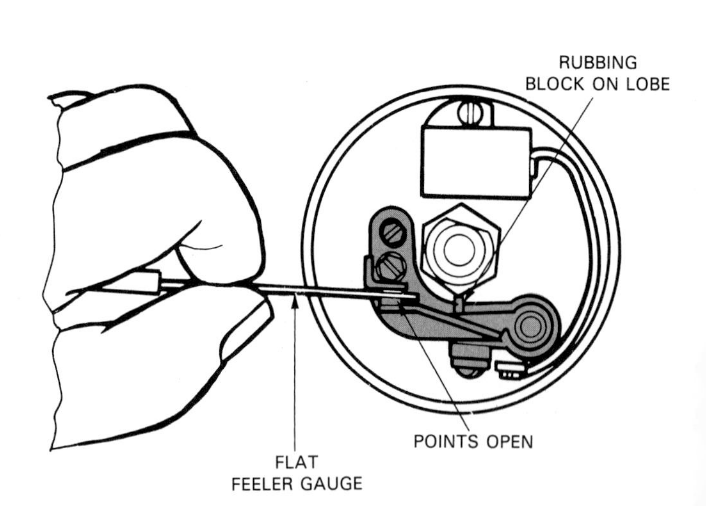

Fig. 14-32. Older contact points are also adjusted with feeler gauge. Crank engine until distributor cam lobe is touching rubbing block on points. Points should be fully open. Adjust points so they just touch correct thickness gauge. (Chrysler)

tighten the hold-down screws and recheck point gap.

CAUTION! Make sure your feeler gauge is clean before inserting it in the points. Oil or grease will reduce the service life of the points.

Using a dwell meter

To use a dwell meter to adjust distributor points, follow the directions provided with the meter. Typically, connect the red lead to the distributor side of the coil (wire going to contact points). Connect the black lead to ground (any metal part on engine), Fig. 14-33.

If an opening is provided in the distributor cap, the points should be set with the engine running. Install the distributor cap and rotor. Start the engine. With the meter controls set properly, adjust the points using an Allen wrench or special screwdriver type tool. Turn the point adjustment screw until the dwell meter reads within specs.

If the distributor cap does NOT have an adjustment window, set the points with the cap removed. Instead of starting the engine, ground the coil wire (connect output end on engine). Crank the engine with the starting motor. This will simulate engine operation and allow point adjustment with the dwell meter.

Dwell specifications (recommended point settings in degrees) vary with the number of cylinders in the engine. An 8-cylinder engine will usually require 30 degrees of dwell. An engine with fewer cylinders will normally require more dwell time. Always obtain exact dwell values from a tune-up chart or shop manual.

Dwell variation (change in dwell meter reading) indicates distributor wear problems. Dwell should remain constant as engine speed is increased or decreased. If dwell varies more than about three degrees, the distributor should be rebuilt or replaced. The distributor shaft, bushings, or advance plate could be worn and loose, allowing a change in dwell. As dwell increases, point gap decreases. As dwell decreases, point gap increases. Also, any change in point gap or dwell will change ignition timing. For this reason, the points should always be adjusted BEFORE the ignition timing.

IGNITION TIMING ADJUSTMENT

Initial ignition timing is the spark timing set by the mechanic with the engine idling (no centrifugal, vacuum, or electronic advance). It must be adjusted anytime the distributor has been removed and reinstalled in an engine. During an engine tune-up, initial timing must be checked and then adjusted if needed.

Initial ignition timing is commonly changed by turning the distributor housing in the engine. This makes the pickup coil and electronic control unit or breaker points fire the ignition coil sooner or later.

Turning the distributor housing against distributor shaft rotation ADVANCES TIMING. Turning the housing with shaft rotation RETARDS TIMING. See Fig. 14-34.

Some computer controlled ignition systems require moving the mounting for the crankshaft position sensor or adjusting a screw or lever on the computer to change timing. In some cases, timing cannot be adjusted and the computer must be replaced if the timing is not within specifications.

When the ignition timing is TOO ADVANCED, the

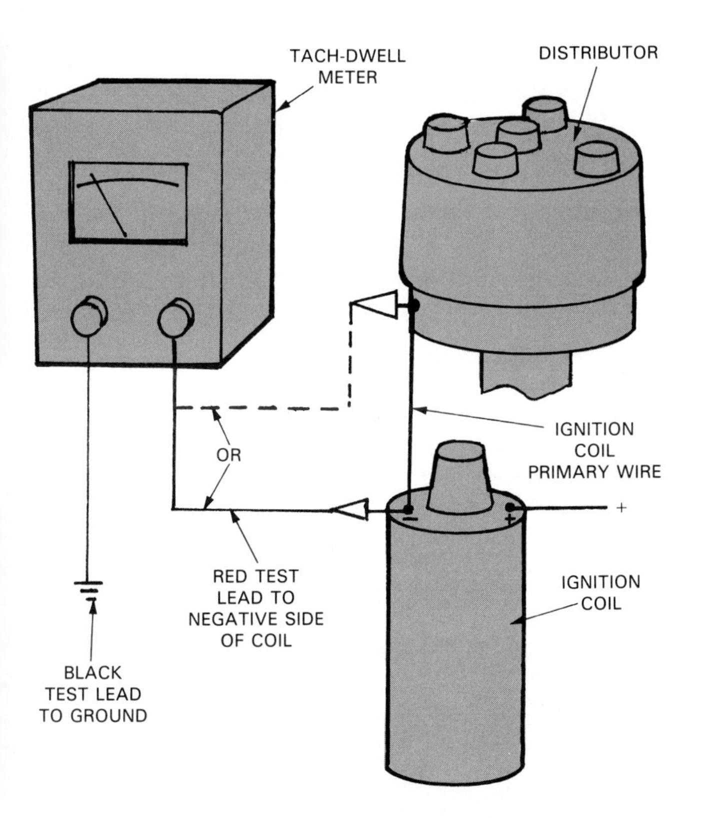

Fig. 14-33. Note basic hookup for tach-dwell meter. Tachometer is used for engine speed adjustment. Dwell meter is for setting contact points.

Fig. 14-34. Engine ignition timing is changed by turning distributor housing in block. A few computer controlled systems cannot be adjusted or provision for adjustment is made on computer itself. (Peerless Instruments)

engine may suffer from spark knock or ping. A light, tapping sound may result when the engine is accelerating or is under a load. The *ping* (abnormal combustion) will sound like a small hammer tapping on the engine.

When timing is TOO RETARDED, the engine will have poor fuel economy, power, and will be very sluggish during acceleration. If extremely retarded, combustion flames blowing out of the opened exhaust valve can overheat the engine and crack the exhaust manifolds.

Measuring ignition timing

A *timing light,* Fig. 14-35, is used to measure ignition timing. A timing light normally has three leads. The two small leads connect to the battery. The larger lead connects to the NUMBER ONE spark plug wire.

Depending upon the type of timing light, the large lead may clip around the plug wire (inductive type). It may also need to be connected directly to the metal terminal of the plug wire (conventional type).

When the engine is running, the timing light will flash ON and OFF like a strobe light. This action can be used to make a moving object appear stationary.

Before measuring engine timing, disconnect and plug the vacuum advance hose going to the distributor. This will prevent the vacuum advance from functioning and upsetting your readings. Check the emission sticker or manual to find out if other devices must also be disconnected. When adjusting timing on vehicles with computer controlled ignitions, you must trigger the base timing. Base timing is simply the ignition timing without computer control. Procedures for triggering base timing vary. Consult the factory service manual for speific instructions.

Start the engine and point the timing light on the timing marks. The timing marks may be on the front cover and harmonic balancer of the engine, Fig. 14-36. The timing marks may also be on the engine flywheel.

The flashing timing light will make the mark or marks on the harmonic balancer or flywheel appear to stand still. This will let you determine whether the engine is timed properly. For example, if there is only one reference line on the harmonic balancer, simply read initial timing by noting the degree marks lined up with the reference line, as in Fig. 14-36.

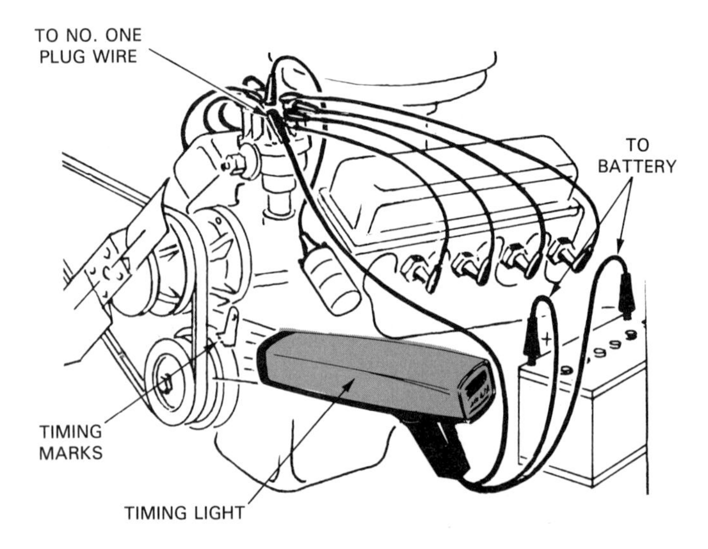

Fig. 14-35. Timing light connects as shown. Two small leads go to battery. Large lead connects to number one spark plug wire. Shine light on engine timing marks following recommended procedure for particular engine.

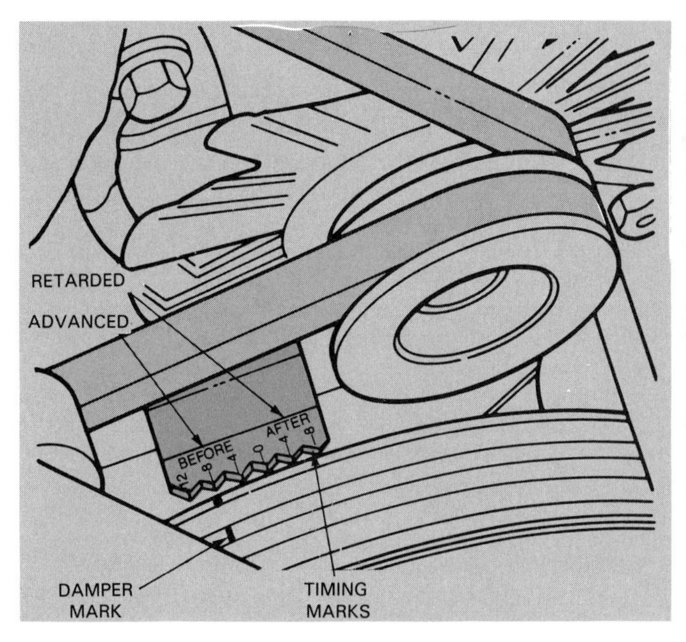

Fig. 14-36. Timing marks are usually on front of engine, near front damper or pulley. However, they can sometimes be on flywheel at rear of engine. BEFORE means advanced and AFTER means retarded. (Chevrolet)

Ignition timing specs

Ignition timing is very critical to the performance of an engine. If the ignition timing is off even two or three degrees, engine fuel economy and power can drop considerably.

Changing ignition timing

Adjust the ignition timing if the timing marks are not lined up correctly. Loosen the distributor hold-down bolt.

A *distributor wrench* (long, special shaped wrench for reaching under distributor housing) is handy. Only loosen the distributor bolt enough to allow distribution rotation. Do NOT remove the bolt, Fig. 14-34.

With the distributor hold-down loosened, shine the timing light on the engine timing marks. Turn the distributor one way or the other until the correct timing marks line up. Tighten the hold-down and double-check the timing. Reconnect the distributor vacuum hose and disconnect the timing light.

DANGER! Keep your hands and the timing light leads away from the engine fan and belts. The spinning fan and belts can damage the light or cause serious injury!

Testing centrifugal advance (in-car)

A timing light can be used to test the general operation of centrifugal advance. Remove the vacuum hose going to the distributor. Start and idle the engine.

While shining the timing light on the engine timing marks, slowly increase engine speed to approximately 3500 rpm. See Fig. 14-37.

If the centrifugal advance is working, the mark should steadily move to a more advanced position with the increase in speed.

If the timing mark jumps around or DOES NOT MOVE advance smoothly, the centrifugal advance is faulty. It may be worn, rusted, have weak springs, or other mechanical problems.

Testing vacuum advance (in-car)

To test the vacuum advance, remove the vacuum advance hose from the distributor. Start the engine and increase rpm to approxiamtley 1500 rpm. Note the location of the timing marks. Then, reconnect the vacuum hose on the distributor diaphragm.

As soon as vacuum is reconnected, engine speed should increase and the timing mark should, as in Fig. 14-37, advance.

If the vacuum advance is NOT working, check the vacuum advance diaphragm and the supply vacuum to the distributor.

Testing a vacuum advance diaphragm

To check the vacuum advance diaphragm, apply a vacuum to the unit using a vacuum pump or your mouth. When suction is applied to the diaphragm, the advance plate in the distributor should swing around. When vacuum is released, the advance plate should snap back into its normal position. See Fig. 14-38. If the ADVANCE DIAPHRAGM LEAKS, and will not hold vacuum, it must be replaced.

Measuring total advance (in-car)

Special timing lights are available that are capable of measuring total advance with the distributor installed in the engine. The timing light has a DEGREE METER built into the back of its case. The meter will register advance quickly and accurately, Fig. 14-37.

A *distributor tester* may also be used to check distributor operation. The distributor is removed and mounted in the tester. The tester will check all distributor functions.

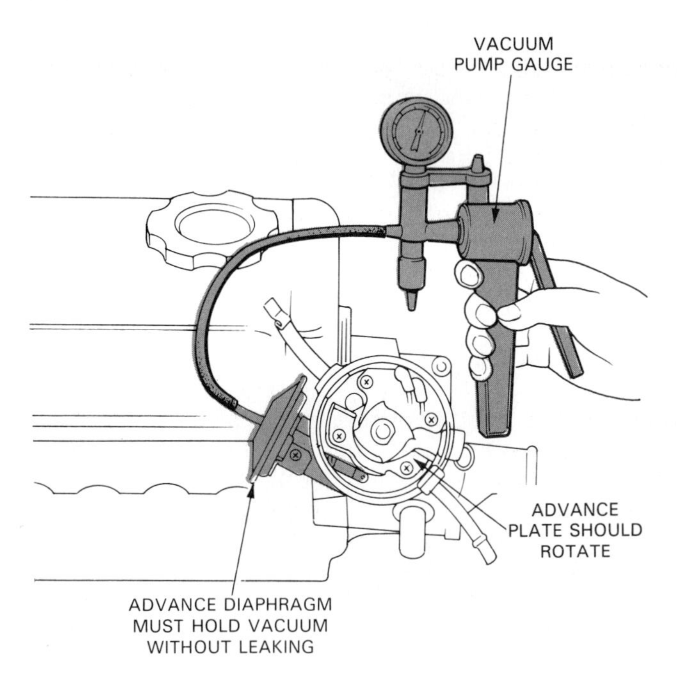

Fig. 14-38. Test vacuum advance diaphragm for operation and leakage with hand vacuum pump. If diaphragm leaks or fails to rotate advance plate, replace diaphragm. (Honda)

Removing ignition distributor

Before removing a distributor, carefully mark the position of the rotor and distributor housing. Place marks on the engine and distributor housing with a scribe or marking pen. Then, if the engine crankshaft is NOT rotated, you will be able to install the distributor by simply lining up your marks. Refer to Fig. 14-39.

To remove the distributor, remove the distributor cap, rotor, primary wires, and distributor hold-down bolt. Pull the distributor upward while rotating it back and forth. If stuck, use a slide hammer puller with a two-prong fork.

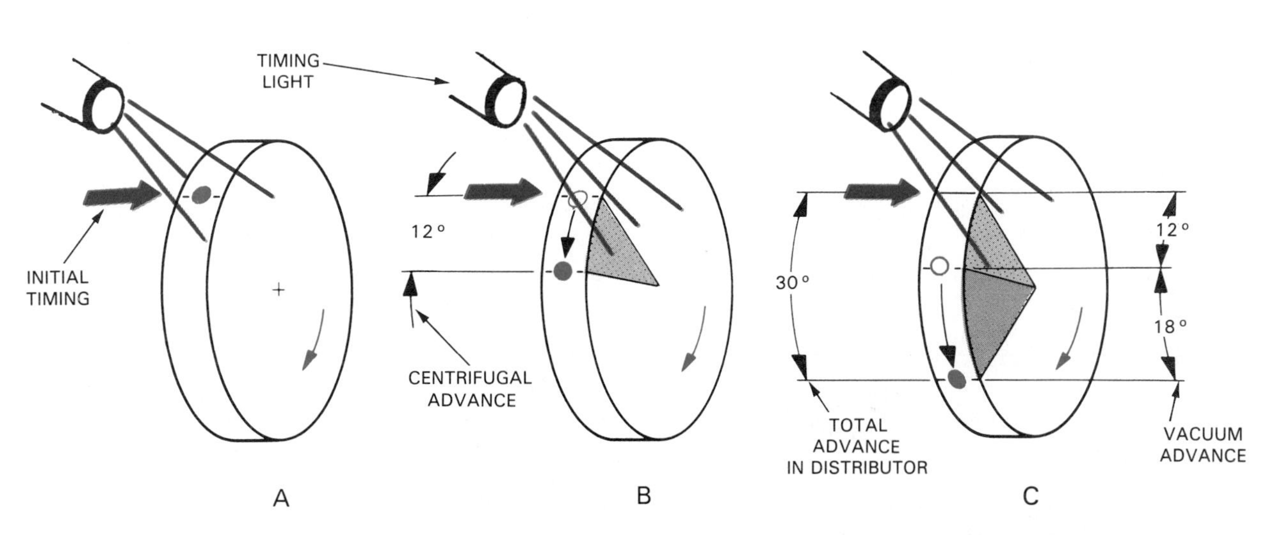

Fig. 14-37. A—Initial timing would be timing set with no distributor advance. B—Centrifugal advance would come in as engine speed increases. C—Vacuum advance would be additional advance when engine is cruising at low load. (Bosch)

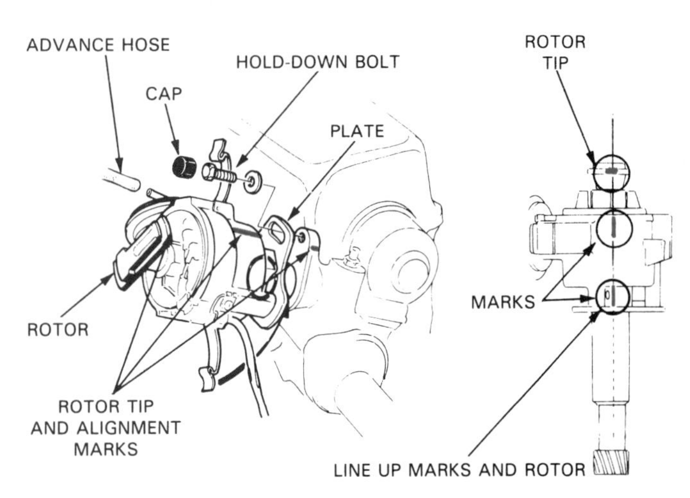

Fig. 14-39. Place marks on engine and distributor as shown before removing distributor. This will let you install distributor correctly upon reassembly. Marks at rotor tip, housing, and engine should align. (Honda)

Installing a distributor

If you made reference marks and the engine crankshaft was NOT turned, install the distributor as it was removed. Align the rotor and housing with the marks on the engine.

Double-check the position of the rotor after installation because the rotor will turn as the distributor gear meshes with its drive gear.

To install a distributor when the engine crankshaft has been rotated, remove the number one spark plug. Bump (crank) the engine until you can feel air blowing out of the spark plug hole. As soon as air blows out, slowly turn the crankshaft until the engine timing marks are on TDC or zero.

With the crankshaft in this position, the distributor rotor should point at the NUMBER ONE SPARK PLUG WIRE. Fit the distributor into the engine so that the rotor points at number one distributor cap tower.

Also, make sure the distributor housing is installed properly. The advance unit should be pointing as it was before removal.

IGNITION MODULE SERVICE

A *faulty ignition module* will produce a wide range of problems: engine stalls when hot, engine cranks but fails to start, engine misses at high or low speeds.

Quite often, an ignition module problem will show up after a period of engine operation. ENGINE HEAT will soak into the module, raising its temperature. The heat will upset the operation of the electronic components in the unit.

Many shop manuals list the ignition module as one of the last components to test. If all of the other components are in good working order, then the problem might be in the electronic control module.

If a specialized tester is available, it may be used to quickly determine the condition of the module. The wires going to the electronic control module are unplugged. The tester is plugged into the circuit. The tester will then indicate whether a fault exists.

The microscopic components inside the module are very sensitive to high temperature. When testing the module, many technicians use a heat gun or light bulb to warm the module. This will simulate the temperature in the engine compartment. The heat may make the control unit act up and allow you to find an erratic problem.

CAUTION! Do not apply too much heat to an electronic control unit or it may be ruined. Only heat the unit to a temperature equal to its normal operating temperature.

COMPUTERIZED IGNITION SERVICE

Many of the components of a computer controlled ignition system are similar to those of electronic or contact point ignition systems. This makes testing about the same for many parts (spark plugs, secondary wires, ignition coil). However, the computerized ignition has engine sensors and a computer which add to the complexity of the system.

Computer self-diagnosis mode

Some computerized systems have a check engine light in the dash that glows when a problem exists. The computer can be activated to produce a number code. The code can be compared to information in the car's service manual to pinpoint the source of a problem. This makes testing and repairing a computerized system much easier.

CAUTION! A computerized ignition system can be seriously damaged if the wrong wire is shorted to ground or if a meter is connected improperly. Always follow manufacturer's testing procedures.

Computer ignition testers

Most auto makers provide specialized testing equipment for their computerized ignition systems. Like an ignition control unit, the computer system tester plugs into the wiring harness. It will then measure internal resistances and voltages in the system to determine where a problem is located.

SUMMARY

The ignition system provides a method of turning the engine on and off. It operates on different supply voltages and steps up battery or alternator voltage to 30,000 volts or more for the spark plugs. It must time these sparks depending upon the operating conditions of the engine.

The primary of the ignition is the low voltage section. The secondary includes all parts operating on high voltage.

The ignition coil is a step-up transformer that changes battery voltage into high voltage. When current flows through the ignition coil, a magnetic field is developed. Then, when the switching device, either breaker points or electronic circuit, opens the circuit supplying current to the coil, the coil fires. High voltage shoots out the coil and to the spark plug.

The spark plug simply provides an air gap inside the engine combustion chamber. When voltage reaches the air gap, an electric arc jumps the gap. This produces enough heat to start combustion. It takes about 4000 to 10,000 volts to fire a spark plug.

Spark plug heat range refers to the operating temperature of the spark plug tip. A hotter plug is needed to burn off deposits. A colder plug would be used in an engine operating at high speeds. Spark plug reach is determined by the length of the plug threads. A resistor plug is commonly used in passenger cars to prevent radio static or interference.

There are two basic types of ignition systems—distributor and distributorless types. The distributor contains contact points or a pickup coil for sensing engine rpm. It may also contain centrifugal and vacuum advance mechanisms for altering ignition timing.

The centrifugal advance simply advances ignition timing as engine speed increses. It uses spring-loaded weights and centrifugal force to turn the distributor cam opposite shaft rotation. The vacuum advance uses engine vacuum to retard timing with high load conditions to prevent ping or knock.

An electronic ignition system uses a module in conjunction with the distributor pickup coil, to operate the ignition coil. The pickup coil produces weak electrical pulses as the trigger wheel on the distributor shaft turns. The module amplifies these on-off pulses for ignition coil operation. The module can be located in the distributor, on a fender panel, firewall, or in the passenger compartment.

The distributor cap is a plastic cover that holds the inner ends of the spark plug wires. The rotor turns inside this cap to distribute high voltage to each spark plug wire.

Electronic spark advance uses various engine sensors and a computer to control ignition timing. Sensors monitor engine speed, engine temperature, intake manifold vacuum, inlet air temperature, etc. The computer uses the electrical information from these sensors to fire the ignition coil at just the right time.

A crankshaft triggered ignition system has the pickup coil mounted next to the engine crankshaft. The pulse ring or trigger wheel is on the crankshaft. This is more accurate because there is no slop or play in the distributor gear or timing chain that could upset timing.

A computer-coil or distributless ignition uses multiple ignition coils, sensors, and a computer. Each coil operates two spark plugs simultaneously. One cylinder is on the power stroke and the others is on the exhaust stroke. This system eliminates the problem of distributor wear.

A spark test should be made when an engine fails to start when cranked. It will make sure the ignition system is providing sparks to the plugs. If there is spark, you would check for fuel next.

Read spark plugs by inspecting the condition of their tips.

Gap spark plugs with a wire feeler gauge to specs. The gauge should drag in the gap slightly. Bend the side electrode to change the gap.

Engine firing order refers to the sequence that combustion occurs in each cylinder. It is needed when installing spark plug wires.

A pickup coil can be tested with an ohmmeter. Its resistance must be within specs. Wiggle the wires going to the coil when testing. They can become broken internally and cause intermittent problems. A low-reading, AC voltmeter will also check the output of a pickup coil.

Initial ignition timing is set by the technician. The engine should be idling and usually the vacuum advance hose disconnected and plugged. The timing light connects to the battery and to number one spark plug wire. Shine the timing light on the timing marks and note how the marks align. Rotate the distributor housing to change ignition timing.

KNOW THESE TERMS

Primary circuit, Secondary circuit, Ignition switch, Ballast resistor, Ignition coil, Primary windings, Secondary windings. Center electrode, Side electrode, Heat range, Reach, Resistor spark plug, Distributor, Contact points, Condenser, Dwell, Point gap, Trigger wheel, Pickup coil, ECU, Coil wire, Plug wire, Resistance plug wire, Distributor cap, Rotor, Timing advance, Timing retard, Centrifugal advance, Vacuum advance, Electronic advance, Engine sensors, Spark control computer, Pulse ring, Distributorless ignition, Detonation sensor, Spark test, Reading plugs, Dead cylinder, Plug gap, Secondary wire resistance, Firing order, Carbon trace, Dwell variation, Initial ignition timing, Timing light, Distributor wrench, Number one plug wire, Computer self-diagnosis.

REVIEW QUESTIONS—CHAPTER 14

1. The ________ section of the ignition system operates on high voltage.
2. How does the ignition coil produce high voltage to operate the spark plugs?
3. A customer complains of oil fouled plugs. The car is very old and the engine needs to be rebuilt because of bad rings.
   Technician A says that colder plugs would help prevent fouling and they should be installed in the engine.
   Technician B says that the plugs recommended by the auto manufacturer should always be installed.
   Who is correct?
   a. Technician A.
   b. Technician B.
   c. Both A and B.
   d. Neither A nor B.
4. How does an electronic ignition system fire the ignition coil?
5. Where can the module for an ignition system be located?
6. ________ spark plugs and plug wires are used to prevent radio interference.
7. It takes about 20,000 volts to jump the rotor-to-cap air gap. True or false?
8. A customer complains of spark knock or ping in her engine. She is using the recommended fuel octane and the engine is in good condition.
   Technician A says that the ignition timing might be too far advanced and it should be checked.

Technician B says that the timing must be retarded and should be adjusted.
Who is correct?
a. Technician A.
b. Technician B.
c. Both A and B.
d. Neither A nor B.

9. Explain the three types of timing advance.
10. List seven engine sensors that can be used by the ECU to control ignition system operation. Briefly explain each sensor.
11. A computer-coil or distributorless ignition system typically fires two spark plug simultaneously. True or false?
12. A _______ _______ is commonly used when an engine cranks but will not start. It quickly checks the ignition system.
13. What is a ''dead cylinder?''
14. How do you gap spark plugs?
15. How do you check spark plug wires with an ohmmeter?
16. A _______ _______ is a small line of carbon that can build up on distributor cap and short out one or more plugs and cylinders.
17. How do you test a distributor pickup coil?
18. How do you adjust engine ignition timing?
19. Rotating the distributor housing in the same direction as rotor rotation would _________ ignition timing.
20. In your own words, how do you install a distributor in an engine?

ASE CERTIFICATION-TYPE QUESTIONS

1. Technician A says that the primary section of an ignition system is a high voltage circuit. Technician B says that the primary section of an ignition system is a low voltage circuit. Who is right?
(A) A only.
(B) B only.
(C) Both A & B.
(D) Neither A nor B.
2. All of the following are basic parts of a distributor type ignition system EXCEPT:
(A) resistance wire.
(B) multi-coil unit.
(C) switching mechanism.
(D) ignition coil.
3. Technician A says that one of the functions of an automotive ignition system is to time the spark so that it occurs as the piston nears TDC on the power stroke. Technician B says that one of the functions of an ignition system is to enable the engine to run on either battery or alternator voltage. Who is right?
(A) A only.
(B) B only.
(C) Both A & B.
(D) Neither A nor B.
4. An ignition system distributor is used to _________.
(A) drive the oil pump on certain engine designs
(B) change spark timing with changes in engine load
(C) change spark timing in response to engine speed
(D) All of the above.
5. Technician A says that an ignition module can be located inside the distributor. Technician B says that an ignition module can be located under a distributorless ignition system's coil pack. Who is right?
(A) A only.
(B) B only.
(C) Both A & B.
(D) Neither A nor B.
6. Technician A says that some computer-controlled ignition systems use a distributor to regulate voltage to the engine's spark plugs. Technician B says that some computer-controlled ignition systems operate without the use of a distributor. Who is right?
(A) A only.
(B) B only.
(C) Both A & B.
(D) Neither A nor B.
7. Technician A says that worn spark plugs can reduce an engine's power output. Technician B says that worn spark plugs can reduce an engine's fuel economy. Who is right?
(A) A only.
(B) B only.
(C) Both A & B.
(D) Neither A nor B.
8. The spark plugs of a particular engine are grayish-tan in color. What does this indicate?
(A) Possible worn piston rings.
(B) The plugs are burning properly.
(C) Oil leaking past the piston rings.
(D) A preignition problem.
9. Technician A says that when using a timing light to test a distributor with centrifugal advance, you should increase engine speed to approximately 2200 rpm. Technician B says that when testing the distributor's centrifugal advance with a timing light, engine speed should be increased to approximately 3500 rpm. Who is right?
(A) A only.
(B) B only.
(C) Both A & B.
(D) Neither A nor B.
10. Technician A says that a faulty ignition module can cause an engine to stall. Technician B says that a faulty ignition module can cause an engine to miss at high or low speeds. Who is right?
(A) A only.
(B) B only.
(C) Both A & B.
(D) Neither A nor B.

Chapter 15
Cooling and Lubrication Systems—Operation and Service

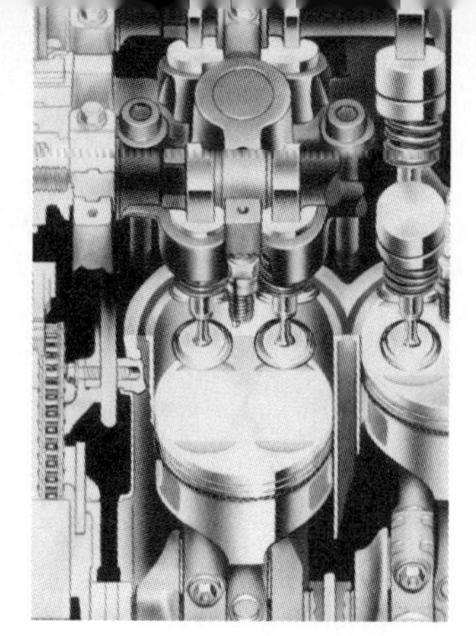

After studying this chapter, you will be able to:
- *Explain the operation of a cooling system.*
- *Compare cooling system types.*
- *Troubleshoot cooling system problems.*
- *Perform basic cooling system tests.*
- *Replace or repair faulty cooling system components.*
- *Explain the operation of an engine lubrication system.*
- *Compare lubrication system design variations.*
- *Diagnose lubrication system problems.*
- *Measure engine oil pressure.*
- *Replace or repair faulty lubrication system parts.*

The cooling and lubrication systems are vitally important to engine performance and service life. Without these systems, an engine can ''self-destruct'' in a matter of minutes.

As an engine technician, you must be fully prepared to work on the cooling system and the lubrication system. They are common causes of engine failure and sources of trouble. Study this chapter carefully!

Note! The first section of the chapter will summarize cooling system operation and service. Then, the second section of the chapter will explain lubrication system operation and service.

COOLING SYSTEM FUNCTIONS

The engine's COOLING SYSTEM has four basic functions:
1. Remove excess heat from engine.
2. Maintain constant engine operating temperature.
3. Increase temperature of cold engine quickly.
4. Provide for heater operation.

Removing engine heat

The burning air-fuel mixture produces a tremendous amount of heat. Flame temperatures can reach 4500 °F (2 484 °C). This is enough heat to melt metal parts.

Some combustion heat is used to produce expansion and pressure for piston movement. However, most of the combustion heat flows out the exhaust and into the metal parts of the engine. Without removal of this excess heat, the engine would be seriously damaged.

Maintain operating temperature

Engine operating temperature is the temperature the engine coolant (water and antifreeze solution) reaches under running conditions. Typically, an engine's operating temperature is between 180 and 195 °F (82 and 91 °C).

When an engine warms to operating temperature, its parts expand. This assures that all part clearances are correct. It also assures proper combustion, emission output levels, and engine performance.

Reaching temperature quickly

An engine must warm up rapidly to prevent poor combustion, part wear, oil contamination, reduced fuel economy, and other problems. A cold engine suffers from several problems.

For instance, the aluminum pistons in a cold engine will not be heat expanded (size increases from heat). This can cause too much clearance between the pistons and cylinder walls. The oil in a cold engine will be very thick. This can reduce lubrication protection and increase engine wear. The fuel mixture will also not vaporize and burn as efficiently in a cold engine.

Heater operation

A cooling system commonly circulates coolant to the car heater. Since the engine coolant is warm, its heat can be used to warm the passenger compartment.

COOLING SYSTEMS TYPES

There are two major types of engine cooling systems: liquid and air.

An *air cooling system* uses cylinder cooling fins and outside air to remove excess heat from the engine.

The *cooling fins* increase the surface area of the metal around the cylinder. This allows enough heat to transfer from the cylinder to prevent engine damage.

An air cooling system commonly uses plastic or sheet metal ducts and shrouds (enclosures) to route air over the cylinder fins. Thermostatically controlled flaps regulate airflow and engine operating temperature. This type does not use a radiator, water pump, etc.

Air cooled automotive engines have been almost totally replaced by liquid (water) cooled engines.

A *liquid cooling system* circulates a solution of water and antifreeze through the water jackets (internal passages in engine). The coolant then collects and carries heat out of the engine.

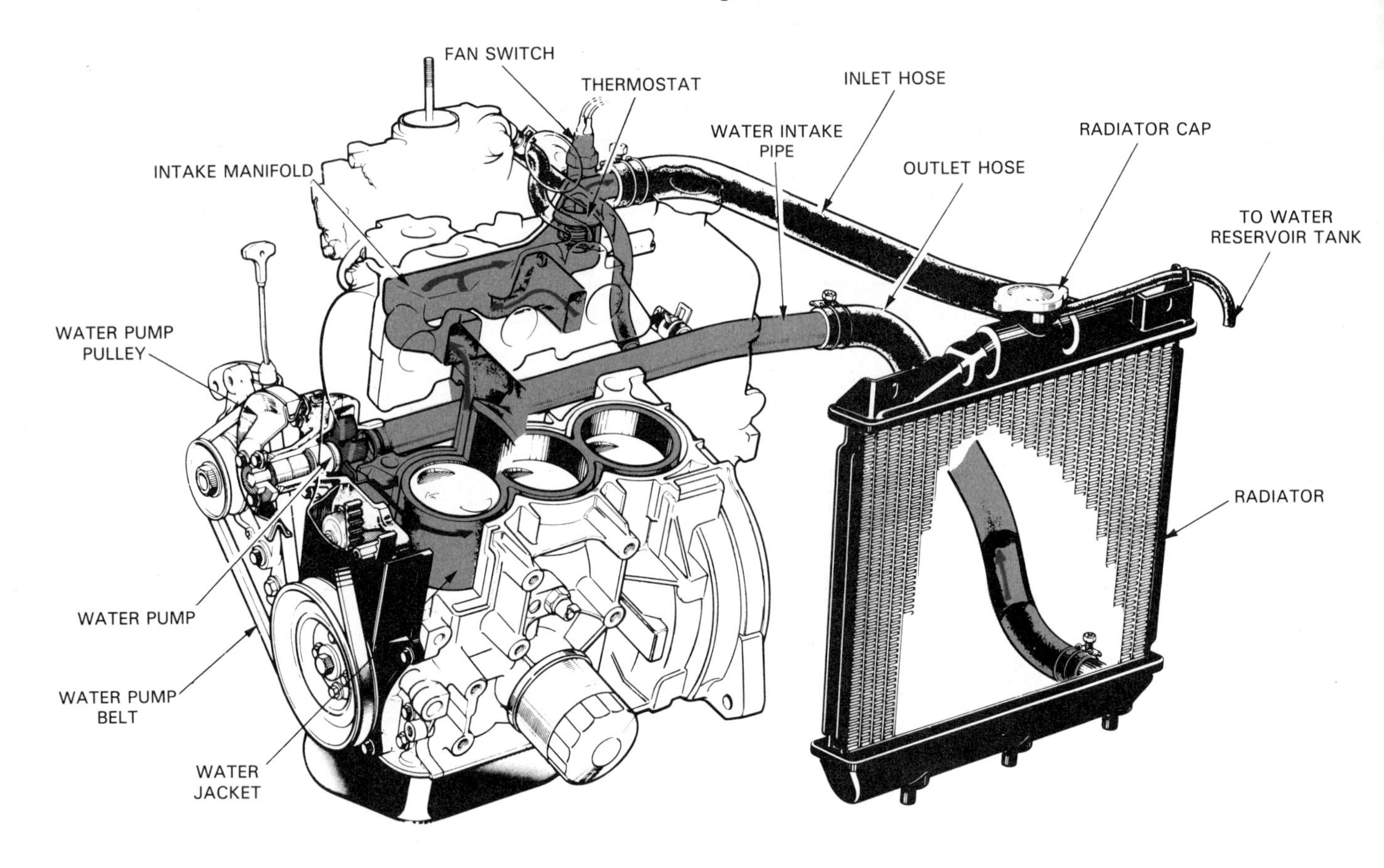

Fig. 15-1. Study parts and flow through cooling system. (General Motors)

A liquid cooling system has several advantages over an air type system:

1. More precise control of enigne temperature.
2. Less temperature variation inside engine.
3. Reduced exhaust emissions because of better temperature control.
4. Improved heater operation to warm passengers.

Note! An *oil cooled engine,* a liquid type system, uses the lubrication system to maintain the engine's operating temperature. Oil is circulated through the head and block instead of antifreeze and water. At present, this is an experimental system.

OPERATION OF A LIQUID COOLING SYSTEM

A fundamental cooling system consists of the components in Fig. 15-1. These parts include:

1. WATER PUMP (forces coolant through engine and other system parts).
2. RADIATOR HOSES (connect engine to radiator).
3. RADIATOR (transfers coolant heat to outside air).
4. FAN (draws air through radiator).
5. THERMOSTAT (controls coolant flow and engine operating temperature).

With the engine running, a belt powers the water pump. The pump forces coolant to circulate through the engine water jackets.

When the engine is cold, the thermostat remains closed. This prevents coolant from going to the radiator. Instead, it circulates around inside the engine. This helps warm the engine quickly.

When the engine reaches operating temperature, the thermostat opens. Hot coolant then flows through the radiator. The fan pulls air through the radiator. Excess heat is then transferred into the air flowing through the radiator. This maintains a proper engine temperature.

Generally, coolant circulates through the engine block, travels to the head, and then returns to the radiator (assuming the thermostat is open). In a *reverse-flow cooling system,* however, coolant circulates through the head before moving through the block and into the radiator. This arrangement helps produce a more uniform temperature throughout the engine, especially around the hot exhaust valves. As a result, the engine is less likely to suffer spark knock or ping. A reverse-flow cooling system allows the use of higher compression ratios and greater spark advance to improve power and efficiency.

WATER PUMP CONSTRUCTION

The *water pump* uses centrifugal force to circulate coolant through the engine water jackets, hoses, and radiator. Normally, it is powered by a belt running off of the crankshaft pulley, Fig. 15-2.

Illustrated in Fig. 15-3, the parts of a typical water pump are:

1. IMPELLER (disc with fan-like blades that spins and produces pressure and flow).
2. PUMP SHAFT (steel shaft that transfers turning force from hub to impeller).

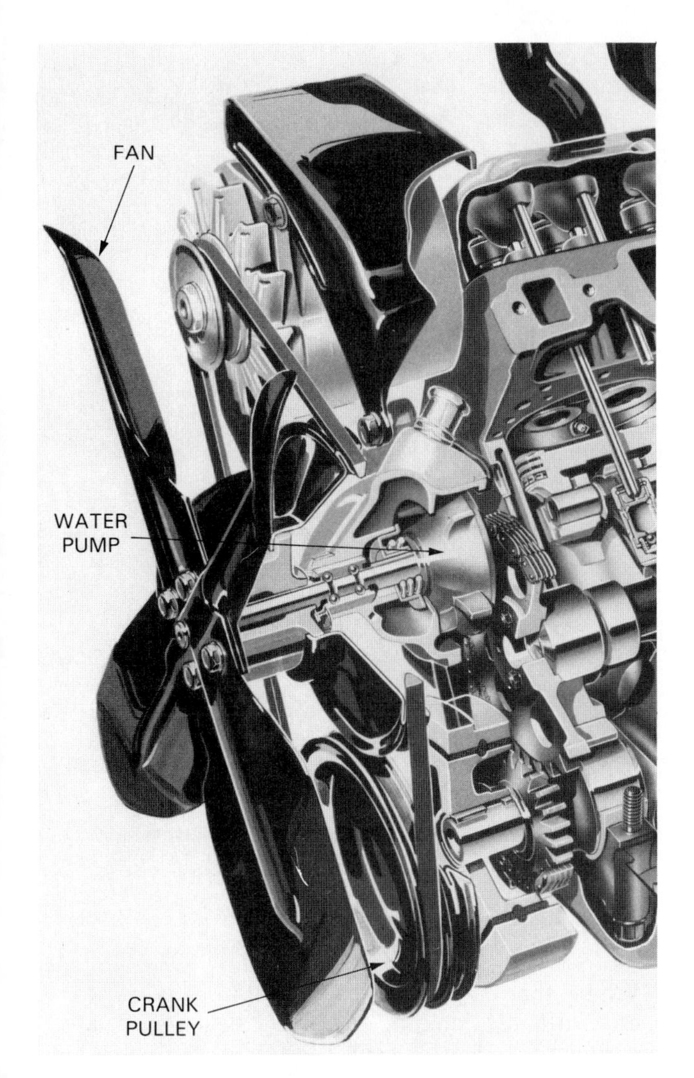

Fig. 15-2. This water pump bolts to front of engine. It is powered by belt running off of crankshaft pulley. (Chevrolet)

3. PUMP SEAL (prevents coolant leakage between water pump shaft and pump housing).
4. PUMP BEARINGS (plain or ball bearings that allows pump shaft to spin freely in housing).
5. PUMP HUB (provides mounting place for belt pulley and fan).
6. PUMP HOUSING (iron or aluminum casting that forms main body of pump).

The water pump normally mounts on the front of the engine. With some transverse mounted engines, it may bolt to the side of the engine. With other designs, the water pump is cast into the front of the engine block. The core of the pump bolts onto the block housing.

Water pump impellers can be metal or plastic. Impeller blades can be curved or straight like paddle wheels. Straight blades are sometimes used to reduce engine power consumption. A *water pump gasket* fits between the engine and pump housing to prevent coolant leakage. RTV sealer may be used instead of a gasket, Fig. 15-4.

RADIATOR CONSTRUCTION

The *radiator* transfers coolant heat into the outside air. The radiator is normally mounted in front of the engine so cool outside air can flow through it.

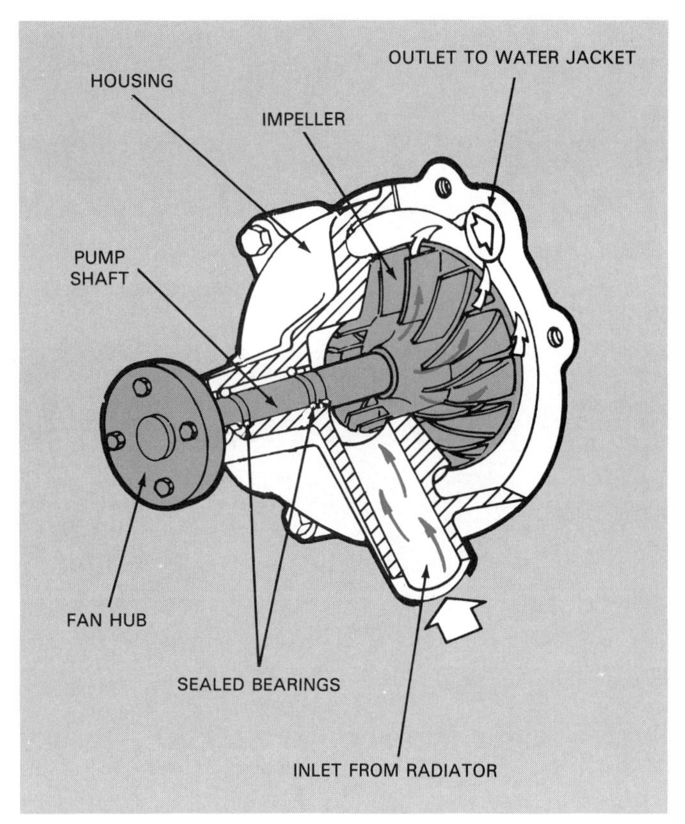

Fig. 15-3. Study parts of typical water pump. (Chrysler)

A radiator consists of:

1. CORE (center section of radiator made up of tubes and cooling fins), Fig. 15-5.
2. TANKS (metal or plastic ends that fit over core ends to provide storage for coolant and fittings for hoses).
3. FILLER NECK (opening for adding coolant; also holds radiator cap and overflow tube).
4. OIL COOLER (inner tank for cooling automatic transmission or transaxle fluid).
5. DRAINCOCK (fitting on bottom of tank for draining coolant from system).

Radiator action

Under normal operating conditions, hot engine coolant circulates through the radiator tanks and core tubes. Heat transfers into the core's tubes and fins. Since cooler air is flowing over and through the radiator fins, heat is removed from the radiator. This reduces the temperature of the coolant before it flows back into the engine.

Radiator cap

The *radiator cap* performs several functions:

1. Seals top of radiator filler neck to prevent leakage.
2. Pressurizes system to raise boiling point of coolant.
3. Relieves excess pressure and vacuum to protect against system damage.
4. In a closed system, it allows coolant flow into and from coolant reservoir.

The radiator cap locks onto the radiator tank filler neck. Rubber or metal seals make the cap-to-neck joint airtight.

The *radiator cap pressure valve* has a spring-loaded disc that contacts the filler neck. The spring pushes the valve into the neck.

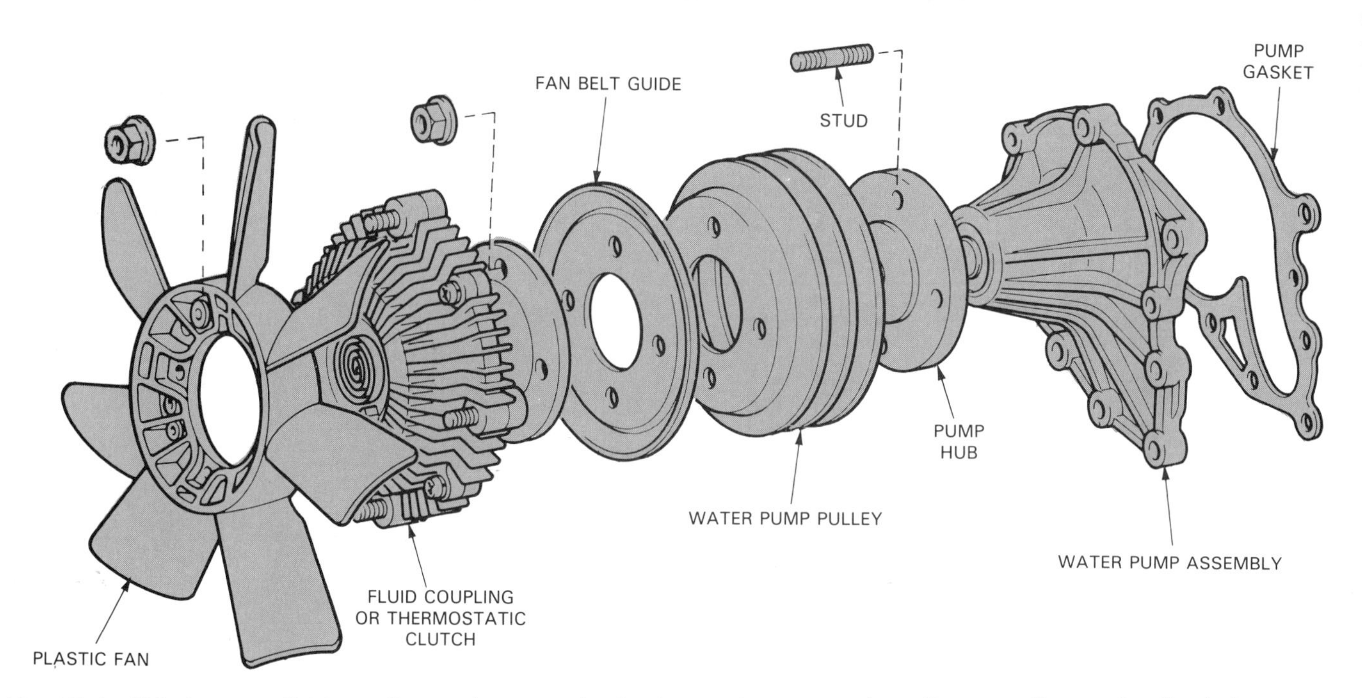

Fig. 15-4. This fan uses fluid coupling or thermostatic clutch to reduce power loss. Fan, coupling, and pulley bolt to pump hub. Gasket fits between engine and pump. (Toyota)

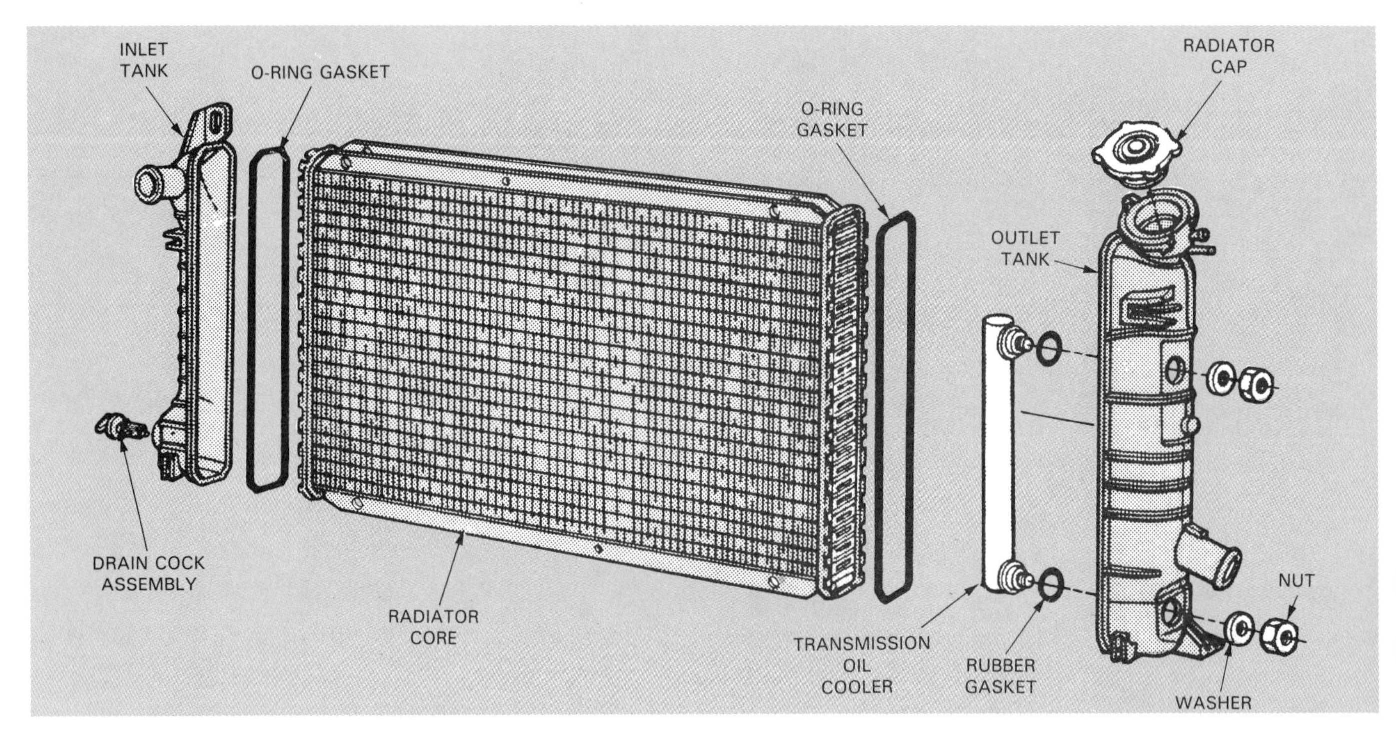

Fig. 15-5. Exploded view of radiator. This type uses rubber seals between core and tanks. (Buick)

Normally, water boils at 212 °F (100 °C). However, for every pound of pressure increase, the boiling point goes up about 3 °F. The radiator cap works on this principle, Fig. 15-6.

Typical *radiator cap pressure* is 12 to 16 psi (83 to 110 kPa). This raises the boiling point of the engine coolant to about 250 to 260 °F (121 to 127 °C). Many of the metal surfaces inside the water jackets can be above 212 °F (100 °C).

If the engine overheats and pressure exceeds the cap rating, the pressure valve opens. Excess pressure forces coolant out the overflow tube and into the reservoir or onto the ground. This prevents high pressure from rupturing the radiator, gaskets, seals, or hoses, Fig. 15-6.

The *radiator cap vacuum valve* opens to allow reverse flow back into the radiator when the coolant temperature

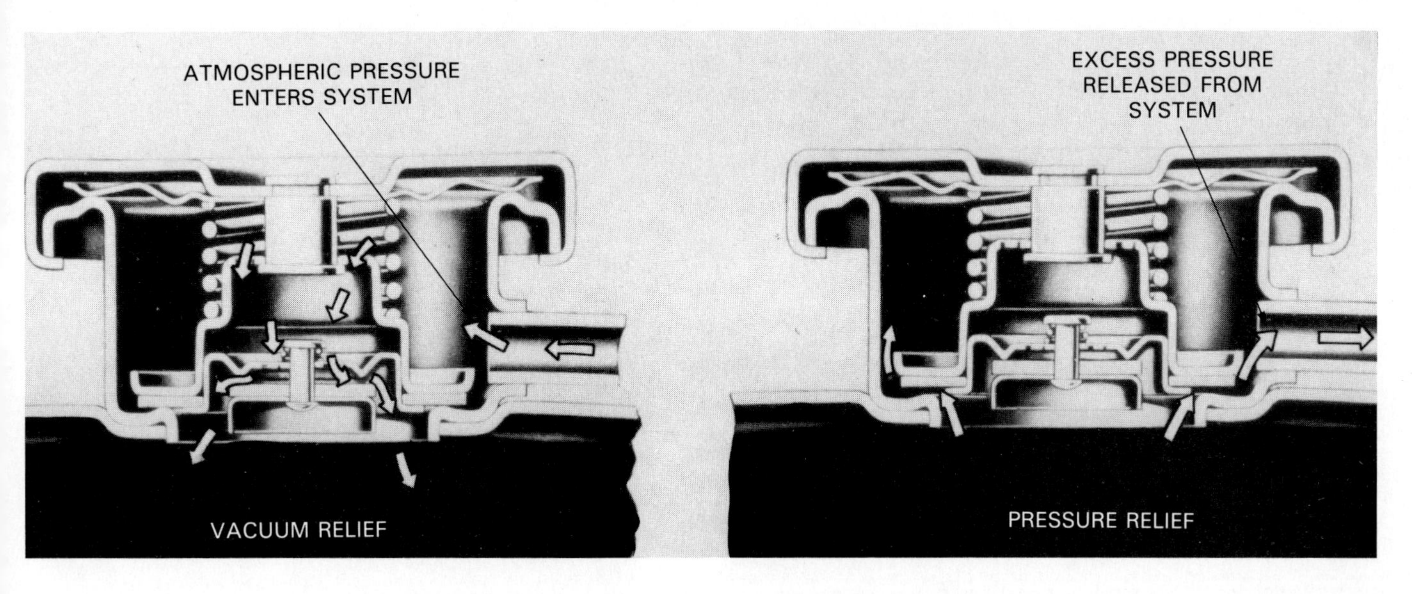

Fig. 15-6. Radiator cap vacuum valve keeps hoses and other parts from collapsing when system cools. Pressure valve releases excess pressure to keep parts from rupturing as system warms. (Pontiac)

drops after engine operation. It is a smaller valve located in the center, bottom of the cap, Fig. 15-6.

Without a cap vacuum valve, the radiator hoses and radiator tanks could collapse when the engine cools. The cooling and contraction of the coolant and air in the system could decrease coolant volume and pressure. Outside atmospheric pressure could then crush inward on the hoses and radiator.

Closed cooling system

A *closed cooling system* uses an expansion tank or reservoir and a special closed radiator cap. The overflow tube is routed into the bottom of the reservoir tank. Pressure and vacuum valve action pull coolant in and out of the reservoir tank as needed. This keeps the cooling system filled at all times.

Open cooling system

An *open cooling system* does NOT use a coolant reservoir. The overflow tube allows excess coolant to leak onto the ground. Also, it does not provide a means of adding fluid automatically as needed. The open cooling system is no longer used on modern automobiles.

Radiator shroud

The *radiator shroud* helps assure that the fan pulls air through the radiator. It fastens to the rear of the radiator and surrounds the area around the fan.

The plastic shroud keeps air from circulating between the back of the radiator and the front of the fan. A huge volume of air flows through the radiator core. Without a fan shroud, the engine could overheat, Fig. 15-7.

COOLING FANS

A *cooling system fan* pulls air through the core of the radiator and over the engine to help remove heat. It increases the volume of air flowing through the radiator, especially when the car is standing still.

Engine powered fans

An *engine powered fan* bolts to the water pump hub and pulley. Sometimes, a spacer fits between the fan and pulley to move the fan closer to the radiator. Refer back to Fig. 15-4.

A *flex fan* has thin, flexible blades that alter airflow with engine speed. At low speeds, the fan blades remain curved and pull air through the radiator. At higher engine speeds, the blades flex until they are almost straight. This reduces fan action and saves engine power.

A *fluid coupling fan clutch* is designed to slip at higher engine speeds. It performs the same function as a flexible fan. The clutch is filled with silicone-based oil. At a specific fan speed, there is enough load to make the clutch slip, Fig. 15-4.

A *thermostatic fan clutch* has a temperature sensitive, bimetallic spring that controls fan action. The spring controls oil flow in the fan clutch. When cold, the spring causes the clutch to slip, speeding engine warmup. After reaching operating temperature, it locks the clutch, providing forced air circulation.

Electric engine fans

An *electric engine fan* uses an electric motor and a thermostatic switch to provide cooling action. An electric fan is needed on front-wheel drive cars having transverse (sideways) mounted engines. The water pump is located away from the radiator, Fig. 15-7.

The *fan motor* is a small, DC motor. It mounts on a bracket secured to the radiator. A metal or plastic fan blade mounts on the end of the motor shaft.

The *fan switch* or *thermo switch* is a temperature sensitive switch that controls fan motor operation. When the engine is cold, the switch is open. This keeps the fan from spinning and speeds engine warmup. After warmup, the switch closes to operate the fan. Some electric fans are computer-controlled.

Fig. 15-8 illustrates an electric cooling fan. Note how the thermo switch controls its operation.

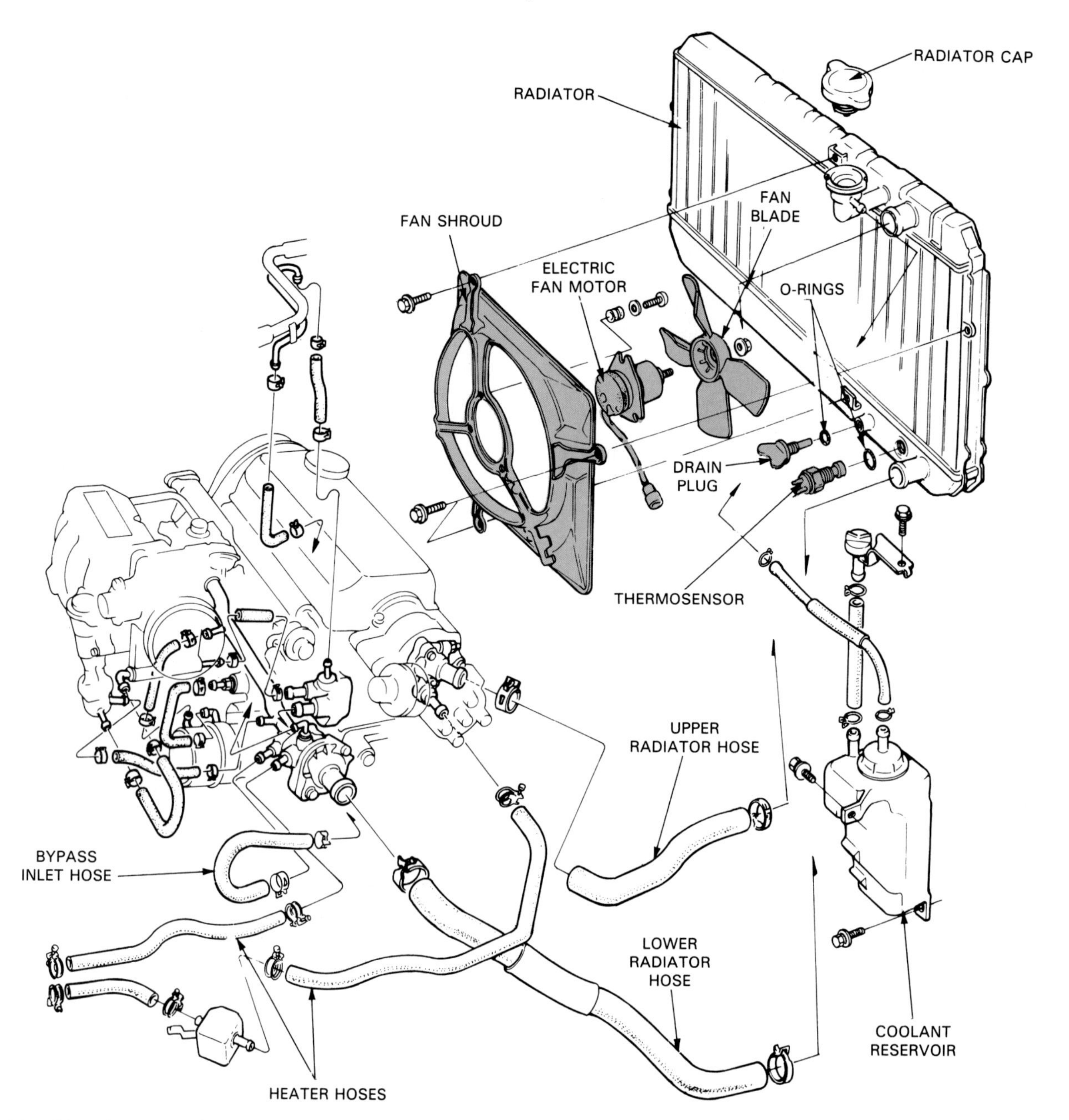

Fig. 15-7. Electric cooling fan is common on late model, front-wheel drive cars. Note thermosensor or fan switch that turns fan on after engine warms up. (Honda)

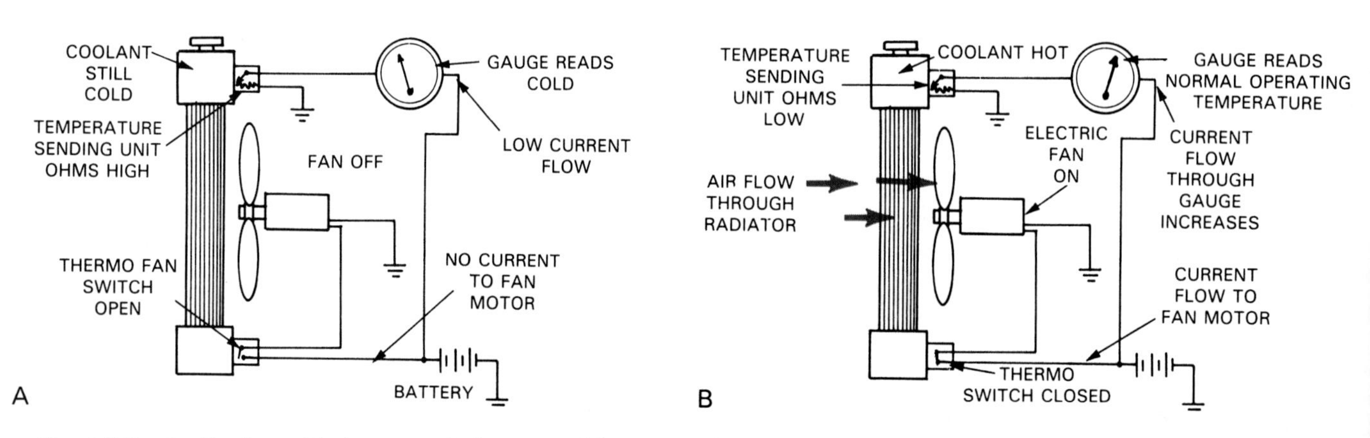

Fig. 15-8. A—Engine cold, thermo switch open and fan stays off to speed warmup. B—Engine hot, switch closes and connects fan motor to source of voltage.

Radiator and heater hoses

Radiator hoses carry coolant between the engine water jackets and the radiator. Being flexible, hoses can withstand the vibrating and rocking of the engine without breakage. Look at Fig. 15-7.

The upper radiator hose normally connects to the thermostat housing on the engine intake manifold or cylinder head. Its other end fits on the radiator. The lower hose connects the water pump inlet and the radiator.

A *hose spring* is frequently used in the lower radiator hose to prevent its collapse. The lower hose is exposed to suction from the water pump. The spring assures that the inner lining of the hose does NOT tear away, close up, and stop circulation.

Heater hoses are small diameter hoses that carry coolant to the heater core (small radiator-like device under car dash). *Hose clamps* hold the radiator and heater hoses on their fittings.

THERMOSTAT

The *thermostat,* Fig. 15-9, senses engine temperature and controls coolant flow through the radiator. It reduces flow when the engine is cold and increases flow when the engine is hot. The thermostat normally fits under a thermostat housing between the engine and the end of the upper radiator hose.

The thermostat has a wax-filled pellet. The pellet is contained in a cylinder and piston assembly. A spring holds the piston and valve in a normally closed position.

When the thermostat is heated, the pellet expands and pushes the valve open. As the pellet and thermostat cool, spring tension closes the valve.

Thermostat rating is stamped on the thermostat to indicate the operating (opening) temperature of the thermostat, Fig. 15-9. Normal ratings are between 180 and 195 °F (82 and 91 °C).

High thermostat heat ranges are used in modern automobiles because they reduce exhaust emissions and increase combustion efficiency. Fig. 15-10 illustrates thermostat operation.

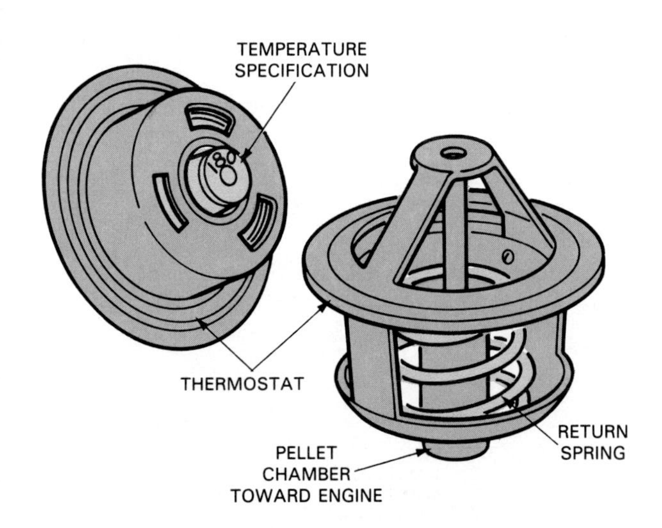

Fig. 15-9. Thermostat controls flow of coolant through engine and radiator. Note temperature rating stamped on bottom of thermostat. (Chrysler)

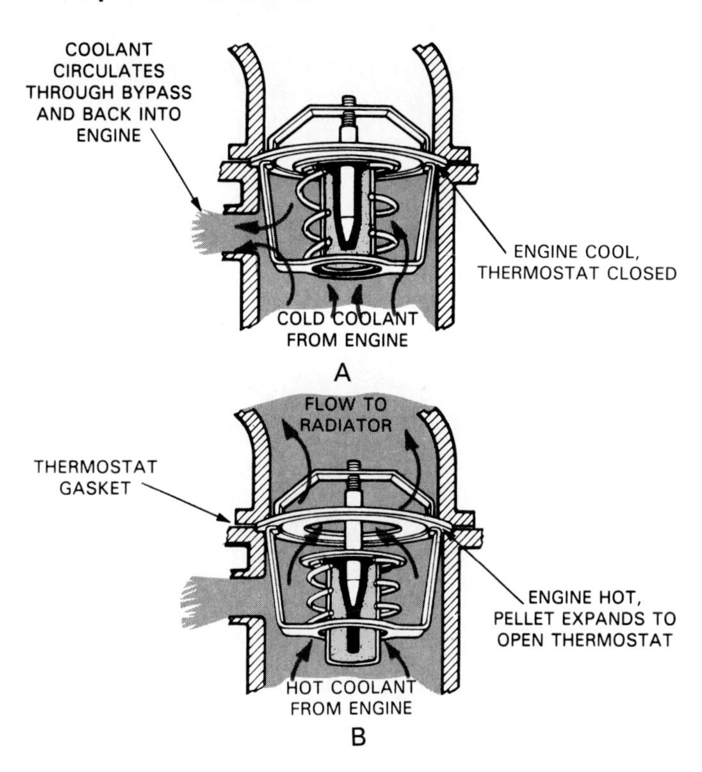

Fig. 15-10. A—Coolant is cold so thermostat pellet remains contracted to keep thermostat closed. Lack of circulation through radiator speeds engine warmup. B—Coolant temperature has increased enough to expand pellet and open thermostat. Circulation through radiator maintains engine temperature. (Chrysler)

A *bypass valve* permits coolant circulation through the engine when the thermostat is closed. If the coolant could NOT circulate, hot spots could develop inside the engine.

ANTIFREEZE

Antifreeze, usually ethylene glycol, is mixed with water to produce the engine coolant. Antifreeze has several functions.

Antifreeze keeps the coolant from freezing in very cold weather (outside temperature below 32 °F or 0 °C).

Coolant freezing can cause serious cooling system or engine damage. As ice forms, it expands. This expansion can produce tons of force. The water pump housing, cylinder head, engine block, radiator, or other parts could be cracked and ruined.

Antifreeze also prevents rust and corrosion inside the cooling system. It provides a protective film on part surfaces.

Antifreeze acts as a lubricant for the water pump. It increases the service life of the water pump, bearings, and seals.

Antifreeze conducts heat better than plain water and, therefore, cools the engine better. It is normally recommended in hot weather.

Engine block heater

A *block heater* may be used on an engine to aid engine starting in cold weather. It is simply a 120 V heating element mounted in the block water jacket. They are commonly found on diesel engines.

COOLING SYSTEM SERVICE

This section of the chapter will review the most common cooling system service procedures done by an engine technician. Coolant system failures can crack cylinder blocks, warp cylinder heads, "blow" head gaskets, burn valves, melt pistons, and do other expensive damage to engines. It is important that you know how to correctly test, service, and repair this important engine system.

COOLING SYSTEM TROUBLESHOOTING

An inspection will frequently let you find the source of the cooling system problem. Obvious troubles include:

1. Coolant leaks, Fig. 15-11.
2. Loose or missing fan belts, Fig. 15-12.
3. Low coolant level.
4. Abnormal water pump noises.
5. Leaves and debris covering outside of radiator.
6. Coolant in oil (oil looks like milk).
7. Combustion leakage into coolant (air bubbles in coolant).

CAUTION! Keep your hands and tools away from a spinning engine fan. Wear eye protection and stand behind, not over, the spinning fan blade. Then, if tools are dropped into the fan or a fan blade breaks, you are not likely to be hit and injured by flying parts.

Cooling system problems can be grouped into three general categories: leaks, overheating, and overcooling.

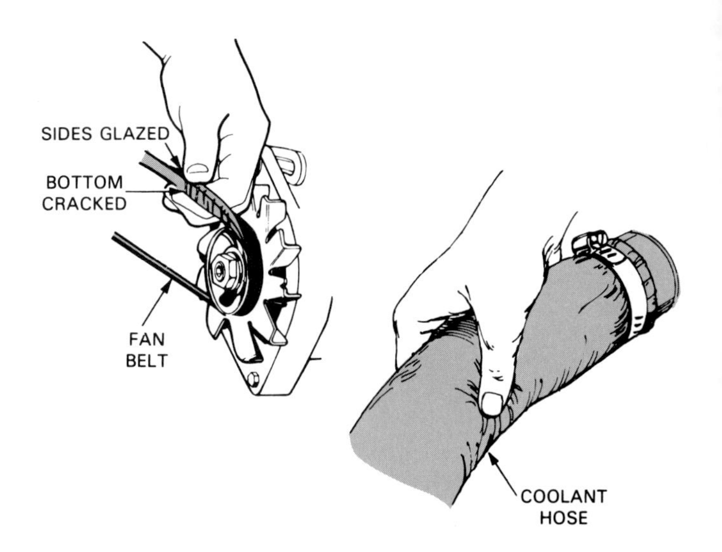

Fig. 15-12. Inspect cooling system and check condition of belts and hoses. Check for loose or deteriorated belts. Check hoses for leaks, hardening, or softening. (Gates Rubber Co.)

Coolant leaks

Coolant leaks show up as a wet, discolored (darkened green, orange, or rust colored) areas in the engine compartment or on the ground. The leaking fluid will smell like antifreeze and have the same general color. Leaks can occur almost anywhere in the system.

A low coolant level may indicate a leak. If not visible, the leak may be internal (cracked engine part, blown head gasket).

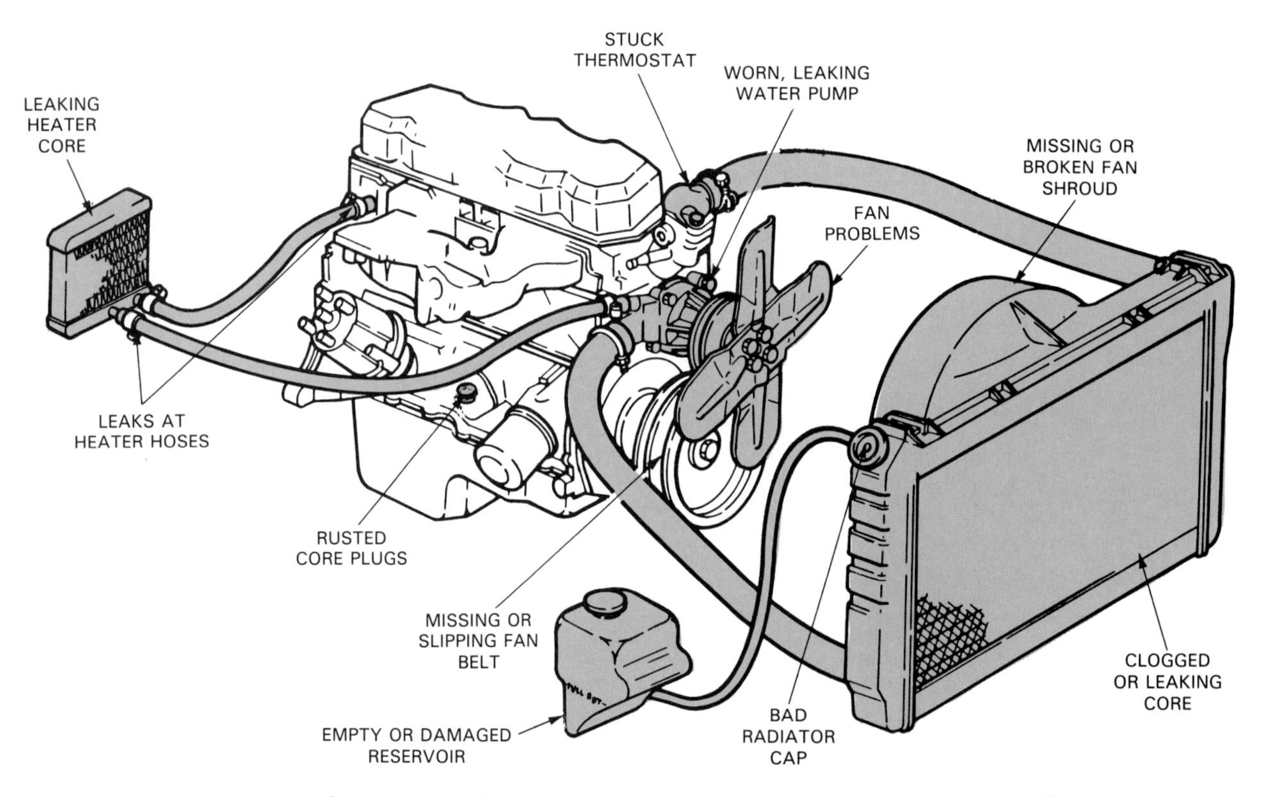

Fig. 15-11. Study types of troubles that can be found in a cooling system. (Pontiac)

WARNING! Never remove a radiator cap when the engine is hot. The pressure release can make the coolant begin to boil and expand. Boiling coolant could shoot out—causing SEVERE BURNS! On closed system, check the coolant level at the reservoir.

Engine overheating

Engine overheating is a serious problem that can cause major engine damage. The driver may notice the engine temperature light glowing, temperature gauge reading high, or the coolant boiling.

Common causes of engine overheating are:

1. LOW COOLANT LEVEL (leak or lack of maintenance has allowed coolant level to drop too low).
2. RUST OR SCALE (mineral accumulations in system have clogged radiator core or built up in water jackets).
3. STUCK THERMOSTAT (thermostat fails to open normally, restricts coolant flow).
4. RETARDED IGNITION TIMING (late ignition timing transferring too much heat into exhaust valves, ports, and exhaust manifold).
5. LOOSE FAN BELT (water pump drive belt slips under load and reduces circulation), Fig. 15-13.

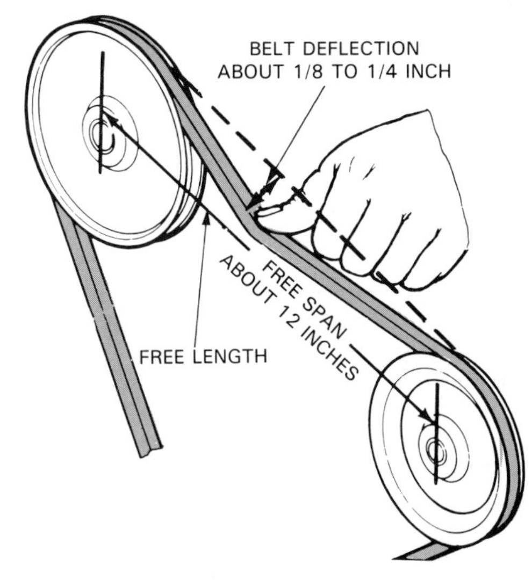

Fig. 15-13. An easy way to check belt tension is to push in on belt with your thumb. Belt should not deflect over about 1/8 to 1/4 inch. (Chrysler)

6. BAD WATER PUMP (leaking seal, broken pump shaft, or damaged impeller blades prevent normal pumping action).
7. COLLAPSED LOWER HOSE (suction from water pump may flatten hose if spring is missing or hose is badly deteriorated).
8. MISSING FAN SHROUD (air circulates between fan and back of radiator, reducing airflow through radiator)
9. ICE IN COOLANT (coolant frozen from lack of antifreeze can block circulation and cause overheating).
10. ENGINE FAN PROBLEMS (fan clutch or electric fan troubles can prevent adequate airflow through radiator).

Some engines are protected from overheating damage from a loss of coolant by the ECU. The ECU monitors the coolant temperature sensor to detect overheating. If overheating is detected, the computer cuts off spark to one cylinder at a time in a controlled sequence and retards ignition timing to reduce top speed. The outside air pulled into the ''dead cylinders'' cools the engine and prevents overheating damage for a time.

Engine overcooling

Overcooling causes slow engine warmup, insufficient warmth from the heater, and sluggish engine performance. Overcooling can cause increased part wear. Overcooling also reduces fuel economy because more combustion heat transfers into the metal engine parts.

Overcooling may be caused by:

1. STUCK THERMOSTAT (thermostat stuck open, allowing too much circulation).
2. LOCKED FAN CLUTCH (fan operates all the time to cause excess airflow through radiator).
3. SHORTED FAN SWITCH (electric fan runs all the time, increasing warmup time).

FAN BELT SERVICE

A *loose fan belt* may slip and not rotate the water pump and fan properly. The engine may even overheat. Always inspect the condition and tension (tightness) of fan belts when servicing a cooling system, Fig. 15-13. If a fan belt is cracked, frayed, glazed (hard, shiny surface), or oil soaked, it should be replaced.

WARNING! Keep your hands away from engine belts. A belt can pull your fingers into the pulleys, causing severe hand injuries.

ENGINE FAN SERVICE

A *faulty engine fan* can cause overheating, overcooling, vibration, and water pump wear or damage. Always check the fan for bent blades, cracks, and other problems. A flexible fan is especially prone to these problems. If any troubles are found, replace the fan.

WARNING! A fan with cracked or bent blades is extremely dangerous. Broken blades can be thrown out with great force causing severe lacerations or death.

Testing a fan clutch

To test a thermostatiac fan clutch, start the engine. The fan should slip when cold. When the engine warms, the clutch engage. Air should begin to flow through the radiator and over the engine. You will be able to hear and feel the air when the fan clutch locks up.

If the fan clutch is locked all the time, it is defective and must be replaced. Excessive play or oil leakage also indicates fan clutch failure.

Electric cooling fan service

Most *electric cooling fans* are controlled by a heat sensitive switch located somewhere in the cooling system (radiator, engine block, thermostat housing). When the engine is cold, the switch keeps the electric fan motor

OFF to speed engine warmup. Then when a predetermined temperature is reached, the switch closes and the fan begins to cool the engine.

Testing an electric cooling fan

To test an electric cooling fan, observe whether the fan turns ON when the engine is warm. Make sure the fan motor is spinning at a NORMAL SPEED and is forcing enough air through the radiator.

If the fan does NOT function, check the fuse, electrical connections, and supply voltage to the motor. If voltage is applied to the fan, yet the motor fails to operate, the fan motor should be replaced.

If the engine is warm and voltage is not supplied to the fan motor, check the action of the fan switch. Use either a VOM or test light. The switch should have almost zero resistance (pass current and voltage) when engine is warm. Resistance should be infinite (stop current and voltage) when the engine is cold.

If these tests do not locate the trouble, refer to a factory service manual for instructions. There may be a defective relay, connection, or other problem.

ANTIFREEZE SERVICE

Antifreeze should be checked and changed at regular intervals. After prolonged use, antifreeze will break down and become very corrosive. It can lose its rust preventative properties and the cooling system can rapidly fill with rust.

A visual inspection of the antifreeze will help determine its condition. Rub your finger inside the radiator filler neck. Check for rust, oil (internal engine leak), scale, or transmission fluid (leaking oil cooler). Also, find out how long the antifreeze solution has been in service.

If contaminated or too old, replace the antifreeze. If badly rusted, you may also need to flush (clean) the system, as described shortly.

Changing antifreeze

Antifreeze should be changed when contaminated or when two years old. Some coolants can last up to five years between service. Check in a service manual for exact change schedules.
on the bottom of the radiator or remove one end of the lower radiator hose. Allow the old coolant to drain.

If the antifreeze is not contaminated with rust, you may then refill the system. Tighten the petcock or hose. Pour in the needed amount of antifreeze (about two gallons or 7.6 liters).

Start and warm the engine. The coolant level may DROP when the thermostat opens. Add more water, if needed. Then install the radiator cap. Remember to dispose of used antifreeze properly by storing and sending it to a recycling center.

Antifreeze/water mixture

For ideal cooling and winter protection, a 50/50 mixture of water and antifreeze is usually recommended, Fig. 15-14. It will provide protection from ice formation to about $-34\,^{\circ}F$ ($-36.7\,^{\circ}C$). Higher ratios of antifreeze may produce even lower freezing temperatures but this much protection is not normally needed.

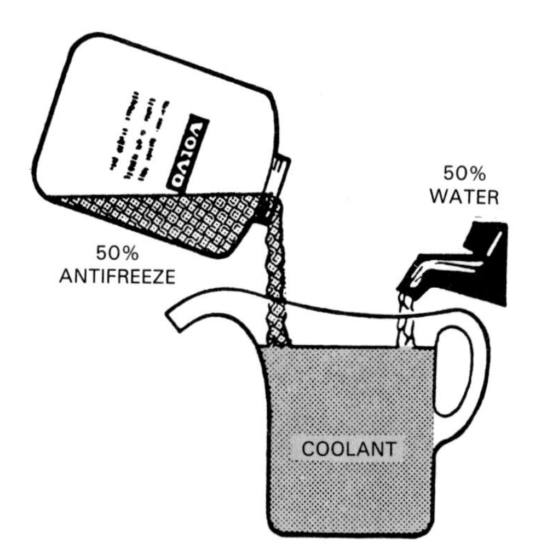

Fig. 15-14. Coolant is commonly mix of 50 percent water and 50 percent antifreeze. (Volvo)

NOTE! Plain water should NEVER be used in a cooling system or the four antifreeze functions just discussed will NOT be adequately provided.

Aluminum components

Many late model cars use aluminum cooling system and engine parts. Only use antifreeze designed for aluminum components.

Aluminum can be corroded by some types of antifreeze. Check the car's service manual or the antifreeze label for details.

Testing antifreeze strength

Antifreeze strength is a measurement of the concentration of antifreeze compared to water. It determines the freeze up protection of the solution.

A *cooling system hydrometer* can be used to measure the freezing point of the cooling system antifreeze solution. See Fig. 15-15.

Minimum antifreeze strength should be several degrees lower than the lowest possible temperature for the climate of the area. For example, if the lowest normal temperature for the area is $-10\,^{\circ}F$ ($-23\,^{\circ}C$), the antifreeze should test to $-20\,^{\circ}F$ ($-29\,^{\circ}C$).

COOLING SYSTEM HOSE SERVICE

Old *radiator hoses* and *heater hoses* are frequent causes of cooling system problems. After a few years of use, hoses deteriorate. They may become soft and mushy or hard and brittle. Cooling system pressure can rupture the hoses and result in coolant loss.

A softened lower radiator hose can be collapsed by the suction of the water pump. The collapsed hose will restrict coolant circulation and cause overheating. The spring inside the lower radiator hose normally prevents hose collapse and it should never be removed.

Checking cooling system hoses

Inspect the radiator and heater hoses for cracks, bulges, cuts, or any other sign of failure.

SQUEEZE the hoses to check whether they are hardened or softened and faulty. Flex or bend the heater

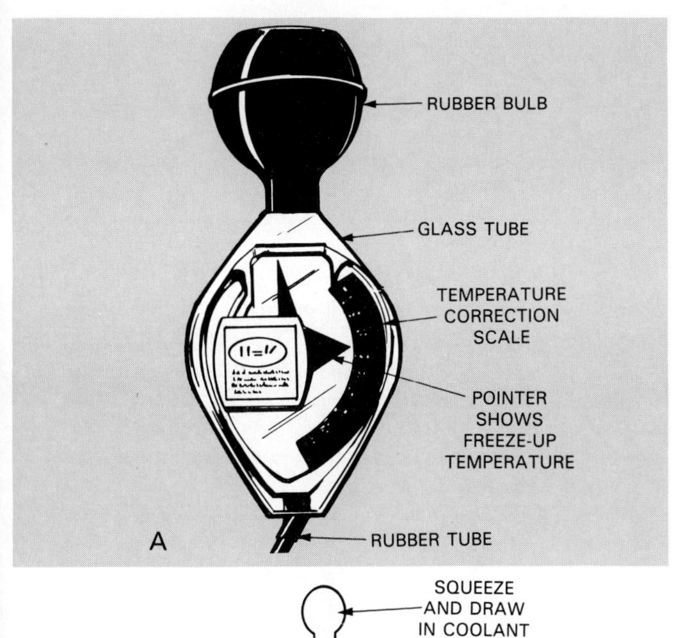

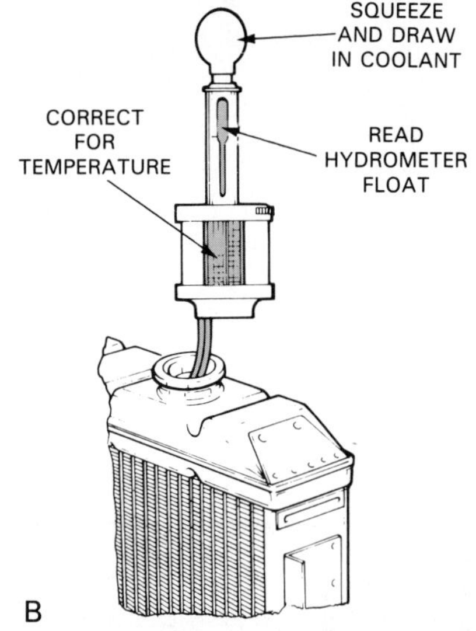

Fig. 15-15. A—Hydrometer will measure antifreeze strength. B—Draw coolant into hydrometer. Read pointer or float. Then correct for temperature. Antifreeze must be strong enough to prevent winter ice formation. (Florida Dept. of Voc. Ed.)

hoses and watch for surface cracks. If any problem is detected, the affected hoses should be replaced.

Hose replacement

To remove a hose, loosen the hose clamps. Twist the hose while pulling. If a new hose is to be installed, you may cut a slit in the end of the old hose to aid removal.

Clean the metal hose fittings. Coat them with a nonhardening sealer. Install the new hose.

Position the hose clamps so that they are fully over the metal hose fitting. Then tighten the clamps.

Install coolant and pressure test the system. Check all fittings for leaks. Refer to Fig. 15-16.

COOLING SYSTEM PRESSURE TEST

A *cooling system pressure test* is used to quickly locate leaks. Low air pressure is forced into the system. This will cause coolant to pour or drip from any leak in the system.

A *pressure tester* is a hand-operated air pump used to pressurize the cooling system for leak detection. Install the pressure tester on the radiator filler neck. Then pump the tester until the pressure gauge reads radiator cap pressure (around 14 psi or 95 kPa), Fig. 15-17.

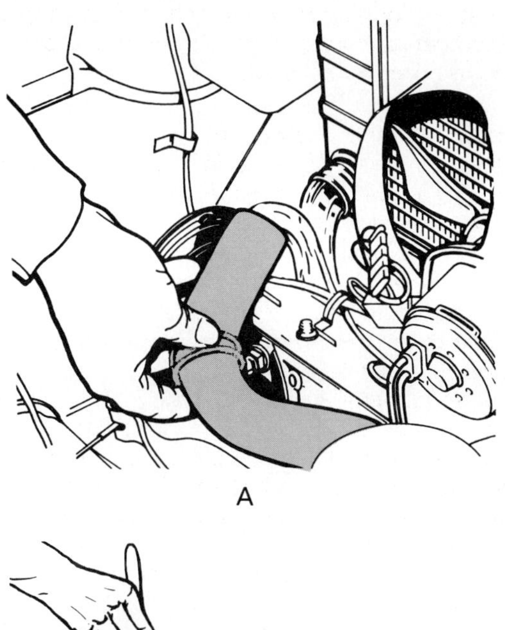

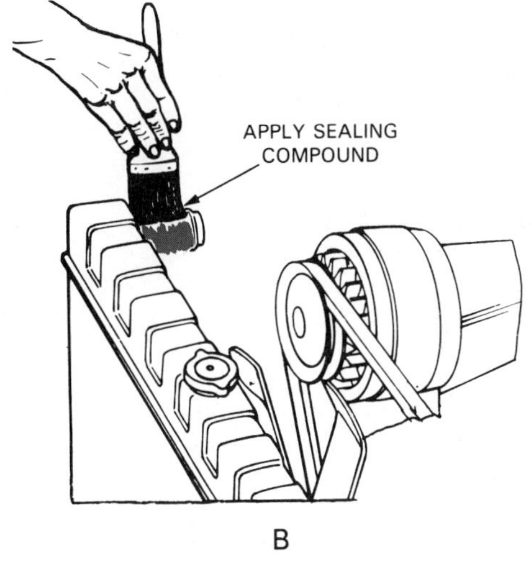

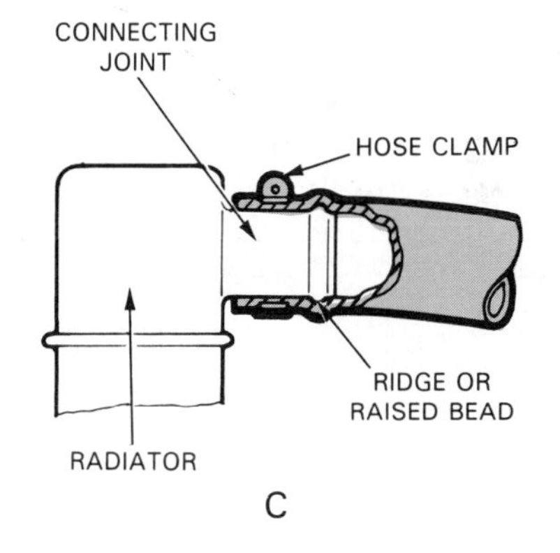

Fig. 15-16. Hose service. A—Loosen clamp. Twist hose and pull off. B—Clean hose fitting. Coat with nonhardening sealer. C—Install new hose. Position clamp properly before tightening. (Ford and Mopar)

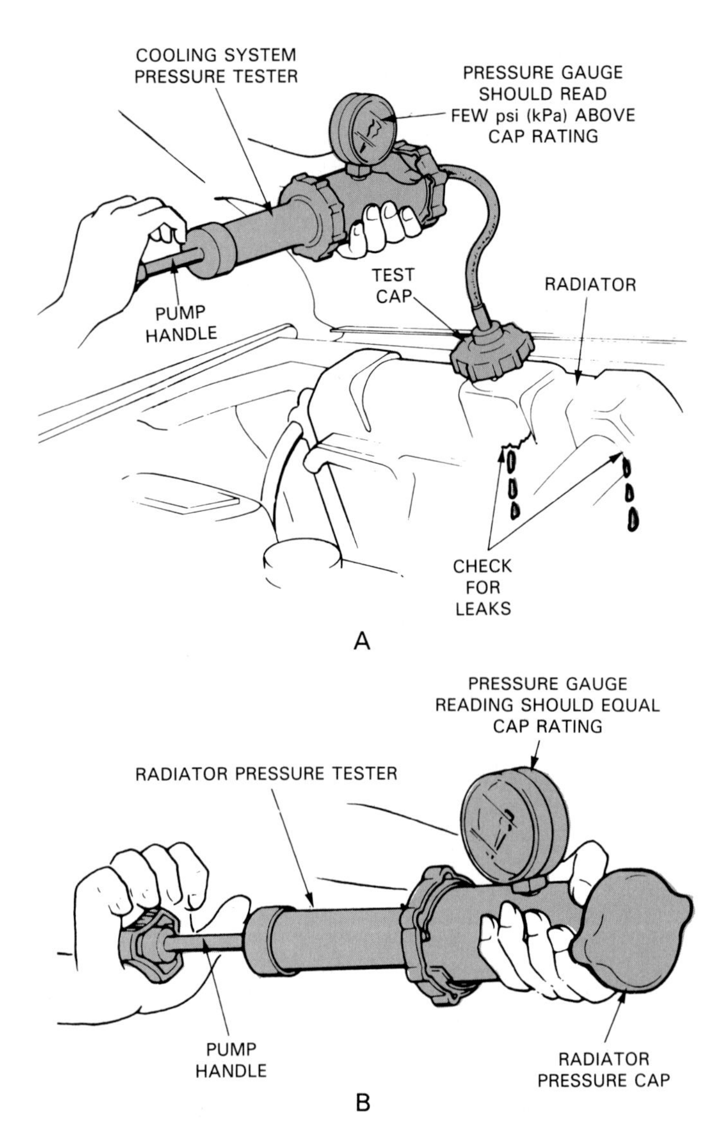

Fig. 15-17. A—To pressure test cooling system, install tester on radiator filler neck. Pump handle until gauge reads a few psi above cap pressure. Then, check for leaks in system. B—To test pressure cap, install cap on tester using adapter. Pump handle until gauge stops increasing. Needle should stop going up at cap rated pressure. It should also hold pressure without leaking. (Honda)

CAUTION! Do not pump too much pressure into the cooling system or part (radiator, hose, or gasket) damage may result. Only equal radiator cap pressure when testing.

With pressure in the system, inspect all parts for leakage. Check at all hose fittings, at gaskets, under the water pump, around the radiator, and at engine freeze (core) plugs. If a leak is found, tighten, repair, or replace parts as needed.

RADIATOR AND PRESSURE CAP SERVICE

When overheating problems occur and the system is NOT leaking, check the radiator and the pressure cap. They are common sources of overheating. The pressure cap could have bad seals, allowing pressure loss. The radiator may be clogged and not permitting adequate air or coolant flow.

Inspecting radiator and pressure cap

Inspect the outside of the radiator for leaves and road dirt. Also, make sure the radiator shroud is in place and unbroken. These troubles could limit air circulation through the core.

If needed, use a water hose to wash debris out of the core. Spray water from the back to push debris out the front of the radiator. You may also use compressed air if pressure is low enough not to damage the core.

Inspect the radiator cap and filler neck. Check for cracks or tears in the cap seal. Check the filler neck sealing surfaces for nicks and dents. Replace the cap or have the neck repaired as needed.

Pressure testing radiator cap

A *radiator cap pressure test* measures cap opening pressure and checks the condition of the sealing washer. The cap is installed on a cooling system pressure tester. The procedure is shown in Fig. 15-17.

Pump the tester to pressurize the cap. Watch the pressure gauge. The cap should release air at its rated pressure (pressure stamped on cap.) It should also hold that pressure for at least one minute.

Radiator repair

A *radiator shop* specializes in radiator repair. It has the facilities to properly disassemble, rod out (clean), solder (reassemble or repair), and pressure test a radiator. Few mechanics try to fix a radiator in-shop.

Some late model radiators have plastic components. Special sealers, seals, and repair techniques are needed on these radiators. Refer to the service manual for details.

WATER PUMP SERVICE

A *bad water pump* may leak coolant (worn seal), fail to circulate coolant (broken shaft or damaged impeller), or it may produce a grinding sound (faulty bearings).

Rust in the cooling system or lack of antifreeze are common reasons for pump failure. These conditions could speed seal, shaft, and bearing wear. An overtightened fan belt is another common cause for premature water pump failure.

Checking water pump

To check for a *bad water pump seal,* pressure test the system and watch for leakage at the pump. Coolant will leak out of the small drain hole at the bottom of the pump or at the end of the pump shaft. Replace or rebuild a leaking pump.

To check for *worn water pump bearings,* try to wiggle the fan or pump pulley up and down. If the pump shaft is loose in its housing, the pump bearings are badly worn. Pump replacement would be necessary. A stethoscope can also be used to listen for worn, noisy water pump bearings.

To check *water pump action,* warm the engine. Squeeze the top radiator hose while someone starts the engine. You should feel a pressure surge (hose swelling) if the pump is working. If not, pump shaft or impeller problems are indicated. You can also watch for coolant circulation in the radiator with the engine at operating temperature.

Water pump removal

To remove the water pump, unbolt all brackets and other components (air conditioning compressor, power steering pump, alternator) preventing pump removal. Then, unscrew the bolts holding the pump to the engine. Keep all bolts organized to aid assembly.

Scrape off old gasket or sealer material. The engine-to-pump mating surfaces must be perfectly clean to prevent coolant leakage. On soft aluminum parts, be careful not to gouge or scratch the sealing surfaces.

Installing water pump

To install a *water pump gasket,* use approved sealer to stick the new gasket to the pump. This will keep the gasket in alignment over the bolt holes during pump installation, Fig. 15-18.

To use a chemical gasket, (sealer used in place of fiber gasket), squeeze out a bead of approved sealer around the pump sealing surface. Form a continuous bead of consistent width (about 1/8 in. or 3 mm).

Fit the pump onto the engine. Move it straight into place. Do not shift the gasket or break the sealant bead. Start ALL of the bolts by hand. Screw them in about two turns. Check that all bolt lengths are correct.

Torque all of the fasteners a little at a time in a crisscross pattern. Go over the bolts several times to assure correct tightening. Install the other components and tighten the belt properly.

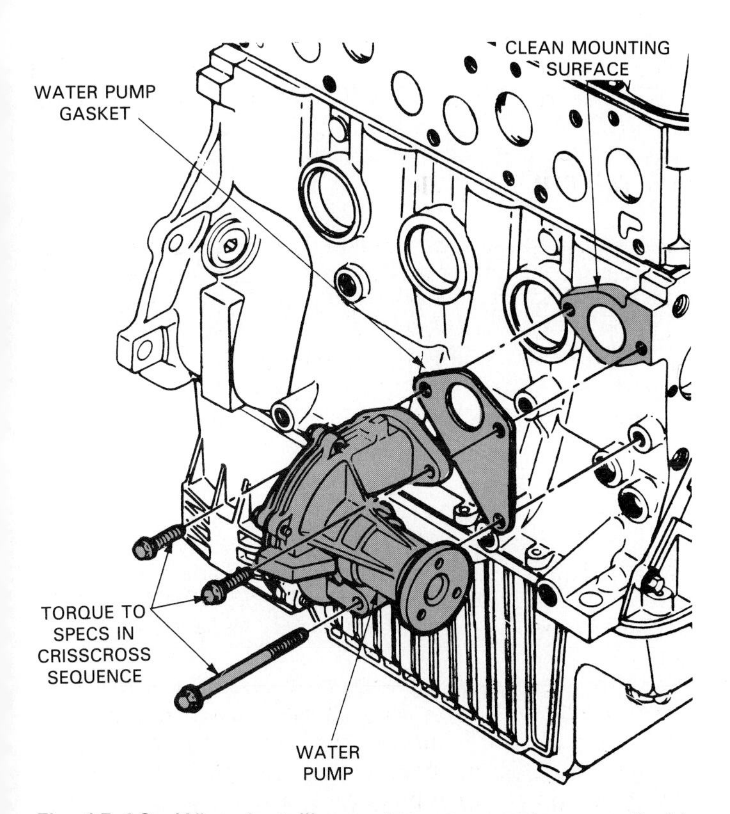

Fig. 15-18. When installing water pump, make sure all old gasket material is off of mating surfaces. Use approved sealer to hold gasket in place. Start bolts by hand and check lengths. Then torque to specs in crisscross pattern. (Ford Motor Co.)

If needed, refer to a factory shop manual. It will give detailed directions on pump service for the exact make and model of car. You may need to remove the radiator in some cars, but not with others. It will give this kind of essential information.

Water pump rebuild

A *water pump rebuild* involves pump disassembly, cleaning, part inspection, worn part replacement, and reassembly. Few mechanics rebuild water pumps. They purchase new or factory rebuilt pumps. Rebuilding takes too much time and would not usually be cost effective.

THERMOSTAT SERVICE

A *stuck thermostat* can either cause engine overheating or engine overcooling. If the thermostat is stuck shut, coolant will NOT circulate through the radiator. As a result, overheating could make the coolant boil at ALL engine speeds. If a thermostat is stuck open, too much coolant may circulate through the radiator. The engine may not reach proper operating temperature. The engine may run poorly for extended periods in cold weather. Engine efficiency (power, gas mileage, and driveability) will be reduced.

Thermostat testing

To check thermostat action, watch the coolant through the radiator neck. When the engine is cold, coolant should NOT flow through the radiator. When the engine warms, the thermostat should open. Coolant should then begin to circulate through the radiator. If this action does not occur, the thermostat may be defective.

In some instances, the thermostat may have to be removed from the engine for testing. The thermostat is placed in a container of water and heated on a hot plate or stove. The opening temperature of the thermostat is observed and compared to specs.

If the thermostat does NOT open at the correct temperature, it is defective and should be replaced.

Replacing thermostat

The thermostat is normally located under the *thermostat housing* (fitting for upper radiator hose). The thermostat housing can be located at either the coolant inlet or outlet on the engine.

To remove the thermostat, drain the coolant and remove the upper radiator hose. Unscrew the bolts holding the thermostat housing to the engine. Tap the housing free with a rubber hammer. Lift off the housing and thermostat.

Scrape all of the old gasket material off the thermostat housing and the sealing surface on the engine.

Make sure the thermostat housing is not warped. Place it on a flat surface and check for gaps between the housing and surface. If warped, file the surface flat. This will prevent coolant leakage.

Make sure the temperature rating is correct. Then place the new thermostat into the engine. Normally, the rod (pointed end) on the thermostat should face the radiator hose. The pellet chamber should face the inside of the engine.

Position the new gasket with approved sealer, Fig. 15-19. Start the fasteners by hand. Then torque them to specs in a crisscross pattern. Do NOT overtighten the thermostat housing bolts or warpage may result. Most housings are made of soft aluminum or "pot metal."

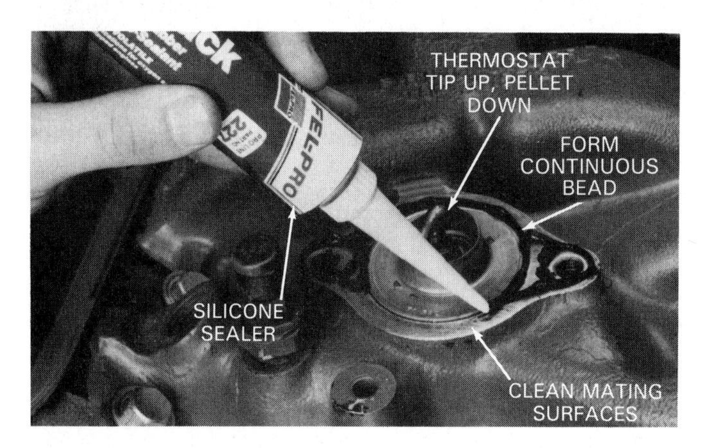

Fig. 15-19. When installing thermostat, clean housing and engine gasket surfaces. Make sure thermostat is pointing in right direction, pellet toward engine. Use fiber gasket or silicone sealer. Tighten bolts equally to specs. (Fel-Pro)

Some thermostat housings use O-rings versus a gasket. In many cases, this O-ring can be reused if it is not damaged. If it shows any signs of damage, replace the O-ring. Some thermostat kits come with a new housing O-ring.

Cooling system bleeder screw

A *cooling system bleed screw* is sometimes provided to help remove trapped air when refilling the system. Some systems have more than one bleed screw or valve. Many late model vehicles with low hood lines require a bleed valve to empty out air pockets formed in areas of the system. Air trapped in the cooling system can quickly cause engine overheating or damage to the part near the air pocket and hot spot. Part cracking or warpage could result.

To bleed a cooling system, fill the system with coolant. Then, start and warm the engine to full operating temperature. Open the bleed screw until all air is purged out and coolant leaks from the valve. Refill the reservoir as needed after bleeding.

WARNING! Never fully remove a cooling system bleed screw or any system connection (hose, fitting, plug, sensor) with the engine at full operating temperature. Steaming hot coolant could squirt out. Wear safety glasses and gloves when working with a hot cooling system.

FLUSHING A COOLING SYSTEM

Flushing (cleaning) of a cooling system should be done when rust or scale is found in the system. Flushing involves running water or a cleaning chemical through the cooling system. This washes out contaminants.

Fast flushing is a common method of cleaning a cooling system because the thermostat does not have to be removed from the engine.

A water hose is connected to a heater hose fitting. The radiator cap is removed and the drain cock is opened. When the water hose is ON and water flows into the system, rust and loose scale are removed.

Reverse flushing of a radiator requires a special adapter that is connected to the radiator outlet tank by a piece of hose. Another hose is attached to the inlet tank. Compressed air, under low pressure, is used to force water through the core backwards. This can be done on the engine block as well.

Chemical flushing is needed when a scale buildup in the system is causing engine overheating. A chemical cleaner is added to the coolant. The engine is operated for a specific amount of time to allow the chemical to act on the scale. Then the system is flushed with water to remove the chemical, Fig. 15-20.

WARNING! Always follow manufacturer's instructions when using a cooling system cleaning agent. The chemical may cause eye and skin burns.

LUBRICATION SYSTEM

This section of the chapter will summarize the operation and construction of an engine lubrication system. As with the cooling system, the lubrication system prevents engine wear and damage.

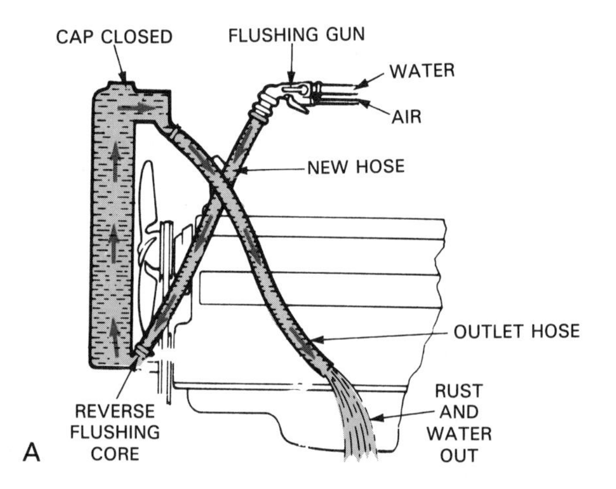

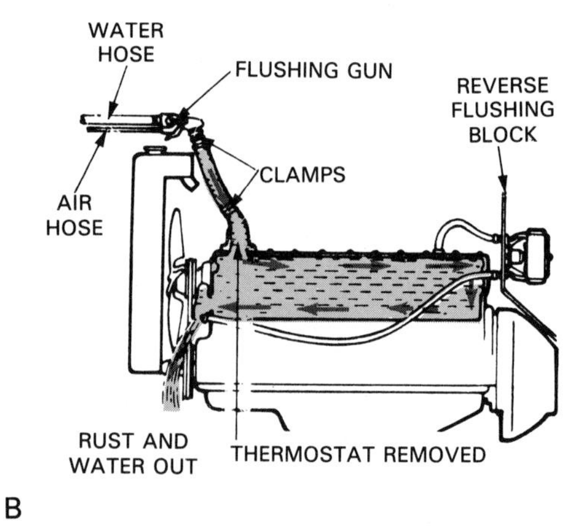

Fig. 15-20. Cooling system flushing is needed to remove rust and other deposits. A—Reverse flushing radiator. B—Reverse flushing engine block. (Chrysler)

FUNCTIONS OF LUBRICATION SYSTEM

An engine lubrication system has several functions. These include:

1. Reduces friction and wear between moving parts.
2. Helps transfer heat and cool engine parts.
3. Cleans the inside of the engine by removing contaminants (metal, dirt, and other particles).
4. Cuts power loss and increases fuel economy.
5. Absorbs shocks between moving parts to quiet engine operation and increase engine life.

ENGINE OIL

Engine oil, also called *motor oil,* is used to produce a lubricating film on the moving parts in an engine. It is commonly refined from crude oil or petroleum.

Synthetic oils (manufactured oils) are also available. They can be made from substances other than crude oil.

An *oil film* (thin layer of oil) separates engine parts to prevent metal-on-metal contact. Without the oil film, the parts would rub together and wear rapidly, Fig. 15-21.

Oil clearance

Oil clearance is the small space between moving engine parts for the lubricating oil film. The clearance allows oil to enter to prevent part contact, Fig. 15-21.

One example, a connecting rod bearing typically has a bearing or oil clearance of about .002 in. (0.05 mm). This clearance is large enough to allow oil entry. However, it is also small enough to keep the parts from hammering together during engine operation.

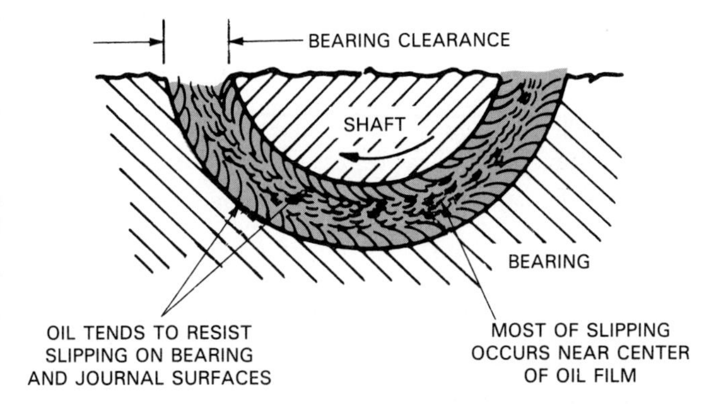

Fig. 15-21. Quality motor oil is essential to engine service life. Exaggerated view of oil clearance shows how oil holds parts away from each other to prevent friction and wear. (Chrysler)

Oil viscosity (weight)

Oil viscosity, also called *oil weight,* is the thickness of fluidity (flow ability) of the motor oil. A high viscosity oil would be very thick and would resist flow, like HONEY. A low viscosity oil would be thin and running, more like WATER.

A *viscosity numbering system* is used to rate the thickness of engine oil. A high number would indicate thicker oil. A lower number would denote a thinner oil.

The oil's viscosity number is printed on the oil bottle. The SAE (Society of Automotive Engineers) standardized this numbering system. For this reason, oil viscosity is written SAE 30, SAE 40, etc., Fig. 15-22.

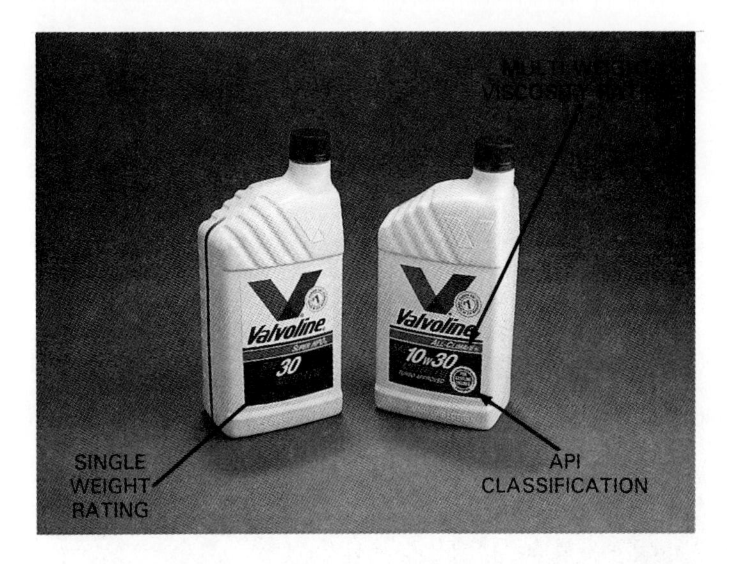

Fig. 15-22. Motor oil viscosity is printed on bottle as numbering system. Larger number means oil is thicker. Multi-viscosity oil holds its weight with a change in temperature and is good in cold weather.

Engine oil viscosities commonly range from a thin SAE 10 weight to a thick SAE 50 weight. Auto manufacturers specify an SAE number for their engines.

Temperature effects on oil

When cold, oil thickens and resists flow. When heated, oil thins and becomes runny. This can pose a problem.

The oil in a cold engine may be so thick that engine starting is difficult. The oil will not pump through the engine properly. This may increase starter drag and result in poor lubrication.

When the engine warms up, the oil film thins out. If it becomes too hot and thin, the oil film can break down and part contact can result.

Multi-viscosity oil

Multi-viscosity oil or *multi-weight oil* will exhibit operating characteristics of a thin, light oil when cold and a thicker, heavy oil when hot. It is a high grade oil that maintains the same viscosity with temperature changes. A multi-weight oil can be numbered SAE 10w30, 10w-40, 20w-50, 10w-20w-50, etc.

For example, a 10w-30 weight oil will flow easily (like a 10w oil) when starting a cold engine. It will then act as a thicker oil (like 30 weight) when the engine warms to operating temperature. This will make the engine start more easily in cold weather. It will also provide adequate film strength (thickness) when the engine is at full operating temperature.

Selecting oil viscosity

Normally, always use the oil viscosity recommended by the auto maker. However, in a very old, high mileage worn engine, a higher viscosity oil may be beneficial. Thicker oil will tend to seal the rings and provide better bearing protection. It also may help cut engine oil consumption and smoking.

Fig. 15-23 is one auto maker's chart showing recommended SAE viscosity number.

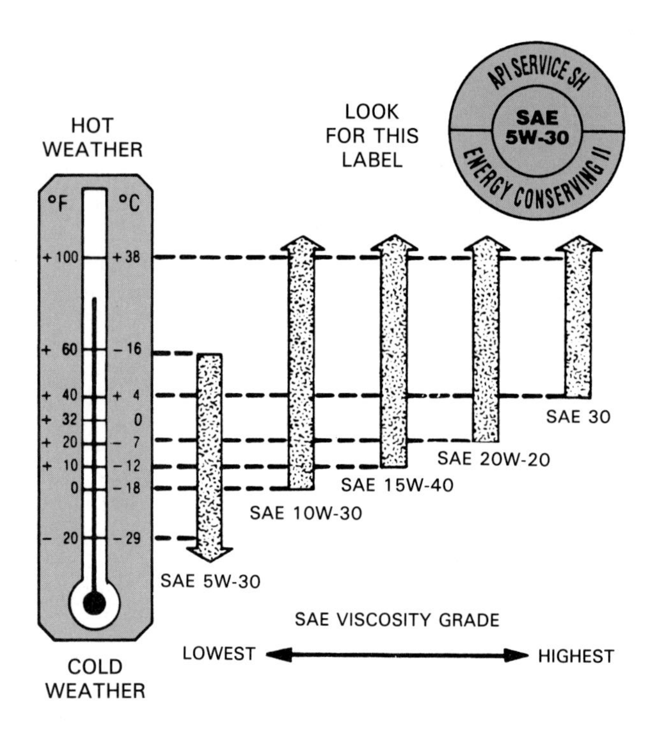

Fig. 15-23. This is a sample graph showing recommended oil weights for various outside temperatures. Note how heavier oils are suggested in hotter weather to maintain adequate oil film. (General Motors)

Oil service rating

An *oil service rating* is a set of letters printed on the oil can to denote how well the oil will perform under operating conditions. This is a performance standard set by the American Petroleum Institute, abbreviated API. The service rating categories are:

1. SA (lowest quality oil that should NOT be used in automotive engines).
2. SB (minimum quality oil for automotive gasoline engines under mild service conditions, not normally recommended).
3. SC (meets oil warranty requirements for 1964 through 1967 automotive gasoline engines).
4. SD (meets oil warranty requirements for 1968 through 1970 automotive gasoline engines).
5. SE (meets oil warranty requirements for 1972 through 1979 automotive gasoline engines).
6. SF (meets oil warranty requirements for 1980 and earlier engines).
7. SG (meets oil warranty requirements for 1987 and earlier engines).
8. SH (current top grade oil recommended for modern gasoline engines).
9. CA through CG4 (oil recommended for diesel engines).

A car owner's manual will give the service rating recommended for a specific vehicle. You can use a better service rating than recommended, but NEVER a lower service rating! A high service rating (SH for example) can withstand higher temperatures and loads while still maintaing a lubricating film.

LUBRICATION SYSTEM PARTS

A *lubrication system* consists of:

1. MOTOR OIL (lubricant for moving parts in engine).
2. OIL PAN (storage area for motor oil), Fig. 15-24.

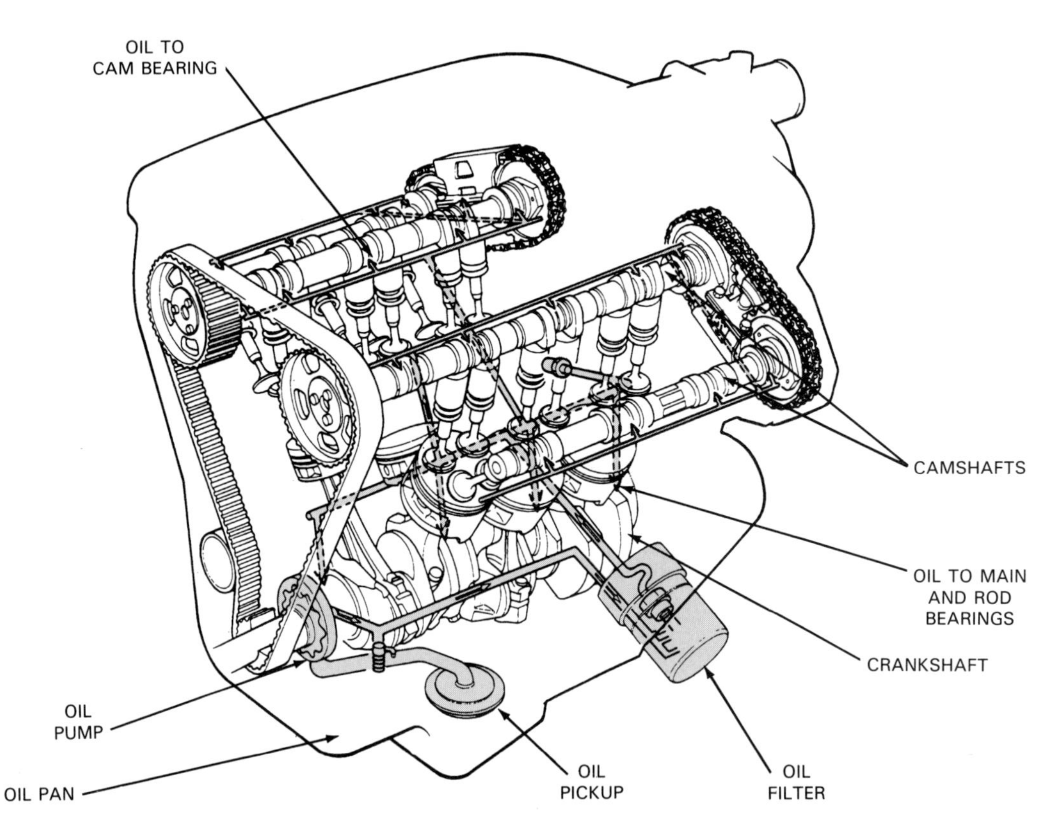

Fig. 15-24. Memorize basic parts of lubrication system and trace flow through engine. (Ford)

3. OIL PUMP (forces oil throughout inside of engine).
4. OIL FILTER (strains out impurities in oil).
5. OIL GALLERIES (oil passages through engine).

As you will learn, other parts are added to increase system efficiency.

Lubrication system operation

With the engine running, the oil pump pulls motor oil out of the oil pan, Fig. 15-24. A screen on the pickup tube removes large particles from the oil before oil enters the pump. The pump then pushes the oil through the oil filter and oil galleries. The oil filter cleans the oil. The filtered oil then flows to the camshaft, crankshaft, lifters, rocker arms, and other moving parts.

When oil leaks out of the engine bearings, it sprays on the outside of internal engine parts. For example, when oil leaks out of the connecting rod bearings, it sprays on the cylinder walls. This lubricates the piston rings, pistons, wrist pins, and cylinders. Oil finally drains back into the oil pan for recirculation, Fig. 15-25.

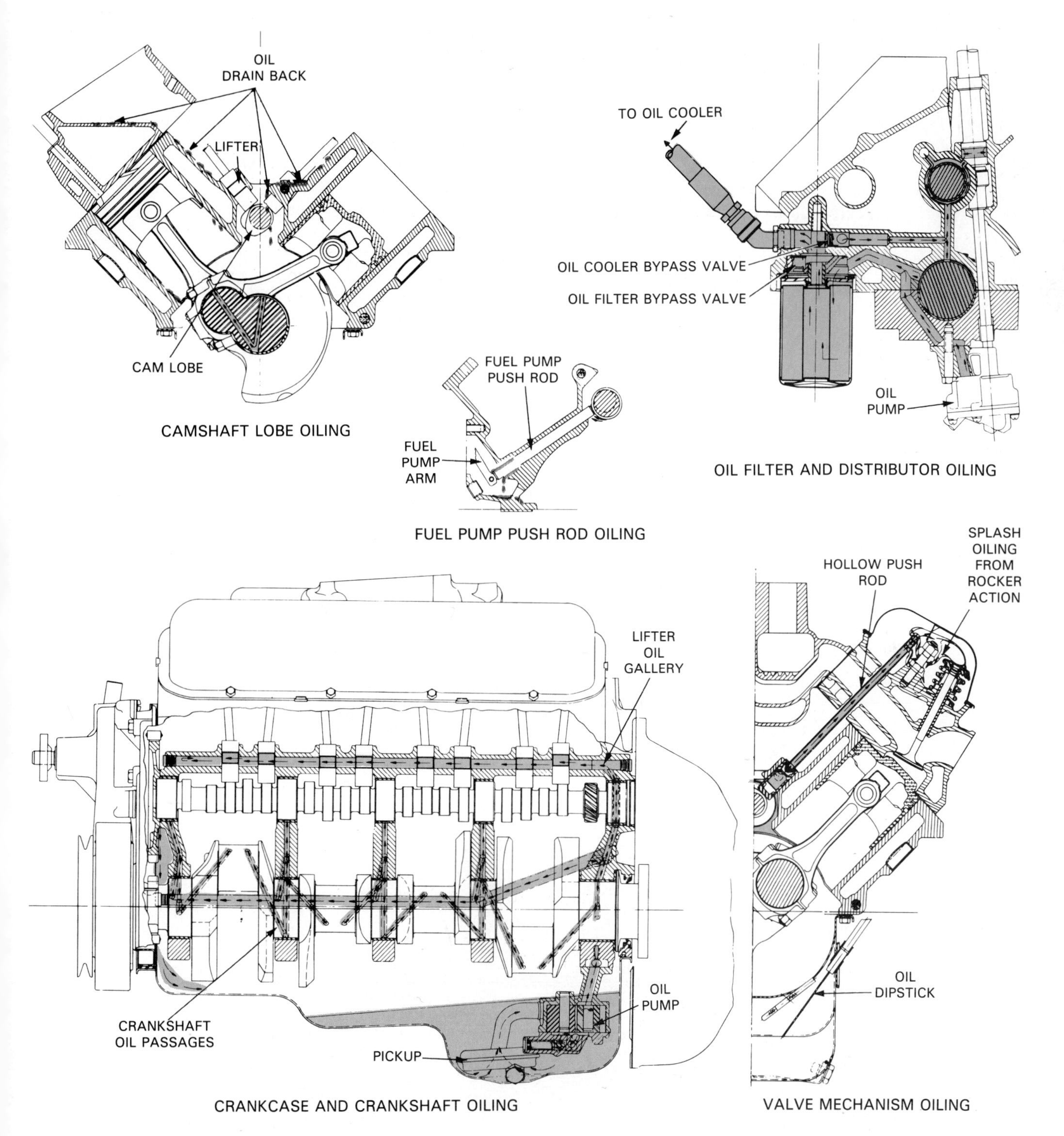

Fig. 15-25. Note how some parts receive pressure-fed lubrication and others get splash oiling. (GM Trucks)

ENGINE OILING METHODS

There are two methods for lubricating engine components: pressure fed oiling and splash oiling. See Fig. 15-25 for an oil flow diagram.

FULL FLOW AND BYPASS SYSTEMS

The *full flow lubrication system* forces all of the oil through the oil filter before the oil reaches the parts of the engine. It is the most common type.

The *bypass lubrication system* does NOT filter all of the oil that enters the engine bearings. It filters some of the extra oil not needed by the bearings. The bypass lubrication system is not very common.

OIL PUMPS

The *oil pump* is the ''heart'' of the engine lubrication system; it forces oil out of the pan, through the engine filter, galleries, and to the engine bearings. The oil pump is frequently driven by a gear on the engine camshaft. It may also be driven by a cogged belt or by a direct connection with the end of the crankshaft, Fig. 15-26. Some oil pumps are gear-driven off of an oil pan-mounted balancer shafts.

There are two basic types of engine oil pumps: rotary and gear.

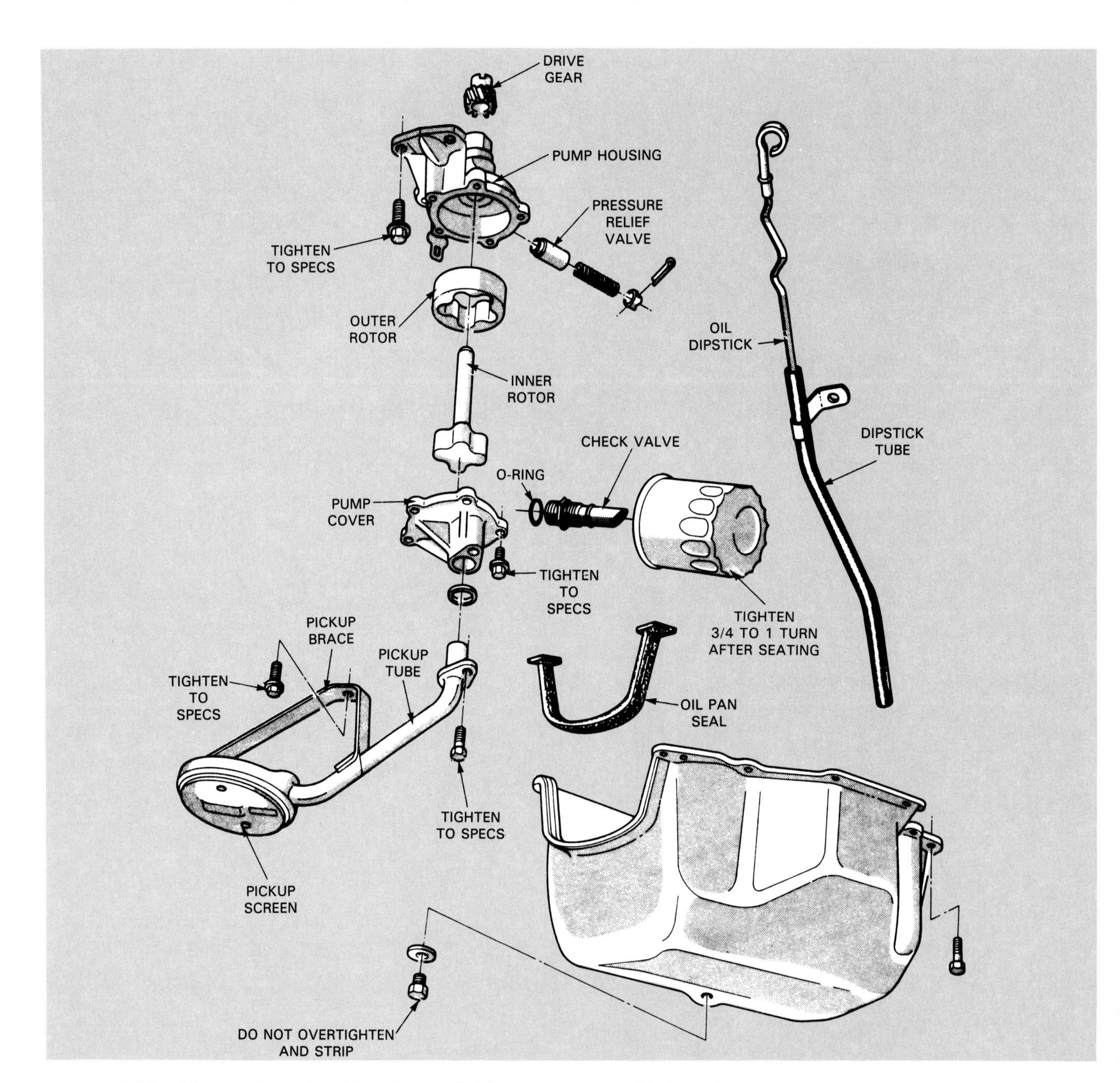

Fig. 15-26. Oil pump is engine driven by small drive gear, a cogged belt, or by being mounted around crankshaft. This is a rotary pump driven by gear on camshaft. Pickup bolts to pump and extends near bottom of oil pan. (Plymouth)

Rotary oil pumps

A *rotary oil pump* uses a set of star shaped rotors in a housing to pressurize the motor oil. As the oil pump shaft turns, the inner rotor causes the outer rotor to spin. The eccentric action of the two rotors form pockets that change in size, Fig. 15-27.

A large pocket is formed on the inlet side of the pump. As the rotors turn, the oil filled pocket becomes smaller as it nears the outlet of the pump. This squeezes the oil and makes it squirt out under pressure. As the pump spins, this action is repeated over and over to produce a relatively smooth flow of oil.

Gear oil pumps

A *gear oil pump* uses a set of gears to produce lubrication system pressure. A shaft, usually turned by the distributor, crankshaft, or accessory shaft, rotates one of the pump gears. This gear turns the other gear which is supported on a very short shaft inside the pump housing, Fig. 15-28.

Oil on the inlet side of the pump is caught in the gear teeth and carried around the outer wall inside the pump housing. When the oil reaches the outlet side of the pump, the gear teeth mesh and seal.

Oil caught in each gear tooth is forced into the pocket at the pump outlet and pressure is formed. Oil flows out of the pump and to the engine bearings, Fig. 15-29.

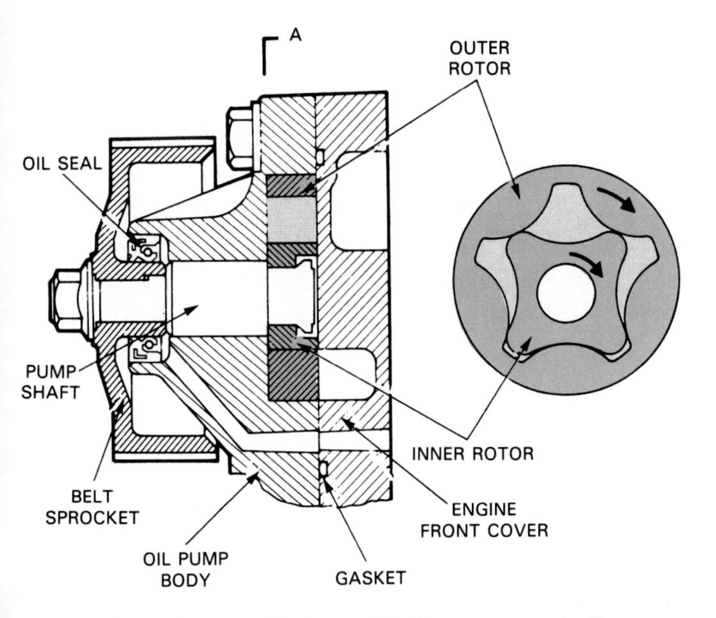

Fig. 15-27. Timing belt drives this front-mounted oil pump. Note action of rotors. (Chrysler)

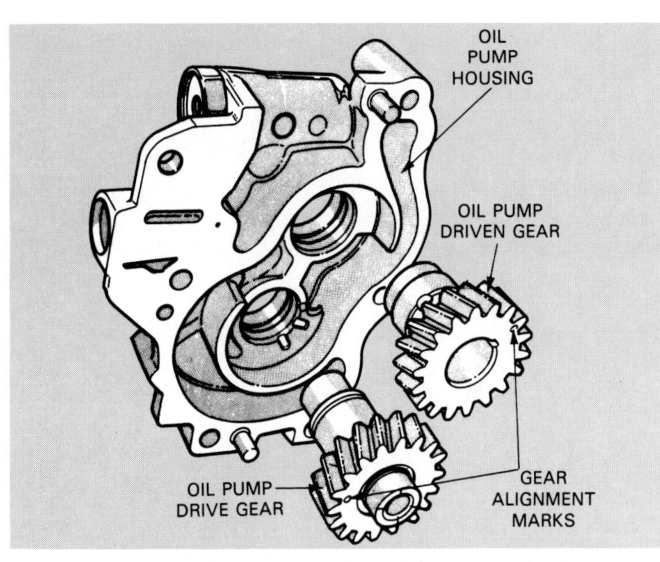

Fig. 15-28. Gear oil pump traps motor oil in gear teeth. As gears turn, oil is carried around housing and pressurized.

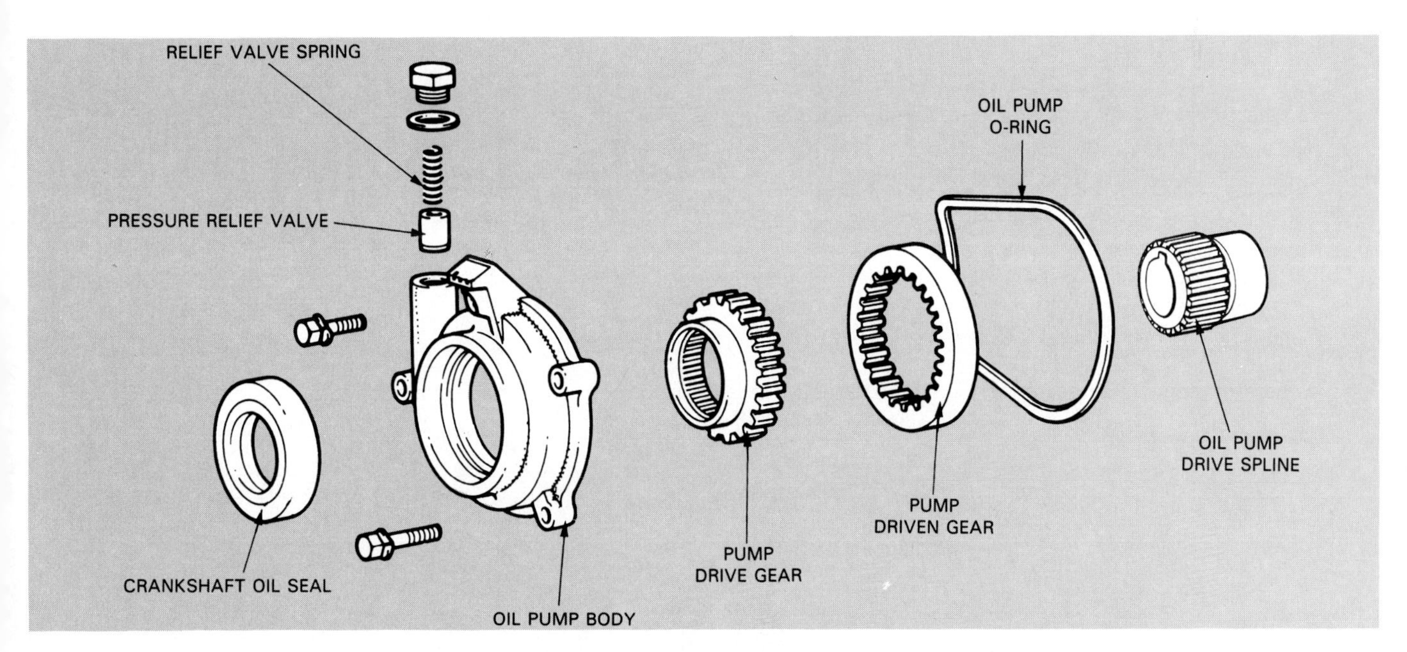

Fig. 15-29. This oil pump mounts around front snout of crankshaft. Spline turns inner drive gear. Drive gear rotates other pump gear. (Toyota)

OIL PICKUP AND SCREEN

The *oil pickup* is a tube extending from the oil pump to the bottom of the oil pan. One end of the pickup bolts or screws into the oil pump or to the engine block. The other end holds the pickup screen, Fig. 15-26.

The *pickup screen* prevents large particles from entering the pickup tube and oil pump. The screen is usually part of the pickup tube. Without the screen, the oil pump could be damaged by bits of valve stem seals and other debris flushed into the pan.

OIL FILTERS

An *oil filter* removes small metal, carbon, rust, and dirt particles from the motor oil. It protects the moving engine parts from abrasive wear, Fig. 15-30.

An *element* is a paper or cotton filtering substance mounted inside the filter housing. It will allow oil flow but will block and trap small debris.

A *filter bypass valve* is commonly used to protect the engine from oil starvation if the filter element becomes clogged. The valve will open if too much pressure is formed in the filter. This allows unfiltered oil to flow to the engine bearings, preventing major part damage.

Oil filter types

The two classifications of engine oil filters are: spin-on filter and cartridge filter.

The *spin-on oil filter* is a sealed unit having the element permanently enclosed in the filter body. When it must be serviced, a new filter is simply screwed into place. This is the most common type, Fig. 15-31.

The *cartridge oil filter* has a separate element and canister, Fig. 15-32. To service this type oil filter, the canister is removed. Then, a new element is installed inside the existing canister. A cartridge type oil filter is sometimes used on heavy duty or diesel applications.

Oil filter housing

The *oil filter housing* is a metal part that bolts to the engine and provides a mounting place for the oil filter.

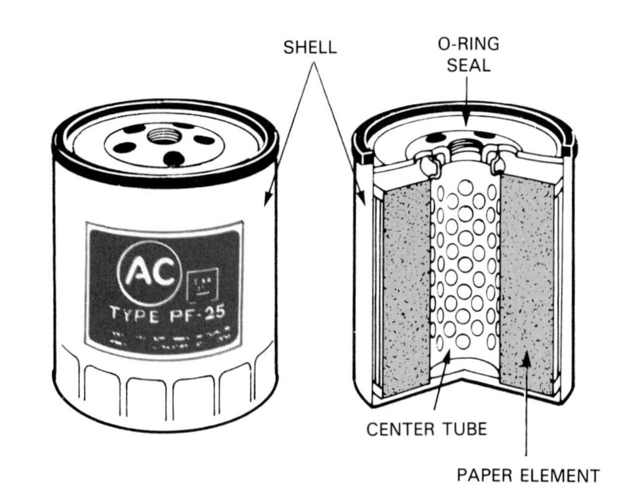

Fig. 15-31. Spin-on filter has metal housing around paper element. Always make sure O-ring diameters are the same when installing new filter. (AC-Delco)

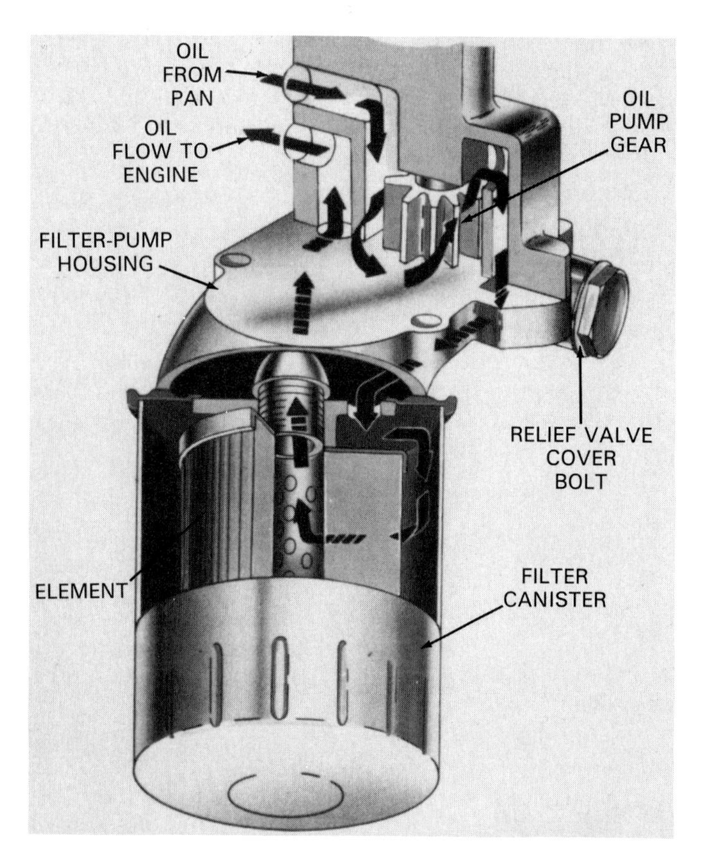

Fig. 15-30. Study gear oil pump and oil filter action. (Oldsmobile)

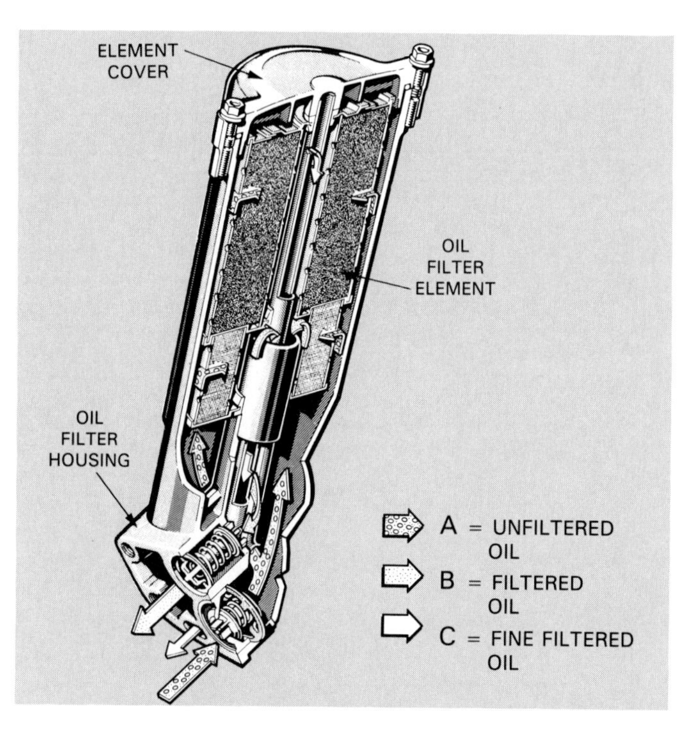

Fig. 15-32. This is a cartridge type oil filter. Element is removed from housing during filter change. Cover unbolts so old element can be pulled out and replaced. (Mercedes Benz)

The housing may also have a fitting for the oil pressure sending unit.

A gasket normally fits between the engine and oil filter housing to prevent leakage. Sometimes, the pressure relief valve, filter bypass valve, or oil pump are inside this housing, Fig. 15-30.

PRESSURE RELIEF VALVE

A *pressure relief valve* limits maximum oil pressure. It is a spring-loaded, bypass valve in the oil pump, engine block, or oil filter housing.

Under high oil pressure conditions (cold, thick oil for example), the pressure relief valve opens, Fig. 15-33. Oil pressure pushes the small piston back in its cylinder by overcoming spring tension. This allows some oil to bypass the main oil galleries and pour back into the oil pan. Most of the oil still flows to the bearings and a preset pressure is maintained, Fig. 15-33.

Some pressure relief valves are adjustable. By turning a bolt or screw or by changing spring shim thickness, the pressure setting can be altered.

OIL PAN AND SUMP

The *oil pan,* normally made of thin sheet metal or aluminum, bolts to the bottom of the engine block. It holds an extra supply of oil for the lubrication system.

The oil pan is fitted with a screw-in drain plug for oil changes. *Baffles* may be used to keep the oil from splashing around in the pan, Fig. 15-34.

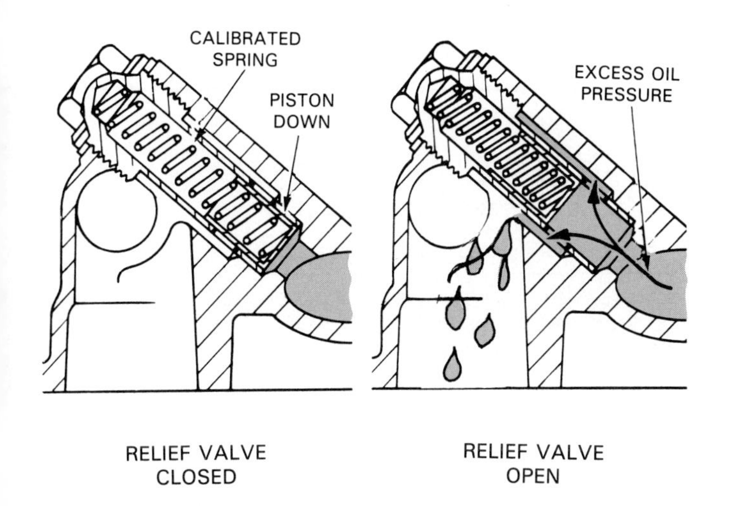

Fig. 15-33. Relief valve prevents excess engine oil pressure that could rupture oil filter. When oil pressure is below specs, spring holds valve closed so all oil goes to bearings. When pressure is too high, as with cold, thick oil, relief valve opens and bleeds some oil back into pan. (AMC)

The *sump* is the lowest area in the oil pan where oil collects. As oil drains out of the engine, it fills the sump. Then the oil pump can pull oil out of the pan for recirculation.

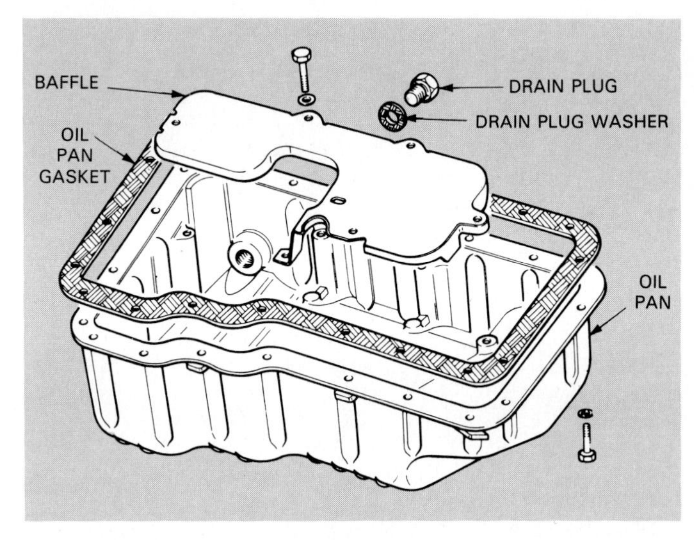

Fig. 15-34. A sump is used to collect oil as it drains out of block. Fiber or chemical gasket is used to prevent leakage.

A *structural oil pan* is designed to add strength to the engine bottom end and cylinder block. Most structural oil pans are made of lightweight, ribbed aluminum or metal construction. Some designs have the main cap bolts going through the oil pan.

OIL GALLERIES

Oil galleries are small passages through the cylinder block and head for lubricating oil. They are cast or machined passages that allow oil to flow to the engine bearings and other moving parts, Fig. 15-35.

The *main oil galleries* are large passages through the center of the block. They feed oil to the crankshaft bearings, camshaft bearings, cylinder heads, and lifters.

OIL COOLER

An *oil cooler* may be used to help lower and control the operating temperature of the engine oil. It is a radiator-like device connected to the lubrication system. Oil is pumped through the cooler and back to the engine, Fig. 15-35. Airflow through the cooler removes heat and lowers the temperature of the oil. Oil coolers are frequently used on turbochanged engines or heavy duty applications.

OIL LEVEL INDICATOR

An *oil lever indicator* senses the amount of oil in the oil pan and signals an indicator in the dash. This alerts the driver of a dangerously low oil level condition.

As shown in Fig. 15-36, a float mechanism is located in the oil pan. The float rides on top of the motor oil in the pan. When the oil level drops too low, the float closes the circuit and current flows to the indicator.

OIL PRESSURE SENDING UNIT

The *oil pressure sending unit* is a diaphragm type device that screws into one of the oil galleries. It changes

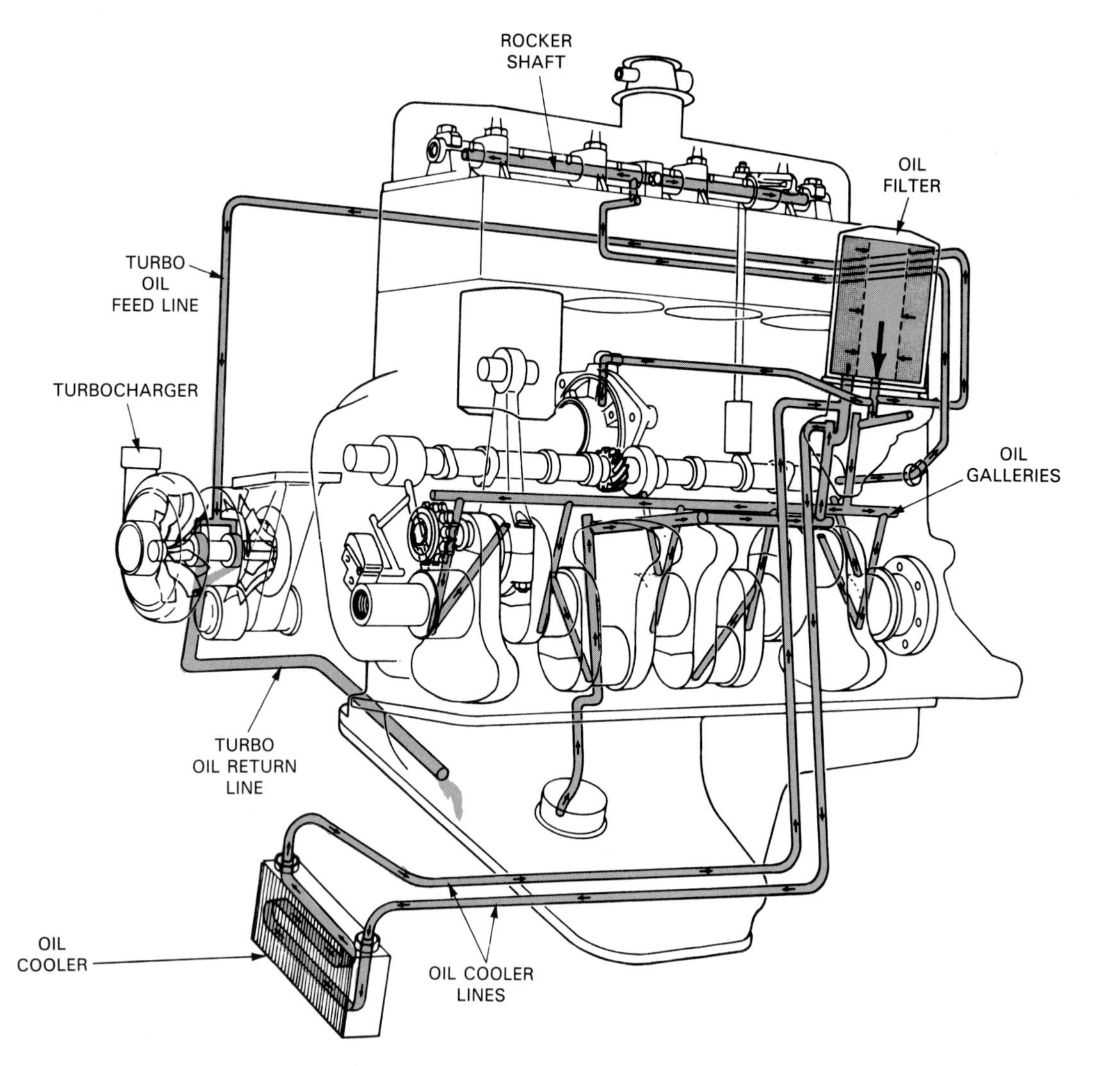

Fig. 15-35. Note how oil lines feed to turbocharger and oil cooler. Oil cooler is needed on high performance or heavy duty applications to help maintain oil viscosity. (Peugeot)

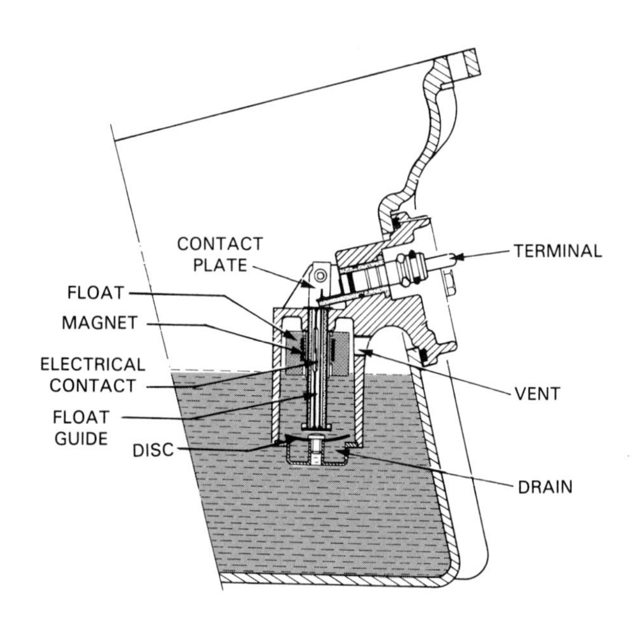

Fig. 15-36. Oil level indicator uses simple float that rides on top of motor oil. As oil level drops, sensor signals dash indicator in driver's compartment. (Mercedes Benz)

resistance with a change in oil pressure. The oil pressure sending unit is used to operate the low oil pressure warning light in the dash.

When oil pressure is normal, the sending unit has infinite resistance and the engine warning light stays out. However, if oil pressure drops too low (below about 5 or 10 psi), the contacts in the unit close and the warning light glows.

LUBRICATION SYSTEM SERVICE

This section of the chapter reviews the types of service tasks you might encounter when working on an engine lubrication system. As mentioned, this is a very important system that affects engine service life.

LUBRICATION SYSTEM TROUBLES

The problems found in a lubrication system are limited in number. They include:

1. High oil consumption (oil must be added to engine frequently).
2. Low oil pressure (gauge reads low, indicator light

glows, or abnormal engine noises).

3. High oil pressure (gauge reads high, oil filter swelled).
4. Defective indicator or gauge circuit (inaccurate operation or readings).

When diagnosing these troubles, make a visual inspection of the engine for obvious problems. Check for oil leakage, disconnected sending unit wire, low oil level, smashed oil pan, or other trouble that would relate to the symptoms, Fig. 15-37.

High oil consumption

High oil consumption is caused by external oil leakage out of the engine or by internal leakage of oil into the combustion chambers.

External oil leakage is easily detected as darkened, oil wet areas on or around the engine. Oil will be dark brown or black if used for an extended period of time. Oil may be found in small puddles under the car. Leaking gaskets or seals are usually the source of external engine oil leakage.

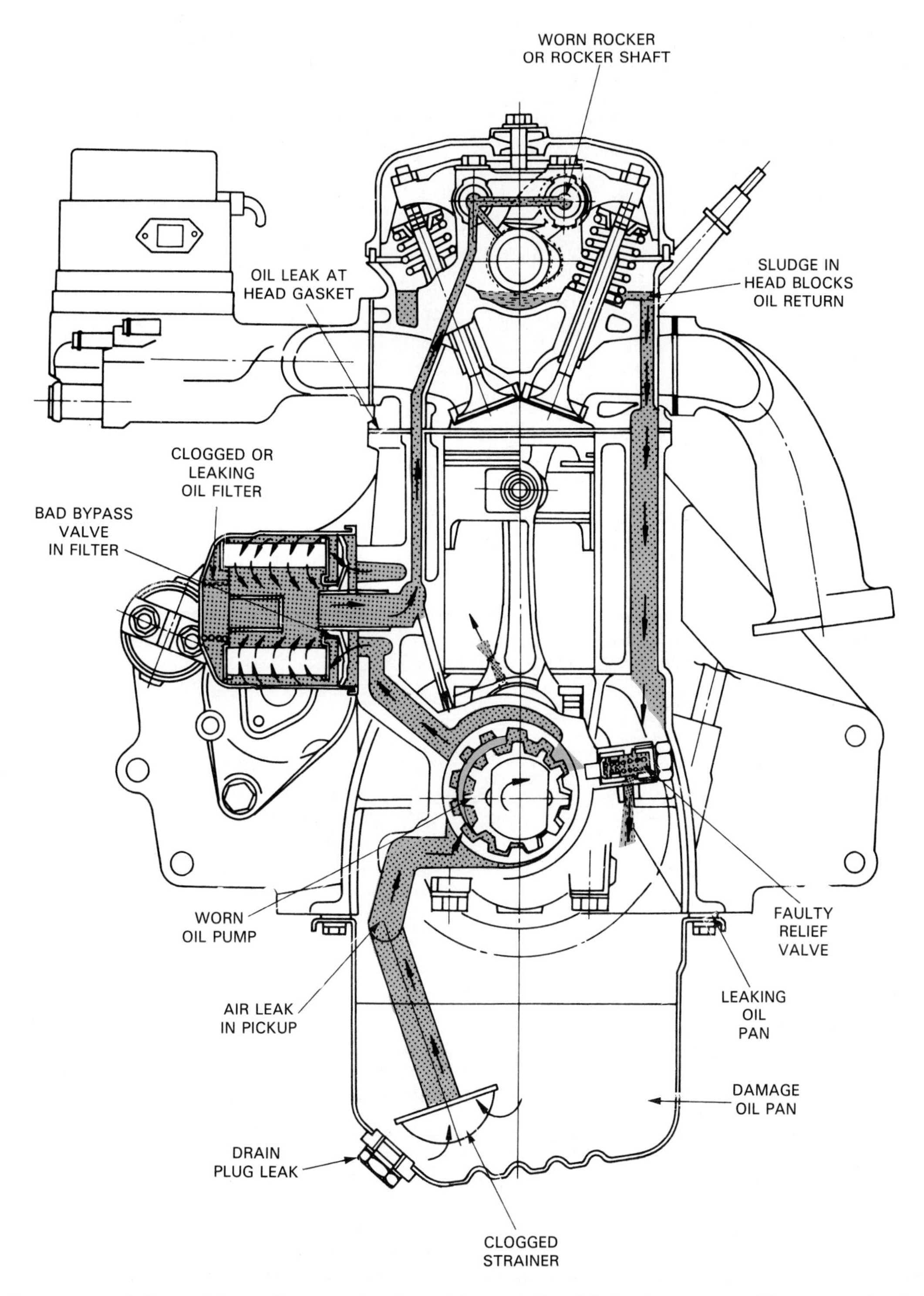

Fig. 15-37. Note some of the problems that can develop with an engine lubrication system. You must check for these troubles during diagnosis.

To locate oil leakage, you may need to raise the car on a lift and visually look for leaks under the engine. Trace the oil leakage to its highest point. The parts around the point of leakage will be WASHED CLEAN by the constant dripping or flow of oil.

Internal oil leakage shows up as BLUE SMOKE coming out of the car's exhaust. For example, if the engine piston rings and cylinders are badly worn, oil can enter the combustion chambers and be burned.

Do not confuse black smoke (excess fuel in cylinder) and white smoke (water leakage into gasoline engine cylinder) with the blue smoke caused by motor oil.

NOTE! Engine oil consumption and smoking is covered fully in the chapter on engine mechanical problems.

Oil pressure problems

Low oil pressure is indicated when the oil indicator light glows, oil gauge reads low, or when the engine lifters or bearings rattle.

Common causes of low oil pressure are:

1. Low oil level (oil not high enough in pan to cover oil pickup).
2. Worn connecting rod or main bearings (pump cannot provide enough oil volume).
3. Thin or diluted oil (low viscosity or gasoline in oil).
4. Weak or broken pressure relief valve spring (valve opening too easily).
5. Cracked or loose oil pump pickup tube (air being pulled into oil pump).

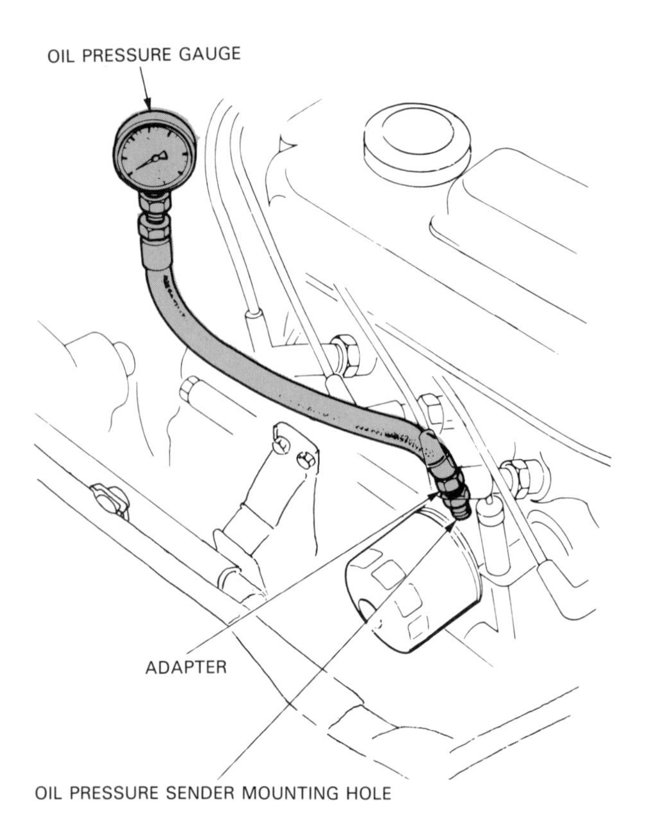

Fig. 15-38. If symptoms point to low oil pressure, measure actual pressure. Remove oil sending unit and install test gauge. Start and warm engine to operating temperature. Test gauge reading should be up to specs at idle and at cruising rpm. (Honda)

6. Worn oil pump (excess clearance between rotor or gears and housing).
7. Clogged oil pickup screen (reduce amount of oil entering pump).

High oil pressure is seldom a problem. The most frequent causes of high oil pressure are:

1. Pressure relief valve stuck closed (not opening at specified pressure).
2. High relief valve spring tension (wrong spring or spring has been improperly shimmed).
3. High oil viscosity (excessively thick oil or use of oil additive that increases viscosity).
4. Restricted oil gallery (defective block casting or debris in oil passage).

A *defective indicator* or *gauge* may appear to be a low or high oil pressure problem. The sending unit, circuit wiring, or gauge may be at fault.

Oil pressure test

An *oil pressure test* uses a gauge to measure actual pressure in the engine. A pressure gauge is screwed into the hole for the oil pressure sending unit, Fig. 15-38.

Start and warm the engine. Run the engine at the service manual recommended rpm. Read the test pressure gauge and compare to specs. If oil pressure is too low or high, you must make repairs as needed.

Depending upon the number of miles on and type of engine, oil pressure should be at least 20 to 30 psi (138-207 kPa) at idle and 40 to 60 psi (276-413 kPa) at crusing speeds. Check service manual specs when testing.

ENGINE OIL AND FILTER SERVICE

It is extremely critical that the engine's oil and oil filter are serviced regularly. Lack of maintenance can greatly shorten engine service life, Fig. 15-39.

Used oil will be contaminated with dirt, metal particles, carbon, gasoline, ash, acids, and other harmful substances. Some of the smallest particles and corrosive chemicals are not trapped in the oil filter. They will circulate through the engine, increasing part wear and corrosion.

Oil and filter change intervals

Auto makers give a maximum number of miles (kilometers) a car can be driven between oil changes. If the oil is not changed at this interval, the car's warranty will become void.

New cars can generally be driven about 6000 miles (10 000 km) between oil changes. Older model cars should have their oil changed more often.

An older, worn engine will contaminate the oil more quickly than a new engine. More combustion byproducts will blow past the rings and enter the oil. Also, engine bearing clearance will be larger, requiring more of the oil and lubrication system.

Also, if a car is only driven for short periods and then parked, its oil should be changed more often. Since the engine may not be reaching full operating temperature, the oil can be contaminated with fuel, moisture, and other substances more quickly.

Diesel engines and turbocharged engines usually re-

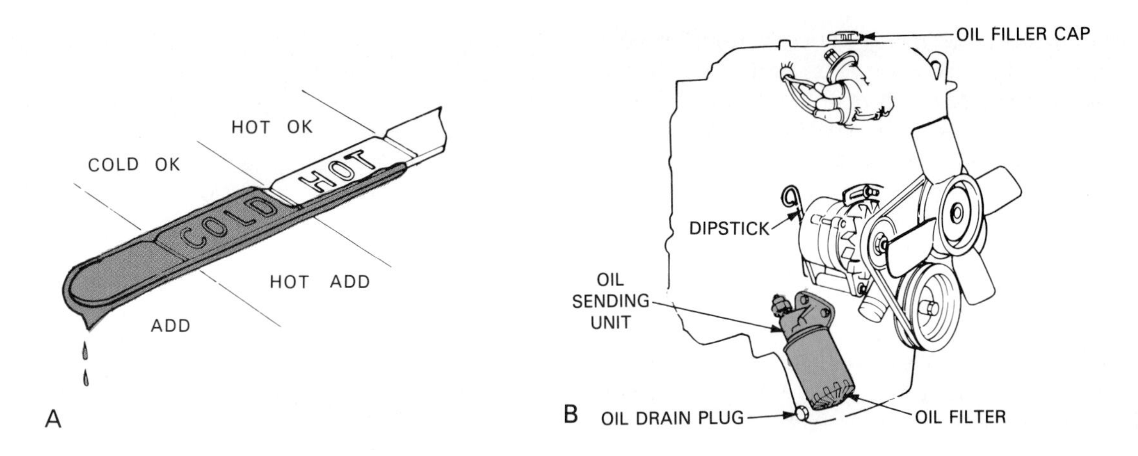

Fig. 15-39. A—When checking oil level, make sure you use correct markings on dipstick. B—Typical locations for filler cap, dipstick, and oil filter. Also note oil sending unit. (Mazda)

quire more frequent oil and filter service than naturally aspirated (non-supercharged) engines.

Changing engine oil and filter

To change the engine oil, warm the engine to full operating temperature. This will help suspend debris in the oil and make the oil drain more thoroughly. Then, follow the basic steps given in Fig. 15-40.

A few rules to remember when changing engine oil and the oil filter are:

1. Keep the car relatively level so all oil drains from the pan.
2. Do NOT let hot oil pour out on your hand or arm!
3. Check the condition of the drain plug threads and the O-ring washer. Replace them if needed.
4. Do NOT overtighten the oil pan drain plug. It will strip very easily.

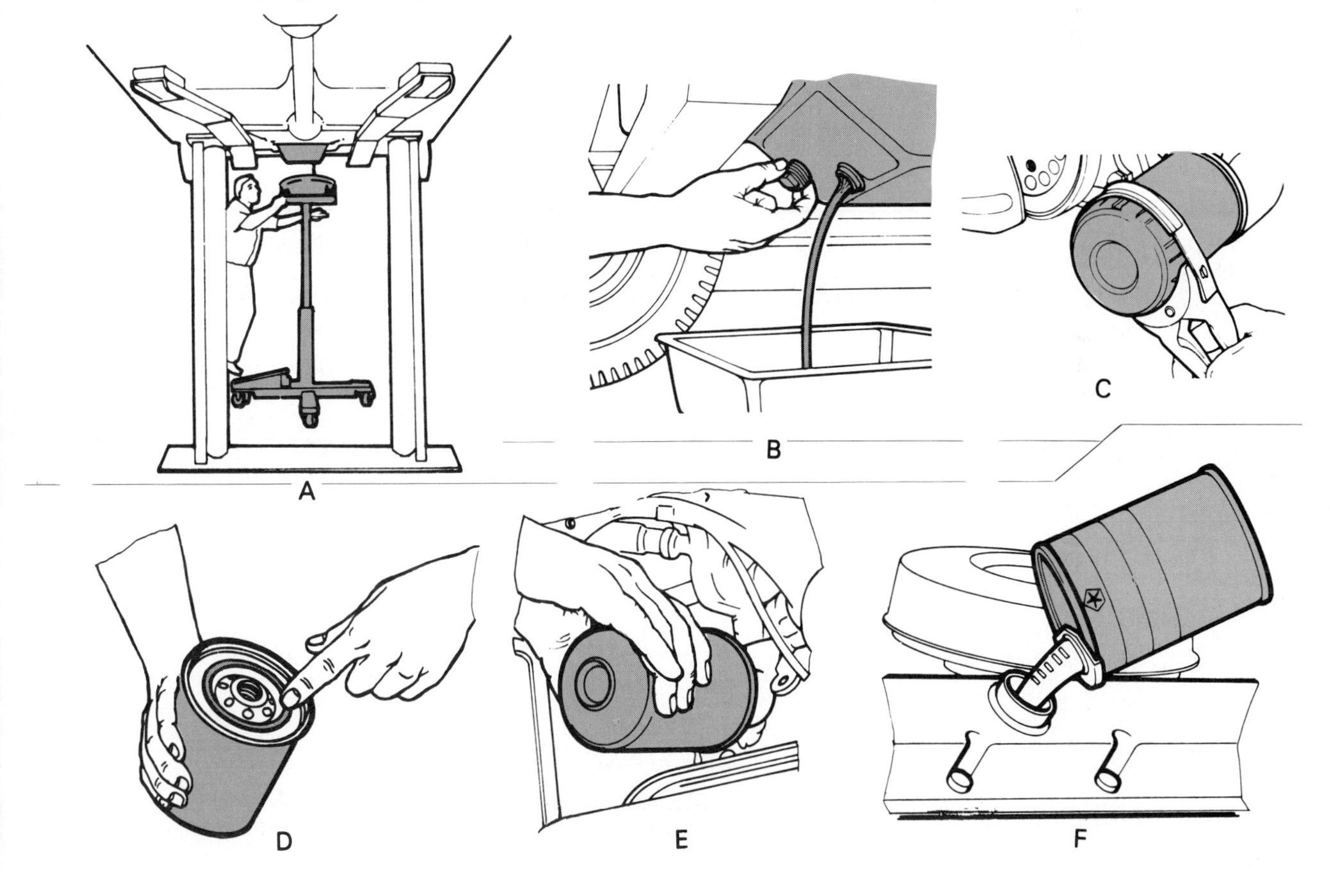

Fig. 15-40. Changing engine oil and filter. A—Use lift or floor jack and jack stands to raise car in level position. Place catch pan under drain plug. B—Unscrew drain plug and allow oil to pour into pan. Be careful of hot oil. It can cause painful burns. C—Use filter wrench to remove old filter. Turn counterclockwise. D—Wipe clean oil on new filter O-ring. This will assure proper tightening. E—Install and tighten oil filter by hand. Hands and filter should be clean and dry. Use a rag if needed. F—Install correct type and quantity of oil. Pour oil into filler or breather opening in valve cover or intake manifold. (Mopar)

5. Wipe oil on the filter O-ring before installation.
6. Hand tighten the oil filter as the filter canister could distort and leak.
7. Fill the engine with the correct amount and type of oil.
8. Check for oil leaks with the engine running before releasing the car to the customer.

Motor oil disposal

Used motor oil is considered hazardous waste. One gallon of used motor oil can be refined into two and one-half quarts of high quality like-new motor oil. It takes about 40 gallons of crude oil to produce this much motor oil. The old oil should be stored in an approved container for recycling.

Recycling old oil not only saves our environment from possible pollution, it also helps save our natural resources. Always send used motor oil to a recycling center! Some recycling companies provide a pick-up service while others require you to take the old oil to their facility. Some shop heaters can burn used motor oil as a fuel.

OIL PUMP SERVICE

A *bad oil pump* will cause low or no oil pressure and possibly severe engine damage. When inner parts wear, the pump may have reduced output. The pump drive shaft can also strip in the pump or distributor, preventing pump operation. See Fig. 15-41.

If tests point to a faulty oil pump, remove and replace or rebuild the pump.

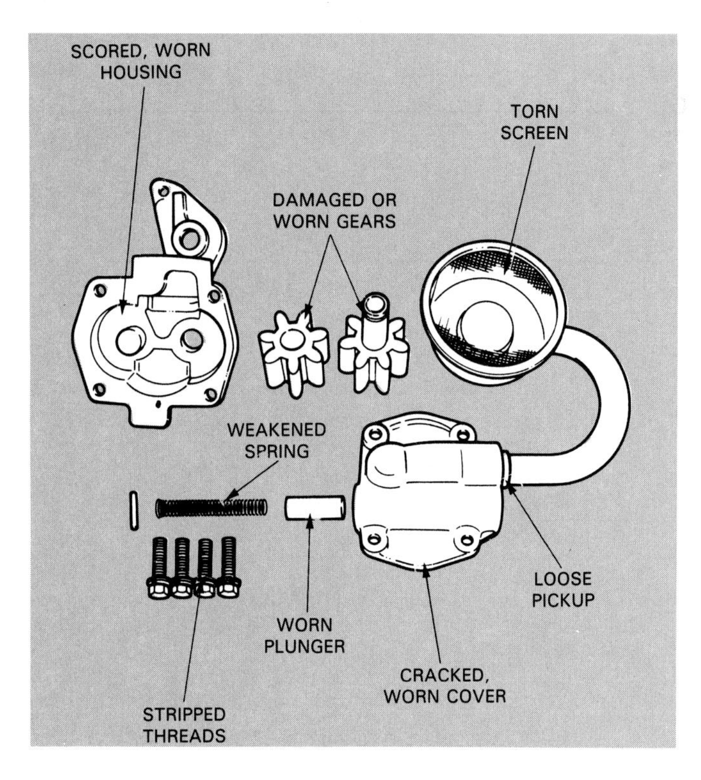

Fig. 15-41. When servicing oil pump, inspect condition of gears or rotors and inside of housing. Slightest signs of wear or damage normally require pump replacement. (Pontiac)

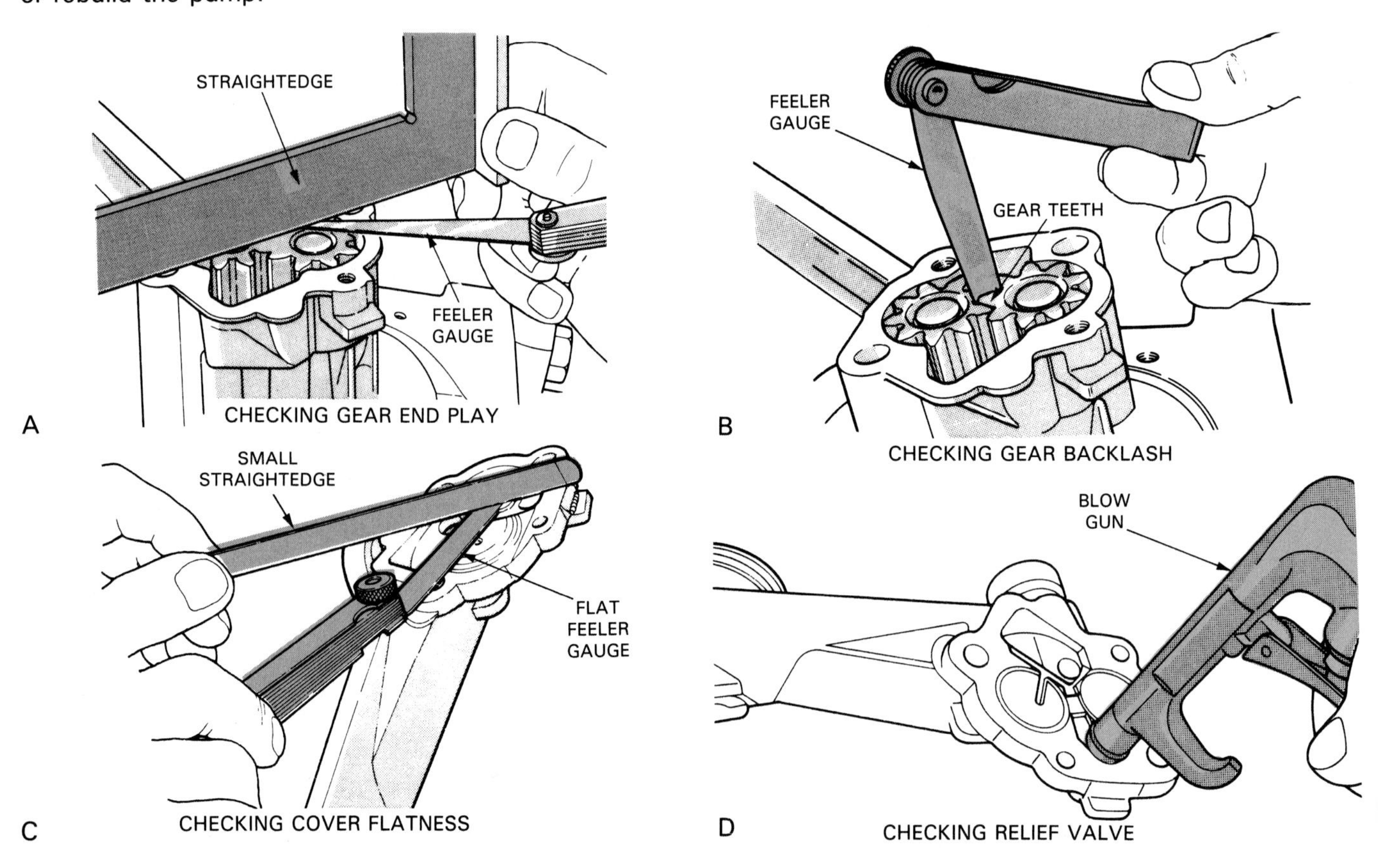

Fig. 15-42. Rebuilding in-pan gear type pump. A—Use straightedge and flat feeler to check gear end play. Sand parts with 400 grit paper on perfectly flat surface if needed. B—Check gear backlash and replace parts if not within specs. C—Straightedge and feeler should be used to check cover flatness. If only slightly warped, true by rubbing on 400 grit laying on flat surface. D—Relief valve action can be checked with air nozzle. (Dodge)

Oil pump removal

Some oil pumps are located inside the engine oil pan. Others are on the front of the engine under a front cover or on the side of the engine. Since removal procedures vary, refer to a shop manual.

Oil pump rebuild

Most technicians install a new or factory rebuilt pump when needed. It is usually too costly to completely rebuild an oil pump in-shop. However, you should have a general understanding of how to overhaul oil pumps.

Figs. 15-42 through 15-44 summarize how to service typical oil pump designs.

Oil pump installation

Before installation, *prime* (fill) *the pump* with motor oil. This will assure proper initial operation upon engine starting.

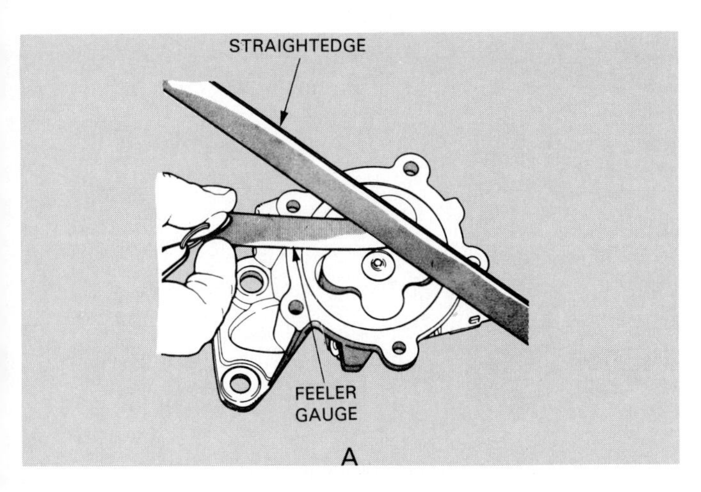

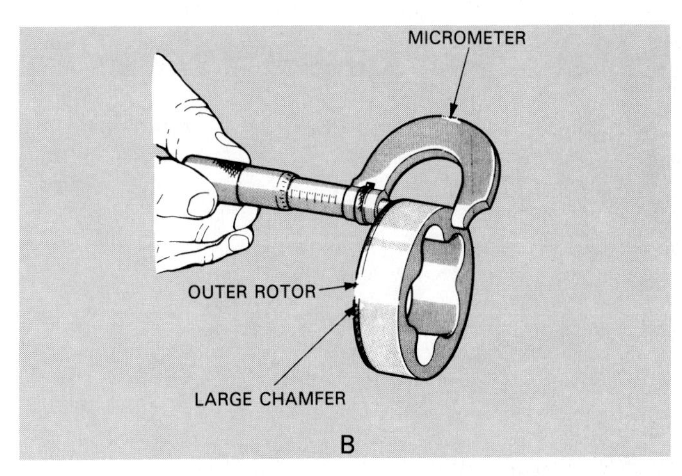

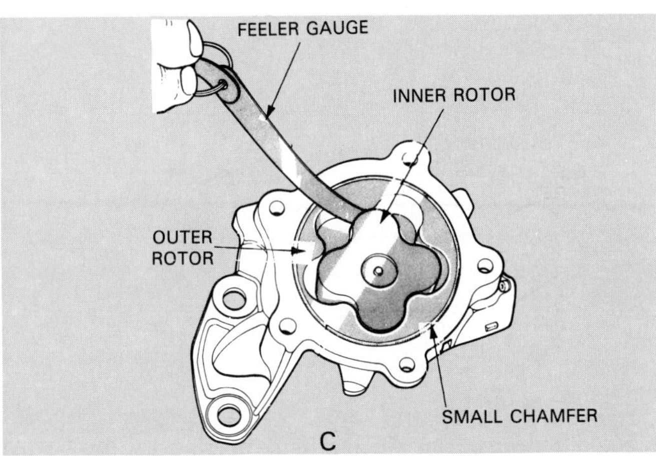

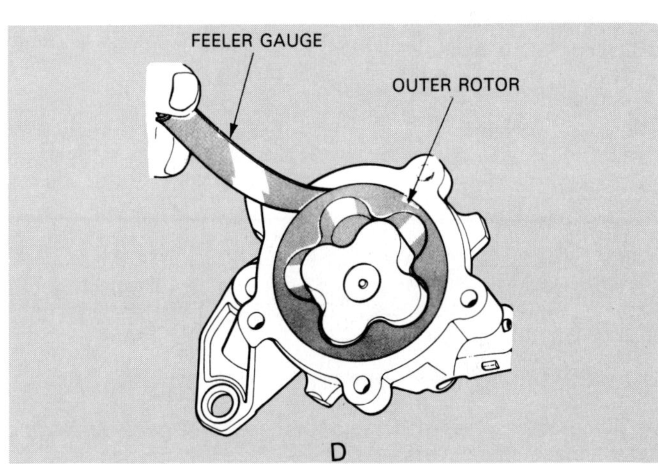

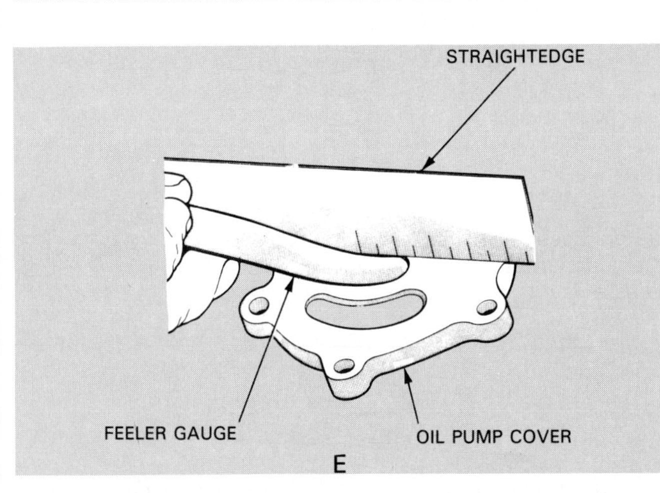

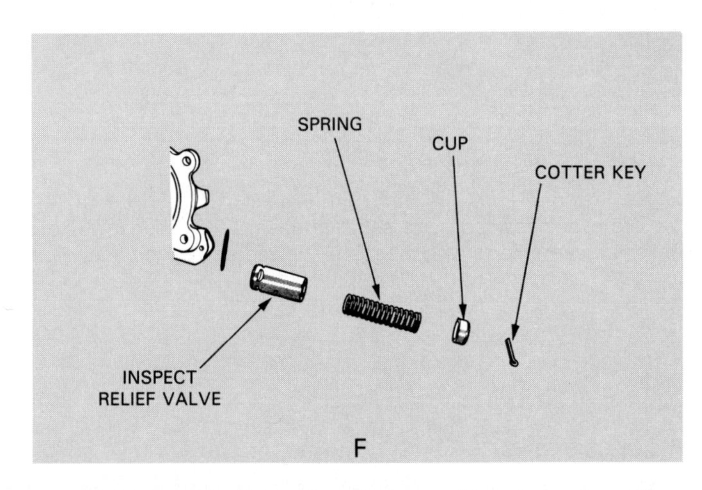

Fig. 15-43. A—Checking rotor end play. B—Measuring outer rotor thickness. C—Measuring clearance between rotors. D—Checking outer rotor clearance. E—Using straightedge and feeler to check cover flatness. F—Inspect parts of oil pressure relief valve carefully. (Chrysler)

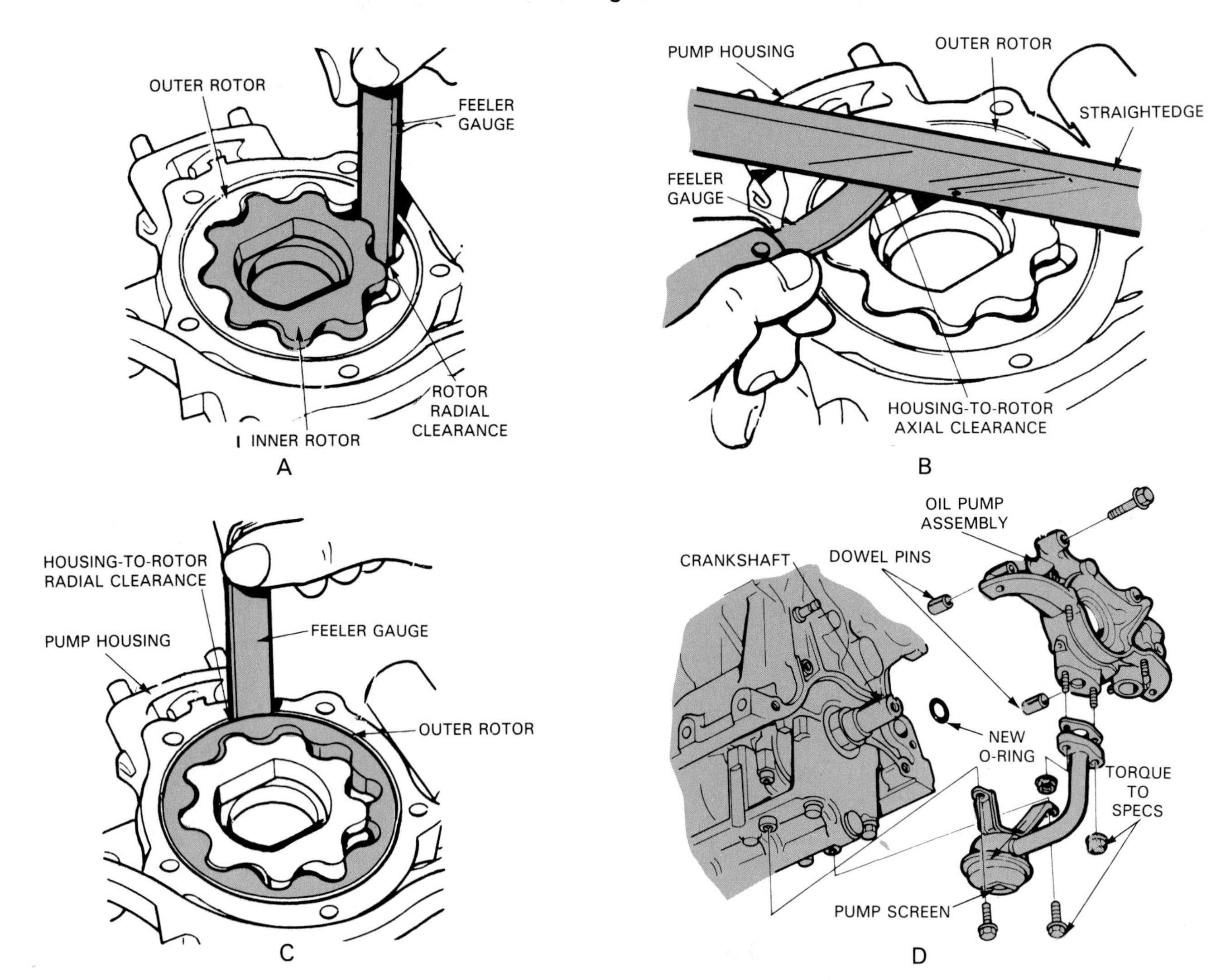

Fig. 15-44. Servicing crank-mounted oil pump. A—Measuring inner rotor radial clearance. B—Measuring housing-to-rotor axial clearance. C—Measuring housing-to-rotor radial clearance. D—Install assembled pump on front of engine using approved serevice manual methods. Dowel pins assure that pump aligns around crankshaft properly. (Honda)

Install the pump in reverse order of removal. Double-check gasket position. Torque all bolts to exact specs. If an oil pickup is used, make sure it is installed properly and tightened, Fig. 15-45.

Figs. 15-46 and 15-47 show how to install oil pumps mounted on the front of an engine.

Pressure relief valve service

A *faulty pressure relief valve* can produce oil pressure problems. The valve may be located in the oil pump, filter housing, or engine block.

If symptoms point to the pressure relief valve, it should be disassembled and serviced. It is also serviced during an engine overhaul.

Remove the cup or cap holding the pressure relief valve. Then, slide the spring and piston out of their bore.

Measure *spring free length* (length extended spring) and compare to specs. If the spring is too short or long, install a new spring. One auto maker recommends checking spring tension on a spring tester. Another provides shims for increasing spring tension or an adjustment for changing opening pressure.

Use a micrometer and small hole gauge to check valve and valve bore wear. Also, check the sides of the valve for scratches or scoring. Slide the plunger in and out of its bore. Replace parts if any problems are found.

OIL PAN SERVICE

An engine oil pan may need to be removed for various reasons: to service engine bearings, oil pump, a damaged pan, a stripped pan drain plug, or during an engine overhaul.

Some engine oil pans can be removed with the engine in the car. After removing any bolt-on crossmembers, steering components in the way, and other obstructions, raise the engine off its motor mounts. This may be needed to give enough clearance to unbolt and slide off the oil pan.

Other vehicle designs do NOT allow in-car oil pan removal. The engine must be lifted from the car before the pan will come off. Check in a service manual for details.

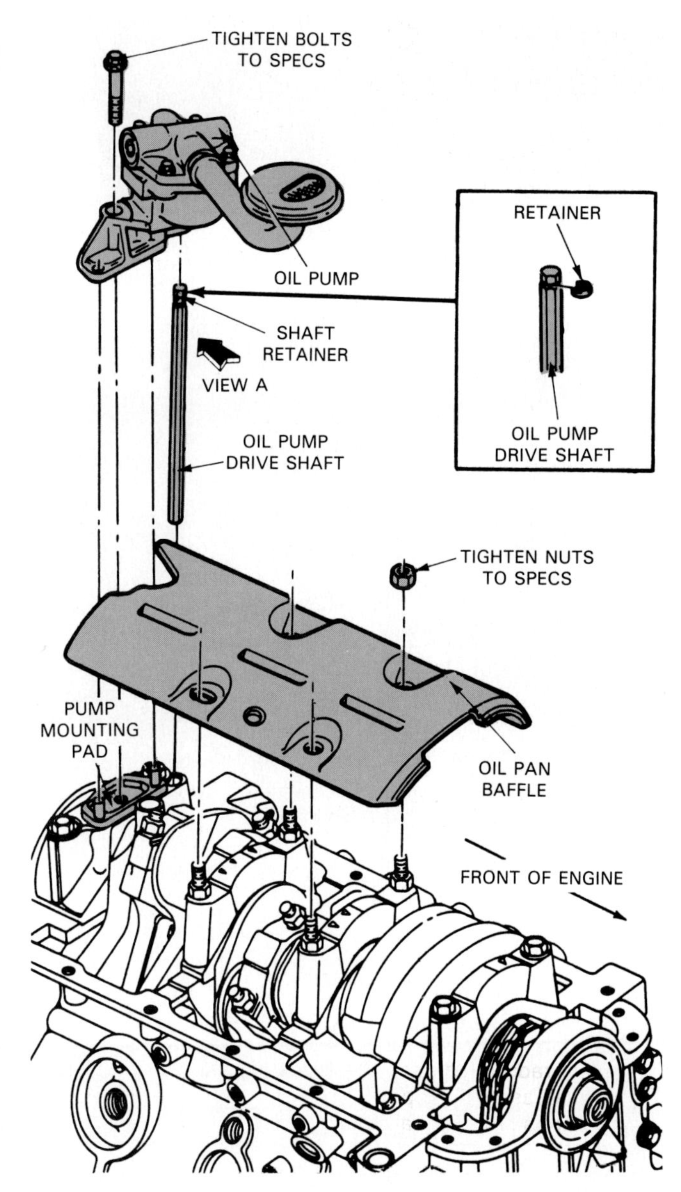

Fig. 15-45. When installing oil pump inside oil pan, oil pump drive shaft installation is critical. Most shafts have small clip that keeps shaft from pulling or falling out during distributor removal. Make sure this clip is positioned correctly. Tighten all oil pump bolts to exact specs. Double-check pickup tube. (Ford)

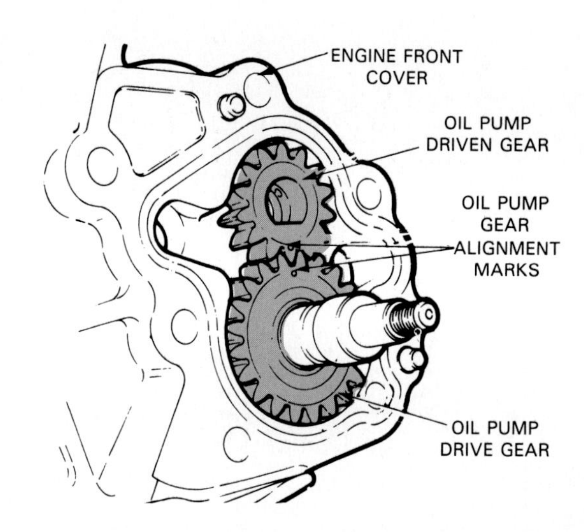

Fig. 15-46. This gear pump, mounted on front of engine, has alignment marks. Gear teeth have been lapped together at factory and must be timed together for proper meshing. (Chrysler)

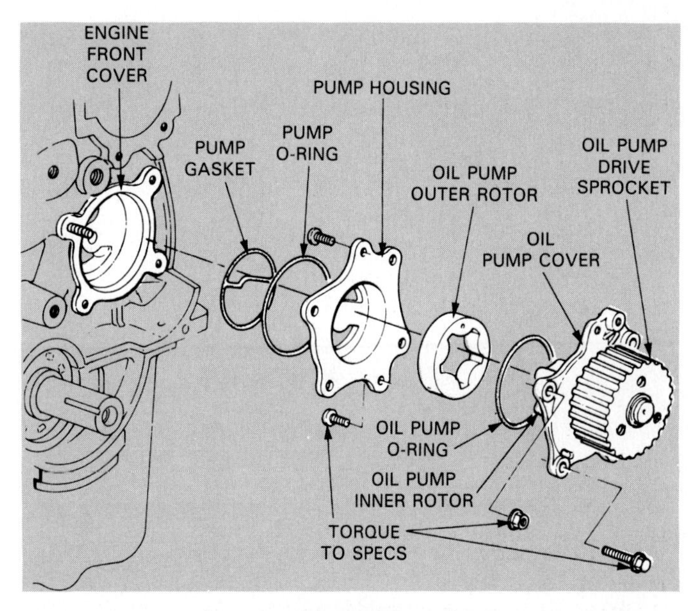

Fig. 15-47. Note how seals install on this front-mounted oil pump. Tighten bolts to specs to prevent part warpage. (Honda)

Removing an engine oil pan

To remove an engine oil pan, first drain out the motor oil. Reinstall the drain plug. Unscrew the bolts around the outside of the pan flange.

To free the pan from the cylinder block, tap on it lightly with a rubber hammer. If stuck tight, carefully pry between the pan flange and the block. Do NOT bend the oil pan flange or leakage may result upon assembly.

Using a gasket scraper, remove all old gasket or silicone material from the pan and engine block. With an aluminum pan, do not gouge the sealing surface.

Check the inside of the pan for debris. Metal bits (bearing particles), plastic bits (timing gear teeth), rubber particles (valve stem seals) indicate engine mechanical problems.

Installing an engine oil pan

To install an engine oil pan, wash the pan thoroughly in cold soak cleaner. Check the drain plug hole for stripped threads. Also, lay the pan upside down on a flat workbench. Make sure that the pan flange is NOT bent. Straighten the flange, if needed, with light hammer blows.

Either a gasket or chemical sealer may be recommended for the oil pan, Fig. 15-48. Use approved gasket adhesive to position a new gasket. If rubber seals are used on each end, press them into their grooves. Place silicone sealer where the gaskets meet the rubber seals.

To use a chemical or silicone gasket, clean the pan and block mating surfaces with a suitable solvent. Then, run

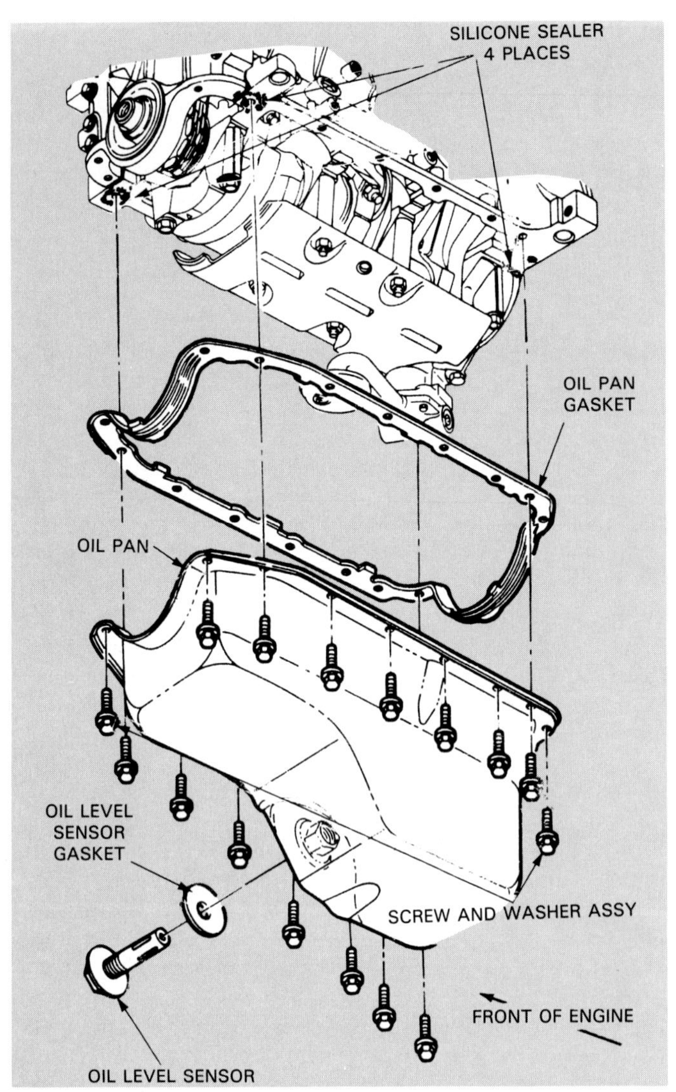

Fig. 15-48. When installing oil pan, use new fiber gasket or silicone sealer. All parts should be perfectly clean and dry. Start all bolts by hand and then tighten to specs in several steps. Do not overtighten or gasket will split and leak. (Ford)

Fig. 15-49. This technician is installing oil pan. Adhere gasket to block. Then place bead of silicone sealer where seal touches gasket. (Fel-Pro)

a uniform bead (about 1/8 in. or 3 mm) of sealer around the pan or block flange as recommended. Form a continuous bead and place extra sealer at part or gasket-seal joints.

Fit the pan carefully into place on the block. Start all of the pan bolts by hand. Check that the gasket or sealer has not been shifted or smeared, Fig. 15-49. Then, tighten each bolt a little at a time in a crisscross pattern. Torque the bolts to specs.

Double-check drain plug torque. Fill the engine with oil. Inspect for oil leaks with the engine running.

OIL PRESSURE INDICATOR AND GAUGE SERVICE

A *bad oil pressure indicator* or *gauge* may scare the customer into thinking there are major engine problems. The indicator light may flicker or stay on, pointing to a low pressure problem. The gauge can read low or high, also indicating a lubrication system problem.

Inspect the indicator or gauge circuit for problems. The wire going to the sending unit may have fallen off (no light, flickering light, or improper gauge readings). The sending unit wire may also be shorted to ground (light stays on or gauge always reads high).

To check the action of the indicator or gauge, remove the wire from the sending unit, Fig. 15-50. Touch it on a metal part of the engine. This should make the indicator light glow or the oil pressure gauge read maximum. If it does, the sending unit may be bad. If it does not, then

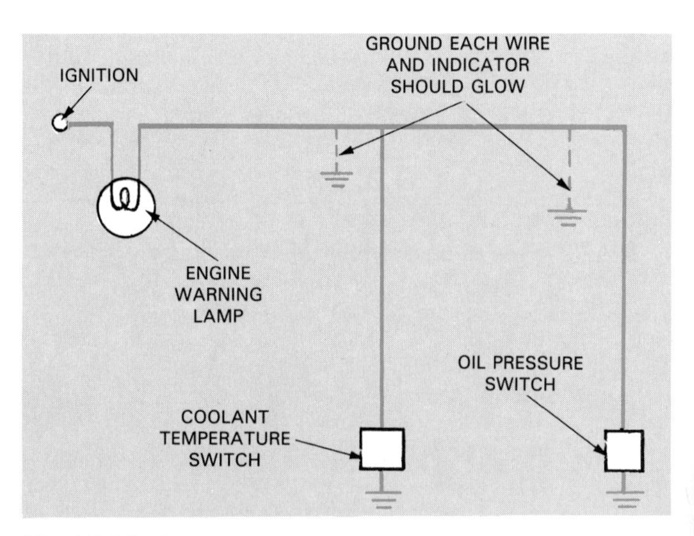

Fig. 15-50. Usually, you can check oil sending unit or temperature sending unit by grounding wire going to unit. When lead is grounded, indicator light should glow. With a gauge, do not ground too long or damage could result. (Ford)

the circuit, indicator, or gauge may be faulty.

Always check the service manual before testing an indicator or gauge circuit. Some auto makers recommend a special gauge tester. This is especially important with some computer controlled sytems. The tester will place a specific resistance in the circuit to avoid circuit damage. Refer to a manual for instructions.

Normally, a low oil pressure light will glow when oil pressure is below 5-10 psi (34-69 kPa).

SUMMARY

An engine technician must know how to service and maintain an engine's cooling and lubrication systems. Both are vitally important to the service life and dependability of an engine.

The cooling system must remove excess heat from the engine. It must also maintain a constant engine operating temperature, speed engine warmup, and provide for heater operation.

The water pump is the "heart" of the system and it circulates coolant through the water jackets, hoses, and radiator. The water pump is driven by a belt running off of the engine crankshaft.

The radiator transfers coolant heat into the outside air. A fan draws air through the radiator. The radiator cap increases pressure to prevent boiling and coolant loss. Typical cap pressure is 12 to 16 psi (83 to 110 kPa).

An electric engine cooling fan uses an electric motor and a fan switch. The switch is thermostatic and only operates the fan motor when the engine temperature is above a specific point.

The thermostat controls coolant flow through the system. When cold, the thermostat blocks circulation to speed warmup. When hot, the thermostat opens to allow cooling action.

Antifreeze is a mixture of water and ethylene glycol. It helps cool the engine better than water, prevent rust formation, and lubricates the water pump. A 50 percent mix of water and antifreeze is common.

Engine overheating is a serious problem that can cause major damage. It can be caused by low coolant level, stuck thermostat, loose fan belt, missing fan shroud, engine fan problems, etc.

Antifreeze should be changed at least every two years. After prolonged use, it turns corrosive and can accelerate rust formation in system.

Antifreeze strength should be checked with a hydrometer. Freeze-up protection should be about 10 degrees below the coldest possible temperature for the climate.

A cooling system pressure test will find system leaks. Pump a few pounds above cap pressure into the system. Then, inspect the radiator, around hoses, the water pump, and other potential leakage points. The pressure tester can also be used to check the radiator cap opening pressure.

Bad water pumps are normally replaced with new or factory rebuilt units. It is too time consuming to rebuild pumps in-shop. When installing a pump, check bolt lengths. Clean all mating surfaces. Use sealer on the new gasket. Tighten bolts to specs in a crisscross pattern.

The lubrication system uses the oil pump to force motor oil to high friction points in the engine. An oil film separates parts to prevent wear.

Oil viscosity is the thickness of the oil. A multi-viscosity oil is for cold weather. Use the type and grade oil recommended by the engine manufacturer.

Oil pumps can be rotor or gear types. Both use engine power to send oil through the galleries in the block. Oil is also forced through passages in the head for valve train lubrication.

The oil filter removes impurities from the oil. It can be a spin-on or cartridge type. It should be changed with the engine oil at prescribed intervals, about 6000 miles or 10 000 km.

An oil pressure sending unit screws into the engine and is exposed to engine oil pressure. It is a switch or variable resistor that operates the dash oil light or gauge.

A pressure gauge can be screwed into the sending unit hole. It will show actual engine oil pressure for problem diagnosis.

The engine oil pan can sometimes be removed without engine removal. Check in a service manual for details. The pan may have to be removed to replace the pan gasket, oil pump, engine bearings, etc. Do not over-tighten the oil drain plug; it will strip very easily.

KNOW THESE TERMS

Water pump, Radiator hoses, Heater hoses, Fan, Thermostat, Impeller, Radiator core, Radiator tanks, Filler neck, Drain cock, Radiator cap valves, Closed system, Open system, Shroud, Thermostatic clutch, Fan switch, Hose spring, Bypass, Antifreeze, Coolant leaks, Stuck thermostat, Collapsed hose, Antifreeze mixture, Hydrometer, Refractometer, Pressure test, Flushing, Oil film, Viscosity, Multi-weight, Service rating, Oil pan, Oil pump, Oil filter, Oil galleries, Oil pickup, Pressure relief valve, Oil cooler, Oil level indicator, Pressure sending unit, Oil pressure test.

REVIEW QUESTIONS—CHAPTER 15

1. List four functions of a cooling system.
2. Air-cooled engines are still used on many makes of cars. True or false?
3. The water pump uses __________ __________ to circulate coolant through the engine __________ __________.
4. The __________ is the center section of the radiator that contains tubes and fins.
5. This is NOT a purpose of the radiator cap:
   a. Seal top of filler neck.
   b. Relieve excess pressure.
   c. Lower boiling point of coolant.
   d. Relieve vacuum in system.
6. A customer complains of engine overheating, only at idle. The system is full of coolant and belts are properly adjusted.
   Technician A notices a missing fan shroud and says that could be the cause of overheating.
   Technician B says a thermostat stuck closed is a more likely cause of the problem.

Who is correct?
a. Technician A.
b. Technician B.
c. Both A and B.
d. Neither A nor B.

7. How does an electric cooling fan system operate?
8. A missing lower hose spring can cause engine overheating. True or false?
9. An electric engine fan fails to turn on when the engine warms up.
Technician A says the first step is to remove and test the fan motor.
Technician B says simply apply battery voltage directly to the motor at the fan switch.
Who is correct?
a. Technician A.
b. Technician B.
c. Both A and B.
d. Neither A nor B.
10. What are two ways of checking antifreeze?
11. How and why do you perform a cooling system pressure test?
12. How do you check for worn water pump bearings?
13. It is very difficult to break a thermostat housing. True or false?
14. Explain how multi-weight oil could be desirable.
15. List and explain the five major parts of a lubrication system.
16. A _________ _________ _________ limits maximum oil pressure.
17. Internal oil leakage shows up as _________ _________ coming out the car's exhaust pipe.
18. How do you check actual engine oil pressure?
19. Diesel and turbocharged engines should have their oil changed more often than naturally aspirated gasoline engines. True or false?
20. What are some of the reasons you may have to remove the engine oil pan?

ASE CERTIFICATION-TYPE QUESTIONS

1. Technician A says that an automotive engine's operating temperature is normally between 220-250 °F (104-121 °C). Technician B says that an engine's operating temperature is normally between 180-195 °F (82.2-90.5 °C). Who is right?
(A) A only.
(B) B only.
(C) Both A & B.
(D) Neither A nor B.
2. Technician A says that a liquid cooling system is better than an air type cooling system because it improves the efficiency of the heater. Technician B says that a liquid cooling system is better than an air type cooling system because engine temperature is more precisely controlled. Who is right?
(A) A only.
(B) B only.
(C) Both A & B.
(D) Neither A nor B.
3. An open cooling system utilizes a(n) __________ to allow excess coolant to drain onto the ground.
(A) coolant reservoir
(B) overflow tube
(C) drain petcock
(D) None of the above.
4. All of the following will cause an automotive engine to overheat EXCEPT:
(A) retarded ignition timing.
(B) a loose fan belt.
(C) a locked fan clutch.
(D) an inoperative electric cooling fan.
5. Technician A says that when the engine is cold, coolant should flow through the radiator. Technician B says that when the engine is cold, coolant should not flow through the radiator. Who is right?
(A) A only.
(B) B only.
(C) Both A & B.
(D) Neither A nor B.
6. Technician A says that a low viscosity motor oil is very thick and resists flow. Technician B says that a low viscosity motor oil is thin and flows easily. Who is right?
(A) A only.
(B) B only.
(C) Both A & B.
(D) Neither A nor B.
7. All of the following are causes of low engine oil pressure EXCEPT:
(A) a clogged oil pump pickup screen.
(B) worn connecting rod bearings.
(C) restricted oil gallery.
(D) main bearings.
8. When performing an oil pressure test, the pressure gauge should be attached to __________.
(A) the oil filter fitting
(B) the oil pressure sending unit hole
(C) the oil pump
(D) None of the above.
9. All of the following are basic rules to follow when changing the engine's oil and oil filter EXCEPT:
(A) apply oil to the oil filter's O-ring before installing filter.
(B) do not overtighten oil pan drain plug.
(C) use a filter wrench to tighten engine oil filter.
(D) keep the car relatively level when draining the oil.
10. Technician A says that before installing an oil pump, you should apply grease to the pickup screen. Technician B says that before installing an engine's oil pump, you should prime the pump with motor oil. Who is right?
(A) A only.
(B) B only.
(C) Both A & B.
(D) Neither A nor B.

Chapter 16

Filters, Fuel Pumps, Carburetors—Operation and Service

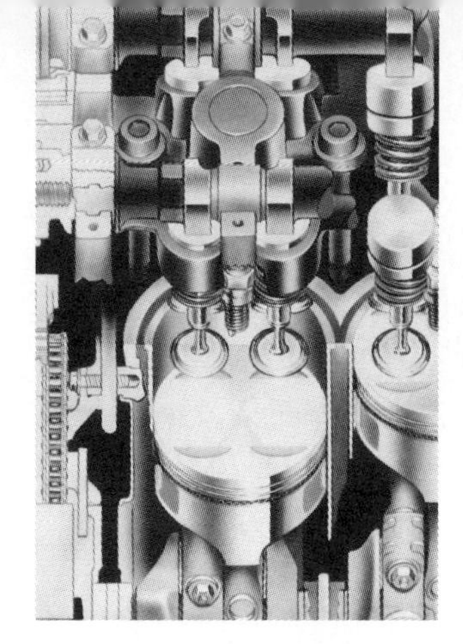

After studying this chapter, you will be able to:
- *Describe the components of a fuel supply system.*
- *Summarize the service of air and fuel filters.*
- *Explain the operation of mechanical and electric fuel pumps.*
- *Describe how to troubleshoot and replace fuel pumps.*
- *Explain the operation of a carburetor.*
- *List the basic carburetor circuits.*
- *Perform typical carburetor service tasks.*
- *Summarize the operation and testing of computer controlled carburetor systems.*

An engine is dependent upon clean air and fuel for efficient operation. The air around us contains a tremendous amount of dirt that can wear out an engine in a very short period of time. Also, without the correct mixture of fuel and air, an engine will not perform properly.

This chapter will summarize the most important information on air filters, fuel filters, mechanical fuel pumps, electric fuel pumps, and carburetors. It will provide the information most commonly used by an engine technician. Study carefully!

FUEL SYSTEM CONSTRUCTION

A fuel system can be divided into three subsystems:

1. *Fuel supply system* (provides filtered fuel, under pressure, to carburetor, or injection system).
2. *Air supply system* (removes dust and dirt from air entering engine).
3. *Fuel metering system* (controls amount of fuel that mixes with air).

It is important that you understand the components included in each system when trying to diagnose and repair engine problems.

Fuel tanks

An automotive *fuel tank* must safely hold an adequate supply of fuel for prolonged engine operation. The size of a fuel tank partially determines a car's *driving range* or the greatest distance the car can be driven without stopping for fuel. See Fig. 16-1.

Fuel tank capacity is the rating of how much fuel a tank can hold when full. An average fuel tank capacity is about 12 to 25 gallons (45 to 95 liters).

The *fuel tank filler neck* is the extension for filling the tank with fuel. The *filler cap* fits on the filler neck.

A *tank pickup unit* extends down into the tank to draw out fuel and to operate the fuel gauge. A coarse filter is usually placed on the end of the pickup tube to strain out larger debris.

Fuel lines and hoses

Fuel lines and fuel hoses carry fuel from the tank to the engine. A main fuel line connects the tank to the

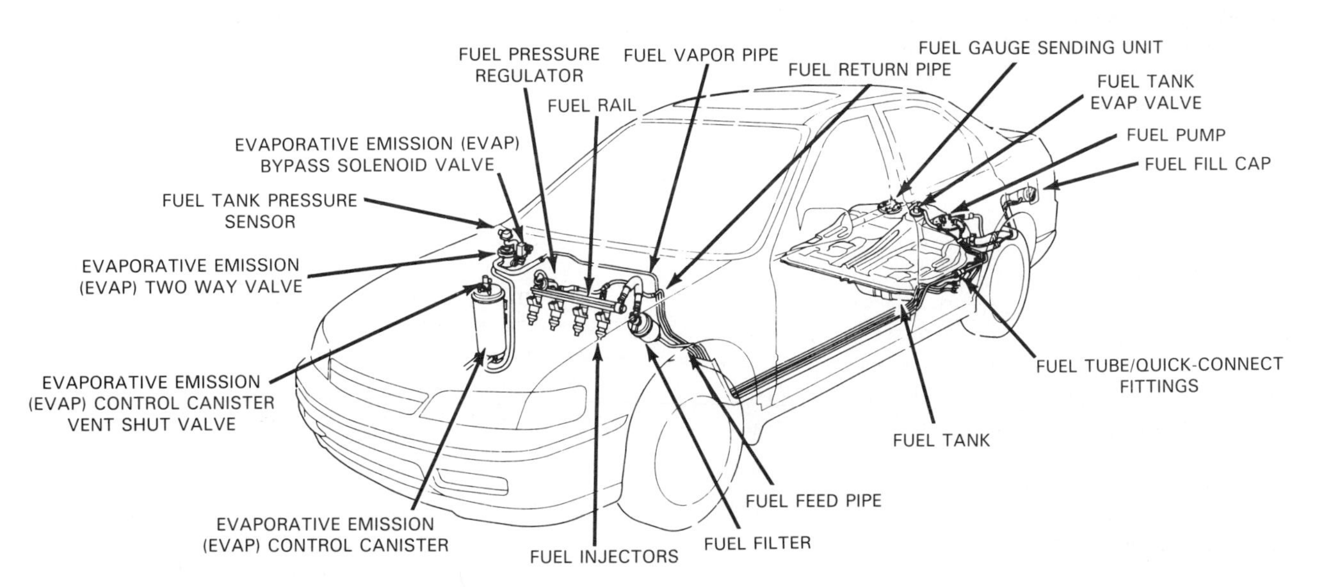

Fig. 16-1. Note the locations of fuel supply system components. The fuel rail is steel while the fuel lines going to and from the tank are made of nylon tubing. (Honda)

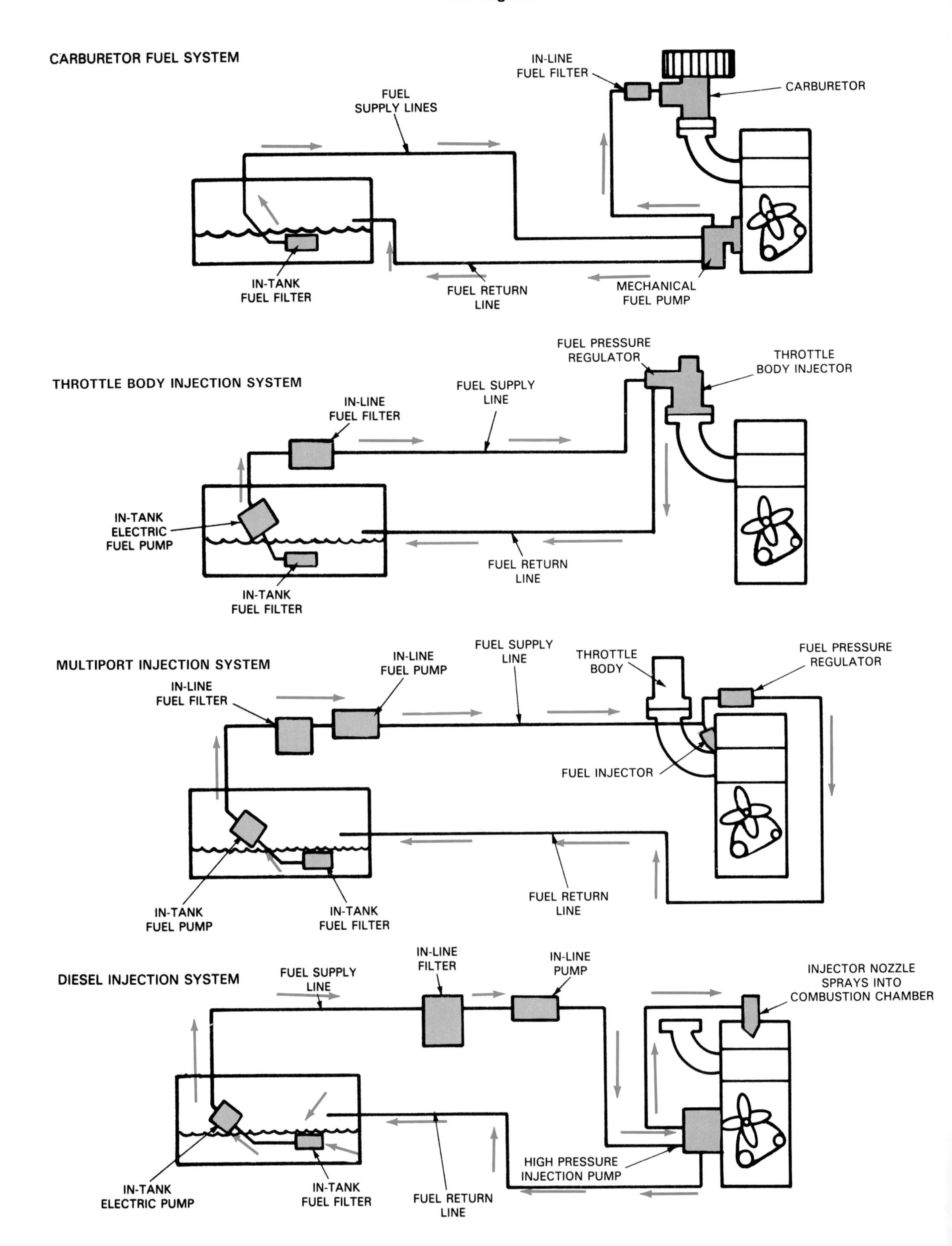

Fig. 16-2. Study locations of fuel filters and other parts for four types of fuel delivery systems.

engine. Fuel is pulled through this line to the carburetor or metering section of the injection system, Fig. 16-1.

Fuel lines are made from double-wall steel, plastic, or nylon tubing. Plastic and nylon tubing is used in modern fuel systems to reduce the possibility of rust and corrosion in the fuel system. A fuel line is strong enough to withstand the constant and severe vibrations produced by normal engine operation.

Fuel hoses can be made of synthetic rubber or nylon covered with a plastic insulator. These hoses are used when movement between two or more parts occur, such as between the body and the engine, since the engine moves on its motor mounts.

A *fuel return system* helps cool the fuel and prevent *vapor lock* (bubbles form in overheated fuel and upset fuel flow). A second return fuel line is used to carry excess fuel back to the tank. This keeps cool fuel constantly circulating through the system, Fig. 16-2.

Fuel filters

Fuel filters stop contaminants (rust, water, corrosion, dirt) from entering the carburetor, throttle body, injectors, injection pump, and any other part that could be damaged by foreign matter.

A fuel filter is normally located on the fuel tank pickup tube. A second fuel filter is located in the main fuel line or inside the carburetor or fuel pump, Fig. 16-2.

FUEL PUMPS

A *fuel pump* develops pressure to force fuel out of the tank and to the engine. There are two basic types: mechanical fuel pump and electric fuel pump.

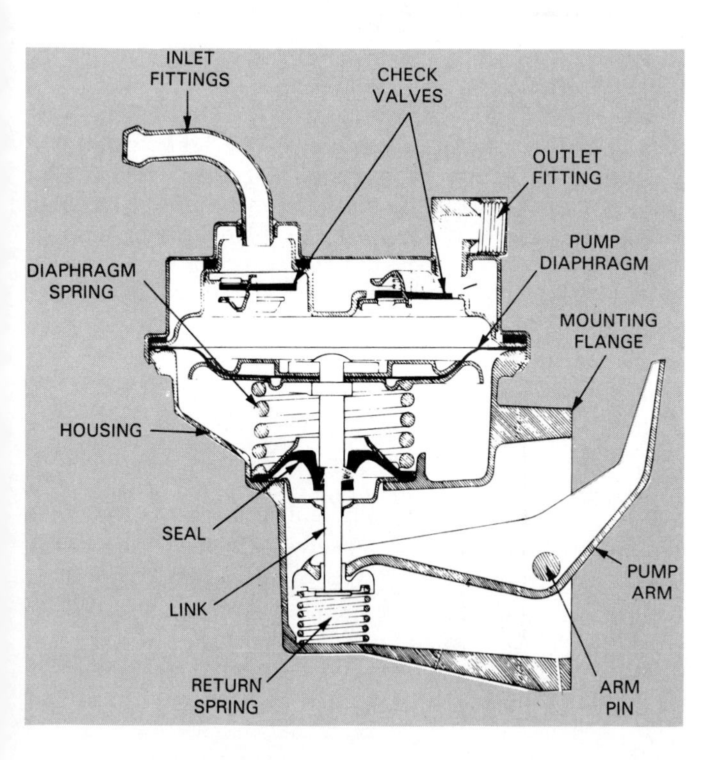

Fig. 16-3. Mechanical fuel pump uses rocker arm action, check valves, rubber diaphragm, and spring to pressurize fuel.

Mechanical fuel pumps

A *mechanical fuel pump* is usually powered by an eccentric (egg shaped lobe) on the engine camshaft. The mechanical fuel pump bolts to the side of the engine block. A gasket prevents oil leakage between the pump and engine. Look at Fig. 16-3.

Mechanical fuel pumps are commonly used with carburetor type fuel systems. They are the oldest type of fuel pump, but they are still found on many cars.

Basically, a mechanical fuel pump consists of:

1. FUEL PUMP BODY (foundation for other pump components).
2. ROCKER ARM (uses engine power to produce pumping action).
3. RETURN SPRING (keeps rocker arm in contact with eccentric).
4. DIAPHRAGM (rubber disc that acts on fuel in pump body).
5. DIAPHRAGM SPRING (pushes on diaphragm to develop fuel pressure).
6. CHECK VALVES (control flow of fuel through pump body).

Fig. 16-4 illustrates the operation of a mechanical fuel pump.

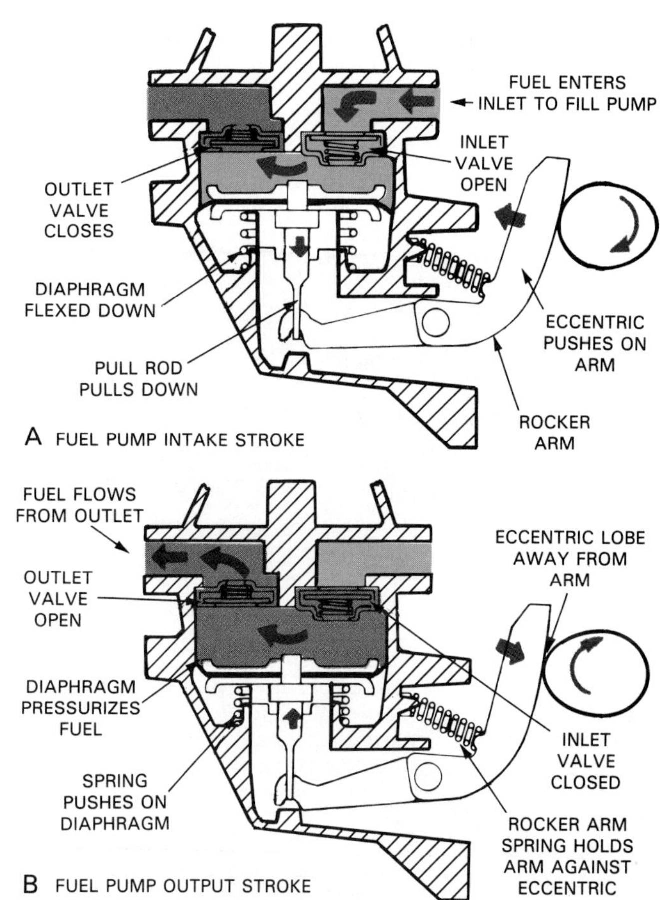

Fig. 16-4. Operation of mechanical fuel pump. A—Eccentric pushes on rocker arm. Pull rod stretches diaphragm down and compresses spring. Fuel is pulled into pump through inlet check valve. B—Eccentric rotates away from pump arm. Spring then acts on diaphragm to pressurize fuel in pump chamber. This closes inlet check and opens outlet check so fuel flows to engine.

Electric fuel pumps

An *electric fuel pump* also produces fuel pressure and flow. Electric fuel pumps are commonly used on late model cars. Refer to Fig. 16-5.

An electric fuel pump can be located inside the fuel tank as part of the fuel pickup unit, Fig. 16-6. It can also be located in the fuel line between the tank and engine.

Many modern electric fuel pumps are controlled by an on-board computer.

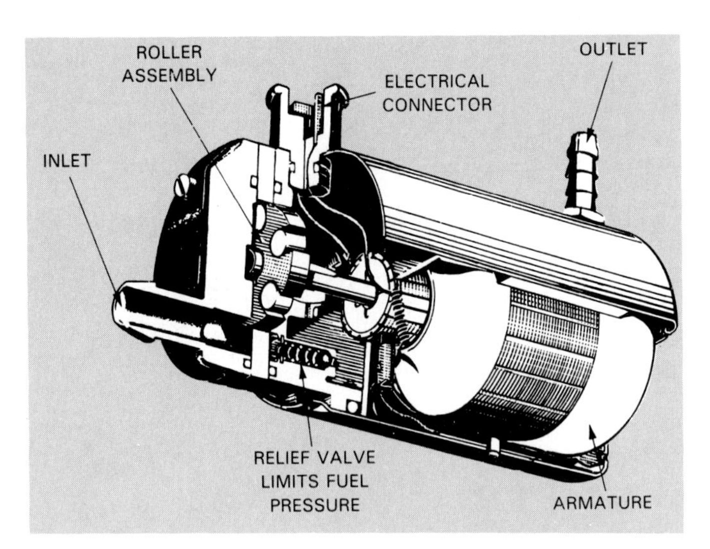

Fig. 16-5. Electric fuel pump uses dc motor to turn pumping element or rotor. Note parts. (Volvo)

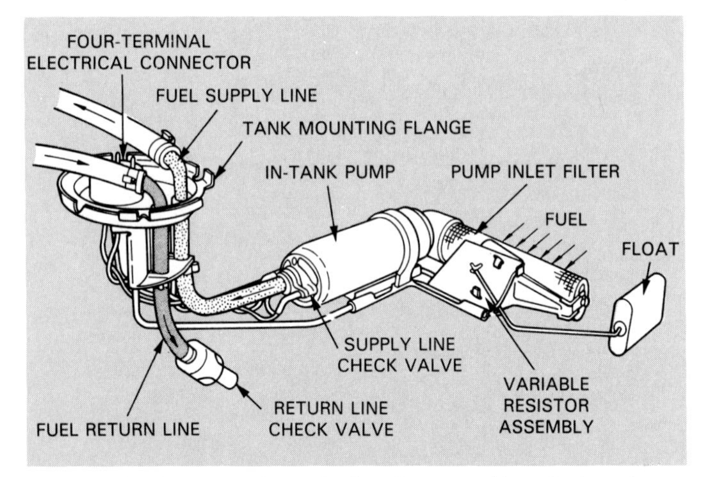

Fig. 16-6. This is an in-tank electric pump. Note fuel strainer, pump motor, fuel return, float, and variable resistor for operating fuel gauge. (Chrysler)

AIR FILTERS

An *air filter,* also called *air cleaner,* removes foreign matter (dirt and dust) from the air flowing into the engine intake manifold. Most air filters use a paper element (filter material). A few cars use a polyurethane (foam) filtering element. The element fits inside a metal or plastic housing, Fig. 16-7.

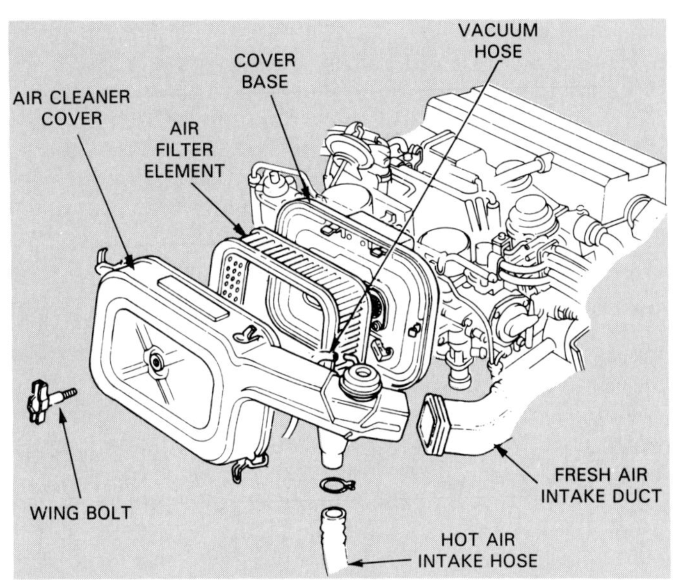

Fig. 16-7. Air cleaner element mounts in air cleaner housing. (Honda)

An *air filter sensor* is used on some vehicles to measure the pressure drop across the element to warn the driver of a clogged or dirty filter. It is often a mechanical device on the filter housing. A window in the sensor shows when the element should be changed.

FUEL SUPPLY SYSTEM SERVICE

Now that you have an understanding of a fuel supply system, you are ready to learn about problems, tests, and repairs.

Always keep a fire extinguisher handy when working on a car's fuel system. During a fire, a few minutes time can be a LIFETIME!

Fuel line and hose service

Faulty fuel lines and *hoses* are a common source of fuel leaks. Fuel hoses can become hard and brittle after being exposed to engine heat and the elements. Engine oil can soften and swell fuel hoses. Always inspect hoses closely and replace any in poor condition.

Metal fuel lines seldom cause problems. However, they should be replaced when smashed, kicked, rusted, or leaking.

Remember these rules when working with fuel lines and hoses:

1. Place a rag around the fuel line fitting during removal. This will keep fuel from spraying on you or the hot engine.
2. Use a tubing wrench on fuel system fittings.
3. Only use approved double-wall steel tubing for fuel lines. Never use copper tubing.
4. Make smooth bends when forming a new fuel line. Use a bending spring or bending tool.
5. Form only double-lap or ISO flares on the ends of the fuel line.
6. Reinstall fuel line hold-down clamps and brackets. If not properly supported, the fuel line can vibrate and fail.
7. Route all fuel lines and hoses away from hot

or moving parts. Double-check clearance after installation.

8. Only use approved synthetic rubber hoses in a fuel system. If vacuum type rubber hose is accidentally used, the fuel can chemically attack the hose. A dangerous leak could result.
9. Make sure a fuel hose slides fully over its fitting before installing the clamps. Pressure in the fuel system could force the hose off.
10. Double-check all fittings for leaks. Start the engine and inspect the connections closely.

Do not bend or mishandle plastic or nylon fuel lines. If the line becomes kinked, it will create a permanent restriction. If a plastic or nylon fuel line is damaged in any way, replace it. Do not attempt to repair a damaged plastic fuel line.

Some late model fuel lines have a special snap type fitting, called a *push-on fitting.* Do not pry the fitting apart or it will be damaged. In some cases, you may need a special tool to release the fuel line fitting for service. This tool can be purchased at most auto parts stores or tool suppliers.

When servicing push-on fuel line fittings, you may also need to replace the O-ring seal to prevent fuel leakage. Make sure you purchase the right material, shape, and size seal for the fuel line and make/model vehicle. If you install the wrong seal, a fire can result.

Fuel filter service

Fuel filter service involves periodic replacement or cleaning of system filters. It may also include locating clogged fuel filters.

A *clogged fuel filter* can restrict the flow of fuel to the carburetor, gasoline injectors, or diesel injection pump. Engine performance problems will usually show up at higher cruising speeds. On older cars, a *clogged in-tank strainer* is a common and hard to diagnose problem.

Fuel filters can be located in the following places:

1. In the fuel line before the carburetor, fuel injectors, or diesel injection pump.
2. Inside the fuel pump.
3. In the fuel line right after the electric fuel pump.
4. Under the fuel line fitting in the carburetor.
5. A fuel strainer is also located in the fuel tank on the end of the fuel pickup tube.
6. Inside the injector body (strainer type).

When in doubt, refer to a service manual.

FUEL PUMP SERVICE

Fuel pump problems usually show up as low fuel pressure, inadequate fuel flow, abnormal pump noise, or fuel leakage from the pump. Both mechanical and electric fuel pumps can fail after prolonged operation.

Low fuel pump pressure can be caused by a weak diaphragm spring, ruptured diaphragm, leaking check valves, or physical wear of moving parts.

Mechanical fuel pump noise (clacking sound from inside pump) is caused by a weak or broken rocker arm return spring or by wear of the rocker arm pin or arm itself. Mechanical fuel pump noise can be easily confused with valve or tappet clatter. To verify mechanical fuel pump noise, use a stethoscope.

Note! Most electric fuel pumps make some noise (buzz or humming sound) when running. Only when the pump noise is abnormally loud should an electric fuel pump be considered faulty.

A clogged tank strainer can cause excess electric fuel pump noise. Pump speed can increase because fuel is not entering the pump properly.

Fuel pump leaks are caused by physical damage to the pump body or deterioration of the diaphragm or gaskets. Most mechanical fuel pumps have a small vent hole in the pump body. When the diaphragm is ruptured, fuel will leak out of this hole.

It is possible for a ruptured mechanical fuel pump diaphragm to contaminate the engine oil with gasoline. Fuel can leak through the diaphragm, pump, into the side of the block, and down into the oil pan.

FUEL PUMP TESTS

Fuel pump testing commonly involves measuring fuel pump pressure and volume. Since exact procedures vary, refer to a manual for exact testing methods. Sometimes, fuel pump vacuum is measured as another means of determining pump and supply line condition.

WARNING! Most fuel injection systems operate on very high fuel pressure. Make sure you tighten all test connections and follow prescribed procedures when testing fuel pump output.

Measuring fuel pump pressure

To measure fuel pump pressure, connect a pressure gauge to the output line of the fuel pump, Fig. 16-8A.

Typically, you would start and idle the engine at spec rpm with a mechanical fuel pump. With an electric fuel pump, you may only need to activate (supply voltage to) the pump motor. Compare your pressure readings to specifications.

Depending upon the type of fuel system, the *fuel pressure test* should show a:

1. Gauge reading of 4 to 6 psi (30 to 40 kPa) for CARBURETOR type fuel system.
2. Gauge reading of 15 to 40 psi (100 to 275 kPa) for GASOLINE INJECTION type fuel system.
3. Gauge reading of 6 to 10 psi (40 to 70 kPa) for DIESEL ENGINE fuel supply system.

Always use factory values when determining fuel pump condition. Pressures vary slightly for each make and model car.

If fuel pump pressure is NOT within specs, check pump volume and the lines and filters before replacing the fuel pump.

Measuring fuel pump volume

Fuel pump volume, also called *capacity,* is the amount of fuel the pump can deliver in a specific amount of time. It is measured by allowing fuel to pour into a graduated (marked) container for a certain time period.

Route the output line from the fuel pump into a special container. For safety, a valve or clip should be used to control fuel flow into the container, Fig. 16-8B.

With the engine idling at a set speed, allow fuel to pour into the container for the prescribed time (normally 30 seconds). Close off the clip or valve. Compare volume output to specs.

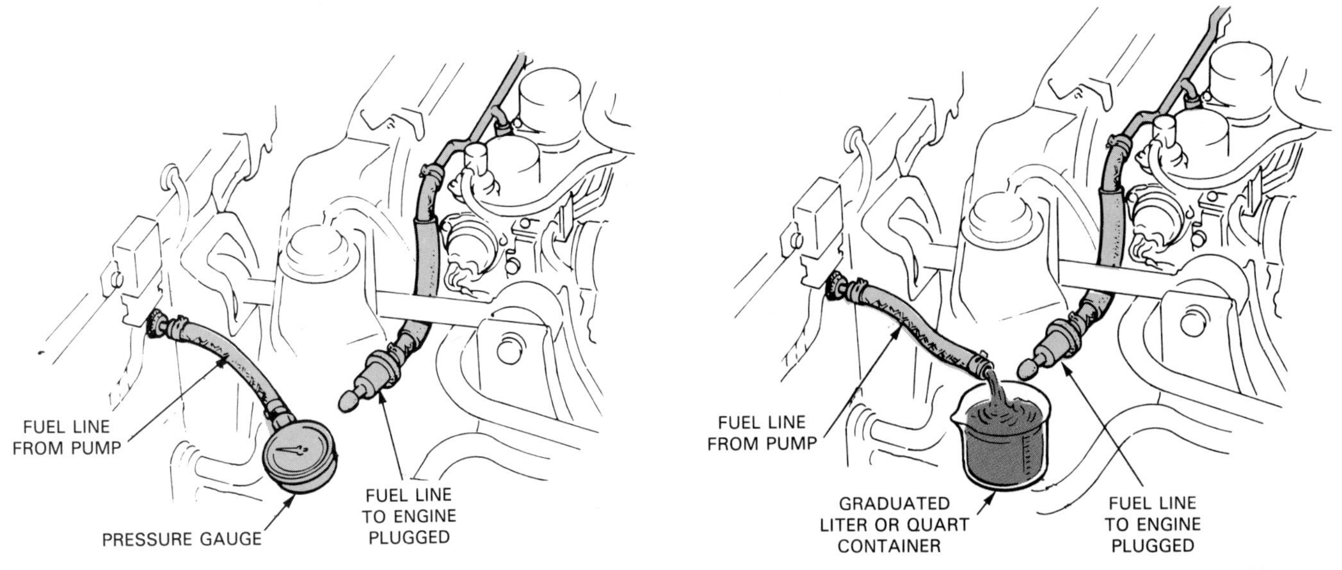

Fig. 16-8. Performance testing fuel pump. A—To check pump pressure, disconnect fuel line after pump and install pressure gauge. Start engine or activate electric pump. Read gauge and compare to specs. B—To check pump volume, insert fuel line hose into graduated container. Start engine or activate pump and allow fuel to flow into container for specific time span. Compare output to specs. Always check for clogged in-tank strainer or fuel filter before condemning pump. (Honda)

Fuel pump volume output should be a minimum of about ONE PINT (0.47 liters) in 30 seconds for carburetor systems. Fuel injection systems typically have a slightly higher volume output from the supply pump. Refer to factory values for the particular fuel pump and automobile.

Measuring fuel pump vacuum

Fuel pump vacuum should be checked when a fuel pump fails pressure and volume tests. A vacuum test will eliminate possible problems in the fuel lines, hoses, filters, and pickup screen in the tank.

Typically, fuel pump vacuum should be about 7 to 10 in./hg. A good vacuum reading indicates a good fuel pump. If the pump failed the pressure or volume test but passed the vacuum test, the fuel supply lines and filters may be at fault.

Fuel pump replacement

Replacing a fuel pump is a relatively simple task. Basically, you must remember to:

1. Relieve system fuel pressure if needed.
2. Place a rag around the pump fittings when loosening, Fig. 16-9.
3. Seal off fuel lines when pump is removed to prevent leakage and a potential fire.
4. Clean mating surface on engine properly before installing new mechanical pump. Use gasket sealer to hold mechanical pump gasket in place.
5. Make sure you have correct replacement pump, as shown in Fig. 16-10.
6. Position mechanical pump eccentric in the released location by cranking engine. This will let you start bolts in engine block more easily.
7. Check supply voltage to electric pump before condemning and replacing pump.
8. Torque all fasteners to specs.
9. Check filter for clogging before condemning pump.
10. Use service manual as needed for specific instructions on specific pump type.
11. Tighten hose clamps, fittings, and fasteners properly.
12. Check for fuel leaks before releasing car to customer.

CARBURETORS

This section of the chapter reviews the principles of carburetion. Many engines entering the shop will still

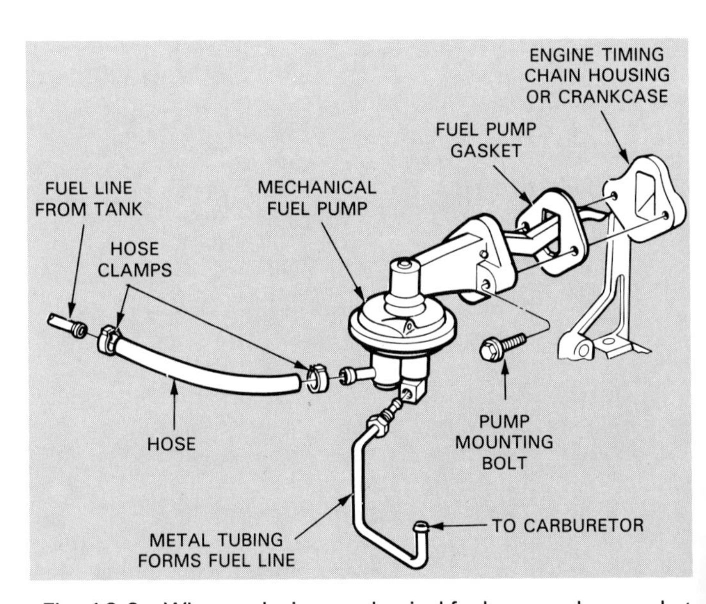

Fig. 16-9. When replacing mechanical fuel pump, clean gasket surfaces. Position eccentric away from pump arm to ease starting of bolts. Torque bolts and tighten clamps properly. Check for leaks after replacement. (Chrysler)

Fig. 16-10. Most technicians do not rebuild fuel pumps in-shop. They install new or factory rebuilt pumps to save time and money. Make sure new pump is same as old unit. (AC-Delco)

have carburetors and an engine technician must know how to find and correct problems efficiently.

PURPOSE OF CARBURETOR

A *carburetor* is a device for mixing air and fuel in the correct proportions (amounts). The carburetor bolts to the engine intake manifold, Fig. 16-11. The air cleaner fits over the top of the carburetor.

Carburetor parts

A basic carburetor, Fig. 16-11, consists of:

1. CARBURETOR BODY (main carburetor housing).
2. AIR HORN (air passage containing venturi, throttle valve, and end of main discharge tube).
3. THROTTLE VALVE (airflow control valve in air horn).
4. VENTURI (restriction or narrowed area in air horn).
5. MAIN DISCHARGE TUBE (fuel passage between fuel bowl and air horn), Fig. 16-11.
6. FUEL BOWL (fuel storage area in body).
7. FLOAT (device for controlling amount of fuel entering bowl), Fig. 16-11.

Carburetor operation

With the engine running, air is pulled into the engine by the downard moving pistons. Air flows through the air cleaner and through the carburetor air horn. Airflow over the venturi produces a high vacuum at the upper end of the main discharge tube. This pulls fuel through the tube and into the air horn. Air mixes with the fuel and the mixture is drawn into the engine intake manifold, Fig. 16-11.

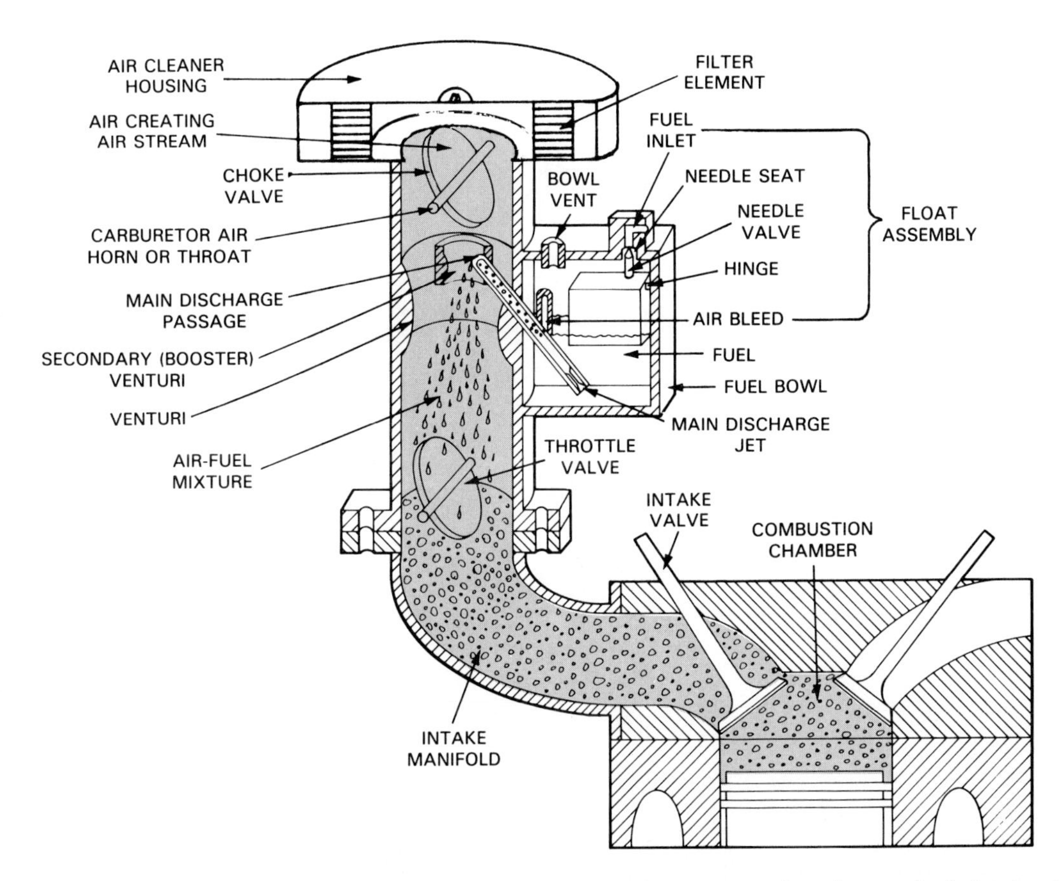

Fig. 16-11. Study basic parts of simple carburetor. Vacuum produced on engine's intake strokes pull air and fuel into cylinders. Throttle valve controls engine speed.

The throttle valve is connected to the driver's gas pedal. When the gas pedal is pressed, the throttle is swung open. This increases airflow, fuel entry, and engine power. Releasing the gas pedal closes the throttle to reduce engine speed and power.

The fuel pump forces fuel into the fuel bowl. As fuel is consumed, the float drops and opens the needle valve. Fuel then flows into the bowl. When enough fuel has entered, the float pushes the needle valve closed. This stops entry of fuel and prevents *engine flooding* (too much fuel entry into carburetor and engine).

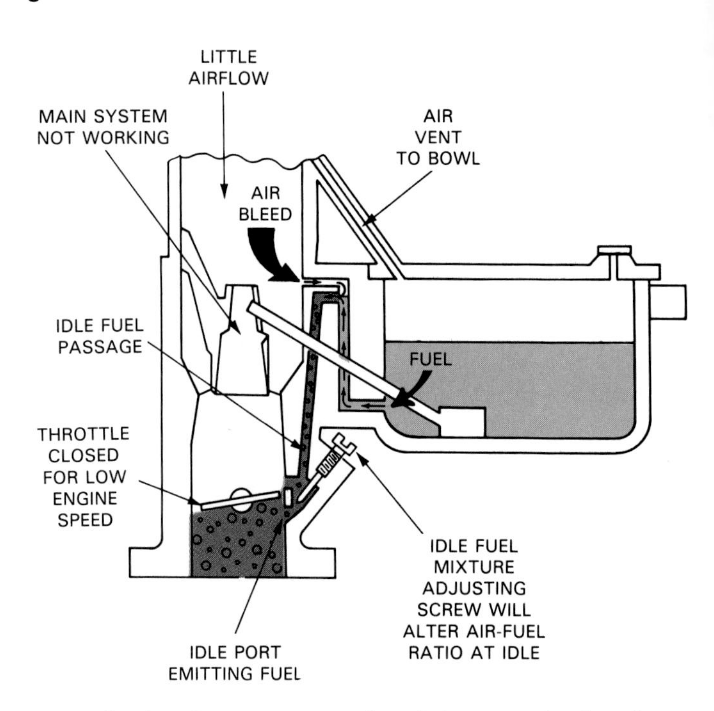

Fig. 16-13. Idle system operation. At low speeds, there is not enough airflow to pull fuel out of main discharge. With throttle closed, high vacuum forms below throttle plate. Fuel is then pulled through idle circuit. Air mixes with fuel in passage to help atomization. Idle mixture screw can be turned in or out to lean or richen mixture at idle. (Ford Motor Co.)

Carburetor systems

A *carburetor system* is a network of passages and related parts that help control the air-fuel ratio under a specific engine condition. Also called a *carburetor circuit,* each system supplies a predetermied air-fuel mixture as engine temperature, speed, and load change.

For example, a gasoline engine's air-fuel mixture may vary from a rich 8:1 ratio to a lean 18:1 ratio. An automotive carburetor, using its various systems, must provide air-fuel ratios of approximately:

1. 8:1 for cold engine starting.
2. 16:1 for idling.
3. 15:1 for part throttle.
4. 13:1 for full acceleration.
5. 18:1 for normal cruising at highway speeds.

Older cars, not subject to strict emission control regulations, have a slightly richer air-fuel ratio. Late model cars have leaner carburetor settings to help reduce exhaust pollution.

The major carburetor systems are:

1. FLOAT SYSTEM (maintains supply of fuel in carburetor bowl), Fig. 16-12.

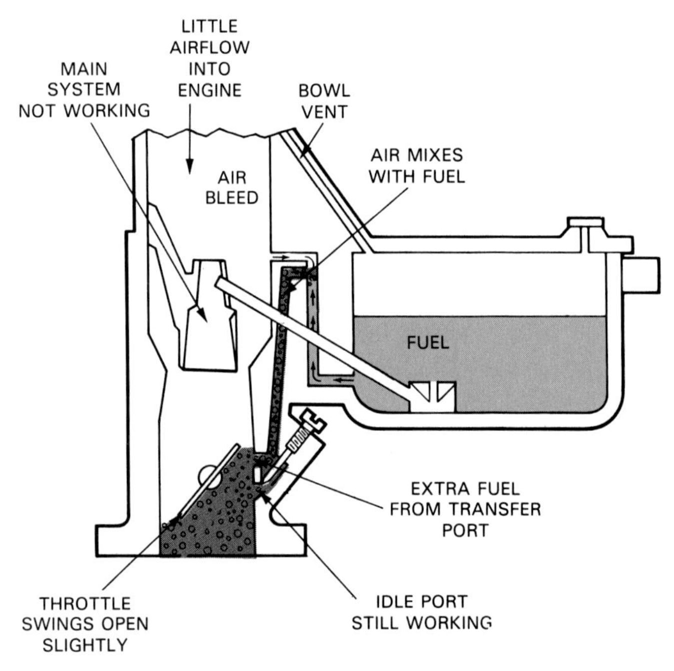

Fig. 16-14. Off-idle system operation. As the throttle is swung open for initial acceleration, more fuel is needed to mix with increased air. Extra opening above idle screw port provides this fuel.

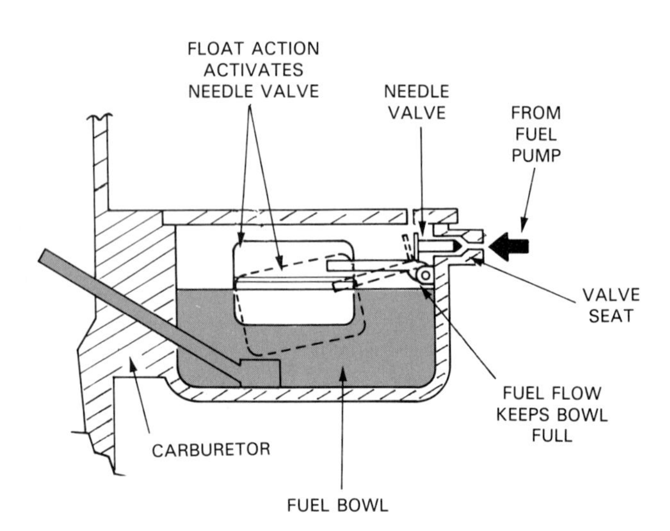

Fig. 16-12. Float rises and falls as fuel is pulled from or enters bowl. Float operates needle valve at fuel inlet to carburetor. (Ford)

2. IDLE SYSTEM (provides a small amount of fuel for low speed engine operation), Fig. 16-13.
3. OFF-IDLE SYSTEM (provides correct air-fuel mixture slightly above idle speeds), Fig. 16-14.
4. ACCELERATION SYSTEM (squirts fuel into air horn when throttle valve opens and engine speed increases), Fig. 16-15.
5. HIGH SPEED SYSTEM (supplies lean air-fuel mixture

at cruising speeds), Fig. 16-16.

6. FULL POWER SYSTEM (enriches fuel mixture slightly when engine power demands are high), Fig. 16-17.

7. CHOKE SYSTEM (provides extremely rich air-fuel

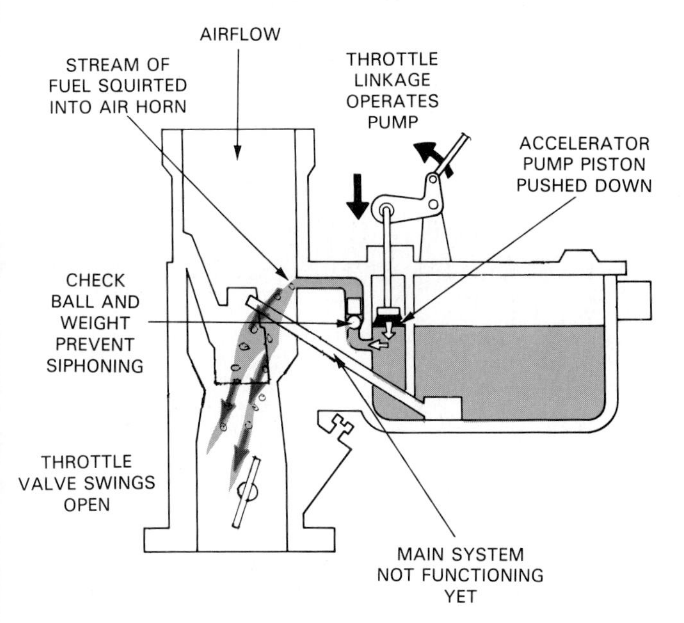

Fig. 16-15. Acceleration system operation. When gas pedal is pressed and throttle swings open, huge amount of air rushes through carburetor. It takes time for fuel to be pulled into air horn through main system. Therefore, accelerator pump is needed to squirt fuel into air horn to prevent engine hesitation caused by temporary lean condition. Note how pump squirts fuel anytime throttle linkage is activated. (Ford)

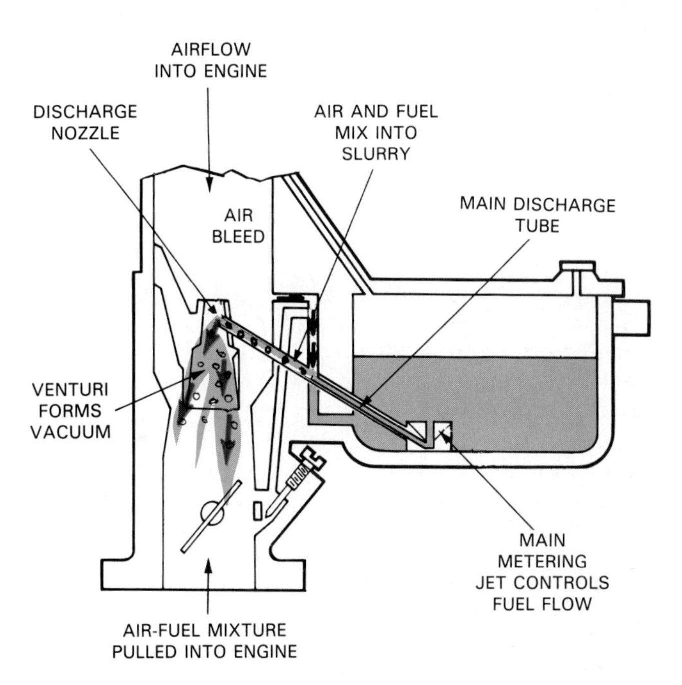

Fig. 16-16. Main system operation. Main system operates at higher engine rpm or at cruising speeds. Main jet has specific size hole that meters right amount of fuel to main discharge tube. Airflow through venturi produces vacuum at tip of tube to pull fuel out of bowl and into airstream.

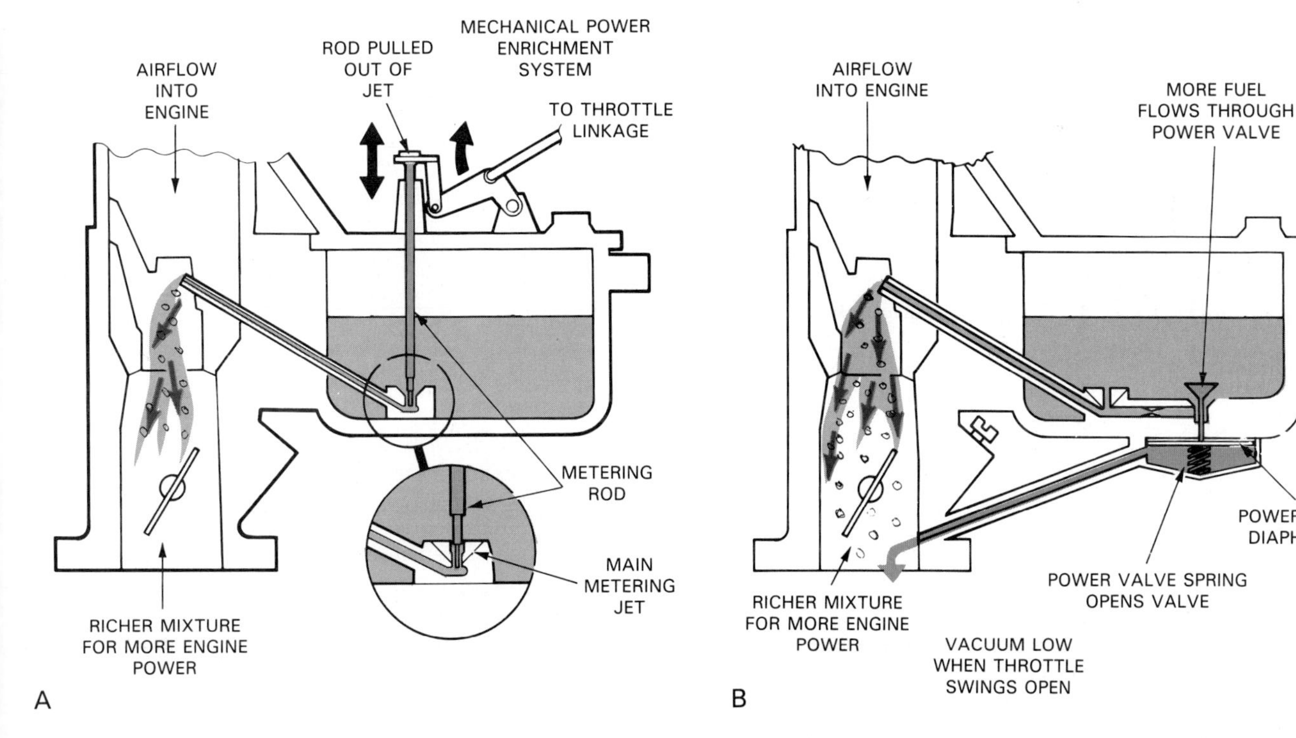

Fig. 16-17. Full power system operation. A—Metering rod is pulled in and out of jet to alter amount of fuel entering main system. Wider throttle opening would pull metering rod out and vice versa. B—Power valve can be used instead of metering rod. Engine vacuum drops with high power output of engine. Therefore, vacuum can be used to open and close power valve as needed. Power valve alters fuel flow through main system to increase power and efficiency of engine. (Ford)

mixture for cold engine starting), Fig. 16-18.

CARBURETOR DEVICES

Some other important carburetor devices include:

1. BOWL VENT (releases pressure and vacuum from fuel bowl), Fig. 16-19.
2. IDLE MIXTURE SCREW (needle valve that can be adjusted to alter air-fuel ratio at idle).
3. LIMITER CAPS (plastic caps that prevent too much adjustment of mixture screws).
4. MIXTURE SCREW PLUGS (metal plugs that prevent tampering of mixture adjustment which could increase emissions).
5. CHECK BALL (valve that only allows fuel or airflow in one direction).
6. AIR BLEED (device for mixing air into the fuel for improved atomization).
7. CHOKE UNLOADER OR BREAK (linkage mechanism and/or vacuum diaphragm that cracks choke plate open to prevent engine flooding), Fig. 16-19.
8. FAST IDLE CAM (linkage-cam mechanism that holds throttle plate open when engine is cold and choke is activated), Fig. 16-20.
9. FAST IDLE SOLENOID (electric solenoid that opens throttle plate to increase engine speed as needed), Fig. 16-21.
10. THROTTLE RETURN DASHPOT (diaphragm assembly that slows throttle closing to prevent engine stalling), Fig. 16-22.
11. AUXILIARY AIR VALVE (secondary air valve that helps keep air speed high through air horn), Fig. 16-19.
12. IDLE SPEED SCREW (screw that can be turned to open or close throttle to adjust engine idle rpm).

Note! Many other specialized devices can be found on carburetors. Refer to the shop manual for details of the specific make and model carburetor.

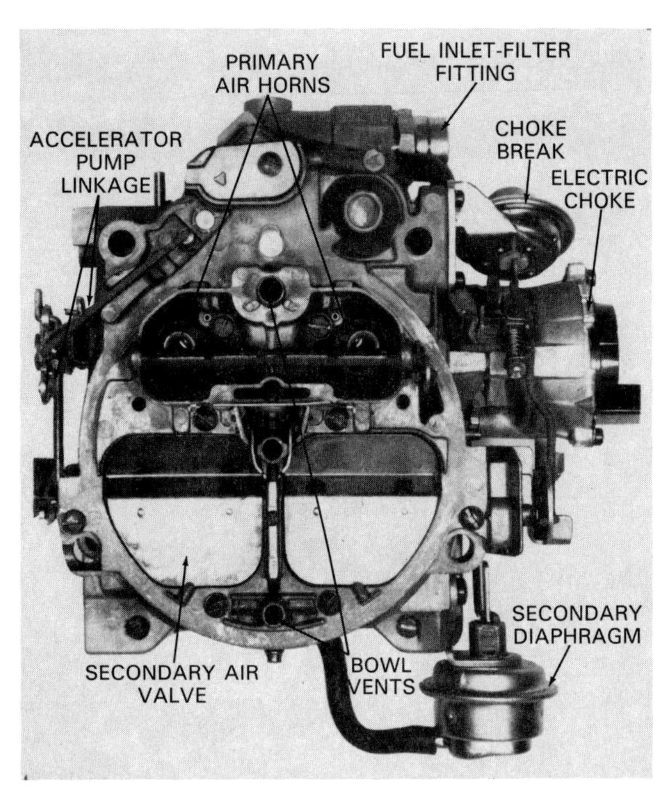

Fig. 16-19. Top view of 4-barrel carburetor shows major components. (Cadillac)

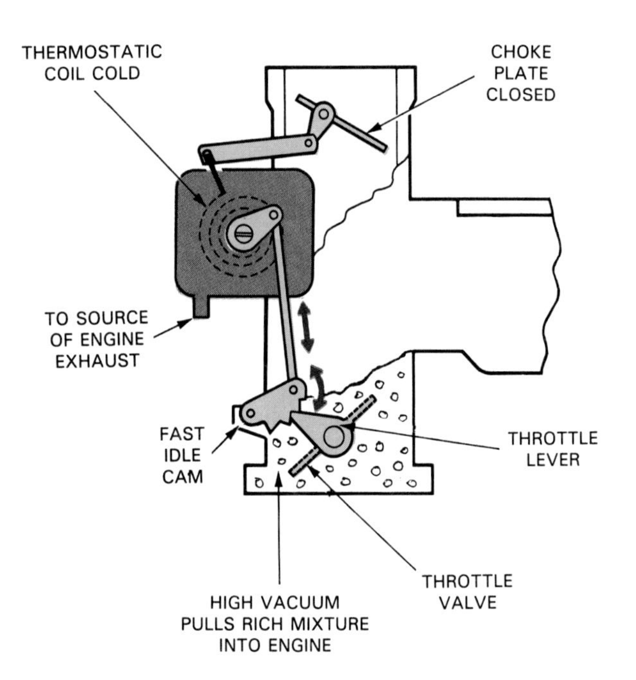

Fig. 16-18. Choke system operation. Thermostatic spring closes choke when cold. This causes huge amount of fuel to be pulled into air horn because of vacuum below choke plate. Fast idle cam is linked to choke so engine speed is also increased with engine cold. Exhaust gases or electric heating element warm spring after engine start-up to open choke. (Ford)

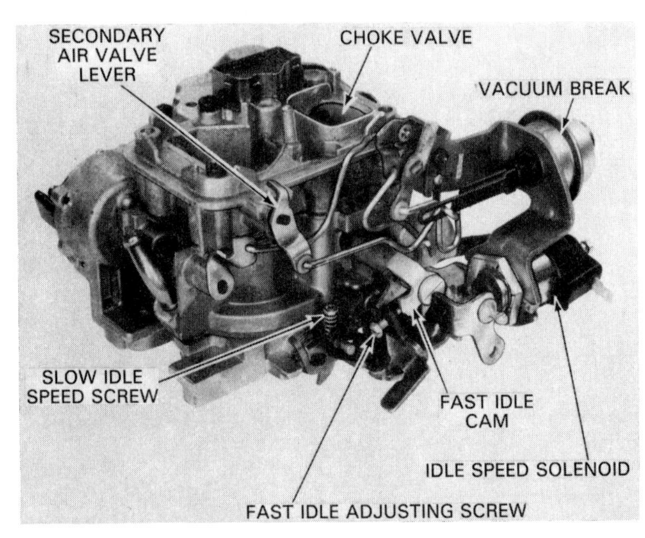

Fig. 16-20. Side view of in-line, 2-barrel carburetor. Note components. (Oldsmobile)

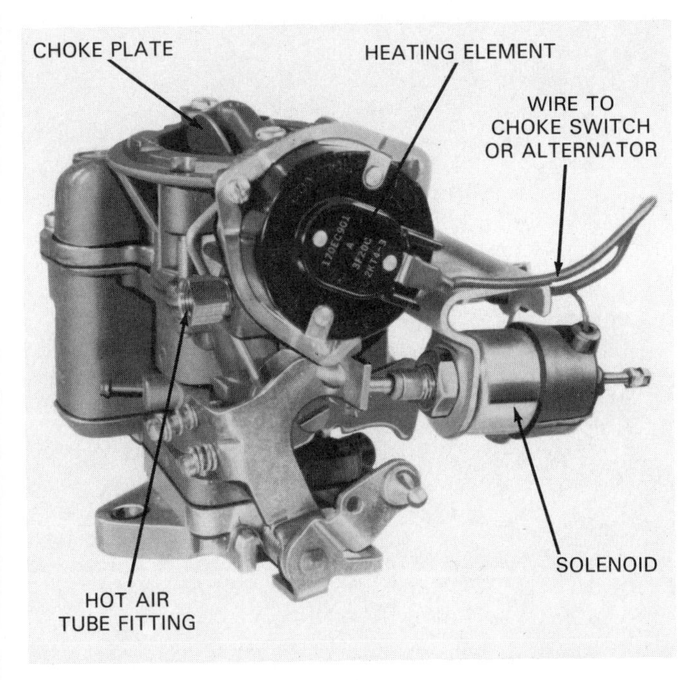

Fig. 16-21. Side view of this single barrel carb shows electric choke, idle speed solenoid, and choke plate. (Ford)

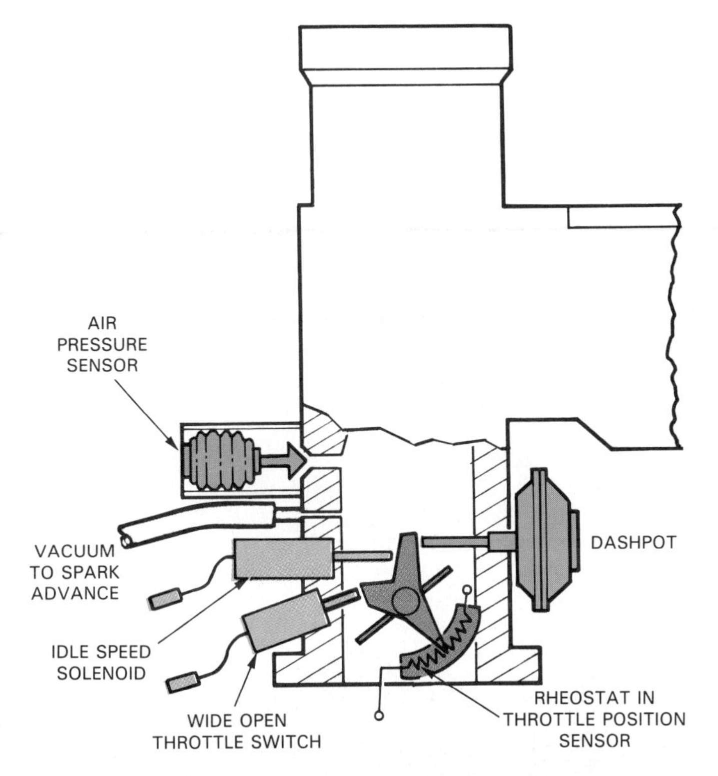

Fig. 16-22. Note other devices that can be found on carburetors or throttle body. Throttle position sensor reports to computer about throttle opening. Dashpot prevents engine stalling when engine is quickly returned to idle speed. WOT switch also reports wide open throttle position for shutting off air conditioning compressor and other power robbing units. Idle speed solenoid controls throttle opening at idle. Vacuum connections provide suction to various devices—distributor vacuum advance for instance. Air or barometric pressure sensor can alter air-fuel mixture with changes in altitude and air density. (Ford)

Carburetor vacuum connections

A modern carburetor has numerous vacuum connections. When the vacuum connections or port is BELOW the carburetor throttle plate, the port ALWAYS receives full intake manifold vacuum. However, when the vacuum port is ABOVE the throttle plate, vacuum is only present at the port when the THROTTLE IS OPENED.

Typical components operated off carburetor vacuum connections are:

1. EGR VALVE (exhaust emission control device).
2. DISTRIBUTOR VACUUM ADVANCE (diaphragm for advancing ignition timing).
3. CHARCOAL CANISTER (emission control container for storing fuel vapors).
4. CHOKE BREAK (diaphragm for partially opening choke when engine is running).

TWO AND FOUR-BARREL CARBURETORS

So far, this chapter has discussed mainly the *one-barrel carburetor* (carburetor body with one air horn). Automotive carburetors also are available in *two-barrel* (two air horns in single body) and *four-barrel* (four air horns in one body) designs.

Multiple barrel carburetors are used to provide increased "engine breathing" (air intake). The amount of fuel and air that enters the engine is a factor limiting engine horsepower.

In-line, 2-barrel, and all 4-barrel carbuaretors are divided into two sections: the primary and the secondary, Figs. 16-19 and 16-20.

The *primary* of a carburetor includes the components that operate under normal driving conditions. In a 4-barrel carburetor for example, it consists of the two front throttle plates and related components.

The *secondary* of a carburetor consists of the components or circuits that function under high engine power output conditions.

COMPUTER CONTROLLED CARBURETORS

A *computer controlled carburetor* normally use a solenoid operated carburetor valve to respond to commands from a microcomputer (electronic control unit). The system uses various sensors to send information to the computer. Then, the computer calculates how to set the carburetor's air-fuel mixture, Fig. 16-23.

A computer controlled carburetor system, sometimes termed a computer controlled emission system, consists of:

An *electromechanical carburetor* is a carburetor having both electrical and mechanical control devices. It is commonly used with a computer control system.

A *mixture control solenoid* in the computer controlled carburetor alters the air-fuel ratio. Electrical signals from the computer activate the solenoid to open and close air and fuel passages in the carburetor, Fig. 16-24.

The system *computer,* also called the *electronic control unit* (ECU), uses sensor information to operate the carburetor's mixture control solenoid.

An *idle speed actuator* may also be used to allow the computer to change engine idle speed. It is usually a tiny electric motor and gear mechanism that holds the car-

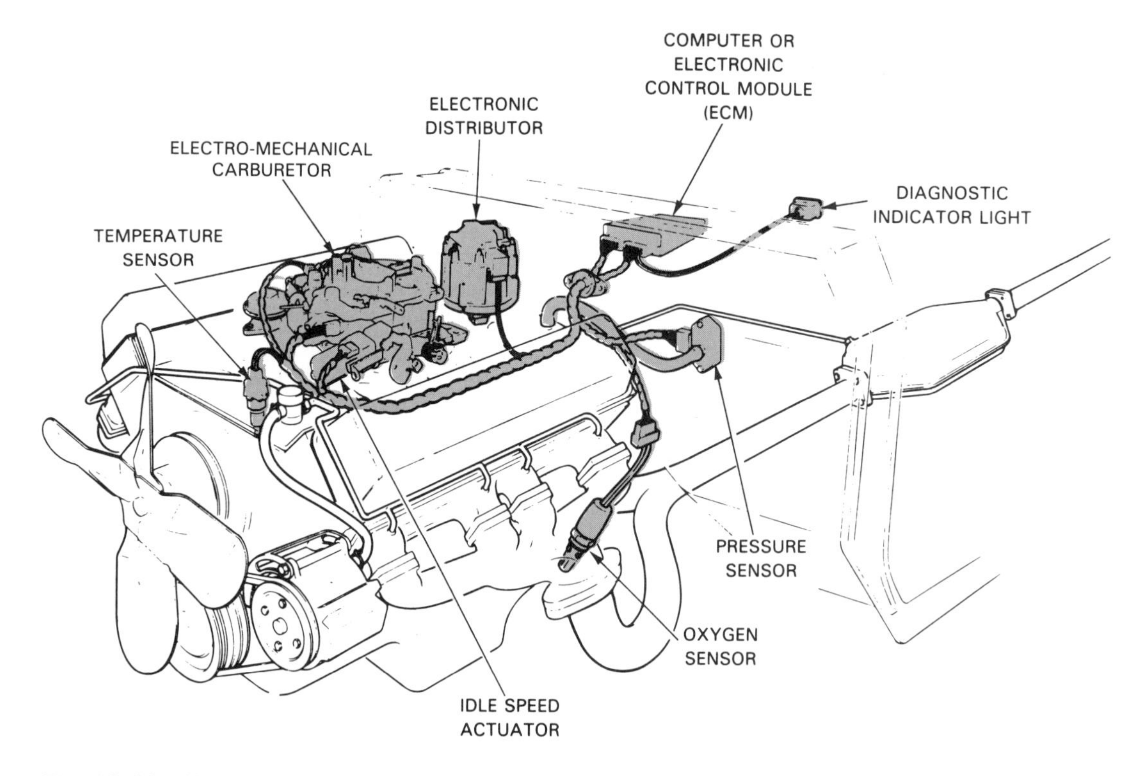

Fig. 16-23. Note basic parts and locations of computerized carburetor system. (General Motors)

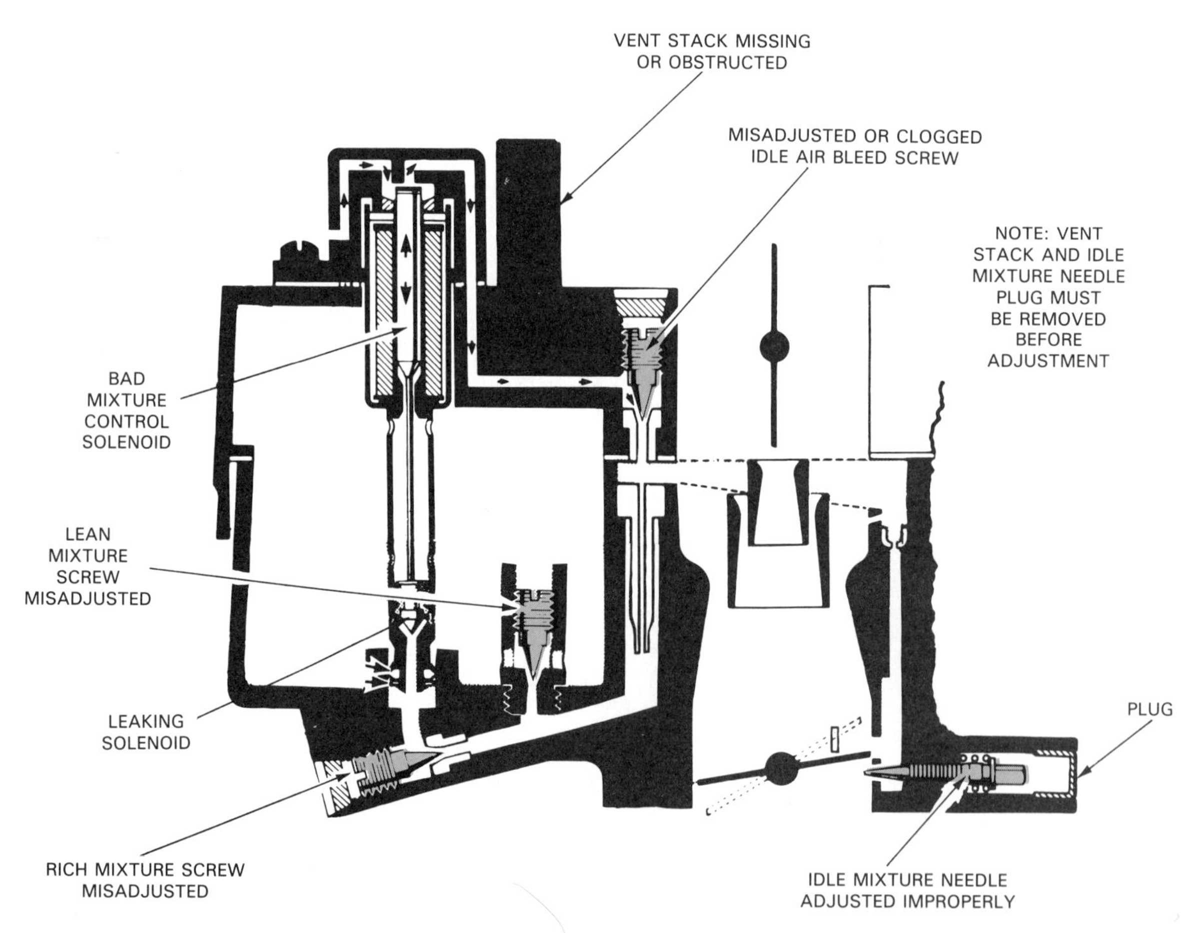

Fig. 16-24. Computer controlled carburetor uses mixture control solenoid. Solenoid reacts to computer signals to open or restrict fuel and air passages in carburetor. This allows carburetor to provide more efficient air-fuel mix for different engine conditions. (American Motors)

buretor throttle lever in the desired position.

An *oxygen sensor* or *exhaust gas sensor* monitors the oxygen content in the engine exhaust. The amount of oxygen in the exhaust indicates the richness (low oxygen content) or leanness (high oxygen content) of the air-fuel mixture. The sensor's voltage output changes with any change in oxygen content in the exhaust, Fig. 16-23.

The *temperature sensor* detects the operating temperatures of the engine. Its resistance changes with the temperature. This allows the computer to enrich the fuel mixture during cold engine operation.

A *pressure sensor* measures intake manifold vacuum and engine load. High engine load or power output causes intake manifold vacuum to drop. The pressure sensor, in this way, can signal the computer with a change in resistance and current flow.

Other sensors may also be used. These sensors are discussed in the chapter on fuel injection.

Open and closed loop

The term *open loop* means that the system is operating on present values in the computer. For example, right after cold engine starting, the computer will operate open loop. This is because the oxygen sensor CANNOT provide accurate information.

Closed loop means that the computer is using information from the oxygen sensor and other sensors. The information forms an imaginary loop (circle) from the computer, through the engine fuel system, into the exhaust, and back to the computer through the oxygen sensor information.

A computer controlled carburetor system normally operates in the closed loop mode. Using ''info'' from the engine sensors, a very precise air-fuel ratio can be maintained.

Computer controlled carburetor operation

In a computer controlled carburetor, the air-fuel ratio is maintained by *cycling* the mixture solenoid ON and OFF several times a second. Fig. 16-25 shows how the control signals from the computer can be used to meter different amounts of fuel out of the carburetor.

When the computer sends a rich command to the solenoid, the signal voltage to the mixture solenoid is

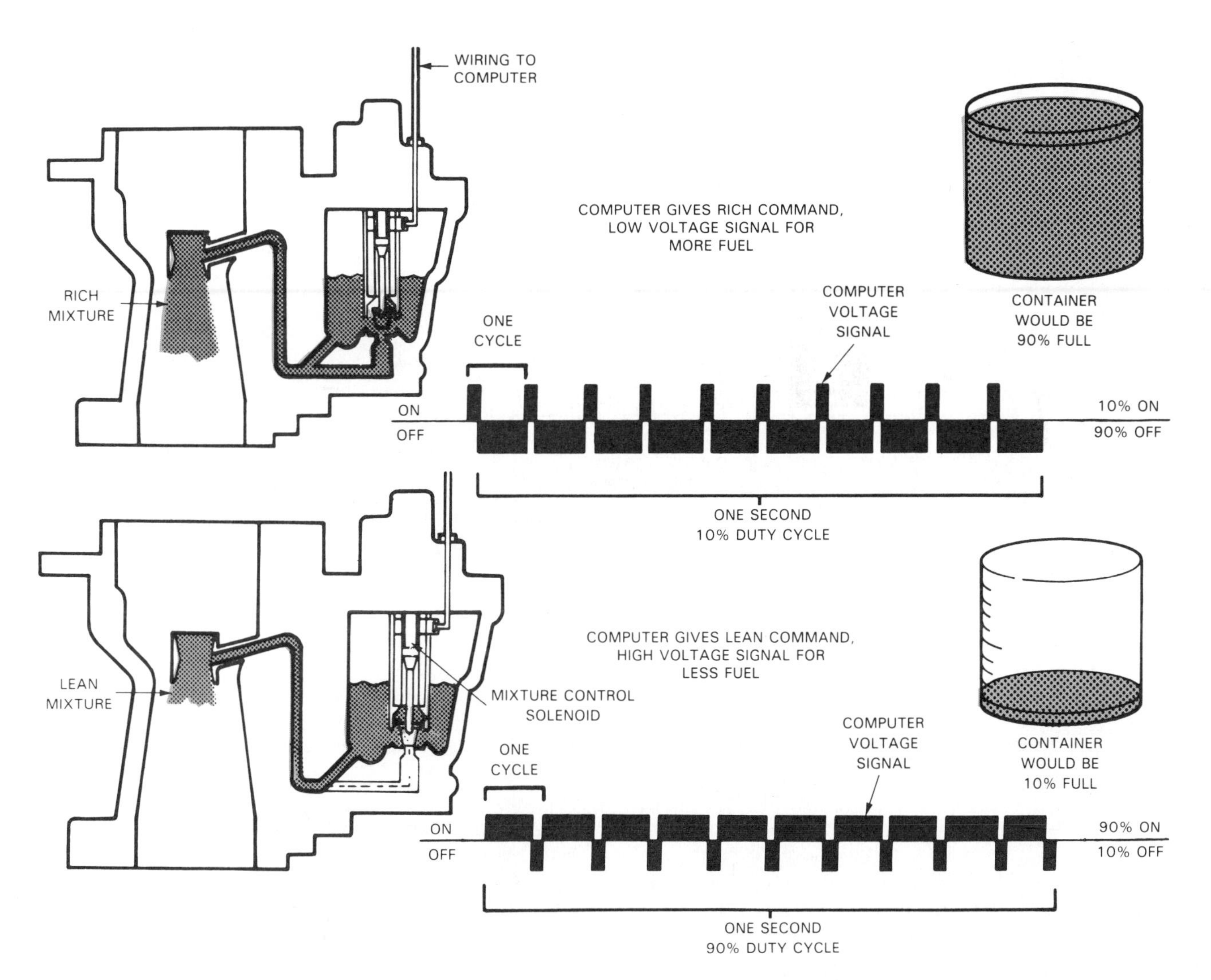

Fig. 16-25. Computer sends electrical pulses to carburetor mixture control solenoid to alter fuel ratio. Note difference in rich and lean commands from computer for this carburetor. (Chrysler)

usually OFF more than it is ON. This causes the solenoid to stay open more. During a lean signal from the computer, the signal usually has more ON time. This causes less fuel to pass through the solenoid valve.

As you will learn in later chapters, a computer controlled carburetor system is very similar to ELECTRONIC FUEL INJECTION. Both systems commonly use engine sensors, a computer, and solenoid action to control the amount of fuel entering the engine.

CARBURETOR SERVICE

This part of the chapter summarizes the more important information needed to service carburetors. A carburetor is vital to efficient engine operation.

Inspecting carburetors

A visual inspection of the carburetor may provide clues to the carburetor problem. Remove the air cleaner. Look for fuel leakage, sticking choke, binding linkage, missing or disconnected vacuum hoses, or any other troubles.

A heavy covering of road dirt usually indicates the carburetor has been in service for a long time. Adjustments or repairs may be needed.

While inspecting the carburetor, also check the rest of the engine compartment. Look for disconnected wires and hoses. Listen for the HISSING SOUND of a vacuum leak. Try to locate anything that could upset engine operation.

Incorrect air-fuel mixture

Many internal carburetor problems show up as an air-fuel mixture that is too rich or too lean.

A *lean air-fuel mixture* is caused by any condition that allows too much air and too little fuel to enter the engine. A lean mixture will cause the engine to MISS ERRATICALLY (every once in a while). The cause may be a vacuum leak, incorrect mixture screw adjustment, clogged fuel passage, low float level, or other problems.

A *rich air-fuel mixture* results from too much fuel entering the engine. A very rich mixture will make the engine roll, lope, and emit black smoke. A clogged air bleed, high float level, incorrect choke setting, or other carburetor troubles can produce a rich mixture.

Carburetor problem diagnosis

When diagnosing carburetor problems, try to determine which carburetor system is at fault. For example, if the engine only runs poorly when cold, check the choke and fast idle systems.

If the engine only misses at idle, suspect the idle or low speed carburetor systems and check them first. Use this type of logic to narrow down the possible sources.

Use a factory or service manual diagnosis chart when you have difficulty locating a carburetor problem. Such charts are designed for each type of carburetor.

CARBURETOR REMOVAL

Carburetor removal is needed during engine repairs or carburetor service. Begin by removing the air cleaner and all wires, hoses, and linkages that would prevent removal of the carburetor.

Fig. 16-26. When removing modern carburetor, it is wise to label all wires and vacuum hoses. This will prevent confusion when reinstalling carburetor. (Fel-Pro)

You may want to label the vacuum hoses and wires that connect to the carburetor. Late model cars have a large number of hoses and wires. Refer to Fig. 16-26.

Remove the four nuts or bolts that secure the carburetor to the engine intake manifold, Fig. 16-27. Carefully lift the carburetor off the engine while holding it level. Do not splash the fuel in the bowl. This would stir up any dirt in the bottom of the bowl. To prevent damage to the throttle plates, mount the carburetor on a stand.

Carburetor rebuild

A carburetor rebuild is needed when carburetor passages become clogged, gaskets or seals leak, rubber parts deteriorate, or components wear and fail.

A *carburetor rebuild* generally involves:

1. Disassembly and cleaning of major parts.
2. Inspection for wear or damage.
3. Installation of a CARBURETOR REBUILD KIT (new gaskets, seals, needle valve and seat, pump diaphragm or cup, and other nonmetal parts).
4. Reassembly of the carburetor.
5. Major adjustments.

Carburetor disassembly

When disassembling a carburetor, follow the detailed directions in a service manual. Generally, you must remove ALL PLASTIC and RUBBER parts. Also, remove any part that will prevent thorough cleaning of the carburetor passages. If not worn or damaged, the throttle plates and shaft can be left in place, Fig. 6-28.

To prevent part damage, always use proper diassembly techniques. When unscrewing jets, use a special jet tool or the correct size screwdriver.

Keep parts organized and note the location of each part. Jet sizes are sometimes different and must be reinstalled in the same location.

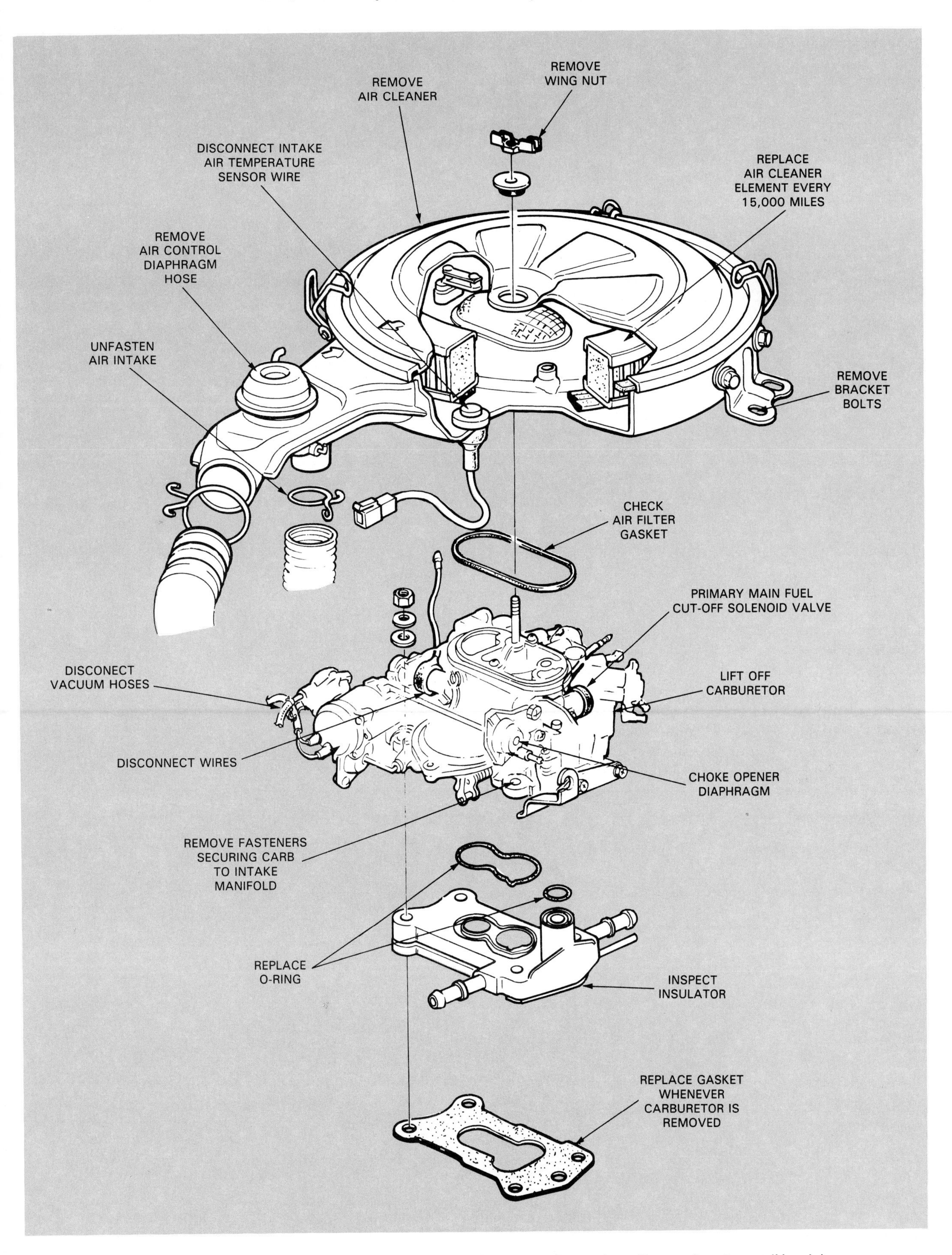

Fig. 16-27. Note parts that must be serviced when removing or installing carburetor. (Honda)

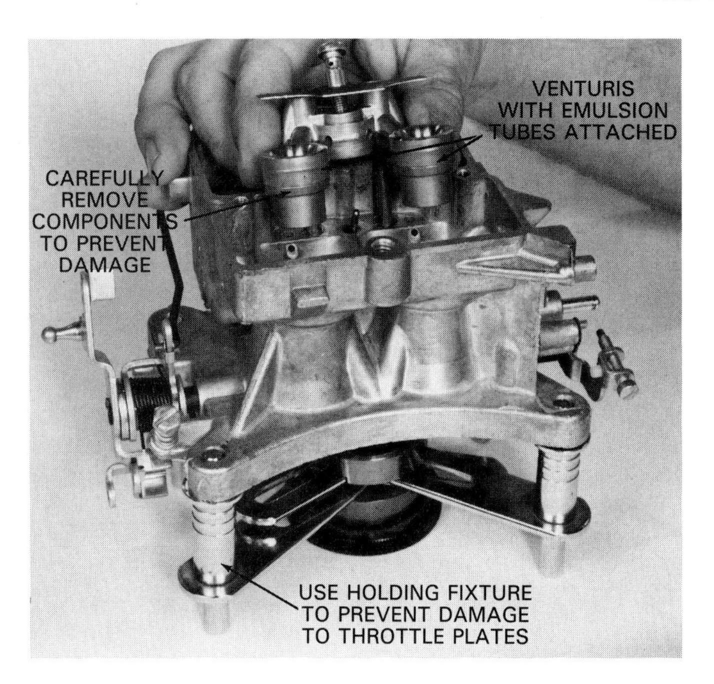

Fig. 16-28. Take your time when disassembling a carburetor. Note how all parts fit together and be careful not to damage delicate components. Note use of stand to protect throttle plates from damage.

Cleaning carburetor parts

Carburetor cleaner, also known as decarbonizing cleaner, removes deposits from a carburetor. A very powerful cleaning agent, it will remove gum, carbon, oil, grease, and other deposits inside air and fuel passages and on external parts.

Do not allow carburetor cleaner to contact plastic and rubber parts. Carburetor cleaner will dissolve and ruin nonmetal parts. Place metal components in a tray and lower them into the cleaner. Allow the carburetor parts to soak for the recommended time, Fig. 16-29.

WARNING! Carburetor cleaner can cause serious chemical burns to the skin and eyes. Wear rubber gloves and eye protection when working with carburetor cleaning solution.

After soaking, rinse all carburetor parts with water, clean kerosene, or clean cold soak cleaning solution. Then, dry the parts with compressed air. FORCE AIR through all passages to make sure they are clear.

NOTE! Do NOT use wire or a drill to clean out carburetor passages. This could scratch and enlarge the passages, upsetting carburetor operation.

Inspecting carburetor parts

1. Inspect for wear and excessive play between the throttle shaft and throttle body.
2. Check for binding of the choke plate and linkage.
3. Inspect carburetor bodies for warpage, cracks, and other problems.
4. Check float for leakage or bent hinge arm.
5. Inspect all gasket mating surfaces for nicks, burrs, and cleanliness.
6. Inspect for damaged or weakened springs and stripped fasteners.
7. Check pointed ends of idle mixture screws for damage.

Replace parts that show wear or damage. Also, discard all of the old parts included in your carburetor rebuild kit.

Carburetor reassembly

If not experienced, follow the directions in a service manual to reassemble a modern carburetor. There can be as many as 20 different carburetors for just one model year. This results in thousands of different reassembly procedures and adjustments. Each is critical to proper engine performance.

Basically, the carburetor is assembled in reverse order of disassembly. The manual will give procedures for installing each part. An exploded view of the carburetor shows how parts fit together. One is given in Fig. 16-30.

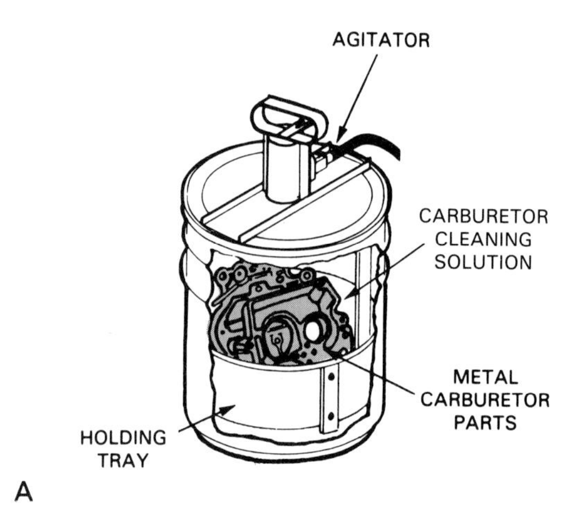

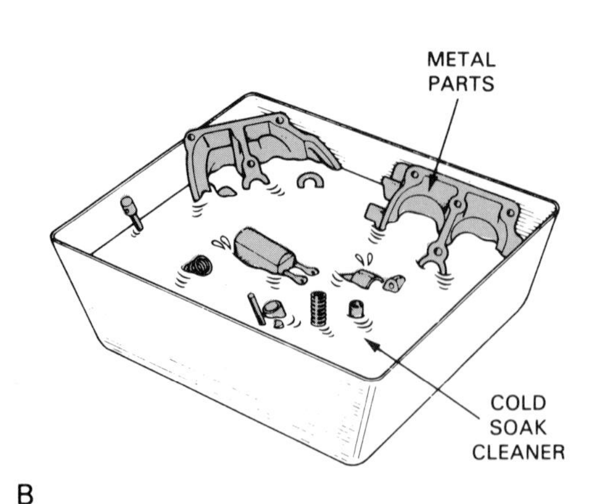

Fig. 16-29. After completely disassembled, all metal parts of carburetor should be cleaned. A—Submerse metal parts in decarbonizing cleaner, often called carburetor cleaner. It is very powerful so do not let cleaner splash into your eyes or get on your hands. Soak parts for a period recommended by solution manufacturer. B—After soaking in carburetor cleaner, wash parts in cold soak cleaner and blow dry. (Tomco and Toyota)

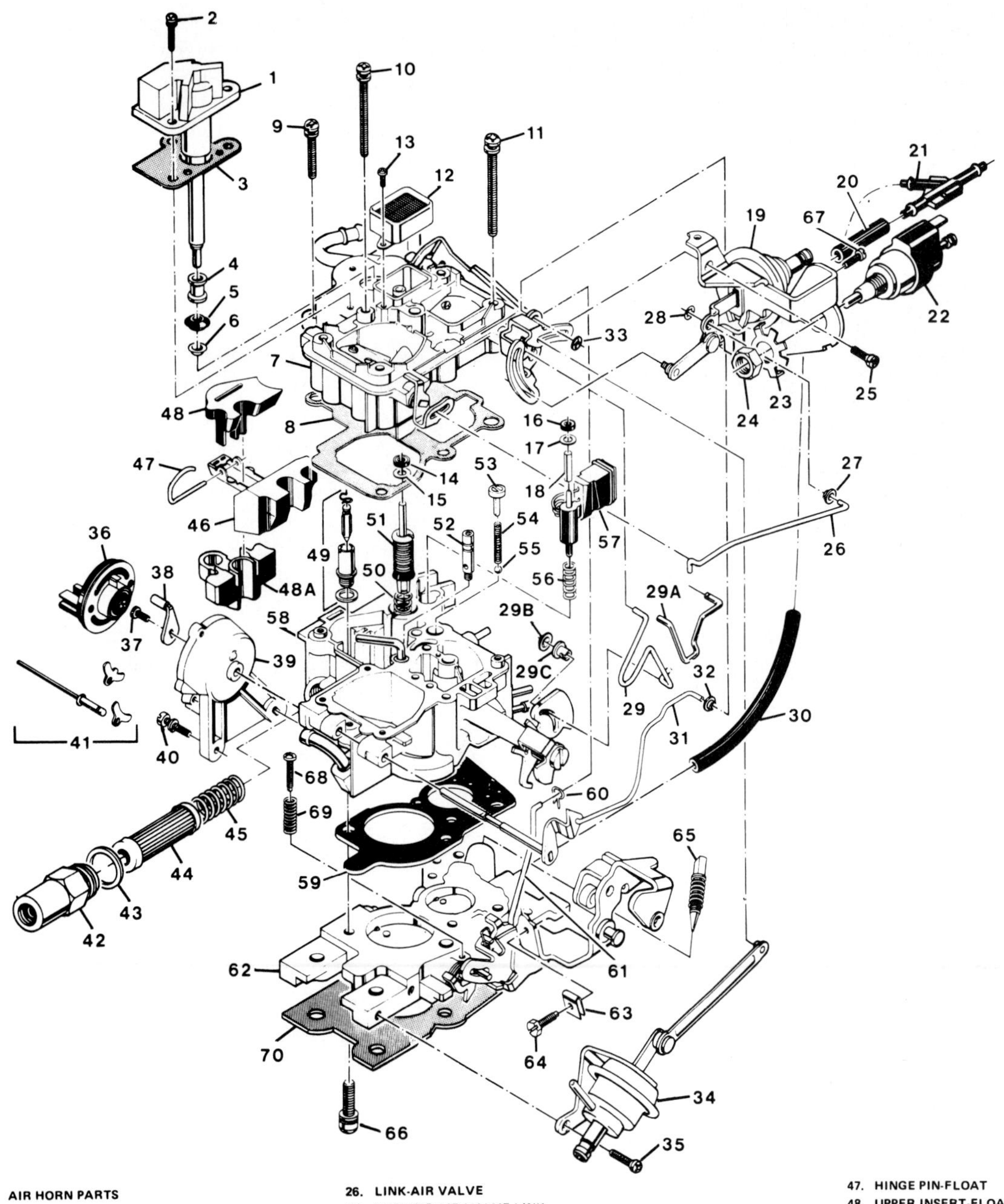

AIR HORN PARTS

1. MIXTURE CONTROL (M/C) SOLENOID
2. SCREW ASSEMBLY-SOLENOID ATTACHING
3. GASKET-M/C SOLENOID TO AIR HORN
4. SPACER-M/C SOLENOID
5. SEAL-M/C SOLENOID TO FLOAT BOWL
6. RETAINER-M/C SOLENOID SEAL
7. AIR HORN ASSEMBLY
8. GASKET-AIR HORN TO FLOAT BOWL
9. SCREW-AIR HORN TO FLOAT BOWL (SHORT)
10. SCREW-AIR HORN TO FLOAT BOWL (LONG)
11. SCREW-AIR HORN TO FLOAT BOWL (LARGE)
12. VENT STACK AND SCREEN ASSEMBLY
13. SCREW-VENT STACK ATTACHING
14. SEAL-PUMP STEM
15. RETAINER-PUMP STEM SEAL
16. SEAL-T.P.S. PLUNGER
17. RETAINER-T.P.S. PLUNGER SEAL
18. PLUNGER-T.P.S. ACTUATOR

CHOKE PARTS

19. VACUUM BREAK AND BRACKET ASSEMBLY- PRIMARY
20. HOSE-VACUUM BREAK PRIMARY
21. TEE-VACUUM BREAK
22. SOLENOID-IDLE SPEED
23. RETAINER-IDLE SPEED SOLENOID
24. NUT-IDLE SPEED SOLENOID ATTACHING
25. SCREW-VACUUM BREAK BRACKET ATTACHING
26. LINK-AIR VALVE
27. BUSHING-AIR VALVE LINK
28. RETAINER-AIR VALVE LINK
29. LINK-FAST IDLE CAM
29A LINK-FAST IDLE CAM
29B RETAINER-LINK
29C BUSHING-LINK
30. HOSE-VACUUM BREAK
31. INTERMEDIATE CHOKE SHAFT/LEVER/LINK ASSEMBLY
32. BUSHING-INTERMEDIATE CHOKE LINK
33. RETAINER-INTERMEDIATE CHOKE LINK
34. VACUUM BREAK AND LINK ASSEMBLY-SECONDARY
35. SCREW-VACUUM BREAK ATTACHING
36. ELECTRIC CHOKE-COVER AND COIL ASSEMBLY
37. SCREW-CHOKE LEVER ATTACHING
38. CHOKE COIL LEVER ASSEMBLY
39. CHOKE HOUSING
40. SCREW-CHOKE HOUSING ATTACHING
41. CHOKE COVER RETAINER KIT
67. SCREW-VACUUM BREAK BRACKET ATTACHING

FLOAT BOWL PARTS

42. NUT-FUEL INLET
43. GASKET-FUEL INLET NUT
44. FILTER-FUEL INLET
45. SPRING-FUEL FILTER
46. FLOAT AND LEVER ASSEMBLY
47. HINGE PIN-FLOAT
48. UPPER INSERT-FLOAT BOWL
48A LOWER INSERT-FLOAT BOWL
49. NEEDLE AND SEAT ASSEMBLY
50. SPRING-PUMP RETURN
51. PUMP PLUNGER ASSEMBLY
52. PRIMARY METERING JET ASSEMBLY
53. RETAINER-PUMP DISCHARGE BALL
54. SPRING-PUMP DISCHARGE
55. BALL-PUMP DISCHARGE
56. SPRING-T.P.S. ADJUSTING
57. SENSOR-THROTTLE POSITION (TPS)
58. FLOAT BOWL ASSEMBLY
59. GASKET-FLOAT BOWL

THROTTLE BODY PARTS

60. RETAINER-PUMP LINK
61. LINK-PUMP
62. THROTTLE BODY ASSEMBLY
63. CLIP-CAM SCREW
64. SCREW-FAST IDLE CAM
65. IDLE NEEDLE AND SPRING ASSEMBLY
66. SCREW-THROTTLE BODY TO FLOAT BOWL
68. SCREW-IDLE STOP
69. SPRING-IDLE STOP SCREW
70. GASKET-INSULATOR FLANGE

Fig. 16-30. Service manual will give exploded view of carb like this one. It will help you make sure everything is installed properly. (Buick)

Carburetor preinstallation adjustments

As you assemble the carburetor, several adjustments must be made. Adjustments typically include:

1. Float drop adjustment, Fig. 16-31.
2. Float level adjustment, Fig. 16-31.
3. Idle mixture screw "rough" adjustment, Fig. 16-32.
4. Choke linkage and spring adjustments, Fig. 16-33.
5. Choke spring adjustment, Fig. 16-34.
6. Anti-stall dashpot adjustment, Fig. 16-34.
7. Accelerator pump adjustment, Fig. 16-35.

There are several other adjustments, as you will learn, that must be made AFTER installing the carburetor on the engine.

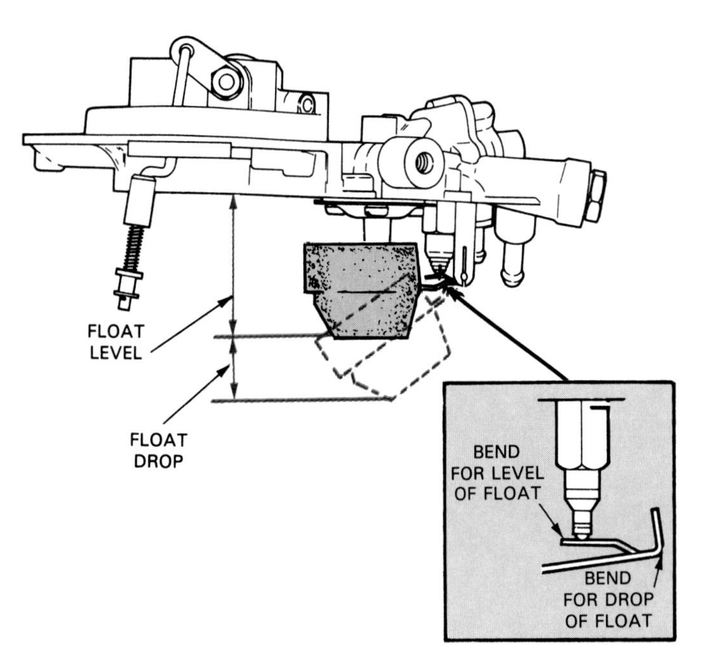

Fig. 16-31. During assembly, adjust float drop by measuring from float to specific location on carburetor. Bend float hinge tang as needed. You must also adjust float level. Float is in up position when measuring. Again, bend correct tang on hinge as needed. (Renault)

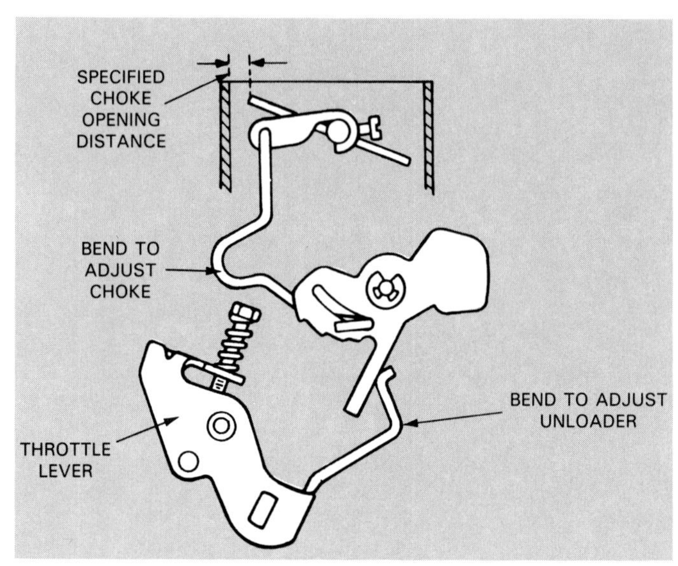

Fig. 16-33. Mechanical choke unloaded cracks choke plate open when throttle opens. This prevents engine flooding with fuel. It should be checked during carburetor service. (Ford)

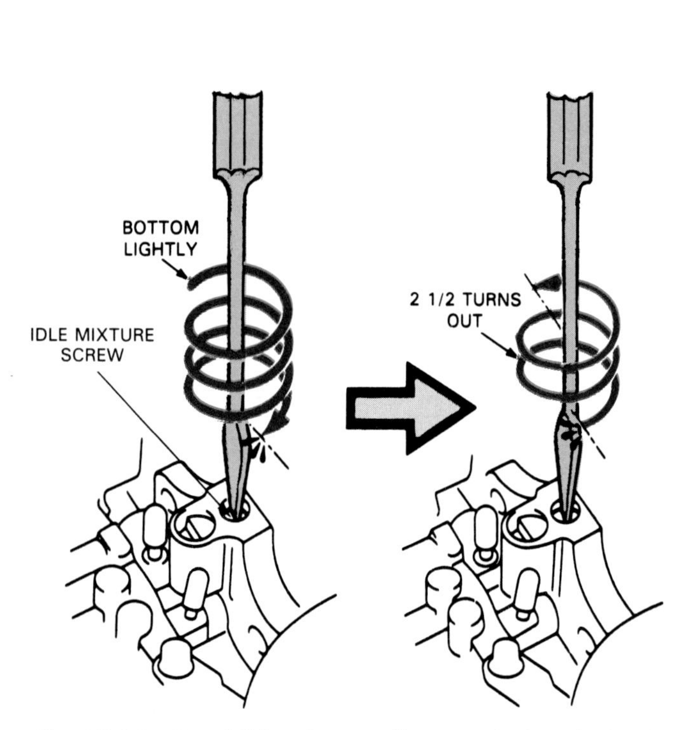

Fig. 16-32. Rough idle mixture adjustment is done before installing carb. Bottom screws and then back them out about 2 1/2 turns. Readjust mixture after installation. (Toyota)

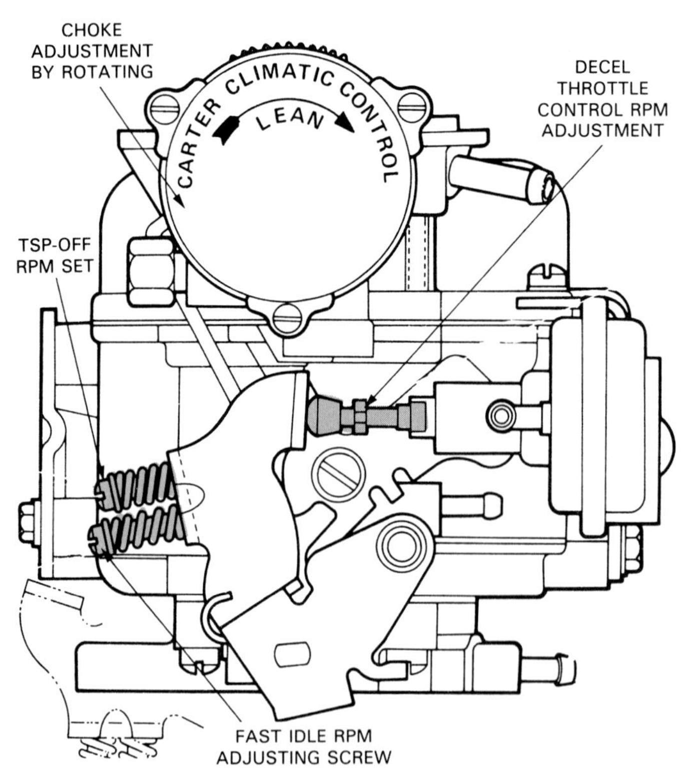

Fig. 16-34. Note speed adjustment screws and choke adjustment. Locations will vary with different carburetors. (Ford)

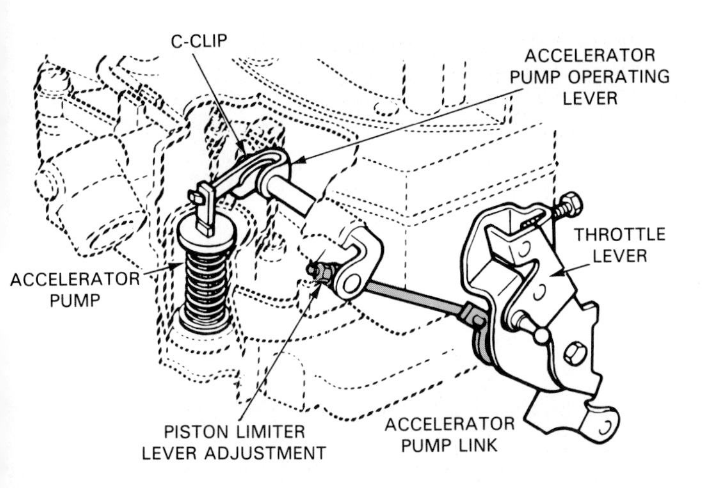

Fig. 16-35. Accelerator pump must be adjusted to spray fuel into air horn as soon as throttle begins to open. Adjust by threaded nut, bending linkage, moving linkage to different hole, etc. (Ford)

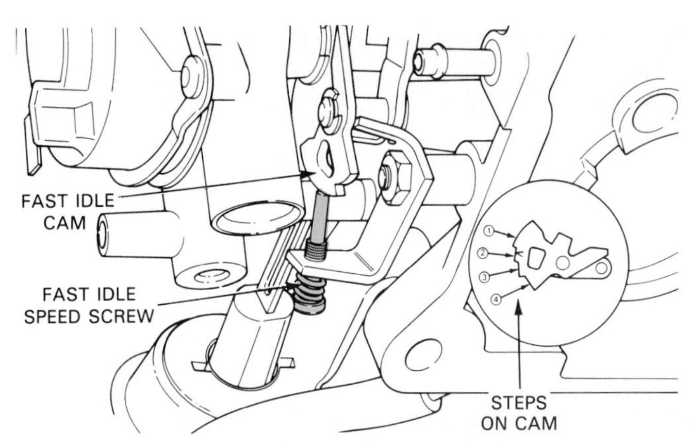

Fig. 16-36. Fast idle adjustment is sometimes done with cam mechanism. Position fast idle screw on correct cam step. Then use tach to measure engine rpm while turning screw.

CARBURETOR INSTALLATION

Place a new base plate gasket on the intake manifold. Then, fit the carburetor over the mounting studs, if used. Install and properly torque the fasteners that secure the carburetor to the intake manifold, Fig. 16-27.

WARNING! Do NOT overtighten the carburetor hold-down nuts or bolts. Overtightening can easily snap off the carburetor base flange.

Reinstall all of the hoses, lines, wires, and linkage rods or cables. Double-check that all of the vacuum hoses have been installed correctly. Hand operate the throttle to make sure the throttle plates are not binding or hitting on the base plate gasket.

COLD IDLE SPEED ADJUSTMENT

To set the cold fast idle or fast idle cam, connect a tachometer to the engine. Warm the engine to operating temperature. Set the emergency brake and block the wheels of the car.

While following service manual instructions, set the fast idle cam to hold the throttle open for a fast idle. The fast idle screw must contact a specified step on the fast idle cam, Fig. 16-36.

Turn the fast idle screw until the tachometer reads within specifications. Cold fast idle speeds vary from approximately 750 to 950 rpm. Some specs are given with the automatic transmission in drive. This will reduce the cold fast idle value.

HOT IDLE SPEED ADJUSTMENT

Many carburetors use a solenoid to control the hot idle speed of the engine. When the engine is running, the solenoid acts on the throttle lever to hold the throttle open for the correct idle speed.

Sometimes, a solenoid is used to maintain idle speed when the air conditioner is turned on. In any case, you must use factory recommendations to adjust the solenoid.

To set hot idle speed, turn the solenoid adjusting screw until the tachometer reads correctly, Fig. 16-37. Sometimes the adjustment is on the solenoid mounting bracket or it may be on the solenoid plunger itself.

Typically, hot idle speed is from about 650 to 850 rpm. It is lower than cold idle speed but higher than curb idle speed.

CURB IDLE SPEED ADJUSTMENT

Curb idle speed is the lowest idle speed setting. On older carburetors, it is the idle speed setting that controls engine speed under normal, warm engine conditions, Fig. 16-37.

On late model carburetors using a fast idle solenoid, curb idle can also be termed *idle drop* or *low idle speed adjustment.* In this case, it is a very low idle speed that only occurs when the engine is shut off. When the solenoid is deenergized, the throttle plates drop to the curb idle setting to keep the engine from dieseling (engine running with ignition key off).

To adjust curb idle, unplug the wires going to the idle

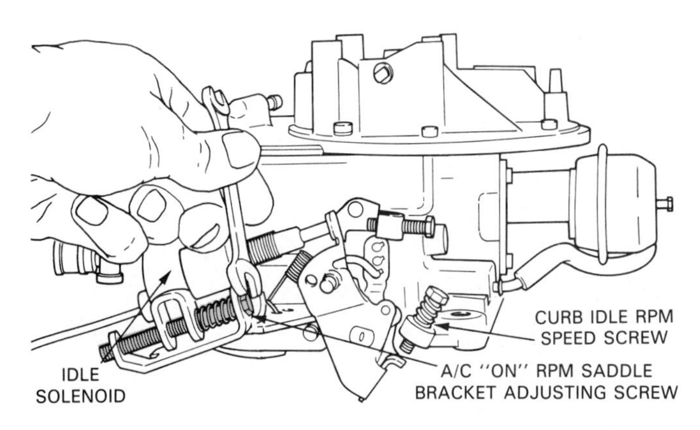

Fig. 16-37. Note how solenoid is moved into or away from throttle lever to change hot idle speed. Curb idle screw is set with solenoid deenergized.

speed solenoid. Then, turn the curb idle speed screw until the tachometer reads within specifications.

Curb idle speed is usually very low, approximately 500 to 750 rpm.

Idle mixture adjustment

After setting idle speed, you may also need to adjust the carburetor idle mixture. There are several methods for doing this.

With older, pre-emission carburetors, the idle mixture screw is turned in and out until the smoothest idle mixture is obtained. Basically, the mixture screw is turned in until the engine misses (lean miss). Then, it is turned out until the engine rolls (rich miss). This mixture screw is then set halfway between the lean miss and rich roll settings. This is done to both mixture screws if needed (two or four-barrel carburetor).

After setting the mixture, check and adjust the idle speed as needed. If you must reset idle speed, then readjust the idle mixture again.

Propane idle mixture adjustment is required on some later model carburetors. A bottle of propane is connected to the air cleaner or intake manifold. The propane enriches the fuel mixture during carburetor adjustment, Fig. 16-38. Following service manual procedures, you must meter propane into the engine while adjusting the mixture screws. This will provide a leaner (less exhaust emissions) setting.

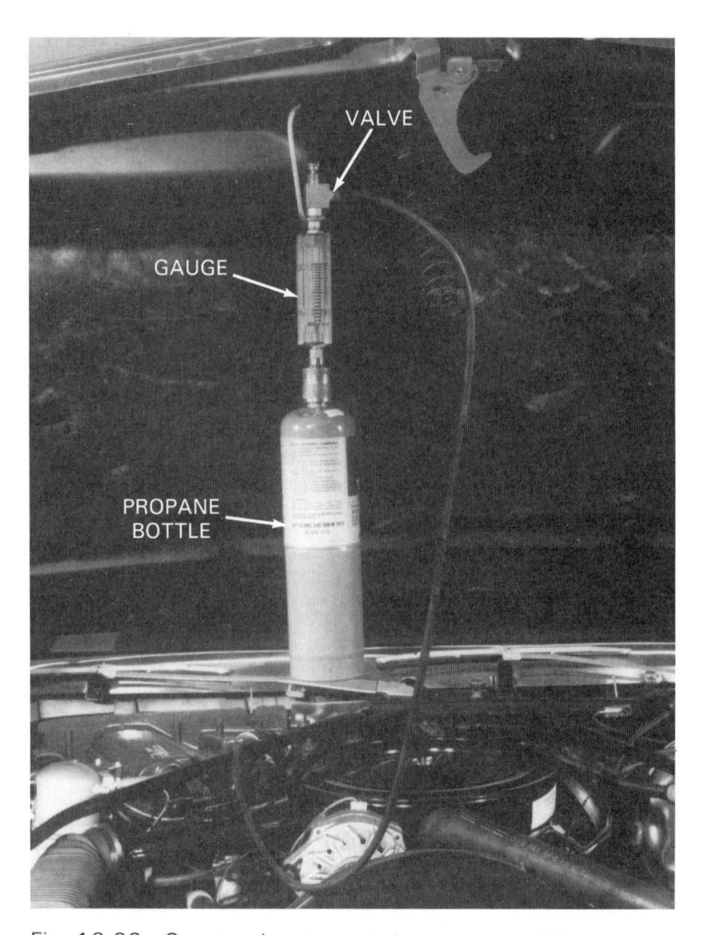

Fig. 16-38. Some carburetors require propane enrichment during idle mixture adjustment. Propane richens mixture so when bottle is disconnected from engine, adjustment will provide clean burning, low emissions setting of idle mixture screws. Check service manual for exact procedures. (General Motors)

WARNING! Most late model carburetors have *sealed idle mixture screws* preset at the factory. Unless major carburetor repairs have been made, do NOT remove and tamper with these screws. To avoid a violation of federal law, follow manufacturer's prescribed procedures when making carburetor adjustments.

An *exhaust gas analyzer* may be used to adjust the idle mixture on late model cars. The mixture screws are adjusted until the exhaust sampling is within specified limits.

COMPUTER CONTROLLED CARBURETOR SERVICE

A computer controlled carburetor system requires specialized service techniques. Most systems have a self-diagnostic mode that allows easy location of system problems. Some computer systems flash a light on and off in a Morse (number) type code. The technician can use the code chart in the service manual to pinpoint the system trouble.

Computer contolled carburetor problems

Computer controlled carburetors can suffer the same problems as conventional carburetors. Gaskets can split and leak. Seals can harden and split. Passages become clogged.

Before suspecting the computer system or any of its components, check all other possible causes. This includes the ignition system (ignition timing, spark plug wires, spark plugs, pick-up coil, etc.). Also check the operation of the air cleaner door and all emission controls (EGR system, evaporative system, PCV system, EFE system, etc.) as well as engine compression (piston rings and valves). Also, inspect for vacuum leaks at the intake manifold gasket, vacuum hoses, and carburetor. A malfunction of any one of these parts could upset engine operation and appear to be a computer system problem.

If none of these checks reveal the cause of the performance problem, the computer control system may be defective and should be tested. The following symptoms could indicate a possible computer system problem:

1. Detonation.
2. Poor idling.
3. Missing.
4. Hesitation.
5. Surging.
6. Poor gas mileage.
7. Sluggish engine.
8. Hard starting.
9. Exhaust odor or smoke.
10. Engine cut-out.
11. Engine (computer) light flashing.

Computer controlled carburetor system testing

In most cases, a built-in diagnostic system catches and indicates computer system problems. When a fault exists in the computer network, a "check engine" light will glow in the instrument panel. This warns the driver and mechanic of a possible problem.

Reading trouble codes

To isolate which part of the computer system is bad, connect a scan tool to, or ground the proper terminals on, the data link connector. Grounding the connector will cause the "check engine" light to flash a code. On other systems, an analog (needle type) voltmeter can be connected to a specified point. Each pulse (sweep) of the voltmeter needle would signal the numbers of the code.

For example, the light or voltmeter may flash or sweep two times and then pause, indicating the number two. Then, after a few seconds, it may flash or sweep three times to indicate the number three. This would mean a trouble code of 23 (2, pause, 3) as in Fig. 16-39.

The technician can then pinpoint the trouble by referring to a code chart in the service manual. For detailed information, refer to a service manual for the computer system being checked. Systems codes and test procedures vary.

Other tests

A system performance test is used to find problems not included in the system code or diagnostic warning system. The check is needed when the warning light does not indicate a system problem and no cause for a problem can be found. It is the last test to be made when an engine performance problem exists. The service manual will give step-by-step directions.

Measuring dwell signal to carburetor

Several computer controlled carburetor systems require a dwell meter and tachometer to check the computer output. The dwell meter measures the duration of the electrical pulses to the carburetor mixture control solenoid. The tach measures engine rpm.

Generally, connect the dwell meter and tach as shown in Fig. 16-40. The dwell meter connects to a specified connector in the wiring harness. The tach connects to the ignition system.

Fig. 16-41 shows the relationship between the dwell meter reading and the carburetor's air-fuel mixture.

When testing the system, compare dwell meter readings to specs. If the dwell is too high or low for engine conditions, there is a problem in the computer system. Normally, the dwell reading will rise and fall within a certain range (around 35 degress plus or minus 5 degrees) with the engine at full operating temperature and at a constant speed.

Interpreting dwell signal from computer

If the computer dwell signal is not within specs, use common sense, basic testing methods, and a service manual to find the problem.

Your service manual will normally give a logical step-by-step procedure for isolating specific problems using a digital VOM or special testers, Figs. 16-41 and 16-42.

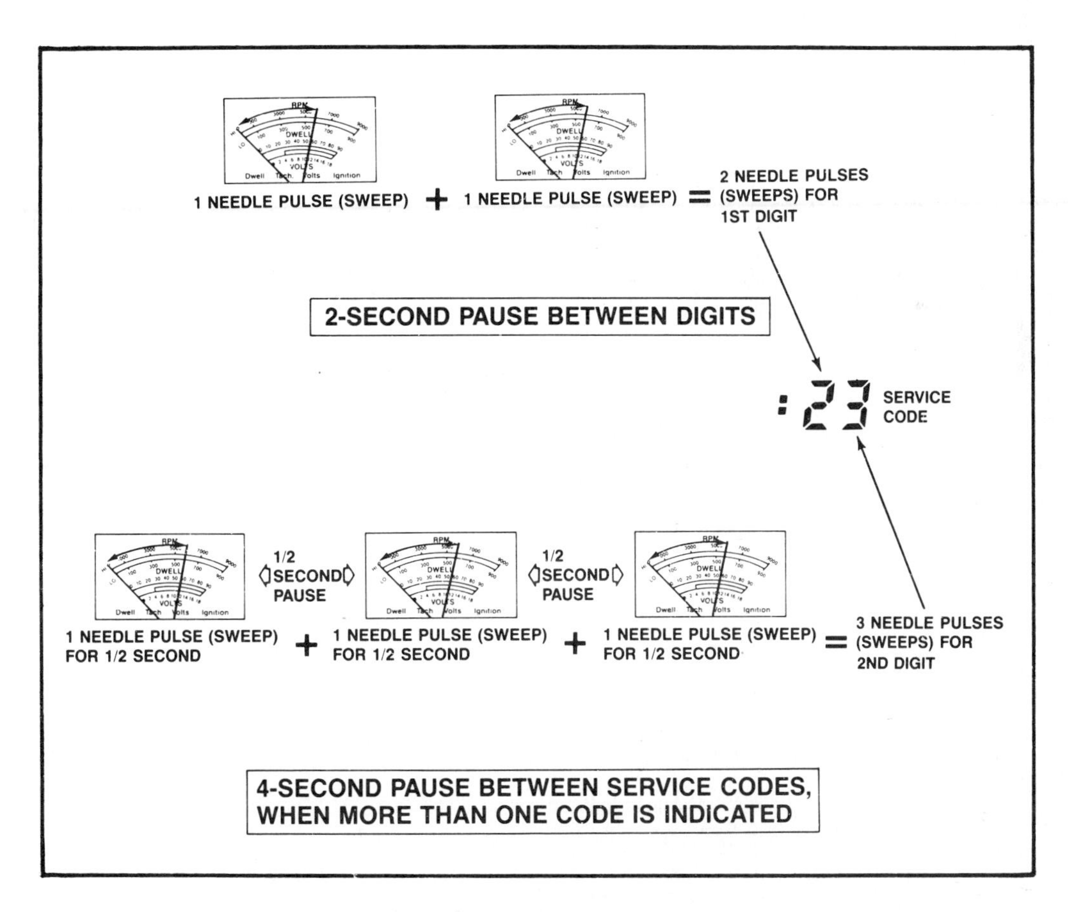

Fig. 16-39. Analog or needle type voltmeter is needed to read the self-diagnosis code of many computer controlled carburetor systems. When activated by jumper wire, computer will send voltage pulses to test socket and meter. Two meter deflections, a short pause, and three more deflections would represent a twenty-three code. Your manual will tell what problem twenty-three represents. (Ford)

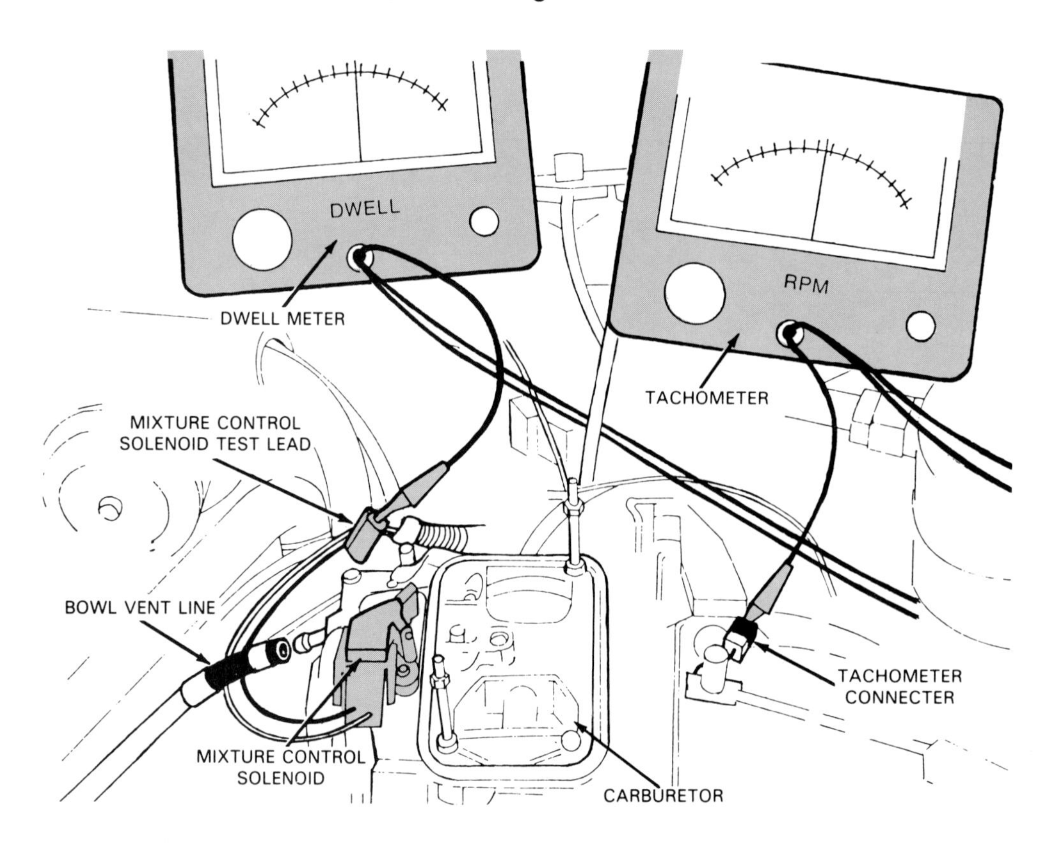

Fig. 16-40. Note basic connections for using a dwell meter to check computer signals to carburetor. Dwell meter will measure pulse width going to mixture control solenoid. This will help you determine whether there is a problem in the control system, sensor system, or carburetor itself.

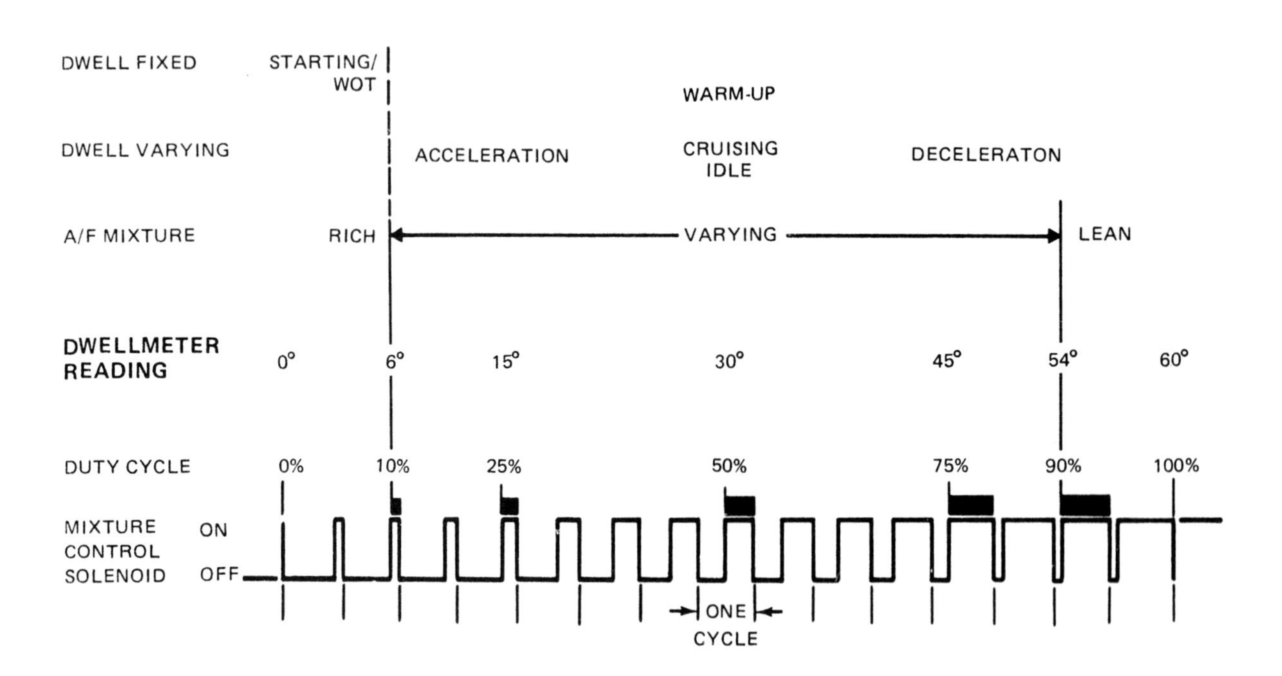

Fig. 16-41. Chart shows relationship between dwell and mixture control solenoid operation. Note how low dwell richens mixture and high dwell leans mixture. (Buick)

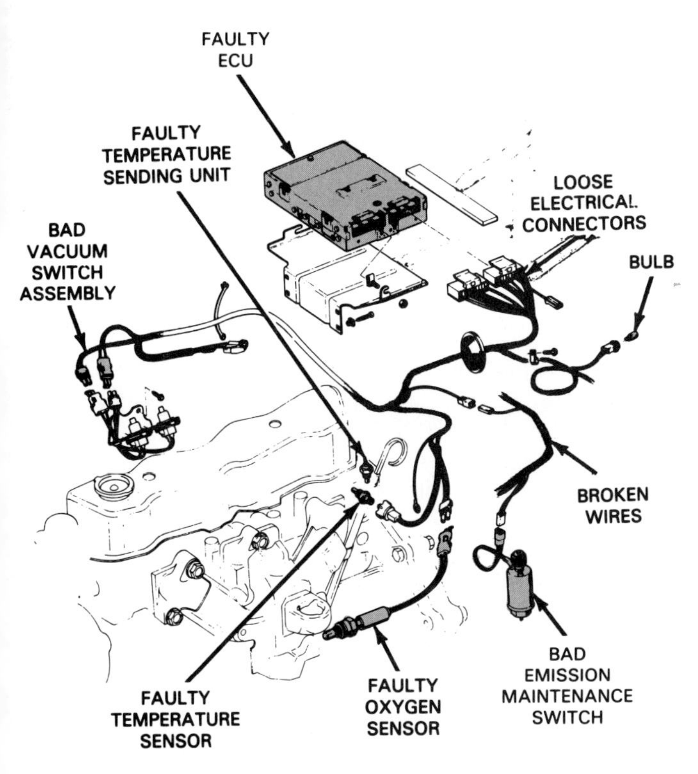

Fig. 16-42. Study types of problems that can occur in a computer controlled carburetor system.

SUMMARY

An engine relies on clean air and fuel. Unfiltered air can cause rapid part wear. Unfiltered fuel can clog injectors, carburetor passages, and cause numerous troubles.

Double-wall steel fuel lines and synthetic rubber hoses carry fuel from the tank to the engine. Fuel filters can be located in the tank, in-line, at the carburetor, or strainers can be inside the injectors.

Either a mechanical or electric fuel pump can force fuel through the lines and to the engine. A mechanical pump is driven by an eccentric inside the engine. This type is commonly found with carburetor systems. An electric pump is more modern and can be found on all fuel system types.

Always check fuel filters for clogging before testing fuel pump output. Measure fuel pump pressure and volume to check its performance. Normally, replace a pump if it fails a performance test.

A carburetor is a mixing device for air and fuel. It must alter the mixture ratio with changes in engine operation. The seven carburetor systems are the float, idle, off-idle, acceleration, high speed, full power, and choke systems. You should be able to visualize the operation of each to be good at carburetor troubleshooting.

A computer controlled carburetor uses engine sensors, an electronic control unit, and electro-mechanical carburetor. The computer can alter carburetor mixture output very quickly to increase combustion efficiency.

When trouble occurs, always inspect the carburetor for problems: fuel leaks, vacuum leaks, missing wires, etc. When removing a carburetor, label wires and hoses. Remove the linkage and four fasteners that secure the carb to the intake manifold.

Mount a carburetor on a repair stand to prevent damage to the throttle plates. During a carburetor rebuild, remove all rubber and plastic components. Then you can soak the metal parts in carburetor cleaner. The rebuild kit will have new gaskets, seals, etc. They must be installed as described in the service manual. Adjustments that must be made during assembly will also be detailed in the manual.

After installing the carburetor, adjust fast idle, hot idle, and idle mixture. Do this by connecting a tach to the engine to measure rpm. Adjust engine speeds to within specifications.

Computerized carburetor systems usually produce a trouble code to help locate problems. A dash light will glow as a sign of trouble. Then the computer can be activated to produce a trouble code. Either the dash light will flash the code or you may have to connect an analog voltmeter to the system to read the code. The service manual will give a chart showing what each number code means.

KNOW THESE TERMS

Fuel supply system, Air supply system, Fuel metering system, Fuel tank capacity, Tank pickup unit, Fuel return system, Mechanical pump, Electric pump, Fuel pump pressure, Fuel pump volume, Fuel pump vacuum, Carburetor, Air horn, Throttle valve, Venturi, Main discharge, Fuel bowl, Float, Carburetor circuit, Air-fuel ratio, Float system, Idle system, Off-idle system, Acceleration system, High speed system, Full power system. Choke system, Bowl vent, Idle mixture screw, Check ball, Air bleed, Choke unloader, Fast idle cam, Fast idle solenoid, Throttle dashpot, Idle speed screw, Two-barrel, Four-barrel, Primary, Secondary, Electro-mechanical carburetor, Mixture control solenoid, Electronic control unit, Idle speed actuator, Oxygen sensor, Open loop, Closed loop, Lean mixture, Rich mixture, Carburetor rebuild, Carburetor cleaner, Float drop, Float level, Idle mixture adjustment, Accelerator pump adjustment, Cold idle adjustment, Hot idle adjustment, Curb or idle drop adjustment, Propane enrichment, Computer self-diagnosis, Trouble code, Computer dwell signal.

REVIEW QUESTIONS—CHAPTER 16

1. A ________ ________ ________ helps cool the fuel going to the engine and prevents ________ ________.
2. A mechanical fuel pump is normally powered by a:
   a. Cogged belt.
   b. Chain sprocket.
   c. Cam eccentric.
   d. Camshaft lobe.
3. An electric fuel pump can be located:
   a. In fuel tank and in-line.
   b. In-line and in carburetor.
   c. In fuel tank and at fuel rail.
   d. On engine and in-line.
4. Why is the air filter important to engine service life?
5. What are the symptòms of a clogged fuel filter?
6. List typical fuel pressures for carburetor, gasoline

injection, and diesel supply systems.

7. A typical volume spec for a fuel pump would be:
   a. Pint per minute.
   b. Quart per second.
   c. Pint per 30 seconds.
   d. Quart per hour.
8. When installing a mechanical fuel pump, the eccentric lobe should be positioned toward the pump arm. True or false?
9. In your own woıds, list and explain the operation of the seven carburetor circuits.
10. How does a computerized carburetor system work?
11. A computer controlled carburetor system normally functions in this mode:
    a. Open loop.
    b. Repeating loop.
    c. Cycling loop.
    d. Closed loop.
12. A very rich fuel mixture will make the engine emit ________ ________.
13. Explain the five steps for a carburetor rebuild.
14. Wear eye protection and rubber gloves when using carburetor cleaner. True or false?
15. How do you adjust the various engine idle speeds?

ASE CERTIFICATION-TYPE QUESTIONS

1. Technician A says that an automobile's fuel return system is used to regulate fuel pressure in a gasoline injection system. Technician B says that an automobile's fuel return system is used to cool the fuel in order to prevent vapor lock. Who is right?
   (A) A only.
   (B) B only.
   (C) Both A & B.
   (D) Neither A nor B.
2. All of the following are basic components of a mechanical fuel pump EXCEPT:
   (A) check valves.
   (B) diaphragm.
   (C) roller assembly.
   (D) return spring.
3. Technician A says that a fuel filter can be located inside the car's fuel pump. Technician B says that a fuel filter can be located inside an injector body. Who is right?
   (A) A only.
   (B) B only.
   (C) Both A & B.
   (D) Neither A nor B.
4. The fuel pump volume output for a carburetor should be at least ________ in thirty seconds.
   (A) one pint
   (B) two pints
   (C) one-half gallon
   (D) one gallon
5. Technician A says that when installing a mechanical fuel pump, the pump eccentric should be placed in the released position. Technician B says that when installing a mechanical fuel pump, the pump eccentric should be in the engaged position. Who is right?
   (A) A only.
   (B) B only.
   (C) Both A & B.
   (D) Neither A nor B.
6. Technician A says that a carburetor's main discharge tube is a fuel passage between the fuel bowl and air horn. Technician B says that a carburetor's main discharge tube is a fuel passage between the throttle valve and intake manifold. Who is right?
   (A) A only.
   (B) B only.
   (C) Both A & B.
   (D) Neither A nor B.
7. Technician A says that a carburetor normally provides an 18:1 air-fuel ratio for cold engine starting. Technician B says that a carburetor normally provides an 8:1 air-fuel ratio for cold engine starting. Who is right?
   (A) A only.
   (B) B only.
   (C) Both A & B.
   (D) Neither A nor B.
8. Which of the following engine components can be operated off carburetor vacuum?
   (A) Choke break.
   (B) EGR valve.
   (C) Charcoal canister.
   (D) All of the above.
9. The carburetor's acceleration system ________.
   (A) supplies a lean air-fuel mixture at cruising speeds
   (B) squirts fuel into the air horn when the throttle valve opens
   (C) provides correct air-fuel mixture slightly above idle speed
   (D) enriches fuel mixture slightly when engine power demands are high
10. Technician A says that several computer-controlled feedback carburetors require the use of a tachometer to help test computer output. Technician B says that several computer-controlled feedback carburetors require the use of a dwell meter to check computer output. Who is right?
    (A) A only.
    (B) B only.
    (C) Both A & B.
    (D) Neither A nor B.

Chapter 17

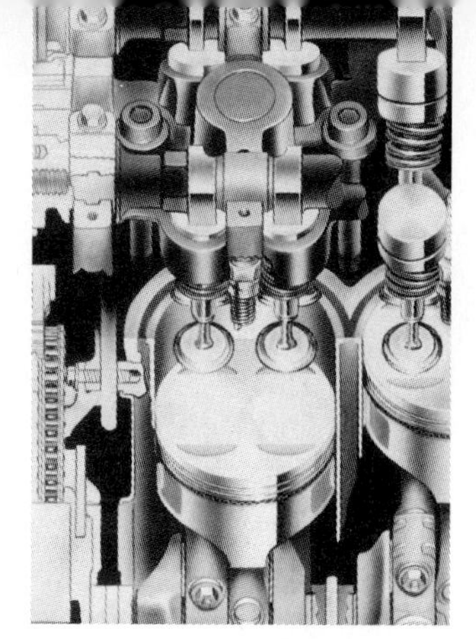

Gasoline Injection—Operation and Service

After studying this chapter, you will be able to:
- *Describe the major parts of an electronic fuel injection system.*
- *Compare different gasoline injection system types.*
- *Explain the operation and construction of an electronic fuel injection system.*
- *Troubleshoot typical fuel injection system problems.*
- *Activate and utilize the self-diagnosis mode found on most EFI systems.*
- *Pressure test a fuel injection system.*
- *Adjust engine idle speed on a fuel injected engine.*

In recent years, auto makers have converted over to gasoline injection. Federal regulations require that cars get a specified fuel economy, or a ''gas guzzler'' charge must be added to the cost of the vehicles. For this reason, today's engines are equipped with electronic fuel injection because it is more efficient than carburetion.

As an engine technician, you must understand the operation, construction, and basic service of gasoline injection systems. For instance, when a car enters the shop with a ''sick'' running engine, the engine techician must be able to find the problem quickly and accurately, whether these problems are in the fuel system, ignition system, emission control systems, or engine itself. This chapter will help you master these important skills.

GASOLINE INJECTION OPERATION

Modern *gasoline injection systems* use engine sensors, a computer, and solenoid-operated fuel injectors to meter the right amount of fuel into the engine. Termed *electronic fuel injection,* (abbreviated EFI), these systems use electric-electronic devices to monitor and control engine operation. Other non-electronic injection systems are used but they will not be covered since they are being replaced by electronic systems.

An ECU (electronic control unit) or computer receives electrical signals (different current or voltage values) from the various sensors. It can then use stored data to operate the injectors, ignition system, and other engine related devices. As a result, less unburned fuel leaves the engine as emissions and the car gets better gas mileage.

Throttle body and multiport injection

Two common types of electronic fuel injection are multiport and throttle body injection, Figs. 17-1 and 17-2.

Throttle body injection mounts one or two injectors inside a throttle body assembly. Fuel is sprayed into one point or location at the center inlet of the engine intake manifold.

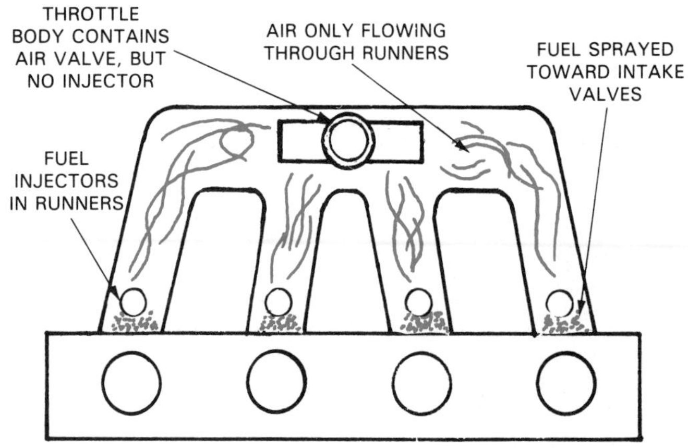

Fig. 17-1. Parts layout for multiport injection system.

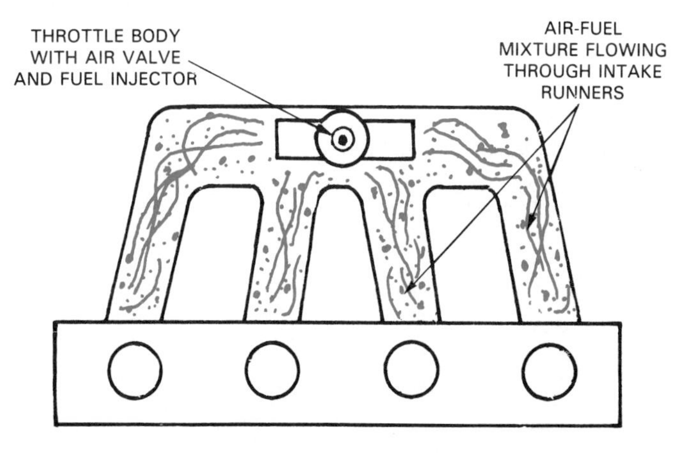

Fig. 17-2. Throttle body injection (TBI) is the other basic type of fuel injection system.

Multiport injection has one injector for each engine cylinder. Fuel is injected in more than one location. This is the most common type and is often called *port injection.*

Gasoline injection components

The basic parts of a multiport, electronic fuel injection system are:

1. ELECTRONIC FUEL PUMP (dc motor driven pump that provides fuel pressure and flow from tank to engine), Fig. 17-3.
2. FUEL RAIL (metal tubing or casting that feeds fuel to pressure regulator and injectors and usually has a service fitting), Fig. 17-3.
3. PRESSURE REGULATOR (diaphragm-operated pressure relief valve that maintains a constant pressure at the injectors), Fig. 17-4.
4. FUEL RETURN LINE (line-hose assembly that carries

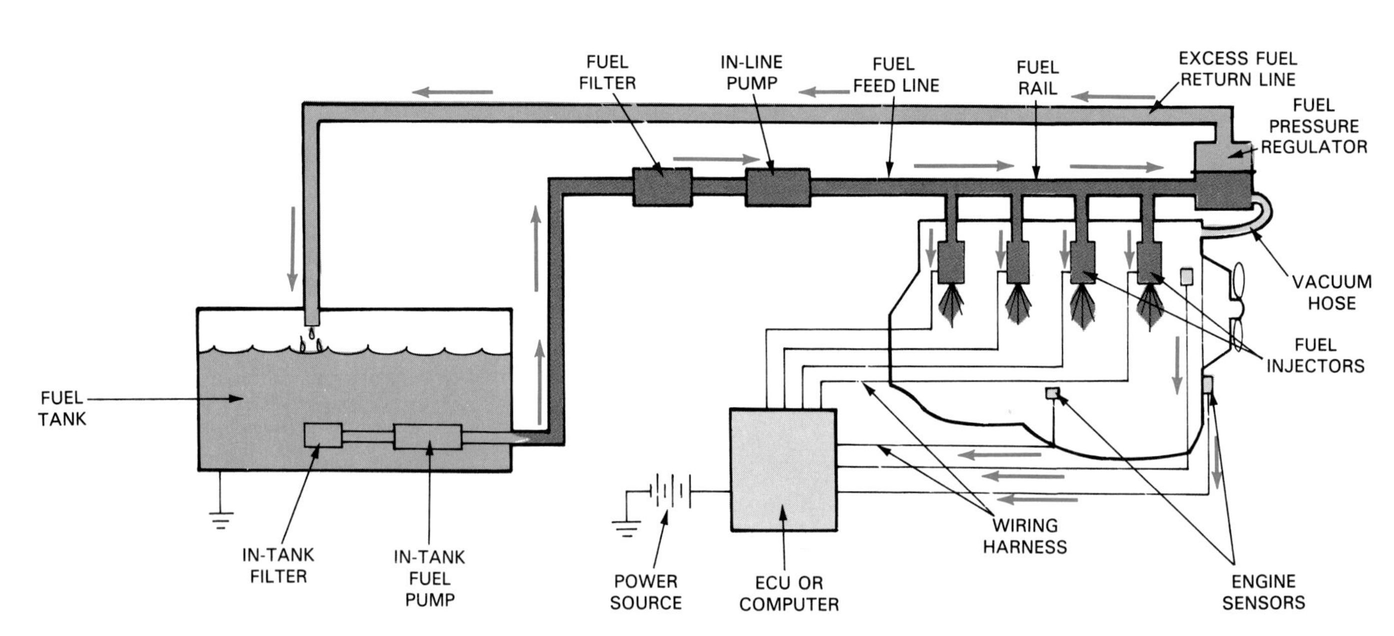

Fig. 17-3. Study basic operation of EFI system. Electric pump(s) force fuel to fuel rail and injectors. Pressure regulator limits pressure and bleeds excess fuel into return line. Fuel constantly circulates through system. Engine sensors report engine conditions to computer. Computer can then open injectors as needed.

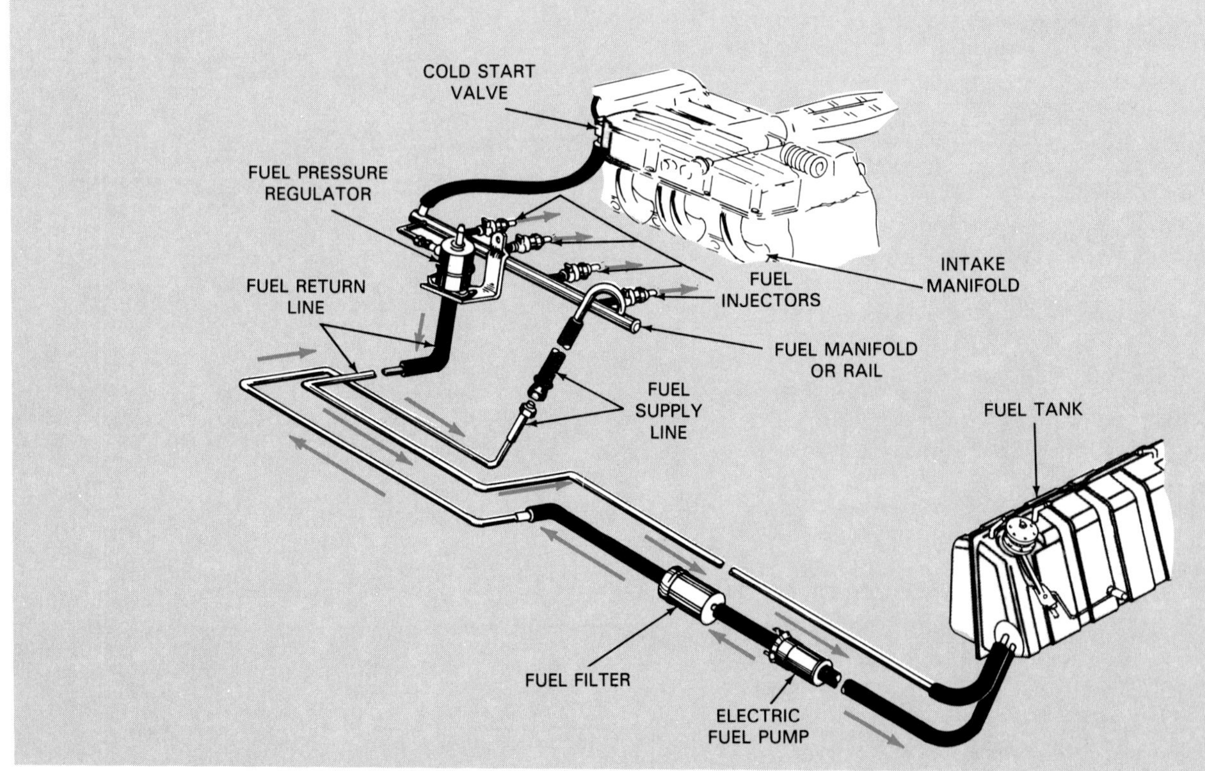

Fig. 17-4. Fuel pressure regulator mounts on or next to fuel rail with multiport systems. It mounts inside throttle body with single-point systems.

excess fuel back to tank from pressure regulator).
5. INJECTORS (with electronic injection, they are solenoid-operated fuel valves that open when energized by computer).
6. SENSORS (devices capable of changing internal resistance or voltage output with a change in a condition—temperature, pressure, position, etc.). Refer to Fig. 17-3.
7. COMPUTER (electronic control unit that is programmed to use sensor data to determine when and how long injectors should be opened for specific conditions).

Look at Fig. 17-3. Imagine electrical signals flowing from the sensors to the computer. The computer can then analyze this data and send electrical pulses to the injectors so that they open and spray fuel toward the engine intake valves.

EFI injector construction

A *fuel injector* for an EFI system is simply a fuel valve. When not energized, spring pressure makes the injector remain closed keeping fuel from entering the engine. When the computer sends current through the injector coil, the magnetic field attracts the injector armature. Fuel then squirts into the intake manifold, Fig. 17-5.

An injector typically consists of:

1. ELECTRIC TERMINALS (electrical connection for completing circuit between injector coil and computer).
2. INJECTOR SOLENOID (armature and coil that opens and closes valve).
3. INJECTOR SCREEN (screen filter for trapping debris before it can enter injector nozzle).
4. NEEDLE VALVE (end of armature that seals on needle seat), Fig. 17-5.
5. NEEDLE SEAT (round hole in end of injector that seals against needle valve tip).
6. INJECTOR SPRING (small spring that returns needle valve to closed position).
7. O-RING SEAL (rubber seal that fits around outside of injector body and seals in intake manifold).
8. INJECTOR NOZZLE (outlet of injector that produces fuel spray pattern).

Injector pulse width

The *injector pulse width* indicates the amount of time each injector is energized and kept open. The computer controls the injector pulse width.

Under full acceleration, the computer would sense a wide open throttle, high intake manifold pressure, and high inlet airflow. The computer would then increase injector pulse width to richen the mixture for more power.

Under low load conditions, the computer would shorten the injector pulse width. With the injectors closed a larger percentage of time, the air-fuel mixture would be leaner for better fuel economy.

Cold start injector

The *cold start injector* is an extra injector that sprays fuel into the center of the engine intake manifold when the engine is cold. It serves the same purpose as the carburetor choke. The cold start injector helps assure engine startup in cold weather. It is used on some multi-point systems, Fig. 17-6.

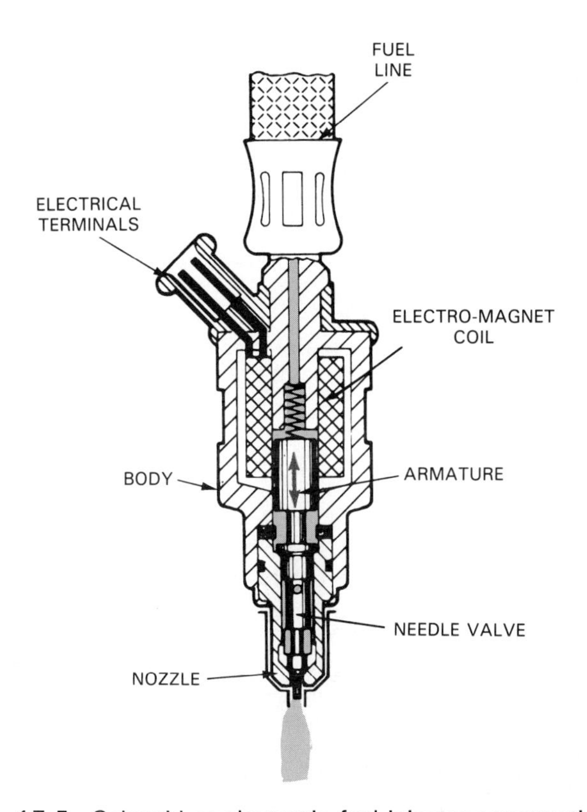

Fig. 17-5. Solenoid or electronic fuel injector construction. Coil produces magnetic field when current flows through it. Magnetic field pulls armature up to open needle valve. Fuel can then spray into engine. (VW)

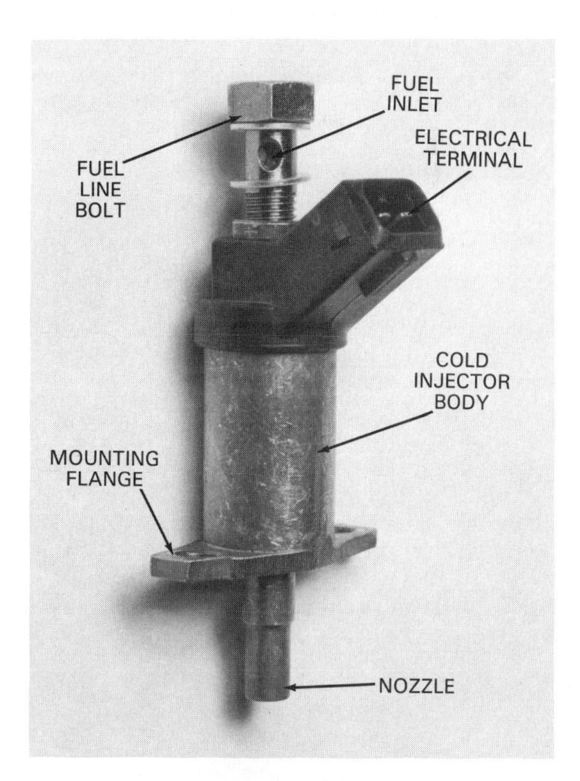

Fig. 17-6. A few systems use a cold start injector. Its construction is similar to the main injectors. (Saab)

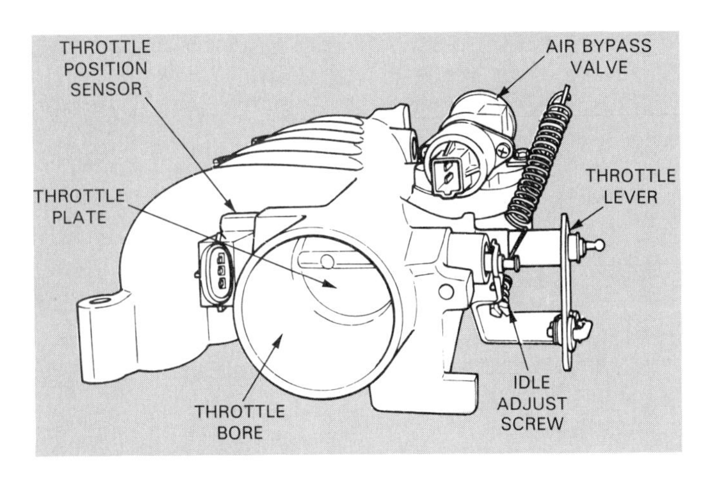

Fig. 17-7. This is a throttle body for a multiport injection system. It allows driver to control airflow into engine, which, in turn, controls engine speed. Air bypass valve allows computer to control idle speed. (Ford)

Throttle body (multiport injection)

A *throttle body assembly* for multiport injection is primarily used to control airflow into the engine. A throttle valve is linked to the driver's gas pedal. This allows the driver to open and close the throttle valve to control engine speed and power. See Fig. 17-7.

An *idle air valve* is frequently mounted on the throttle body to control engine idle speed. The computer can energize the bypass as needed to increase or decrease idle rpm. When the valve bypasses more air, engine speed goes up and vice versa.

Throttle body injection (TBI) unit

A *throttle body injection unit* also contains the fuel injector(s) and pressure regulator. It should not be confused with a throttle body for multiport injection. Look at Figs. 17-8 and 17-9.

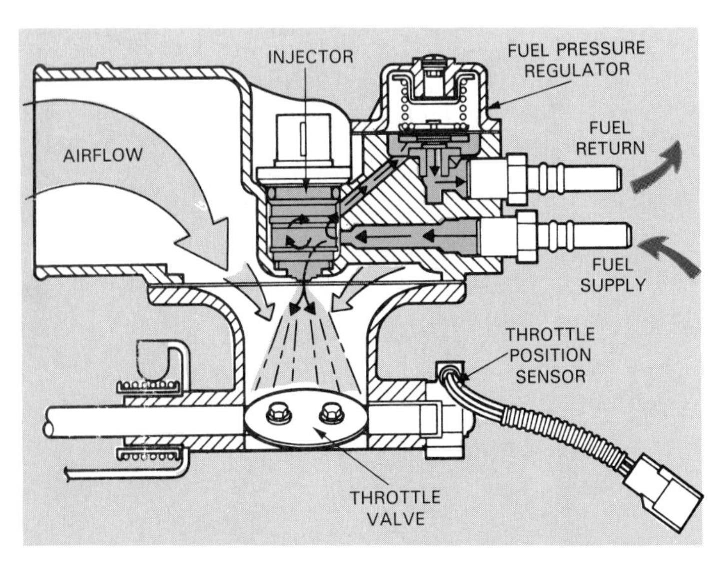

Fig. 17-8. Throttle body injection unit controls fuel and air. Pressure regulator and injector mount inside throttle body. Note fuel and air flow. (Ford)

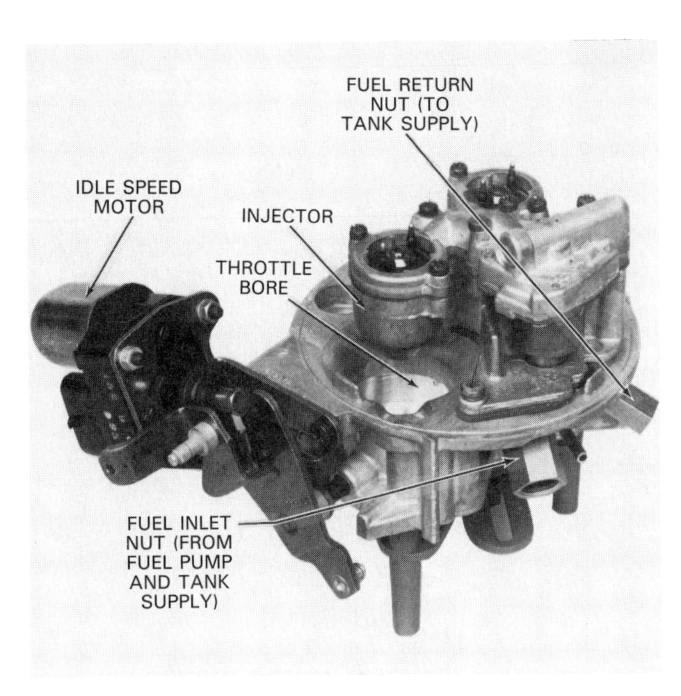

Fig. 17-9. Photo of actual TBI unit shows injector, idle speed motor, and other parts. (Cadillac)

The fuel injector is mounted in the top of the TBI unit. Fuel is injected down at the throttle valve and into the center of the engine intake manifold. The air-fuel mixture is then drawn into the engine cylinders.

Fuel pressure regulator

The *fuel pressure regulator* must maintain a preset pressure inside the fuel rail. In this way, when the computer alters the injector pulse width, a predetermined amount of fuel is injected into the engine. This provides a precise control of the engine's air-fuel ratio.

As shown in Fig. 17-10, a pressure regulator consists of:

1. INLET FITTING (opening for fuel from fuel pump).
2. PRESSURE CHAMBER (area inside regulator so fuel can act upon flexible diaphragm).
3. VACUUM CHAMBER (area on opposite side of diaphragm so engine vacuum can affect fuel pressure slightly).
4. DIAPHRAGM (flexible membrane separating two chambers and linked to pressure relief valve.).
5. PRESSURE RELIEF VALVE (disc shaped valve that can open or close the fuel return outlet).
6. OUTLET FITTING (opening so fuel can enter return line).
7. RELIEF VALVE SPRING (spring that holds valve closed until fuel pressure reaches specs).

As an example, during engine starting, the pressure regulator valve remains closed. This lets fuel pressure rise quickly. Then, when spec pressure is obtained, the diaphragm is flexed upward, opening the relief valve. Fuel pours into the outlet and return and pressure is held constant in the rail and at the injector inlets.

If the engine is crusing at a constant speed engine

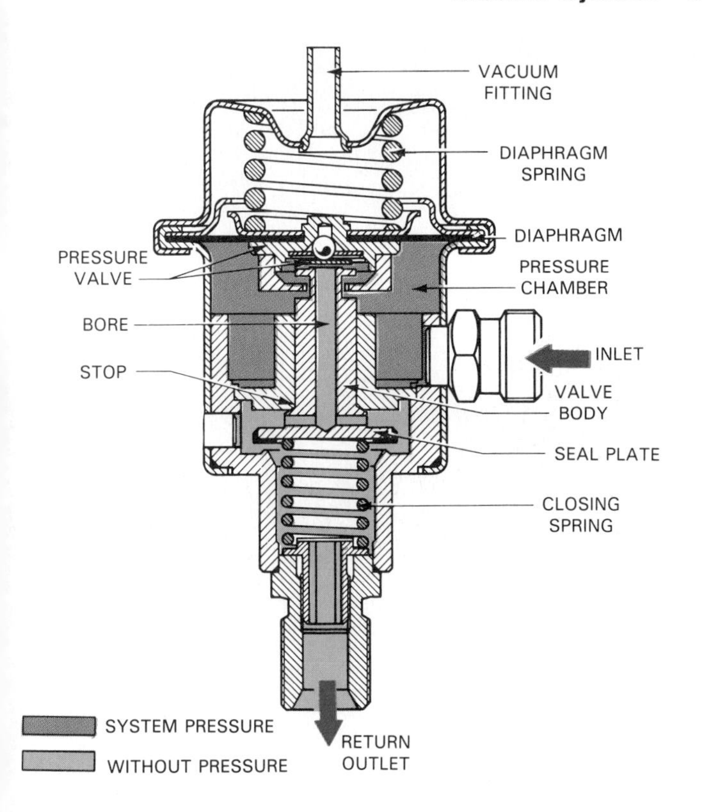

Fig. 17-10. Fuel pressure regulator uses flexible diaphragm, spring, and relief valve to control fuel pressure. When pressure reaches specs, diaphragm pushes up to compress spring, fuel can then flow out and into return line to tank. (Mercedes Benz)

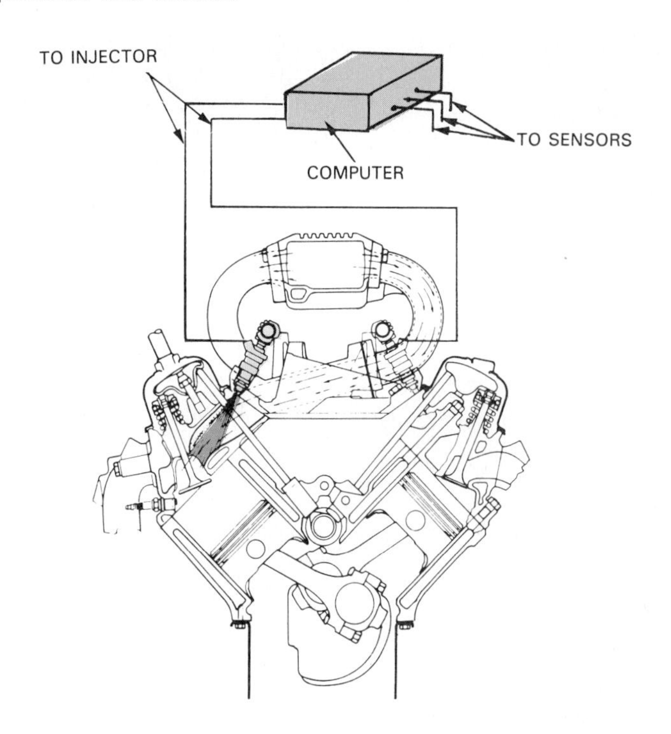

Fig. 17-11. Sensor inputs are fed to computer. Computer can then use power transistors to open injectors when needed. (Chevrolet)

vacuum is high. This pulls up on the diaphragm and fuel pressure is lowered slightly. This leans the mixture for good fuel economy.

However, if the engine is accelerated, as when passing another car, engine vacuum drops. This lets the diaphragm flex down, closing the relief valve slightly. Fuel pressure then increases to richen the mixture for more power.

Idle speed motor

An *idle speed motor* is sometimes used on throttle body assemblies to control engine idle speed, Fig. 17-9. The computer actuates the positioner to open or close the throttle plates. In this way, the computer can maintain a precise idle speed with changes in engine temperature, load (air conditioning ON for example), and other conditions.

Computer

Mentioned briefly, the *computer* or ECU is the ''intelligence'' of an electronic fuel injection system. It can use sensor data (various voltage or current signals) to calculate when and how long each injector should be opened, Fig. 17-11.

The computer's IC's (integrated circuits) can store information on how the injection system should react to different driving conditions. A photo of the inside of a car's computer is given in Fig. 17-12.

Basically, to open an injector, power transistors in the computer connect the injector coil to battery voltage. To

Fig. 17-12. Photo of inside of computer shows integrated circuits, power transistors, and other components that form ''brain'' of electronic fuel injection system. (Ford Motor Co.)

close an injector, the power transistor changes into a resistor to block current flow to the injector.

Engine sensors

An *engine sensor* is an electrical device that changes circuit resistance or voltage with a change in a condition. For example, a temperature sensor's resistance may decrease as temperature increases. The computer can use the increased current flow through the sensor to

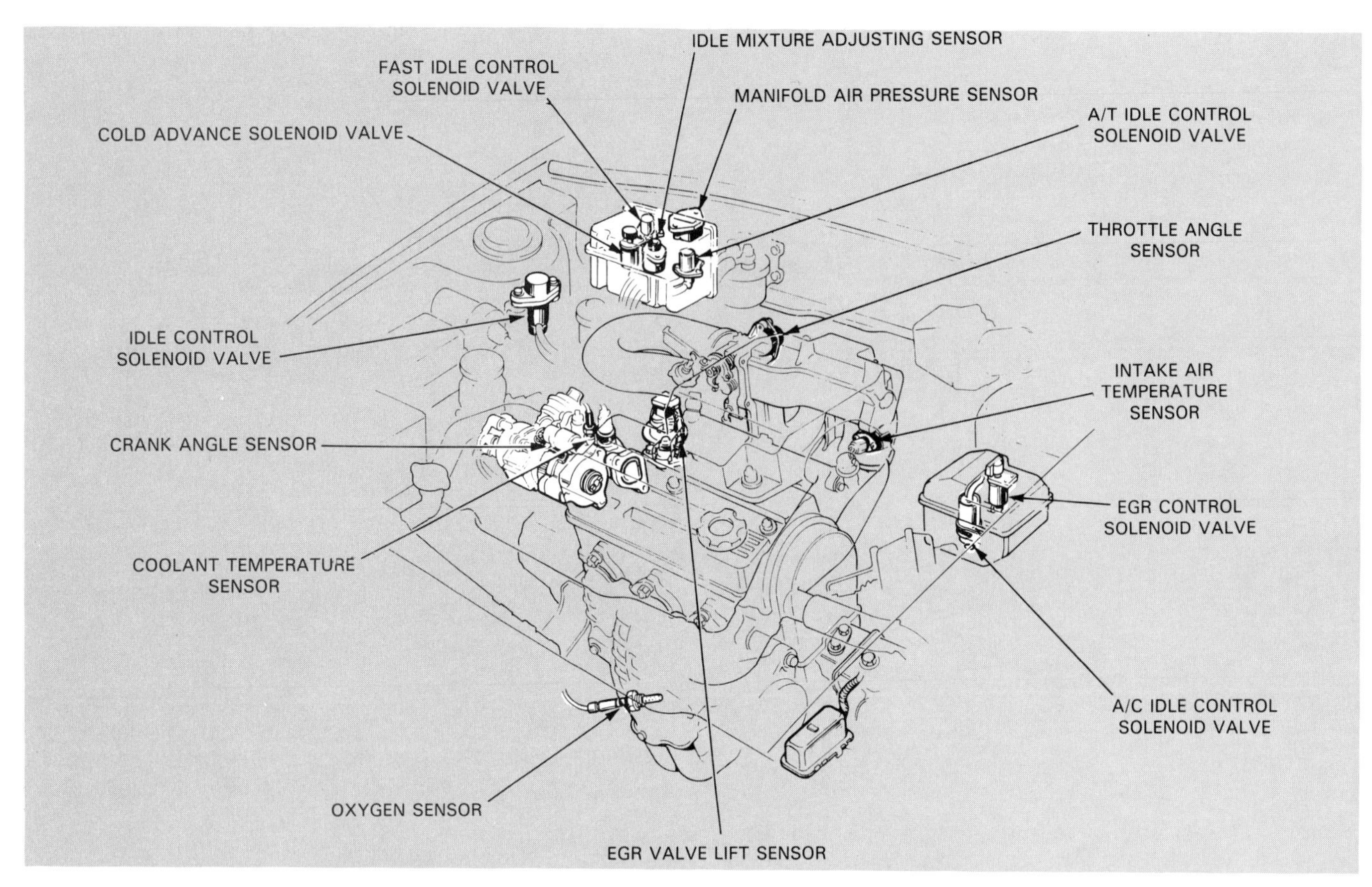

Fig. 17-13. Engine sensors can be located in various locations. This is a typical example. Refer to manual for exact locations on car being serviced. (Honda)

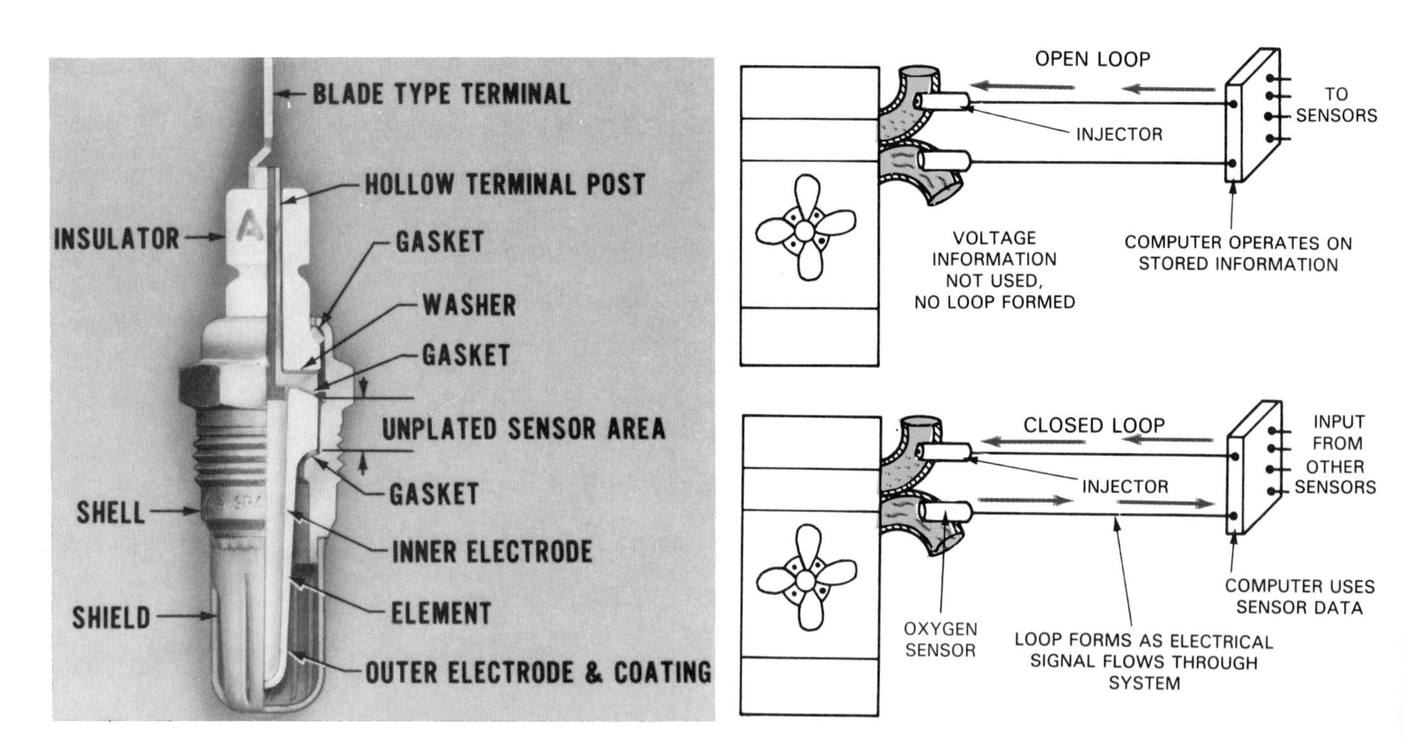

Fig. 17-14. Oxygen sensor is primary sensor because it checks actual air-fuel ratio by ''sniffing'' amount of oxygen in engine exhaust. High oxygen content would show lean mixture. Low oxygen content would show rich mixture. Sensor produces voltage output of about .5 volts for computer. (AC-Delco)

Fig. 17-15. When engine is cold, system is in open loop. Computer operates on stored data, not on information from sensors. After engine warms, oxygen sensor can function and send data to computer. Computer then goes into closed loop and information flows in imaginary loop or circle.

calculate any needed change in injector opening. Look at Fig. 17-13.

Typical sensors for an EFI system include the following:

1. OXYGEN SENSOR (senses amount of oxygen in engine exhaust to check air-fuel ratio, voltage output changes in proportion to air-fuel ratio), Figs. 17-14 and 17-15.
2. ENGINE COOLANT TEMPERATURE SENSOR (senses temperature of engine coolant, allows computer to richen mixture for cold starting).
3. MASS AIRFLOW SENSOR (monitors mass or volume of air flowing into intake manifold for adjusting fuel mixture).
4. INTAKE AIR TEMPERATURE SENSOR (checks temperature of outside air entering engine for fine mixture adjustments).
5. THROTTLE POSITION SENSOR (senses movement of throttle plates so mixture can be adjusted for engine speed or acceleration).
6. MANIFOLD ABSOLUTE PRESSURE SENSOR (monitors vacuum in engine intake manifold so mixture can be richened or leaned with changes in engine load).
7. CRANKSHAFT POSITION SENSOR (senses rotation of engine crankshaft for speed and timing of injection).
8. KNOCK SENSOR (microphone type sensor that detects ping or preignition so ignition timing can be retarded).

Other sensors can also be used in the computer system. Refer to the shop manual for more specific details when needed.

GASOLINE INJECTION SERVICE

When a car enters the shop with an engine performance problem, the engine technician may have to diagnose and repair the trouble. He or she must find out whether the problem is a mechanical problem (burned valve, worn rings, etc.) or a problem in an engine support system (fuel injection, ignition, emission). This section of the chapter will help you find and correct common troubles with EFI systems.

Diagnosing EFI problems

To diagnose problems in a gasoline injection system, you must use:

1. Your knowlege of system operation.
2. Basic troubleshooting skills.
3. Service manual.

As you try to locate problems, visualize the operation of the system. Relate the function of each component to the problem. This will let you eliminate several possible sources and concentrate on others, Fig. 17-16.

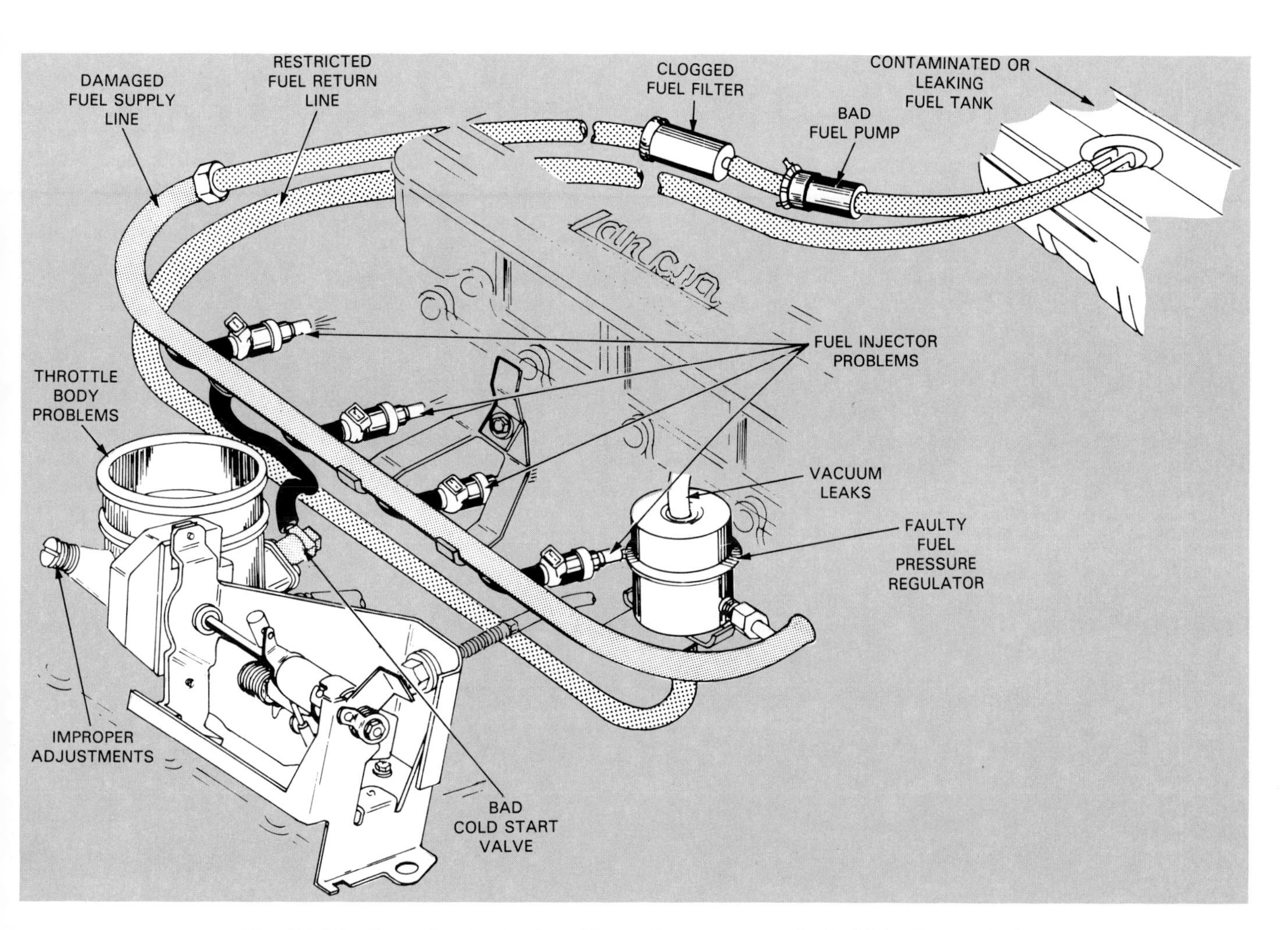

Fig. 17-16. Note the types of problems that can occur in fuel injection system.

Fig. 17-17. Inspect engine compartment when you suspect fuel injection problems. Look for loose wires, leaking hoses, vacuum leaks, and other similar types of trouble. (Ford)

Inspection of the engine compartment will sometimes locate gasoline injection troubles. Check the condiiton of all hoses, wires, and other parts. Look for fuel leaks, vacuum leaks, kinked lines, loose electrical connections, and other troubles, Fig. 17-17.

With EFI, you may need to disconnect and check the terminals of the wiring harness. Inspect them for rust, corrosion, or burning. High resistance at terminal connections is a frequent cause of problems.

ECU self-diagnostics

Modern computer control systems test themselves and can indicate the location of a problem. There are two types of computer control systems in use today. The first has been in use for years and is referred to as *on-board diagnostics generation one* or *OBD I.*

The second system was first installed in 1994 on a few vehicles and was implemented on all new vehicles starting with the 1996 model year. This system can detect sensor malfunctions before they become noticeable to the driver or even the technician. This system is referred to as *on-board diagnostics generation two* or *OBD II.*

While each manufacturer's systems have their differences, they are similar in many ways. Both diagnostic systems work as follows:

1. ECU illuminates a *malfunction indicator lamp (MIL)* or check engine light on the dash.
2. ECU displays a digital number code in the dash climate control or driver information center.
3. ECU lights LEDs on the computer itself to indicate problem codes.

OBD II systems will flash the MIL if a problem exists that could damage the vehicle's emission system. When the MIL light illuminates, it tells the driver and technician that a problem exists, Fig. 17-18.

Data link connector

A test terminal, also called a *data link connector,* is used to access the computer system. Each terminal on

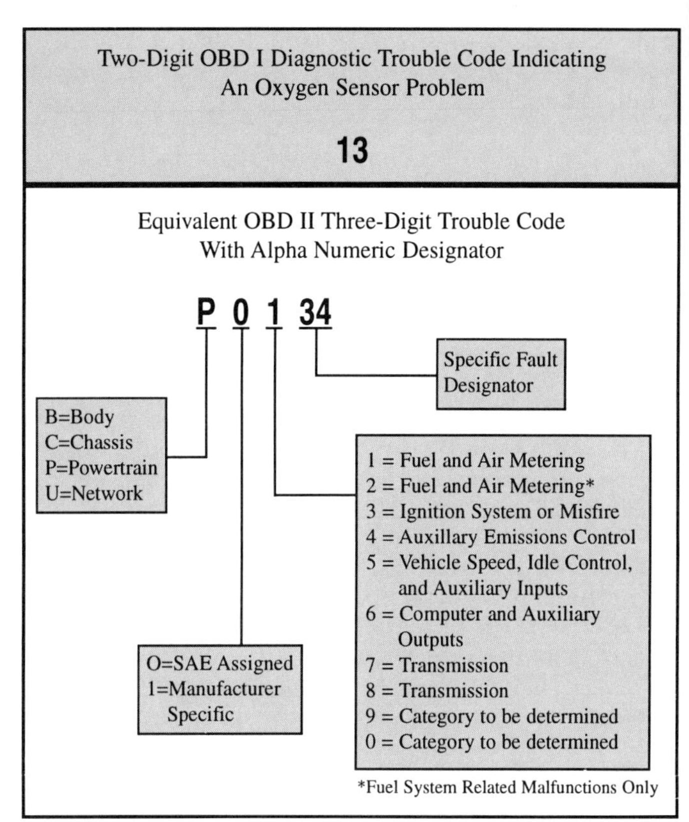

Fig. 17-18. Most modern computer systems will test themselves. Codes may be read on a vehicle's digital display, grounding the DTC to cause the MIL light to flash, or via a scan tool. OBD I ECU's will produce a two number code. OBD II systems will produce a three number code with an alpha numeric designator. Do not try to correlate the older fault codes with the OBD II fault designator.

the connector is wired to an ECU, battery power, or ground. OBD I data link connectors can be accessed by grounding terminals or with an analog meter or scan tool. OBD II data link connectors must be accessed using a scan tool.

The data link connector can be located almost anywhere on OBD I systems, so check the service manuals. OBD II systems mount the connector in a location accessible from the driver's seat.

Activating self-diagnosis

Warm the engine to operating temperature. This will ensure that the system has the proper conditions to enter closed loop operation. Shut off the engine and turn the ignition key to on or run.

To activate the self-diagnostics on OBD I equipped vehicles, depending on the system, you may have to use a jumper wire or test connector to ground terminals on the data link connector or read the pulse sweeps of an analog meter as you did in Chapter 16.

However, the recommended method of reading codes on OBD I and OBD II equipped vehicles is to use a scan tool. The other methods described earlier cannot be used on OBD II systems.

Reading codes

If the output is a digital display, simply read the number and compare it with the chart in the service manual.

If the output is flashed from the dashboard MIL or ECU mounted LED lamp, count how many flashes occur between each pause. Write down the number and compare it to the service manual.

If an analog meter is used, count the number of needle deflections between pauses. Write down the number and compare it to the chart in the service manual.

Each system may vary. The service manual will give you instructions for each make and model of vehicle.

Interpreting trouble codes

When looking up trouble codes in the service manual, start from the lowest number and work up. If multiple codes were stored, you can often eliminate one or more codes by using this method. If you had codes 13 and 45 stored in an ECU, you would look up code 13 first and then 45.

OBD I codes will be two digit numbers, such as 32. These numbers correspond with a particular problem listed in the service manual. OBD II codes are three digit numbers with an *alpha-numeric designator.* The alpha numeric designator indicates in what system the problem is located and whether the code is SAE or manufacturer specific. The first digit in the three number code indicates the nature of the code. The last two digits of the number indicate the specific fault, Fig. 17-18.

Do not attempt to correlate OBD I codes with the OBD II fault designator. In most OBD I systems, only one or two codes are used to indicate a problem with a specific component. OBD II systems can use up to 20 OR MORE SAE and manufacturer specific codes for problems with the same sensor.

All codes, no matter the system used, should be looked up in a *trouble code chart.* These charts give the meaning for the codes present in the ECU. Additional information on both of these diagnostic systems is given in Chapter 21.

Clearing trouble codes

To *clear* or remove the trouble codes from an OBD I computer system, you must usually disconnect the battery or the fuse to the computer. Disconnect them for about 10 seconds. You must use a scan tool to clear codes on vehicles with OBD II systems.

After clearing the codes, check to see if the engine light comes back on after engine operation. This would indicate there is still a problem in the system. Further tests would be needed.

RELIEVING EFI SYSTEM PRESSURE

DANGER! Always relieve fuel pressure before disconnecting any EFI fuel line. Many systems retain fuel pressure (as high as 50 psi or 345 kPa), even when the engine is NOT running.

Some EFI systems have a special *relief valve* (test fitting). Or, the pressure regulator may allow pressure relief for bleeding pressure back to the fuel tank.

FUEL PRESSURE REGULATOR SERVICE

A *faulty fuel pressure regulator* can cause an extremely rich or lean mixture. If the output pressure is high, too much fuel will spray out of the injectors, causing a rich mixture. If the regulator bypasses too much fuel to the tank (low pressure), not enough fuel will spray out each injector. This causes a lean mixture.

To test a fuel pressure regulator, install a pressure gauge at the test fitting on the fuel rail or before the regulator. Start and idle the engine while reading the gauge. Then, apply the recommended vacuum to the pressure regulator and note the pressure drop. This is shown in Fig. 17-19.

If either reading is not within specs, the regulator may be faulty. Check other possible problems (bad fuel pump, clogged filter, smashed return line) before replacing the regulator, however.

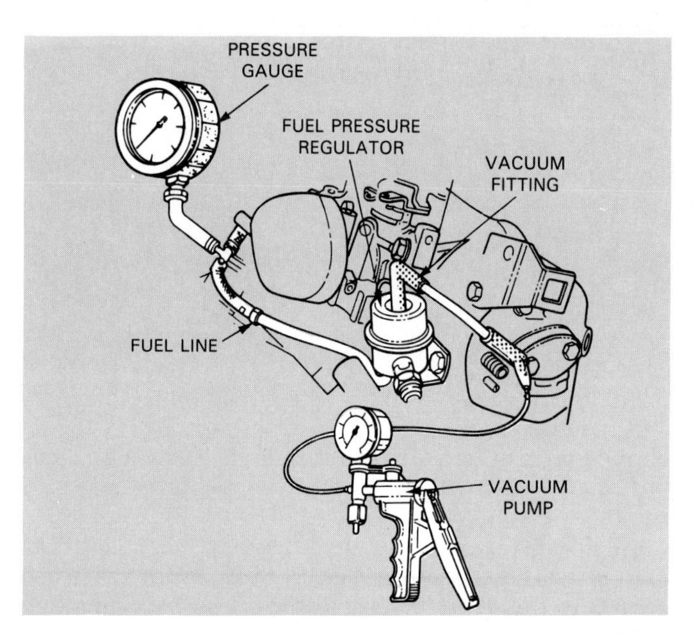

Fig. 17-19. Fuel pressure regulator is common source of engine performance problems. To test regulator, connect pressure gauge to system and vacuum pump to regulator diaphragm. Check pressure with and without vacuum applied to diaphragm. Compare to specs. Check return line, filters, and pump before condemning regulator. (Renault)

INJECTOR SERVICE

A *bad injector* can cause a wide range of problems: rough idle, hard starting, poor fuel economy, and engine miss. It is very important that each fuel injector provides the correct fuel spray pattern.

A *leaking injector* richens the fuel mixture by allowing extra fuel to drip out the closed injector valve. The injector valve may be worn or dirty, or the return spring may be weak or broken.

A *dirty injector* can restrict fuel flow and make the air-fuel ratio too lean. If foreign matter collects in the valve, a poor spray pattern and inadequate fuel delivery can result.

An *inoperative EFI injector* normally has shorted or opened coil windings. Current is reaching the injector, but since the coil is bad, a magnetic field cannot form and open the injector valve.

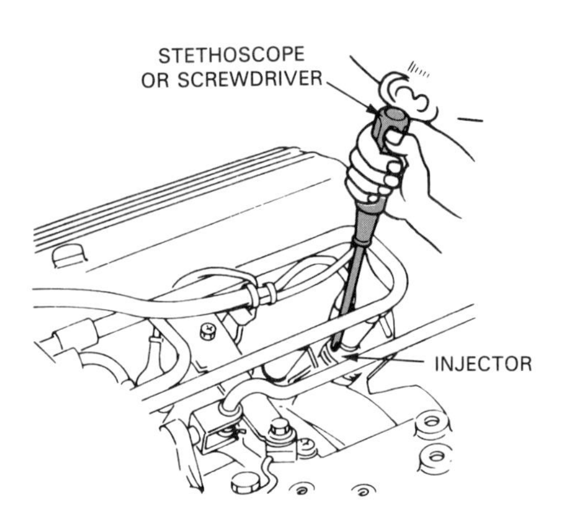

A

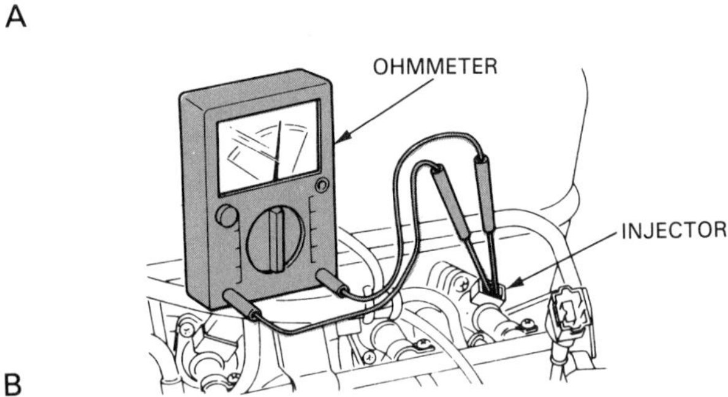

B

Fig. 17-20. Two methods of checking for faulty injector. A—Use stethoscope or screwdriver to listen for clicking noise that shows injector is working. B—Use ohmmeter to check coil for opens or shorts.

Checking injector operation

To quickly make sure each EFI injector is opening and closing, place a STETHOSCOPE (listening device) against each injector. A clicking sound means the injector is opening and closing. If you do NOT hear a clicking sound, the injector is not working. The injector solenoid, wiring harness, or computer control circuit may be bad. See Fig. 17-20A.

With the engine off, you can check the condition of the coils on the inoperative injectors. Use an ohmmeter. Measure the resistance across the injector coil and check for shorts to ground. If the coil is open (infinite resistance) or shorted (zero resistance to ground), you must replace the injector. Refer to Fig. 17-20B.

If the injector tests good, you may need to check the wiring going to that injector. Following the service manual, check supply voltage to the inoperative injector. You may also need to measure the resistance in the circuit between the injector solenoid and the computer. A high resistance would indicate a frayed wire, broken wire, or poor electrical connection.

Refer to a wiring diagram when solving complex fuel injection electrical problems. The diagram will show all electrical connections and components.

Replacing EFI multiport injectors

An EFI multiport injector is easy to replace. After bleeding off fuel pressure, simply remove the hose from the injector and fuel manifold. Unplug the electrical connection and remove any fasteners holding the injector. Pull the injector out of the engine.

Inspect the boot and other rubber parts closely. Some manufacturers suggest that you replace the boot, seals, and hose if the injector is removed for service.

Install the new or serviced injector in reverse order. Follow shop manual directions when servicing a throttle body injection unit or multiport injector.

ENGINE SENSOR SERVICE

Most EFI engine sensors can be tested with self-diagnosis, a scan tool, or a special tester (analyzer), as covered earlier in the chapter. However, an engine sensor can also be tested with a digital meter or sometimes a test light. Refer to a shop manual for specifics.

Throttle position sensor service

A throttle position sensor should produce a given amount of resistance for different throttle openings. For example, the sensor might have high resistance with the throttle plates closed and a lower resistance when the throttles are open. Compare your ohmmeter readings to specs to determine the condition of a throttle position sensor.

To replace a throttle position sensor, you must usually unstake the attaching screws. They may be soldered in place to prevent tampering. Use a file to grind off the stake and remove the sensor.

After installing the new throttle position sensor, adjust the sensor. Refer to your manual for details. For most systems, the sensors should read a prescribed resistance with the throttle plates in specified positions.

Oxygen sensor service

Most auto manufacturers recommend that you test the oxygen sensor with a specialized system tester or by the self-diagnosis mode. A digital voltmeter can also be used to check oxygen sensor voltage. Normally, oxygen sensor output voltage should be about .5 volt or more.

To replace an oxygen sensor, unplug the wire connection at the sensor. Unscrew the old sensor.

Coat the threads with anti-seize compound. Start the sensor by hand. Then screw in and tighten the sensor with a wrench. Do not over-tighten the sensor or it may be damaged. Check fuel injection system operation after sensor installation.

Temperature sensor service

EFI systems can use both a coolant temperature sensor and an intake air temperature sensor. If these sensors are bad, they will make the engine run rich or lean. The internal resistance of these sensors changes with temperature.

A digital ohmmeter is often recommended for testing a temperature sensor. The service manual will give resistance values for various temperatures. If the ohmmeter test readings are not within specs for each temperature value, the sensor is bad.

Servicing other EFI sensors

The other sensors in an electronic fuel injection system are tested using the same general procedures just discussed. You would use the self-diagnosis mode, a special analyzer, or a digital meter to check each sensor. Refer to a service manual for exact procedures.

COMPUTER SERVICE

If your system tests find a bad ECU, you may need to replace the complete unit or a section of the unit. Again, because of system variations, follow manufacturer procedures.

A PROM (programmable read only memory) is a computer chip used in some on-board computers. It is a computer chip calibrated for the particular engine. When a computer tests faulty, you must normally reuse the old prom in the new computer. Some ECUs use an eraseable PROM that is affixed to the board. Depending on the system, it may be necessary to program or "burn in" the vehicle operating information from a scan tool or computerized analyzer to the new ECU.

An EFI computer is normally mounted under the dash. This keeps the electronic circuits away from engine heat and vibration. In a few cars, however, the computer is in the engine compartment.

GASOLINE INJECTION ADJUSTMENTS

As with a carburetor, there are several adjustments on some gasoline injection systems. These include:

1. Engine idle speed adjustment, Fig. 17-21.
2. Throttle plate stop adjustment.
3. Minimum idle air rate adjustment.
4. Throttle cable adjustment, Fig. 17-21.

Not all vehicles have adjustments, so refer to a service manual for adjustment procedures.

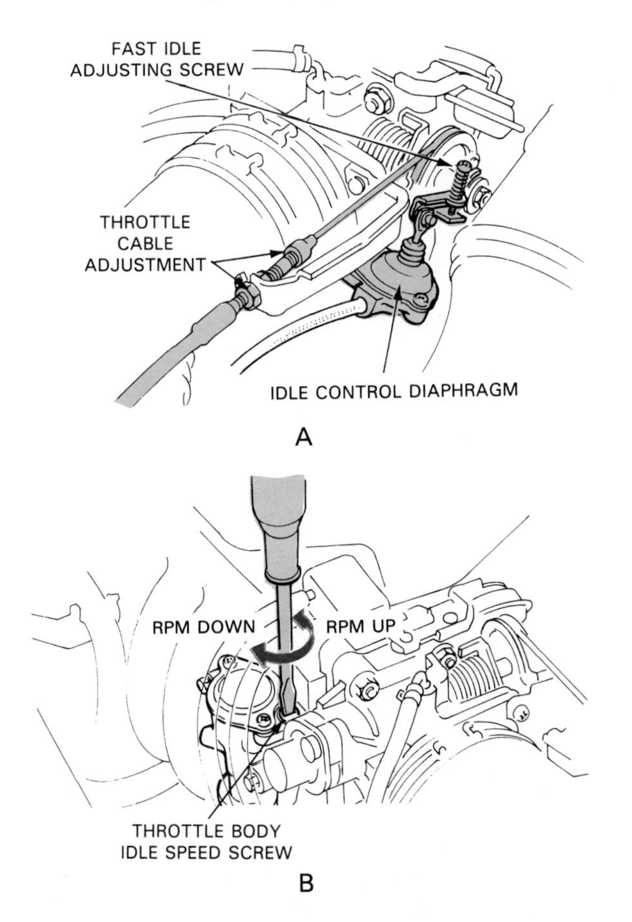

Fig. 17-21. Refer to service manual when doing adjustments on fuel injection system. A—Throttle cable and fast idle adjustment. B—Curb idle speed adjustment. (Honda)

SUMMARY

Most new engines come equipped with electronic fuel injection. For this reason, today's engine mechanic must know how to do the more common tests and repairs done on injection systems.

EFI systems use engine sensors, a computer, and electrically-operated injectors. The sensors monitor engine temperature, air-fuel ratio, airflow, throttle position, and other variables. The sensors produce different voltage or current values with changes in conditions. The computer uses these signals from the sensors to control engine functions.

A throttle body injection system has the injector(s) inside the throttle body assembly. Multiport injection has one injector for each cylinder. Multiport is replacing single-point injection because it is more efficient.

An electric fuel pump pressurizes the system. The pressure regulator maintains a preset pressure in the fuel rail and at the injectors. In this way, the computer can control the engine's air-fuel ratio by varying injector pulse width or open time.

A throttle body for multiport injection is used to control airflow and engine speed.

The computer or electronic control unit is the "brain" for the system. It contains IC's or computer chips that store information electronically. After analyzing data from the sensors, it uses power transistors to energize the fuel injectors.

To find EFI problems, begin by inspecting the engine compartment. Look for loose wires, leaking hoses, vacuum leaks, etc. Most new cars have a computer self-diagnosis mode. After energizing the computer, a number code will be produced by the computer. Follow manual procedures to activate and interpret the computer self-diagnosis. A trouble code chart in the manual will give the names of the faulty components. A data link connector is often found on late model cars.

A pressure gauge can be used to check the fuel pressure regulator. A special test fitting is normally located on the fuel rail. Pressures must be within specs.

A bad fuel injector can cause a wide range of problems. A stethoscope can be used to listen for injector opening. An ohmmeter will measure injector coil resistance to check for opens or shorts.

Engine sensors can be checked with a digital VOM. Again, compare test values with specs. Replace any sensor not within specified ohms or voltage specifications.

Follow manual instructions when adjusting a fuel injection system. Typical adjustments are idle speed, throttle plate stop, air-fuel mixture, and throttle cable.

KNOW THESE TERMS

EFI, ECU, Multiport injection, TBI, Fuel rail, Pressure regulator, Fuel return, Injectors, Computer, Sensor, Pulse width, Cold start injector, Throttle body, Air bypass valve, Throttle positioner, Oxygen sensor, Temperature sensor, Airflow sensor, Throttle position sensor, Manifold pressure sensor, Crank position sensor, Knock sensor, Computer self-diagnosis, Trouble code, Analog meter, Digital meter, Test connector.

REVIEW QUESTIONS—CHAPTER 17

1. What is the difference between throttle body and multiport injection?
2. List and explain the seven basic parts of an electronic fuel injection system.
3. How does an electronic injector open?
4. _______ _______ _______ is the amount of time computer holds injector open and it controls the engine's _________ ratio.
5. This device is used to control engine idle speed on a fuel injected engine:
   a. Dashpot.
   b. Aneroid bellows.
   c. Air bypass valve.
   d. Pressure regulator.
6. A TBI assembly and a multiport injection throttle body perform the same functions of controlling air and fuel flow. True or false?
7. How does a fuel pressure regulator work?
8. An engine _______ is a device that can change resistance or voltage output with a change in condition.
9. List and explain eight engine sensors.
10. How do you use voltage pulses or light flashes to determine a computer self-diagnosis code?
11. A customer car has the computer check engine light glowing.
    Technician A says that a scan tool should be connected to the data link connector for reading the code.
    Technician B says only a digital meter should be used on a computer system to avoid part damage.
    Who is correct?
    a. Technician A. c. Both A and B.
    b. Technician B. d. Neither A nor B.
12. During self-diagnosis, the engine light flashes once, pauses, and flashes six more times. This would represent a trouble code of _________.
13. What is a service manual trouble code chart?
14. How do you check the operation of a fuel pressure regulator?
15. An engine problem is difficult to diagnose. The engine seems to lack power and pings or knocks. Ignition timing, self-diagnosis, and other electrical tests fail to find a problem.
    Technician A says that the fuel pressure regulator could be keeping fuel pressure too low, making the air-fuel ratio too lean.
    Technician B says the pressure regulator could be producing too much pressure.
    Who is correct?
    a. Technician A. c. Both A and B.
    b. Technician B. d. Neither A nor B.

ASE CERTIFICATION-TYPE QUESTIONS

1. Technician A says that gasoline injection systems use a computer to help regulate the correct amount of fuel sprayed into the engine. Technician B says that gasoline injection systems use engine sensors to help regulate the correct amount of fuel sprayed into the engine. Who is right?
   (A) A only. (C) Both A & B.
   (B) B only. (D) Neither A nor B.
2. All of the following are basic components of a multiport fuel injection system EXCEPT:
   (A) a pressure regulator.
   (B) a throttle body.
   (C) a fuel rail.
   (D) a centrally located fuel injector.
3. Technician A says that the electronic fuel pump controls injector pulse width in a multiport fuel injection system. Technician B says that the computer controls injector pulse width in a multiport fuel injection system. Who is right?
   (A) A only. (C) Both A & B.
   (B) B only. (D) Neither A nor B.
4. Technician A says that a throttle body assembly for a multiport injection system is primarily used to control airflow into the engine. Technician B says that a throttle body assembly for a multiport injection system is primarily used to control fuel flow into the engine. Who is right?
   (A) A only. (C) Both A & B.
   (B) B only. (D) Neither A nor B.
5. All of the following are typical sensors used in a modern throttle body fuel injection system EXCEPT:
   (A) knock sensor.
   (B) mass airflow sensor.
   (C) engine coolant temperature sensor.
   (D) oxygen sensor.
6. Technician A says that a throttle body for a TBI injection system normally contains the system's injector(s) and fuel pressure regulator. Technician B says that the throttle body for a TBI injection system normally contains only the injector(s). Who is right?
   (A) A only. (C) Both A & B.
   (B) B only. (D) Neither A nor B.
7. Technician A says that some computer control systems activate a light emitting diode on the computer itself to indicate a diagnostic trouble code. Technician B says that most computer control systems activate a light in the dash to indicate a diagnostic trouble code. Who is right?
   (A) A only. (C) Both A & B.
   (B) B only. (D) Neither A nor B.
8. Technician A says that to clear the trouble codes from most computer systems, you should disconnect the ECU from battery power. Technician B says that to clear the trouble codes from an OBD II computer system, you should disconnect the fuse to the computer. Who is right?
   (A) A only. (C) Both A & B.
   (B) B only. (D) Neither A nor B.
9. All of the following are possible adjustments on fuel injection systems EXCEPT:
   (A) throttle plate stop.
   (B) minimum idle air rate.
   (C) throttle cable.
   (D) air-fuel mixture.
10. Technician A says that a fuel injection system's oxygen sensor output voltage should normally be 4.5 volts or more. Technician B says that a fuel injection system's oxygen sensor output voltage should normally be .5 volts or more. Who is right?
    (A) A only. (C) Both A & B.
    (B) B only. (D) Neither A nor B.

Chapter 18

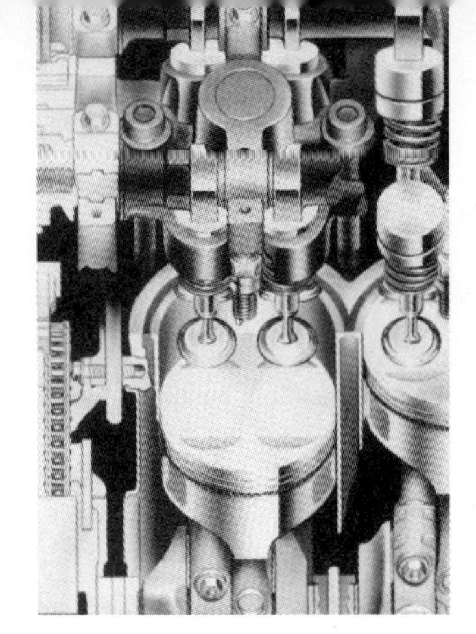

Diesel Injection—Operation and Service

After studying this chapter, you will be able to:

- *Explain the major parts of a diesel injection system.*
- *Summarize the operation of a diesel injection system.*
- *Diagnose common diesel injection problems.*
- *Perform basic tests on a diesel injection system.*
- *Remove and replace major diesel injection components.*
- *Adjust injection timing and diesel engine speed settings.*

Today's engine technician must understand the operation of diesel engines. Millions of diesels are on the road and require service and repair.

As you will learn, a diesel injection system is fairly simple and easy to repair. See Fig. 18-1.

DIESEL INJECTION SYSTEM OPERATION

Modern *diesel injection systems* use a high pressure mechanical pump to force diesel oil directly into the engine combustion chambers. Many late model systems use a computer, sensors, and actuators (motors or solenoids) to control the mechanical injection pump.

As shown in Fig. 18-2, the basic parts of a diesel in-

Fig. 18-1. Other than the fuel system, a diesel engine is very similar to a gasoline engine. An engine technician must understand the operation and service of diesel injection to survive in today's service facility. This is a V-6 diesel. (Oldsmobile)

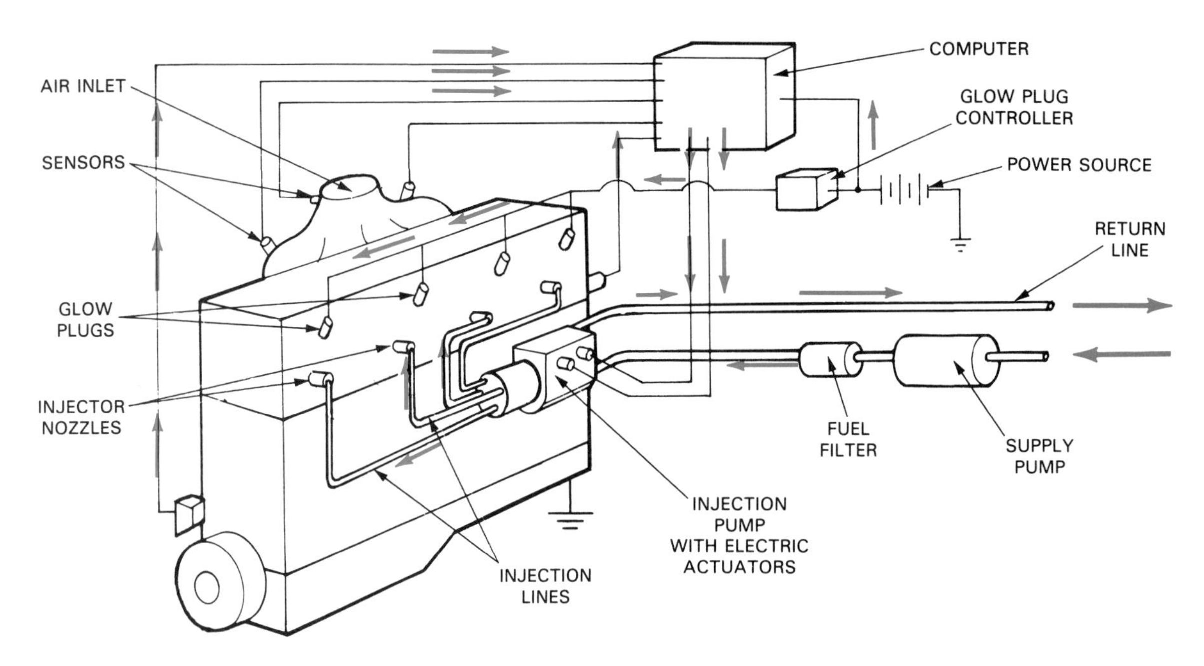

Fig. 18-2. Study basic parts of modern electronically controlled diesel injection system. Older, all-mechanical systems did not have computer, sensors, or injection pump actuators.

jection system are:

1. SUPPLY PUMP (feeds fuel from tank to injection pump).
2. INJECTION PUMP (high pressure, mechanical-electrical pump for forcing fuel to injector nozzles).
3. INJECTION LINES (high pressure, steel lines that feed fuel from injection pump to nozzles).
4. INJECTOR NOZZLES (spring-loaded fuel valves that spray fuel into engine precombustion chambers).
5. FUEL RETURN LINE (fuel line that carries excess fuel back to tank).
6. COMPUTER (used on new cars to increase efficiency by closely controlling injection pump operation).
7. SENSORS (monitor diesel engine conditions and report to computer).
8. ACTUATORS (motors or solenoids on injection pump that control injection timing and quantity).
9. GLOW PLUGS (electric heating elements in prechambers that warm air to aid cold engine starting).
10. GLOW PLUG CONTROLLER (electronic circuit that turns glow plugs on when diesel engine is cold).

With the diesel running, the supply pump forces fuel into the injection pump and through the return line. The injection pump is powered directly off of the engine. It has mechanical pumping plungers that force fuel out the injection lines and to the nozzles. Fuel sprays into the cylinders to produce each power stroke, Fig. 18-3.

The glow plugs are only energized when the engine is cold. Being a compression ignition engine, the glow plugs heat the air so the fuel will ignite upon injection. The controller only sends current to the glow plugs when engine temperature is below a specific point.

Today's diesel injection systems are also equipped with a computer control system. Engine sensors monitor engine intake manifold vacuum, throttle position, crank speed, air temperature, etc. They send electrical signals to the computer. Then, the computer can send current to the actuators on the injection pump to alter injection timing and duration as needed.

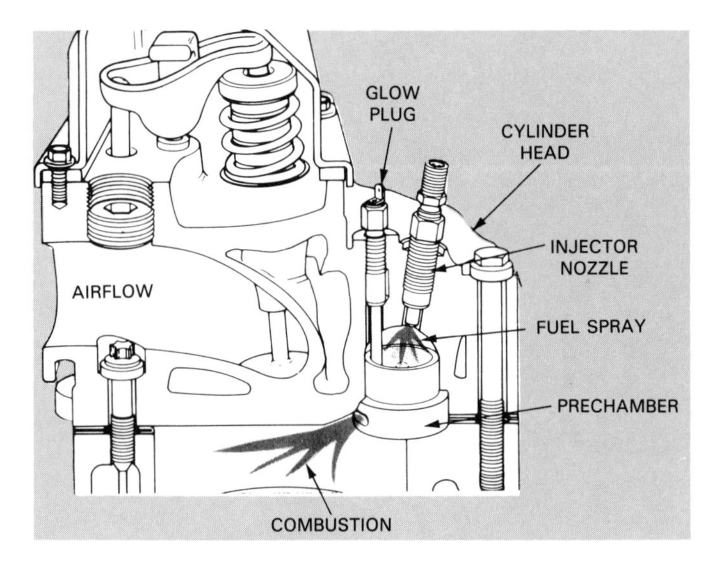

Fig. 18-3. Injectors and glow plugs screw into cylinder head. Tips protrude into precombustion chambers that are press-fit in head. Combustion flame starts in prechamber and blows out into main chamber. (Fel-Pro Gaskets)

Diesel injection pump

An *injection pump,* Fig. 8-4, for a diesel engine has several critical functions. These include:

1. Meter fuel to each injector.
2. Circulate fuel through fuel lines and nozzles.
3. Produce extremely high fuel pressure.
4. Time fuel injection to meet speed and load of engine.
5. Allow driver to control engine power output.

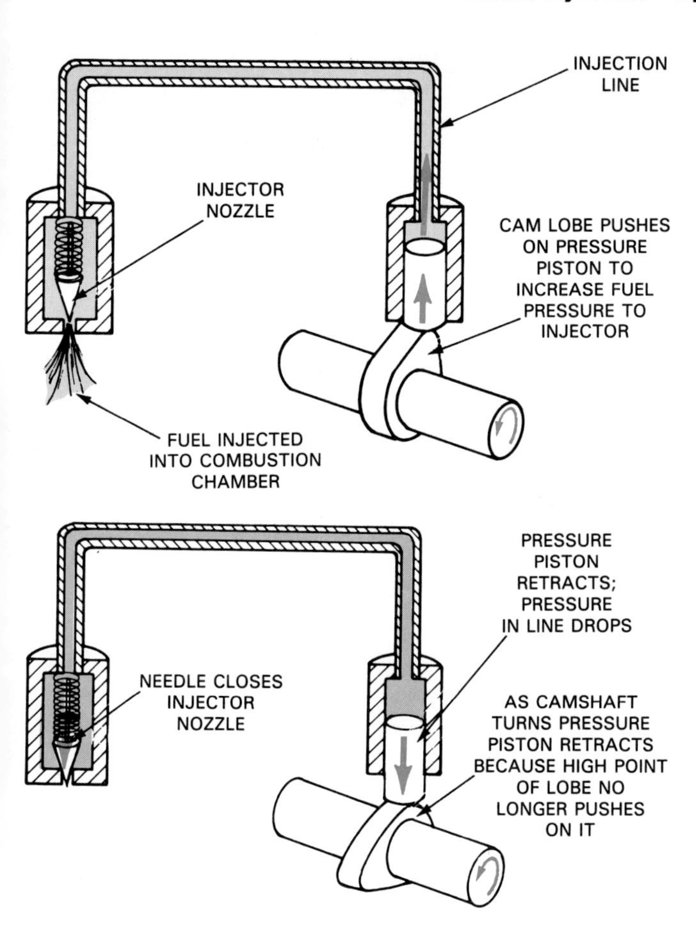

Fig. 18-4. Basic principle of diesel injection pump. Cam rotates and pushes on small pumping plunger. Plunger compresses fuel and forces it out injector nozzle.

6. Control engine idle speed and maximum engine speed.
7. Help close injector nozzles after injection.
8. Provide a means of shutting engine off.

The injection pump bolts to the side or top of the engine. It is powered by the engine using a chain, gears, or cogged belt.

The injection pump uses the principle pictured in Fig. 18-4. A camshaft acts on plunger. The plunger slides up in its barrel (cylinder), compressing and pressurizing the fuel. The fuel then flows through the injection line and out the injector nozzle.

Injection pump components

The major parts of a diesel injection pump are pictured in Fig. 18-5. They include:

1. PUMPING PLUNGER (small piston that is moved by cam action to produce high injection pressure).
2. METERING SLEEVE (fits over pumping plunger and can be moved to control amount of fuel injected).
3. DELIVERY VALVES (spring-loaded valves that help nozzles close quickly, without leakage).
4. CAM (lobed part that moves pumping plunger).
5. TRANSFER PUMP (pump inside injection pump that fills pumping plunger area and circulates fuel through pump).
6. GOVERNOR (mechanism for controlling maximum engine speed by moving metering sleeve).
7. INJECTION TIMING ADVANCE (mechanism for changing when injection occurs as engine speed changes).
8. THROTTLE LEVER (lever connected to throttle cable or linkage that is moved by car's gas pedal to control engine speed and power).

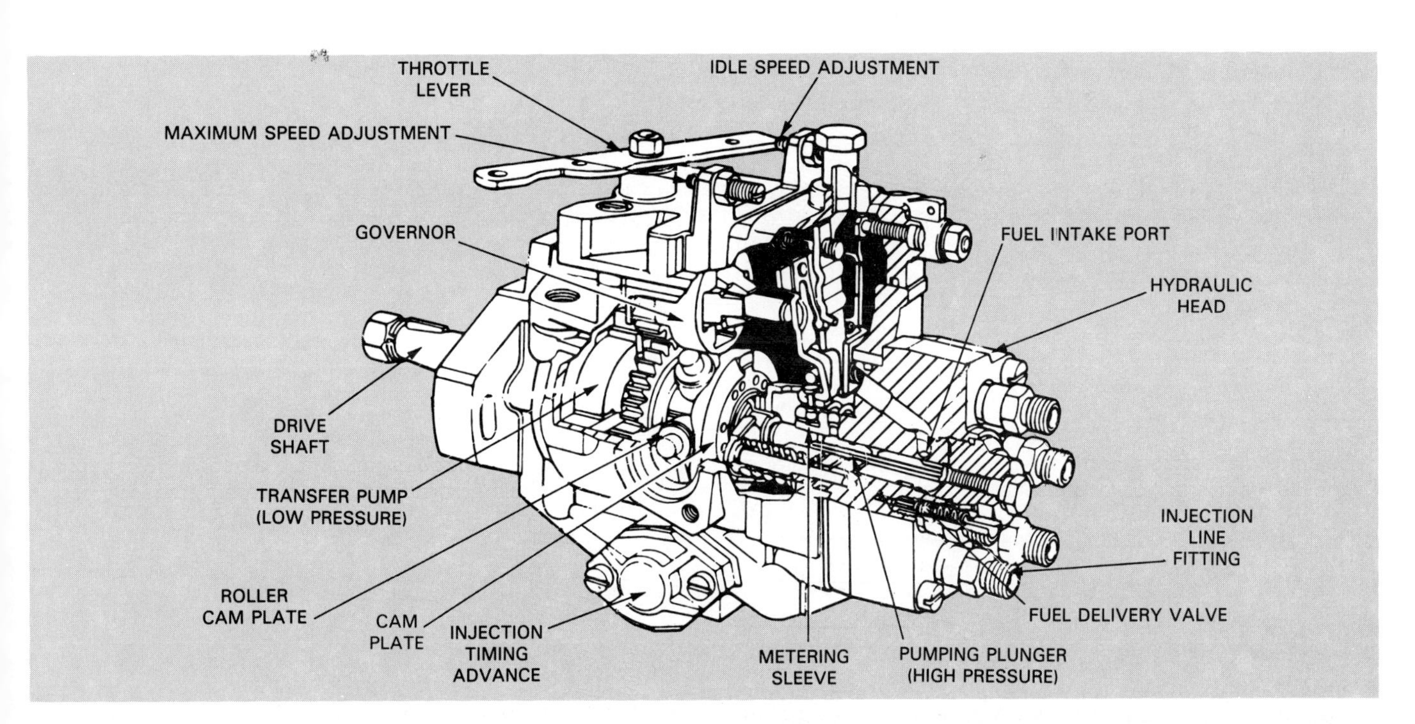

Fig. 18-5. Today's distributor injection pump uses one or two pumping plungers to feed fuel to all nozzles. Note major components: metering sleeve, pumping plunger, advance unit, transfer pump, throttle lever, governor, and fuel delivery valves. (Chrysler)

9. DRIVE SHAFT (shaft that transfers engine power into injection pump).
10. HYDRAULIC HEAD (mechanism for feeding fuel to each injection line).
11. SPEED SCREWS (screw adjustments for setting idle speed and maximum speed of diesel).

Plunger stroke

The *effective plunger stroke* is the amount of pumping plunger movement that pressurizes fuel. It controls the amount of fuel delivered to the injectors. This lets the pump control engine speed by controlling the amount of fuel injected into the chambers. Plunger stroke is altered by moving a sleeve that surrounds the pumping plunger to open or close fuel passages in the injection pump.

Governor

A *governor* is used on an injection pump to control engine idle speed and also to limit maximum engine speed. A diesel engine can be DAMAGED if allowed to run too fast.

The governor uses centrifugal (spinning) weights, springs, and levers. The levers are connected to the control sleeve. If engine speed increases too much, the governor weights are thrown outward. This moves the levers and sleeve to reduce the effective stroke of the plungers. Engine speed and power output are limited.

Injection timing

Injection timing refers to when fuel is injected into the combustion chambers in relation to the engine's piston position. It is similar to spark timing in a gasoline engine.

Fuel shutoff solenoid

A *fuel shutoff solenoid* blocks fuel flow through the injection pump so the diesel engine can be shut off. It is mounted on the injection pump as shown in Fig. 18-6. A wire from the driver's ignition switch connects to the shutoff solenoid.

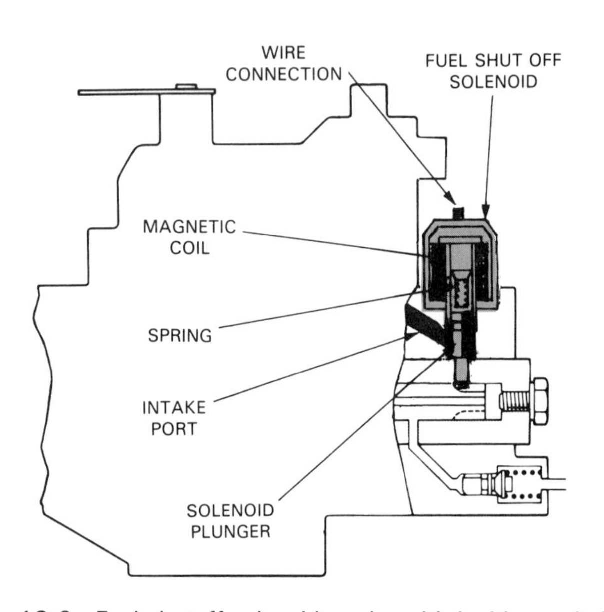

Fig. 18-6. Fuel shutoff solenoid works with ignition switch to allow driver to start or shutoff diesel engine. Current flow to solenoid makes solenoid open fuel passage for diesel operation. When current is broken, solenoid closes fuel passage and diesel shuts off.

Injection pump types

There are two basic types of diesel injection pumps common on late model cars: in-line and distributor. Other types of injection pumps are used on large trucks.

An *in-line injection pump* has one pumping plunger for each engine cylinder. The plungers are arranged one after the other in a straight line, like an in-line engine.

A *distributor injection pump* uses one or two pumping plungers that provide fuel for all of the engine's cylinders. Fuel is distributed, using aligning passages, to each nozzle. This is a more modern type pump.

DIESEL INJECTOR NOZZLES

Injector nozzles are spring-loaded valves that spray fuel directly into the diesel engine precombustion chambers. They are threaded or clamped into the cylinder head. One injector is provided for each engine cylinder, Fig. 18-7.

Nozzle components

The parts of a diesel injector include:

1. HEAT SHIELD (protects injector from engine heat and forms seal between injector and cylinder head).
2. INJECTOR BODY (main section of injector that holds other parts), Fig. 18-8.
3. NEEDLE VALVE (opens and closes nozzle or fuel opening).
4. INJECTOR SPRING (holds injector needle in normally closed position and controls opening pressure).
5. PRESSURE CHAMBER (inner cavity in injector body where injection pump pressure forces needle valve open).
6. NOZZLE (opening or orifice that forms cone spray pattern during injection).

Nozzle operation

When the injection pump produces high pressure, fuel flows through the injection line and into the inlet of the

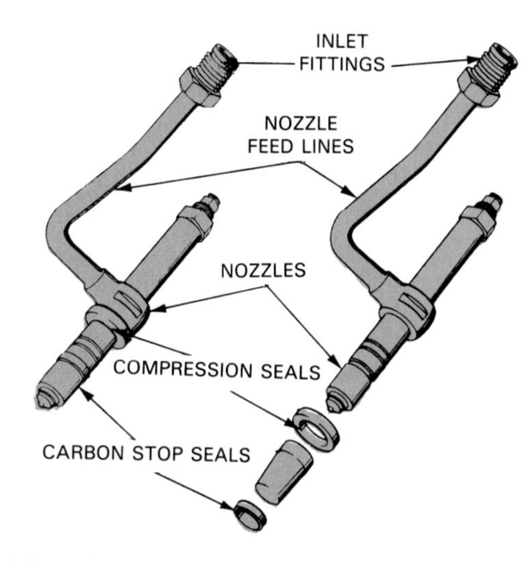

Fig. 18-7. Injector nozzles either clamp or screw into head. Note seals that prevent leakage. (Oldsmobile)

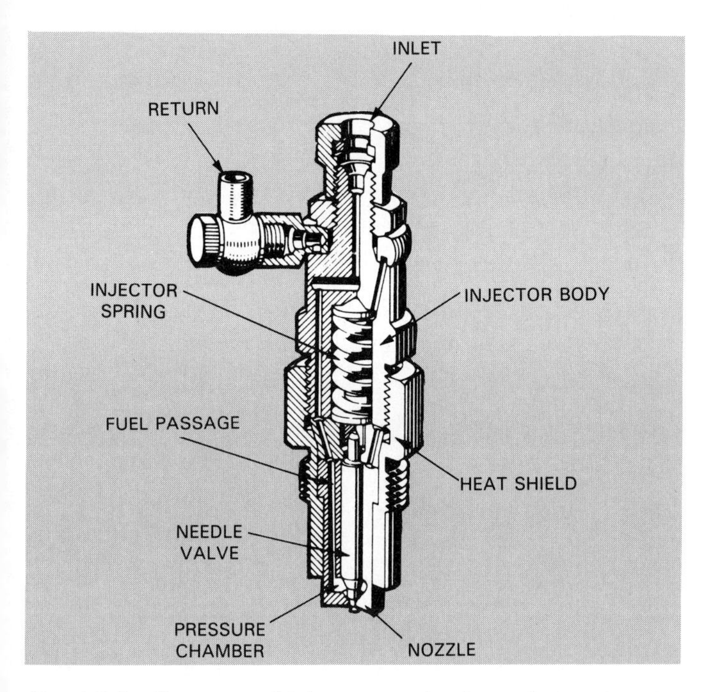

Fig. 18-8. Cutaway of injector nozzle shows internal parts. Spring holds needle in normally closed position. When injection pump forces fuel into nozzle, high pressure is formed in chamber. Pressure then pushes up on needle, compressing spring so fuel can spray into engine. (Robert Bosch)

injector valve. Fuel then flows down through the fuel passage in the injector body and into the pressure chamber.

The high fuel pressure in the chamber forces the needle upward, compressing the spring. This allows diesel fuel to spray out.

COMPUTER CONTROLLED DIESEL INJECTION

Computer contolled diesel injection, also called *electronic diesel injection,* uses engine sensors, injection pump mounted actuators, and an electronic control unit (computer), to increase efficiency.

Basically, the computer helps control injection timing and the amount of fuel delivered to each injector nozzle. This increases fuel economy and reduces emissions. It also improves engine power, acceleration, and cuts down engine smoke (particulates), especially after system parts begin to wear from high mileage, Fig. 18-9.

Diesel engine sensors

A typical computer controlled diesel system has *sensors* that monitor engine speed, manifold pressure, crankshaft position, nozzle needle position, throttle lever position, and EGR operation. Diesel sensors function the same way as sensors in a gasoline injection system operate. They monitor a condition and convert it into an electrical signal that the computer can measure. This lets the computer alter injection pump operation for maximum efficiency.

Computer controlled injection pump

A *computer controlled injection pump* is similar to a conventional all-mechanical pump. However, electric control motors are used to improve fuel metering in the

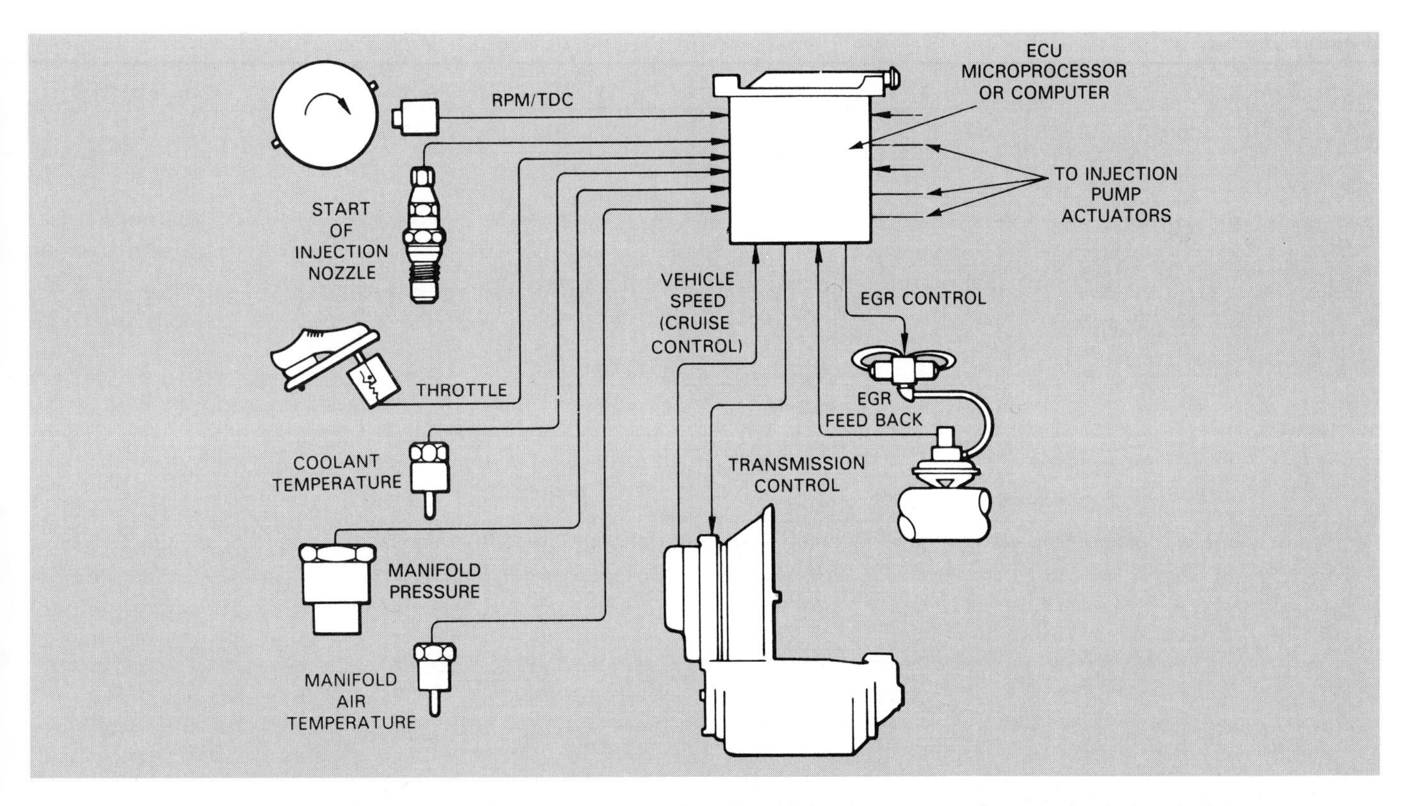

Fig. 18-9. Diagram shows basic parts of computer controlled diesel injection system. Sensors feed electrical data to computer concerning engine operating conditions. Computer can then produce electrical signals that are sent to special diesel injection pump. (Stanadyne Inc.)

injection pump. The actuators or motors serve as an interface or communication link between the electronics of the computer and the mechanical parts in the injection pump. The computer can send an electrical signal (voltage) to an actuator and the actuator converts it into movement to change the output of the pump.

An electronic-mechanical injection pump is shown in Fig. 18-10. Note the actuators on top of the pump.

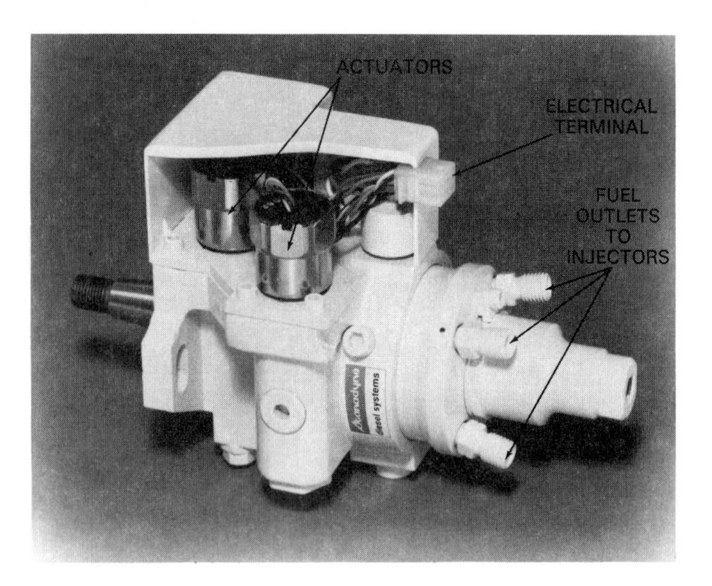

Fig. 18-10. This modern diesel injection pump has actuators or electric motors that can respond to signals from computer. It is very small and lightweight. (Stanadyne Inc.)

The *injection pump actuators* are small, reversible DC motors that operate a screw mechanism. In this way, the computer can make the motors turn a rod outward or inward to change injection volume and timing. The computer does this by changing the electrical polarity to the motors. With one polarity, the small motor spins in one direction to lengthen its output rod and with the other polarity, it reverses direction to shorten its rod.

The *fuel control motor* alters the effective plunger stroke and amount of fuel injection. The computer, after almost instantaneous gathering of data from the sensors, can energize the fuel control motor to move the fuel metering sleeve to increase or decrease plunger stroke, Fig. 18-11.

For example, if the throttle position sensor detects wide open throttle, the computer would energize the fuel control motor to move the control sleeve for a longer effective plunger stroke. This would cause more fuel to spray out the injection nozzles to increase engine power.

The *injection advance control motor* is used to alter injection timing with changes in engine speed, load, temperature, and other variables. It works just like the fuel control motor, but is used to bleed off pressure to the hydraulic advance piston. Injection timing is then more closely matched to engine needs.

Fig. 8-11 shows a cutaway view of a computer controlled diesel injection pump.

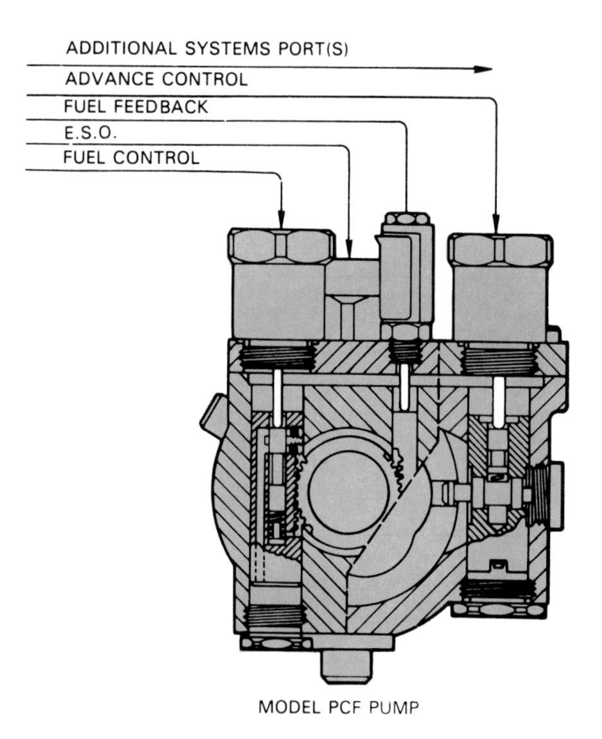

Fig. 18-11. Cutaway shows internal parts of computer controlled diesel injection pump. Actuator motors turn screw mechanisms that convert electrical signals into mechanical movement. This lets computer control mechanical parts in injection pump for improved efficiency. (Stanadyne Inc.)

Electronic governor

An *electronic governor* uses the engine speed sensor, computer, and fuel control devices in the pump to limit maximum engine rpm. Unlike an all-mechanical injection pump that uses flyweight action, electronic pulses control fuel metering at high speed. This reduces the number of components in the injection pump and protects the engine from high rpm damage. An electronic governor system can also check idle speed and change it as needed.

Diesel system computer

The *diesel system computer* is similar in construction to the computer for an electronic gasoline injection system. Tiny electronic chips called *integrated circuits* process the large amount of information from all of the sensors and then send electrical signals to the various actuators (motors or solenoids), Fig. 18-9.

As the components in the injection pump wear, the computer, by monitoring nozzle opening and fuel metering valve action, can compensate for part wear and keep injection timing and volume set for maximum efficiency. This keeps the engine "in-tune" for longer periods.

The computer sends a reference voltage to most of the engine sensors. This reference voltage is usually between 5 and 12 volts. When the resistance in a sensor changes, a different current level returns to the computer. The computer can then alter injection pump operation for the changed condition.

Fail safe system

The *fail safe system* allows the diesel injection system

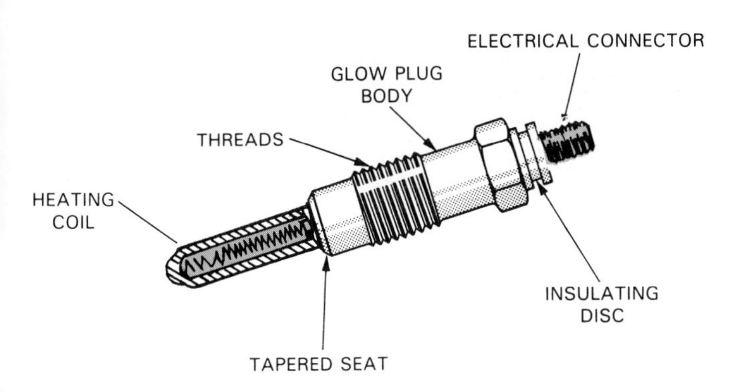

Fig. 18-12. Glow plug is heating element that aids cold starting of diesel. It has small electric heating element that glows red hot when energized. (Mercedes Benz)

to provide engine operation with failure of the computer or engine sensors. This will let the car be driven to a service center for repairs.

GLOW PLUGS

Glow plugs are heating elements that warm the air in the precombustion chambers to help start a cold diesel engine. The glow plugs are screwed into threaded holes in the cylinder head. See Fig. 18-12.

Glow plug control circuit

A *glow plug control circuit* automatically turns the glow plugs OFF after a few seconds of operation.

A temperature sensor or thermo switch checks the temperature of the engine coolant. It feeds electrical data to a control unit, Fig. 18-13. This lets the control unit keep the glow plugs OFF if the engine is already warm.

DIESEL INJECTION SERVICE

An as engine technician, you may be required to diagnose and repair diesel engines. For example, if a diesel enters the shop with a miss, you must be able to find the problem. The trouble could be an internal engine mechanical problem or a problem with the injection system. This section of the chapter will help you "build" your diesel engine service skills.

Diesel injection diagnosis

Start diagnosis by checking the operation of the engine. Check for:

1. Abnormal exhaust smoke.
2. Excessive knock.
3. Engine miss.
4. Engine no start condition.
5. Lack of engine power.
6. Poor fuel economy.

If needed, refer to the troubleshooting chart in the car's

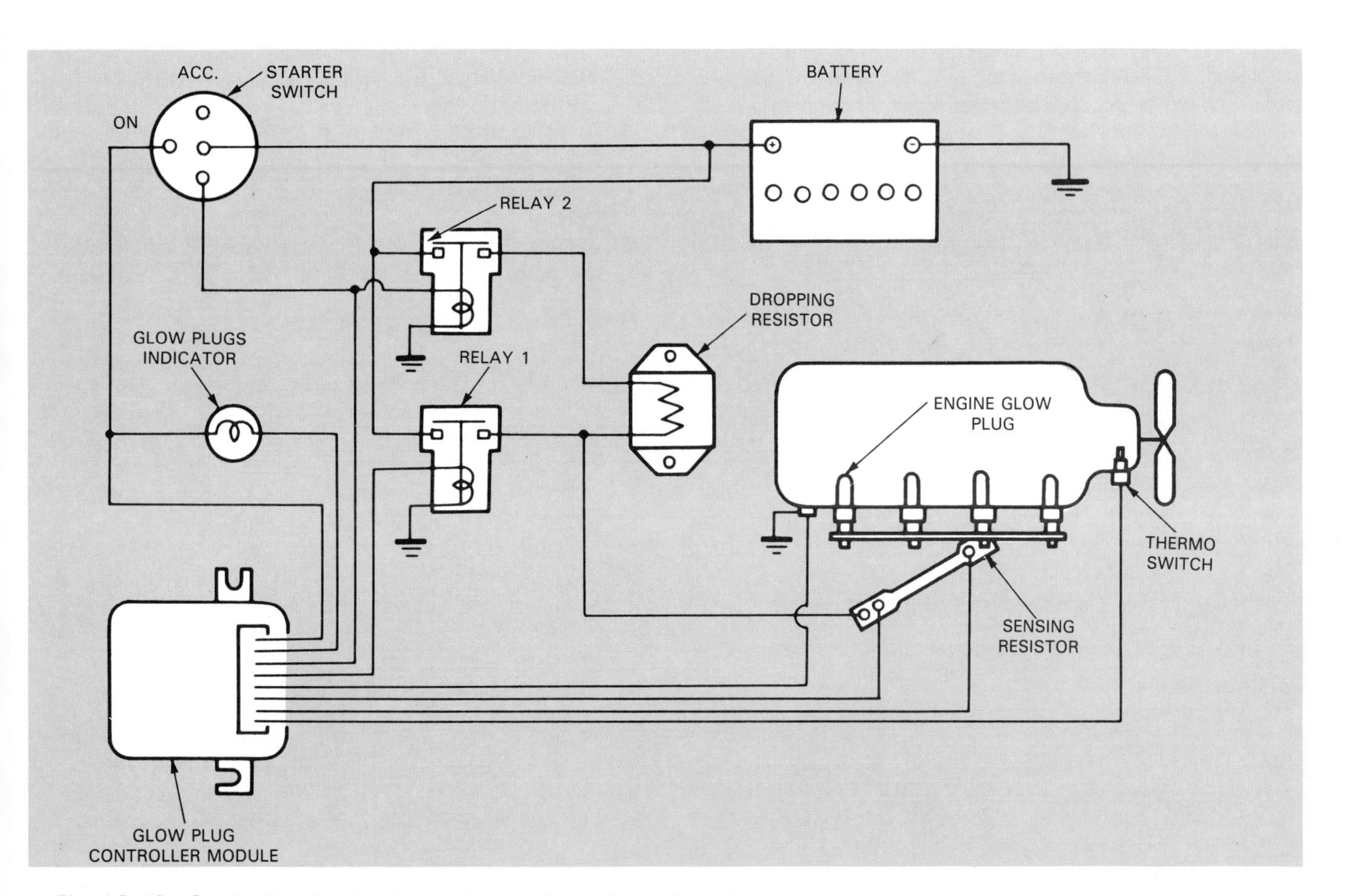

Fig. 18-13. Study this electric circuit diagram for a diesel glow plug system. Controller only turns glow plugs on when temperature sensor is below specific value. (General Motors)

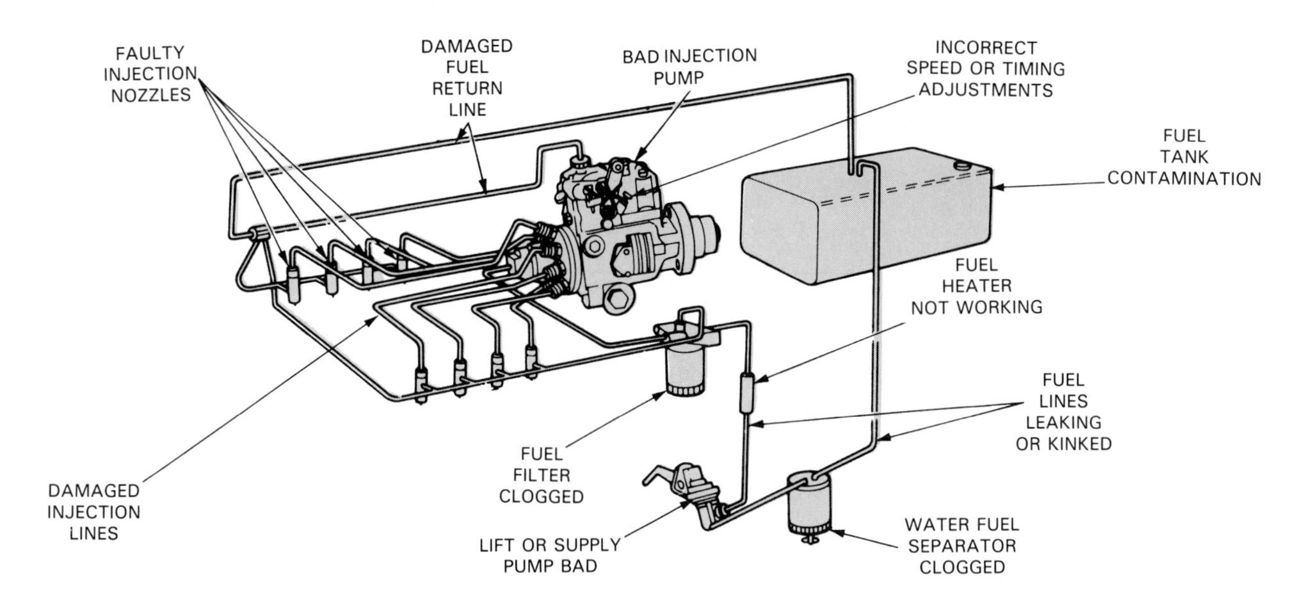

Fig. 18-14. Note types of troubles that you will find with a diesel injection system.

service manual. It will list the possible causes for each condition. The chart will be accurate because it is designed for the particular type of injection system.

Fig. 18-14 shows some typical diesel injection system problems.

Diesel engine miss

A *diesel engine miss* results from one or more cylinders not firing (burning fuel) properly. Just as a gasoline engine will miss if a spark plug does not produce a spark, a diesel engine can also run roughly because of injection system problems.

A miss in a diesel engine can be due to: faulty injectors, clogged fuel filters, incorrect injection timing, low cylinder compression, injection system leak, air leak, or faulty injection pump.

Diesel engine will not start

When a diesel engine *does not start,* it may be due to inoperative glow plugs, restricted air or fuel flow, bad fuel shutoff solenoid, contaminated fuel, slow cranking speed, or injection pump problems.

Lack of engine power

When a diesel engine *lacks power,* check the throttle cable adjustment, governor setting, fuel filters, air filter, engine compression, and other factors affecting combustion. Keep in mind that a diesel engine does NOT produce as much power as a gasoline engine of equal size.

Poor fuel economy

Poor fuel economy may be due to a fuel leak, clogged air filter, incorrect injection timing, or leaking injectors. Normally, a diesel engine will get better fuel economy than a gasoline engine.

DIESEL INJECTION SERVICE RULES

Remember these rules when servicing diesel injection systems:

1. Cap lines and fittings to prevent foreign matter from entering the fuel system.
2. Treat the diesel injector or injection pump gently. Do not pry on them or drop them.
3. High pressure inside a diesel injection system can cause serious injury. Use care.
4. Remember, diesel injection systems must be bled (air removed) after repairs.
5. Clean dirt around fittings before disconnecting them.
6. Follow all torque specifications. This is extremely important on a diesel engine.
7. Replace bent, kinked, or damaged fuel injection lines.
8. Place a piece of screen mesh over the air inlet when the engine is to be operated without the air cleaner. This will keep other objects from being sucked into the engine. Also, do not cover the air inlet with your hand while the engine is running. Injury may result.
9. Check fuel filters and water separators periodically. Water can cause expensive corrosion of injection system parts.
10. Observe safe practices concerning ventilation and fire hazards when operating a diesel engine.
11. Wear safety glasses when working on a diesel injection system.
12. When in doubt, refer to instructions in a service manual for the make and model of vehicle being serviced. Avoid mistakes which could upset engine performance or cause engine damage.
13. Use service manual procedures when working on an electronic or computer controlled diesel injection system. The slightest mistake could ruin expensive components (computer, sensors, actuators, etc.).

DIESEL INJECTION MAINTENANCE

Refer to a service manual for details on periodic maintenance of a diesel injection system. You will need to change or clean filters periodically.

If you detect signs of fuel leakage, use a piece of cardboard around each fitting. If there is a serious leak, it will strike the cardboard and not your hand. Replace any injection line or return hose that is not is perfect condition.

Fuel filters are normally located in the fuel tank (sock filter), in the fuel line (main filter), and sometimes in the injector assemblies (final filter screens). These filters must be kept clean.

The main fuel filter may have a drain. The drain can be used to bleed off trapped water. When mixed with diesel oil, WATER causes rapid corrosion and pitting of components.

TESTING DIESEL INJECTION OPERATION

There are several ways to check the operation of a diesel injection system. We will briefly explain the most common methods.

Cylinder balance test

A *cylinder balance test* on a diesel engine involves disabling one injector at a time to check the firing of that cylinder. Just as you can remove a spark plug wire on a gasoline engine to check for combustion, you can loosen the injector line to disable the cylinder. See Fig. 18-15A.

To perform a cylinder balance test on a diesel, wrap a rag around the injector and loosen one injector fitting. When the fitting is loosened, fuel should slowly leak out of the connection. Fuel leakage will prevent the injector from opening.

As the injector line is cracked, engine RPM SHOULD DROP and the engine should idle roughly. If ''killing'' the cylinder does NOT affect engine operation, that cylinder has NOT been firing. There may be a bad injector, low compression, or injection pump problem. Further tests will be needed. Check all injectors.

DANGER! When loosening an injection line, only unscrew the fitting enough to allow fuel to drip from the connection. Wear safety glasses, leather gloves, and obtain instructor approval before completing this test.

Glow plug resistance-balance test

A *glow plug resistance-balance test* provides a safe way of finding out if each cylinder is firing. When combustion is occurring in a cylinder, it will raise the temperature and internal resistance of the glow plug in that cylinder.

To do this test, allow the engine to cool. Unplug the wires to all of the glow plugs. Connect a digital ohmmeter across each glow plug and ground. Write down the ohms reading of each glow plug. Start the engine and let is run for a few minutes. Shut the engine off and recheck glow plug resistance.

If a cylinder is NOT firing, the resistance (ohms value) of its glow plug will NOT increase as much as the others. See Fig. 18-15B.

Digital pyrometer balance test

A *digital pyrometer* is an electronic device for making very accurate temperature measurements. It can sometimes be used to check the operation of a diesel engine.

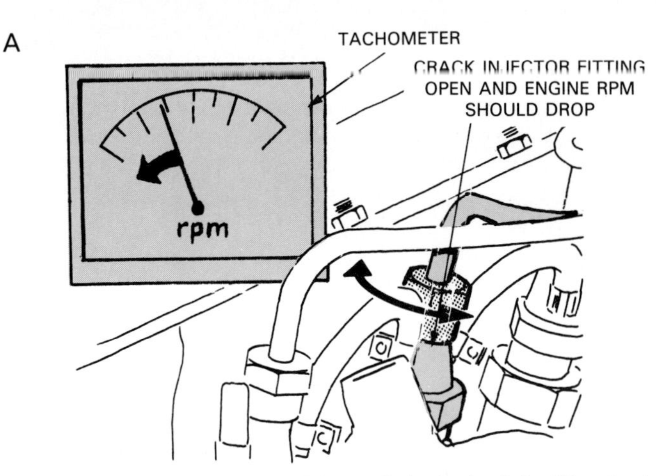

A—Cracking open injector fitting will let you find a dead cylinder. When line is cracked open, injector will stop functioning. This should make engine rpm drop and engine should idle more roughly. If idle stays same, that cylinder is not producing power. Wear gloves and eye protection.

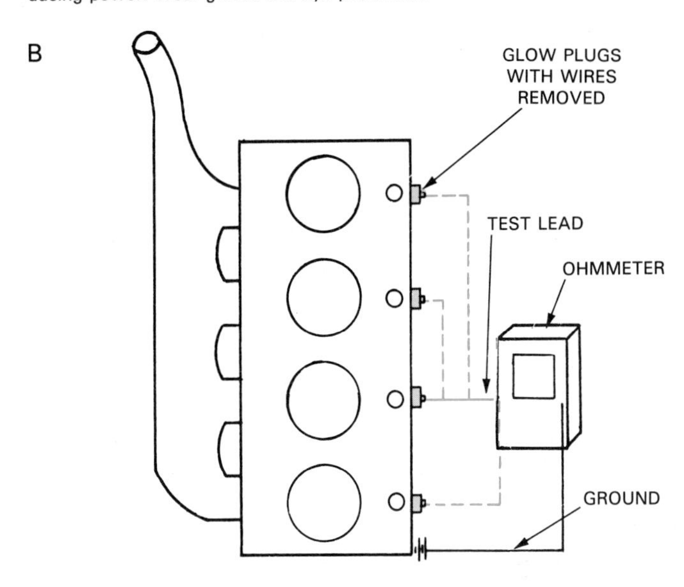

B—Glow plug resistance-balance test will also check for dead cylinder. Measure resistance of each glow plug with engine cold. Start and warm engine. Then recheck glow plug resistance. If all do not have same resistance change, something is wrong in unequal cylinder.

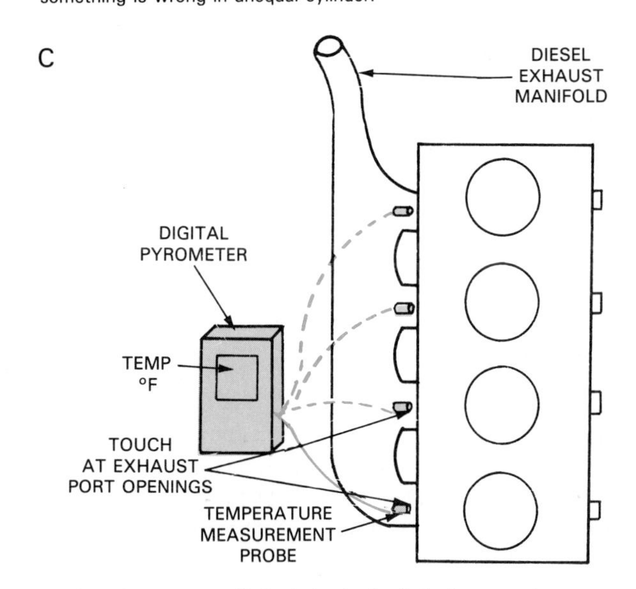

C—Digital pyrometer will check for dead cylinder by measuring temperature of outside of exhaust manifold. If area next to one cylinder head port is cooler, that cylinder is not producing combustion power.

Fig. 18-15. Three ways to check whether each cylinder in diesel is producing power or combustion. (Volvo)

Touch the digital pyrometer on the exhaust manifold at each exhaust port. The temperature of the exhaust manifold with the engine running should be almost equal. If the manifold reading is cooler next to one exhaust port, that cylinder is not firing, Fig. 18-15C.

DIESEL INJECTOR SERVICE

A *bad diesel injector* can cause the engine to miss. It may also reduce engine power or cause smoking and knocking.

The injector nozzles are exposed to the direct heat and by-products of combustion. They can wear, become clogged with carbon, or be damaged. This can result in an incorrect opening pressure, incorrect spray pattern, fuel leakage, and other problems.

Injector substitution

An easy way to verify an injector problem is to *substitute* a good injector for the possible bad injector. If the cylinder fires with the good injector, then the old injector is faulty. If the cylinder still misses with the good injector, then other engine or injection problems exist.

Injector removal

If your tests indicate a faulty injector, remove the injector for service. Following the shop manual, disconnect the battery to prevent engine cranking. Using the appropriate tools, remove the injection line. Be careful not to bend or kink the high pressure line. Unscrew or unclamp and remove the injector.

Pop testing diesel injector

A *pop tester* is a device for checking a diesel injection nozzle while it is out of the engine. One type tester is pictured in Fig. 18-16A.

To use a pop tester, fill the tester reservoir with the correct *calibration fluid* (test liquid). Do NOT use diesel oil because it is too flammable and test results may not be reliable.

Open the tester valve and pump the tester handle. As soon as solid fluid (no air bubbles) sprays out the tester, close the valve. Connect nozzle to tester, Fig. 18-16B.

DANGER! Extremely high pressures are developed when pop testing a diesel injector nozzle. Wear eye protection and keep your hands away from the fuel spray.

To check *injector opening pressure,* purge any remaining air from the nozzle by pumping the tester lever up and down. Then, pump the handle slowly while watching the pressure gauge. Note the reading when the injection nozzle opens. Repeat the test until you are sure you have an accurate reading.

Typical diesel injector opening pressure is approximately 1700 to 2200 psi (11 713 to 15 158 kPa). If opening pressure is NOT within service manual specs, rebuild or replace the injector.

To check the *injector's spray pattern,* operate the pop tester handle while watching the fluid spray out of the nozzle. As in Fig. 18-16C, there should be a narrow, cone-shaped mist of fluid. A solid stream of fuel, uneven spray, excessively wide spray, or spray filled with liquid droplets indicates that the injector requires service or replacement.

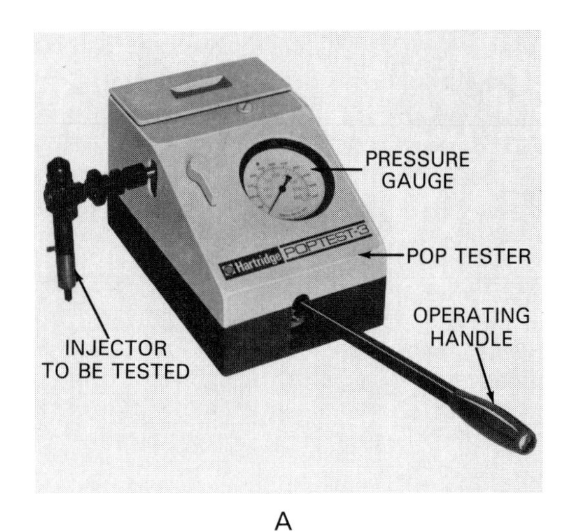

A

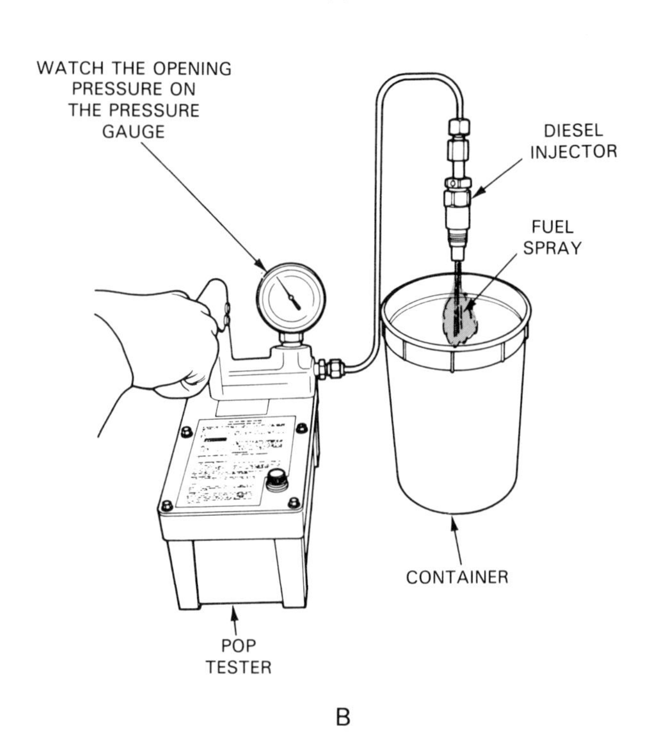

B

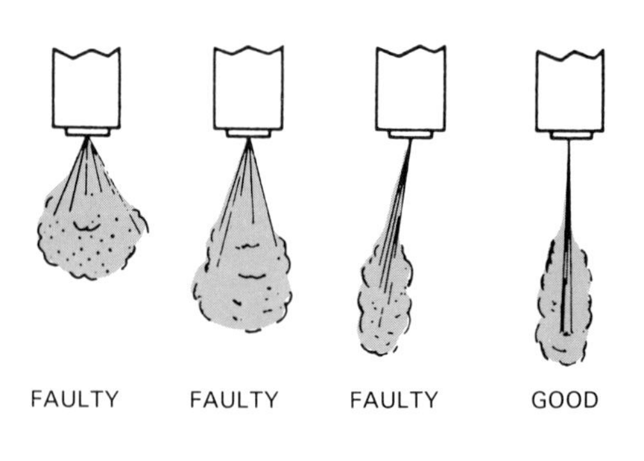

C

Fig. 18-16. Testing diesel injector. A—Pop tester will allow you to see and measure injector performance. B—Mount nozzle on pop tester. Pump handle while watching pressure gauge and spray pattern. C—Good and bad spray patterns. (Toyota, Ford, and Hartridge)

To check *injector leakage,* slowly operate the tester handle to maintain a pressure LOWER than the nozzle opening pressure. Many auto manufacturers recommend a pressure of about 300 psi (2060 kPa) below opening pressure. With this pressure, the diesel injector should NOT leak or drip for 10 seconds. Leakage would point to a dirty injector nozzle or worn components.

Diesel injector rebuild

An *injector nozzle rebuild* involves disassembly, cleaning, inspecting, replacing bad parts, reassembly, and testing the injector. Since injector designs vary, always follow the detailed directions in a service manual.

Installing diesel injectors

When installing a diesel injector, coat the threads with anti-seize compound. Use a new heat shield or seal to prevent compression leakage. Screw the injector into the cylinder head by hand. Then, torque it to specs. Without bending it, reconnect the injector line. Tighten the connection properly, Fig. 18-17.

Reconnect the wire on the fuel shutoff solenoid and start the engine. Check the system for leaks and proper operation. A few systems may require you to BLEED AIR out of the system.

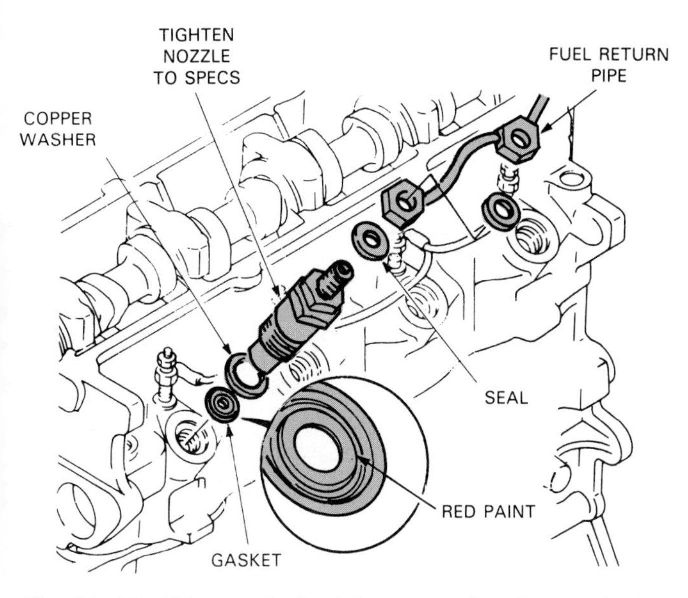

Fig. 18-17. When replacing injector nozzle, place anti-seize compund on threads. Use new gaskets or seals if needed. Torque to specs. Be careful not to kink or bend injection line when disconnecting or connecting to nozzle. (Ford)

GLOW PLUG SERVICE

Inoperative glow plugs will make a diesel engine hard to start when cold. If only one or two glow plugs are inoperative, the engine may miss while still cold.

If glow plug problems are indicated, use a test light or VOM to check for voltage to the glow plugs. Touch the probe on the feed wires to the glow plugs. A clip-on ammeter can also be used to check current flow to the glow plugs.

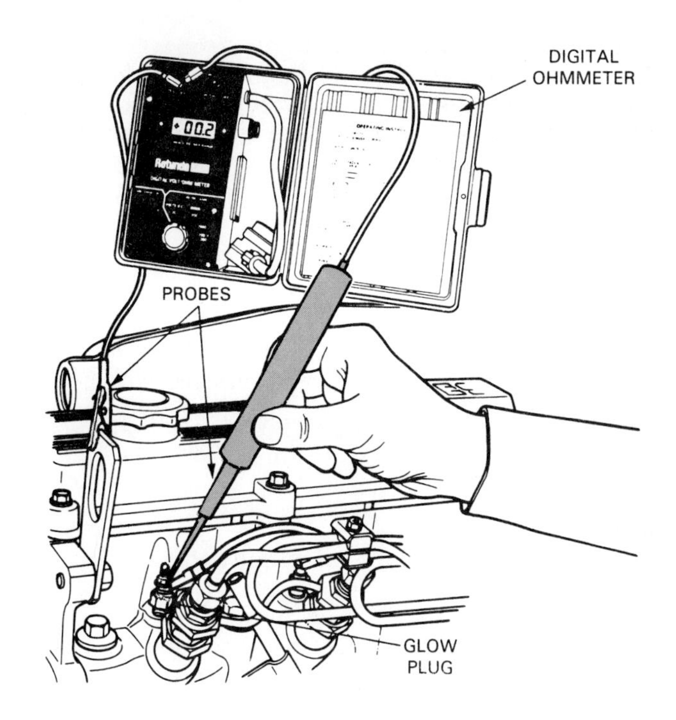

Fig. 18-18. Ohmmeter will check for opened or shorted glow plug. Resistance must be within specs. Plugs can be checking while in or out of engine. (Ford)

An ohmmeter can be used to determine the condition of each glow plug. Connect the ohmmeter across each glow plug terminal and ground. The resistance should be within specs. If the ohms value is too high or low, replace the glow plug, Fig. 18-18.

Glow plug replacement

To remove glow plugs, disconnect the wires going to them. Then, use a deep well socket and ratchet to unscrew the glow plugs.

DANGER! A glow plug can reach temperatures above 1000 °F (538 °C). This can cause serious burns to your hand.

When the glow plugs are removed, inspect each one closely. Look for damage or a heavy coating of carbon. A carbon buildup can insulate the plug and make the engine hard to start. Clean any plugs that are to be reused. Replace all that are faulty.

When installing the glow plugs, coat the threads with anti-seize compound. Start the plugs by hand and then tighten to specs.

INJECTION PUMP SERVICE

A *bad injection pump* can keep the engine from running, or cause engine missing, smoking, and other problems. The injection pump is usually a very trouble-free unit. However, water contamination, prolonged service, leaking seals, and physical damage from accidents may require replacement, repairs, or adjustment.

Most garages remove and install a NEW or FACTORY REBUILT injection pump when internal parts are faulty. The diesel injection pump is a very precise mechanism. Specialized tools and equipment are needed for repair.

Injection pump timing

Diesel *injection pump timing* is adjusted by rotating the injection pump on its mounting. To advance injection timing, turn the pump OPPOSITE drive shaft rotation. To retard injection timing, turn the pump housing WITH drive shaft rotation.

Injection pump timing must be set whenever the pump is removed from the engine or when an incorrect adjustment is discovered.

There are several methods for adjusting injection timing. Basically, injection pump timing can be adjusted by:

1. Aligning timing marks on engine and injection pump with No. 1 piston at TDC (rough adjustment).
2. Using a dial indicator to measure injection pump plunger stroke in relation to engine piston position, Fig. 18-19.

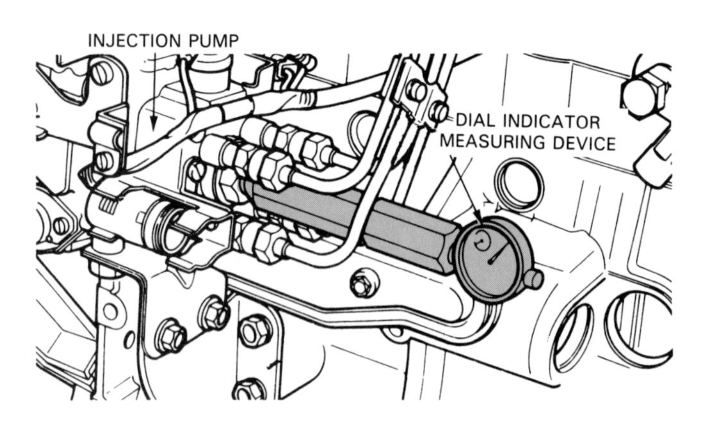

Fig. 18-19. Some manufacturers recommend dial indicator for measuring pumping plunger movement to set injection timing. Plunger movement is compared to timing marks on crank damper so injection can be timed with engine piston position. Follow manual directions. (Ford)

3. Using a luminosity (light) device to detect combustion flame in No. 1 cylinder, Fig. 18-20.
4. Using a fuel pressure (injection) detector.

Refer to a factory shop manual for details. The manual will describe which adjustment method should be used. Also, follow the operating instructions for the specific type of timing device. Injection timing is very critical to diesel engine performance.

Injection pump throttle cable adjustment

Basically, injection pump *throttle cable* or *linkage adjustment* involves making sure the pump's throttle lever opens and closes freely. As parts wear, the cable or linkage may need to be reset. During adjustment, check for binding. Lubricate if needed, Fig. 18-21A.

INJECTION PUMP SPEED ADJUSTMENTS

There are three speed adjustments commonly used on a diesel injection pump: curb idle speed, maximum speed, and fast (cold) idle speed.

Diesel idle speed adjustments

Diesel idle speed is usually set using a special diesel tachometer. An idle stop screw on the injection pump is turned to vary engine speed.

A *diesel tachometer* may be similar to or part of the injection timing device. With all designs, the diesel tach will read engine speed in rpm (revolutions per minute).

Diesel curb idle speed

To adjust diesel *curb idle speed,* place the transmission in neutral or park. Start the engine. Allow the engine to run until it reaches full operating temperature. Connect the tachometer to the engine.

Compare the tach reading with specs. Curb idle speed will be given on the engine compartment emission con-

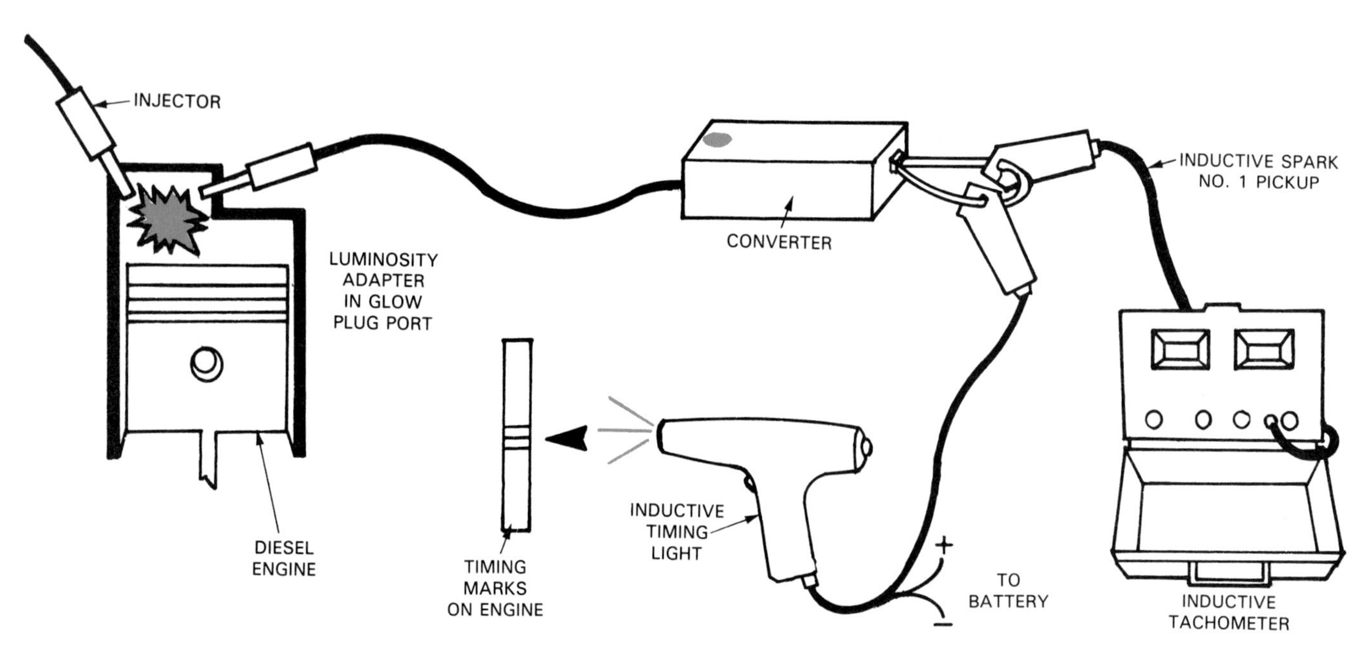

Fig. 18-20. Luminosity probe will detect combustion flame. Converter amplifies signal from probe so conventional timing light or tach can be used on diesel. A diesel does not have an electrical ignition system for powering a timing light or tach.

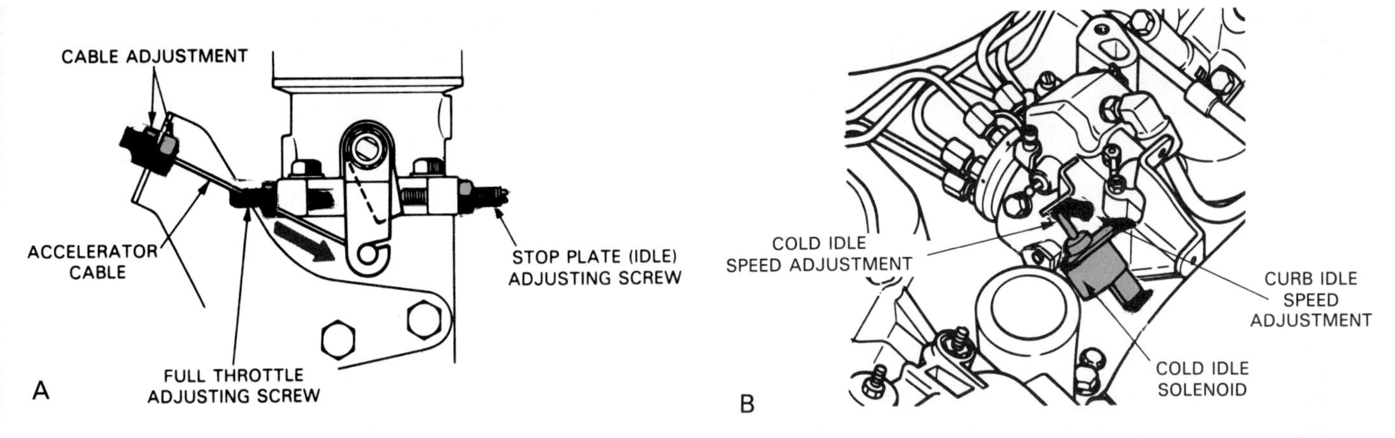

Fig. 18-21. Speed adjustments on diesel are fairly simple. Turn adjustment screw until tach reads with specs. A—Cable adjustment, idle adjustment, and maximum speed adjustment. B—This view shows cold idle solenoid adjustment and curb idle screw. Follow service manual procedures when adjusting. (Ford)

trol decal or in the service manual.

If needed, turn the curb idle speed adjusting screw on the injection pump to raise or lower idle rpm, Fig. 18-21B. You must usually loosen a locknut before the screw will turn. Hold the screw and retighten the locknut when the tach reads correctly.

Diesel cold idle speed adjustment

To set the *cold idle speed* of a diesel, connect a jumper wire from the battery positive terminal and the fast idle solenoid terminal. This will activate the fast idle, even when the engine is warm.

Raise the engine speed momentarily to release the solenoid plunger. Again, compare your tachometer readings to specs. Adjust the solenoid if needed.

Sometimes a cold start lever replaces a cold idle solenoid. Refer to manufacturer manuals for details of adjustment.

Diesel maximum speed adjustment

A *maximum speed adjustment* is used to limit the highest possible diesel engine rpm. If maximum governor rpm is set too high, internal engine damage may result. If diesel maximum rpm is too low, the engine will not produce enough power.

To adjust the maximum speed of a diesel, position your tachometer so that it can be read from within the driver's compartment. With the transmission in neutral or park, hold your left foot on the brake. Press the accelerator pedal slowly to the floor.

CAUTION! When performing a maximum speed test, be ready to release the accelerator pedal at any time to prevent overspeed damage.

Once the tachometer reads the maximum speed specs, engine speed should NO LONGER INCREASE. If the maximum speed is not within specs, turn the maximum speed adjusting screw on the injection pump. Tighten the locknut on the screw after adjustment. Recheck maximum rpm.

FUEL CUTOFF SOLENOID SERVICE

A *bad fuel cutoff solenoid* will usually keep the diesel engine from running. Windings may fail so it can no longer open to allow fuel entry to the pumping chambers in the pump. It is also possible for the solenoid to fail in the open position, keeping the diesel engine from shutting off.

Always check the supply circuit to the fuel cutoff solenoid. Wires feeding current to the solenoid could be opened or shorted, causing faulty operation of the solenoid.

Testing fuel cutoff solenoid

Check for power to the solenoid first. Use a test light. The test light should glow with the ignition key on or the engine cranking. If not, disconnect the wire from the solenoid. Check for power on the incoming wire alone. If you do not have voltage, the circuit is bad. If you do have voltage, the solenoid is probably faulty.

An ohmmeter can also be used to measure the resistance of the windings in the solenoid. If open or shorted, replace the unit.

Replacing fuel cutoff solenoid

To replace the fuel cutoff solenoid, simply unscrew the old unit. Be careful not to let dirt or other debris fall into the opening in the injection pump. After making sure it is clean, screw the new solenoid into the pump. Torque it to specs. Connect its feed wires and check engine operation.

COMPUTER CONTROLLED DIESEL INJECTION SERVICE

Most computer controlled diesel injection systems have self-diagnosis. The computer will produce a number code indicating possible troubles with components in the system. This is similar to the self-diagnosis discussed for gasoline injection.

Always refer to a factory shop manual for details of activating and testing a computer controlled diesel injection system. Procedures vary and the slightest mistake could cause part damage.

Sensor and actuator problems

Like a gasoline injection system, engine sensors and actuators on diesel injection systems can fail and upset engine operation.

You must use the same basic techniques to check the internal resistance of components, voltage output at various wires, and current levels. If a voltage, resistance, or current value is not within specs, corrective action must be taken.

SUMMARY

Diesel injection is a high pressure system that forces fuel into the engine combustion chambers. An injection pump produces this high pressure and times injection with the engine's power strokes.

Injector nozzles are spring-loaded fuel valves. Pressure from the injection pump opens the valves for injection into the diesel precombustion chambers.

The pumping plunger(s) in the injection pump use cam action to pressurize the fuel. A metering sleeve in the pump can be shifted to control effective plunger stroke and injection quantity. Delivery valves in the pump make the nozzles close quickly without leakage.

A governor in the injection pump limits top engine rpm. Too much speed could damage engine parts.

An injection timing advance alters injection timing with changes in engine speed. Injection should occur sooner as engine speed increases.

A throttle lever on the injection pump connects to the driver's gas pedal. Movement of the pedal rotates the lever to change injection quantity and engine power output.

A fuel shutoff solenoid is used to prevent fuel flow into the pumping plunger area of the injection pump. This lets the driver turn the engine off with the key switch.

The injector nozzles screw into threaded holes in the diesel cylinder head. Glow plugs screw into the head next to the nozzles.

The glow plugs are small electric heating elements. They help heat the air in the prechambers so a cold diesel engine will start.

Some later model systems use a computer to control injection pump operation. Engine sensors report various conditions to the computer. The computer can then send electrical signals to actuators on the pump. The actuators are small electric motors that operate screw mechanisms. In this way, computer signals are converted into mechanical movement for controlling the injection timing and injection quantity.

When diagnosing diesel engine problems, you must determine whether the cause of the problem is in the engine itself or the fuel system.

A cylinder balance test is done to find a "dead cylinder" (cylinder not firing). Crack open an injection line and note whether engine rpm drops. If rpm drops, that cylinder is firing and producing power. If rpm stays the same, that cylinder is dead. Something is preventing combustion.

A glow plug resistance-balance test will also check for a dead cylinder. Measure glow plug resistance with an ohmmeter with the engine cold and then with it warmed up. If the resistance does not change equally at all glow plugs when warm, something is wrong in that cylinder.

A digital pyrometer will check for a dead cylinder by accurately measuring the temperature of the exhaust manifold. Check temperature at the exhaust ports. If the area around one port is cooler, that cylinder may not be producing power.

If you suspect a bad injector, substitute a known good injector. If this corrects the diesel miss, you have found the problem.

A pop tester will check the operation of a diesel injector. It will force test fluid into the nozzle so you can check opening pressure, leakage, and spray pattern.

Injection pump problems usually require a specialist. Pumps are very complicated and precise. Most shops intall a new or rebuilt injection pump when needed.

Injection pump timing is critical to diesel engine performance. Injection timing can be set using several techniques: aligning marks on pump and engine, using dial indicator to measure pump stroke, and luminosity device. Always follow service manual procedures.

Diesel speed adjustments require a special diesel tachometer since a diesel does not have an electric ignition system. Speed adjustment screws for curb idle, fast idle, and maximum speed are usually on the injection pump.

KNOW THESE TERMS

Supply pump, Injection pump, Injection line, Nozzle, Return line, Computer, Sensors, Actuators, Glow plugs, Glow plug controller, Pumping plunger, Metering sleeve, Delivery valves, Transfer pump, Governor, Throttle lever, Drive shaft, Hydraulic head, Effective plunger stroke, Injection timing, Fuel shutoff solenoid, In-line pump, Distributor pump, Needle valve, Computer controlled diesel, Electronic governor, Fail safe system, Cylinder balance test, Digital pyrometer, Calibration fluid, Injector rebuild, Diesel tach.

REVIEW QUESTIONS—CHAPTER 18

1. List and explain the ten basic parts of a modern diesel injection system.
2. This component controls injection quantity.
   a. Transfer pump.
   b. Delivery valves.
   c. Solenoid.
   d. Metering sleeve.
3. The __________ __________ __________ is the amount of plunger movement that pressurizes fuel.
4. How does a mechanical governor work?
5. __________ __________ refers to when fuel is sprayed into the combustion chambers.
6. What makes a diesel engine shut off?
7. What are the two major types of injection pumps?
8. How does a diesel injector operate?
9. How does a computer controlled diesel system work?
10. Glow plugs are __________ __________ that warm the air in the prechamber and aid __________ starting of a diesel.
11. A diesel engine comes into the shop with a miss at idle.
    Technician A says that the shop's digital pyrometer should be used to check for a cold section next to a port on the exhaust manifolds.

Technician B says that cracking open the injection lines is a safer and faster way of finding a dead cylinder.
Who is correct?
a. Technician A.
b. Technician B.
c. Both A and B.
d. Neither A nor B.

12. Diesel oil is commonly used inside a pop tester. True or false?
13. Typical diesel injector opening pressure is about ________ to ________ psi.
14. What is the most common symptom of inoperative glow plugs?
15. Tests find a faulty diesel injection pump. Technician A says the pump should be disassembled to find and repair the source of the problem. Technician B says it is better to install a new or rebuilt injection pump.
Who is correct?
a. Technician A.
b. Technician B.
c. Both A and B.
d. Neither A nor B.

ASE CERTIFICATION-TYPE QUESTIONS

1. Technician A says that modern diesel injection systems use a high pressure electrical pump to force diesel fuel into the engine's combustion chambers. Technician B says that modern diesel injection systems use a high pressure mechanical pump to force diesel fuel into the engine's combustion chambers. Who is right?
(A) A only.
(B) B only.
(C) Both A & B.
(D) Neither A nor B.
2. All of the following are basic parts of a diesel injection system EXCEPT:
(A) glow plug controller.
(B) actuators.
(C) fuel rail.
(D) injector nozzles.
3. Technician A says that one of the functions of a diesel injection pump is to provide a means of shutting off the engine. Technician B says that one of the functions of a diesel injection pump is to allow the driver to control engine power output. Who is right?
(A) A only.
(B) B only.
(C) Both A & B.
(D) Neither A nor B.
4. Technician A says that a diesel injection pump's governor is used to control engine idle speed. Technician B says that a governor is used on a diesel injection pump to limit maximum engine speed. Who is right?
(A) A only.
(B) B only.
(C) Both A & B.
(D) Neither A nor B.
5. Technician A says that one of the sensors in a computer-controlled diesel injection system is used to monitor manifold pressure. Technician B says that one of the sensors in a computer-controlled diesel injection system is used to monitor injector nozzle needle position. Who is right?
(A) A only.
(B) B only.
(C) Both A & B.
(D) Neither A nor B.
6. Technician A says that an inline injection pump uses one or two pumping plungers to provide fuel for all the engine's cylinders. Technician B says that an inline injection pump has one pumping plunger for each engine cylinder. Who is right?
(A) A only.
(B) B only.
(C) Both A & B.
(D) Neither A nor B.
7. Technician A says that during injection system diagnosis, you should check the engine for excessive knocking. Technician B says that during diesel injection system diagnosis, you should check for abnormal exhaust smoke. Who is right?
(A) A only.
(B) B only.
(C) Both A & B.
(D) Neither A nor B.
8. If a diesel engine's cylinder is not firing, the resistance of its glow plug will not ____________ the other glow plugs.
(A) increase more than
(B) increase as much as
(C) decrease more than
(D) decrease as much as
9. All of the following are various methods used to check the operation of a diesel injection system EXCEPT:
(A) cylinder balance test.
(B) glow plug resistance-balance test.
(C) spark test.
(D) digital pyrometer balance test.
10. Technician A says that a dial indicator can be used to adjust diesel injection pump timing. Technician B says that a luminosity device can be used to adjust diesel injection pump timing. Who is right?
(A) A only.
(B) B only.
(C) Both A & B.
(D) Neither A nor B.

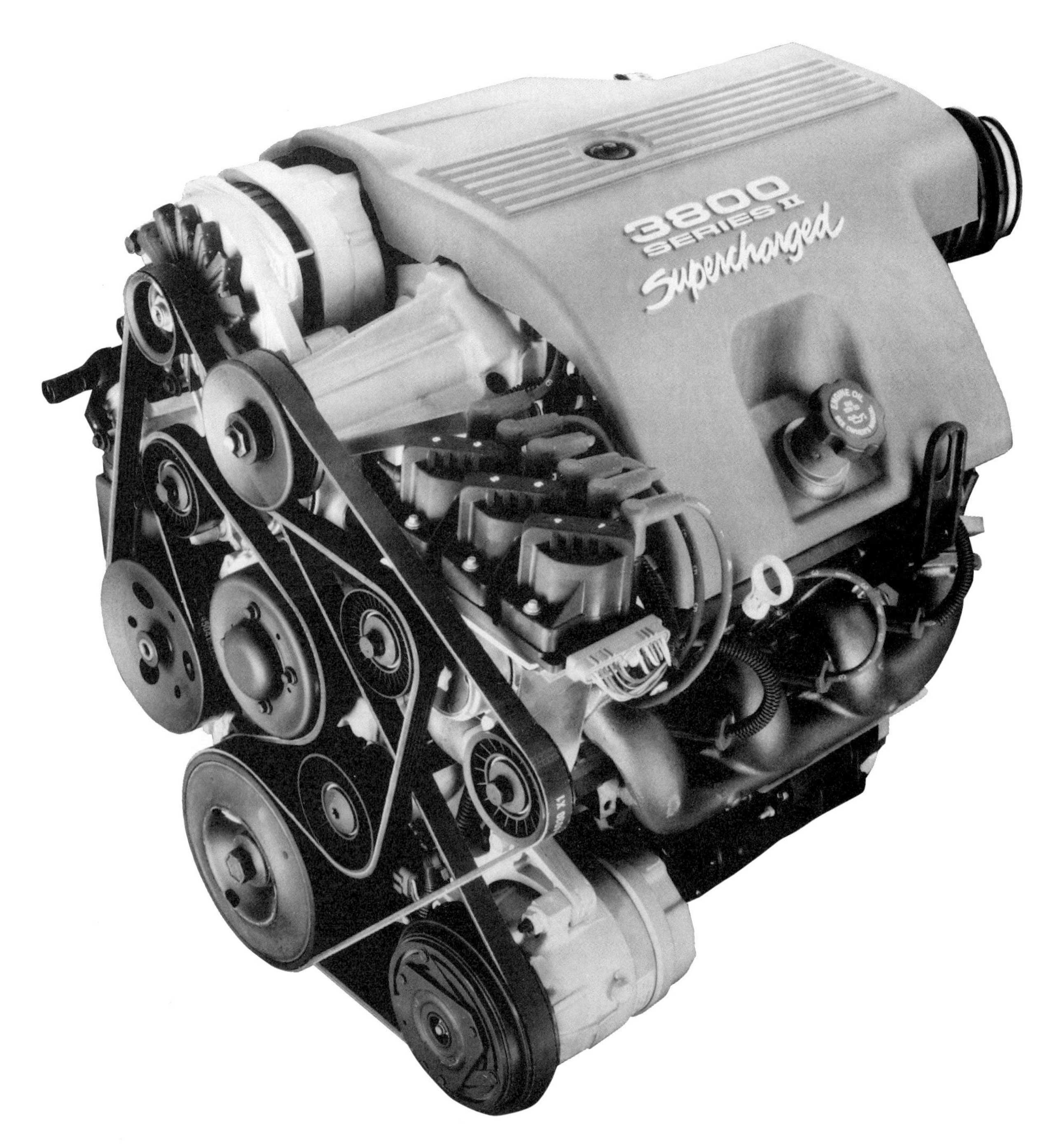

Superchargers have been used on race cars for years and are now being used on production vehicles. They have the ability to produce tremendous horsepower while placing little drag on the engine. (General Motors)

Chapter 19

Turbocharging and Supercharging Systems

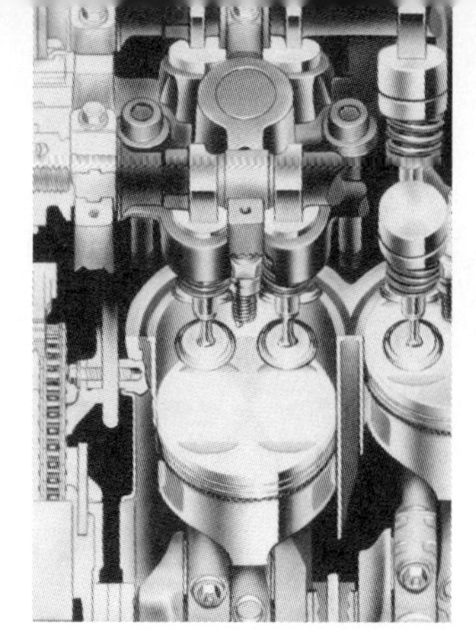

After studying this chapter, you will be able to:

- *Explain the construction of a supercharger.*
- *Explain the construction of a turbocharger.*
- *Describe the construction of a waste gate.*
- *Summarize turbocharging system operation.*
- *List the reasons why turbochargers and superchargers are used on today's engines.*
- *Explain the operation of an intercooler.*
- *Diagnose common turbocharger problems.*
- *Remove and replace a turbocharger and waste gate.*
- *Remove and replace a supercharger.*

Turbochargers are now used by almost all automobile manufacturers. A "turbo" has the ability to increase the horsepower of an engine by as much as 50 percent. For this reason, they are commonly used on small in-line four-cylinder and V-six cylinder engines to increase performance. Turbochargers, as you will learn, can increase power under a load and still have little affect on fuel economy at cruising speeds.

Superchargers have been used on race engines for many years. Manufacturers began installing superchargers on production engines a few years ago and are beginning to see increased use.

To be a competent technician, you must have some knowledge of turbochargers and superchargers because so many engines are equipped with them. Study carefully!

NORMAL ASPIRATION

Normal aspiration means the engine uses only outside air pressure or atmospheric pressure to cause airflow into the combustion chambers. At sea level, this would be 14.7 psi (101 kPa).

Normal aspiration refers to an engine that is NOT supercharged or turbocharged. With only outside air pressure to carry oxygen into the engine cylinders, engine power is limited by the engine's low volumetric efficiency.

Volumetric efficiency is a measure of how much air or air-fuel mixture the engine can draw in on its intake strokes. A high volumetric efficiency means the engine "breathes" easily because of good intake port, valve, combustion chamber, and camshaft design.

Because of *pumping losses* (restriction to airflow through intake manifold and cylinder head), most engines do not have high volumetric efficienty. As a result, they do not produce much horsepower for their size. Turbocharging and supercharging is used on many gasoline and diesel engines to improve efficiency and power output.

SUPERCHARGING AND TURBOCHARGING

Supercharging is a term that refers to any method of increasing intake manifold pressure. A *supercharger* is a mechanically driven blower that forces air into the engine under pressure. A belt is used to spin the blower and force air (diesel) or air-fuel mixture (gas engine) into the combustion chambers. They are used on some passenger car engines and can be found on many racing engines. See Fig. 19-1A.

A *turbocharger* is a blower or special type fan assembly that uses engine exhaust gases to turn the blower. A mechanical linkage with the engine is not needed. Look at Fig. 19-1B.

Passenger car supercharging

A supercharger for a passenger car is a small unit mounted near the front of the engine. It is usually belt driven off the engine crankshaft damper, as in Fig. 19-1.

Some superchargers have an electromagnetic clutch, similar to the clutch on an air conditioning system compressor. This allows the supercharger to be turned on and off. One system uses a wide open throttle switch to activate the supercharger, Fig. 19-1. When the gas pedal is pushed to the floor for passing or getting up to highway speed, the switch sends power to the clutch. This engages the supercharger and engine power is instantly increased. However, most superchargers are direct drive.

Most superchargers used on production vehicles are small units that provide 4-10 pounds (27.58-68.95 kPa) of boost. Larger units that provide much more boost are used on racing engines and diesel trucks. The supercharger uses two or three rotors that compress air entering the engine. This increased air is mixed with fuel and pulled into the combustion chamber.

Supercharging has the advantage of instant air pressure increase on demand. With a turbocharger, there is a slight delay in pressure and power increase as the unit has to build up rpm (speed). This will be discussed in detail shortly.

Turbocharger construction

Pictured in Fig. 19-2, the major parts of a turbocharger are:

1. TURBINE WHEEL (exhaust driven fan that turns turbo

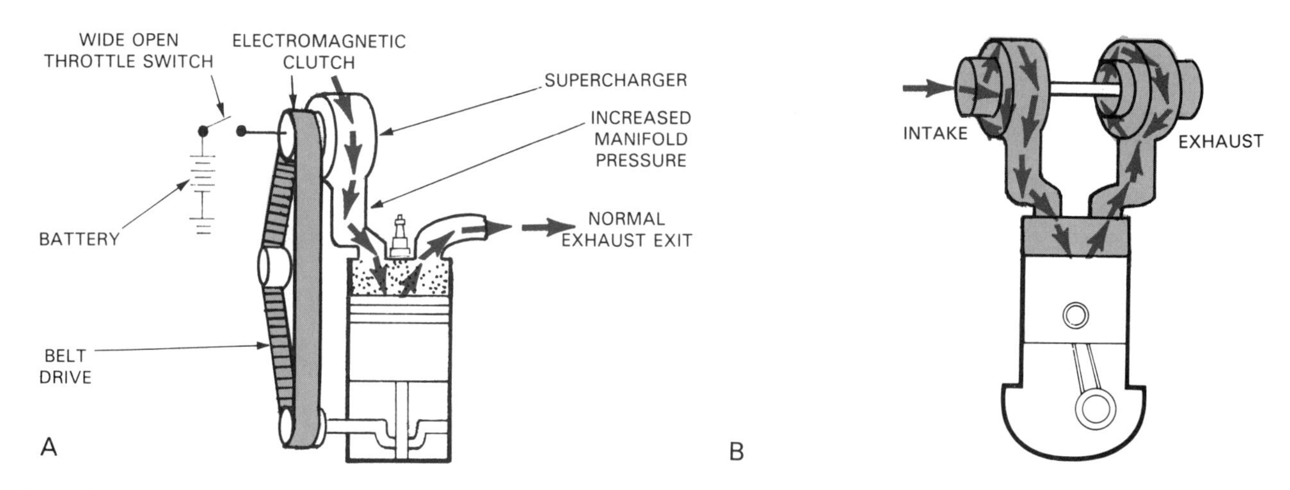

Fig. 19-1. Comparison shows basic differences between supercharging and turbocharging. A—Automotive supercharger normally uses an engine driven belt to spin blower. Air-fuel mix is forced into cylinders under pressure. B—Turbocharger does not have mechanical link with engine. Burning exhaust gases spin turbo. Other end of turbo can then pressurize intake charge for more engine power.

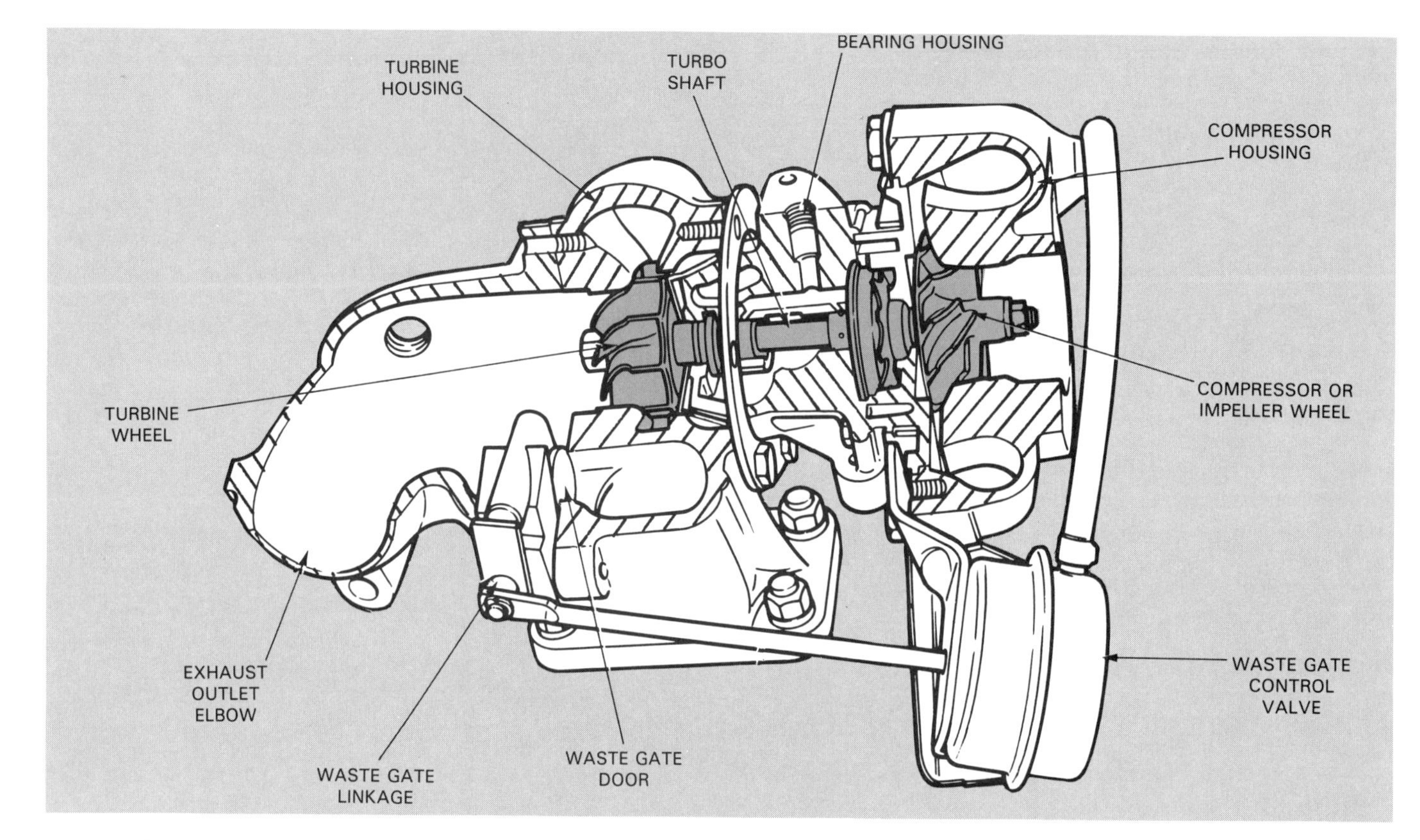

Fig. 19-2. Study basic parts of turbocharger. Turbine is driven by exhaust gases. Turbo shaft connects turbine to compressor wheel. Compressor wheel is simply a fan that blows charge into engine. (Chrysler)

shaft and compressor wheel).
2. TURBINE HOUSING (outer enclosure that routes exhaust gases over turbine wheel).
3. TURBO SHAFT (steel shaft that connects turbine and compressor wheels. It passes through center of turbo housing).
4. COMPRESSOR WHEEL (driven fan that forces air into engine intake manifold under pressure).
5. COMPRESSOR HOUSING (part of turbo housing that surrounds compressor wheel. Its shape helps pump air into engine).
6. BEARING HOUSING (enclosure around turbo shaft that contains bearings, seals, and oil passages).

Turbocharger operation

During engine operation, hot exhaust gases blow out the open exhaust valves and into the exhaust manifold, Fig. 19-3. The exhaust manifold and connecting tubing route these gases into the turbine housing. As the gases pass through the turbine housing, they strike the fins or blades on the turbine wheel. When engine load is high enough, there is enough exhaust gas flow to rapidly spin the turbine wheel.

Since the turbine wheel is connected to the compressor wheel by the turbo shaft, the compressor wheel rotates with the turbine. Compressor wheel rotation pulls air into the compressor housing. Centrifugal force throws the air outward. This causes air to flow out of the turbocharger and into the engine cylinder under pressure.

With more air and fuel in the cylinder on the engine's intake stroke, more pressure and combustion force result during the engine's power stroke.

Turbocharger location

Turbochargers normally bolt to the side, the top, or the front of the engine. Exhaust tubing routes engine exhaust gases into the inlet or turbine housing. Tubing or hose also connects the impeller or outlet end of the turbo tothe engine intake manifold, Fig. 19-3.

Theoretically, the turbocharger should be located as close to the engine exhaust manifold as possible. Then, a maximum amount of exhaust heat will enter the turbine housing. When the hot gases blow onto the spinning turbine wheel, they are still burning and expanding to help rotate the turbine.

Turbo lag

Turbo lag is a short delay period before the turbo develops sufficient boost. *Boost* is any pressure above atmospheric pressure in the intake manifold.

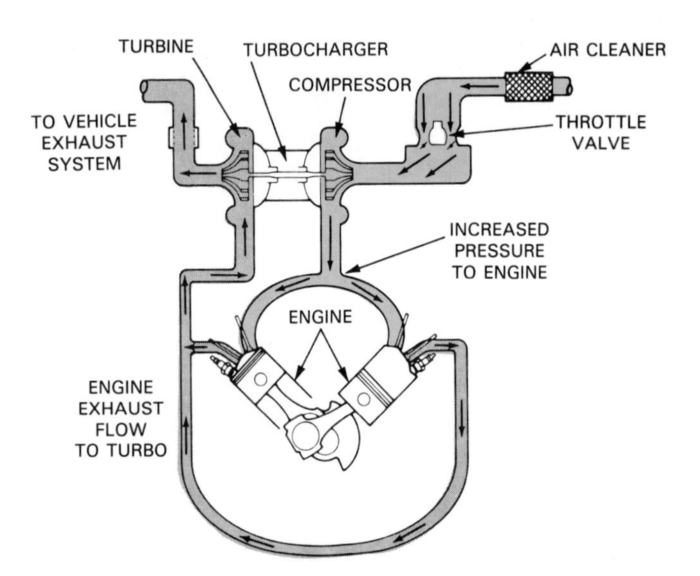

Fig. 19-3. Exhaust gases blow out of running engine and are directed into turbine housing. Housing routes flow over blades on turbine wheel. This spins turbine and compressor wheels at high speed. Centrifugal force produced at compressor wheels blades throws air or air-fuel mixture outward and into combustion chambers under pressure. (General Motors)

When the car's accelerator pedal is pressed down for rapid acceleration, the engine may lack power for a few seconds. This is caused by the compressor and turbine wheels not spinning fast enough. It takes time for the exhaust gases to bring the turbo up to operating speed.

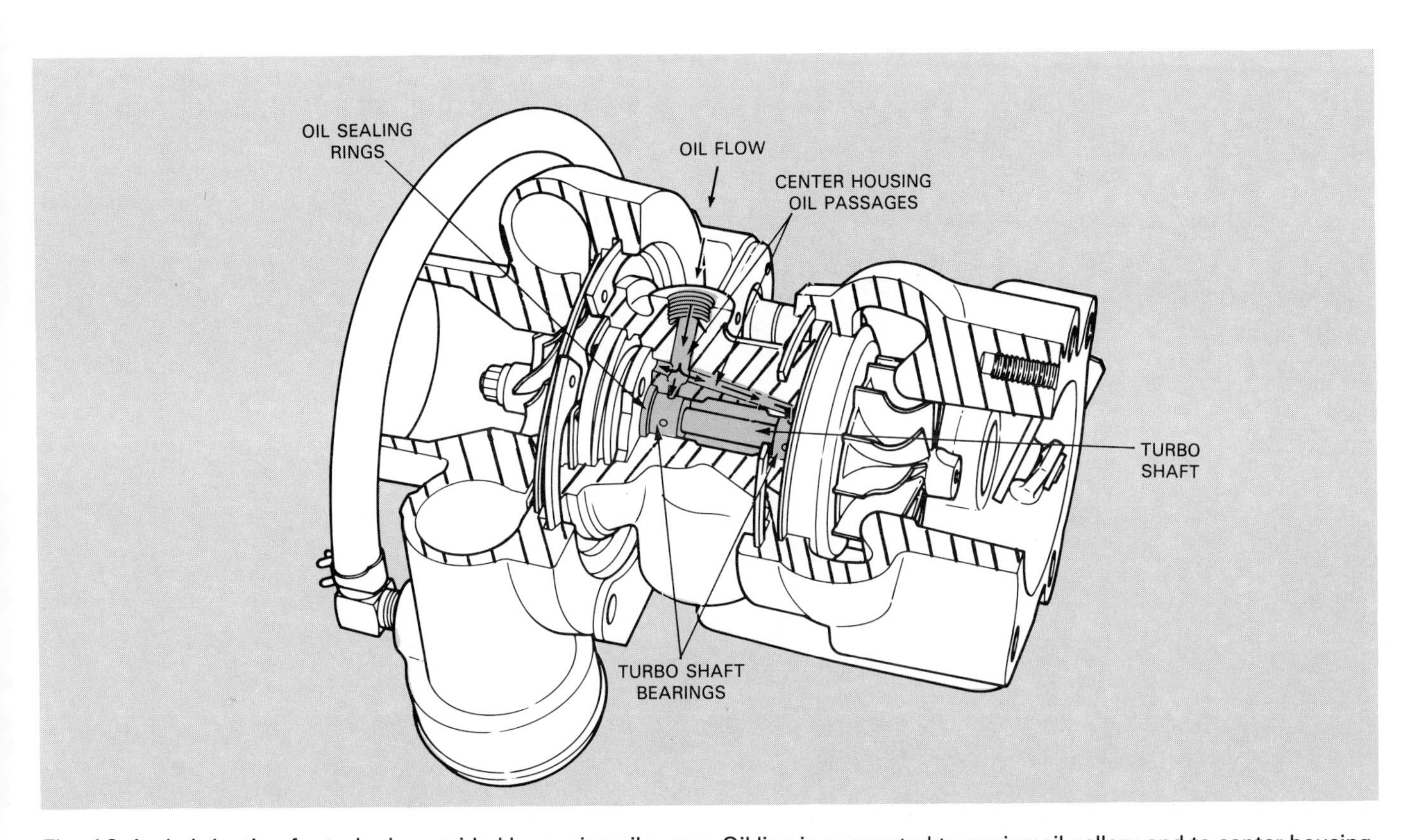

Fig. 19-4. Lubrication for turbo is provided by engine oil pump. Oil line is connected to engine oil gallery and to center housing of turbo. Passages in housing feed oil to turbo shaft and bearings. Oil drains into another line and back to engine oil pan. (Chrysler)

Modern turbo systems suffer from very little turbo lag. Their turbine and compressor wheels are very light so that they can accelerate up to rpm quickly.

Some late model turbocharger impellers are made of carbon-fiber reinforced plastic. This reduces impeller weight and the problem of turbo lag considerably.

Turbocharger lubrication

Turbocharger lubrication is needed to protect the turbo shaft and bearings from damage. A turbocharger can operate at speeds up to 100,000 rpm. For this reason, the engine lubrication system forces motor oil into the turbo shaft bearings, as shown in Fig. 19-4.

Oil passages are provided in the turbo housing and bearings. An oil supply line runs from the engine to the turbo. With the engine running, oil enters the turbo under pressure. This causes the turbo shaft to ride on a thin film of oil, avoiding metal-to-metal contact.

Sealing rings (piston type rings) are placed around the turbo shaft, at each end of the turbo housing. They prevent oil leakage into the compressor and turbine housings.

A drain passage and drain line allow oil to return to the engine oil pan after passing through the turbo bearings.

WASTE GATE

A *waste gate* limits maximum boost pressure developed by the turbocharger. It is a butterfly or poppet valve that allows exhaust to bypass the turbine wheel.

Without a waste gate, the turbo could produce too much pressure in the combustion chambers. This could lead to detonation (spontaneous combustion) and engine damage. See Fig. 19-5.

A waste gate consists of:

1. DIAPHRAGM (flexible membrane that reacts to different amounts of boost pressure).
2. DIAPHRAGM SPRING (coil spring that holds the waste gate valve in the normally closed position).
3. WASTE GATE VALVE (poppet or flap type valve that can open to bypass exhaust gases away from turbine wheel), Fig. 19-5.
4. HOUSING (air tight metal container that encloses parts and provides fitting for pressure hose).
5. PRESSURE LINE (hose that connects waste gate with source of intake manifold pressure).

Waste gate operation

The operation of a turbocharger waste gate is shown in Fig. 19-6. Under partial load, the system routes all of the exhaust gases through the turbine housing. The waste gate is closed by the diaphragm spring. This assures that there is adequate boost to increase engine power.

Under full load, boost may become high enough to overcome diaphragm spring pressure. Manifold pressure compresses the spring and opens the waste gate valve. This permits some of the exhaust gases to flow through the waste gate passage and into the exhaust system. Less exhaust is left to spin the turbine. Boost pressure is limited to a preset value (about 5 to 7 psi).

INTERCOOLER

A *turbocharger intercooler* is an air-to-air or liquid heat exchanger that cools the air entering the engine. It is a radiator-like device mounted at the pressure outlet of the turbocharger. See Fig. 19-7.

Outside air flows over and cools the fins and tubes of the intercooler. Then, when the air or coolant flows through the inside of intercooler, heat is removed.

By cooling the air entering the engine, engine power is increased because the air is more dense and contains more oxygen by volume. Cooling also reduces the tendency for engine detonation.

COMPUTER CONTROLLED TURBOCHARGING

A *computer controlled turbocharging system* uses the electronic control unit to operate the waste gate con-

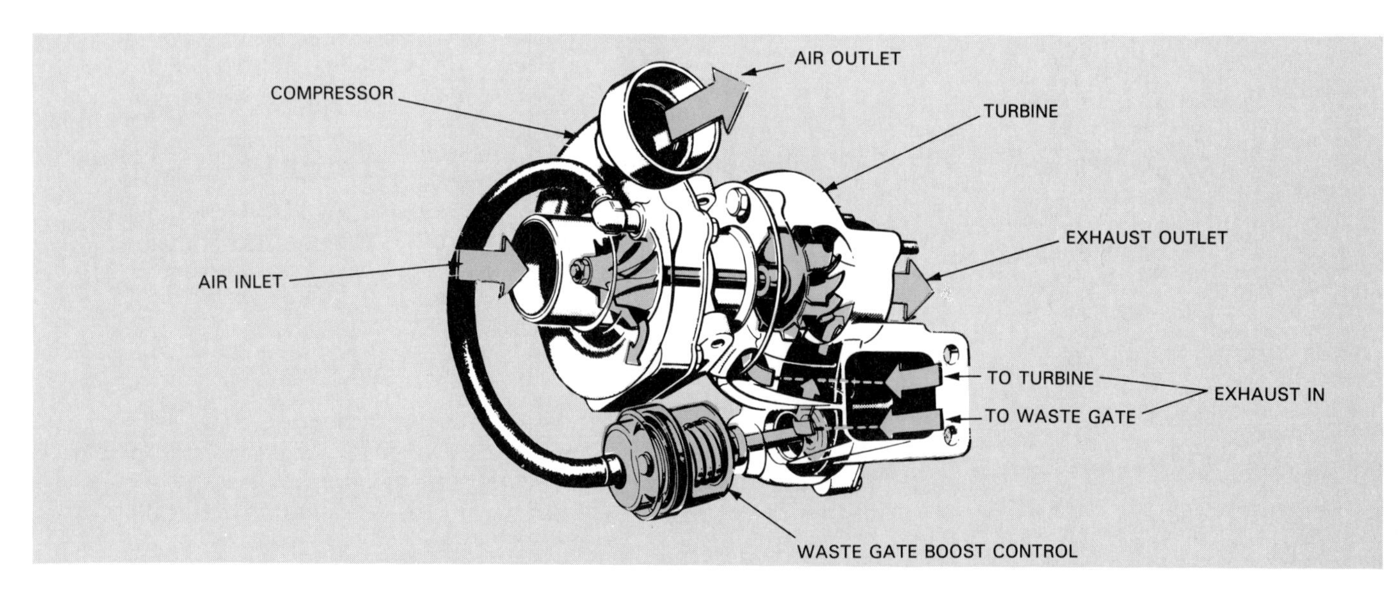

Fig. 19-5. Waste gate is diaphragm operated valve that limits maximum boost pressure developed by turbo. It usually fastens to turbo. Hose connects turbine housing or manifold to diaphragm chamber. Note valve inside housing. (Mercedes Benz)

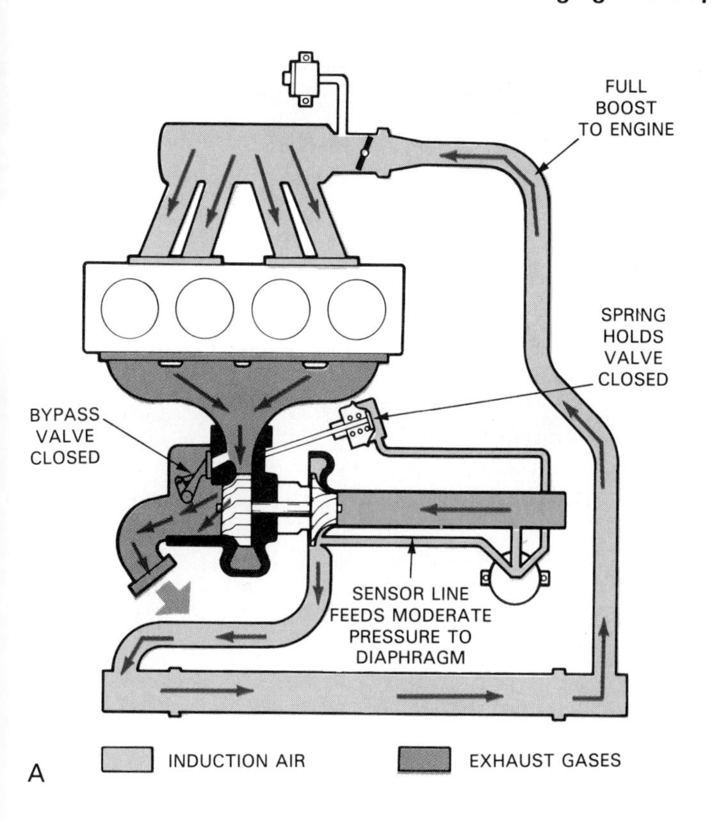

A—Waste gate remains closed so long as boost pressure is not too high. Spring in waste gate holds valve closed so all exhaust gases are directed against turbine wheel blades. This produces full boost.

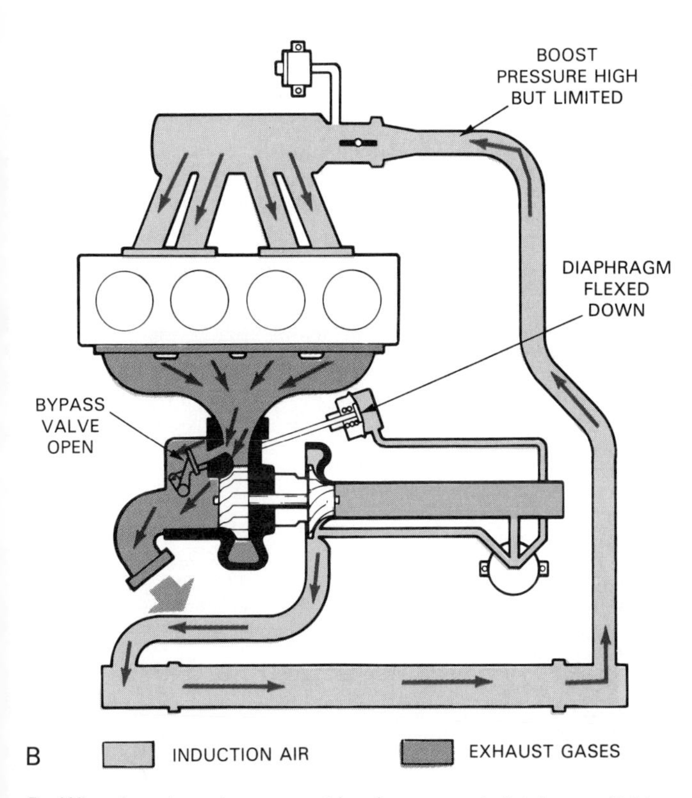

B—When boost reaches present level, pressure in intake manifold or compressor housing acts on diaphragm in waste gate. It pushes diaphragm down, compresses spring, and forces valve open. This bypasses some exhaust gases away from turbine wheel. Turbo rpm stops increasing and boost is limited to protect engine from damage.

Fig. 19-6. Study waste gate operation. (Saab)

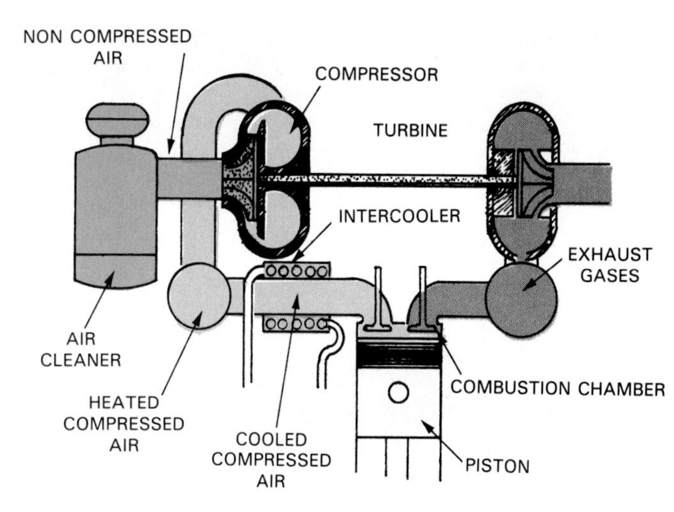

Fig. 19-7. Intercooler is sometimes used to increase engine power even more. It lowers temperature of charge entering combustion chambers. Cooler charge is more dense and can produce more power. Most auto intercoolers are air-to-air, meaning outside air cools air flowing through intercooler. (Waukesha)

trol valve through sensor signals, Fig. 19-8.

The computer output is sent to a vacuum control solenoid. The vacuum solenoid controls engine vacuum going to the waste gate diaphragm. Several sensors, especially the knock sensor, speed sensor, and oxygen sensor, also provide data to the computer.

The computer can then control vacuum to the waste gate diaphragm as needed. For example, if the knock sensor detects pinging, the computer can shut off current to the vacuum solenoid. This will let the solenoid open the waste gate to prevent detonation and damage.

The computer normally energizes the solenoid all of the time. When it does not, the waste gate opens. Other systems vary, however.

ENGINE MODIFICATIONS

Turbocharged and supercharged engines normally have several modifications to make them withstand the increased horsepower.

Shown in Fig. 19-9, a few of these modifications are:

1. Lower compression ratio.
2. Stronger rods, pistons, crankshaft, and bearings.
3. Higher volume oil pump and an oil cooler.
4. Larger radiator.
5. O-ring type head gasket.
6. Heat resistant valves.

TURBOCHARGING SYSTEM SERVICE

Turbocharging system problems usually show up as:

1. Inadequate boost pressure and lack of engine power.
2. Leaking shaft seals and oil consumption.
3. Damaged turbine or compressor wheels and vibration and noise.
4. Excess boost and detonation.

Refer to a factory service manual for a detailed troubleshooting chart if needed. It will list the common troubles for the particular turbo system.

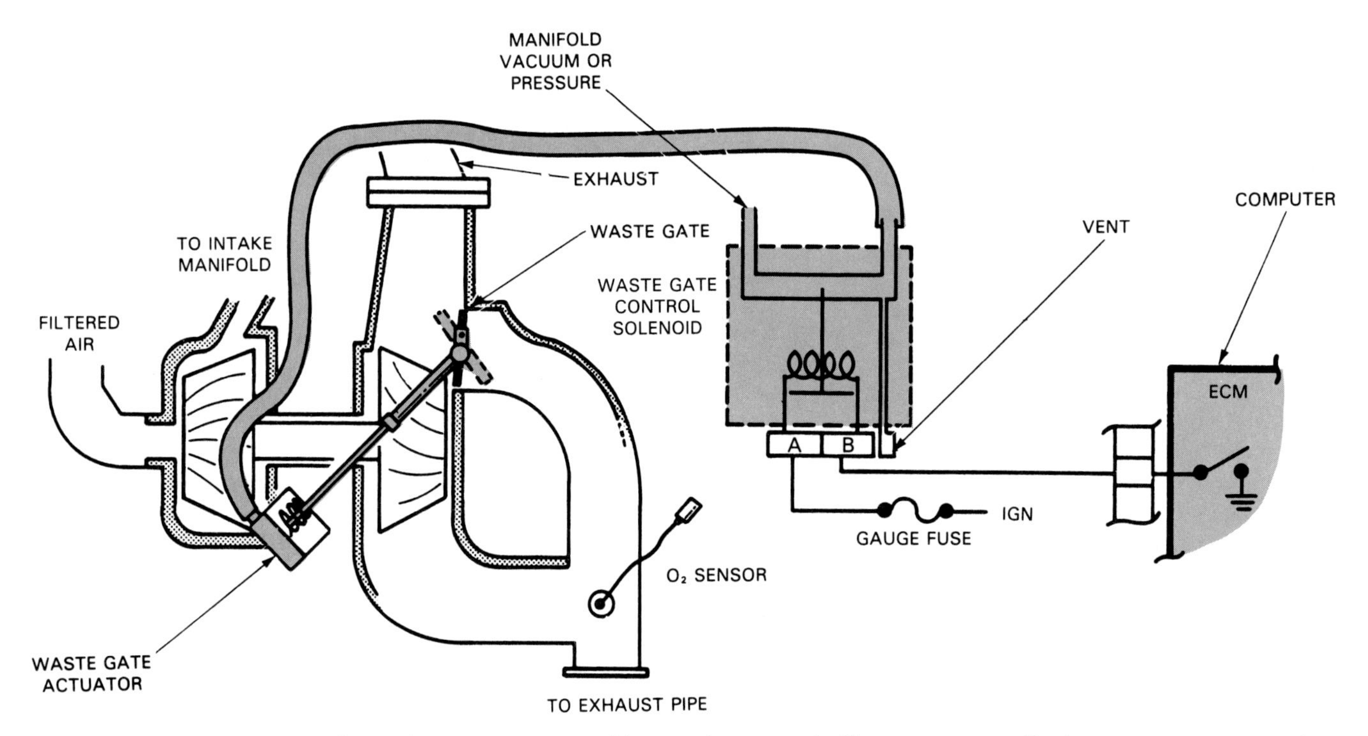

Fig. 19-8. Computer controlled turbo waste gate provides precise control of boost pressure. Engine sensors report engine operating conditions to computer. Computer can then calculate how much boost can be provided for these conditions. It opens and closes waste gate using solenoid. Solenoid, in this example, operates vacuum valve. Vacuum valve can feed or block flow to diaphragm. (Buick)

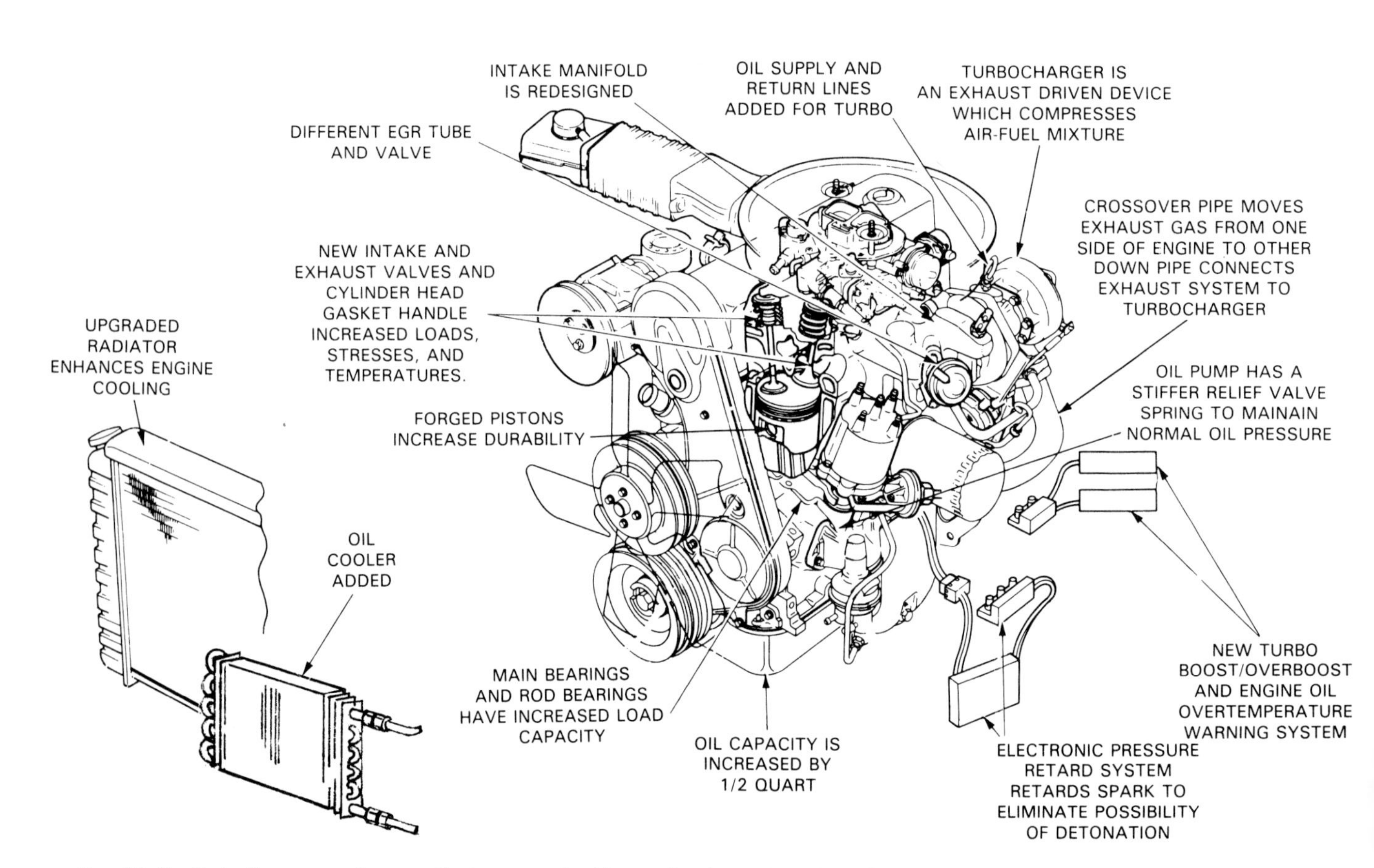

Fig. 19-9. Note the many changes that are used with a turbocharged engine. Generally, the parts must be made stronger to withstand increased loads. Adequate cooling is also critical. Supercharged engines use similar modifications. (Ford)

Increasing turbocharger life

The average service life of a turbocharger is approximately 50,000 miles (about 80 500 km). However, this figure can be considerably lower if certain rules are not followed. Some turbos last over 100,000 miles when properly maintained.

With the high replacement cost of a turbo, it is wise to pass on to your customers some basic tips on driving and servicing a turbocharged engine:

1. Change the engine oil and filter at regular intervals. Both turbocharged and supercharged engines should have their oil changed at no more than 3500 miles.
2. Allow the engine and engine oil to reach full operating temperature before ''kicking in'' or ''boosting'' the turbo. An oil temperature gauge may help.
3. If oil temperature is too high, it can also cause turbo charger failure. An oil cooler may be installed to help protect the turbocharger.
4. After driving the car at highway speeds, allow the engine to idle for a few moments before shutting the engine off. This will let the turbo cool down from its potential internal temperature of about 1000 °F (538 °C). At idle, exhaust gases are much cooler and will help prevent heat damage as the unit cools. Idling may also possibly prevent oil starvation damage. It allows the turbo to slow down before the engine and oil supply stop.

Checking turbocharging system

There are several checks that can be made to determine turbocharging system condition. These include:

1. Check connection of all vacuum lines to waste gate and oil lines to turbo. See Fig. 19-10.
2. Use a regulated low pressure air hose to check for waste gate diaphragm leakage and operation.
3. Use a dash gauge or a test gauge to measure boost pressure (pressure developed by turbo under a load). If needed, connect the pressure gauge to an intake manifold fitting. Compare boost to specs.
4. Use a stethoscope to listen for bad turbocharger bearings.

Servicing turbocharger

Checking out the condition of the inside of the turbocharger requires removing it from the engine. Disconnect the oil lines, and unbolt the unit from its mountings. Fig. 19-11 summarizes the procedures for removing a turbocharger. Move the unit to a workbench.

Before disassembling the unit, scribe an alignment mark on the housing. Open the housing and inspect the interior for oil contamination. Also check the turbine and compressor wheels. They should be clean and free of damage. Even the slightest imperfection could throw the wheels out of balance and cause severe vibration or disintegration. Make sure the turbine assembly spins

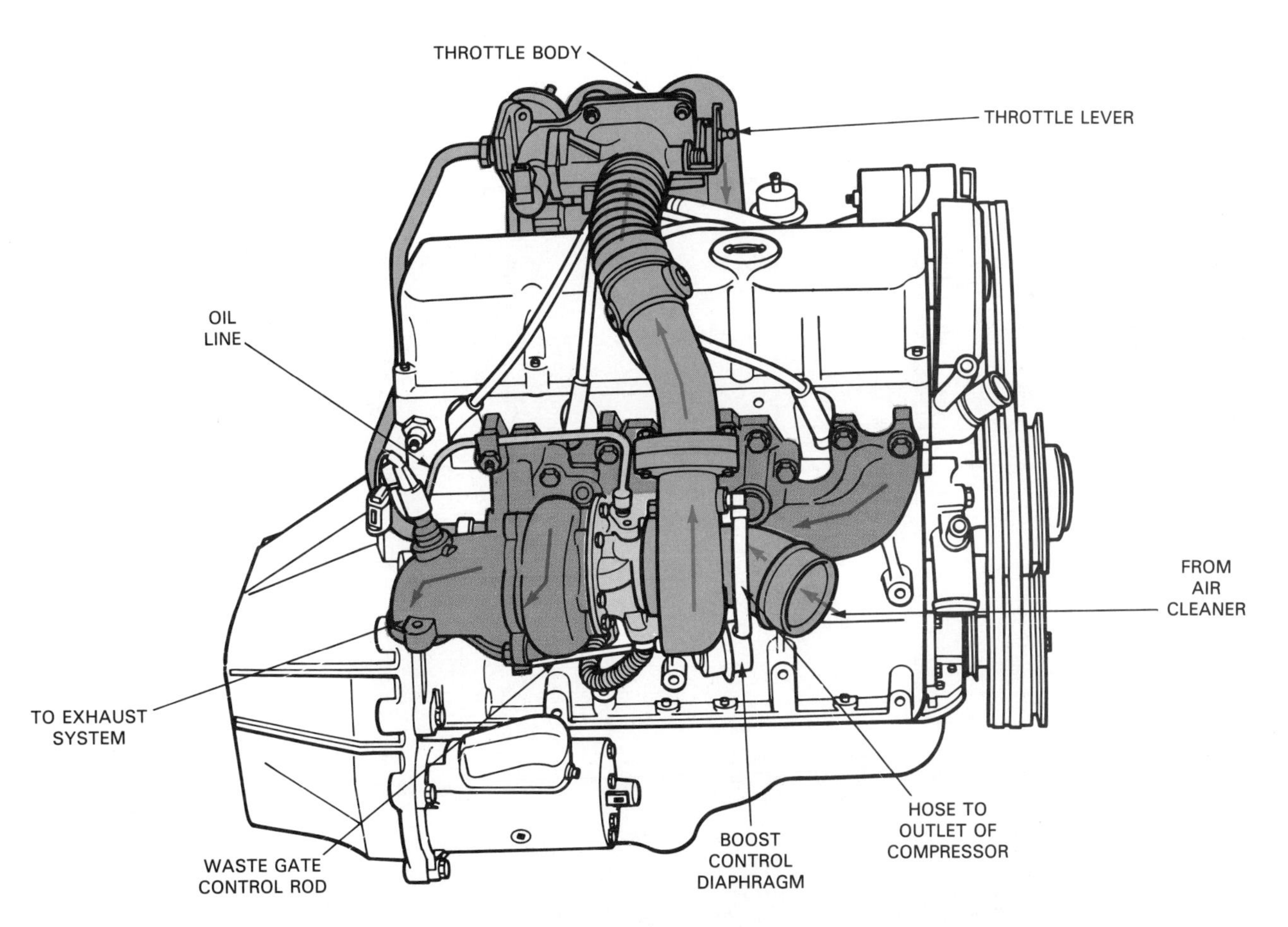

Fig. 19-10. This turbo is mounted right next to exhaust manifold for high efficiency. Always check for damage to feed lines, hoses, linkages, etc. Leaks are common source of turbo malfunctions. (Ford)

1. Remove air cleaner.
2. Remove turbo heat shield (not shown).
3. Remove exhaust outlet pipe.
4. Remove exhaust down pipe.
5. Remove boost control tube.
6. Remove oil supply line.
7. Remove throttle bracket.
8. Remove rear waste gate actuator vacuum line.
9. Remove EGR tube.
10. Remove dipstick and tube if necessary.
11. Remove necessary vacuum hoses.
12. Remove bolts attaching turbocharger to intake manifold.
13. Remove bolts from turbocharger rear brace and remove turbocharger from engine.
14. For installation of turbocharger, replace O-rings, grease the compressor outlet O-ring, and reverse the above procedure.

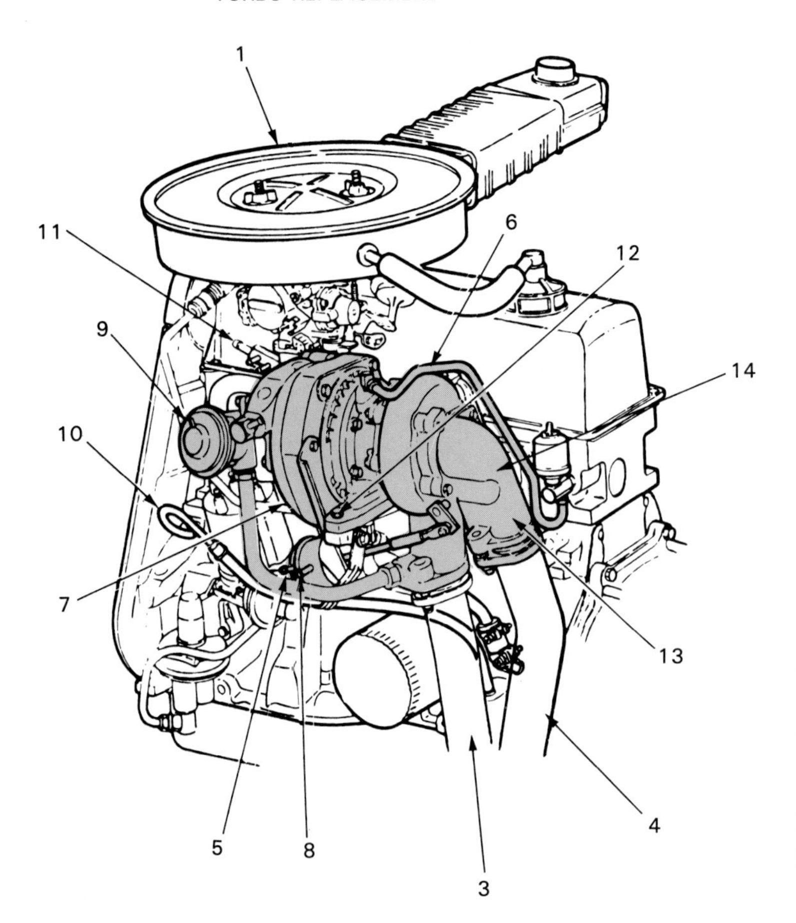

Fig. 19-11. These are typical steps for replacement of a turbocharger. Follow manual directions, when needed, with an actual repair. (Ford)

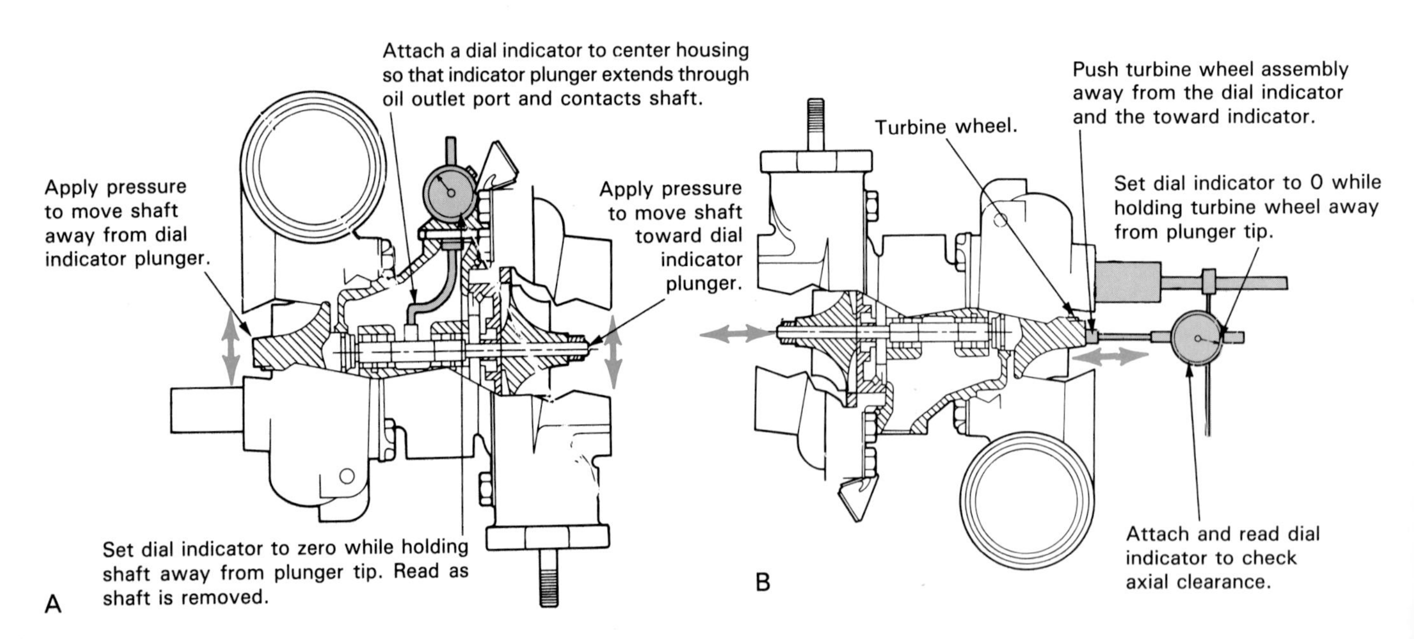

A—To measure bearing radial wear, mount indicator as shown. Wiggle wheels and shaft up and down while reading indicator. Generally, clearance should not exceed .003 to .006 in. (.08 to .15 mm).

B—To measure axial clearance, mount indicator on end of turbo shaft. Slide shaft back and forth in housing while reading indicator. Typical specs call for no more than .001 to .003 in. (.03 to .08 mm) clearance.

Fig. 19-12. Common problem with high mileage turbo is worn shaft or bearings. Note how to check wear. (Ford)

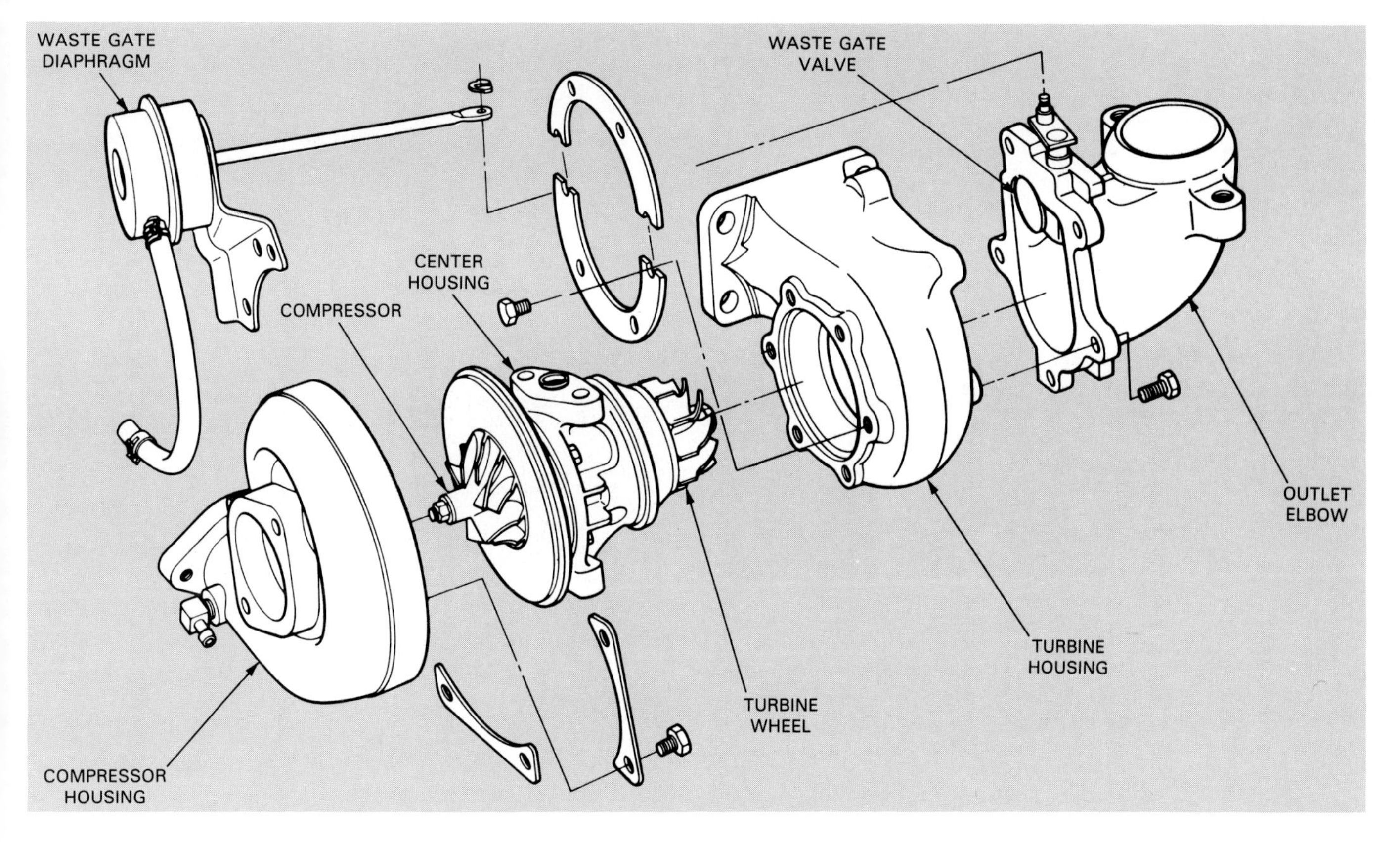

Fig. 19-13. Exploded view shows how internal parts of turbo fit together. Damage to turbine or compressor wheels requires replacement of unit. Leaking housing gaskets, bad seals, damaged housings, and other nonmoving parts of turbo can be serviced in-shop. Waste gate can also be replaced in-shop if bad. Follow manual procedures for exact type unit. (Ford)

freely and does not rub on the housing when rotated by hand. Fig. 19-12 shows how to use a dial indicator to check for turbo bearing wear.

Turbocharger repairs

Most problems with a turbo cannot be repaired in the field. If the wheels, shaft, bearings, or center section of the turbo fail, a new or factory rebuilt unit is commonly installed.

Minor problems, however, such as a bad waste gate control diaphragm, leaking seals or hoses, damaged housings, etc., can be done in-shop. Always refer to a service manual for the exact make and model engine or turbo being serviced, Fig. 19-13.

Never use a hard metal object or sandpaper to remove carbon deposits from the turbine wheel. If you gouge or remove metal, the wheel can vibrate and destroy the turbo. Only use a soft wire brush and solvent to clean the turbo wheels.

Installing turbocharger

Many turbocharger problems are NOT repaired in the field. Most mechanics install a new or rebuilt unit. When installing a turbo, you should:

1. Make sure the new turbo is the correct type.
2. Use new gaskets and seals.
3. Torque all fasteners to specs.
4. If needed, change engine oil and flush oil lines before starting engine.
5. If the failure was oil related, check oil supply pressure in feed line to turbo.
6. Check waste gate operation.

Waste gate service

An *inoperative waste gate* can either cause too much or too little boost pressure.

Before condemning the waste gate, always check other parts. Check the knock sensor (spark retard system, if used) and the ignition timing. Make sure the vacuum-pressure lines are all connected properly.

Follow service manual instructions when testing or replacing a waste gate. Pictured in Fig. 19-14, waste gate removal is simple. Unbolt the fasteners. Remove the lines and lift the unit off of the engine. Many manuals recommend waste gate replacement, rather than in-shop repairs.

Supercharger Service

Many of the problems and tests associated with turbochargers relates to superchargers as well. Generally, supercharger failure can include drive belt failure, engine smoking, lack of boost pressure, vacuum and fluid leaks, and loud bearing noises from the ends of the supercharger.

Visually inspect the unit for obvious problems first. Are there signs of drive belt slippage? Does the belt squeal when the engine is started? This could indicate mechanical failure or bearing seizure. Are there any visible fluid leaks? Excessive fluid leaks indicate a loss of supercharger lubricant.

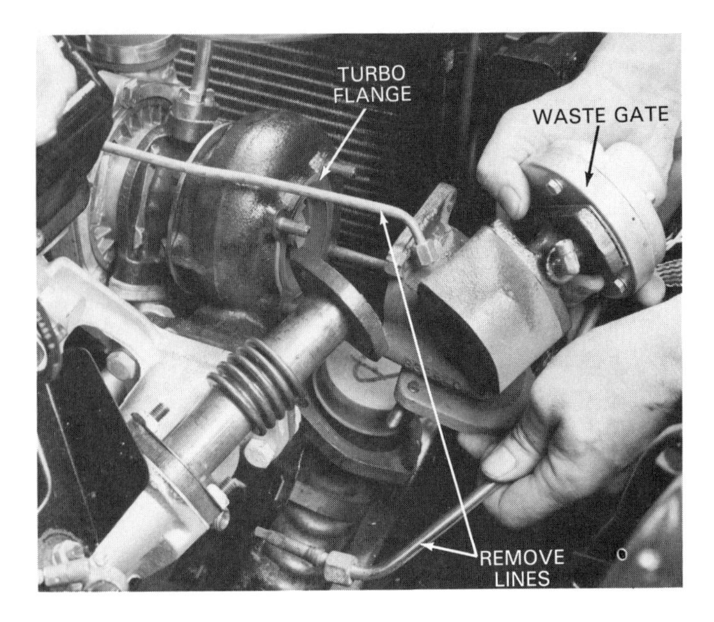

Fig. 19-14. Technician is removing waste gate. Pressure test showed that waste gate had leaking diaphragm, a common problem. (Saab)

Place a stethoscope against the ends of the supercharger housing to isolate any bearing noises. Failed supercharger bearings will make a loud rumble noise when the stethoscope probe is placed over the bearings on the housing.

Install a pressure gauge to measure actual boost pressure. Compare actual boost pressure to specifications for engine speed. If boost is low, check the pressure regulator or you might have internal rotor or housing damage.

If supercharger removal is needed, follow service manual instructions, Fig. 19-15. Procedures vary with engine and supercharger design. Generally, remove any parts that prohibit supercharger removal. This would include the drive belt, wires, hoses, lines, and cables. Remove the bolts that secure the supercharger to the engine. Free the gasket with light prying. Then, lift the unit off the engine.

Because of the special tools and cleanliness needed, superchargers are not normally serviced in the field and are simply replaced as a unit. To install the supercharger, use a new base gasket and apply sealer as directed in the service manual. Slowly lower the supercharger onto the engine without damaging the gasket. Start all bolts by hand. Then, use a torque wrench to tighten them in the factory recommended sequence. Reinstall any accessory parts removed earlier.

Finally, reinstall the drive belt and related parts. Start and warm the engine. Test drive the vehicle to make sure you have corrected the problem. Make sure the engine has normal power, no abnormal noises, and does not smoke.

SUMMARY

A supercharger refers to a mechanically driven blower. Turbocharger refers to an exhaust drive blower. Both can increase engine horsepower by up to 50 percent.

A turbo consists of a turbine wheel that is rotated by engine exhaust gases. It turns a shaft connected to the compressor wheel. The compressor wheel is another fan

Fig. 19-15. Superchargers are replaced as a unit, do not attempt to service one in the field without the proper tools and training. (Buick)

that blows air or air-fuel mixture into the engine cylinders under pressure.

A center bearing housing supports the turbo shaft. Oil is fed into this housing from the engine for the shaft bearings. Doughnut-shaped housings are formed around the turbine and compressor wheels to aid flow.

With the engine running, hot exhaust blows out of the engine and is routed into the turbine housing. The gases strike the blades on the turbine and make it spin at very high speed. Since the compressor wheel is connected to the turbine, it also spins.

The spinning action of the blades on the compressor wheel make the air inside the turbo circulate. Centrifugal force throws the air outward and pressurizes it. The air is then routed to the engine intake manifold.

Turbochargers and superchargers normally bolt to the side, top, or in front of the engine.

Turbo lag refers to the short delay period before boost or a manifold pressure increase. This is due to the time it takes the exhaust gases to accelerate the turbine and compressor wheels up to speed.

A waste gate limits maximum boost pressure to prevent detonation and engine damage. It is a diaphragm operated valve that can bypass exhaust gases around the turbine wheel when manifold pressure is too high.

An intercooler is a radiator-like device that cools the air entering the engine from the turbocharger. When the air is compressed, it tends to heat up. By cooling the air back down, engine power is improved and the tendency for detonation is reduced.

A computer controlled turbo system uses engine sensors, a microprocessor, and a waste gate solenoid. The solenoid, when energized or deenergized by the computer, can open or close the waste gate to closely control boost pressure.

To withstand the increased horsepower, an engine requires several modifications. A few of these are a lower compression ratio, stronger internal components, high volume oil pump, larger radiator, special heat resistant valves, and oil cooler.

Turbocharging system troubles can produce various symptoms: lack of engine power, engine smoking, vibration, noise, detonation, etc. Supercharger problems are similar.

You should check all hose and linkage connections. Check actual boost pressure with a test gauge. Use a stethoscope to listen for noise.

Most shops install a new or factory rebuilt turbochargers and superchargers. Both are precisely balanced and in-shop repairs can be time consuming and sometimes, impossible.

When installing, make sure you have the correct type or model unit. Use new gaskets and seals. Torque all fasteners to specs. If needed, change the engine oil and filter and flush the feed line. Make sure oil is being fed to new turbos.

High or low boost would indicate waste gate problems. A pressure gauge can be connected to a source of intake manifold pressure.

KNOW THESE TERMS

Supercharger, Turbocharger, Turbine, Compressor, Turbo lag, Boost, Waste gate, Intercooler, Waste gate solenoid, Boost pressure test.

REVIEW QUESTIONS—CHAPTER 19

1. What is the difference between an automotive supercharger and a turbocharger?
2. List and explain the six major parts of a turbocharger.
3. In your own words, how does a turbocharger work?
4. Turbochargers are normally located near the engine exhaust manifold to increase efficiency. True or false?
5. ________ ________ is the short delay before the turbo develops sufficient ________ or increased pressure.
6. How is a turbocharger lubricated?
7. This is NOT part of a waste gate.
   a. Diaphragm.
   b. Valve.
   c. Diaphragm spring.
   d. Turbine.
8. Explain four ways of increasing turbocharger service life.
9. A car enters the shop with a severe detonation problem. Engine power is good and initial ignition timing is within specs.
   Technician A says that the trouble could be in the knock sensor.
   Technician B says waste gate could be stuck closed.
   Who is correct?
   a. Technician A.
   b. Technician B.
   c. Both A and B.
   d. Neither A nor B.
10. When a turbocharger is found to be bad, most shops install a new or rebuilt unit. True or false?

ASE CERTIFICATION-TYPE QUESTIONS

1. Technician A says that a normally aspirated engine is equipped with a supercharger. Technician B says that a normally aspirated engine is equipped with a turbocharger. Who is right?
   (A) A only.
   (B) B only.
   (C) Both A & B.
   (D) Neither A nor B.
2. All of the following are basic components of a turbocharger EXCEPT:
   (A) bearing housing.
   (B) turbine wheel.
   (C) transfer pump.
   (D) turbo shaft.
3. Technician A says that with a supercharged engine, there is a slight delay in pressure and power increase because the unit must build up rpm. Technician B says that with a supercharged engine air pressure increases immediately. Who is right?
   (A) A only.
   (B) B only.
   (C) Both A & B.
   (D) Neither A nor B.
4. Technician A says that volumetric efficiency refers to how much air or air-fuel mixture an engine can draw in on its power strokes. Technician B says that volumetric efficiency refers to how much air or air-fuel mixture an engine can draw in on its intake strokes. Who is right?
   (A) A only.
   (B) B only.
   (C) Both A & B.
   (D) Neither A nor B.
5. All of the following are locations a turbocharger can be mounted on an engine EXCEPT:
   (A) on top.
   (B) on the side.
   (C) in front.
   (D) inside the engine.
6. Technician A says that boost refers to any pressure above atmospheric pressure in an engine's combustion chambers. Technician B says that boost refers to any pressure above atmospheric pressure in an engine's intake manifold. Who is right?
   (A) A only.
   (B) B only.
   (C) Both A & B.
   (D) Neither A nor B.
7. Technician A says that a turbocharger is lubricated by oil from its own lubrication supply system. Technician B says that a turbocharger is lubricated by an oil supply line that runs from the engine to the turbo. Who is right?
   (A) A only.
   (B) B only.
   (C) Both A & B.
   (D) Neither A nor B.

8. All of the following are modifications needed on a turbocharged engine EXCEPT:
   (A) heat resistance valves.
   (B) smaller cooling system radiator.
   (C) higher volume oil pump and an oil cooler.
   (D) ignition retard system.
9. Technician A says that after driving a turbocharged engine at highway speeds you should allow the engine to idle for a few minutes before shutting it off. Technician B says that you should allow a turbocharged engine to reach full operating temperature before kicking in the turbo. Who is right?
   (A) A only.
   (B) B only.
   (C) Both A & B.
   (D) Neither A nor B.
10. Technician A says that a faulty waste gate can cause too much boost pressure in a turbocharger. Technician B says that a faulty waste gate can reduce turbo boost pressure below specs. Who is right?
   (A) A only.
   (B) B only.
   (C) Both A & B.
   (D) Neither A nor B.

Late model engine design improvements have not only been increasing engine power and fuel economy, they have helped reduce emissions. This engine uses sequential, multipoint fuel injection, turned runner intake manifold, and other innovations to produce very fuel efficient, clean running power plant. (Ford Motor Company)

Chapter 20

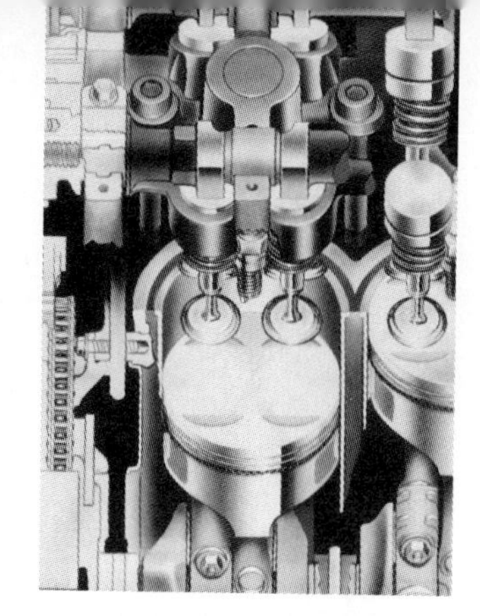

Emission Control Systems—Operation and Service

After studying this chapter, you will be able to:
- *Explain the sources of air pollution.*
- *Explain the four types of auto related emissions.*
- *Describe the operation of emission control systems.*
- *Describe how a computer control system helps prevent engine emissions.*
- *Troubleshoot common emission control system problems.*
- *Inspect emission control systems.*
- *Use an emission control information sticker.*
- *Test and service emission control systems.*
- *Use a service manual and self-test mode to service a computerized emission control system.*

Emission control systems are used on engines to reduce the amount of harmful chemicals or emissions released into the atmosphere. These systems help keep our air clean, Fig. 20-1.

This chapter introduces the terminology, parts, and systems that control emissions. It will prepare you for later chapters on engine diagnosis and tune-up. Study carefully!

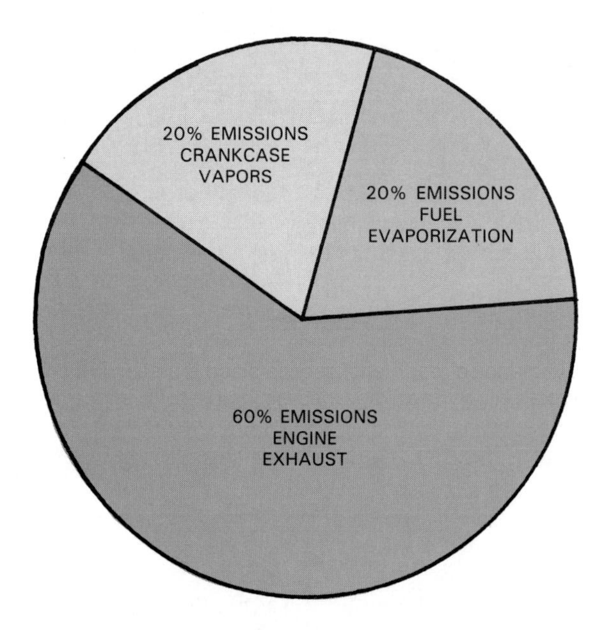

Fig. 20-1. These are basic proportions of emissions produced by automobile. Note high percentage from engine exhaust.

EMISSIONS

Emissions must be controlled to protect our environment. There are four basic types of emissions: hydrocarbons, carbon monoxide, oxides of nitrogen, and particulates.

Emissions are from three sources:

1. ENGINE CRANKCASE BLOWBY FUMES (20 percent of emissions).
2. FUEL VAPORS (20 percent of emissions).
3. ENGINE EXHAUST GASES (60 percent of emissions).

Various engine modifications and emission control systems help reduce pollution from these sources.

Hydrocarbons (HC)

Hydrocarbons, abbreviated HC, is an emission resulting from the release of UNBURNED FUEL into the atmosphere. Hydrocarbon emission can be caused by incomplete combustion or by fuel evaporization. For example, HC is produced when unburned fuel blows out an untuned engine's exhaust system. It can also be caused by fuel vapors escaping from a car's fuel system. Look at Fig. 20-2.

Hydrocarbon emissions can contribute to eye, throat, and lung irritation, other illnesses, and possibly cancer.

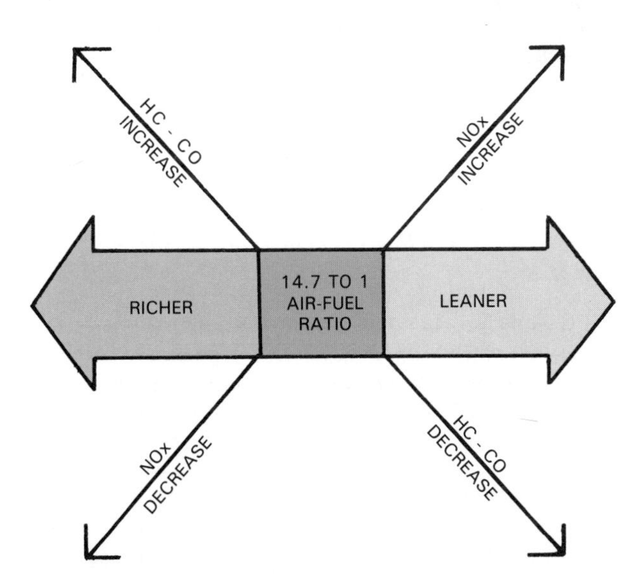

Fig. 20-2. Generally, rich mixture increases HC and CO emissions. However, lean mixture increases NOx emissions. (GM)

Carbon monoxide (CO)

Carbon monoxide, abbreviated CO, is an extremely toxic emission resulting from the release of PARTIALLY BURNED FUEL. It is a colorless, odorless, but deadly gas. It can cause headaches, nausea, respiratory (breathing) problems, and even death if inhaled in large quantities. CO prevents human blood cells from carrying oxygen to body tissues.

Any factor that reduces the amount of oxygen present during combustion increases carbon monoxide emissions. For example, a rich air-fuel mixture (high ratio of fuel to air) would increase CO. As the mixture is leaned, CO emissions are reduced, Fig. 20-2.

Oxides of nitrogen (NOx)

Oxides of nitrogen, abbreviated NOx, are emissions produced by HIGH TEMPERATURES during combustion. Air consists of about 80 percent nitrogen and 20 percent oxygen. With enough heat (above approximately 2500 °F or 1 370 °C), nitrogen and oxygen in the air-fuel mixture combine to form NOx emissions.

ENGINE MODIFICATIONS FOR EMISSION CONTROL

The best way to reduce exhaust emissions is to burn all of the fuel inside the engine. For this reason, several engine modifications have been introduced to improve efficiency.

Today's engines can have the following modifications to lower emissions:

1. LOWER COMPRESSION RATIOS allow the use of unleaded gasoline. Unleaded fuel burns quickly to lower HC emissions. It also does not contain lead additives that cause particulate emissions. Lower compression stroke pressure also reduces combustion temperatures and NOx emissions.
2. LEANER AIR-FUEL MIXTURES lower HC and CO emissions because more air is present to help all of the fuel burn.
3. HEATED INTAKE MANIFOLDS speed warm-up and permit the use of leaner mixtures during initial start-up.
4. SMALLER COMBUSTION CHAMBER SURFACE VOLUMES reduce HC emissions. A smaller chamber increases combustion efficiency by lowering the amount of heat dissipated out of the fuel mixture. Less combustion heat enters the cylinder head and more heat is left to burn the fuel.
5. INCREASED VALVE OVERLAP cuts NOx emissions. A camshaft with more overlap dilutes the incoming air-fuel mixture with inert exhaust gases. This reduces peak combustion temperatures and NOx.
6. HARDENED VALVES AND SEATS can withstand unleaded fuel. Lead additives in fuel, besides increasing octane, act as high temperature lubricants. They reduce wear at the valve faces and seats. Hardened valves and seats are needed to prevent excessive wear when using unleaded fuels.
7. WIDER SPARK PLUG GAPS will properly ignite "super-lean," clean burning air-fuel mixtures. Wider gaps produce hotter sparks that can ignite hard to burn, lean air-fuel mixtures.
8. REDUCED QUENCH AREAS in the combustion chambers lower HC and CO emissions. A *quench area* is produced when the engine pistons move too close to the cylinder head. When these metal parts are too close, it tends to quench (put out) combustion and increase emissions due to unburned fuel. Modern engines have redesigned cylinder heads and pistons which prevent high quench areas.
9. HIGHER OPERATING TEMPERATURES improve HC and CO emissions; therefore, today's engines have "hotter" thermostats. If the metal parts in an engine are hotter, less combustion heat will dissipate out of the burning fuel.

VEHICLE EMISSION CONTROL SYSTEMS

Several systems are used to reduce the pollution produced by the engine and its fuel system. These include:

1. PCV SYSTEM (positive crankcase ventilation system to keep engine crankcase fumes out of atmosphere).
2. HEATED AIR INLET SYSTEM (thermostatic controlled air cleaner maintains constant air temperature entering engine for improved combustion).
3. FUEL EVAPORATION CONTROL SYSTEM (closed vent system prevents fuel vapors from entering atmosphere).
4. EGR SYSTEM (exhaust gas recirculation system injects burned exhaust gases into engine to lower combustion temperature and prevent NOx pollution).
5. AIR INJECTION SYSTEM (air pump forces outside air into exhaust system to help burn unburned fuel).
6. CATALYTIC CONVERTER (thermal reactor for burning and chemically changing exhaust byproducts into harmless substances).

Variations of these systems and computer control systems are all used to make the modern engine very efficient.

Positive crankcase ventilation (PCV)

A *positive crankcase ventilation system* uses engine vacuum to draw toxic blowby gases into the intake manifold for reburning in the combustion chambers.

Detailed in earlier chapters, engine blowby is caused by pressure leakage past the piston rings on the power strokes. A small percentage of combustion gases can flow through the ring end gaps, back of the piston ring grooves, and into the crankcase, Fig. 20-3.

Blowby gases can cause:

1. Air pollution, if released into the atmosphere (HC and CO).
2. Corrosion of engine parts (sulfur and acid).
3. Dilution of engine oil (HC, water, sulfur, and acid).
4. Formation of sludge (chocolate pudding like substance that can clog oil passages).

A PCV system is designed to prevent these problems. It helps keep the inside of the engine clean and also reduces air pollution.

A *PCV valve* is commonly used to control the flow of air through the crankcase ventilation system, Fig. 20-4. It may be located in a rubber grommet in a valve cover or in a breather opening in the intake manifold. The PCV valve changes airflow for idle, cruise, acceleration, wide open throttle, and engine-off conditions. This is

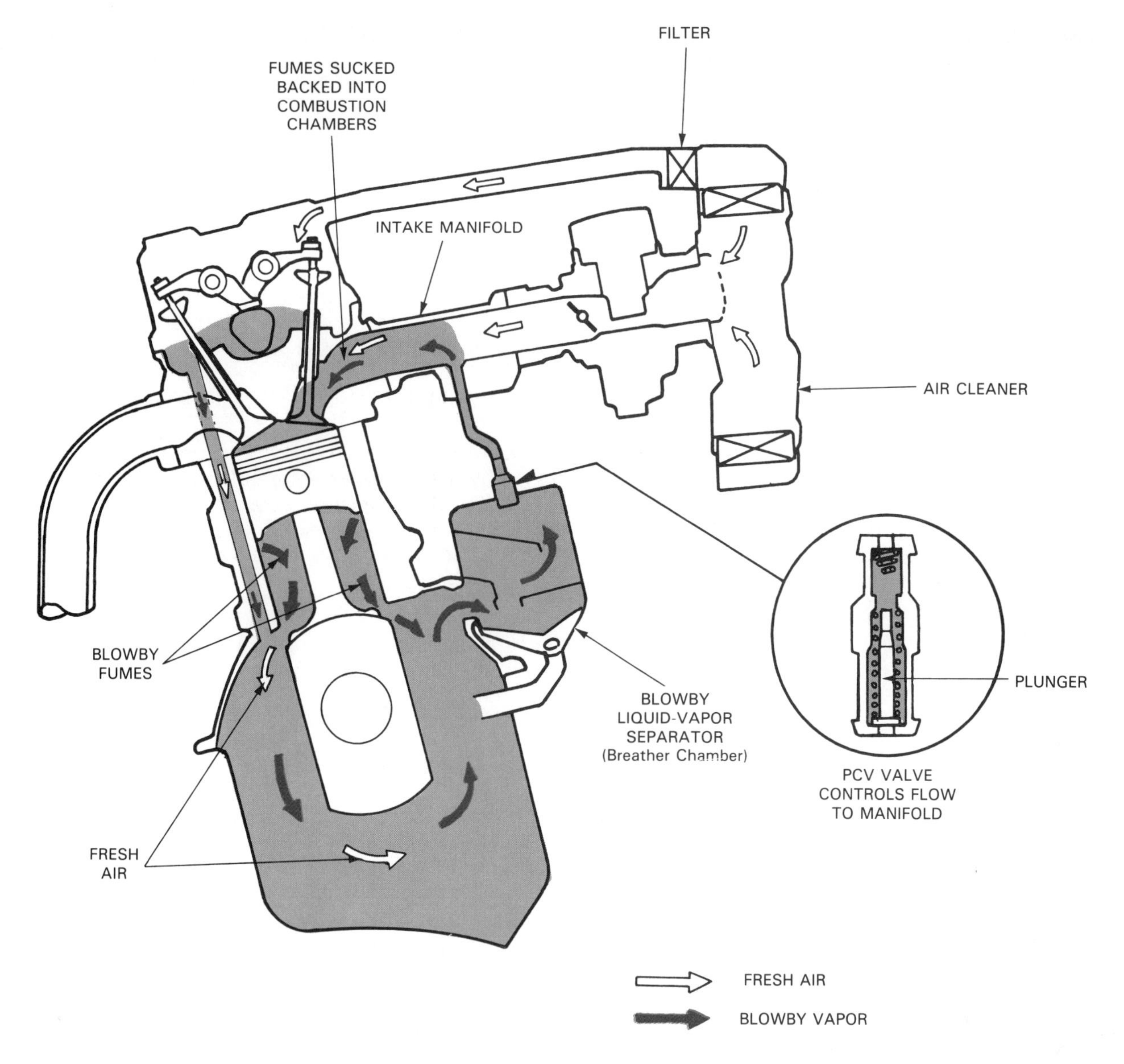

Fig. 20-3. Positive crankcase ventilation system uses engine vacuum to draw toxic crankcase fumes into intake manifold for burning. PCV valve controls flow so idle mixture is not too lean. (Honda)

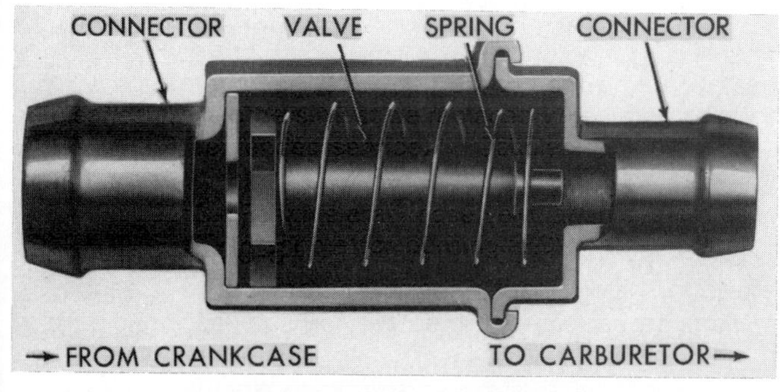

Fig. 20-4. Note construction of PCV valve. Some units can be cleaned while others must be replaced when dirty or after extended service. (Cadillac)

shown in Fig. 20-5.

In case of an *engine backfire* (air-fuel mixture in intake manifold ignites), the PCV valve plunger is seated against the body of the valve. This keeps the backfire (burning) from entering and igniting the fumes in the crankcase.

Heated air inlet system

The *heated air inlet system* speeds engine warm-up and keeps the temperature of the air entering the engine constant. Late model gasoline engines commonly use this system.

By maintaining a more constant inlet air temperature, the carburetor or fuel injection system can be calibrated leaner to reduce emissions. This system, also called *thermostatic air cleaner system,* also helps prevent carburetor icing during warm-up. Fig. 20-6 shows thermostatic air cleaner operation.

Evaporative emissions system

The *fuel evaporative emissions system* prevents vapors from the fuel systems from entering the atmosphere. Older, pre-emission cars used vented gas tank caps. Car-

ENGINE NOT RUNNING OR BACKFIRE

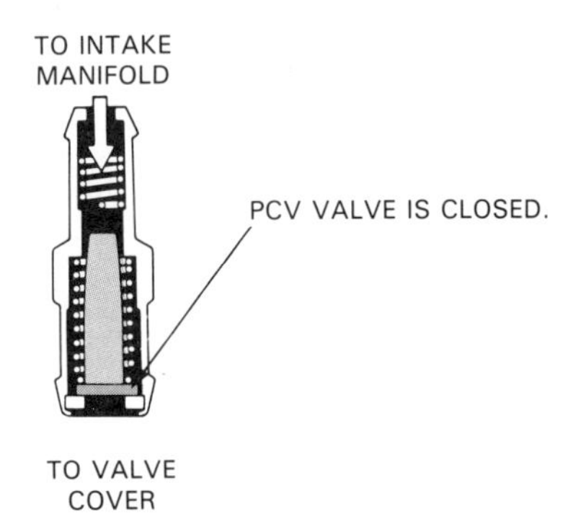

NORMAL OPERATION

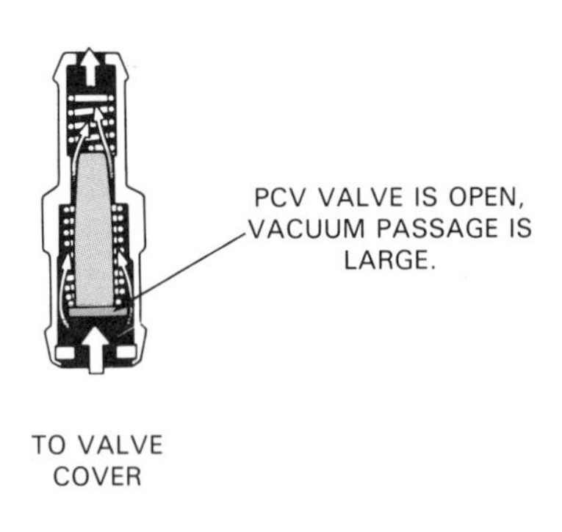

IDLING OR DECELERATING

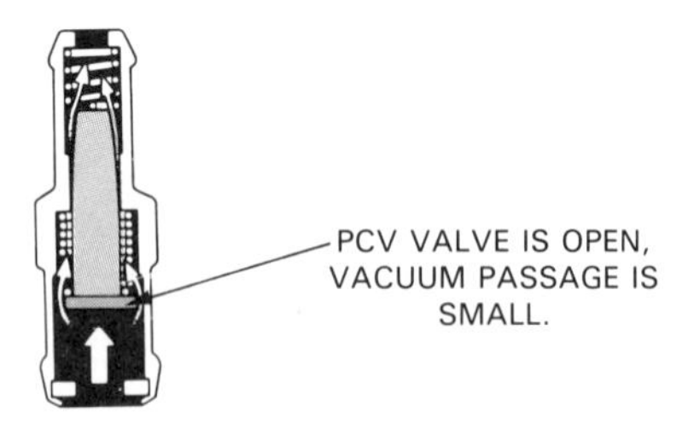

ACCELERATION OR HIGH LOAD

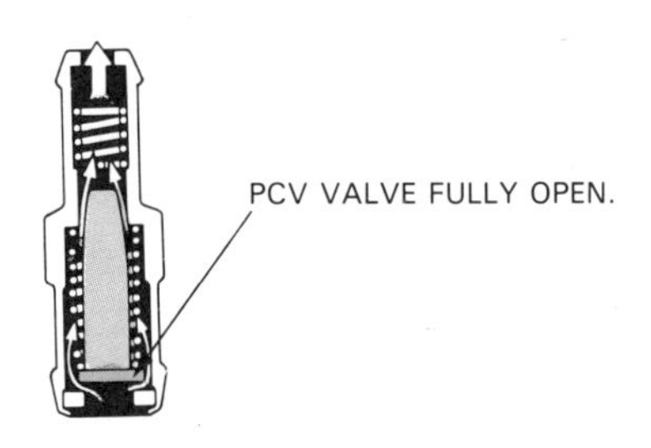

Fig. 20-5. Study operation of PCV valve in different modes. (Toyota)

HOT AIR POSITION FOR WARMUP

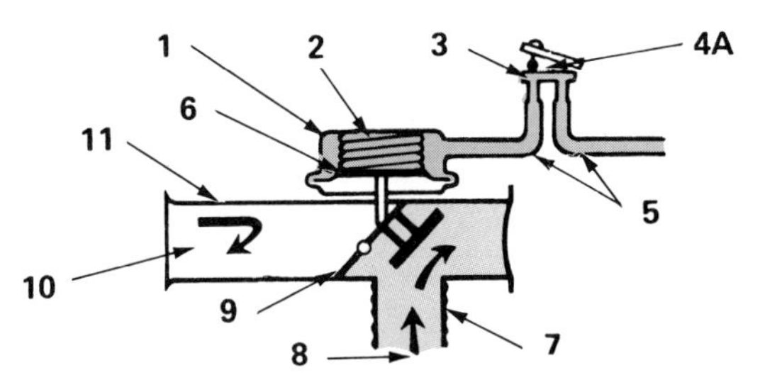

TEMPERATURE CONTROL OR BLEND

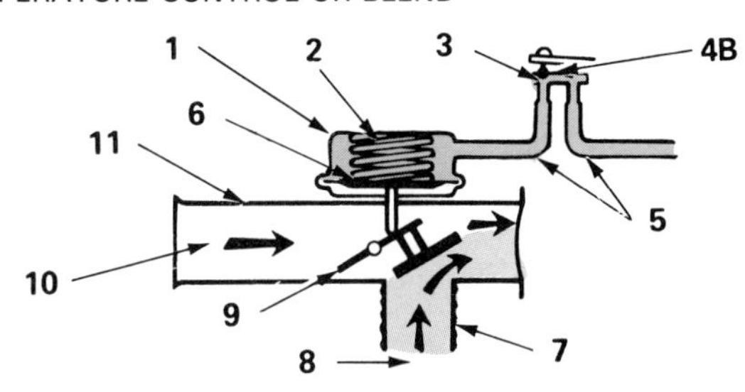

COOL AIR POSITION

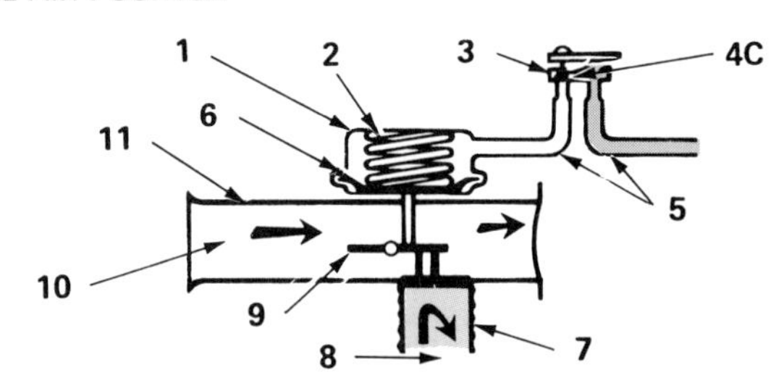

1 VACUUM DIAPHRAGM MOTOR
2 DIAPHRAGM SPRING
3 TEMPERATURE SENSOR
4A AIR BLEED VALVE—CLOSED
4B AIR BLEED VALVE—PARTIALLY OPEN
4C AIR BLEED VALVE—OPEN
5 VACUUM HOSES
6 DIAPHRAGM
7 HEAT STOVE
8 HOT AIR (EXHAUST MANIFOLD)
9 DAMPER DOOR
10 OUTSIDE INLET AIR
11 SNORKEL

Fig. 20-6. Thermostatic air cleaner maintains relatively constant air inlet temperature to more closely control air-fuel mixture. Hot air position is used when engine or outside air are cold. Cool air position results in very hot weather. Note action of temperature sensor and vacuum motor. (Oldsmobile)

buretor bowls were also vented to atmosphere. This caused a considerable amount of emissions. Modern cars commonly use a fuel evaporization control system to prevent this source of air pollution, Fig. 20-7.

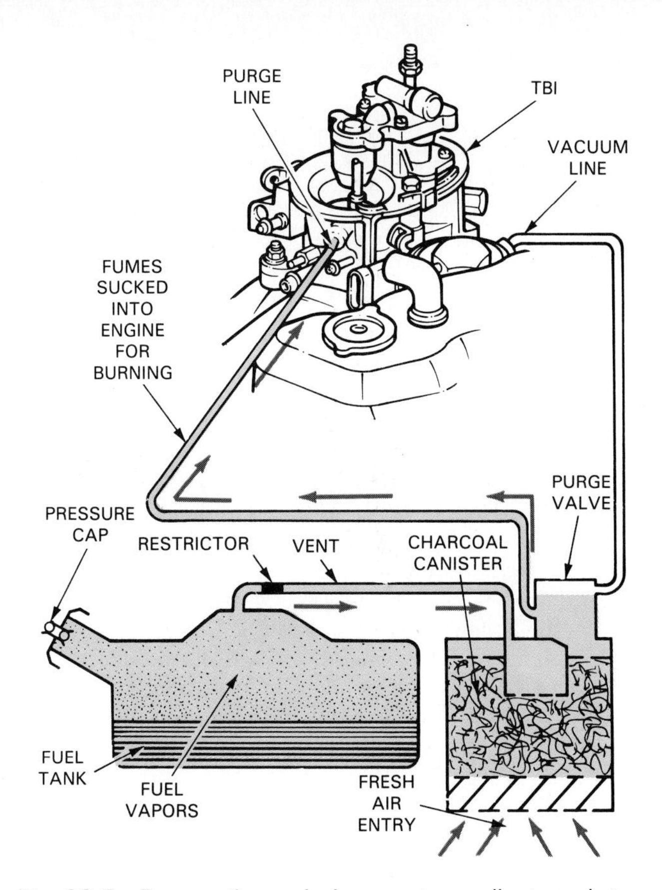

Fig. 20-7. Evaporative emissions system collects and stores fuel vapors in charcoal canister. When engine is started, purge line pulls vapors out of charcoal and into intake manifold for burning in engine. (Buick)

A *liquid-vapor separator* is frequently used to keep liquid fuel from entering the evaporation control system. It is simply a metal tank located above the main fuel tank. Liquid fuel condenses on the walls of the liquid-vapor separator. The liquid fuel then flows back into the fuel tank, Fig. 20-8.

A *roll-over valve* is sometimes used in the vent line from the fuel tank. It keeps liquid fuel from entering the vent line after an auto accident where the car rolls upside-down, Fig. 20-8.

A *fuel tank vent line* carries fuel tank vapors up to a charcoal canister in the engine compartment, Fig. 20-8.

The *charcoal canister* stores fuel vapors when the engine is NOT running. The metal or plastic canister is filled with activated charcoal granules. The charcoal is capable of absorbing fuel vapors.

The top of the canister has fittings for the fuel tank vent line, carburetor vent line, and the purge (cleaning) line. The bottom of the canister has an air filter that cleans outside air entering the canister, Fig. 20-8.

A *purge line* is used for removing or cleaning the stored vapors out of the charcoal canister. It connects the canister and the engine intake manifold. When the engine is running, engine vacuum draws the vapors out of the canister and through the purge line, Fig. 20-8.

Exhaust gas recirculation system

The *exhaust gas recirculation (EGR) system* allows burned exhaust gases to enter the engine intake manifold to help reduce NOx emissions. When exhaust gases are

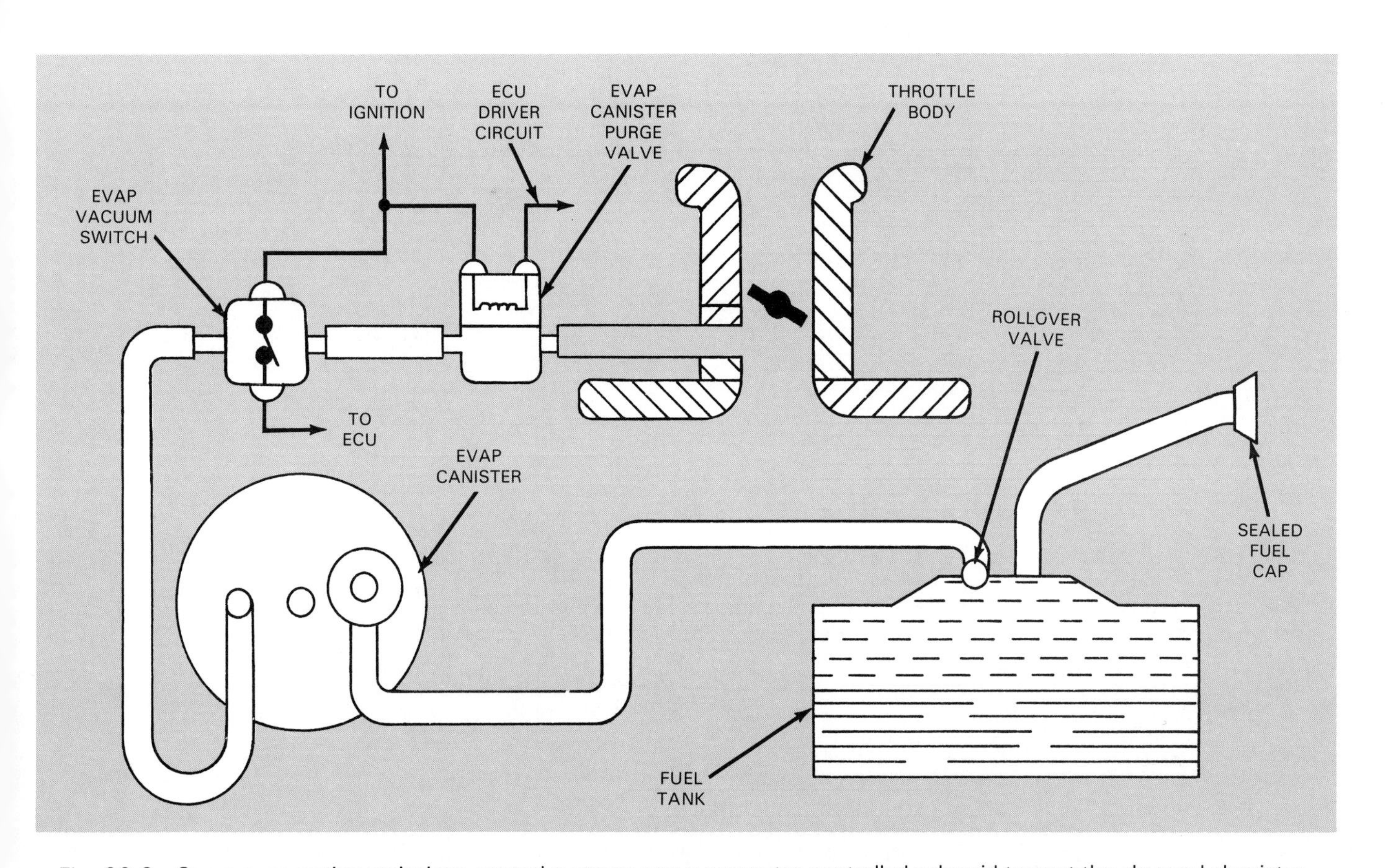

Fig. 20-8. Some evaporative emissions control systems use a computer-controlled solenoid to vent the charcoal chanister. This system is referred to as an "enhanced" evaporative emissions control system.

added to the mixture, they decrease peak combustion temperatures and lower NOx.

The EGR valve usually bolts to the engine intake manifold or an adapter plate. Exhaust gases are routed through the cylinder head and manifold to the EGR valve.

The *EGR valve* consists of a vacuum diaphragm, spring, plunger, exhaust gas valve, and a diaphragm housing, Fig. 20-9.

At idle, the throttle plate in the carburetor or fuel injection throttle body is closed. This blocks engine vacuum so it cannot act on the EGR valve. The EGR spring holds the valve shut and exhaust gases do NOT enter the intake manifold. If the EGR valve were to open at idle, it could upset the air-fuel mixture and the engine could stall.

When the throttle plate is swung open to increase speed, engine vacuum is applied to the EGR hose. Vacuum pulls the EGR diaphragm up. In turn, the diaphram pulls the valve open. Engine exhaust can then enter the intake manifold and combustion chambers, Fig. 20-9. Some EGR valves are equipped with pintle position and flow sensors that monitor system performance.

Air injection system

An *air injection system* forces fresh air into the exhaust ports of the engine to reduce HC and CO emissions. The exhaust gases leaving an engine can contain unburned and partially burned fuel. Oxygen from the air injection systems causes this fuel to continue to burn, Fig. 20-10.

The *air pump* is belt driven and forces air at low pressure into the system. A rubber hose or metal line connects the output of the pump to a diverter valve. Emission system air pumps can also be driven by a small electric motor. This reduces emissions because a more constant flow of air is produced by the electric motor since air pump speed does not change with engine speed.

The *diverter valve* keeps air from entering the exhaust system during deceleration. This prevents backfiring in the exhaust system. The diverter valve also limits maximum system air pressure when needed. It releases excess pressure through a silencer or muffler, Fig. 20-10.

An *air manifold* directs a stream of air toward each engine exhaust valve. Fittings on the air distribution manifold screw into threaded holes in the exhaust manifold or head, Fig. 20-10.

An *air check valve* is usually located in the line between the diverter valve and the air distribution manifold. It keeps exhaust from entering the air injection system.

Pulse air system

A pulse air system performs the same function as an air injection system. However, it uses pressure pulses in the exhaust system which operate check valves.

The *aspirator valves,* called check valves, gulp valves, or reed valves, block airflow in one direction and allow airflow in the other direction. Pulse air systems are no longer used on modern vehicles. Fig. 20-11 shows how a typical system operates.

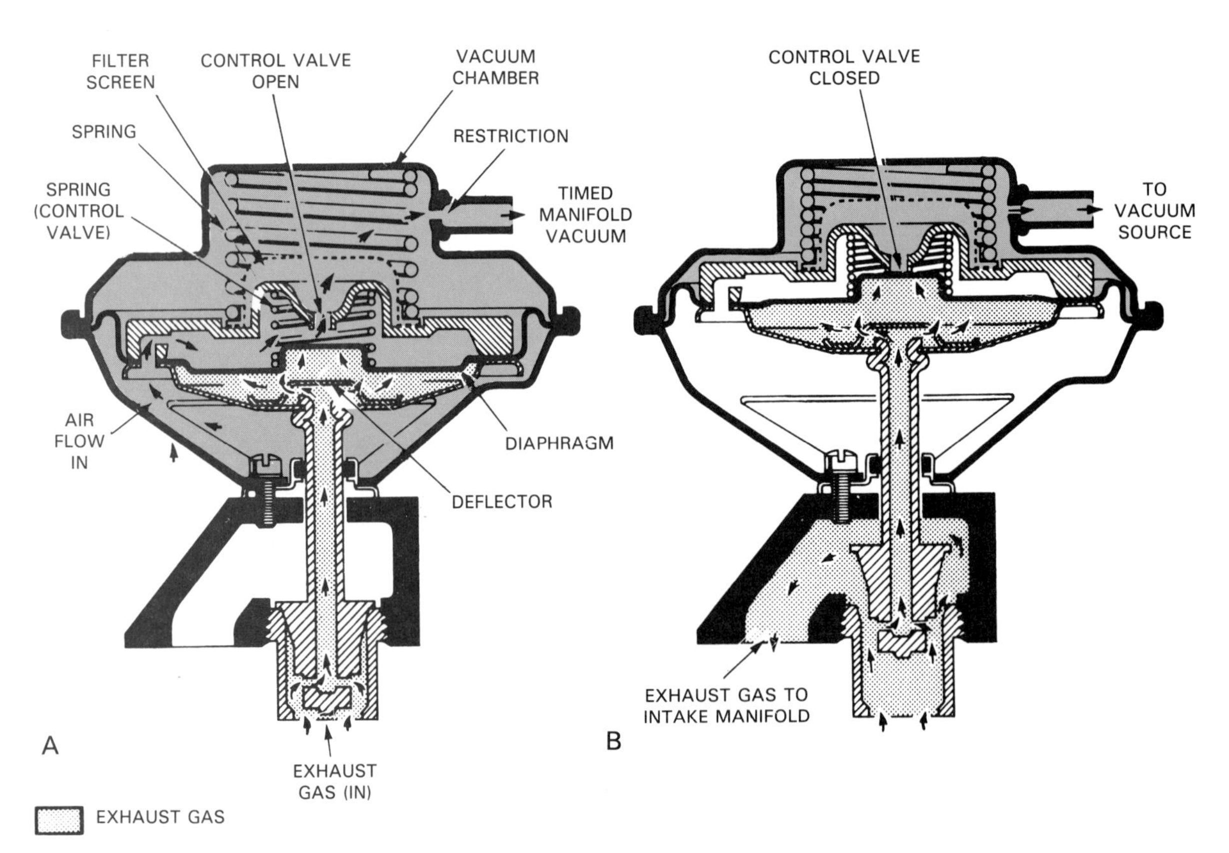

Fig. 20-9. EGR valve allows exhaust gases to enter intake manifold so combustion flame is not too hot. This prevents too much NOx pollution. A—Engine idling. EGR closed so idle mixture is not too lean. B—Engine speed increased. Intake vacuum applied to EGR. This pulls diaphragm up and opens valve so exhaust gases can flow to intake manifold and into combustion chambers. (Buick)

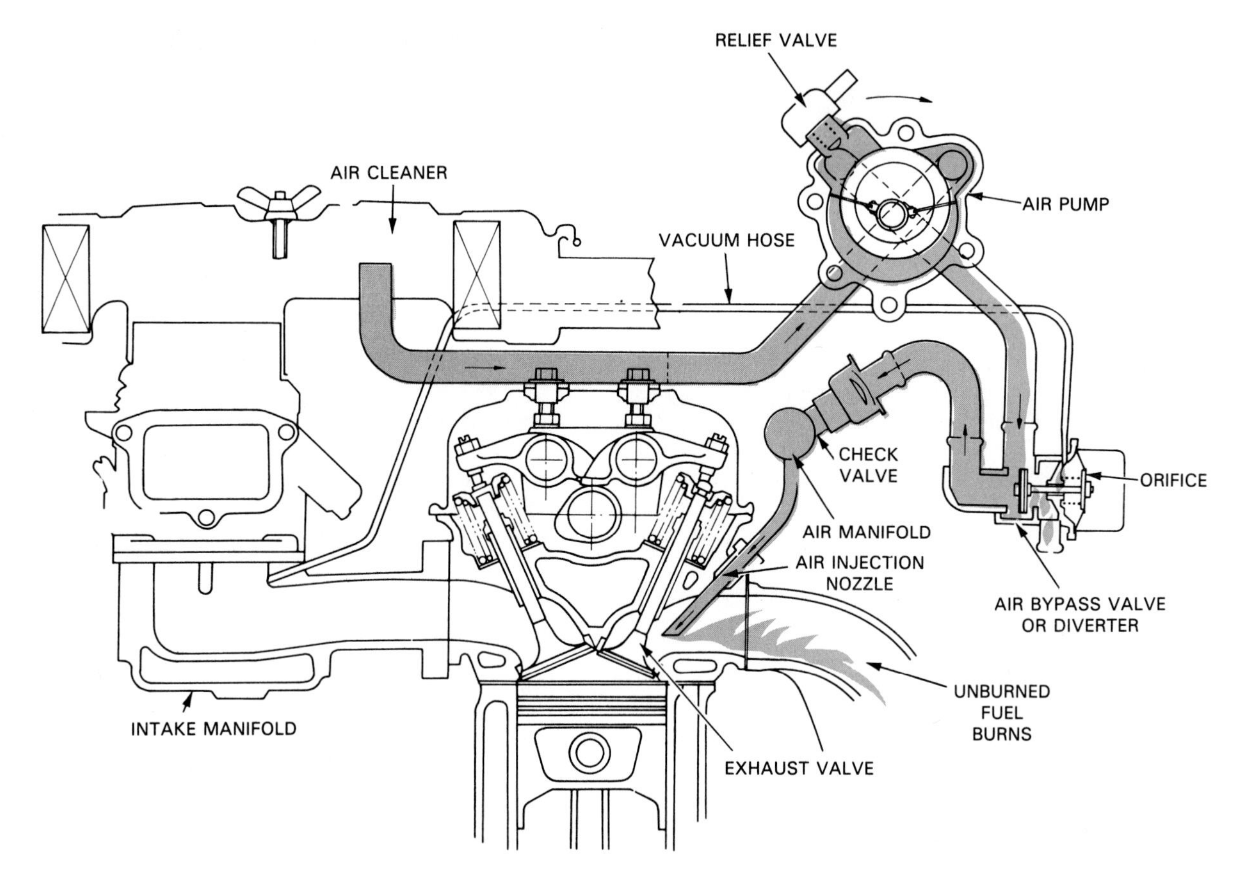

Fig. 20-10. Air injection system uses low pressure air pump to force fresh air into exhaust ports or exhaust manifold. This makes any fuel burn to reduce emissions. Diverter or antibackfire valve blocks air injection upon deceleration. Check valve keeps exhaust from entering air line. Study system. (Chevrolet)

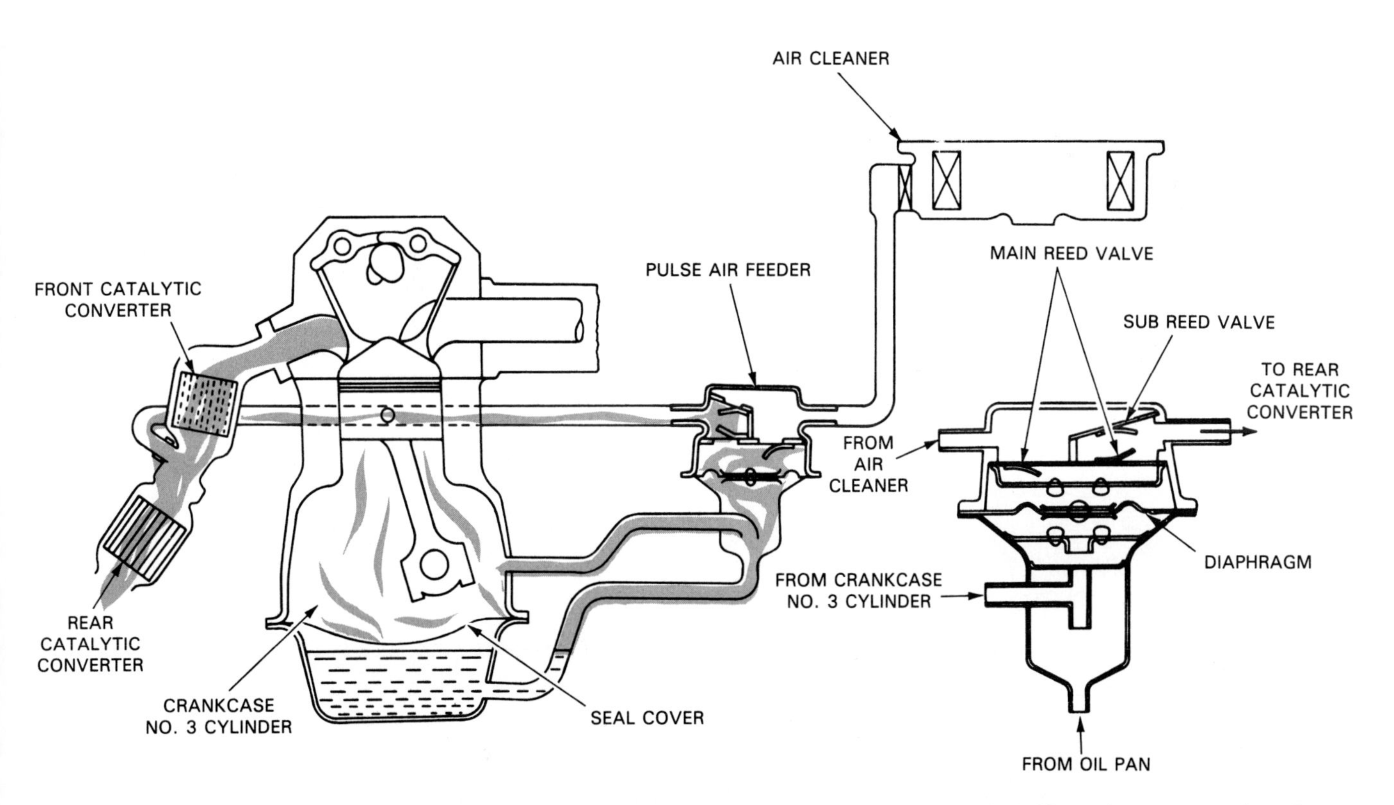

Fig. 20-11. Pulse air system does same thing as air injection, only air pump is not needed. Natural pressure pulses from opening and closing exhaust valves operate system. Aspirator valves are simply one-way valves that cause air to flow toward and into exhaust manifold to aid combustion in exhaust system. (Honda)

Heat riser

A *heat riser* can be used at the outlet of the engine exhaust manifold to speed engine warmup. It is simply a thermostatic spring or vacuum diaphragm-operated flat or valve in the exhaust system. See Fig. 20-12.

When the engine is cold, the heat riser closes to block or restrict exhaust flow. Hot exhaust gases are then forced to flow up into the intake manifold and/or cylinder head. Exhaust passages in the manifold or head allow the gases to speed part warmup. As the engine warms, the heat riser opens to permit normal exhaust flow away from the engine.

Catalytic converter

A *catalytic converter* oxidizes (burns) the remaining HC and CO emissions that pass into the exhaust system.

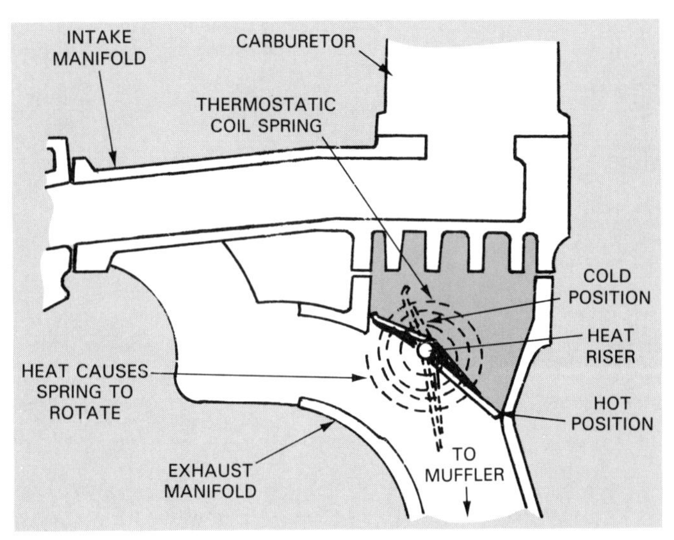

Fig. 20-12. Heat riser is flap mechanism in exhaust manifold. When engine is cold, thermostatic spring or vacuum switch and diaphragm close valve to cause high backpressure. Exhaust gases are forced to intake manifold to speed engine warmup. When engine warms, riser opens. An engine is very inefficient and produces high emission levels when cold. (Deere & Co.)

Extreme heat of approximately 1400 °F or 760 °C ignite these emissions and change them into harmless carbon dioxide (CO_2) and water (H_2O), Fig. 20-13.

The *catalyst operating temperature* is attained when the catalyst agents are hot enough (above approx. 300 °F or 149 °C) to start treating emissions.

A catalytic converter contains a catalyst substance, usually platinum, palladium, rhodium, or a mixture of these substances. Platinum and palladium treat the HC and CO emissions. Rhodium acts on the NOx emissions. Newer converters contain a base metal called cerium. This metal has the ability to stabilize the catalyst operation by attracting and releasing excess oxygen in the exhaust.

A catalytic converter using a ceramic block type catalyst is often termed a *monolithic* type converter.

When small ceramic beads are used, it is called a *pellet* type catalytic converter.

A *dual-bed catalytic converter* contains two separate catalyst units enclosed in a single housing. A mixing chamber is provided between the two. Air is forced into the chamber to help burn the emissions, Fig. 20-14. Most modern converters are *three-way converters.*

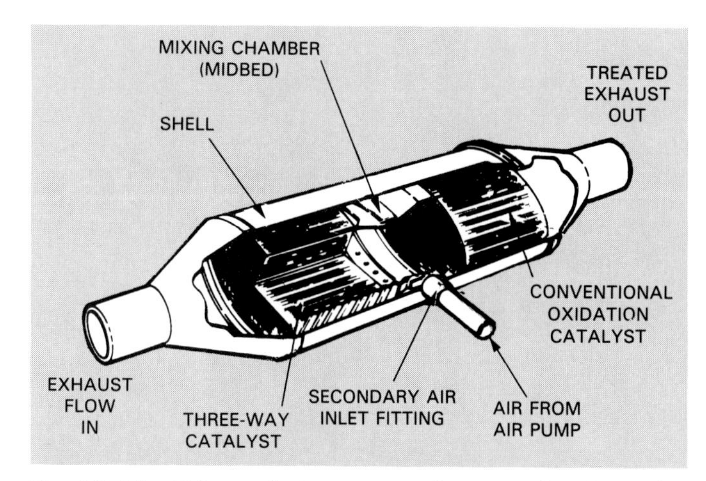

Fig. 20-14. This catalytic converter has two chambers. Air injection pump forces outside air into center of chambers to air thermal reaction or burning. (Ford)

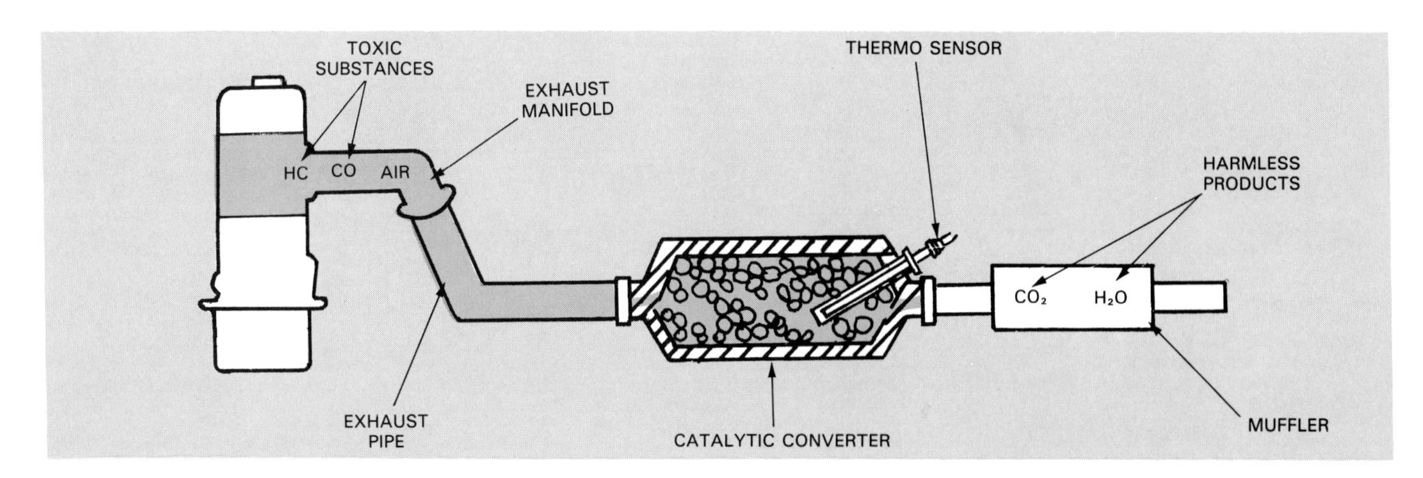

Fig. 20-13. Catalytic converter is thermal or heat reactor that burns any emissions that enter exhaust system. Hopefully, harmless CO_2 and water come out exhaust pipe.

COMPUTERIZED EMISSION CONTROL SYSTEMS

A *computerized emission control system* uses engine sensors, an on-board computer, and electric actuators (solenoid, motors, switches, etc.). See Fig. 20-15.

Fig. 20-16 shows a computerized engine control system. Note how it controls the air management or air injection system and the charcoal canister purge, which are both emission control systems.

By more closely controlling the engine idle speed, the fuel mixture, spark timing, transmission shift points, etc., a computer increases efficiency and reduces emissions considerably.

For more information on computer control, refer to the text index. Computers, sensors, and actuators are discussed in several chapters.

EMISSION CONTROL SYSTEM SERVICE

This section of the chapter will summarize the most common tests and repairs for emission control systems. A malfunction in an emission system can upset normal engine operation. As an engine technician, you should have a working knowledge of these engine related systems.

Exhaust gas analyzer

An *exhaust gas analyzer* is a testing instrument that measures the chemical content of engine exhaust gases. It can measure actual engine emissions produced by the engine, Fig. 20-17.

The analyzer probe (sensor) is placed in the car's tailpipe. With the engine running, the exhaust analyzer

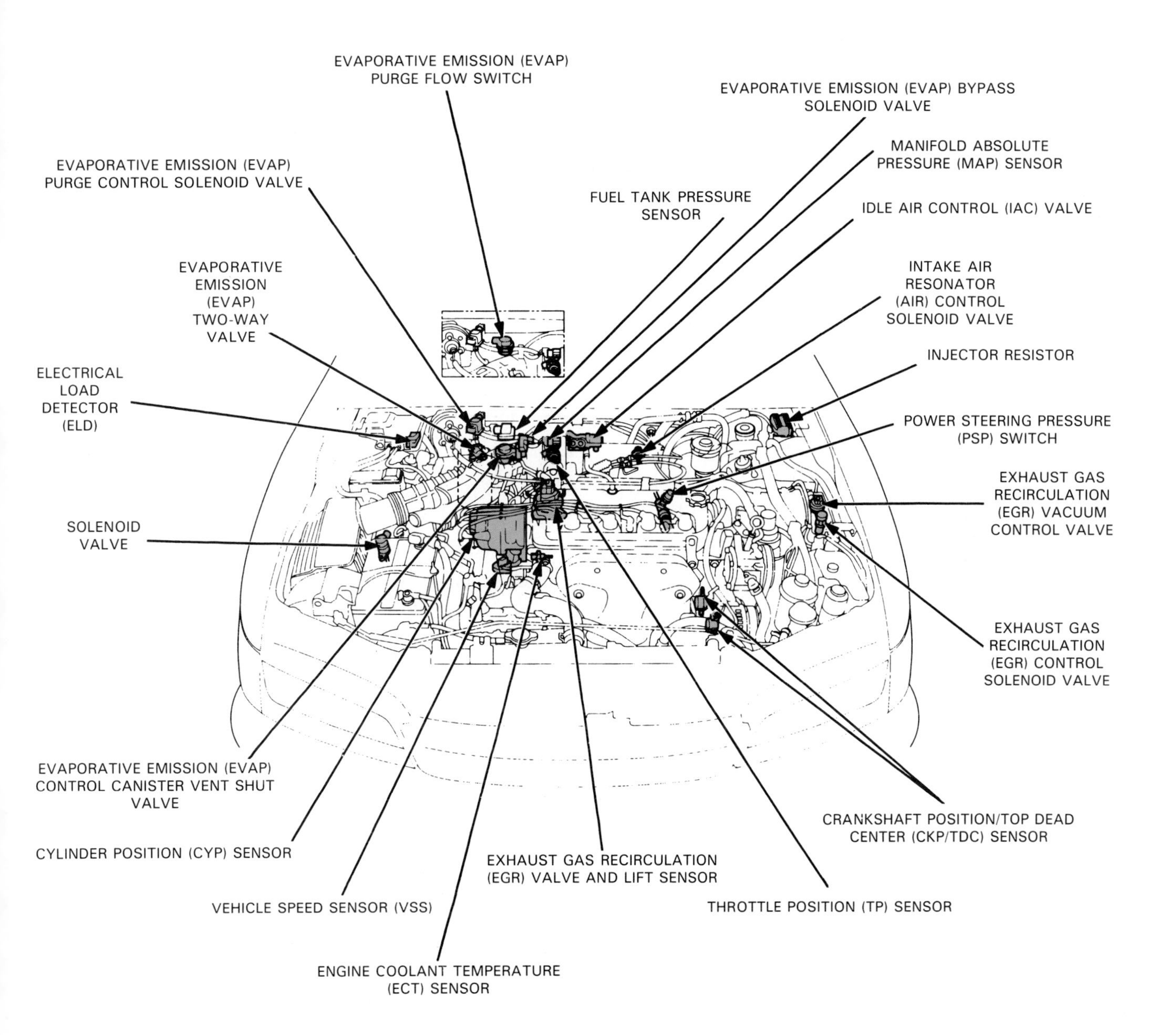

Fig. 20-15. A computer or ECU is now used to monitor and control today's emission control systems. Illustration shows general location and types of sensors and actuators used to reduce emissions. (Honda Motor Co.)

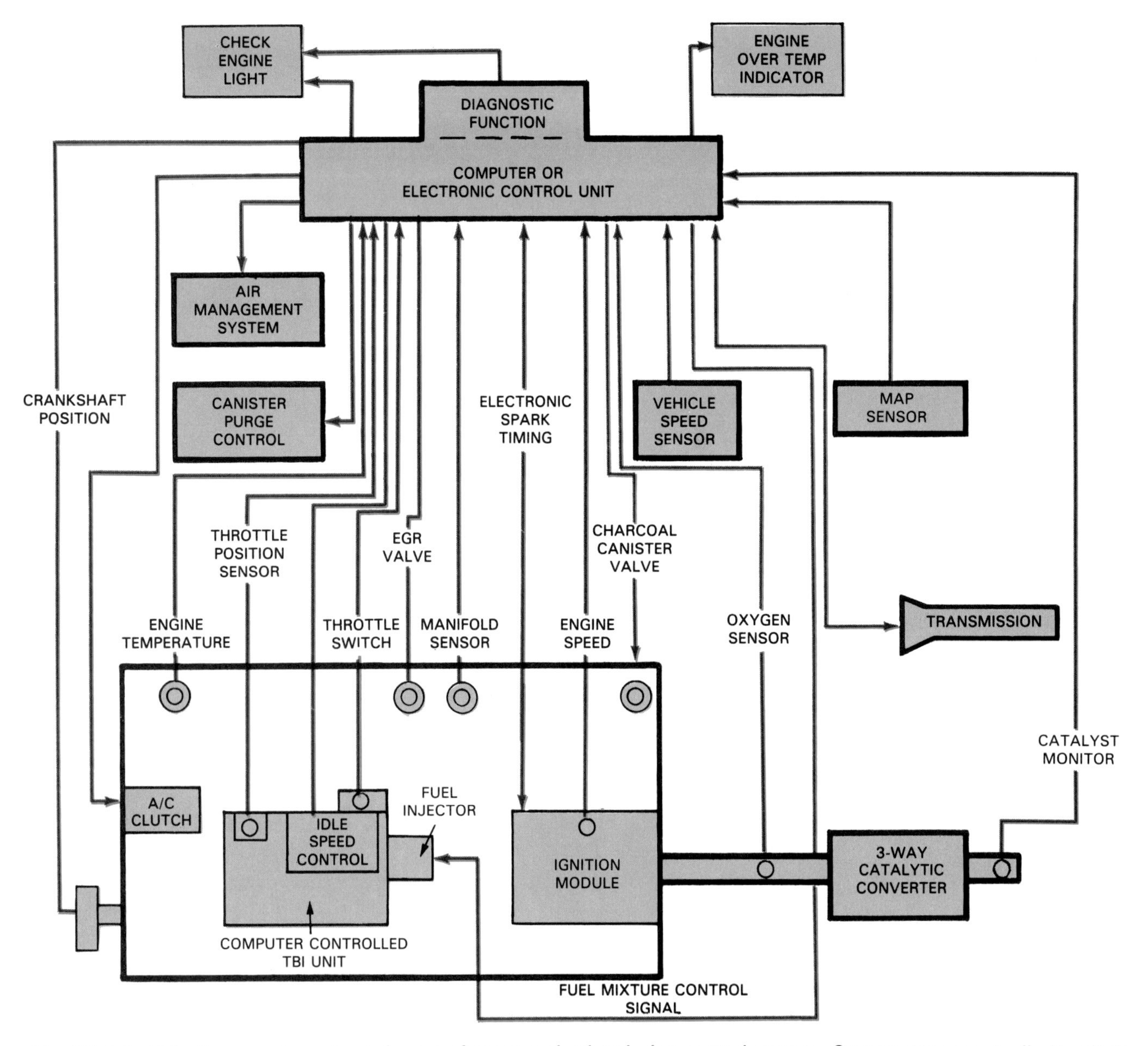

Fig. 20-16. This diagram shows basic layout of computerized emission control system. Oxygen sensor actually "sniffs" exhaust for oxygen which is an indicator of air-fuel ratio. Other sensors check engine temperature, speed, throttle opening, etc. Note other emission devices also controlled by ECU. (Buick)

will indicate the amount of pollutants and other gases in the exhaust. The technician can use this information to determine the condition of the engine and other systems.

An exhaust gas analyzer is an excellent diagnostic tool that will indicate:

1. Carburetor or fuel injection problems.
2. Engine mechanical problems.
3. Vacuum leaks.
4. Ignition system problems.
5. PCV troubles.
6. Clogged air filter.
7. Faulty air injection system.
8. Evaporative control system problems.
9. Computer control system troubles.
10. Catalytic converter condition.

Engine exhaust gases contain chemical substances that change with combustion (engine) efficiency. Some of these substances are harmful hydrocarbons (HC), carbon monoxide (CO), and oxides of nitrogen (NOx). Other by-products of combustion are NOT harmful: carbon dioxide (CO_2), oxygen (O_2), and water (H_2O). By measuring these substances, you can find out if the engine and emission systems are working properly.

The *two-gas exhaust analyzer* measures the amount of HC and CO in the engine exhaust. It has been used for a number of years.

The *four-gas exhaust analyzer* measures the quantity of HC, CO, CO_2, and O_2 in the exhaust. Although CO_2 and O_2 are not toxic, they provide useful data about combustion efficiency. Late model engines are so clean burning, a four-gas exhaust analyzer is needed to accurately

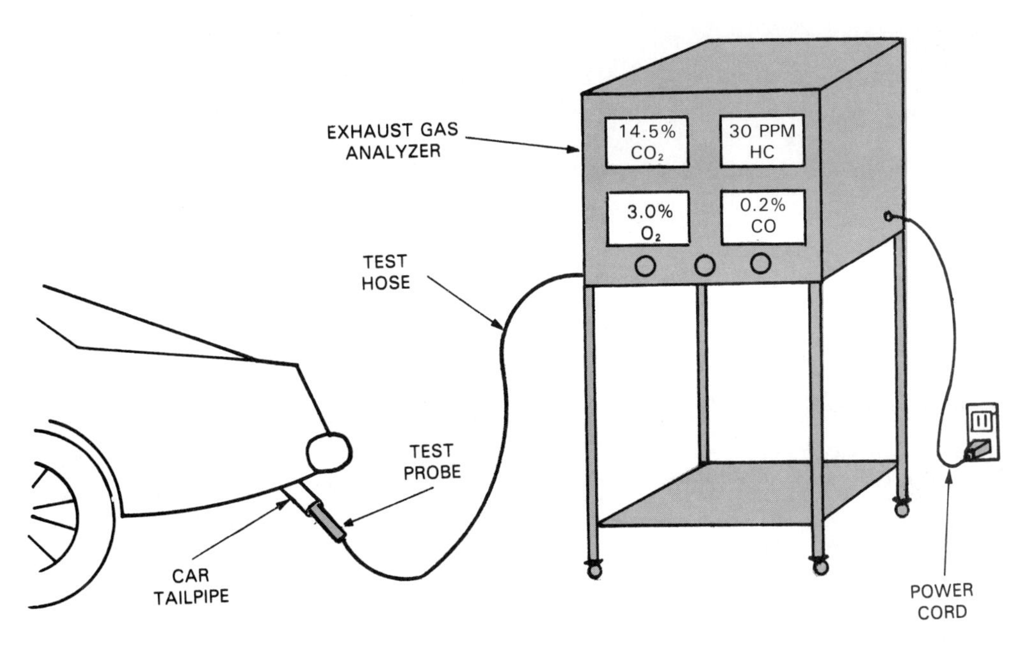

Fig. 20-17. Exhaust analyzer will measure pollutants and other substances in exhaust for checking engine efficiency. If readings are above specs for particular year of car, adjustments or repairs would be needed. Follow equipment operating instructions when using an exhaust analyzer.

evaluate the exhaust gases. It provides extra information for diagnosing problems and making adjustment, Fig. 20-17.

HC readings

An exhaust analyzer measures *hydrocarbons* (HC) in parts per million (ppm) by volume. An analyzer reading of 100 ppm means there are 100 parts of HC for every million parts of exhaust gas. An older engine might have an HC spec of 400 ppm. A newer car, having stricter requirements, could have a 100 ppm HC spec. If HC is higher, the car's HC emissions (unburned fuel) are too high. An adjustment or repair would be needed.

NOTE! Always refer to the emission control sticker in the engine compartment or a service manual for emission level specs. Values vary year by year.

Higher-than-normal HC readings can be caused by one or more of the following conditions:

1. Rich or lean air-fuel mixture (carburetor or fuel injection system problem).
2. Improper ignition timing (distributor, computer, or adjustment problem).
3. Engine problems (blow-by, worn rings, burned valve, blown head gasket).
4. Faulty emission control system (bad PCV, catalytic converter, evaporative control system).
5. Ignition system troubles (fouled spark plug, cracked distributor cap, open spark plug wire).

CO readings

An exhaust analyzer measures *carbon monoxide* (CO) in percentage by volume. A one percent reading would mean that one percent of the engine exhaust is made up of CO. High CO is basically caused by incomplete burning and a lack of air (oxygen) during combustion.

If the analyzer reading is higher than specs, you would need to locate and correct the cause of the problem.

The exhaust analyzer's CO reading is related to the air-fuel ratio. A HIGH CO reading would indicate an over-rich mixture. A LOW CO reading would indicate a lean air-fuel mixture.

Typical causes of high CO readings are:

1. Fuel system problems (bad injector, high float setting, clogged carburetor air bleed, restricted air cleaner, choke out of adjustment, bad engine sensor, computer troubles).
2. Emission control system troubles (almost any emission control system problem can upset CO).
3. Incorrect ignition timing (timing too advanced or advance mechanism problem).
4. Low idle speed (carburetor or injection system setting wrong).

O_2 readings

Four-gas exhaust analyzers measure *oxygen* (O_2) in percentage by volume. Typically, O_2 readings should be between .1 and 7 percent. Oxygen is needed for the catalytic converter to burn HC and CO emissions. Without O_2 in the engine exhaust, emissions can pass through the converter and out the tailpipe.

The air injection system or air pulse system adds O_2 to the exhaust. As air is added, CO and HC emissions decrease. As a result, O_2 readings can be used to check the operation of the carburetor, fuel injection system, air injection system, catalytic converter, and control computer.

Oxygen in the engine exhaust is an accurate indicator of a LEAN air-fuel mixture. As the air-fuel mixture becomes lean enough to cause a LEAN MISFIRE (engine miss), O_2 readings rise dramatically. This provides a very accurate method of measuring lean air-fuel ratios.

Today's fuel systems must be adjusted almost to the lean misfire point to reduce exhaust emissions and increase fuel economy.

CO_2 readings

The four-gas exhaust analyzer also measures carbon dioxide (CO_2) in percent by volume. Typically, CO_2 readings should be above 8 percent.

Carbon dioxide is a by-product of combustion. CO_2 is not toxic at low levels. When you breathe, for example, you exhale carbon dioxide.

Normally, oxygen and carbon monoxide levels are compared when evaluating the content of the engine exhaust. For example, if the percent of CO_2 exceeds the percent of O_2, the air-fuel ratio is on the rich side of the *stoichiometric* (theoretically perfect) fuel mixture.

Using an exhaust gas analyzer

Drive the vehicle to bring it to operating temperature before testing. Never test a vehicle with a cold engine or inaccurate readings will result. Warm the analyzer as described by the manufacturer. Then, zero and calibrate the analyzer while sampling clean air (no exhaust gases in room), Fig. 20-17.

Generally, measure the vehicle at idle and at approximately 2500 rpm. Compare the analyzer readings with specifications.

When testing some electronic (computer) fuel injection systems without a load, only idle readings on the exhaust analyzer will be accurate. A dynamometer must be used to load the engine, simulating actual driving conditions.

Inspecting emission control systems

After studying your exhaust analyzer readings, you must find the source of any problems. Start out by inspecting all engine vacuum hoses and wires. A leaking vacuum hose or disconnected wire could upset the operation of the engine and emission control systems. See Fig. 20-18.

Inspect the air cleaner for clogging. Check that the air pump belt is properly adjusted. Try to locate any obvious problems. If nothing is found during inspection, each system should be checked and tested.

Emission control information sticker

The *emission control information sticker* gives instructions, diagrams, and specs for complying with EPA regulations. The information given on the sticker must be used when doing a tune-up, for example. The emission sticker is normally located in the engine compartment, on the radiator support or valve cover, Fig. 20-19.

PCV SYSTEM SERVICE

An *inopertive PCV system* can increase exhaust emissions, cause engine sludging, engine wear, a rough engine idle, and other problems. A leaking PCV system can cause a vacuum leak and produce a lean air-fuel mix-

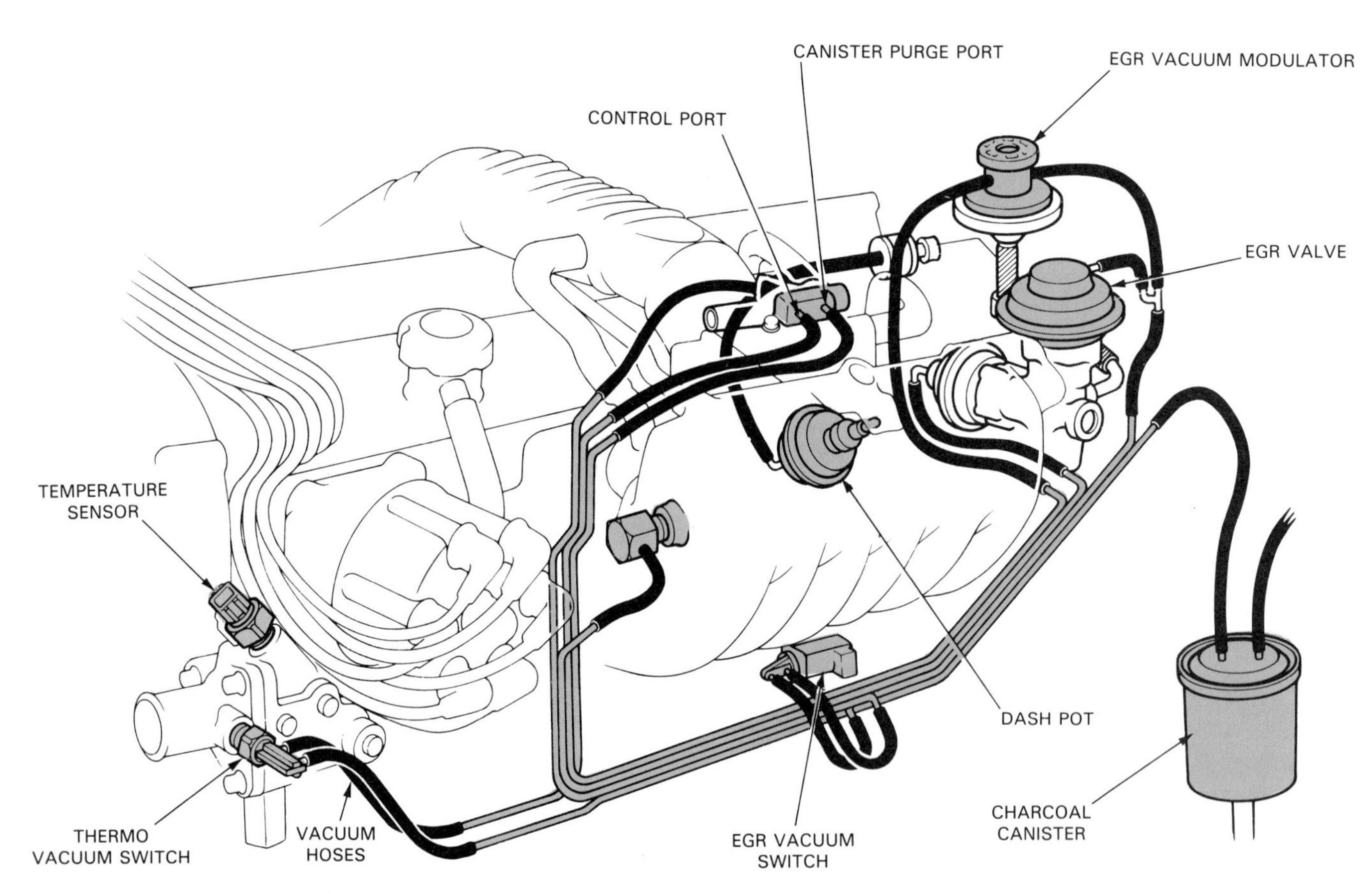

Fig. 20-18. During inspection, always check for leaking vacuum hoses, disconnected wires, loose air pump belt, or other troubles that could increase air pollution. (Toyota)

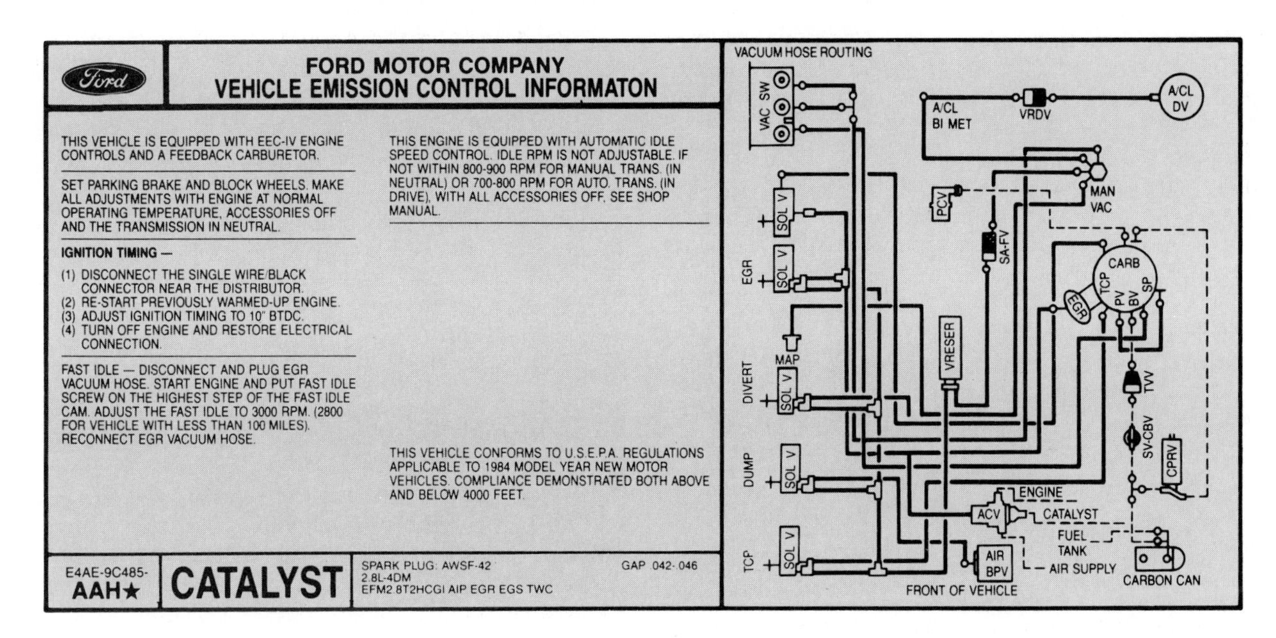

Fig. 20-19. Emission control sticker will be in engine compartment. Like this one, it will give data on setting ignition timing, adjust carburetor or injection system, routing vacuum hoses, etc. Always check this sticker when making tune-up type adjustments. (Ford)

ture at idle. A restricted PCV system can enrich the fuel mixture, also affecting engine idle.

Most auto makers recommend periodic maintenance of the PCV system. Inspect the condition of the PCV hoses, grommets, fittings, and breather hoses. Replace any hose that shows signs of deterioration. Clean or replace the breather filter if needed. Also, check or replace the PCV valve as needed.

To test a PCV system, pull the PCV valve out of the engine. With the engine idling, place your finger over the end of the valve. You should feel suction on your finger and the engine idle speed should drop about 40-80 rpm. If you cannot feel vacuum, the PCV valve or hose might be plugged with sludge. Look at Fig. 20-20.

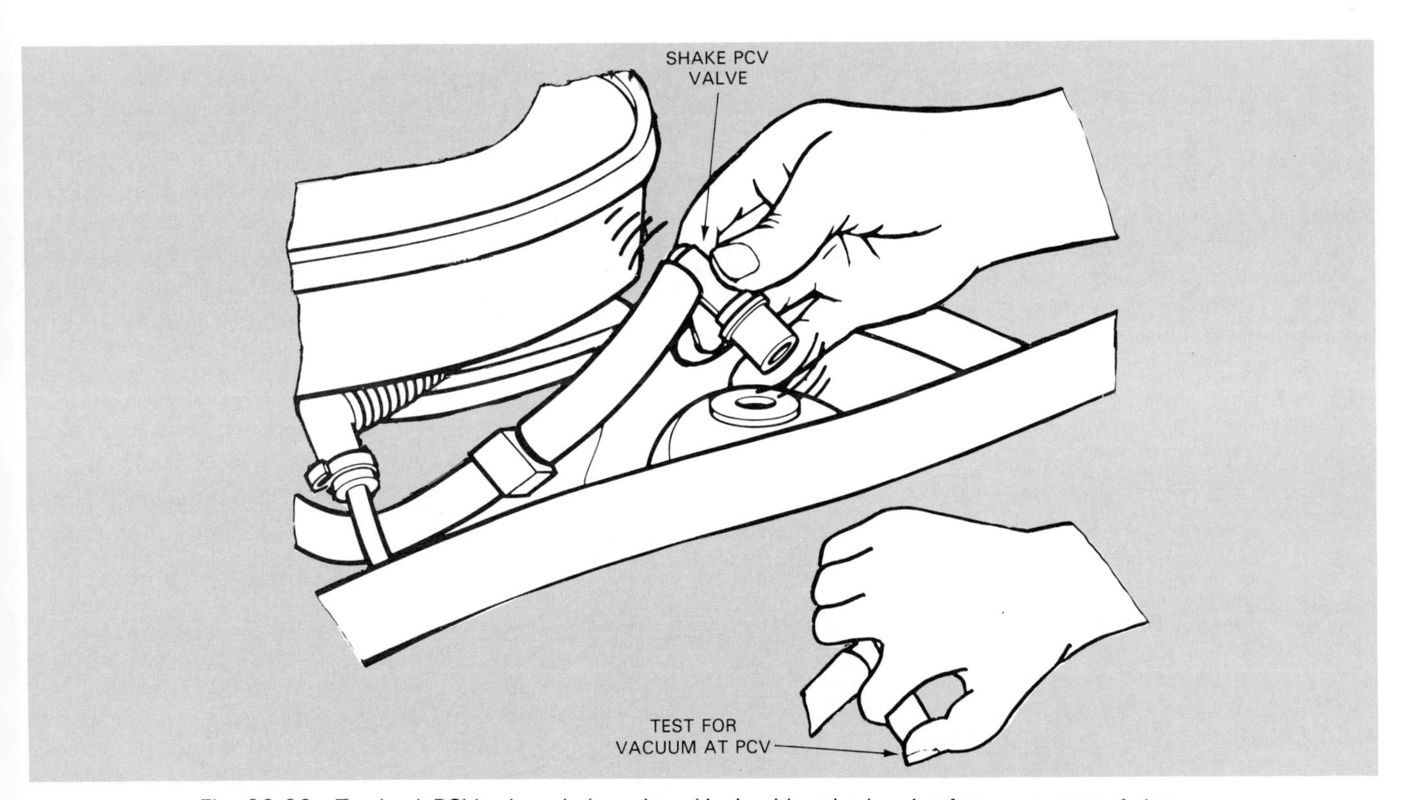

Fig. 20-20. To check PCV valve, shake unit and it should rattle showing free movement of plunger. Also, feel for vacuum with engine running to make sure hose is not stopped up. (Chrysler)

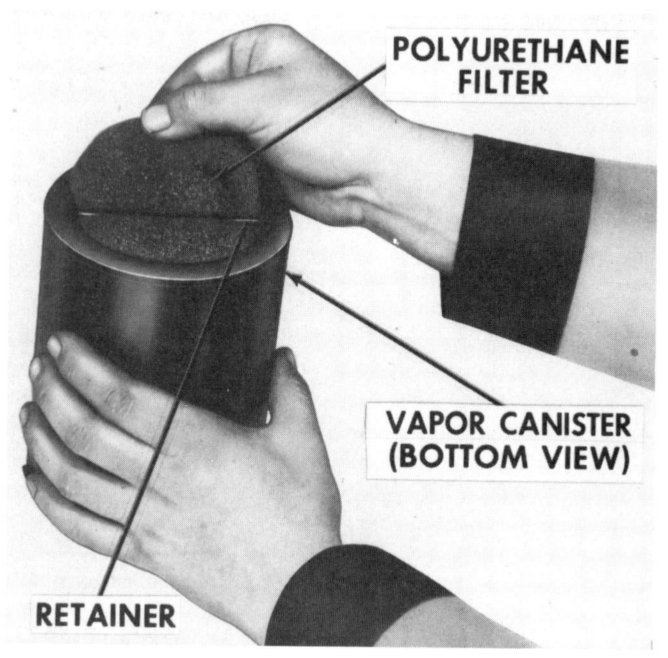

Fig. 20-21. Charcoal canister usually has a filter that should be serviced at regular intervals. Check manual for details. (Cadillac)

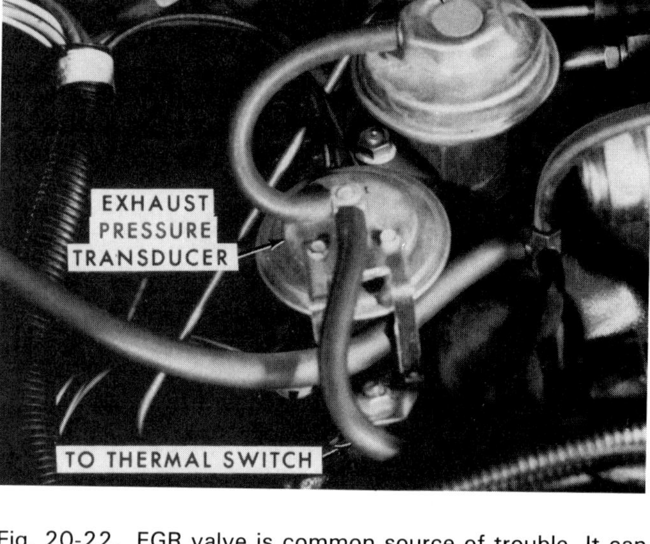

Fig. 20-22. EGR valve is common source of trouble. It can become fouled with carbon. If stuck open, this will cause a vacuum leak that upsets engine operation at idle. If stuck closed, this can cause higher combustion temperatures, and resulting ping and knock. (Cadillac)

A four-gas exhaust analyzer can also be used to check the general condition of a PCV system. Measure and note the analyzer's readings with the engine idling. Then, pull the PCV valve out of the engine, but not off the hose. Compare the readings after the PCV valve is removed.

A *plugged PCV system* will show up on the exhaust analyzer when O_2 and CO do NOT CHANGE. *Crankcase dilution* (excessive blowby or fuel in oil) will usually show up as excessive (one percent or more) increase in O_2 or a one percent or more decrease in CO.

EVAPORATIVE EMISSIONS SYSTEM SERVICE

A *faulty evaporative emissions system* can cause fuel odors, fuel leakage, fuel tank collapse (vacuum buildup), excess pressure in the fuel tank, or a rough engine idle. These problems usually stem from a defective fuel tank pressure-vacuum cap, leaking charcoal canister valves, deteriorated hoses, or incorrect hose routing.

Maintenance on an evaporative control system typically involves inspecting, cleaning, or replacing the filter in the charcoal canister. Service intervals vary, Fig. 20-21.

EGR SYSTEM SERVICE

EGR system malfunctions can cause engine stalling at idle, rough engine idling, detonation (knock), and poor fuel economy. Look at Fig. 20-22.

If the EGR valve sticks open, it will act as a large vacuum leak, causing a lean air-fuel mixture. The engine will run rough at idle or stall. If the EGR fails to open or the exhaust passage is clogged, higher combustion temperatures can cause pinging or knocking.

To test an EGR system, start, idle, and warm the engine. Operating the throttle linkage by hand, increase engine speed to 2000-3000 rpm. If visible, observe the movement of the EGR valve stem. The stem should move as the engine is accelerated. If it does not move, the EGR system is not functioning.

Sometimes the EGR valve stem is not visible. You then need to test each EGR system component separately. Follow the procedures described in a service manual.

To test the EGR valve, idle the engine. Connect a hand vacuum pump to the EGR valve. Plug the supply vacuum line to the EGR valve. When vacuum is applied to the EGR valve, the engine should begin to MISS or STALL. This lets you know that the EGR valve is opening and that exhaust gases are entering the intake manifold. See Fig. 20-23.

If EGR valve does NOT affect engine idling, remove the valve. The valve or the exhaust manifold passage could be clogged with carbon. If needed, clean the EGR valve and exhaust passage. When the EGR valve does not open and close properly, replace the valve.

EGR valves which provide electrical data to a computer control system require special testing procedures. Refer to a shop manual covering the specific system.

HEATED AIR INLET SYSTEM SERVICE

An inoperative thermostatic air cleaner can cause several engine performance problems. If the air cleaner flap remains in the OPEN POSITION (cold air position), the engine could miss, stumble, stall, and warm up slowly. If the air cleaner flap stays in the CLOSED POSITION (hot air position), the engine could perform poorly when at full operating temperature.

You should inspect the condition of the vacuum hoses and hot air hose from the exhaust manifold heat shroud. The hot air tube is frequently made of heat resistant

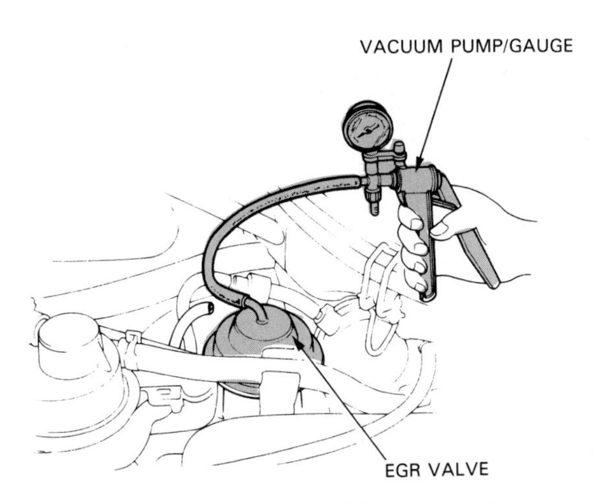

Fig. 20-23. To check operation of EGR valve, apply vacuum to unit with engine idling. When valve opens, engine should begin to miss or stall. If nothing happens, unit is clogged or diaphragm is ruptured. (Honda)

paper and metal foil. It will tear very easily.

To test the system, watch the action of the air flap in the air cleaner snorkel. Start and idle the engine. When the air cleaner temperature sensor is cold, the air flap should be closed. Place an ice cube on the sensor if needed. Then, when the engine and sensor warm, the flap should swing open.

If the air cleaner flap does not function, test the vacuum motor and sensor. Use manual procedures.

AIR INJECTION SYSTEM SERVICE

Air injection system problems can cause engine backfiring (loud popping sound), other noises, and increased HC and CO emissions. Inadequate air from the air injection system could also prevent the catalytic converter from functioning properly.

A four-gas exhaust analyzer provides a quick and easy method of testing an air injection system. Run the engine at idle and record the readings. Then, disable the air injection system. Remove the air pump belt or use pliers to pinch the hoses to the air distribution manifold. Compare the exhaust analyzer readings before and after disabling the air injection system.

Without air injection, the exhaust analyzer O_2 reading should drop approximately 2-5%. HC and CO readings should increase. This would show that the air injection system is forcing air (oxygen) into the exhaust system.

If the analyzer readings do NOT change, the air injection system is not functioning. Test each component until the source of the problem is found.

PULSE AIR SYSTEM SERVICE

Many of the maintenance and testing methods discussed for an air injection apply to the pulse air system. Inspect all hoses and lines. Measure O_2 with a four-gas analyzer to test the system. Analyzer oxygen readings should drop when the pulse air system is disabled. If readings do not drop, check the action of the aspirator (reed) valves.

Fig. 20-24. To check air injection aspirator valves, place fingers over units. You should feel suction pulses but no pressure pulses. Replace units if needed. (Saab)

Fig. 20-24 shows a technician testing the operation of the aspirator valves. With the engine running, you should be able to feel vacuum pulses on your fingers. However, you should NOT feel exhaust pressure pulses trying to blow back through the valves. Replace the valves if needed.

CATALYTIC CONVERTER SERVICE

Catalytic converter problems are commonly caused by lead contamination, overheating, and extended service.

A *clogged catalytic converter,* resulting from lead deposits or overheating, can increase exhaust system backpressure. High backpressure decreases engine performance because gases cannot flow freely away from the engine. A clogged catalytic converter is a fairly common problem. The increased back pressure will reduce engine power considerably. You may smell rotten eggs at the tailpipe. A clogged converter can also overheat, possibly causing a fire.

After extended service, the catalyst in the converter can become coated with deposits. These deposits can keep the catalyst from acting on the HC, CO, and NOx emissions. Increased air pollution can result. The inner baffles and shell can also deteriorate. With a pellet type catalytic converter, this can allow BB-size particles to blow out the tailpipe.

Catalytic converters must be replaced when the catalyst becomes damaged or contaminated.

CAUTION! Remember that the operating temperature of a catalytic converter can be over 1400 °F (760 °C). This is enough heat to cause serious burns.

COMPUTERIZED EMISSION SYSTEM SERVICE

Computerized engine and emission systems can cause a wide range of problems. The computer may control the carburetor, fuel injection system, EGR valve, evaporative control system, and other components. If not controlled properly, these parts can upset engine operation.

Testing computerized emission systems

Most computerized emission control systems have a built-in diagnostic system. If a problem develops, a malfunction indicator light will flash on and off, indicating a problem. To pinpoint the trouble, follow service manual test procedures.

Access the computer's self-diagnostics through the vehicle's data link connector, Fig. 20-25. The procedures for this were given in Chapters 16 and 17. If any codes are present, write them down and look them up in the service manual, which contains charts listing all of the possible codes, Fig. 20-26.

Another method of checking out a computerized emission control system is to measure voltages at the ECU terminals. The service manual will have an illustration like the one in Fig. 20-27. If test voltages are not within specs at a terminal, you would know which section of the computer system has problems.

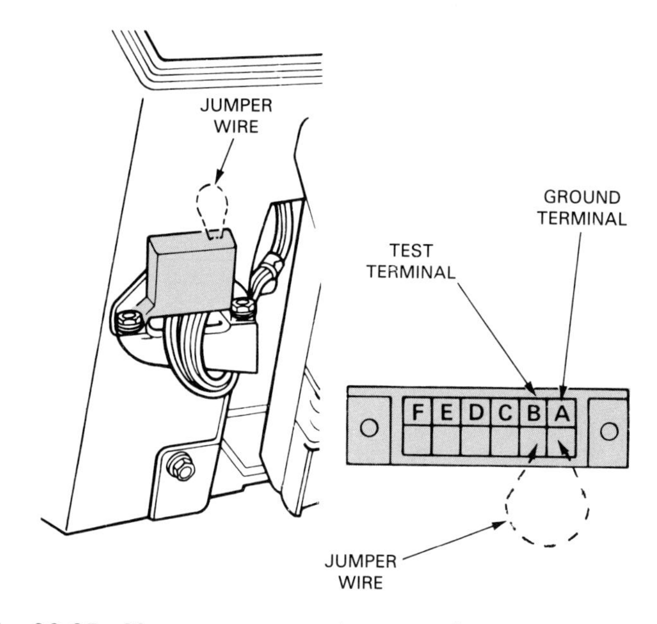

Fig. 20-25. Many systems require you to jump across two connectors at test terminal. This will place computer in self-diagnosis mode. Computer will produce number code that can be compared to service manual chart giving what problem might be in system. Follow service manual directions closely to prevent component damage. (Oldsmobile)

Replacing computer system components

Before replacing any computer system part, make sure all electrical connections are clean and tight. A corroded connection could be affecting system or component operation.

When replacing the computer, you may need to remove and reuse the prom. The *prom* (programmable read only memory) is a computerized chip (miniaturized electronic circuit) designed for use with the specific vehicle. Usually, it is not faulty and can be used over. Only the control section of the computer usually fails.

TROUBLE CODE IDENTIFICATION

The "CHECK ENGINE" light will only be "ON" if the malfunction exists under the conditions listed below. It takes up to five seconds minimum for the light to come on when a problem occurs. If the malfunction clears, the light will go out and a trouble code will be set in the ECM. Code 12 does not store in memory. If the light comes "on" intermittently, but no code is stored, go to the "Driver Comments" section. Any codes stored will be erased if no problem reoccurs within 50 engine starts.

The trouble codes indicate problems as follows:

TROUBLE CODE 12	No distributor reference pulses to the ECM. This code is not stored in memory and will only flash while the fault is present. Normal code with ignition "on," engine not running.
TROUBLE CODE 13	Oxygen Sensor Circuit — The engine must run up to five minutes at part throttle, under road load, before this code will set.
TROUBLE CODE 14	Shorted coolant sensor circuit — The engine must run up to five minutes before this code will set.
TROUBLE CODE 35	Idle speed control (ISC) switch circuit shorted. (Over 50% throttle for over 2 seconds.)
TROUBLE CODE 41	No distributor reference pulses to the ECM at specified engine vacuum. This code will store in memory.
TROUBLE CODE 42	Electronic spark timing (EST) bypass circuit or EST circuit grounded or open.
TROUBLE CODE 43	Electronic Spark Control (ESC) retard signal for too long a time; causes retard in EST signal.

Fig. 20-26. This is partial page from actual service manual showing self-diagnosis code chart. Read what some of the codes mean for this system. (Cadillac)

ECM TERMINAL VOLTAGE

THIS ECM VOLTAGE CHART IS FOR USE WITH A DIGITAL VOLTMETER TO FURTHER AID IN DIAGNOSIS. THESE VOLTAGES WERE DERIVED FROM A KNOWN GOOD CAR. THE VOLTAGES YOU GET MAY VARY DUE TO LOW BATTERY CHARGE OR OTHER REASONS, BUT THEY SHOULD BE VERY CLOSE.

THE FOLLOWING CONDITIONS MUST BE MET BEFORE TESTING:

- ENGINE AT OPERATING TEMPERATURE
- CLOSED LOOP
- ENGINE IDLING (FOR "ENGINE RUN" COLUMN)
- TEST TERMINAL NOT GROUNDED
- SCANNER NOT INSTALLED

Voltage Key "ON"	Voltage Engine Run	Voltage Circuit Open	VIEW – TOP OF BOX	Pin	Pin	VIEW – BOTTOM OF BOX	Voltage Key "ON"	Voltage Engine Run	Voltage Circuit Open
0	0	0	Sensor Return	22	1	BARO Sensor Output	4.75	4.75	*.5
5	5	5	5V Reference	21	2	TPS Sensor Output	*1.0	*1.0	5.0 †WOT
4.75	1.5	*.5	MAP or Vacuum Sensor Output	20	3	Coolant Temp. Sensor	*2.5	*2.5	5.0
			Not Used	19	4	Air Control Solenoid	12.5	*1.0	*.5
12	4–7 (var.)	*.5	M/C Solenoid	18	5	Trouble Code Test Term	5	5	5
*.5	*.5	12	3rd Gear Switch	17	6	EFE Relay - Fed. EFE - Calif.	12 12.1	12 hot *1 cold 14.1	*.5 *.5
8	8	8	VSS Signal	16	7	Coolant Temp. Sensor Return	0	0	0
			E Cell	15	8	ISC Switch	0 (C.T.)	0 hot 12 cold	12 12
*.5	*.5	.90	Oxygen Sensor - Lo	14	9	Oxygen Sensor - Hi	.45	.2—.8 (var.)	.45
*.5	*.5	.75	Dist. Ref. Pulse - Lo	13	10	Dist. Ref. Pulse - Hi	*.5	1–2 (var.)	*.5
*.5	1–2 (var.)	*.5	EST	12	11	Ign. Module By-Pass	*.5	4.2	*.5

* = Value Shown or Less Than that Value
† = Wide Open Throttle
C.T. = Closed Throttle
(var.) = variable
P/N = Park or Neutral
D/R = Drive or Reverse

Fig. 20-27. Another way of finding trouble in computer system is to measure voltages. This service manual diagram shows what voltages should be at ECM test terminal. If voltages are too high or low, you would know that there is a problem in that section of computer system. Read through this example. (General Motors)

Most sensors can be tested with a digital VOM following manual directions. Test them properly before replacement.

Protecting electronic circuits

When handling computers or electronic control units, keep one hand on frame ground and use the other to remove the component. This will prevent any static electrical surge from entering and damaging the electronic circuitry. If available, wear an anti-static wrist strap when working on computer circuits. Static electricity may not instantly ruin an electronic part but it can reduce its useful service life from years to days.

SUMMARY

An engine technician must know how to service emission control systems because they affect engine performance and service life.

HC, CO, NOx, and particulates are the four types of emissions. They are produced by engine blowby fumes, fuel vapors, and engine exhaust gases. Emission control systems are designed to reduce these pollutants.

Various engine modifications are made to assist emission control systems: lower compression ratios, leaner mixtures, increased valve overlap, hardened valves and seats, wider spark plug gaps, higher operating temperatures, etc.

The PCV system draws toxic fumes out of the crankcase and into the intake manifold for burning.

Evaporative emissions system collects and stores fuel vapors. It also causes them to be burned in the engine upon starting.

The EGR system admits burned exhaust gases into the intake manifold. This lowers peak combustion temperatures and NOx pollution.

The air injection system forces outside air into the exhaust manifold. This helps burn any fuel leaving the combustion chambers.

The catalytic converter also burns any fuel flowing through the exhaust system.

A computerized emission control system uses sensor information so the computer can more closely control some of the emission control devices.

An exhaust analyzer is commonly used to diagnose engine and emission control system troubles. It measures the HC, CO, and sometimes CO_2 and O_2.

Always inspect emission control systems for obvious problems. Use the service manual for specific how-to directions for the particular make and model car when doing complicated repairs.

Use the service manual for specific how-to directions for the particular make and model car when doing complicated repairs.

KNOW THESE TERMS

Emissions, Blowby, Hydrocarbons, Carbon monoxide, Oxides of nitrogen, Particulates, PCV, Thermostatic air cleaner, Fuel evaporation control, EGR, Air injection, Catalytic converter, PCV valve, Liquid-vapor separator, Charcoal canister, Purge line, Air pump, Diverter valve, Air manifold, Aspirator valves, Heat riser, Exhaust analyzer, Lean misfire, Stoichiometric, Prom.

REVIEW QUESTIONS—CHAPTER 20

1. What are the three sources of emissions?
2. Define the four types of emissions.
3. Explain nine engine modifications to reduce emissions.
4. A PCV system uses _________ _________ to draw blowby gases out of the _________ for reburning in the _________ _________ of the engine.
5. Explain some causes of higher than normal HC and CO readings with an exhaust analyzer.
6. Technician A says a PCV system can be quickly checked by placing a finger over the PCV valve to check for suction.
   Technician B says that you should also shake the PCV valve to make sure it is free.
   Who is correct?
   a. Technician A.
   b. Technician B.
   c. Both A and B.
   d. Neither A nor B.
7. When vacuum is applied to an EGR valve at idle, the engine should _________ or _________.
8. How do you check the operation of aspirator valves?
9. A catalytic converter can become clogged from _________ _________ or _________.
10. Many computer controlled emission systems have a self-diagnosis mode. True or false?

ASE CERTIFICATION-TYPE QUESTIONS

1. Technician A says that hydrocarbon emission can be caused by incomplete burning of the fuel in the engine's combustion chambers. Technician B says that hydrocarbon emission can be caused by fuel evaporation. Who is right?
   (A) A only.
   (B) B only.
   (C) Both A & B.
   (D) Neither A nor B.
2. Technician A says that a rich air-fuel mixture would increase CO emissions. Technician B says that a rich air-fuel mixture reduces CO emissions. Who is right?
   (A) A only.
   (B) B only.
   (C) Both A & B.
   (D) Neither A nor B.
3. Technician A says that wider spark plug gaps are now utilized on modern vehicles to burn leaner fuel mixtures. Technician B says that today's automotive engines are equipped with hotter thermostats to reduce HC and CO emissions. Who is right?
   (A) A only.
   (B) B only.
   (C) Both A & B.
   (D) Neither A nor B.
4. All of the following are modern vehicle emission control systems EXCEPT:
   (A) evaporative emissions control system.
   (B) pulse air system.
   (C) air injection system.
   (D) heated air inlet system.
5. Technician A says that an engine's PCV valve can be located in a valve cover. Technician B says that an engine's PCV valve can be located in the intake manifold. Who is right?
   (A) A only.
   (B) B only.
   (C) Both A & B.
   (D) Neither A nor B.
6. Technician A says that some PCV valves can be cleaned and reused. Technician B says that certain engine PCV valves must be discarded and replaced. Who is right?
   (A) A only.
   (B) B only.
   (C) Both A & B.
   (D) Neither A nor B.
7. Technician A says that an engine's air injection system is operated by an air pump. Technician B says that an air injection system operates off engine exhaust system pressure. Who is right?
   (A) A only.
   (B) B only.
   (C) Both A & B.
   (D) Neither A nor B.
8. Which type of catalytic converter is used on modern automotive exhaust systems?
   (A) Pellet type converter.
   (B) Three-way converter.
   (C) Dual-bed converter.
   (D) All of the above.
9. Technician A says that an exhaust gas analyzer can detect possible malfunctions in an engine's ignition system. Technician B says that an exhaust gas analyzer can detect possible engine vacuum leaks. Who is right?
   (A) A only.
   (B) B only.
   (C) Both A & B.
   (D) Neither A nor B.
10. Technician A says that if an engine's exhaust produces a higher-than-normal HC reading, an ignition system problem may exist. Technician B says that if an engine's exhaust produces a higher-than-normal HC reading, the engine may have a blown head gasket. Who is right?
   (A) A only.
   (B) B only.
   (C) Both A & B.
   (D) Neither A nor B.

Chapter 21

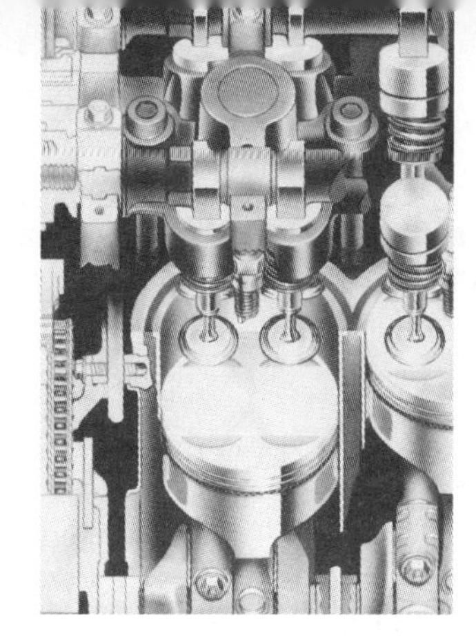

Engine Performance Problems

After studying this chapter, you will be able to:
- *Describe the most common engine performance problems.*
- *Explain the symptoms of common engine performance problems.*
- *Summarize how engine support systems might cause many engine performance problems.*
- *Use a systematic approach when correcting engine performance problems.*

An *engine performance problem* is any trouble that lowers engine power, fuel economy, drivability, or dependability. Performance problems can result from troubles in the fuel system, ignition system, emission control systems, or engine itself. This makes troubleshooting very challenging. You must use your knowledge of system operation and basic testing techniques to quickly and accurately locate and correct the source of problems.

This chapter summarizes the typical performance problems that are discussed in other text chapters. As a result, you will be better prepared to become employed as an engine technician.

LOCATING ENGINE PERFORMANCE PROBLEMS

Use a systematic approach when trying to locate performance problems. Do NOT use hit and miss repairs. A *systematic approach* involves using your knowledge of auto technology and a logical process of elimination. Think of all of the possible systems and components that could upset engine operation. Then, one by one, mentally throw out the parts that could NOT produce the symptoms.

To troubleshoot properly, ask yourself these kinds of questions:

1. What are the symptoms (noise, miss, smoke)?
2. What system could be producing the symptoms (ignition, fuel, engine)?
3. Where is the most logical place to start testing?

For example, suppose an engine with electronic fuel (gasoline) injection misses and emits black smoke after running for a few minutes. The black smoke would tell you that too much fuel is entering the engine. Since the engine does not run poorly when cold, engine temperature relates to the problem. Through simple deduction, you might think of the cold start injector. It richens the fuel mixture and is affected by temperature. Possibly, the cold start injector is staying on too long, allowing an excessively rich air-fuel mixture to enter the engine. The thermostatic switch that operates the injector could be shorted to the on position.

As you can see, logical thought will help you check the most likely problems. If your first idea is incorrect, rethink the problem and check the next most likely trouble source. See Fig. 21-1.

Fig. 21-1. By using logical processes and close inspection, you can often find the sources of many engine performance problems without complex testing procedures. Even with computer control systems, loose wires, vacuum leaks, arcing spark plug wires, engine overheating, and clogged filters are still common. (Mitsubishi)

Performance problem diagnosis charts

If you have trouble locating an engine performance problem, refer to a service manual diagnosis chart. It will list problems, causes, and corrections. The chart will be written for the particular make and model of car.

Fig. 21-2 shows an engine performance troubleshooting chart from a service manual.

Reading spark plugs

Reading spark plugs involves inspecting the condition and color of the spark plug tips. This will show you whether combustion is normal or abnormal (rich mixture,

Problem	Possible cause	Remedy	Page
Rough idle, stalls or misses (cont'd)	EGR valve faulty	Check valve	3-59
	Valve clearance incorrect	Adjust valves	2-7
	Compression low	Check compression	4-2
Engine hesitates/ Poor acceleration	Spark plugs faulty	Inspect plugs	3-6
	High tension wires faulty	Inspect wires	3-6
	Vacuum leaks • PCV hoses • EGR hoses • Intake manifold • Carburetor hoses • Air cleaner hoses	Repair as necessary	
	Incorrect ignition timing	Reset timing	2-26
	Air cleaner clogged	Check air cleaner	2-22
	Fuel system clogged	Check fuel system	5-2

Fig. 21-2. This is a partial page from a service manual. Note how troubleshooting chart gives problem, possible cause, remedy, and page number for service manual coverage. (Toyota)

detonation, lean mixture, oil entering cylinder, etc.). As a result you will have a better idea as to the condition of the engine and its related systems.

A *brown* to *grey-tan* spark plug tip is burning normally. Combustion is clean and good. See Fig. 21-3.

A *black* or *wet plug* would indicate a dead cylinder or

Fig. 21-3. Reading spark plugs will tell you about condition of engine and related systems. A—Normal burning plug has light deposits and is grey-tan. B—This plug is oil fouled from bad rings or oil seals. C—Carbon fouled plug from bad plug wire, rich mixture, etc. D—High mileage plug has heavy deposits and corroded center electrode. Rounded electrode will make it difficult for spark to jump gap. (Champion Spark Plug Co.)

engine mechanical problem allowing oil to enter the cylinder. It could also be caused by an extremely rich air-fuel mixture.

To read plugs, remove each plug and place it in order on your fender cover or tool cart. Then you will know which plug came from which cylinder. Closely inspect the plugs.

For example, if one plug is fouled with carbon and the others are clean, you would know that there is a problem in THAT cylinder. It could have a burned valve, bad plug wire, broken piston ring, etc. You might do a compression test in that cylinder.

Vacuum gauge

A *vacuum gauge* can be used to find some engine performance and mechanical troubles. The gauge is connected to a source of intake manifold vacuum. Gauge readings are then compared to known good readings. Fig. 21-4 show some sample vacuum gauge readings with their potential problem indications.

Test light

A *test light* will quickly check for power in an electric circuit and can be used to find the source of several engine performance problems.

One example, if an engine will not start and you discover no spark at a spark plug wire, use the test light to check for power at the coil. As shown in Fig. 21-5, simply touch the test light on the coil positive. If the light glows, then you would know that the ignition circuit before the coil (ignition switch, wiring, wiring connections, etc.), is functioning.

This type of testing can be used on many other electrical components. However, do NOT connect a test light to some electronic components. The light could draw too much current and ruin the electronic part. Refer to a service manual when in doubt.

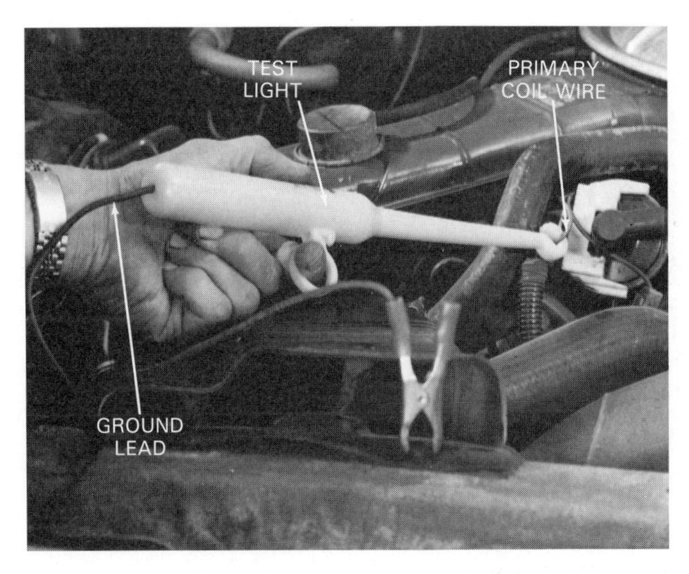

Fig. 21-5. Test light will quickly check for power at various components. Technician is checking for voltage going to coil when diagnosing no-start condition. (Peerless)

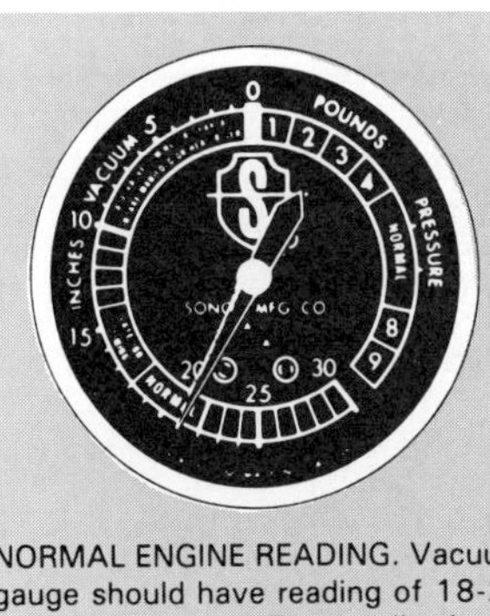

NORMAL ENGINE READING. Vacuum gauge should have reading of 18-22 inches of vacuum. The needle should remain steady.

BURNED OR LEAKY VALVES. Burned valve will cause pointer to drop every time burned valve opens.

WEAK VALVE SPRINGS. Vacuum will be normal at idle but pointer will fluctuate excessively at higher speeds.

WORN VALVE GUIDES. If pointer fluctuates excessively at idle but steadies at higher speeds, valves may be worn allowing air to upset fuel mixture.

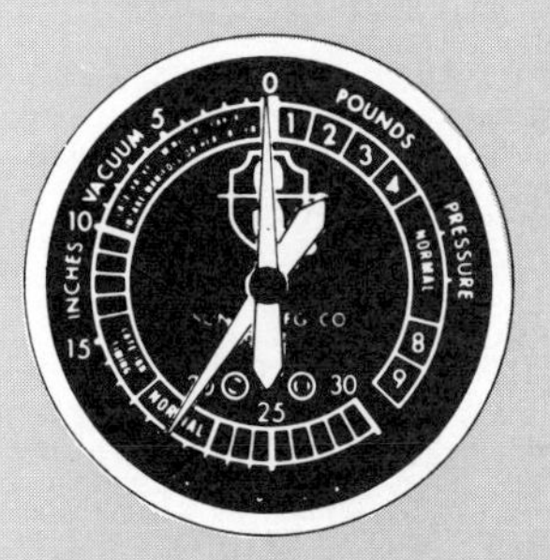

CHOKED MUFFLER. Vacuum will slowly drop to zero when engine speed is high.

INTAKE MANIFOLD AIR LEAK. If pointer is down 3 to 9 inches from normal at idle throttle valve is not closing or intake gaskets are leaking.

CARBURETOR OR FUEL INJECTION PROBLEM. A poor air-fuel mixture at idle can cause needle to slowly drift back and forth.

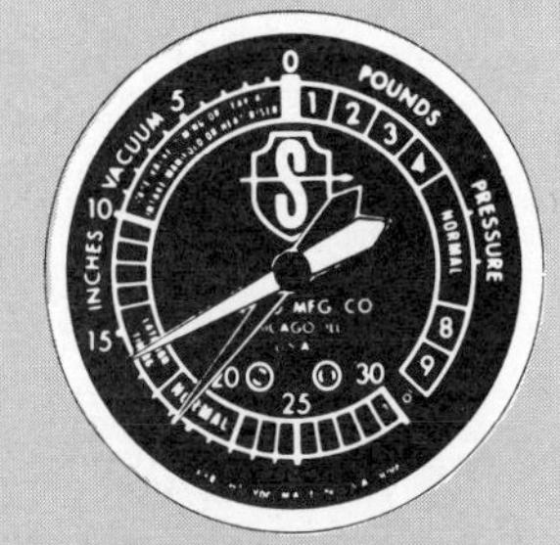

STICKING VALVES. A sticking valve will cause pointer to drop intermittently.

Fig. 21-4. Study vacuum gauge readings and their indications. (Sonoco)

Digital multimeter

A digital multimeter is another handy tool for finding performance problems on today's engines. It will accurately measure the voltage, current, or resistance in a circuit or component. You can compare your test values to specs to find problems, Fig. 21-6.

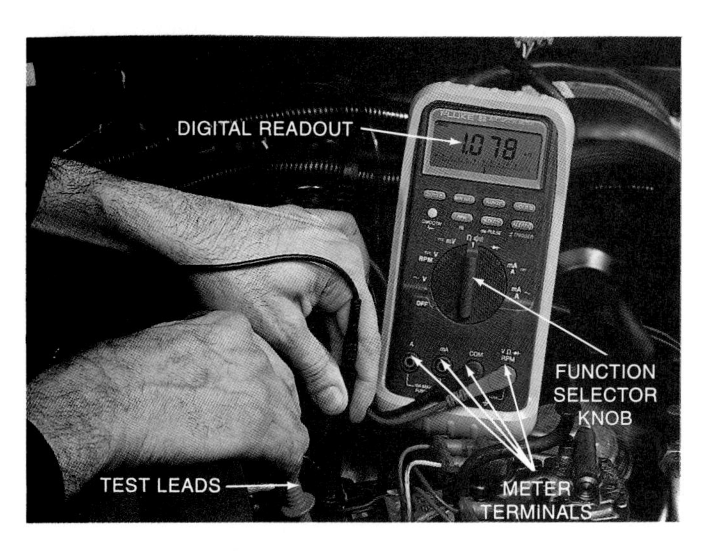

Fig. 21-6. A digital multimeter can be used to help find numerous engine performance problems. Current, voltage, and resistance can be compared to specs. Many electric-electronic components can upset engine performance. (Fluke)

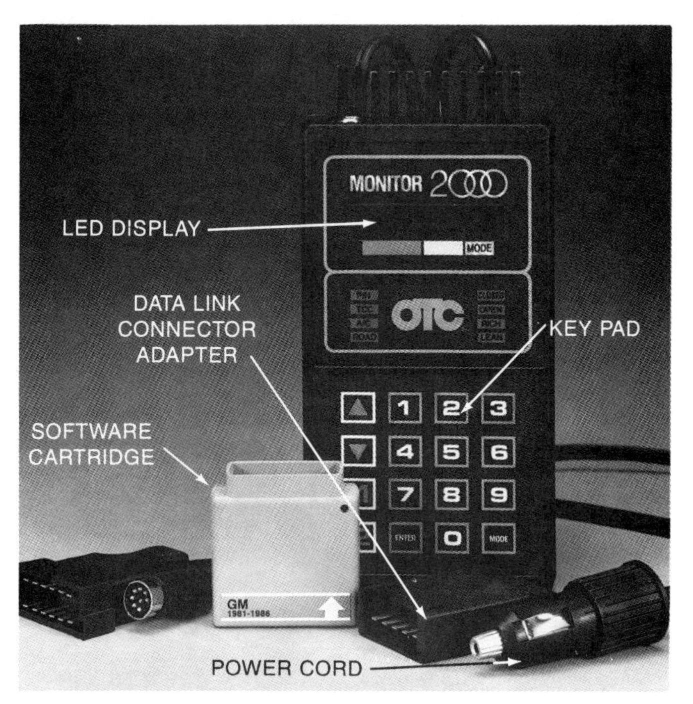

Fig. 21-7. Scan tool will communicate with the vehicle's on-board computer to help find problems. This tester will actually give directions and convert trouble code numbers into actual words describing problems so you do have to refer to service manual trouble code chart. (OTC Tools)

Scan tools

A *scan tool* is a testing device to help you quickly read and interpret computer system trouble codes. The tool lead is connected to the vehicle's data link connector. This allows it to retrieve and display trouble codes. Scan tools will also let you test sensors, actuators, and the computer itself, Fig. 21-7.

Most scan tools use data cartridges that are plugged into the tool. However, the latest scan tools contain fixed memory boards that can be programmed by downloading information from a computer or computerized engine analyzer.

Engine analyzer

An *engine analyzer* is a group of test instruments mounted in a roll-around cabinet, Fig. 21-8. It is commonly used by the engine technician to find engine performance problems.

The engine analyzer will usually contain:

1. Oscilloscope (TV tube type voltmeter for checking operating condition of ignition and other systems).
2. Tach-dwell (digital or analog meter for checking engine speed, the ignition, and computer system).
3. Exhaust gas analyzer (instrument for measuring chemical content in engine exhaust).
4. VOM (for measuring electrical values).
5. Compression gauge (pressure gauge for checking engine compression stroke pressure and engine condition).
6. Vacuum gauge (negative pressure gauge for check-

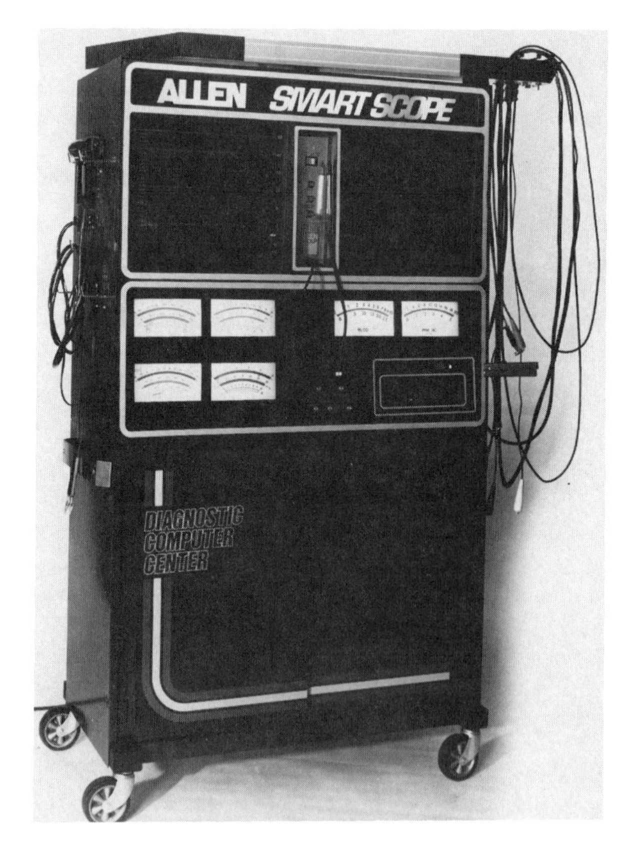

Fig. 21-8. Engine analyzer is also used when finding engine performance problems. Scope will measure instantaneous voltages in ignition system. Other conventional testers are also included. (Allen)

ing engine condition and various vacuum-operated devices).
7. Cylinder balance tester (mechanism for shorting cylinders electronically for finding ''dead'' or ''weak'' cylinder).
8. Cranking current balance tester (device for checking engine compression by monitoring amount of current needed to crank each cylinder through compression stroke—low compression would require less current).
9. Timing light (strobe light for checking and adjusting ignition timing).

As an engine technician in today's shop, you should know how to use an engine analyzer.

Pressure gauge

A *pressure gauge* will measure engine fuel pressure and can be used to find engine performance problems. See Fig. 21-9.

The pressure gauge can be connected to the fuel line, usually at a test fitting on the fuel rail, to check the operation of the fuel pump, pressure regulator, fuel filters, and fuel lines. If pressure is within specs, you have eliminated these parts as a source of the performance problem.

TYPICAL PERFORMANCE PROBLEMS

You must understand the most common engine performance problems. The following will help you use test equipment, troubleshooting charts, and a service manual during diagnosis.

No-start problem

A *no-start problem* occurs when the engine ''cranks'' or is turned over by starting motor but fails to ''fire'' and run on its own power. This is the most obvious and severe performance problem.

With a no-start problem, CHECK FOR SPARK, first. Pull one spark plug wire (gasoline engine). Install a spark tester on the wire end. Attach it to ground. A bright spark should jump the gap when the engine is cranked, Fig. 21-9. If not, the problem is in the ignition system.

If you have spark, CHECK FOR FUEL. With a carburetor fuel system, operate the throttle level. Watch for fuel squirting out the accelerator pump discharge. With throttle body fuel injection, watch the injector outlet. With multiport injection, install a pressure gauge on the fuel rail. If you do not have fuel, then something is wrong with the fuel system, Fig. 21-9.

If a fuel injection system has fuel pressure, check the injector harness for the presence of a pulse. Pull a harness connector loose and install an injector ''noid light'' in the harness connector. The noid light should flash when the engine is cranked. If the light does not flash, there is a problem with the computer, input sensor, or electrical system. If you have both fuel and spark, check engine compression. A jumped timing chain or belt could be keeping the engine from starting. With a diesel, a slow cranking speed can prevent starting.

Hard starting

Engine *hard starting* has excessive cranking time and is due to partial failure of a system. The carburetor choke may be inoperative. The fuel injection coolant temperature sensor may be bad.

Stalling (dying)

Engine stalling is a condition where the engine stops running. This may occur at idle, after cold starting, or

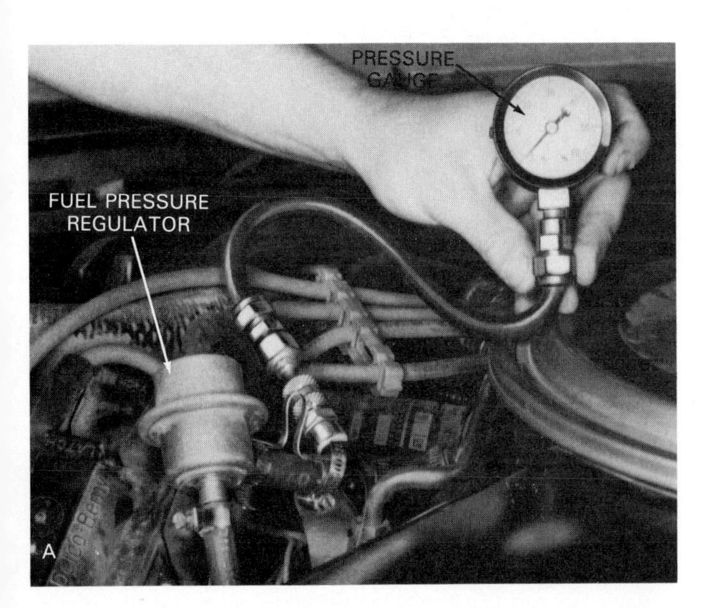

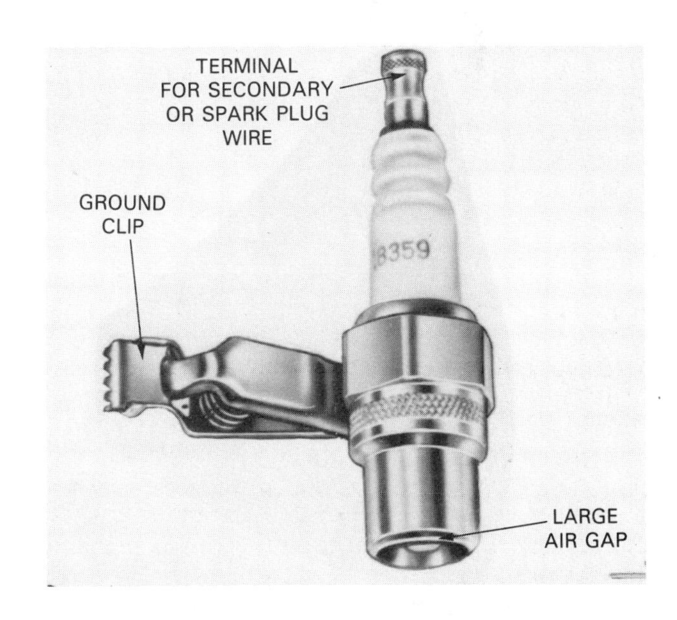

Fig. 21-9. When an engine fails to start, remember to check for spark and for fuel. A—With multiport fuel injection, a pressure gauge is an efficient way of checking the fuel system. B—Spark tester can be connected to spark plug wire. When engine is cranked, you should have strong spark. This can also be done by holding wire next to ground. (OTC and Kent-Moore)

after warm-up.

There are many causes of stalling: low idle speed, carburetor or injection system problem, ignition system trouble, severe vacuum leak, or inoperative thermostatic air cleaner.

Engine miss

An *engine miss* results from one or more cylinders failing to fire (produce normal combustion). The engine will run rough. The engine may miss at idle, under acceleration, or at cruising speeds.

If an engine only misses at idle, for example, check the components that affect idle. If the engine has a carburetor, the idle circuit may be clogged with debris. With fuel injection, possibly an injector is not working properly. A fouled spark plug, open plug wire, cracked distributor cap, corroded terminals, and vacuum leak are a few other possible causes for an engine miss.

Vacuum leak

A *vacuum leak* is a common cause of rough idling. If a vacuum hose hardens and cracks, it will allow outside air to enter the engine intake manifold, bypassing the carburetor, airflow sensor, or throttle body. This will cause an incorrect air-fuel mixture, preventing normal combustion. See Fig. 21-10.

Usually, a vacuum leak will produce a HISSING SOUND. The engine roughness will smooth out when rpm is increased.

A section of vacuum hose can be used to locate vacuum leaks. Place one end of the hose next to your ear. Move the other end around the engine. When the hiss becomes very loud, you have found the leak.

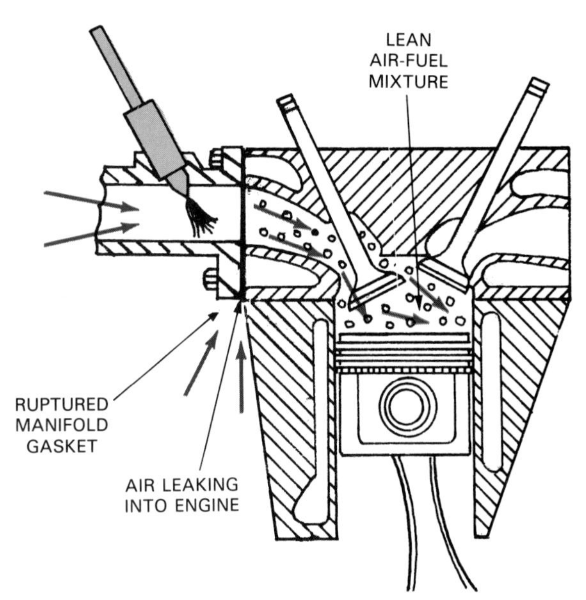

Fig. 21-10. Vacuum leak allows extra air to enter engine after throttle valves or airflow sensor. This will usually lean mixture and cause a rough idle.

Hesitation (stumble)

A *hesitation,* also called *stumble,* is a condition where the engine does not accelerate normally when the gas pedal is pressed. The engine may also stall before developing power.

A hesitation is usually caused by a temporary lean air-fuel mixture. With a carburetor, the accelerator pump may not be functioning. With fuel injection, the throttle position switch may be bad. Check the parts that aid engine acceleration.

Surging

Surging is a condition where engine power fluctuates up and down. When driving at a steady speed, the engine seems to speed up and slow down, without movement of the gas pedal.

Surging is sometimes caused by an extremely lean carburetor or fuel injection setting. Surging can also be caused by ignition or computer control system problems.

Backfiring

Backfiring is caused by the air-fuel mixture igniting in the intake manifold or exhaust system. A loud BANG or POP sound can be heard when the mixture ignites.

Backfiring can be caused by incorrect ignition timing, crossed spark plug wires, cracked distributor cap, bad carburetor accelerator pump, exhaust system leakage, faulty air injection system, or other system fault.

Dieseling (after-running, run-on)

Dieseling, also called *after-running* or *run-on,* occurs when the engine fails to shut off. The engine keeps firing, coughing, and producing power. Dieseling is usually caused by a high idle speed, carbon buildup in the combustion chambers, low octane rating of fuel, or overheated engine.

Pinging (spark knock)

Pinging or *spark knock* is a metallic tapping or light knocking sound, usually when the engine accelerates under load. Pinging is caused by abnormal combustion (preignition or detonation). Pinging is normally caused by low octane fuel, advanced ignition timing, carbon buildup in combustion chambers, or engine overheating.

Vapor lock

Vapor lock occurs when the fuel is overheated, forming air bubbles that upset the air-fuel mixture. Vapor lock can cause engine stalling, lack of power, hard starting, and no starting. Vapor lock is caused by too much engine heat transferring into the fuel.

Gas line freeze

Gas line freeze results when moisture in the fuel turns to ice. The ice will block fuel filters and prevent engine operation. With diesel fuel, the overcooled fuel can form wax that blocks the filters.

To correct gas line freeze, you may need to place the car in a warm garage until the fuel is thawed.

Poor fuel economy

Poor fuel economy is a condition causing the engine to use too much fuel for the miles driven. Fuel economy can be measured by comparing the miles that can be driven on one gallon (3.79 L) of fuel.

Poor fuel economy can be caused by a wide range of problems: rich air-fuel mixture, engine miss, incorrect ignition timing, or leakage.

Lack of engine power

Lack of engine power, also termed a *sluggish engine,* causes the vehicle to accelerate slowly. When the gas pedal is pressed, the car does not gain speed properly. The engine ''hits'' on all cylinders and runs smoothly but does not produce normal power.

As with poor fuel economy, there are many troubles that can reduce engine power: fuel system problems, ignition system problems, emission control system problems, and engine mechanical problems.

OTHER PERFORMANCE PROBLEMS

There are many other more specific engine performance problems (flooding, carburetor icing, leaking fuel injector, incorrect ignition timing). Many of these are covered in other chapters of this book.

On-Board Diagnostics Generation Two (OBD II)

The main source of vehicle air pollution is the poorly tuned engine or malfunctioning engine system. Early on-board diagnostic systems simply illuminated a ''check engine'' indicator lamp in the dash if a sensor or circuit stopped working. This system is referred to as On-Board Diagnostics Generation One (OBD I).

A newer on-board diagnostic system goes a step further by monitoring how efficiently each part of the system is operating. This new system is called *On-Board Diagnostics Generation Two,* abbreviated *OBD II.* OBD II is designed to monitor the condition of hardware and software that affect emissions. OBD II diagnostics can detect and compensate for part ''wear'' (change in performance) and not just detect complete part failure.

OBD II system components

OBD II systems use many of the sensors present in most OBD I systems. However, OBD II systems have several additional features. These include:

- A sixteen-pin data link connector located within easy reach of the driver's seat. This connector is standardized across all manufacturers vehicles.
- The ''check engine light'' is now known as a *malfunction indicator lamp (MIL).*
- At least one pre-converter and a post-converter heated oxygen sensor. The post-converter oxygen sensor is called the *catalyst monitor.*
- A *misfire* and *fuel trim* (air-fuel mixture) *monitor.* This is part of the ECU programming.
- EGR valves with flow and pintle position sensors.
- An enhanced evaporative emissions system equipped with a vent solenoid, diagnostic test fitting, and a fuel tank pressure sensor.
- A refrigerant monitor that will disable the air conditioning system if a low refrigerant level is detected.

OBD II diagnostic codes

OBD II systems generate a three digit diagnostic code or *fault designator,* with an *alpha-numeric designator.* A typical OBD II code might be P0115. The *P* indicates that the problem is powertrain related. The *0* indicates the code is a Society of Automotive Engineers (SAE) diagnostic code. A *1* would indicate that the code is a manufacturer assigned diagnostic code. The first *1* in the fault designator indicates the system, in this case, the code is fuel and air metering related. The *15* indicates the specific fault, which is a engine coolant temperature sensor malfunction in this example, Fig. 21-11.

OBD II systems have four levels of diagnostic trouble codes (DTC):

- *Type A* codes are emissions related and will illuminate the MIL, or in the case of misfire or fuel trim DTCs, flash the MIL when present.
- *Type B* code is also emissions related. However, the ECU will illuminate the MIL only when this type of code appears on two consecutive keystarts or ''trips.''
- *Type C* codes are non-emissions related. Will not illuminate the MIL, however, it will store a DTC and illuminate a ''service lamp'' or the service message on vehicle equipped with a driver's information center.
- *Type D* codes are also non-emissions related, will store a DTC, and will not illuminate any lamps.

When clearing OBD II codes, a scan tool should be used. Disconnecting the battery in most cases will not work as some ECUs will retain any stored codes for several days without battery power.

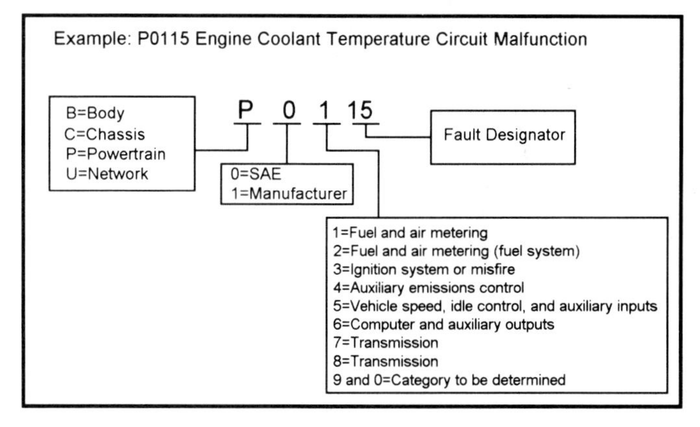

Fig. 21-11. OBD II diagnostic code.

SUMMARY

Use a systematic approach when trying to find the source of engine performance problems. Use your understanding of system operation and basic testing methods to check the most logical sources of trouble.

A service manual diagnosis chart should be used when a problem is difficult to locate and correct.

Reading spark plugs is a common method of finding the source of performance problems. The condition and color of the spark plug tips can indicate air-fuel ratio, engine compression, condition of rings, cylinders, etc. A properly firing cylinder should have a plug tip grey-tan to brown in color.

A vacuum gauge is a handy tool for finding performance problems. You can compare intake manifold vacuum readings to known normal readings.

A test light and multimeter are useful when diagnosing electrical troubles that affect engine performance.

Specialized testers are available for analyzing engine and engine system problems. Follow the operating instructions for the particular tester.

An engine analyzer is a group of test instruments in a single cabinet. It is commonly used by the engine technician to find problems. You should enroll in an automotive electronics course that explains analyzer use.

A pressure gauge is commonly used to check fuel pressure. With EFI engines, this will check the fuel pump, pressure regulator, fuel line, and return line.

KNOW THESE TERMS

Systematic approach, Diagnosis chart, Reading plugs, Digital multimeter, Oscilloscope, No-start problem, Hard starting, Stalling, Missing, Vacuum leak, Hesitation, Surging, Backfiring, Dieseling, Pinging, Vapor lock, Gas line freeze, Poor fuel economy, Sluggish performance.

REVIEW QUESTIONS—CHAPTER 21

1. An engine performance problem is any trouble that lowers engine __________, __________ __________, __________, or __________.
2. What are three questions you should ask yourself when troubleshooting engine performance problems?
3. A gasoline injected engine runs poorly. Light black smoke blows out the tailpipe. The engine has electronic multiport injection.
   Technician A says the problem could be a stuck pressure regulator that is allowing too much fuel to enter the return line.
   Technician B says cold start injector could be on and the thermo-time switch should be checked.
   Who is correct?
   a. Technician A.
   b. Technician B.
   c. Both A and B.
   d. Neither A nor B.
4. How do you read spark plugs?
5. Explain nine test instruments usually found on an engine analyzer.
6. When an engine cranks, but fails to start, you should check for __________ and for __________ to narrow down the possible sources.
7. This trouble will cause an engine to idle roughly and make a hissing sound.
   a. Hesitation.
   b. Pressure leak.
   c. Vacuum leak.
   d. Dieseling.
8. __________ is caused by the air-fuel mixture igniting in the intake manifold or exhaust system.
9. An engine diesels when shut off.
   Technician A says that a high idle speed setting is a common cause.
   Technician B says that a low idle speed setting is the cause.
   Who is correct?
   a. Technician A.
   b. Technician B.
   c. Both A and B.
   d. Neither A nor B.
10. Define the term "vapor lock."

ASE CERTIFICATION-TYPE QUESTIONS

1. Technician A says that a black or wet plug may indicate engine mechanical problems. Technician B says that a black or wet plug may indicate an extremely rich air-fuel mixture. Who is right?
   (A) A only. (C) Both A & B.
   (B) B only. (D) Neither A nor B.
2. Technician A says that a vacuum gauge can detect an intake manifold air leak. Technician B says that a vacuum gauge can detect worn valve guides. Who is right?
   (A) A only. (C) Both A & B.
   (B) B only. (D) Neither A nor B.
3. Technician A says that a test light can be used to test all electronic engine components. Technician B says that a test light can be used to check the operation of specified electronic engine components. Who is right?
   (A) A only. (C) Both A & B.
   (B) B only. (D) Neither A nor B.
4. An engine analyzer normally contains all of the following EXCEPT:
   (A) compression gauge
   (B) oscilloscope
   (C) exhaust gas analyzer
   (D) cylinder leak tester
5. To test the operation of a multiport injection system's pressure regulator, the pressure gauge should be connected to the __________.
   (A) PCV valve vacuum port
   (B) fuel rail fitting
   (C) injector wiring harness
   (D) evaporative emissions purge valve
6. Technician A says that with a no-start engine problem, you should first check the engine's ignition system. Technician B says that with a no-start engine problem, you should first check for engine mechanical problems. Who is right?
   (A) A only. (C) Both A & B.
   (B) B only. (D) Neither A nor B.
7. All of the following can cause an engine to stall EXCEPT:
   (A) a severe vacuum leak.
   (B) an inoperative thermostatic air cleaner.
   (C) a lean air-fuel mixture.
   (D) low idle speed.
8. Technician A says that a cracked distributor cap may cause an engine to miss. Technician B says that a fuel injector malfunction may cause an engine to miss. Who is right?
   (A) A only. (C) Both A & B.
   (B) B only. (D) Neither A nor B.
9. All of the following are components of an OBD II system EXCEPT:
   (A) an eight-pin data link connector.
   (B) a misfire monitor.
   (C) a post-converter catalyst monitor.
   (D) an EGR valve with a pintle position and flow sensor.
10. Technician A says that engine surging is normally the result of a rich air-fuel mixture. Technician B says that engine surging is usually caused by a low idle speed setting. Who is right?
   (A) A only. (C) Both A & B.
   (B) B only. (D) Neither A nor B.

Chapter 22

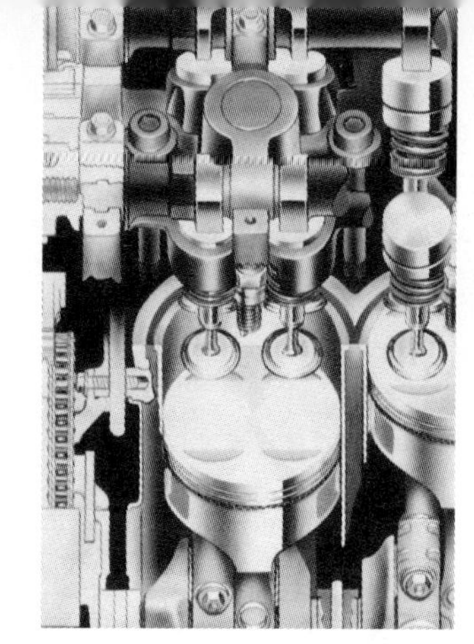

Engine Mechanical Problems

After studying this chapter, you will be able to:
- *Describe typical mechanical failures in an engine.*
- *Explain the causes of engine mechanical failures.*
- *Summarize the symptoms for common engine mechanical breakdowns.*
- *Diagnose engine mechanical problems.*
- *Be better prepared to service an engine.*

This chapter will summarize the causes, symptoms, and results of mechanical part failures in an engine. One of the most difficult aspects of being an engine technician is troubleshooting before engine teardown. You must be able to quickly and accurately locate mechanical problems and correct their source so that the same problem does not happen again.

In previous chapters you learned about engine operation, construction, design, and engine system principles and service. Now you will learn about burned valves, blown head gaskets, and other similar troubles. This will help prepare you to diagnose engine problems and understand later chapters.

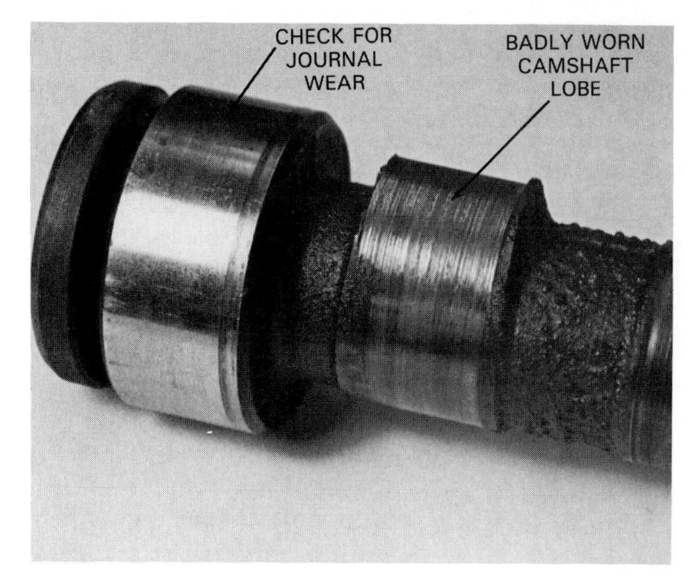

Fig. 22-1. This camshaft had badly worn cam lobes. This is due to extended service and lack of oil changes. (Midwest Crankshaft)

VALVE TRAIN PROBLEMS

The engine valve train is a common cause of mechanical problems in an engine. Friction between parts is high and most of the parts are splash oiled. Wear can occur between parts after extended service or from lack of maintenance.

Valve train problems can cause engine noise, power loss, missing, rough idling, and even piston, cylinder, and head damage.

Camshaft problems

Camshaft problems include lobe wear, journal wear, cam breakage, drive gear failure, etc.

Cam lobe wear is a condition where friction from the lifter has worn off some of the cam lobe. This usually happens in a high mileage engine or one that has not had its oil changed regularly. See Fig. 22-1.

With excess lobe wear, valve train clatter, reduced valve lift, and poor engine performance can result. Measure rocker arm motion with a dial indicator to find a worn cam lobe.

Cam journal wear and *cam bearing wear* will increase part clearance and can reduce engine oil pressure. Again, this normally only happens after extended service.

Cam thrust surface wear can allow the cam to move too far forward and rearward in the engine. This can sometimes produce a light knocking sound, timing chain or gear problems, and increased lifter and lobe wear.

Camshaft breakage, though not common, will keep some of the valves from operating. The lobes on one end of the broken cam will not rotate and open their valves. Severe performance problems or valve damage can result.

Removal of a valve cover will let you check valve train and camshaft action.

Lifter or tappet problems

Lifter or *tappet problems* can produce valve train noise or a light tapping sound from inside the valve cover. Look at Fig. 22-2.

External lifter wear has the bottom of the lifter concaved (sunk in middle) from cam lobe friction. A new lifter is actually convex (crowned or humped) on the bottom so it rotates during engine operation. High mileage can wear off this crown and speed cam lobe wear. This can add to valve train clearance. Engine disassembly is needed to verify external lifter wear.

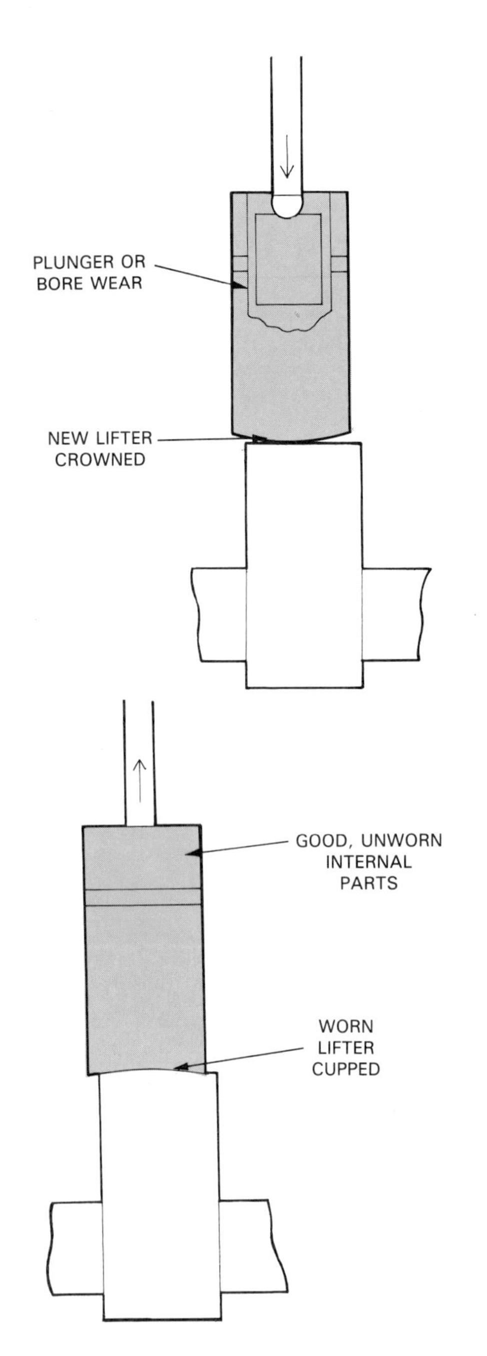

Fig. 22-2. Lifters can wear externally and internally. A—Outside wear occurs where lifter rides on cam lobe. Inside wear can cause too much leak down and valve train clatter. B—Actual worn lifters. (Dana Corp.)

Internal lifter wear, also called just a *bad lifter,* can prevent the lifter from pumping up to take up valve train clearance, Fig. 22-2. This can cause the rocker arm to clatter as it strikes the valve stem tip. The noise will be near the top of the engine, as if a small ball peen hammer were hitting the engine.

With the engine idling and the valve cover removed, push down on the push rod end of the rocker arm. This should easily increase noise so you can find the bad lifter.

Low engine oil pressure can cause hydraulic lifter clatter. Check the oil level and oil pressure before condemning the lifter or lifters. Contaminated or dirty oil can also cause lifter noise.

Push rod problems

A *bent push rod* is not straight and has been bent by a stuck valve, spring bind, spring breakage, or other type of mechanical problem. If severely bent, the push rod can fall out of the rocker arm pocket. This will stop valve action and severe missing and tapping will result. Replace, do not straighten, bent push rods.

A *broken push rod* usually has the small ball on the end broken off. Like a bent push rod, this will cause loud valve train noise and usually push rod disengagement from the rocker arm, Fig. 22-3.

A *clogged push rod* has the oil hole in the middle stopped up from hardened oil and sludge deposits. This will speed rocker arm wear because of reduced lubrication. Always check or clean the hole in push rods during an overhaul or valve job.

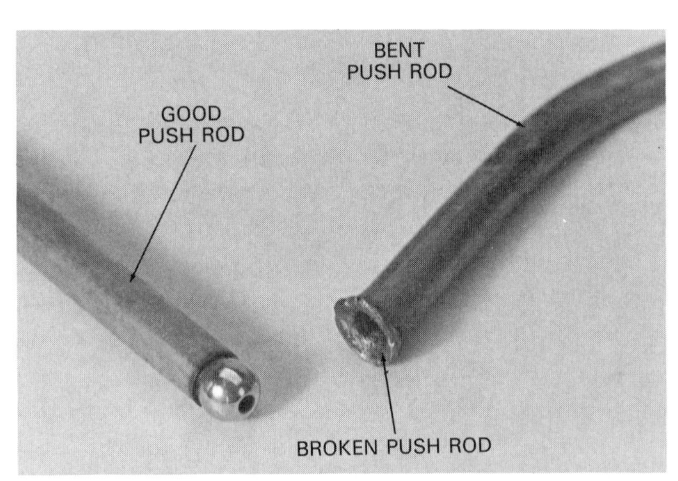

Fig. 22-3. Compare good and failed push rods. One on right is bent and broken. Failure resulted from driver overrevving engine. (Delcotto)

Rocker arm problems

A *worn rocker shaft* results from friction with the rocker arm. This can happen because of inadequate lubrication or high mileage. As shown in Fig. 22-4, the clearance between the rocker arm and shaft will increase. This can also cause valve train clatter.

A *worn rocker arm* can wear on its inside diameter or where it contacts the valve or push rod. This too can

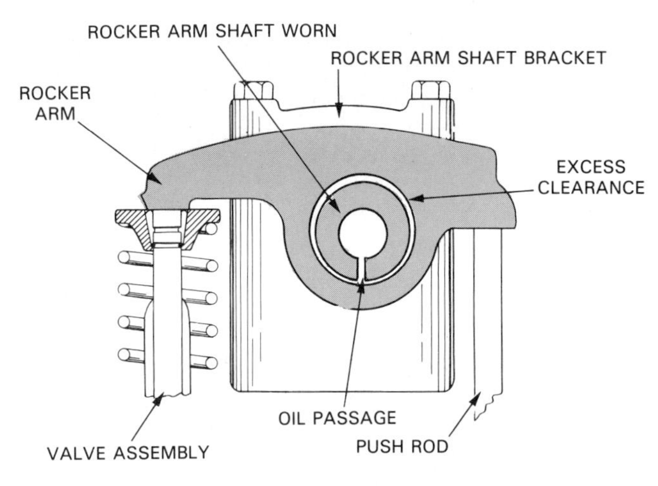

Fig. 22-4. Worn rocker arm or rocker shaft can extend lifter plunger too far up. Valve train noise can result. (Perfect Circle)

make the lifter plunger move too far out and a light tapping noise can result. With an overhead cam engine, the rocker can wear at its contact point with the camshaft, Fig. 22-5. Inspection is needed to verify rocker wear problems.

A *worn rocker stand* is a common problem because the stands can be made of soft aluminum. The stand will wear where it rubs on the rocker arm. This will also cause valve train noise.

Rocker arm or *stand breakage* can occur if the valve strikes the piston from overrevving or if the valve is stuck in the head after sitting unused. This will prevent valve action and cause a severe engine miss.

Fig. 22-5. This is a worn, scored overhead cam rocker arm. Lack of lubrication from deteriorated oil was the cause. (Midwest Crankshaft)

VALVE PROBLEMS

Valves open and close millions of times during the life of an engine. They must do this while sealing the tremendous heat of combustion. After extended service, various valve problems can develop. You must be able to diagnose and locate these problems efficiently.

Burned valve

A *burned valve* has had a portion of the valve margin and face melted and burned away. Pressure can leak past the valve, reducing power in that cylinder. The engine will miss or run rough, especially at idle. See Fig. 22-6A.

With a burned valve, you may be able to hear a puffing sound as pressure blows past the valve. There may be a popping or puffing sound at the carburetor or throttle body (bad intake valve) or at the exhaust system tailpipe (burned exhaust valve).

To fix a burned valve, you must remove the cylinder head from the engine.

Valve face erosion

Valve face erosion is the wearing away of the face where it contacts the seat in the cylinder head. This is more common of a problem with unleaded fuels. They do not have lead type fuel additives that reduce high temperature friction. Look at Fig. 22-6B.

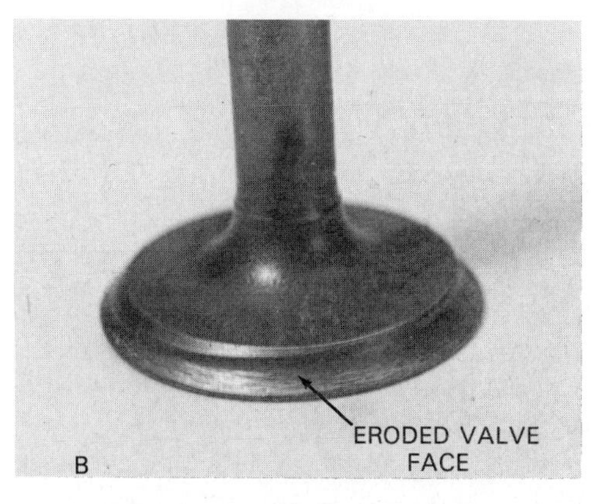

Fig. 22-6. A—This burned valve came out of high mileage engine. Exhaust heat actually melted and burned away portion of face and margin. Engine lost compression in affected cylinder. B—This valve has eroded valve face. Note how it is dished from riding on seat. Unleaded fuel, when valve was designed for leaded fuel, was cause.

Valve stem problems

Valve stem wear results from friction with the cylinder head guide. Only a small amount of oil enters the guide, which makes stem wear common, Fig. 22-7.

Severe stem or guide wear can produce a light knocking or tapping sound. The valve can flop sideways when acted upon by the rocker arm or valve spring. Head removal and valve replacement is needed.

Valve stem breakage is a severe failure that usually results when the piston hits the valve head. This will bend and break off the valve head at the stem. A broken valve spring, weak spring, popped out keepers, etc. can cause this problem.

When the valve head breaks off, it will usually be hit by the piston and smashed up into the head. Piston and head damage normally results.

A *scored valve stem* has been overheated and scarred by severe friction. The rough surface of the stem will accelerate valve guide wear.

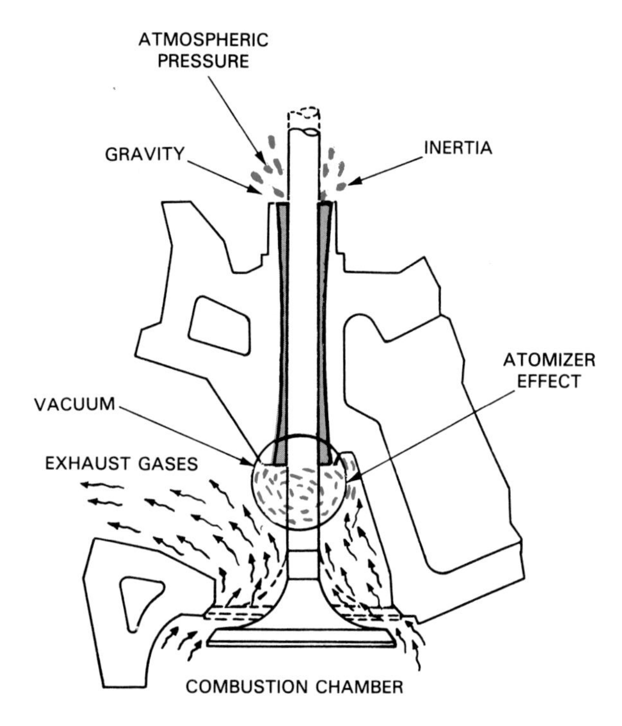

Fig. 22-8. Bad valve seals or worn valve guides will let oil flow down into port and be pulled into combustion chamber. Worn guide or stem can also let valve flop sideways in guide and make tapping type noise.

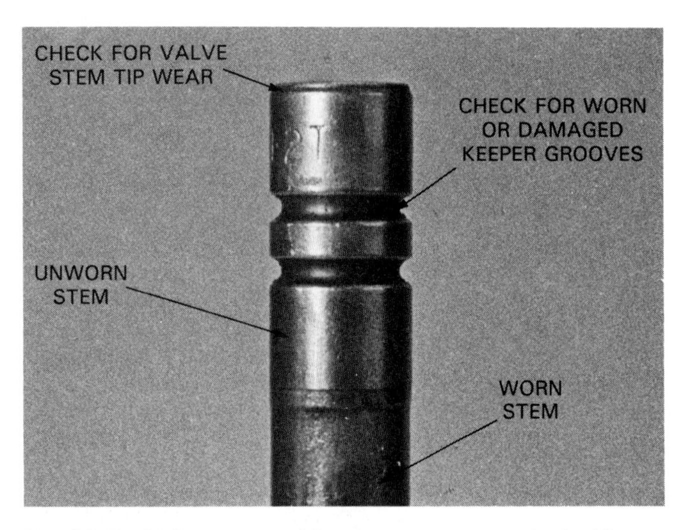

Fig. 22-7. Valve stem problems are easy to overlook. Always inspect stem, keeper grooves, and tip closely.

Fig. 22-9. Bad valve stem seals will make engine smoke right after startup. Seals will harden after prolonged use. Seal on right has cracked and would break up and fall down into oil pan if not replaced. (Fel-Pro Gaskets)

Valve guide problems

Valve guide wear will produce the same symptoms as valve stem wear. The valve will be free to shift sideways in the guide, as shown in Fig. 22-8. Oil can be drawn down through the excess clearance and burned in the combustion chamber. A tapping noise can also result.

A *dropped valve guide* has broken free from its press-fit in the head. This will let the guide fall down around the head of the valve. It can wedge the valve open and cause severe part damage. This is more of a problem with aluminum heads.

Bad valve seals

Bad valve seals have hardened or broken from extended service or engine overheating. The synthetic rubber seals are no longer soft and pliable. With severe failure, the seal can disintegrate and fall down into the oil pan. Look at Fig. 22-9.

Bad oil seals will make the engine produce blue smoke, especially right after engine startup. Oil will drip down into the guides with the engine sitting. Then, upon start-up, the oil will be drawn into the cylinders and burned.

Valve seals can usually be replaced WITHOUT removing the cylinder head. Air pressure is used to hold the valves up in the head. Then, a special tool is used to compress the valve spring. The keepers, springs, and seals can then be removed for service.

Stuck valve

A *stuck valve* results when the valve stem rusts or corrodes and locks in the valve guide. This can happen when the engine sits in storage for an extended period.

A stuck or frozen valve can sometimes be fixed by squirting rust penetrant around the top of the valve guide. If this fails to correct the problem, remove the cylinder head for service.

Valve spring problems, valve float

Valve float is a condition where weakened valve springs, hydraulic lifter problems, or excess engine rpm cause the valves to remain partially open. This problem usually occurs at higher engine speeds. The engine may begin to miss, pop, or backfire as the valves float.

Weak valve springs result after prolonged use. The springs lose some of their tension. The springs may become too weak to close the valves properly.

A *broken valve spring* will frequently let the valve hang partially open. Excess valve-to-rocker clearance may cause valve train clatter (light tapping noise). Popping or backfiring can also result.

Valve springs can be replaced WITHOUT cylinder head removal. As with valve seal replacement, air pressure and a special tool will permit spring replacement.

Inspection is often needed to verify a broken valve spring. The cylinder may pass a compression test.

TIMING MECHANISM PROBLEMS

Timing mechanism problems can affect when and if the camshaft rotates properly to open the valves at the right times. This relates to timing gears, a timing chain, or timing belt.

Timing sprocket problems are common when the sprocket has plastic teeth. The plastic teeth can break off and cause timing chain slack. The broken teeth can also let the timing chain jump over teeth, upsetting valve timing. Severe performance problems and even valve damage can result because the valves can then open at the wrong times, Fig. 22-10A.

Timing gear problems are not very common but can happen after extended service or breakage of other parts. The most common problem is broken gear teeth, as shown in Fig. 22-10B. Broken timing gear teeth can cause a loud knocking noise from inside the front cover. If enough teeth break off, the timing can jump and the valves will open incorrectly.

A *worn timing chain* will upset valve timing, reducing compression stroke pressure and engine power. Wear can cause slack between the crankshaft sprocket and camshaft sprocket. The cam and valves will no longer be kept in time with pistons.

To check for a worn timing chain, rotate the crankshaft back and forth while watching the distributor rotor or rocker arms. If you can turn the crank several degrees without rotor or valve movement, the timing chain is worn. The timing chain would need to be replaced.

A compression test would show lowered compression in ALL cylinders. Incorrect valve timing and a worn chain would affect all cylinders equally. Erratic timing mark motion during a tune-up could also be due to a worn timing chain.

A *worn timing belt* will usually break, jump off its sprockets, or skip over a few sprocket teeth. Severe performance problems or valve damage can result. With some engine designs, the pistons can move up and slam into the open valves, bending or breaking them.

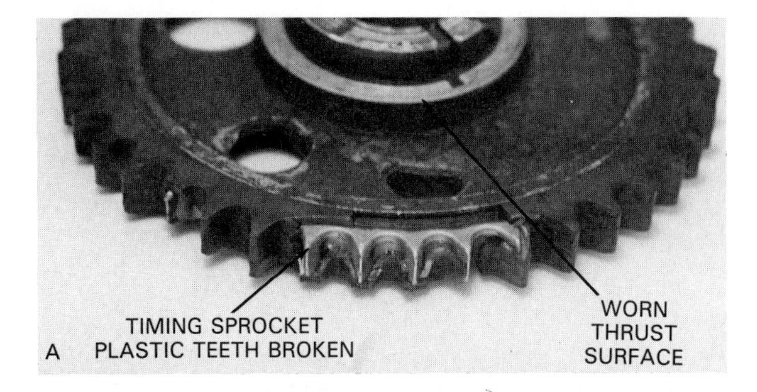

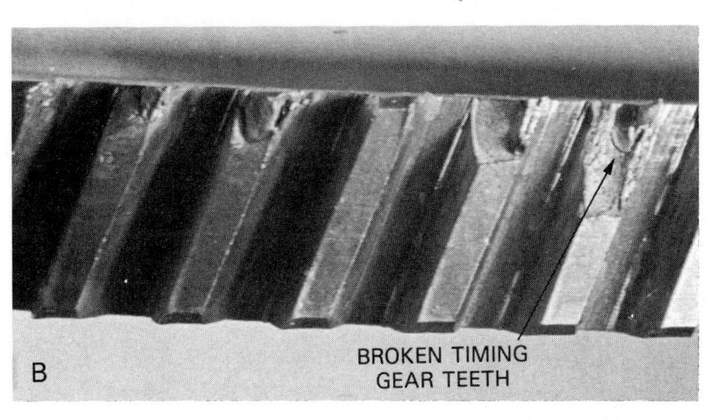

Fig. 22-10. A—This timing sprocket has broken plastic teeth. Timing chain would have excess slack and could jump over teeth, throwing off valve timing. A worn chain will reduce compression in all cylinders. B—This timing gear has broken teeth. It made clunking type sound with engine running.

ENGINE GASKET PROBLEMS

A *blown head gasket* can cause a wide range of problems: overheating, missing, coolant or oil leakage, engine smoking, even head or block damage (burned mating surfaces). Quite often, a blown head gasket will show up during a compression test. Two adjacent cylinders, usually the two center ones, will have low pressure, Figs. 22-11 and 22-12.

A *leaking intake manifold gasket* can cause a vacuum leak, with resulting rough idle. To check for an intake gasket leak, squirt oil along the edge of the gasket. The oil may temporarily seal the leak, showing an intake gasket rupture. A low vacuum gauge reading can also indicate an intake leak.

A *leaking exhaust manifold gasket* will show up as a clicking type sound. As combustion gases blow into the manifold and out the bad gasket, an almost metallic-like rap is produced.

Part warpage (sealing surface on part not flat) is a common cause of gasket failure. Always check for warpage when servicing a bad gasket, Fig. 22-12.

CYLINDER HEAD PROBLEMS

A *warped cylinder head* has been overheated and is no longer true. Usually, the deck surface will warp and

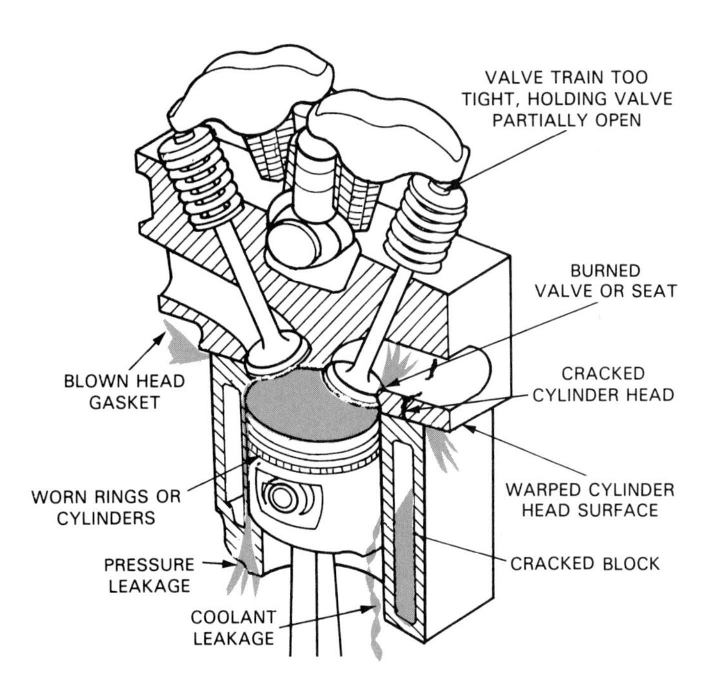

Fig. 22-11. Note some of the problems that can develop with part warpage and cracking.

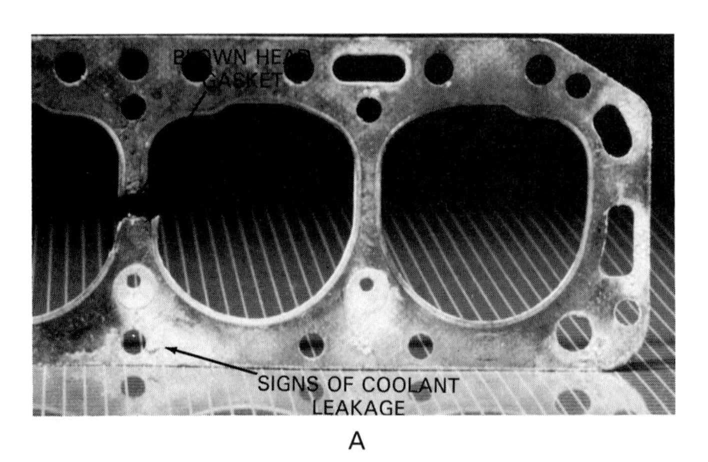

A

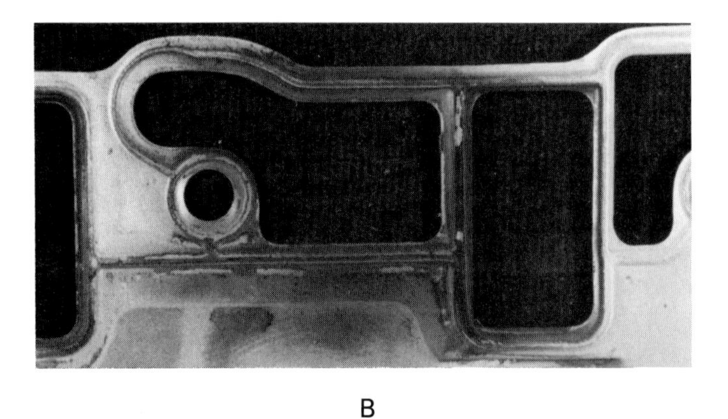

B

Fig. 22-12. Gasket problems are common. A—This is a blown head gasket that resulted from warped head. Signs of coolant leakage are also evident. B—This leaking intake manifold gasket allowed oil to be drawn into cylinders. Engine smoked and used quart of oil every 300 miles.

cause head gasket leakage. Coolant may leak into the engine oil, producing a milk-like substance. This is more of a problem with aluminum cylinder heads.

A *cracked cylinder head* usually results from overheating. Unequal heat expansion will break the head, usually near an exhaust port or on the deck between cylinders. Weak coolant can also freeze in cold weather and crack a cylinder head. A cylinder head crack is shown in Fig. 22-13.

A cracked head can leak coolant or combustion pressure. Coolant leakage can enter the combustion chamber or run down into the pan. Pressure leakage will allow combustion gases to enter the cooling system. Coolant on a spark plug, milky oil, white exhaust smoke, and overheating are symptoms of a cracked head. The head would have to be removed and welded or replaced.

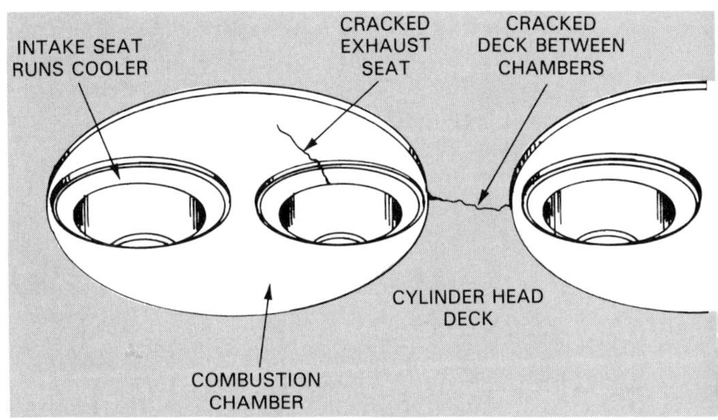

Fig. 22-13. Cylinder head will usually crack at exhaust seat and between combustion chambers on deck surface. This can lead to coolant, oil, and pressure leakage.

CYLINDER BLOCK PROBLEMS

Cylinder block problems include cracks, warpage, worn cylinders, mineral deposits in coolant passages, misaligned main bores, etc. Being a large casting, it suffers from the same troubles as a cylinder head.

Deck warpage results from overheating and can cause head gasket leakage. This is not a very common problem.

Mineral deposits in the water jackets can cause engine overheating. Lime can build up on the inner walls in the block. This can restrict coolant flow and cause localized hot spots, Fig. 22-14.

Crankcase warpage will throw the main bores out of alignment. This can make the crank lock in the block during engine reassembly with new bearings. The block would need line boring to correct this problem, as shown in Fig. 22-15.

A *cracked block* can have cracks in the cylinder, deck surface, lifter gallery, etc. This usually results from overheating, ice formation in cold weather, or breakage of pistons or connecting rods.

Fig. 22-16 shows a badly damaged cylinder block. The connecting rod broke and knocked a large hole in the cylinder. Sleeving would be needed to salvage the block.

A *worn cylinder block* has worn cylinders from piston

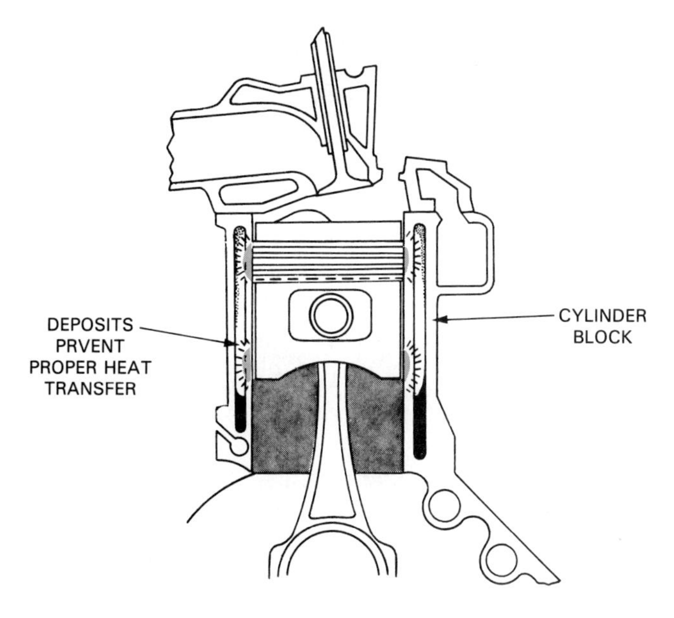

Fig. 22-14. Mineral deposits in block can cause overheating. (Deere and Co.)

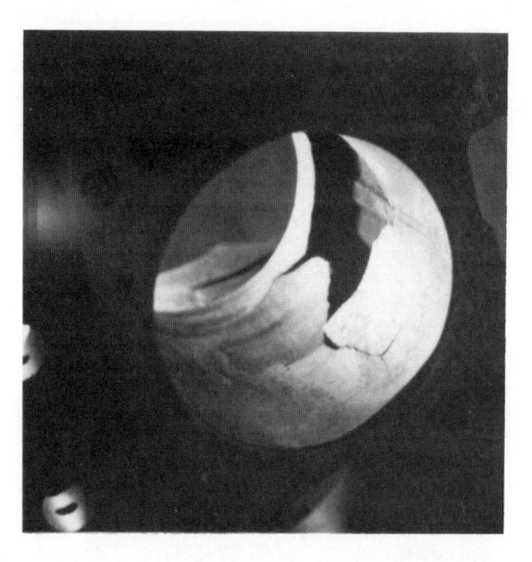

Fig. 22-16. This cylinder has been cracked and broken when connecting rod bolts broke off. Crankshaft smashed rod into side of cylinder wall. Sleeving would be needed to salvage block. Ideally, the block should be replaced.

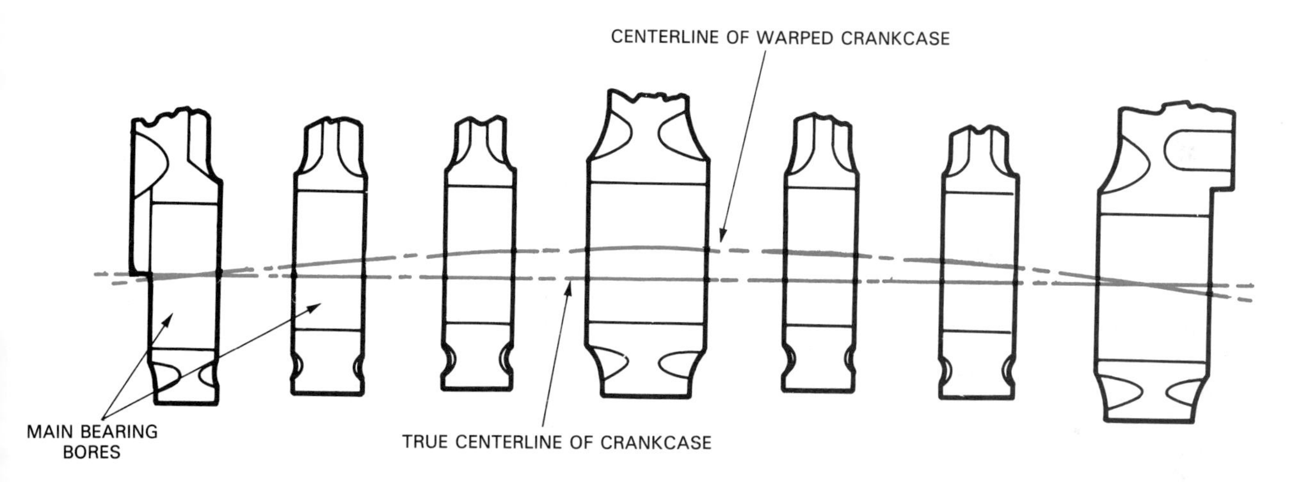

Fig. 22-15. Block crankcase warpage can throw main bores out of alignment. New bearings would make crank lock up when installed in block. Line boring would be needed to fix misalignment. (Federal Mogul Bearings)

ring friction. Because there is more oil lubricating the bottom of the cylinders, the top of the cylinder will wear more. Since the piston rings do NOT rub at the very top of the cylinder, a ring ridge is formed.

A *ring ridge* is a small lip formed where the rings do not wear the cylinder, Fig. 22-17. It is an obvious sign of cylinder block wear. It must be cut out of the cylinder before piston removal, as you will learn later.

Discussed earlier, you can read plugs as a way of finding a worn cylinder. The rings will not seal well on a bad cylinder and oil-fouled plugs can result, Fig. 22-18.

Cylinder problems are major and require engine removal so the block can be bored or replaced.

Piston problems

Piston problems are also major and they include skirt breakage, crown failure, worn ring grooves, etc. Piston failure normally results in oil consumption, engine smoking, and sometimes knocking.

Piston breakage

Piston breakage usually occurs when the skirt breaks off. The piston will then flop sideways in its cylinder and cause major cylinder damage. The affected cylinder will lose compression, burn oil, and make a knocking sound. Skirt breakage can result from too much piston and cylinder clearance, overrevving, or inadequate lubrication. Fig. 22-19 shows an example.

Piston knock (slap)

Piston knock or slap is a loud metallic knocking sound produced when the piston flops back and forth inside its cylinder. It is caused by excess piston skirt or cylinder wear, clearance, and possibly damage.

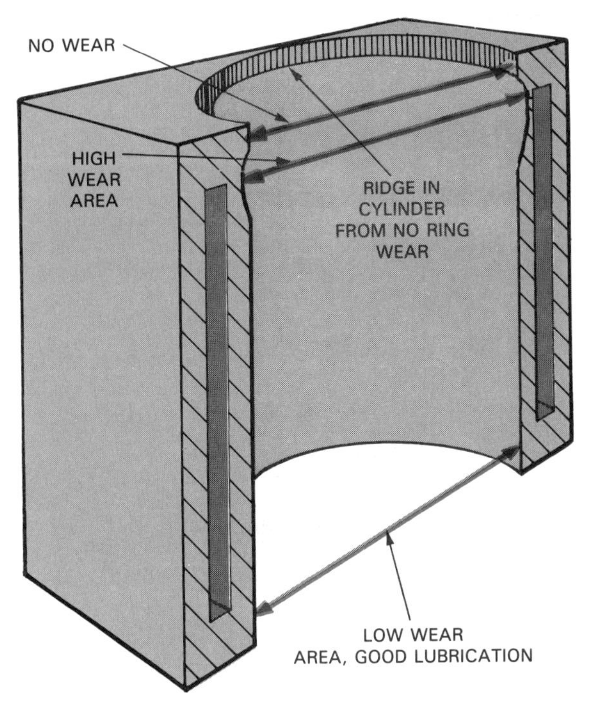

Fig. 22-17. Cylinder will wear more near top because of less splash oiling. Ring ridge is formed where rings do not rub on cylinder. Ring ridge would quickly show cylinder wear.

Fig. 22-19. Broken piston skirt resulted from too much clearance and low oil level in pan. Note scoring of skirt that indicates oiling problem.

Fig. 22-18. Oil-fouled plug would indicate bad rings, valve seals, turbocharger, etc. Reading plugs is important. (Champion Spark Plug Co.)

Piston slap is normally LOUDER when the engine is COLD and tends to quiet down as the engine warms to operating temperature. Heat expansion of the aluminum piston takes up clearance to quiet the piston knock.

Piston pin knock

Piston pin knock occurs when too much clearance exists between the piston pin and piston pin bore or connecting rod bushing. Excessive clearance allows the pin to hammer against the rod or piston as the piston changes direction in the cylinder.

Piston pin knock will usually make a DOUBLE KNOCK (two rapid knocks and then a short pause).

Burned piston

A *burned piston,* discussed in the chapter on combustion, is often a result of preignition or detonation damage. Abnormal combustion, excessive pressure, and heat actually melt and blow a hole in the piston crown or area around the ring lands. The engine may smoke, knock, have excessive blowby, or other symptoms, Figs. 22-20 and 22-21.

A compression test or cylinder leakage test may indicate a burned piston. However, engine disassembly is usually needed to verify the problem. If the cylinder wall is not damaged, the repair can be an in-car operation.

Damaged spark plugs can indicate piston problems. As mentioned, an oil-fouled plug would indicate bad rings or cylinders. However, melted aluminum deposits point to a melted or partially failed piston. A smashed electrode can result from being hit by the piston. Also, when the tip is damaged from detonation, you would want to correct the cause of detonation, possibly preventing piston damage. See Fig. 22-22.

Worn piston rings and cylinders

Worn piston rings or *cylinders* result in blowby, blue-grey engine smoke, low engine power, spark plug fouling, and other problems caused by poor ring sealing.

To check for ring and cylinder problems, increase engine speed while watching the tailpipe and valve cover

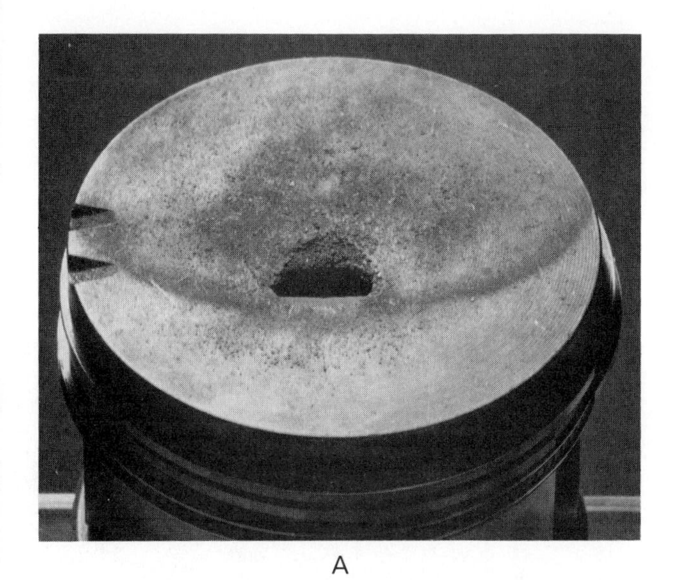

A

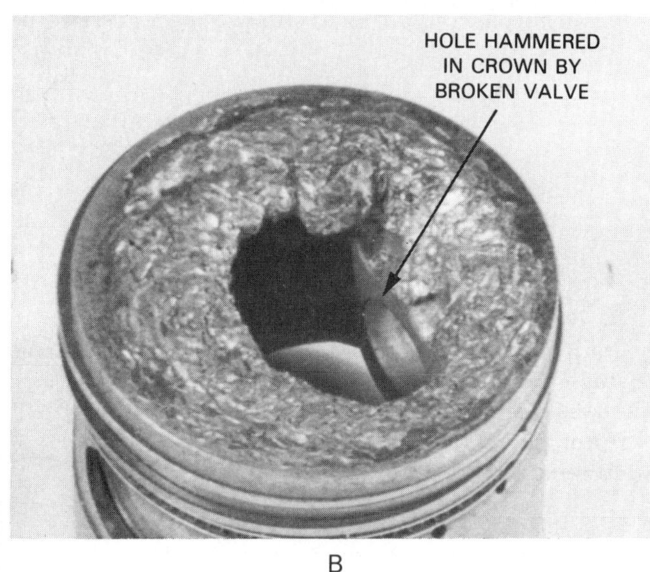

B

Fig. 22-20. A—Burned piston from prolonged detonation. Heat and pressure melt and blow hole in head. Engine cylinder lost compression and started pumping oil, producing heavy blue smoke out exhaust pipe. B—Holed piston happened when valve broke and hammered hole in crown. (Champion Spark Plug Co.)

Fig. 22-21. This burned piston has damage on one side of crown and ring land area. Cylinder damage also resulted. (Fel-Pro Gaskets)

Fig. 22-22. When technician removed this spark plug, burned electrodes were discovered. Stuck turbocharger waste gate was allowing too much boost. Technician corrected problem before pistons were burned. (Champion Spark Plug Co.)

breather opening. If blue-grey smoke pours out of the car's exhaust under load, the oil rings and cylinders may need service. If excessive oil vapors and air blows out the valve cover breather, blowby is entering the crankcase. Piston, compression ring, or cylinder problems are indicated. Worn ring grooves can also cause oil consumption.

If after disassembly, the engine cylinders are found to be worn, the engine must be removed from the vehicle. The block must be sent to a machine shop for *boring* (cylinders machined oversize) or *sleeving* (liners installed to restore cylinders).

If just the rings are found to be worn, NOT the cylinders, the engine may be rebuilt while still installed in the car. Check a manual for details.

CONNECTING ROD PROBLEMS

A *bent connecting rod* has been damaged by detonation or excessive engine rpm. The rod is no longer straight. As shown in Fig. 22-23, it will load one side of the rod bearing and cause the piston to cock at an angle in the cylinder.

A bent connecting rod is usually not discovered until engine disassembly. You will see an uneven wear pattern on the piston skirt and rod bearing. Fixtures are

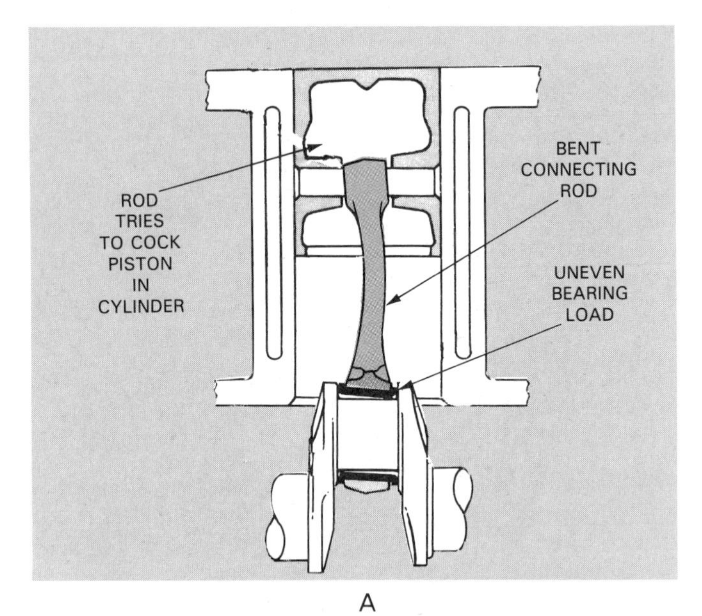

A

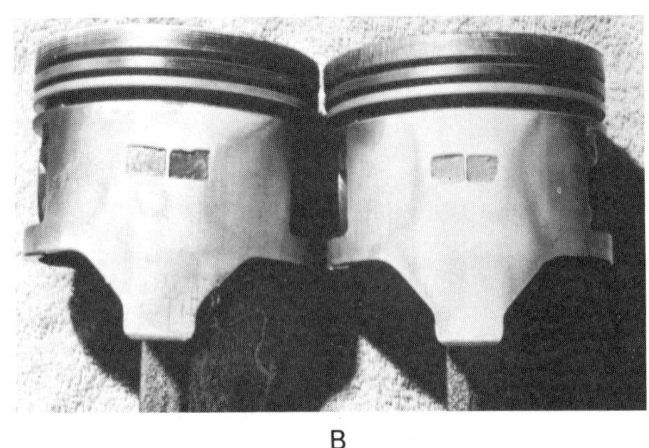

B

Fig. 22-23. A—Bent connecting rod will load one side of piston skirt and rod bearing. B—Note uneven piston wear pattern from bent rod.

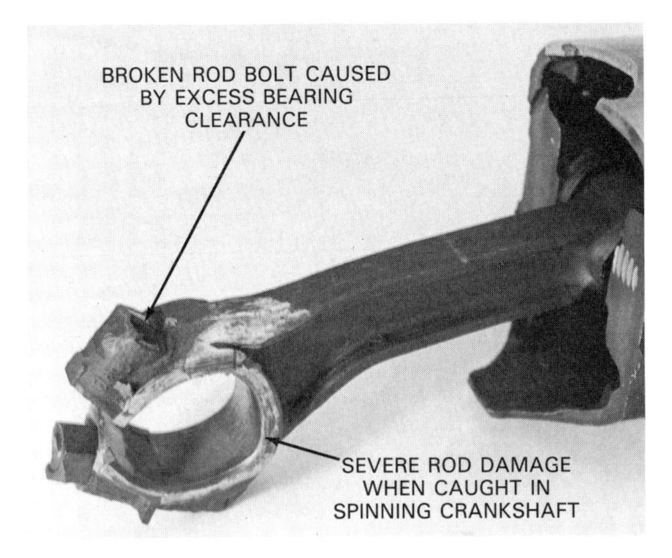

Fig. 22-24. High engine rpm on old engine caused rod bolt to snap off. Cap then came off rod and rod was crushed into block by spinning crankshaft. Driver ignored or did not notice rod bearing knock which usually precedes complete bearing and rod failure. (Midwest Crankshaft)

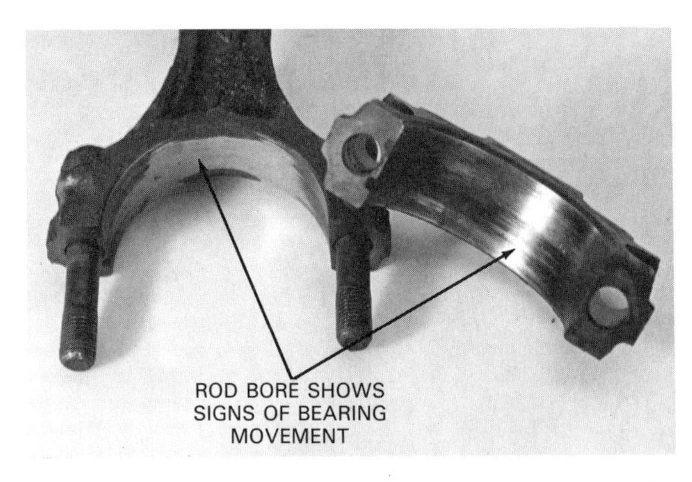

Fig. 22-25. An out-of-round rod bore or stretched rod bolts can let bearing move or shift inside rod. This can wear back of bearing and rod until finally bearing spins in rod. A spun rod bearing will damage the connecting rod and crankshaft. (Midwest Crankshaft)

available for checking rod straightness.

A *broken connecting rod* has suffered a complete failure of the I-beam or rod bolts and cap. Rod breakage can result from excessive engine speed, running an engine after rod bearing failure, loss of lubrication, improper assembly, etc. See Fig. 22-24.

When a connecting rod breaks, severe engine damage usually results. The spinning crankshaft can slam into the disengaged rod, crushing the rod into the block, sometimes ruining the block. The piston can also fly upward and smash into the cylinder head, causing head and valve damage. Normally, complete engine disassembly, crank turning, block boring, etc. are needed to repair a broken connecting rod.

Connecting rod bore problems result from extended service, rod bearing failure, oil pump problems, etc. The big end of the rod can become out-of-round and can no longer support the rod bearing properly.

As shown in Fig. 22-25, always inspect for rod bore problems during engine service. Marks on the bore can show rod bearing movement in the rod. This would tell you that the rod should be rebuilt (bore remachined by machine shop) or replaced.

CRANKSHAFT AND BEARING PROBLEMS

The crankshaft and its bearings must withstand extreme loads and friction during engine operation. The slightest mistake during assembly, a lubrication problem, or driver abuse can lead to bearing or crankshaft troubles.

Rod bearing knock

Connecting rod bearing knock is caused by wear and excessive rod bearing-to-crankshaft clearance. It will produce a light, regular, rapping noise with the engine floating (point at which throttle is held constant and engine is not accelerating). It is loudest after engine warmup. In a cold engine, thickened oil tends to cushion

Fig. 22-26. A bad rod bearing allowed this piston to hit cylinder head, making loud knocking noise. Note shiny area that shows contact.

Fig. 22-27. Main bearing wear will lower oil pressure considerably and can cause knocking under load. This is badly worn main thrust bearing.

and quiet rod knock.

To locate a bad rod bearing, short out or disconnect each spark plug wire one at a time. The loose, knocking rod bearing may quiet down or change pitch when its spark plug is disabled.

With some engines, excessive rod bearing clearance will let the piston head strike the cylinder head. This will make a loud knocking sound, similar to a failed piston. Fig. 22-26 shows a piston that has been hitting the head due to rod bearing failure.

Spun rod bearing

A *spun rod bearing* is a bearing that has been hammered out flat and then rotated inside the connecting rod bore. It will make a rod bearing type knock, but much louder. When a rod bearing spins, it will normally damage the rod and crankshaft journal. The driver did not notice or ignored the slight rod knock that would precede total rod bearing failure.

Worn main bearings

Worn main bearings have too much clearance and can let the crank shift up and down in the block with load changes. A thin layer of metal has been worn off the bearing face, as shown in Fig. 22-27.

Main bearing knock

Main bearing knock is similar to rod bearing knock but is slightly deeper or duller in pitch. It is usually more pronounced when the engine is pulling or lugging under a load. Bearing and possibly journal wear are letting the crankshaft move up and down inside the cylinder block.

Main bearing wear will usually reduce oil pressure significantly. To verify main bearing noise, remove the oil pan. Pressure test the lubrication system. Excessive oil flow out of one or more of the main bearings implies too much bearing clearance. If a pressure tester is not available, remove and inspect each of the main bearings. If the crankshaft is not worn, bearing insert replacement should correct the problem.

Excess crankshaft end play

Excess crankshaft end play is caused by a worn main thrust bearing. Thrust bearing wear can produce a deep knock, usually when applying and releasing the clutch (manual transmission). With an automatic transmission, the end play problem may only show up as a single thud or knock upon acceleration or deceleration.

A knock occurring with clutch or torque converter action could also be caused by loose flywheel bolts or other drive train problems. Check out all possible causes before beginning repairs.

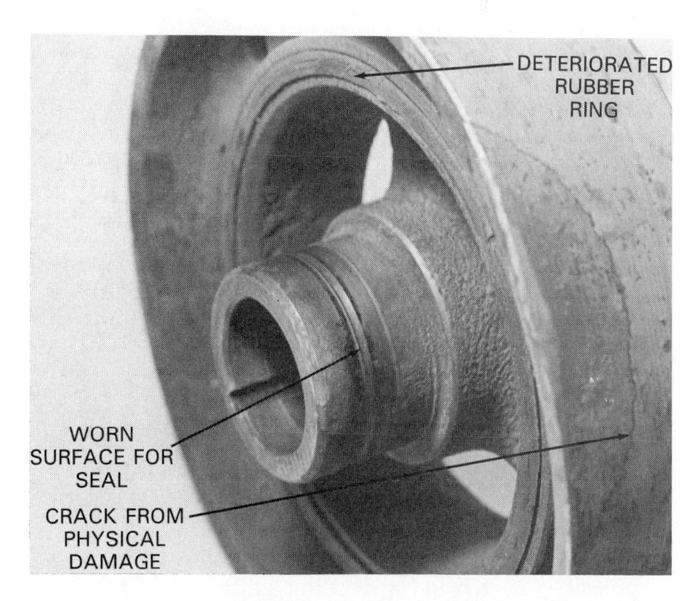

Fig. 22-28. Note problems with this harmonic balancer. Keep these troubles in mind during engine service.

CRANKSHAFT DAMPER PROBLEMS

Crankshaft damper problems include physical damage, deteriorated rubber ring, seal journal wear, slipped or sheared key, etc. Engine vibration can result from damage. Oil leakage can occur if the seal surface is worn or scored. If the outer metal ring slips on its rubber mount, the timing marks can be out of position. This can cause confusion when trying to set ignition timing.

Fig. 22-28 illustrates a few of the problems you can encounter with a crankshaft damper.

FLYWHEEL PROBLEMS

Flywheel problems include loose bolts, ring gear teeth failure, friction surface scoring, warpage, etc. A loose flywheel can make a knocking sound similar to main bearing knock. The teeth on the flywheel ring gear can wear or break off and cause starter engagement troubles. Keep these problems in mind during diagnosis.

INTAKE AND EXHAUST MANIFOLD PROBLEMS

Exhaust manifold problems usually consist of a cracked or warped manifold. Engine exhaust heat can crack an exhaust manifold, allow leakage and exhaust noise. Heat can also warp a manifold, allowing it to leak next to the cylinder head mating surface. See Fig. 22-29.

Intake manifold problems are not very common because the manifold is not exposed to as much heat as other parts. An intake manifold can sometimes warp and cause a vacuum leak. Plugs can also loosen or fall out, allowing vacuum or coolant leakage.

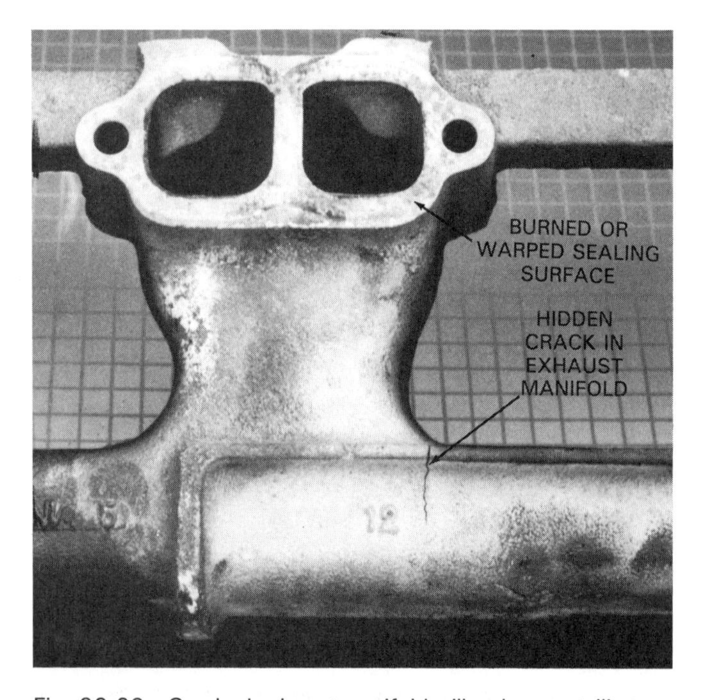

Fig. 22-29. Cracked exhaust manifold will make a metallic-type ticking sound as exhaust pressure blows out crack. This crack was on backside of manifold and was difficult to locate. Using section of vacuum hose as listening device found leak. (Fel-Pro)

MECHANICAL PROBLEM DIAGNOSIS

If a technician does not know how to properly diagnose (locate) engine problems, a great deal of time, effort, and money will be wasted. In fact, an untrained mechanic may unknowingly rebuild an engine when a minor repair might have corrected the fault.

For example, a worn or stretched timing chain can cause the engine valves to open and close at the wrong time in relation to piston movement. This could cause low compression stroke pressure in all cylinders. The engine would have low power output and low compression readings.

If not diagnosed correctly, the technician could mistake the trouble (worn timing chain) as more major problems (worn piston rings, scored cylinders, etc.). The engine could be overhauled when a new timing chain would have corrected the problem. Unfortunately, the customer would have to pay for the technician's lack of training.

Symptoms of engine mechanical problems

Symptoms (signs) that result from engine mechanical problems include:

1. EXCESSIVE OIL CONSUMPTION (engine oil must be added too often).
2. CRANKCASE BLOWBY (combustion pressure blows past piston rings, into crankcase, and out breather).
3. ABNORMAL ENGINE NOISES (knocking, tapping, hissing, rumbling).
4. ENGINE SMOKING (blue-grey, black, or white smoke blows out tailpipe).
5. POOR ENGINE PERFORMANCE (rough idle, vehicle accelerates slowly, engine vibrates).
6. COOLANT IN ENGINE OIL (oil has white, milky appearance).
7. ENGINE LOCKED UP (crankshaft will not rotate).

With any of these troubles, inspect and test the engine to determine the exact problem source. You must find out what repair is needed. Then, determine whether the engine can be repaired in the car or if it must be removed for more major repairs.

Engine pre-teardown inspection

After gathering information from the customer or service writer, inspect the engine using all of your senses (sight, smell, hearing, touch). Look for external problems (oil leaks, vacuum leaks, part damage, contaminated oil).

If a leak is found, smell the fluid to determine if it is oil, coolant, or other type of fluid. Listen for unusual noises that indicate part wear or damage.

Increase engine speed while listening and watching for problems. The engine may run fine at idle but act up at higher speeds.

A few engine problems that may be located through inspection are:

Coolant in oil will show up as white or milky looking oil. It is caused by a mechanical problem that allows engine coolant to leak into the engine crankcase. There may be a cracked block or head, leaking head gasket, leaking intake manifold gasket (V-type engine), or similar troubles.

Oil fouled spark plugs point to internal oil leakage into the engine combustion chambers. They are an indication that the engine has worn rings, worn or scored cylinder walls, or bad valve seals. You will need to perform additional tests to find the source of the problem.

Oil in the coolant is usually NOT an engine problem. It is caused by a leak in the radiator (transmission) oil cooler.

Engine oil leaks occur when gaskets harden and crack, when seals wear, when fasteners work loose, or when there is part damage (warped surfaces, cracked parts). To find oil leaks, clean the affected area on the outside of the engine. Then, trace the leak upward to its source. Oil will usually flow down and to the rear of the engine because of cooling fan and airflow action.

External coolant leaks will show up as a puddle of coolant under the engine. Leaks can be caused by hose problems, rusted out freeze (core) plugs, or warped, worn, or damaged parts. Use a cooling system pressure tester to locate external coolant leaks.

Engine blowby occurs when combustion pressure blows past the piston rings into the lower block and oil pan area of the engine. Pressure then flows up to the valve covers and out the breather. Excessive blowby will show up as smoke or oil around the breather. With the valve cover breather removed, oil vapors may blow out when engine speed is increased.

Engine vacuum leaks show up during inspection as a hissing sound, like air leaking out of a tire. Vacuum leaks are loudest at idle and temporarily quiet down as the engine is accelerated (manifold vacuum drops with engine acceleration). Very ROUGH ENGINE IDLING usually accompanies a vacuum leak.

Engine smoke

Engine smoking is normally noticed at the tailpipe when the engine is accelerated or decelerated. The color of the smoke can be used to help diagnose the source of the problem.

With a GASOLINE ENGINE, the exhaust smoke may be:

1. BLUE-GREY SMOKE (indicates motor oil is entering combustion chambers, may be due to worn rings, worn cylinders, leaking valve stem seals, or other troubles).
2. BLACK SMOKE (caused by extremely rich air-fuel mixture, not engine mechanical problem).
3. WHITE SMOKE (if not water condensation on cool day, may be due to internal coolant leak into cylinders).

With a DIESEL ENGINE, exhaust smoke may be:

1. BLUE SMOKE (oil entering combustion chambers and being burned because of ring, cylinder, or valve seal problems).
2. BLACK SMOKE (injection system problem or low compression keeping fuel from burning).
3. WHITE SMOKE (unburned fuel, cold engine, or coolant leaking into combustion chambers).

Fig. 22-30 shows engine exhaust smoke problems for an engine. Refer to the index for more information on exhaust smoke.

Finding abnormal engine noises

Abnormal engine noises (hisses, knocks, rattles, clunks, popping) may indicate part wear or damage. You must be able to quickly locate and interpret engine noises and decide what repairs are needed.

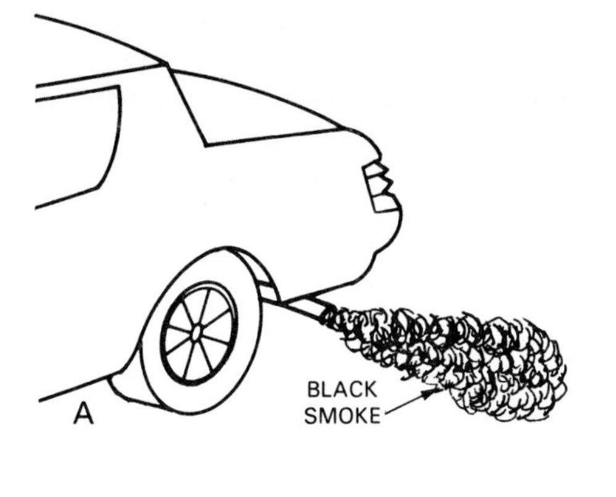

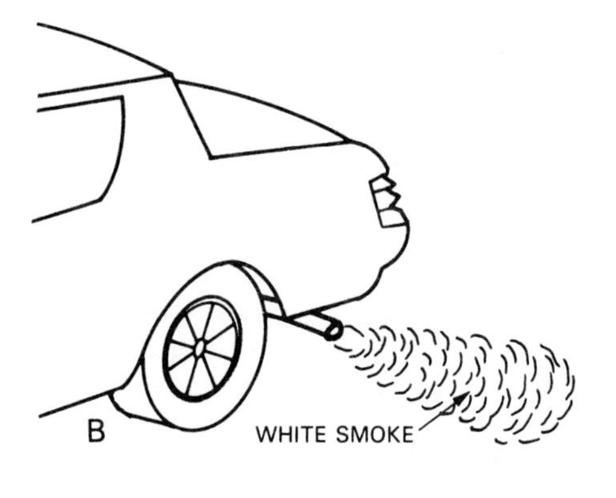

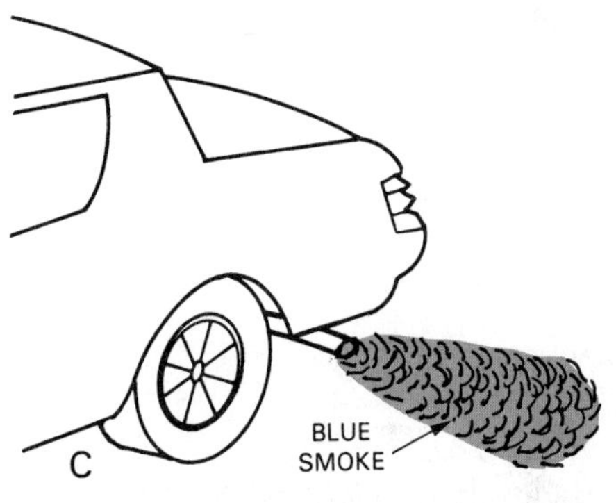

Fig. 22-30. Engine exhaust smoke is good indicator of engine mechanical condition. A—Black smoke indicates rich fuel mixture and carburetor or injection system trouble. B—White smoke can be from coolant leakage into combustion chambers or unburned fuel with diesel. C—Blue smoke indicates oil consumption. Oil is entering and being burned in cylinders.

A *stethoscope* is a listening devide for finding internal sounds in parts. Like a doctor's stethoscope for listening to your heart, it will amplify (increase) the loudness of noises.

To use a stethoscope, place the head set in your ears. Then, touch the probe on different parts around the noise. When the sound becomes the loudest, you have touched near the part producing abnormal noise.

A *long screwdriver* can be used when a stethoscope is NOT available. Sounds will travel through the screwdriver, like with a stethoscope.

A *section of hose* can be used to locate vacuum leaks

and air pressure leaks. Place one end of the hose next to your ear. Move the other end around the engine compartment. The hiss will become loudest when the hose is near the leaking part.

CAUTION! When using a listening device, keep it away from the spinning engine fan or belts. Severe injury could result if the stethoscope, screwdriver, or hose were hit or pulled into the fan or belts!

Decide what repair is needed

After performing all of the necessary inspections and tests, decide what part or parts must be repaired or replaced to correct the engine mechanical problem. Evaluate all data from your pre-teardown diagnosis. If you still cannot determine the exact problem, the engine may have to be partially disassembled for further inspection.

Before you can properly repair engine problems, you must be able to:

1. Explain the function of each engine part.
2. Describe the construction of each engine part.
3. Explain the cause of engine problems.
4. Describe the symptoms of major engine problems.
5. Select specific methods to pinpoint specific problems.
6. Know which parts must be removed for certain repairs.
7. Know whether the engine must be removed from the car during the repair.

These skills can only be developed through study and work experience.

SUMMARY

Part of being an engine technician involves diagnosing engine mechanical problems. By visualizing what happens during an engine failure, you will be better prepared to troubleshoot an engine.

Valve train problems can produce light tapping noises, engine missing, power loss, and major engine damage.

A worn camshaft has its lobe or lobes worn down by friction. This can prevent the valves from opening fully. Cam journal or bearing wear can reduce engine oil pressure.

A bent push rod results from a stuck valve, valve hitting piston, etc. It will make a tapping noise or can disengage from the rocker and prevent valve action.

A bad lifter will also cause valve train clatter. External wear shows up as a concaved bottom on the lifter. Internal wear will make the lifter leak down too quickly and noise will result.

A burned valve has had some of its face and margin blown away. A loss of cylinder compression results. Valve erosion is a wearing away of some of the valve face, and is more common with unleaded fuel.

Worn valve guides can produce a tapping noise and oil consumption. Bad valve seals will also cause engine smoking, especially upon initial engine startup.

Valve float is caused by weak valve springs; the valves fail to close fully at high engine rpm. A broken valve spring will also cause valve train noise.

A worn timing chain will reduce engine compression in all cylinders because valve timing will be off. A worn timing belt can fly off its sprockets and cause valve damage.

A blown head gasket can cause overheating, missing, coolant or oil leakage, etc. It usually occurs when the cylinder head is warped.

Cylinder heads and blocks can be cracked and warped by engine overheating or ice formation in cold weather. A warped head would have to be milled.

A worn cylinder will wear more at the top and a ring ridge will be formed. A worn cylinder will cause oil consumption and smoking.

A broken piston usually breaks at the skirt. Excess cylinder clearance lets the piston flop sideways, breaking off the bottom of the skirt. Smoking, knocking, and cylinder damage are consequences.

Piston knock or slap is a result of too much clearance. It is usually loudest when the piston is cold and quiets down as the engine warms.

A burned piston has a hole melted or blown in the head or land area. It is usually caused by detonation.

Worn piston rings also cause engine oil consumption and blue smoke. Oil can be drawn past the rings and into the combustion chambers. Smoking is more pronounced with the engine warm and with a load.

When a connecting rod breaks, it is usually hit by the spinning crankshaft and jammed into the side of the block. Parts are severely damaged.

Rod bearing knock is from too much crank-to-bearing clearance. It will make a light, regular rapping noise with a constant engine speed. Shorting out cylinders will affect the knock and help find the bad cylinder number.

A spun rod bearing has been hammered out and rotated inside the rod bore. Severe rod and crank damage normally results. The excess clearance can let the piston hit the cylinder head.

Worn main bearings will make a deeper pitch knock and lower engine oil pressure considerably. Main bearing knock is loudest when lugging the engine.

The signs or symptoms of engine mechanical problems include oil consumption, blowby, abnormal noises, smoking, poor performance, coolant in oil, and an engine that is locked up.

Blue-grey smoke indicates oil is entering the combustion chambers. Black smoke is a sign that the fuel mixture is too rich. White engine smoke is from coolant leakage or from poor combustion in a diesel engine.

KNOW THESE TERMS

Lobe wear, Journal wear, Bent push rod, Broken push rod, Bad lifter, Burned valve, Valve erosion, Valve stem wear, Valve breakage, Valve guide wear, Dropped valve guide, Bad valve seals, Stuck valve, Valve float, Weak valve spring, Broken valve spring, Worn timing chain, Worn timing belt, Blown head gasket, Leaking intake manifold gasket, Leaking exhaust manifold, Warped cylinder head, Cracked head, Crankcase warpage, Ring ridge, Piston breakage, Piston knock, Piston pin knock, Burned piston, Worn piston rings, Bent connecting rod, Broken connecting rod, Rod bearing knock, Spun rod bearing, Worn main bearings, Main bearing knock, Milky oil, Blowby, Vacuum leak, Blue-grey smoke, Black smoke, White smoke, Stethoscope.

REVIEW QUESTIONS—CHAPTER 22

1. A worn cam lobe will reduce valve lift and engine performance. True or false?
2. Worn cam bearings can reduce engine ________ ________.
3. A push rod can be bent from a ________ ________, ________ ________, ________ ________, or other type mechanical problem.
4. Explain the difference between external and internal lifter wear.
5. An engine has valve train clatter. The engine has high mileage. Inspection found sludge or hardened oil in the engine.
   Technician A says to inspect components (rockers, push rods, stands) for wear and replace them if worn.
   Technician B says you should check the hole in the center of the push rods for clogging.
   Who is correct?
   a. Technician A.
   b. Technician B.
   c. Both A and B.
   d. Neither A nor B.
6. A car has a dead miss and a puffing or popping noise can be heard at the throttle body.
   Technician A says there could be a leak at an intake valve—bad spring, bent valve, etc.
   Technician B says the engine could have a burned piston.
   Who is correct?
   a. Technician A.
   b. Technician B.
   c. Both A and B.
   d. Neither A nor B.
7. Worn valve guides can produce a light tapping noise. True or false?
8. Define the term "valve float."
9. Which is the most common timing mechanism problem?
   a. Broken timing gear teeth.
   b. Spun timing gear key.
   c. Broken timing sprocket teeth.
   d. Jumped timing chain.
10. How can you detect a worn timing chain?
11. This problem will lower compression, especially in two adjacent cylinders.
    a. Burned valve.
    b. Cracked head.
    c. Leaking intake gasket.
    d. Blown head gasket.
12. A ring ridge is a sign of a worn ________.
13. In your own words, what is rod bearing knock and main bearing knock?
14. List seven symptoms that result from engine mechanical problems.
15. What is blowby?

ASE CERTIFICATION-TYPE QUESTIONS

1. Technician A says that engine valve train problems can produce a rough idling condition. Technician B says that valve train malfunctions can cause engine cylinder damage. Who is right?
   (A) A only.
   (B) B only.
   (C) Both A & B.
   (D) Neither A nor B.
2. All of the following are common symptoms caused by cam lobe wear EXCEPT:
   (A) valve train noise.
   (B) excessive combustion pressure.
   (C) engine performance problems.
   (D) reduced valve lift.
3. Technician A says that a burned valve will cause an engine to run rough especially at high speeds. Technician B says that a burned valve will cause an engine to run rough especially at idle. Who is right?
   (A) A only.
   (B) B only.
   (C) Both A & B.
   (D) Neither A nor B.
4. Technician A says that a blown head gasket can cause engine coolant leakage. Technician B says that a blown head gasket can damage the engine block. Who is right?
   (A) A only.
   (B) B only.
   (C) Both A & B.
   (D) Neither A nor B.
5. Technician A says that a crack in a cylinder head usually occurs near an intake port. Technician B says that a cylinder head crack normally occurs near an exhaust port. Who is right?
   (A) A only.
   (B) B only.
   (C) Both A & B.
   (D) Neither A nor B.
6. Technician A says that piston skirt breakage can result from overrevving an engine. Technician B says that piston skirt breakage can be caused by a malfunction in the engine's lubrication system. Who is right?
   (A) A only.
   (B) B only.
   (C) Both A & B.
   (D) Neither A nor B.
7. Technician A says that a burned piston will normally cause a no-start engine problem. Technician B says that a burned engine piston will sometimes produce excessive blowby. Who is right?
   (A) A only.
   (B) B only.
   (C) Both A & B.
   (D) Neither A nor B.
8. Which of the following is a symptom of worn piston rings?
   (A) Blue-grey engine smoke.
   (B) Low engine power.
   (C) Fouled spark plugs.
   (D) All of the above.
9. Technician A says that a bent connecting rod is sometimes caused by engine detonation. Technician B says that a bent connecting rod is normally caused by using high octane fuel. Who is right?
   (A) A only.
   (B) B only.
   (C) Both A & B.
   (D) Neither A nor B.
10. Technician A says that black engine exhaust smoke is usually caused by a lean air-fuel mixture. Technician B says that black engine exhaust smoke is normally the result of an extremely rich air-fuel mixture. Who is right?
    (A) A only.
    (B) B only.
    (C) Both A & B.
    (D) Neither A nor B.

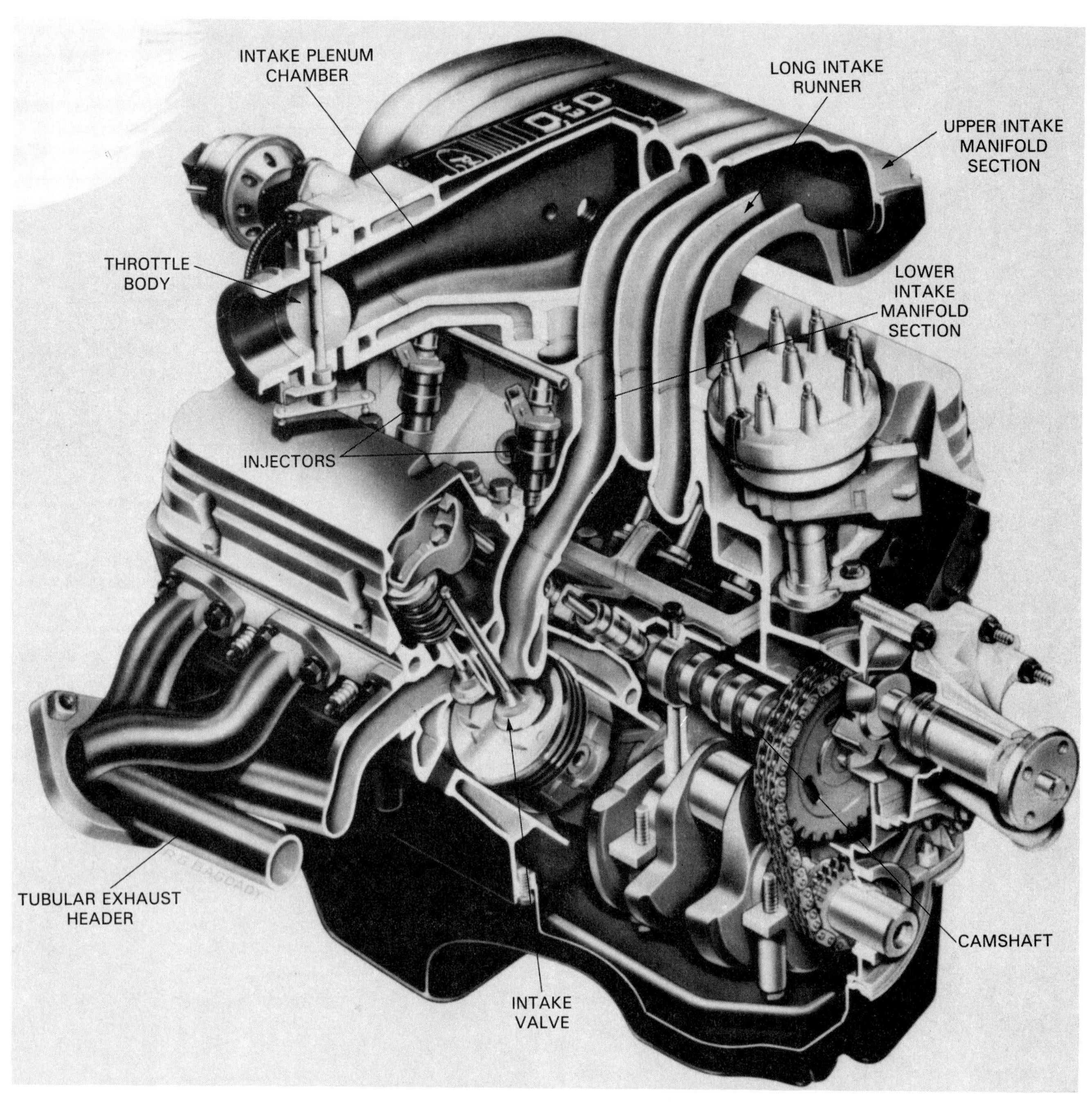

This is a cutaway of a high performance V-8 engine. Note induction or intake system. Very long intake runners are used to produce a ''ram effect.'' Inertia of moving air tends to force or ram more air and fuel into combustion chambers on intake strokes. This increases engine power, especially at low rpm. Also note tubular exhaust headers. (Ford)

Chapter 23

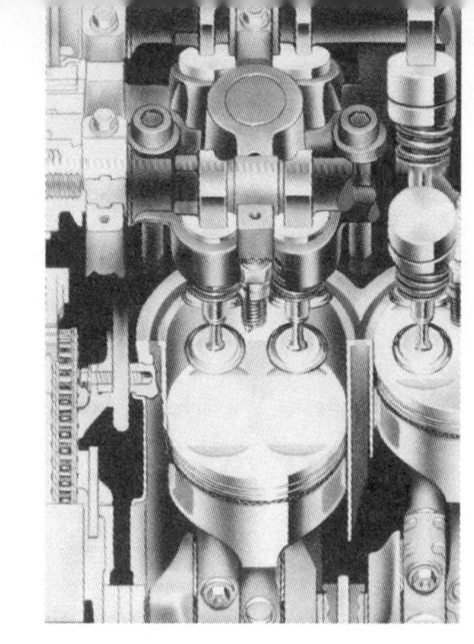

Engine Tune-Up

After studying this chapter, you will be able to:
- *Compare a minor tune-up and a major tune-up.*
- *List the fundamental steps for a tune-up.*
- *Summarize the hazards of doing a tune-up.*
- *Describe the maintenance operations for a diesel engine.*

An engine tune-up is not only important to engine power and fuel economy, it is equally critical to air quality and conservation of our natural resources. An untuned engine consumes a tremendous amount of fuel and produces high levels of air pollution. Both are critical to our world as a whole. As an engine technician, you must be able to keep an engine running efficiently.

The chapter overviews subject matter given in several other text chapters. For this reason, you may want to refer to the index to find additional information on specific subjects: installing distributor, carburetor or injection system service, using an exhaust analyzer, checking spark plug wire resistance, and reading spark plugs. This chapter will add to your study.

WHAT IS A TUNE-UP?

An *engine tune-up* is a maintenance type operation that returns the engine to peak performance after part wear and deterioration. It involves the replacement of worn parts, basic service tasks, making adjustments, and, sometimes, minor repairs.

A tune-up assures that the engine, ignition system, fuel system, and emission control systems are within factory specs. Manufacturers normally recommend a tune-up after a specific period of engine operation. See Fig. 23-1.

The procedures for a tune-up vary from shop to shop. In one garage, a tune-up may include only the most routine tasks. In another shop, it may involve a long list of tests, adjustments, and repairs.

Minor tune-up

A minor tune-up is a basic tune-up done when the engine is in good operating condition. For example, a new vehicle only driven 20,000-30,000 miles (32 187-48 280 km) may only require a minor tune-up. Some new engines do not require plug replacement for 100,000 miles (160 000 km). Most of the engine systems are in satisfactory condition with little or no wear.

A minor tune-up typically involves these tasks:

1. Visual inspection for obvious problems affecting engine performance (loose wires, missing vacuum hoses, oil leaks, coolant leaks, etc.).
2. Check for tripped trouble codes on computer-controlled vehicles, Fig. 23-1. On older, high mileage vehicles, it would include distributor service (points and condenser replacement) and spark plug replacement.
3. Check and adjustment of ignition timing if adjustable.
4. Replacement of air and fuel filters.
5. Fuel system adjustments (idle speed, for example) if any adjustments are available.
6. Clean throttle body and idle air bypass on fuel injected engines.
7. Emission control system tests and service (vapor canister filter replacement, for example).
8. Filling of all fluid levels.
9. Road test.

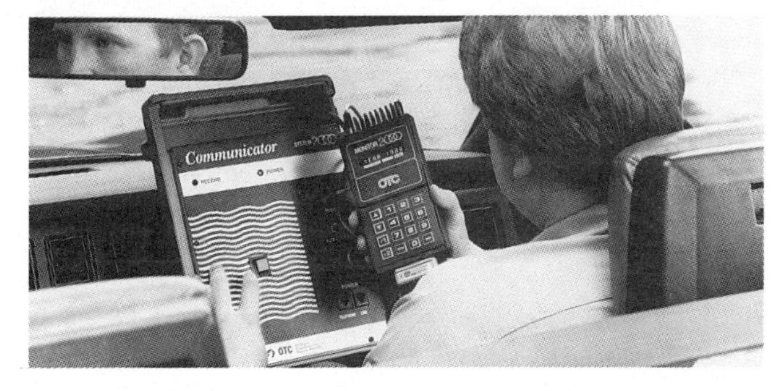

Fig. 23-1. The modern tune-up is performed as maintenance, rather than as a repair for a performance problem. Here, a technician is scanning the ECU for potential diagnostic trouble codes. (OTC)

Major tune-up

A *major tune-up* is more thorough and is done when the engine systems (ignition, fuel, emission, and engine itself) are in poor condition. After prolonged use, parts can wear or deteriorate, requiring more attention.

Besides the steps listed for a minor tune-up, a major tune-up typically includes more diagnosis type tests (scope analysis, compression test, vacuum test, etc.) to determine what should be done during the tune-up. It may also include carburetor or injection system repairs, throttle body or distributor rebuild, and other more time consuming jobs.

Fig. 23-2 gives a few problems that should be located and corrected during a major tune-up.

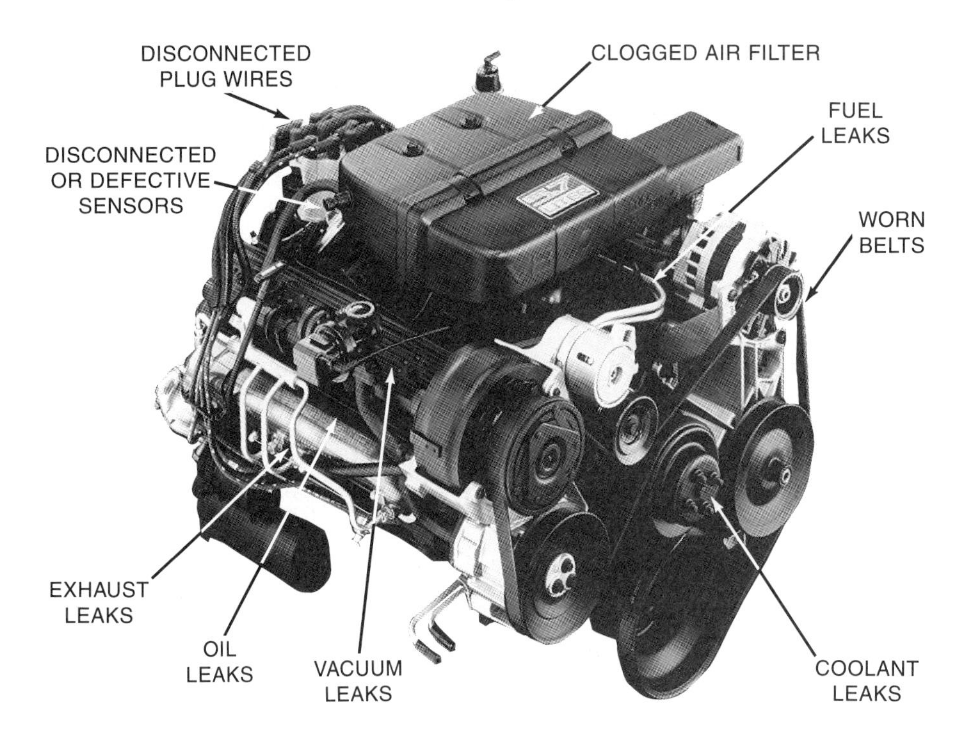

Fig. 23-2. During inspection, look for any troubles that could upset engine performance and dependability. (General Motors)

Reasons for a tune-up

A tune-up is critical to the operation of an engine. it can affect:

1. Engine power and acceleration.
2. Fuel consumption.
3. Exhaust pollution.
4. Smoothness of engine operation.
5. Ease of starting.
6. Engine service life.

You must make sure you return every engine related system to peak operating condition during a tune-up. If you overlook just one problem, the engine will NOT perform properly and the car owner will NOT be happy with your work.

TUNE-UP SAFETY

When doing a tune-up, remember these safety rules:

1. Engage the emergency brake and block the wheels when the engine is to be running.
2. Place an exhaust hose over the tailpipe when running the engine in an enclosed chop.
3. Keep clothing, your hands, tools, and equipment away from a spinning engine fan and belts.
4. Disconnect the battery when recommended in the service manual. This will help avoid accidental engine cranking or an "electrical fire."
5. Be careful not to touch the hot exhaust manifold when removing old spark plugs. Keep test equipment leads away from the engine exhaust manifolds, fan, and belts.
6. Wear eye protection when blowing debris away from spark plugs or when working near the spinning engine fan.
7. Keep a fire extinguisher handy when performing fuel system tests and repairs.
8. With a diesel engine, disable the injection pump when removing an injection line. System pressure is high enough to puncture your skin and eyes.

TYPICAL TUNE-UP PROCEDURES

The following is a summary of the most common procedures for an engine tune-up. They are typical and apply to most makes and models of cars.

Preliminary inspection

To begin a tune-up, inspect the engine compartment. Try to find signs of trouble:

1. BATTERY PROBLEMS (dirty case top, corroded terminals, physical damage).
2. AIR CLEANER PROBLEMS (clogged filter, inoperative air flap, disconnected vacuum hoses).
3. FUEL SYSTEM PROBLEMS (leaks, clogged fuel filter, etc.), Figs. 23-3 and 23-4.
4. BELT TROUBLES (loose belts, fraying, slippage).
5. DETERIORATED HOSES (hardened or softened cooling system, fuel system, and vacuum hoses).
6. POOR ELECTRICAL CONNECTIONS (loose or corroded connections; frayed or burned wiring).

If any problems are found, correct them before beginning the tune-up. Many of these problems could affect engine performance.

EVALUATING ENGINE AND ENGINE SYSTEMS

During a tune-up, it is very important that you test and evaluate the condition of the engine and its systems. Various test instruments are used for this purpose. A compression gauge, leakdown tester, ignition scope, vacuum gauge, exhaust analyzer, timing light, and VOM are a few of the test instruments used during a tune-up.

Fig. 23-3. This engine had binding throttle linkage. Cleaning, lubrication, and adjustment corrected erratic idle speed. (Fel-Pro Gaskets)

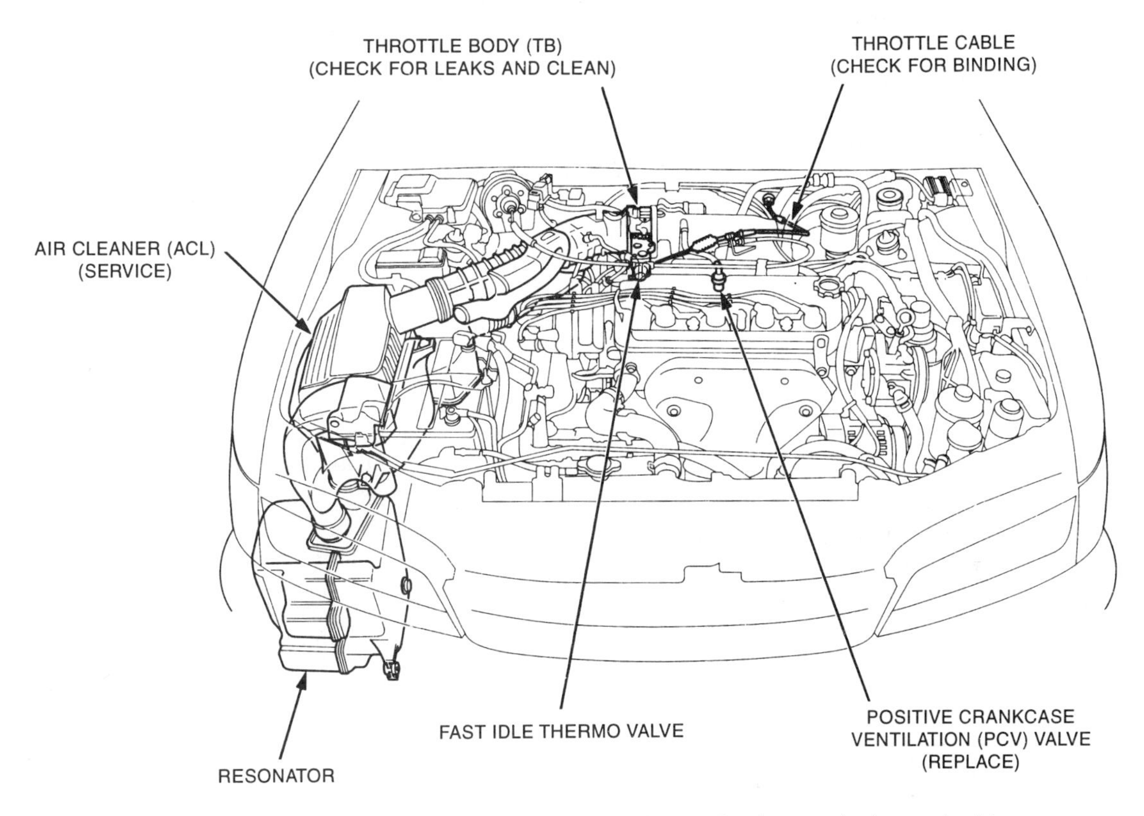

Fig. 23-4. Air filter and throttle body must be serviced at regular intervals. Also check for bad hoses, belts, and other problems. (Honda)

Compression testing

A *compression test* is a common method of determining ENGINE MECHANICAL CONDITION. It should be done anytime symptoms point to cylinder pressure leakage. An extremely rough idle, popping noise from air inlet or exhaust, excessive blue smoke, or blowby are all reasons to consider a compression test.

A compression test is frequently made during a tune-up. It is impossible to tune an engine that is not sound mechanically. If the engine fails the compression test, mechanical repairs must be made BEFORE the tune-up.

A *compression gauge* is used to measure compression stroke pressure during this test. If gauge pressure is lower than normal, pressure is leaking out of the engine combustion chamber.

LOW ENGINE COMPRESSION can be caused by:

1. BURNED VALVE (valve face damaged by combustion heat).
2. BURNED VALVE SEAT (cylinder head seat damaged by combustion).
3. PHYSICAL ENGINE DAMAGE (hole in piston, broken valve, etc.)
4. BLOWN HEAD GASKET (head gasket ruptured).
5. WORN RINGS OR CYLINDERS (part wear prevents

ring-to-cylinder seal).

6. VALVE TRAIN TROUBLES (valves adjusted with insufficient clearance which keep valves from fully closing, broken valve spring, etc.).
7. JUMPED TIMING CHAIN OR BELT (loose or worn chain or belt has jumped over teeth, upsetting valve timing).

For other less common sources of low compression, refer to a service manual troubleshooting chart.

Compression test (gasoline engine)

To do a compression test on a gasoline engine, remove all of the spark plugs so that the engine will crank easily. Block open the carburetor or fuel injection throttle plates. This will prevent a restriction of airflow into the engine.

Disable the ignition system to prevent sparks from arcing out of the disconnected spark plug wires. Some electronic ignition systems may be damaged if operated with the spark plug wires disconnected. Usually, the feed wire going to the ignition coil can be removed to disable the system.

If electronic fuel injection is used, it should also be disabled so that fuel will not spray into the engine. You may need to unplug the wires to the injectors. Check your service manual for specific directions.

Screw the compression gauge into one of the spark plug holes. See Fig. 23-5. Crank the engine and let the engine rotate for about six compression strokes (compression gauge needle moves six times). Write down the gauge readings for each cylinder and compare to specs.

Compression test (diesel engine)

A diesel engine compression test is similar to a compression test for a gasoline engine. However, do not use a compression gauge intended for a gasoline engine. It can be damaged by the high compression stroke pressure. A diesel compression gauge must read up to approximately 600 psi (4 134 kPa).

To perform a diesel compression test, remove either the injectors or the glow plugs. Refer to a service manual for instructions but most suggest glow plug removal. Install the compression gauge in the recommended hole. Disconnect the fuel shut-off solenoid to disable the injection pump. Crank the engine and note the highest reading on the gauge. Crank the engine about six ''puffs'' (compression strokes).

Note! Cranking rpm must be up to specs (about 200 rpm minimum) and the engine should be warm for accurate test results.

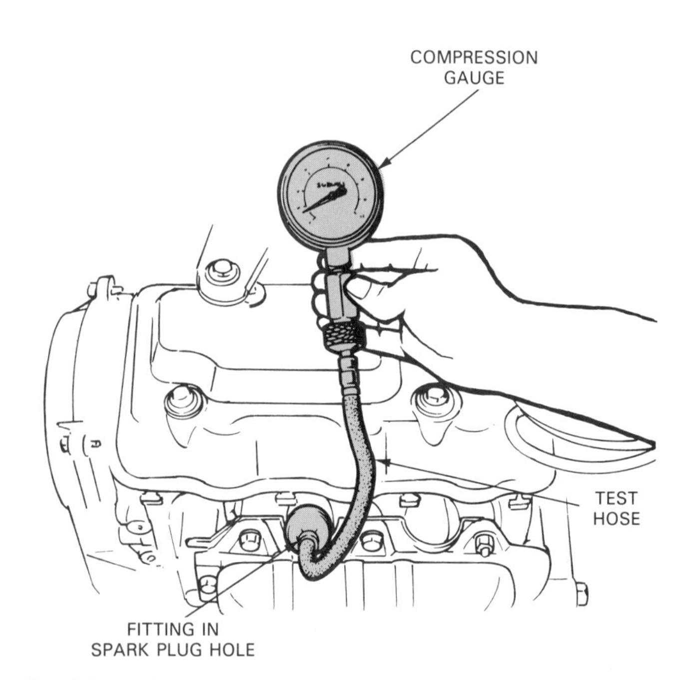

Fig. 23-5. Compression gauge installs in spark plug hole on gasoline engine and glow plug hole on diesel. Crank engine for about six ''puffs'' and then write down gauge reading. If below specs, something is causing leakage. (General Motors)

Compression test results

GASOLINE ENGINE COMPRESSION READINGS should run around 125 to 175 psi (861 to 1 206 kPa). Generally, the compression pressure should not vary over 15 to 20 psi (102 to 138 kPa) from the highest to the lowest cylinder. Readings must be within about 10 to 15 percent of each other.

DIESEL ENGINE COMPRESSION READINGS will average 275 to 400 psi (1 895 to 2 756 kPa), depending upon engine design and compression ratio. Compression levels must not vary more than about 10 to 15 percent (30 to 50 psi or 206 to 344 kPa).

Look for cylinder PRESSURE VARIATION during an engine compression check. If some cylinders have normal pressure and one or two have low readings, engine performance will be reduced. The engine will have a rough idle and lack power.

If ALL of the cylinders are low (worn timing chain for example), the engine may run smoothly but lack power and gas mileage.

If TWO ADJACENT CYLINDERS read low, it might point to a blown head gasket between the two cylinders. A blown head gasket will sometimes produce a louder than normal puffing noise from the spark plug or glow plug holes with the gauge removed.

Wet compression testing

A *wet compression test* should be completed if cylinder pressure reads below specs. It will help you determine what engine parts are causing the problem. See Fig. 23-6A.

As in Fig. 23-6B, squirt a tablespoon of 30 weight motor oil into the cylinder having the low pressure reading. Install the compression gauge and recheck cylinder pressure.

If the compression gauge reading GOES UP with oil in the cylinder, the piston rings and cylinders may be worn and leaking pressure. The oil will temporarily coat and seal bad compression rings to increase pressure. Look at Fig. 23-6C.

If the low pressure reading STAYS ABOUT THE SAME, then the engine valves or head gasket may be leaking. The engine oil will seal the rings but NOT a burned valve or blown head gasket. In this way, a wet compression test will help diagnose the cause of low compression.

Do NOT squirt too much oil into the cylinder during a wet compression test or a false reading will result. With excessive oil in the cylinder, compression readings will

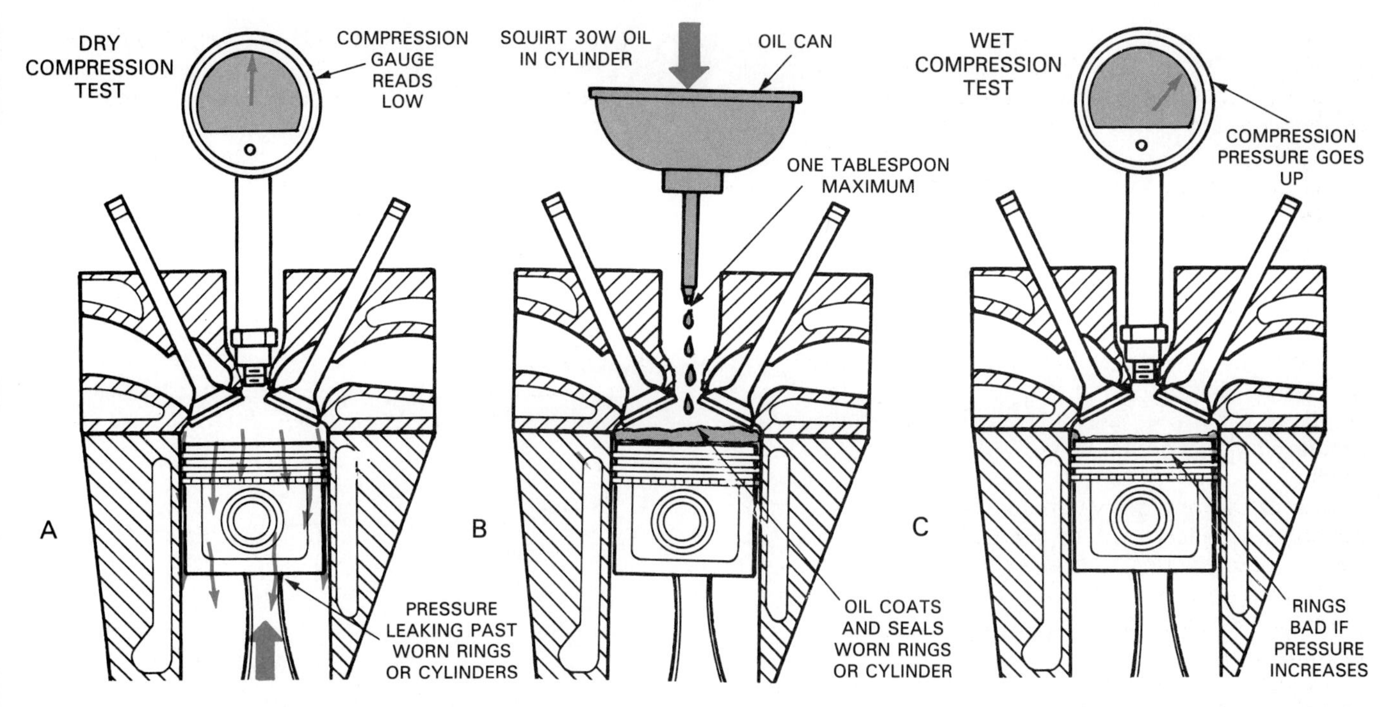

Fig. 23-6. Basic method of making dry and wet compression test. A—Complete conventional, dry compression test with pressure gauge. Record all readings and compare to specs. B—If compression is low, squirt a tablespoon of oil into cylinder. This will temporarily seal rings. C—Measure compression pressure again. If pressure reading goes up, that cylinder may have bad rings or worn cylinder. Same pressure reading might point to burned valve or blown head gasket.

go up even if the piston rings and cylinders are not in good condition. Oil, like any liquid, WILL NOT COMPRESS. It will take up space in the cylinder, raising the compression ratio and gauge readings.

Note! Some auto makers warn against performing a wet compression test on a diesel. If too much oil is squirted into the cylinder, hydraulic lock and part damage could result because the oil will be highly pressurized, like a solid, in the small cylinder volume.

CYLINDER LEAKAGE TEST

A *cylinder leakage test* also measures air leakage out of the engine combustion chamber to find mechanical problems. It is a more accurate way of finding the specific source of the trouble because a shop air source feeds air into the cylinder. A pressure gauge and valve assembly is used to measure actual percent of cylinder leakage.

As shown in Fig. 23-7, the gauge-valve unit is installed in a spark plug or glow plug hole. Shop air pressure is connected to the other gauge line. The crankshaft must be rotated so the cylinder being tested has its piston at TDC with both valves closed.

First, air pressure is adjusted so that the gauge reads zero leakage with the test hose disconnected or shut off. Then, air is allowed to enter the cylinder and the pressure gauge will drop. The amount of pressure drop depends upon how well the rings, valves, head gasket, etc. seal pressure in the cylinder.

If the pressure gauge shows only a 10 percent or less pressure drop, the engine is in fair condition. A HIGHER PERCENTAGE pressure drop indicates excess leakage

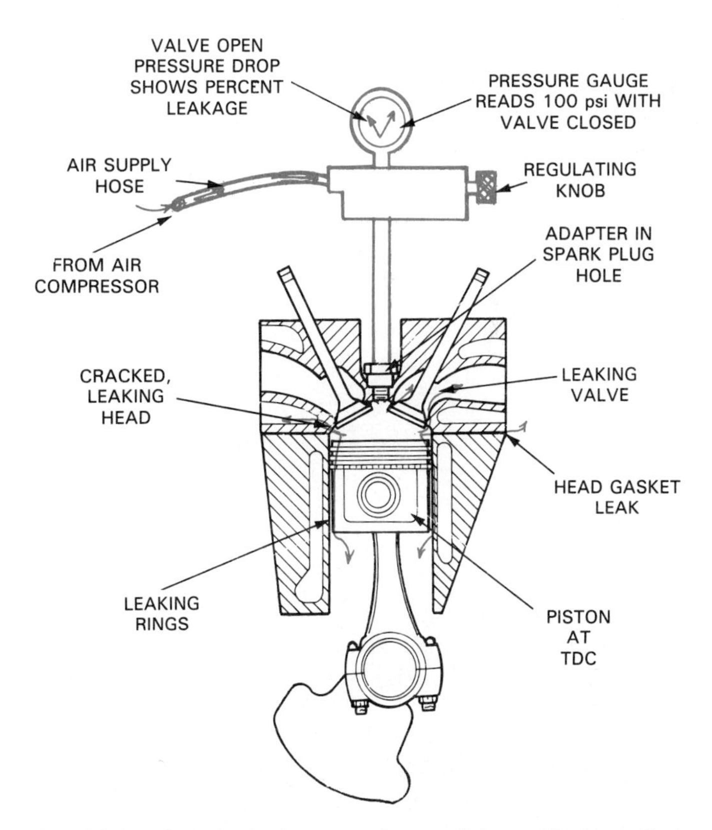

Fig. 23-7. Cylinder leakage test forces air into cylinder to find leakage. First adjust air pressure so gauge reads zero. Then, allow air to flow into cylinder with piston at TDC on compression stroke. Gauge pressure will drop and show percent of cylinder leakage. If there is a problem, you may be able to hear hiss (air leak) that indicates source of problem.

and a mechanical engine problem.

To find the source of leakage, listen for a hissing sound at the intake manifold inlet, exhaust pipe, or valve cover breather. Depending upon where you hear leakage, the problems could include the following:

1. HISS AT INTAKE MANIFOLD INLET (leaking intake valve, intake valve adjusted too tight, broken intake valve spring, damaged valve).
2. HISS AT TAILPIPE (burned exhaust valve, lash adjustment too tight, broken exhaust valve spring, bent valve).
3. HISS AT VALVE COVER BREATHER (worn piston rings, worn cylinder walls, blown head gasket, burned piston).
4. HISS AROUND OUTSIDE OF HEAD (blown head gasket, warped head, warped block deck, cracked head or block, etc.).
5. HISS FROM ADJACENT SPARK PLUG HOLE (blown head gasket, cracked head or block).

Note! During a cylinder leakdown test, make sure you have the piston at TDC on the compression stroke. If not, a valve could be open and cause false indications.

VACUUM GAUGE TESTING

A *vacuum gauge* will measure intake manifold or engine vacuum and indicate engine condition. Detailed in other chapters, the vacuum gauge is connected to an intake manifold vacuum fitting, as in Fig. 23-8.

Warm the engine. Run the engine at the correct idle speed and read the vacuum gauge. Vacuum should be steady and read within specs (about 15 to 17 in/Hg or 40 to 45 cm/Hg).

A burned valve, late ignition timing, worn timing chain, restricted exhaust system, etc. will show up as abnormal vacuum gauge readings. Refer to the text index and a service manual for more information.

Cylinder balance test

A *cylinder balance test* ''kills'' one cylinder at a time to find out if that cylinder is firing properly. For instance, if an engine has a ''dead miss'' at idle, you can pull one spark plug wire at a time. If engine rpm drops and the engine idles more roughly, then that cylinder is firing. If you pull off a plug wire and idle speed and smoothness stay the same, that cylinder is ''dead.'' Something is preventing normal combustion.

There are several ways to do a cylinder balance test: using a simple tachometer, an engine analyzer, or by simply watching and listening as each wire is disconnected.

Fig. 23-9 shows how a tach is used to complete a cylinder balance test. As you remove each wire, watch the tachometer needle. It should drop equally as each cylinder is made inoperative. If the needle fails to drop enough on any cylinders, a compression test or leakage test may be done to find the source of the problem. A fouled spark plug or bad wire could also be the cause.

With an engine analyzer or scan tool, there may be cylinder balance buttons that will short out cylinders electronically. Push a button and it will keep the coil from firing on that cylinder only. Some will short cylinders automatically and record the results on a paper printout. Follow equipment operating instructions since they vary.

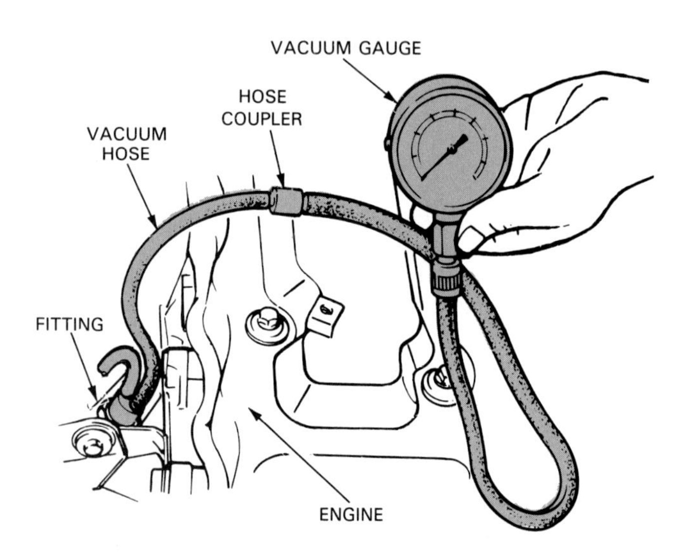

Fig. 23-8. Vacuum gauge will also help find problems during a tune-up. At idle, engine vacuum should be steady and up to specs. (Oldsmobile)

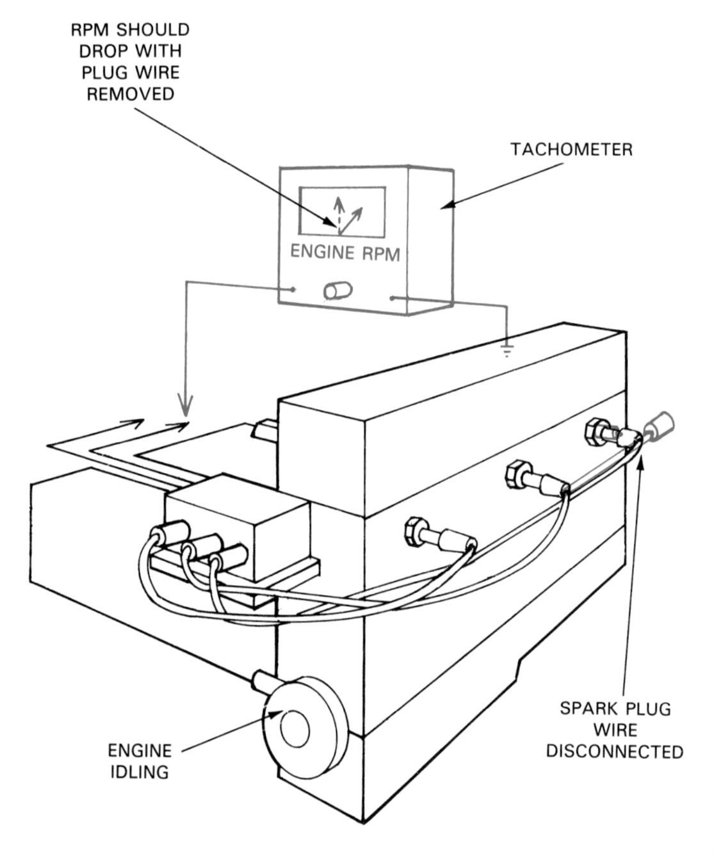

Fig. 23-9. Cylinder balance test can be done with tachometer. Connect tach and fast idle engine. Pull off one spark plug wire at a time while watching tach. If rpm does not drop equally in one cylinder, that cylinder is dead. You could have bad wire, fouled plug, low compression, etc.

VALVE ADJUSTMENT

When mechanical (solid) lifters are used, the valves require periodic adjustment. Proper valve adjustment is important to the performance of the engine. It is sometimes done during a major tune-up.

If a valve opens too much or not enough, it will upset the amount of air-fuel mixture pulled into the cylinder. It also affects valve lift and duration. This will affect combustion and reduce engine efficiency.

Valve adjustment can be changed by turning rocker arm adjusting screws on some engines, Fig. 23-10. On some OHC engines, you must change shim thickness to alter valve lash or valve clearance, Fig. 23-11.

Basically, remove the valve cover. Crank the engine until a cam lobe is pointing away from the rocker arm or until the valve is fully closed. Adjust the valve until the correct thickness feeler gauge fits over the valve or under the rocker arm, Figs. 23-10 and 23-11. If clearance is incorrect, turn the adjusting screw or install a shim of different thickness. Repeat this procedure on all the valves.

Some engine manufacturers recommend cold valve adjustment while others require that the engine be warmed to full operating temperature. Always refer to the service manual for specific procedures and specifications.

Note! Usually the exhaust valve will require more lash or clearance than the intake valve. It runs hotter and will expand or lengthen more than the intake valve.

Tune-up parts replacement

Depending upon the age and condition of the engine, any number of parts may need replacement during a tune-up. With a new, late model car, you may only need to replace the spark plugs and filters. With an older, high mileage car, you may have to replace the spark plugs, injectors, distributor components, spark plug wires, or any other parts reducing engine efficiency.

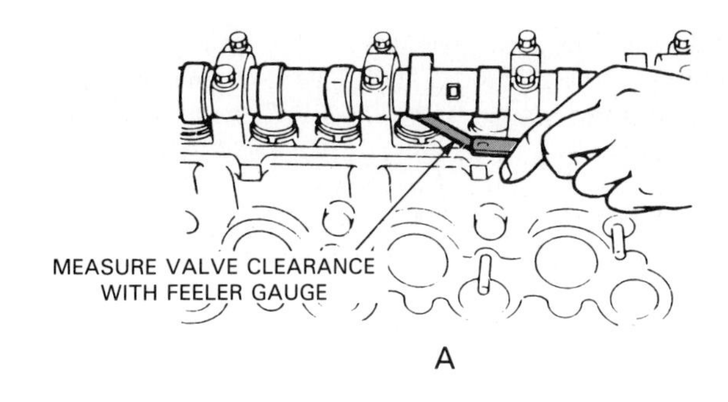

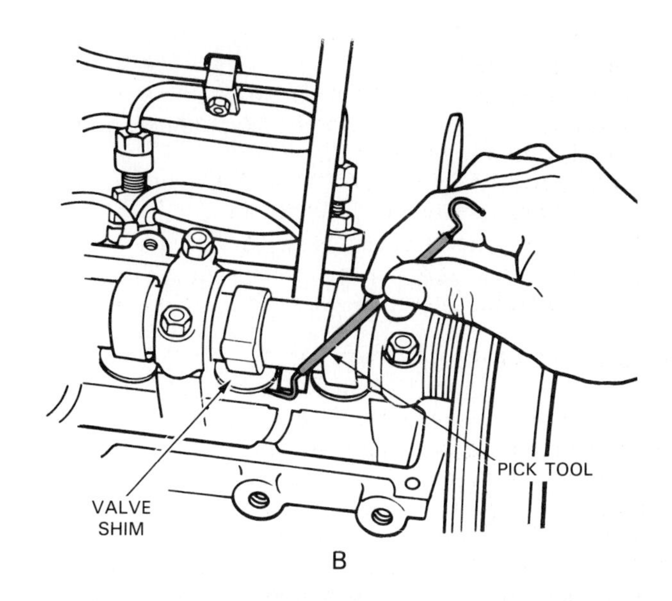

Fig. 23-11. This OHC engine has valve adjusting shims. A—Checking valve clearance with feeler gauge. B—Installing shim of different thickness to correct lash. (Ford)

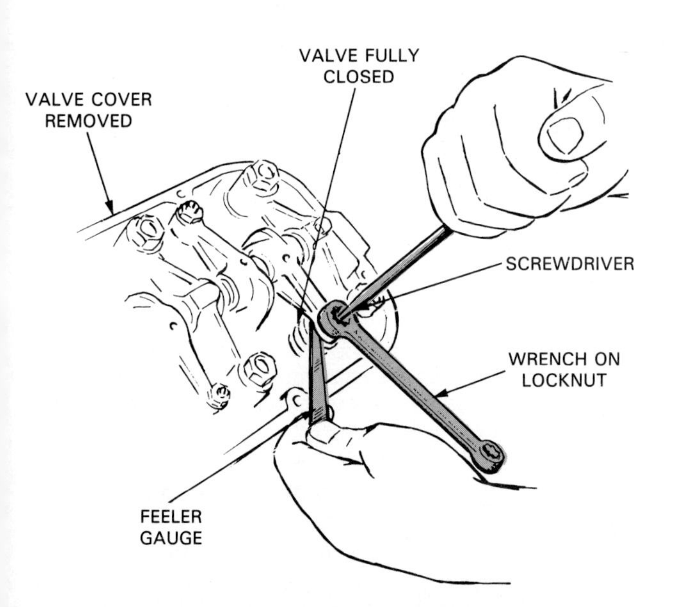

Fig. 23-10. Valve adjustment is sometimes needed during tune-up. Crank engine until valve is fully closed. Slide correct size feeler gauge between valve and rocker. Turn adjusting screw to change lash if needed. Repeat on all valves. Check manual for specifics. (Volvo)

During a tune-up, you may have to replace or service:

1. SPARK PLUGS (spark plugs are almost always serviced or replaced during tune-up), Fig. 23-12.
2. SPARK PLUG WIRES (wires should be checked for high resistance or electrical leakage through insulation), Fig. 23-13. Refer to Chapter 14 on Ignition System Service for details.
3. DISTRIBUTOR PARTS (may have to replace cap and rotor, set pickup coil air gap or contact point gap, adjust ignition timing, etc.), Fig. 23-14.
4. FUEL SYSTEM PARTS (must check and sometimes adjust idle speed, choke, idle mixture, change filters, etc.).
5. EMISSION CONTROL PARTS (changing canister filter, PCV valve, etc.).

TUNE-UP ADJUSTMENTS

Again, the adjustments needed during a tune-up will vary with the make, model, and condition of the car. A few of the most common tune-up adjustments include:

1. Spark plug gap, Fig. 23-12.
2. Pickup coil air gap (electronic ignition) or breaker

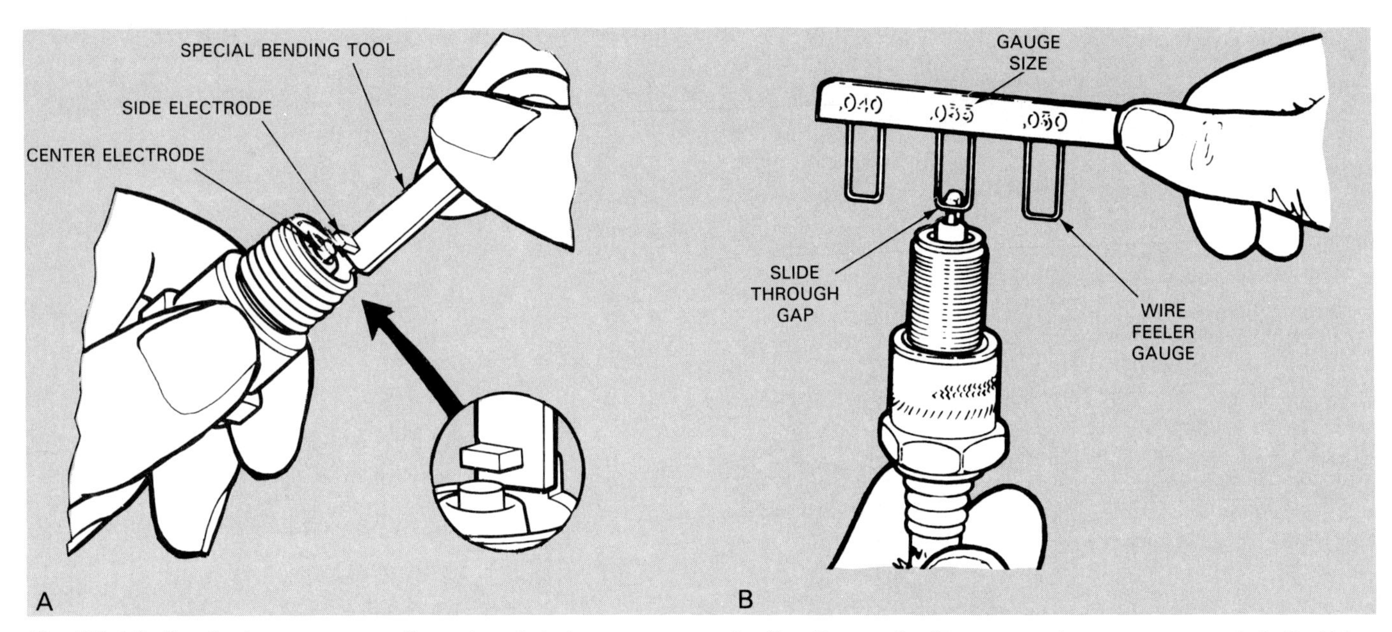

Fig. 23-12. Spark plugs are normally replaced during a tune-up. A—Bending tool will open or close gap as needed. B—Wire feeler gauge is slid inside gap to check electrode spacing. Use wire gauge of correct thickness. (Chrysler)

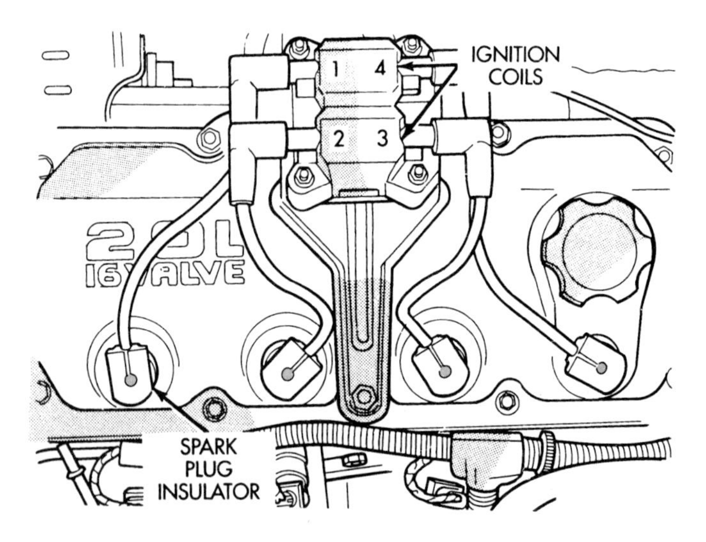

Fig. 23-13. Spark plug wires are still a common source of engine performance problems. They can break internally, their insulation can leak high voltage to ground, voltage can be induced into another wire, or they can simply come off the coil or spark plug. (Chrysler)

point gap or dwell (contact point ignition).

3. Ignition timing (gasoline engine) or injection timing (diesel engine), Fig. 23-14.
4. Idle speed.

Other adjustments may also be needed. Follow the specific directions and specs in a service manual. This will assure a thorough and long lasting tune-up. Refer to the index for more text information.

General tune-up rules

Keep these general rules in mind when doing a tune-up:

1. Gather information about the performance of the engine. Ask the customer about the car. This may give you ideas about what components should be tested and replaced.
2. Make sure the engine has warmed to full operating temperature. Usually, you cannot evaluate engine operation and make tune-up adjustments with a cold engine.
3. Use professional, high quality tools and equipment. Make sure the equipment is accurate and will give precise readings.
4. Refer to the car service manual or emission sticker for specs and procedures. Today's cars are so complex, the slightest mistake could ruin the tune-up.
5. Use quality parts. Quality parts will assure that your

Fig. 23-14. Timing light, as explained in Chapter 14, is used to check and adjust ignition timing. With a diesel, you would have to check injection timing. (Peerless Instruments)

tune-up lasts a long time. Cheap, bargain parts are usually inferior and can fail quickly.

6. Keep service records. You should write down all of the operations performed on the car. This will give you and the customer a record for future reference. If a problem develops, you can check your records to help correct the trouble.
7. Complete basic maintenance service. Most good garages will lubricate door, hood, and trunk latches and hinges during a tune-up. They will also check all fluid levels, belts, and hoses. This will build good customer relations and assure vehicle safety and dependability.
8. *Tune-up intervals* are the recommended time spans in months or miles when an engine should be tuned-up. They are given by the auto maker. One maker may require tune-up type operations every 36 months or 36,000 miles, while another requires more or less frequent service. Check the manual for details. The vehicle warranty may be void if tune-up intervals are missed.

DIESEL ENGINE TUNE-UP (MAINTENANCE)

Diesel engines do NOT require tune-ups like gasoline engines. A diesel does NOT have spark plugs to replace or an ignition system to fail. The diesel injection system is very dependable and only requires major service when problems develop.

A *diesel engine tune-up,* more accurately called *diesel maintenance,* typically involves:

1. Replacing air filter element.
2. Cleaning, draining, or replacing fuel filters.
3. Adjusting engine idle speed.
4. Adjustment of throttle cable.
5. Inspecting engine and related systems.
6. Check of injection timing.
7. Changing engine oil and oil filter.
8. Periodic service of emission control systems.

Refer to Chapter 18 and a shop manual for more information on diesel engine maintenance.

SUMMARY

An engine tune-up involves keeping all engine systems working at peak performance. It is a maintenance type operation that must be done at regular intervals.

A minor tune-up is done on an engine in good condition and only involves spark plug replacement, filter service, and minor adjustments. A major tune-up can involve more time-consuming tasks—carburetor rebuilt, throttle body service, replacing spark plug wires, etc.

Keep safety in mind during a tune-up because you will be working around a running engine.

Start your tune-up with an inspection of the engine compartment. Look for battery problems, fuel system and ignition system troubles, bad hoses, loose wires, low fluid levels, etc.

A compression test is done when engine missing or other symptoms point to mechanical problems. A compression test measures actual compression stroke pressure. If low, a mechanical trouble is allowing air to leak out of the engine combustion chamber through a burned valve, blown head gasket, worn piston rings, etc.

A cylinder leakage test is like a compression test but uses outside air pressure to check for combustion chamber leakage. Air is pumped into the cylinder with the piston at TDC on the compression stroke. Any leak will show up as an excessive gauge pressure drop and hissing sound.

A cylinder balance test shorts out spark plugs to see if the cylinder is firing. If a plug wire is disconnected, engine rpm should drop. If engine idle speed stays the same, that cylinder is not producing power.

Valve adjustment is sometimes done during a tune-up when the engine has mechanical lifters or followers. Either adjustment screws or adjustment shims are provided to change valve lash or clearance.

During a tune-up, various parts may need replacement or adjustment. Usually, spark plugs are replaced and gapped. Sometimes, you must also test and replace spark plug wires, points, fuel filters, air filters, PCV valve, and other components. You must check out engine operation and refer to a shop manual and service records to determine what must be done during the tune-up.

Always check ignition timing and idle speeds during a tune-up. Road test the car after the tune-up to check your work.

KNOW THESE TERMS

Minor tune-up, Major tune-up, Compression test, Burned valve, Blown head gasket, Worn rings, Insufficient valve clearance, Jumped timing chain, Diesel compression gauge, Wet compression test, Cylinder leakage test, Vacuum gauge test, Cylinder balance test, Valve adjustment, Gapping spark plugs, Pickup coil and point gap, Ignition timing, Idle speed, Tune-up interval.

REVIEW QUESTIONS—CHAPTER 23

1. An engine tune-up is important to:
   a. Engine power.
   b. Fuel economy.
   c. Air pollution.
   d. All of the above.
   e. None of the above.
2. An engine tune-up is a _________ operation that returns the engine to _________ _________ after part _________ and deterioration.
3. In your own words, what is the difference between a minor tune-up and a major tune-up?
4. Explain eight tune-up safety rules.
5. A _________ _________ is a common method of checking engine mechanical condition.
6. List problems that cause low engine compression.
7. How do you do a compression test?
8. Most auto manufacturers suggest a wet compression test on diesels. True or false?
9. A cylinder leakage test shows a 30 percent pressure drop in one cylinder. A hissing sound can be heard from the car's tailpipe.
   Technician A says an intake valve could be burned or damaged.
   Technician B says an exhaust valve could be too tight or burned.
   Who is correct?
   a. Technician A.
   b. Technician B.
   c. Both A and B.
   d. Neither A nor B.
10. A cylinder balance test shows that one cylinder is not causing an rpm drop.
    Technician A says that there could be pressure leakage and a mechanical problem.
    Technician B says that there could be a fouled spark plug or bad spark plug wire.
    Who is correct?
    a. Technician A.
    b. Technician B.
    c. Both A and B.
    d. Neither A nor B.

ASE CERTIFICATION-TYPE QUESTIONS

1. Technician A says that a minor tune-up is performed when the engine is in poor running condition. Technician B says that a minor tune-up is performed when the engine is generally running in good condition. Who is right?
   (A) A only.
   (B) B only.
   (C) Both A & B.
   (D) Neither A nor B.
2. Technician A says that during a minor engine tune-up, the ignition timing should be checked and adjusted if needed. Technician B says that during a minor engine tune-up, the emission control system should be tested and serviced if needed. Who is right?
   (A) A only.
   (B) B only.
   (C) Both A & B.
   (D) Neither A nor B.
3. Technician A says that you should check for fuel leaks when doing a major engine tune-up. Technician B says that checking for oil leaks should be performed during this type of tune-up. Who is right?
   (A) A only.
   (B) B only.
   (C) Both A & B.
   (D) Neither A nor B.
4. Technician A says that a tune-up can affect the service life of an engine. Technician B says that a tune-up can affect fuel economy. Who is right?
   (A) A only.
   (B) B only.
   (C) Both A & B.
   (D) Neither A nor B.
5. Low engine compression can be caused by all of the following EXCEPT:
   (A) faulty spark plugs.
   (B) a burned valve seat.
   (C) a blown head gasket.
   (D) worn rings.
6. Technician A says that gasoline engine compression readings should be approximately 275-400 psi. Technician B says that gasoline engine compression readings should be approximately 125-175 psi. Who is right?
   (A) A only.
   (B) B only.
   (C) Both A & B.
   (D) Neither A nor B.
7. If a cylinder leakage test shows a 30% pressure drop in each cylinder _________.
   (A) the leakage gauge is out of calibration.
   (B) the engine is in fair condition.
   (C) a sensor problem exists.
   (D) an engine mechanical problem exists.
8. Technician A says that a tachometer can be used to perform a cylinder balance test. Technician B says that an engine analyzer can be used to perform a cylinder balance test. Who is right?
   (A) A only.
   (B) B only.
   (C) Both A & B.
   (D) Neither A nor B.
9. When adjusting valve lash, the engine's exhaust valve normally requires _________ clearance than the intake valve.
   (A) less
   (B) more
   (C) similar
   (D) Any of the above, depending on the engine.
10. Which of the following is usually performed during a diesel engine tune-up?
    (A) Cleaning or replacing fuel filters.
    (B) Adjusting throttle cable.
    (C) Check injection timing.
    (D) All of the above.

Chapter 24

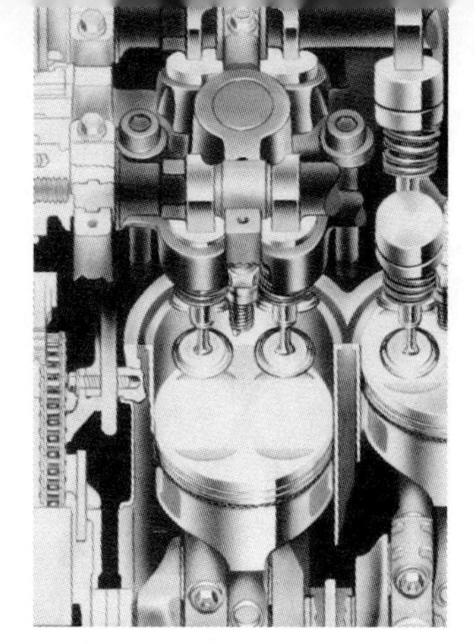

Engine Removal, Disassembly, Cleaning

After studying this chapter, you will be able to:
- *List common engine repairs that do not require engine removal.*
- *Use a vehicle identification number.*
- *Explain why engine removal would be necessary.*
- *Summarize the steps to prepare for engine removal.*
- *List the safety rules to follow during engine removal.*
- *Describe how to keep parts organized during engine removal and teardown.*
- *Summarize how to attach and use engine lifting devices.*
- *Explain the major steps for engine disassembly.*
- *Describe the methods used to clean engine parts.*

In previous chapters, you learned about engine design, part construction, performance problems, engine mechanical failures, lubrication and cooling system service, and engine troubleshooting. This chapter will expand your knowledge of engine service by giving the most important or typical steps for engine removal, teardown, and cleaning. Keep in mind that exact steps vary from engine to engine. However, there are many general methods that apply to all engine makes and models.

ENGINE IDENTIFICATION

Identifying the type of engine to be serviced is needed when ordering parts. There are several ways of doing this: knowledge from work experience, studying engine construction, using the vehicle identification number, casting numbers, and measurement after disassembly.

Identify engine type

By looking at the engine, you can determine:

1. Number of cylinders.
2. Cylinder arrangement (V-type, in-line, etc.).
3. Type fuel system (carburetor, electronic injection, diesel).
4. Mounting (longitudinal, transverse).
5. Accessories used (air conditioning, power steering, etc.).
6. Other facts (horsepower given on tag, etc.)
7. Make or manufacturer's name.

The parts person can usually use this type information along with the year, make, and model of car to help find out what engine is installed in the car.

Vehicle identification number

The *vehicle identification number,* abbreviated VIN, is used to identify what equipment is installed on the automobile. It is a number code that gives the engine type, transmission type, differential type, color code, etc.

As pictured in Fig. 24-1, the VIN can be located in several locations: on dash, fender, engine, door, or in the engine compartment. The service manual will tell where to find the VIN.

An engine technician will sometimes use the VIN. An example is given in Fig. 24-2. Note how the letter "D" represents the engine type. You would go to another chart in the manual to find what the letter means, Fig. 24-3.

Before ordering parts, make sure the car has its original engine. If another shop changed engines, the VIN may not be correct and you could order the wrong parts.

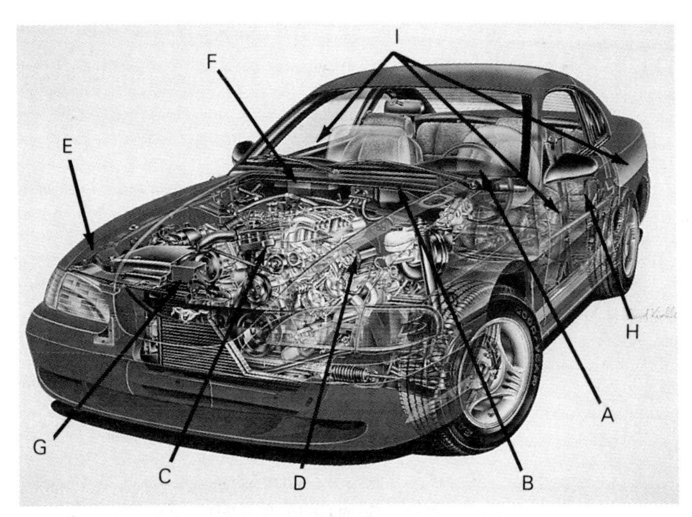

A—Left side of dashpad.
B—Upper left corner of firewall.
C—Stamped on front of engine.
D—Stamped or printed on side of engine.
E—Right front wheel house.
F—Upper right corner of firewall.
G—On radiator support.
H—Lockface of left front door.
I—Body panels (varies by manufacturer).

Fig. 24-1. VIN plate can be located in these locations depending upon make and model car. VIN will have code identifying type engine. VIN is also stamped on most body panels. (Ford)

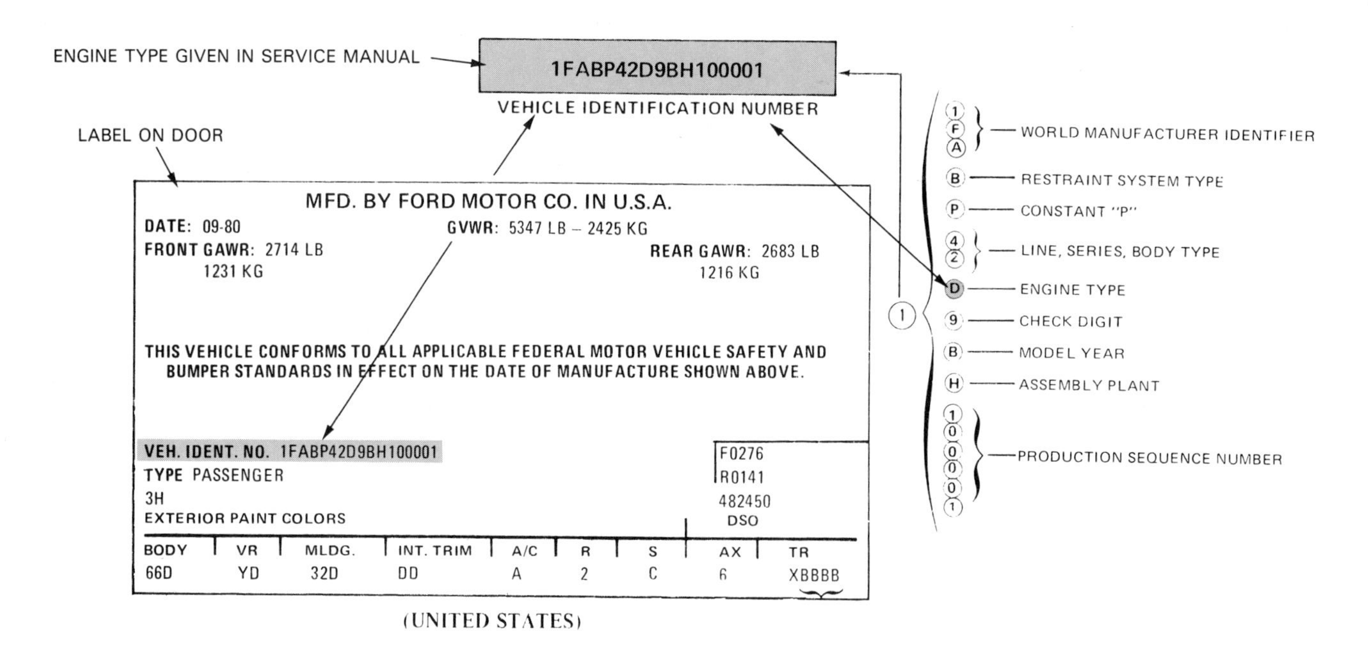

Fig. 24-2. This car had VIN number on label at edge of door. You would have to then go to service manual to find out what letter "D" means. (Ford)

Casting numbers

Casting numbers are numbers formed into parts during manufacturing. They can be found on major components: block, camshaft, heads. These numbers can be used to find out information when ordering parts. Books or charts are available that decode casting numbers.

Part measurement.

Part measurement is another way of finding out what engine you are working on. You can measure cylinder bore, valve sizes, head port sizes, clutch diameter, etc. This will help you and the parts person determine block, head, and engine type.

Note! For more information on engine identification, refer to the text chapters on engine construction and types. Knowing how to tell a "hemi head" from a "wedge head" or an OHC from a push rod engine, is essential to the engine technician.

IS ENGINE REMOVAL NECESSARY?

Many engine repairs can be made with the engine block in the chassis. Repairs limited to the cylinder head, valve train, and other external parts are commonly in-car operations.

Engines are removed when the cylinder block or crankshaft is badly damaged. Depending upon the year, make, and model of the car, engine removal can also be required for other repairs.

For example, one car might need engine removal simply to replace a damaged oil pan. Another car may allow quick and easy oil pan removal, permitting in-car replacement of piston rings, bearings, and other major components.

When in doubt, always refer to a manufacturer's service manual. It will give directions for the exact car and engine.

A few of the engine repairs that can be done with the engine still installed in the chassis include these jobs:

1. A VALVE JOB involves reconditioning the valve faces, seats, and other valve train or cylinder head parts. Normally, the intake manifold, exhaust manifold, and cylinder heads can be removed with the short block bolted in the chassis.
2. VALVE SEALS and SPRINGS can usually be replaced without removing the cylinder heads nor the valves. Shop air pressure can be injected into the cylinders to hold the valves closed. Then a small compressor can be used to remove the springs, retainers, keepers, and seals, Fig. 24-4.
3. OIL PUMPS can often be replaced without engine removal. However, if the oil pan cannot be removed by simply raising the engine off its mounts, engine removal may be needed for oil pump service.
4. ROD and MAIN BEARINGS can sometimes be replaced with the block in the car. After removing the pan, you can unbolt the rod and main caps to service the bearings. The upper main inserts can be rotated out of the block with a thin shank screwdriver or with a special cotter pin type tool that fits into a crank oil hole. Pushing on the screwdriver or turning the crank will make the tool push the inserts out of the block, Fig. 24-5.
5. EXTERNAL ENGINE PARTS (exhaust manifolds, intake manifolds, valve covers, etc.) can normally be serviced without engine removal. With exhaust manifolds, it may help to unbolt the motor mounts. Then you can lift the engine several inches with a floor jack.
6. REAR MAIN OIL SEAL REPLACEMENT, if the oil

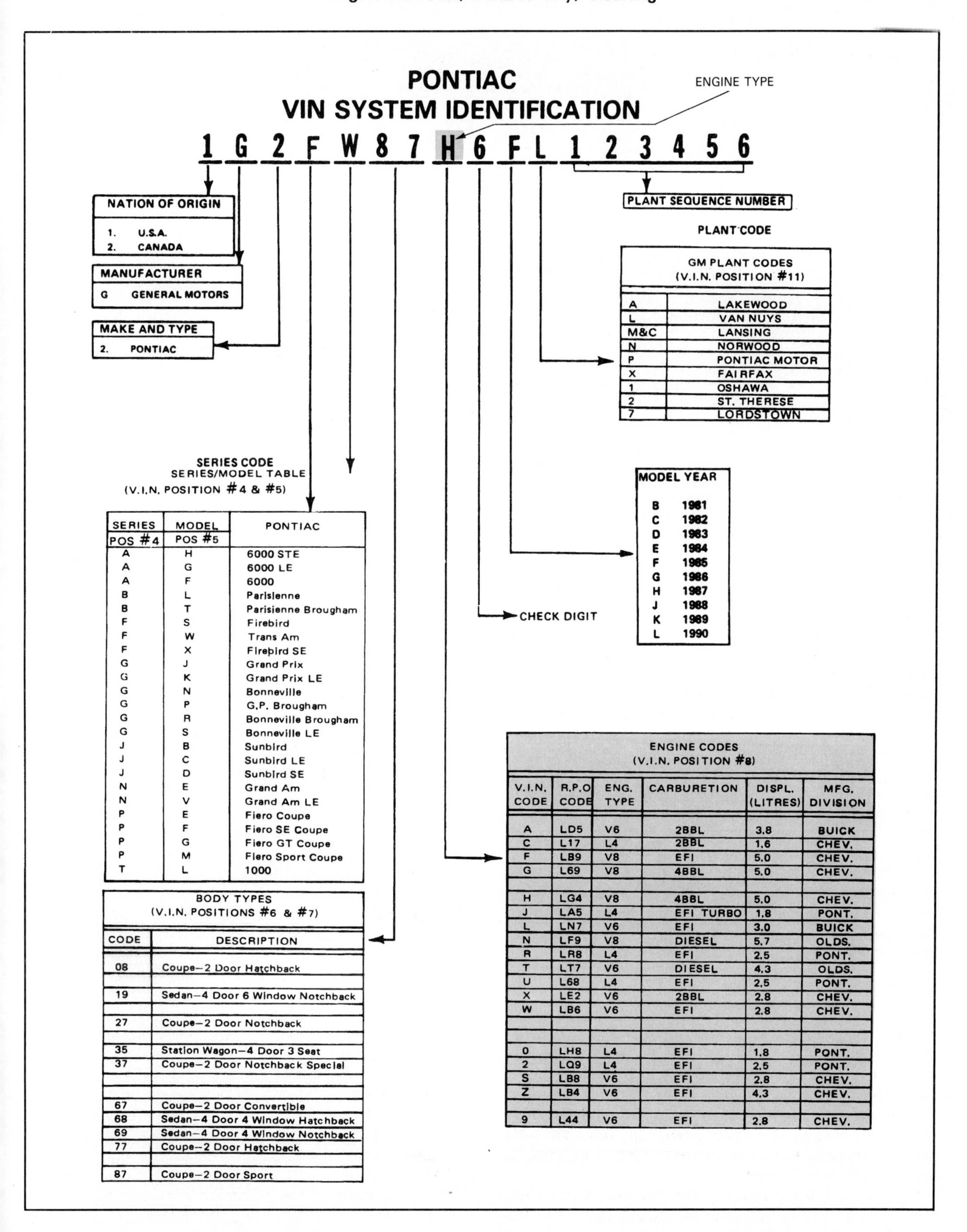

GM PLANT CODES (V.I.N. POSITION #11)

A	LAKEWOOD
L	VAN NUYS
M&C	LANSING
N	NORWOOD
P	PONTIAC MOTOR
X	FAIRFAX
1	OSHAWA
2	ST. THERESE
7	LORDSTOWN

SERIES POS #4	MODEL POS #5	PONTIAC
A	H	6000 STE
A	G	6000 LE
A	F	6000
B	L	Parisienne
B	T	Parisienne Brougham
F	S	Firebird
F	W	Trans Am
F	X	Firebird SE
G	J	Grand Prix
G	K	Grand Prix LE
G	N	Bonneville
G	P	G.P. Brougham
G	R	Bonneville Brougham
G	S	Bonneville LE
J	B	Sunbird
J	C	Sunbird LE
J	D	Sunbird SE
N	E	Grand Am
N	V	Grand Am LE
P	E	Fiero Coupe
P	F	Fiero SE Coupe
P	G	Fiero GT Coupe
P	M	Fiero Sport Coupe
T	L	1000

BODY TYPES (V.I.N. POSITIONS #6 & #7)

CODE	DESCRIPTION
08	Coupe—2 Door Hatchback
19	Sedan—4 Door 6 Window Notchback
27	Coupe—2 Door Notchback
35	Station Wagon—4 Door 3 Seat
37	Coupe—2 Door Notchback Special
67	Coupe—2 Door Convertible
68	Sedan—4 Door 4 Window Hatchback
69	Sedan—4 Door 4 Window Notchback
77	Coupe—2 Door Hatchback
87	Coupe—2 Door Sport

ENGINE CODES (V.I.N. POSITION #8)

V.I.N. CODE	R.P.O CODE	ENG. TYPE	CARBURETION	DISPL. (LITRES)	MFG. DIVISION
A	LD5	V6	2BBL	3.8	BUICK
C	L17	L4	2BBL	1.6	CHEV.
F	LB9	V8	EFI	5.0	CHEV.
G	L69	V8	4BBL	5.0	CHEV.
H	LG4	V8	4BBL	5.0	CHEV.
J	LA5	L4	EFI TURBO	1.8	PONT.
L	LN7	V6	EFI	3.0	BUICK
N	LF9	V8	DIESEL	5.7	OLDS.
R	LR8	L4	EFI	2.5	PONT.
T	LT7	V6	DIESEL	4.3	OLDS.
U	L68	L4	EFI	2.5	PONT.
X	LE2	V6	2BBL	2.8	CHEV.
W	LB6	V6	EFI	2.8	CHEV.
0	LH8	L4	EFI	1.8	PONT.
2	LQ9	L4	EFI	2.5	PONT.
S	LB8	V6	EFI	2.8	CHEV.
Z	LB4	V6	EFI	4.3	CHEV.
9	L44	V6	EFI	2.8	CHEV.

Fig. 24-3. This is actual service manual chart explaining VIN. What engine is in this car? (Pontiac)

Fig. 24-4. Valve seals and valve springs are examples of parts that can be replaced without cylinder head or engine removal. Install air hose in spark plug hole. Air pressure from compressor will hold valves up in head when removing springs. Compress spring and remove keepers to free spring and seal from valve. (Fel-Pro)

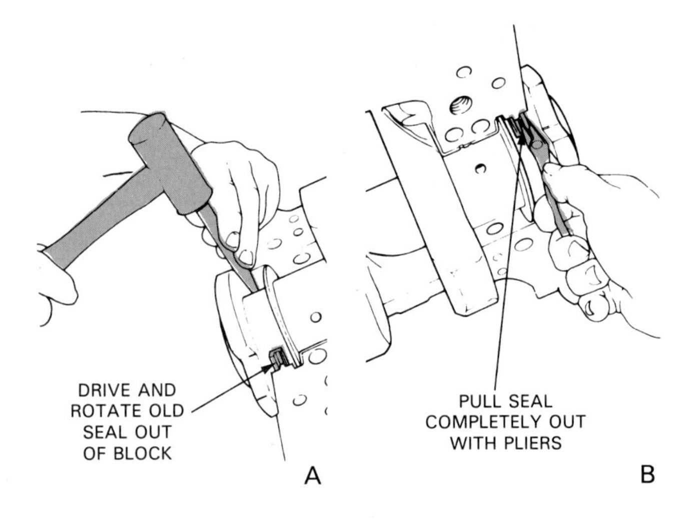

Fig. 24-6. Rear main oil seal can also be replaced without removing engine from car. Remove oil pan and rear main cap. Then service rear seal. A—Drive upper half of neoprene seal out with driver. If engine has rope type seal, turn screw into rope. B—Use pliers to pull seal out of block. Install new seal in cap and block. Make sure new neoprene seal lip is facing in right direction. (Buick)

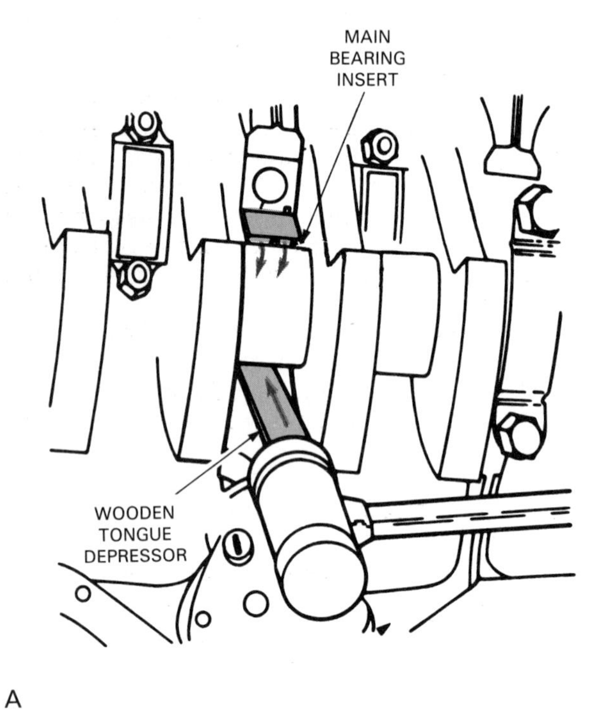

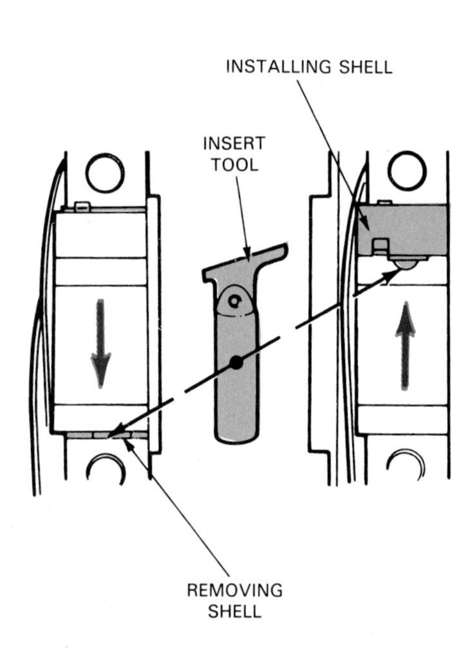

Fig. 24-5. Engine rod and main bearings can be replaced without removing engine from car if oil pan will come off. A—Using wooden tongue depressor to rotate old bearing out of block. B—Using special tool that fits into oil hole in crank to remove upper bearing insert. Bearings in caps can be removed by hand. (Renault and General Motors)

pan can be removed, is an in-car operation. An old neoprene type oil seal can be pushed out of the block with a small screwdriver. With an old rope type rear oil seal, turn a screw into the seal. Then pull the screw and seal out of the block with a pair of pliers, Fig. 24-6. The other half of the seal fits in the main cap. The new seal should be a neoprene type!

7. PISTON SERVICE, so long as the block is in good condition, can sometimes be done without pulling the engine from the car. For example, if a piston pin is knocking in a late model V-6 engine, find which piston pin is knocking with a stethoscope. Then, pull

the pan and cylinder head on the correct side of the block. Drive out that piston. Replace the piston and pin. Then reassemble the engine. This would save time over pulling the engine out of the car and servicing other parts.

A few of the engine repairs that normally require engine removal are the following:

1. CRANKSHAFT SERVICE because of wear or damage requires you to pull the engine out of the car. However, if the connecting rods and other parts are NOT damaged, you do NOT have to remove the cylinder heads, valve covers, rods, and other engine components, Fig. 24-7.

   By removing the pan and unbolting all of the rod caps, the crank will normally lift out of the bottom of the block for repairs or replacement. This saves hours of work!

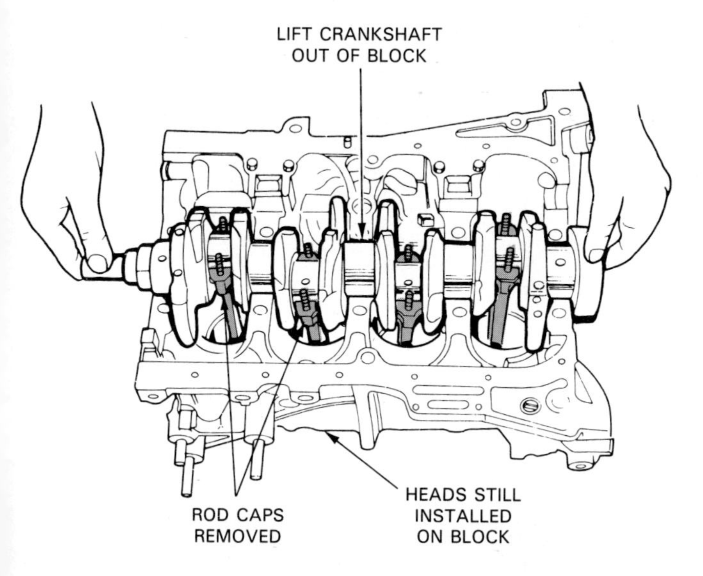

Fig. 24-7. Engine removal from car is needed for crankshaft service. However, you do not always have to remove cylinder heads and other top end parts. Remove pan and unbolt all rod caps. This will let you lift crank out of block. If rod is damaged, you will have to remove head to service rod. (Honda)

2. CYLINDER BLOCK WEAR or DAMAGE is another out-of-car operation. Seldom can any repairs be done to the block unless major parts are removed and the block is sent to a machine shop.

Remember that there are exceptions to almost any rule! Machines are available for turning crankshaft journals in-car but they are not common. Some engines must be lifted halfway out of the car just to change spark plugs. This makes it important for you to refer to a manual when in doubt.

ENGINE OVERHAUL

The term *engine overhaul* or *engine rebuild* is understood to mean:

1. Complete disassembly of the engine.
2. Every part is cleaned, inspected, and measured (if needed).
3. Worn or damaged parts are replaced or serviced.
4. Then the engine is reassembled.

Basically, the following parts are replaced or reconditioned during a major engine rebuild:

1. Piston rings and sometimes the pistons (if worn or damaged).
2. Rod, main, and camshaft bearings.
3. Oil pump.
4. Timing chain, gears, or belt.
5. Core plugs.
6. Camshaft and lifters (if worn).
7. Valve train components, especially valve springs (if needed).
8. Valve guides or seats.
9. All gaskets and seals are replaced.
10. Other parts (if not within specs).

As you will learn later, sometimes the block must be bored and oversize pistons installed. You may need to have the crankshaft turned and install undersize main and rod bearings. Cam bearings could be worn and lowering oil pressure. Later study will make this more clear to you.

In-car engine overhaul

An *in-car engine overhaul* is a major engine rebuild with the cylinder block still bolted to the chassis. It can only be done when cylinder wall and crankshaft wear is within specs and the oil pan and heads can be easily removed in-car. The general condition of the crank can be determined by measuring oil pressure, listening for bearing knock, and/or inspection.

Generally, remove the top end from the engine. Check cylinder wear to make sure the cylinders will take new rings without boring. Remove the oil pan and rod caps. Drive the pistons out the top of the block. Hone the cylinders and replace the main bearings. Then, reinstall the pistons and rods with new rings and bearings in place. Install the new oil pump, reconditioned heads, and other engine parts.

Note! An in-car engine overhaul is only efficient when there is a large engine compartment and plenty of room to work under the engine. A pickup truck is a good example. In most cases, it is just as easy to pull the engine during major service.

PREPARING FOR ENGINE REMOVAL

To prepare for engine removal, most technicians disconnect everything under the engine first: motor mounts, bell housing bolts, torque converter bolts, exhaust system. See Fig. 24-8.

Disconnecting parts under the engine is done first so oil, fuel, and coolant are NOT leaking out and dripping in your face because of top end disassembly. Also, you can still crank the engine over with a remote starter switch to remove the torque converter bolts (automatic transmission or transaxle). Then, you can drain the fluids, remove the starter, motor mounts, and disconnect everything on top of the engine.

Note! Depending upon whether the car is rear or front-wheel drive, you must modify your procedures for removing an engine.

Basically, follow these steps to get ready for engine removal:

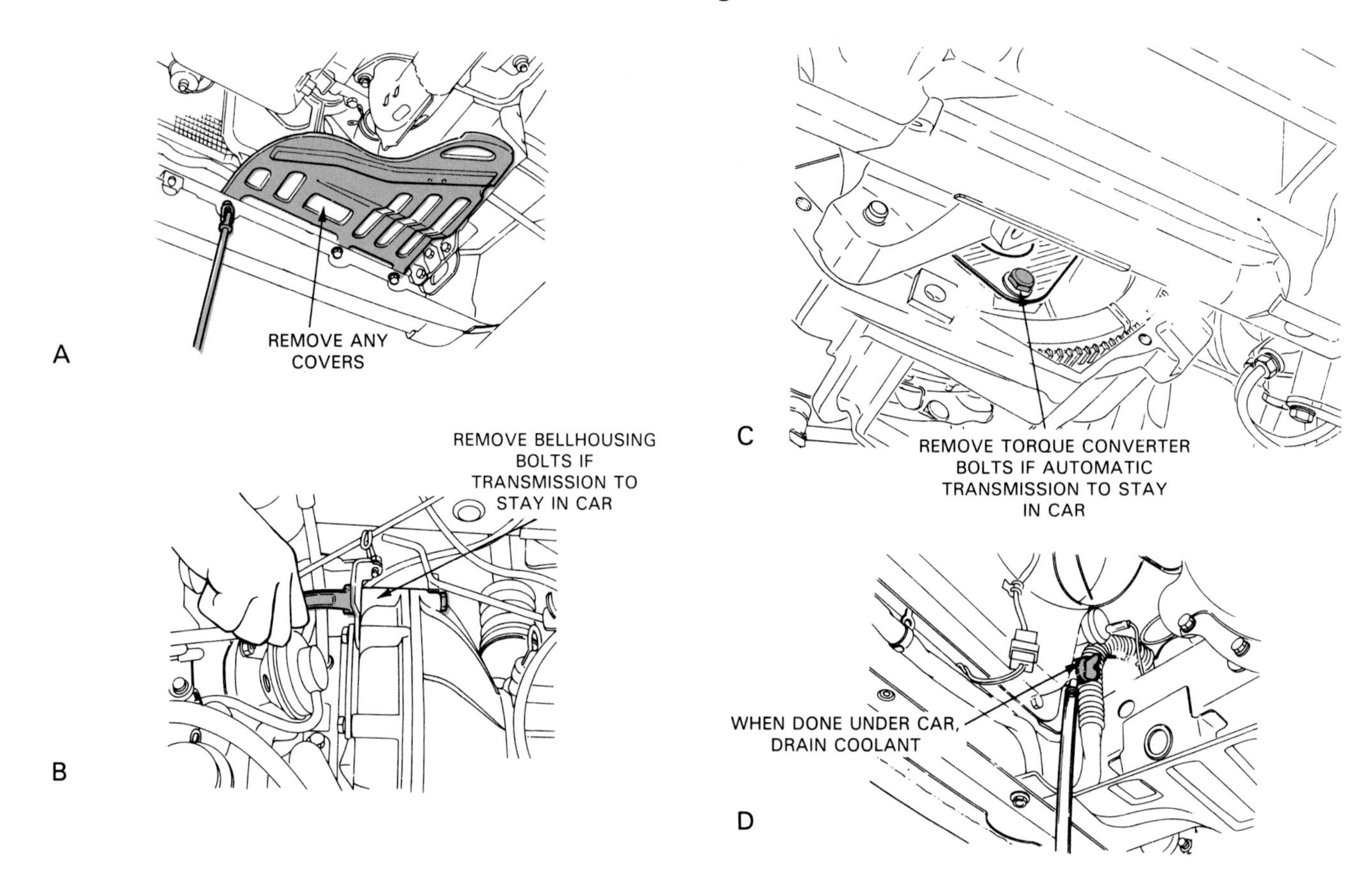

Fig. 24-8. Most technicians begin engine removal under car. A few of the parts that must be disconnected include the following: A—Remove any air shrouds or shields that are in the way. B—Unbolt bell housing if transmission or transaxle is to remain in car. C—Also, unbolt flex plate or flywheel if automatic is to remain in car. D—Drain coolant and oil last so you do not get dripped on by fluids. (Subaru and Peugeot)

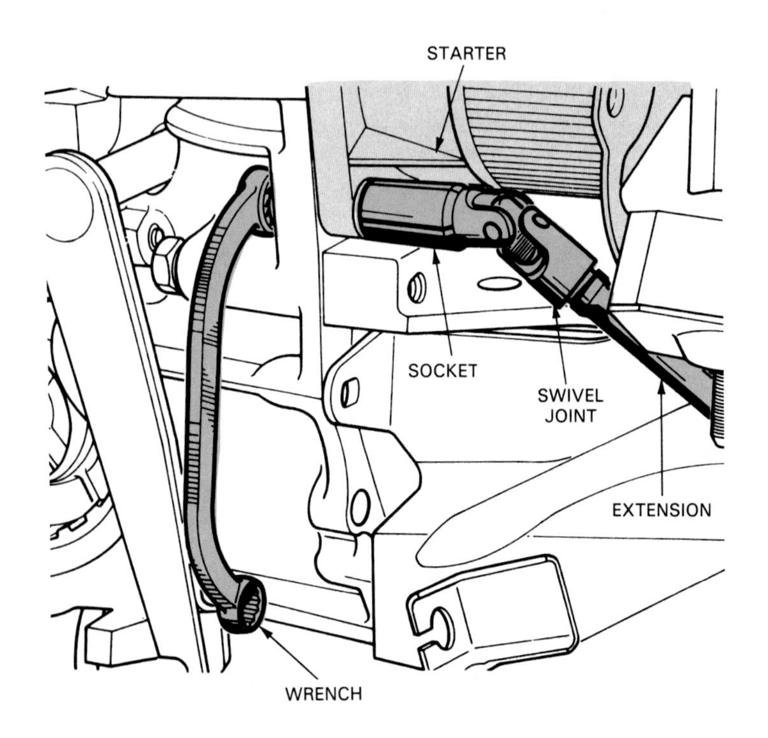

Fig. 24-9. These two wrenches can be handy in tight quarters. Long extension and swivel will let you reach hidden bolts. Bent box wrench will also reach difficult nuts or bolts.

1. Park the car so there is plenty of work space on both sides and in front of the engine compartment.
2. Place the car on stands and unfasten everything preventing engine removal UNDER THE CAR.

   See Figs. 24-9 and 10. The exhaust system can be rusted and the header pipes can be difficult to unbolt. Use rust penetrant and a six-point socket if the bolts or nuts are hard to unscrew.
3. Drain the engine oil and coolant.
4. Lower the car and place covers over the fenders to protect the paint.
5. Scribe lines around the hood hinges to aid realignment. Then have someone help you remove the hood. Store the hood in a safe place.
6. Disconnect the battery to prevent electrical shorts. Remove the battery if in the way, Fig. 24-11.
7. Disconnect or remove the components on the top of the engine: air cleaner, throttle linkage, distributor, injectors. Keep all fasteners in a separate container.
8. Unplug all electrical wires between the engine and chassis. If needed, use masking tape to label or identify the wires. This will simplify reconnection, Figs. 24-12 and 13.
9. Remove all coolant, fuel, and vacuum hoses or lines that prevent engine removal. Label vacuum hoses if confusing. If needed, bleed fuel pressure before

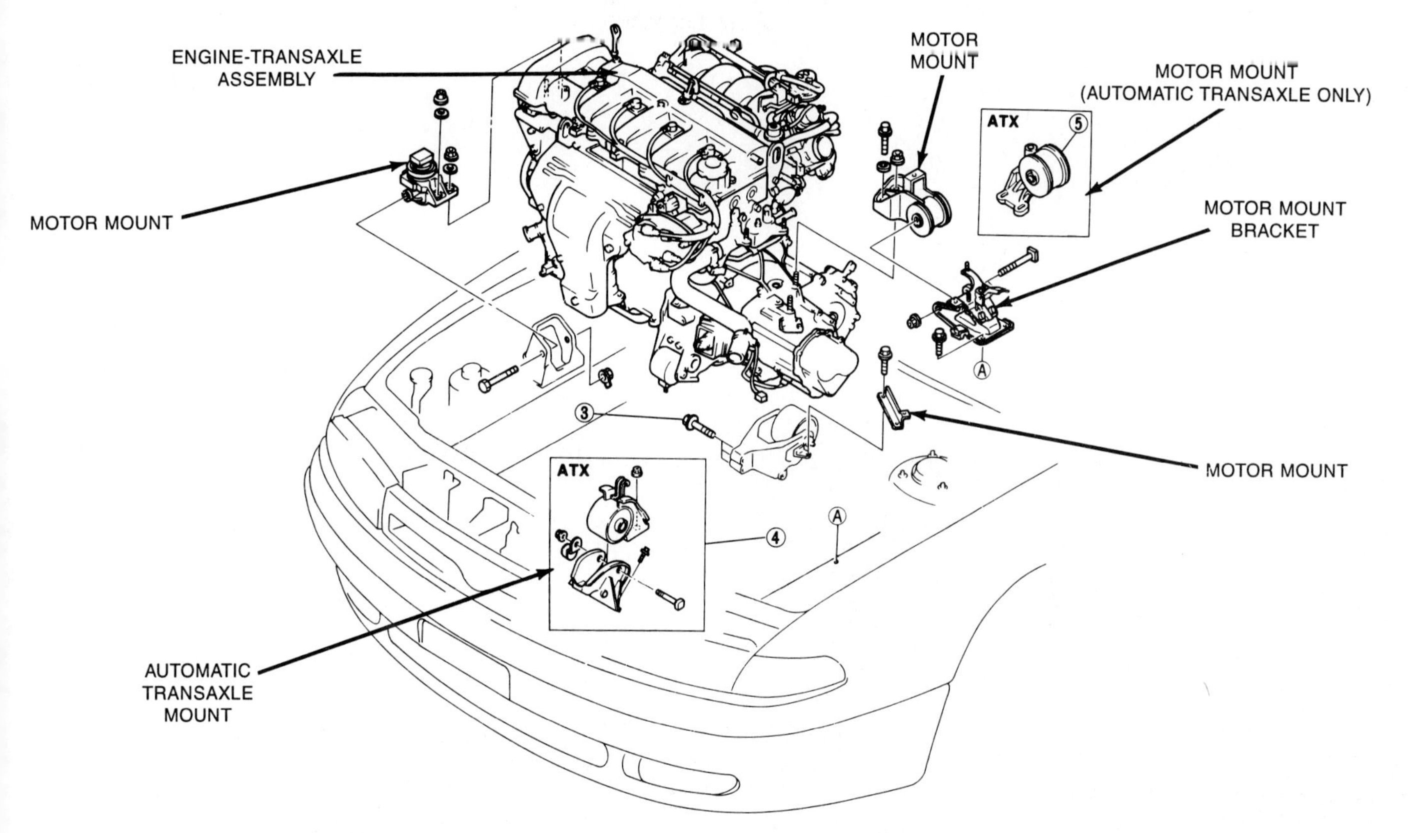

Fig. 24-10. Engine mount designs vary. Sometimes through bolts are used to secure the engine to the frame. Refer to service manual for specific directions when in doubt. (Mazda Motor Co.)

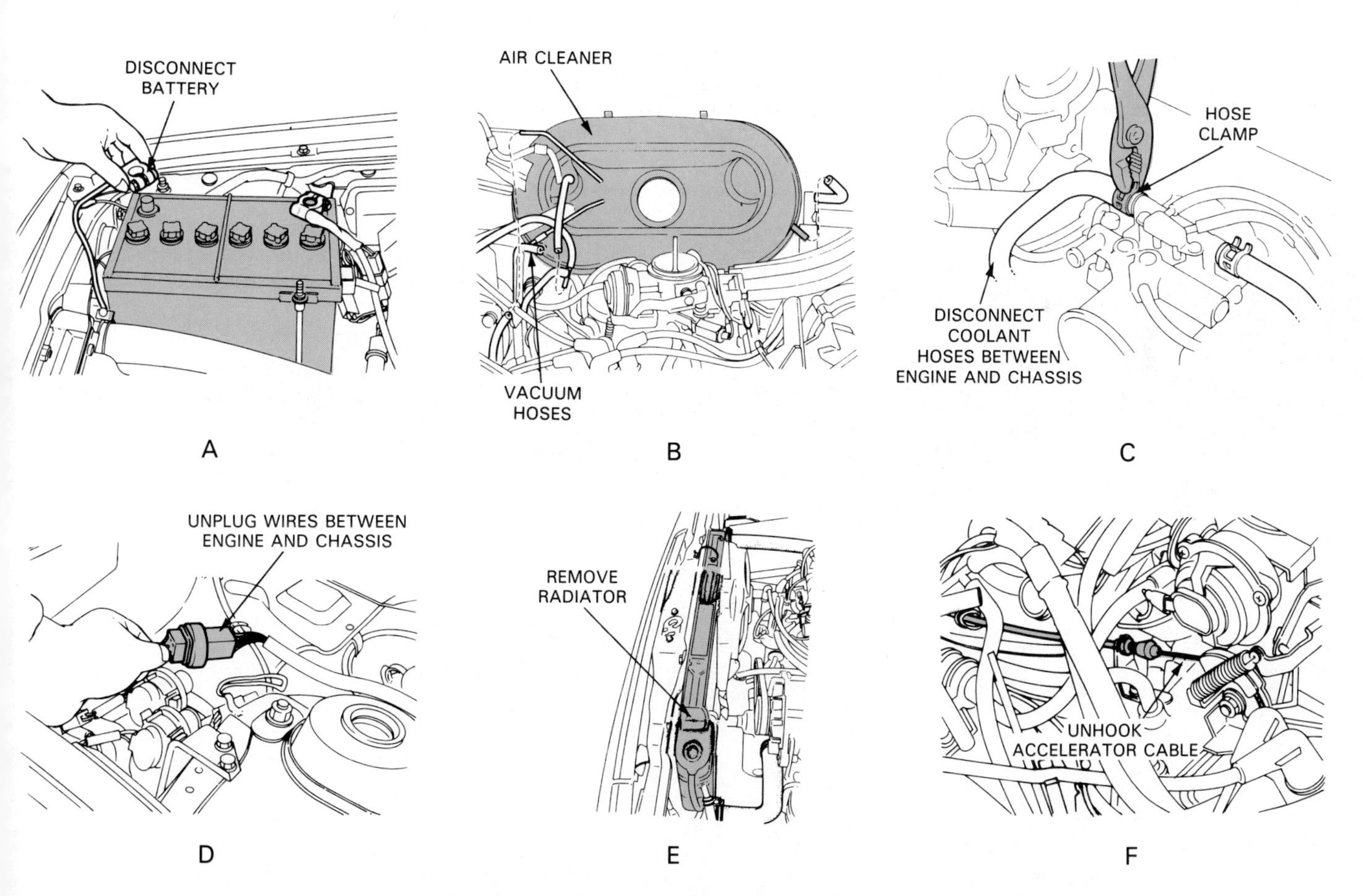

Fig. 24-11. Major components on top of engine compartment that must be disconnected. A—Disconnect battery to prevent electrical fire. B—Air cleaner and vacuum hoses. C—Remove hose clamps and hoses as needed. D—Unplug wiring harnesses and wires preventing engine removal. E—You must usually remove radiator. F—Throttle cable is normally disconnected at throttle body or carburetor. (Subaru)

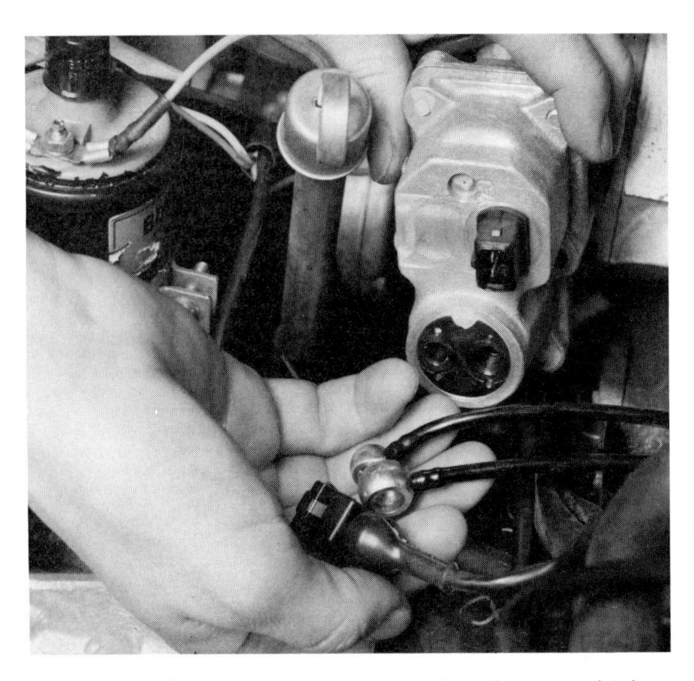

Fig. 24-12. Make sure you remember how hoses and wires reconnect. If you forget, performance problems can result after reinstalling the engine. Label hoses and wires when needed. (Saab)

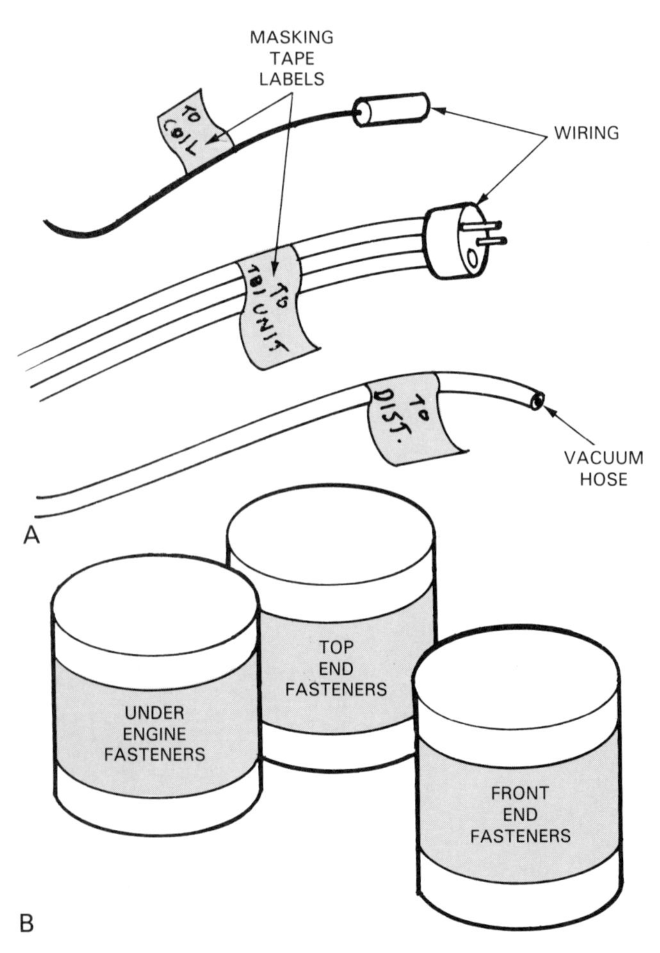

Fig. 24-13. A—Masking tape can be made into labels for hoses and wires. B—Use several containers to keep bolts, nuts, brackets, and other parts organized. Then, when reinstalling the engine, you can go to each can when working on one area of engine compartment.

disconnecting fuel lines, Fig. 24-14.

10. Do NOT disconnect any power steering or air conditioning lines unless absolutely needed. Usually, the power steering pump or air conditioning compressor can be unbolted and placed on one side of the engine compartment. This will save fluid and prevent loss of refrigerant.
11. Remove the radiator, fan, and other accessory units in front of the engine. Be careful not to hit or drop the radiator.
12. Again, keep fasteners organized in several different containers. For instance, keep all of the bolts and nuts from the front of the engine in one container. Keep engine top end and bottom end fasteners separate in two more containers. This will speed reassembly, Fig. 24-13.
13. Remove any other part that prevents engine removal. Check behind the engine for hidden ground wires, dipstick tube, etc., Fig. 24-14.

REMOVING TRANSMISSION WITH ENGINE

It is sometimes necessary to remove the engine and transmission together. Some front-wheel drive cars, using a transaxle, require that the two be removed as a unit. It may be easier to remove the two together. Check a manual for details.

You may want to drain the fluid from the transmission or transaxle if it is to be removed. With rear-wheel drive, the drive shaft, transmission and clutch linkage, speedometer cable, rear motor mount, and other parts must also be removed. With a transaxle, the axle shafts must be disconnected.

ENGINE REMOVAL

Before engine removal, DOUBLE-CHECK that everything is disconnected or removed. For example, check:

1. Behind and under engine for hidden wires or ground straps.
2. That all bell housing bolts are out.
3. That you have removed the torque converter bolts (automatic transmission/transaxle to stay in car).
4. That all fuel lines are disconnected and plugged.
5. That motor mounts are unbolted.
6. That a floor jack or holding bar is supporting the transmission.

Installing lifting fixture or chain

Connect the lifting fixture or chain to the engine. Position the fixture at recommended lifting points. Sometimes, brackets are on the engine, Fig. 24-15.

If a lifting chain is to be used, fasten it to the engine on opposite sides and ends. If a single length of chain is used, install a bolt, nut, and washer on the chain to keep it from slipping and dropping the engine. Be careful!

If bolts are used to secure the chain to the engine, make sure they are large enough in diameter and that they are fully installed. The bolts must not be too long (stick out from chain) or too short (must thread into hole distance equal to one and one-half diameter of

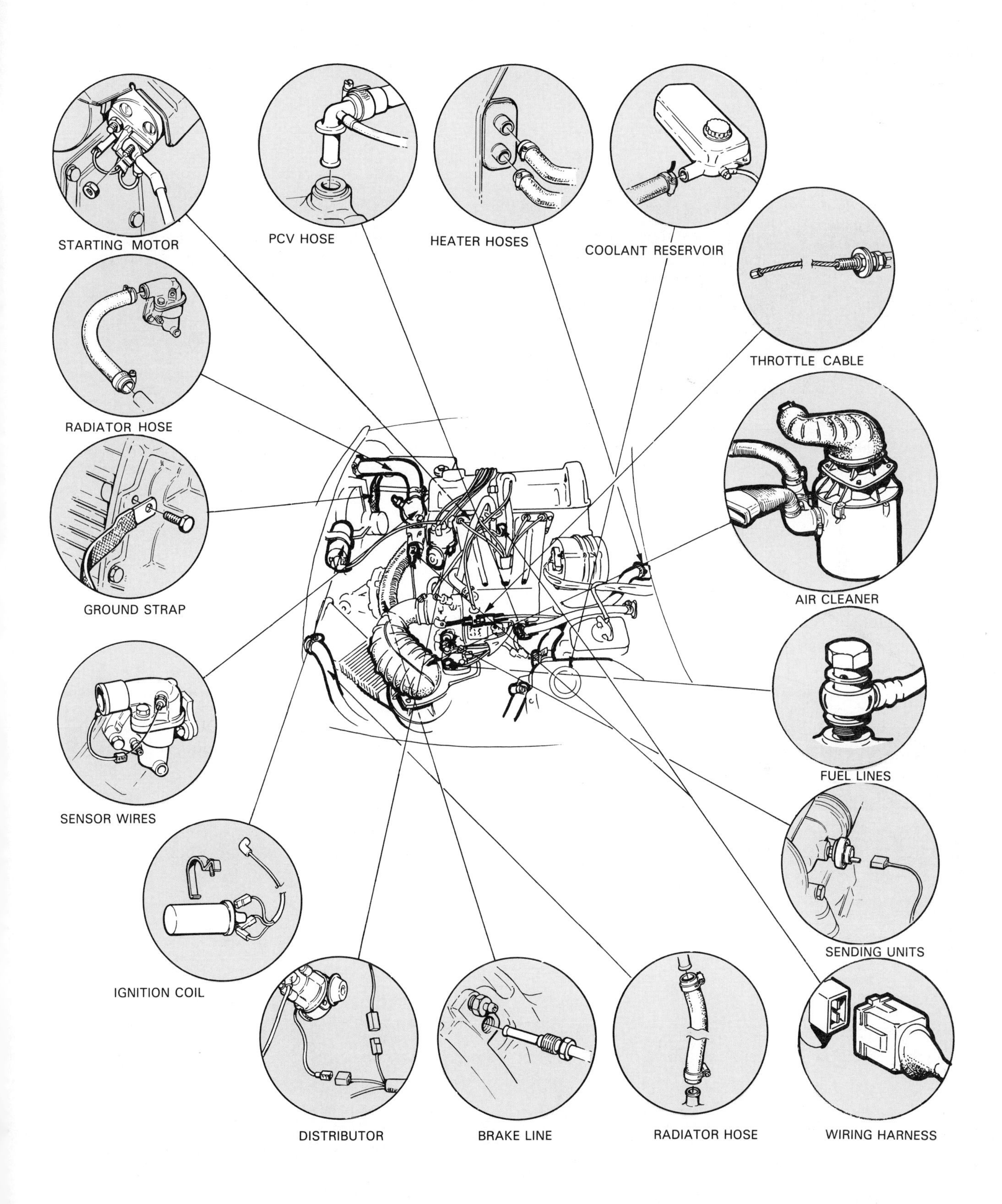

Fig. 24-14. This is service manual illustration showing many of the parts that must be disconnected before engine removal. (Saab)

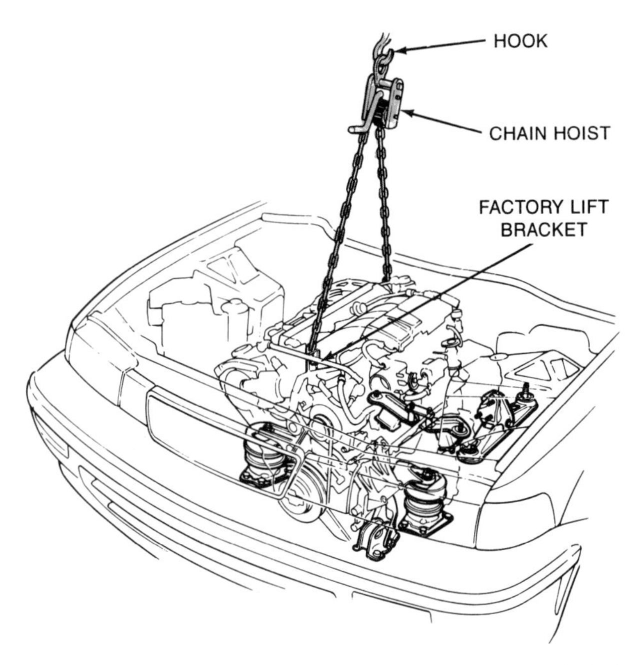

Fig. 24-15. When lifting an engine, use a secure chain hoist, preferably connected to factory lift brackets on the engine. (Honda)

bolt). This is illustrated in Fig. 24-16.

Generally, position the fixture or chain so that it will raise the engine in a level manner or at the correct angle if needed. If one lifting point is at the right-front of the cylinder head, the other should be on the left-rear of the head. Use common sense and follow manufacturer's instructions, Fig. 24-17.

CAUTION: Avoid damage to motor mounts during engine removal. Some engine motor mounts are liquid-filled or computer-controlled, hydraulic units to reduce engine noise. The computer can alter the stiffness of the mounts by using solenoid valves that control hydraulic flow in the mounts. If you damage any of these mounts, it can be an expensive mistake.

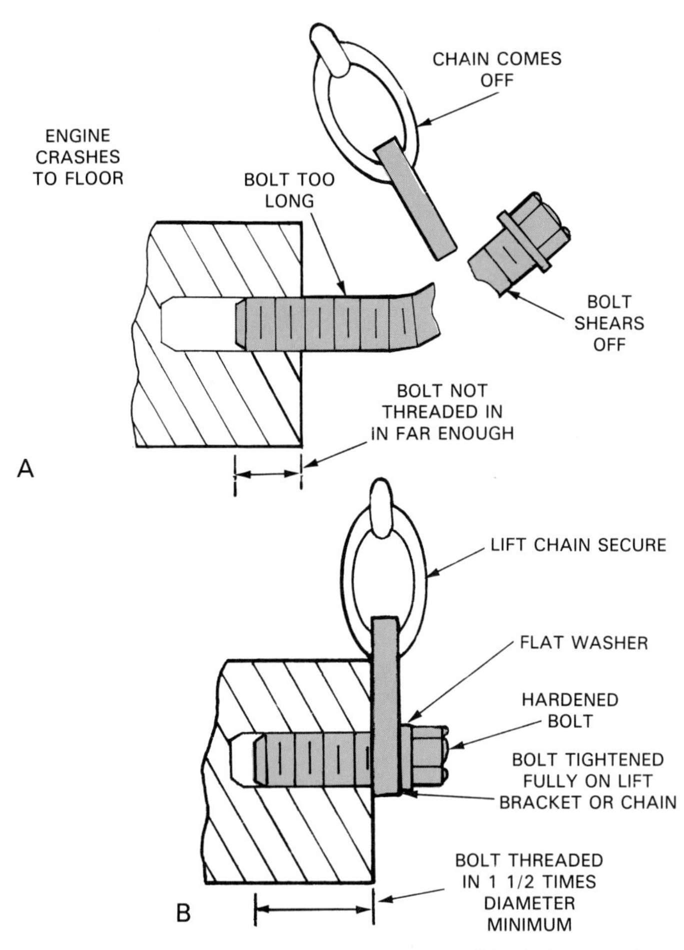

Fig. 24-16. Make sure bolts used to fasten lift chain to engine are large enough in diameter. Bolt should be fully tightened against chain and should thread into part 1 1/2 times its diameter. A—Bolt too long was bent and snapped off. Engine crashed to floor causing serious part damage. B—Bolt correctly installed. It is tightened down and threaded in far enough.

Lifting engine out of car

Attach the lifting device (crane or hoist) to the fixture or chain on the engine. Fig. 24-18 shows the use of a portable crane. Make sure the crane boom or hoist is

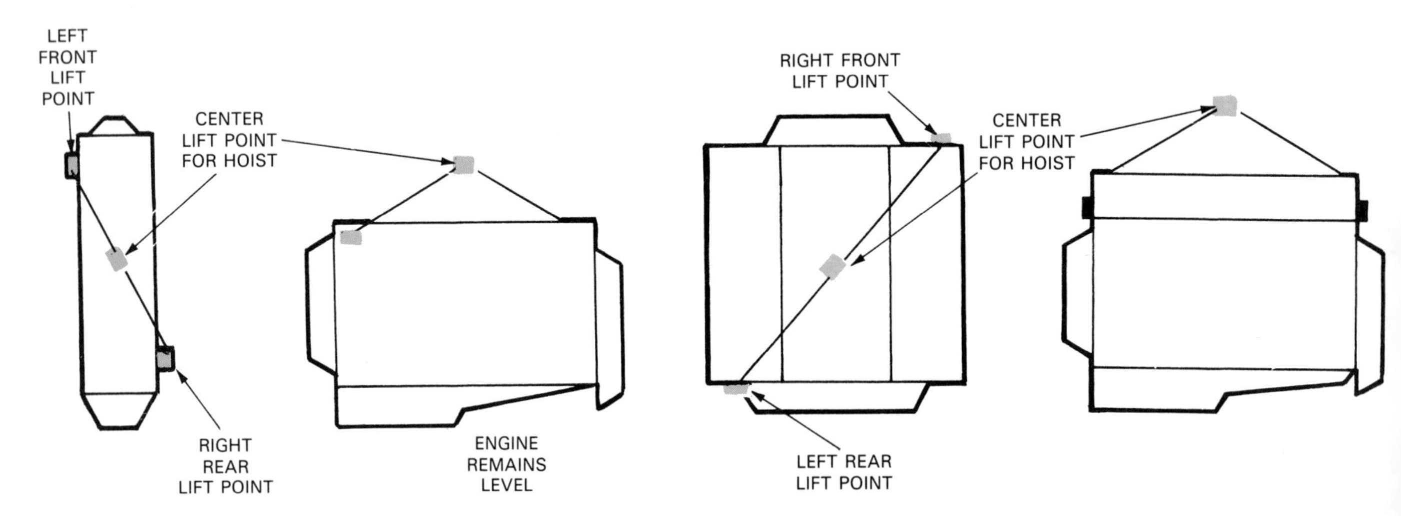

Fig. 24-17. Lift points should be on opposite sides and ends of engine. A center lift point will raise engine in level manner. If you need to tilt engine, as when removing engine and transmission together, move center lift point one way or other.

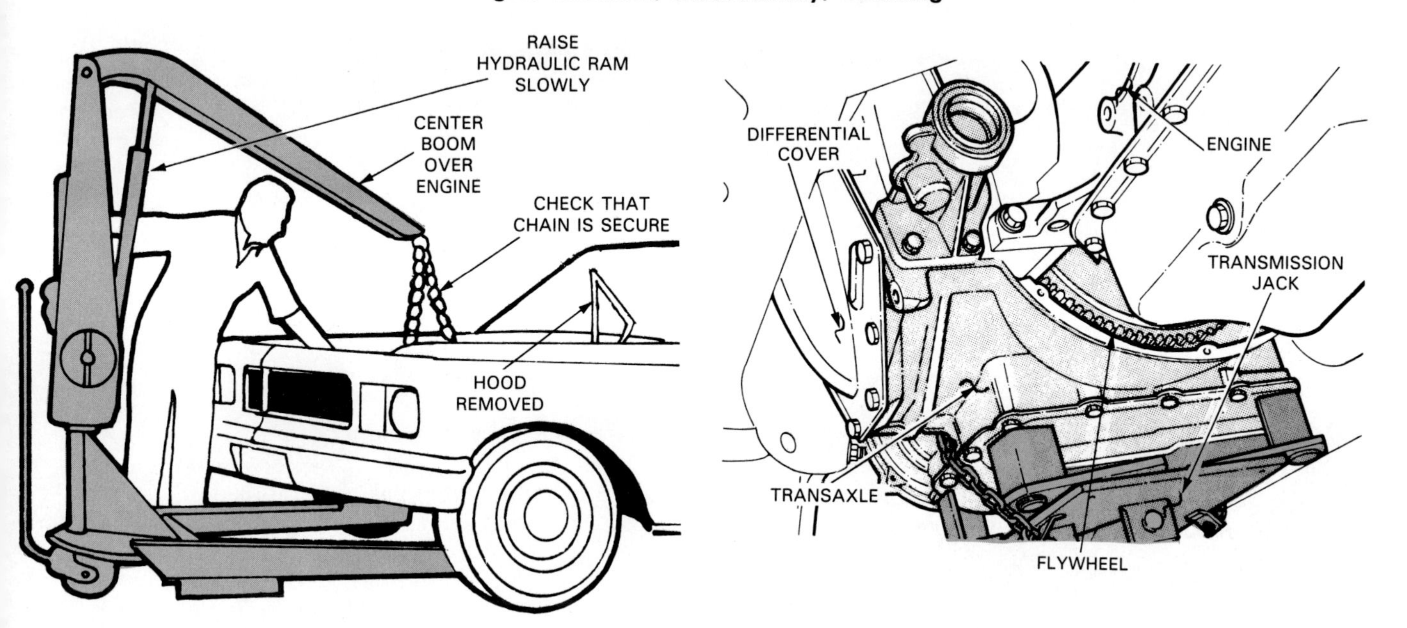

Fig. 24-18. Crane boom should be centered directly over lift point to keep from shifting and binding engine. Raise slowly while watching for clearance all around engine. (Florida Dept. of Voc. Ed.)

Fig. 24-19. If transmission or transaxle is to stay in car, support transmission or transaxle. Do not let it hang unsupported! Here technician has placed transmission jack under transaxle. (Chrysler)

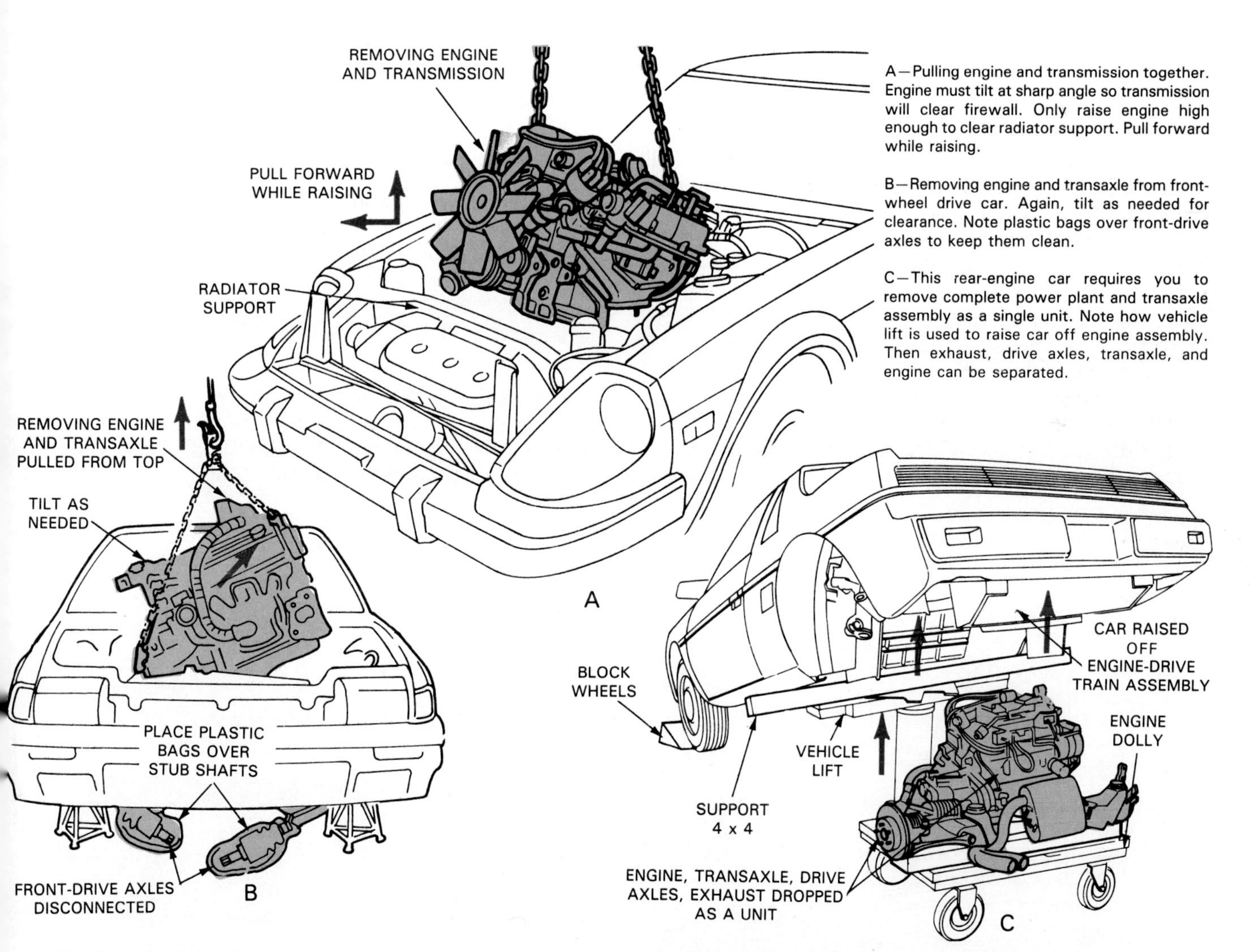

Fig. 24-20. Typical methods for removing engine or engine and transmission. Methods vary with design of vehicle. (Nissan, Honda, Pontiac)

centered over the engine. Place a floor jack under the transmission if needed, Fig. 24-19.

Raise the engine slowly about an inch or two. Then check that everything is out of the way and disconnected, Fig. 24-20A.

DANGER! Never place any part of your body under an engine held in the air. A heavy engine can chop off fingers, crush hands or feet, cripple, or even kill when dropped!

Continue slowly raising the engine while pulling. This will separate the engine from the transmission or slide the transmission out from under the firewall. Do not let the engine bind or damage parts.

When the engine is high enough to clear the radiator support, roll the crane and engine straight out and away from the car. With a stationary hoist, roll the car out from under the engine. Look at Fig. 24-20B.

Do NOT let a transmission or transaxle hang unsupported after engine removal. This could damage the mounts, drive shaft, drive axles, and other parts.

Note! With some front-wheel drive cars, the engine and transaxle are removed from underneath the car, Fig. 24-20C. With some vans, the engine can be removed through the large door in the side of the body. Again, check a service manual for details.

As soon as you can, lower the engine to the ground or mount it on an engine stand.

DANGER! NEVER work on an engine that is held by a crane or hoist. The engine could shift and fall, damaging the engine or causing serious injury!

Keep your work area clean

Coolant and engine oil will usually drip onto the shop floor during engine removal. To prevent an accident, wipe up spills as soon as they occur. There is nothing professional about trying to work in a "grease pit."

Separating engine from transmission/transaxle

If you pulled the engine and transmission/transaxle together, you will need to separate them on the shop floor. Place blocks of wood under the engine or leave the hoist and chain on the engine to keep it from flipping over while sitting on the ground.

With an automatic, remove the bell housing-to-engine bolts. Check closely all the way around the engine because some can be hard to find. Remove the starting motor and any other parts preventing transmission removal.

Some technicians prefer to leave the torque converter bolted to the flywheel. Others unbolt the converter so it will come out with the transmission. Lift up, wiggle, and pull the transmission or transaxle off the back of the engine.

With a manual transmission or transaxle, follow about the same procedures. However, you must unbolt the transmission first. Then remove the bell housing, clutch, and flywheel.

There are many design variations with rear and front-wheel drive. If you are not sure about something, check the manufacturer's manual.

WARNING: Manual clutches contain asbestos, which can cause cancer. Wear appropriate respiratory protection.

ENGINE DISASSEMBLY

With the engine bolted to an engine stand or sitting on blocks, you are ready to begin teardown. Go slowly and inspect each part for signs of trouble. Look for wear, cracks, damage, seal leakage, or gasket leakage.

Remember! If you overlook one problem, your engine repair may fail in service. All of your work could be for nothing.

Engine teardown methods vary somewhat from engine to engine. However, general procedures are similar and apply to all engines. The following is a guide.

Engine front end disassembly

Engine front end disassembly is simple if a few basic rules and service manual instructions are followed. The engine front end should be removed first, especially with OHC engines.

1. Remove the water pump and any other parts bolted in front of the engine timing cover. If a timing belt is used, remove the belt cover. Loosen the tensioner and slip off the belt. (This would have to be done before cylinder head removal.)
2. NOTE! Do not attempt to rotate the crankshaft of an OHC engine with the timing belt off (cylinder head still in place). The pistons could slide up and bend the valves.
3. A WHEEL PULLER is normally needed to remove the harmonic balancer. The balancer is commonly press-fitted onto the crankshaft. Fig. 24-21 shows how to use a puller when removing a front oil seal.
4. Unbolt and remove the timing chain or gear cover. If prying is necessary, do it lightly while tapping with a rubber or plastic hammer. Do not bend or scar mating surfaces.
5. Remove the old slinger and timing mechanism. Usually, the timing gears or sprockets will slide off after

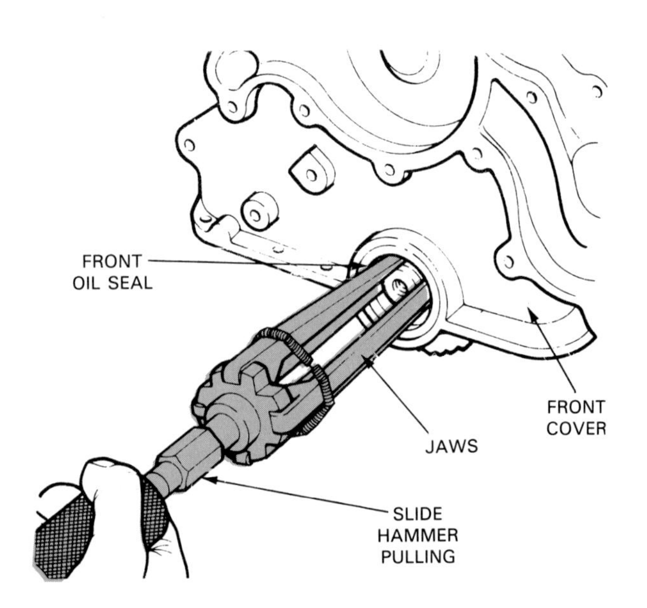

Fig. 24-21. Most technicians start engine teardown with front end parts. This is especially true with OHC engines. Here technician is using slide hammer puller to remove front oil seal from front cover. Note that water pump is already off. (Ford)

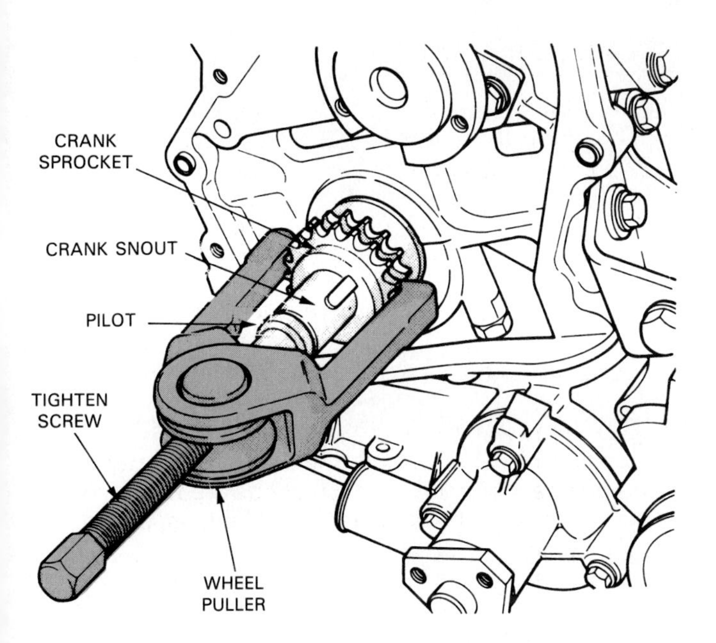

Fig. 24-22. Front damper and crank sprocket can require use of screw-type wheel puller. Tighten screw and jaws will force part off crank snout. (Chrysler)

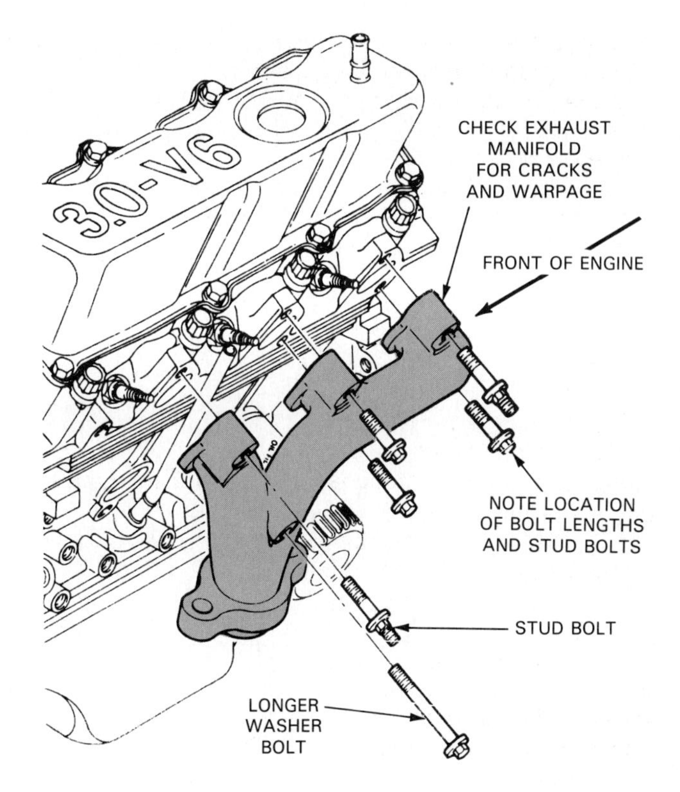

Fig. 24-23. When removing exhaust manifold, note odd bolt lengths and location of stud bolts. They must be installed in same holes. (Ford)

light taps with a brass hammer. If not, use a wheel puller, Fig. 24-22.

6. If the oil pump or other components are in the front cover, slip them out while inspecting for wear.

Engine top end disassembly

The engine top end generally includes the valve train and cylinder head related components. These are normally the second set of parts of the engine to be disassembled.

1. Remove external engine parts (carburetor, fuel rail, or throttle body unit, spark plug wires, distributor, etc.) Take off all parts that could be damaged or that prevent removal of the cylinder head(s).
2. Remove the exhaust manifold(s). Note the location of any stud bolts. They must be replaced in the same hole, Fig. 24-23.
3. Unbolt the valve cover(s). If light prying is needed, be careful not to damage the mating surfaces. Light taps with a rubber hammer will also free a valve cover, Fig. 24-24.
4. Remove the rocker arm assemblies. If rocker shafts are used, loosen the fasteners using the sequence in Fig. 24-25. This will prevent the end fasteners from being bent by spring tension.
5. With V-type, push rod engines, you may need to remove the valve train components before the intake manifold. The push rods can pass through the bottom of the intake. See Fig. 24-26.
6. If the lifters, push rods, and rocker arms are to be reused, keep them in exact order. Use an ORGANIZING TRAY (tray or board with holes in it) or label these parts with masking tape. Wear patterns and select-fit parts require that most components be installed in their original locations.
7. WARNING! Do not unbolt an aluminum cylinder head when it is still hot from engine operation. This could cause head warpage. Allow the head to cool before removal.
8. Unscrew the cylinder head bolts, Fig. 24-27. Most technicians use an air impact or a breaker bar and six-point socket, Fig. 24-28. With a V-type engine, punch mark the heads right and left.
9. Frequently, the head will be stuck to the block. To free the head, insert a breaker bar in one of the ports and pry up. If stuck tight, install the spark plugs or glow plugs and crank the engine with the starting motor. Compression stroke pressure will pop the head up and free it from the block. Avoid prying on mating surfaces!
10. Inspect the head gasket and deck surfaces for signs of leakage. Also look for oil in the combustion chambers, indicating seal or ring problems.

Cylinder head disassembly

Disassembly procedures for a cylinder head can vary for overhead valve and overhead cam heads.

Keep these rules in mind during head teardown:

1. Inspect for signs of trouble before disassembly. Look for oil-fouled combustion chambers indicating bad rings, valve seals, and valve guides. Inspect for a blown head gasket, signs of coolant leakage, rust, and cracks. Close inspection may find other troubles.
2. Use a valve spring compressor to remove the valve keepers, retainers, and spring. Before compressing the springs, strike the valve retainers, not the valve tips, with a brass hammer. This will free them from

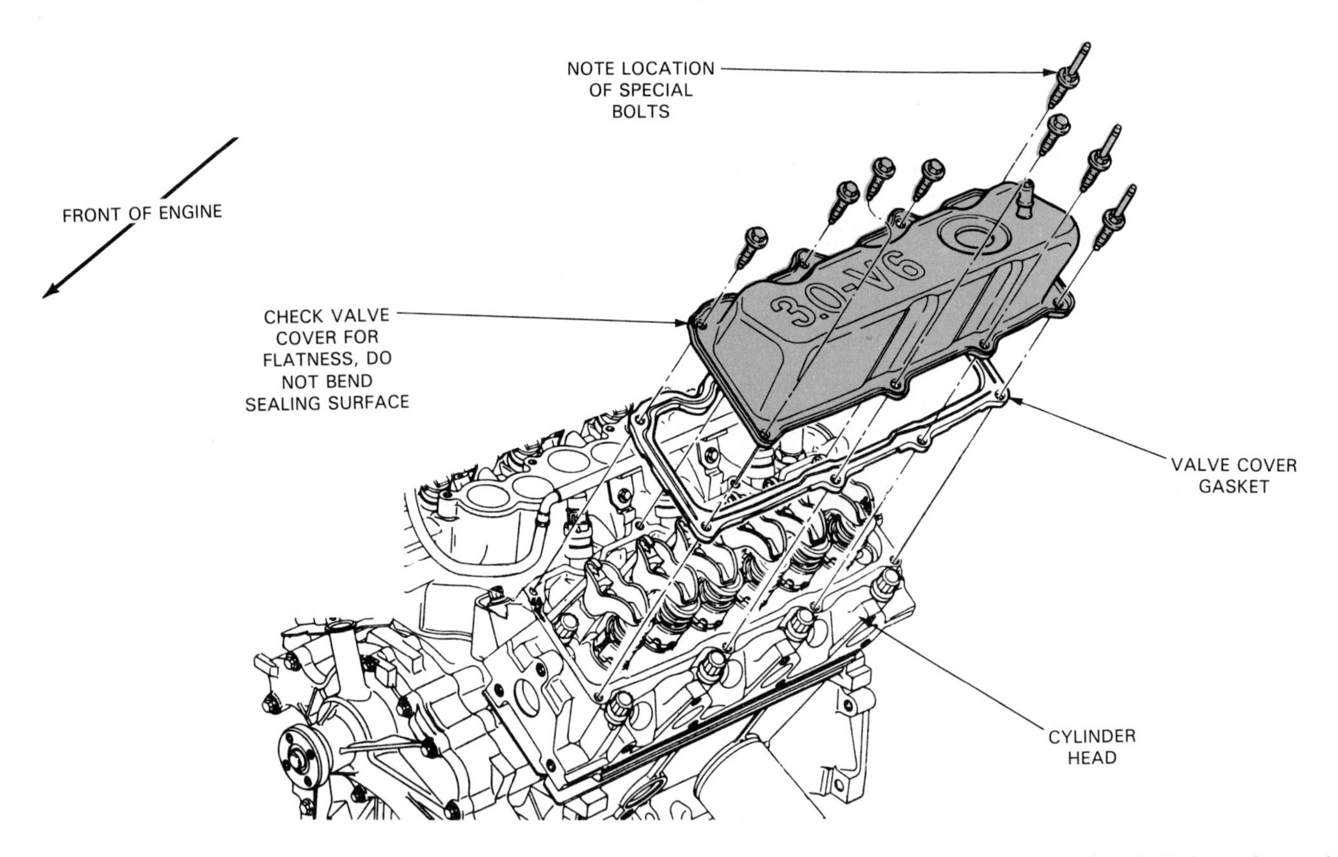

Fig. 24-24. Do not bend valve covers during removal. Rubber mallet will sometimes free cover from head. Light prying under cover may be needed but do not bend lip. Inspect valve train after cover is off.

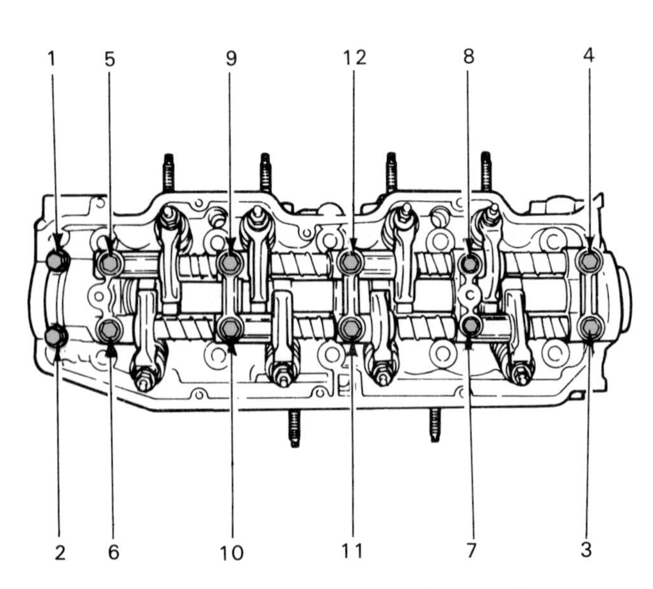

Fig. 24-25. When unbolting rocker shafts, use this sequence. If you start on one end and work across, spring tension could bend rocker shafts or break caps. (General Motors)

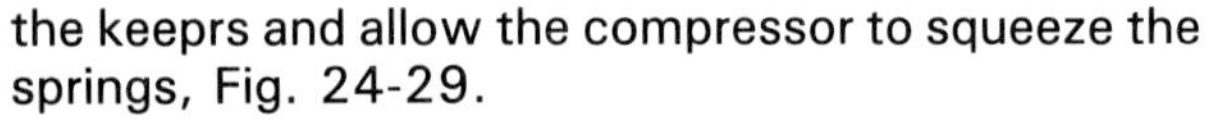
the keeprs and allow the compressor to squeeze the springs, Fig. 24-29.

3. A special valve spring compressor is needed for some overhead cam cylinder heads. One is shown in Fig.

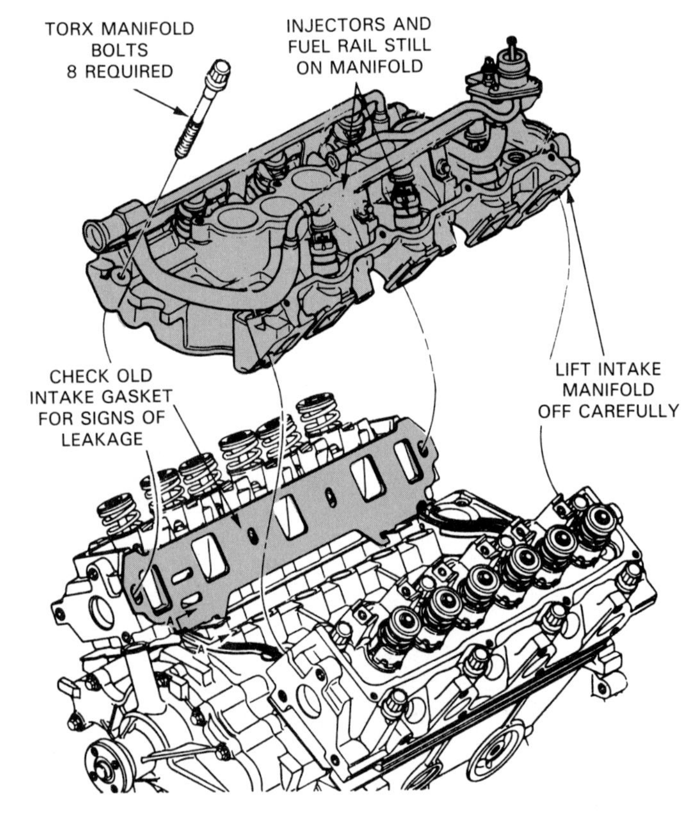

Fig. 24-26. Intake can be removed with injectors and fuel rail attached. Pry up lightly to free manifold and lift off. Inspect old gasket for signs of leakage.

Fig. 24-27. Air impact will save time during engine disassembly. You may need to crack larger bolts loose with breaker bar before using impact.

Fig. 24-29. Valve spring compressor is needed for cylinder head disassembly. Place one end over valve head and other over spring. Pull on lever arm to compress spring. Remove keepers and release compressor to free spring. Wear safety glasses because parts can fly if compressor slips off spring. (Fel-Pro Gaskets)

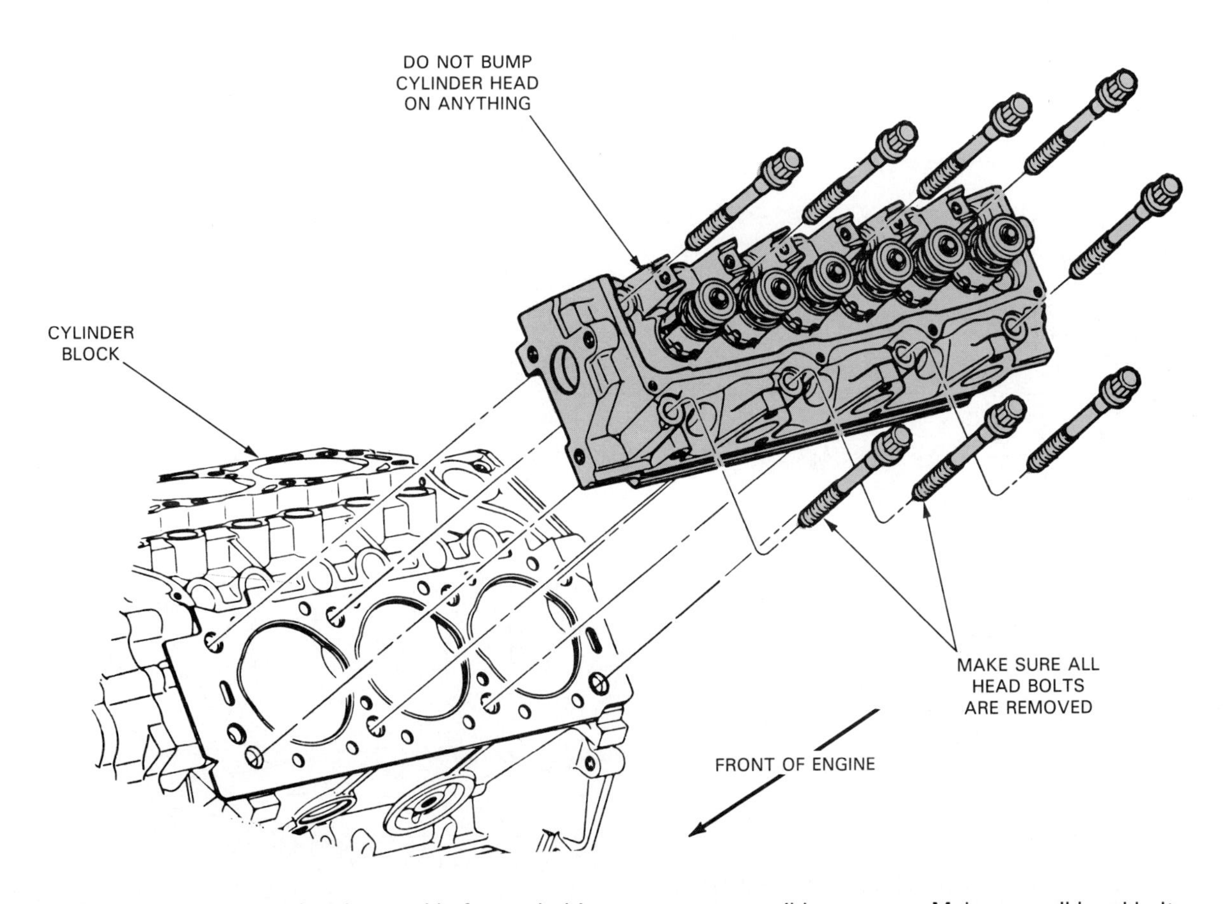

Fig. 24-28. Allow aluminum head to cool before unbolting to prevent possible warpage. Make sure all head bolts are out. Free head by inserting breaker bar in intake port or crank engine and let pressure free head from gasket. Inspect head and gasket for trouble right after removal. (Ford)

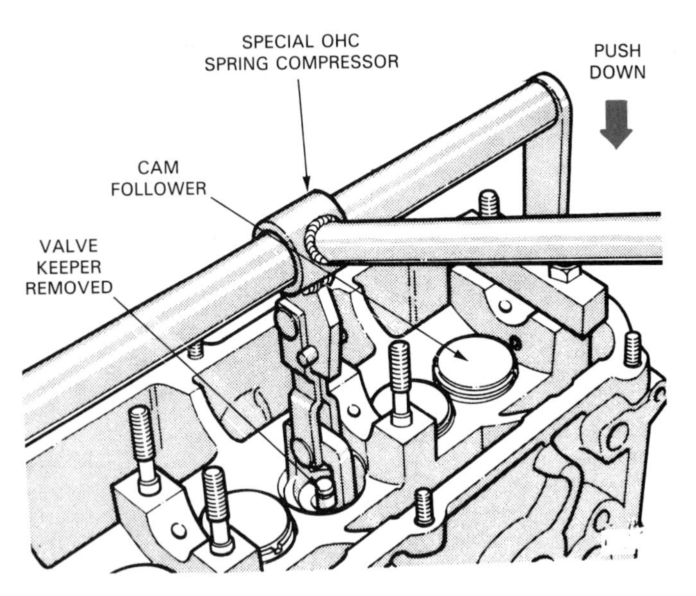

Fig. 24-30. Special spring compressor is needed on some OHC engines. Note how valve springs are recessed inside head. (Chrysler)

24-30. It will reach down inside the tappet bore to compress the springs.

4. Before removing the valves, check their tips for *mushrooming* (tips hammered larger by rocker action). File a chamber on the valve tips if needed. Never try to drive mushroomed valves out of a head. This can score or split the valve guides or head. Look at Fig. 24-31.

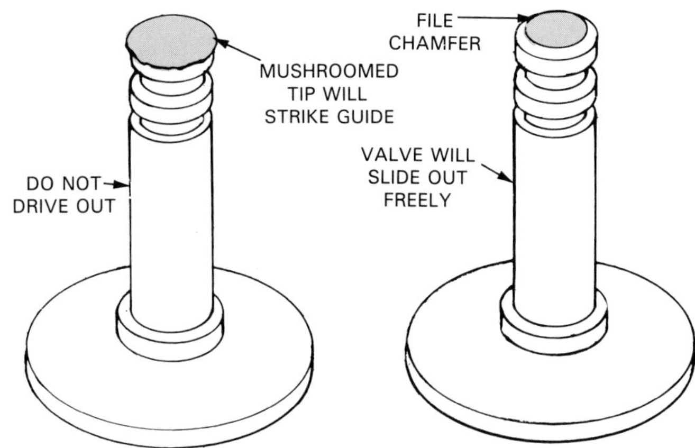

Fig. 24-31. Before removing valves, check them for mushroomed tip. Enlarged tip end must be filed before valve removal to prevent valve guide damage.

5. Keep all parts organized. It is best to return parts to their same locations in the engine. Parts can be

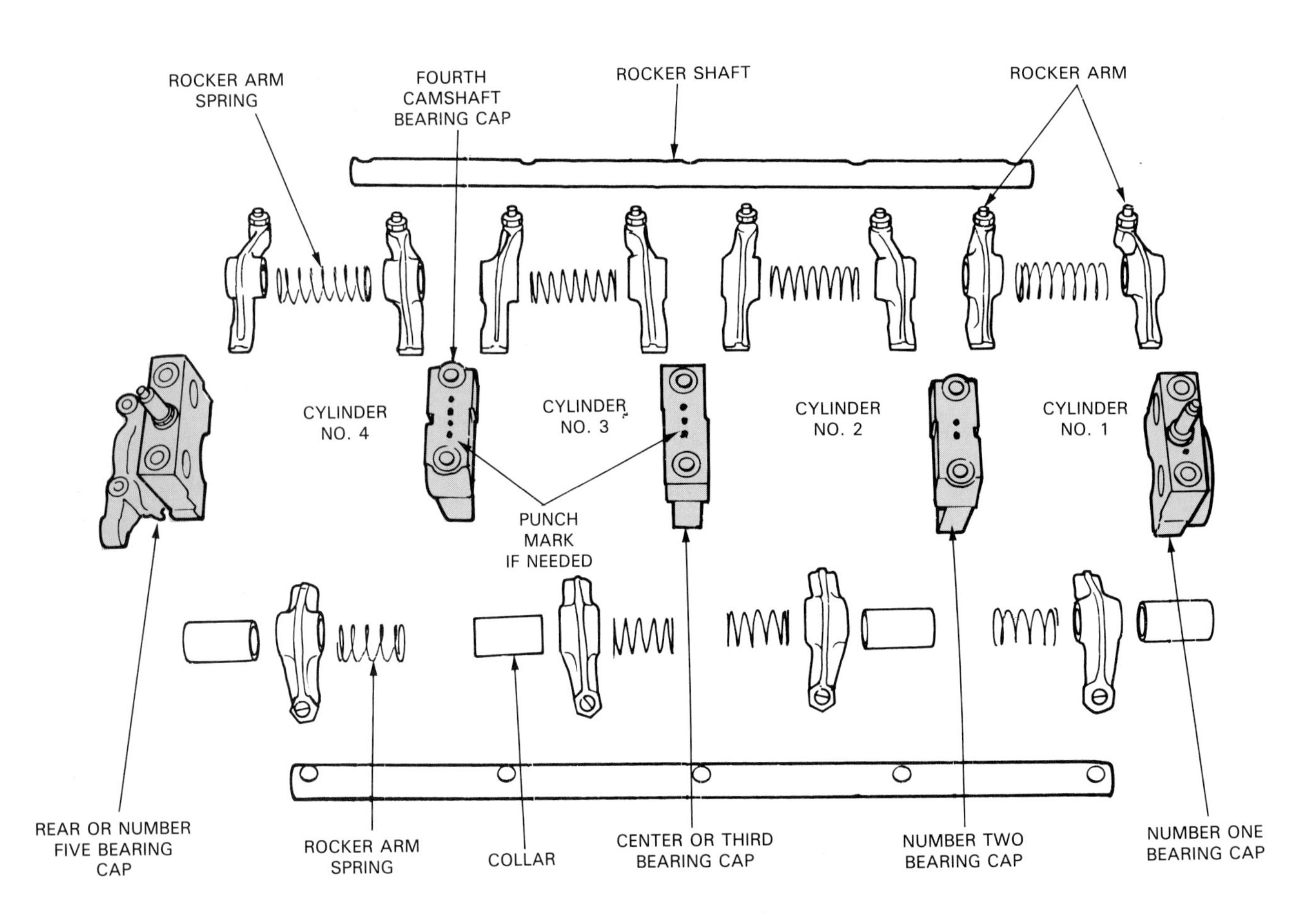

Fig. 24-32. Make sure cylinder head camshaft caps are marked before removal. They cannot be mixed up or cam bore can become out of round, locking cam in head during reassembly. Remove rockers from shafts for inspection. (Honda)

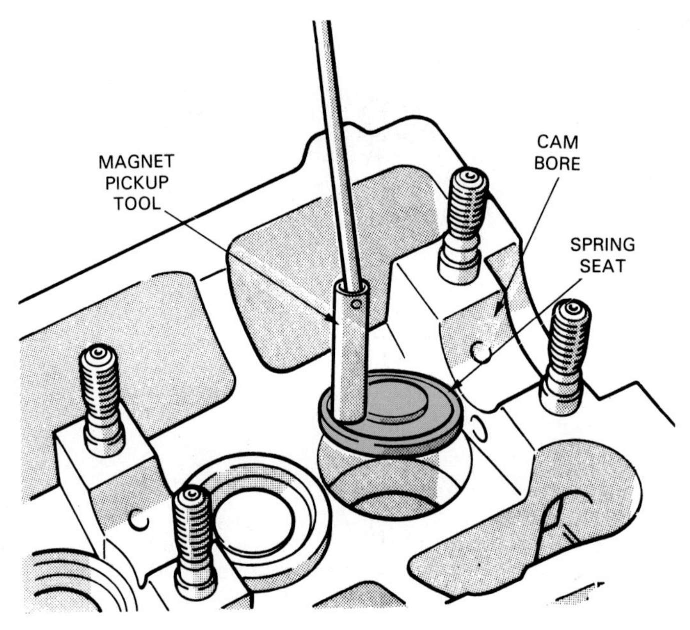

Fig. 24-33. Technician is using magnetic pickup tool to remove spring seat from OHC head. (Chrysler)

select-fit from the factory and their sizes may be matched for more accurate clearances, Fig. 24-32.

6. Remove any other parts that would prevent head service, Fig. 24-33.

Engine bottom end disassembly

After top end and front end disassembly, you are ready to take the bottom end apart. The bottom end typically includes the pistons, rods, crankshaft, and related bearings.

1. Inspect the cylinders for signs of excess wear. Use your fingernail to feel for a lip or ridge at the top of the cylinder wall, Fig. 24-34A.

   A *cylinder* or *ring ridge* may be formed at the top of the cylinder walls, where ring friction does not wear the cylinder.

2. A *ridge reaming tool* is needed to cut out and remove a ridge at the top of a worn cylinder, Fig. 24-34B. Use a wrench to rotate the reamer and cut away the metal lip. Cut until flush with the rest of

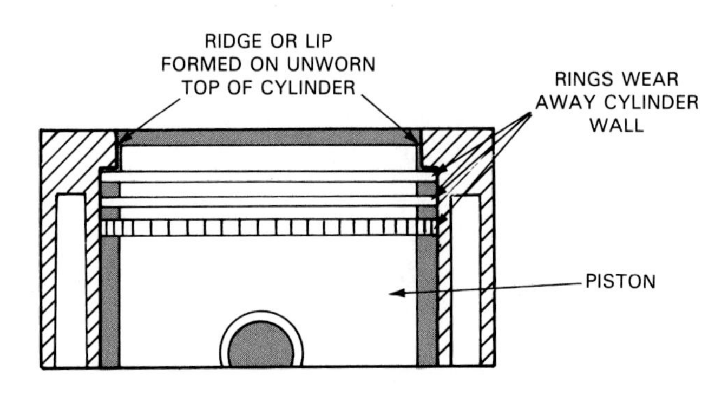

A—Ring ridge is formed at top of cylinder where rings do not rub on and wear cylinder wall. Rings would bind on ridge if you tried to force piston out of cylinder.

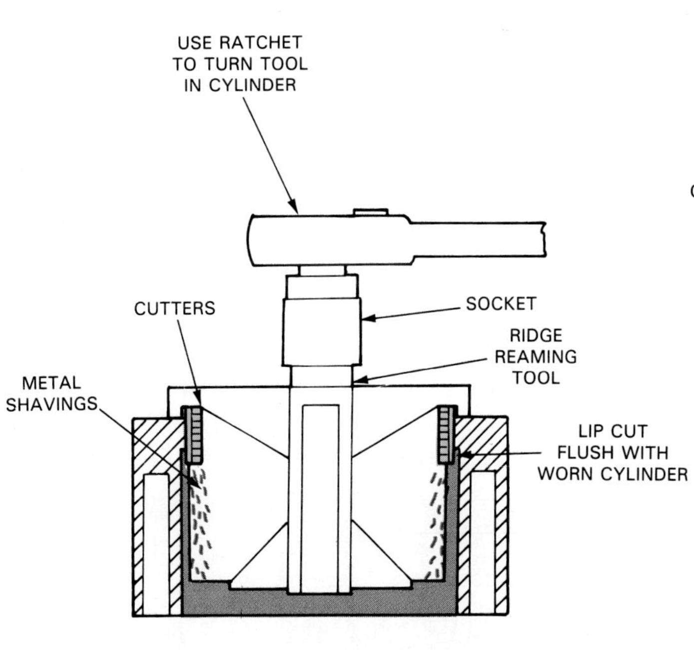

B—Ridge reaming tool will cut off ridge. Only rotate tool until ridge is flush with worn part of cylinder.

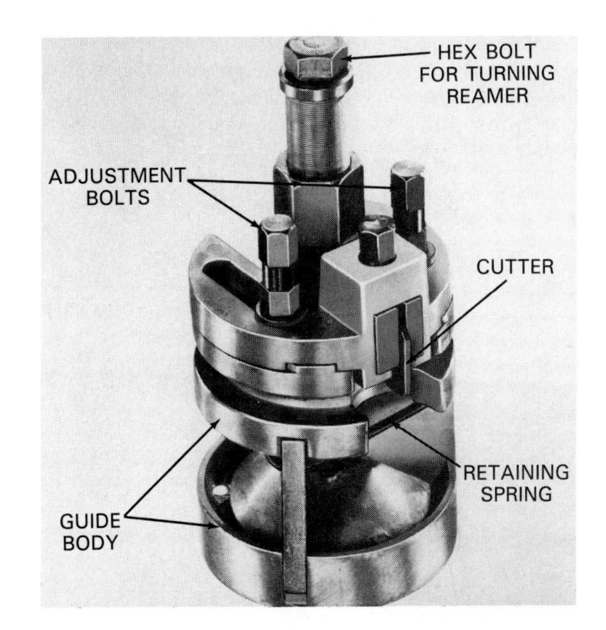

C—Note parts of ridge reamer.

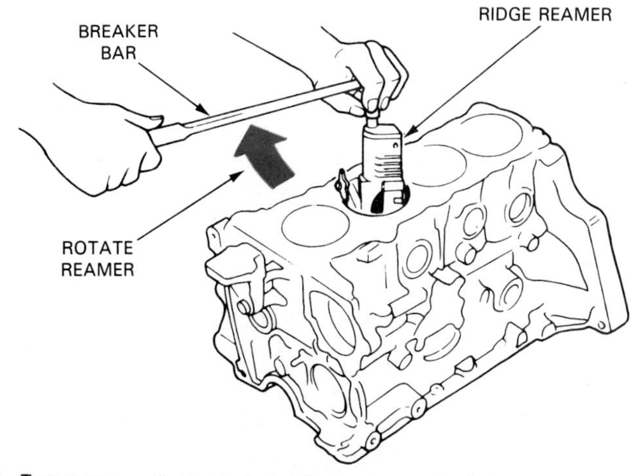

D—Turn reamer until ridge is flush with worn part of cylinder.

Fig. 24-34. Ring ridge must be removed before driving pistons out of block. You can feel ridge with fingernail.

the cylinder wall. This will prevent piston damage during piston removal, Figs. 24-34C and 34D.

3. Use compressed air to blow metal shavings out of the cylinder after ridge reaming. This will prevent cylinder or piston scoring.
4. Unbolt and remove the oil pan and oil pump. Inspect the bottom of the pan for debris. Metal chips and plastic bits may help you diagnose and find engine problems, Fig. 24-35.

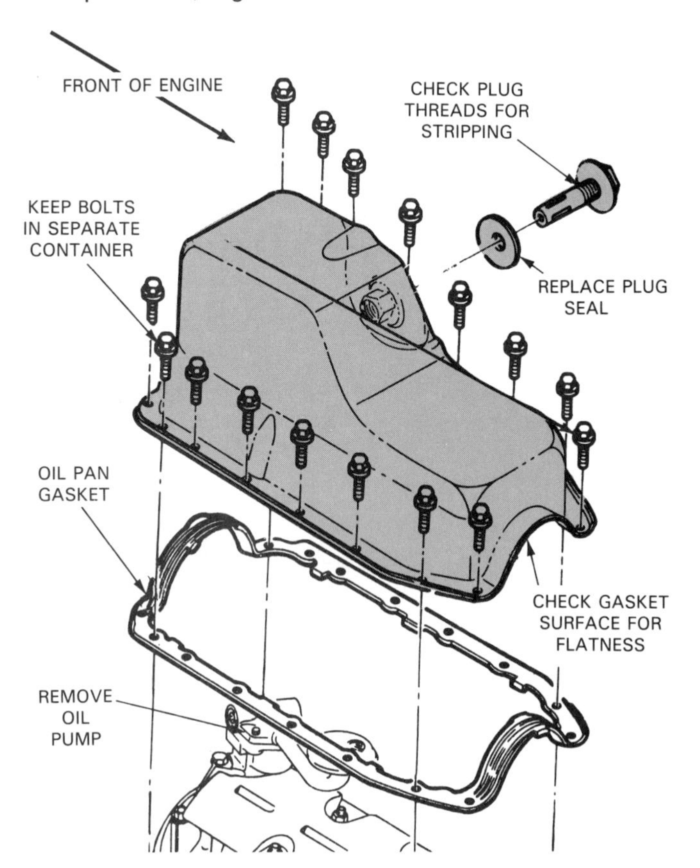

Fig. 24-35. To begin short block disassembly, turn engine over on engine stand and remove oil pan. Make sure you have drained oil first or you will have a mess on shop floor. Check inside of pan for debris that could help find other problems. (Ford)

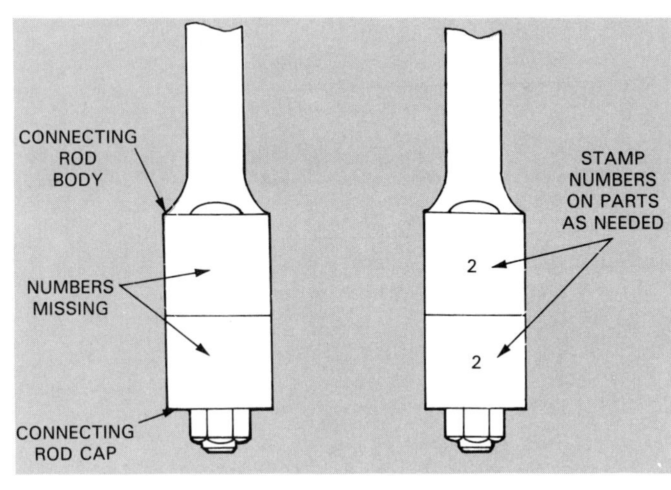

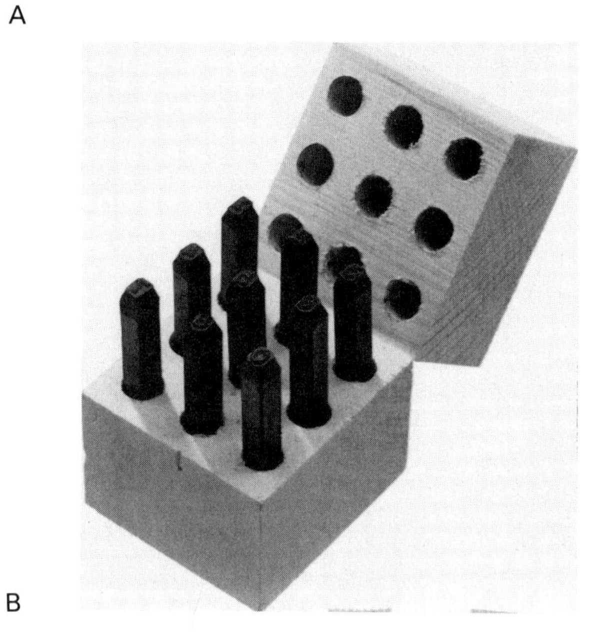

Fig. 24-37. As each piston and rod is removed, make sure it has numbers on cap. If needed, punch mark rod and cap so caps will not get mixed up. Caps must be kept with same rod or bore will not be round. A—New rods sometimes have numbers missing. B—Number stamp set or center punch can be used to mark rods.

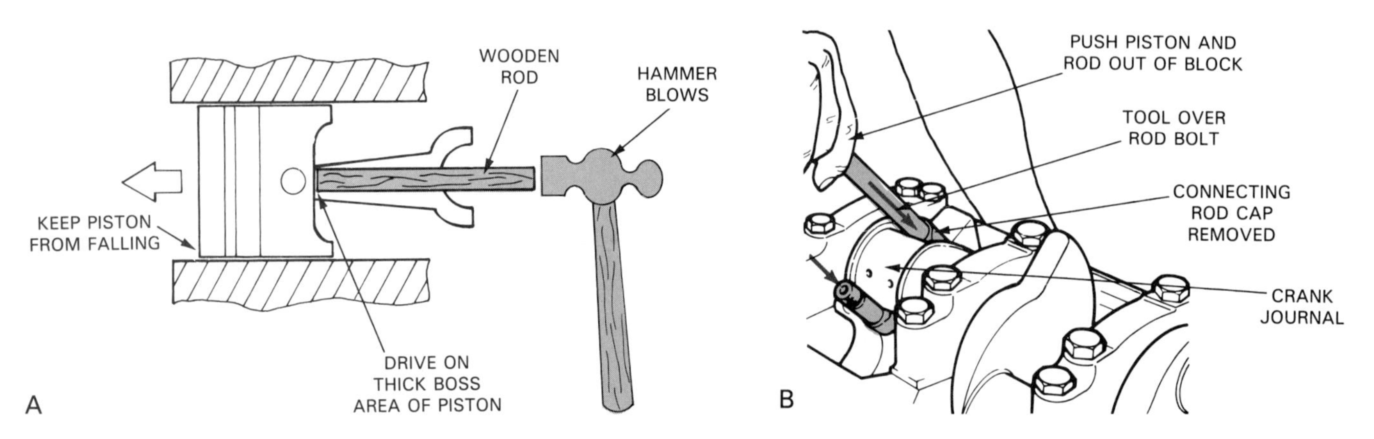

Fig. 24-36. A—Wooden rod or old oak handle is one way to drive out pistons. Place end of handle on piston boss. Light hammer blows should push piston and rod out of block. Do not let piston fall. B—This technician is using threaded rod that fits over rod bolt to hand-force piston out of block.

5. Unbolt one of the connecting rod caps. Then, use a wooden hammer handle or wooden rod to tap the piston and rod out of the cylinder, Fig. 24-36.
6. As soon as the piston is out, replace the rod cap. Also, check the piston head and connecting rod for identification markings. The piston will usually have an arrow pointing to the front of the engine. The connecting rod and rod cap should have numbers matching the cylinder number, Fig. 24-37A.
7. If needed, mark the piston heads with arrows or numbers. Also, if needed, number the connecting rods. If you mix up the pistons or rod caps, severe problems can develop when trying to reassemble the engine, Fig. 24-37B.
8. Remove the other piston and rod assemblies one at a time. Reinstall each cap on its rod. Mark them if needed.

   Note! Do not do anything to dent or mar the connecting rod I-beams. For example, never clamp the I-beam of the rod in unprotected vise jaws. The dent or indentation in the metal could cause a stress point. The rod could crack and then break during engine operation. Only mark the rods on the side of the cap and rod body.
9. Remove all of the old rings from their pistons. Spiral the rings off with your fingers or use a ring expander.
10. If the car has a manual transmission, check for flywheel warpage. Use a dial indicator setup. Turn the crankshaft while noting the indicator reading. The crank will turn easily with all of the piston out of the block. If runout is beyond specs, send flywheel to a machine shop for resurfacing.
11. Before removing the main bearing caps, check that they are numbered. Normally, numbers and arrows are cast on each cap. Number one cap is at the front of the engine.
12. If needed, use a *number set* (punch set for indenting numbers in metal parts) or a center punch to mark the main caps. If the caps are mixed up, the crank bore may be misaligned and the crank can lock in the bore and main bearings during reassembly.
13. Unbolt the main caps. To remove the caps, wiggle them back and forth while pulling with channel locks or with the main cap bolts. Then, lift the crankshaft carefully out of the block. Do NOT hit and nick the journals, Fig. 24-38.

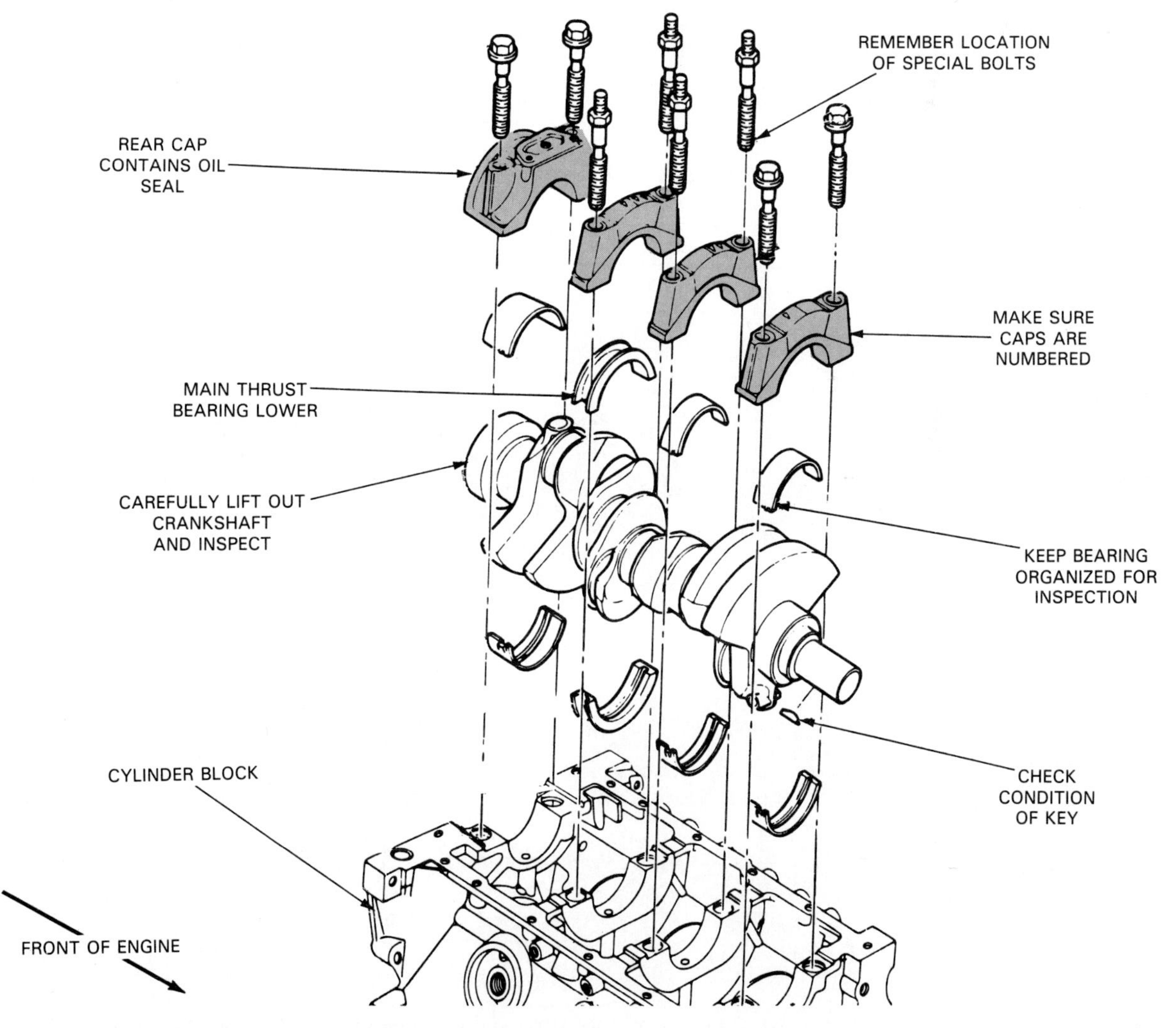

Fig. 24-38. Make sure main caps are numbered before removal. They too must be repositioned in same locations to keep main bore round. Carefully lift crank out of block without hitting journals. (Ford)

14. If the engine is old, pry out the block and head core or freeze plugs. They rust out and will leak after prolonged service. This is also needed if the block is going to be *boiled* (cleaned in strong chemicals at machine shop to remove mineral deposits in water jackets).
15. If the cylinders have deep ridges, the block must be sent to a machine shop for boring and hot tank cleaning. Make sure all external hardware (rubber motor mounts, oil and coolant temperature sending units) are removed.

CLEANING ENGINE PARTS

After you have removed all of the parts from the engine block and cylinder head(s), everything should be cleaned. Different cleaning techniques are needed depending upon part construction and type of material.

Parts can be coated with carbon (valves and combustion chambers), sludge (valve covers, block valley, bottom on intake) and an oil film (other engine parts). See Fig. 24-39.

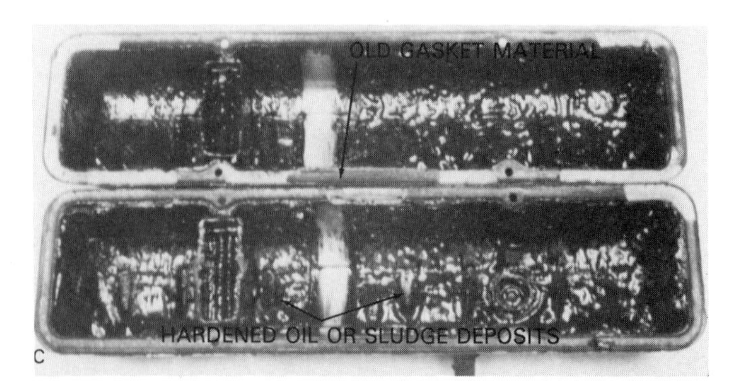

Fig. 24-39. Three common types of engine deposits. A—Light oil film. B—Carbon deposits. C—Sludge. (Texaco)

Closer part inspection can be done during and after part cleaning. Some problems are difficult to see when a part is covered with oil, grease, or carbon deposits.

Scrape off old gaskets and hard deposits

Begin engine parts cleaning by scraping off all old gasket material and hard deposits. Scrape off the gaskets for the valve covers, head, front cover, intake manifold, oil pan, and other components. Also scrape off as much sludge and carbon as you can, Fig. 24-40.

Use a dull scraper and work carefully when cleaning soft aluminum parts. The slightest nick, in a cylinder head for instance, could cause head gasket leakage when returned to service.

WARNING! When using a gasket scraper, push away from your body, NOT toward your body. A scraper can inflict serious cuts!

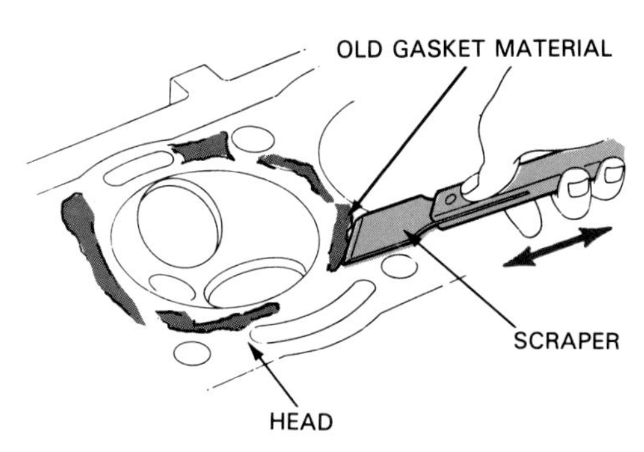

Fig. 24-40. Hand scraper is commonly used to remove heavy deposits and old gasket material. Do not nick soft aluminum parts. Scrape away from body. (Toyota)

Use power cleaning tools

There are several power cleaning tools used by the professional technician. If used properly, they can speed up and ease engine repairs.

A *power brush* is driven by an air or electric drill to remove hard carbon. It is especially handy inside hard-to-reach areas—combustion chambers, for example, Fig. 24-41A. Fig. 24-41B shows a power brush being used inside a valve guide.

DANGER! Always wear eye protection when cleaning parts with power tools. Metal bristles, bits of carbon, or metal chunks from tool or part breakage can fly into your face.

A *wire wheel* on a grinder is another common method of cleaning engine parts. Frequently it is used to clean carbon off valves. Keep safety shield and tool rest in place. See Fig. 24-42.

A *scuff pad* is a tough plastic mesh material mounted on a rubber backing and arbor. The arbor's end is placed in an air drill or similar tool. The scuff pad will cut and lift off gasket material and silicone but is too soft to damage metal surfaces.

Use cleaning solvents

After scraping off the gaskets, use cleaning solvent to remove hard-to-reach deposits.

A *cold soak tank* is a cleaning machine for removing oil and grease from parts, Fig. 24-43. It will NOT remove hard carbon or mineral deposits. Most auto shops have

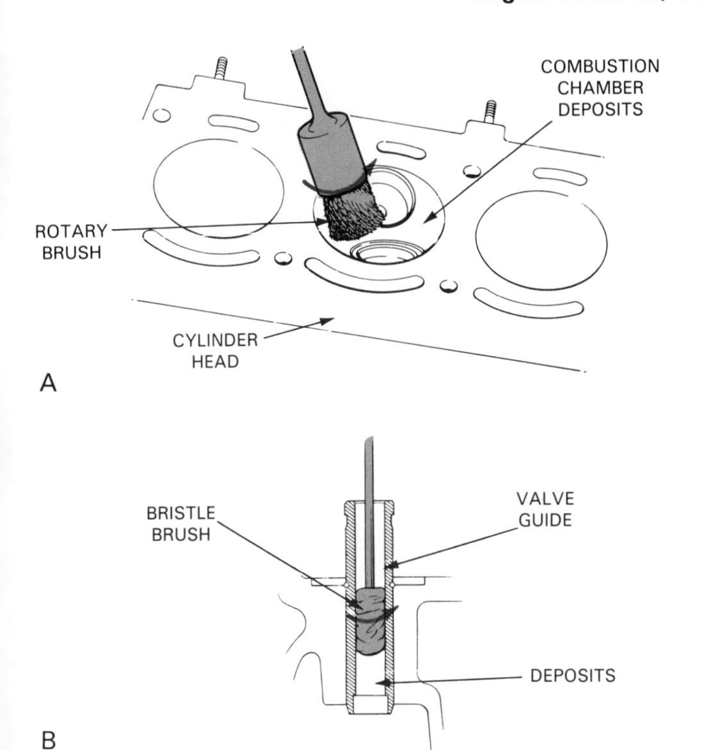

Fig. 24-41. A—Power brush in electric or air drill will clean combustion chambers. B—Small brush on drill will clean out valve guides. (Toyota)

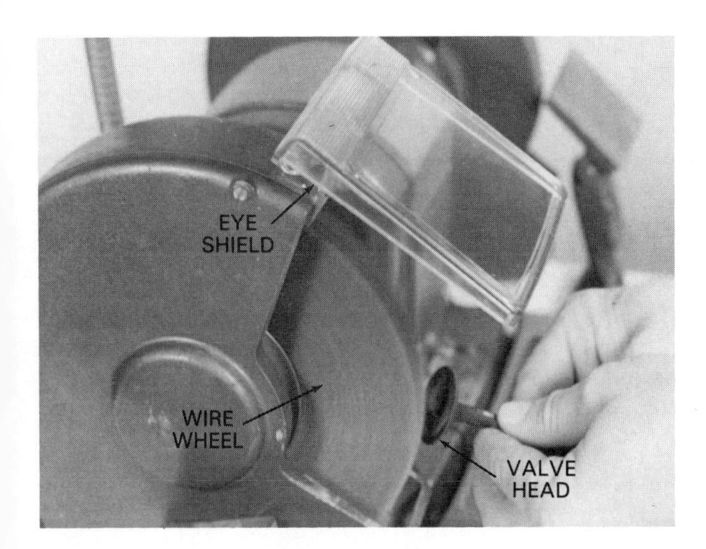

Fig. 24-42. Wire wheel on bench grinder is commonly used to remove hard carbon from engine valves. Wear eye protection and keep shields in place.

a cold cleaning tank or machine. It has a pump and filter that circulates clean solvent out of a spout. To wash off parts, direct the stream of solvent on the part while rubbing with a soft bristle brush.

DANGER! Never use gasoline to clean parts. The slightest spark or flame could ignite the fumes, causing a deadly fire!

Spray-on gasket solvents are also available. Sometimes gaskets can be very difficult to scrape off.

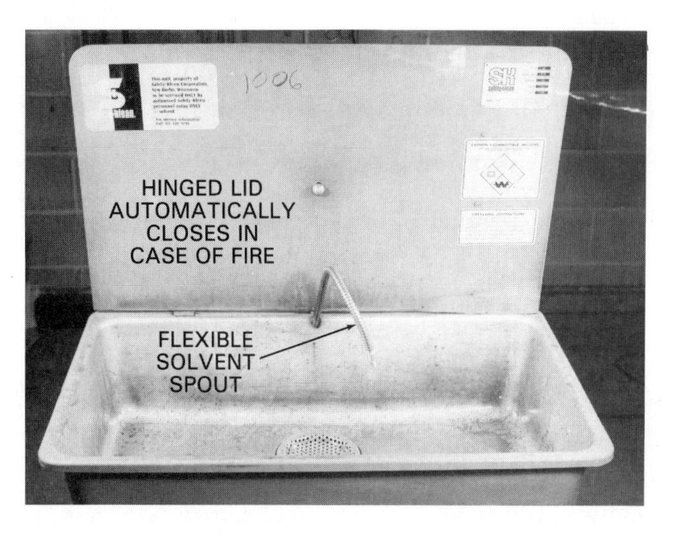

Fig. 24-43. Cold soak tank is for removing oil film. If parts are covered with sludge, scrape off before cleaning in tank.

Fig. 24-44. Sometimes gaskets can become very hard to remove after years of service. Chemical sprays are available for dissolving gaskets and speeding removal. Softened gaskets can then be easily scraped off. (Fel-Pro)

Gasket removing chemicals can be helpful. Fig. 24-44 shows a technician using a spray-on chemical for head gasket and valve cover gasket removal. After spraying, let the chemical work for a few minutes. Then, scrape off the softened gasket with a scraper.

A *hot tank* is a large cleaning machine filled with strong, corrosive chemicals. One is pictured in Fig. 24-45. It will remove mineral deposits in the water jackets, hard carbon deposits, oil, grease, and even paint. Automotive machine shops normally have a hot tank.

Note! Aluminum components can be corroded or etched by soaking in a hot tank. Clean only cast iron and steel parts in a hot tank.

Carburetor cleaner or *decarbonizing cleaner* is another very powerful chemical. It will remove carbon, paint, gum, and most other deposits. An agitation tank containing this type cleaner is shown in Fig. 24-46.

Carburetor cleaner can be used on smaller engine parts or in special situations. It is an expensive solvent and is not commonly used unless necessary. Also, it should NOT be used on soft aluminum parts because it could etch them. Steel engine valves, rocker arms, or push rods would be examples of parts cleaned this way.

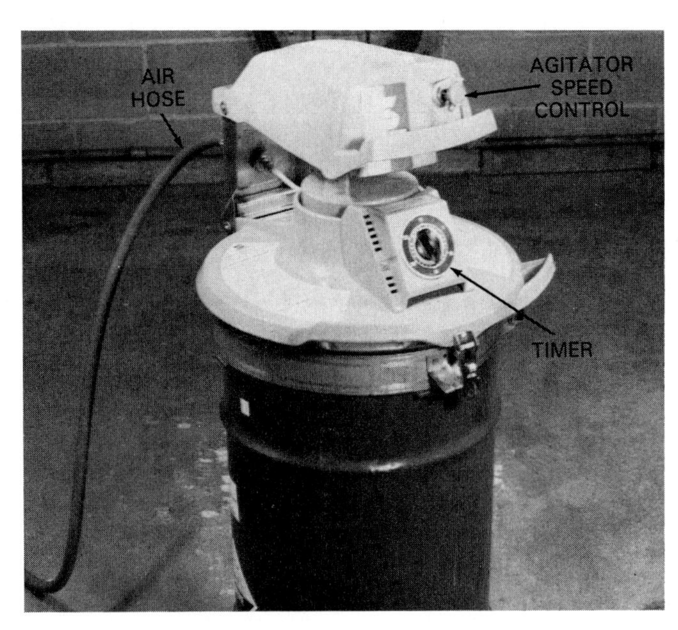

Fig. 24-46. Carburetor cleaner is sometimes used to clean engine parts. For example, if rocker arms are coated with gumlike oil, soak them in carburetor cleaner. Do not submerse aluminum parts in cleaner because they can be etched and ruined.

Shot blast cleaning

Shot blast cleaning uses air pressure to blow glass beads or sand (not as desirable because of more erosion) onto the parts. This will remove paint, carbon, and other dry deposits. The parts must be free of oil and sludge to prevent machine clogging. Shot blast cleaning will make used parts look like new.

Fig. 24-47A shows a shot blast cabinet. Place the part to be cleaned inside the cabinet. Close the door. Insert your arms into the holes and rubber gloves. Then, direct the blast nozzle over the part to be cleaned, Fig. 24-47B.

Fig. 24-45. Machine shop will have hot tank for cleaning blocks, heads, and other ferrous parts. (Storm Vulcan)

Cylinder heads, intake manifolds, valve covers, etc. can all be cleaned and ready for painting with shot blast cleaning. However, be careful when cleaning soft aluminum parts. Only use enough blast action to clean off deposits. Too much blasting can wear off aluminum and ruin the part.

Carbon coated piston heads are very hard to clean. You can clean piston heads with a soft bristle brush (slow) or quickly with careful use of the shot blaster. Only clean the heads of the piston, not the skirt because of the danger of skirt erosion.

Air blow gun

An *air blow gun* is normally the last method of cleaning parts. It uses pressure from the shop's air compressor to blow off small bits of dirt, solvent, water, and other debris. Look at Fig. 24-48.

WARNING! Use extreme care when using an air gun. Wear goggles and avoid aiming the gun at your body. If air enters your blood stream, a blood clot could cause death!

Special cleaning tools

Other special cleaning tools are also needed for engine components. These tools (piston ring groove cleaner for example) will be explained in later chapters.

SUMMARY

Many engine repairs can be done with the engine block still bolted to the chassis. This includes front end and top end service. Valve jobs, camshaft service, and valve train repairs, can all be done without removing the engine from the car.

Crankshaft and cylinder block service requires engine

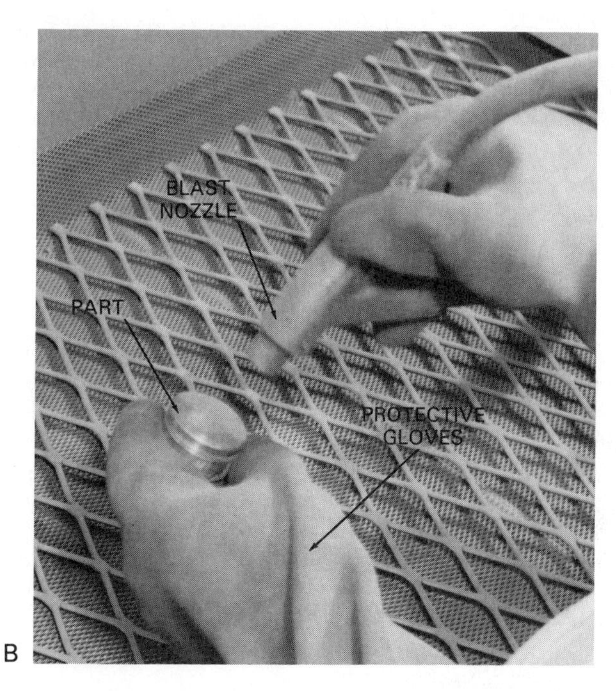

Fig. 24-47. Blast cleaning booth will make used parts look like new. A—Note parts of blaster. B—Only blast parts enough to remove deposits.

removal. In some cases, however, you can service pistons and rings by removing the heads and oil pan. The engine can be rebuilt in-car. Check a manual for details. Usually, it is just as easy to remove the engine for major service.

An engine overhaul involves cleaning, inspecting, measuring, and replacing parts as needed. When part measurements are not within specs, the parts must be machined, repaired, or replaced. Usually, piston rings, bearings, seals, gaskets, core plugs, timing chain, belt or gears, oil pump, and any other worn parts are replaced during an overhaul.

During engine removal, you should try to keep everything organized. Label wires and hoses. Place fasteners in cans. Use one container for bolts off the front of the engine; another for lower end fasteners, etc. This will save time and confusion upon reassembly.

When attaching the lift chain, make sure it is positioned to lift the engine level or at the correct angle. Also, bolts should be large enough in diameter and should be fully threaded into holes at least 1 1/2 times their diameter.

When raising the engine, double-check that everything is disconnected. Keep hands out from under the engine. Only raise the engine high enough to clear the radiator support. As soon as the engine is out, lower it to the ground. Never work on an engine held in the air with a crane or hoist. Mount the engine on a stand.

Generally, to disassemble the engine, start with the engine front end. Then, remove the top end: intake, exhaust manifolds, valve train, heads, and finally the bottom end or short block.

Wheel pullers are needed to remove some dampers, seals, and sprockets. A ridge reamer is needed to cut out any lip formed at the top of worn cylinders before piston removal. Check for numbers on the rods as they are removed. Number them as needed. This also applies to block main caps and head camshaft caps.

Fig. 24-48. Use compressed air and blow gun to final clean parts. Air blast will blow off tiny particles and dry solvent. Wear eye protection and do not direct air blast toward body.

After disassembly, clean all parts using approved methods. Inspect parts closely for problems before and after cleaning. Look for signs of leakage, cracks, stripped threads, etc.

Scrapers are used to remove old gaskets and sludge. Power wire wheels will clean off carbon in combustion chambers, valve guides, and on valves. Cold soak cleaner is for removing an oil film. Hot tank cleaning and carburetor cleaner will remove more stubborn deposits: hard carbon, paint, etc. Do not submerse aluminum parts (pistons, rocker stands, etc.) in these solvents because they will etch and remove aluminum, ruining the parts.

KNOW THESE TERMS

Valve job, Engine overhaul, In-car overhaul, VIN, Casting numbers, Engine crane, Wheel puller, Organizing tray, Mushroomed valve, Ring ridge, Ridge reaming tool, Power brush, Wire wheel, Cold soak cleaner, Hot tank, Carburetor cleaner, Shot blasting, Air blow gun.

REVIEW QUESTIONS—CHAPTER 24

1. Tests find a burned engine valve. The engine is not smoking and has good oil pressure.
   Technician A says that this should be an in-car operation and the block can remain bolted to the chassis.
   Technician B says it is just as easy to pull the engine to service the valves.
   Who is correct?
   a. Technician A. c. Both A and B.
   b. Technician B. d. Neither A nor B.
2. Which of these jobs does NOT require head removal?
   a. Crankshaft replacement.
   b. Piston replacement.
   c. Connecting rod replacement.
   d. Diesel prechamber replacement.
3. Define the term "engine overhaul."
4. An in-car engine overhaul is more common than an out-of-car engine overhaul. True or false?
5. How can you identify the type engine in a car?
6. Which is the most common general sequence for preparing to remove an engine from a car?
   a. Remove top end parts, front end parts, and then parts under engine.
   b. Remove parts under car, then on top and front of engine.
   c. Remove front end parts, top end parts, and then bottom end parts.
7. Why should you label wires and vacuum hoses?
8. It is a waste of time to use several cans to keep fasteners organized. True or false?
9. When using a chain to remove an engine, the bolts holding the chain should be tightened or threaded __________ and threaded in at least __________ the diameter of the bolt.
10. Describe three safety rules to follow when lifting an engine out of a car.
11. A __________ __________ is sometimes needed to remove a crank damper, seal, or crank sprocket.
12. Why should most parts be returned to their original locations in an engine?
13. What can happen if you drive out a valve that has a mushroomed tip?
14. A __________ __________ __________ is needed to cut out a lip at the top of a worn cylinder to prevent __________ damage.
15. Explain six ways of cleaning engine parts.

ASE CERTIFICATION-TYPE QUESTIONS

1. Technician A says that the vehicle identification number (VIN) can be used to identify the automobile's engine type. Technician B says that a vehicle identification number can be used to identify the automobile's transmission type. Who is right?
   (A) A only. (C) Both A & B.
   (B) B only. (D) Neither A nor B.
2. All of the following engine repairs can be performed with the engine still installed in the vehicle EXCEPT:
   (A) valve job.
   (B) intake manifold replacement.
   (C) crankshaft service.
   (D) valve seal removal and replacement.
3. Technician A says that the vehicle identification number can be located on the front of the engine near the fuel pump. Technician B says that the vehicle identification number can be located on the upper left corner of the firewall. Who is right?
   (A) A only. (C) Both A & B.
   (B) B only. (D) Neither A nor B.
4. Technician A says that an engine's timing belt should be replaced during a major engine rebuild. Technician B says that an engine's cylinder head(s) should be reconditioned during a major engine rebuild. Who is right?
   (A) A only. (C) Both A & B.
   (B) B only. (D) Neither A nor B.
5. Technician A says that when preparing to remove an engine from a vehicle, you do not need to disconnect the car's battery. Technician B says that when preparing for engine removal, you should remove the automobile's radiator. Who is right?
   (A) A only. (C) Both A & B.
   (B) B only. (D) Neither A nor B.
6. Technician A says that when removing the engine on certain front-wheel drive cars equipped with a transaxle, you must remove the engine and transaxle as a unit. Technician B says that during engine removal on all front-wheel drive cars equipped with a transaxle, the engine and transaxle must be removed separately. Who is right?
   (A) A only. (C) Both A & B.
   (B) B only. (D) Neither A nor B.
7. Teardown on an OHC engine commonly begins with __________ disassembly.
   (A) front end (C) bottom end
   (B) top end (D) Any of the above.
8. You should place a breaker bar __________ to free the cylinder head from the block.
   (A) between the head and block deck surface
   (B) between the head and a solid, stationary object
   (C) between the head and exhaust manifold
   (D) in one of the ports
9. An engine block has deep ridges in its cylinders. Technician A says that a brush hone should be used to recondition the cylinders. Technician B says that the block should be sent to a machine shop to have the cylinders bored. Who is right?
   (A) A only. (C) Both A & B.
   (B) B only. (D) Neither A nor B.
10. Technician A says that aluminum engine components can be cleaned in a hot tank. Technician B says that aluminum engine components can be corroded if cleaned in a hot tank. Who is right?
    (A) A only. (C) Both A & B.
    (B) B only. (D) Neither A nor B.

Chapter 25

Short Block Service

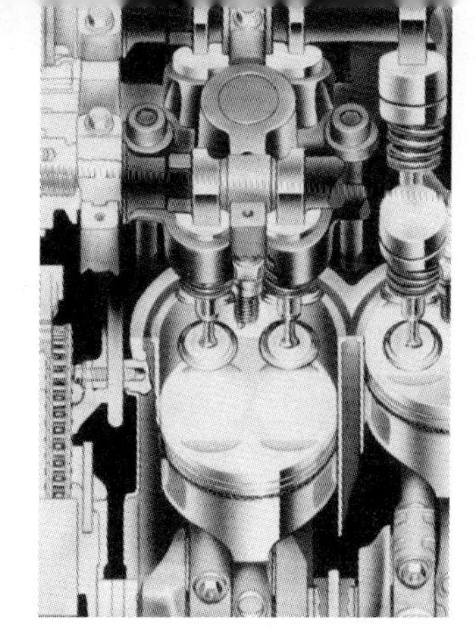

After studying this chapter, you will be able to:

- *Check for cylinder block damage, cracks, and wear.*
- *Explain how to have a machine shop repair cracks or flaws in a cylinder block.*
- *Check for main bore and deck warpage.*
- *Measure cylinder bore wear and compare it to specs.*
- *Describe how a machine shop bores or sleeves a worn or damaged cylinder.*
- *Hone or deglaze a cylinder.*
- *Properly clean a cylinder after honing.*
- *Install camshaft and balancer shaft bearings.*
- *Check a crankshaft for wear, cracks, and runout.*
- *Install a rear main oil seal.*
- *Install main bearings and the crankshaft.*
- *Plastigage rod and main bearings to check clearance.*
- *Check crankshaft end play.*
- *Service pistons and piston rings.*
- *Service connecting rods and wrist pins.*
- *Measure ring and piston clearances.*
- *Properly assemble a short block.*

You learned about engine problems, teardown, part cleaning, and other aspects of engine service in earlier chapters. This chapter will summarize the overhaul of an engine short block. You will learn how to properly inspect, measure, and install the crankshaft, main bearings, pistons, rings, rods, piston pins, rod bearings, and cam bearings into the cylinder block.

If needed, you may want to refer back to earlier text chapters. These chapters explained engine removal, disassembly, and part cleaning. These operations must be understood before you can properly assemble an engine short block.

CYLINDER BLOCK SERVICE

Cylinder block service typically includes:

1. Inspection for cracks, excess wear, scored cylinders, and other block damage.
2. Checking for main bore and deck warpage; having block machined if needed.
3. Measuring cylinder wear and having block bored if cylinders are worn beyond specs.
4. Honing or deglazing cylinders to prepare cylinder walls for new rings.
5. Installing cam or balancer shaft bearings.
6. Final cleaning of block before assembly.

As you will learn, most of the jobs can be done in the average auto shop. However, milling and boring operations usually require the help of a machine shop.

Inspecting cylinder block

Begin short block reassembly by closely inspecting the block for problems. Look for the kind of troubles illustrated in Fig. 25-1. In particular, check the cylinder walls for scoring, heavy wear, cracks, etc.

Using a drop light, closely inspect the surface of each cylinder wall. Rub your fingernail around in the cylinder. This will help you feel and locate problems.

Finding cracks in block

If symptoms point to possible cracks in the block, you should magnaflux the block or use dye penetrant.

Magnafluxing uses a magnet and a ferrous (iron) powder to highlight cracks and pores in cast iron blocks. Generally, place the magnet over the area to be tested.

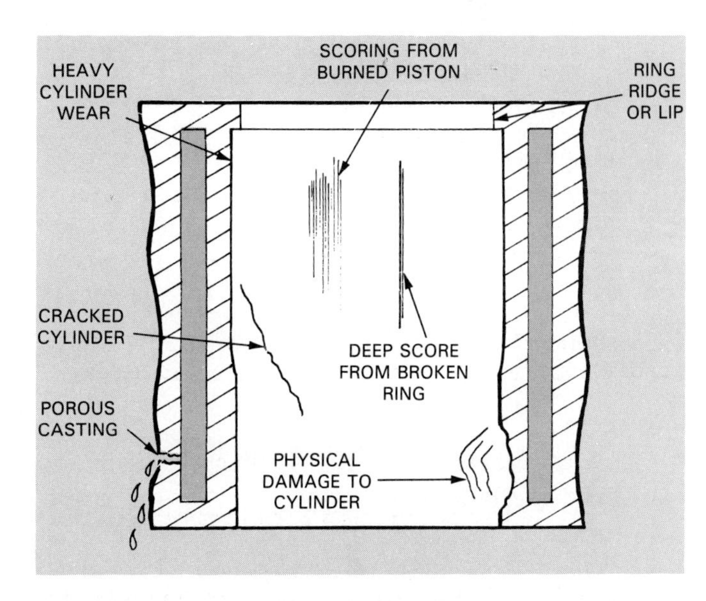

Fig. 25-1. Closely inspect cylinder walls for these kinds of trouble. Use your fingernail to feel for cracks and scoring.

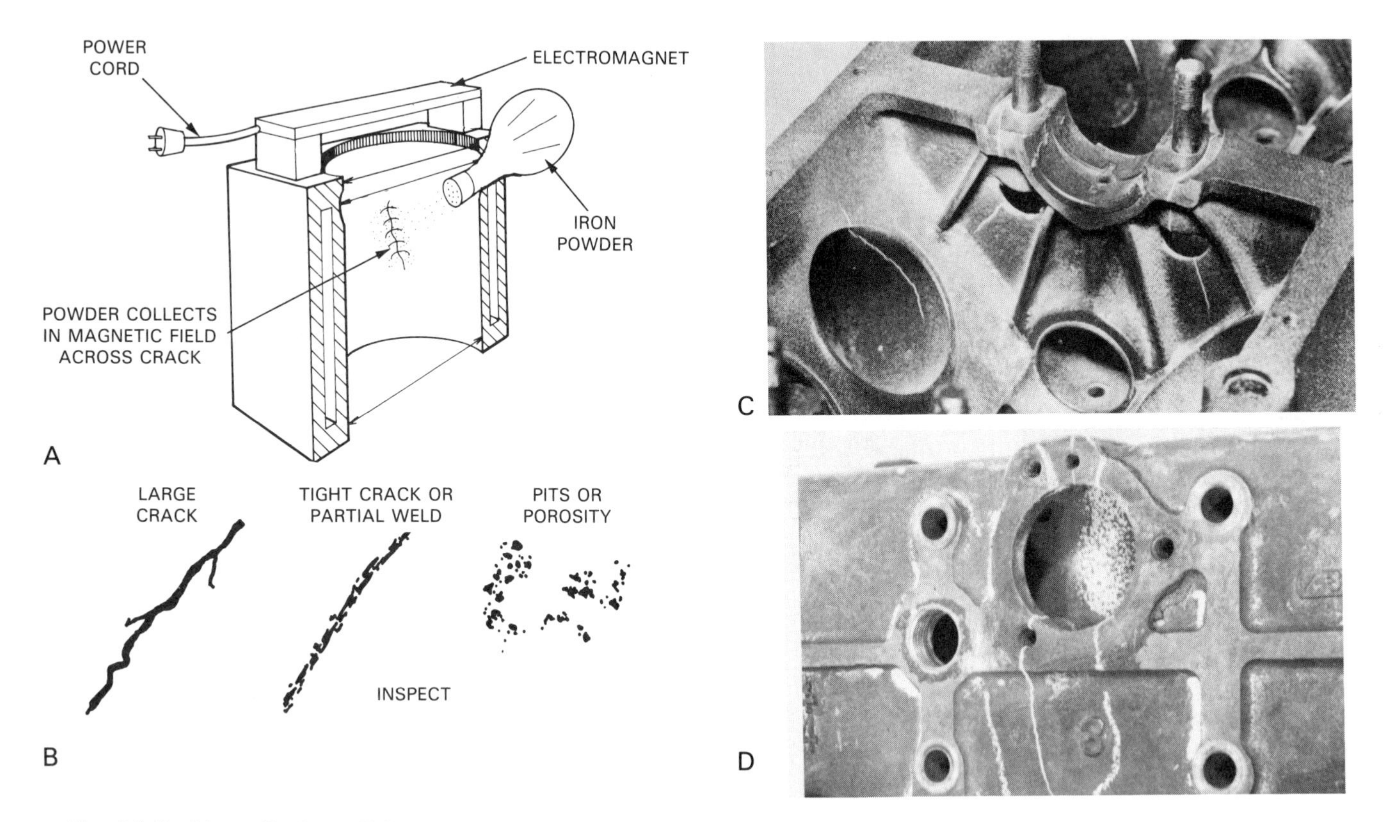

Fig. 25-2. Magnafluxing will find cracks in iron parts. A—Magnafluxing equipment found cracked cylinder. B—Type of problems magnetic inspection will locate. C—Actual cracked block caused by part breakage. D—Coolant freeze damage to block. (K-line and Magnaflux Corp.)

Spread the powder over the block. Any crack will make the powder collect in the flaw. A break in the block metal sets up a magnetic field that pulls the powder to the crack or flaw. This will make the crack or pore easy to detect, Fig. 25-2.

Dye penetrant is commonly used to find cracks and casting imperfections in aluminum cylinder blocks. The special dye is sprayed over the block. Next, a powderlike developer is spread over the penetrant. This will make the penetrant turn red if it has collected in any crack or casting flaw. You will be able to then find any block cracks or flaws more easily.

Zyglo or *fluorescent penetrant* is similar to dye penetrant but it requires a black light to show up cracks. The chemical is sprayed over the part (block, head, manifold) and then placed under a black light. The chemical will collect in any crack. The concentration in the crack will glow under the light, Fig. 25-3.

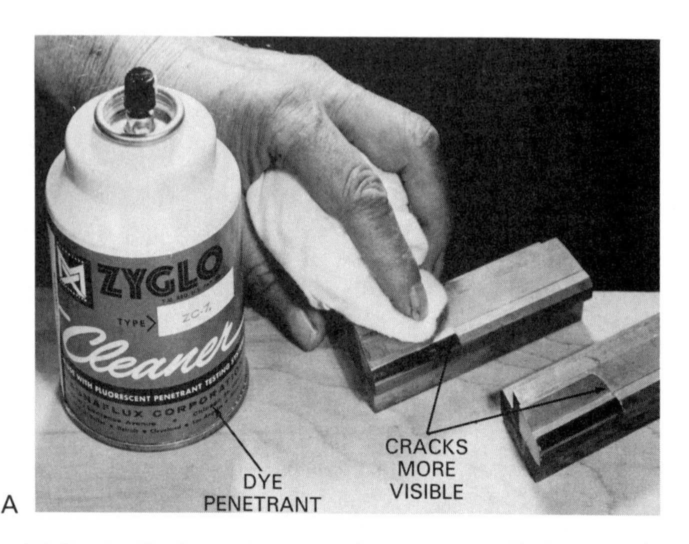

Fig. 25-3. A—Zyglo can be sprayed over parts to find cracks. Chemical collects in cracks and then will show up when placed under black light. B—Another dye type crack finding system. Follow specific directions. (Magnaflux Corporation)

Pressure testing involves filling the block or head water jackets with air pressure to find cracks or flaws. All water jacket openings are sealed using rubber expander plugs. Sheets of rubber are bolted over the deck surface with the metal plates. This makes the inside of the block air tight. An air hose is connected to the water jackets and the block is lowered into a large tank of water. Any crack or casting pour will show up as air bubbles rising to the top of the water.

Repairing block cracks

Cracks and pores in a cylinder block can be corrected by:

1. Sleeving cylinder if crack is in cylinder wall area.
2. Welding cast iron with NI (high nickel) rod after heating in furnace or welding aluminum block with a heli-arc welder and aluminum rod.
3. Plugging which involves drilling series of holes along crack and threading in repair plugs.
4. Using special epoxy or metallic plastic to seal casting flaw or pore if in certain areas of block, Fig. 25-4. Should NOT be used on cracks.
5. Replacing block if cost of repair is higher than new or good used block.

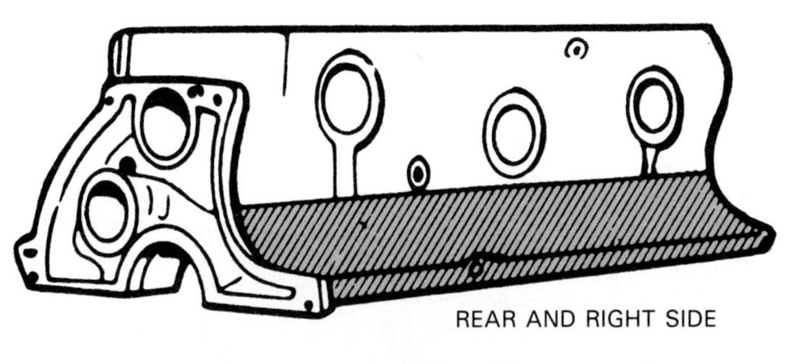

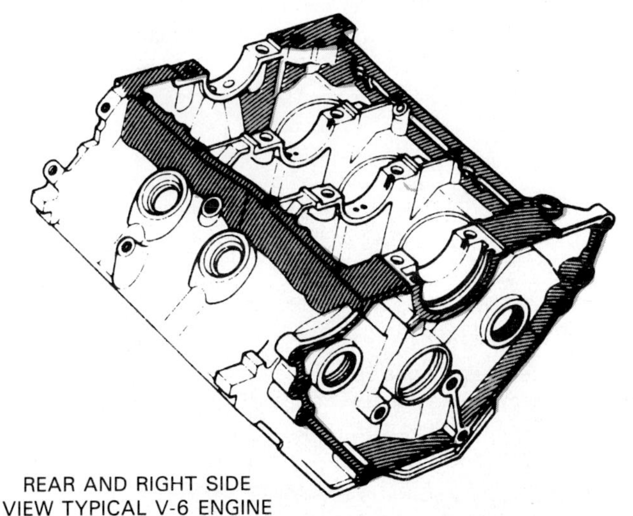

Fig. 25-4. Casting flaws or pores can sometimes be repaired with special epoxy or plastic. This is service manual illustration showing areas of block that will seal with epoxy. (Ford Motor Co.)

Block repairs are usually done by a specialized machine shop. They have the special equipment and skills needed to repair blocks. As an engine technician, you should know that block repairs can be made by some shops.

Checking main bearing bores

After repeated heating and cooling or overheating, the main bearing bores in the cylinder block can warp or twist. This will affect main bearing insert alignment and crankshaft fit in the block. Under severe cases of main bearing bore misalignment, the crankshaft can lock up when the main caps are torqued.

To measure main bearing bore alignment, lay a straightedge on the bores. Insert a .0015 in. (0.038 mm) feeler gauge under the straightedge. If the blade fits, the main bearing bores are misaligned.

Note! ALWAYS check bore misalignment after main bearing failure, Fig. 25-5.

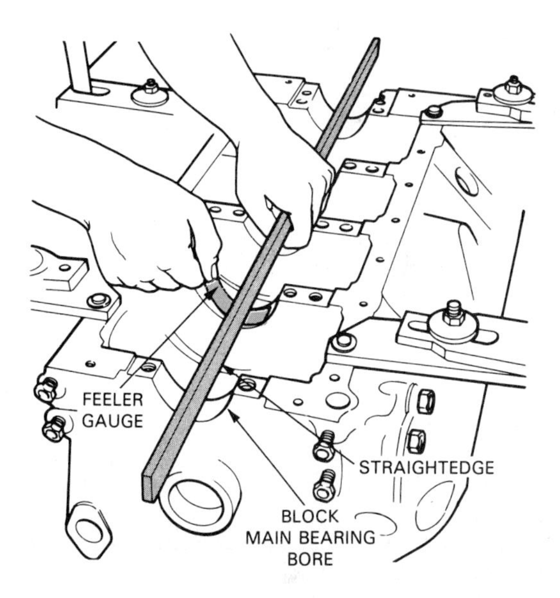

Fig. 25-5. Check main bearing bore alignment if main bearings were badly worn or spun. Try to fit a .0015 inch feeler gauge between the straightedge and block bore. If it fits, block should be line bored by machine shop. (Federal Mogul)

Line boring block

Block line boring (machine tool) or *line honing* (rigid stones) is used to straighten or true misaligned main bearing bores. A block should be line bored when misalignment is over about .0015 in. (0.038 mm).

A machine shop will have a boring or honing bar for machining the bore back into alignment. Line boring is becoming more common with today's thin-wall, lightly constructed blocks. Look at Fig. 25-6.

Measuring deck warpage

Deck warpage is measured with a straightedge and feeler gauge on the head gasket sealing surface of the block. It should be checked when the old head gasket was blown and leaking. It should also be done on late model aluminum blocks since they distort easily.

To check for block deck warpage, lay a straightedge

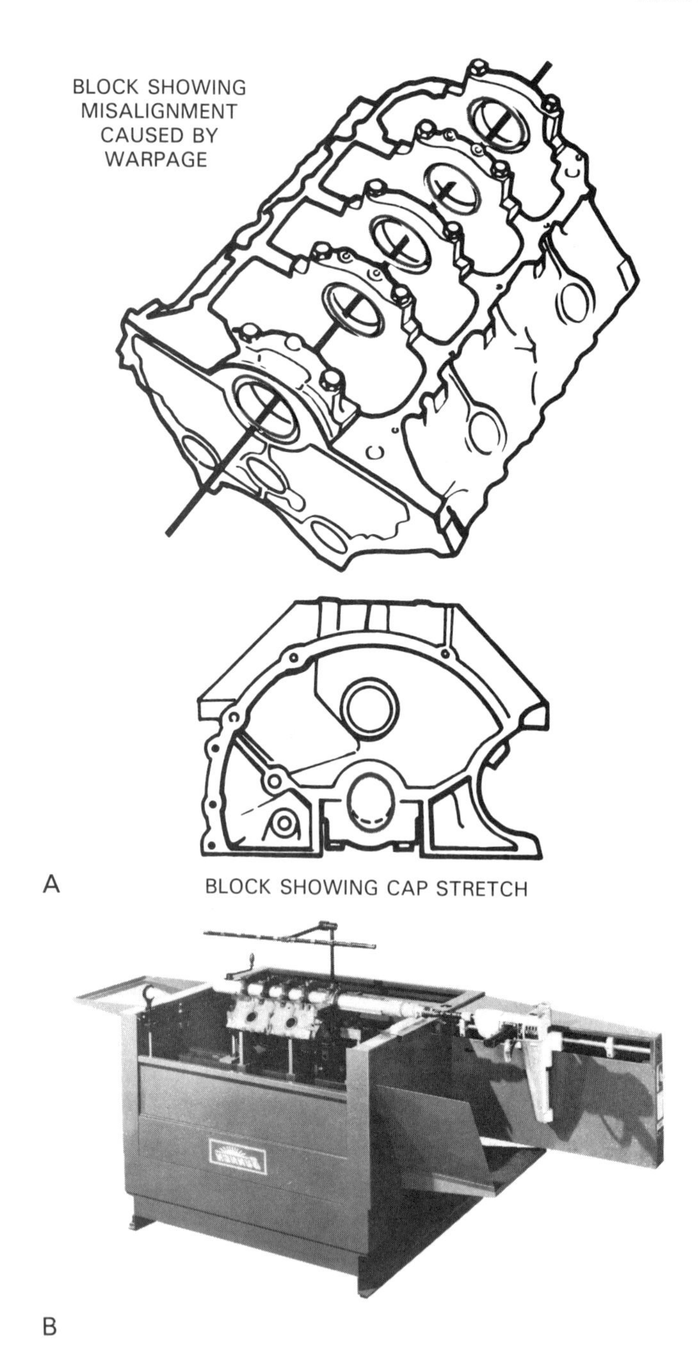

Fig. 25-6. A—Note how warpage, cap stretch, or a spun bearing could cause problems. If misaligned, crank could lock in block. B—Modern line honing or boring machine. (Sunnen)

on the clean block surface, Fig. 25-7. Try to slip different thickness feeler gauge blades between the block and straightedge. The thickest blade that fits shows warpage. Check in different locations on the block.

If beyond specs (about .003 to .005 in. or 0.08 to 0.13 mm), replace the block or send it to a machine shop for surface milling. See Fig. 25-8. *Decking a block* involves machining the cylinder head mounting surfaces until they are parallel and equal distance from the main bore. This is also called ''squaring the block.''

Cylinder wear

Discussed earlier, a worn cylinder is NOT round or true. It will be larger in diameter near the top and be wider

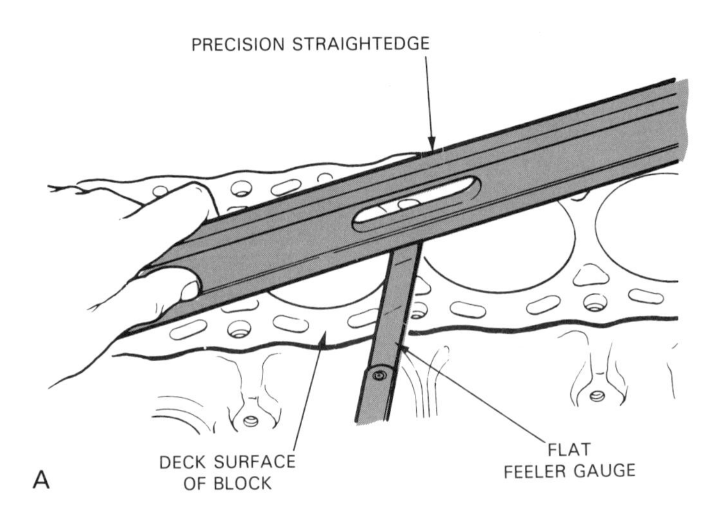

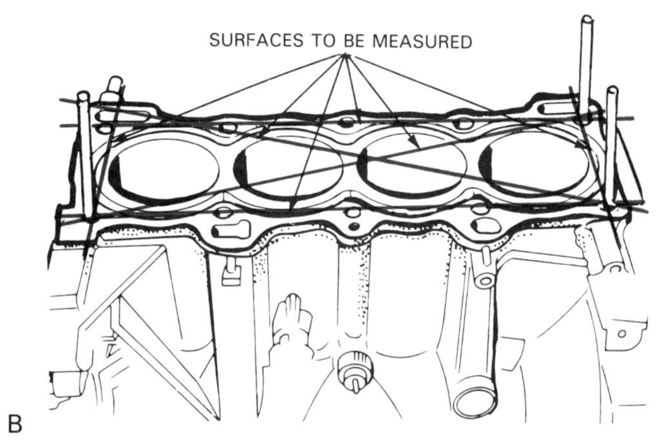

Fig. 25-7. Block deck warpage can happen after overheating of engine, especially with thin wall or aluminum blocks. A—Checking warpage. B—Check at these locations on block. (Honda)

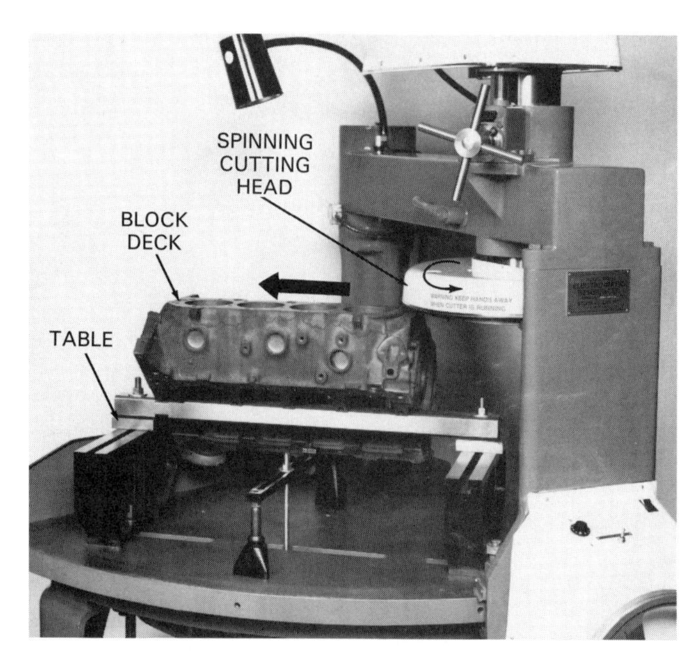

Fig. 25-8. Block deck should be milled if warped beyond specs. This is a machine shop milling machine being used to true block deck. (Storm Vulcan)

across its thrust surface, Fig. 25-9.

Cylinder taper is a difference in the diameter at the top and bottom of the cylinder. It is caused by less lubricating oil at the top of the cylinder. More oil splashes on the lower area of the cylinder. As a result, the top of the cylinder wears faster (larger) than the bottom, producing a taper, Fig. 25-9A.

Cylinder out-of-roundness is a difference in cylinder diameter when measured front-to-rear and side-to-side in the block. Piston thrust action normally makes the cylinders wear more at right angles to the centerline of the crankshaft or piston pins, Fig. 25-9B.

If new rings are installed in an out-of-round cylinder, blowby will result. There will be a gap where the round ring does not contact the cylinder. Pressure and oil can leak past this gap.

If new rings are installed in a tapered cylinder, they will be expanded and contracted every time they slide up and down in the cylinder. The rings can then, at high engine rpm, fail to stay in contact with the cylinder properly. A loss of power, blowby, and oil consumption can also result.

Measuring cylinder wear

If the cylinder is not badly scratched or scored, you must measure cylinder wear to assure that the new rings will seal properly. Obviously, new, round rings cannot seal in worn, out-of-round, or tapered cylinders. Also, cylinder measurement will let you determine piston-to-cylinder clearance, Fig. 25-9C.

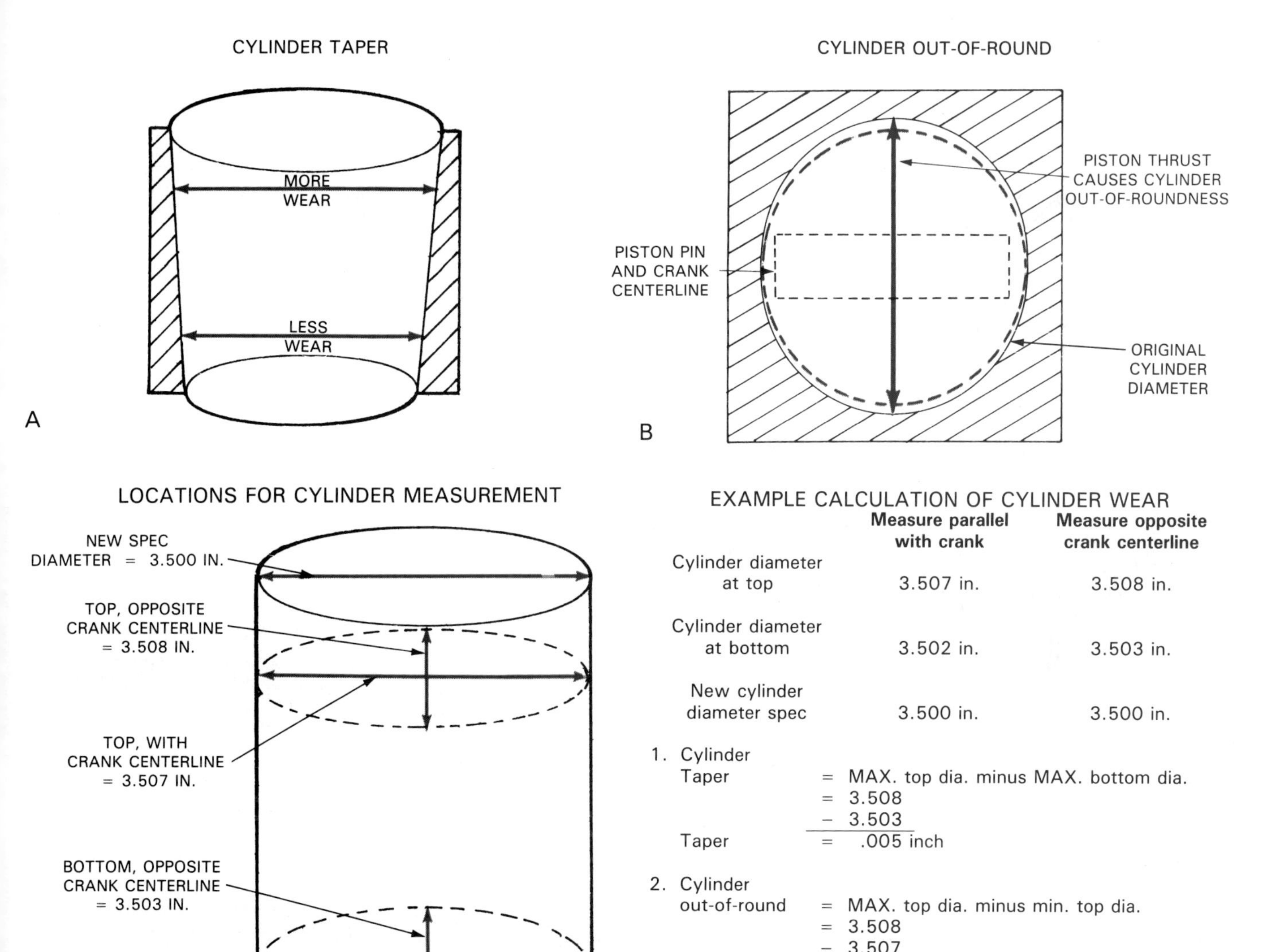

Fig. 25-9. Measuring cylinder wear. A—Cylinder will wear more at top than bottom, forming taper. B—Cylinder will also wear more on piston thrust surface or opposite crankshaft and wrist pin, forming out-of-round. C—Measure cylinder in these locations and write down your readings. D—Example measurements and calculations of cylinder wear from part C of this illustration.

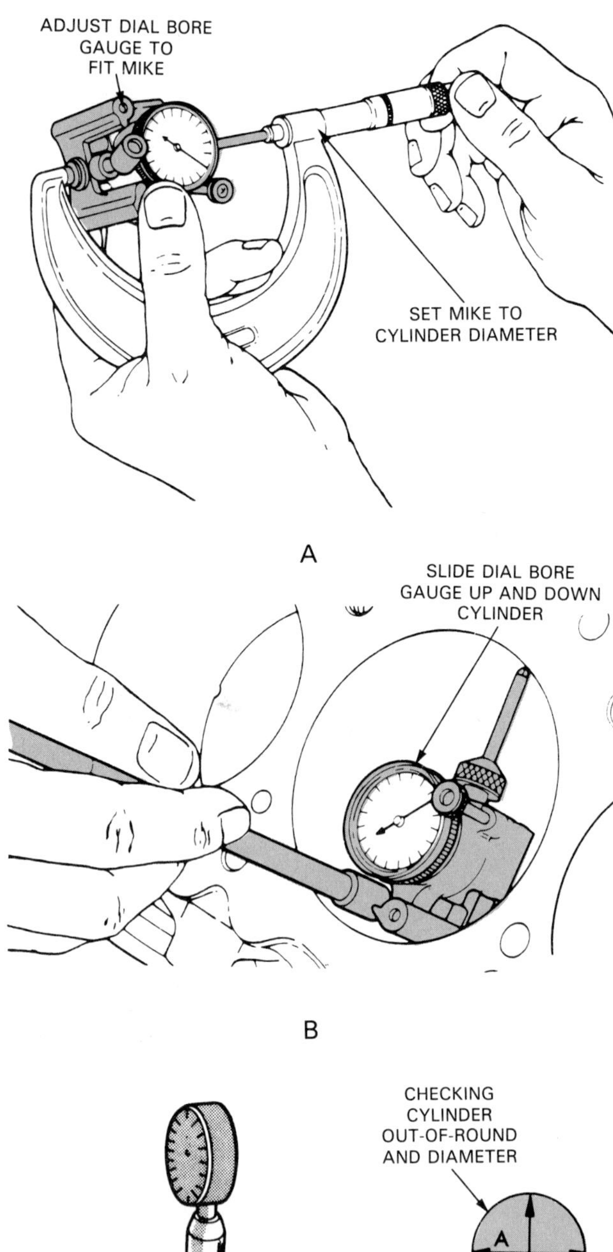

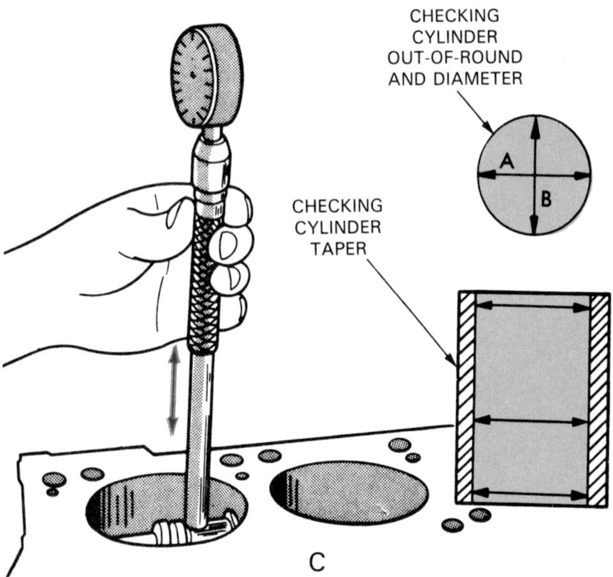

Fig. 25-10. Using dial bore gauge to check cylinder wear. A—Set mike to diameter of new cylinder. Then, adjust dial bore gauge to fit inside micrometer. B—Slide dial bore gauge up and down in cylinder to read taper. Rotate gauge around cylinder to read out-of-round. Cylinder wear equals difference between new spec and actual readings. C—Locations for measurements. (General Motors)

A *dial bore gauge* is a fast and precise tool for measuring cylinder wall wear. Use an outside micrometer to adjust the dial bore gauge to the correct diameter. Then, slide the bore gauge up and down in the cylinder. Indicator movement equals cylinder taper. Check both parallel and perpendicular to the bore centerline to determine out-of-round. Look at Fig. 25-10.

A large inside micrometer or a telescoping gauge and an outside micrometer can also be used to measure cylinder dimensions. They are slightly more time consuming to use, however. Refer to Chapter 2 for more information on measuring tools.

Note! WRITE DOWN your cylinder measurements carefully, as was shown in Fig. 25-9. The slightest mistake in math could be critical.

Cylinder wear limits

Generally, cylinder boring is needed when:

1. CYLINDER OUT-OF-ROUND is more than about .005 inch (0.13 mm).
2. CYLINDER TAPER is more than approximately .008 inch (0.20 mm).
3. MAXIMUM CYLINDER DIAMETER, at its widest point, exceeds .010 inch (0.25 mm) over standard bore specifications.

Some engine manufacturers do NOT allow even this much wear. Refer to engine specs for more accurate wear values.

If the cylinders in the block are worn more than specs, you will have to bore the block or replace it. If bored, oversize pistons will be needed.

Fig. 25-9D gives an example of how to measure and find cylinder wear.

Measuring cylinder diameter

When measuring cylinder wear, you must also check cylinder bore diameter. It is possible that the block was bored earlier and is not a *standard diameter* (original diameter for new block).

To check cylinder diameter:

1. Check the service manual for standard bore specifications.
2. Use your dial bore gauge, inside mike, or telescoping gauge and outside mike to measure cylinder diameter at its widest point, near the top.
3. Measure the cylinder diameter at the bottom, in an unworn section.
4. Compare your measurements to specs.

As an example, say your measurements show the cylinder bore to be 4.011 in. near the unworn bottom area and specs call for a 4.000 in. cylinder diameter. You would then know that the cylinder has been bored .010 inch oversize (.010 overbore plus .001 wear). This would allow you to order the right size piston rings, know correct overbore limits, etc.

CYLINDER HONING

Cylinder honing, also called *deglazing,* is used to break the *glaze* (polished surface) on a used cylinder. It must also be used to smooth a very rough cylinder surface after boring. Most ring manufacturers recommend deglazing or honing before new ring installation.

A *cylinder hone* produces a precisely textured, crosshatched pattern on the cylinder to aid ring seating and sealing. Tiny scratches from the hone cause initial ring and cylinder wall break-in wear. This makes the ring fit in the cylinder after only a few minutes of engine operation.

Hone types

As shown in Fig. 25-11, there are several types of engine cylinder hones:

1. A *brush hone* has small balls of abrasive material formed on the ends of round metal brush bristles, Fig. 25-12A. It is desirable when the cylinder is in good condition and requires very little honing. It will also smooth a bored cylinder wall.
2. A *flex hone* has hard, flat, abrasive stones attached to spring-loaded, movable arms, Fig. 25-11. It is used when the cylinder wear is slight and a moderate amount of honing is needed.
3. A *ridged* or *sizing hone* has stationary, but adjustable, stones that lock into a preset position, Figs. 25-11 and 25-12B. It will remove a small amount of cylinder taper or out-of-roundness. A ridged hone can be used like a boring bar to true a cylinder when wear is within specs.

Fig. 25-11. Hones are needed to deglaze shiny, worn cylinder and to smooth rough bored cylinder walls. Rigid hone will help remove taper and out-of-round. Flex hone, if used more at bottom of cylinder will help true cylinder. Brush hone will produce very accurate surface finish to help ring sealing. (Snap-On Tools)

Power honing

Power honing is a machine shop operation that rotates rigidly mounted stones through a worn cylinder. Like boring, it will enlarge and true the walls of the cylinder. See Figs. 25-12B and 25-13.

Power honing can be used to true and clean up a worn cylinder and allow the use of standard size rings and pistons. The pistons must be knurled however to take

A

B

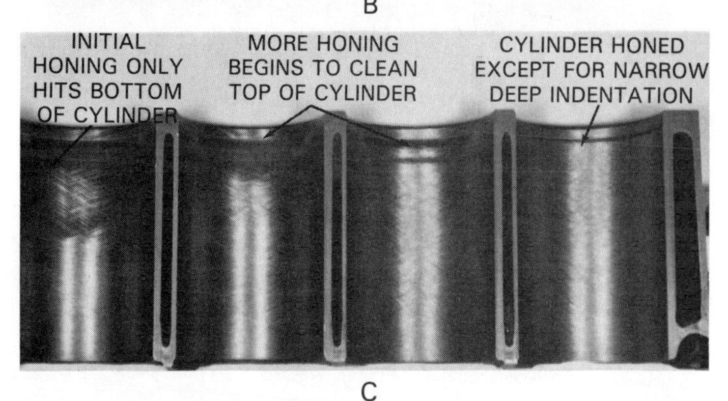

C

Fig. 25-12. Honing a cylinder. A—Technician is using electric drill to spin brush hone to remove glaze in cylinder with little wear. B—Machine shop is power honing cylinder to remove more severe wear. C—Note stages of power honing and how it helps true cylinder. Small dark ring at top of right cyinder is only .001 inch deep and would not affect ring sealing in passenger car engine. (Goodson and Sunnen)

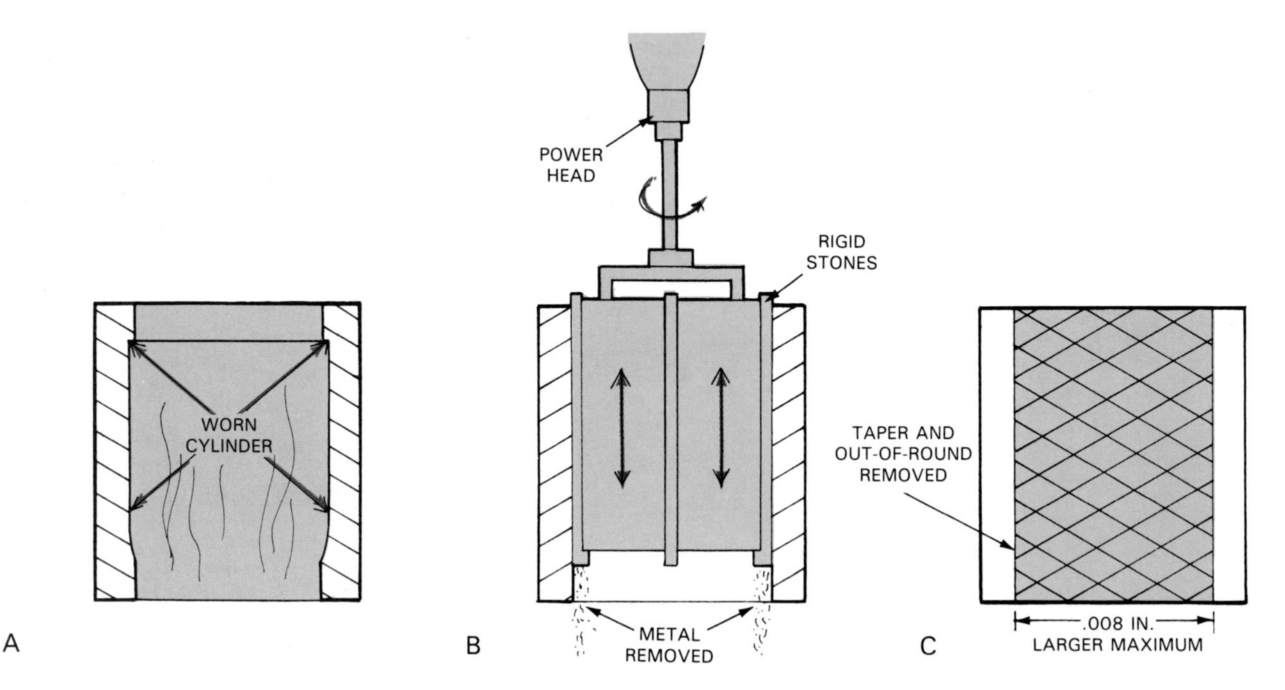

Fig. 25-13. Rigid power hone will remove cylinder taper and out-of-round, making cylinder equal size in all areas. A—Cylinder has minor scratches and small ring ridge. B—Power hone is spun inside cylinder. C—Power hone can be used to true cylinder up to .008 inch oversize. Standard rings and knurled pistons will normally keep everything within specs.

up the excess clearance, Fig. 25-12C.

Generally, standard rings and pistons can be used with a power hone cylinder enlargement of .008 in. (0.20 mm). Most new standard piston ring sets will seal in a cylinder that is .010 inch oversize. This will save you from having to bore the cylinder .010 in. larger and install oversize pistons and rings.

Note! For new rings to seal, the most critical cylinder aspect is roundness. The second more critical dimension is taper. Round rings must have a round cylinder to prevent blowby. Rings can still seal a slightly oversize cylinder wall if it is round and untapered.

Hone grit selection

As mentioned, it is critical that the cylinder wall have the proper smoothness or texture before installing new rings. A hone will place tiny scratches in the surface of glazed (shiny or used) cylinder wall. It will also smooth a bored (rough) cylinder wall.

The pocket in these hone marks or scratches hold oil to prevent ring scoring. The peaks or high points in the honed metal surface contact the pistons rings. They quickly wear down from ring friction so that the ring can match the shape of the cylinder wall.

If the cylinder surface is too coarse, ring and cylinder scoring could result. If the cylinder wall surface is too smooth, the ring may not seal properly and lead to oil consumption.

Surface smoothness is measured in *microns* or millionths of an inch. Special testers are available that can be passed over a surface to measure RMS (root mean square) which is a measurement of surface smoothness.

Cylinder smoothness or RMS should be matched to the type of piston ring. Generally, softer ring coatings require a coarser stone and surface. Remember these rules:

1. Use 220 grit stones with CAST IRON RINGS to produce a textured 25 to 30 RMS surface texture.
2. Use 280 grit stones with CHROME RINGS to produce a medium 20 to 25 RMS surface.
3. Use 320 grit stones with MOLY RINGS to produce a very fine 10 to 15 RMS surface.

This is not extremely critical but is an ideal recommendation. However, if hard moly rings are installed in a coarse finished cylinder, oil consumption and smoking can result.

Using a hone

To hone a cylinder, follow equipment manufacturer's instructions. Install the hone in a larger, low-speed electric drill. Compress the stones (squeeze inward) and slide them into the cylinder. Be careful not to scratch the cylinder. Turn on the electric drill and move the spinning hone up and down in the cylinder.

WARNING! Make sure you do NOT pull the hone out of the cylinder while honing. The hone could break and bits of stone may fly out.

Move the hone up and down in the cylinder fast enough to produce a 50 to 60 degree CROSS-HATCH PATTERN, illustrated in Fig. 25-14. Moving the hone up and down faster or slower will change the pattern.

If the cylinder has NOT been bored or power honed true, hone more at the BOTTOM OF THE CYLINDER. This will remove taper and enlarge the bottom slightly to make the cylinder equal in diameter at the top and bottom. Hone at the bottom for a few seconds and then move up and down to produce the 50 to 60 degree cross-hatch.

Note! It is usually acceptable if a small band in the cylinder, under the ring ridge, fails to polish up when honing. This will have little effect on ring sealing.

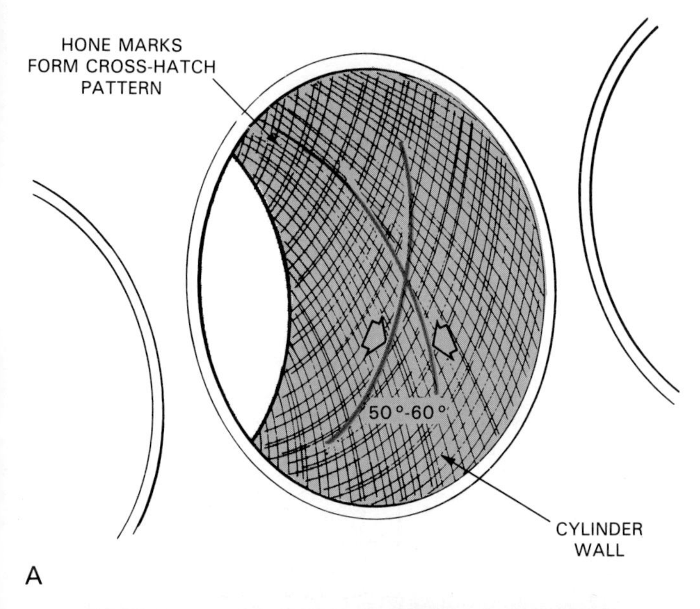

Fig. 25-14. A—When honing, try to produce a 50 to 60 degree crosshatch pattern. B—Actual hone marks on inside of cylinder.

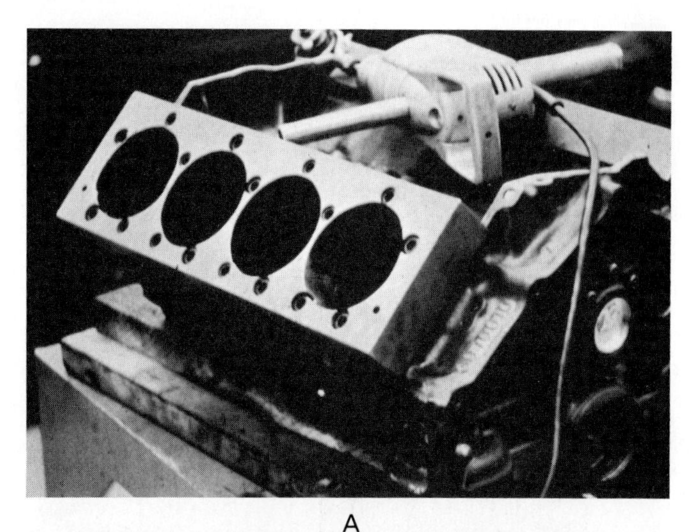

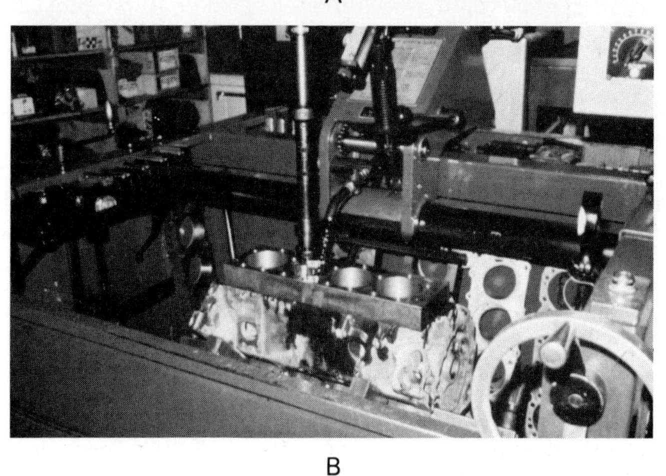

Fig. 25-15. A—This machine shop used a deck plate while power honing. The deck plate simulates the cylinder head being bolted to the block. Any cylinder distortion from head bolt torque is then corrected when honing. B—This block is being bored and honed for oversize pistons. (Sealed Power)

BLOCK BORING

Block boring is needed to correct excess cylinder wall wear, scoring, or scratching. Boring will true the cylinder walls so the rings and pistons will fit properly. It is also done when installing a repair sleeve.

Boring is needed when your measurements show cylinder wear is beyond specs. If .008 in. power honing will not clean up the cylinder, boring would be needed. If taper is more than .012 in. and out-of-round is more than .005 in., boring is also needed. After boring, you would have to install oversize pistons and rings.

A *boring bar* is a machine shop tool that will cut a thin layer of metal out of the cylinders. As shown in Fig. 25-15, the boring bar is rotated down through each cylinder

CAUTION: Always find and use a reputable machine shop. Your work is only as good as the machine work done on your parts. Do not trust the machine shop's work; always check any machined part before assembly.

Overbore sizes

Normally, a block is bored in increments of .010 inch. Depending upon the amount of wear, the block cylinders could be bored .010, .020, .030, .040, .050, or .060 inch oversize. Most shops bore cylinders either .030 or .060 inch oversize because these piston oversizes are the most available.

Overbore limit

The *overbore limit* is the largest possible diameter a block should be bored oversize. It is a spec given by the engine manufacturer. Some blocks can be bored .120 inch oversize while others can only be bored .020 inch oversize. The overbore limit depends upon how thick the cylinder walls are cast during manufacturing. A thin wall block could become too thin if its overbore limit were exceeded. The cylinder wall could crack from the pressure of combustion. Always check the overbore limit of the specific block before recommending an overbore diameter. The machine shop should also be able to determine the overbore limit.

Note! Boring increases engine compression slightly. This can be critical with some engines.

Oversize pistons and rings

Oversize pistons and *rings* are required to fit a cylinder block that has had its cylinders bored out. The pistons must be purchased to match cylinder oversize.

Boring and oversize pistons will make the engine like new. New pistons and rings will operate on a freshly machined cylinder surface, providing excellent ring sealing and service life.

If the block is bored .030 inch oversize, then you would have to order .030 inch oversize pistons and rings. The proper clearances are made into the oversize pistons.

Note! Make sure the correct piston oversize is available before having the block bored. If .020 inch oversize pistons are not available, you do NOT want to bore the block that oversize diameter. Many shops will not bore a block until after obtaining the new, oversize pistons.

Cylinder sleeving

Cylinder sleeving involves machining or boring one or more of the cylinders oversize and pressing in a cylinder liner. Sleeving is needed, for example, after part breakage has severely damaged (gouged, nicked, chipped, or cracked) the cylinder wall. The damage is too deep to clean up with power honing or boring.

Sleeving also allows the bad cylinder to be restored to its original diameter. The same size pistons can be reused. For example, if only one piston and cylinder are damaged, all of the other pistons and cylinders may be good and usable! Sleeving would save the customer money on the repair.

Sleeve installation is a machine shop operation. If you have an aluminum block with cast iron sleeves, sleeve replacement is simple. The machine shop can drive out the old sleeve and press in a new one. Boring is not needed. This can be done for a badly damaged cylinder, for example.

Sleeving a cast iron block is more complex. The old cylinder must be bored oversize to a diameter slightly smaller than the outside diameter of the sleeve. The sleeve is then pressed into the block. Then, the sleeve is bored to the correct bore diameter.

Sleeve or *liner protrusion* is the distance the sleeve sticks up above the deck of the block. This measurement is critical to head gasket sealing. Fig. 25-16 shows how to check sleeve protrusion.

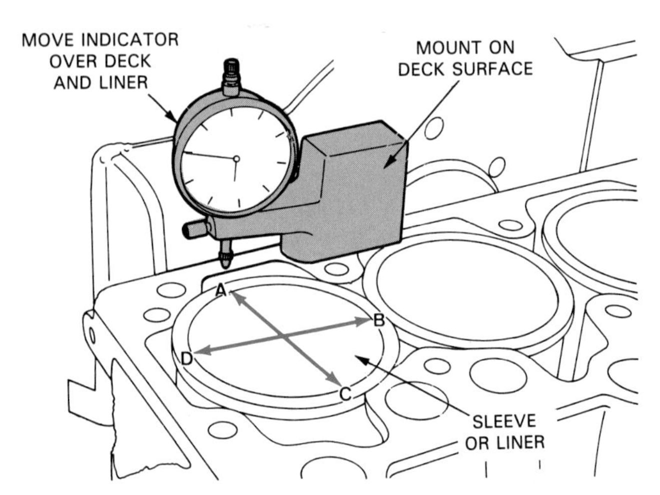

Fig. 25-16. Proper cylinder liner protrusion is critical to head gasket sealing. Note how dial indicator is used to measure height of liner above block deck surface. (Renault)

Cleaning cylinder walls

After boring, sleeving, and honing, it is very important to clean and remove all *grit* (bits of stone and metal) from inside the engine. This grit, if not removed, will act like grinding compound circulating through the engine. It can wear bearings, rings, and other vital engine parts.

Shown in Fig. 25-17, a pressure washer is the most efficient way of cleaning a cylinder block. The high pressure soap and water will blow all grit out of the cylinders and crankcase. After pressure washing, blow the block dry and wipe the cylinders down with motor oil.

Fig. 25-17. Pressure washer is quick and efficient way of cleaning cylinder block after boring or honing. It will blow all debris out of hone marks and inside of crankcase. Wash block by hand if you do not have pressure washer. (Sealed Power)

If a pressure washer is not available, hand wash the cylinders with warm water and soap (detergent). A soft bristle brush, NOT a wire brush, will quickly loosen and remove particles inside the hone marks. Blow the block dry with compressed air.

Next, place motor oil on a clean shop rag. Wipe the cylinder down thoroughly with the OIL-SOAKED RAG. The heavy oil will pick up any remaining stone grit embedded in the cylinder honing marks. Wipe the cylinders down until the rag comes out clean.

After cleaning, recheck the cylinder for scoring or scratches. If honing did NOT clean up all of the vertical scratches in the cylinder, cylinder boring or sleeving may be needed.

Check all block threads for damage and cleanliness. Clean all threads and repair any damaged threads found. Thread repair was discussed in Chapter 3.

CAM BEARING INSTALLATION

New cam bearings should be installed during major engine service. Worn cam bearings are common and can lower engine oil pressure.

Many technicians have the machine shop intall cam bearings. However, some technicians have the special

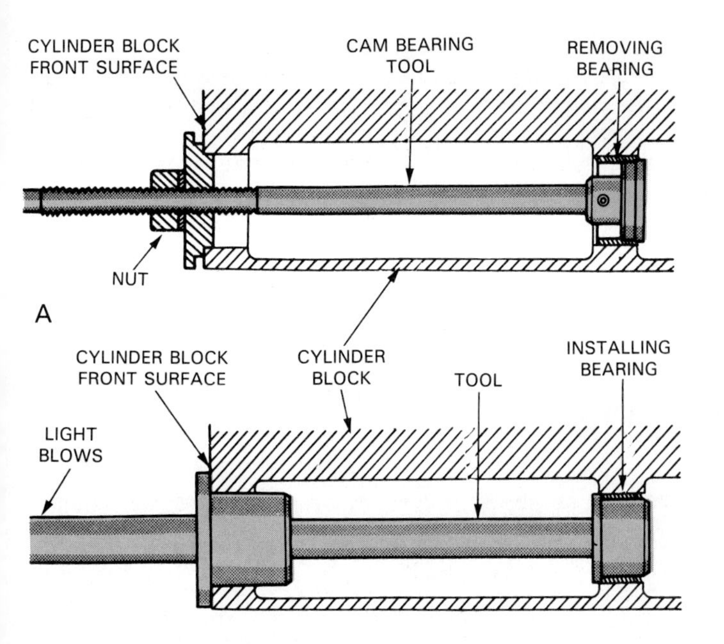

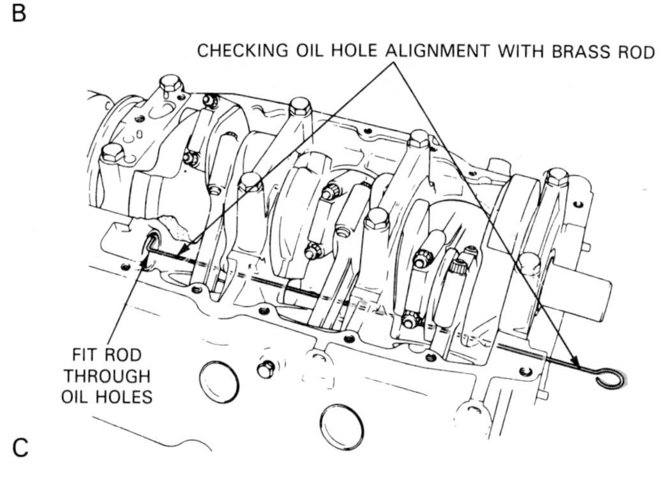

Fig. 25-18. A—Driver is being used to pull old cam bearings out of block. B—Driver being used to push new cam bearings into place. C—Make sure cam bearing oil holes are aligned after installation or severe valve train damage can result.

drivers for pressing cam bearings in and out of the block.

Fig. 25-18 summarizes how camshaft or balancer shaft bearings are removed and installed.

Basically, a driver is used to force the used bearings out of their bores. The new bearings are then forced into place using a similar bearing driver.

It is critical that all oil holes in the block line up properly with the oil holes in the cam bearings. If they do not align, severe valve train wear can result from lack of lubrication.

Before engine reassembly, double-check to make sure the block is in acceptable condition. Remember that the block is the foundation of the engine rebuild. If the block has problems, you are "building your house on quicksand."

CRANKSHAFT SERVICE

Crankshaft service involves measuring crank journal wear, checking crank straightness (if symptoms or damage require it), checking for cracks (if severely damaged from rod breakage), cleaning crank, and proper installation. Quite often, the crankshaft will be in good condition and nothing must be done to the crank other than checking for wear. However, in a high mileage engine, oil pump wear and reduced lubrication can cause excess journal wear requiring crank grinding.

Look at each connecting rod and main journal surface closely. Look for scratching, scoring, and any signs of wear. The slightest nick or groove is VERY SERIOUS.

Very fine crocus cloth may be used to clean up minor burrs or marks on journals. Polish around the journal, NOT across it.

Measuring journal taper

If one side of a crankshaft journal is worn more than the other, the crank journal is *tapered.* To measure journal taper, use an outside micrometer, as shown in Fig. 25-19. Measure both ends of each journal, Fig. 25-20. Any difference indicates taper. Taper beyond recommended limits requires crankshaft turning.

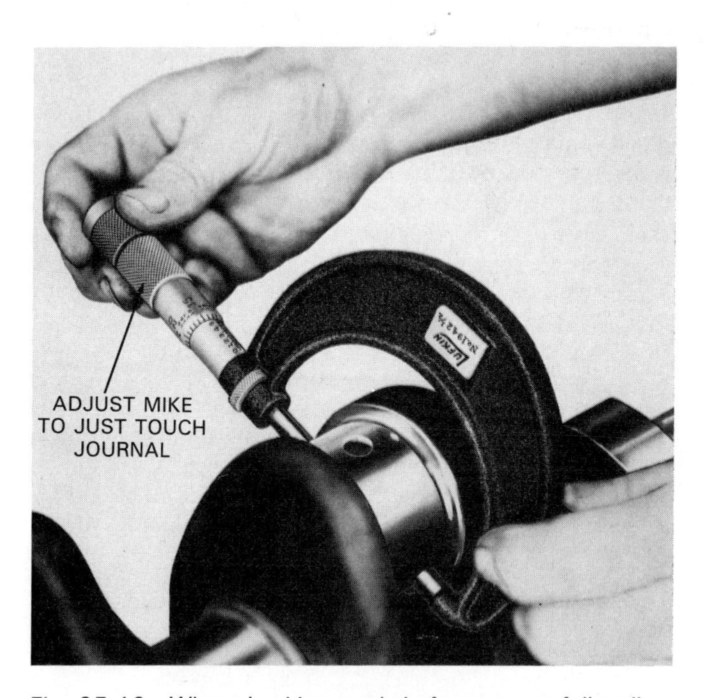

Fig. 25-19. When checking crankshaft wear, carefully adjust mike to just drag over journals. (Lufkin)

Measuring journal out-of-round

When you measure for journal taper, also measure *journal out-of-roundness* (journal worn more on top than bottom). As in Fig. 25-20, measure across the journal from side to side then from top to bottom. If not within spec limits, send the crankshaft to a machine shop for grinding.

Generally, crankshaft journal taper and journal out-of-round must not exceed .0005 to .001 inch (0.13 to 0.025 mm). If worn more, you should have the crankshaft ground undersize. Check service manual specs for the exact engine.

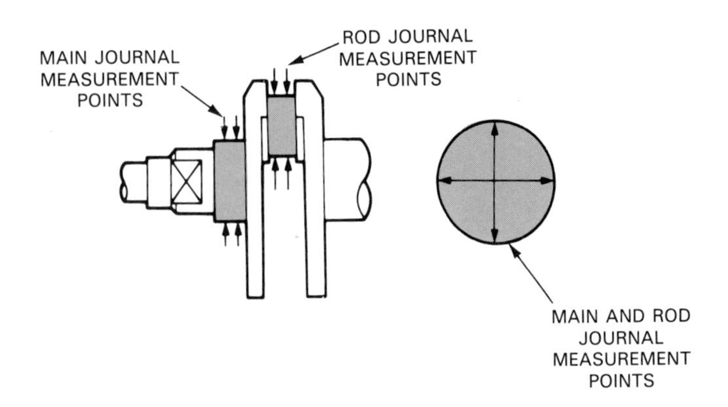

Fig. 25-20. Measure crank and main journals at these locations. This will check journal taper and out-of-round. (Buick)

Checking for cracked crankshaft

A *cracked crankshaft* can result from part breakage, harmonic balancer failure, flywheel failure, or driver abuse. Cracks can be hard to find on a crankshaft because they can be so small.

As described earlier, Magnafluxing will help locate cracks. Sprinkle the metal powder over the crank. The magnetic field will make the powder collect in any crack. Cracks usually form on the side of the journals. Replace the crankshaft if cracked.

Checking crankshaft straightness

A *bent crankshaft* can ruin new main bearings or cause the engine to lock up when the main caps are tightened.

To measure crankshaft straightness, mount a dial indicator against the center main journal. The crank can be mounted on V-blocks, Fig. 25-21. The crank can also be placed in the block main bearings.

Slowly turn the crank while watching the indicator. Indicator movement equals crankshaft bend. If runout is not within spec limits (about .001 inch or 0.03 mm), replace the crank or have it straightened and turned by a machine shop.

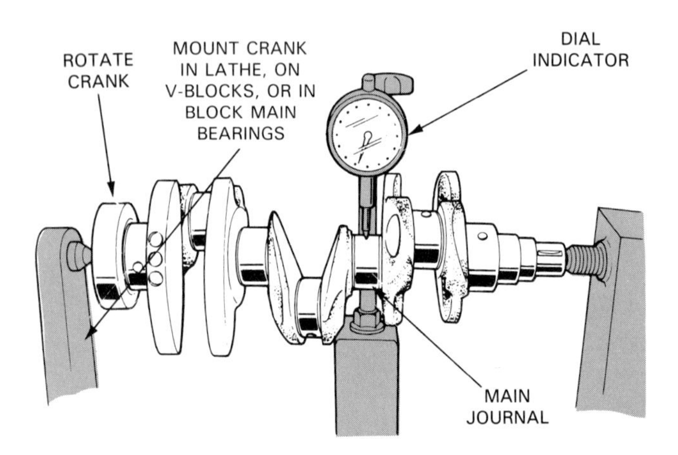

Fig. 25-21. If parts were broken or main bearings were badly worn, check crankshaft runout. Mount crank in lathe, V-blocks, or in block main bearings. Mount dial indicator on main journal. Turn crank and read indicator showing runout. (Pontiac)

Crankshaft turning or grinding

Crankshaft turning involves grinding the rod and main journals smaller in diameter to fix journal wear or damage. This is done by a machine shop, Fig. 25-22.

Undersize bearings are needed after the crankshaft has been turned. Since the crank journals are smaller in diameter, the bearings must be thicker to provide the correct bearing-to-journal fit.

Fig. 25-22. Machine shop is grinding crankshaft main and rod journals undersize to correct wear or damage. You would have to purchase undersize bearings to fit crank. (Landis Tool)

Undersize bearing markings

Always look at the old crankshaft bearings to determine whether they are undersize. Undersize bearings may be used on rebuilt engines or even in a few new engines.

Inspect the back of the old bearings for an UNDERSIZE NUMBER, as in Fig. 25-23. The number stamped on the bearing back denotes bearing undersize (.010 in. for example). A letter code can also be used to give bearing and undersize.

If during the engine repair, you find that the old crankshaft is in good condition, install the same size bearings as the old ones. However, if the crank is badly worn or damaged, exchange the crankshaft for a new or reconditioned (ground undersize) crank. Then, install the appropriate size main and rod bearings.

Cleaning crankshaft

The crankshaft must be perfectly clean before installation into the block. This is especially critical if the crank has been ground and debris (metal shavings, stone grit, etc.) are inside the oil passages.

Remove any gallery plugs, as shown in Fig. 25-24. Wash the crank carefully in parts cleaner. Then, use a blow gun to blow out all passages and dry off the journals. Reinstall the core or galley plugs. Wipe the journals down with an oil-soaked rag.

Note! Freshly overhauled engines have been ruined when technicians failed to remove crankshaft core plugs

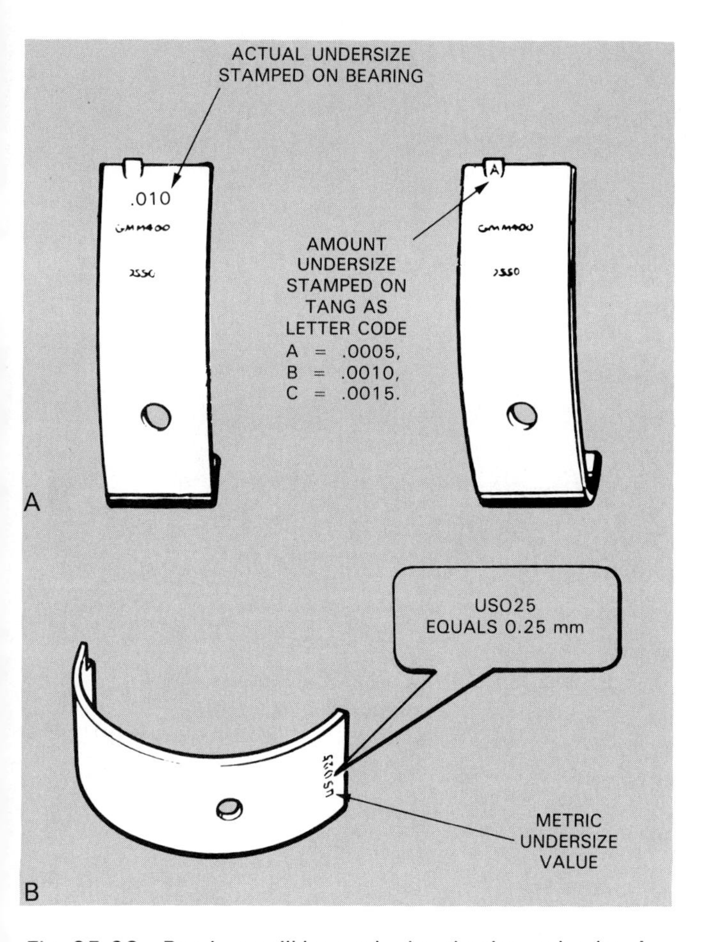

Fig. 25-23. Bearings will be marked undersize on backs. A—Usually, number is given for .010, .020, etc., inch undersize. Letter code requiring manual can also be given. B—Do not confuse metric undersize markings with conventional. This 025 marking means the bearing is for a 0.25 mm undersize crank journal. This would equal about a .010 inch undersize. (General Motors)

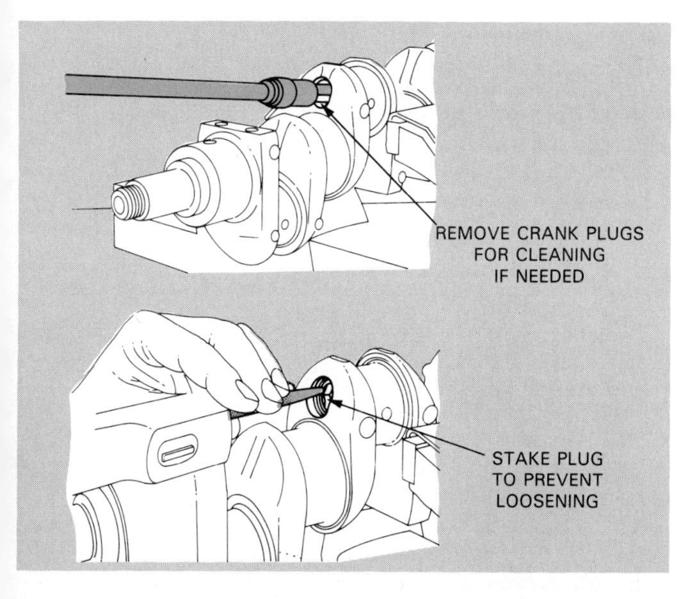

Fig. 25-24. After crank grinding, make sure you remove any plugs. Debris can collect inside hollow crankshafts. Inside of crank must be cleaned to prevent bearing damage upon engine operation.

for cleaning. When a crank is ground, stone grit and metal particles can enter and fill the pockets inside the crank. Bits from bearing failure can also collect inside the crank. If not removed, this debris can flow out and onto the bearings during engine operation. Bearing damage and failure can result.

INSTALLING REAR MAIN OIL SEAL

There are three basic types of rear main oil seals: two-piece synthetic rubber seal, two-piece rope (wick) seal, and one-piece synthetic rubber seal. Each requires a different installation technique.

Two-piece rubber seal

The two-piece synthetic rubber rear seal is very easy to install. Simply press it into place in the block and rear main cap.

The sealing lip on the rear main seal must point towards the inside of the engine. If installed backwards, oil will pour out of the back of the seal upon engine starting. Lubricate the sealing lip with motor oil, Fig. 25-25.

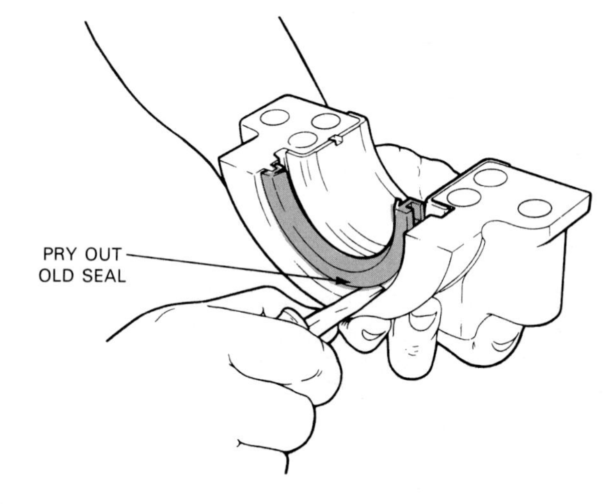

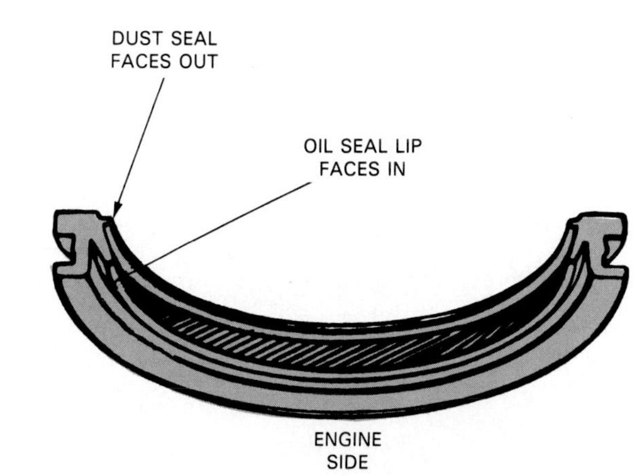

Fig. 25-25. A—Use small screwdriver to pop out old two-piece rubber seal from cap and block. B—Look at new seal closely to make sure oil sealing lip faces inside of block. (Oldsmobile)

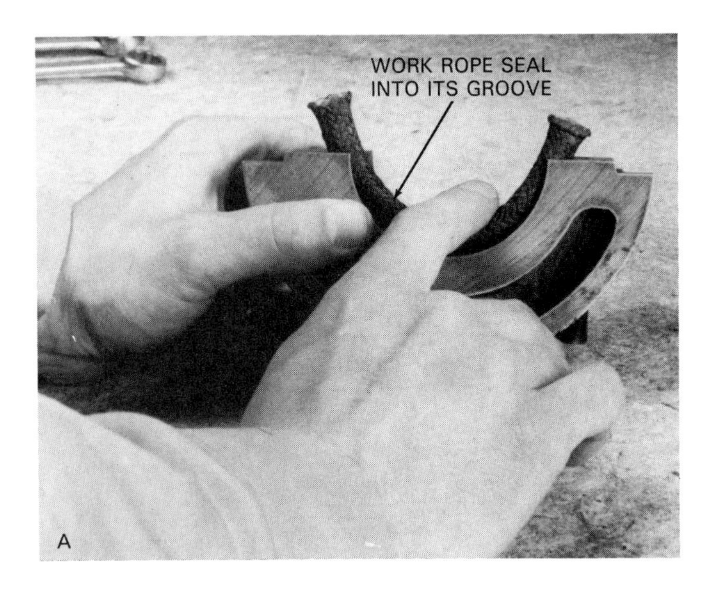

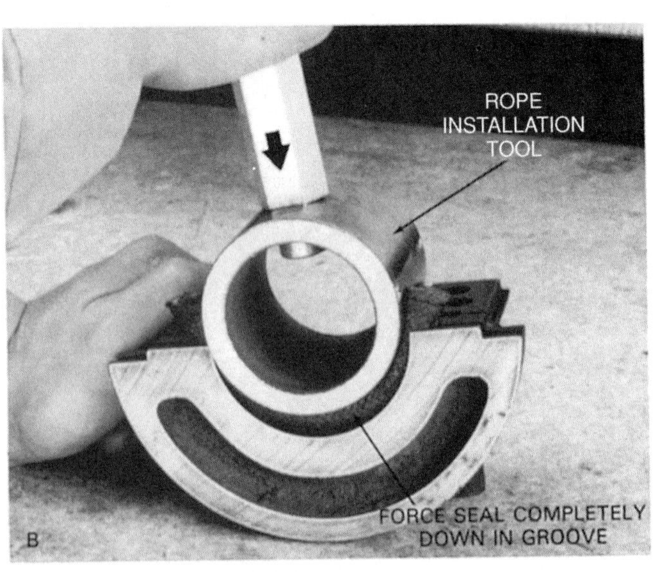

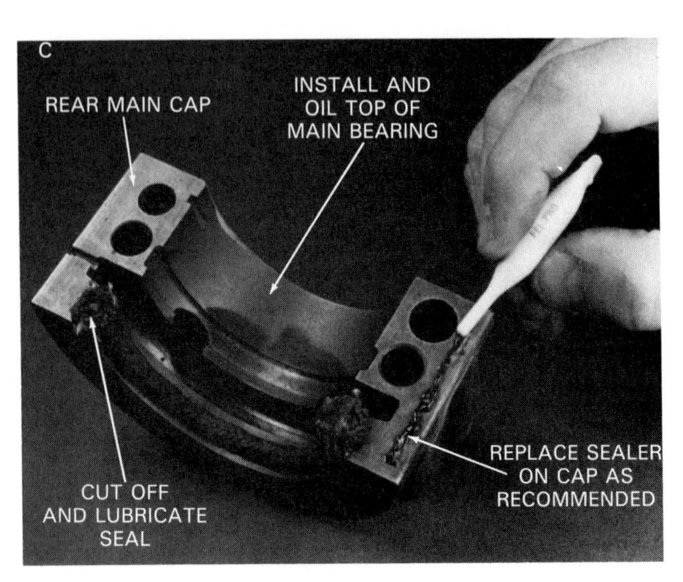

Fig. 25-26. Installing rope type oil seal. A—Start rope into its groove in cap and block. B—Use special driver or round bar to force seal completely into groove. C—Cut ends of rope flush with single-edge razor blade and apply recommended sealer to cap. (Fel-Pro Gaskets)

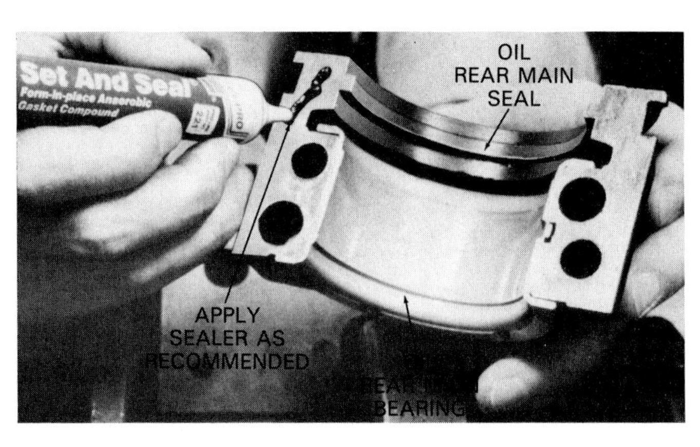

Fig. 25-27. Do not forget to seal main cap when needed. Anaerobic sealer is needed where there is tight metal-on-metal contact. RTV or silicone sealer is for a more loose fit, as on sides of cap. (Fel-Pro)

Rope seal

A two-piece rope or wick rear main oil seal must be worked down into its cap and block groove carefully. Use a special seal installing tool and light hammer blows or hand pressure and a smooth steel bar, as in Fig. 25-26. Use a single-edge razor blade to cut the rope seal flush with the cap and block parting line, Fig. 25-26.

Note! *Silicone sealer* or anaerobic sealer are commonly recommended on the rear main cap to prevent oil leakage. Check the manual on which type to use. The sealer keeps oil from seeping between the main cap and block mating surfaces. Fig. 25-27 gives an example.

If additional side seals are provided for the cap, follow the instructions with the gasket set. Sealer may be recommended on main cap side seals.

One-piece rubber seal

A one-piece synthetic rubber seal is usually installed after the rear main cap has been bolted to the block. It is driven into position from the rear of the engine using a seal driver. After installation, make sure the seal is square and undamaged.

Fig. 25-28 summarizes the steps for one-piece rear main seal installation.

MAIN BEARING AND CRANK INSTALLATION

With the block honed and cleaned, you are ready to install the main bearings and crankshaft. Discussed earlier, make sure you have the correct size main bearings to go with the crankshaft.

Installing main bearings

With a clean, dry rag, wipe the main bores free of any dust or oil. Check main bearing sizes on the boxes and backs of the bearings.

Without touching the front of the bearing inserts, snap each main bearing into place in the block. The bearing tab fits into a notch in the main bore. Make sure the oil holes in the bearings line up with oil holes in the cylinder block, Fig. 25-29A. Snap the other bearing halves into the main caps. Check that the main thrust bearing is located properly.

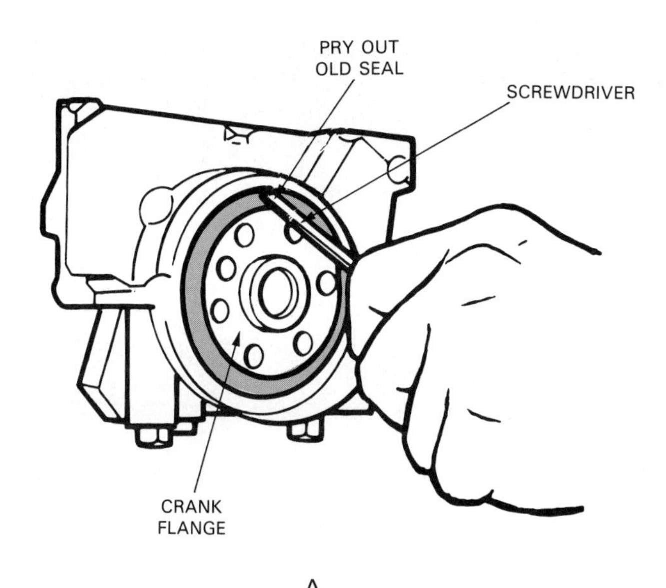

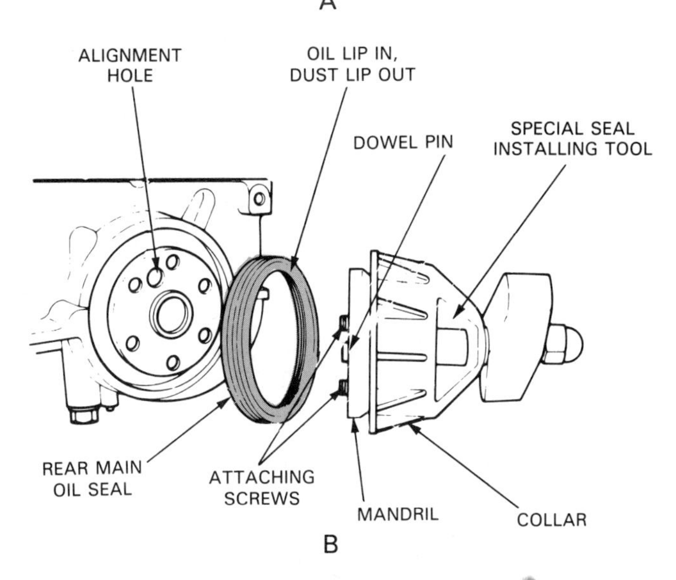

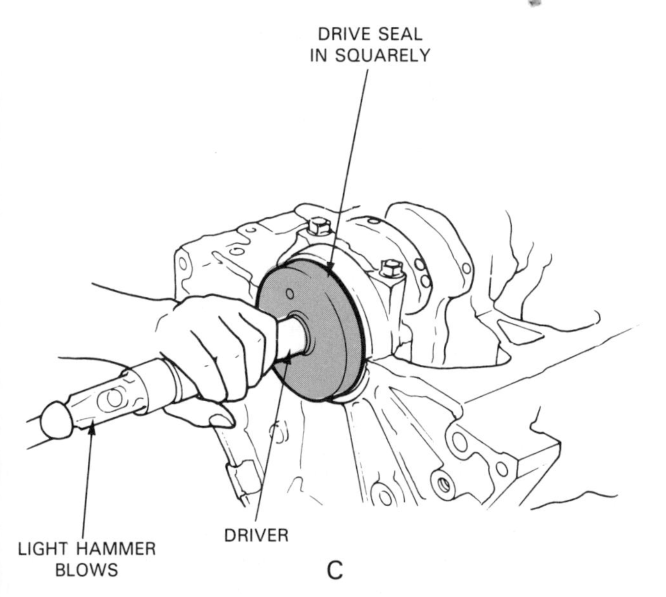

Fig. 25-28. Installing modern one-piece rear main seal. A—Pry out old seal without scratching block. B—This special seal installing tool is tightened with wrench. C—Light hammer blows are used to push seal squarely into place. (Buick and Pontiac)

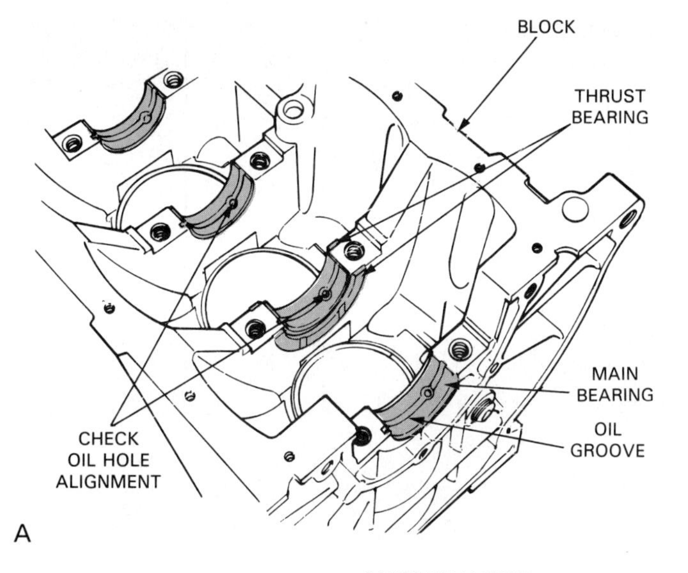

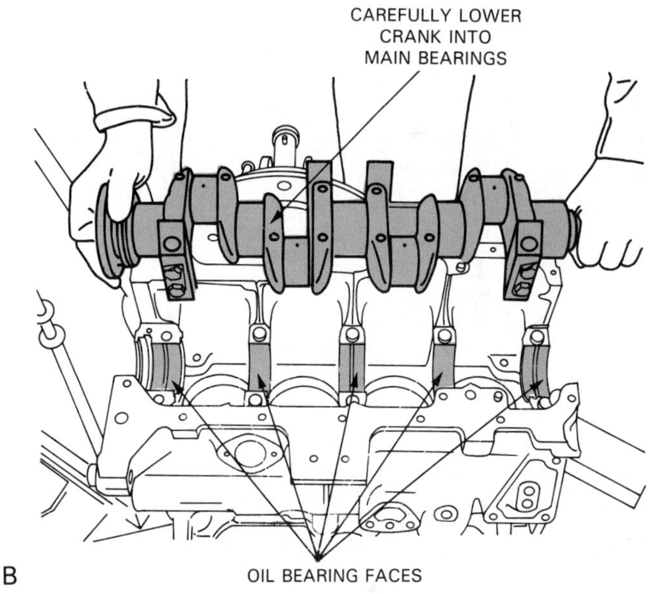

Fig. 25-29. A—Install main bearings clean and dry. Check that all oil holes align. Then, coat faces of bearings with heavy engine oil or white grease. B—Holding snout and flange, carefully lower crank into main bearings. Do not hit journals or bearings. (GM and Peugeot)

WARNING! Never oil the backs of bearings or bearing bores. This could decrease bearing clearance and make the bearings spin in their bores upon engine starting.

Installing crankshaft

With the main bearings and rear main seal installed, coat the tops of the bearings and rear seal lip with heavy engine oil or white grease. Then, as in Fig. 25-29B, carefully place the crank into the block. Be careful not to hit the bearings or journals. Do not turn the crank without the main caps in place or the bearings could shift out of place.

Checking main bearing clearance

To check the oil clearance between the crank journal and main bearing, place a small bead of PLASTIGAGE (clearance measuring device) on the unoiled crankshaft.

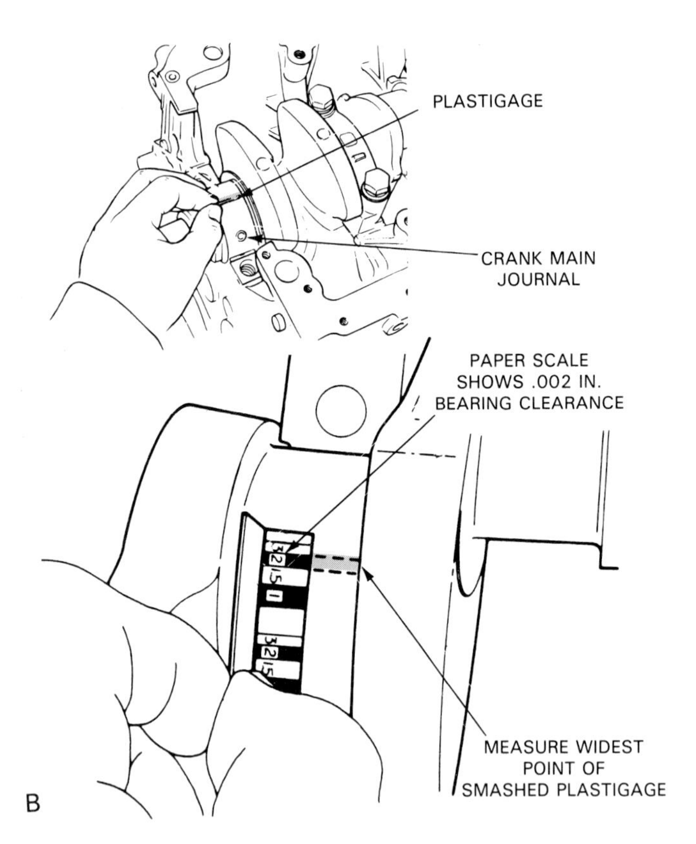

Fig. 25-30. Checking main bearing clearance. A—Place strip of Plastigage on journal, parallel with crank centerline. Install and torque cap with bearing installed. Remove cap. B—Use paper scale to find clearance. Match width of smashed plastic with scale. If plastic is same as two, then the oil clearance is .002 inch. (Toyota and Buick)

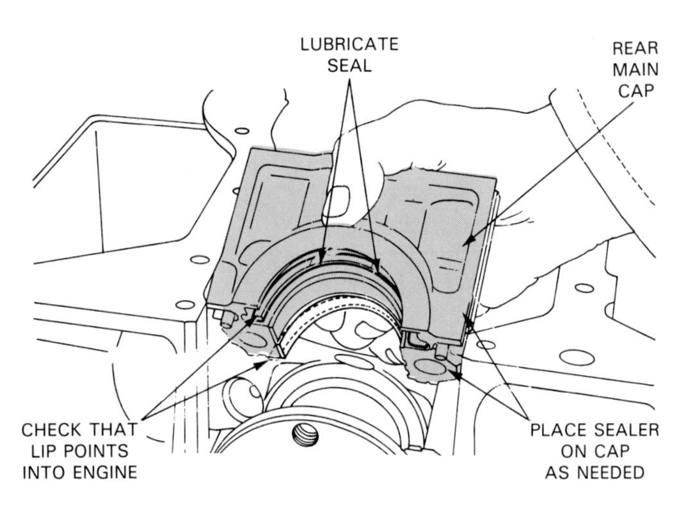

Fig. 25-31. When installing rear main cap, check installation of rear main seal. Use sealer as recommended. (Peugeot)

Install and torque the main cap, Fig. 25-30A.

Remove the main bearing cap and compare the smashed Plastigage to the paper scale, Fig. 25-30B. If clearance is NOT correct, check bearing sizes and crankshaft journal measurements.

An average main bearing clearance would be .001 to .002 inch or 0.025 to 0.05 mm. This can vary up or down slightly so check engine specs. Some technicians only check one main bearing clearance; others check all of the bearings.

Torquing main caps

Place each main cap into the block. Check main cap NUMBERS and ARROWS to be sure they are positioned correctly. Each cap must be facing the right direction and in its original location. Spin the bolts down but do not tighten them.

When installing the rear cap, double-check the rear main oil seal. If it is a two-piece rubber type, make sure the sealing lip is pointing toward the inside of the engine, Fig. 25-31. Remember to place sealer on the rear cap if needed.

Look up the torque specs for the main bearing cap bolts in a service manual. Then, use a torque wrench to tighten the bolts to about one-half torque. Go over all of the main cap bolts, as in Fig. 25-32. Tighten the bolts to three-fourths torque. Then, tighten the main cap bolts to their full torque at least TWO TIMES. This will assure proper torque and that no bolts have been skipped.

Note! It is possible for a main cap to become cocked in the block. The outer edge of the cap could catch on

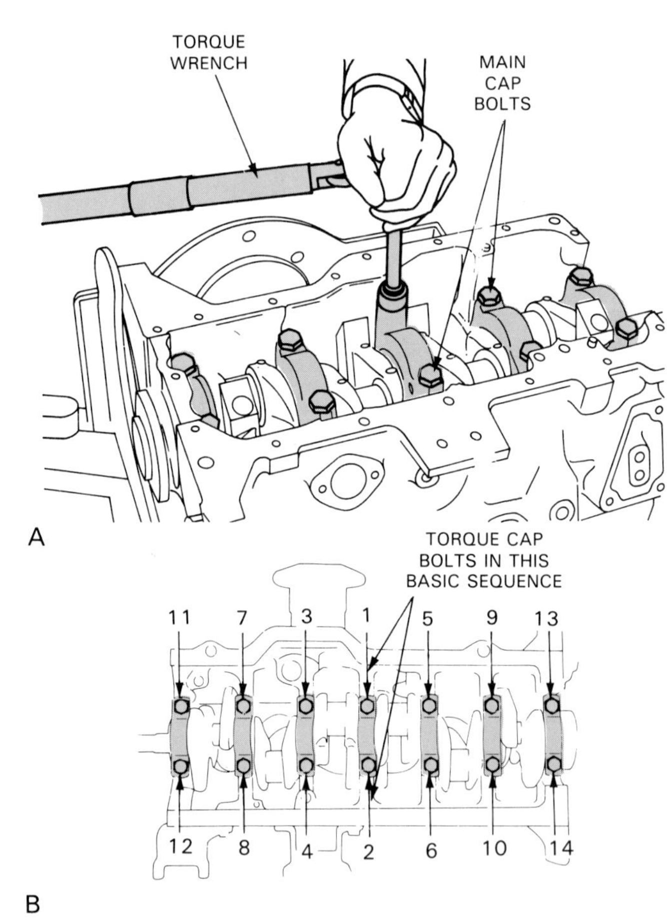

Fig. 25-32. Torquing block main cap bolts. A—Make sure all caps are fully seated. Torque each bolt a little at a time up to partial specs. Use an even pull on the wrench. B—Torque each cap in sequence as shown. This will pull each cap down squarely and help avoid forgetting any bolt.

the side of the block. Make sure the caps are square and fully seating before torquing them.

Checking crankshaft end play

Crankshaft end play is the amount of front-rear movement of the crankshaft in the block. It is controlled by the clearance between the main thrust bearing and the crankshaft thrust surface or journal.

To measure crankshaft end play, mount a dial indicator on the block. Position the dial indicator against the crankshaft so that the indicator stem is parallel with the crank centerline. This is illustrated in Fig. 25-33.

Zero the indicator. Then, pry back and forth on a counterweight with a small pry bar. Indicator needed movement equals crankchaft end play.

Compare your measurements to specs. If not within specs or about .006 inch (0.15 mm), check the main thrust bearing and width of the crank journal thrust surfaces.

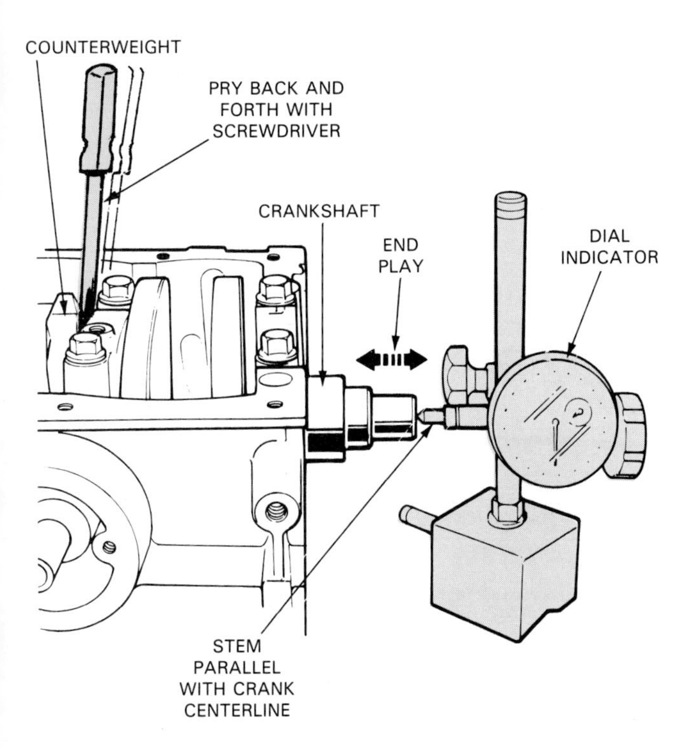

Fig. 25-33. To check crankshaft end play, use dial indicator as shown. Make sure indicator plunger stem is parallel with crank. Pry back and forth on crank counterweight while reading indicator. (Buick)

Replacing pilot bearing

If the car has a manual transmission or transaxle, you should replace the pilot bearing in the rear of the crankshaft. It supports the end of the transmission input shaft. If worn, clutch and transmission problems can develop.

To remove the pilot bearing, fill the cavity in the crank and pilot with heavy grease. Then, use an old input shaft or driver of the correct diameter to drive out the old pilot bearing.

Use hammer blows to force the shaft or driver into the grease. The grease will build pressure on the back of the pilot bearing, pushing it out of the crank.

To install the new pilot bearing, drive the pilot squarely into the crank, as shown in Fig. 25-34. Be careful not to damage the inside diameter of the bearing. Place a small amount of grease in the pilot bearing.

PISTON SERVICE

Pistons are made of aluminum which is very prone to wear and damage. It is very critical that each piston be checked thoroughly. Look for cracked or collapsed skirts, worn ring grooves, cracked ring lands, pin bore wear, or other problems. You must find any trouble that could affect piston performance and engine service life. Discard any damaged pistons.

Cleaning piston ring grooves

Detailed in the previous chapter, make sure you use a ring groove cleaner or old broken ring to clean the piston grooves. All hard deposits must be removed from inside the piston ring grooves, Fig. 25-35.

If not cleaned, the hard deposits could force the new rings out against the cylinder walls upon piston heating and expansion. This could score the rings and cylinder walls, possibly causing the engine to lock up (rings melt and weld to cylinder walls).

Measuring piston wear

A large outside micrometer is used to measure piston wear. Mike readings are compared to specs to find wear.

Piston size is measured on the skirts, just below the piston pin hole, as in Fig. 25-36. Measure perpendicular to the piston pin, Fig. 25-37. Adjust the micrometer for a slight drag as it is pulled over the piston. If worn more than specs (about .005 inch or 0.13 mm), replace or knurl the piston(s).

If in doubt, check in the service manual for directions to properly measure piston wear. Some engine manufacturers require measurement even with the pin hole.

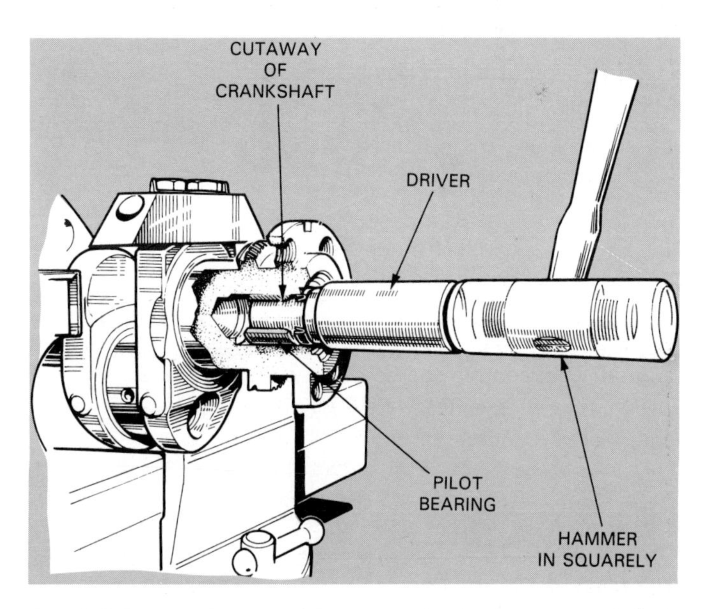

Fig. 25-34. If needed, install new pilot bearing in crank. Use heavy grease and driver to force out old bearing. Then use driver to push new bearing squarely into position. (Renault)

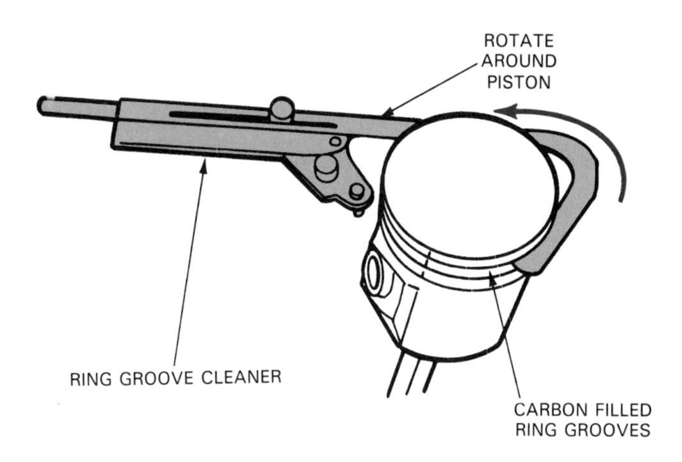

Fig. 25-35. Make sure you clean carbon out of ring grooves or new rings could be scored upon piston heating and expansion. Ring groove cleaner is fast method of removing carbon from inside piston grooves. (Ford)

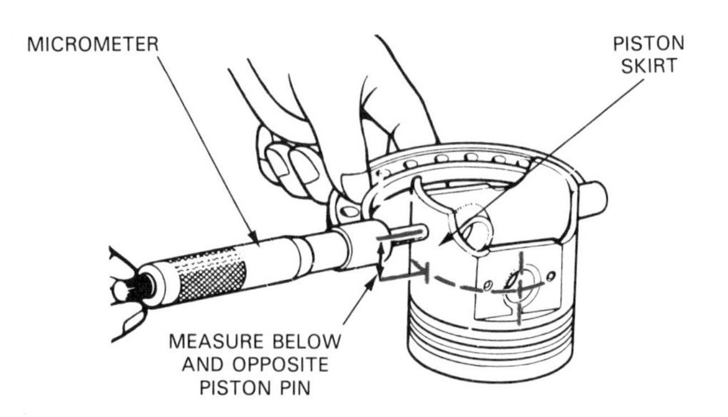

Fig. 25-336. Usually, measure piston wear just below piston pin on piston skirt. Adjust mike for slight drag when pulled across skirt. Compare your measurements to specs. (Toyota)

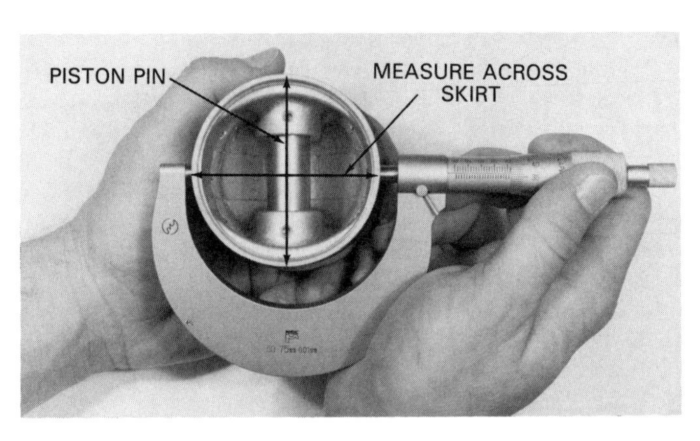

Fig. 25-37. Note how piston measurements should be opposite piston pin on skirt. Hold mike properly when measuring for accurate readings.

Others require measurement lower on the skirt. This will make their piston diameter specs inaccurate if you measure in the wrong location. Most give specs for just below the piston pin, perpendicular to the pin, however.

Knurling a piston

Piston knurling can be used to increase the diameter of the skirt a few thousandths of an inch (hundredths of a mm). A machine shop can usually knurl pistons and increse piston diameter up to .005 inch (0.13 mm).

Knurling makes grooves in the skirts, pushing up metal next to the grooves. This increases piston diameter and also provides grooves for motor oil to reduce friction.

Fig. 25-38 shows a piston knurler and a knurling operation. This is an easy way to salvage used pistons that are slightly worn. It will also take up some of the clearance in a slightly worn cylinder to prevent piston slap or knock.

Fig. 25-38. Pistons with slight wear can be knurled to increase their diameter and provide proper cylinder clearance. A—One type of piston knurling machine. B—Knurling roller smashes grooves in soft aluminum and displaces metal up around grooves to increase piston size. (K-Line and Deere & Co.)

Measuring piston clearance

To find *piston clearance,* subtract piston diameter from cylinder diameter. The difference between the cylinder bore measurement and the piston diameter measurement will equal piston clearance. Cylinder and piston measurement were explained earlier.

Average piston-to-cylinder clearance is about .001 to .003 inch (0.025 to 0.080 mm). Since specs vary, always refer to the service manual.

Another way of measuring piston clearance is with a feeler gauge strip. A long, flat feeler gauge strip is placed on the piston skirt. Then the gauge and piston are pushed into a cylinder. A spring scale is used to pull the feeler gauge strip out of the cylinder. When the spring scale reading equals specs, the size of the feeler gauge equals piston clearance.

When piston-to-cylinder clearance is excessive, you must either:

1. Knurl the pistons.
2. Install new standard size pistons (providing cylinders are not worn beyond specs).
3. Bore the cylinders and purchase oversize pistons.
4. Sleeve the cylinders.

Measuring piston ring gap

Discussed in earlier chapters, *piston ring gap* (clearance between ends of ring when installed in cylinder) is very important. If the gap is too small, the ring could lock up or score the cylinder upon heating and expanding. If the ring gap is too large, ring tension against the cylinder wall may be low, causing blowby.

To check ring gap, compress and place a compression ring in its cylinder. Then, push the ring to the bottom of normal ring travel with the head of a piston. This will square the ring in the cylinder and locate it at the smallest cylinder diameter, Fig. 25-39A.

Measure ring gap with a flat feeler gauge. Compare your measurements to specs. If not correct (between .010 to .020 inch or 0.25 to 0.51 mm), you may have the wrong piston ring set or cylinder dimmensions may be off, Fig. 25-39B.

Some manufacturers allow *ring filing* (special, thin grinding wheel removes metal from ends of ring) to increse piston ring gap. Others do not!

Checking ring groove depth

To check ring groove depth, fit a new compression ring into the piston groove, as shown in Fig. 25-40. Make sure the ring will fit below the surface of the piston. This makes sure you have to correct ring set. If the ring wall thickness were too large, the rings could be forced out against the cylinder walls after piston heat expansion. Part damage could result.

Measuring piston ring side clearance

Piston ring side clearance is the space between the side of a compression ring and the inside of the piston groove. Ring groove wear tends to increase this clearance. If worn too much, the ring will not be held square against the cylinder wall. Oil consumption and smoking can result.

To measure ring side clearance, obtain the new piston rings to be used during the overhaul. Insert or install the

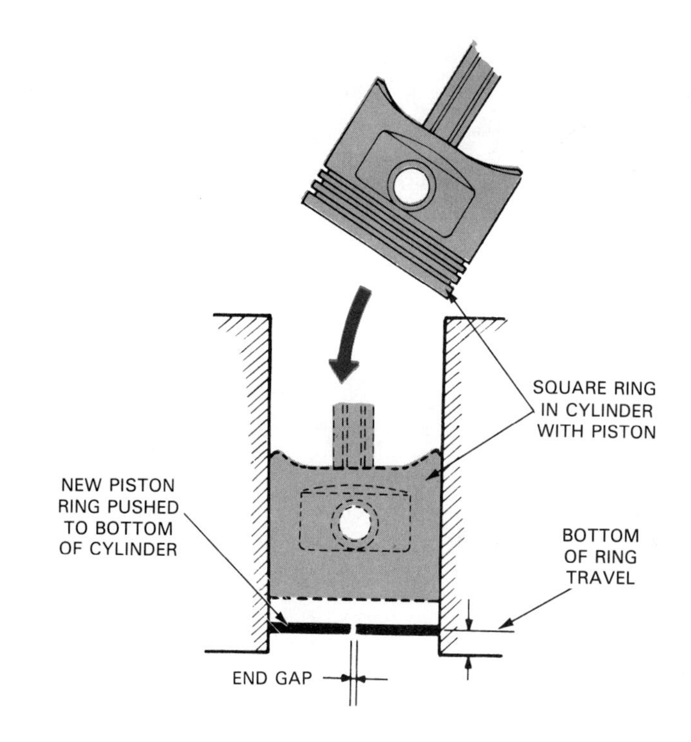

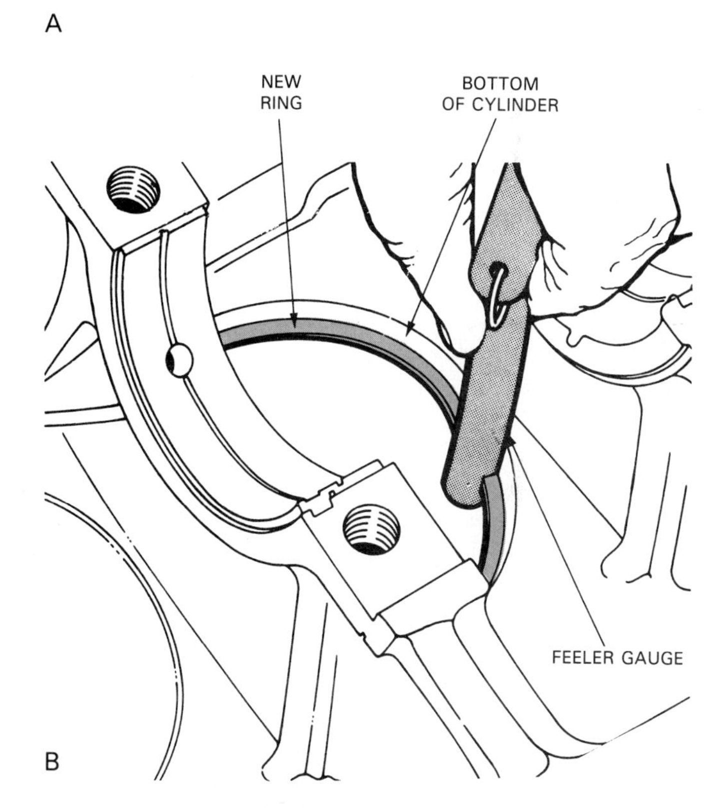

Fig. 25-39. Checking ring gap in cylinder. A—Use piston to push new ring to bottom of ring travel in cylinder. This will square ring in cylinder. B—Use feeler gauge to check ring gap. If too small, file ring ends so ring does not lock against cylinder wall upon heat expansion. (Toyota and Chrysler)

new ring into its groove. Then slide a feeler gauge between the ring and groove, Fig. 25-41.

The largest feeler gauge that fits between the ring and groove indicates ring side clearance. The top ring groove is usually checked because it suffers from more com-

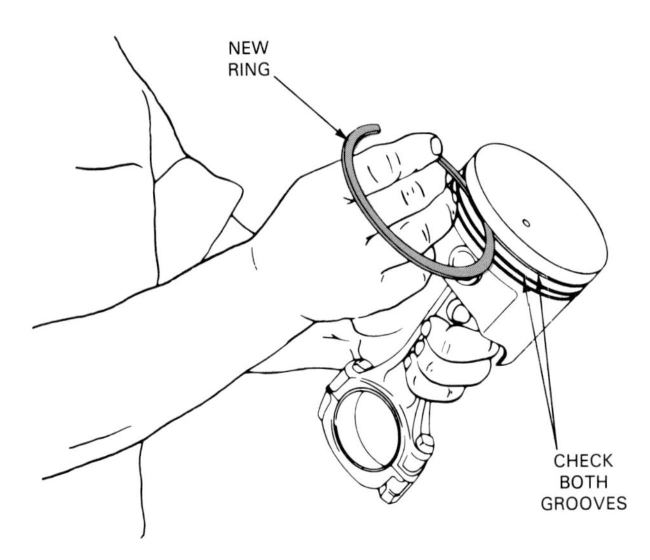

Fig. 25-40. Fit ring into groove. Make sure ring will go below surface of piston. This checks groove depth with ring. (Pontiac)

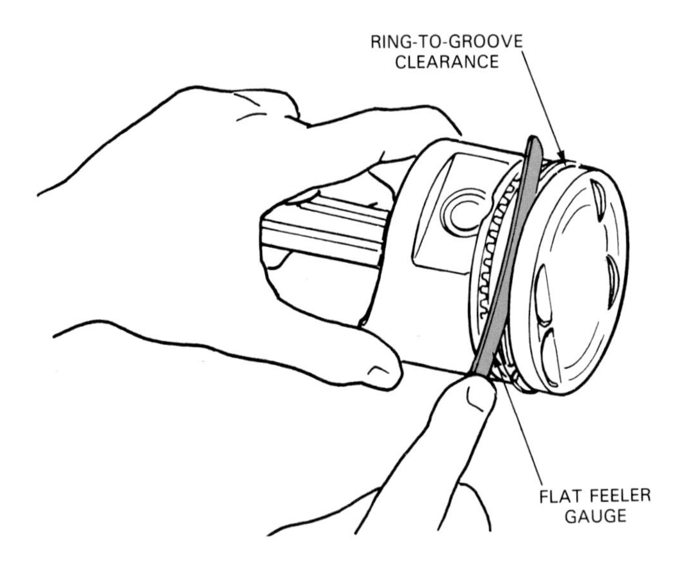

Fig. 25-41. To check ring side clearance, insert feeler gauge between ring and groove. Thickest gauge that fits equals clearance. If too large, replace pistons, check ring dimensions, or have groove cut for ring side spacers. (Toyota)

bustion heat and wear than the second groove.

If ring side clearance is beyond specs (about .002 to .005 inch or 0.05 to 0.13 mm), either repace the pistons or have a machine shop fit ring spacers in the grooves.

Ring spacers are thin steel rings that fit next to the compression rings. The piston groove is machined wider to accept the spacer. This will restore ring side clearance to desired limits. Fig. 25-42 shows a machine shop tool for cutting ring grooves wider for spacers.

PISTON PIN SERVICE

Depending upon the type and make of engine, the piston pin may either be *free-floating* (pin will turn in both rod and piston) or *press-fit* (pin force fitted in rod or

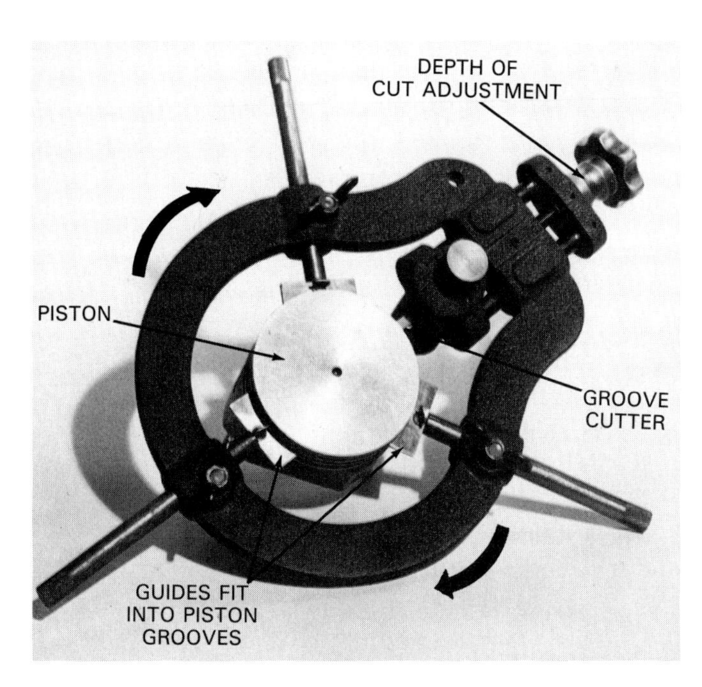

Fig. 25-42. This tool can be used to cut ring grooves wider when worn. Then, thin steel spacers can be placed next to rings to correct ring side clearance. Tool is rotated around piston. (K-Line Tools)

piston). Other setups have been manufactured but are not common.

During piston and rod service, check the pin clearance on both floating and pressed-in pins.

To check for excessive piston pin clearance, clamp the connecting rod I-beam slightly in a vise. Use vise caps to protect the rod. Pull and push up and down on the piston and try to rock the piston against normal pin movement, Fig. 25-43A.

If play can be detected, the pin, rod bushing, or piston bore is worn. A small telescoping gauge and outside micrometer should be used to measure exact part wear after pin removal.

Free-floating pin service

To remove a free-floating pin from the piston, use snap ring pliers to compress and lift out the snap rings on each end of the pin, Fig. 25-43B. Then, push the pin out of the piston with your thumb. A brass drift and light hammer blows may be needed. Make sure the piston and rod are marked and kept organized, Fig. 25-43C.

When the pin is worn, it should be replaced. If the pin bore in the pistons measures larger than specs, replace the pistons. The pin bore may also be reamed larger. Oversize piston pins can then be used. Pin bore reaming is usually done by a machine shop.

Pressed-in pin service

To remove a pressed-in piston pin, you will need to use a press and a driver setup similar to the one shown in Fig. 25-44. Wear eye protection and make sure the piston is mounted properly, Fig. 25-45.

Some technicians use a torch to heat and expand the piston or rod small end. This will make the piston pin slide out easier. See Fig. 25-46.

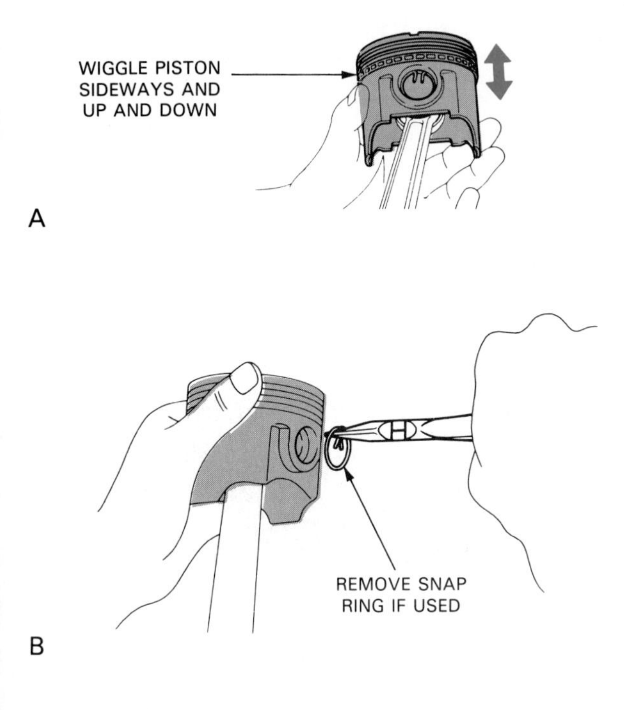

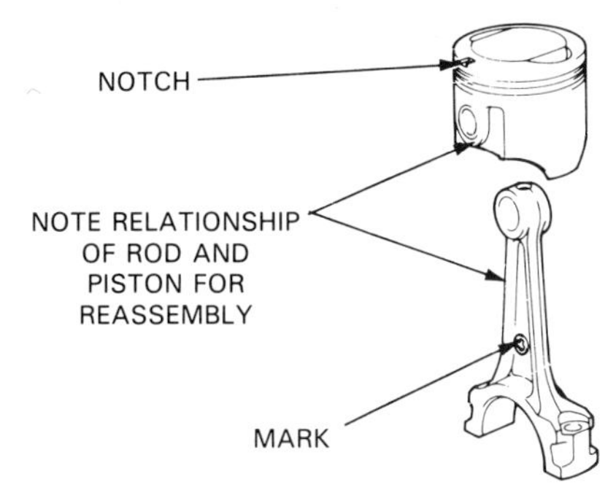

Fig. 25-43. A—Wiggle piston against normal swivel to check for loose wrist pins. B—If floating type, remove snap rings and push out pins. If press-fit, you will need to use a press to remove pins. C—Note how rods go on pistons for reassembly. (Toyota)

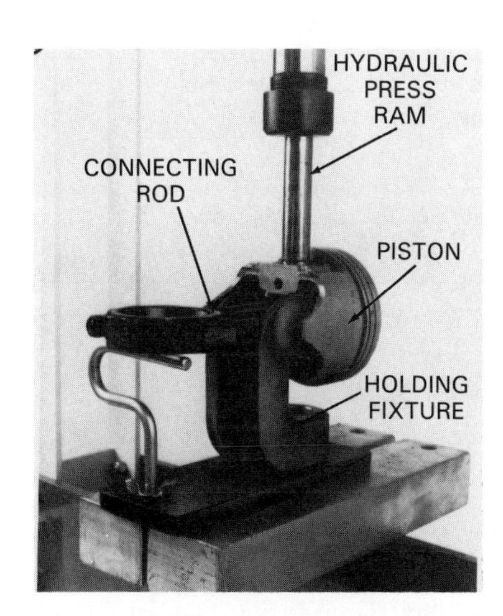

Fig. 25-44. Holding fixture and hydraulic press are being used to drive out press-fit piston pin. (OTC Tools)

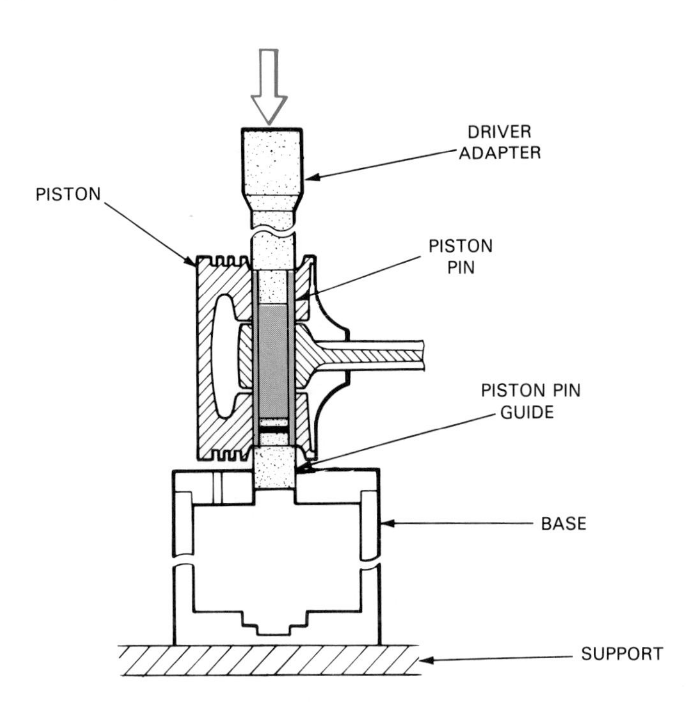

Fig. 25-45. Note how correct size adapters are needed to drive pin out of piston. (General Motors)

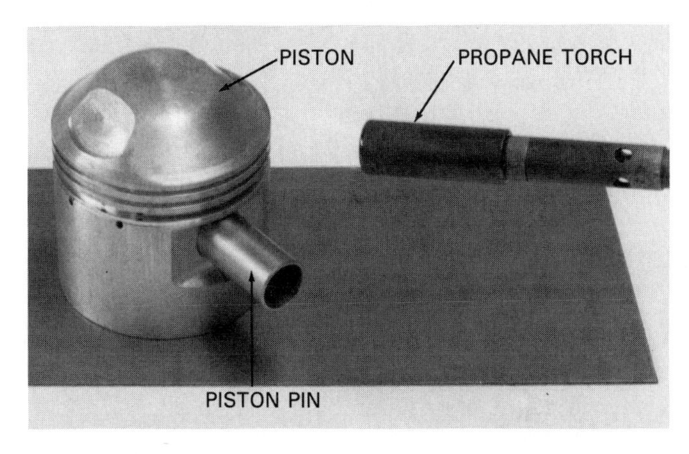

Fig. 25-46. Some technicians use heat from torch to expand parts so pin can be pushed out more easily.

Measuring piston pin wear

Measure piston pin wear with an outside micrometer. Measure in the middle and at both ends of the pin (floating types only). Compare wear to specs. If worn too much, replace the piston pins. Look at Fig. 25-47.

A typical piston pin clearance is only .0001 to .0003 inch or .0025 to .0075 mm. This is in ten-thousandths of an inch, not thousandths. Keep this in mind while measuring piston pin wear, piston bore wear, and rod small end wear.

Piston pin installation

Before installing a piston pin, make sure the piston is facing in the right direction in relation to the connecting rod. Normally, a piston will have some form of marking on its head which should point towards the front of the engine.

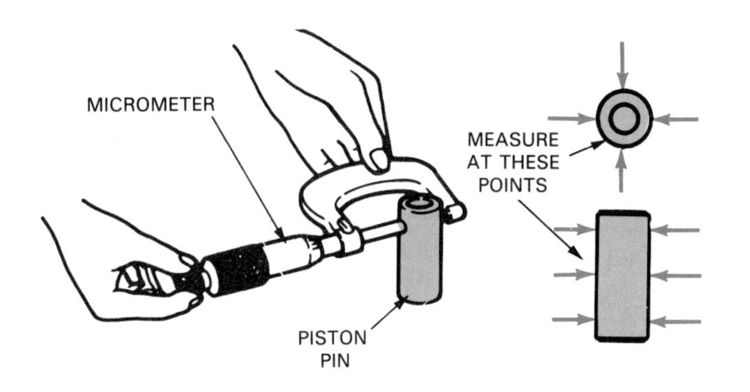

Fig. 25-47. Measure piston pin wear in locations shown. Specs are often given in ten-thousandths of an inch so measure carefully.

The connecting rod may have one edge of the big end bore chamfered (faces outside of journal on V-type engines). The rod may also have an oil spray hole or rod numbers that must face in only one direction. If needed, check the vehicle's shop manual for directions.

To start a pressed-in type piston pin, tap it into the bore with a brass hammer. Then, use a press to force the pin into the piston. The connecting rod small end must be centered on the pin.

After pushing a floating type piston pin into the piston, install the snap rings to secure the pin. Double-check that the snap rings are fully seated in their grooves.

CONNECTING ROD SERVICE

Connecting rods are subjected to tons of force during engine operation. As a result, they can wear, bend, or even break, Fig. 25-48.

The old piston and the bearing inserts will indicate the condition of the connecting rod. If any piston or bearing wear abnormalities are found, there may be a problem with the connecting rod.

For example, if one side of the bearing is worn, the connecting rod may be bent. If the back of the bearing insert has marks on it, the rod big end may be distorted, allowing the bearing insert to shift inside the rod.

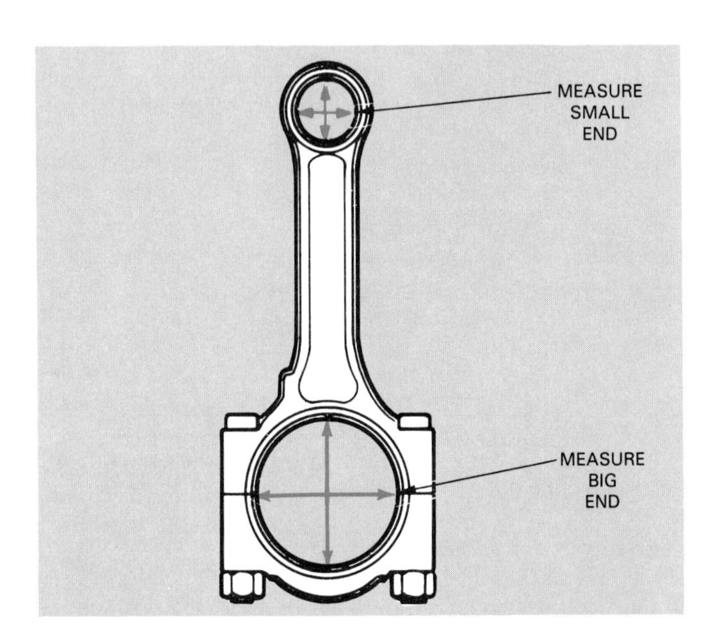

Fig. 25-48. Connecting rod small end and big end must be measured as shown. Check for out-of-round and taper in bores. If beyond specs, have the rod rebuilt. (Buick)

Rod small end service

Measure the rod small end with a telescoping gauge and a micrometer, Fig. 25-49A. If worn beyond specs, have a machine shop replace the rod bushing, Fig. 25-49B and 25-49C. The pin will have to be *fitted* (bushings reamed for proper clearance) in the rod.

Note! Oversize piston pins may be available (.0015 to .003 in. oversize). They can be used in reamed pistons and rods to correct wear.

Rod big end service

To check the connecting rod big end for problems, remove the bearing insert and bolt the rod cap to the rod. Torque to specs. Then, measure the rod bore diameter on both edges and in both directions, Fig. 25-48.

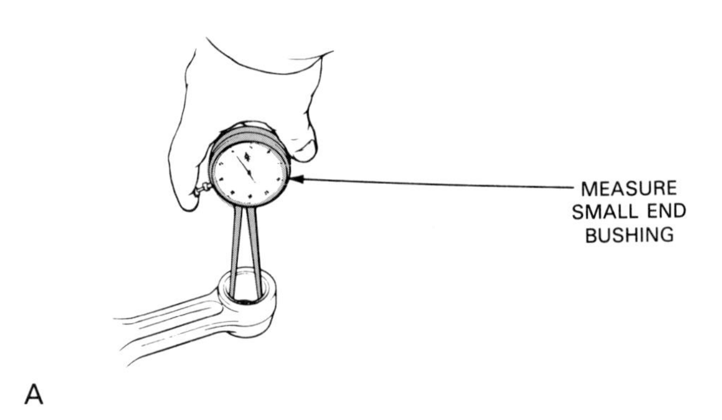

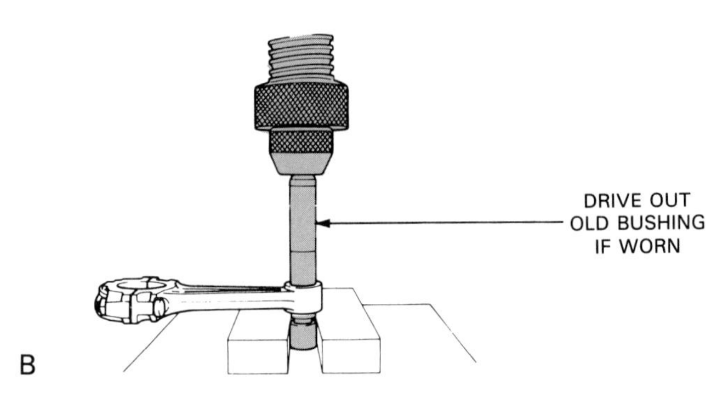

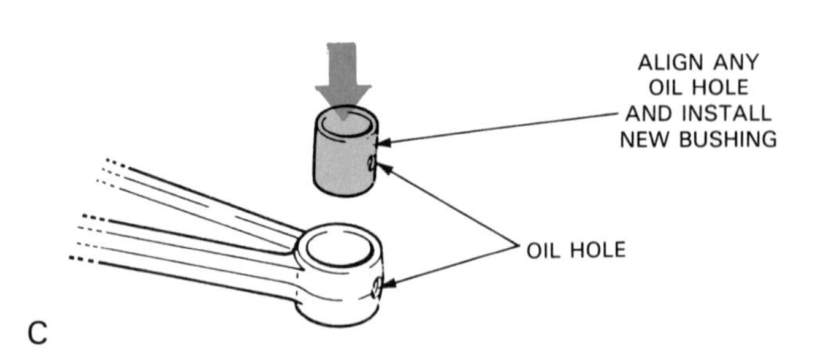

Fig. 25-49. A—Special dial gauge or telescoping gauge can be used to check rod small end wear. B—Using press to push out old worn bushing. B—When driving in new bushing, make sure any oil holes align. (Toyota)

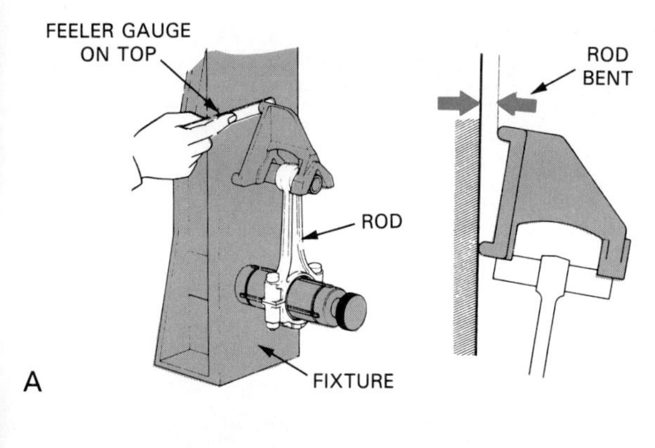

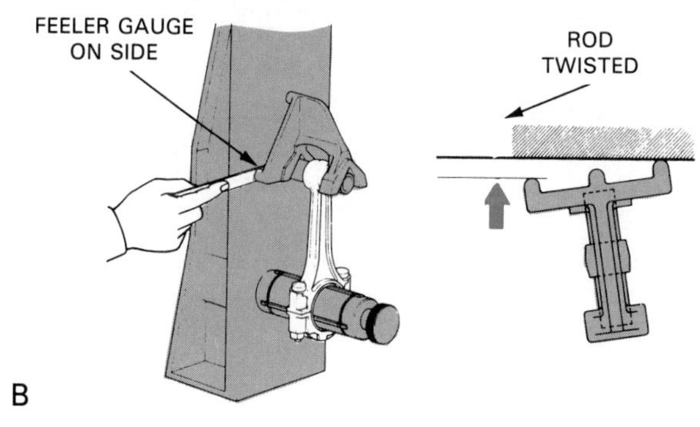

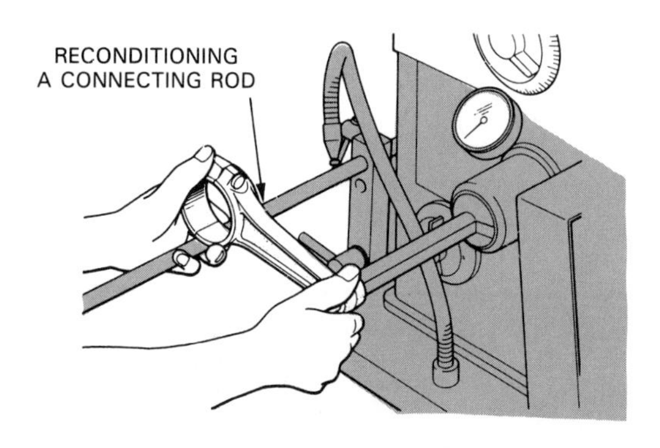

Fig. 25-50. Machine shop should have equipment for rebuilding and checking connecting rods. A—Fixture being used to measure rod bend. B—Check rod twist. C—Machine tool for honing rod bores. (Toyota)

Any difference in diameter equals rod big end taper. Any difference in the cross diameters equals rod big end out-of-roundness. If taper or out-of-round are greater than specs, have a machine shop rebuild the rod or purchase a new rod.

Checking rod straightness

To determine if a rod is bent, a special *rod alignment fixture* is needed. It will check whether the rod small end and big end are perfectly parallel. Use the operating instructions provided with the particular fixture or send the rods to a machine shop. See Fig. 25-50. Replace any bent rods.

Rebuilding connecting rods

Connecting rods can be rebuilt by most machine shops. This should be done when inspection or measurements show major problems. Sometimes, only one rod needs to be rebuilt, as when a bearing spins and damages the rod big end bore. In a high mileage engine, all of the connecting rods may require rebuilding.

A connecting rod is rebuilt by:

1. Machining cap and big end bore so bore diameter meets specs.
2. Replacing small end bushing.
3. Checking or correcting rod bore alignment.
4. Installing new rod bolts, if needed.

As shown in Fig. 25-50C, special machine shop equipment is available for accurately machining rods.

INSTALLING PISTON RINGS

With the cylinders, pistons, and connecting rods all checked, you are ready to install the new piston rings on the pistons. Make sure you checked ring gap as explained earlier.

Installing oil rings

Install the oil ring first. To hold the piston, clamp the rod lightly in a vise. Use wooden blocks or brass vise caps to prevent rod damage. The bottom of the piston should touch the top of the vise to keep the piston from swiveling.

Wrap the expander-spacer around its groove. BUTT

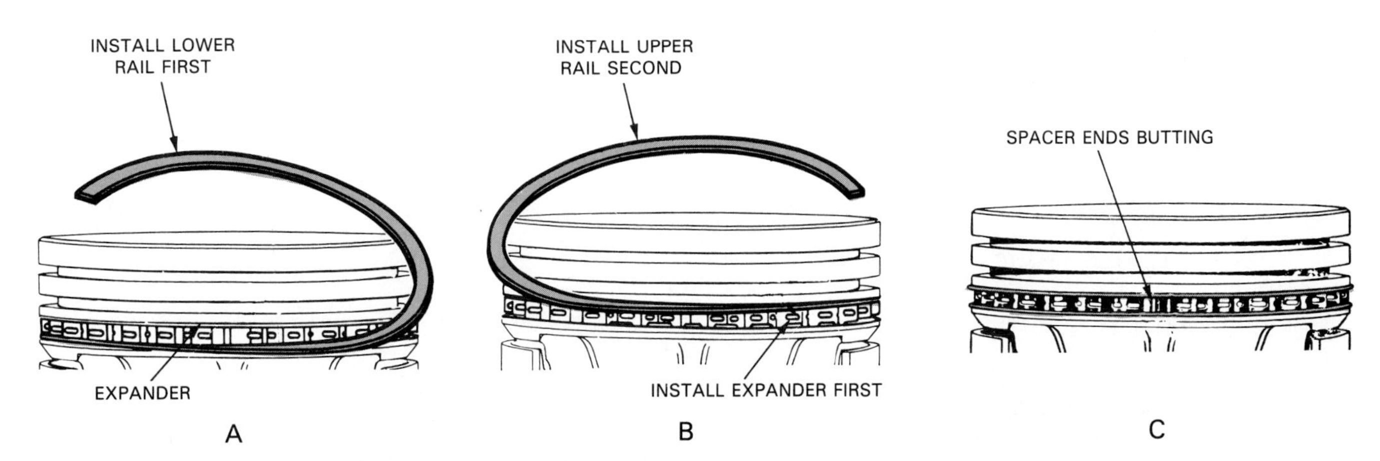

Fig. 25-51. Installing oil ring on piston. A—Install expander-spacer and butt ends together. Then spiral lower rail next to spacer in groove. B—Spiral on top rail and fit it next to spacer. C—Check that ring gaps are opposite each other and to one side of piston pin. Double-check that expander-spacer ends are not overlapped; they must butt together.

the ends together.

Spiral the bottom rail next to the bottom of the expander-spacer, Fig. 25-51A. Then, spiral in the top oil ring rail, Fig. 25-51B. Double-check that the expander-spacer is NOT overlapped. Its ends must butt together, Fig. 25-51C.

Make sure the oil ring assembly will rotate on the piston. There will be a moderate drag as the oil ring is turned. One ring gap should be almost aligned with the end of the piston pin. The other gap should be at the opposite end of the pin.

Installing compression rings

Note the instructions with the new piston ring set. There will usually be hints on proper ring installation. Usually, compression rings have a top and a bottom. If installed upside down, ring failure or leakage may result.

Ring markings are usually provided to show how compression rings should be installed. The markings show the top of each ring and which ring goes into the top or second piston groove. Refer to Fig. 25-52.

CAUTION! Compression rings, unlike steel rings, are made of very brittle cast iron. They will BREAK EASILY if expanded or twisted too much.

Using a *piston ring expander* (special tool for spreading and installing rings), slip the compression rings into their grooves. If a ring expander is not available, use your fingers to carefully spiral the compression rings onto the piston. See Fig. 25-53.

Piston ring gap spacing

A specific *piston ring gap spacing* is normally recommended to reduce blowby and ring wear. Fig. 25-54 shows a typical method.

Note that each gap is directly opposite the one next

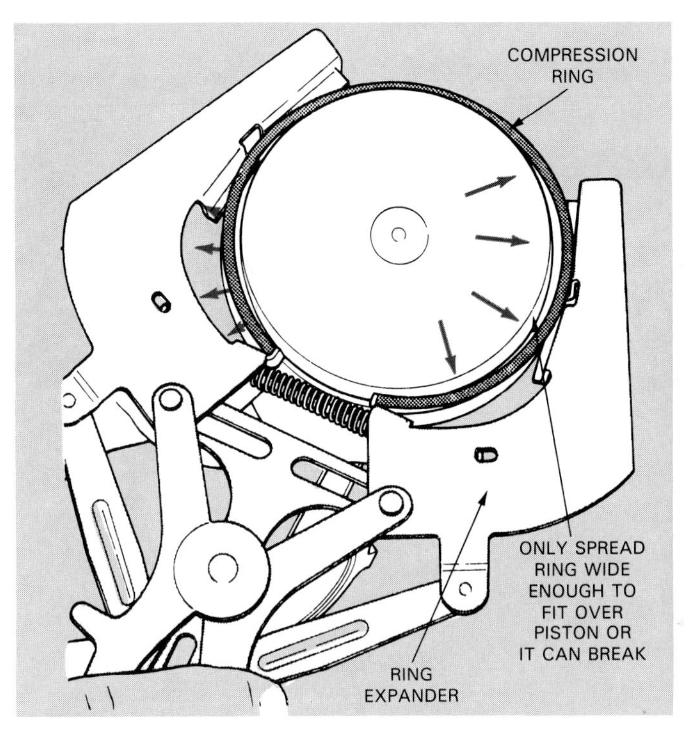

Fig. 25-53. If available, use ring expander to spread rings and install them on pistons. You can use your fingers if careful not to break rings. (Chrysler)

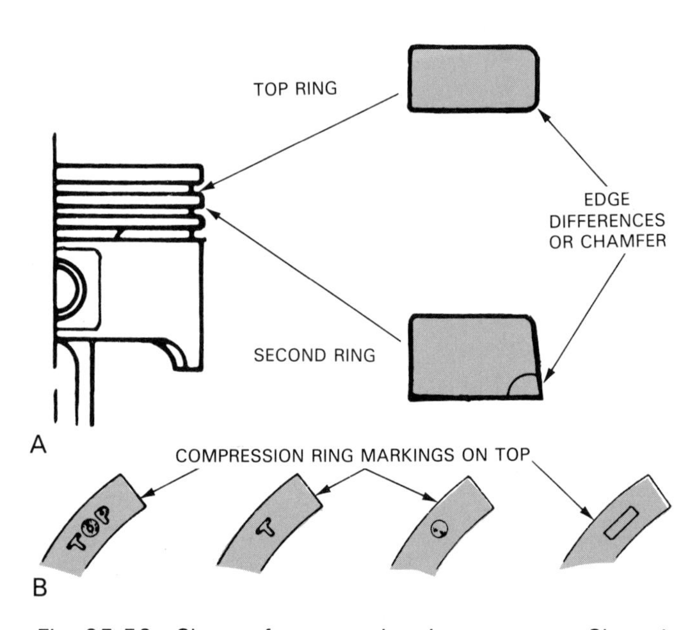

Fig. 25-52. Shape of compression rings can vary. Shape is used to help hold ring against cylinder wall and seal properly. A—Example of different ring shapes. Ring installation instructions will tell you which ring goes in which groove. B—Marks will also denote top of compression rings when needed. (Honda and Ford)

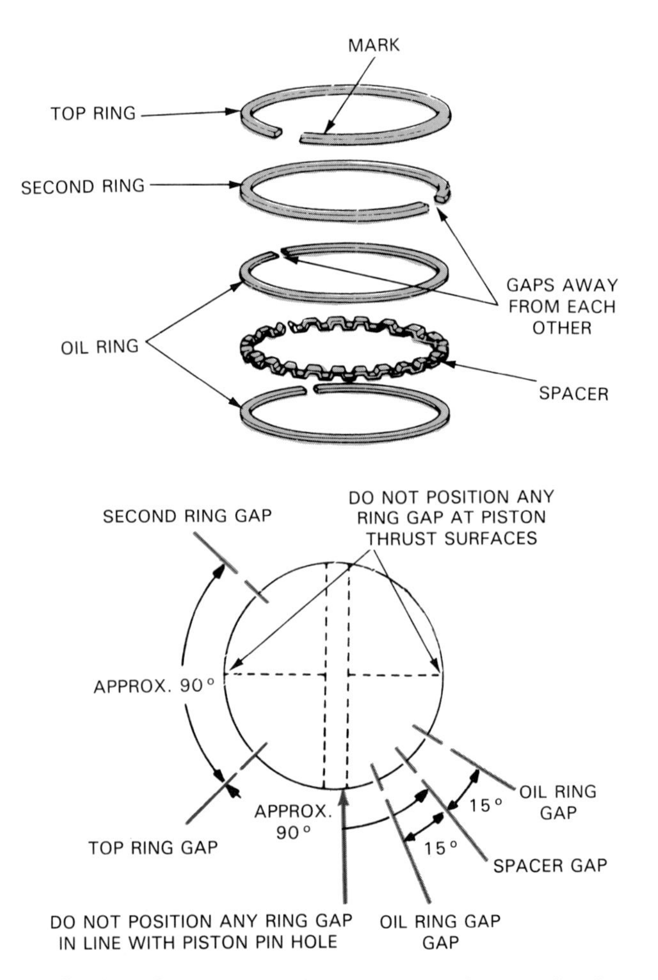

Fig. 25-54. Space piston ring gaps away from each other as shown. (Toyota)

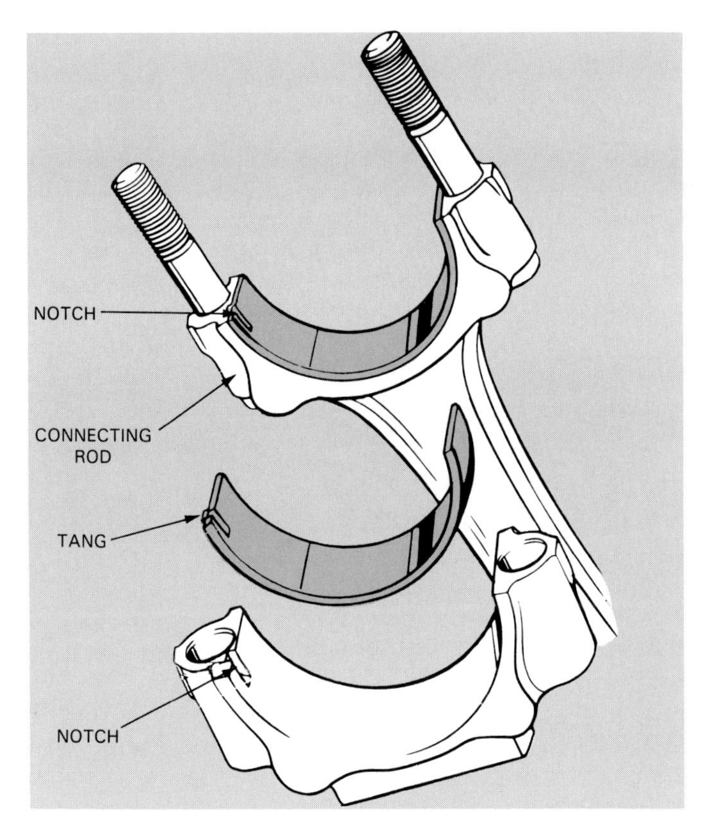

Fig. 25-55. Fit clean, dry rod bearings into connecting rod body and cap. Make sure bearing is square in bore. (Oldsmobile)

to it. The gaps are also almost lined up with the piston pin hole. This provides maximum distance between each gap for minimum pressure leakage. Being next to the pin hole reduces ring wear because the gap is not on a major thrust surface.

INSTALLING PISTON AND ROD ASSEMBLY

Fit the rod bearing inserts into the connecting rods, Fig. 25-55. Fit the matching inserts into the correct caps. Then, wipe a generous layer of oil on the face of the bearings, as in Fig. 25-56. The backs of the bearings must be clean and DRY (not oiled).

To install a piston assembly in the block, dip the head of the piston and new rings into clean engine oil. Double-check that the ring gaps are still spaced properly.

Clamp a ring compressor around the rings, as in Fig. 25-57. While tightening the compressor, hold the compressor square on the piston. The small dents or indentations around the edge of the compressor should face the bottom of the piston.

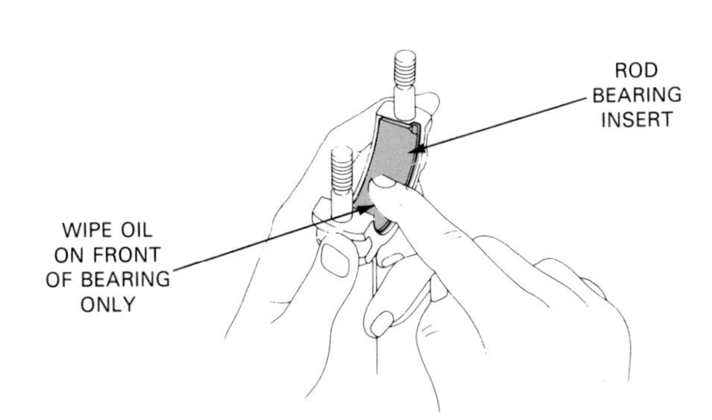

Fig. 25-56. Wipe thick engine oil or white grease on top of rod bearings. (Toyota)

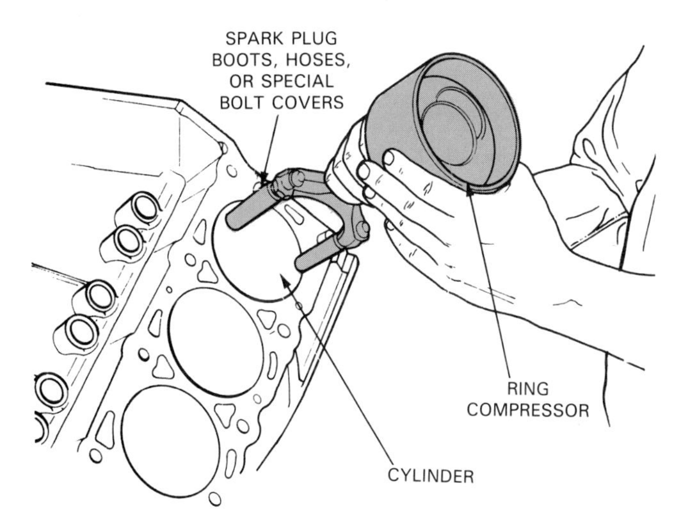

Fig. 25-57. Install ring compressor over piston and rings. Tighten it down moderately. Check rod and piston alignment marks. Slide protectors over rod bolts. Fit piston and rod assembly into block. (Ford)

Protecting the crankshaft

Slide *plastic* or *rubber hoses* or special *rod bolt covers* over the connecting rod bolts, Fig. 25-57. This will prevent the rod bolts from possibly scratching the crankshaft journals. Turn the crankshaft until the corresponding rod journal is at BDC.

Double-check the markings on rod and piston. Make sure the rod is facing the right direction and that the cap number matches the rod number. Also check that the piston notch or arrow is facing the front of the engine.

For example, if the rod is number one, it would normally go in the very front cylinder with the piston marking to the front. Check the service manual if in doubt.

Installing piston and rod in block

Place the piston and rod into its cylinder. While guiding the rod bolts over the crank with one hand, tap the piston into the engine with a WOODEN HAMMER HANDLE. A soft wooden hammer handle will not mark or dent the head of the piston. Hold the ring compressor squarely against the block deck. Keep tapping until the rod bearing bottoms around the crank journal. Look at Fig. 25-58.

If a piston ring pops out of the compressor, do NOT try to force the piston down into the cylinder. This would break or damage the piston rings or piston, Fig. 25-59. Instead, loosen the ring compressor and start over.

Double-check that the rod is facing in the right direction. See Fig. 25-60.

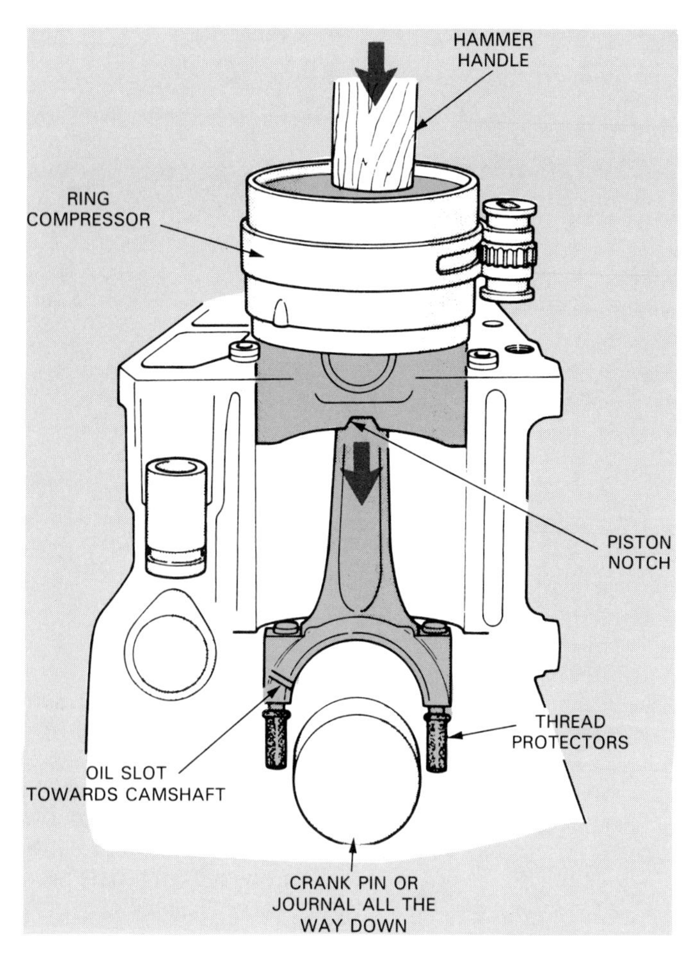

Fig. 25-58. Gently tap piston into block. Use one hand to guide rod bolts over crank pin. Tap the piston down until bearing is tight around crank. (Chrysler)

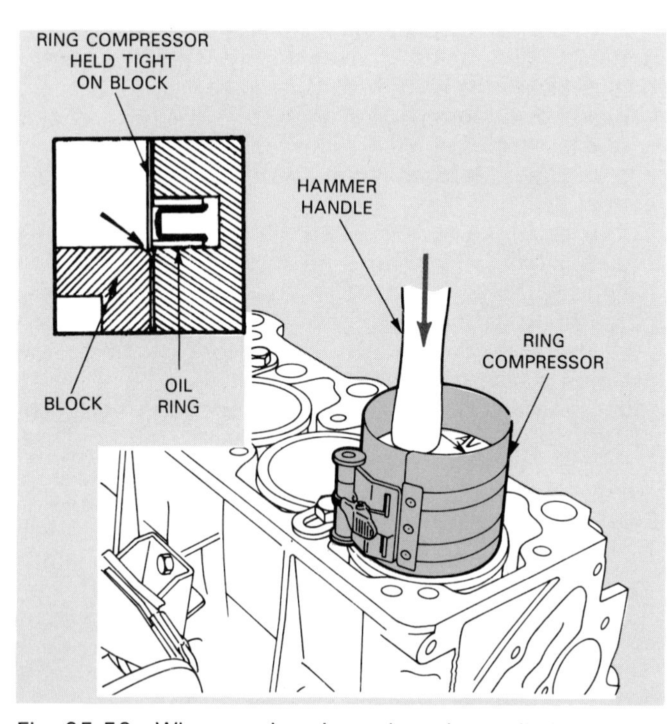

Fig. 25-59. When tapping piston down into cylinder, stop if oil ring pops out and catches on block. Reinstall compressor and start over to avoid ring damage.

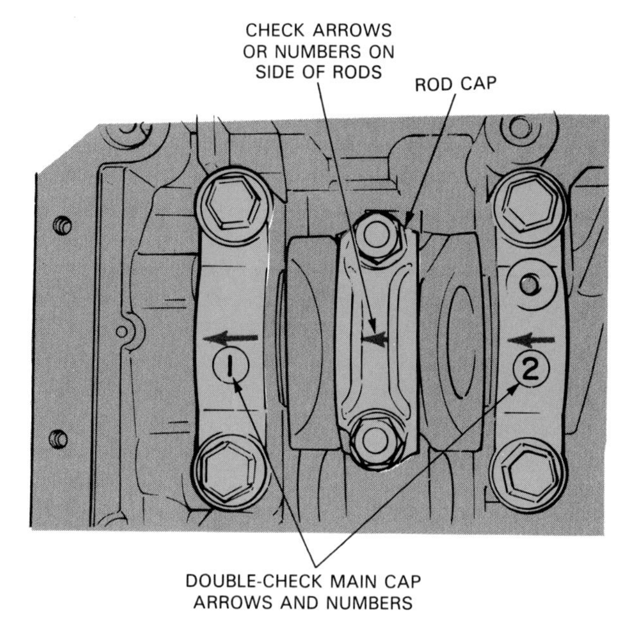

Fig. 25-60. Recheck that the rod caps and main caps are installed properly; numbers must match and point in correct direction. (General Motors)

Checking rod bearing clearance

To measure connecting rod bearing clearance, use Plastigage. Place a bead of Plastigage across the crank rod journal. Assemble and torque the rod cap bolts. Then, remove the cap and compare the smashed Plastigage to the paper scale. The width of the smashed Plastigage will let you determine bearing clearance, as in Fig. 25-61. Typical rod bearing clearance is about .0015 inch (0.038 mm). This would equal a 1.5 reading on the Plstigage paper scale.

When installing a connecting rod cap, make sure the rod and cap NUMBERS ARE THE SAME. If the rod is numbered with a five, then the cap should also have five stamped on it. Mixing up rod caps can damage the bearings or crankshaft.

Torquing connecting rods

It is very important for you to properly torque each rod nut or bolt to specifications.

If a rod is over-tightened, the rod bolt could snap off during engine operation. Severe block, crank, piston, and cylinder head damage could result as parts fly around inside the engine.

If a rod is under-tightened, the bolts could stretch under load allowing the bearing to spin or hammer against the crank. Again, serious crankshaft and rod damage could result.

Install the rod cap. If the bolts are used, place locking compound on the threads, Fig. 25-62. Using a TORQUE WRENCH, tighten each rod fastener a little at a time. This will pull the rod cap down squarely. Keep increasing torque until full specs are reached. Double-check each rod bolt or nut torque several times, Fig. 25-63.

Checking rod side clearance

Connecting rod side clearance is the distance between the side of the connecting rod and the side of the

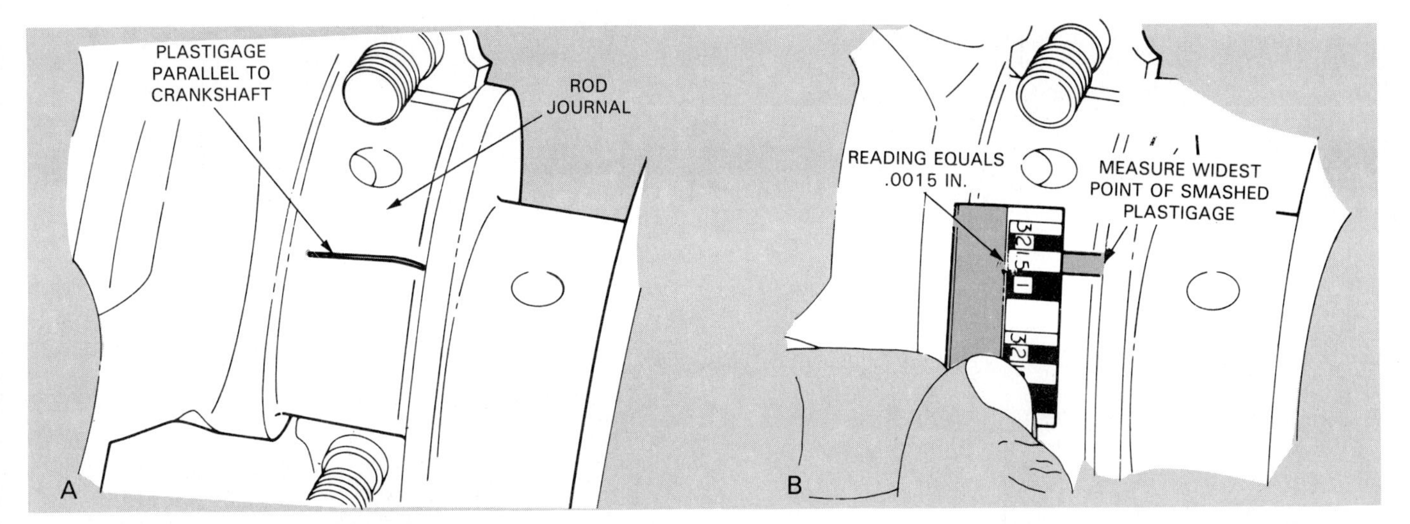

Fig. 25-61. Plastigage rod bearings as you did main bearings. A—Place piece of gauge on journal. Install and torque cap and remove. B—Compare paper scale with smashed plastic bead. If Plastigage was as wide as the three, what would that tell you? Ask your instructor if needed! (Buick)

Fig. 25-62. On used rod bolts, you might want to use thread locking compound. It will prevent potential nut loosening and rod failure after extended service.

crankshaft journal or other rod. To measure rod side clearance, insert different size feeler gauge blades into the gap between the rod and the crank, as in Fig. 25-64.

The largest feeler blade that slides between the rod indicates side clearance. Compare your measurements to specs which vary from .005 to .020 inch (0.127 to 0.51 mm). If not within specifications, crankshaft journal or connecting rod width is incorrect.

ENGINE BALANCING

Engine balancing may be needed when the weight of the pistons, connecting rods, or crankshaft is altered. Engine balancing is done to prevent engine vibration.

For example, if new, oversize pistons are installed and they weight more than the old, standard pistons, engine (crankshaft) balancing may be required.

Most large, automotive machine shops have engine

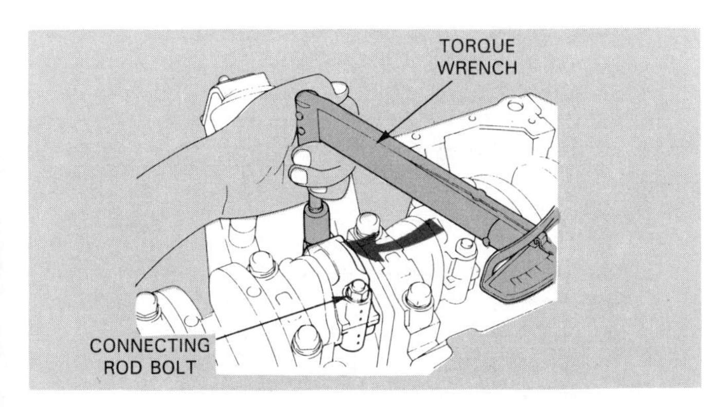

Fig. 25-63. As with the main caps, torque the rods to specs. Tighten one bolt and then the other until up to specs. Go over each bolt a couple of times. Incorrect torque can lead to rod breakage and severe part damage. (Toyota)

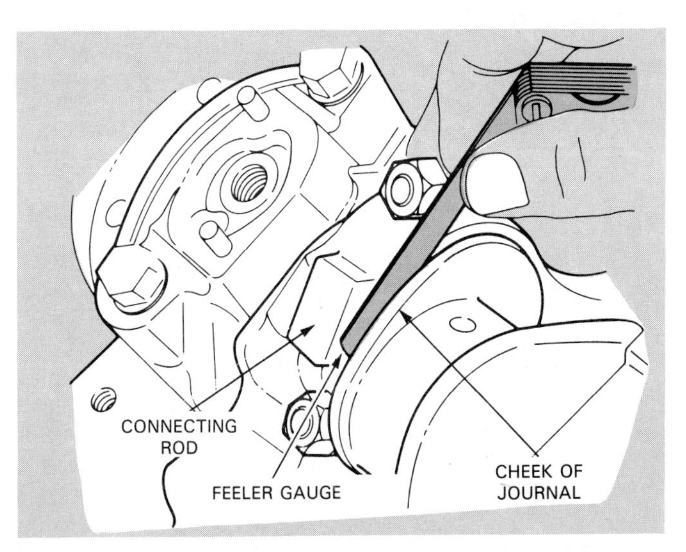

Fig. 25-64. To measure rod side play, use flat feeler gauge. The largest blade that fits equals side play. (Pontiac)

balancing equipment. Basically, the pistons, rings, piston pins, connecting rods, and rod bearings are weighed on an accurate scale. Material is machined or ground off the pistons and rods until each assembly weighs the same. All rod big ends and small ends should also be equal in weight.

Then, bob weights comparable to the weight of each piston and rod assembly are bolted to the crankshaft rod journals. The crankshaft, front damper, and flywheel are bolted together and rotated on the engine balancing machine. It will show where weight should be added (metal welded on) or removed (metal drilled out) from the crankshaft counterweights, damper, and flywheel.

Proper engine balance is very critical with today's small, high rpm, economy car engines. Keep engine balancing in mind with major engine modifications.

Balance shaft service

Some modern engines are equipped with balance shafts that are crankshaft or camshaft driven. These shafts are normally located in the block or below the crankshaft, depending on the engine design. This shaft would normally be removed during engine disassembly.

To remove the balance shaft, begin by removing any retainers holding the shaft to the engine. In some cases, it may be necessary to use a slide hammer to remove the shaft. After the shaft has been removed, be sure to remove the balance shaft bearings for inspection.

Inspect the shaft for nicks, burrs, and wear. Check the bearings used with the balance shaft for wear. The balance shaft and bearings should be replaced as a set if damaged or worn.

Often, special tools are needed to service the balance shaft in most engines. Check the service manual for installation and other service procedures.

SUMMARY

Inspect the cylinder block closely for problems before assembly. Magnafluxing is a quick way of finding cracks in cast iron blocks. Dye penetrant will find cracks in aluminum blocks. Cracked blocks can sometimes be repaired by welding, special epoxy, plugging, and sleeving.

Main bore alignment and deck warpage are checked with a straightedge and feeler gauge. Machining is needed to correct misalignment or warpage of the block.

Cylinder wear is best checked with a dial bore gauge. However, a telescoping gauge and outside mike will also work. You could also use an inside micrometer. Check for cylinder out-of-round, taper, and diameter.

Cylinder honing is needed on used cylinders or cylinders that have been bored oversize. Honing will produce the correct surface texture for proper ring break-in. Power honing will help remove cylinder taper and out-of-round. Flex or brush honing will simply prepare the cylinder surface. A finer grit stone is needed with hard chrome or moly rings.

Do not pull a spinning hone too far out of the cylinder or tool breakage can result.

A block should be bored when cylinder wear is beyond specifications. A machine shop can bore the block cylinders oversize. Then, oversize pistons would have to be purchased and installed. Some blocks can be bored up to .120 inch oversize.

Cylinder sleeving can be done to repair a badly worn or damaged cylinder wall. The sleeve can be installed so that the same size piston will fit.

Always clean the block carefully after boring or honing. You must remove all grit so that it cannot circulate through the engine during engine operation. Use a pressure washer or hand wash the block with soap and water. Wipe the block down with an oil-soaked rag to final clean the grit.

Measure the crankshaft journals for wear. Check for general diameter, taper, and out-of-round. If worn too much, have a machine shop grind the crank journals undersize. Then, undersize bearings would be needed. Bearing undersize is usually stamped on the back of the bearing inserts.

Make sure you install the rear main oil seal correctly. The sealing lip must face inside the engine. A rope seal must be worked down into its groove and then cut off flush. Use a driver to install a one-piece rear seal to prevent seal damage.

Never oil the back of engine bearings. They should be installed clean and dry. Check oil hole alignment. Coat the bearing faces with heavy engine oil or white grease. Check bearing clearance with Plastigage. Torque main and rod bearing caps to specs. Start out at about one-half torque and work up to full torque. Recheck final torque at least twice.

Clean piston ring grooves of carbon with a ring groove cleaner or broken ring. Measure piston wear with an outside micrometer. Usually, measure just below the piston pin on the skirt. Piston clearance is found by subtracting piston diameter from cylinder diameter. Clearance must not be too big or piston slap can result.

Check piston ring gap by sliding the ring to the bottom of the cylinder bore with a piston. A feeler gauge will check end gap. If too small, file the ring ends until within specs.

Ring side clearance is checked by fitting a compression ring in its groove. Then, slide the right size feeler gauge in next to the ring. If the groove is too wide, have the grooves machined and install spacers or replace the pistons.

Measure piston pins and the connecting rod bores. Make sure their dimensions are within specs. Replace parts or have the rods rebuilt if needed.

Install the oil ring on the piston by hand first. Make sure the expander-spacer ends butt together and do not overlap. Use your fingers or a ring expander to install the compression rings. Space the end gaps away from each other and slightly to one side of the piston pin ends.

Use a ring compressor to install the piston and rod into the block. Protect the crank and drive the piston down into the block with a hammer handle. Torque rod and main cap bolts properly.

Check rod side clearance with a feeler gauge. Engine balancing may be needed if components are altered or changed. Factory engines have the weights of the crank counterweights, rods, pistons, etc., all matched to prevent engine vibration.

KNOW THESE TERMS

Cylinder boring, Overbore limit, Oversize pistons and rings, Cylinder sleeving, Cylinder taper, Cylinder out-of-roundness, Dial bore gauge, Cylinder hone, Honing grit, Block line boring, Deck warpage, Piston size, Piston taper, Piston knurling, Piston clearance, Ring-to-groove clearance, Ring spacers, Piston ring gap, Ring markings, Ring expander, Ring gap spacer, Crankshaft turning, Undersize bearings, Journal taper, Journal out-of-roundness, Plastigage, Main bearing cap torque, Crankshaft end play, Piston and rod markings, Rod bolt covers, Rod bolt torque, Connecting rod side clearance.

REVIEW QUESTIONS—CHAPTER 25

1. _________ uses a ferrous powder and magnet to find cracks in parts.
2. Explain five ways to repair cracks in blocks.
3. How do you check for main bore alignment?
4. How do you measure cylinder taper and out-of-round?
5. A block should be bored if cylinder out-of-round is more than _______ inch or taper is more than _______ inch.
6. All blocks can be bored .060 inch without problems. True or false?
7. How can power honing be beneficial?
8. In your own words, how dc you hone or deglaze a cylinder?
9. A car iron V-6 block is almost new and has very low mileage. However, a piston skirt broke off and gouged one cylinder. All of the other pistons are in good condition.
   Technician A says to bore the block and install new pistons to clean up the cylinder damage.
   Technician B says to sleeve the damaged cylinder and buy a new piston.
   Who is correct?
   a. Technician A.
   b. Technician B.
   c. Both A and B.
   d. Neither A nor B.
10. Why is proper cleaning of a cylinder block so important after boring or honing?
11. How do you measure crankshaft main and rod journal wear?
12. What is an undersize bearing?
13. On a two-piece rubber rear main oil seal, the oil sealing lip must face the _________ of the engine.
14. Check bearing clearance with:
    a. Flat feeler gauge.
    b. Dial indicator.
    c. Plastigage.
    d. Micrometer.
15. Main caps can be installed in different locations on the block without problems. True or false?
16. How do you measure piston wear?
17. To find piston clearance, subtract _________ _________ from _________ _________.
18. When should you have pistons knurled?
19. Technician A says you should torque each connecting rod bolt a little at a time until up to full specs and then go over each bolt two more times to check full torque.
    Technician B says to torque the first rod bolt to full torque and then the other rod bolt to full torque and then check torque one more time.
    Who is correct?
    a. Technician A.
    b. Technician B.
    c. Both A and B.
    d. Neither A nor B.
20. When should you consider engine balancing?

ASE CERTIFICATION-TYPE QUESTIONS

1. Technician A says that checking for main bore and deck warpage is a typical procedure performed during short block service. Technician B says that cylinder head milling is a common procedure performed during short block service. Who is right?
   (A) A only.
   (B) B only.
   (C) Both A & B.
   (D) Neither A nor B.
2. Which of the following problems should you look for when inspecting a cylinder block for problems?
   (A) Cylinder ring ridge.
   (B) Cylinder scoring.
   (C) Cracked cylinder.
   (D) All of the above.
3. Technician A says that block deck warpage is measured with a straightedge and feeler gauge. Technician B says that block deck warpage is measured with an inside micrometer and dial indicator. Who is right?
   (A) A only.
   (B) B only.
   (C) Both A & B.
   (D) Neither A nor B.
4. Technician A says that a short block should be lined bored when misalignment is over approximately .0015" (.004 mm). Technician B says that a short block should be line bored when misalignment is over approximately .0001" (.002 mm). Who is right?
   (A) A only.
   (B) B only.
   (C) Both A & B.
   (D) Neither A nor B.
5. Cylinder taper is the difference in cylinder diameter when measured _________.
   (A) from front-to-rear and side-to-side
   (B) at the top and bottom
   (C) at the ridge
   (D) on the deck
6. Technician A says that a brush hone should be used to true a cylinder when wear is within specs. Technician B says that a ridged hone should be used to true a cylinder when wear is within specs. Who

is right?
(A) A only.
(B) B only.
(C) Both A & B.
(D) Neither A nor B.

7. Technician A says that many machine shops commonly bore cylinders .030″ (.762 mm) oversize. Technician B says that several machine shops commonly bore cylinders .060″ (1.5 mm) oversize. Who is right?
(A) A only.
(B) B only.
(C) Both A & B.
(D) Neither A nor B.

8. Technician A says that block boring reduces engine compression. Technician B says that block boring increases engine compression. Who is right?
(A) A only.
(B) B only.
(C) Both A & B.
(D) Neither A nor B.

9. Technician A says that piston knurling can be used to increase the diameter of the piston skirt. Technician B says that piston knurling helps reduce piston-to-cylinder friction. Who is right?
(A) A only.
(B) B only.
(C) Both A & B.
(D) Neither A nor B.

10. Technician A says that an engine's compression rings are usually made of steel. Technician B says that an engine's compression rings are normally made of cast iron. Who is right?
(A) A only.
(B) B only.
(C) Both A & B.
(D) Neither A nor B.

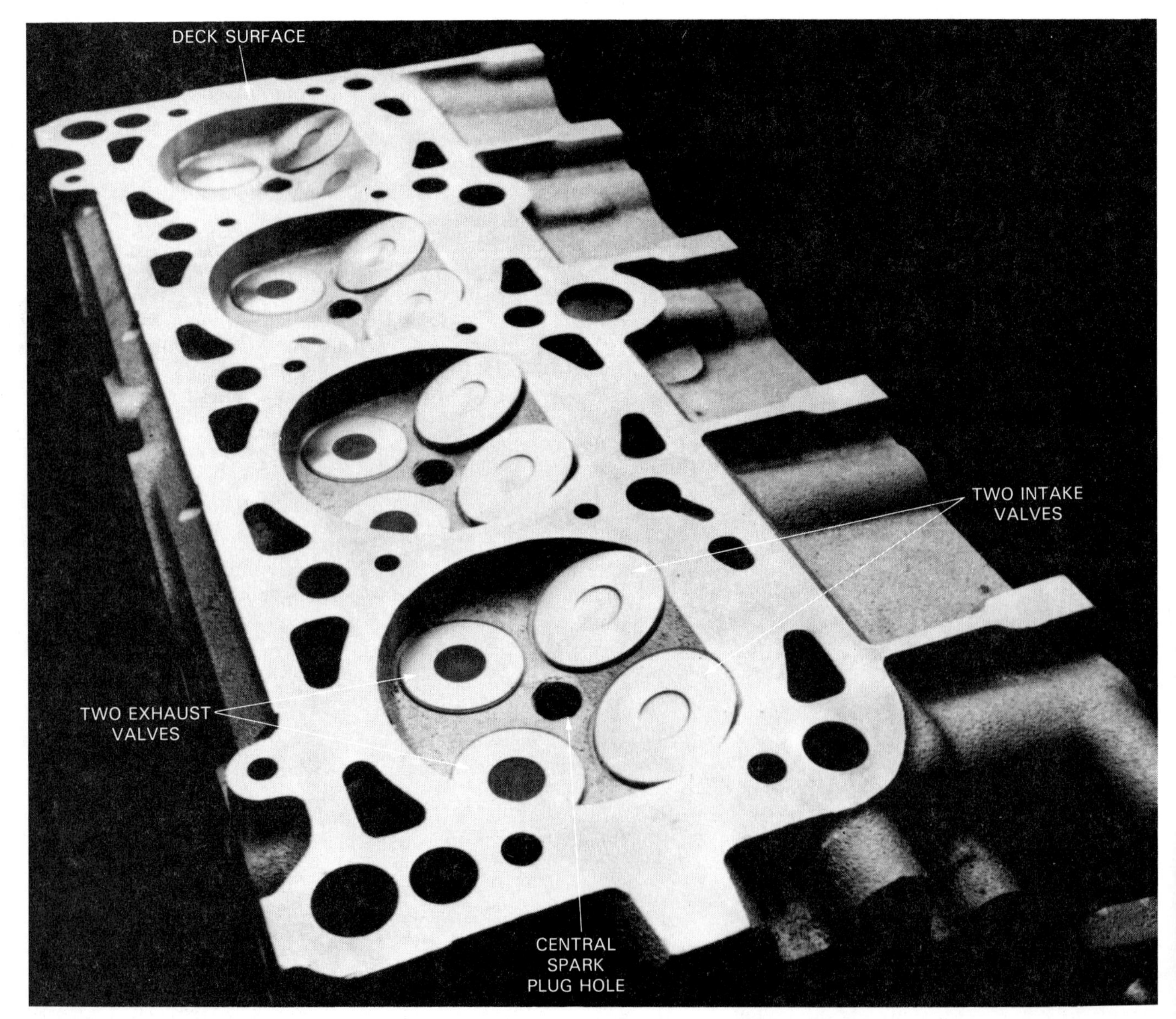

Photo shows deck side of the late model, high performance cylinder head. With four valves per cylinder, reconditioning of head is more time consuming than with more conventional two-valve per cylinder head. (Volkswagen)

Chapter 26

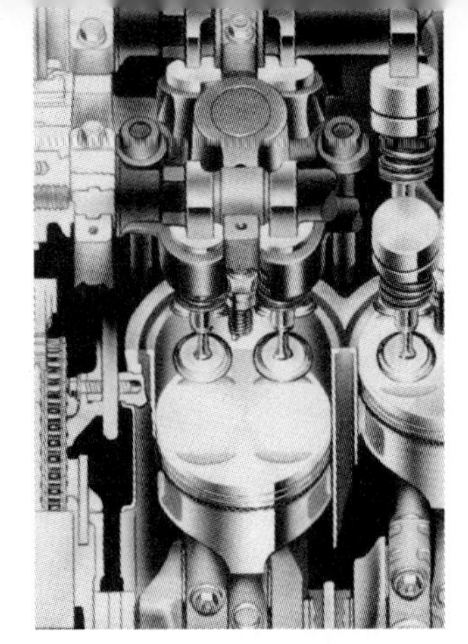

Top End Service

After studying this chapter, you will be able to:
- *Check a cylinder head for cracks, warpage, and other problems.*
- *Service diesel precombustion chambers.*
- *Measure valve guide and valve stem wear.*
- *Knurl, ream, or replace valve guides.*
- *Replace valve seats.*
- *Recondition valve seats.*
- *Grind engine valves.*
- *Test, shim, and install valve springs.*
- *Properly install both O-ring and umbrella type valve stem seals.*
- *Reassemble a head using a valve spring compressor.*
- *Measure camshaft wear, camshaft end play, and bearing clearance.*
- *Service rocker arms and rocker arm shafts.*
- *Service push rods and lifters.*

This chapter outlines the most important steps for reconditioning a cylinder head and its related parts. The head has to seal the top of the block and control flow in and out of the cylinder. With the tremendous heat of combustion, this is a difficult task. After prolonged service, heads can warp, valves can burn, valve train parts can wear, and engine performance can suffer. You will learn how to return these parts to specifications for maximum engine power, smoothness, and economy.

Fig. 26-1 shows some of the things you will learn to check and correct in this chapter.

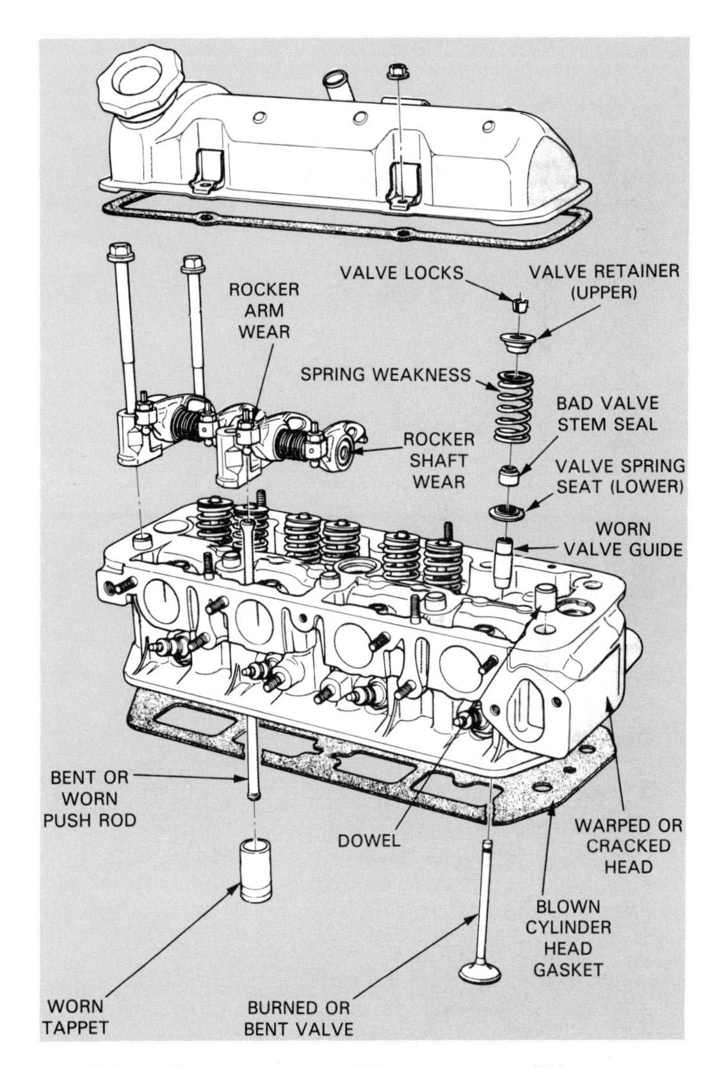

Fig. 26-1. These are some of the parts you will learn to service in this chapter. Cylinder head installation is covered in Chapter 28, however. (Chrysler)

CYLINDER HEAD PROBLEMS

With the head disassembled, inspect closely for problems. Keep in mind that if any trouble is NOT found, your repair could fail in service.

Mount the head on a stand and check for cracks, burned seats, burned deck surface between chambers, cracked valve guides, and other problems.

Checking for cracked head

A *cracked head* can result from engine overheating or physical damage from other broken parts. Usually, the crack will happen at an exhaust seat or between chambers, allowing coolant or pressure leakage. This can lead to further engine overheating, coolant in oil, and other symptoms, described in Chapter 22.

Magnafluxing is a common way of finding cracks in cast iron cylinder heads. Pictured in Fig. 26-2, magnet(s) are placed on the head where a crack might occur. Then, a squeeze bulb is used to spread fine iron powder over

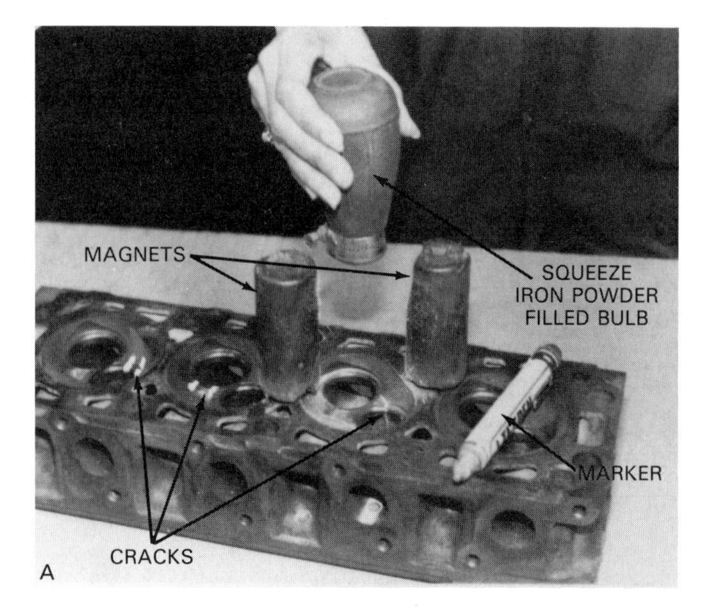

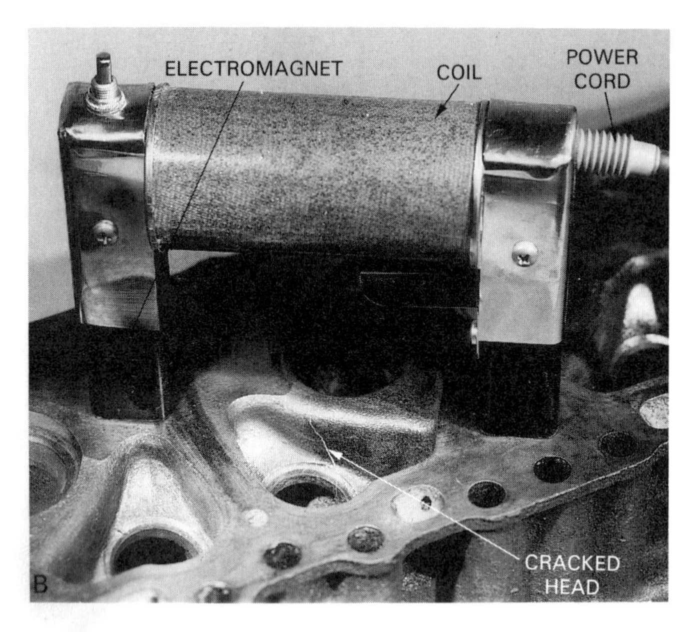

Fig. 26-2. A—Technician is Magnafluxing head. Note that cracks have been found in several places. B—Electromagnet has found crack across combustion chamber. C—Close-up of head crack. (Midwest Crankshaft, K-Line Tools, and Magnaflux)

the head. A crack will set up a small field that will attract the powder. The powder will collect in the crack and make it visible.

Dye penetrant can be used to find cracks in aluminum cylinder heads because Magnafluxing will not work with nonferrous aluminum. Detailed in the previous chapter, there are two processes. You must follow manufacturer instructions since procedures vary. One process requires a black light and the other requires a powder-like developer to make the crack more visible.

If you do not have the equipment and the engine has been overheated, you can send the head to a machine shop for crack detection. They will check the head and assure it can be reused on the engine.

Repairing cracked heads

When a cylinder head is cracked, it can either be welded, plugged (series of holes and metal plugs used to fix crack), or replaced. With most heads, replacement is more cost effective. With more exotic or expensive heads, welding may be desirable.

The welding of cast iron heads normally requires a specialty shop. The head should be heated in a furnace. Then, an NI (high nickel content welding rod) is used to repair the crack.

Aluminum heads can be welded using a heli-arc welder. Preheating in a furnace is usually NOT needed.

Note! Crack repair in heads and blocks should be done by an expert. Special welding skills are essential.

Measuring head warpage

A *warped head* has actually become curved slightly from overheating. This is more common with modern thin cast heads, especially aluminum heads. A warped head can lead to head gasket failure because of uneven clamping pressure on the head gasket. This can cause coolant leakage, compression loss, and other troubles.

A straightedge and flat feeler gauge are used to measure head warpage, as in Fig. 26-3. Place the

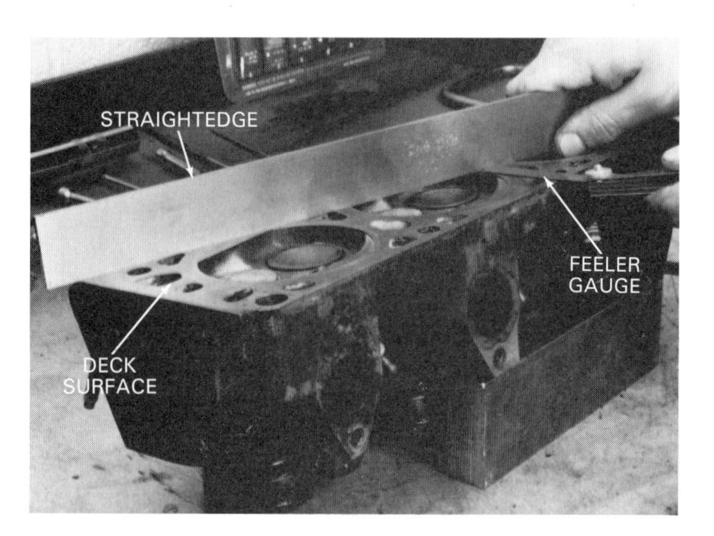

Fig. 26-3. To check for head warpage, place straightedge on deck surface. Then try to slide feeler gauge under straightedge. Thickness of gauge equals warpage. Have machine shop mill head to correct warpage, if not excessive.
(Fel-Pro Gaskets)

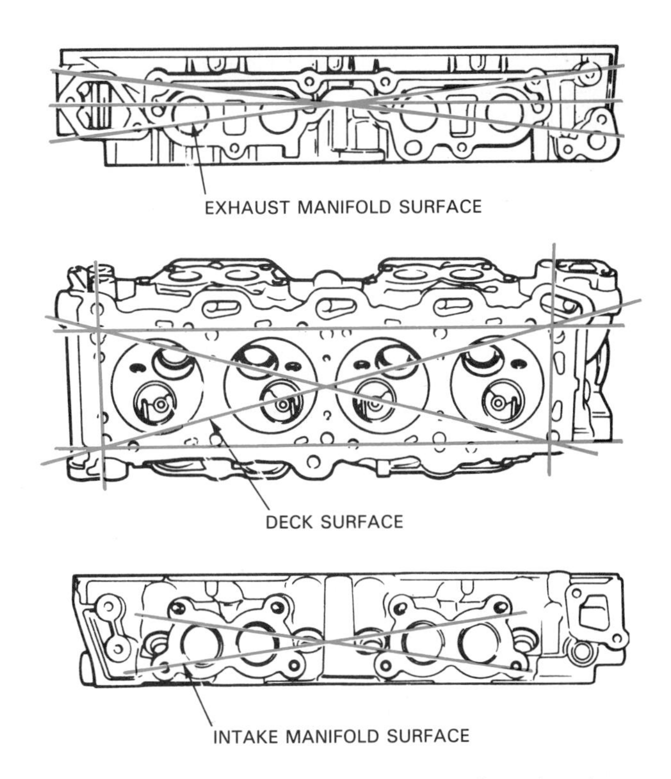

Fig. 26-4. Check head for warpage at these locations. (General Motors)

straightedge on the head. Try to slip different feeler gauge blade thicknesses under the straightedge. The thickest blade size that fits equals head warpage.

Check warpage in different positions across the head surfaces, Fig. 26-4. The most common place warpage shows up is between the two center combustion chambers. Generally, head warpage should NOT exceed about .003 to .005 inch (0.08 to 0.13 mm).

With an overhead cam head, you should also use the straightedge and feeler gauge to check the cam bore. Place the straightedge in the bore and try to slip blades under it. If warped beyond specs (over approximately .001 inch or 0.03 mm), have a machine shop line bore the head cam bores.

Cylinder head milling

Cylinder head milling is a machine shop operation where a thin layer of metal is cut or machined off the head gasket (deck) surface of the cylinder head. It is done to correct head warpage.

Check manufacture specs to see how much metal can be milled from the specific head. Milling increases the engine's compression ratio and can also affect valve train geometry (relationship or angles between parts).

On V-type engines, the intake manifold will require milling if an excess amount of metal is milled off the cylinder head.

With a diesel engine, cylinder head milling is very critical. A diesel engine compression ratio is so high, any milling can reduce clearance volume and increase compression pressures beyond acceptable limits.

Some diesel engine cylinder heads are case hardened and cannot be milled. They must be replaced when warped. Check auto maker directions when in doubt.

DIESEL PRECHAMBER SERVICE

Auto diesel engines use precombustion chambers. They are small chambers pressed into the cylinder head The tips of the diesel injectors and glow plugs extend into the precombustion chambers.

After prolonged use, the precombustion chambers may require removal for cleaning or replacement because of damage.

Prechamber removal

A *brass drift* and *hammer* are commonly used to remove a diesel engine precombustion chamber, Fig. 26-5A. The drift may be inserted through the glow plug or the injector hole in the head. Light blows with the hammer will drive out the chamber. Compare new prechamber dimensions to the old one, Fig. 26-5B.

Prechamber installation

When installing a precombustion chamber, be careful not to damage the chamber. Use a special driver or brass hammer to tap the unit back into the head, Fig. 26-5C. Hammer only on the outer edge of the chamber. The center area of the precombustion chamber could cave in or dent from hammering.

Make sure the precombustion chamber is perfectly FLUSH with the deck surface of the cylinder head. If it is not, head gasket leakage can result, Fig. 26-5D.

HEAD DISASSEMBLY

Cylinder head disassembly was explained in Chapter 24. Basically, you must:

1. Hit spring retainers with plastic hammer to free keepers.
2. Compress springs with valve spring compressor, as shown in Fig. 26-6.
3. Use fingers or small magnet to lift keepers out of valve stem grooves.
4. Release compressor and remove spring assembly from valve.
5. Repeat this procedure on the other valve assemblies.
6. Keep all parts organized so that they can be reinstalled in the same locations. Valve stems may be select-fit from the factory.

WARNING! Wear eye protection when using a valve spring compressor. If it were to slip off the retainer, spring pressure could make parts fly into your face!

VALVE GUIDE SERVICE

Valve guide wear is a common problem; it allows the valve to move sideways in its guide during operation. This can cause oil consumption (oil leaks past valve seal and through guide), a light knocking or tapping sound, burned valves (poor seat-to-valve face seal), or valve breakage. Look at Fig. 26-7.

Measuring valve guide and stem wear

To check for valve guide wear, slide the valve into its guide. Pull it open about one-half inch (12.7 mm) Then, try to wiggle the valve sideways. If the valve moves sideways in any direction, the guide or stem is worn.

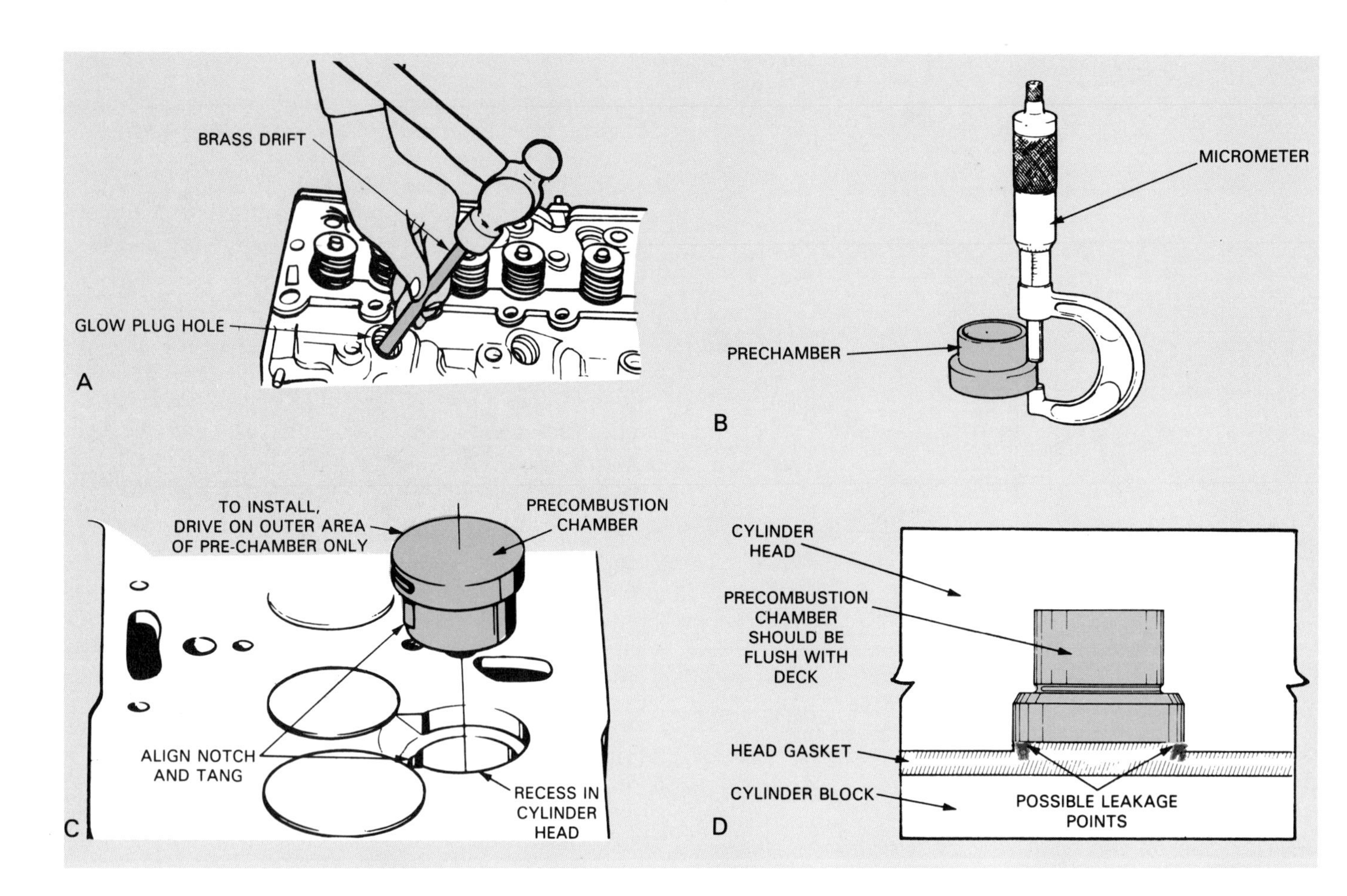

Fig. 26-5. Servicing diesel prechamber. A—Drive out old chamber through glow plug or injector hole. B—Make sure new prechamber has correct dimensions. C—Align prechamber and drive in without damaging it. D—Prechamber should be flush with cylinder head deck. If not, head gasket leakage could result. (Oldsmobile and Fel-Pro)

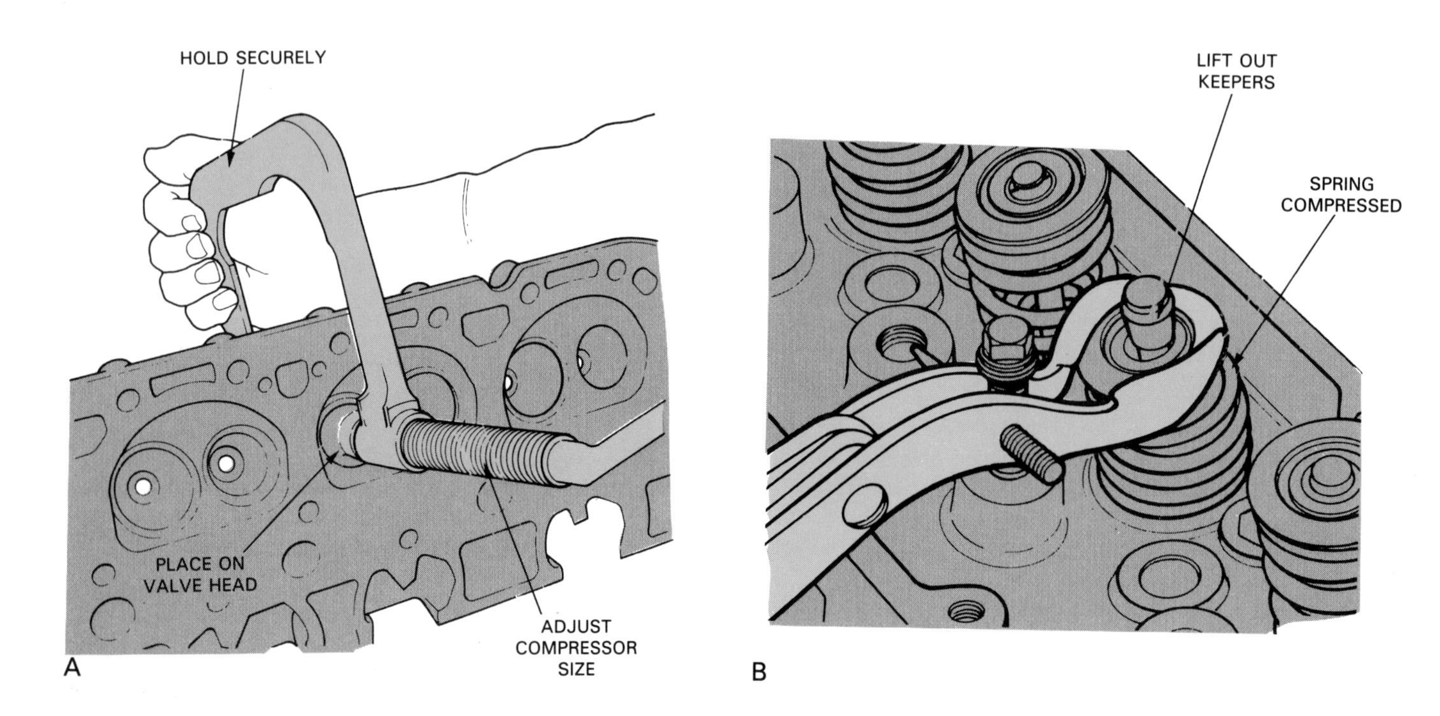

Fig. 26-6 A—Fit valve spring compressor over valve head and valve retainer. Then, compress spring. B—With spring compressed, use magnet or fingers to lift keepers out of stem grooves. (Buick and Ford)

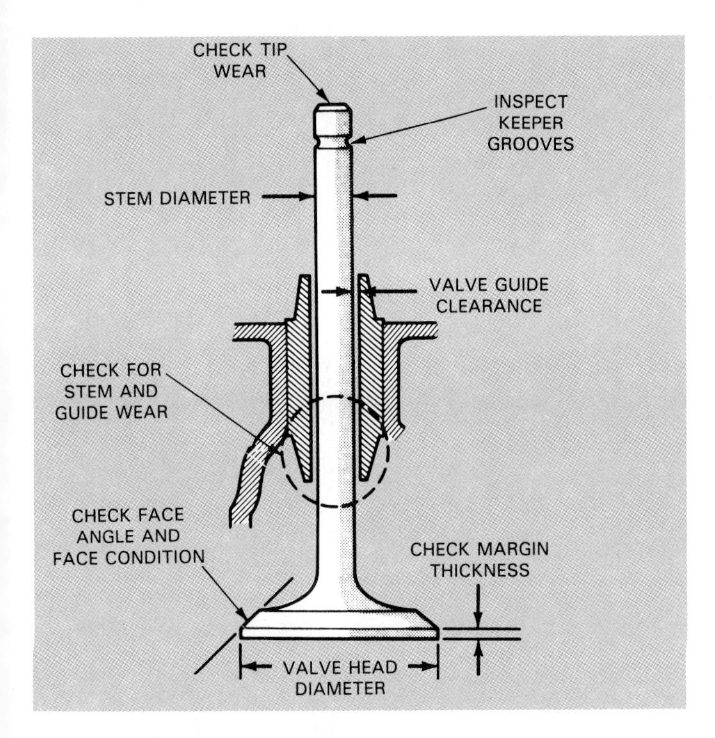

Fig. 26-7. These are the kinds of problems you must look for when servicing valves and valve guides. (Chrysler)

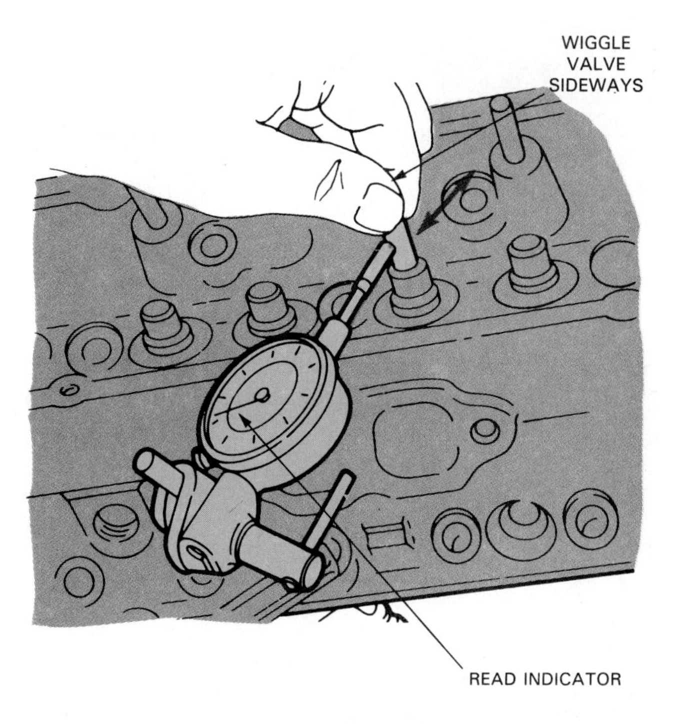

Fig. 26-9. Valve guide wear and clearance can also be measured with dial indicator at stem or at valve head. Here technician is wiggling valve stem to measure clearance. (Pontiac)

Fig. 26-8 shows how a small hole gauge and outside micrometer are used to measure guide and valve stem wear. Fig. 26-9 pictures how a dial indicator is used to measure valve stem clearance at the stem. If not within specs, part replacement or repair is needed.

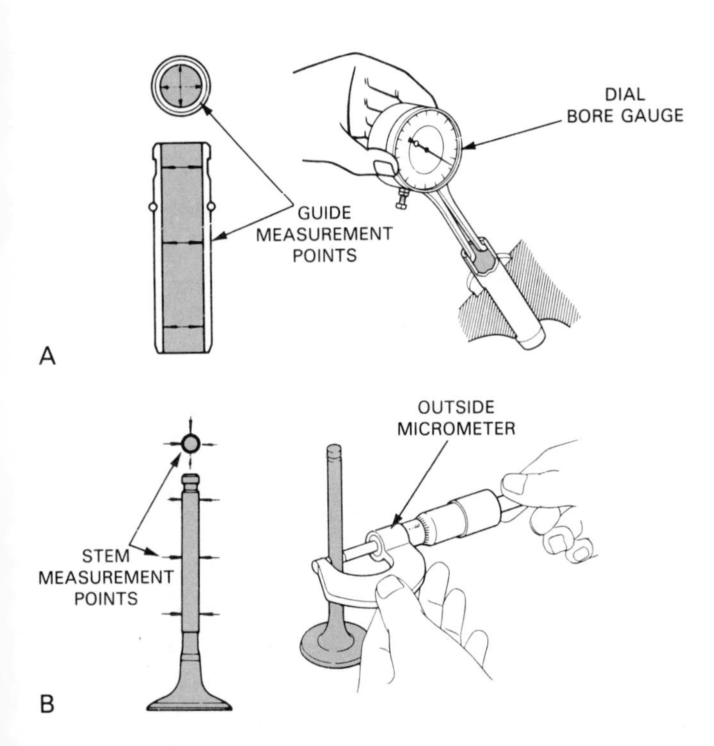

Fig. 26-8. Measuring valve guide or valve stem clearance. A—Use hole gauge to measure in these locations. B—Mike valve stem at these points. Subtract valve stem diameter from valve guide diameter to get clearance. (Toyota)

Repairing valve guide wear

There are four common methods used to repair worn valve guides. These include:

1. *Knurling valve guide* (machine shop tool used to press indentations in guide to reduce its inside diameter), Fig. 26-10.
2. *Reaming valve guide* (guide reamed to larger diameter and new valves with oversize stems installed). Fig. 26-11 shows a technician reaming a guide oversize.
3. *Installing valve guide insert* (old guide pressed or machined out and new guide pressed into head). Fig. 26-12 shows the basic steps for guide replacement.
4. *Installing bronze guide liner* (old guide reamed and thin guide liner installed), Fig. 26-13.

VALVE SEAT RECONDITIONING

Valve seat reconditioning involves grinding (using a stone) or cutting (using carbide cutter) to resurface the cylinder head valve seats. Like a valve, the seats are exposed to tremendous heat, pressure, and wear.

Checking valve seats

Inspect the valve seats closely for problems: burning, cracks, excess width, and retrusion (sinking). Measure seat width and retrusion as shown in Fig. 26-14. If the seat is badly worn or burned, too much grinding may be needed to recondition the seat.

Valve seat retrusion results from heat erosion or seat grinding that causes the valve head to sink into the combustion chamber. The valve stem would also protrude too far out the other side of the head, Fig. 26-14. Seat replacement is needed to correct retrusion. Seat retrusion causes lowered compression, reduced gas flow

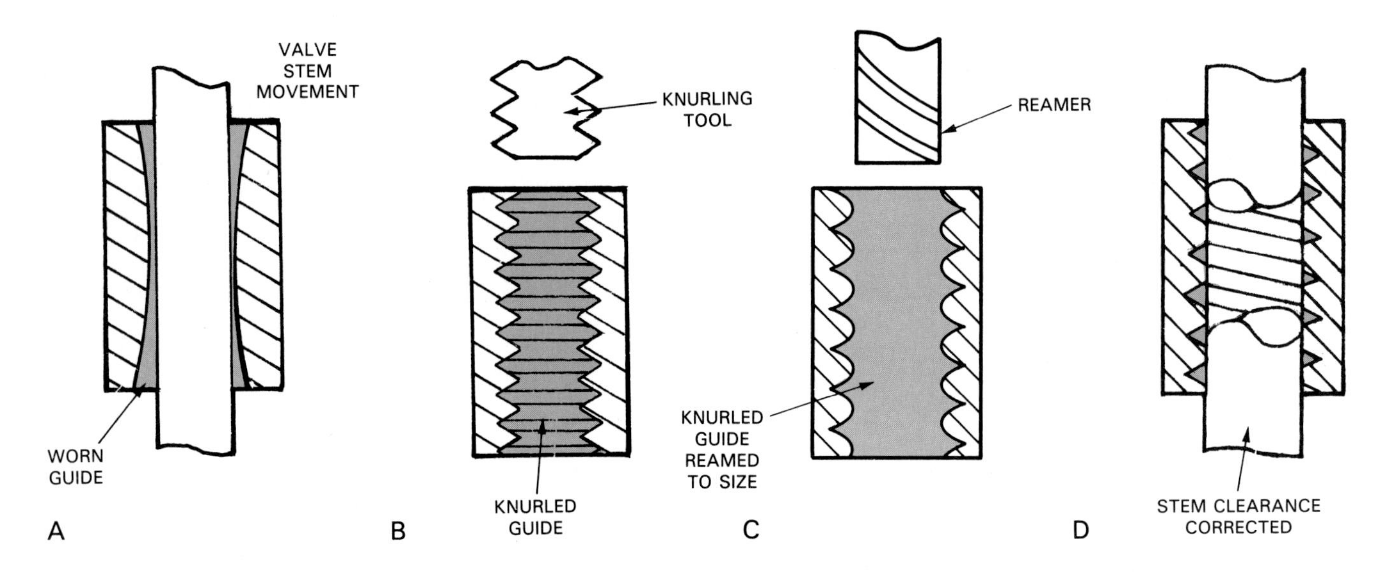

Fig. 26-10. Knurling valve guides. A—Old guide is worn and valve is free to flop sideways. B—Knurling tool is run down through guide, pushing metal outward. C—Guide is reamed to original diameter. D—Valve stem now fits in guide correctly.

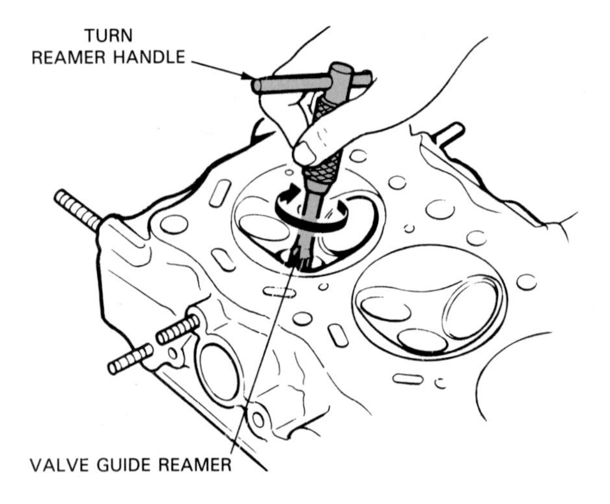

Fig. 26-11. Worn valve guides can also be reamed oversize. Then new valves, with oversize valve stems, can be installed. (Honda)

through port, incorrect valve train geometry, and decreased performance.

Valve seat replacement

Valve seat replacement is only needed when seat wear or damage is severe. Normally, valves seats can be ground or cut and returned to service.

Most technicians send the cylinder head to an automotive machine shop for seat replacement when the seat is part of the cylinder head. Most auto shops do NOT have the specialized tools required for seat removal and installation, especially with integral seats. A new valve seat insert should have an interference fit of around .002 to .004 inch (0.05 to 0.10 mm). This locks the insert in the head.

Press-in valve seat replacement is not as difficult. You do not have to machine a pocket in the combustion chamber to install a new seat. The old seat can be removed and a new one installed.

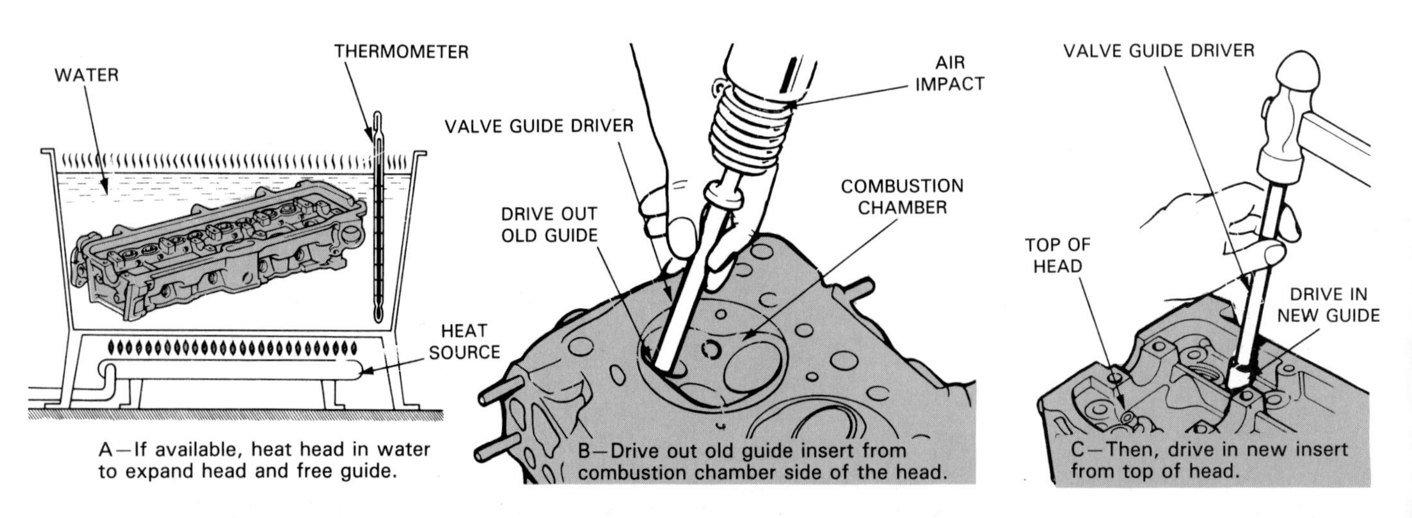

Fig. 26-12. This aluminum head has pressed-in valve guide inserts. They can be replaced easily. (Toyota)

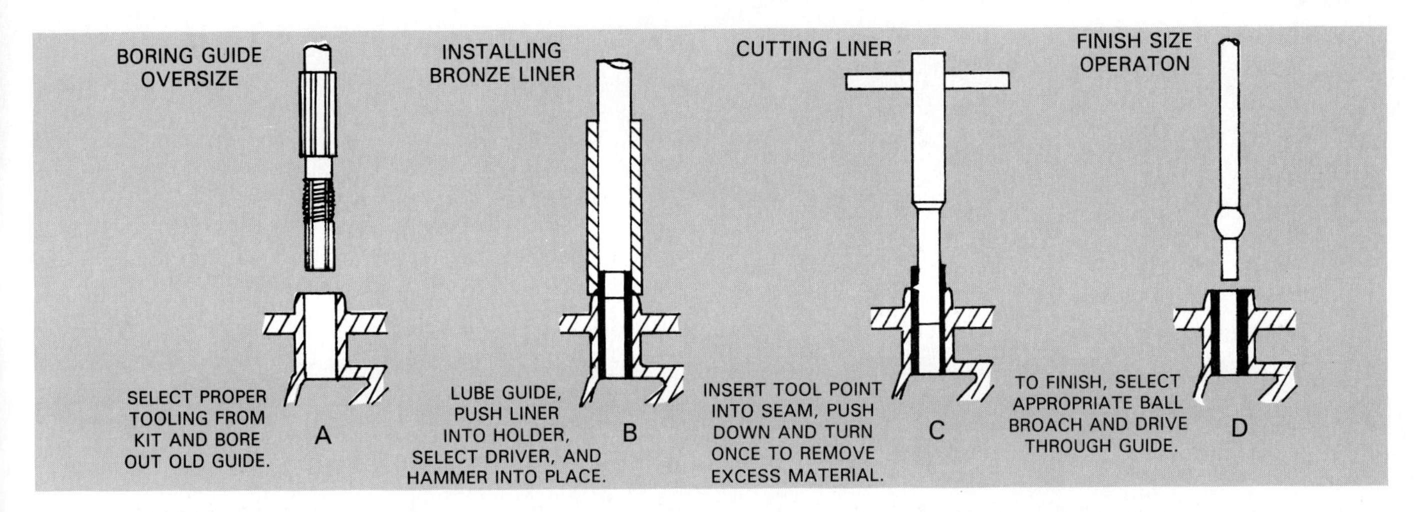

Fig. 26-13. Installing bronze wall liner. A—Ream or bore guide slightly oversize using tool in kit. B—Drive in bronze guide liner. C—Cut liner off flush with old guide. D—Broach inside diameter of liner to size. (K-Line Tools)

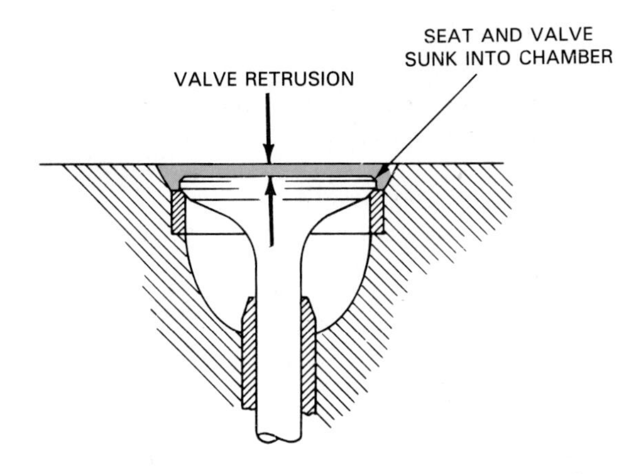

Fig. 26-14. Valve retrusion occurs as seat and valve face wear. This causes valve to sink in combustion chamber and stem to move further through head. (Ford)

Fig. 26-15 shows the tools recommended by one manufacturer for pressed-in valve seat service.

Another way to remove a pressed-in seat is to split the old seat with a sharp chisel. Then, pry out the seat with a hook nose type chisel. Extreme care must be taken not to damage the cylinder head.

To install a seat, some machinists chill and shrink the seat in dry ice; others heat and expand the head in a furnace. This helps lock the seat in the head upon reaching room temperature.

Use a driving tool to force the seat into the recess in the head. Seat installation tools vary. Follow equipment directions.

Staking the seat involves placing small dents in the cylinder head next to the seat, Fig. 26-16. The stakes will swell the head metal over the seat and keep the seat from falling out.

SEAT INSERT REMOVAL

TURN SCREW

EXHAUST SEAT REMOVAL TOOL

A

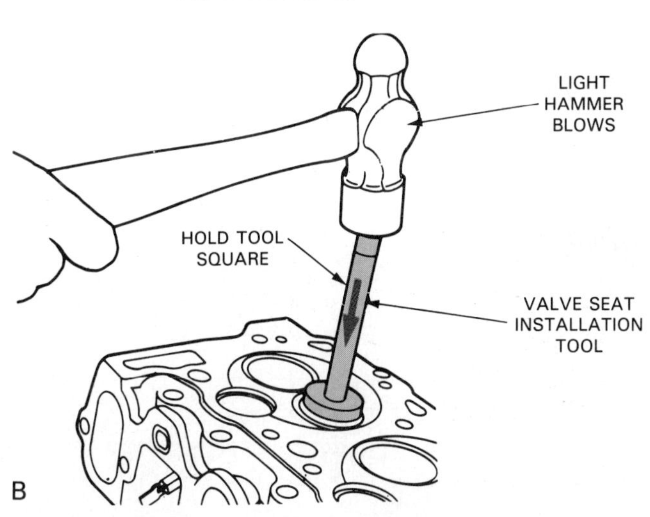

Fig. 26-15. If the head already has pressed-in valve seat insert, service is made easier. However, most technicians still send head out to machine shop for seat replacement. A—Puller is being used to remove old, damaged seat. B—Driver is forcing new seat into head pocket. Stake seat to keep it from falling out. (Ford)

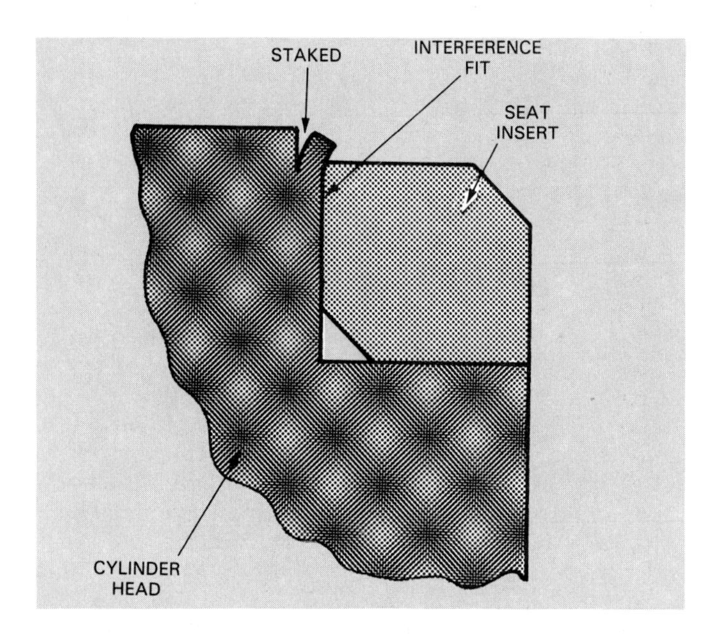

Fig. 26-16. A seat is staked by using chisel, punch, or similar tool to smash metal over seat. This forms a lip that keeps seat in place during engine operation. (TRW)

Top cutting may be needed to make the top of the seat flush with the surface of the combustion chamber.

VALVE SEAT RESURFACING

After new seat installation or when the old valve seats are in serviceable condition, grind or cut the face of the valve seats. The equipment needed to grind valve seats is shown in Fig. 26-17.

Selecting pilot and stone

The *pilot* is a steel shaft that installs in the valve guide to support the grinding stone. Since valve guide diameters vary, you must find the correct size pilot. Find one that slides down into the guide far enough, yet fits the guide snugly.

A T-handle type tool is usually provided for installing the pilot, as pictured in Fig. 26-18. Push down and twist as you fit the pilot into the guide.

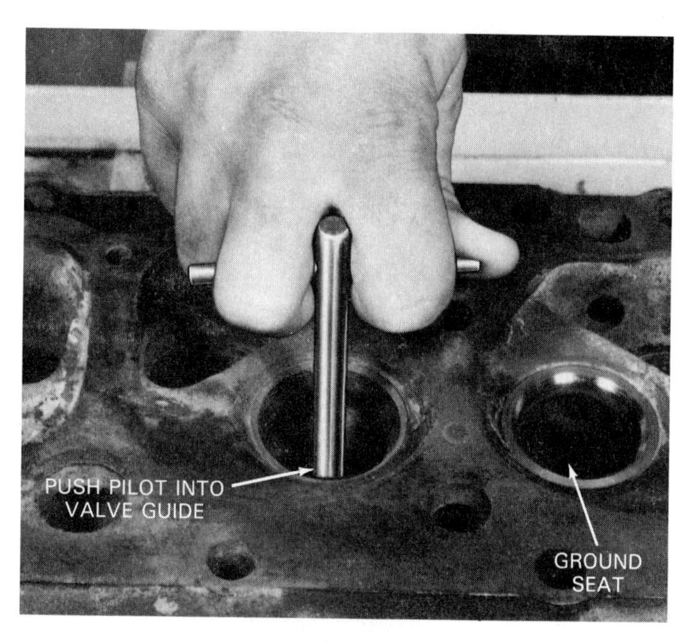

Fig. 26-18. Select right size pilot and push it into guide. Pilot should fit almost to end of taper but not be loose in guide. (Sioux Tools)

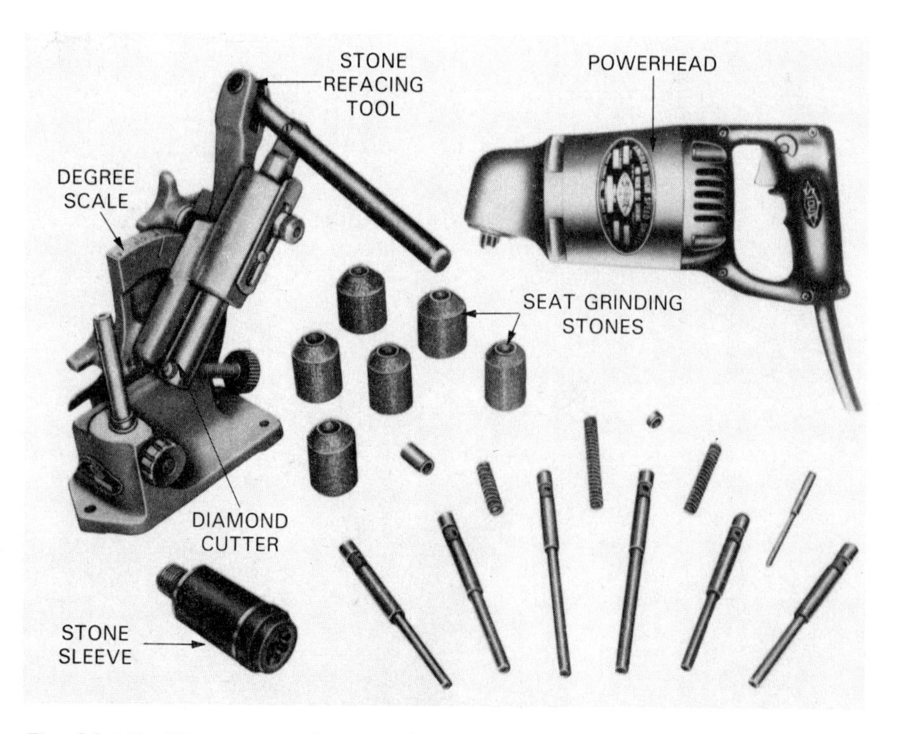

Fig. 26-17. These are tools needed to grind valve seats with stones. Study names of parts. (Sioux Tools)

The *seat stone* is used to cut a fresh surface on the face of the valve seat. It is a grinding stone mounted on a threaded drive attachment.

Find the correct diameter stone for the seat, Fig. 26-19. The stone should be slightly LARGER than the seat but it must not hit the side of the combustion chamber. The stone must also have the correct angle, usually 45 degrees or sometimes 30 degrees.

Screw the stone onto its drive head and check its fit on the seat. If the stone is too small, the seat can be ruined by even momentary grinding!

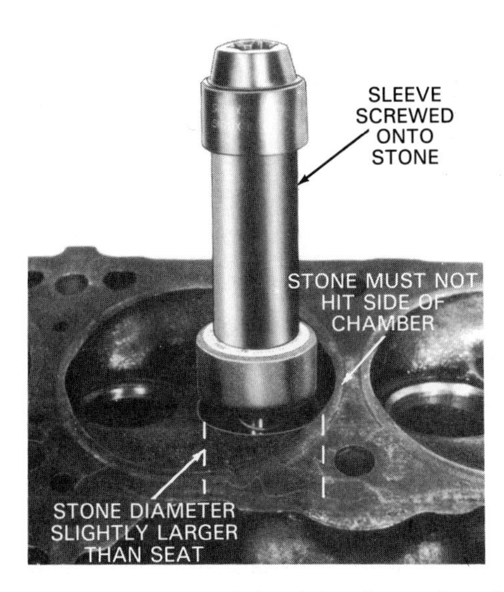

Fig. 26-19. Find a stone of the right size and angle. Stone should be larger than seat but not big enough that it hits side of combustion chamber. (Sioux Tools)

Dressing stone

Dressing a stone involves using a diamond cutter to true the surface of the stone to the correct angle. A stone dressing tool, like the one in Fig. 26-20A, is set to the angle of the valve seat. Then, the drive motor is used to spin the stone.

When dressing the stone, lubricate the pilot with a drop of oil. Set the diamond angle correctly. A scale is provided on the tool that reads in degrees. Lock the tool at the correct angle.

Adjust the diamond cutter up next to the stone. Make sure it will clear the stone slightly before spinning the stone with the power head. See Fig. 26-20B.

CAUTION! The diamond cutter can be ruined if it is adjusted too far into the stone. The stone can grind metal away from the diamond and the diamond can fall out. They are expensive, so adjust the diamond correctly first!

With everything adjusted properly, spin the stone with the power head. Slowly feed the diamond into the stone until it just touches the stone. Then, move the lever up and down to true the entire face of the stone. Only cut as little off the stone as needed.

Grinding valve seats

After you are sure you have the correct stone and it is dressed properly, you are ready to grind the valve seat. Place a drop of oil on the pilot. Slide the stone down onto the seat.

As shown in Fig. 26-21, install the power head into the stone driver. Use one hand to support the power head and the other to operate the trigger.

Note! Do NOT push down on the power head when grinding seats. The weight of the tool is enough downward pressure. Simply hold the tool steady.

Hit the trigger and grind the seat for a second or two.

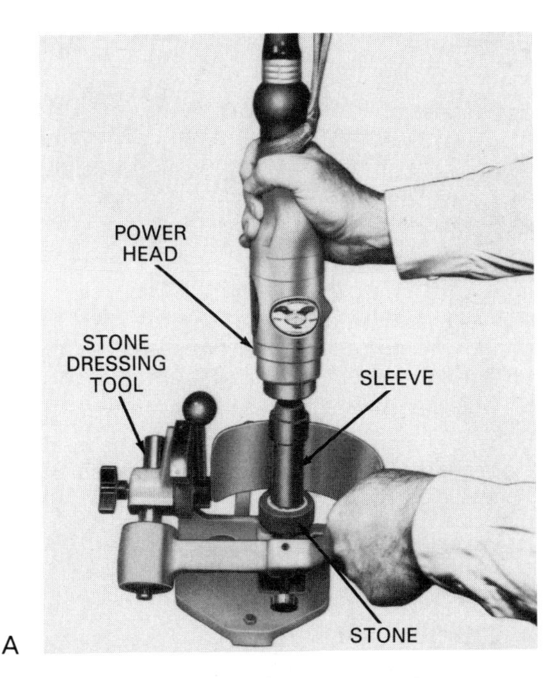

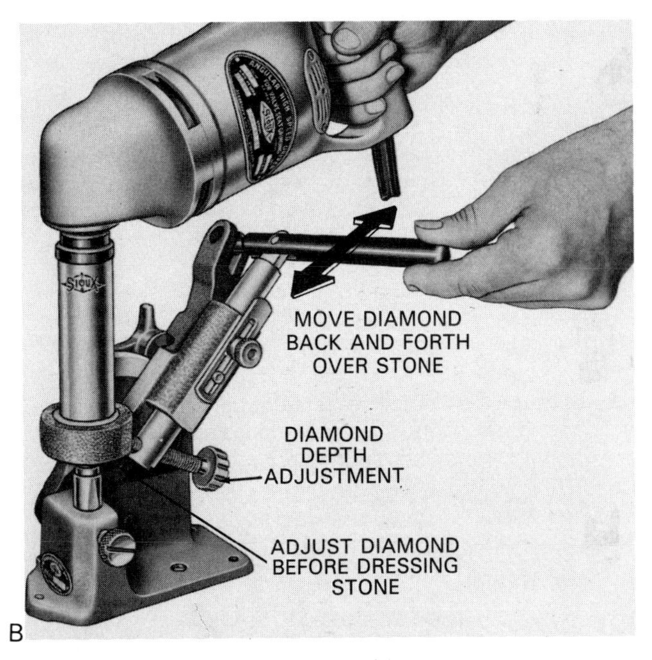

Fig. 26-20. Dressing the stone. A—Stone and sleeve are mounted on stone dressing tool. Tool must be set for correct seat angle. B—After setting angle, adjust diamond so it almost touches stone. Turn on motor and slowly feed diamond into stone while moving lever up and down. Stone should be touched lightly with diamond to avoid rapid stone consumption. (Sioux Tools)

Fig. 26-21. Slide stone and sleeve over pilot. Use power head to spin stone on seat for a second or two. Do not push down on power head. Simply support it and hold it square. (Sioux Tools)

Remove the stone and check the seat. If it is shiny all the way around, no more grinding is needed. You want to grind as little as possible to reduce valve seat retrusion and widening.

Go to the next seat and grind it. As shown in Fig. 26-22, grind all of the intake seats first. Then, install the other stone and grind the exhaust seats. This will keep you from having to change the stone each time. Dress the stone every couple of seats or as needed.

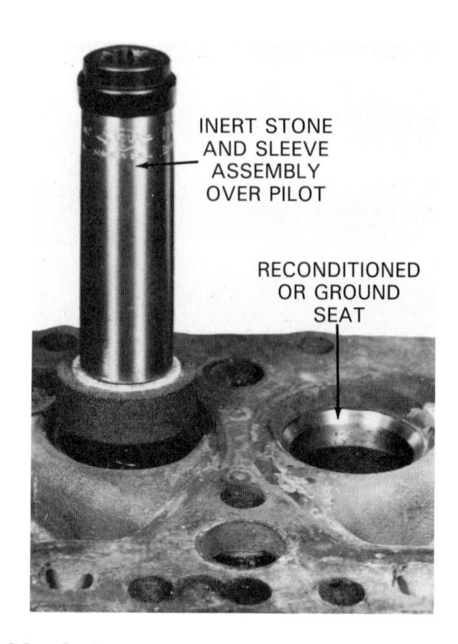

Fig. 26-22. Grind all of the seats the same amount, if possible. Inspect each seat closely after grinding. You must remove all pits or black marks. (Sioux Tools)

Cutting valve seats

The term *cutting valve seats* is used when a carbide cutter, not stones, is used to recondition the cylinder head seats. The carbide cutters are very hard and will machine off a layer of metal when turned against the seats.

Fig. 26-23 shows a carbide valve seat cutting tool. Note how the small carbide blades fit into the tool. This particular tool has a 15 degree angle on the top and 45 degree angle on the bottom.

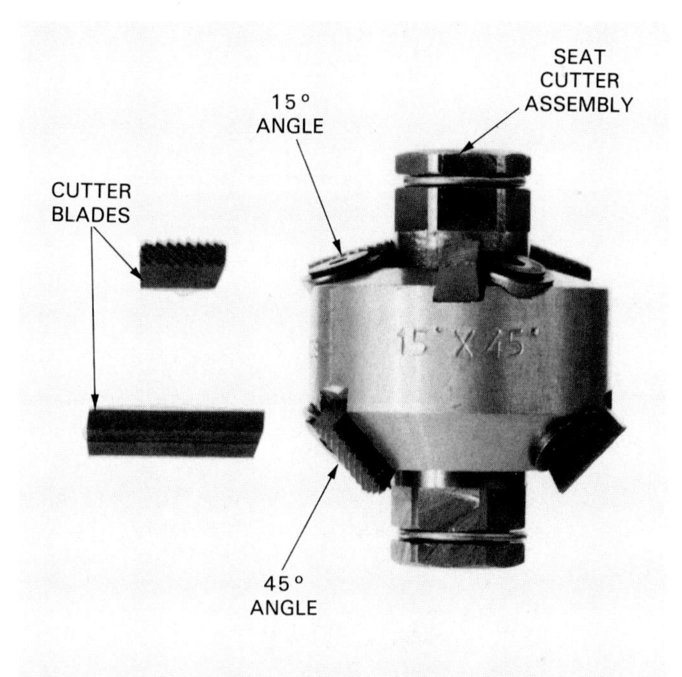

Fig. 26-23. Carbide cutter is another excellent way of reconditioning valve seats. Cutters are made of very hard metal and will remove metal accurately.

To use the cutter, make sure the blades are installed properly. Select the correct angle for the seat. Install the pilot. Slide the cutter over the pilot. Then, use the T-handle to rotate the cutter on the seat. Only cut long enough and clean up all dark areas on the seat. Repeat on the other seats, Fig. 26-24.

Checking seat concentricity and width

After grinding or cutting the seats, some mechanics like to check concentricity or seat runout. A special runout gauge, Fig. 26-25A, is mounted and turned on the seat. The indicator will read seat runout. If not correct, something was wrong when grinding or installing the seat.

You must also check seat width. Grinding or cutting widens the seat. Measure seat width as shown in Fig. 26-25B. If wider than specs, you will have to narrow and position the seat.

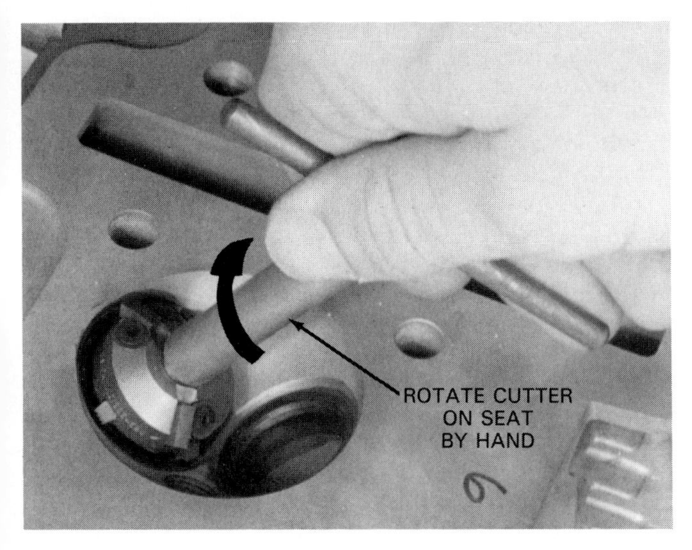

Fig. 26-24. Select correct size and angle cutter. Mount it on T-handle. Rotate cutter against seat by hand until cleaned up.

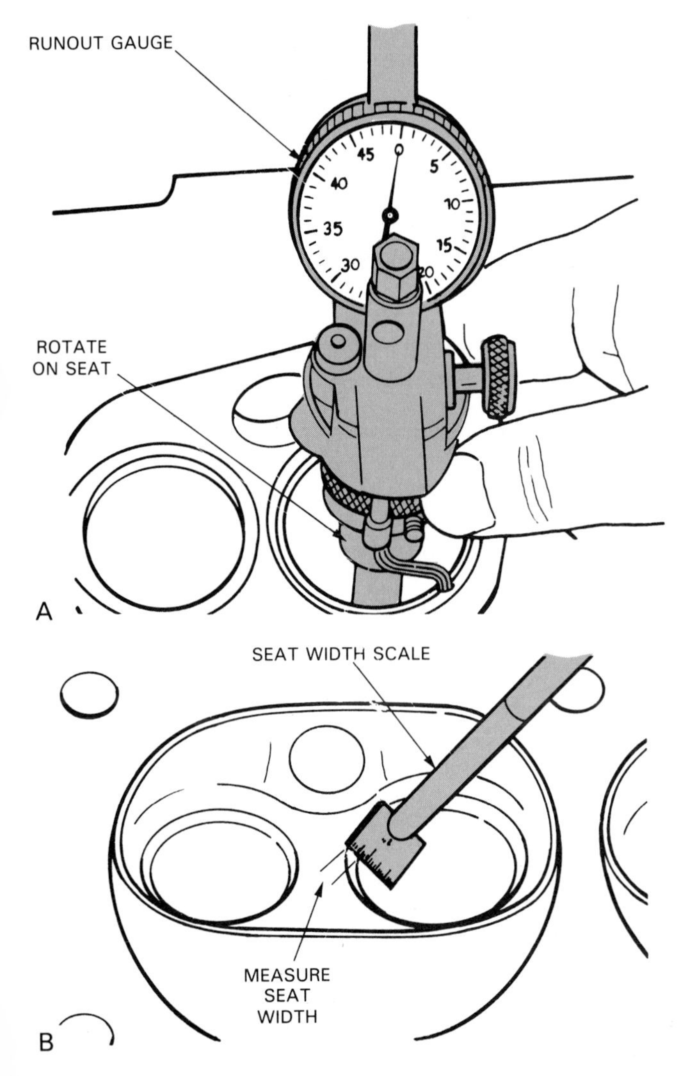

Fig. 26-25. After grinding or cutting, check concentricity and seat width. A—Measure concentricity or runout as shown. B—Use small scale to check seat width. If too wide, you will have to narrow seat with other stone or cutter angles. (Ford)

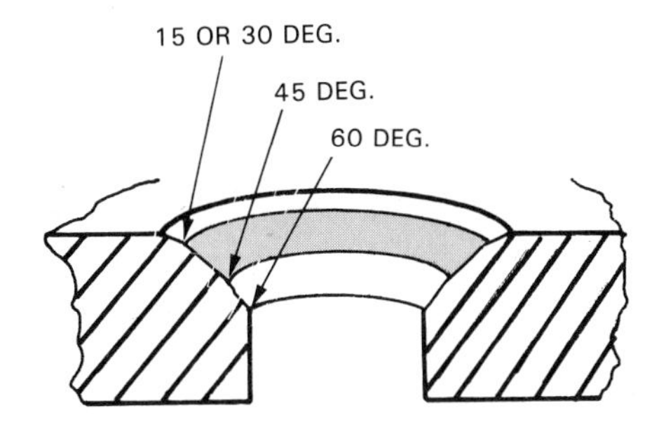

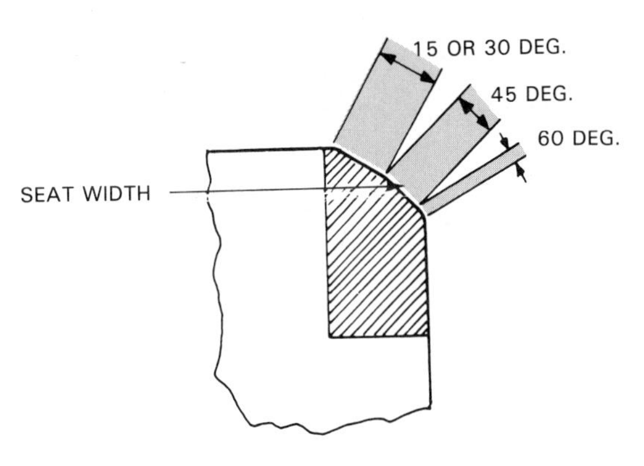

Fig. 26-26. Often called a three-angle valve job, either 15 or 30 degree stone or cutter removes metal from above seat face. A 60 degree stone is used to remove metal below seat. This can be done to narrow and move seat. It can also increase engine power.

Valve seat narrowing

After prolonged service, heat and friction from valve action can wear away the seat. Then, after grinding, the seat can be too wide or it can contact the valve too far out on the face.

Seat narrowing, also called *seat positioning,* is needed so that the seat contacts the valve face correctly. For example, if the seat angle is 45 degrees, use 30 and 60 degree stones to move the seat and narrow it. This is sometimes called a *three-angled valve job* because three different stone angles are used, Fig. 26-26.

Fig. 26-27 shows how a carbide cutter is used to position and narrow a valve seat. First, use a 60 degree cutter to remove material from the inside diameter of the seat. Then, use a 15 degree cutter to remove metal from the outside diameter of the seat as needed. Finally, use the correct cutter angle (45 degree for example) to resurface the face of the seat.

Checking valve-to-seat contact

To check valve-to-seat contact, you can use valve lapping compound or Prussian Blue. Both require the valve to be ground first, as will be explained shortly.

Valve lapping compound is a paste material filled with abrasives. A small amount is placed on the valve face.

Then, a *lapping stick* (stick with suction cup on end) is used to spin the valve on its seat. To turn the stick, rub the palms of your hands together on each side of the

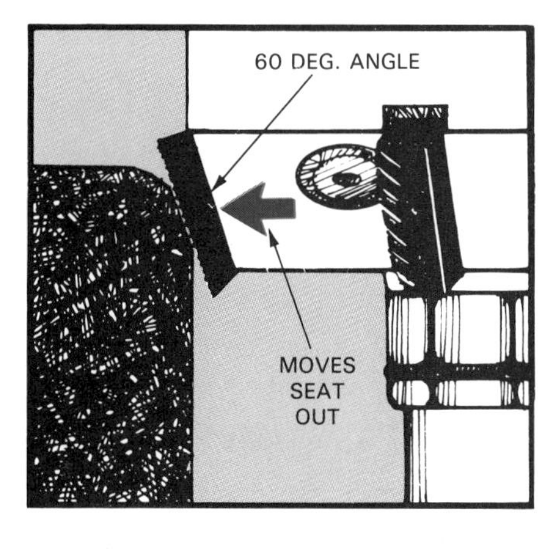

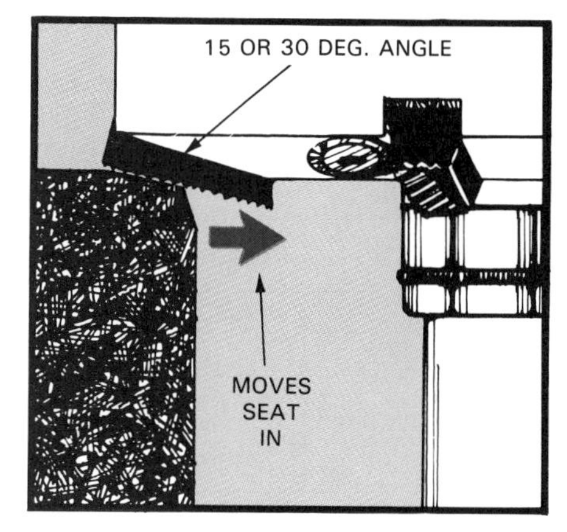

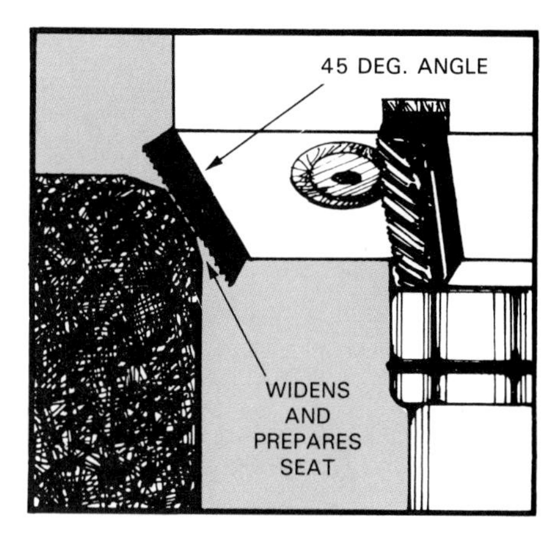

Fig. 26-27. Note basic steps for using carbide cutter to do a three-angle valve job. Estimating seat width and location before cutting will help you know how much to cut with each angle. (Neway)

stick, while picking up and repositioning the valve slightly. Look at Fig. 26-28A.

Remove the valve and check the contact point. A DULL GRAY STRIPE around the seat and face indicates the valve-to-seat contact point. This will help you find out how to move the valve seat if needed, Fig. 26-28B.

A few manufacturers do NOT recommend valve lapping. Refer to a service manual for details.

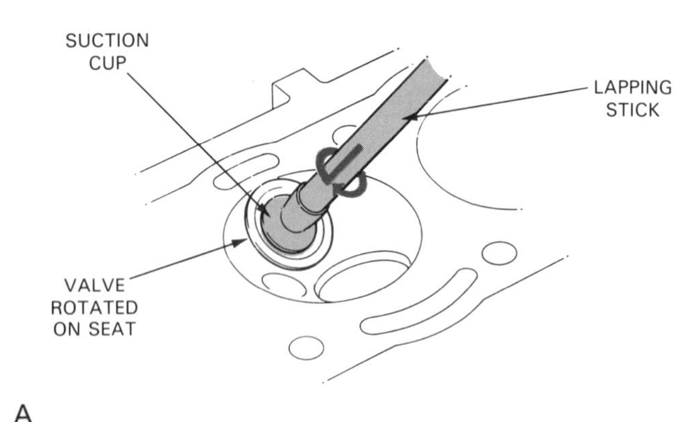

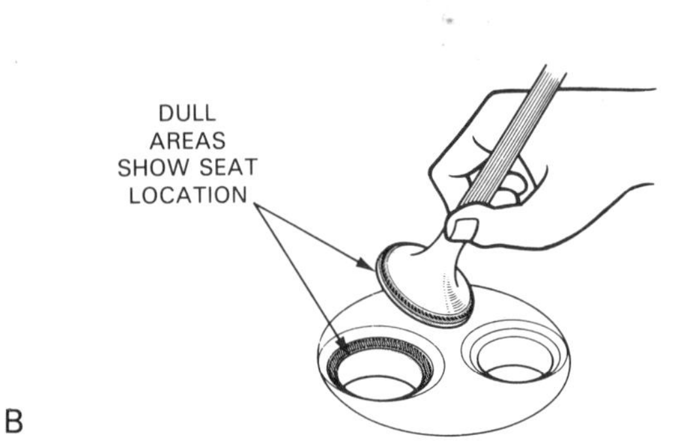

Fig. 26-28. Lapping or grinding compound is an easy way to check seat location on valve. A—Wipe small amount of compound on valve. Then use lapping stick to rotate valve on seat. B—Remove valve and check dull stripe on valve and seat. This will show how seat touches valve face. Clean off all abrasive compound from parts before head assembly. (Toyota)

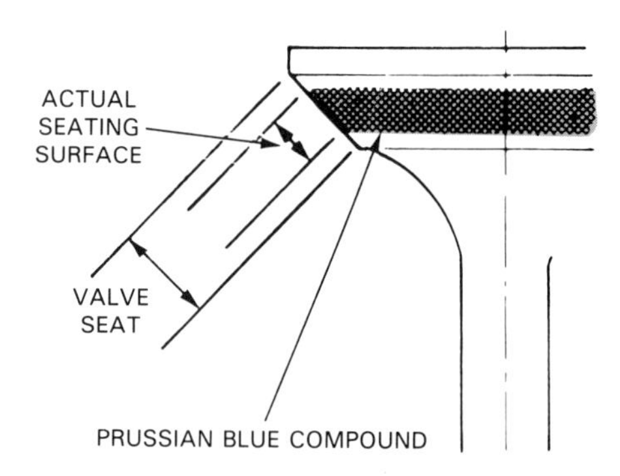

Fig. 26-29. Prussian Blue is another way to check valve-to-seat contact. Wipe dye on valve all the way around face. Insert valve into guide. Then, use your hand to slap valve closed several times. Remove valve and shiny area shows contact. This will tell you if you need to do more grinding or cutting to move seat location. (Honda)

WARNING! Make sure you clean all of the valve grinding compound off of the valve and cylinder head. The compound can cause rapid part wear.

To use Prussian Blue to check the seat, wipe the dye on the valve face. Then, snap the valve closed on its seat several times by hand. The contact with the seat will wipe off some of the Prussian Blue, making the valve face shiny. This will show you where the seat and valve touch. Refer to Fig. 26-29.

Fig. 26-30 shows some good and bad seat contact patterns. You can check this with the procedures just described.

Fig. 26-31 shows typical seat specs for both intake and exhaust valve seats. Note how the exhaust seat is wider so it can dissipate heat better.

An INTAKE VALVE should generally have a valve-to-seat contact width of about 1/16 in. (1.6 mm). An EXHAUST VALVE should have a valve-to-seat contact width of about 3/32 in. (2.4 mm) Check manual specs for exact values.

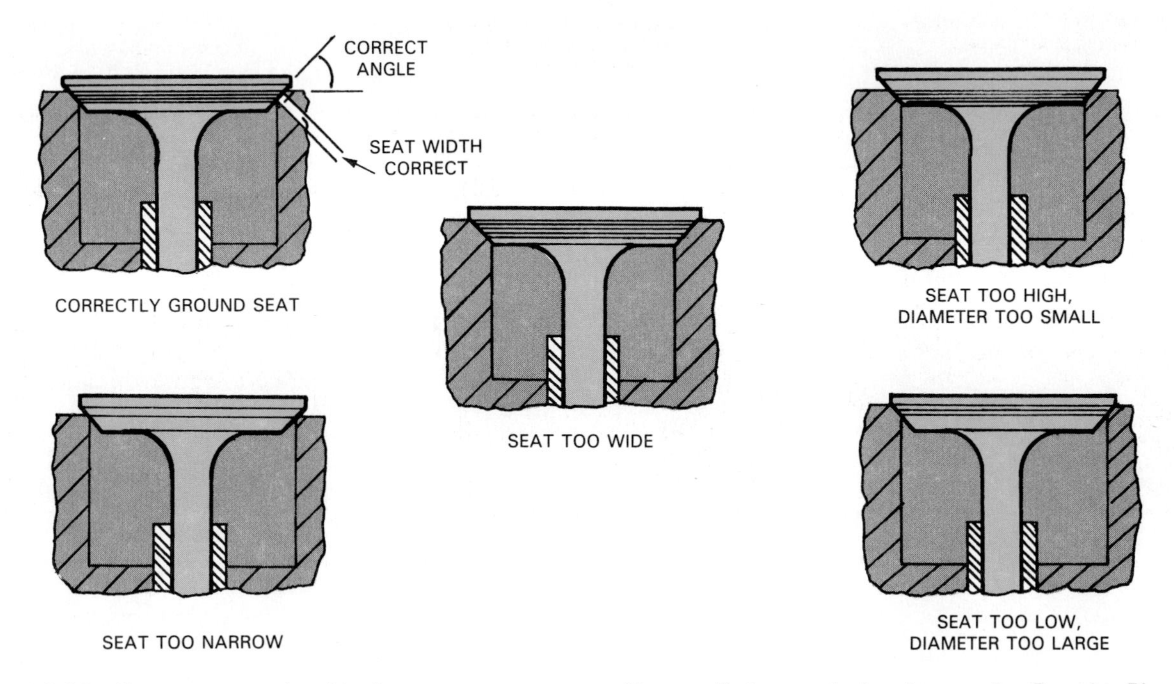

Fig. 26-30. Note some good and bad seat contact patterns. These will show up by lapping or using Prussian Blue. (Ford)

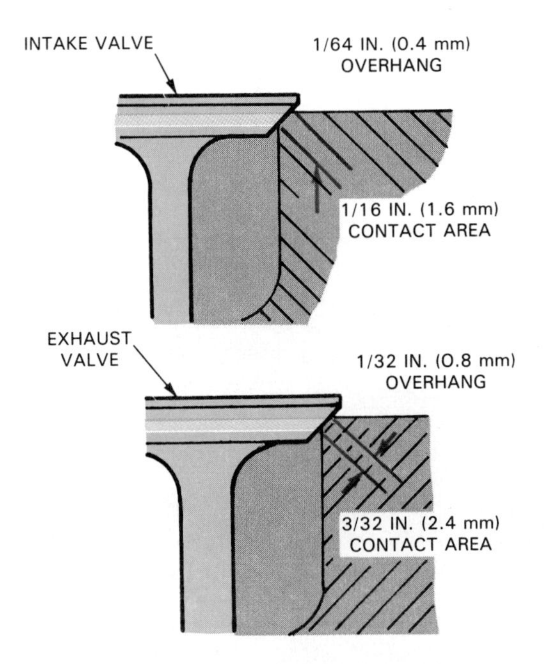

Fig. 26-31. These are typical specs of intake and exhaust valves. Exhaust seat must be slightly wider to dissipate heat more quickly. When possible, use engine manufacturer specifications. (TRW)

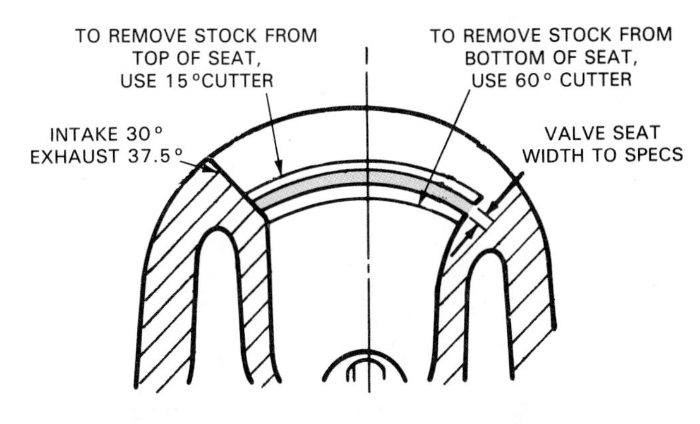

Fig. 26-32. Note again different angles used to grind seat properly.

Moving seat contact point

If the valve seat does NOT touch the valve face properly (wrong width or location on valve), regrind the seat using different stone angles. As mentioned, use 15 or 30 and 60 degree stones, Fig. 26-32.

To MOVE THE SEAT IN and narrow it, grind the valve seal with a 15 or 30 degree stone. Use a 30 degree stone with a 45 degree seat and 15 degree stone with a 30 degree seat angle. This will remove metal from around the top of the seat. The seat face (valve contact surface) will move closer to the valve stem.

To MOVE THE SEAT OUT and narrow it, machine the valve seat with a 60 degree stone or cutter. This will cut metal away from the inner edge of the seat. The seat contact point will move toward the margin or outer edge of the valve.

VALVE GRINDING

Valve grinding machines a fresh, smooth surface on the valve faces and stem tips. Valve faces suffer from burning, pitting, and wear caused by opening and closing millions of times during the service life of an engine. Valve stem tips wear because of friction from the rocker arms.

Before grinding, inspect each valve face for burning

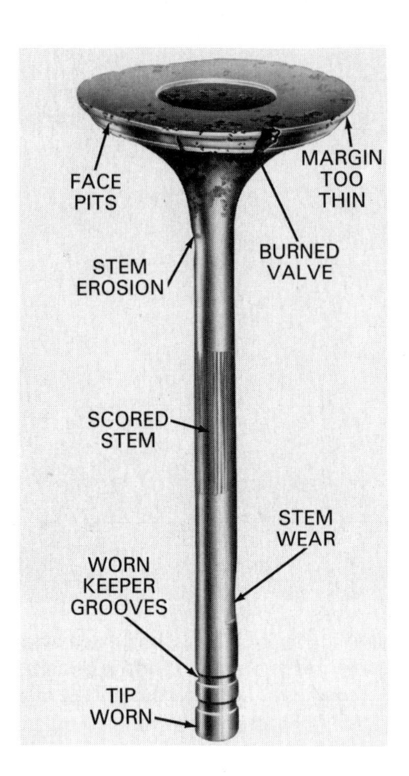

Fig. 26-33. These are some of the troubles you will find with used valves. (Sioux Tools)

and each stem tip for wear. Replace any that are badly burned or worn. You must also grind a new valve as you do the old, used valves. Fig. 26-33 shows valve problems.

DANGER! Wear a face shield when grinding valves. The stone could shatter, throwing debris into your face.

Valve grind machine

A *valve grind machine* will resurface the valve faces and stems. Although there are some variations in design, most valve grind machines are basically the same. They use a grinding stone and precision chuck to remove a thin layer of metal from the valve face and tip, Fig. 26-34.

Valve grinder setup

Dress the stone by using a diamond cutter to true the stone surface. Do this before grinding the valves. A diamond tipped cutting attachment will be provided with the machine for truing the stone. Follow equipment manufacturer's instructions.

CAUTION! Be very careful when using a diamond tool to dress a stone. Wear eye protection and feed the diamond into the stone SLOWLY. If fed in too fast, tool or stone breakage may result.

Set the chuck angle by rotating the valve grinding machine chuck assembly. A degree scale or digital readout is provided so that the cutting angle can be set precisely. Normally, you must loosen a locknut and swivel the chuck assembly into the desired angle. This is illustrated in Fig. 26-35A.

An *interference angle* is normally a one degree difference in the face valve angle and the valve seat angle. It is set on the valve grind machine. If the valve seat angle is 45 degrees, the chuck is set to 44 degrees. This produces an interference angle between the valve face and seat. The break in and sealing time of the valve is reduced. See Fig. 26-35B.

Chuck the valve in the valve grind machine by inserting the valve stem into the chuck. Make sure the stem is inserted so that the chuck grasps the MACHINED SURFACE of the stem near the valve head. The chuck must NOT clamp onto an unmachined surface or runout will result, Fig. 26-35.

Fig. 26-34. Valve grind machine will resurface valve faces and stem tips. Note names of machine parts. (Sioux Tools)

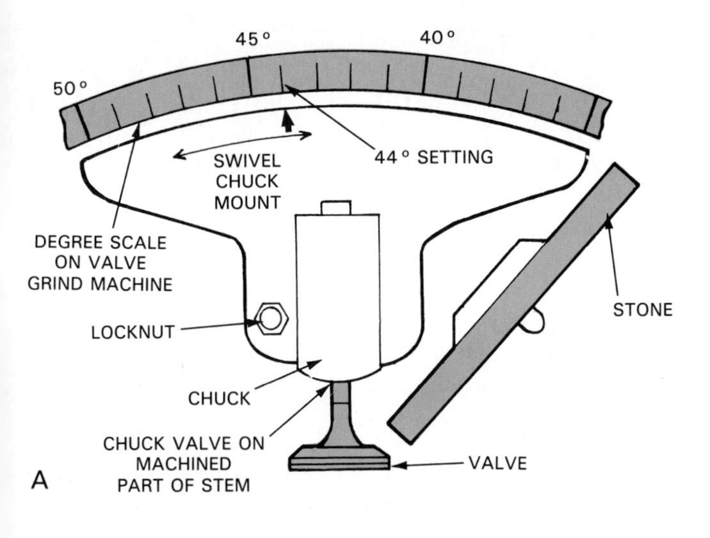

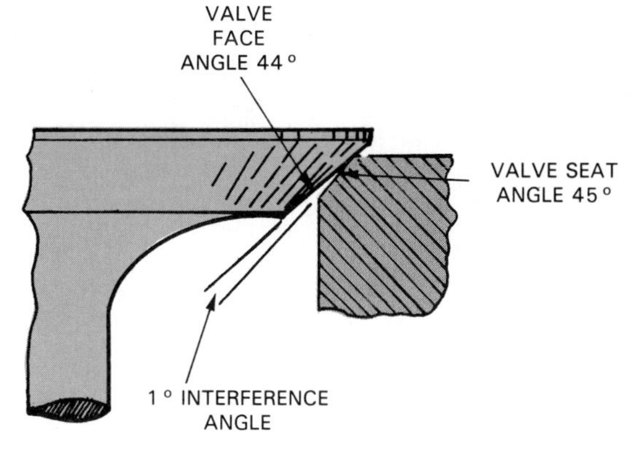

Fig. 26-35. Before grinding valves, adjust stone angle and dress stone. A—If valve has 45 degree face, adjust grinder to 44 degrees. Scale or digital readout will be provided. Also, chuck valve so it is resting on machined portion of stem near valve head. B—Note how interference angle makes valve and seat touch so contact area does not widen rapidly in service and to speed initial valve sealing.

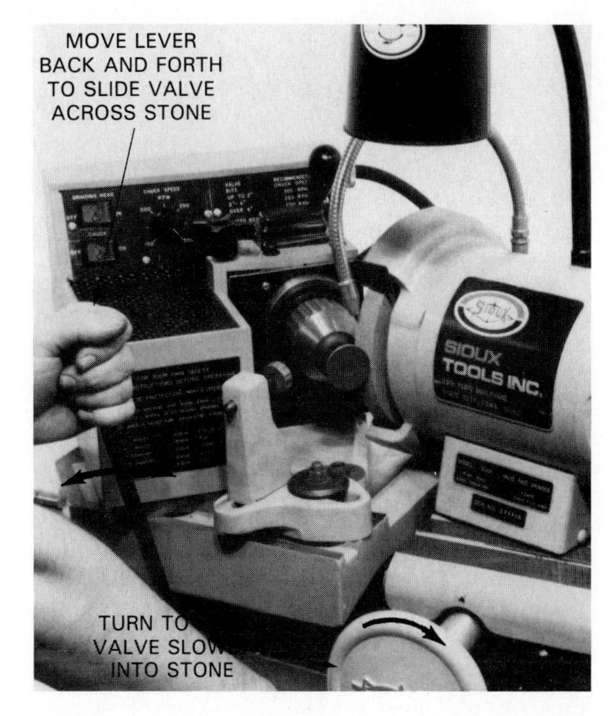

Fig. 26-36. To grind valve face, turn wheel to slowly feed valve into spinning stone. Make sure coolant is adjusted to flow over valve. As soon as stone touches valve, begin using lever to move valve back and forth over stone. Grind until valve face is shiny and free of pits. (Sioux Tools)

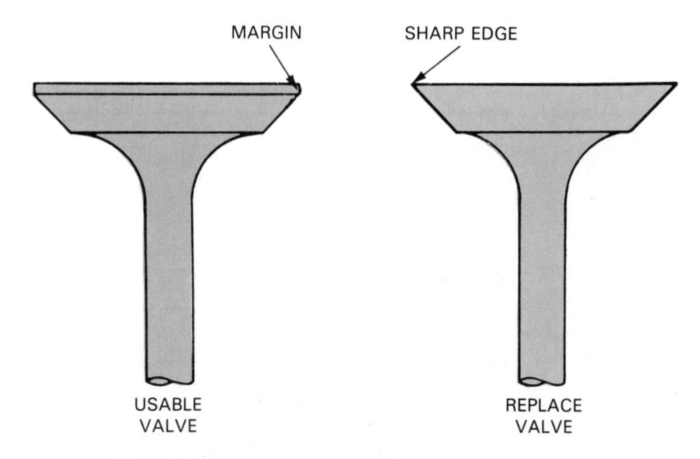

Fig. 26-37. After valve grinding, check margins on valves. If thinner than specs, replace the valve. If margin is too thin, valve could burn easily during engine operation. (TRW)

Grinding valve faces

Turn on the valve grind machine and the cooling fluid. SLOWLY feed the valve face into the stone. While feeding, slowly move the valve back and forth in front of the stone. Use the full face of the stone but do NOT let the face move out of contact with the stone while cutting. Look at Fig. 26-36.

Note! Only grind the valve long enough to ''clean up'' its face. When the full face looks shiny, with no darkened pits, shut off the machine and inspect the face. Look carefully for pits or grooves.

Grinding, by removing metal from the face, will make the valve stem extend through the head more. This will affect spring tension and rocker arm geometry. Grind the face of each valve as little as possible.

A *sharp valve margin* indicates excess valve face removal and requires valve replacement. Auto makers give a spec for minimum valve margin thickness.

If the margin is too thin, the valve can burn when returned to service. It may not be thick enough to dissipate heat fast enough. The head of the valve can actually begin to melt, burn, and blow out the exhaust port. Look at Fig. 26-37.

A *burned valve,* if not noticed during initial inspection, will show up when excess grinding is needed to clean up the valve face. A normal amount of grinding will not remove a deep pit or groove. Replace the valve if burned.

A *bent valve* will show up when the valve head wobbles as it turns in the valve grind machine. If the valve is chucked improperly, it can appear like a bent valve. Shut off the machine and find the cause, Fig. 26-38.

Repeat the grinding and inspecting operation on the other valves. Return each ground valve to its place in an

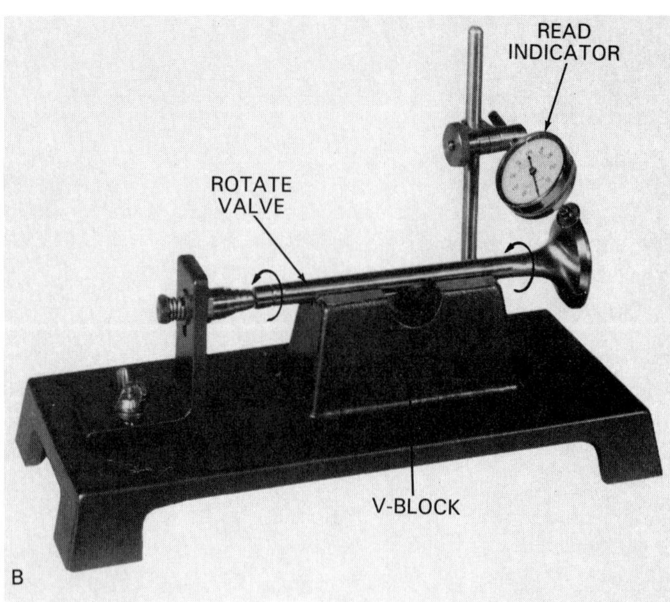

Fig. 26-38. A—This valve had to be replaced. It was bent slightly and eroded. Only one side of face cleaned up when being ground. B—V-blocks and indicator is more accurate way to check valve runout, but it is time consuming. (K-Line Tools)

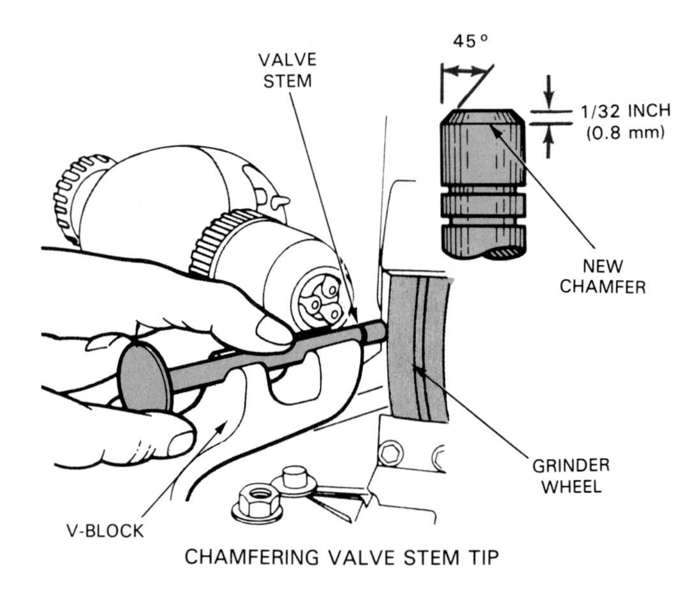

Fig. 26-39. After grinding valve faces, you should also true valve stem tips. Here technician is putting small chamfer on tip. However, if too much material has been worn off tip, valve should be replaced. Check manual for specs. (Ford)

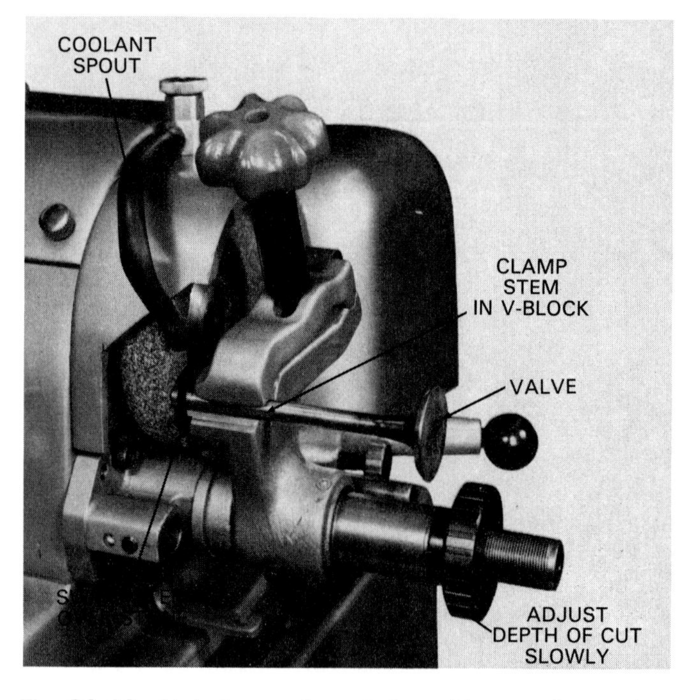

Fig. 26-40. Note how valve grind machine can be used to grind valve tip, making it flat and smooth. Mount valve in V-chuck. Slowly feed tip into spinning stone. Only grind enough to true tip. (Sioux Tools)

organizing tray. Used valves should be returned to the same valve guide in the cylinder head. The stems may have been select fit.

Grinding valve stem tip

Another stone on the valve grind machine is provided for truing and chamfering the valve stem tips. Figs. 26-39 and 26-40 show how to grind and chamfer valve stems.

Grind as little off the stem as possible. Many stems are hardened. Too much grinding will cut through the hardened layer and result in rapid wear when the valve is returned to service.

An indicator is provided on the valve grind machine to show the depth of cut for both the valve face and valve stem tip. Generally, CUT THE SAME amount of metal off of the face and the stem. This will help keep valve train geometry correct.

Valve spring service

Valve springs tend to weaken, lose tension, or even break after extended service. Always test valve springs to make sure they are usable.

Valve spring free height can also be measured with the combination square or with a valve spring tester. Simply measure the length of each spring in a normal, uncompressed condition. If too long or too short, replace the spring, Fig. 26-41A.

Valve spring tension or pressure is measured on a spring tester. Compress the spring to specification or installed height and read the scale. Spring pressure must be within specs. If too low, the spring has weakened and needs replacement or shimming, Fig. 26-41B.

Valve spring squareness is easily checked with a combination square. Place each spring next to the square on a flat work surface. Rotate the spring while checking for a gap between the side of the spring and the square. Replace the spring(s) if not square, Fig. 26-41C.

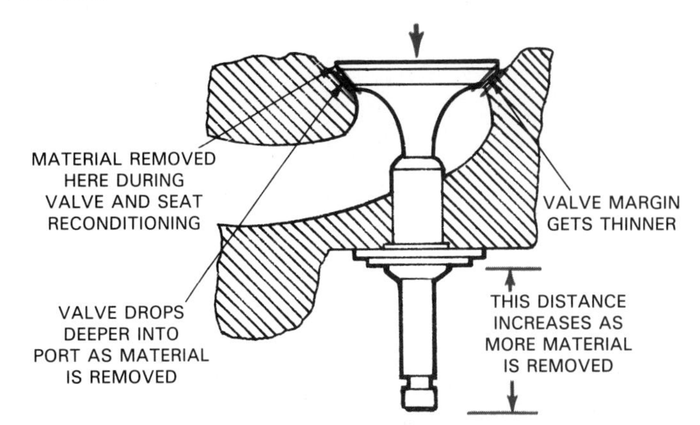

Fig. 26-42. When valve and seat are ground, valve stem extends through head farther. This increases spring installed height and reduces spring pressure.

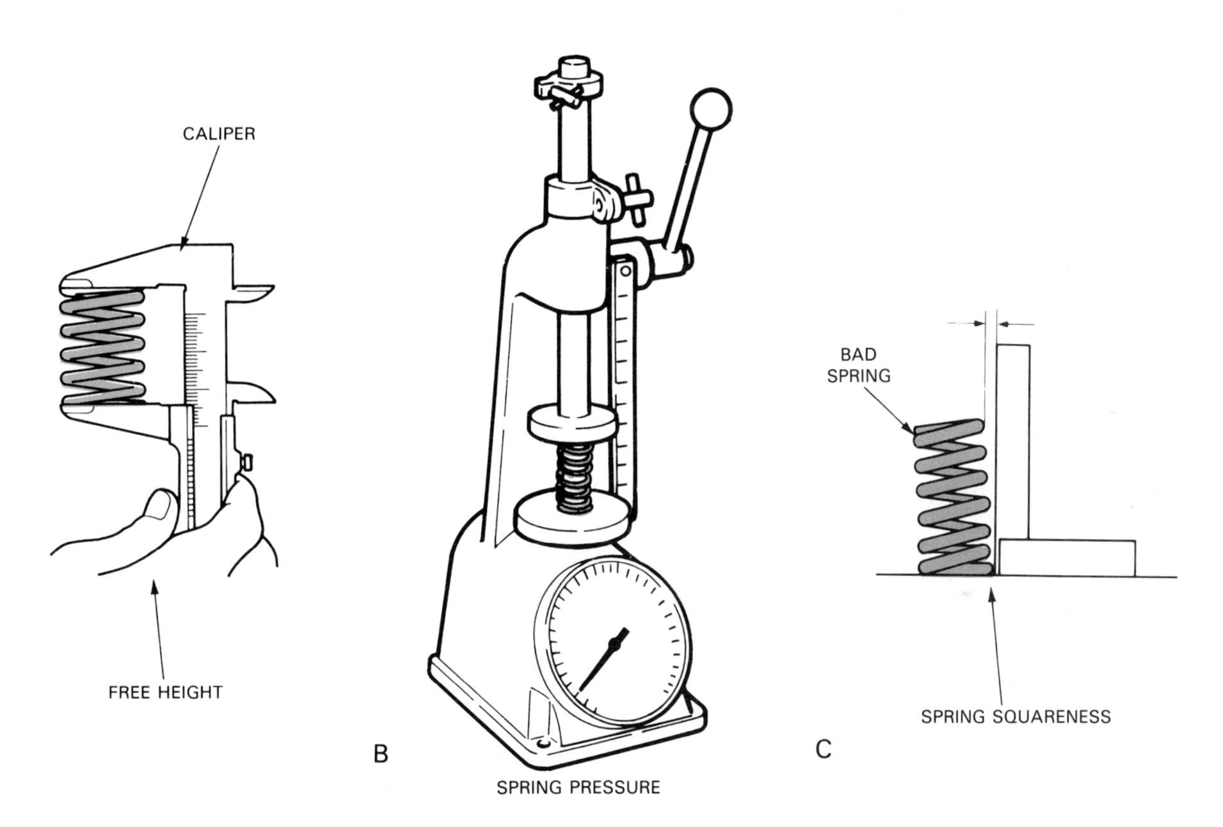

Fig. 26-41. Valve springs must be checked closely before reuse in engine. They can weaken, lose tension, and break. A—Measure free height or length. B—Check spring pressure. C—Make sure spring is square. (Buick)

Shimming valve springs

Shimming valve springs is done to maintain correct tension when the springs are installed on the cylinder heads. When valves and seats are ground, the valve stems stick through the head more. This increases valve spring installed height and reduces spring pressure. This is illustrated in Fig. 26-42.

Valve spring installed height is the distance from the top and the bottom of the valve spring with the spring installed on the cylinder head. It can be measured, as in Fig. 26-43. If greater than specs, add enough shims to reduce installed height and return spring pressure to normal.

Mentioned briefly, a *valve spring tester* is used to measure spring pressure at spring installed height. This allows you to determine how much spring pressure will be present after cylinder head assembly. See Fig. 26-44.

To use a valve spring compressor, place a valve spring on the compressor. Press down on the lever until the scale equals spec pressure for the specific engine springs. Then read the pointer scale to find how far the spring has been compressed. This will let you calculate

Fig. 26-43. After grinding valves and seats, install valve, retainer, and keepers as shown. Then measure valve spring installed height. This will let you determine if you need valve spring shims. (Oldsmobile)

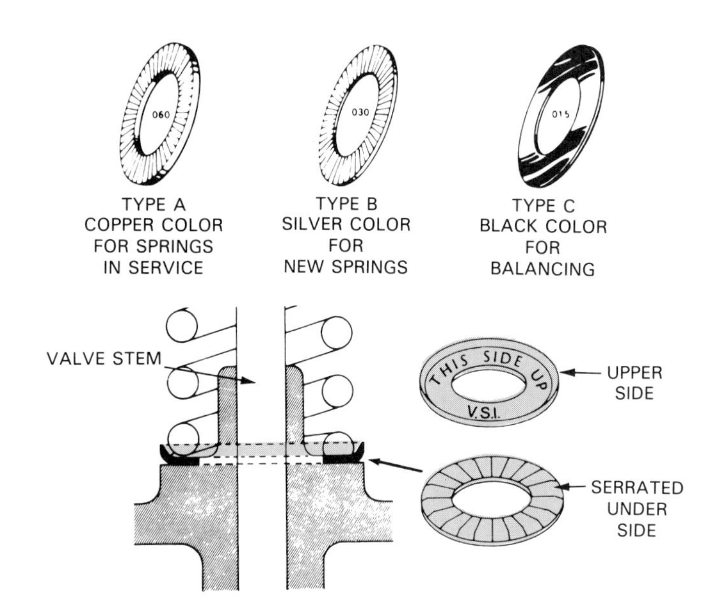

Fig. 26-45. Note types of spring shims and how serrated side should be down. (K-Line Tools)

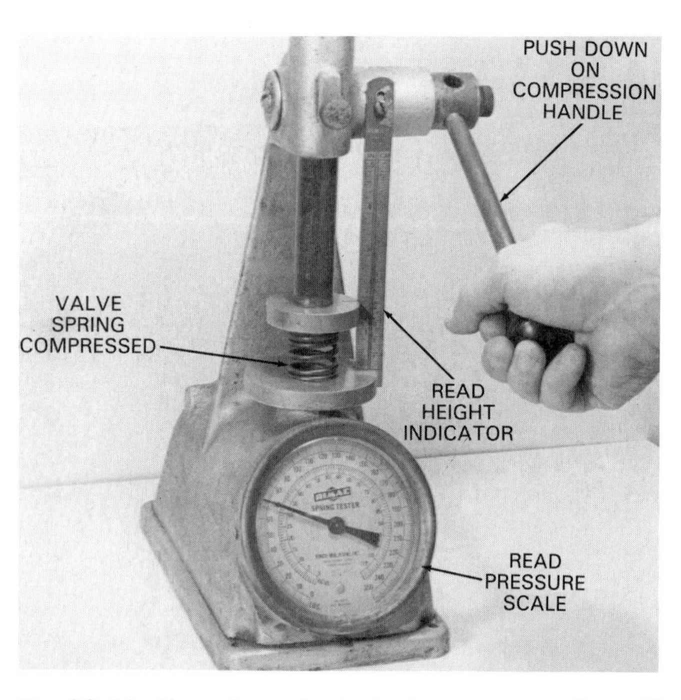

Fig. 26-44. Use valve spring tester to compress spring until scale reads spec pressure. Then read height indicator. Difference in tester height and spring installed height would equal the thickness of shim needed to attain correct spring pressure.

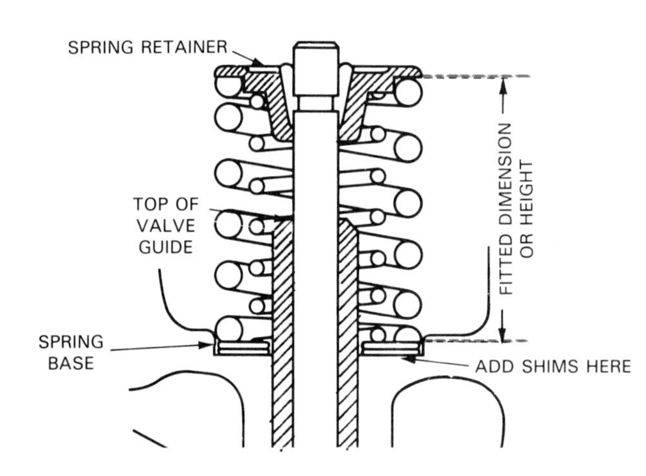

Fig. 26-46. Do not use shims that are too thick, .060 is a typical maximum. (TRW)

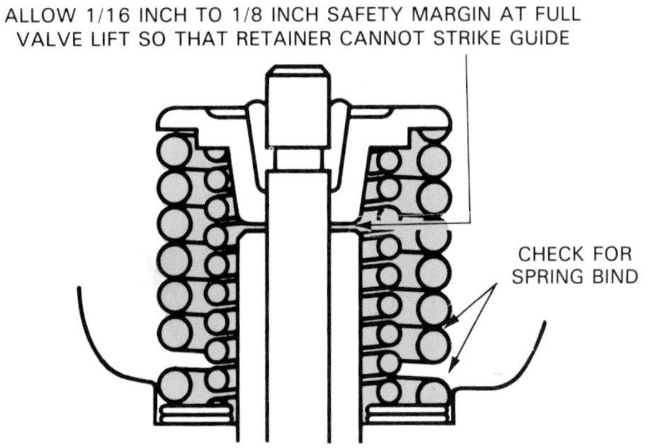

ALLOW 1/16 INCH TO 1/8 INCH SAFETY MARGIN AT FULL VALVE LIFT SO THAT RETAINER CANNOT STRIKE GUIDE

CHECK FOR SPRING BIND

Fig. 26-47. Spring bind is a result of too much cam lift or shims that are too thick. New guide must also be short enough to clear spring retainer. (TRW)

how thick the shims should be to attain the correct spring pressure.

For example, if you had to compress the spring .010 inch more than the installed spring height to attain spec pressure, shim the spring, .010 inch, Fig. 26-45.

Generally, never shim a valve spring over .060 in. (1.5 mm). Thicker shimming could cause spring bind. *Spring bind* causes the spring to fully compress and lock the valve train. This could possibly damage components. Look at Figs. 26-46 and 26-47.

CYLINDER HEAD REASSEMBLY

After cleaning, inspection, measurement, valve guide service, seat grinding, and valve grinding, you are ready to assemble the head. Place a drop of oil on the valve stems and slide them into their guides.

Installing valve seals

Depending upon whether the head has umbrella or O-ring type valve seals, you will have to alter assembly procedures slightly.

With UMBRELLA VALVE SEALS, simply slide the seals over the valve stems. Then install the valve spring assembly. Refer to Fig. 26-48A.

With O-RING VALVE SEALS, compress the valve spring BEFORE fitting the seal on the valve stem. If you install the seal first, it will be cut, split, or pushed out of its groove. Engine oil consumption and smoking will result, Fig. 26-48B.

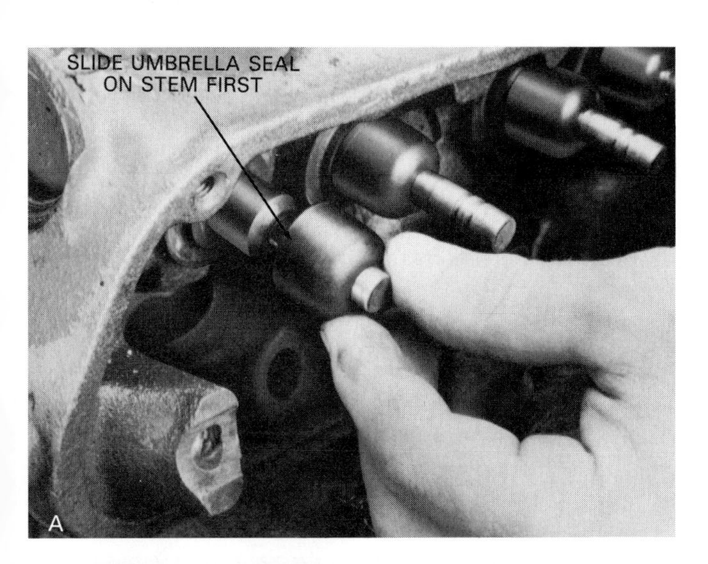

Fig. 26-48. Sequence for valve seal installation varies. A—With umbrella type seal, slide seal over valve first. B—With O-ring seal, compress the spring first. Then fit seal into groove in valve stem. If you forget and install O-ring first, it can be cut or torn, causing engine to smoke. (Fel-Pro Gaskets)

Fig. 26-49 shows how to install a POSITIVE TYPE VALVE SEAL. The top of the valve guide has been machined so that the seal locks onto the guide. First, a protective cap is placed over the valve stem. Then, an installation tool is used to force the valve seal into position.

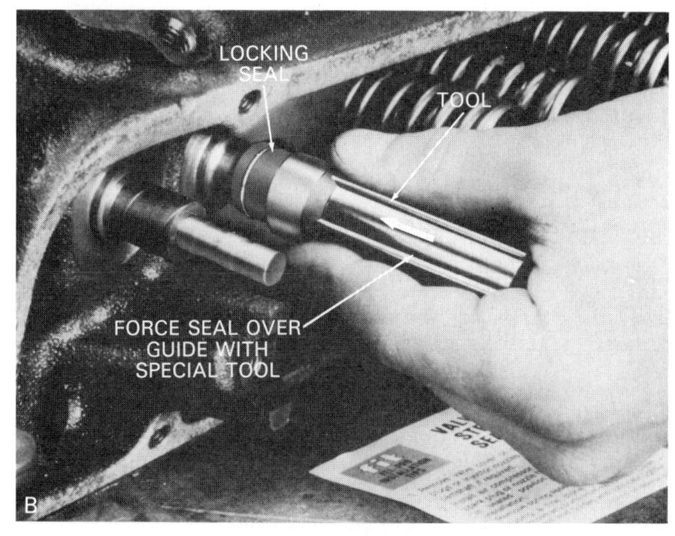

Fig. 26-49. Positive valve seals lock around valve guides. Guides must be machined for these seals. A—Protective cap should be placed over stem to protect rubber seal. B—Special tool is needed to force seal over guide. (Fel-Pro Gaskets)

Installing valve springs

With the spring compressed, fit the retainer and keepers into place on the valve. Install all of the valve assemblies. Tap on the valve stems with a brass hammer to seat the keepers in their grooves, Fig. 26-50.

A valve spring compressor for installing the springs on an OHC engine is pictured in Fig. 26-51. It is needed to reach down into the valve spring pockets.

Checking for valve leakage

To check for valve leakage after head reconditioning, lay the cylinder head on its side. Pour clean soak solvent

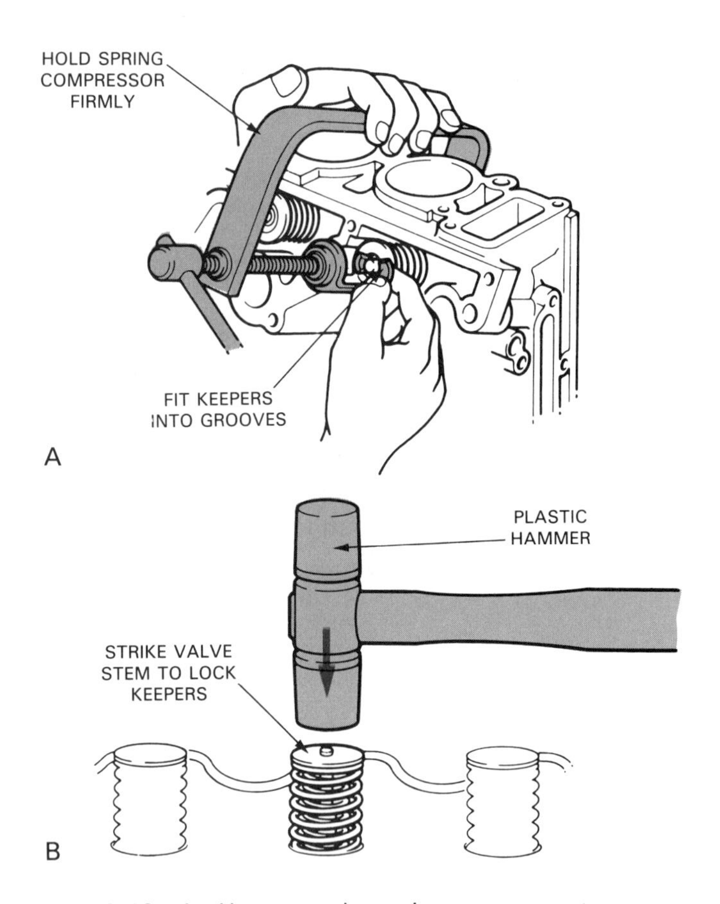

Fig. 26-50. A—Use your valve spring compressor to assemble head. Hold tool securely while fitting keepers into grooves in valve stem. B—Use plastic hammer to hit stem tips. This will set keepers in grooves. (Toyota)

Fig. 26-51. This is a special valve spring compressor needed on this overhead cam cylinder head. Note how it fastens to camshaft cap stud. (AMC)

or water into the intake and exhaust ports. With the fluid in the ports, watch for leakage around the valve heads. If solvent or water drips from around a valve, that valve is leaking. Remove the valve from the head and check for problems. See Fig. 26-52.

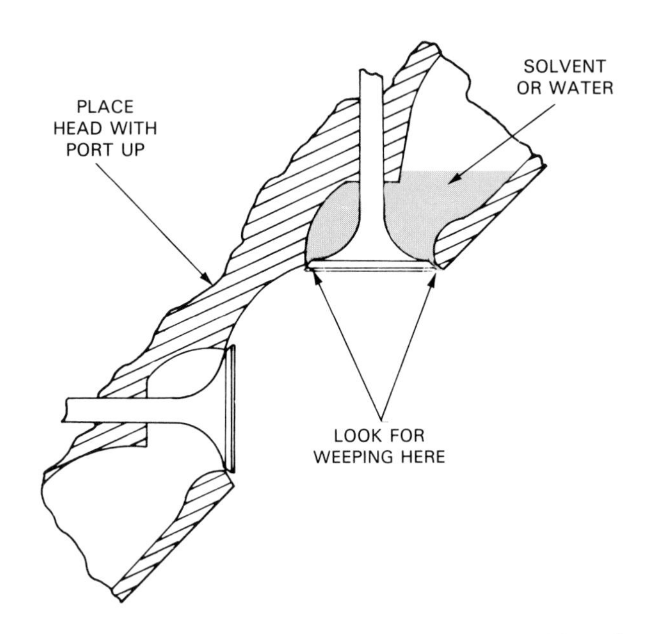

Fig. 26-52. To check for valve leakage before head installation, lay head with ports up. Pour clean solvent or water into ports. Watch for leakage or weeping at valve heads. If any leak, disassemble and find problem.

ROCKER ARM STUD SERVICE

Rocker arm studs can require replacement after thread damage or if they pull out of the head. Some are press-fit in the head; others screw into the head.

Fig. 26-53 summarizes how to replace pressed-in cylinder head rocker arm studs.

IN-CAR VALVE SEAL SERVICE

Discussed in Chapter 24, valve seals and springs can be serviced without cylinder head removal. Use a shop air hose and a special fitting to inject air into one of the cylinders. The air will hold the valves in that cylinder up against their seats. Then, use a special pry bar type compressor to remove the keepers and springs for that cylinder. This will allow in-car seal or spring replacement.

CAMSHAFT SERVICE

Camshaft service involves measuring cam lobe and journal wear, camshaft end play, and cam bearing clearance. It also includes distributor-oil pump gear inspection and cam bearing replacement.

Measuring camshaft wear

Cam lobe wear can be measured with a dial indicator with the camshaft installed in the engine. This is pictured in Fig. 26-54.

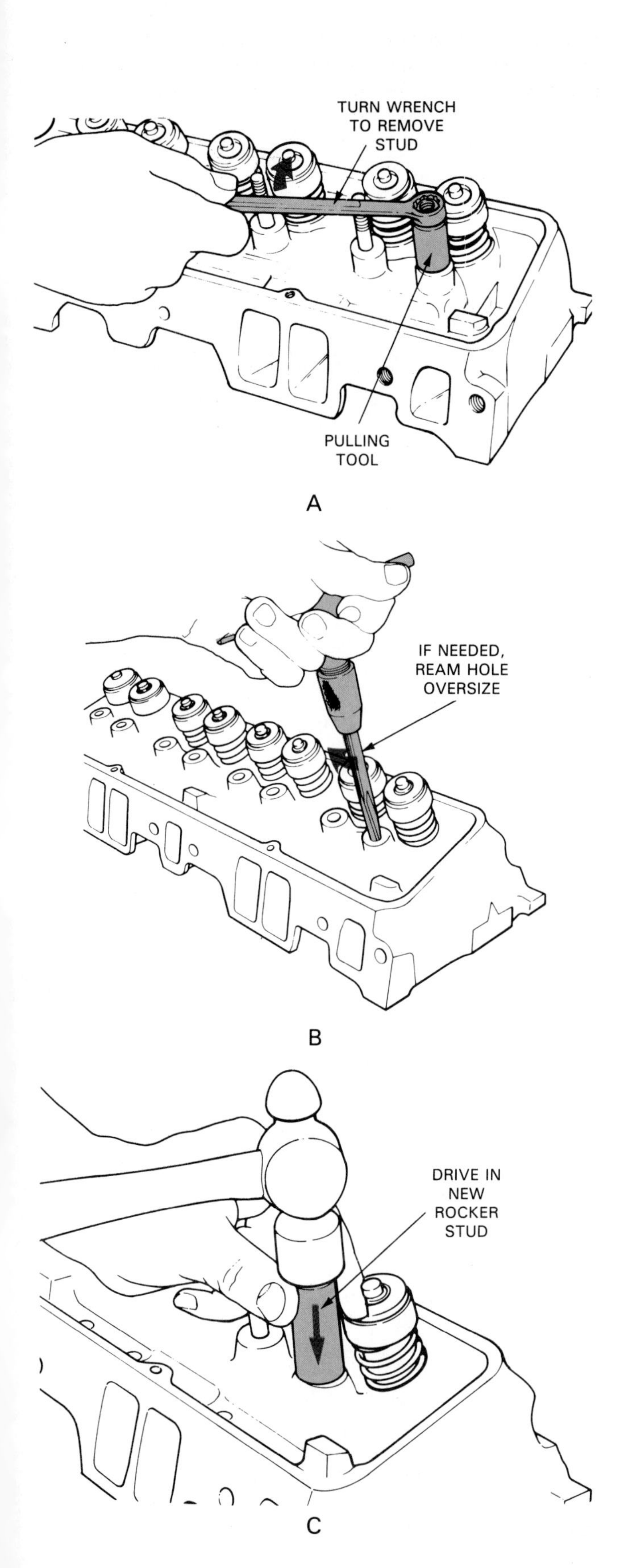

Fig. 26-53. Rocker arm stud replacement. A—If not screwed in, use puller to remove old stud. B—If needed, ream hole for oversize stud, as when old stud pulled out of head. C—Drive new stud in place. (Chevy)

Fig. 26-54. Cam lobe wear can be checked before engine teardown as shown. Mount dial indicator against lobe. Rotate cam and measure lift. If below specs, lobe is worn and cam replacement or grinding would be needed. (Central Tool)

With the camshaft out of the engine, an outside micrometer can be used, as in Fig. 26-55A.

If lobe lift or lobe dimensions are too small to meet

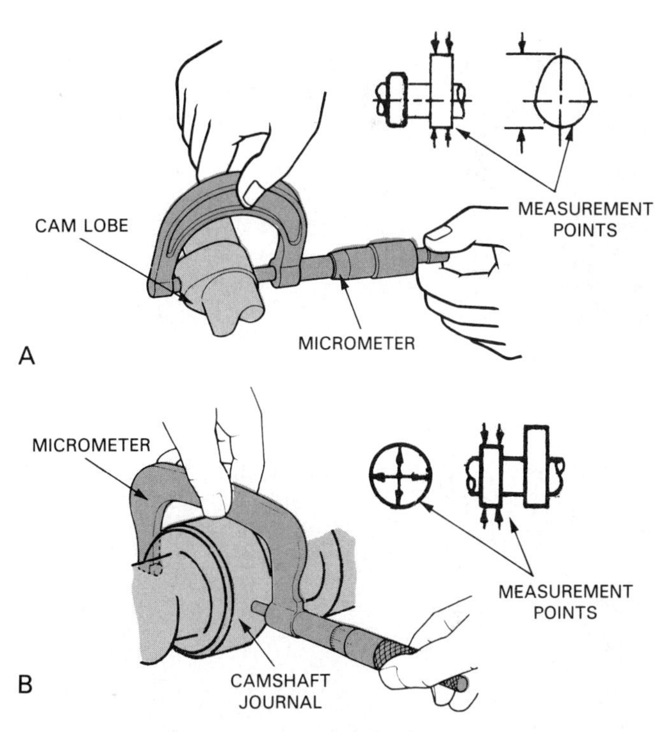

Fig. 26-55. With cam out of engine, you can use "mike" to check camshaft condition. A—Measure lobe size and compare to specs. B—Measure journal diameter and compare to specs. (Toyota)

specs (worn over about .005 inch or 0.13 mm), the cam is worn and should be reground or replaced.

Cam journal wear is measured with an outside micrometer, as in Fig. 26-55B. If worn more than specifications (approximately .002 inch or 0.05 mm) the cam is usually replaced. Journal wear will lower engine oil pressure. Fig. 26-56 shows how to slide a mike over a cam journal.

Camshaft straightness is checked with V-blocks and a dial indicator. This is shown in Fig. 26-57. If the dial indicator reads more than specs (about .002 inch or 0.05 mm), the cam is bent and must be replaced.

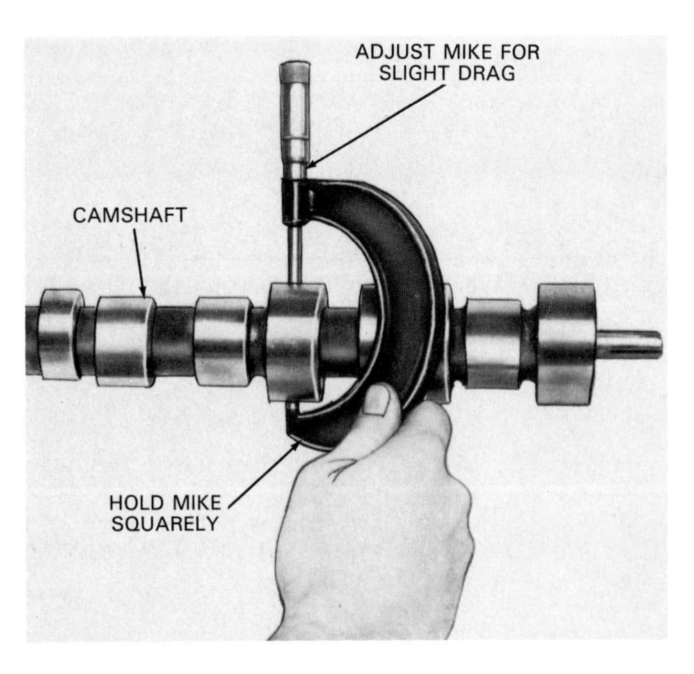

Fig. 26-56. When using micrometer, hold it squarely and adjust it for slight drag when pulled over cam. (Deere & Co.)

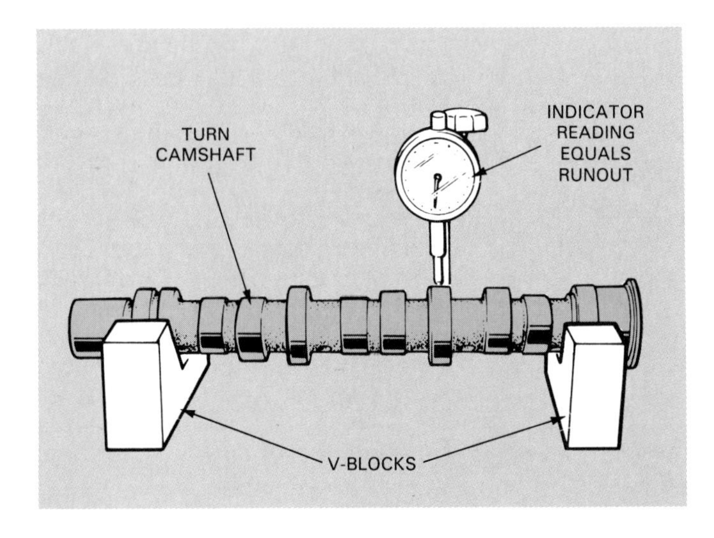

Fig. 26-57. Though not common, camshaft runout can be measured with V-blocks and indicator. Turn cam and read indicator. If more than specs, replace or regrind camshaft. (General Motors)

Camshaft grinding

Camshaft grinding is done to restore cam lobe and journal dimensions to within specifications. Pictured in Fig. 26-58, a cam grinding machine regrinds all of the cam lobes. If a journal is badly damaged, it can be welded up and then ground back down to size.

A ground camshaft is usually much cheaper than a new camshaft. So long as the lobes are hardened properly, it can give good service life.

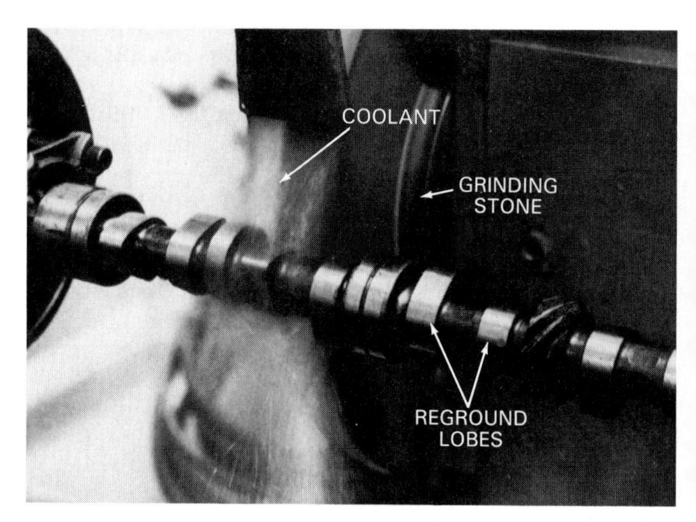

Fig. 26-58. This camshaft is being reground by machine shop. This will bring lobes to within specs. (Landis Tool)

Camshaft end play

Camshaft end play is the front to rear movement of the cam in the cylinder head or block. It is commonly measured with a dial indicator. See Fig. 26-59.

Mount the indicator so its stem is parallel with the cam centerline. Zero the indicator and then pry back and forth on the camshaft. Indicator needle movement equals camshaft end play.

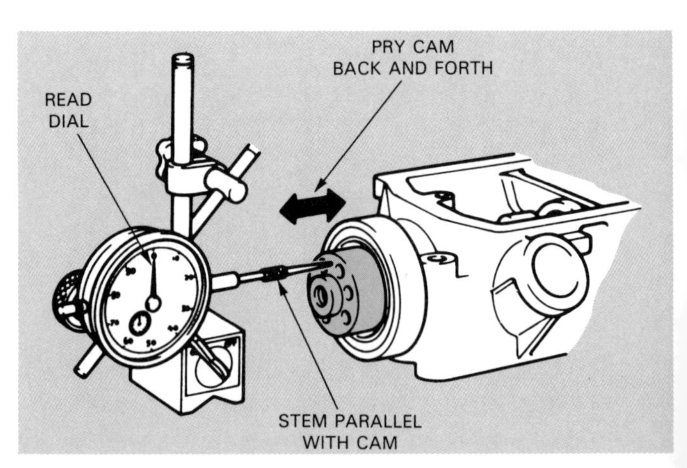

Fig. 26-59. Too much camshaft end play can result with part wear. Measure it with indicator by shifting cam back and forth in head or block. (Toyota)

If end play is not within specs, replace or repair parts as needed. The flange on the cam could be worn or the thrust surface on the head or block could be damaged.

OHC BEARING SERVICE

Cylinder head cam bearings are serviced in much the same way as in-block cam bearings. You must check them for wear and damage. Worn cam bearings can reduce engine oil pressure.

Cam bearing wear is checked by measuring cam bearing diameter. Measure bearing diameter with a telescoping gauge or dial bore gauge, Fig. 26-60A. If worn more than specs (about .002 inch or 0.05 mm), replace the bearings.

Most technicians replace cam bearing during major engine service. They are not very expensive and are important to engine service life.

Cylinder head cam bearings can be one or two-piece designs. If one-piece, you must drive them in and out with a bearing driver or send them to a machine shop for replacement. If two-piece, they simply snap into place like rod or main bearings.

Note! Always make sure the cylinder head or cam bearing caps are installed in their correct locations. If mixed up or turned backwards, the cam bearing bore will not be round. This is shown in Fig. 26-60B.

When installing cam bearings, make sure the oil holes align. Do not dent or mar the bearing surfaces. With two-piece cam bearings, you can measure bearing clearance with Plastigage, Fig. 26-61. With one-piece bearings, subtract cam journal diameter from cam bearing diameter to get cam bearing clearance.

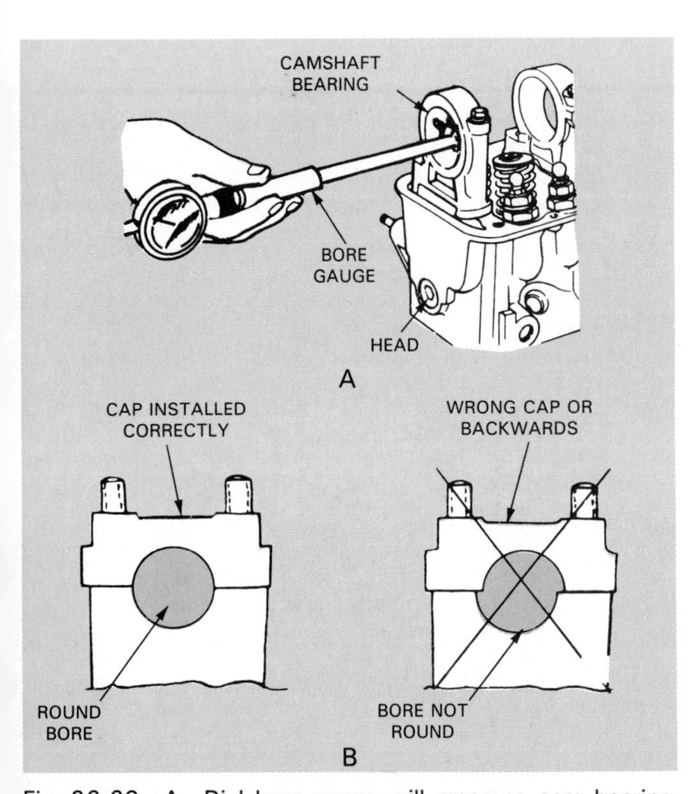

Fig. 26-60. A—Dial bore gauge will measure cam bearing wear. B—Make sure camshaft caps are installed correctly. If mixed up or turned backwards, cam bore will not be round. Cam will lock in head when caps are torqued. (VW and Nissan)

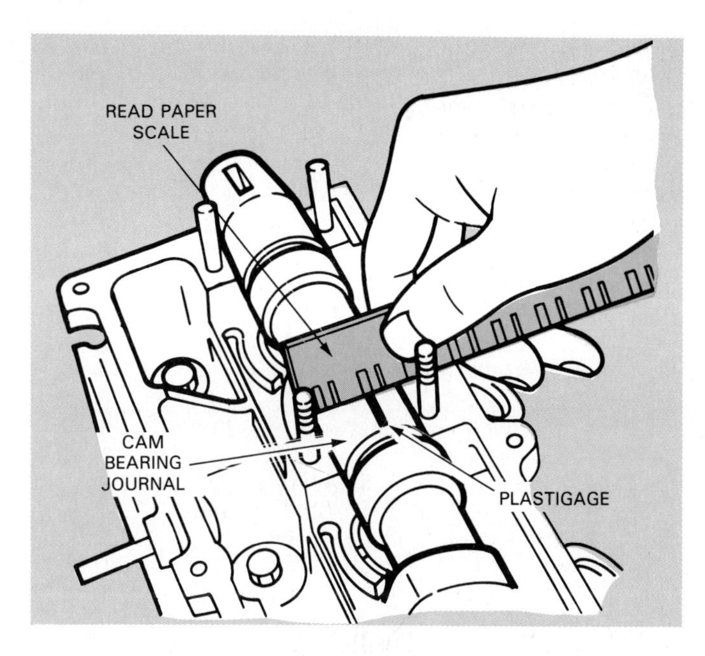

Fig. 26-61. Cam bearing clearance on OHC engines with two-piece bearings can be measured with Plastigage. Place plastic on journal. Torque and remove cap. Compare smashed plastic with paper scale to find clearance. If not within specs, check bearing and journal sizes. (Ford)

Note! Cam bearing installation was also explained in Chapter 25. Refer to this chapter for more information as needed.

ROCKER ASSEMBLY SERVICE

Rocker assembly service involves cleaning, inspection, and measurement of the rocker arms, rocker shafts, and related components. A typical assembly is in Fig. 26-62.

Inspect all wear points closely. Check the rocker shafts for signs of heavy wear or scoring. Check the rocker arms for wear at the locations illustrated in Fig. 26-63.

A telescoping gauge or dial bore gauge is needed to measure rocker arm bushing or inside diameter wear, Fig. 26-64. An outside mike will check rocker shaft wear. If more than specs, replace parts or install new rocker arm bushings, if used.

During rocker reassembly, make sure the rocker shaft oil holes point down. This is the loaded side of the shaft and oil is needed at this high friction area. If the oil holes are installed up, rapid shaft and rocker arm wear will result.

With a ball or stand type valve mechanism, inspect the ball, pivot, or stand for wear. If the rocker stand or pivot is aluminum, inspect it closely. The soft aluminum is prone to wear and frequently requires replacement. Replace any rocker part worn beyond safe limits.

PUSH ROD SERVICE

Servicing push rods involves checking them for tip wear and straightness. When a push rod tip is worn, the

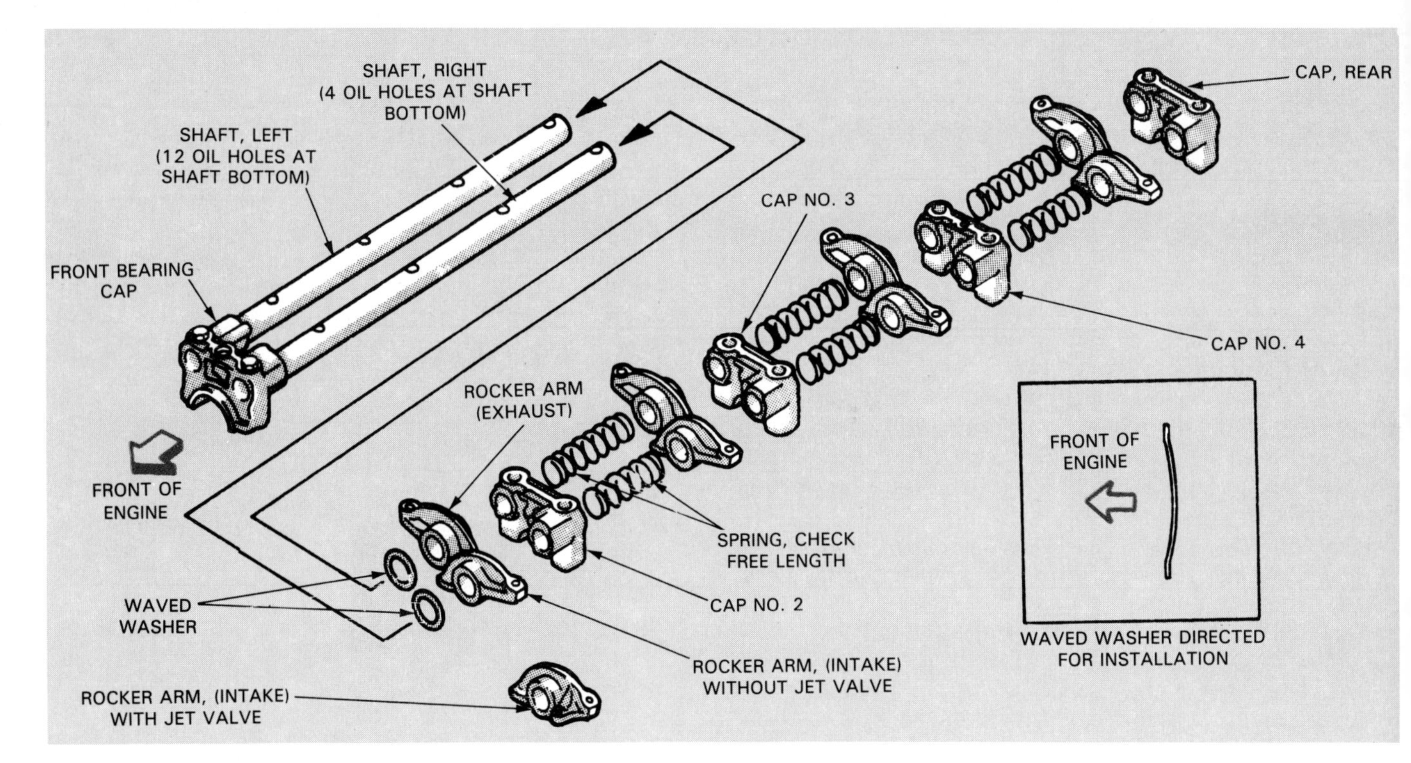

Fig. 26-62. Close inspection of all rocker arm parts is important. Rockers suffer from poor lubrication and can wear at high friction points. (Chrysler)

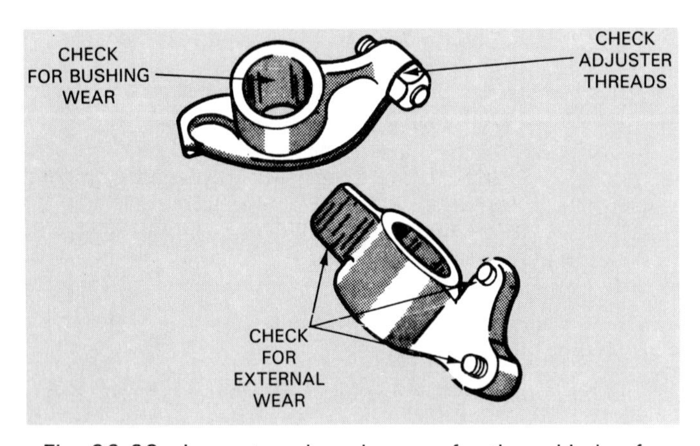

Fig. 26-63. Inspect each rocker arm for these kinds of problems. (Chrysler)

center tends to become pointed where it fits into the rocker arm. A push rod does not wear as much in the lifter end.

A bent push rod can result from a valve hitting a piston, stuck valve, and other similar problems. You can usually check push rod straightness by rolling the push rods on a smooth, flat work surface. A more accurate, but more time consuming, method is to mount them in V-blocks and measure runout with a dial indicator. This is shown in Fig. 26-65.

Note! Push rod straightness can also be checked by mounting each in a drill press. Turn the drill press on and watch push rod movement. If the push rod wobbles in the drill press, it is bent and must be replaced.

Do NOT attempt to straighten a push rod because it will not be as strong as the others.

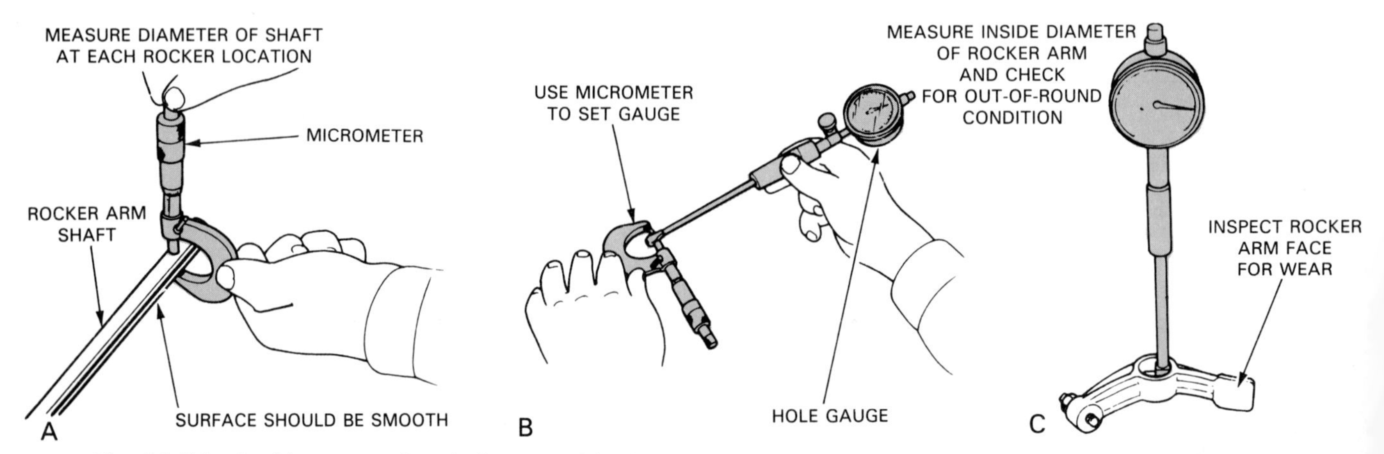

Fig. 26-64. A—Measure rocker shaft wear with micrometer. B—Adjust dial bore gauge to correct size. C—Insert gauge into rocker arm bushing to measure wear. Replace parts or install new rocker bushings as needed. (Honda)

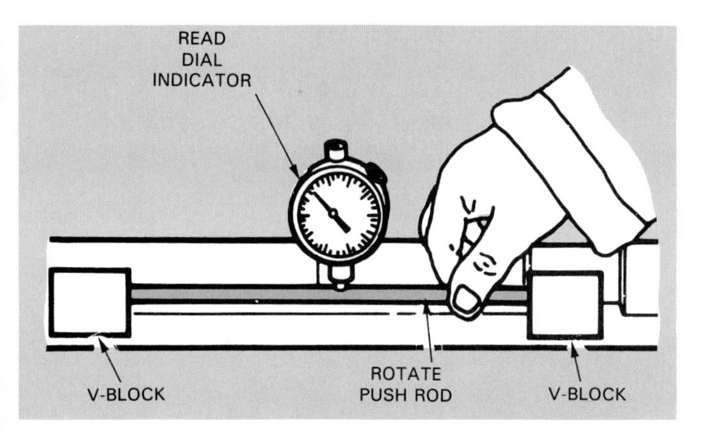

Fig. 26-65. Push rods should be checked for straightness. Use V-blocks and indicator as shown or spin push rods in drill press. (Ford)

If hollow, also look down through the push rods to make sure they are not clogged with debris. Some push rods feed oil up to the rocker arms. Run wire through the push rods if needed to clean them out.

LIFTER (TAPPET) SERVICE

The contact surface between a lifter and cam lobe is one of the highest friction and wear points in an engine. Hydraulic lifters can also wear internally, causing valve train clatter.

Inspecting lifters

Inspect the bottoms of the lifters for wear. A good, unworn lifter will have a slight hump or convex shape on the bottom. Worn lifters will be flat or concave on the bottom. Replace the lifters if the bottoms are worn, Fig. 26-66.

NEVER install used, worn lifters on a new camshaft. Worn lifters will cause rapid cam lobe wear and additional lifter wear. Normally, install new lifters when a camshaft is replaced.

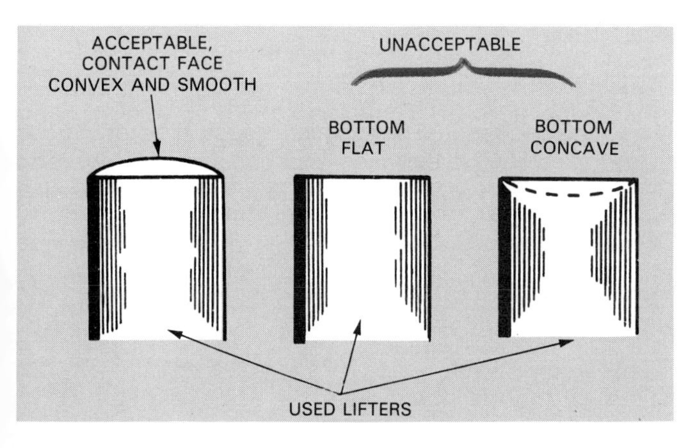

Fig. 26-66. Good lifters should be humped or crowned slightly on bottom. If not, lifters are worn and should be replaced. (Ford)

Testing lifter leak down rate

Lifter leak down rate is measured by timing how long it takes to push the lifter plunger to the bottom of its stroke under controlled conditions. A lifter tester is pictured in Fig. 26-67.

Generally, fill the tester with a special test fluid. Place the lifter in the tester. Then, follow specific instructions to determine lifter leak down rate. If leak down is too fast or slow, replace or rebuild the lifter.

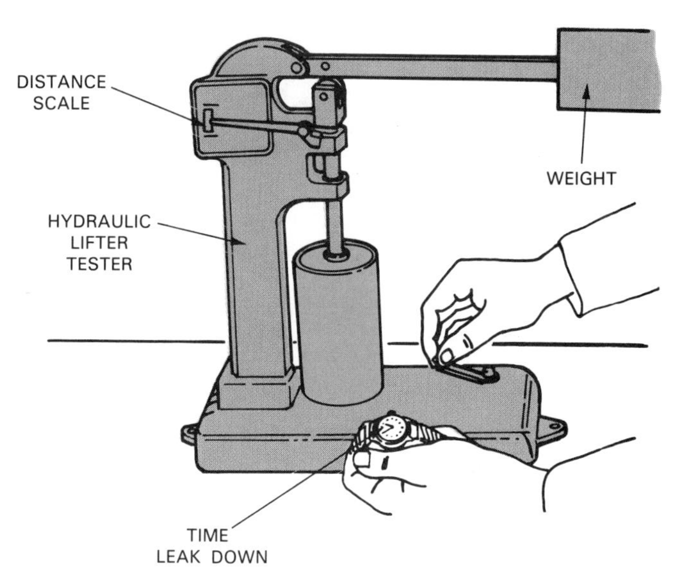

Fig. 26-67. Mechanic is checking lifter leakdown rate. Time leakdown while following equipment instructions. If not in specs, rebuild or replace lifters. (Pontiac)

Rebuilding hydraulic lifters

Rebuilding hydraulic lifters typically involves complete disassembly, cleaning, and measurement of lifter components. All worn or scored parts must be replaced following manufacturer instructions, Fig. 26-68.

Many auto shops do NOT rebuild hydraulic lifters. If the lifters are defective, new ones are installed. This can save time and money under most circumstances.

Warning! Some lifters can be oversize from the factory. This makes it important for you to keep all lifters in order so they can be reinstalled in the same location in the engine. Fig. 26-69 shows how the lifter bore is marked, denoting an oversize lifter and lifter bore.

MORE INFORMATION

The next two chapters will give you more information relating to cylinder head service. The next chapter will detail timing mechanism service. Chapter 28 discusses engine reassembly. These chapters will summarize subjects like: torquing cylinder heads and manifolds, installing timing covers, timing belts, valve covers, front covers, etc.

Fig. 26-70 shows how the parts of a typical gasoline engine cylinder head fit together.

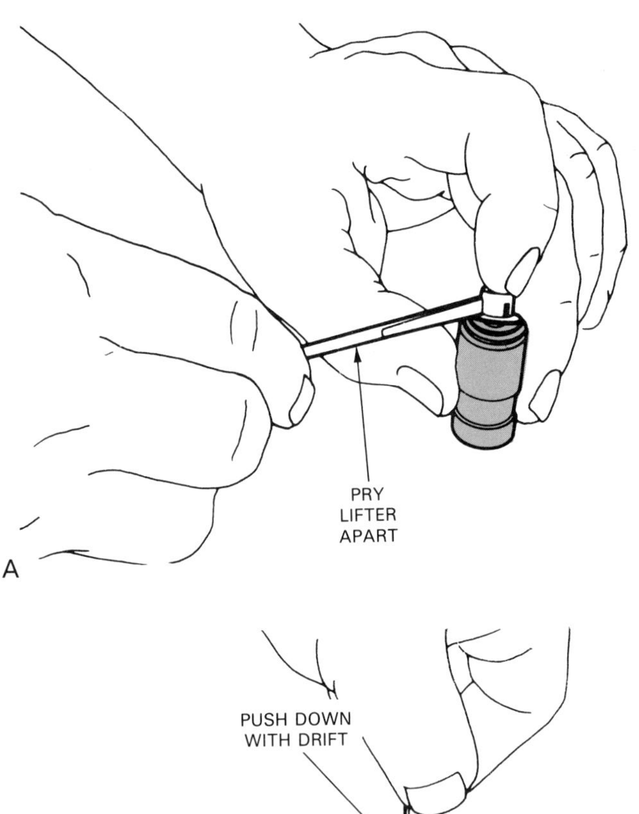

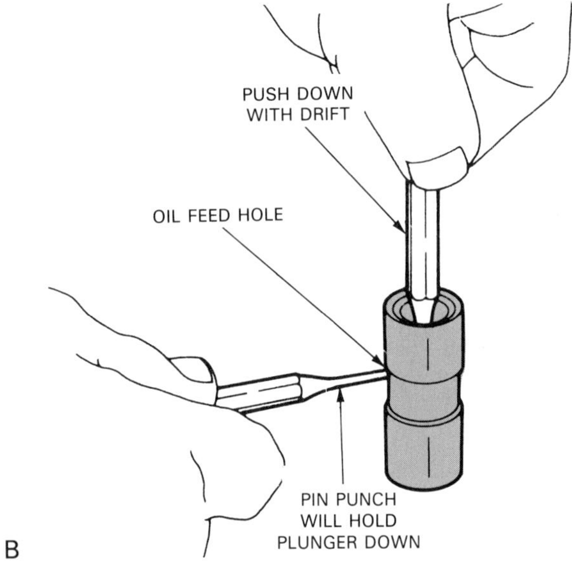

Fig. 26-68. Most technicians replace, rather than rebuild, lifters. A—Disassemble lifter and inspect all parts for wear. B—If in good condition, clean parts and reassemble. Here technician is forcing plunger down into lifter. Then pin in used to hold plunger while installing snap ring. Follow manual directions. (Buick)

SUMMARY

Cylinder heads can suffer from a wide range of problems: warpage, cracks, burned seats, burned valves, worn valve guides, etc. As an engine mechanic, it is important for you to know how to correct these problems.

Magnafluxing is commonly used to find cracks in cast iron heads. Dye penetrant is used to find cracks in aluminum heads. Cracks can sometimes be repaired by welding. If badly damaged, it might be better to install a good used or new head.

Head warpage is checked with a straightedge and feeler gauge. Check at different angles across the head. Warpage is very common with today's thin wall heads, especially aluminum heads.

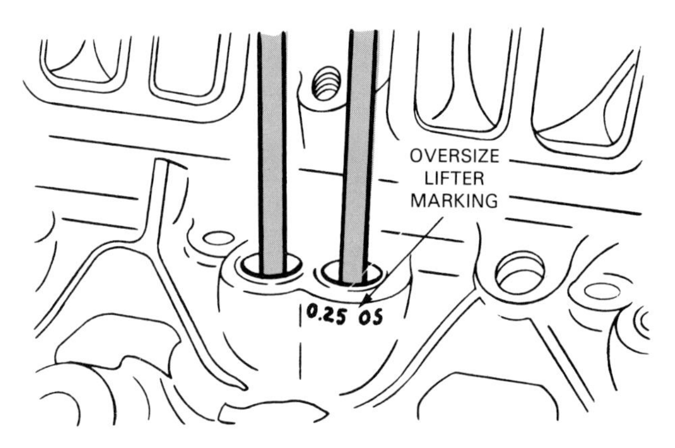

Fig. 26-69. Sometimes, lifters can be oversize. This will normally be marked on block. This is one reason why used lifters should be kept in order.

Milling is used to correct head warpage. The machine shop will true the head deck so the head gasket does not fail in service. Do not mill too much off of a head or it can affect the compression ratio and fit of V-type intake manifolds. Milling diesel cylinder heads is critical so follow manual recommendations.

Diesel prechambers can usually be driven out of the head with a drift and brass hammer. When installing a prechamber, be careful not to dent the center of the chamber. The chamber must be the correct size and fit flush with the head deck after installation.

Valve guide wear is common and can be checked with a dial indicator or small telescoping gauge. Valve guide wear can be corrected by knurling, reaming oversize, installing a bronze liner, or installing a new guide insert.

Valve seat reconditioning involves grinding or cutting a new face on the seat. Valve retrusion is a problem caused by wear or grinding when the valve sinks into the combustion chamber. Valve seat replacement is commonly done by a machine shop.

Valve seats can be resurfaced with a grinding stone or carbide cutter. Seats are usually 45 degrees, but a few high performance heads use a 30 degree intake seat. Only grind as little as needed off of the seat.

You may have to move the seat in or out so it touches the valve properly. Use 15 or 30 and 60 degree stones or cutters to reposition the valve-to-seat contact point.

Valve grinding compound or Prussian Blue can be used to check seat location on the valve. Make sure you clean off all grinding compound to avoid part wear.

An intake valve should have a seat contact width of about 1/16 inch (1.6 mm). The exhaust valve needs a little more contact area, 3/32 inch or 2.4 mm being typical.

Valves are reconditioned on a valve grind machine. The valve faces and stem tips are ground smooth and true. Only grind as little as possible off the valves. Wear eye protection when grinding! Follow equipment operating instructions!

Valve springs should be replaced or checked closely for problems. Check spring squareness, spring tension and spring free length.

Valve spring shimming may be needed to produce the correct spring pressure. Measure spring installed height

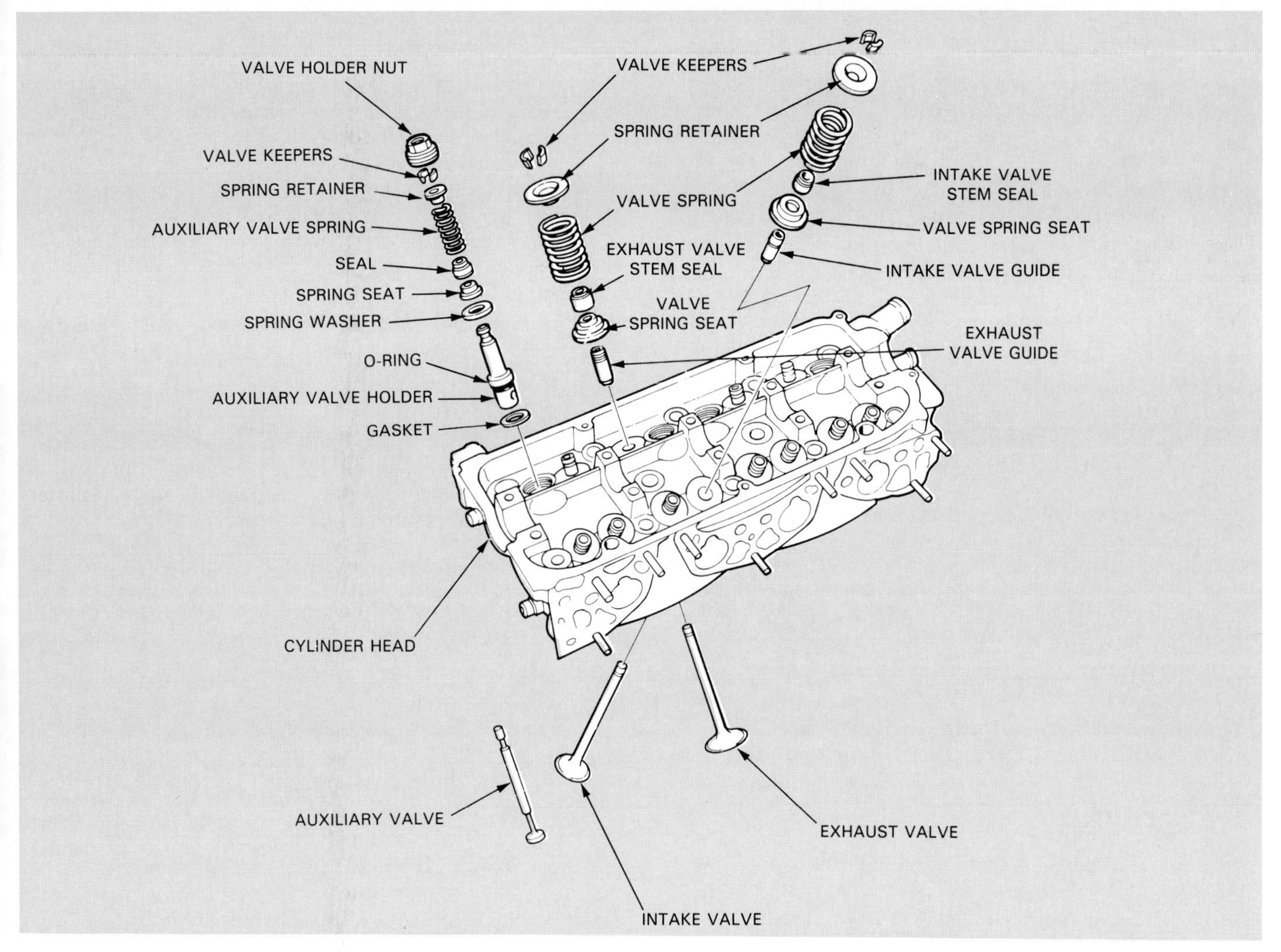

Fig. 26-70. Exploded view of gasoline engine cylinder head assembly. This head uses extra auxiliary valve. (Honda)

Use a valve spring tester to measure spring pressure. Use spring shims to reduce installed height to equal the tester value and meet spring pressure specs.

Install valve seals carefully. Umbrella seals simply slide off the valves before spring installation. However, you must compress the valve springs before installing O-ring type valve seals. This will prevent O-ring seal damage.

Check the camshaft closely for wear or damage. Measure lobe height, journal diameter, and runout. Replace or have a machine shop grind the camshaft if needed.

Inspect all valve train parts closely. Replace any components showing wear or damage. Measure lifter leak down rate, cam bearing wear, push rod straightness, etc.

KNOW THESE TERMS

Magnafluxing, Dye penetrant, Warped head, Warped cam bore, Head milling, Valve guide knurling, Valve guide insert, Valve guide liner, Oversize valve stem, Seat or valve retrusion, Seat staking, Seat grinding stone, Carbide seat cutter, Concentricity, Seat narrowing, Valve lapping compound, Valve-to-seat contact width, Valve grind machine, Interference angle, Sharp valve margin, Burned valve, Bent valve, Valve spring squareness, Valve spring free height, Valve spring tension, Valve spring shimming, Valve spring tester, Spring bind, Positive valve seal, Cam lobe wear, Cam journal wear, Camshaft grinding, Camshaft end play, Lifter leak down rate.

REVIEW QUESTIONS—CHAPTER 26

1. An engine has coolant in the oil, producing a milk-like substance in the crankcase.
   Technician A says the engine could have a warped head and blown head gasket.
   Technician B says that the engine could have a cracked head or block.
   Who is correct?
   a. Technician A.
   b. Technician B.
   c. Both A and B.
   d. Neither A nor B.
2. ________ ________ ________ is a machine shop operation that removes a thin layer of metal from the head deck surface.
3. Technician A says most diesel prechambers must be flush with the head deck surface.
   Technician B says most prechambers can be driven

out with a drift and light hammer blows.
Who is correct?
a. Technician A.
b. Technician B.
c. Both A and B.
d. Neither A nor B.

4. Explain four ways of correcting valve guide wear.
5. Valve seat retrusion can cause lowered ________, reduced ________ ________, incorrect valve train ________, and decreased engine ________.
6. A seat grinding stone should be slightly smaller in diameter than the valve seat. True or false?
7. When dressing a seat stone, how can you damage the diamond cutter?
8. When grinding a seat, do NOT push down on the power head. True or false?
9. The most common valve seat angle is ______ degrees, but some are also ______ degrees.
10. In your own words, how do you check and position a valve seat so it touches the valve properly?
11. What are typical contact widths for intake and exhaust valves?
12. How do you move a seat in or out?
13. How would you adjust the valve grind machine for a 45 degree angle seat?
14. A ________ ________ ________ indicates too much valve face removal and requires valve replacement.
15. How do you determine valve spring shim thickness?
16. What is spring bind?
17. Technician A says that umbrella and O-ring valve seals are installed after compressing the valve spring.
Technician B says to install umbrella or O-ring valve seals and then compress the valve spring.
Who is correct?
a. Technician A.
b. Technician B.
c. Both A and B.
d. Neither A nor B.
18. How can you check for valve leakage before head installation?
19. When installing a rocker arm shaft, the oil holes should be ________.
20. How can a drill press be used to check for push rod straightness?

ASE CERTIFICATION-TYPE QUESTIONS

1. Technician A says that magnafluxing should be used to detect cracks in most aluminum cylinder heads. Technician B says that dye penetrant should be used to find cracks in aluminum cylinder heads. Who is right?
(A) A only.
(B) B only.
(C) Both A & B.
(D) Neither A nor B.
2. Technician A says that some cracked cylinder heads can be welded. Technician B says that many cracked cylinder heads can be plugged. Who is right?
(A) A only.
(B) B only.
(C) Both A & B.
(D) Neither A nor B.
3. Technician A says that certain diesel cylinder heads must be replaced when warped. Technician B says that any diesel cylinder head can be milled when warped. Who is right?
(A) A only.
(B) B only.
(C) Both A & B.
(D) Neither A nor B.
4. Technician A says that valve guide wear can cause engine coolant leakage. Technician B says that valve guide wear can cause engine oil consumption. Who is right?
(A) A only.
(B) B only.
(C) Both A & B.
(D) Neither A nor B.
5. Technician A says that knurling is a common method used to repair worn valve guides. Technician B says that reaming is a common method used to repair worn valve guides. Who is right?
(A) A only.
(B) B only.
(C) Both A & B.
(D) Neither A nor B.
6. Valve seat replacement should be performed ________.
(A) any time the cylinder head is removed from the engine
(B) only when seat wear is severe
(C) when replacing the valves
(D) when replacing the valve springs
7. Technician A says that an interference angle is normally a 4° difference in the valve face angle and the valve seat angle. Technician B says that an interference angle is usually a 1° difference in the valve face angle and the valve seat angle. Who is right?
(A) A only.
(B) B only.
(C) Both A & B.
(D) Neither A nor B.
8. Technician A says that you should never shim a valve spring over 1.20'' (30.5 mm). Technician B says that you should never shim a valve spring over 1.45'' (36. 8 mm). Who is right?
(A) A only.
(B) B only.
(C) Both A & B.
(D) Neither A nor B.
9. Technician A says that cam lobe wear should be measured with a telescoping gauge. Technician B says that cam lobe wear should be measured with a dial indicator. Who is right?
(A) A only.
(B) B only.
(C) Both A & B.
(D) Neither A nor B.
10. Technician A says that a bent push rod can result from a valve hitting a piston. Technician B says that a bent push rod can result from a stuck valve. Who is right?
(A) A only.
(B) B only.
(C) Both A & B.
(D) Neither A nor B.

Chapter 27

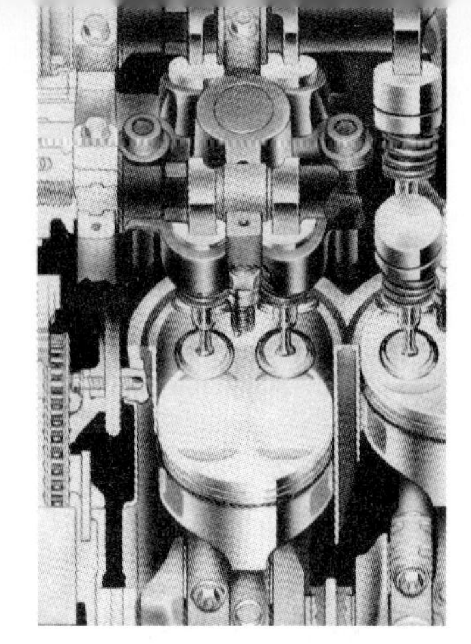

Timing Mechanism Service

After studying this chapter, you will be able to:
- *Check a timing chain for wear.*
- *Install a timing chain assembly.*
- *Install timing gears.*
- *Check timing gear runout and backlash.*
- *Install a timing sprocket and belt.*
- *Adjust timing belt tension.*
- *Replace a front cover oil seal.*
- *Install an engine front cover.*
- *Properly torque timing gear or sprocket fasteners.*
- *Service timing chain tensioners.*
- *Install a timing belt cover.*

Today's engines, as you have learned in earlier chapters, are much more complex than the ones used in cars just a few years ago. Engine manufacturers are now using more overhead cam engines with either a long timing chain or timing belt. More conventional cam-in-block engines with shorter timing chains and timing gears are also common. This makes the service of an engine timing mechanism or front end even more challenging.

This chapter will summarize the most critical information concerning the service and repair of timing chains, timing belts, and timing gears. This will help prepare you to work on any make or model of engine.

Remember! Timing mechanism service is critical. If a mistake is made, severe engine damage or performance problems may result. Study carefully!

IN-CAR TIMING MECHANISM SERVICE

Usually, a timing chain, gears, or belt can be serviced WITHOUT removing the engine from the car. Usually, the parts on the front of the engine and timing cover can be removed with the engine intact. Some OHC engines require the camshaft(s) to be secured before timing mechanism removal. If this is not done on these engines, the camshaft(s) will slip out of time and can cause engine damage. Refer to a service manual for specific direction and specifications if needed.

TIMING MECHANISM DISASSEMBLY

Earlier chapters covered information useful to your full understanding of this chapter. For instance, Chapter 24 explained engine disassembly, including front end or timing mechanism removal. Chapter 11 explained the construction and types of camshaft drives.

TIMING CHAIN SERVICE

Timing chains transfer power from the crankshaft to the camshaft. Sometimes they also power other devices. They are used in both overhead valve and overhead cam engines.

Timing chain problems

Timing chains suffer from wear after prolonged service. The chain links can wear and stretch. The sprocket teeth can also wear down. This can cause excess *chain slack* or the chain can be loose on its sprockets.

Timing chain wear and slack can allow the camshaft to go out of time with the crankshaft. As a result, valve timing can be thrown off. The valves can open too late during the four-stroke cycle. This can reduce compression, power, fuel economy, and exhaust cleanliness.

Inspecting timing chain

You should inspect a timing chain before removal. Fig. 27-1 shows one method of checking OHC timing chain wear. Use a wrench to rotate the cam sprocket back and forth while measuring sprocket movement. Compare your measurement to specifications.

Generally, if you can wiggle one side of a timing chain over about one-half inch (12.7 mm), replace the chain and sprockets.

To check timing chain wear with an OHV engine, you can remove the distributor cap or valve cover. Then use a large wrench on the crankshaft snout. Rotate the crankshaft one way and then the other while watching the distributor rotor or valve train. The rotor or valves should operate with little crankshaft movement. If you can turn the crank over about 5 to 10 degrees without valve or rotor motion, the timing chain is loose.

If the timing chain is worn more than specs, it should be replaced. Many manufacturers recommend replacement of the chain and both sprockets as a set. Others allow you to reuse steel sprockets if not worn excessively.

If the engine has a cam sprocket with plastic teeth, it should normally be replaced. The plastic teeth are prone to wear and failure. Since not very expensive, this type sprocket should be replaced whenever it is removed or disassembled for service.

If the timing chain has a tensioner, check it for wear. The fiber rubbing block can wear down. Replace all parts as needed.

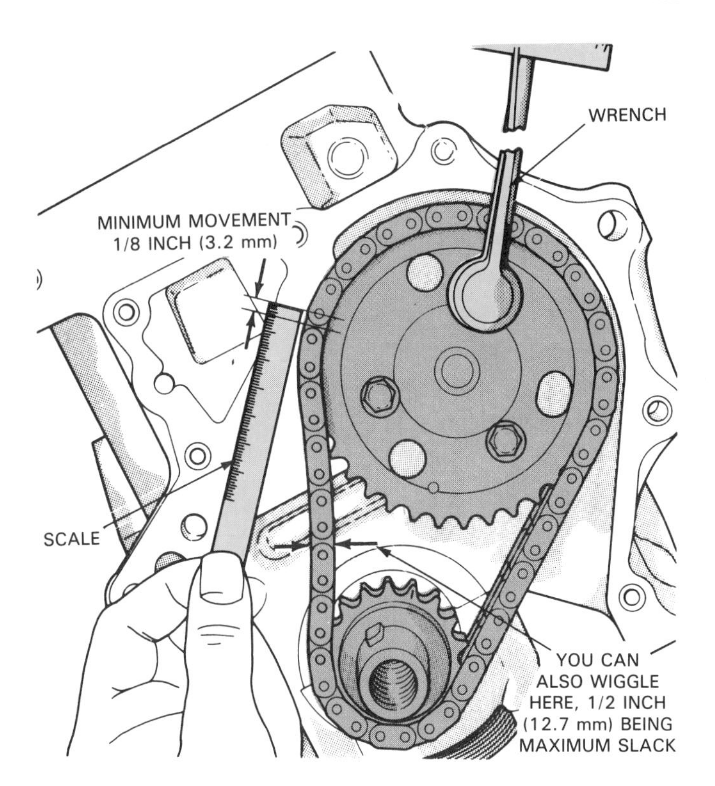

Fig. 27-1. Timing chain wear can be checked by rotating crankshaft back and forth. If sprocket rotation is excessive, chain and usually sprockets would have to be replaced. You can also check for wear by just wiggling chain on slack side. (Chrysler)

Installing timing chain

Make sure the crankshaft key is the right size, unworn, and properly installed in its groove. Also, check that the camshaft is installed properly. If needed, measure camshaft end play and check any fasteners holding a thrust plate on the front of the engine.

Turn the crankshaft and camshaft so that the sprocket timing marks align, Fig. 27-2. Then, fit the timing chain and both sprockets in place as a unit, Fig. 27-3.

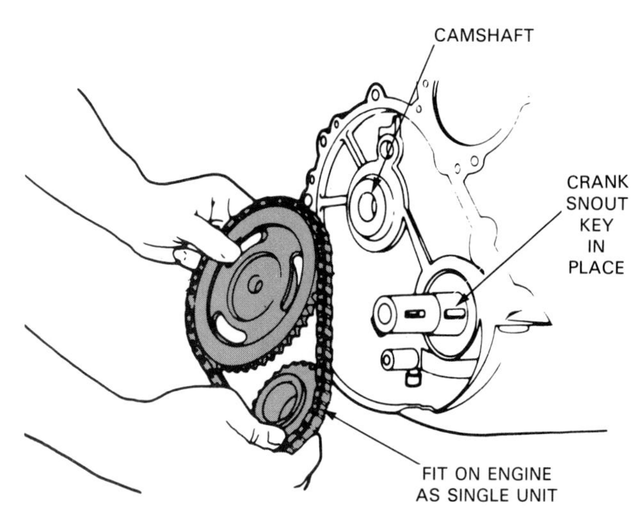

Fig. 27-3. With the cam and crank turned into position, fit the chain and sprockets into place together. Make sure the key and dowel are securely in place. (Renault)

Do NOT hammer on one side of the sprockets to force them in place. Usually, tightening the fasteners or hand pressure will slide them over the cam and crank. If you must use a hammer, use a large socket or driver to exert equal pressure in the center. This will drive the sprockets on squarely, Fig. 27-4.

Timing marks will either be dots, lines, circles, or other shapes indented or cast into the timing sprockets (also used on timing gears and belt sprockets). These markings must be pointing in the correct direction to time the camshaft (valves) with the crankshaft (pistons).

After chain installation, double-check the timing marks. Turn the crankshaft by hand until the marks realign. If needed, refer to the service manual. Sometimes, the key ways or a dowel can be considered timing marks. Usually, the timing marks are dots, like the ones shown in Fig. 27-5.

Fig. 27-6 shows the timing chain for an overhead cam

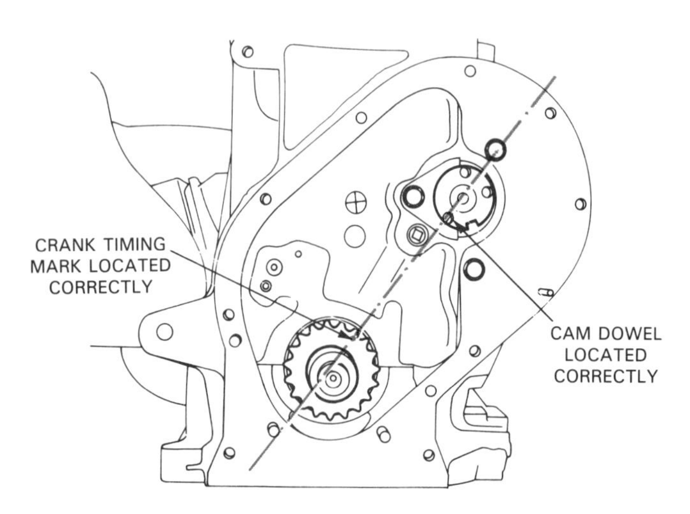

Fig. 27-2. To begin installation of timing chain, rotate camshaft and crankshaft until their key ways or dowel are in the general position to align the timing marks. This will let you fit the sprockets in place. (Peugeot)

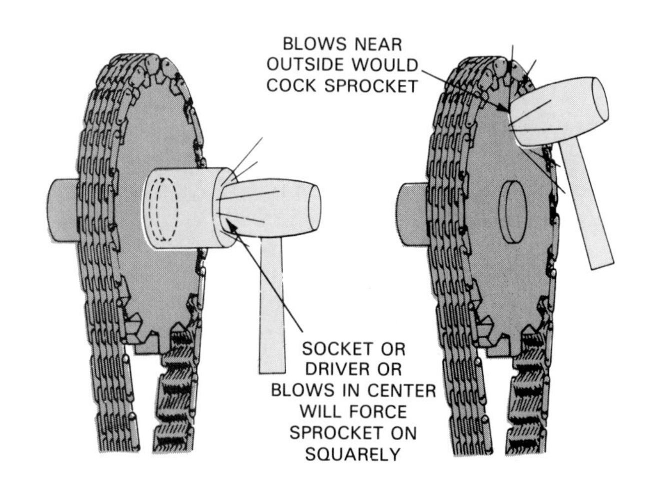

Fig. 27-4. When driving sprockets, drive in the center. Never drive around the outside because this could cock or break sprocket.

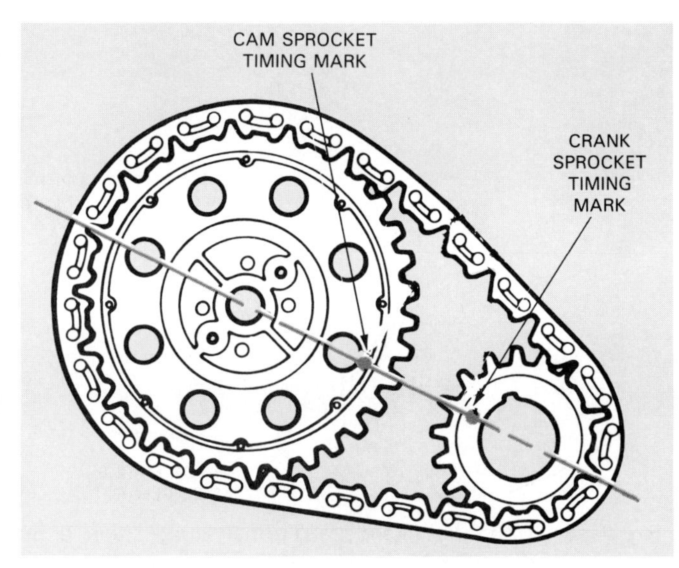

Fig. 27-5. With overhead valve engines, mark on cam sprocket should normally point at mark on crank sprocket. An imaginary line should pass through marks and sprockets as shown. (Renault)

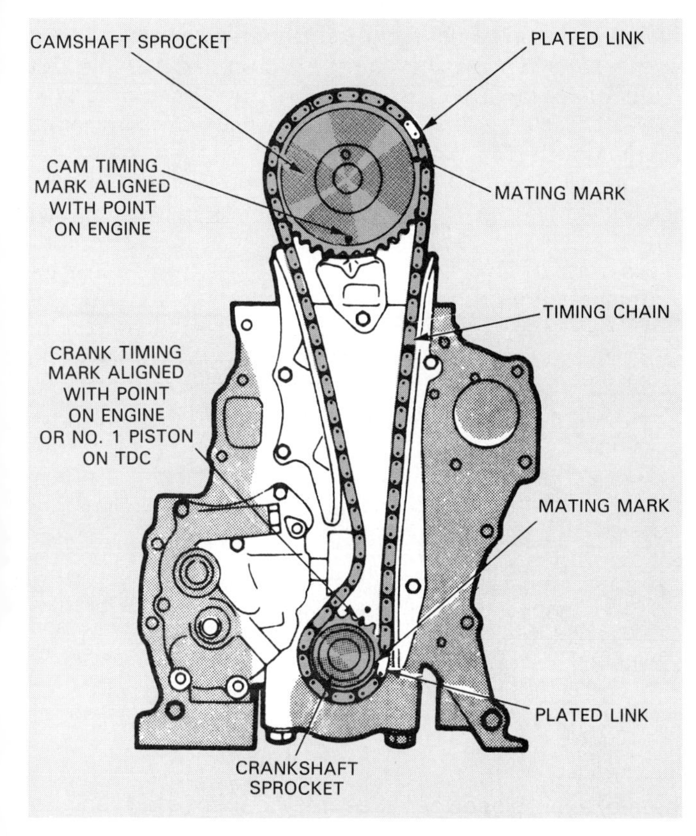

Fig. 27-6. Aligning timing marks on an overhead camshaft engine can be more challenging. This engine also has special chain links that should be aligned with sprocket marks. Refer to the service manual for details of the specific type engine. (Chrysler)

engine. Note that sometimes it has special links (plated or color coded links) that should be aligned with marks on the chain sprockets. The number one piston must be at TDC on the compression stroke with these marks aligned.

CAUTION! Do NOT use the starting motor to turn the crankshaft on some overhead cam engines with the timing chain or belt removed. It is possible, with some designs, for the pistons to hit the valves. The valves could be bent. Turn the cam and crank very slowly by hand.

If a fuel pump eccentric mounts on the cam sprocket, make sure it is positioned properly. Usually, the camshaft dowel must fit through a hole in the eccentric. The cam sprocket bolt installs through the center.

Torque the cam sprocket fasteners to specs. Use a torque wrench to assure the bolts do not loosen in service. See Fig. 27-7.

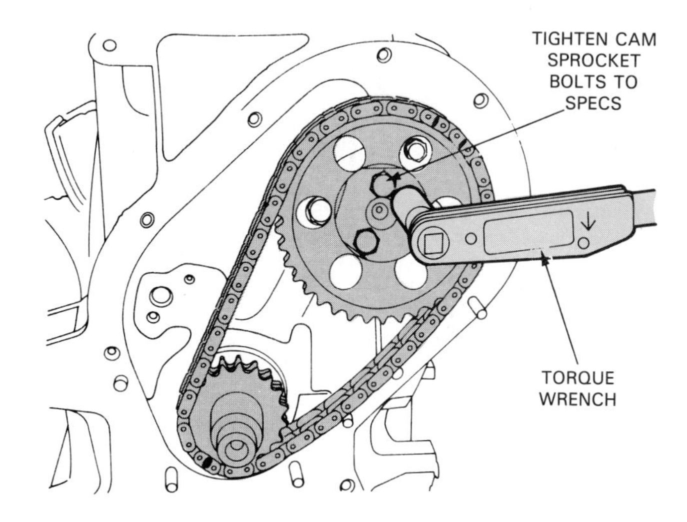

Fig. 27-7. Torque sprocket fasteners to specs. This could prevent problems later. (Peugeot)

Installing chain tensioner

A chain tensioner is used on most OHC timing chains. Inspect it closely. The fiber rubbing block can wear off. Replace it if needed, Fig. 27-8.

Fit the tensioner onto the engine. Then, pull on the chain to make sure it moves properly. Some plunger types can bind if worn and extended too much.

Installing oil slinger

The oil slinger fits in front of the crankshaft sprocket. It slings oil out of the chain and also prevents front seal leakage. Look at Fig. 27-9.

It is a troublefree part. However, if bent or damaged, it should be straightened or replaced.

Make sure you have the slinger facing in the right direction. If installed backwards, it can rub on the front cover. This mistake would require front cover removal—a time consuming procedure.

TIMING BELT SERVICE

Many OHC engines use a synthetic rubber belt to operate the engine camshaft. The cogged (toothed) belt

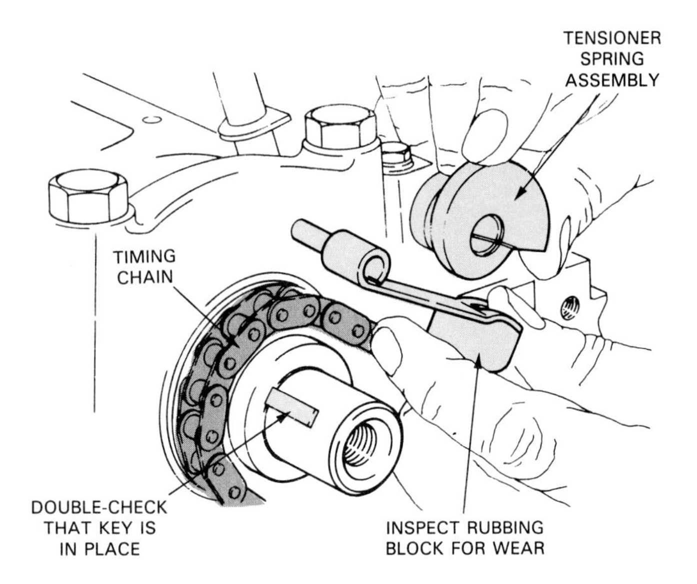

Fig. 27-8. If used, inspect chain tensioner closely before installation. Fiber rubbing block can wear. Compare to wear specs if needed. (Ford)

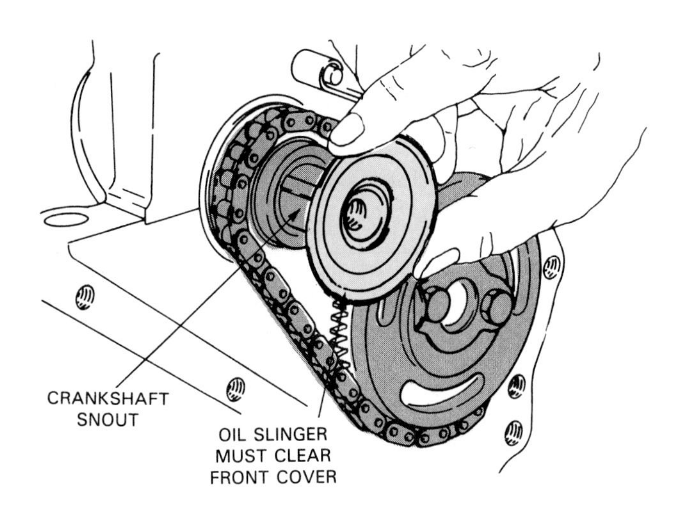

Fig. 27-9. Make sure you do not install oil slinger backwards or it will rub on timing cover and make terrible noise. Partial teardown simply to turn slinger around would be frustrating. (Ford)

provides an accurate, quiet, light, and dependable means of turning the camshaft, Fig. 27-10.

Timing belt service is very important. If the timing belt were to break or be timed improperly, engine valves, pistons, and other parts could be damaged.

Inspecting timing belt mechanism

During timing belt service, always inspect the belt, sprockets, tensioner, and other parts for trouble. Most technicians replace the timing belt when it is removed for service.

Most auto manufacturers recommend belt replacement about every 50,000 miles (80 450 km). More fre-

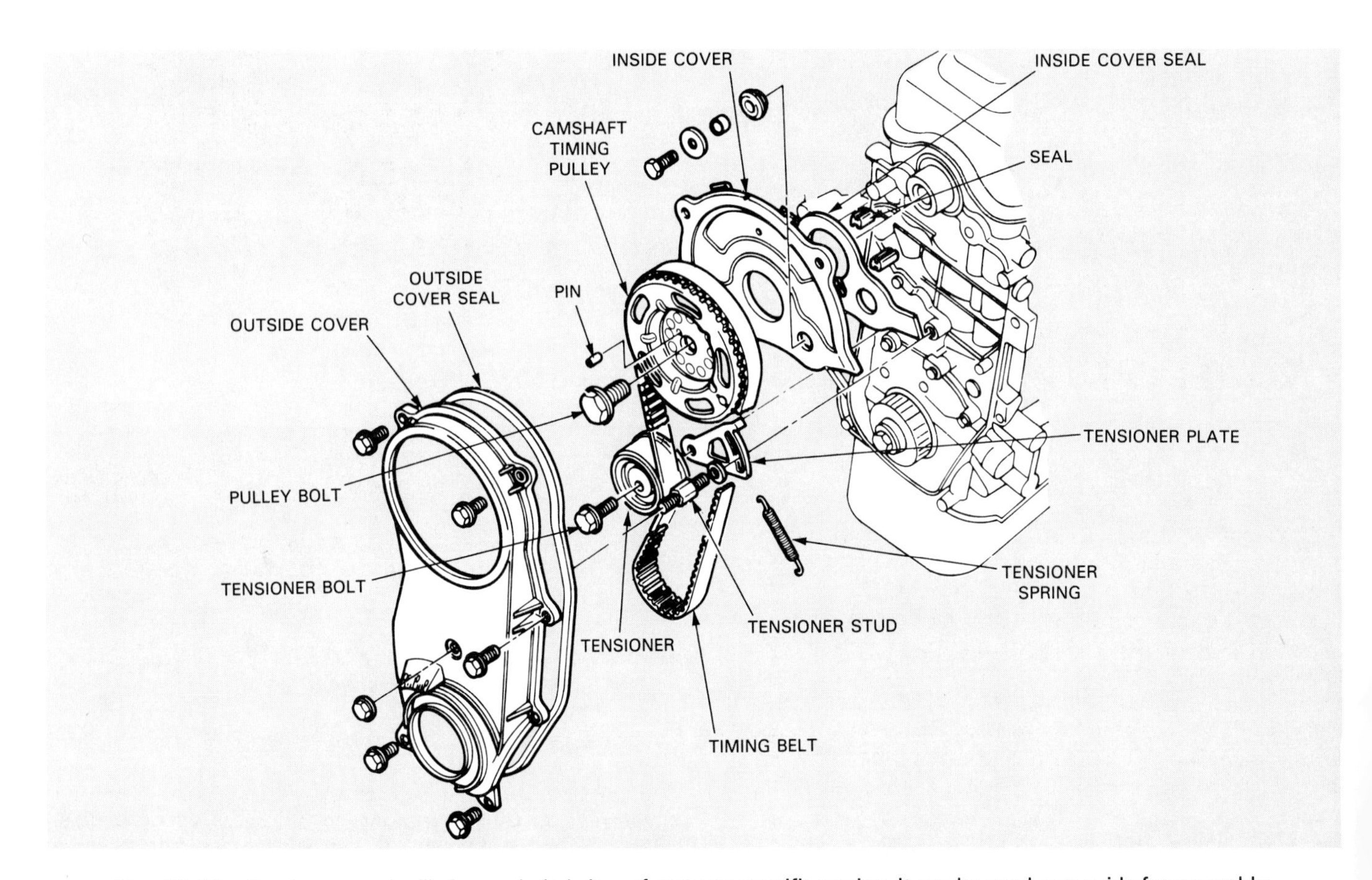

Fig. 27-10. Service manual will give exploded view of parts on specific engine. It can be used as a guide for assembly of components. (Ford)

quent replacement is needed after oil or coolant contamination of the timing belt. Some late model engine timing belts are designed to last the life of the engine without adjustment or replacement. Refer to the owner's manual or service manual for belt service intervals if in doubt.

Since belt removal can be very time consuming, it is normally replaced during major engine repairs. Timing belts are relatively inexpensive.

Inspect the timing sprockets for rounded teeth or physical damage. Spin the tensioner wheel by hand to check for a bad bearing. The wheel should spin freely and quietly. Replace the tensioner bearing if you can feel any roughness.

If any oil seals are leaking (only partial engine disassembly), replace the seals. Oil will cause rapid timing belt failure.

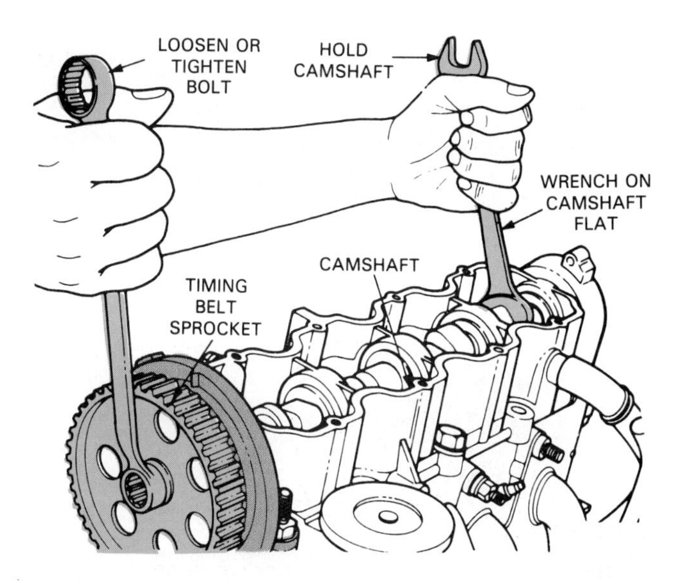

Fig. 27-12. A flat surface is frequently ground on camshaft so it can be held stationary when removing or installing cam sprocket. (Buick)

Installing timing belt sprockets

Start installation by installing the crankshaft and camshaft sprockets. The camshaft sprocket usually fits over a small steel dowl, as in Fig. 27-11. A bolt or bolts then secure the sprocket.

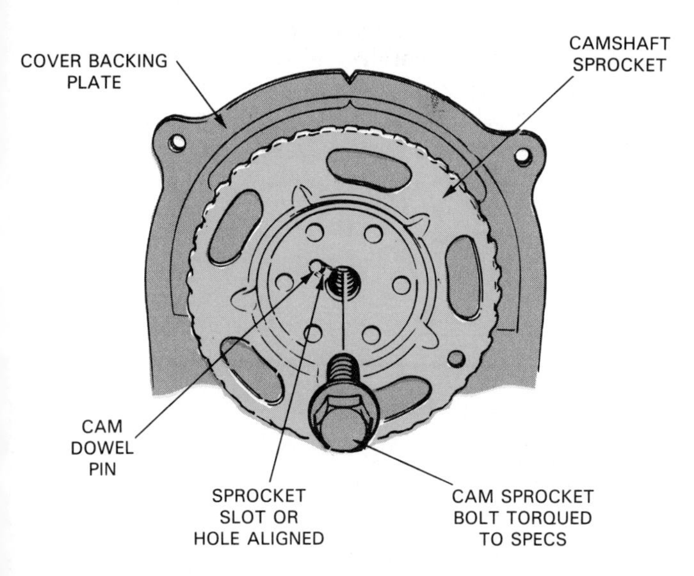

Fig. 27-11. A timing belt sprocket, as with a chain sprocket, is frequently positioned with a small dowel pin. Torque bolt to specs and align marks on sprocket and engine.

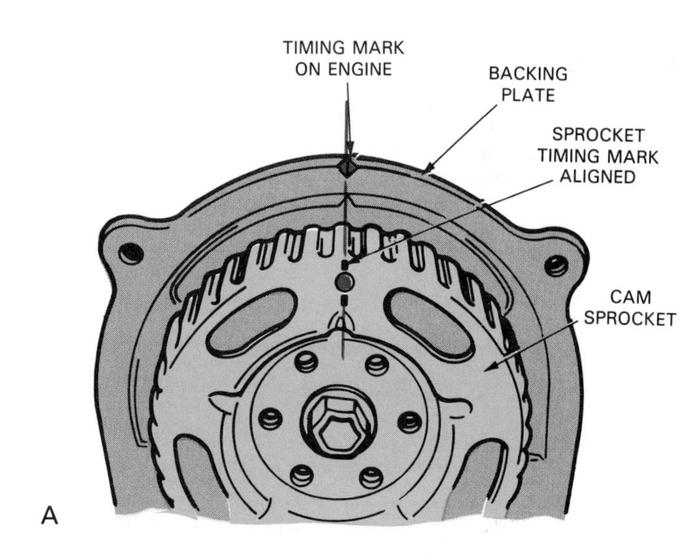

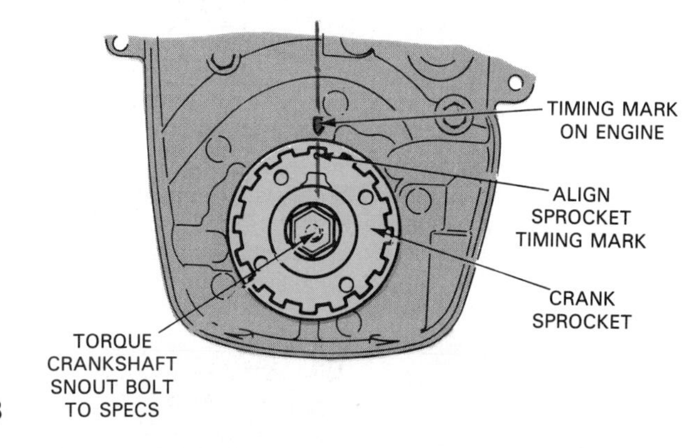

Fig. 27-13. Aligning timing belt sprocket marks. A—Camshaft sprocket must usually be aligned with mark on front of engine. B—Crankshaft sprocket will also have mark that must be aligned with specified point on engine. Check the service manual if in doubt. (Pontiac)

Torque the cam sprocket fasteners to specs. Usually, a flat is provided on the camshaft so you can hold the cam while tightening the bolt, Fig. 27-12.

Make sure you do not damage a camshaft lobe or the sprocket teeth when trying to hold the camshaft! Grasp an unmachined surface if needed.

As was described for a chain sprocket, fit the crankshaft sprocket into position. Turn the sprockets until the timing marks align or until number one piston is at TDC on the compression stroke, Fig. 27-13.

Installing timing belt

If the old timing belt is to be reused, it should have been marked with a directional arrow before removal.

A used belt must be installed so that it rotates in the same direction. This will prevent premature failure. Refer to Fig. 27-14.

Since exact belt installation procedures vary, refer to a shop manual if needed. It will explain special methods for aligning timing marks, positioning the cam and crank, and installing the belt tensioner.

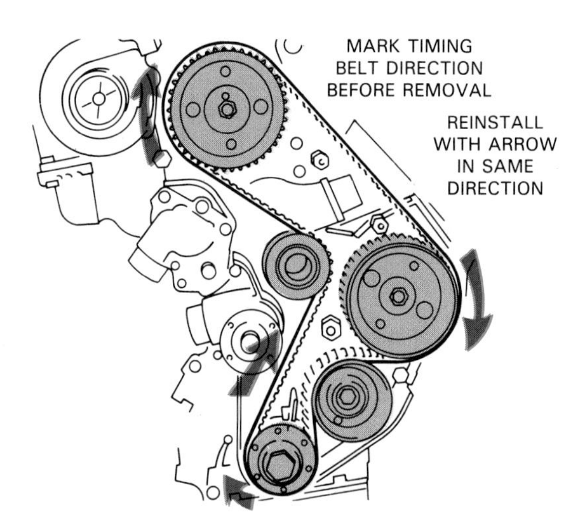

Fig. 27-14. When reusing a fairly new timing belt, it must be installed so that it turns in the same direction. If rotation is reversed, belt can fail quickly. (Ford)

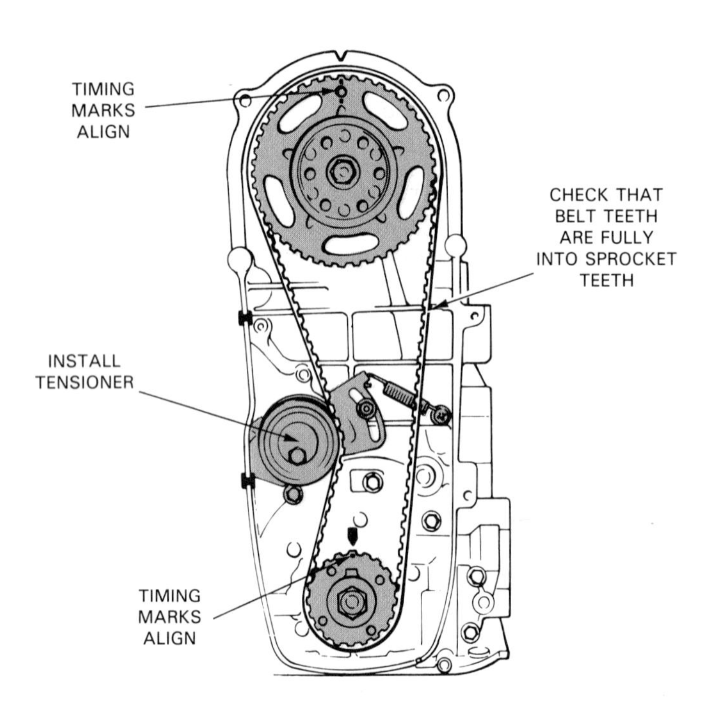

Fig. 27-15. Here is a front view of a typical OHC engine with the timing belt installed.

Fig. 27-15 shows the front view of an OHC engine with the belt in place. Note how the timing marks align.

If the engine has a belt driven distributor or diesel injection pump, its sprocket would also have to be in time with the other two.

Adjusting belt tension

It is important that timing belt tension is adjusted properly. If too loose, the belt could flap or vibrate in service and fly off its sprockets. If too tight, the timing belt could snap. Both could lead to severe valve damage.

Again, if you are not familiar with the specific procedure, check in a service manual. It will detail how to adjust tension properly.

Fig. 27-16 shows a belt tensioner. Note how you can

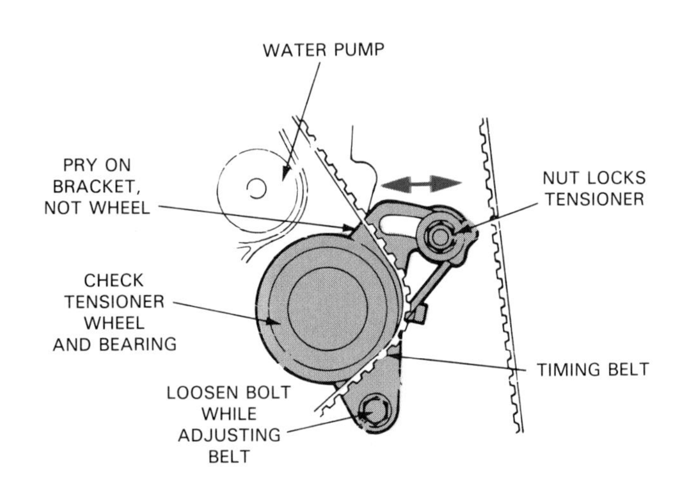

Fig. 27-16. This timing belt adjuster has a slot that lets you swivel adjuster to the side. Pry on adjuster and tighten fastener. Then check for proper tension on belt.

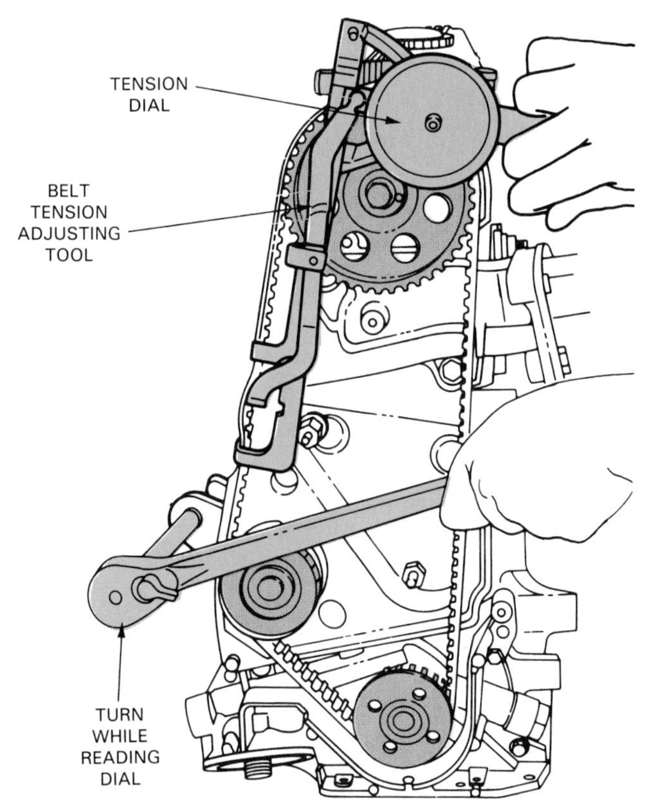

Fig. 27-17. This is a special tool for adjusting belt tension. It measures deflection of straight section of belt. If it is not available, adjust until moderate finger pressure is needed to twist belt one-quarter turn. (Chevrolet)

loosen the attaching bolts and swivel the tensioner. Pry on the tensioner bracket to preload the belt.

Fig. 27-17 shows a special tool being used to set belt tension correctly. An adjusting tool is turned with a wrench until the belt gauge reads within specs. Tighten the tensioner bracket bolts to secure the adjustment.

There is a simpler way to check timing belt tension if a gauge or tool is not available. Adjust the belt until moderate finger and thumb pressure is needed to TWIST THE BELT about one-quarter turn. This should provide good belt service.

After installing the belt, it is wise to rotate the engine by hand. Use a wrench on the crankshaft snout bolt. Turn the crank in the direction of normal rotation until the timing marks come around and realign. Then, double-check your timing marks. See Fig. 27-18.

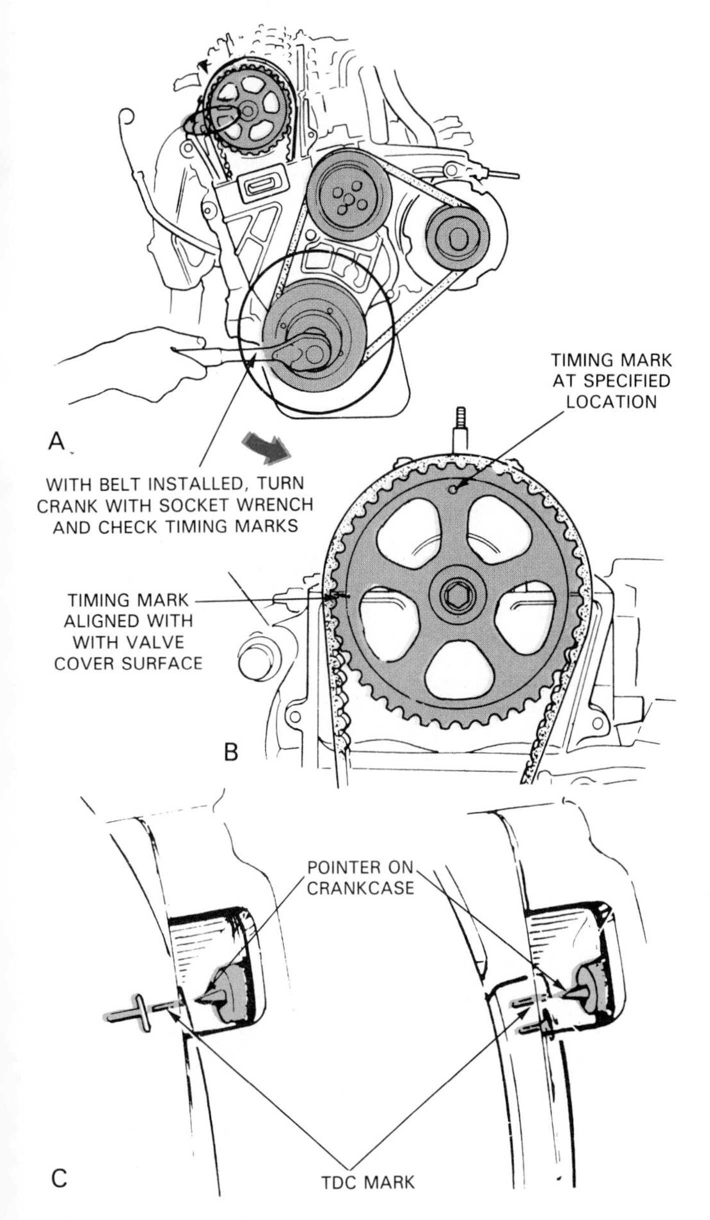

Fig. 27-18. A—To check belt installation, rotate crankshaft by hand. B—Make sure sprocket and engine marks align. C—Also check that engine timing mark is on zero or number one piston is at TDC on its compression stroke. (Honda)

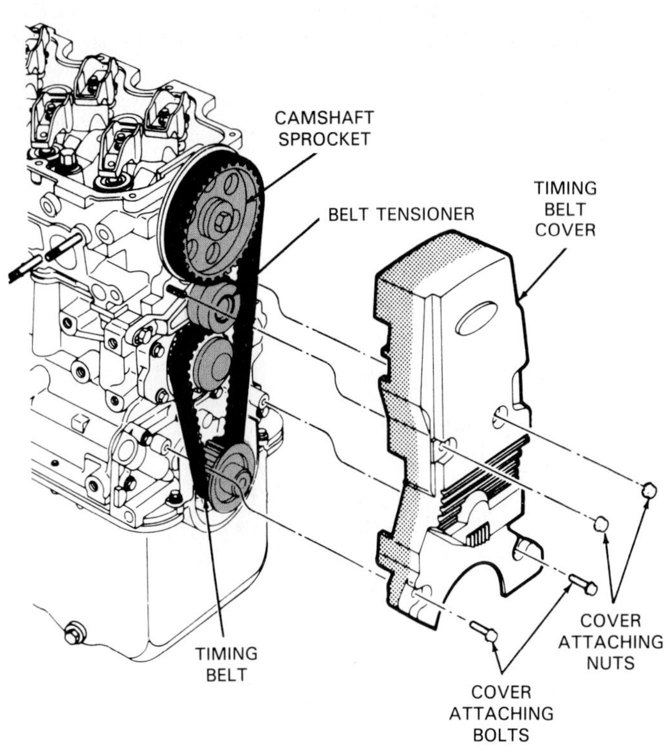

Fig. 27-19. When installing timing belt cover, make sure you use the correct bolts and that all rubber grommets or spacers are in place. The rubber washers prevent rattling and breakage of cover. It can be made of thin sheet metal or plastic. (Ford)

Installing timing belt cover

The *timing belt cover* is simply a plastic or sheet metal shroud around the belt. It keeps debris off the belt and protects your hands when working in the engine compartment. It does NOT contain an oil seal, Fig. 27-19.

To install the cover, simply fit it into place on the front of the engine. If rubber washers are used, make sure you use them. They prevent cover vibration, noise, and cracking of the cover. Check that all fasteners are in place and that the belt and sprockets do not rub on the timing belt cover.

Fig. 27-20 shows an exploded view of a timing belt mechanism and its related parts. Note how this belt cover is two-piece and uses a lower dust seal.

TIMING GEAR SERVICE

Timing gear service is similar to timing chain service. Timing gears are normally more dependable than timing chains. They will provide thousands of miles of trouble-free engine operation. However, after prolonged use, timing gear teeth can wear or become chipped and damaged, requiring replacement.

Inspect the old timing gears carefully. Look for any signs of wear or other problems. Replace the gears AS A SET if needed. See Fig. 27-21.

Installing timing gears

Timing gears are usually press-fit on the crankshaft and camshaft. A wheel puller is normally needed to remove the crankshaft gear.

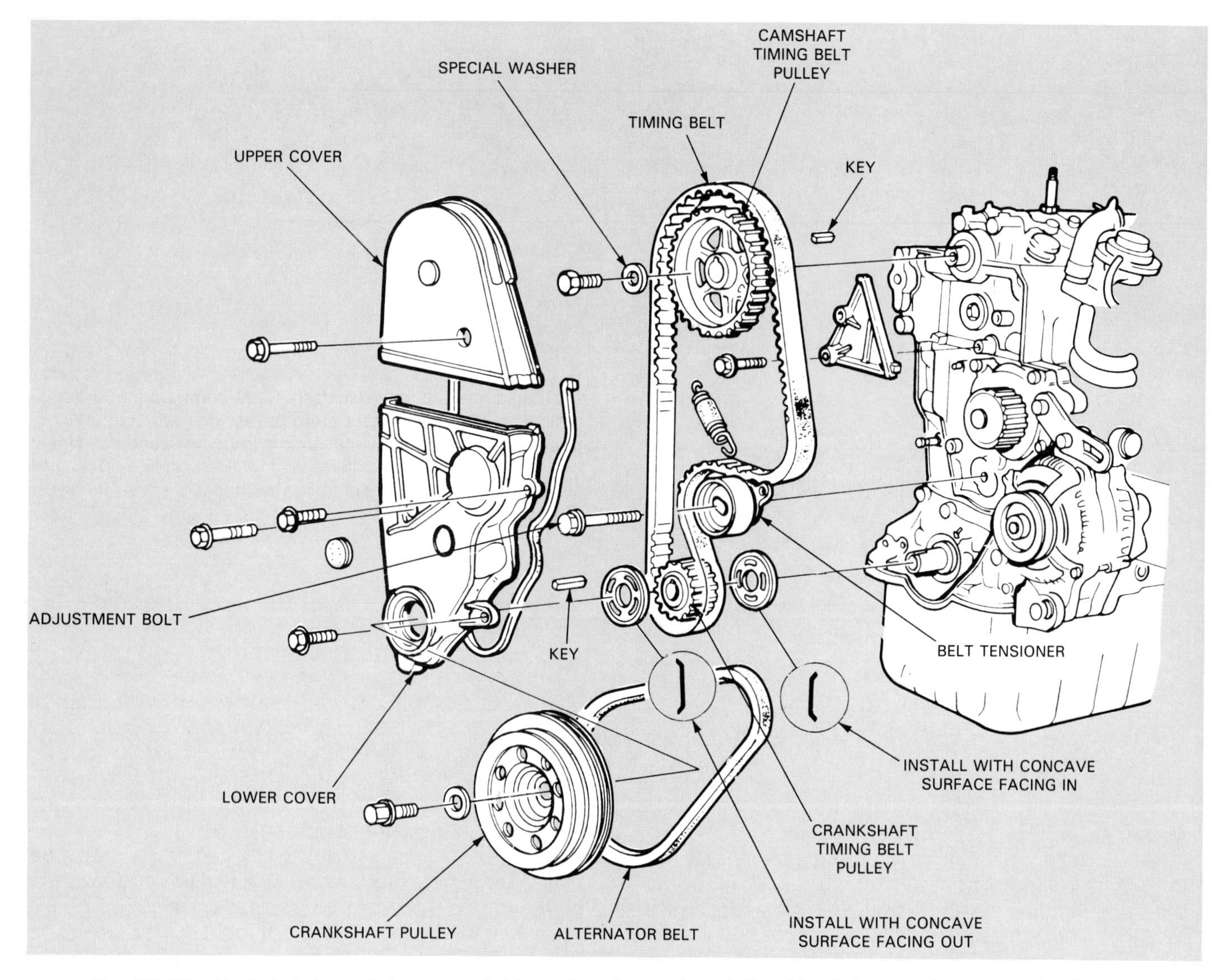

Fig. 27-20. Exploded view of the front of this engine shows the relationship of front end components. Torque all fasteners properly. (Honda)

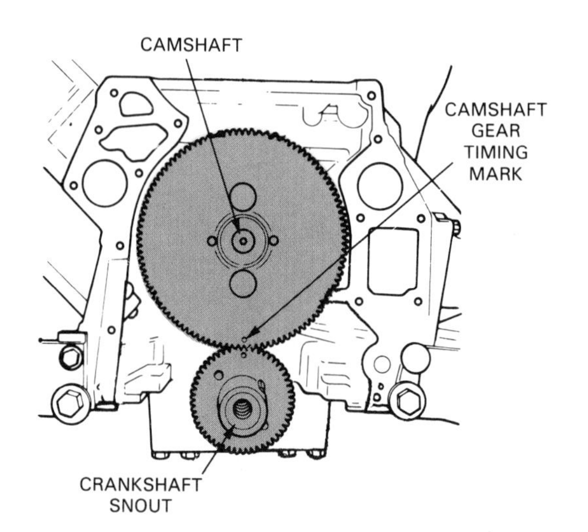

Fig. 27-21. Timing gears are very dependable. However, after prolonged service, teeth can wear or break. When installing new timing gears, position crankshaft and camshaft in relatively correct positions. Then, fit gears in place so that timing marks align. (Ford Motor Co.)

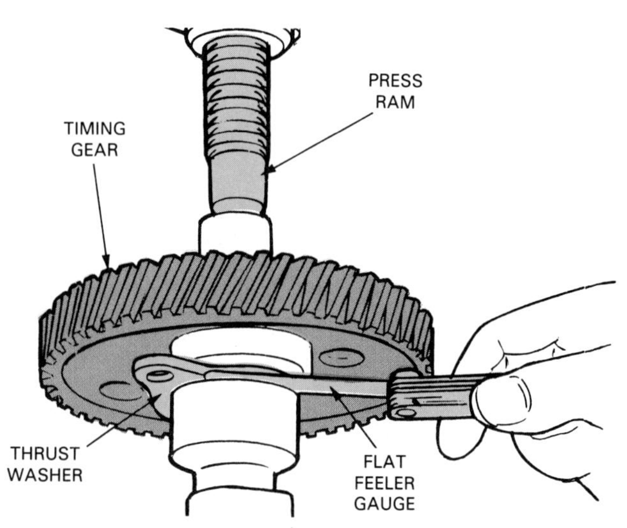

Fig. 27-22. This particular timing gear must be pressed onto camshaft snout. Clearance must be maintained between thrust plate and cheek on camshaft. This clearance allows for proper end play of camshaft. (Ford)

Fig. 27-22 shows how a press is used to service the cam gear on the camshaft. With this engine design, the camshaft must be out for timing gear replacement.

Timing gears can usually be installed with light blows from a brass hammer. Make sure the key and key ways are aligned. Tap in a circular motion to force the gears squarely into position. Do not hit and damage the gear teeth. A press may also be needed to install the cam gear on the camshaft.

As with a timing chain and sprockets, double-check the alignment of the timing marks, Fig. 27-23. The timing marks must be positioned properly to time the camshaft with the crankshaft.

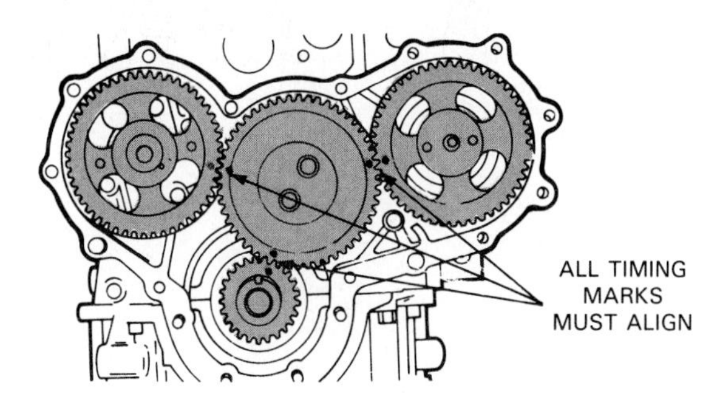

Fig. 27-23. This large truck diesel engine uses four timing gears. Note how marks must be aligned.

Measuring timing gear runout

Timing gear runout or wobble is measured with a dial indicator. Position the indicator stand on the engine block. Place the indicator stem on the outer edge of the camshaft timing gear. The stem should be parallel with the camshaft centerline, as shown in Fig. 27-24.

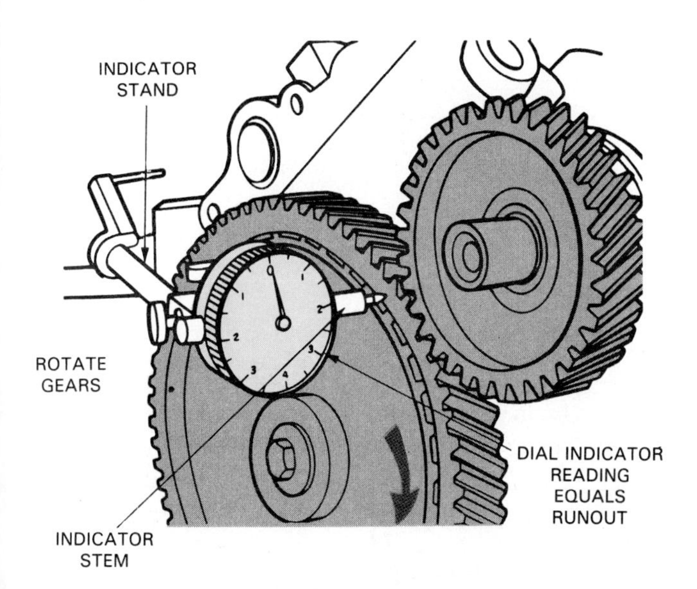

Fig. 27-24. To check timing gear runout or wobble, position dial indicator as shown. Rotate engine crankshaft and read indicator. Needle movement equals runout. If beyond specs, check your installation or the gears might have machining problems. The camshaft could also be bent. (Ford)

Fig. 27-25. Timing gear teeth clearance can be measured with small feeler gauge or dial indicator. Size feeler gauge blade that fits in gear teeth equals backlash. More common method is dial indicator. Mount indicator stem to rest on gear tooth. Wiggle gear back and forth while noting indicator reading. Needle movement equals backlash or teeth clearance. Too little clearance can make gears lock up when expanded by engine heat. Too much clearance can reduce engine power and increase gear noise.

To measure gear runout, turn the engine crankshaft while noting indicator needle movement. No indicator reading equals no runout. If runout is greater than specs, remove the timing gears and check for problems. The gear may not be fully seated or it may be machined improperly. Also check camshaft straightness.

Measuring timing gear backlash

Timing gear backlash is the amount of clearance between the timing gear teeth. Backlash can be measured to determine timing gear teeth wear. Look at Fig. 27-25.

A dial indicator is an accurate tool for measuring timing gear backlash. Set the indicator stand on the engine. Locate the indicator stem on one of the cam gear teeth. The stem must be parallel with gear tooth travel.

Wiggle the cam gear one way and then the other, without turning the crankshaft. Read the indicator by noting needle travel.

If timing gear backlash is greater than specs, the gears are worn. They should be replaced. Refer to the service manual for backlash specs.

Fig. 27-26 shows a technician getting ready to install the front cover after timing gear service.

CRANKSHAFT FRONT SEAL SERVICE

The *crankshaft front seal* keeps engine oil from leaking out from between the crankshaft snout and the engine front cover. Replace the front seal whenever it is leaking or when the front cover is removed.

Seal replacement in the car requires only partial engine disassembly. Typically, you must remove the radiator and other accessory units on the front of the engine. Use a wheel puller to remove the crankshaft damper.

Sometimes, the front crankshaft seal can be replaced without front cover removal. Shown in Fig. 27-27A, a special seal puller may be used to remove the old seal. Then, use a seal driver to squarely seat the new seal into its bore, Fig. 27-27B.

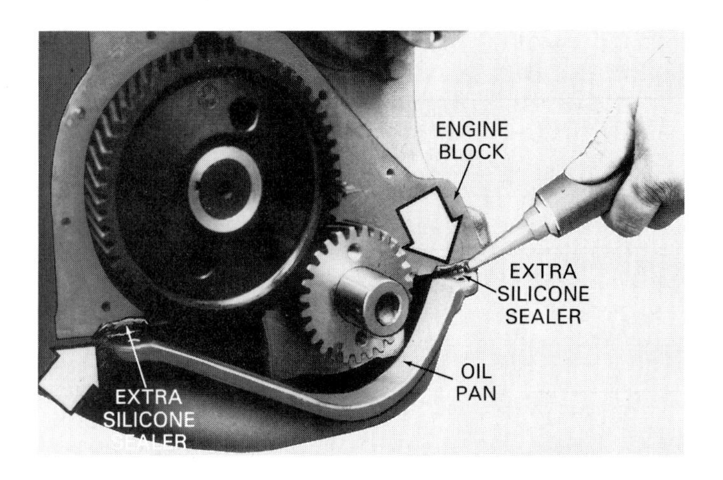

Fig. 27-26. Here timing gears have been serviced with engine and oil pan installed in car. Oil pan bolts must be loosened in front to give clearance for installing front cover. Technician is installing silicone sealer where parts and gaskets come together. (Chevrolet)

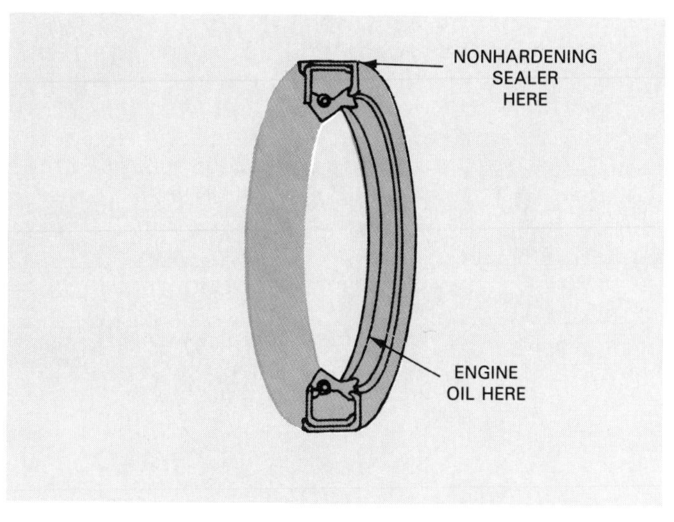

Fig. 27-28. When installing crankshaft seal, coat outside diameter with nonhardening sealer. Coat inside sealing lip with motor oil.

When the crankshaft front seal drives in from the rear, timing cover removal is needed for seal replacement. You must remove the cover to drive the old seal out. Then drive in the new seal, Fig. 27-27.

Always compare the old seal with the new one. Their outside and inside diameters must be the same. You may also be able to match part numbers stamped on the metal flange of the seals.

Before installing a front seal, coat the outside diameter of the seal with nonhardening sealer. This will prevent oil seepage between the seal body and the front cover. Coat the rubber sealing lip with engine oil. This will lubricate the seal during initial engine startup, Fig. 27-28.

When available, seal driver should be used to install an engine front cover seal. However, if you do not have one, you can use a block of wood, as pictured in Fig. 27-29. Drive the seal in squarely without damaging the metal housing of the seal.

Fig. 27-30 shows a technician installing a front seal in a cylinder head. It seals the front camshaft bearing of an OHC engine. Its outside diameter should also be

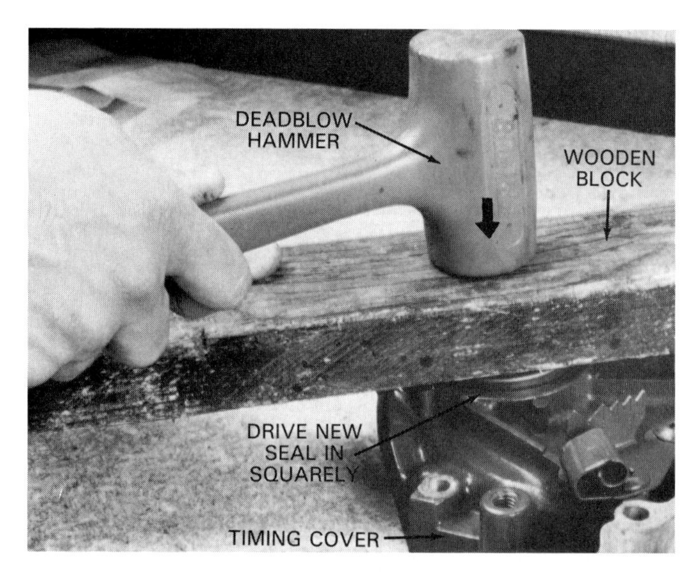

Fig. 27-29. If a seal driver is not available, a block of wood will help you install seal without damage. (Fel-Pro Gaskets)

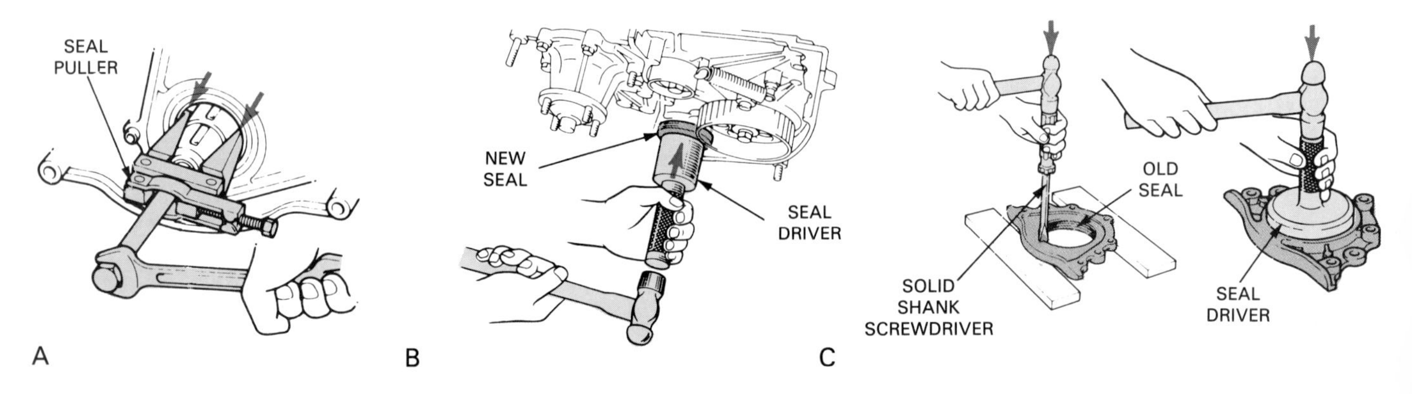

Fig. 27-27. Servicing front oil seal. A—Sometimes front seal can be replaced without removing front cover. It can be pulled out from the front and a new one pressed in. B—Driving in new seal with cover installed. C—Most engines require front cover removal for front seal service. Oil seal is driven out back. Then seal driver can be used to install new seal. (Toyota)

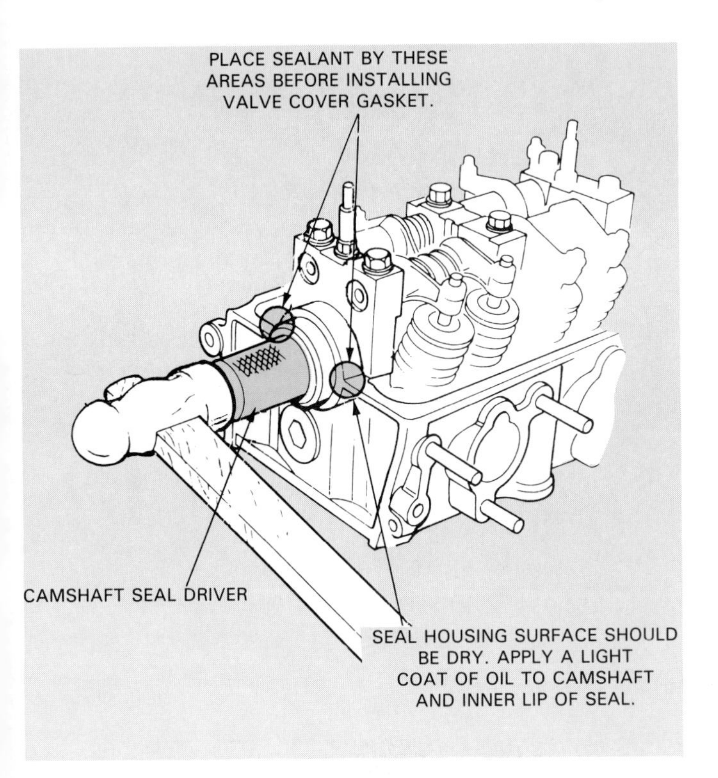

Fig. 27-30. With an overhead cam engine, seal may be mounted on cylinder head as shown. Seal driver is best way to install seal without bending, denting, or damaging seal. (Honda)

coated with nonhardening sealer. Place engine oil or grease on its inside diameter.

ENGINE FRONT COVER SERVICE

The *engine front cover,* also termed *timing cover,* holds the front oil seal and encloses the timing gears or chain.

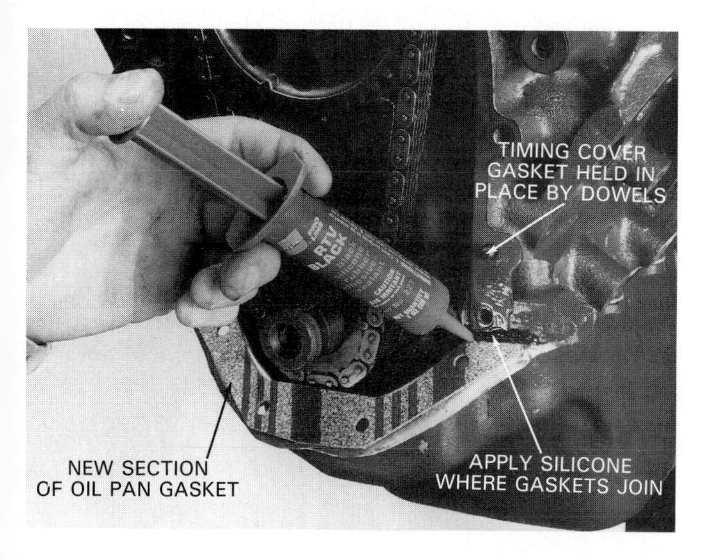

Fig. 27-31. When installing front cover, make sure gasket holes align properly. Use dowel, adhesive, or sealer to hold gaskets in place on engine. Use extra silicone sealer where gaskets meet to prevent weepage of oil out of engine. (Fel-Pro Gaskets)

It can be made of thin stamped, sheet metal or cast aluminum.

Scrape off all gasket material and install the new oil seal. Make sure the cover sealing surface is true. Lay it on a flat work surface and check for gaps under the cover. Straighten the cover if needed.

Sealing front cover

Either a conventional gasket or chemical sealants can be used during front cover installation. Usually, both a gasket and sealer are used, as is shown in Fig. 27-31.

Fig. 27-32 shows how anaerobic and silicone sealers are placed on an aluminum front cover. Use a small bead of sealer. Make sure you have a continuous bead to prevent oil or coolant leakage. Circle all holes that carry oil or coolant.

When using a conventional gasket, check that all holes align on the engine and in the gasket. Gasket sealer or adhesive is commonly used to hold the gasket in place during part assembly.

Make sure you use a bead of silicone sealer where two gaskets or parts come together, Fig. 27-31. This is a common leakage point.

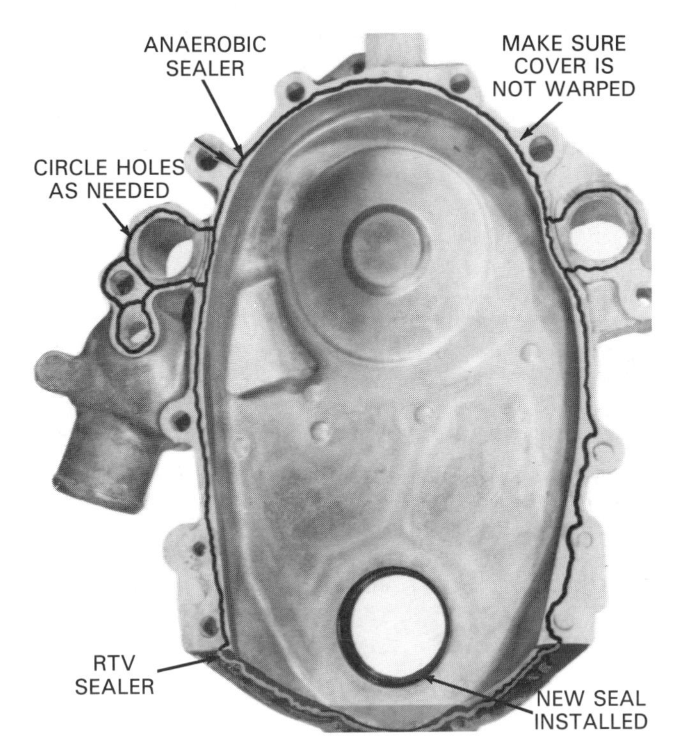

Fig. 27-32. Here is a service manual illustration showing recommended sealing method for a front cover made of cast aluminum. Anaerobic sealer is used where two thick or solid pieces mate up tightly. RTV or silicone sealer is recommended where flexible oil pan mates with front cover. (General Motors Corp.)

Installing front cover

As shown in Fig. 27-33, carefully fit the front cover onto the engine. Do not shift the gasket or smear the sealer.

If the oil pan is on the engine (in-car cover service), loosen the oil pan bolts. Pry down on the pan so the front

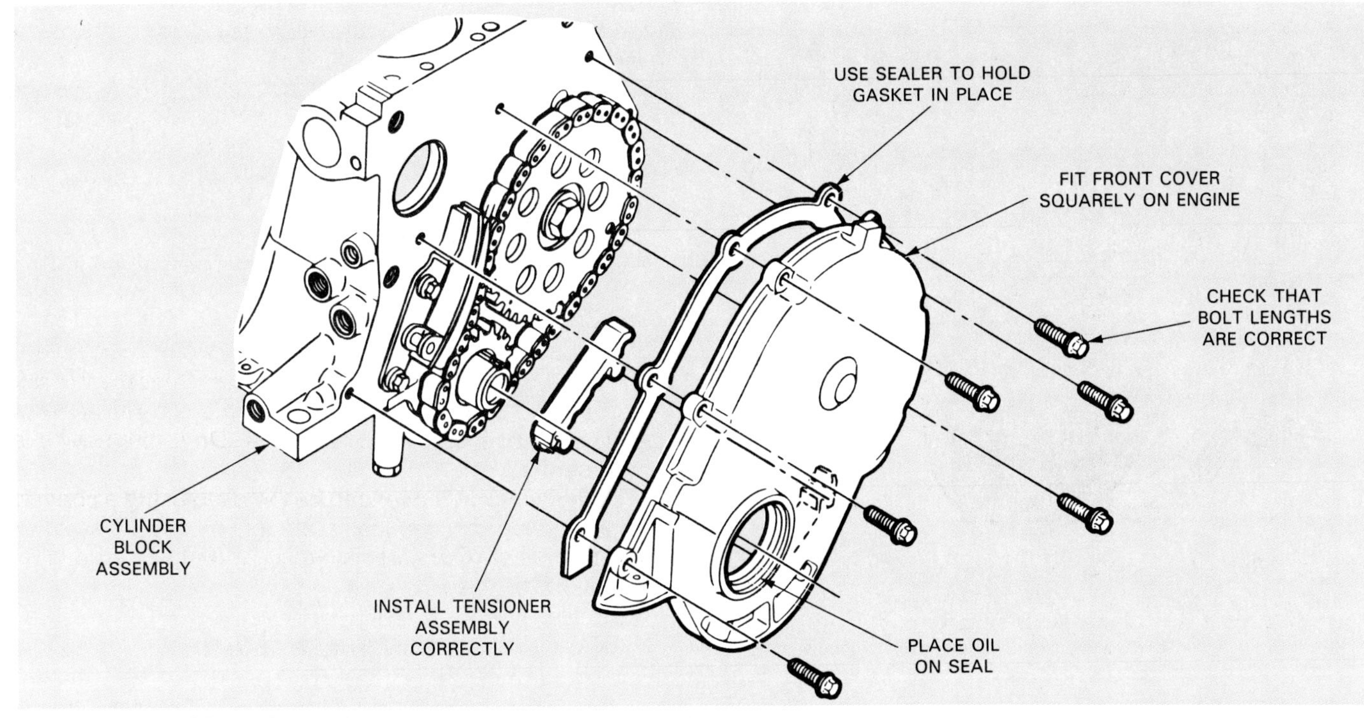

Fig. 27-33. Before installing front cover, check that all fasteners under cover are torqued properly. If used, check installation of chain tensioner. Double-check timing mark alignment. Carefully position front cover without disturbing gasket. Make sure bolt lengths are correct. Seal any bolt threads entering water or oil passages. Torque fasteners to specs. (Ford)

cover will fit into place.

You may need to use a *seal centering tool* that fits over the crankshaft snout. It positions or centers the front cover seal correctly around the crankshaft.

Install the front cover bolts. Check bolt lengths and locations.

Any bolt entering a water jacket or oil passage should be sealed. Wipe a small amount of nonhardening sealer on these bolt threads to prevent leakage. The service manual will explain which bolts require sealer if in doubt.

Tighten the bolts to specs in gradual steps following a crisscross pattern. If used, remove the seal alignment tool. Double-check that all bolts are installed properly. It is easy to miss a hidden hole under the lower section of the timing cover.

Install the front oil pan bolts if needed. Then, finish by assembling other external components: water pump, brackets, front damper, etc.

CRANK DAMPER INSTALLATION

To install the crankshaft damper, check the crank snout key. It should be properly installed and tight in its groove. Fit the damper onto the crankshaft without pushing the key out of its groove.

Usually, the snout bolt will tighten down and force the damper into position, Fig. 27-34. However, sometimes a special driver is needed to install the damper, Fig. 27-35.

In any case, make sure the damper pulls down squarely on the snout. If it cocks sideways, stop and align the damper to prevent part damage.

Do NOT hammer on the outer edge of a crank damper during removal or installation. Remember that the outer ring is mounted on rubber. Hammering could shift the outer ring and ruin the damper. Install any other parts: crank pulleys, timing pointer, brackets, as needed.

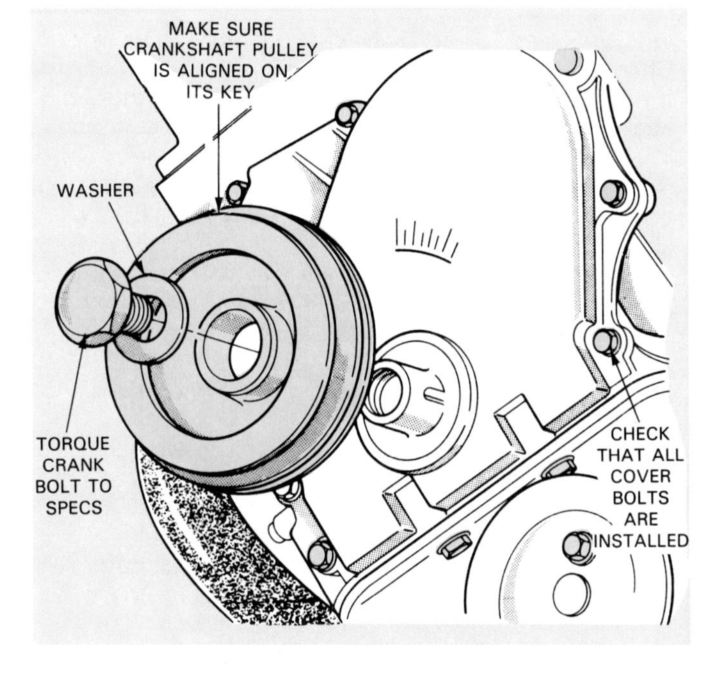

Fig. 27-34. Crankshaft damper will usually slide partially over crank snout. Then, large snout bolt will pull damper into position. If light hammering is necessary, hammer in the center of the damper. Never hammer on its outside diameter or you can damage rubber ring and ruin the harmonic balancer. (Honda)

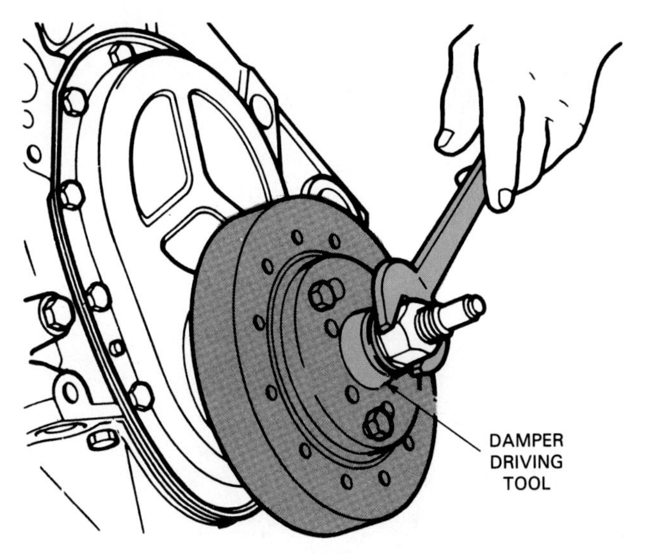

Fig. 27-35. This damper requires a special tool for installation. Only a press-fit holds the damper in place. It is not held by crank snout bolt. Again, never hammer on outer ring of vibration damper. (Chevrolet)

SUMMARY

Modern engine front ends are more complex. Overhead cam engines are the main reason more skill is needed to service today's engines.

Timing chain wear can let the camshaft go out of time with the crankshaft. This can reduce engine power. Check chain slack by turning the crank back and forth. If it can be turned excessively without distributor, camshaft, or valve train movement, the chain is worn.

Timing sprockets with plastic gear teeth are a common problem source. The plastic teeth can wear or break off. A bad chain tensioner can also cause excess chain slack.

To begin timing chain installation, check any keys or dowel. Turn the crank and cam to align the timing marks. Fit the chain and sprockets in place together.

Do not hammer on the outer edges of the sprockets to force them into place. Use light blows in the center of the sprockets.

Make the timing marks on the sprockets align. The marks can be small dents, circles, or other figures on the sprockets. The service manual will usually illustrate them.

Do NOT crank an engine with the starting motor with the head installed and timing chain or belt removed. The pistons could slide up and bend the valves.

An oil slinger fits in front of the crankshaft sprocket to prevent oil leakage and to lubricate the timing chain or gears. Make sure it is not installed backwards.

Timing belt service is very important. If a belt were to break, severe engine damage could result. Inspect the belt for signs of deterioration. Most mechanics replace the belt whenever it is removed for service. Most manufacturers recommend timing belt replacement every 50,000 miles or 80 450 km.

Install the timing belt as you would a timing chain. Align the sprocket timing marks. The number one piston should usually be at TDC on the compression stroke. If the belt is reused, it should be installed to rotate in the same direction.

Adjust timing belt tension correctly to prevent belt breakage or looseness. Special belt tension gauges are available. However, you can adjust the belt until moderate finger pressure is needed to twist the belt about one-quarter turn.

Timing gears are very dependable. Gear teeth can break and wear, however, after extended service. During timing gear service, measure gear backlash and runout with a dial indicator.

When installing a new front oil seal, drive out the old seal without damaging the front cover. Coat the outside diameter of the new seal with nonhardening sealer. Wipe engine oil on the inside of the seal. Drive the new seal in squarely.

An engine front cover can be installed using a conventional gasket or chemical sealants, or both. Use extra silicone where two seals come together. Torque all fasteners to specs.

KNOW THESE TERMS

Chain slack, Sprocket timing marks, Chain tensioner, Timing belt tension, Timing belt cover, Timing gear runout, Timing gear backlash, Front oil seal, Seal alignment tool, Timing cover.

REVIEW QUESTIONS—Chapter 27

1. Timing chains are used on both ________ ________ and ________ ________ engines.
2. How do you check OHV engine timing chain wear without timing cover removal?
3. In your own words, explain how to fit the timing chain and sprockets on an engine.
4. Drive the chain sprockets on with hammer blows around the outside diameter of the sprockets. True or false?
5. Timing marks will be ________, ________, ________, or other shapes on the sprockets.
6. Why should you never crank an engine with the starting motor when the timing chain or belt is removed?
7. This part keeps oil from spraying out the front seal and also helps lubricate the timing chain or gears.
    a. Tensioner.
    b. Shroud.
    c. Slinger.
    d. Front cover.
8. An engine is in for service. Worn camshaft lobes are found and the camshaft is being replaced. The car has 35,000 miles (56 300 km) on it.
    Technician A says that the timing belt should be replaced to protect from engine failure.
    Technician B says that the belt is good for 50,000 miles or 80 450 km, and it should be reused.
    Who is correct?
    a. Technician A.
    b. Technician B.
    c. Both A and B.
    d. Neither A nor B.
9. If you do not have a belt tension gauge, how can

you adjust belt tension properly?

10. This is measured with a dial indicator with the indicator plunger resting on one of the timing gear teeth.
    a. Timing gear runout.
    b. Timing gear diameter.
    c. Timing gear out-of-round.
    d. Timing gear backlash.

ASE CERTIFICATION-TYPE QUESTIONS

1. Technician A says that most engines must be removed from the vehicle in order to replace the timing chain. Technician B says that normally, a timing chain can be replaced without removing the engine from the car. Who is right?
   (A) A only.
   (B) B only.
   (C) Both A & B.
   (D) Neither A nor B.
2. Technician A says that timing chain wear can reduce engine compression. Technician B says that timing chain wear can affect exhaust cleanliness. Who is right?
   (A) A only.
   (B) B only.
   (C) Both A & B.
   (D) Neither A nor B.
3. Most manufacturers recommend replacing an engine's timing belt about every ________.
   (A) 70,000 miles (120 000 km)
   (B) 100,000 miles (160 000 km)
   (C) 200,000 miles (320 000 km)
   (D) 50,000 miles (80 000 km).
4. Technician A says that the camshaft sprocket mark should normally be aligned with a mark on the side of the engine. Technician B says that the cam sprocket mark should normally be aligned with a mark on the front of the engine. Who is right?
   (A) A only.
   (B) B only.
   (C) Both A & B.
   (D) Neither A nor B.
5. Technician A says that a loose timing belt could lead to severe valve damage. Technician B says that if a timing belt is too tight, severe valve damage can occur. Who is right?
   (A) A only.
   (B) B only.
   (C) Both A & B.
   (D) Neither A nor B.
6. Technician A says that the function of a timing belt cover is to protect the oil seal and belt from debris. Technician B says that the function of a timing belt cover is to keep debris off the belt and protect your hands when working in the engine compartment. Who is right?
   (A) A only.
   (B) B only.
   (C) Both A & B.
   (D) Neither A nor B.
7. Timing gear runout should be measured with a ________.
   (A) dial indicator
   (B) inside micrometer
   (C) outside micrometer
   (D) feeler gauge
8. Technician A says that timing gear backlash is the amount of clearance between the crankshaft snout and timing gear. Technician B says that timing gear backlash is the amount of clearance between the timing gear teeth. Who is right?
   (A) A only.
   (B) B only.
   (C) Both A & B.
   (D) Neither A nor B.
9. Technician A says that the engine front cover is also called the timing belt cover. Technician B says that another name for the engine front cover is the timing cover. Who is right?
   (A) A only.
   (B) B only.
   (C) Both A & B.
   (D) Neither A nor B.
10. Technician A says that normally, the snout bolt will tighten down and force a crank damper into position. Technician B says that you must normally hammer on the outer edge of a crank damper to force it into place on the crankshaft. Who is right?
    (A) A only.
    (B) B only.
    (C) Both A & B.
    (D) Neither A nor B.

Chapter 28

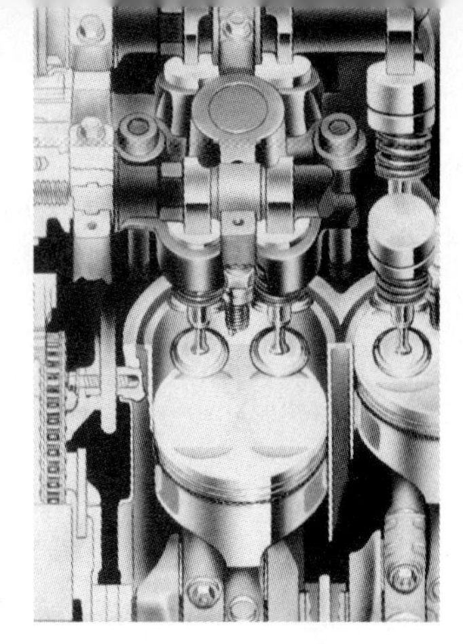

Engine Reassembly, Installation, Break-In

After studying this chapter, you will be able to:
- *List the general steps to prepare for engine reassembly.*
- *Properly install an oil pump and oil pan.*
- *Install and torque a cylinder head assembly.*
- *Install and torque exhaust and intake manifolds.*
- *Correctly install engine gaskets.*
- *Adjust engine valves.*
- *Properly install valve covers.*
- *Install engine accessory units.*
- *Check and install a clutch assembly.*
- *Refit the engine into the vehicle.*
- *Prepare for engine start-up.*
- *Use a good engine break-in procedure.*

This chapter will summarize the basic methods for reassembling the major components of an engine. Previous chapters explained how to assemble the short block and cylinder head. This chapter will build upon this knowledge by explaining how to install the cylinder head, manifolds, and covers on the short block. Then, you will learn about valve adjustment, engine installation, start-up, and break-in.

This is an important chapter that will supplement and complete what you have learned earlier about engine service. Study carefully!

GENERAL ENGINE ASSEMBLY RULES

There are several basic rules you must remember when assembling and installing a car engine. These include:

1. *All parts and tools should be perfectly clean and organized!* Remember that the slightest bit of dirt, metal, plastic, gasket material, etc. could cause major engine damage, Fig. 28-1.
2. *Parts that might be* SELECT FIT (part size matches

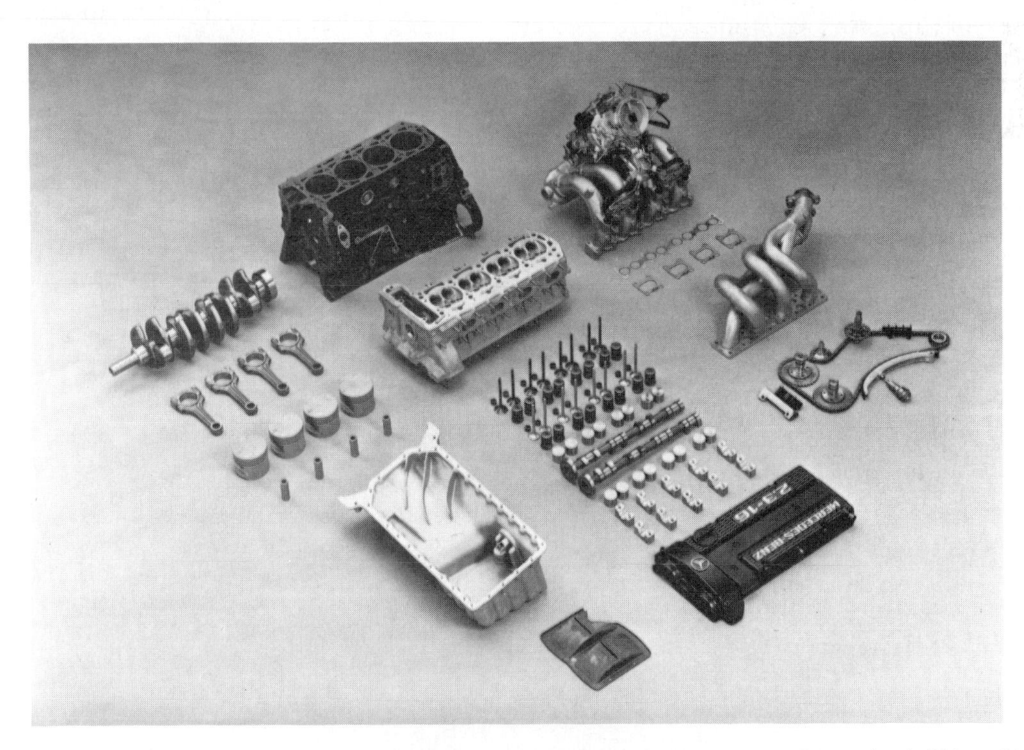

Fig. 28-1. Before starting engine reassembly, all components should be perfectly clean and organized. Sides of block, outside of valve cover, etc., may also require repainting with engine enamel. (Mercedes-Benz)

size of corresponding part) *should be reinstalled in the same location!* For example, if lifters on a low mileage engine are going to be reused on the same camshaft, install them in their exact same lifter bores. Lifter diameters and their wear patterns will vary slightly.

3. *Lubricate parts as needed!* All moving parts should be coated with engine oil or white grease. This will provide protection during initial engine start up.
4. *Make sure bolts and nuts are correct!* It is best to make sure you have exactly the same fastener as used by the manufacturer. Some engines use both conventional and metric fasteners; do not mix them or thread damage will result. Also, check bolt lengths. A bolt that is too long could damage parts by bottoming out. A bolt that is too short could strip its threads and fail in service.
5. *Keep components covered until you are ready to assemble them!* For example, keep a cloth over a short block sitting on an engine stand. This will help keep debris out of the cylinders and off of other areas in the block. Keep the intake manifold inlet covered, especially when it is installed on the head. Anything dropped into the manifold could cause severe engine damage to the pistons, valves, head, and block.
6. *Torque all components properly!* An air impact does not provide the control needed to properly tighten many engine fasteners. Do not try to save time by impacting down, and NOT torquing, critical fasteners. This applies to head bolts, manifold bolts, oil pump bolts, etc. Modern engines are cast very thin and light. Improper torque can cause fluid leakage, part warpage, or even part breakage.
7. *Inspect everything one last time as you install engine components!* Look for signs of trouble: unusual wear, cracks, scoring, etc. You do NOT want to find something wrong AFTER the engine is assembled and installed in the car!
8. *Do not get too excited and rush!* It is very easy to become excited about finishing an engine overhaul, wanting to see the engine "fire up" and run. You must have control and work at a steady pace.
9. *Avoid talking to anyone while working on an engine!* If distracted, it is very easy to forget to do something. Forgetting to torque even one bolt could result in major engine damage, or a serious leak, at your expense.
10. *Refer to a service manual if in doubt about any service task!* The manual will answer any questions you might have concerning an assembly method, torque value, valve adjustment specification, etc.

ENGINE REASSEMBLY

This section of the chapter will describe typical steps for the reassembly of an engine. All parts should already be clean, as described in Chapter 24. The cylinder head should be reconditioned, as explained in Chapter 26. The short block should also be together, discussed in Chapter 25. Refer back to these chapters for more information as needed.

The methods given in this chapter are general and they apply to most makes and models of engines. However, always follow the more specific procedures given in the service manual. The manual will give information that might be unique to the particular engine.

The short block should be placed on an engine stand. All connecting rods should be torqued and all surfaces perfectly clean. Appropriate areas on the outside of the engine should be painted. The front cover and timing chain or gears should also be installed, as was summarized in Chapter 27. Refer to the index if you want to review any of these operations.

Installing oil pump

With the front cover and timing mechanism (not belt however) in place, you can usually begin by installing the engine oil pump. As explained in Chapter 15, make sure the used pump is in good condition. Many technicians prefer to use a new or factory rebuilt oil pump. They are not extremely expensive and are very important to engine service life.

When possible, always prime an oil pump. This can be done by pouring oil or petroleum jelly into the pumping cavity during assembly. You can also submerge the inlet into a container of oil and turn the pump by hand. This will draw oil into the pump. Priming will help protect the pump and ensure quicker engine oil pressure upon initial starting of the engine.

If you have a front cover-mounted oil pump, position the new gaskets, seals, rotors, and the other parts in the engine. Torque all pump fasteners to specs. Turn the pump by hand to check for smoothness. Make sure the pump is well oiled, Fig. 28-2A.

If your engine has a more conventional oil pump that installs in the oil pan, fit the oil pump drive into position. This is shown in Fig. 28-2B. Make sure the oil pump drive rod or shaft is located in its guide hole in the block. As you fit the oil pump to the block, the drive rod must fit into the pump and the hole in the block.

Make sure the oil pump gasket holes align properly. Avoid using sealer on the oil pump gasket because a gob of sealer could lock the oil pump. Usually, install the oil pump gasket clean and dry. Start the pump bolts by hand and then torque them to specs.

Note! Oil pump mounting bolts must be tightened properly. It might be wise to use thread locking compound on the oil pump bolt threads. If these bolts were to loosen or fail in service, the engine will lose oil pressure and major bearing damage can result.

If support brackets are provided on the oil pump, make sure they are installed properly. They prevent pump vibration that could break the pump housing or cause bolt failure. The oil pump pickup tube and screen must also be in good condition and properly installed.

After pump installation, double-check everything. The pump gasket should be in place. The pump drive rod must be located in the block correctly. The pickup tube must not leak at the pump or block.

Installing oil pan

With the pump and front cover in place, you are ready to install the oil pan. Lay the pan on a flat work surface.

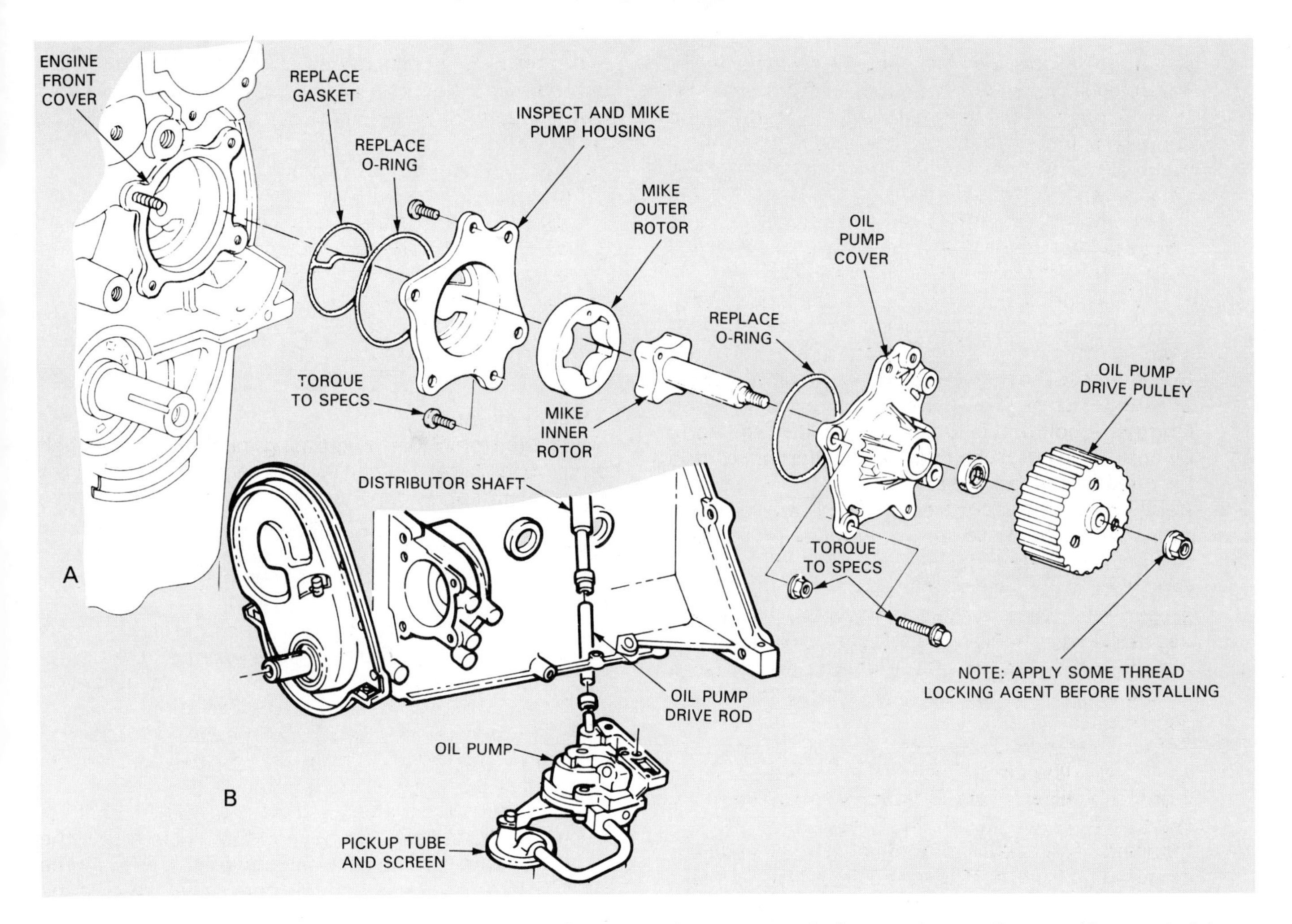

Fig. 28-2. Measure oil pump internal components and compare them to specs before reusing an oil pump. Many technicians install a new oil pump during major engine service. A—Use new gaskets and seals when installing a front-cover mounted oil pump. B—Make sure drive rod is up in hole in block with conventional oil pump. Fit gasket in place and start bolts by hand. Tighten bolts to specs. Make sure pickup tube is secure.

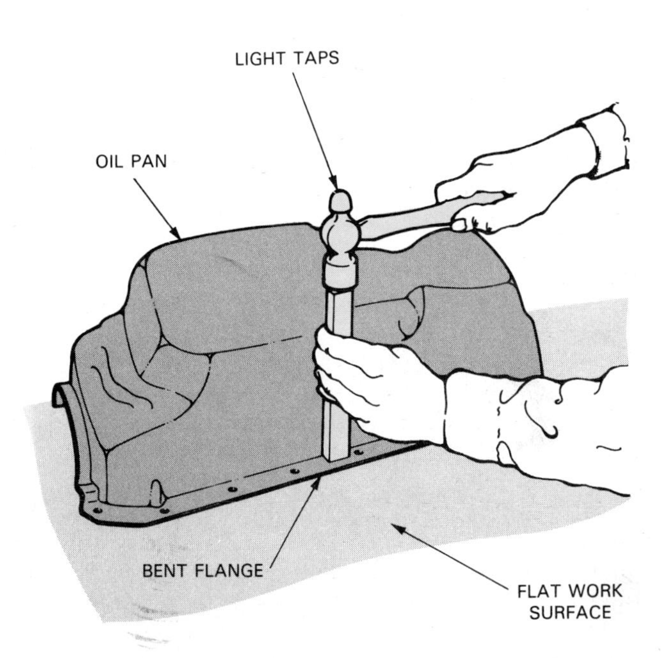

Fig. 28-3. Lay oil pan on workbench and check for a bent flange. If needed, straighten flange. Also check condition of oil drain plug and threads in pan. (Ford)

If the pan flange is bent or untrue, straighten it, as shown in Fig. 28-3. A bent flange can cause oil pan gasket leakage and a unwelcomed return of the customer with a valid complaint.

Install the oil pan gasket. First, lay it in place on the pan. Check that all bolt holes align and that you have the correct gasket. You can either adhere the gasket to the pan or to the engine block.

If the engine is installed in the car, it might be easier to glue the gasket to the oil pan, Fig. 28-4. If you are working on an engine stand, it might be better to bond the gasket to the block, Fig. 28-5. Then, there is less chance of the gasket falling or shifting during assembly.

Sometimes the end seals for the oil pan have to be worked down into grooves in the oil pan or in the engine front cover and block. If this is the case, the pan gasket should be installed next to or under the rubber end seals. RTV sealer should be used to help prevent leakage where the gasket and end seals overlap.

Allow the gasket cement to set up slightly before installing the pan. This will help keep bolt holes aligned. Fit the pan into place on the engine block. Start all bolts by hand before tightening any of the fasteners. If you tighten any of the bolts before others are started, the

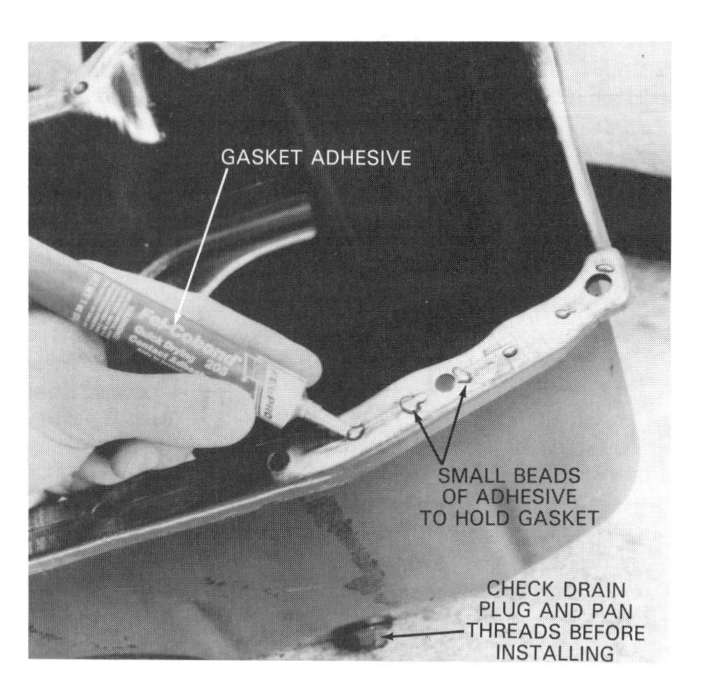

Fig. 28-4. When using a gasket, adhesive should be used to hold oil pan gasket in alignment during pan installation. Depending upon gasket and end seal design, you may need to bond gasket to either oil pan or engine block. (Fel-Pro Gaskets)

Fig. 28-5. With engine on engine stand, gasket can sometimes be positioned on block. This manual illustration recommends sealer to hold gasket in place. (Oldsmobile)

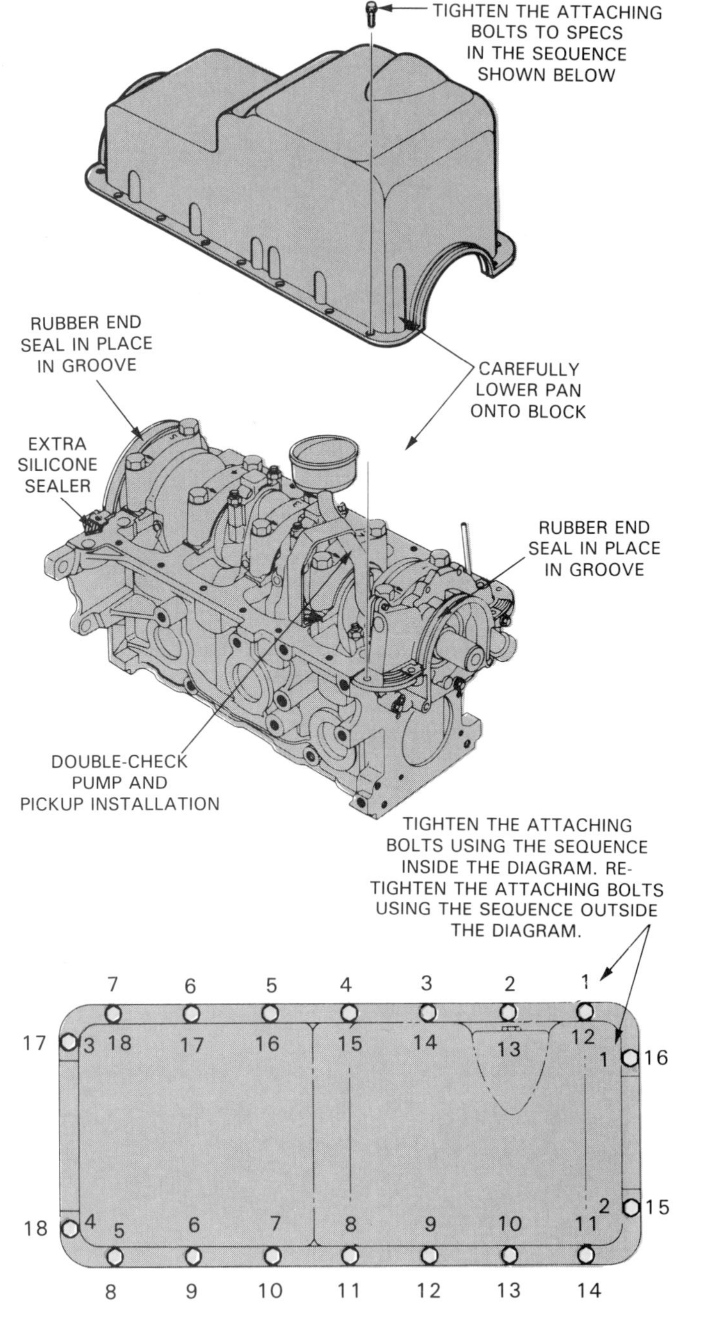

Fig. 28-6. Position oil pan on block. Start all bolts by hand before tightening any bolts. If you tighten bolt, gasket will compress and squeeze out into bolt holes. Other bolts will be difficult to start. Torque bolts to specs in recommended sequence. (Ford Motor Co.)

gasket can compress and block the bolt holes.

With all bolts started, tighten the oil pan bolts to specs. Use a crisscross pattern to assure even gasket compression. One sequence is given in Fig. 28-6.

Fig. 28-7 shows a partial oil pan gasket. It is handy when the oil pan is not removed during a repair. The old section of oil pan under the front cover can be cut off. Then, the new piece of oil pan gasket can be installed and sealed. This is only when servicing some part of the engine front end. A whole new oil pan gasket should be installed when the oil pan is removed for service.

Installing accessory units

With the engine upside down in the stand, install any components that are easily accessible. For example, you might bolt on the engine motor mounts, any sending units, oil filter bracket, etc. These parts are easy to get at with the engine on the stand after oil pan installation.

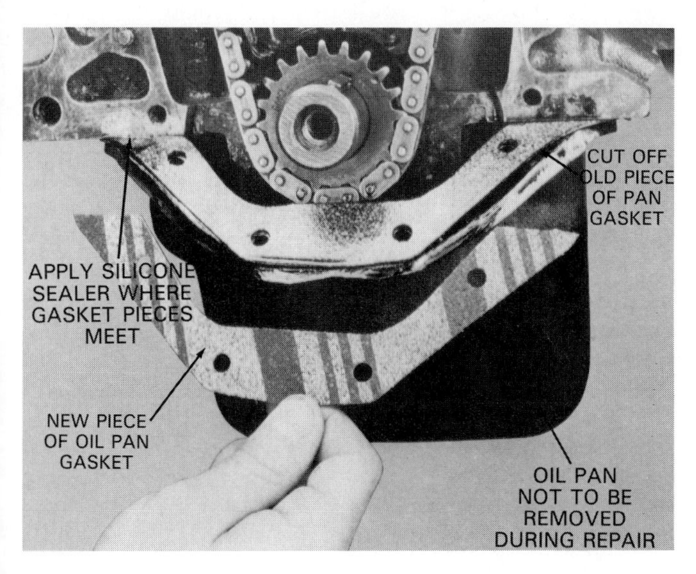

Fig. 28-7. This is a partial oil pan gasket. It is handy if you do not remove an oil pan during a repair, replacing timing chain for example. Section of old oil pan gasket can be cut off. Then new piece installed. Use silicone sealer where pieces mate. (Fel-Pro Gaskets)

When installing the water pump, check the fit of the new gasket. Then, use gasket adhesive to hold the gasket on the pump, Fig. 28-8. If a chemical gasket or RTV is used, form a continuous bead and circle all bolt holes and coolant passages. Fit the pump on the engine and start all bolts by hand. Check bolt lengths and place sealer on threads of any bolts extending into coolant and oil passages in the block.

Fig. 28-8. When installing water pump, check gasket fit on pump first. Then apply adhesive to hold gasket in place on pump. (Fel-Pro Gaskets)

INSTALLING CYLINDER HEAD

Installation methods are very critical when bolting the cylinder head to the block. Late model engines use thin wall, cast iron heads or aluminum heads that can leak, warp, or crack if installed improperly. Late model engines also operate at higher temperatures which can increase the strain on the seal between the head and block. Great care must be taken to assure proper installation techniques for cylinder heads.

Installing head gasket

Cylinder head gasket markings are normally provided to show the front or top of a head gasket. A head gasket can usually only be installed one way. If installed backwards, the gasket will fit but coolant or oil passages may be blocked. This mistake could cause engine overheating or oil starvation to the valve train, Fig. 28-9.

The head gasket may be marked "TOP," "FRONT," or it may have a LINE to show installation direction. Metal dowels are frequently provided to hold the head gasket on the block.

Most modern teflon-coated, permanent torque (do not need retorquing after engine operation) cylinder head gaskets should be installed clean and dry. Sealer is NOT recommended on modern, quality head gaskets. However, some steel shim or low quality fiber-asbestos head gaskets may require retorquing and sealer. Refer to gasket manufacturer's instructions when in doubt.

Do you need new head bolts?

Note! Some engine manufacturers recommend using NEW HEAD BOLTS when reassembling the engine. The used head bolts have been *torqued-to-yield* (tightened until stretched a specific amount) at the factory. New

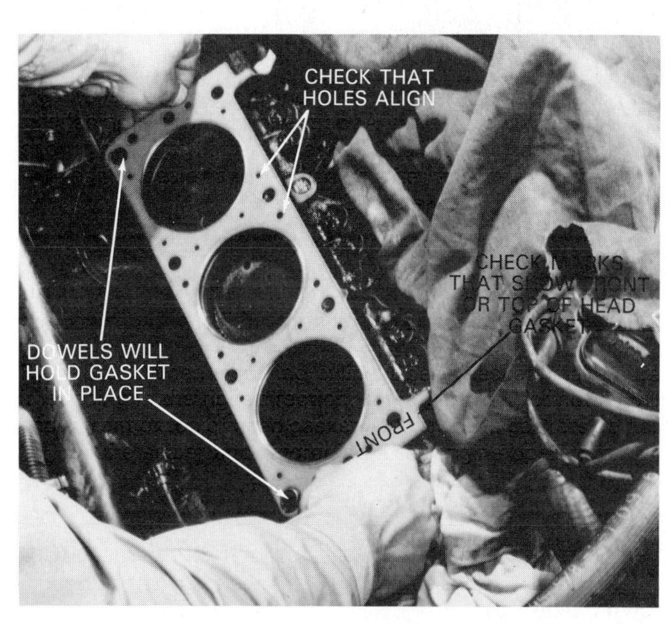

Fig. 28-9. When installing cylinder head gasket, make sure it is facing in the right direction. A word, arrow, or notch will show front or top of head gasket. Quality head gaskets do not require sealer and retorquing after engine operation. (Fel-Pro Gaskets)

bolts are needed to assure the correct gasket compression and bolt clamping action. Refer to the manual to find out if new head bolts are needed, Fig. 28-10.

Diesel cylinder head installation

With diesel engines, head gasket thickness and bore size are very critical. Head gaskets are provided in different thicknesses to allow for cylinder head milling or varying block deck heights. Gasket thickness may be denoted with a color code, series of notches, holes, or other marking system.

When a diesel engine is bored oversize, it also requires a special gasket. You must request an overbore gasket from the parts supplier. A standard bore gasket will usually stick out into the cylinder, causing problems.

Remember! When buying a DIESEL engine head gasket, make sure you have the right one. Gasket THICKNESS and BORE SIZE must be correct for the engine being repaired. Refer to the manual for details.

Fig. 28-10. Always refer to a shop manual for specifications and special procedures when assembling an engine. This engine manufacturer recommends new head bolts because old ones have been stretched by a torque-to-yield process at factory. (Fel-Pro Gaskets)

Oil bolt threads

Make sure all bolt holes in the block are clean. Blow them out if needed. Place a drop of oil on the head bolt threads, Fig. 28-11. This will assure an accurate torque when tightening the head bolts. Do NOT squirt oil in the bolt holes or a hydraulic lock could keep the bolts from torquing down properly. Do NOT oil torque-to-yield bolts.

Use dowels or guide studs

Gently place the cylinder head over the head gasket and block. You must do this without bumping and damaging the gasket.

Make sure the head is over its dowels. If dowels are not provided on the block deck, install stud bolts in the block. They will serve as guides to hold the head gasket and position the cylinder head. See Fig. 28-12. Remove the guide studs if used and start all of the bolts by hand.

Fig. 28-11. Internal and external threads should be perfectly clean. Place a drop of oil on threads to assure accurate head bolt torque. (Fel-Pro Gaskets)

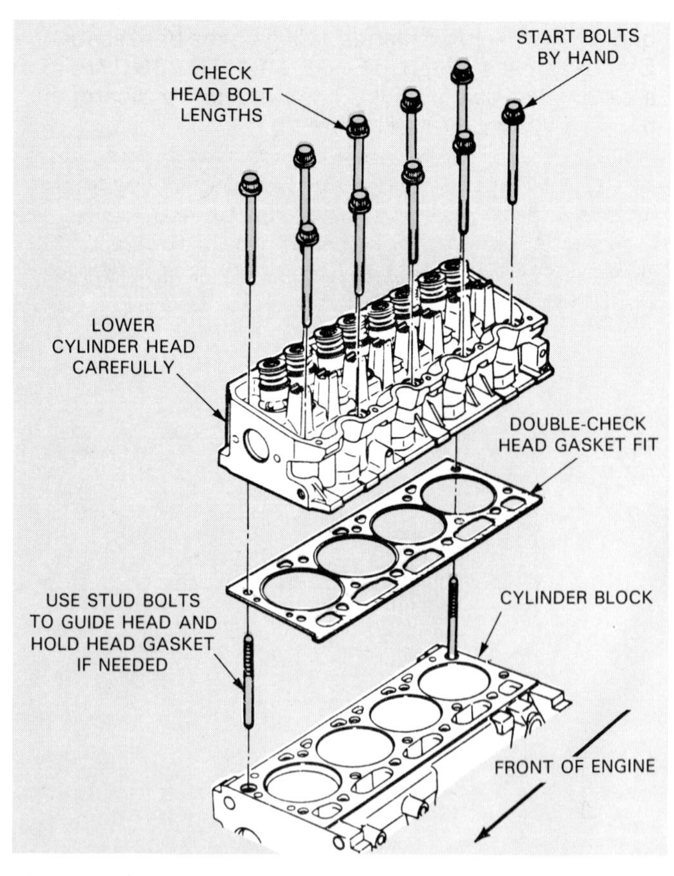

Fig. 28-12. Use studs to guide head over block if dowels are not provided in block. Install head bolts and check their lengths. (Ford)

Fig. 28-13. Head bolts that extend into coolant passages should have sealer on their threads. The manual will point out which bolts require sealer. (Fel-Pro Gaskets)

Sealing bolt threads

Place sealer on any head bolts that extend into water or oil jackets. The service manual will show which bolts must be sealed, Fig. 28-13.

Torquing head bolts

Torque the head bolts to specs using a crisscross sequence. The service manual will give the best sequence. One example is given in Fig. 28-14. Fig. 28-15 shows a technician using a very accurate dial indicator type torque wrench on head bolts.

Generally, tighten the bolts starting in the middle. Then, work your way outward, tightening each bolt a little at a time (1/2, 3/4, then full torque).

For example, if the head bolt torque is 100 ft.-lb. (130 N•m), tighten all of the bolts to 50 ft.,-lb. (68 N•m), then 75 ft.-lb. (101 N•m), then 100 ft.-lb. (130 N•m).

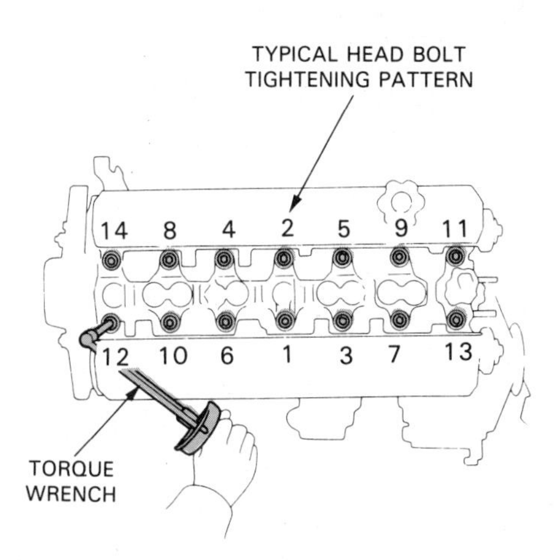

Fig. 28-14. Torque the head bolts to exact specs. Generally, start in the middle and work your way to the outer bolts. Gradually increase torque until up to specs. Then, go over all bolts a couple of times to assure precise tightness. (Toyota)

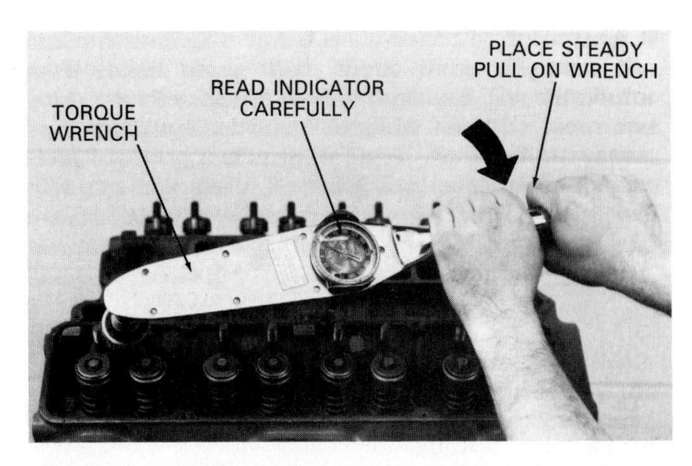

Fig. 28-15. Pull on torque wrench slowly while reading indicator or listening for click sound. A slow, steady pull on the wrench will produce the most accurate bolt torque. (Fel-Pro Gaskets)

Tighten each bolt to full torque several times to assure an accurate torque.

INSTALLING INTAKE MANIFOLD

The installation of an engine intake manifold is also critical. Some manifolds not only seal vacuum; they can also carry coolant and hot exhaust gases. If incorrectly installed, the manifold can allow a vacuum leak or a coolant leak. Partial engine teardown might be needed to correct the problem.

Checking intake manifold warpage

Use a straightedge and flat feeler gauge to check the intake manifold sealing surfaces for warpage. As shown in Fig. 28-16, lay the edge over the port openings. Try to slide different thickness blades under the straightedge. Measure at different points on the manifold.

The thickest blade that will easily slide under the

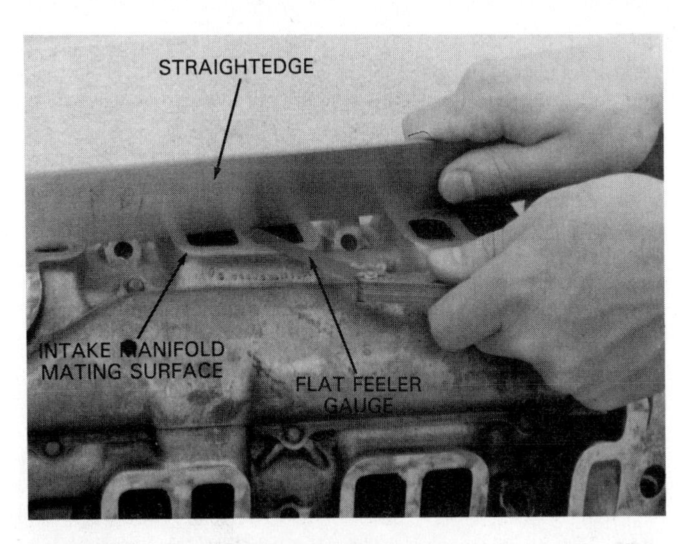

Fig. 28-16. Straightedge and feeler gauge will quickly check for intake manifold warpage. Check closely around the center ports and heat riser passage.

straightedge equals intake manifold warpage. If more than specs, the intake manifold surfaces must be machined or a new manifold purchased.

Checking intake manifold for damage

Inspect the intake manifold for cracks, burned areas, and other problems. Closely inspect the bottom of the manifold and near the heat passage from the cylinder head. Hot exhaust gases for the EGR or for manifold warming can burn and crack the manifold. Also, check the head mating surface near the heat passage. A ruptured gasket could have allowed the hot gases to burn the mating surface.

Installing intake manifold gasket

Place the new intake manifold gasket over the head to check its fit. Make sure all passages align.

Sometimes, ports or passages can vary slightly with the specific head. It is possible for the ports to be slightly larger. A coolant passage might be located differently. Any of these problems could cause a vacuum leak, coolant leakage, or even serious engine overheating damage. Double-check all openings in the gasket and head, as shown in Fig. 28-17.

If you are using a metal valley tray type intake manifold gasket, like the one in Fig. 28-18A, use RTV sealer on it. Silicone or RTV sealer should circle all ports and coolant passages. The thin metal gasket can leak easily and silicone provides added protection against a leak.

If you are using a multi-piece intake manifold gasket made of fiber, you do not have to circle every intake port. Simply use gasket adhesive to hold the gasket in place, Fig. 28-18B. Some technicians still prefer to circle coolant passages with silicone sealer.

With the head sections of the intake manifold gasket in place, install the end seals. Adhesive can be used to hold the seals in place if needed. If large barbs are provided on the seals and they fit properly down into the block, no adhesive is needed.

Place a small bead of silicone sealer where the end seals and main intake gaskets overlap. This will keep engine oil from weeping out during engine operation, Fig. 28-19.

Fig. 28-17. Check intake manifold gasket fit before applying sealer. Make sure ports and other passages align perfectly.

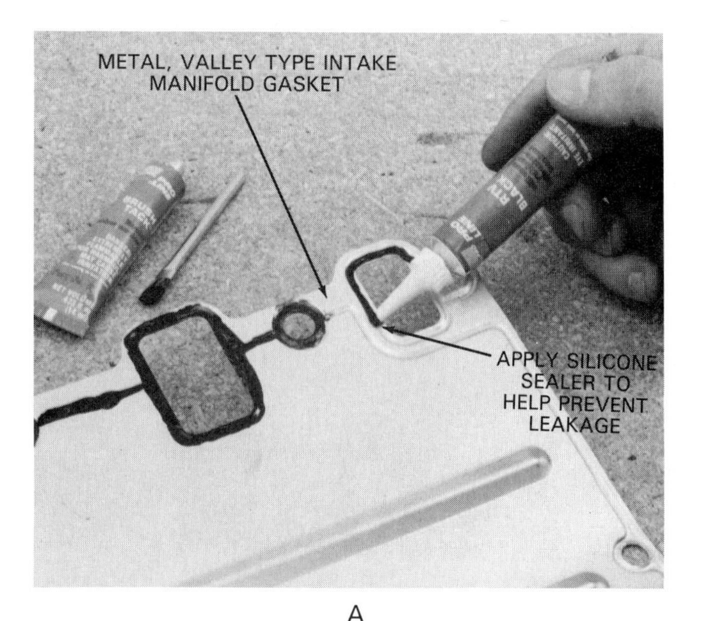

A

B

Fig. 28-18. A—With this valley tray type intake manifold gasket made of steel, silicone sealer will help prevent leakage. Circle all ports and coolant passages. B—This intake gasket already has a small bead of silicone rubber around openings and does not require extra sealer around ports. Adhesive can be helpful to hold gasket in place during assembly however. (Fel-Pro Gaskets)

Carefully position intake manifold

After installing the intake manifold gasket, you should be ready to position the intake manifold. If you wait too long, the sealer can completely cure and affect the manifold installation. Allow the sealer and adhesive to cure slightly but do NOT let them dry totally.

As pictured in Fig. 28-20, carefully lower the intake

Fig. 28-19. After positioning end seals with adhesive or rubber barbs, use silicone sealer where gaskets and seals overlap. (Fel-Pro Gaskets)

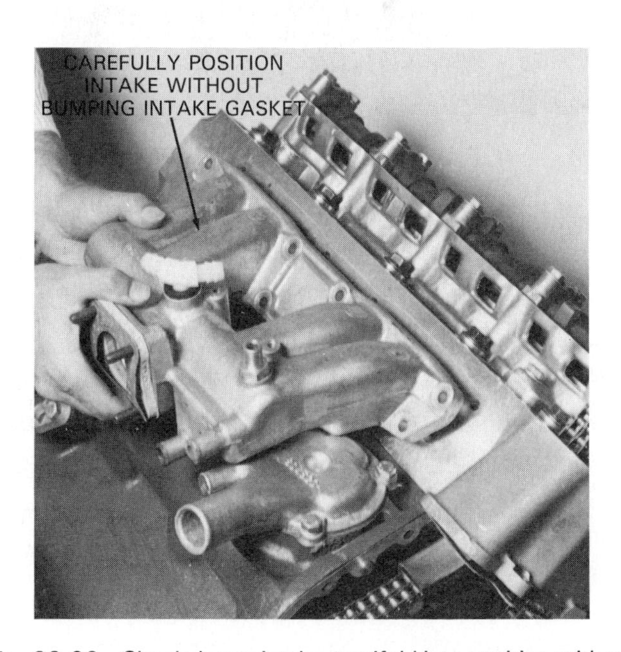

Fig. 28-20. Slowly lower intake manifold into position without shifting new gasket. (Saab)

manifold onto the engine. Be careful not to bump and shift the gasket or smear the sealer. In some applications, you might want to use guide studs so the manifold lowers straight down into place.

Torquing intake manifold

Start all intake manifold bolts by hand before tightening any bolts. If you tighten some of the bolts, you will not be able to shift and start the remaining bolts. With all bolts threaded in a few turns, run them down lightly.

Use a torque wrench to tighten the intake manifold bolts to specs. Use a crisscross pattern to draw the manifold down evenly. See Fig. 28-21.

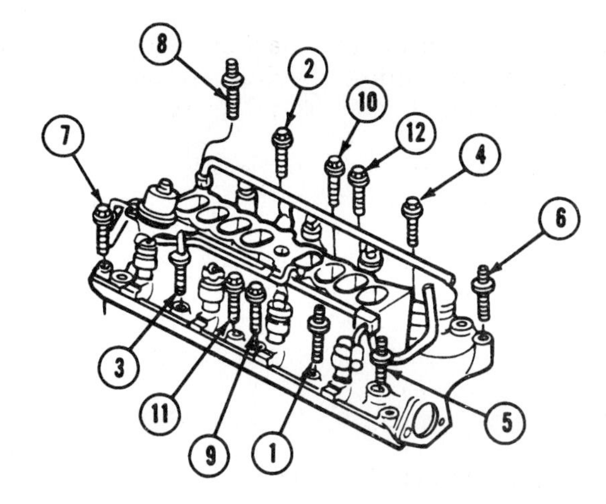

Fig. 28-21. When installing an intake manifold, follow the bolt torque sequence recommended by the manufacturer. Make sure that any stud bolts are in the proper location to hold other parts. (Ford)

Generally, torque the bolts to one-half of their specified torque. Then, tighten them to three-fourths torque, then, to their full torque value. Go over all bolts two or three times. As each bolt is tightened, the ones next to it will loosen.

Fig. 28-22 shows a multi-piece intake manifold. You would use the same basic procedure to install the base of the manifold. Torque it in a crisscross pattern to specifications. Then, you would do the same for the upper sections of the intake manifold.

With plastic engine parts like intake plenums, brass thread inserts are usually pressed into the plastic. This allows additional parts (sensors, vacuum fittings, etc.) to be bolted to the plastic part without thread failure.

INSTALLING EXHAUST MANIFOLDS

Exhaust manifolds route extremely hot gases into the car's exhaust system. As a result, it can suffer from cracks, warpage, rust, and other problems.

Always inspect an exhaust manifold closely before installation. Look for cracks, burned areas, erroded mating surfaces, severe rust damage, etc.

To check the manifold for warpage, use a straightedge and flat feeler gauge as described earlier. This is shown in Fig. 28-23.

Most used exhaust manifolds should have a gasket installed under them. Position the gasket on the head, Fig. 28-24. Then, slide the exhaust manifold into position. Studs are helpful to guide the manifold and gasket in place and hold them while starting the bolts, Fig. 28-25.

Start all bolts by hand and then turn them down until snug. Use a torque wrench to tighten the exhaust manifold bolts to specs, Fig. 28-26. Again, use a crisscross pattern, Fig. 28-27.

Check thermostat

When doing major engine service, you should check the engine thermostat. Make sure one is installed and that it operates properly. Also, check that the thermostat

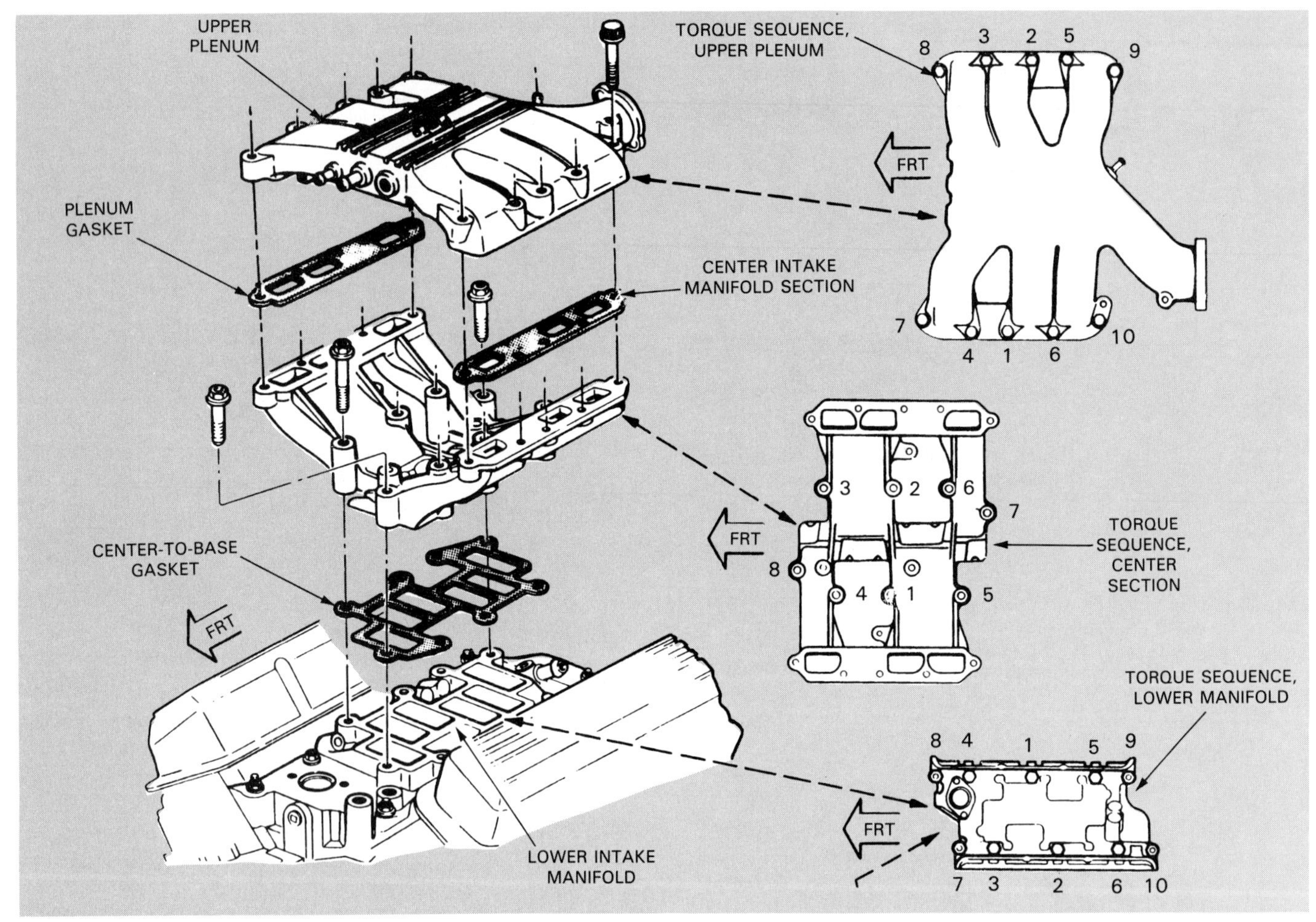

Fig. 28-22. Use the same general techniques just described to install multi-piece intake manifold. Torque all sections to specs. (Buick)

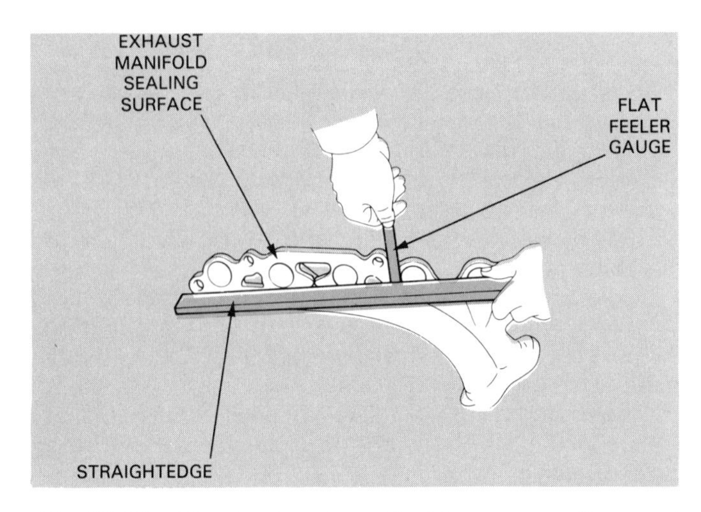

Fig. 28-23. Check exhaust manifold for warpage. If warped slightly, machine shop can usually machine it true. Replace the manifold if warped excessively. Also check it for cracks.

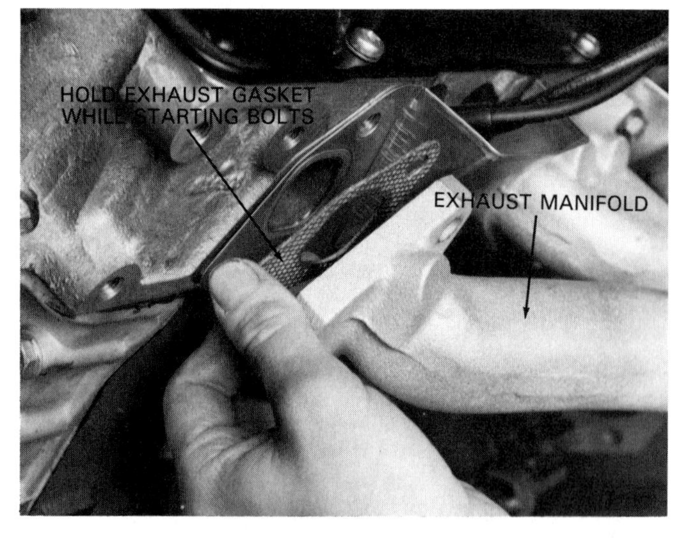

Fig. 28-24. It is a good practice to place a gasket under a used exhaust manifold to prevent leakage. (Saab)

heat range is correct. Someone could have changed it or left it out.

When installing the thermostat housing, make sure the thermostat is centered under the housing. Tighten each bolt a little at a time to specifications. Remember that the housing will warp or crack easily if tightened too much or improperly, Fig. 28-28.

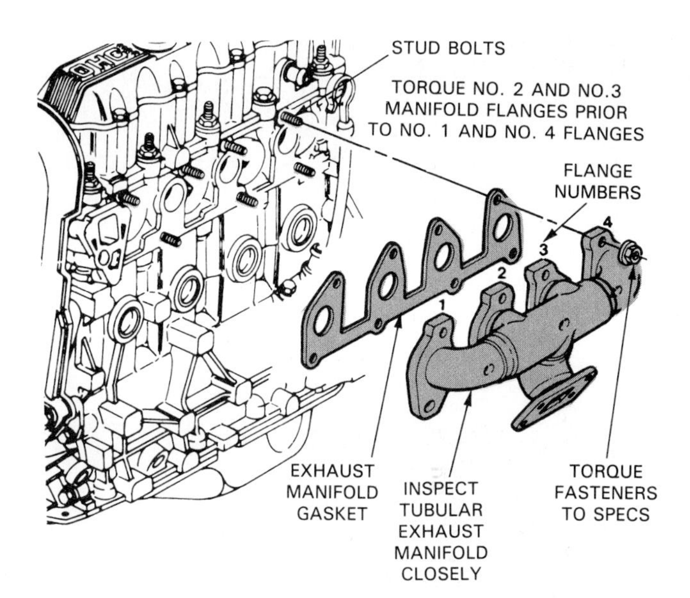

Fig. 28-25. Stud bolts will hold exhaust manifold gasket and manifold in place while starting bolts. Do not use sealer because high heat will harden sealer, possibly causing leakage. (Pontiac)

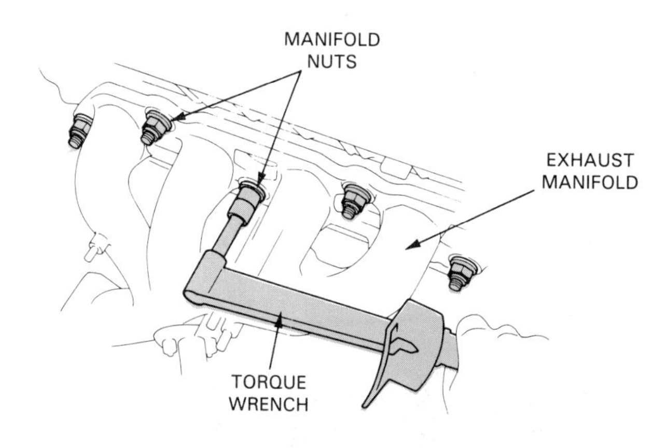

Fig. 28-26. Torque exhaust manifold to specs to avoid warpage, breakage, and leakage. (Toyota)

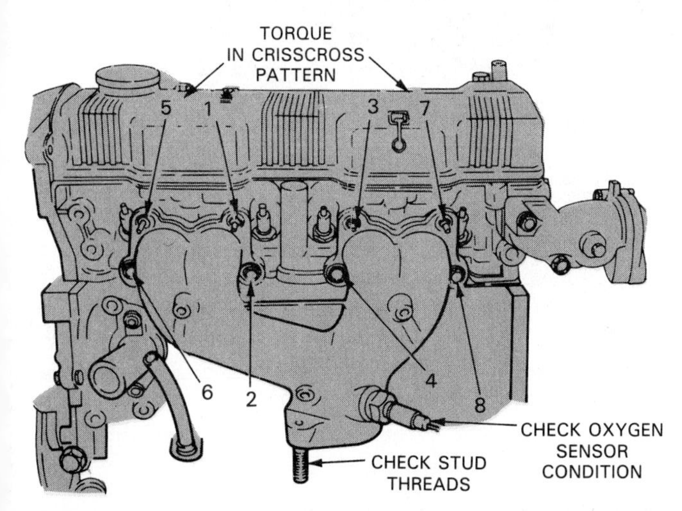

Fig. 28-27. Tighten exhaust manifold bolts in a crisscross pattern to pull it down evenly. Also check installation of oxygen sensor, stud bolts, etc. (General Motors Corp.)

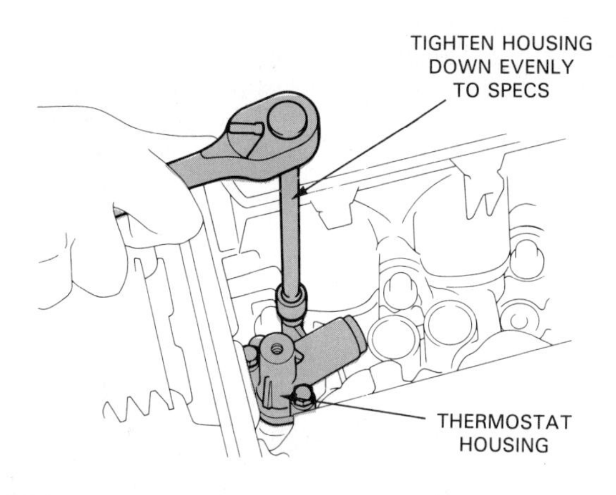

Fig. 28-28. Always check thermostat during major engine service. Make sure it has correct heat range and is in good operating condition. Tighten housing bolts evenly to prevent warpage or breakage.

OVERHEAD CAMSHAFT INSTALLATION

A camshaft for an overhead cam engine is installed in a similar way as was described for a overhead valve (cam-in-block) engine. This was explained in Chapter 25. However, there are a few slightly different procedures. Look at Fig. 28-29.

After making sure the camshaft bearings are in good condition or replaced, lubricate the camshaft journals and cam lobes. White grease is commonly recommended on the lobes to reduce friction and wear during initial start-up.

Slide the camshaft into the head without bumping the cam bearings. Turn the cam as it slides into the head. Look at Fig. 28-30.

Installing cylinder head caps

Some OHC heads use cylinder head caps or camshaft bearing caps. They hold the upper cam bearing shells and bolt over the camshaft journals, Fig. 28-29.

First, install the lower cam bearing shells in the head. Make sure any oil holes align. The back of the bearings should be clean and dry. Wipe engine oil on the face of the bearings.

Lower the camshaft into the bearings. Install the upper half of the cam bearings into the caps. Oil the journals and position the caps on the head.

Just like block main caps, cam bearing caps are numbered and MUST be installed in their original locations. If you mix up caps or install them backwards, the cam can lock in the cylinder head. The bearings and camshaft can be severely damaged upon engine starting.

Torque the camshaft or head caps to specs. Use the general sequence shown in Fig. 28-31. Make sure you can still turn the camshaft after torquing the camshaft bearing caps. If not, something is wrong. Check bearing size, camshaft straightness, cap locations, etc.

ASSEMBLING VALVE TRAIN

Depending upon engine design, procedures for installing valve train components will vary. Refer to a service

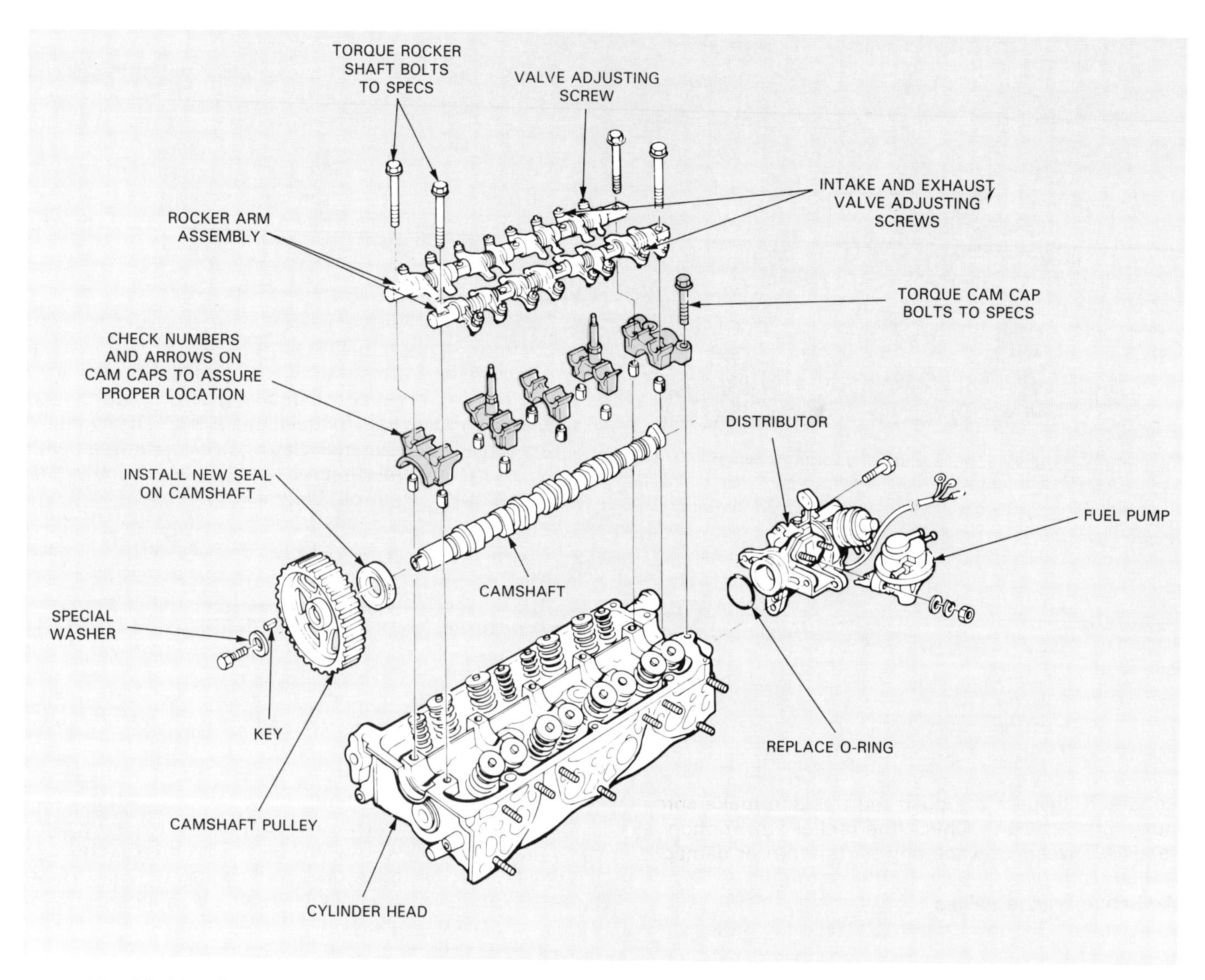

Fig. 28-29. This engine uses caps to secure camshaft bearings and cam into OHC cylinder head. Caps must be installed in right sequence, just like main bearing caps in a cylinder block. (Honda)

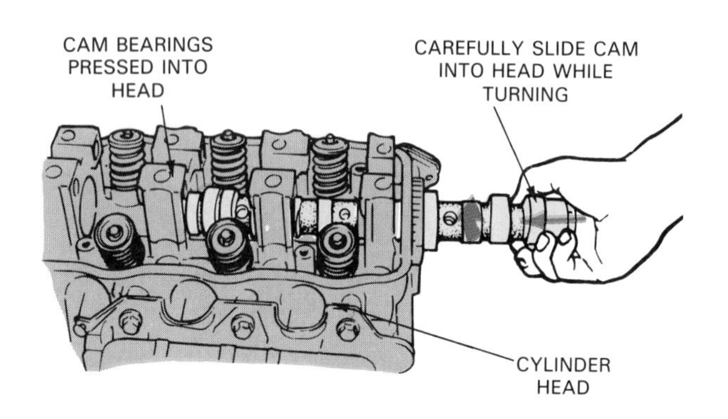

Fig. 28-30. This head does not use removable caps and you must slide camshaft into front of the head. (Oldsmobile)

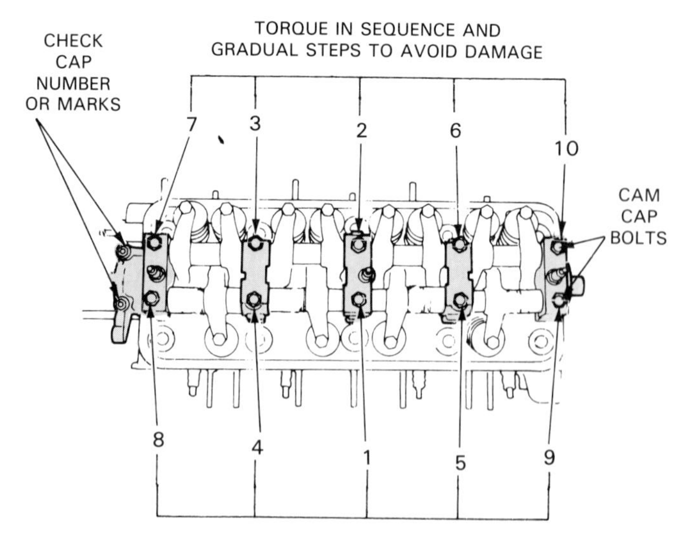

Fig. 28-31. After double-checking cam bearing cap numbers or arrows, torque them to specs. Make sure you can still turn cam after tightening. Note sequence. (Honda)

manual if in doubt.

If you are working on an overhead valve engine, install the lifters in their bores. If the lifters are used, they

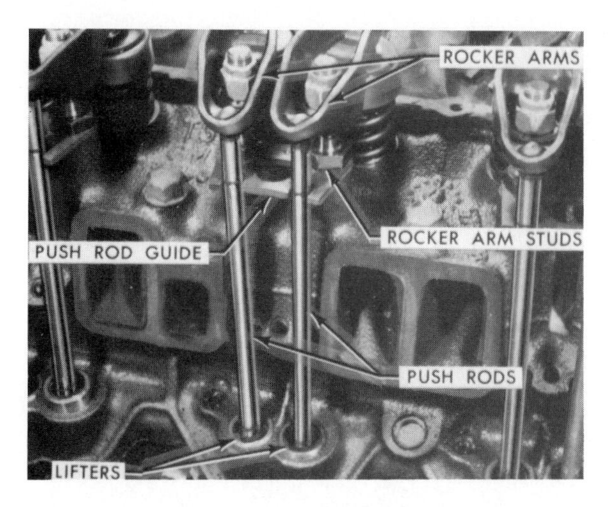

Fig. 28-32. Follow service manual directions when installing valve train. Designs and methods vary. In any case, try to install parts in their original locations. (Chevrolet)

should be installed on the same cam lobes to prevent excessive wear. Then, you can install the push rods and rocker arms. See Fig. 28-32.

Sometimes the intake and exhaust rocker arms are different. Usually, an "E" for exhaust and an "I" for intake will be stamped on the rocker arms if they are not the same. Refer to Fig. 28-33.

Again, make sure all parts are unworn and are in usable condition. Inspect the push rod tips and make sure no push rods are bent. Check the rocker arm friction surfaces for wear. Replace any parts worn or damaged.

Adjusting engine valves

After installation of the valve train components, you may have to adjust the valves. This is done by turning the rocker arm nuts, installing OHC follower shims, or

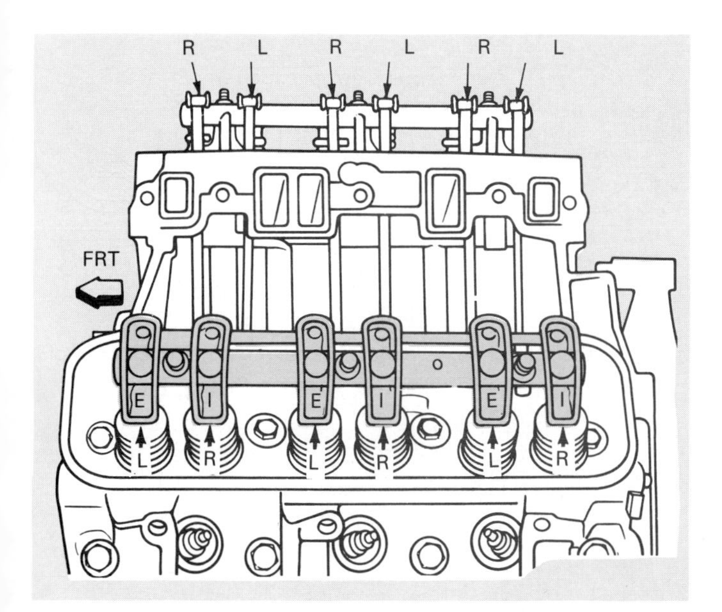

Fig. 28-33. Some rocker arms are marked intake and exhaust and they must be installed on correct valve. (Pontiac)

using oversize push rod lengths.

Valve adjustment is critical to engine performance and service life. If the valves are TOO LOOSE, valve train noise, sometimes termed lifter clatter, will result. A light tapping or clicking sound will be produced from under the valve cover. If the valves are adjusted TOO TIGHT, burned valves can result. The valves may not close tightly and hot exhaust gases can erode away metal from the valve face and valve seat.

Nonadjustable rocker arms

Nonadjustable rocker arms are used on many push rod engines with hydraulic (self-adjusting) lifters. The hydraulic lifters automatically compensate for changes in valve train clearance, maintaining *zero valve lash* (no clearance in valve train for quiet operation). The hydraulic lifters can adjust valve train clearance as parts wear, with changes in temperature (part contraction or expansion), and with changes in oil thickness.

If adjustment is needed because of valve grinding, head milling, or other conditions, different PUSH ROD LENGTHS can sometimes be purchased for use with nonadjustable rocker arms. Refer to the service manual or a part supplier for details. Normally, torquing the rocker fastener (nut or bolt) to specs adjusts the lifters or rocker arms properly.

Warning! Overtightening a nonadjustable rocker arm can cause engine missing and rocker stud damage. Torque them to specs!

Adjusting hydraulic lifters

Hydraulic lifter adjustment is done to center the lifter plunger in its bore. This will let the lifter automatically adjust itself to take up or allow more valve train clearance. Some manuals recommend adjustment with the engine off. However, many technicians adjust hydraulic lifters with the engine running.

To adjust hydraulic lifters with the ENGINE OFF, turn the crankshaft until the lifter is on the camshaft base circle (not on lobe). The valve must be fully closed.

Loosen the adjusting nut until you can wiggle the push rod up and down. See Fig. 28-34. Then, slowly tighten the rocker until all play is out of the valve train (cannot wiggle push rod).

To center the lifter plunger, tighten the adjusting nut about ONE MORE TURN. Refer to a manual for details because this can vary with engine design. Repeat the adjusting procedure on the other rockers.

To adjust hydraulic lifters with the ENGINE RUNNING, install special oil shrouds or another device for catching oil spray off the rockers. Start and warm the engine to operating temperture.

Tighten all of the rockers until they are quiet. One at a time, loosen a rocker until it CLATTERS. Then, tighten the rocker slowly until it QUIETS DOWN. This will be zero valve lash.

To set the lifter plunger halfway down in its bore, tighten the rocker about one-half to one turn MORE. Tighten the rocker slowly to give the lifter time to leak down and prevent engine missing or stalling. Repeat the adjustment on the other rockers.

Other adjustment methods may be recommended. Check the manual for more detailed information.

Fig. 28-34. Valves must be adjusted to center hydraulic lifter plunger in its bore. Here technician is slowly tightening rocker nut until all play is out of valve train. Then, another one-half to one full turn will center plunger. Cam lobe must be rotated away from lifter when adjusting. (Oldsmobile)

Adjusting mechanical lifters

Adjusting mechanical lifters, also called *solid lifters,* is done to assure proper valve train clearance. Since hydraulic lifters are NOT used to maintain zero lash, you must adjust mechanical lifters periodically. Check the car's service manual for adjustment intervals and clearance specs.

Typical valve clearance is approximately .014 in. (0.35 mm) for the intake valves and .016 in. (0.40 mm) for the exhaust valves.

Mechanical lifters make a clattering or pecking sound during engine operation. Unlike hydraulic lifters, this is normal. Mechanical lifters are used on heavy-duty engines (taxi cabs, pickup trucks, diesel engines) and high performance engines.

To adjust mechanical lifters, position the lifter on its base circle (valve fully closed). This can easily be done by cranking the engine until the piston is at TDC on its COMPRESSION STROKE (you can feel air blow out of spark plug hole). At TDC on the compression stroke, both valves in that cylinder can be adjusted.

Slide the correct size flat feeler gauge between the rocker arm and the valve stem, as shown in Fig. 28-35. When properly adjusted, the gauge should slide between the valve and rocker with a slight drag, Fig. 28-36.

If needed, adjust the rocker to obtain valve clearance. You will normally have to loosen a locknut and turn an adjusting screw. Tighten the locknut and recheck adjustment. Repeat the procedure on the other valves.

Note! Some engines must have valves (mechanical lifters) adjusted with the ENGINE COLD. Others require a HOT ENGINE (at operating temperature). Check the manual because a change in temperature will cause part

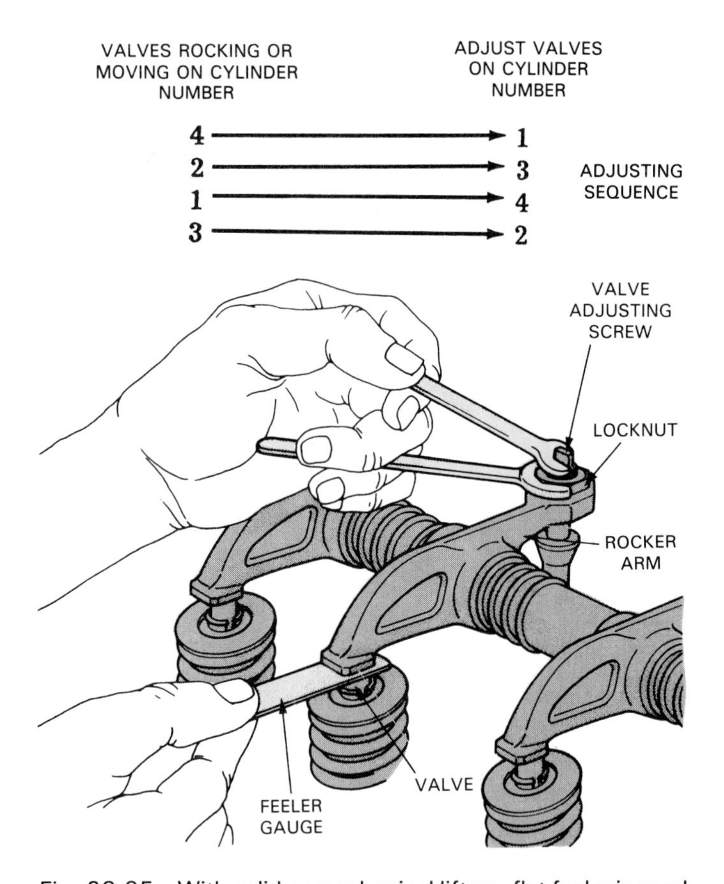

Fig. 28-35. With solid or mechanical lifters, flat feeler is used to measure clearance between rocker and valve. Manual will give correct feeler gauge thickness. When adjusted properly, blade should drag slightly when pulled in and out. Screw provides adjustment. Tighten locknut and then recheck lash. Note valve adjusting sequence given for this particular engine. (Chrysler)

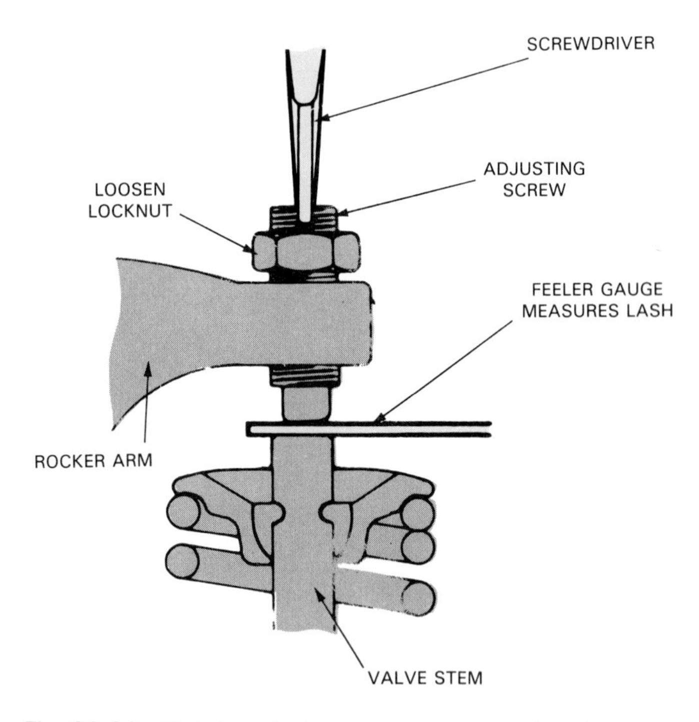

Fig. 28-36. Note how feeler gauge measures valve clearance or lash. (Buick)

expansion or contraction. This, in turn, causes a change in valve train clearance.

OHC engine valve adjustment

There are many methods of adjusting the valves on OHC engines. Some adjust like mechanical lifters in a push rod engine. The rocker arm adjuster is turned until the correct size feeler gauge fits between the rocker or cam lobe and valve stem. Others use hydraulic cam followers and adjustment may not be necessary.

Valve adjusting shims may also be used on modern OHC engines to allow the adjustment of cam-to-valve clearance, Fig. 28-37. Measure valve clearance with a feeler gauge. Then, if needed, remove and change shim thickness as needed.

For example, to calculate new shim thickness, compare measured clearance to specifications. If the engine requires .015 in. (0.38 mm) lash and actual lash is .020 in. (0.51 mm) you need a shim that is .005 in. (0.13 mm) thicker than the old one (.020 minus .015 equals .005). The new shim would provide spec lash.

Other OHC engines have an Allen adjusting screw in the cam followers. Turning the screw changes valve clearance. Refer to the shop manual for detailed directions. Exact procedures for valve adjustment vary with engine design.

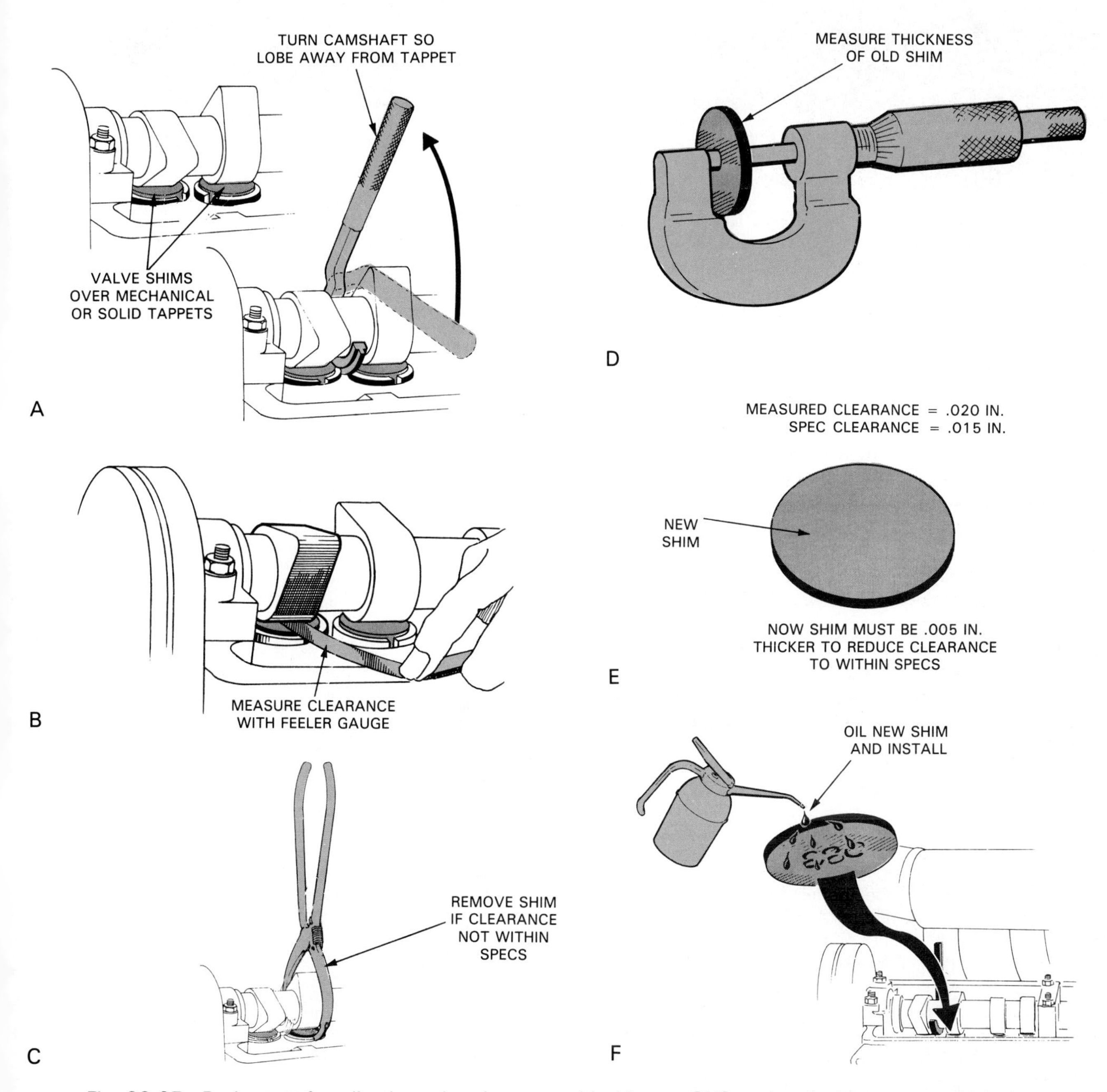

Fig. 28-37. Basic steps for adjusting valve clearance with shims on OHC engine. A—Turn cam until lobe is away from follower or tappet. B—Measure clearance. C—If not within specs, remove old adjusting shim. D—Measure old shim thickness with micrometer. Calculate new shim thickness. E—Obtain new shim that will result in correct clearance. F—Oil and install new shim. (Volvo)

INSTALLING VALVE (ROCKER) COVERS

If NOT installed properly, a valve cover, also termed rocker cover, can leak oil very easily. It is important for you to realize how easily a valve cover will leak. This may help prevent an incorrect installation technique and "comeback" (customer returns to shop after failure).

Some valve covers use a cork or synthetic rubber gasket. A few late model valve covers are factory sealed using silicone sealer. When reinstalling the valve cover, either sealer or gaskets may be used, depending on cover design and accessibility of the cover.

Checking valve cover sealing surface

Before installing a valve cover, make sure the cover flange is NOT bent. Lay the cover on a flat workbench. View between the gasket surface and the workbench to detect gaps (dents, bends, or warpage). You could also use a straightedge, Fig. 28-38.

A thin, sheet metal cover can be straightened with taps from a small, ball peen hammer. A cast aluminum cover can sometimes be sent to a machine shop for resurfacing. Replace a warped plastic valve cover.

Installing valve cover gasket

To install a valve cover gasket, place a very light coating of approved adhesive around the edge of the valve cover. This is mainly needed to hold the gasket in place during assembly, Fig. 28-39. Fit the gasket on the cover and align the bolt holes.

After letting the adhesive cure slightly, place the cover and gasket on the cylinder head and hand start ALL of the bolts. Tighten the valve cover bolts to specs using a crisscross pattern. See Fig. 28-40.

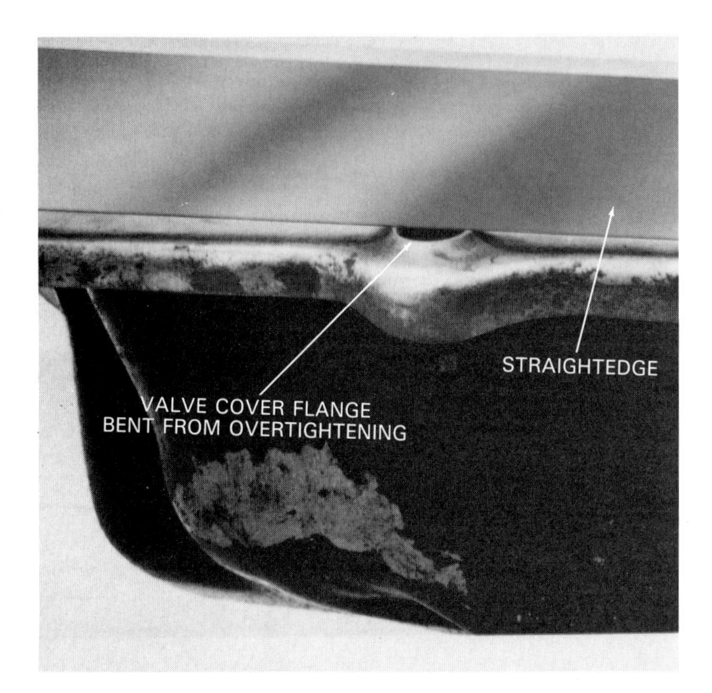

Fig. 28-38. Check valve cover closely for bent or warped flange. Here straightedge is being used. Note how bolt hole is dented from overtightening. This could cause oil leakage. Use hammer to straighten flange. (Fel-Pro Gaskets)

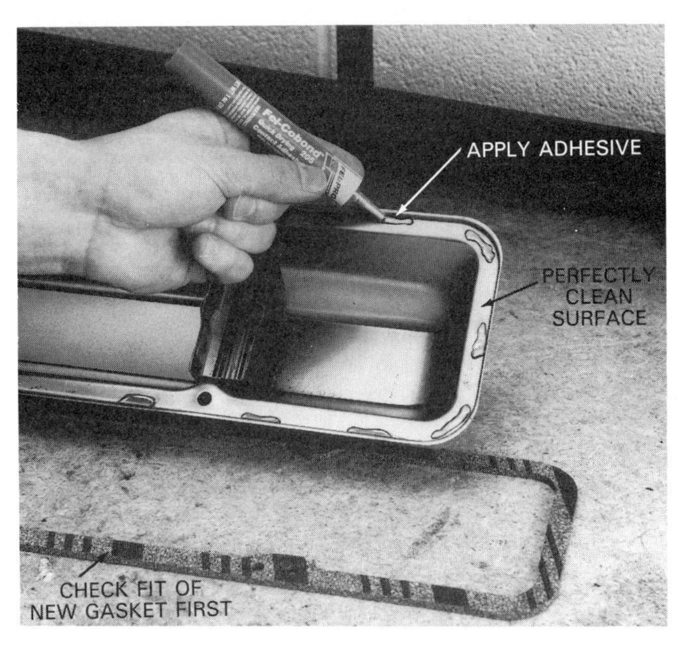

Fig. 28-39. Adhere new gasket to valve cover using approved adhesive. Make sure bolt holes align while allowing adhesive to become tacky. (Fel-Pro Gaskets)

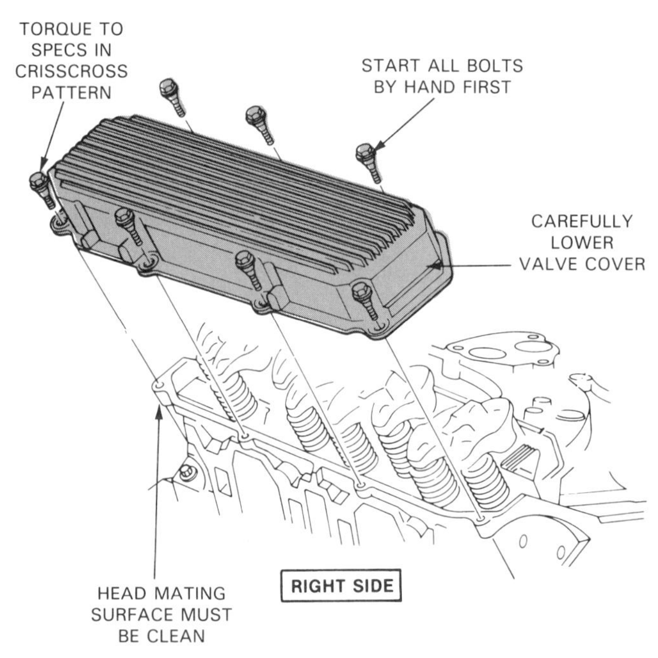

Fig. 28-40. Carefully lower valve cover onto engine without bumping and shifting gasket. Start all bolts by hand before tightening any bolts. (Cadillac)

NOTE! A very common mistake is to OVERTIGHTEN valve cover bolts. Overtightening can smash and split the gasket. It can also bend the valve cover, causing oil leakage. Torque the bolts to specs, generally just enough to lightly compress the gasket.

Installing valve cover with silicone sealer

To use silicone or RTV sealer on the valve cover, double-check that the cover and cylinder head gasket surfaces are PERFECTLY CLEAN. Sealer will NOT bond and seal on a dirty, oily surface.

Apply a continuous bead of sealer all the way around the valve cover sealing surface. Typically, the bead should be about 3/16 in. (1.6 mm) wide. Immediately, lower the valve cover onto the head carefully. You must not smear and break the bead of silicone sealer. Torque the fasteners to specs.

ENGINE INSTALLATION

Installing an engine in a car is just about the opposite as removal (covered in Chapter 24). However, there are a few common steps you should know.

Installing flywheel

Before installing a flywheel, check its ring gear teeth. Make sure none of the teeth are worn or broken off. Ring gear teeth damage will cause problems with starting motor engagement and operation. If needed, replace the ring gear or flywheel.

Unlike automatic transmission flywheels, most manual transmission flywheels have a removable ring gear. Use a chisel to split and remove the damaged ring gear, Fig. 28-41A. Then, heat the new ring with a torch. Heat the gear evenly so that is will expand evenly.

Use a hammer to tap the heat-expanded ring gear over the flywheel lip. Handle the ring gear carefully and do not strike the gear teeth with the hammer. Drive on the inner diameter of the ring gear until it is seated on the flywheel. See Fig. 28-41B.

Install the flywheel on the crankshaft. Do not forget to use any spacer plate that fits between the engine and transmission or transaxle.

The flywheel and crank flange normally have offset bolt holes so the flywheel will only go on in one position. Turn the flywheel on the crank until all bolt holes align. Install and torque the flywheel bolts to specs.

Note! Chapter 24 discussed how to check flywheel runout, etc.

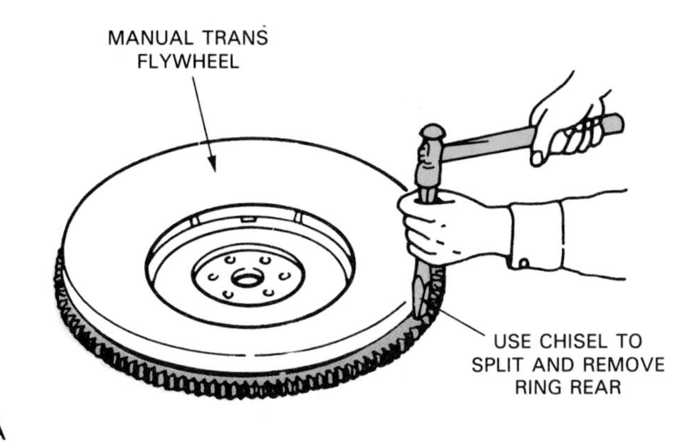

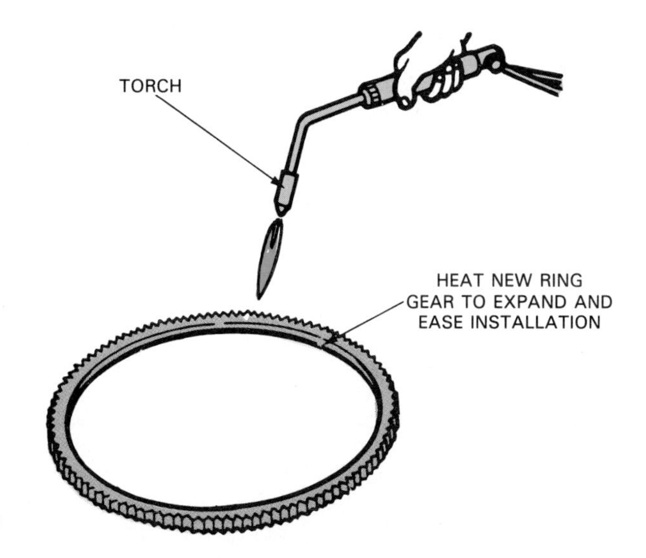

Fig. 28-41. Inspect flywheel ring gear teeth closely. One broken or badly worn tooth could require you to remove the transmission or transaxle. A—Use chisel to break and remove old ring gear if needed. B—Heat new ring gear to expand it larger. Drive hot ring gear over manual flywheel until fully seated. (Buick)

Installing clutch

With a manual transmission or transaxle, you will have to check and install the clutch before engine installation. Most technicians prefer to use a new pilot bearing, clutch disc, and throw-out bearing because these parts wear and fail quickly. Some even like to replace the pressure plate. In any case, check these parts closely before reuse.

Fig. 28-42 shows how to measure clutch disc wear. If the disc is worn more (thinner) than specs, it should be replaced. Also, purchase a new disc if it is contaminated with oil or shows signs of slippage and overheating. Be sure to wear respiratory protection when working with clutches.

Check the pressure plate friction surface for heat checking, signs of overheating, worn release arms, weakened springs, etc. The throw-out bearing must turn freely with no signs of bearing roughness.

Again, it is a good policy to replace the pilot bearing, clutch disc, and throw-out bearing anytime the clutch is disassembled. These parts wear and fail in relatively few miles.

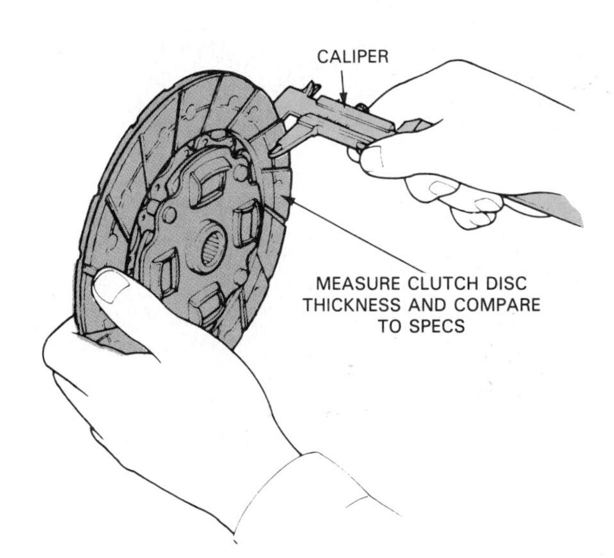

Fig. 28-42. Inspect clutch disc closely and measure thickness before reuse. Most technicians install new disc during major repairs. If worn too thin, it must be replaced. (Honda)

To install the clutch, fit the disc and pressure plate onto the flywheel. Insert a *pilot tool* (alignment shaft) or old transmission input shaft through the disc and into the pilot bearing. This will hold the clutch and center it while torquing the pressure plate bolts. This is illustrated in Fig. 28-43.

Warning! Make sure you use the factory bolts to install a pressure plate. They are high tensile strength and usually are not threaded near the head. If you use other bolts, they can break and lead to serious part damage!

Fit the throw-out bearing onto its release arm. Make sure it is secure because it must stay in place during engine or transmission-transaxle installation, Fig. 28-44.

Bolt the bellhousing to the rear of the engine block. Torque its fasteners to specs. Attach any other components to the bellhousing, clutch bracket for example.

If the transmission or transaxle is out of the car, you can usually install the engine and transmission or transaxle as a single unit. Refer to the manual for details.

Use an engine crane to hold the engine on the shop floor. The crane will keep the engine from falling over. Lift the transmission or transaxle up and slide its input shaft through the clutch. You might have to wiggle the ''trans'' while pushing in.

Make sure to hold the transmission or transaxle perfectly straight with the centerline of the crankshaft or it will not slide fully into the pilot bearing. Once butted up against the clutch housing, install the transmission or transaxle bolts. See Fig. 28-45.

WARNING! Do NOT use the transmission or transaxle bolts to pull the unit up to the bellhousing. The trans should slide freely up against the bellhousing. If you use the bolts to pull them together, the input shaft can smash into the side of the pilot bearing and cause part damage.

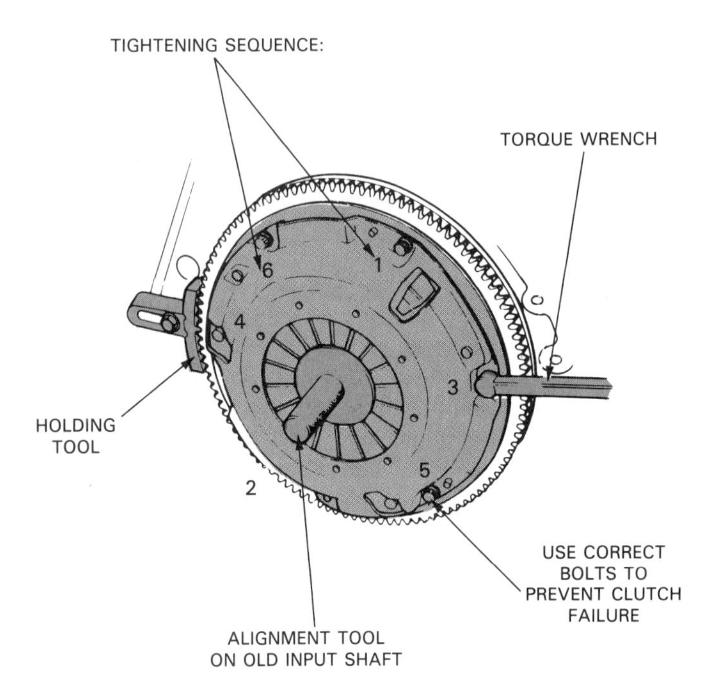

Fig. 28-43. Use pilot shaft to align clutch assembly on flywheel. Torque bolts to specs and use factory bolts. Install new pilot bushing if needed.

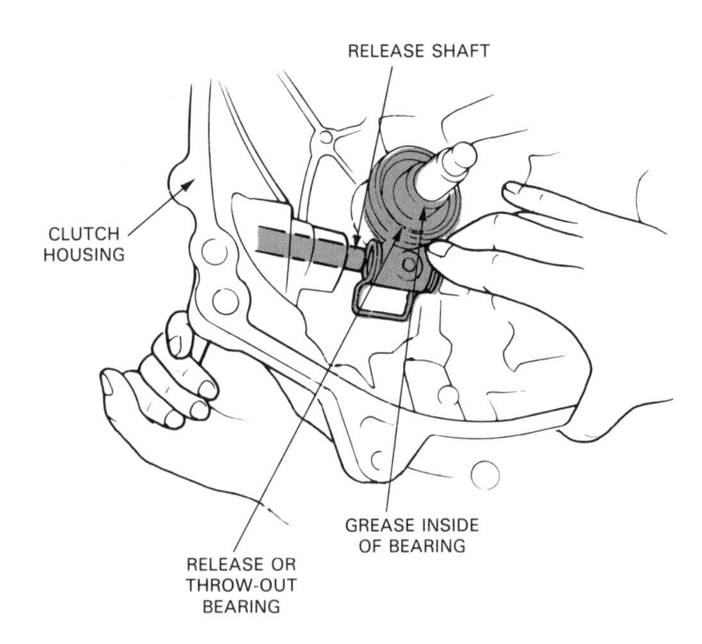

Fig. 28-44. Position new throw-out bearing carefully. It must stay in place when bolting transmission or transaxle to engine. Again, most technicians install new throw-out bearing since it is common wear point. (Honda)

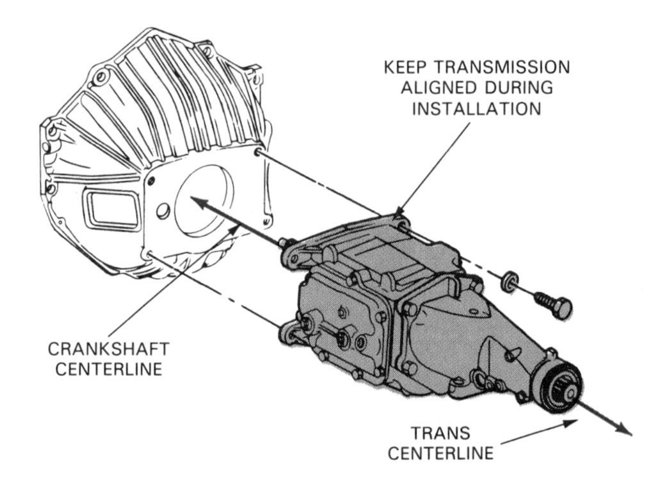

Fig. 28-45. When bolting transmission or transaxle to engine, keep their centerlines in perfect alignment. (GMC)

Installing torque converter

If you are attaching an automatic transmission or transaxle to the engine, make sure the torque converter is completely installed over its shaft. Wiggle the converter up and down while pushing in. View down the side of the transmission or transaxle to make sure the converter is not sticking out too far.

If you bolt the transmission or transaxle to the engine and the converter is NOT completely over its shaft, severe transmission or transaxle damage can occur. The front pump, housing, or converter could be ruined—a time consuming and costly mistake.

If the transmission or transaxle are to be installed with the engine, bolt them together when on the shop floor. As mentioned for a manual transmission or transaxle, make sure the input shaft is in perfect alignment with the engine crankshaft.

If the torque converter has a drain plug, it should be rotated down. The hole for the drain plug in the flywheel must also be rotated down. When the two are bolted up, the converter drain plug will stick through this enlarged hole in the flywheel. This will also help you align everything as you bolt the automatic transmission or transaxle to the engine. Torque all fasteners to specs.

Fig. 28-46. Do not work on engine supported on crane. Only use crane to install or remove engine. Keep feet and hands out from under engine. (OTC Tools)

INSTALLING ENGINE IN THE CAR

Installing an engine in a car is about the opposite as removing the engine. Engine removal was covered in Chapter 24.

With modern computer-controlled vehicles, never re-route wires. Keep all wires in their original locations and reinstall them in their original positions. If you re-route spark plug wires or some high current carrying wires, current can be induced into low current sensor or computer feedback wires, which could upset computer system operation.

A few rules to remember when installing an engine include:

1. Keep your hands and feet out from under the engine, Fig. 28-46.
2. Only raise the engine on the hoist while it is being placed in the vehicle. Never work on an engine raised in the air on a hoist or crane.
3. With a manual transmission, check the condition of the pilot bearing (bushing in end of crankshaft). Replace it if worn.
4. Position the lifting chain or fixture on the engine so that the engine is raised LEVEL. If NOT level, it will be difficult to slide the engine against the transmission. See Fig. 28-47.
5. Slowly lower the engine into the vehicle while watching for clearance all around the engine compartment. Position the engine so that its crankshaft centerline aligns with the transmission input shaft centerline, Fig. 28-48.
6. Push the engine back and align the engine dowel pins with the holes in the transmission. Use a large bar to shift the engine.
7. As soon as the dowel pins slide FULLY into their holes, install an engine-to-transmission bolt, but do NOT tighten it. Start another bolt on the other side of the engine.
8. Check that the torque converter is properly lined up with the holes in the flywheel. Turn

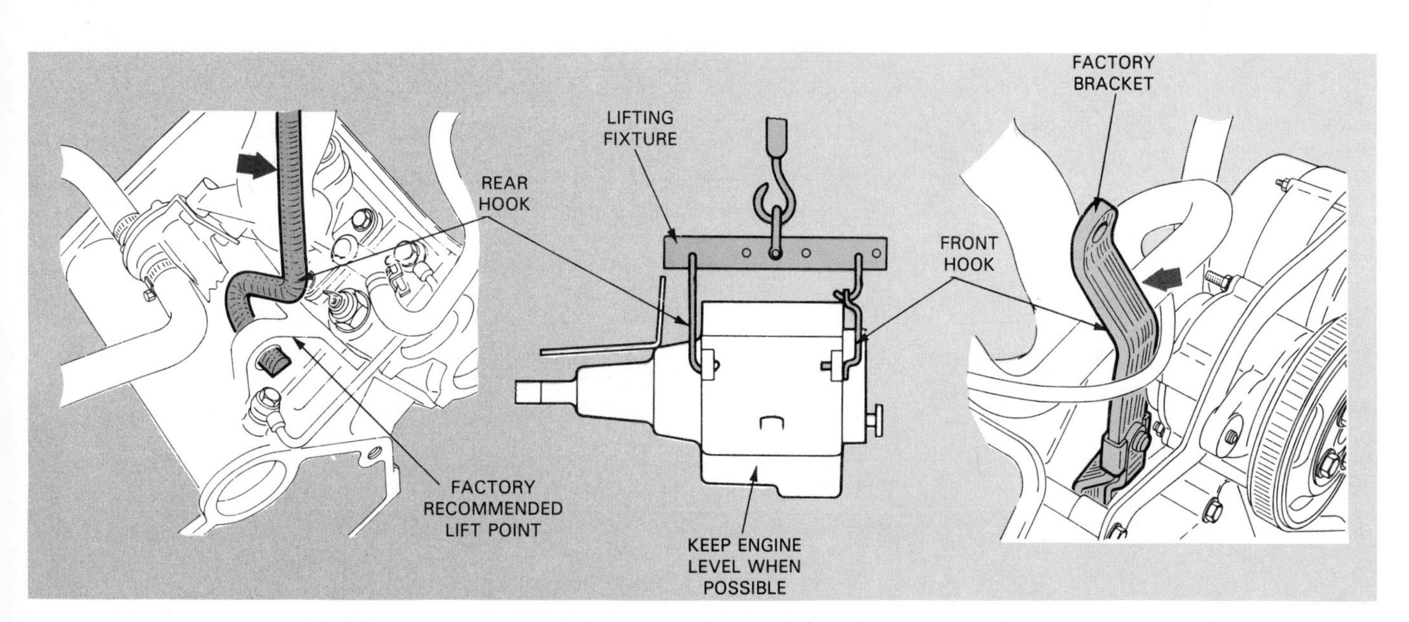

Fig. 28-47. Attach lifting chain or bracket so that engine is lifted squarely. This will help align everything during installation.

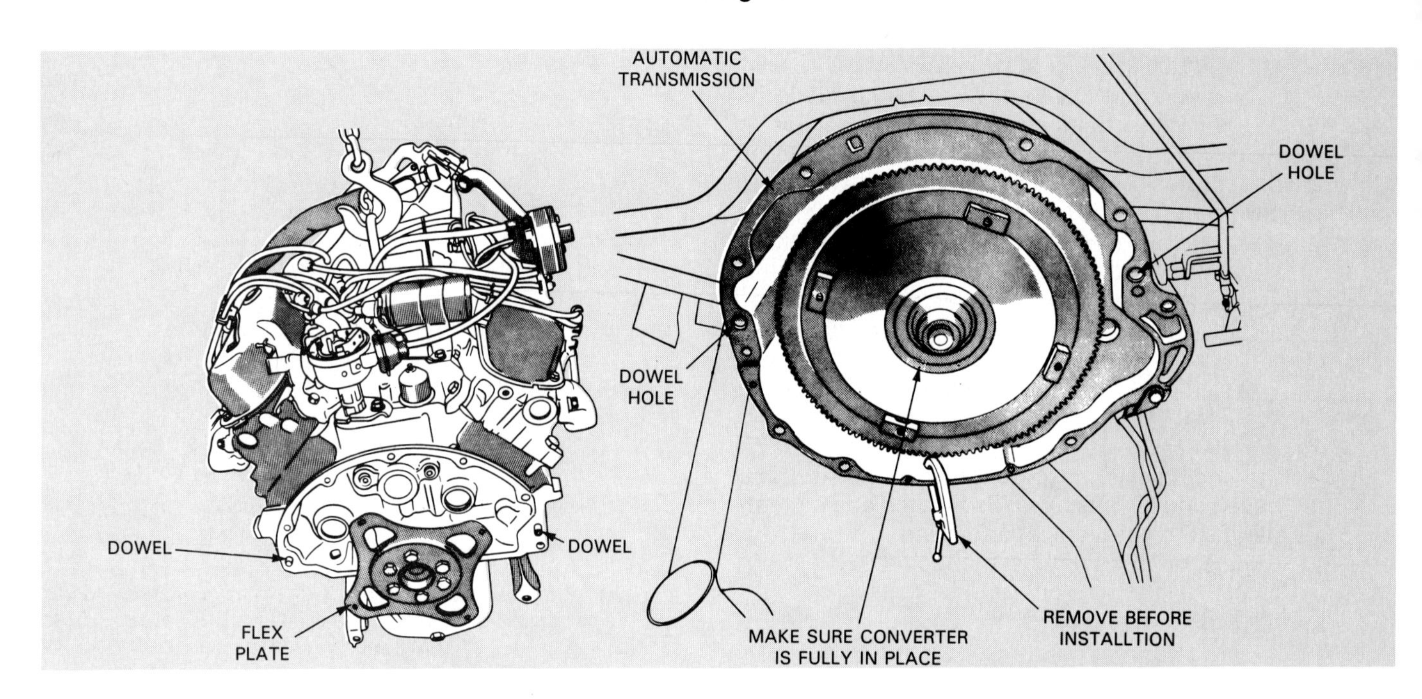

Fig. 28-48. With automatic transmission or tranaxle, make sure torque converter is fully installed over its shaft. Also have everything aligned and out of the way before lowering engine into engine compartment. (Chrysler)

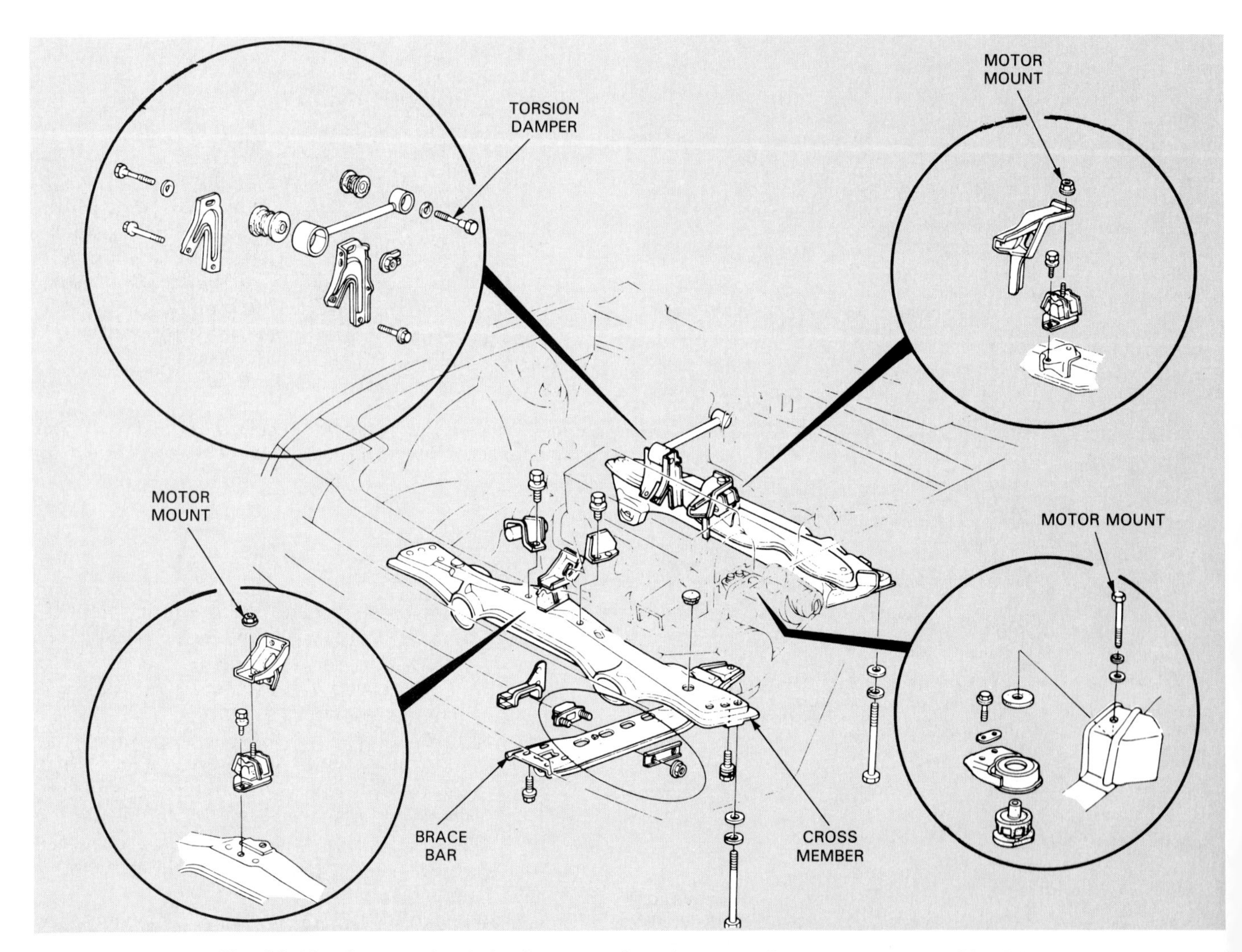

Fig. 28-49. Once engine is in place, attach and secure all motor mounts. (Honda)

the converter if needed.

WARNING! Never use the transmission or bellhousing bolts to force the engine against the transmission or transaxle or part damage can result. The two should slide together freely using your hands, feet, and pry bars.

9. Finish installing the other components: motor mounts, oil filter, fuel lines, wiring, vacuum hoses, throttle cable or linkage, battery cable ground, fan belts, front-drive axles, etc. Refer to Figs. 28-49 through 28-55.
10. Fill the engine with motor oil and the radiator with coolant.
11. Start and fast idle the engine until warm. Watch the engine oil light or gauge to make sure you have good oil pressure. Also watch engine temperature to prevent overheating.
12. If the engine will not start, check the following:
    a. Ignition timing and spark intensity.
    b. Spark plug wire routing.
    c. Primary wire routing to ignition coil and engine sensors.
    d. Fuel supply to carburetor or injection system.

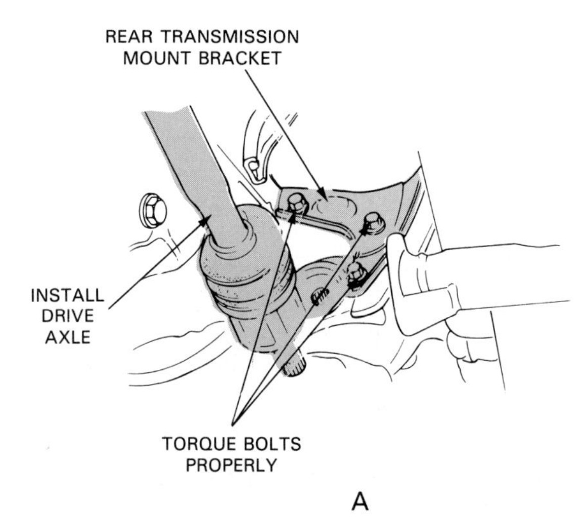

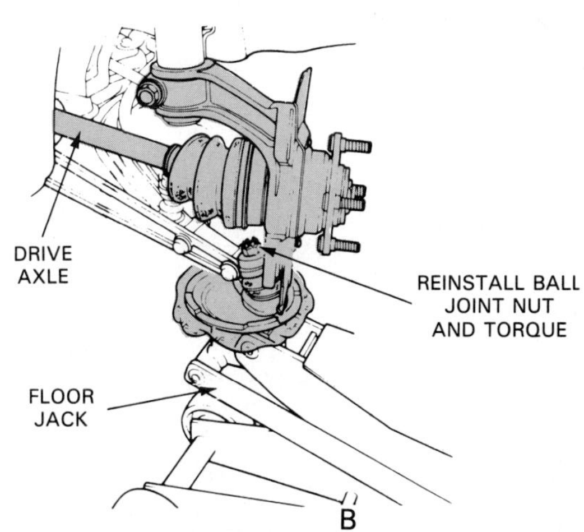

Fig. 28-50. With a front-wheel drive car, you will have to install front drive axles as you install engine and transaxle. Refer to service manual for particular car. A—Install drive axle. B—Reassemble ball joint and related parts.

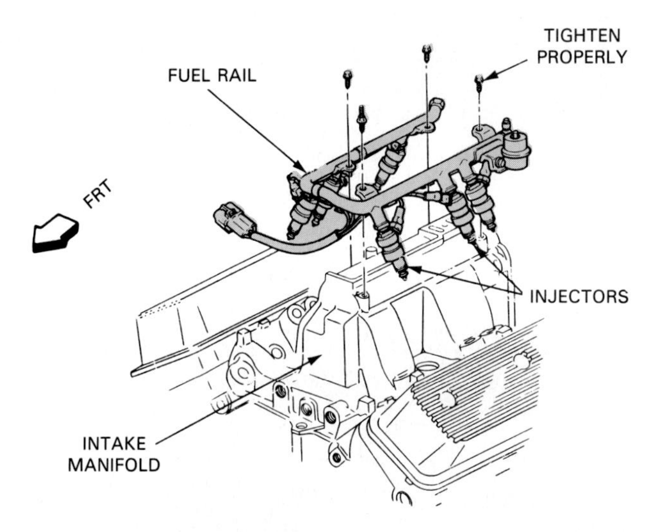

Fig. 28-51. Injectors and carburetor are usually installed after engine is back in car. This protects them from damage during installation. (Oldsmobile)

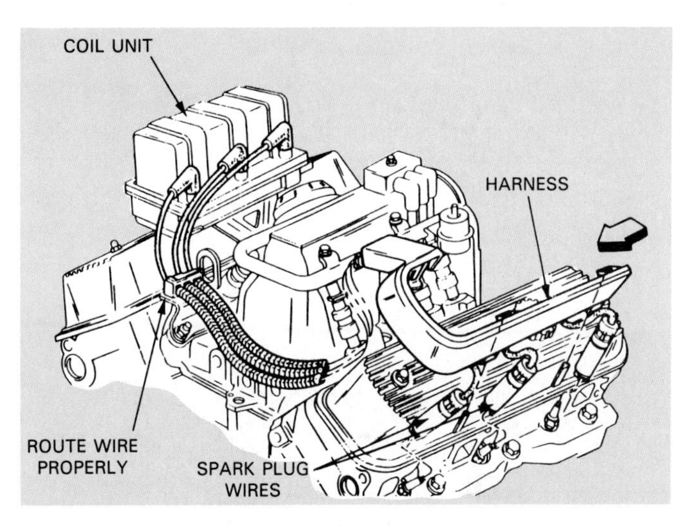

Fig. 28-52. Route all ignition system wires away from hot or moving parts. Double-check that wires feed to correct spark plugs. (Buick)

Refer to the text index for more help if the engine will not start and run properly. For example, Chapter 21 explains engine performance problems which relate to initial engine starting problems.

13. After the engine "fires" and warms up, shut the engine off. Recheck critical tune-up type adjustments: ignition timing, idle speed, etc.
14. Restart the engine and let it run at a fast idle. Watch for leaks under the engine and make sure the engine does not overheat. See Fig. 28-56.
15. Replace and align the hood. Double-check everything visually.

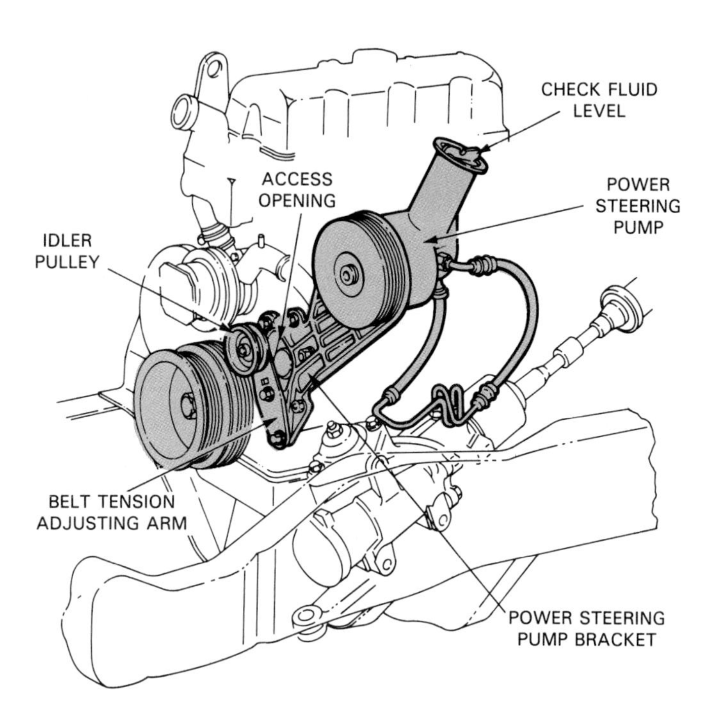

Fig. 28-53. Install all accessory units in correct sequence. Sometimes one bracket will cover and block access to other bolts. (Ford)

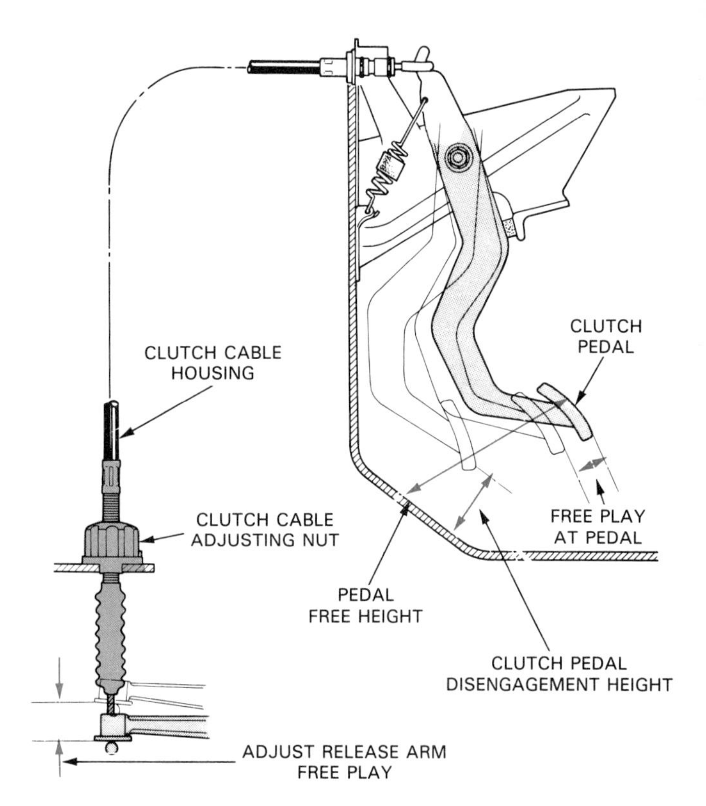

Fig. 28-55. Manual clutch or automatic transmission-transaxle linkage may also require adjustment. Follow manual directions. (Honda)

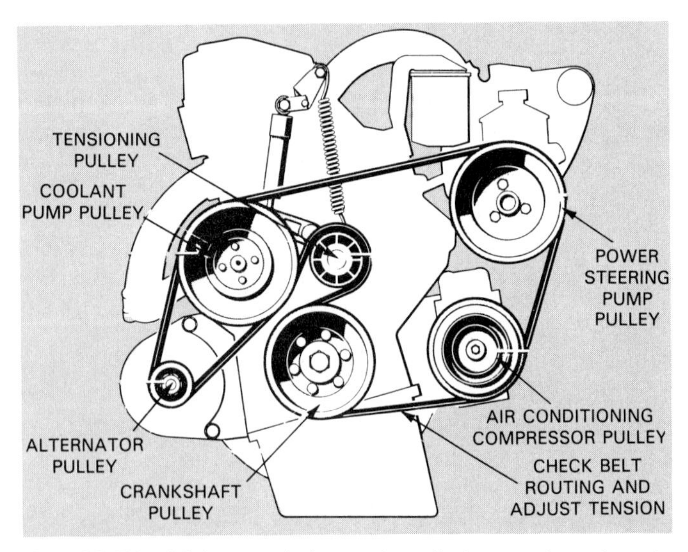

Fig. 28-54. Make sure belts are installed properly and are in good condition during major engine service. (Mercedes-Benz)

Fig. 28-56. Always watch for oil and coolant leaks when running engine for first time. Visually double-check everything to protect engine and yourself. (Ford)

ENGINE BREAK-IN

Engine break-in is done mainly to seat and seal new piston rings. It also aids initial wearing in of other components under controlled conditions.

After warm-up at a fast idle, most mechanics road test the car. At the same time, they use moderate acceleration and deceleration for engine break-in.

Generally, accelerate the car to about 40 mph (65 km/h). Then, release the gas pedal fully and let the car coast down to about 20 mph (32 km/h). Do this several times while carefully watching engine temperature and oil pressure.

Do NOT allow the engine to overheat during break-in; ring and cylinder scoring may result.

CAUTION! When road testing and breaking in an engine, drive the car on a road that is free of traffic. Do NOT exceed posted speed limits nor normal safe driving standards.

Inform the car owner of the following rules concerning the operation of a freshly overhauled engine:

1. Avoid prolonged highway driving during the first 100 to 200 miles (161 to 322 km). This will prevent ring friction from overheating the rings and cylinders, possibly causing damage.
2. Do not worry about oil consumption until after about 2000 miles (3 220 km) of engine operation. It will take this long for full ring seating.
3. Check engine oil and other fluid levels frequently.
4. Change the engine oil and filter after approximately 200 miles (3 220 km) of driving. This will help remove any particles in oil.
5. Inform the customer of any problems not corrected by the engine repair or overhaul.

For example, if the radiator is in poor condition (has been previously repaired or was filled with rust), tell the customer about the consequence of NOT correcting the problem. The problem may upset engine performance or reduce engine service life. Having the customer sign a release form will protect you if the unserviced part fails.

SUMMARY

When assembling an engine, all parts must be perfectly clean. Parts should be reinstalled in their same locations when possible. Lubricate moving parts. Make sure bolt lengths are correct. Keep components covered until assembly. Torque all components properly. Inspect everything as it is installed. Work at a steady pace and do not talk to anyone while you are working.

When installing an oil pump, make sure all parts are in perfect condition. Check gasket fit. If an oil pump drive rod is used, make sure the rod fits up into its hole in the block and into the pump. Torque the pump bolts to specs.

When installing the oil pan, straighten its flange if bent. Adhere the gasket to the pan or block. Use silicone sealer where the end seals overlap the pan gasket.

When installing a head gasket, check its fit on the block. Make sure it is facing in the right direction. Markings are normally provided showing the top or front of a head gasket. If dowels are not provided in the engine, install long stud bolts to hold the gasket and guide the head.

Lower the head onto the block without hitting the gasket. Start all of the bolts and then turn them down lightly. Use a torque wrench to tighten the head bolts in an approved sequence. Tighten them in steps and then go over them several times to assure full torque specifications.

New head bolts are needed with some engines. Some used head bolts are stretched and will not provide proper clamping action on the head gasket.

With a diesel engine, make sure you have the correct head gasket. They come in different thicknesses and different bore diameters. Diesel compression ratio and proper gasket selection is critical.

Before intake manifold installation, check it for warpage with a straightedge. Also check gasket fit. Adhere the gasket in place if needed. With a metal type intake gasket, circle all ports and coolant passages with silicone sealer. If the gasket is high quality fiber with special sealing beads, extra sealer is not needed. However, place a bead of sealer where the gasket and end seals overlap.

Lower the intake manifold into position without shifting the gasket or end seals. Start all bolts and then torque them using a crisscross sequence. Go over all of the bolts several times.

Also check the exhaust manifold for damage or warpage before installation. Check bolt lengths and torque the manifold bolts properly.

Since valve train designs vary, follow the details in the service manual. In any case, lube the lifters and fit them in their bores. Then install the push rods and rockers. With an overhead cam engine, position the cam followers.

Valve adjustment methods also vary. Many engines do not require valve adjustment. Simply torquing the rocker arm nut will center the lifter plunger and provide for quiet engine operation. Adjustable rockers must be turned to center the lifter plunger.

To adjust hydraulic lifters with the engine off, position the cam lobe away from the lifter or rocker. Turn the nut until all play is out of the valve train. Then turn the rocker nut about one more turn. To adjust them with the engine running, back off the rocker nut until the valve clatters. Then tighten the nut slowly until the valve noise stops. One more turn should center the hydraulic lifter.

To adjust mechanical lifters, the engine must be off. Turn the cam lobe away from the lifter or rocker. Slide the correct size feeler gauge between the rocker and valve stem tip. Turn the adjusting nut until the feeler drags slightly when pulled back and forth. Tighten the locknut and recheck the adjustment.

Valve adjusting shims are used on some OHC engines. You must obtain the correct shim thickness to provide proper valve train clearance. An Allen screw is provided on hydraulic tappets in some OHC cylinder heads. Turning the Allen screw will change valve train clearance.

Check the valve cover flange for straightness before installation. Adhere the gasket to the cover or use a continuous bead of silicone. All surfaces should be perfectly clean.

Carefully lower the cover into place. Start all bolts and then tighten them in a crisscross sequence. Do not overtighten valve cover bolts or you can bend the cover or split the gasket.

Engine installation is about the reverse of engine removal. Make sure all external parts are installed: motor mounts, sending units, sensors, flywheel, clutch, etc.

Keep the engine crankshaft and transmission-transaxle input shaft in perfect alignment during installation. Do not use the bellhousing bolts to force the transmission against the engine. The two should mate together by hand before installing the bolts.

Fill the engine and radiator before starting the engine. Start and fast idle the engine. Check oil pressure and do not let the engine overheat. Watch for leaks. Adjust ignition timing, idle speed, etc.

Test drive the car. Accelerate and decelerate several times to help seat the piston rings. Warn the car owner about driving methods after an engine over-haul.

KNOW THESE TERMS

Select fit, Priming oil pump, Gasket alignment, Head gasket markings, Torque-to-yield, Diesel head gasket thickness, Sealing bolt threads, Gasket-seal overlap,

Camshaft bearing cap numbers, Exhaust and intake rockers, Nonadjustable rockers, Adjustable rockers, Zero lash, Valve adjusting shims, Pilot tool, Engine break-in.

REVIEW QUESTIONS—CHAPTER 28

1. What are ten general rules to remember during engine assembly?
2. Where does the upper end of an oil pump drive rod install?
3. The oil pump gasket should be sealed with silicone sealer to prevent leakage. True or false?
4. In your own words, how do you install an oil pan?
5. How do you know how to position a cylinder head gasket?
6. If dowels are NOT provided in the block, what should you do?
7. Head bolts should have a drop of oil on their threads. True or false?
8. What is the general way to torque cylinder head bolts?
9. In your own words, how do you install an intake manifold?
10. Used exhaust manifolds should have a ________ installed under them to prevent ________.
11. Cam bearing ________ are ________ and must be installed in their exact original locations.
12. With nonadjustable rocker arms, you must:
    a. Run the rocker nuts down with an impact switch.
    b. Grind material off valve stems for adjustment.
    c. Torque the rocker nuts or bolts to specifications.
    d. None of the above are correct.
13. Hydraulic lifters can be adjusted with the engine off or with the engine running. True or false?
14. How do you install OHC valve adjusting shims?
15. List 14 rules to follow during engine installation.

ASE CERTIFICATION-TYPE QUESTIONS

1. When assembling and installing an engine, you should ________.
   (A) tighten head bolts as tight as possible
   (B) reuse torque-to-yield bolts
   (C) coat all moving parts with motor oil or white grease
   (D) install the oil pump dry
2. Technician A says that you should use thread locking compound when installing oil pump mounting bolts. Technician B says that thread locking compound should not be used when installing oil pump mounting bolts. Who is right?
   (A) A only.
   (B) B only.
   (C) Both A & B.
   (D) Neither A nor B.
3. Technician A says that the oil pan gasket can be adhered to the engine block mating surface. Technician B says that the oil pan gasket can be adhered to the oil pan itself. Who is right?
   (A) A only.
   (B) B only.
   (C) Both A & B.
   (D) Neither A nor B.
4. Technician A says that some engine manufacturers recommend using new head bolts when installing an engine's cylinder head. Technician B says that to prevent compression leakage, most engine manufacturers recommend using the old head bolts during cylinder head installation. Who is right?
   (A) A only.
   (B) B only.
   (C) Both A & B.
   (D) Neither A nor B.
5. Technician A says that normally when torquing head bolts, you should start at one end of the cylinder head and work inward. Technician B says that when torquing head bolts, you should normally start in the middle and work your way outward. Who is right?
   (A) A only.
   (B) B only.
   (C) Both A & B.
   (D) Neither A nor B.
6. Technician A says that if certain intake manifolds are installed improperly, an engine coolant leak can result. Technician B says that if certain intake manifolds are installed improperly, an exhaust leak can occur. Who is right?
   (A) A only.
   (B) B only.
   (C) Both A & B.
   (D) Neither A nor B.
7. You should use a ________ to check an intake manifold sealing surface for warpage.
   (A) dial indicator and flat feeler gauge
   (B) straightedge and flat feeler gauge
   (C) outside caliper and straightedge
   (D) dial indicator and straightedge
8. Technician A says that an OHC engine's cam bearing caps must be installed in their original location. Technician B says that it is not necessary to install an OHC engine's cam bearing caps in their original location. Who is right?
   (A) A only.
   (B) B only.
   (C) Both A & B.
   (D) Neither A nor B.
9. Technician A says that an engine's hydraulic lifters can be adjusted with the engine running. Technician B says that an engine's hydraulic lifters can be adjusted with the engine off. Who is right?
   (A) A only.
   (B) B only.
   (C) Both A & B.
   (D) Neither A nor B.
10. After an engine has been overhauled, you should avoid prolonged highway driving during the first ________.
    (A) 500 miles (800 km)
    (B) 200 miles (320 km)
    (C) 1000 miles (1600 km)
    (D) 100 miles (160 km)

Chapter 29

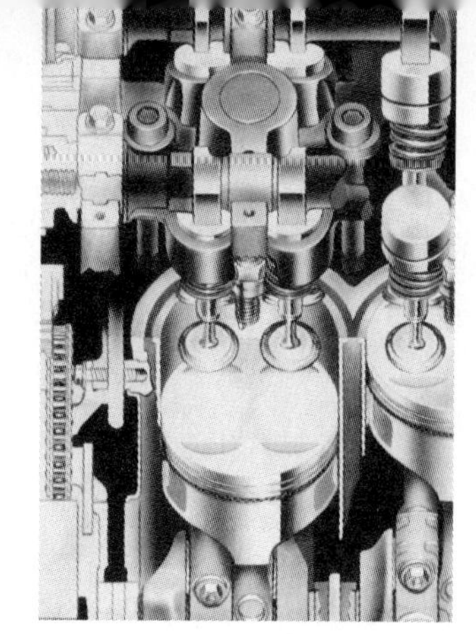

ASE Engine Certification

After studying this chapter, you will be able to:
- *Summarize the ASE testing program.*
- *Explain the Engine Repair Test content.*
- *Explain the Engine Performance Test content.*
- *Describe why ASE certification can be helpful to the mechanic and to the shop owner.*
- *Summarize methods of taking an ASE test successfully.*

This chapter will quickly review the ASE (NIASE) testing program. It will explain the two tests available in engine service and give hints on how to pass them. This textbook has covered information that will be helpful when taking both ASE engine tests.

Remember! Always try to learn more about auto mechanics. Engines, and cars in general, are changing every year at a rapid pace. If you do NOT continue to study and read technical publications, you will fall behind in your knowledge. Your ability to diagnose and repair engines will suffer.

With the precision and complexity of today's cars, the mechanic is called the ''technician.'' The term ''technician'' implies a higher degree of skill. No longer can the ''shade tree mechanic'' survive with modern automobile designs.

ASE (NIASE)

ASE stands for ''Automotive Service Excellence.'' It was shortened from NIASE which is an abbreviation for ''National Institute for Automotive Service Excellence.'' ASE is a nonprofit, nonaffiliated (no ties to industry) organization formed to help assure the highest standards in automotive service.

ASE directs an organized program of self-improvement under the guidance of a 40-member board of directors. These members represent all aspects of the automotive industry—educators, shop owners, consumer groups, government agencies, aftermarket parts companies, and auto manufacturers. This broad group of experts guides the ASE testing program and helps it stay in touch with the needs of the industry.

VOLUNTARY CERTIFICATION

ASE tests are voluntary. They do not have to be taken and they do not license technicians.

Some countries and a few states have made technician certification a requirement, however, technicians take the tests to show their employer and customers that they are fully qualified to work on a system of a car or an engine.

ASE gives statistics stating that over 300,000 technicians have taken certification tests and passed them. Thousands of these technicians have been retested and recertified after five years to maintain their credentials.

TEST CATEGORIES

In auto mechanics, there are eight test categories: Engine Repair, Engine Performance, Automatic Transmission/Transaxle, Manual Drive Train and Axles, Suspension & Steering, Brakes, Electrical/Electronic Systems, and Heating and Air Conditioning.

You can take any one or all of these tests. However, only four tests (200 questions maximum) should be taken at one testing session.

There are also seven tests in medium/heavy-duty truck repair and five tests in collision repair and refinishing.

Fig. 29-2 gives a breakdown of what each automotive service test involves. Study what they cover!

Engine repair test

The *engine repair test* has questions relating to the service of valve train, cylinder head, and block assemblies. The questions are primarily on how to do repairs or find mechanical problems. Questions are also given on lubricating, cooling, ignition, fuel, exhaust, battery, and starting system service, Fig. 29-1.

If you plan on taking this test, review the textbook chapters that explain these topics. In particular, restudy the service chapters in Sections 3, 4, and 5 of this book.

Engine performance test

The *engine performance test* has questions dealing with engine diagnosis and the diagnosis of engine related systems. Troubleshooting tune-up type problems is the general thrust of this test. You can pick between tests that cover either carburetor and feedback fuel injection or import fuel injection, Fig. 29-2.

To study for the performance test, primarily review Sections 3 and 5 in this textbook. Concentrate on Chapters 21, 22, and 23.

WHO CAN TAKE ASE TESTS?

To take ASE certification tests and receive certification, you must either have two years of on-the-job experience or one year of approved educational credit and one year of work experience.

However, you may take the tests even if you do NOT have the required two years experience. You will be sent a score for the test, but NOT certification credentials. After you have gained the mandatory experience, you can notify ASE and they will mail you a certificate.

You will be granted credit for formal training by one, or a combination, of the following types of schooling:

1. High school training for three full years in

Specifications for the Automobile Tests
Content Area/Number of Questions

A1 Engine Repair	**80**	**A2 Automatic Transmission/Transaxle**	**50**
A. General Engine Diagnosis	18	A. General Transmission/Transaxle Diagnosis	19
B. Cylinder Head & Valve Train Diagnosis & Repair	16	B. Transmission/Transaxle Maintenance & Adjustment	6
C. Engine Block Diagnosis & Repair	15	C. In-Vehicle Transmission/Transaxle Repair	11
D. Lubrication & Cooling Systems Diagnosis & Repair	10	D. Off-Vehicle Transmission/Transaxle Repair	14
E. Ignition System Diagnosis & Repair	7	1. Removal, Disassembly, & Assembly (3)	
F. Fuel & Exhaust Systems Diagnosis & Repair	8	2. Oil Pump & Converter (3)	
G. Battery & Starting System Diagnosis & Repair	6	3. Gear Grain, Shafts, Bushings, & Case (3)	
		4. Friction & Reaction Units (5)	
A3 Manual Drive Train & Axles	**40**	**A4 Suspension & Steering**	**40**
A. Clutch Diagnosis & Repair	6	A. Steering Systems Diagnosis & Repair	9
B. Transmission Diagnosis & Repair	7	1. Steering Columns & Manual Steering Gears (2)	
C. Transaxle Diagnosis & Repair	10	2. Power-Assisted Steering Units (4)	
D. Drive (Half) Shaft & Universal Joint Diagnosis & Repair	6	3. Steering Linkage (3)	
E. Rear Axle Diagnosis & Repair	7	B. Suspension Systems Diagnosis & Repair	13
1. Ring & Pinion Gears (3)		1. Front Suspensions (6)	
2. Differential Case Assembly (2)		2. Rear Suspensions (5)	
3. Limited Slip Differential (1)		3. Miscellaneous Service (2)	
4. Axle Shafts (1)		C. Wheel Aligment Diagnosis, Adjustment, & Repair	13
F. Four-Wheel Drive Component Diagnosis & Repair	4	D. Wheel & Tire Diagnosis & Repair	5
A5 Brakes	**55**	**A6 Electrical/Electronic Systems**	**50**
A. Hydraulic System Diagnosis & Repair	16	A. General Electrical/Electronic System Diagnosis	11
1. Master Cylinders (5)		B. Battery Diagnosis & Service	5
2. Fluids, Lines, & Hoses (3)		C. Starting System Diagnosis & Repair	5
3. Valves & Switches (4)		D. Charging System Diagnosis & Repair	6
4. Bleeding, Flushing, & Leak Testing (4)		E. Lighting Systems Diagnosis & Repair	6
B. Drum Brake Diagnosis & Repair	6	1. Headlights, Parking Lights, Taillights, Dash Lights, & Courtesy Lights (3)	
C. Disc Brake Diagnosis & Repair	13	2. Stoplights, Turn Signals, Hazard Lights, & Back-up Lights (3)	
D. Power Assist Units Diagnosis & Repair	4	F. Gauges, Warning Devices, & Driver Information Systems Diagnosis & Repair	6
E. Miscellaneous Diagnosis & Repair	7	G. Horn & Wiper/Washer Diagnosis & Repair	3
F. Anti-Lock Brake System Diagnosis & Repair	9	H. Accessories Diagnosis & Repair	8
		1. Body (4) 2. Miscellaneous (4)	
A7 Heating & Air Conditioning	**50**	**A8 Engine Performance**	**70**
A. A/C System Diagnosis & Repair	12	A. General Engine Diagnosis	10
B. Refrigeration System Component Diagnosis & Repair	11	B. Ignition System Diagnosis & Repair	13
1. Compressor & Clutch (5)		C. Fuel, Air Induction & Exhaust Systems Diagnosis & Repair	14
2. Evaporator, Receiver/Drier, Condenser, etc. (6)		D. Emissions Control Systems Diagnosis & Repair	9
C. Heating & Engine Cooling Systems Diagnosis & Repair	6	1. Positive Crankcase Ventilation (1)	
D. Operating Systems & Related Controls Diagnosis & Repair	16	2. Exhaust Gas Recirculation (3)	
1. Electrical (7)		3. Exhaust Gas Treatment (2)	
2. Vacuum/Mechanical (4)		4. Evaporative Emissions Control (3)	
3. Automatic & Semi-Automatic Temperature Controls (5)		E. Computerized Engine Controls Diagnosis & Repair	17
E. Refrigerant Recovery, Recycling & Handling	5	F. Engine Related Service	3
		G. Engine Electrical Systems Diagnosis & Repair	4
		1. Battery (1)	
		2. Starting System (1)	
		3. Charging System (2)	

Fig. 29-1. These are certification test categories for automobiles. If you pass one test, you will be a Certified Automobile Technician. If you pass all of them, you will be a Master Automobile Technician. (ASE)

automotives may be substituted for one year of work experience.

2. Post-high school training for two years in a public or private facility can be substituted for one year of work experience.
3. Two months of short training courses can be substituted for one month of work experience.
4. Three years of an apprenticeship program, where you work under an experienced mechanic as a form of training, can be substituted for both years of work experience.

To have schooling substituted for work experience, you must send a copy of your transcript (list of courses taken), a statement of training, or certificate to verify your training or apprenticeship. Each should give your length of training and subject area. This should accompany your registration form and fee payment.

TEST LOCATIONS AND DATES

ASE administers tests twice a year in over 700 locations across the country. The test sites are usually community colleges or high schools. Tests are given in May and November of each year. Contact ASE for more specific dates and locations for the tests.

TEST RESULTS

The results of your test will be mailed to your home. Only YOU will find out how you did on the tests. You can then inform your employer if you like.

Test scores will be mailed out a few weeks after you have completed the test. If you pass a test, you can consider taking more tests. If you fail, you will know that more study is needed before retaking the test.

TEST TAKING TECHNIQUES

Each engine test has 80 multiple-choice questions. You must carefully read the question and evaluate it. Then, read through the possible correct answers. You must then select the MOST CORRECT response. One answer will be more correct than the others. Sometimes more than one response is correct. Fig. 29-3 gives two example questions. Try to answer them before reading the caption.

You will NOT be required to recall exact specifications unless they are general and apply to most makes and models of cars. For example, compression test pressure readings all typically about the same with all gasoline engines and so are many engine clearances. This type general information might be needed to answer some questions.

These are a few tips that might help you pass ASE certification tests:

1. Read the statements or questions slowly. You might want to read through them twice to make sure you fully understand the question.
2. Analyze the statement or question. Look for hints that make some of the possible answers wrong.
3. Analyze the questions as if you were the technician trying to fix the car. Think of all possible situations and use common sense to pick the most correct response.
4. When two technicians give statements concerning a problem, try to decide if either one is incorrect. If both are valid statements about a situation, mark both technicians correct. If only one is correct, neither are correct, or just one, mark the answer accordingly. This is one of the most difficult types of questions.
5. If the statement only gives limited information, make sure you do not pick one answer as correct because it may be a more common condition. If the statement does not let you conclude one answer is better than another, both answers are equally correct.
6. Your first thought about which answer is correct is usually the correct response. If you think about a question too much, you will usually read something into the question that is not there. Read the question carefully and make a decision.
7. Do not waste time on any one question. Make sure you have time to answer all of the questions on the test.
8. Visualize how you would perform a test or repair when trying to answer a question. This will help you solve the problem more accurately.

BENEFITS OF ASE CERTIFICATION

When you pass an ASE test, you will be given a shoulder patch for your work uniform. The patch has the

PRACTICE QUESTIONS

1. A compression test shows that one cylinder is too low. A leakage test on that cylinder shows that there is too much leakage. During the test, air could be heard coming out of the tail pipe. Which of these could be the cause?
   (A) Broken piston rings
   (B) A bad head gasket
   (C) A bad exhaust gasket
   (D) An exhaust valve not seating

2. After the compression readings shown below were taken, a wet compression test was made. The second set of readings was almost the same as the first.
   Technician A says that a burned valve could cause these readings.
   Technician B says that a broken piston ring could cause these readings.
   Who is right.?
   (A) A only
   (B) B only
   (C) Both A and B
   (D) Neither A nor B

140
135
5
140
Compression Specifications 140 psi

Fig. 29-2. These are sample questions for ASE. Try to answer them. The answers are 1 (D), 2 (A). (ASE)

Blue Seal insignia with either the words "Automotive Technician" (passed test in one area or more) or "Master Auto Technician" (passed all tests).

The patch will serve as good public relations, showing everyone that you are well trained to work on today's complex vehicles. It will also tell employers that you are someone special that has taken extra effort to prove your value as a technician. It should lead to more rapid advancement and more income as customers indicate their preferences for a certified technician.

MORE INFORMATION

For more information on ASE certification tests, write for a "Registration Booklet" from:

ASE
13505 Dulles Technology Drive, Suite 2
Herndon, VA 22071

The bulletin will give test locations, testing dates, costs, sample questions, and other useful information.

SUMMARY

This textbook has helped prepare you to pass two ASE certification tests: Engine Repair and Engine Performance. The theory given in this book will help you analyze problem symptoms—if you know how an engine should operate, you will be better at finding faults when an engine is NOT operating properly. The diagnosis and service coverage in this book will also help you because ASE questions are primarily troubleshooting and repair type questions.

ASE stands for Automotive Service Excellence. Previously called NIASE (National Institute for Automotive Service Excellence), this is a non-profit organization. It is not affiliated directly with any industry or organization.

The ASE tests are voluntary. They will help show your employer and customers that you are fully competent to work on their engine. The two engine tests contain 80 multiple-choice questions. Other tests are available in Automatic Transmissions/Transaxles, Manual Transmissions and Axles, Suspension & Steering, Brakes, and Electrical/Electronic Systems.

When taking the test, read the questions and statements slowly. Mentally drop the incorrect responses. Try to imagine you are really working on a car when reading the question. Usually, your first idea of the right answer or answers is correct. If you think too much about an answer, you might pick the second most correct response. Do not waste too much time on any question or you may not have time to answer all of the questions.

Write to ASE for a "Registration Booklet." It will give test dates, locations, costs, etc.

KNOW THESE TERMS

ASE, NIASE, nonaffiliated, Engine Repair Test, Engine Performance Test, Test Taking Techniques, Shoulder patch, Bulletin of Information.

REVIEW QUESTIONS—Chapter 29

1. How many questions are there on the two engine tests?
2. ASE tests are mandatory in some states. True or false?
3. How many tests and questions should be taken at one time?
   a. Five tests and 250 questions.
   b. Four tests and 200 questions.
   c. Three tests and 150 questions.
   d. All of the tests and no limit on questions.
4. What is the difference between the Engine Performance Test and the Engine Repair Test?
5. Who can take ASE tests?
6. Explain eight hints for taking ASE tests.
7. What is the address of ASE?

ASE CERTIFICATION-TYPE QUESTIONS

1. Technician A says that some of the members of the ASE board of directors are auto repair shop owners. Technician B says that some of members of the ASE board of directors represent government agencies. Who is right?
   (A) A only.
   (B) B only.
   (C) Both A & B.
   (D) Neither A nor B.
2. Technician A says that questions relating to exhaust system service are part of the ASE certification tests. Technician B says that questions relating to cooling system service are part of the ASE certification tests. Who is right?
   (A) A only.
   (B) B only.
   (C) Both A & B.
   (D) Neither A nor B.
3. Technician A says that in order to take ASE certification tests, you must have two years of on-the-job experience. Technician B says that in order to receive ASE certification, you must have two years of on-the-job experience. Who is right?
   (A) A only.
   (B) B only.
   (C) Both A & B.
   (D) Neither A nor B.
4. Technician A says that ASE tests are given in May and September of each year. Technician B says that ASE tests are given in June and October of each year. Who is right?
   (A) A only.
   (B) B only.
   (C) Both A & B.
   (D) Neither A nor B.
5. Technician A says that each ASE engine test consist of 80 multiple-choice questions. Technician B says that each ASE engine test consist of 40 multiple-choice and 20 True or False questions. Who is right?
   (A) A only.
   (B) B only.
   (C) Both A & B.
   (D) Neither A nor B.

Chapter 30

Career Success

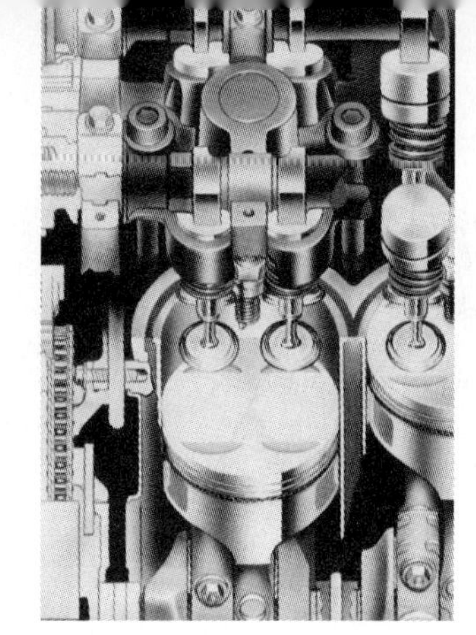

After studying this chapter, you will be able to:

- *Explain why the continuous study of mechanics is essential to job success.*
- *Describe how both work quality and work speed are equally important to job success.*
- *Summarize the different systems used to pay mechanics for their labor.*
- *Explain the types of auto repair facilities.*
- *Cooperate with other technicians as a means of improving working conditions and profits.*
- *Summarize how you can advance in the field of auto mechanics.*

This chapter will briefly discuss factors that can determine your degree of success when employed in auto mechanics. There are many simple things that can help or hurt you as an engine technician. There is more to being a good technician than just being able to work on engines.

DEPENDABILITY

Dependability is critical! If you fail to report to work on time, leave early, or take too many days off, the shop suffers. The shop owner or service writer is responsible for scheduling repair work based on a certain number of technicians. If one or more technicians do not report to work, this schedule is useless and customers may have to wait for their cars longer than planned.

If customers are unhappy about being inconvenienced, they will not return to the shop again. The technicians will have less work and the shop will not be as profitable. Everyone can suffer when a technician is constantly absent or late for work on a busy day.

The number one reason technicians are fired is lack of dependability.

TEAM EFFORT

The technicians in a shop should consider all employees as a "team." They should work together to make the garage a safe, enjoyable, and efficient operation.

If technicians do not get along and do not cooperate, morale and work output will suffer. Keep in mind—a shop is competing with other shops for business. Technicians should never compete with each other.

Many times, tasks are difficult to complete by yourself. You might need help lowering a heavy intake manifold onto an engine or replacing a hood. If you have helped other technicians with similar tasks, they will be glad to help you.

Also, if you do not have a car to work on, offer to assist another technician who is completing a job. He or she should be willing to help or return the favor at a later date. As a result, you will make more money and so will the other employees.

CONSTANTLY LEARN!

To be successful in engine repair, you must constantly try to learn more about engines and engine repair techniques. Engine technology is changing at a frantic pace. If you think you KNOW EVERYTHING, you are very quickly going to be left behind in your knowledge of auto technology.

To be a wise technician, you must always try to learn something new. Engine repair is a very complex and challenging area of employment. You must always try to read service manuals, service bulletins, automotive magazines, and other literature.

Try to learn something new each day and your job will become easier and more profitable each day.

WORK QUALITY

A primary concern of any engine technician, shop owner, or customer is work quality. Repairs must be done correctly, following factory specifications.

It is tempting for a technician to try to work too fast in an attempt to make more money. The excess speed usually "backfires" and costs everyone (technician, customer, shop owner) money. When a technician works too quickly, comebacks (repair must be done over) increase. More customers return unsatisfied with the repairs done on their cars. Then the technician must do the repair over again, free-of-charge. He or she loses money on lost work time, the shop owner loses money on wasted stall time, and the owner of the car loses because he or she must return to the shop a second time.

WORK SPEED

Work speed is still very important to the engine technician. For those working on a percentage basis, income will depend upon the amount of work completed each week. For those on an hourly basis, employers may not give pay raises because the technicians are not making the shop enough money.

To increase work speed without affecting work quality, constantly think about new methods to increase work efficiency. Consider a new tool or technique that will save time without lowering the quality of the repair.

This kind of mental attitude will help you improve your skills. As a result, you will become more productive each day.

EARNINGS

There are three basic methods shops use to pay their technicians: hourly rate, commission, and commission plus parts.

Hourly rate

The *hourly rate* is simple; a stated amount of pay is given for each hour worked. This pay method is found in every type of shop. It may be desirable when you do not like the pressure of producing a quantity of work each week. The hourly rate is usually slightly lower than the commission method of pay.

Commission

When you earn a *commission,* you obtain a set percent of the labor charged the customers. For example, with a 50 percent commission at $32 an hour, you would get $16 dollars an hour for work completed.

If you are fairly quick and can match or exceed the *flat rate manual* (book that states how long each repair should take), you can earn a good wage. However, if you are slow or if there is not enough work to keep you busy, your income could be as low or lower than a technician paid hourly. Some technicians get commission plus parts in which a commissioned technician also gets a small percentage for the parts installed on engines.

TYPES OF SHOPS

The type of shop you work in can affect your success as an engine technician. Some shops may be more desirable than others.

Dealerships

Large *auto dealerships* are good because they usually pay well and have a large volume of work to be done. They also have good benefits. However, they require their technicians to do warranty work.

Warranty work is repairs done on cars or engines still under warranty. The rate of pay is considerably lower under warranty. If you are fairly experienced and fast, a large dealership may be a good place to seek employment.

Private garages

A *privately owned garage* can also be an excellent place to work. All of the work done is nonwarranty and the pay is normally good. As long as the shop is reputable and has plenty of volume to keep you busy, you might want to seek a job in a small garage. The shop owner controls the work atmosphere and conditions for the technicians. Inquire about insurance and other benefits.

Service stations

A *service station,* having a repair area, may be an excellent place to work, especially when you are a beginner. Much of the work will be *quick service* (repairs that only take an hour or two). You will replace belts, hoses, water pumps, thermostats, tire, and other items. Many service stations offer a parts percentage incentive which can increase your earnings. You can learn much about engine repair in a service station with repair stalls.

Specialized engine shops

An *engine shop* specializes in the rebuilding of engines. Many machine shops will rebuild or overhaul engines. The engine must be removed from the car and taken to the engine shop in most instances.

The engine shop will usually have all of the specialized equipment needed to bore blocks, power hone cylinders, resurface decks, magnaflux components, etc.

Small garages, service stations, and other types of facilities may use the service of a specialized engine shop. They will make money on the installation and removal of the engine. The engine shop will make money on the rebuild.

WORK CLOTHES

Work clothes are a vital part of a technician's preparation for work. Many shops provide a standard set of work clothes for their technicians. The technicians usually must pay a small fee for purchasing or renting the work clothes and having them cleaned.

Work clothes are very important. They should be comfortable, well fitting, attractive in appearance, and clean. To the public, your work clothes reflect your mechanical abilities and work attitudes. If you look dirty and sloppy, customers will suspect that your work may be less than desirable. You must always project a PROFESSIONAL IMAGE!

SHOP SUPERVISOR AND SERVICE WRITER

Always maintain good working relations with the shop supervisor and service writer. In larger shops, they have the authority to assign jobs to technicians. If you do not treat them with respect, you may be assigned more than your share of less desirable jobs.

PARTS DEPARTMENT

As just mentioned, keep good relations with everyone in the shop. This includes the workers in the parts department.

You must depend on parts people to get parts quickly and correctly for engine repairs. This also applies to the people working at a parts house. If you do not give them respect, do not expect them to go out of their way to help you obtain a hard-to-get part.

ADVANCEMENT IN AUTOMOTIVE TECHNOLOGY

Keep in mind the many opportunities in automotive technology. A few of these include:
1. Automotive instructor.
2. Service manager.
3. Shop supervisor.
4. Technical representative for an auto maker.
5. Technical representative for an auto aftermarket company.
6. Auto aftermarket sales representative.
7. Automotive engineer.
8. Shop owner.

Some of these positions will require you to take special training. You may have to attend college or take specialized courses. You may also want ASE master technician certification. However, a basic knowledge of auto technology will give you an edge over others seeking the same kind of position.

The auto repair and manufacturing field is one of the largest employment areas in the nation. Your chances for succeeding in this field depend primarily upon your initiative and WILLINGNESS TO LEARN.

ENTREPRENEURSHIP

An *entrepreneur* is someone who starts a business. This might be a muffler shop, tune-up shop, parts house, or similar facility. To be a good entrepreneur, you must be able to organize all aspects of the business: bookkeeping, payroll, facility planning, hiring, etc. After gaining experience, you might want to consider starting your own business.

SUMMARY

Being an engine technician offers many challenges and rewards. Your success will depend upon your ability to develop competent skills, cooperate with employers and fellow workers, and keep up-to-date with automotive technological developments.

As an engine technician, you will be expected to work efficiently and rapidly while still producing quality work. Depending on the shop or employer, you may be paid an hourly rate, a straight commission or percentage on the dollar amount of your work, or a commission plus a percentage of the parts you install.

The type of shop will have some effect on your success as a technician. Large auto dealerships, as a rule, pay well, have good benefits, and have a large volume of work. On the other hand, technicians will be required to do warranty work which is lower paying.

Private garages normally pay well and do no warranty work. Service stations can be a good place for a beginner.

KNOW THESE TERMS

Team effort, Work quality, Work speed, Hourly rate, Commission, Commission plus parts, Warranty work, Quick service, Service bulletins, Shop supervisor, Automotive instructor, Service manager, Technical representative, Auto aftermarket sales representative, Automotive engineer, Shop owner.

REVIEW QUESTIONS—CHAPTER 30

1. As an engine technician, what is meant by the term "team effort?"
2. An undependable technician in a busy auto shop fails to show up for work. This only hurts this technician. True or false?
3. Some experienced technicians know everything they need to know. True or false?
4. Why is work quality equally important as speed?
5. Explain three ways that you can be paid as an engine technician.
6. This would be used to figure your pay for a certain repair.
   a. Service manual.
   b. Service bulletin.
   c. Commission manual.
   d. Flat rate manual.
7. Warranty work usually pays less than customer paid work. True of false?
8. Small garages and service stations may take engines to an ________ ________ for major rebuilding.
9. Why should you show respect to the service writer in a large dealership?
10. List eight positions of advancement in automotive technology.

ASE CERTIFICATION-TYPE QUESTIONS

1. The number one reason an auto technician is usually fired is due to lack of ________.
   (A) mechanical skills
   (B) dependability
   (C) education
   (D) productivity
2. Which of the following determines the success of an auto technician?
   (A) Work quality.
   (B) Work speed.
   (C) Method of earnings.
   (D) All of the above.
3. Technician A says that some auto shops pay their technicians on an hourly rate basis. Technician B says that certain auto shops pay their technicians on a commission basis. Who is right?
   (A) A only. (C) Both A & B.
   (B) B only. (D) Neither A nor B.
4. Which of the following type of work is normally performed at a service station?
   (A) Water pump replacement.
   (B) Alternator belt replacement.
   (C) Thermostat replacement.
   (D) All of the above.
5. Technician A says that most specialized machine shops will include engine removal and installation when performing an engine rebuild. Technician B says that engine removal and installation is not included when a machine shop performs an engine rebuild. Who is right?
   (A) A only. (C) Both A & B.
   (B) B only. (D) Neither A nor B.

CONVERSION CHART

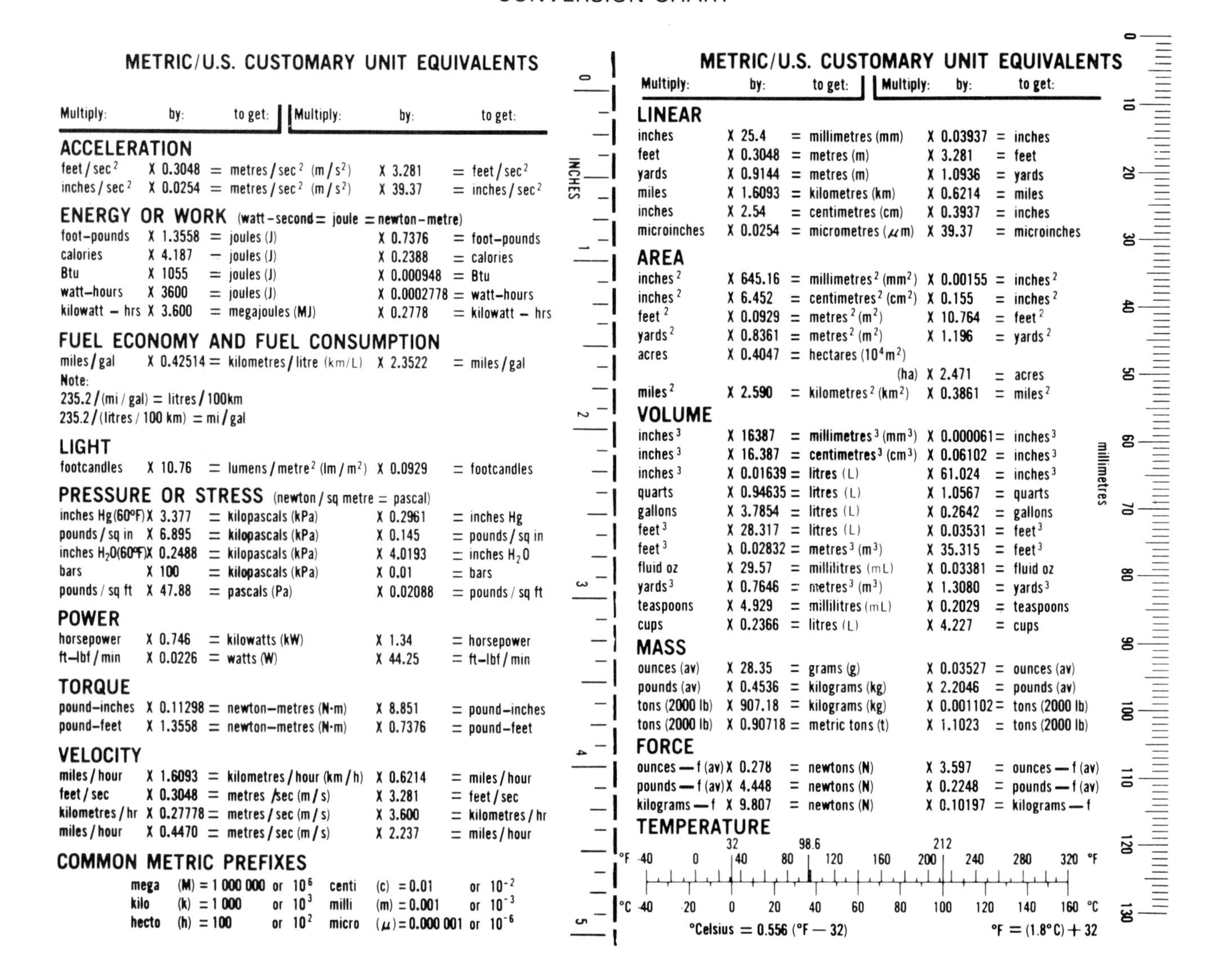

METRIC/U.S. CUSTOMARY UNIT EQUIVALENTS

Multiply:	by:	to get:	Multiply:	by:	to get:
ACCELERATION					
feet/sec²	X 0.3048	= metres/sec² (m/s²)		X 3.281	= feet/sec²
inches/sec²	X 0.0254	= metres/sec² (m/s²)		X 39.37	= inches/sec²
ENERGY OR WORK (watt-second = joule = newton-metre)					
foot-pounds	X 1.3558	= joules (J)		X 0.7376	= foot-pounds
calories	X 4.187	= joules (J)		X 0.2388	= calories
Btu	X 1055	= joules (J)		X 0.000948	= Btu
watt-hours	X 3600	= joules (J)		X 0.0002778	= watt-hours
kilowatt – hrs	X 3.600	= megajoules (MJ)		X 0.2778	= kilowatt – hrs
FUEL ECONOMY AND FUEL CONSUMPTION					
miles/gal	X 0.42514	= kilometres/litre (km/L)		X 2.3522	= miles/gal
Note: 235.2/(mi/gal) = litres/100km; 235.2/(litres/100 km) = mi/gal					
LIGHT					
footcandles	X 10.76	= lumens/metre² (lm/m²)		X 0.0929	= footcandles
PRESSURE OR STRESS (newton/sq metre = pascal)					
inches Hg(60°F)	X 3.377	= kilopascals (kPa)		X 0.2961	= inches Hg
pounds/sq in	X 6.895	= kilopascals (kPa)		X 0.145	= pounds/sq in
inches H_2O(60°F)	X 0.2488	= kilopascals (kPa)		X 4.0193	= inches H_2O
bars	X 100	= kilopascals (kPa)		X 0.01	= bars
pounds/sq ft	X 47.88	= pascals (Pa)		X 0.02088	= pounds/sq ft
POWER					
horsepower	X 0.746	= kilowatts (kW)		X 1.34	= horsepower
ft-lbf/min	X 0.0226	= watts (W)		X 44.25	= ft-lbf/min
TORQUE					
pound-inches	X 0.11298	= newton-metres (N·m)		X 8.851	= pound-inches
pound-feet	X 1.3558	= newton-metres (N·m)		X 0.7376	= pound-feet
VELOCITY					
miles/hour	X 1.6093	= kilometres/hour (km/h)		X 0.6214	= miles/hour
feet/sec	X 0.3048	= metres/sec (m/s)		X 3.281	= feet/sec
kilometres/hr	X 0.27778	= metres/sec (m/s)		X 3.600	= kilometres/hr
miles/hour	X 0.4470	= metres/sec (m/s)		X 2.237	= miles/hour

COMMON METRIC PREFIXES

mega	(M) = 1 000 000	or 10^6	centi	(c) = 0.01	or 10^{-2}
kilo	(k) = 1 000	or 10^3	milli	(m) = 0.001	or 10^{-3}
hecto	(h) = 100	or 10^2	micro	(μ) = 0.000 001	or 10^{-6}

METRIC/U.S. CUSTOMARY UNIT EQUIVALENTS

Multiply:	by:	to get:	Multiply:	by:	to get:
LINEAR					
inches	X 25.4	= millimetres (mm)		X 0.03937	= inches
feet	X 0.3048	= metres (m)		X 3.281	= feet
yards	X 0.9144	= metres (m)		X 1.0936	= yards
miles	X 1.6093	= kilometres (km)		X 0.6214	= miles
inches	X 2.54	= centimetres (cm)		X 0.3937	= inches
microinches	X 0.0254	= micrometres (μm)		X 39.37	= microinches
AREA					
inches²	X 645.16	= millimetres² (mm²)		X 0.00155	= inches²
inches²	X 6.452	= centimetres² (cm²)		X 0.155	= inches²
feet²	X 0.0929	= metres² (m²)		X 10.764	= feet²
yards²	X 0.8361	= metres² (m²)		X 1.196	= yards²
acres	X 0.4047	= hectares (10^4m²) (ha)		X 2.471	= acres
miles²	X 2.590	= kilometres² (km²)		X 0.3861	= miles²
VOLUME					
inches³	X 16387	= millimetres³ (mm³)		X 0.000061	= inches³
inches³	X 16.387	= centimetres³ (cm³)		X 0.06102	= inches³
inches³	X 0.01639	= litres (L)		X 61.024	= inches³
quarts	X 0.94635	= litres (L)		X 1.0567	= quarts
gallons	X 3.7854	= litres (L)		X 0.2642	= gallons
feet³	X 28.317	= litres (L)		X 0.03531	= feet³
feet³	X 0.02832	= metres³ (m³)		X 35.315	= feet³
fluid oz	X 29.57	= millilitres (mL)		X 0.03381	= fluid oz
yards³	X 0.7646	= metres³ (m³)		X 1.3080	= yards³
teaspoons	X 4.929	= millilitres (mL)		X 0.2029	= teaspoons
cups	X 0.2366	= litres (L)		X 4.227	= cups
MASS					
ounces (av)	X 28.35	= grams (g)		X 0.03527	= ounces (av)
pounds (av)	X 0.4536	= kilograms (kg)		X 2.2046	= pounds (av)
tons (2000 lb)	X 907.18	= kilograms (kg)		X 0.001102	= tons (2000 lb)
tons (2000 lb)	X 0.90718	= metric tons (t)		X 1.1023	= tons (2000 lb)
FORCE					
ounces — f (av)	X 0.278	= newtons (N)		X 3.597	= ounces — f (av)
pounds — f (av)	X 4.448	= newtons (N)		X 0.2248	= pounds — f (av)
kilograms — f	X 9.807	= newtons (N)		X 0.10197	= kilograms — f

TEMPERATURE

°F -40 0 32 40 80 98.6 120 160 200 212 240 280 320 °F

°C -40 -20 0 20 40 60 80 100 120 140 160 °C

°Celsius = 0.556 (°F — 32) °F = (1.8°C) + 32

TAP/DRILL CHART

COARSE STANDARD THREAD (N. C.) Formerly U. S. Standard Thread					FINE STANDARD THREAD (N. F.) Formerly S.A.E. Thread				
Sizes	Threads Per Inch	Outside Diameter at Screw	Tap Drill Sizes	Decimal Equivalent of Drill	Sizes	Threads Per Inch	Outside Diameter at Screw	Tap Drill Sizes	Decimal Equivalent of Drill
1	64	.073	53	0.0595	0	80	.060	3/64	0.0469
2	56	.086	50	0.0700	1	72	.073	53	0.0595
3	48	.099	47	0.0785	2	64	.086	50	0.0700
4	40	.112	43	0.0890	3	56	.099	45	0.0820
5	40	.125	38	0.1015	4	48	.112	42	0.0935
6	32	.138	36	0.1065	5	44	.125	37	0.1040
8	32	.164	29	0.1360	6	40	.138	33	0.1130
10	24	.190	25	0.1495	8	36	.164	29	0.1360
12	24	.216	16	0.1770	10	32	.190	21	0.1590
1/4	20	.250	7	0.2010	12	28	.216	14	0.1820
5/16	18	.3125	F	0.2570	1/4	28	.250	3	0.2130
3/8	16	.375	5/16	0.3125	5/16	24	.3125	I	0.2720
7/16	14	.4375	U	0.3680	3/8	24	.375	Q	0.3320
1/2	13	.500	27/64	0.4219	7/16	20	.4375	25/64	0.3906
9/16	12	.5625	31/64	0.4843	1/2	20	.500	29/64	0.4531
5/8	11	.625	17/32	0.5312	9/16	18	.5625	0.5062	0.5062
3/4	10	.750	21/32	0.6562	5/8	18	.625	0.5687	0.5687
7/8	9	.875	49/64	0.7656	3/4	16	.750	11/16	0.6875
1	8	1.000	7/8	0.875	7/8	14	.875	0.8020	0.8020
1 1/8	7	1.125	63/64	0.9843	1	14	1.000	0.9274	0.9274
1 1/4	7	1.250	1 7/64	1.1093	1 1/8	12	1.125	1 3/64	1.0468
					1 1/4	12	1.250	1 11/64	1.1718

BOLT TORQUING CHART

METRIC STANDARD

GRADE OF BOLT		5D	.8G	10K	12K	
MIN. TENSILE STRENGTH		71,160 P.S.I	113,800 P.S.I.	142,200 P.S.I.	170,679 P.S.I.	
GRADE MARKINGS ON HEAD		5D	8G	10K	12K	SIZE OF SOCKET OR WRENCH OPENING
METRIC		FOOT POUNDS				METRIC
BOLT DIA.	U.S. DEC EQUIV.					BOLT HEAD
6mm	.2362	5	6	8	10	10mm
8mm	.3150	10	16	22	27	14mm
10mm	.3937	19	31	40	49	17mm
12mm	.4720	34	54	70	86	19mm
14mm	.5512	55	89	117	137	22mm
16mm	.6299	83	132	175	208	24mm
18mm	.709	111	182	236	283	27mm
22mm	.8661	182	284	394	464	32mm

SAE STANDARD / FOOT POUNDS

GRADE OF BOLT	SAE 1 & 2	SAE 5	SAE 6	SAE 8		
MIN. TEN STRENGTH	64,000 P.S.I.	105,000 P.S.I.	133,000 P.S.I	150,000 P.S.I.		
MARKINGS ON HEAD					SIZE OF SOCKET OR WRENCH OPENING	
U.S. STANDARD	FOOT POUNDS				U.S. REGULAR	
BOLT DIA.					BOLT HEAD	NUT
1/4	5	7	10	10.5	3/8	7/16
5/16	9	14	19	22	1/2	9/16
3/8	15	25	34	37	9/16	5/8
7/16	24	40	55	60	5/8	3/4
1/2	37	60	85	92	3/4	13/16
9/16	53	88	120	132	7/8	7/8
5/8	74	120	167	180	15/16	1.
3/4	120	200	280	296	1-1/8	1-1/8

DECIMAL CONVERSION CHART

FRACTION	INCHES	M/M
1/64	.01563	.397
1/32	.03125	.794
3/64	.04688	1.191
1/16	.06250	1.588
5/64	.07813	1.984
3/32	.09375	2.381
7/64	.10938	2.778
1/8	.12500	3.175
9/64	.14063	3.572
5/32	.15625	3.969
11/64	.17188	4.366
3/16	.18750	4.763
13/64	.20313	5.159
7/32	.21875	5.556
15/64	.23438	5.953
1/4	.25000	6.350
17/64	.26563	6.747
9/32	.28125	7.144
19/64	.29688	7.541
5/16	.31250	7.938
21/64	.32813	8.334
11/32	.34375	8.731
23/64	.35938	9.128
3/8	.37500	9.525
25/64	.39063	9.922
13/32	.40625	10.319
27/64	.42188	10.716
7/16	.43750	11.113
29/64	.45313	11.509
15/32	.46875	11.906
31/64	.48438	12.303
1/2	.50000	12.700

FRACTION	INCHES	M/M
33/64	.51563	13.097
17/32	.53125	13.494
35/64	.54688	13.891
9/16	.56250	14.288
37/64	.57813	14.684
19/32	.59375	15.081
39/64	.60938	15.478
5/8	.62500	15.875
41/64	.64063	16.272
21/32	.65625	16.669
43/64	.67188	17.066
11/16	.68750	17.463
45/64	.70313	17.859
23/32	.71875	18.256
47/64	.73438	18.653
3/4	.75000	19.050
49/64	.76563	19.447
25/32	.78125	19.844
51/64	.79688	20.241
13/16	.81250	20.638
53/64	.82813	21.034
27/32	.84375	21.431
55/64	.85938	21.828
7/8	.87500	22.225
57/64	.89063	22.622
29/32	.90625	23.019
59/64	.92188	23.416
15/16	.93750	23.813
61/64	.95313	24.209
31/32	.96875	24.606
63/64	.98438	25.003
1	1.00000	25.400

INDEX

F

L

M

N

O

P

R

ACKNOWLEDGMENTS

The author would like to thank all of the companies and individuals that helped make this book possible.

AMERICAN AUTO MANUFACTURERS

Buick Motor Car Division; Cadillac Motor Car Division; Chevrolet Motor Division; Chrysler Motor Corp.; Ford Motor Company; GMC Truck & Coach Division; Oldsmobile Division; Pontiac Motor Division.

FOREIGN AUTO MANUFACTURERS

American Honda Motor Co.; Alfa Romeo, Inc.; Aston Martin Lagonda, Inc.; Audi of North America; BMW of North America, Inc.; British Leyland Motors, Inc. (Triumph and MG); Nissan Motor Corp.; Datsun; Fiat Motors of North America, Inc.; Jaguar; Maserati Automobiles, Inc.; Mazda Motors of America, Inc.; Mercedes-Benz of North America, Inc.; Mitsubishi Motor Sales of America; Peugeot, Inc.; Renault USA, Inc.; Rolls-Royce, Inc.; Saab-Scandia of America, Inc.; Subaru of America, Inc.; S.A. Automobiles Citroen; Toyota Motor Sales, USA, Inc., Volkswagen of America, Inc.-Porsche; Volvo of America.

DIESEL AND TRUCK RELATED COMPANIES

Airesearch Industrial Div.; Caterpillar Tractor Co.; Cummins Engine Co., Inc.; Detroit Diesel Allison Div.; GMC Truck and Coach Div.; International Harvester; Motor Vehicle Manufacturer's Assn.; White Diesel Div.

AUTOMOTIVE RELATED COMPANIES

AC-Delco; Airtex Automotive Division; Alloy; American Bosch; American Hammered Automotive Replacement Division; Ammco Tools, Inc.; AP Parts Co.; Armstrong Bros. Tool Co.; AP Parts Co.; Applied Power, Inc.; Automotive Control System Group; Beam Products Mfg. Co.; Bear Automotive; Belden Corp.; Bendix; Binks Mfg. Co.; Black & Decker, Inc.; Blackhawk Mfg. Co.; Bonney Tools; Borg-Warner Corp.; Bosch Power Tools; Carter Div. of AFC Ind.; Champion Spark Plug Co.; C.A. Laboratories, Inc.; Clayton Manufacturing Co.; Cleveland Motive Products; Clevite Corp.; Colt Industries; Chicago Rawhide Mfg. Co.; CRC Chemicals; Cy-lent Timing Gears Corp.; D.A.B. Industries, Inc.; Dana Corp.; Dake; Dayco Corp.; Debcor, Inc.; Deere & Co.; Delco-Remy Div. of GMC; Detroit Art Services, Inc.; The DeVilbiss Co.; Duro-Chrome Hand Tools; The Echlin Mfg. Co.; Edu-Tech-A Division of Commercial Service Co.; H.B. Egan Manufacturing Co.; Ethyl Corp.; Exxon Co. USA; Fairgate Fuel Co., Inc.; Federal Mogul; Fel-Pro Inc.; Firestone Tire and Rubber Co.; Ford Parts and Service Division; Florida Dept. of Vocational Education; FMC Corporation; Fram Corp.; Gates Rubber Co.; General Tire & Rubber Co.; Goodall Manufacturing Co.; The B.F. Goodrich Co.; The Goodyear Tire & Rubber Co.; Gould Inc.; Gunk Chemical Div.; Hartridge Equipment Corp.; Hastings Mfg. Co.; Heli-Coil Products; Helm Inc.; Hennessy Industries; Holley Carburetor Div.; Hunter Engineering Co.; Ingersoll-Rand Co.; International Harvester Co.; Kansas Jack, Inc.; K-D Tools Manufacturing Co.; Keller Crescent Co.; Kem Manufacturing Co., Inc.; Kent-Moore; Killian Corp.; Kline Diesel Acc.; Kwick-Way Mfg. Co.; Lincoln St. Louis, Div. of McNeil Corp.; Lisle Corp.; Lister Diesels, Inc.; Lufkin Instrument Div.-Cooper Industries Inc.; Marquette Corp.; McCord Replacement Products Division; Mac Tools Inc.; Maremont Corp.; Minnesota Curriculum Services Center; Mobile Oil Corp.; Moog Automotive Inc.; Motorola; National Institute for Automotive Service Excellence; NAPA; OTC Tools & Equipment; Owatonna Tool Co.; Parker Hannifin Corp.; Precision Brand Products; Proto Tool Co.; Purolator Filter Division; Quaker State Corp.; Rochester Div. of GM; Roto-Master; Sealed Power Corp-Replacement Products Group; SATCO; Schwitzer Cooling Systems; Sears, Roebuck and Co.; Sellstrom Mfg. Co.; Sem Products, Inc.; Shell Oil Co.; Simpson Electric Co.; Sioux Tools, Inc.; Snap-on Tools Corp.; Speed Clip Sales Co.; Stanadyne, Inc.; The L.S. Starrett Co.; Stewart-Warner; Sun Electric Corp.; Sunnen Product Co.; Test Products Division-The Allen Group, Inc.; Texaco, Inc.; 3-M Company; Tomco (TI) Inc.; TRW Inc.; TWECO Products, Inc.; Uniroyal, Inc.; American Bosch Diesel Products; Vaco Products Co.; Valvoline Oil Co.; Victor Gasket Co.; Waukesha Engine Division, Dresser Industries, Inc.; Weatherhead Co.; a special thanks goes to the Eastwood Co. (1-800-345-1178), TIF, and to my wife (Jeanette) and children (Danielle, Jimmy, and DJ).